D1174912

HANDBOOK OF
CONSTRUCTION
MANAGEMENT
AND
ORGANIZATION

HANDBOOK OF
CONSTRUCTION
MANAGEMENT
AND
ORGANIZATION

J. B. BONNY, *Editor*

JOSEPH P. FREIN, *Associate Editor*

VNR VAN NOSTRAND REINHOLD COMPANY
New York   Cincinnati   Toronto   London   Melbourne

Van Nostrand Reinhold Company Regional Offices:
New York   Cincinnati   Chicago   Millbrae   Dallas

Van Nostrand Reinhold Company International Offices:
London   Toronto   Melbourne

Library of Congress Catalog Card Number: 72-7364
ISBN: 0-442-20898-7

Manufactured in the United States of America

Published by Van Nostrand Reinhold Company
450 West 33rd Street, New York, N.Y. 10001

Published simultaneously in Canada by Van Nostrand Reinhold Ltd.

15 14 13 12 11 10 9 8 7 6 5 4 3 2 1

Library of Congress Cataloging in Publication Data

Bonny, John Bruce, 1903-
    Handbook of construction management and organization.

    1. Construction industry—United States—Management.
I. Frein, Joseph P., 1904-   II. Title.
HD9715.U52B58   658'.99'00973   72-7364

CONTRIBUTORS

JAMES A. ATTWOOD
Senior Vice-President
Equitable Life Assurance Society of the U.S.
New York, N. Y.

DARIO DE BENEDICTIS
Member, California Bar
San Francisco, California

R. E. BERNARD
Vice-President
Kaiser Engineers
Oakland, California

J. B. BONNY
President and Chairman (Retired)
Morrison-Knudsen Company, Inc.
Boise, Idaho

HENRY C. BOSCHEN
Chairman (Retired)
Raymond International, Inc.
New York, New York

DAVID V. BURGETT
Partner
Lybrand, Ross Bros. and Montgomery
San Francisco, California

PAUL W. CANE
Manager of Public Relations
The Bechtel Corporation
San Francisco, California

FRANCIS DURAND
Partner
Ernst and Ernst
New Orleans, Louisiana

O. P. EASTERWOOD, JR.
Partner
McNutt, Dudley and Easterwood
Washington, D. C.

JOHN FONDAHL
Professor of Civil Engineering
Stanford University
Stanford, California

J. P. FREIN
Vice-President and Director (Retired)
Morrison-Knudsen Company, Inc.
Boise, Idaho

EARDLEY W. GLASS
Financial Vice-President (Retired)
Morrison-Knudsen Company, Inc.
Boise, Idaho

DAN GORTON
Vice-President (Retired)
Fidelity and Deposit Company of Maryland
Baltimore, Maryland

JOSEPH J. HYDE
Partner
Lybrand, Ross Bros. and Montgomery
San Francisco, California

GEORGE F. KERNAN
Vice-President
Continental Illinois National Bank and
   Trust Company
Chicago, Illinois

LEE E. KNACK
Director of Labor Relations
Morrison-Knudsen Company, Inc.
Boise, Idaho

WILLIAM S. LAMBIE, JR.
    Vice-President, Parts and Service
    Caterpillar Tractor Co.
    Peoria, Illinois

JOHN LANGBERG
    Procurement Manager
    International Engineering Company, Inc.
    San Francisco, California

JOHN W. LEONARD
    Chief Engineer
    Morrison-Knudsen Company, Inc.
    Boise, Idaho

ROBERT M. McLEOD
    Member, California Bar
    San Francisco, California

W. F. MEYER
    IBM Marketing Manager
    New York, New York

JOHN C. MOORE
    President
    Marsh and McLennan
    Seattle, Washington

DONALD G. PERRY
    Partner
    Lybrand, Ross Bros. and Montgomery
    San Francisco, California

MARK A. ROBINSON
    Director of Procurement
    Morrison-Knudsen Company, Inc.
    Boise, Idaho

FREDERICK P. SLOAT
    National Director of Actuarial and
        Benefits Consulting Services
    Lybrand, Ross Bros. and Montgomery
    San Francisco, California

FOREWORD

The construction industry, like any other, has through evolution and technological advance become increasingly complex. Always a high-risk business, it has thus become even riskier.

But underneath this complexity lie certain truths, certain basic fundamentals. The need has long existed for a "handbook" of these fundamentals that would be of guidance to general contractors, specialty contractors and subcontractors as well as to professors and students in universities or technical institutes.

Such a handbook would, by example, examine the basics of construction management, the organizational structures which in practice have proved to be the most workable, and the capabilities and requisites which experience has shown to be necessary for success.

Obviously, no text can provide all the answers. However, this compendium by proven professionals, edited by leaders of the construction industry, represents a valuable addition to the body of literature on this subject.

In our experience, no other business demands such a broad range of management talents and skills which are adaptable to other industries. Successful construction management requires another quality—the 24-hour approach to problem-solving.

Jack Bonny has spent a lifetime in construction, a lifetime full of challenge and achievement. He and the associates he has gathered together to contribute to this handbook write from experience and with the shared conviction that the book fills a real need.

We, their friends and associates, commend them for this effort.

Edgar F. Kaiser
Chairman of the Board
Kaiser Industries

FOREWORD

PREFACE

The primary purpose of this Handbook is to make available to general contractors, specialty contractors, and subcontractors, as well as to professors and students in universities or technical institutes which offer courses on this subject, the fundamentals of construction management together with the most workable types of organization, and the necessary capabilities they must include to reasonably ensure success and minimize the possibility of failure in this most hazardous profession.

The second and equally important purpose is to furnish to equipment manufacturers, dealers, material suppliers, bankers, surety bondsmen, and others, who traditionally rely on financial statements and general reputation, something more concrete to look for—type of management and organization, their scope and capability—in deciding how far to go along with contractors with whom they deal or wish to deal.

In order to cover the subjects of this Handbook adequately and with the greatest consideration of their practical aspects, the editor called upon an associate of long standing, familiar with both theory and practice of construction management and the structure and functions of the departments, both in the field and in central and district offices, that make up construction organization.

The editor and his associate editor, Mr. Joseph P. Frein, have each written chapters on the specific subjects in which they are most experienced. Together they have chosen and invited a number of executives outstanding in their field to author chapters or sections on the theory and practice of construction management, its basic concepts, and in some detail its important functions.

They have likewise selected leading specialists in their particular fields to write chapters on the vital segments making up the structure of construction management and organization. These fields include job organization by general types of projects, equipment maintenance, preventive maintenance and overhaul, engineering and estimating, scheduling and controls, data processing and the uses of the computer equipment in engineering and accounting techniques, office administration, corporate and cost accounting, payroll, employment and labor relations, safety, public relations and company publications, legal and contractual problems, banking and finance, taxes, surety bonding, insurance, pension and retirement problems, and others.

All of these subjects have been treated from the standpoint of practice as well as theory, since every contributor is an expert whose knowledge came through practical application, and their presentation in a single volume should provide a valuable "tool kit" for anyone concerned with construction. Almost nothing on construction management and organization has been published in the last six years, and the few books still available for reference are often too general and theoretical to be of much practical use. It is hoped, therefore, that this Handbook will serve adequately its useful purpose in this most important but almost entirely neglected field.

Idaho City, Idaho   J. B. BONNY

INTRODUCTION

The construction industry, by its very nature, differs from rather than resembles any other business. It is true that certain basic principles of sound business management apply equally to all enterprises. Management must comply with codes of ethical conduct, it must strive to build a loyal and efficient organization, it must have sound fiscal policies, it must have adequate accounting and cost controls, it must purchase materials wisely and produce its product economically in order to realize a profit. At this point the differences begin to exceed the similarities.

The construction business is to a significant degree made up of a multiplicity of contradictions and anomalies.

Giant in size and ever growing larger, employing fifteen out of every hundred workers, the largest user of steel, aluminum, copper, cement, rubber, lumber, brick and building supplies, fuel, power, and the products of a host of other businesses whose prosperity depends upon it, construction is, in truth, not an industry at all but a service profession.

Construction is above all a "people" profession. Management must first and always be able to communicate with and lead people. Management must develop to a high degree a loyal, dedicated and efficient staff and train its members as well as allow them to make prompt and sound independent decisions in the field, above and beyond usual industrial practice. The "man on the spot" and his staff must constantly deal with new groups of Union officials and workers, since Union practice and labor economics have over the years nearly eliminated the custom of construction workers drifting from job to job for the same employer, and a ready-made organization from top to bottom at each location is today very rare indeed. Rather, the project manager and his permanent people today lack entirely the easy familiarity and knowledge of each individual's capabilities, nor do they have the "Company loyalty" of their skilled and unskilled workers on each job. Instead much time and effort must be spent in overcoming the indifference and frequently the hostility of a constantly changing group of strangers who must be molded into a workable unit in order to complete the project on time and with a profit.

Construction, ever expanding as population and needs increase, lends itself to the development of larger and constantly growing enterprises, capable of handling huge and numerous individual projects and spreading geographically throughout the nation and the world. Yet the smaller contractor with less capital, if he has learned the principles of competent management, can compete on equal terms and in most cases look forward to a much greater return on his investment, due to lower overhead and more personal supervision, than his large competitor. He must, however, be prepared to accept a greater hazard, since with a smaller number of jobs under way he cannot "spread the risk" to protect himself against the one bad contract which may break him.

The contractor, contrary to any other industry or business, is constantly striving to work himself out of a job. He moves to a location, sets up a "factory" (his construction plant), and proceeds to produce only one of its kind, as each project is unique from all others. If the plant functions as it must from the start, and if labor—the men who operate the plant and work with their hands—performs as expected, it is because the contractor and his supervisors successfully lead the workers and properly plan and schedule the work, in which case the job is finished on time and if priced safely, with a profit.

A project completed, the contractor, his supervisors, and his plant move on to other projects and the performance is repeated. Regardless of how large or how small, whether building one project at a time or fifty, he is by the nature of his profession mobile and constantly moving, with each new project presenting a brand new set of problems.

The editor many years ago was blessed with a partner, an elderly and successful retired contractor, who backed him financially in what was then his construction infancy. Bob Muir, famous in his day, was an old country Scotsman who had come to the U.S. as a very young engineer in the late 1870's and worked West with the construction of the Santa Fe railroad. He had a great fund of experience and knowledge of the fundamentals of dealing with men at all levels.

Two quotations from his often salty homilies might be appropriate here as introduction to the now more complicated, more demanding, and sometimes more frustrating, but still basically and fundamentally human theory and science of construction management and organization. One on the philosophy of contracting, as true today as it was then, was: "There are no great problems in the contracting business—only that you are always dependent on what God and your fellow man will do." Another one equally valid today as it was yesterday and will be tomorrow, was: "Jock, never let your head swell—remember always that you can learn *something* from the meanest mucker in your employ!"

Idaho City, Idaho   J. B. BONNY

INTRODUCTION

CONTENTS

4

BID STRATEGY   34
J. B. Bonny

5

CONSTRUCTION CONTRACTS   45
Dario De Benedictis and Robert M. McLeod

6

ARRANGING FOR FINANCING   55
Eardley W. Glass

7

CONSTRUCTION FINANCING   64
George F. Kernan

8

SURETY BONDS   80
Dan Gorton

9

EQUIPMENT MAINTENANCE AND REPAIR   97
William S. Lambie, Jr.

10

METHODS OF DECIDING OVERHAUL OR REPLACEMENT   160
William S. Lambie, Jr.

11

CHARGES FOR USE OF EQUIPMENT   167
Mark A. Robinson

12

PURCHASING, EXPEDITING, TRAFFIC, AND TRANSPORTATION   175
John Langberg

13

**FUNCTIONS AND ORGANIZATION
OF CONTRACTOR'S ENGINEERING
SECTION   192**
J. P. Frein

14

**COST ESTIMATING FOR LUMP-SUM
AND UNIT-PRICE CONTRACTS   256**
J. P. Frein

15

ESTIMATING OTHER THAN FIRM-PRICE CONTRACTS   345
J. P. Frein

16

COST CONTROLS, RELATION, AND COORDINATION WITH ENGINEERING AND ACCOUNTING   406
J. P. Frein

17

NETWORKING TECHNIQUES FOR PROJECT PLANNING, SCHEDULING, AND CONTROL   411
John Fondahl

18

USE OF COMPUTERS IN CONTRACTOR'S ENGINEERING ORGANIZATION   441
John W. Leonard

19

COMPUTER CAPABILITIES IN CONSTRUCTION MANAGEMENT   478
W. F. Meyer

20

OFFICE ADMINISTRATION: HEADQUARTERS AND FIELD   503
Donald G. Perry

21

CORPORATE AND COST ACCOUNTING   515
Francis Durand

22

PAYROLL PROCEDURES   528
Donald G. Perry

23

EMPLOYMENT PRACTICES AND RECORDS   534
R. E. Bernard

24

LABOR RELATIONS AND THEIR EFFECT ON EMPLOYMENT PROCEDURES   555
Lee E. Knack

25

SAFETY PROCEDURES AND PRACTICES   568
Lee E. Knack

26

PUBLIC RELATIONS FOR CONTRACTORS   582
Paul W. Cane

27

LEGAL AND CONTRACTUAL PROBLEMS   591
O. P. Easterwood, Jr.

28

TAXES   609
Joseph J. Hyde

29

CONTRACTORS' INDUSTRIAL INSURANCE 622
John C. Moore

30

GROUP INSURANCE PLANS 630
James A. Attwood

31

FUNDAMENTAL CONCEPTS UNDERLYING PENSION PLAN FINANCING AND COSTS 635
Frederick P. Sloat and David V. Burgett

HANDBOOK OF

CONSTRUCTION
MANAGEMENT
AND
ORGANIZATION

1 BASICS OF CONTRACTING

J. B. BONNY

President and Chairman (Retired)
Morrison-Knudsen Company, Inc.
Boise, Idaho

IT HAS BEEN SAID that a contractor is a man who takes contracts, and a successful contractor is one who makes money on them. This statement, although basically true, is an obvious oversimplification.

For purposes of this Handbook the word "contractor" is taken to mean construction contractor, whether he be a general, specialty, or subcontractor.

It is true that purveyors of thousands of items and types of goods to the government, from airframes and engines to thimbles and ships and shoes, have come to be called "contractors." The misfortunes and malfeasances of some of these manufacturers, when featured by the news media, tend through public misunderstanding to fall on the heads of construction contractors as well, so that "contractor" had become a dirty word in many places through association.

1-1. Characteristics of Contractors

Contractors come from many walks of life, though to be sure the majority of those who start, lacking proper training and experience and a flair for management, have little chance of success.

The lure and romance of contracting, like the lure of finding gold or oil, attract many hardy optimists. When one drops out through lack of finances or training plus experience or natural aptitude, there are two to take his place. Thus the field is always overcrowded. Competition, though more intense at some times than at others, is always keen, and the weak are soon weeded out. Like the old couplet about settling the West a hundred years ago: "The cowards never started and the weak died on the way."

Successful contractors are a breed apart and their histories make colorful monuments to their ingenuity. Many, starting from scratch without money or training, climbed to the top by sheer courage and hard work, learning as they went along. Some were shovel runners, cat skinners, carpenters, steel tiers, plumbers, or concrete workers. Some were civil or mechanical engineers, attracted from a safe profession to its more dangerous and exciting cousin. Even a few lawyers, timekeepers, and accountants took to the construction field and became outstanding contractors. They all, however, shared the same characteristics: optimism, courage, basic good judgment, and willingness to work.

1-2. Changing Conditions

As time passed the rules of the game changed. Taxes and wages went up, up, and up some more. As wages became inflated, individual efficiency went down with the law of diminishing returns. New, larger, and more productive equipment continued to be developed, effectively braking inflationary costs of machine operation. Work that had to be done by a man's hands, however, became more and more expensive—to the point of pricing itself out of the market. Bricklayers, for example, forced wages and fringe benefits up from around three dollars an hour to seven dollars, at the same time limiting the number of bricks permitted to be laid in a day to five hundred instead of the thousand that had been a standard day's work for many years. Wooden forms, where they had to be used in lieu of steel because of limited reuse or no reuse at all, went up in cost five-fold in little more than ten years (1960-1970), and carpenters continued to demand more and more for less and less production. The same trend has applied to all other crafts that could not be fully mechanized—steel erectors, re-bar tiers (rod busters), plumbers, and electricians.

1-3. Problems with Some Owner's Engineers

The other horn of the dilemma to be faced is the problem of the owner's engineers. The vast majority of them are

1

conscientious, clear thinking, and fair. They are making every effort to design the work so the contractor can get maximum reuse of forms, panels, and other expendable materials and to the greatest possible degree eliminate expensive hand labor. Unfortunately there are still many who, contrary to the best interests of both contractor and owner (who in the end must pay the bill), design plants and structures without achieving minimum costs by the use of standard design to permit economical reuse of forms, such as allowing steel panels to be set in place by cranes at a fraction of the cost of special wood forms which must be used once and thrown away.

All too many owner's engineers fail to recognize that they are, in fact, in positions of special trust and should conduct themselves as fair arbiters of the intent of the contract. Instead, some write specifications that charge the contractor with the cost of correcting any and all mistakes in the design and specifications—even though such mistakes may be entirely those of the owner's engineer. Unreasonably harsh interpretation of the specifications, always against the contractor, results in excessive costs, effectively robbing him of the already thin margin of profit he is trying desperately to preserve.

The occasional engineer who attempts to "break" the contractor by a combination of vicious and unreasonable provisions in the specifications, omissions in the plans, or misleading and conflicting statements, may appear for a while to get away with it. Vicious and sadistic enforcement of unreasonable provisions may for a time get the owner a quality and quantity of work he did not pay for. Forcing the contractor to perform work beyond the specifications and then refusing to pay for it may please a stupid or venal owner when reported by the engineer. But eventually the engineer and succeeding owners who employ him pay the price.

The first contractor he "breaks" may never work again for him—or in fact for any other engineer. But the word gets around. The current owner may be successfully sued by a contractor he couldn't "break." Contractors who have been burned will warn others. Before long the engineer will either be unable to get anyone to bid on his work or the bids will be so high and carry so much contingency allowance for the "tough" engineer that owners cannot pay the price and will look elsewhere for engineers—and the engineer will find himself without clients.

On the other hand, the engineer, whether he is chief engineer of a large company requiring many facilites to be constructed or a design consultant, who is consistently fair demands good quality construction and pays for what he orders. He will get the lowest possible bids, with contingency factors low or zero, and in the end will save hundreds of thousands or millions of dollars for his company or his clients.

1-4. Basic Knowledge Required

Courage and optimism and willingness to work are no longer enough to assure success in contracting. The new breed of contractor and all of his staff, as he grows larger and spreads out geographically, must have knowledge and great competence in many fields. He must understand how to choose and organize his staff. He must know how to command and keep the loyalty and enthusiasm of his subordinates.

As entrepreneur, the contractor must know how to get the financing so necessary in his field and to gain and hold the confidence of his banks and his bonding company.

The principles of engineering and estimating must be understood. He must know the strategy of bidding against tough competitors, plus when and where to pick jobs that offer the best margin of profit. A strong and competent field organization that will finish the jobs on time and economically must be developed and trained.

Accurate and detailed costs are a must, with records that cannot be denied kept ready to back up change orders and claims when and if necessary.

The contractor must be familiar with all possible forms of insurance in order to protect his payrolls and other funds, his equipment, his jobs, and his employees. He must have at his fingertips knowledge of a multitude of taxes and be familiar with their ramifications.

Labor relations and public relations, if not in competent hands, will destroy him.

Knowledge of all these things and more must be acquired and retained and used to have a well rounded and successful company. No longer is it possible to run a construction company "by guess and by God" with a little luck.

The strongest candidate today to head a construction company would be a man under forty with at least fifteen years of experience in construction preceded by university training and majors in civil engineering, business administration, and contract Law. Certain universities today are offering courses in contruction management. These can be of immense value in cutting down the learning time necessary before the college graduate is able to contribute significantly in the profession of his choice.

1-5. Branches of Construction—Heavy

To define the construction contractor accurately it is necessary to look first at the major branches of the profession, to determine how each operates, and to compare similarities and differences in the various branches.

The first major branch—the largest in volume of work but not in numbers—is made up of the heavy engineering contractors. This portion of the industry is in turn divided into contractors primarily concerned with:

1. Dams, canals, and hydroelectric work
2. Highways and railroads
3. Tunnels, subways, and tubes
4. Bridges
5. Aqueous and subaqueous, i.e., piers, jetties, dredging, and breakwaters
6. Airports
7. Pipelines and pumping stations
8. Water and sewer lines, sewerage plants, and other and more sophisticated antipollution installations
9. Atomic power, missile and tracking stations, other military installations, power transmission.

1-6. Dam and Canal Construction

Many of the major contractors, particularly in the West, started in business building dams and canals for the U.S. Bureau of Reclamation and various irrigation districts shortly after the turn of the century.

The work was usually bid on unit prices—i.e., prices were submitted on such items of work as common excavation per cubic yard, rock per cubic yard, concrete per cubic yard, and reinforcing steel and structural steel gates and gate guides per pound or per ton. The quantities were listed as estimated by the owner's engineer and frequently varied widely from the actual quantities. The contract was let on the basis of the lowest bid computed on the total price for all units bid upon.

The competition was always keen and the prices low, for it was necessary not only to beat the competition but also to be in line with the engineer's estimate. The Bureau of Reclamation maintained its own construction force until the middle 1920's and preferred to do its own work "force account."

The "force account" work was performed by a rather large and highly efficient construction organization within the Bureau and was headed up by several outstanding "construction superintendents" reporting to the chief engineer. They worked on plans and specifications prepared by the Bureau and the work was inspected by Bureau engineers. They requisitioned equipment and, with the chief engineer's approval, hired their foremen, master mechanics, and workmen, and ran their own cookhouses and camps. Careful costs were maintained and high standards of work were reached.

As the policy shifted, from building major dams and canals by force account and contracting smaller structures and portions of canals, to calling for bids on most projects and awarding contracts when the contractors' figures were at or below the force account "bid," and then to calling for bids on all work and awarding to the lowest offerer, the force account organization was gradually disbanded.

Some construction superintendents joined construction firms where Frank Crowe, the superintendent of Hoover Dam and later Shasta Dam, and Ralph M. (Mickey) Conner and others became famous. Many remained with the Bureau in supervisory positions and ended their careers in high places—men like Harry Bashore, Commissioner of Reclamation, who followed after the late great Elwood Mead, Walker (Brig) Young, and Grant Bloodgood as chief engineer, and Jack Savage as chief design engineer and world-renowned consultant.

These and many others were a tough and knowledgeable breed who knew more about construction methods than most of the contractors of their day. The specifications were enforced with extreme rigidity, and woe betide the contractor who attempted to slight the letter of them or got behind schedule on his work.

Equipment prior to 1930 was all small. Many contractors still used horse teams and scrapers to move dirt. The jobs were of fairly long duration; the retained percentage held from the monthly estimates by the owner to guarantee completion of the job (usually 10 percent) was onerous, and the contractor was always short of funds. As a consequence, in the late 1920's the "joint venture" was conceived.

The joint venture is a limited partnership between two or more contractors covering a single job. In practice usually one partner is the sponsor; he holds a larger percentage interest in the work (usually 40 to 60 percent), and runs the work subject to his partner's concurrence on major policy matters.

The purpose of the joint venture (which remains equally true today) was: (1) to spread the risk (as the jobs became larger, a contractor going alone could easily go broke on a single job); (2) to reduce the amount of financing required from each partner; (3) to get the advantage of each partner's judgment in pricing the work; and (4) to permit each partner to participate in more, hopefully profitable, projects than would be possible if he went alone. The joint venture has proved sound over the years and is in use today on most large jobs.

The dam builders were the first to practice, from sheer necessity, the charting and scheduling of the sequence and timing of all major items of work. Equipment and plant were furnished to meet a preplanned schedule. The schedule itself was set up to coordinate the work with the delivery of embedded permanent materials, frequently all or in part furnished by the owner. Then, too, opening up the riverbed and diverting the stream had to be timed to get the concrete or the fill into the riverbed and up to a safe height between flood seasons. This led to total scheduling in which every concrete pour was numbered and the proposed date of pouring shown at the outset of the work. From the late 1920's this was done laboriously by hand, until the advent of the computer, when it became "critical path" or some other designated computerized control of scheduling. Thus the idea of "critical path" had actually been in use for some time before the coming of the computer—manually produced but basically the same principle.

Canal work, like dam work, required careful planning. The jobs were usually of fairly long duration—eighteen months and up to thirty-six months—but if the canal was unlined very little equipment was required per dollar of contract volume. Therefore the burden of financing was not too onerous. In the case of a concrete lined canal, the equipment load was greater; aggregate and mixing plants, batching plants, cranes, and a multitude of canal lining equipment made the equipment and plant investment larger in proportion to the dollar value of work.

1-7. Hydro Projects

Hydroelectric work, including as it did power houses, usually in conjunction with a dam, and either canal work or a tunnel, required the contractor to have a fairly versatile organization and strong financing.

Hydro work generally is attractive, since in spite of comparatively long-term construction schedules and the hazardous type of work, it is possible to command a higher markup over estimated cost. It becomes particularly attractive if a mix with lower yield but shorter term work, such as pipelines and power transmission, can be worked out.

1-8. Highways and Railroads

Highway and railroad work are listed together since many successful contractors started in this field. Highway work in particular, while not usually as long term as most other heavy construction, is the least profitable of all. Although not usually awarded in large contracts, it requires an inordinate amount of equipment, particularly if grading and surfacing or paving is included in a single contract. The amount of equipment required does not vary with the size of the job and includes tractors with dozers and scrapers or rubber-tired scraper units, trucks, shovels, draglines, cranes, discing and harrowing tools, both sheepsfoot and rubber-tired rollers, perhaps a back hoe for digging structures, steel forms for concrete pavement with pavers and spreading tools, vibrators, finishing machines, aggregate and rock plants, batchers if concrete or paving plants, leveling machines and smooth rollers if black top, etc.

As a result, a highway project may and frequently does require more than a million dollars worth of equipment to perform a contract of a million dollars or less. Obviously a single job cannot pay for the equipment to perform it and the contractor is forced to take another job to keep up his payments—and another and another, for whatever price he can get. This brings the competition of desperation, and little or no profit can be expected on the average. Also the organization required is extensive. The general superintendent must have a number of highly competent superintendents and foremen, since by the nature of the work and the sequences required, they are spread over several miles, and even with a large number of pickup trucks equipped with two-way radio, communication is poor and the job difficult to control.

Many large and well-diversified contractors, however, continue to do highway work. The reason for this, in the writer's own experience and in discussion with many others, can be summed up in one sentence: "It's a great school for superintendents—if they can make it on highway work they can make it anywhere."

Railroad work, on the other hand, has all the advantages but one. In the years following the completion of the first transcontinental link at Promontory Point in Utah, railroads in the United States grew tremendously. Many contractors expanded and became rich and famous on railroad work and nothing else.

On major line changes or extensions, quite commonly the railroad's chief engineer would allocate a new line in sections of a number of miles each to several contractors. Each contractor, in turn, might parcel out one or more miles to "station gangs," furnishing each "gang" with tools, powder, food, and lodging to be charged against their earnings which were computed on a price per yard and paid once a month or at the end of the job, in which case the main contractor might make small advances from time to time. Many of the station gangs, particularly in the Northwest, were Swedes. Rock was drilled by hand, one man turning a drill steel rod or "bull prick" and another swinging a maul. Shooting was usually black powder poured in the hole and fired by a fuse.

The station man's life was a hard one—long hours of back-breaking work for small pay. Occasionally, however, a station gang would get a break; the rock would be softer and break better or the men would be tougher and produce more than the others. In these cases the gang, if it had a smart leader, would accumulate sufficient funds to finance a subcontract on its own, and might even function as a general contractor and secure a section from the railroad direct. Some of the big names in contracting even today hark back to such a beginning.

By the 1930's, however, most of the railroad expansion was completed. The roadbeds had been modernized and laid with heavy rail. The day of the railroad contractor as such was over.

1-9. Underground Construction

Tunnels, shafts, and subways, although performed by many large "diversified" contractors, are fundamentally a special type of work and even carry their own special classifications, each with its own breed of men. The most hazardous of all construction work, with the possible exception of deep bridge piers, underground construction carries with it the constant possibility of a large profit—"if everything goes well."

Only a few major contractors specialize exclusively in underground work and some of them have been highly successful. These few have developed tunnel or shaft organizations complete with entire crews, from general superintendents, walkers, and shifters, to powdermen, dinky skinners, and mucker operators, who follow the general superintendent from job to job.

The large diversified contractor usually carries only the nucleus of an underground organization, and may have a vice-president or "manager of underground construction" who is sufficiently well known to be able to pick up entire crews of hard rock men, soft ground men, and even sandhogs (men who work under air pressure where the ground is too soft or too wet to stand with ordinary supports until concrete lined and must be supported by internal air pressure). These sandhogs are tough and temperamental characters who risk their lives daily against possible cave-ins or collapse. They are exposed to the deadly "bends" which may cripple or kill without warning, but are paid higher wages than any other craft, sometimes for a shift under 40 pounds of air pressure that lasts 2 hours or less.

1-10. Bridge Construction

Bridge construction, too, is a profession by itself. A few companies specialize in it to the exclusion of other work, but most bridges are built by highway or diversified contractors. Since many contracts carry rather simple bridges with shallow piers and little or no hazard along with highway work, the general contractor may elect to subcontract his bridges, culverts, and structure concrete to one or more subcontractors who specialize in this work, and thus avoid organizing for this portion of his contract.

Occasionally a multimillion dollar bridge project develops in the United States or abroad which challenges the imagination and offers the opportunity of a generous profit to the contractor sufficiently well financed and with courage to accept the risks inherent in deep piers and long spans, with probably high tides and rough water.

Frequently these large bridge jobs are let in two contracts—the piers and approaches in one, and the superstructure in another. In this case the heavy construction contractor bids alone or in joint venture with others on the substructure and approaches; and a steel company or its subsidiary bids on furnishing and erecting the superstructure, usually a suspension span or spans in large bridges, because structural spans with fixed members become too expensive and finally impractical as the length of span increases.

In other cases the steel company may submit a bid on the entire job, subcontracting the approaches and substructure to a general contractor in whom he has confidence and who has quoted the steel company a price before the bid, which the steel company has used in turn in its bid. Occasionally the general contractor and the steel company join to bid in joint venture, and if successful each performs his part of the work. Finally, the steel company may quote the general contractor or all general contractors bidding, hoping to sell the superstructure to the successful bidder.

1-11. Waterfront Construction

Aqueous and subaqueous piers, jetties, dredging, and breakwaters are usually done by a special class of contractor. A number, however, content themselves with concentrating on dredging only; others on rock-fill piers and jetties or breakwaters. Some large diversified contractors include all but dredging and may have separate departments to follow "water work." A very few maintain equipment and have a separate organization or own a subsidiary that does dredging. The reasons are clear and lie in the nature of the work.

Piers, jetties, and breakwaters are most frequently jobs of considerable size, often in the five, ten, or twenty million dollar bracket. They are located at home and abroad, on salt-water coasts or bays, and less frequently on large inland lakes. Piers may be of placed rock decked with concrete, or more commonly concrete or treated timber piles decked with wood or concrete.

Jetties and breakwaters are usually of large dumped or placed rock, sometimes capped for a roadway. It follows that the local or regional contractor with sufficient foresight to option or buy a satisfactory quarry close to the proposed site has a tremendous advantage, and for his patience and acumen is in a position to reap a sizable reward. This type of work becomes most attractive if it can be secured, in spite of high insurance costs for "builder's risk," since fierce winter storms may well wash away a whole summer's work in a single day.

1-12. Dredging

This is a different game. The cost of dredges is high. A large bucket or dipper dredge will cost from several hundred thousand to more than a million dollars, whereas a suction dredge with necessary pipe and equipment in the 20-inch to 30-inch class may run from three and a half million up. At the same time the cost of maintaining a captain and skeleton crew while the dredge is idle is high, and unless it operates at least 50 percent of the time can be ruinous. Thus the dredging business can be "feast" or "famine," and although the dredge may well last from twenty to thirty years, it can ruin its owner unless he has luck and the energy to go after business wherever it is, and gets it at a fair price.

The dredging contractor also faces a special kind of competition. By tradition, the Rivers and Harbors Division of the Corps of Engineers maintains its own fleet of dredges and a dredging organization. It performs emergency work and improvements "force account" on the Mississippi River, the Gulf of Mexico, and the East Coast. At times when the government equipment is busy or when larger or special types of dredges are needed, contracts are let for dredging projects, either in competition with the government "bids" or to supplement the Corps of Engineers equipment. This results in stabilizing prices at low minimums and minimizes the chances of the dredging contractors picking up "fat jobs" in those areas. Yet a number of dredging contractors have been highly successful and continue to operate a profitable and very efficient business on the Great Lakes, the Mississippi, the Gulf Coast, and in fact all over the world.

1-13. Airports

These are becoming an increasingly important source of profitable work for both the local and the regional or national contractor. The local contractor is able to bid in runway repairs and resurfacing, lengthening of runways and taxiways, drainage, parking aprons, and improvement of freight and passenger handling facilities.

The major airports, faced with ever-increasing traffic, larger and faster planes, and urgent need for enlarged and more efficient baggage handling, passenger facilities, and freight terminals, are and will be for some time a major source of business for large national diversified contractors. Since virtually no main line or international airport today is capable of handling the present traffic, let alone future needs of bigger jets, development of these airports is a "must," and money for the expansion will of necessity be found. Nearly every type of construction is involved in

these huge terminals; hence they become of prime importance to both the diversified general contractor and the specialty contractor—electrical, plumbing, structural, or what not—as well as to the subcontractor able to handle selected portions of the work.

The decade of the 1970's cannot hope to more than scratch the surface of airport developments that must be made to handle adequately the mass of traffic that is even now crowding airports beyond capacity and creating a hazard of the first magnitude to air travel. It may well be that separate satellite airports may be required at all large centers of population to segregate the long-haul and international traffic from the shorter-haul and commuter traffic. The need is already apparent, and only shortage of funds and prevailing high interest rates prevent expansion to take care of minimum requirements.

All this means construction work and lots of it. The contractor, or the contractor and the airlines with the banks or lending agencies, who is able to work out a reasonable "pay as you go" formula for financing these urgently needed developments, need have no fear of shortage of business—or lack of ample profits.

1-14. Pipelines and Pumping Stations

These constitute a branch of general contracting that is the most volatile, the most "feast or famine," the most "go for broke" of all facets of contracting. The business of cross-country lines, particularly the "big inch," 30 inches or more in diameter, for transmission of gas, oil and petroleum products, is, to say the least, one that never lacks excitement. A single "spread' of trenching, wrapping, laying, and back-filling equipment—equipment sufficient to put in the ground at least a mile a day of "big inch" pipe—costs in the neighborhood of a million dollars or more. The contractor (or pipeliner) must have this equipment in his possession and ready to go when the job is offered, or his better equipped competitor will get the job. A "spread," to "pay out," must do at least $3 million worth of work a year, so the pipeliner who gets in the business buys his equipment before he has the job and gambles on what he may be able to pick up. Since oil companies normally build lines only when they have the market assured, they demand that the work be done in a very short time.

If the contractor is of middle size and possesses, say, three "spreads," he is in fair shape to compete if the business is offered. He must then fly over the line (there is never time to drive or walk it), *guess* at the price including river crossings or get subcontract figures on them, and if his bid is low he goes to work to finish the line on a very tight schedule, both to meet the contract date and to release his spread for other work he hopes to get. This procedure alone would make the business enough of a gamble —but there is more. Because of the speed at which the work is done, it may be half completed before the current costs are in and totaled. Thus the pipeliner hardly knows whether he is making or losing money until so much of the work is finished that he can't do much about it. Pipeliners, however, *do* make money. Many have succeeded performing only this work, and a few diversified contractors operate pipeline divisions or own pipeline subsidiaries with satisfactory results.

1-15. Water Supply and Sewerage

Water and sewer lines, sewerage plants, and all forms of antipollution facilities are growing fields with inherent profit possibilities.

For many years the most necessary utilities, serving first the larger cities and then smaller and smaller urban centers, were most frequently the preserves of specialty contractors. These, particularly in the major centers, were inclined to be local or regional operators and often consisted of two or more brothers and the male members of their immediate families. Often of Italian or middle-European origin, they would band together with individual members acting as superintendents or foremen in the field. This would result in a highly efficient and economical unit, able to operate with low overhead and come up with a profit shared in the family, at prices ruinous to outside competition.

Many cities would tend to offer their utility work in increments and contracts sufficiently small to permit "the locals" to absorb the work—excluding large regional or national "outsiders"—and still obtain a profit for themselves and a margin for "favors" to corrupt city or county officials. As time passed, however, with the rapid growth of urban population, the jobs of necessity grew larger and larger. The administration of public utilities, by popular demand, was frequently turned over to commissions and districts not responsible to the mayors or city councils, and relatively free from political corruption or influence.

1-16. Power Projects

Utility work has become big business. With the increasing demand for environmental control and public awareness of the dangers from water, soil, and air pollution, it has become a rapidly expanding source of profitable business for regional and national as well as for local contractors. This branch of construction is growing and is sure to continue to grow. With the population growth and the tightening regulations that force industry as well as municipalities to take steps, however expensive, to prevent all forms of environmental pollution, it should be of prime concern and interest both to established contractors and to all those who wish to grow and succeed.

The most technical and yet one of the most attractive opportunities in the heavy construction field comes within the general areas of atomic power, missile, and tracking stations, military construction, and finally power transmission.

Atomic power in its second decade is still plagued by two bugaboos: local resistance due to fears of explosion or radiation, both unwarranted; and the relatively higher cost of delivered power compared to hydroelectric or simpler steam plants using coal or gas. Time, however, is on the side of nuclear power. Hydro sites are becoming scarcer and costs of the dams and conventional powerhouses are going up and up. Meanwhile more and more conservation organizations and governmental bodies are working to prevent new dam construction in order to protect recreational areas and farmlands from further flooding. The cost of fuel for conventional steam plants, as well as that of the plants themselves, is constantly rising while the fuel reserves are being depleted.

Nuclear plants, on the other hand, are benefiting from constant research, and thus the cost of the plants in terms of kilowatt output is going down as the operating efficiency goes up. New developments in heat transfer, perhaps the perfection of direct heat transfer eliminating the use of nuclear heated steam, promise to lower future costs still more.

To date the number of contractors who have engaged in atomic power construction is limited. A few pioneers have learned how not only to perform the heavy construction involved but to install the equipment as well. The nuclear and power equipment, however, is manufactured by a handful of major suppliers who in practice have made their arrangements direct with power companies and government bodies or power districts. They have guaranteed both the installed cost and the kilowatt price, and in turn have contracted the "brick and mortar" work and occasionally the equipment installation with large national contractors experienced in the field.

Atomic or nuclear power cannot long remain a mystery; the basic construction involved will become simpler, better understood and cheaper. Normal performance will become standard and accepted. Since a continually increasing need for power is inevitable, this field of construction will certainly open up and become attractive to the contractor who is looking for new fields.

1-17. Military Work

Missile installations, tracking stations, and NASA facilities, as well as military installations of all kinds, are at least for the immediate future slowed to a fraction of their volume in the 1950's and 1960's.

There developed in the 1950's a number of major military construction jobs abroad, in Greenland, and in North Africa awarded on a cost plus a fixed fee (C.P.F.F.) by the U.S. Corps of Engineers and Spanish Bases and port work by the U.S. Navy, Bureau of Yards and Docks (now Facilities Engineering Command). Concurrently Canada awarded two contracts, one for the East Half and one for the West Half of the DEW line (Distant Early Warning system), also on C.P.F.F. contracts, with the West Half going to a Canadian company owned by an American contractor. Soon thereafter the U.S. Air Force awarded a fee contract to Western Electric for design and construction of a reflector telephone communications system in Alaska, known (it was secret for the first few months) as "White Alice."

The impact of these jobs was formidable. There simply were not sufficient accountants, warehousemen—to say nothing of engineers and field supervisors—in any one construction organization or joint venture to keep track of government property and maintain current costs. The situation was made untenable by this, coupled with the fact that all were "crash" programs and required that much work be accomplished and some completed for use during the period when, under ordinary conditions, the jobs would be organized, and that shipments of permanent materials, parts, and supplies and even equipment be inventoried and warehoused prior to release to the proper work site.

Shipments received at the docks were promptly grabbed by sometimes overeager expediters for the various project managers, frequently loaded on trucks, and sent to the various job sites without the central warehouse manager being able to check inventories against ships' manifests, or even having "whiz tickets" with sufficient description to know what went where.

The using agency, usually not the construction agency, insisted constantly on new additions to each facility and a multiplicity of changes in plans. Few of these resulted in "changes in scope" increasing the estimated cost for purposes of computing the contractor's fee, but all tended to increase the cost, heighten the confusion, and force the construction agency to go back to Congress time after time for additional appropriations.

The situation became critical, and the contractor was

forced to reorient his staff. The general manager or resident partner in the field became two-thirds engineer, warehouseman, and detective, and one-third construction manager. It didn't matter so much what the work cost, but it was necessary to know and know quickly what went into it. The chief engineer and his site engineers enlarged their forces, and the revised organization spent their days and most of their nights taking off quantities of materials from "as built" plans and partially completed drawings to determine what had been used. The business manager, the controller, and their staffs dropped everything and all hands took inventory. Warehouses became mad-houses as dozens of people went down each row in the yard and each slot and drawer in the buildings to find what was there and had not been used—or stolen—or never received.

Meanwhile, at home, things were happening. The general accounting office (the comptroller general) had sent an investigating team to the sites fairly early in the game and had released the usual preliminary report, highly critical, to Congress, and threatened to withhold large sums of cost reimbursement from the contractor pending further checking.

On the heels of this came a flurry of investigations by various committees in both the Senate and the House. One or two, basing statements on perjured or highly colored allegations by the contractor's exemployees who had been fired for drunkenness, dereliction of duty or both, tried and condemned the contractor in the news media without a hearing. It was necessary to employ attorneys of outstanding reputation at high cost (reimbursable, however, under the contract) to prepare briefs on information gathered by the contractor's staff at home and in the field for presentation refuting questions raised. All this took months and occupied much of the contractor's top management at home and abroad, to the general disruption of all work underway and a resulting overrun in costs. Eventual clearance of all contractors concerned, not only by the general accounting office but likewise by the congressional committees involved, removed the stigma but did not repair the damage.

A movement was started, however, by the construction agencies, together with top contractor executives, to convert by negotiation the balance of the work to be done on two of the major projects to firm unit price contracts.

The procedure was logical and highly successful. Inventories of all materials to be used in the work, plus fuel, lubricants, tires, parts, and equipment were made and prices assigned to each item by a joint government-contractor board. Estimates were made by the contractor and unit prices were determined to be negotiated with an official board of the contracting agency which would be awarded provided that the firm-price arrangement showed material savings to the government.

The rules provided that the contractor would take over all materials and equipment, paying agreed prices for materials and stocks on hand as well as use charges on equipment to depreciate the same. Additional equipment parts or material were to be purchased by the contractor for his own account. Further rules made the contract subject to redetermination, i.e., complete review from when work first started on unit price plus all work that might be later added by negotiation to project completion, with any excess profit resulting as determined by the final review board to be returned to the contracting agency. Another and final stipulation called for interim review every six months and return to the government, as excess, any profit over 10 percent made during the preceding period.

The contracts, if successfully negotiated, would be true, guaranteed maximum unit-price contracts. Dangerous— yes, under the conditions. However, there was one saving grace. The first increment to be negotiated would only cover the work on which plans were complete. There were bound to be additions of succeeding increments many times the volume of the first and a corresponding chance to raise prices in future negotiations if necessary.

In one case, at least, the sponsor's operating vice-president and a top representative of each of the partners arrived at the project headquarters a week ahead of the proposed negotiation date. The sponsor's vice-president of engineering, each of the partners' chief engineers, and the entire engineering staff on the project had been working for more than a month to arrive at unit prices (and totals) for negotiation in a week's time.

An overall review of the proposed unit prices and comparison on costs to date on like items showed that there would be practically no savings to the government if these prices were used. In this case there would be no conversion of the contract. A hurried conference with the sponsor's and partners' top engineers revealed that, in arriving at cost, they had carefully analyzed costs to date, deducted certain overhead items directly connected with warehousing and accounting procedures necessary in cost-plus-fixed-fee work, and then added back in all estimated costs in connection with converting the operation to unit price into the first increment of work. Not a single engineer would agree or even admit that the cost of conversion should only be charged partially to the first increment. Nor would any of them admit that the price should be arrived at by analyzing what it should be if they were bidding a unit-price job in the area with no previous cost-plus experience.

The partners then had a meeting alone with the resident partner who agreed to the principle that it was a "new ball game." There was no time to make a new estimate—the engineering staff was adamant. There remained nothing to do unless the partners took the matter into their own hands.

Accordingly the project chief engineer (under duress) and the resident partner simply went through the list of nearly three hundred units and cut them each 20 percent, which resulted in an apparent saving to the government of some $5 million on the first increment. The offer was made to the Military Board, was accepted, and the job was converted to unit price on the designated date.

The sponsor's vice-president–operations was in the dog house; his president was worried but recognized that a contracting opportunity existed. The VP–enginering was courteous but coldly formal. The resident partner reduced his engineering staff and accounting and warehouse staffs to bare minimums and released nearly half of his superintendents and general foremen; those remaining, scenting a bonus if it worked, doubled their efforts.

When the first six months went by and the costs were totaled there was an excess profit. The excess of $3 million was returned to the government, and future prices reduced. This continued until the end of the job some two years later. The construction agency had saved the government some $20 million. The joint venture had distributed a handsome profit to the partners, and some very competent engineers had learned that neither the slide rule nor the calculator were, in themselves, equipped with brains.

The next development in the "cold war" was the great missile race, Atlas, Titan, and finally the solid fueled Minute Man. The using agency, the Air Force, awarded the hardware, the systems components, and the "birds" themselves to various airframe and space-oriented manufacturers. The brick and mortar work was awarded by the Corps of Engineers in a series of large contracts to various joint ventures on a combination of unit price and lump-sum items competitively bid. The plans were sketchy and based on the initial criteria furnished by the Air Force who would inspect, together with the manufacturers' representatives, the installation of most of the hardware; some of the critical items of wiring and piping control mechanisms and consoles would be installed by the manufacturers themselves. Combinations of the larger contractors banded together to bid, and the time frame was such that a number of joint ventures were awarded the work in different locations over a period of a few months so that they were all starting at nearly the same time.

The contractors who had been through the bases abroad and conversion to unit price felt quite secure. Their top field organizations were virtually intact and trained to handle large unit-price jobs spread over a fairly large area. The missile jobs, although spread over a large number of square miles, were easy of access or soon would be made so, and the work was largely repetitive, much of it being subcontracted. Their organization should do well.

They reckoned, however, without all the facts. No sooner had the roads been laid out, equipment brought in to excavate at the sites and drill, shoot, and muck the missile shafts than the changes started. As tests continued to be run on missiles with dummy heads or conventional explosives in the Pacific, the changes multiplied. Everything except the number of missiles at the sites changed. Shaft collars were enlarged, shaft gates were redesigned, and electrical, mechanical, and embedded piping had to be torn out and replaced. Shock doors were modified and collars excavated out and reset in the connecting tunnels of each flight, and a thousand other items were changed, some major, some minor, but all holding up the work and costing money.

The contractor transferred his assistant chief engineer, claims and change orders, to full time at the site and added more enginering help by robbing other jobs. This wasn't enough. The changes continued to pour in. No price that could be put on them even remotely covered the cost. Changing one bulkhead door in a connecting flight tunnel (a tunnel connecting two silos in a flight of perhaps three silos) by excavating the collar sunk in concrete, drilling out a larger collar and setting in the new and heavier door might take four days with five men and the direct cost be at or near a thousand dollars—but the tunnel with all the work in it was tied up, as well as direct access between the two silos. Crews had to be transferred or laid off; then overtime and Sunday work were used to get back on schedule. The real cost of this bulkhead door might well be fifteen to twenty thousand dollars with no chance to collect it.

The contractor appealed to his Washington attorneys for help. A senior partner came to the job and brought a young lawyer with training in claims and some understanding of the work. They worked on the problem for nearly three weeks.

Meanwhile, gossip had it, nearly all the missile contractors were going broke; the changes were wrecking them and no reasonable price put on changes would help. Some formula had to be worked out that would relieve all the contractors, since their problems were from the same cause. The partners were called in, the sponsor's VP–engineering arrived, the executive VP–operations outlined the problem, and the senior law partner took over.

"I think we may have a solution. Several other missile contractors have approached us but we have hesitated to represent them for fear of possible conflict of interest. In every case the *impact* of this multitude of changes has caused a *ripple effect* touching every part of the work. In turn the ripple effect has forced *acceleration* all over the jobs. The government has in the past recognized acceleration under certain conditions as a factor in cost for change orders. If they will accept *impact* and *ripple effect* as proper causes, and the comptroller general will approve, it may save a disaster."

So the executive VP–operations and the senior legal partner went to Washington. The chief of engineers and his general counsel listened and promised a prompt answer, since they were well aware of the situation facing a considerable number of the nation's leading contractors.

In less than two weeks the chief had approved the policy, the comptroller general had agreed, and *Impact, Ripple Effect,* and *Acceleration* had become a part of construction language. A crisis had been averted and all concerned could return to the relatively simple job of performing the work.

1-18.   Construction in Battle Zones

In 1961 the decision was made to furnish aid to the South Vietnamese in their struggle with Communist North Viet Nam. This resulted in the award by the Bureau of Yards and Docks of the Navy Department (now Facilities Engineering Command) of contracts for airfields and other facilities amounting to "about $20 million" to a joint venture formed by two large contractors.

Since plans had to be completed during construction, the work was of necessity on a cost plus fixed-fee contract. As the United States involvement grew, so did the volume of work required. The contractors, with Navy approval, added two large and competent members to the joint venture, to provide a supplemental reservoir of skilled supervisors. The work continued to expand and so did the problems. Attempting to account for government property and equipment in a combat zone became a nightmare. Often the Viet Cong would cut a section of highway between Saigon and one of the many bases, holding up convoys of cement, lumber, or what not. Frequently the Cong officer in charge of the raiders would hand the South Vietnamese driver of each truck a receipt for his confiscated load and permit the driver to proceed on his way. This system satisfied all concerned, including the accountants, and since the amount of material hijacked was relatively insignificant, no one was greatly concerned.

This was not true, however, of the raids made on central warehouse stocks by the "Arvan" forces (our allies, the South Vietnamese army). In these cases, since the Vietnamese guards were not permitted to be armed, a convoy of Arvan trucks would drive through the main gate, load up whatever they chose, and depart without even a "thank you," much less a receipt. Woe betide the guard who reported the incident together with identification of the trucks! The office of the South Viet Nam District General (who controlled security clearances for Vietnamese employees) would send a note that the "offending guard" was a "security risk" and must be discharged.

At the peak of this contract, Americans, third-country nationals, and South Vietnamese employees numbered fifty-three thousand and accomplished $56 million of installed work in one thirty-day period. As the work was being phased out and the computers and accountants had long ago caught up, the very small percentage of material, supplies, and equipment lost or unaccounted for has become a source of pride to both the Navy and the contractors.

The moral of this is: when major military type contracts are offered, they *may* be a source of considerable revenue, if not profit, but should be approached with great care by young and growing concerns, preferably in joint venture with older and more experienced partners.

1-19. Power Transmission

This is a growing field and will continue to be a source of profitable business for some time to come.

Most major as well as local and urban transmission systems have traditionally been built by electrical contractors, who in turn subcontracted line and feeder approach roads and tower foundations to heavy construction contractors. The development of longer systems, made possible by using extremely high voltage in a-c (alternating current) lines and even higher voltage in d-c (direct current) lines, has brought a greater proportion of the work within the ability of the heavy contractor. This has led to many joint-venture agreements between heavy and electrical contractors on major jobs, a trend which will no doubt continue in the future and provide much profitable work for both.

With the trend to longer lines and huge regional and international grids and the development of the "solid state" concept which requires very elaborate "step up" and "step down" facilities at the terminals, it appears certain that more and more partnerships between electrical and heavy constructors will become necessary. The high cost of these systems, whether carrying hydro, steam plant, or nuclear generated power or all three, offers a bright future for diversified heavy constructors and the larger electrical organizations.

1-20. Building Construction

In the building fields are to be found the largest number of contractors. The various classifications attract many of the smaller and medium-sized companies who tend to specialize in a particular segment of building work. Others larger and better financed may cover two or more specialties, and some of the larger diversified firms either maintain divisions or own subsidiaries who cover all the fields classified as building construction.

The general divisions of the building industry include:

1. High-rise urban building, offices, hotels, apartments, air-space structures, etc., contracted on fee or firm price; promotional high-rise where the contractor participates in equity or finances alone or with lending institutions and operates as rental property.
2. Industrial building construction, cost plus, bid or negotiated lump sum, guaranteed maximum or lease-back, construction management, or package contracts.
3. Commercial developments, shopping centers, supermarkets, recreational developments, branch banks, etc. Cost plus, firm price, or lease-back.
4. Hospitals, post offices, schools, etc. Cost plus, firm price, or lease-back.
5. Housing, single and complexes, multiunit, duplexes, condominiums, promotional housing developments, etc.

1-21. High-Rise Construction

High-rise buildings in urban areas are the traditional field of the large and medium-sized builder. The volume, although somewhat restricted in a period of high interest rates and shortage of mortgage money, is always considerable. Population grows, and even though more and more city dwellers move to the suburbs to live, ground values in the cities continue to increase, taxes and utilities go up, and the suburban dwellers continue to conduct their business and to be employed in the city. The demand for office space, apartment space, and hotel space in new, modern, high-rise buildings seems to have no end. Construction costs continue to rise, and with them leasehold prices and rentals.

Architects and engineers with great ingenuity devise panel shells of glass, stainless steel, or aluminum, all of which reduce hand labor of erection. Sectional assembly of utilities and automation of elevators help to slow, but only to slow, skyrocketing costs of construction and operation of new buildings.

The high-rise builder who depends on securing work on cost plus or firm price faces some momentous problems. To be sure, such work requires a relatively small percentage of his total contract for financing. He subcontracts perhaps as much as 90 percent of his work to specialty contractors who get paid as he is paid. He does, however, have to carry his home-office expense in any case, and, if his contract is firm price, he must also carry his job supervision and job accounting costs plus taxes and insurance. If his contract is on cost plus a fee, he may well have for home-office expense and profit 2 percent or less; if on firm price, his markup for job supervision, job accounting, and home office, plus profit and interest on borrowed money, may be from 3 to 5 percent.

No wonder, then, that even in normal times high-rise builders of national reputation would have to perform a $100 million of contract work to net $1 million or 1 percent on volume. On the basis of interest rates at 8 percent, his final earnings would be paltry indeed.

The high-rise contractor who is extremely well financed or who is able to find a partner to finance the construction and real estate is most fortunate, even though the real estate may be only the air space over an existing low-rise building or structure. He is then able to conclude leases in the building during construction, or perhaps lease it to one major tenant who will then pay for the building plus a profit over a period of years. The partner, if he finds one, may be an insurance company, a pension fund, a conglomerate, a bank holding company, or some other institution anxious to have "a piece of the action." This type of arrangement, where possible, can result in far more profit than a mere fee or 5 percent markup on cost.

1-22. Industrial Buildings

Industrial building construction is in itself a whole series of speciality fields. Some types of industrial construction are open to medium-sized contractors of local or regional scope. Large warehouses, airline mechanical bases, and even tire plants and automobile assembly plants often are awarded on competitive bids. As a general rule these and other types of plants are contracted on plans prepared by the owner's architect-engineer. Sometimes the architect-engineer or a large firm of national engineering constructors is designated as construction manager, and for a fee prepares plans and specifications, contracts the work to one or a number of different contractors, and acts as inspector as well as designer. Occasionally the owner may have his own engineering department, act as his own contractor, and subcontract out various parts of the project, coordinating the efforts of the various contractors with his staff project or construction manager forces.

The latter case, that of the owner acting as his own general contractor, is the least desirable. Often the owner's staff is lacking in broad experience in the building field, is likely to usurp the functions of the actual building contractors employed, fails to coordinate their efforts properly, signs onerous labor contracts on their behalf to gain advantage later in the owner's labor agreements in operating the plant, and generally destroys the actual builder's chances of profit and even of finishing on time. Many wiser and more experienced builders, having been burnt, refuse to submit proposals to the owner who acts as contractor-engineer and parcels out the work.

The architect-engineer or national constructor who acts as designer, inspector, and contract manager is, by the nature of his job, making every effort to get the project completed within the original estimate he furnished the owner for financing or budget purposes. Any savings he can make under the estimate improves his image in the owner's eyes and helps advertise his ability for references in securing work from other owners. His tendency is to be tough in enforcing specifications which he wrote to protect the owner and himself.

It is true, however, that the architect-engineer or national constructor construction manager is, in most cases, experienced and likely to state clearly in the specifications what the builder contractor is required to do. Also the architect-engineer or constructor serving as construction manager expects to stay in business. He knows that if he seriously hurts or breaks too many contractors, his reputation will precede him, and before long he will have difficulty in getting contractors to bid on his work. This tends to temper his treatment, and the contractor who reads the specifications and plans well and thoroughly before he bids may have a fair chance of emerging with even a little profit.

The majority of industrial companies in the more technical "process" fields—steel mills and ferrous plants, oil refineries, copper, nickel, aluminum and other nonferrous plants, cement, pulp and paper, heavy machinery, food, textile, beverage and many other technical installations—tend to employ by bidding or negotiation regional or national industrial builders with staffs skilled and oriented in the particular field.

The successful contractor is chosen because of his ability and available organization in that special type of plant, and is usually expected, starting with a flow sheet furnished by the owner (how much in terms of barrels or tons or gallons or whatever quantity per day or month or year must be produced), to design and build the plant ready to operate.

The contractor either has built a similar installation or many of them, or has a "subject matter director or specialist" in the field plus a "skills inventory" showing the names of engineering, executive, and supervisory employees in his organization with experience in similar installations.

The owner then negotiates with the chosen builder a cost-plus, firm lump-sum price, or guaranteed maximum contract, based on the contractor's original bid or estimate proposed for detailed design and construction. In practice, the cost-plus-a-fee contract is rare and firm price is uncommon because of incomplete or nonexistent plans.

The most common arrangement is the so-called "guaranteed max." This presupposes that the contractor has previous experience (or has a subject matter specialist on his staff with such experience) in the design and building of a similar or identical installation.

Should the design and capacity be identical, the procedure is fairly simple. The contractor takes the proper file which contains plans, equipment and material costs, labor rates, labor efficiency, data on precipitation and temperature, and finally total cost. He adjusts these figures for conditions at the new site—weather, labor supply and rates, labor efficiency, material and installed equipment, etc. This, when contingency and fee are added, then becomes the "guaranteed max." Should the cost exceed this figure, the excess is for the account of the contractor. On the other hand, if the costs are less than the guaranteed max (minus the fee which the contractor collects), the balance of saving is shared by the owner and the contractor, perhaps 75 percent to the owner and 25 percent to the contractor, or in some cases fifty-fifty.

Should the new plant be materially different due to process changes, size and capacity, or other factors, the contractor must weigh these changes and evaluate them as additions to or subtractions from the original plan and *then* weigh all the factors up or down as when he adjusted the cost in the case of an identical plan. Obviously, under the "guaranteed max," the contractor with previous successful experience in a given specialty field has a built-in advantage over the newcomer who lacks the previous experience.

In one case, at least, a national contractor considering the possibility of getting into the oil refinery business did a careful analysis of the cost of acquiring the personnel and designing plans to let him compete in this field, which did not appear to be overcrowded. The answer after six months of investigation was: "Yes, if you want to spend three and a half million dollars just to get in position to submit a proposal."

There are, however, other roads to industrial construction that are not so badly blocked. One is the lease-back, if the contractor can find a financial partner willing to put up the money for a half interest in the action. In this case the contractor and his financial partner build a plant by prearrangement with the industrial operator, who in turn agrees to lease for a period of years sufficient to write off the plant. The lease may carry options to extend a stated number of years or perhaps an agreed purchase price at the end of the period.

Well-financed industries able to use their own surplus funds for new plant construction on occasion employ a national firm as construction manager-engineer, or they may make a package deal with an architect-engineer, who in turn contracts all or major portions of the work to local or area contractors, thereby giving them a chance to gain know-how in the field.

1-23. Commercial Developments

Shopping centers, supermarkets, recreational facilities, branch banks, etc., are the bread and meat of the small to medium-sized local or regional contractor.

These developments are normally rather simple construction, are usually (with the exception of recreational developments) awarded in small to medium-sized contracts, and in most cases do not attract outside national contractors. They are usually let on a lump-sum or unit-price basis by architects for the owners; hence plans are complete or nearly so. Since chain stores are the most likely to be the landowners, and since they tend to encourage chain stores in other lines to purchase and build on portions not needed by the owner for building or parking and thus develop a "mall" or shopping center, the contractor who is successful in getting an award on the first job may very well be in a position to build a major portion or the entire complex.

Branch banks frequently establish offices at developing shopping centers, and the contractor on the grounds has a chance to bid on this work also. The branch bank construction, although tending to be somewhat more ornate than a grocery or drugstore and of heavier construction—concrete or brick—particularly in the vault area, is still rather straightforward construction. The contractor who succeeds in getting this type of work may, if he performs successfully, look forward to future work in other areas of the state for the same or other expanding banks. Occasionally this type of work may be awarded on a cost-plus basis to contractors who have performed well in the past.

More frequently the market or drug or department store chain may elect to go the lease-back route and often makes its own arrangements for financing through pension funds or insurance companies, who in turn own the facility and "lease-back" to the operator. In such a case the contractor may find himself with a contract paid for by the owner but administered by the lessee-operator or his architect. This should not concern him too much, as he will no doubt receive the same treatment as he would if the lessee-operator had remained the owner.

1-24. Recreational Developments

These are usually found in country or mountain areas, frequently remote from large centers, on lakes or streams, and occasionally on seacoasts or islands.

They are traditionally promotions of sizable investment complexes and are prone to be started by construction of roads, water systems, and sewerage disposal systems. They are usually large in size, in the multimillion dollar bracket, and tend to attract major national diversified builders who may well be themselves the promoters and owners.

As the development proceeds, if successful, there will be increments including not only housing but apartments, hotels, golf clubs, and the like. It is in this latter stage that small and medium-sized regional contractors have a chance to participate, usually in housing or apartment work let by the architect for the purchasers of lots and small tracts. This is attractive work and may keep the regional contractor busy for some years.

1-25. Public Buildings

Builders of hospitals, post offices, and schools tend with some exceptions to be fairly large regional or national builders. They are in most cases specialists in these lines, and the field is rarely invaded by the large diversified constructors. On occasion, however, smaller school buildings and now and then a smaller post office on straight contract or lease-back goes to the local builder who has been constructing supermarkets and perhaps apartments. The reasons are clear. There are not enough hospitals or post offices or even schools built in a given area to provide steady volume for the local builders.

The architectural and equipment requirements tend to be somewhat difficult and costly. The manufacturers of the installed equipment and the subcontract trades who follow the field are also specialists, and lacking sufficient volume in one area they follow where the work is.

Thus a contractor as well as the architect who specializes in hospitals keeps up on the latest developments in medical equipment, which is expensive, difficult to install, and frequently requires that special areas or space be provided.

Too often a new or improved item of equipment is developed during the construction of the building and insisted upon by the doctors on the board of the hospital. The experienced architect and builder have been through this before and have allowed for it in the figure so that the always slender margin between the bid and the funds is not too badly disturbed. Had the contractor been a neophyte in the field and the doctors intransigent, a long delay in the work and high additional cost would have resulted.

A post office in a sense has a similar problem. Should the experienced contractor have the job on a firm-price basis, he makes every effort to limit changes in design, equipment, or both. If he is successful and if visiting architects from the Post Office Department are not too ingenious in inventing demands, he may realize his anticipated profit.

If the post office happens to be a lease-back and the usual clause requires the lessor to repair or replace any part of the building or grounds that shows any fault or failure within a year or more of operation, the experienced post office contractor will have someone nearby during the guaranty period to explain, excuse, and sometimes fix minor imperfections that would probably never have shown up if the facility had been properly maintained by the department.

A school building or a school complex also needs experienced hands and much watching. School boards, usually made up of sincere, well-meaning people with little practical knowledge of what goes into a schoolhouse, are prone to make new demands and changes at every meeting of the board. An experienced builder who knows what to look for will be available and ready with logical and cogent answers to all possible demands and questions as well as complaints. He must always remember that this school is not only the first but the only one that *this* Board will ever contract. To them it is the most important one ever built, and each member of the board is eager to inject his or her ideas of improvement in design (without, of course, adding to the cost of the contract).

Fortunate indeed is the builder when the board has selected an architect with wide experience and competence in this field. He can and will protect the builder from most of the complaints, and will ward off most suggested changes—particularly those that are not only expensive and wasteful but also can disrupt the builder's schedule to the point of complete frustration.

1-26. Residential Construction

Finally, the backbone of the building construction profession is, and always will be, the house builder. He starts out with the smallest unit in the building field, a house. He may have learned his trade working for a house builder, through carpenter to foreman to superintendent. He may have been an accountant, an architect, or an engineer. He may even have been a baseball player who retired early with a little money saved.

Traditionally the house builder acquires a lot, employs an architect (unless he is one), and builds a house. If he is successful he sells the house at a profit and becomes a building contractor. He may then team up with the architect and build the houses he designs, sometimes for a fee, sometimes for a firm price. If still successful, he purchases enough land for a subdivision and with his architect friend designs houses and builds them.

When prices are going up the subdivision builder reaps a harvest, and if his judgment of land is good and his subdivisions sell, he is in a position to take the mortgages to the bank and collect his money, or he may establish relations with a bank or mortgage loan company that will lend sufficient money direct to the purchaser so that with a down payment he can pay off the contractor.

The contractor builder, his credit established, is now able with his own funds and bank credit to branch out very rapidly. He may elect to stay with housing developments in new areas surrounding the city. He may go into more expensive residential areas that are centrally located, build multiunit apartments and condominiums (where the buyer purchases a floor or apartment in a multistory building and becomes a stockholder or owner of his part, with others in turn owning their floors or apartments).

As long as demand for housing keeps up, interest rates remain stable, and the builder has used reasonably good judgment in his selection of areas to develop, he prospers and, depending on his courage and timing, becomes wealthy.

Should the builder and developer, however, get too extended, prices and building costs rise too fast, interest rates go up rapidly, and credit become scarce, he may suddenly find that his loans are eating him up. Costs of construction and interest have dried up his market.

The prospective purchaser can no longer afford the down payment or the high interest. If he buys at all he will look for an older house with a mortgage at a lower interest rate. The speculative builder then is overextended. He may have to sacrifice houses and apartments at a loss, saving what land and assets he can manage until the turn comes, credit eases, and money becomes cheaper again.

1-27. Summary of Requirements

Contracting is and always will be a demanding and rewarding but an unforgiving profession. It demands experience, sometimes acquired at heartbreaking cost. It demands the full attention of an alert and imaginative mind. It demands the ability to choose and train competent associates and in turn, with them, to build a sound and loyal organization, for no man can succeed in this profession alone. It demands all these, plus a sense of timing and the ability to weigh each opportunity and promptly decide which one has more chance of gaining than of losing.

If all these demands are met and none are faulted, the rewards are great and the satisfaction of accomplishment most gratifying.

2 MANAGEMENT FUNCTIONS, PROBLEMS, AND TYPES OF ORGANIZATION

J. B. BONNY

President and Chairman (Retired)
Morrison-Knudsen Company, Inc.
Boise, Idaho

AFTER DECIDING on the type of construction work to be pursued, management engages in its prime functions: the problems of getting work, financing the work until reimbursed, and performing the work properly and on time.

2-1. Sales Functions

Getting the work (sales) is the first responsibility of the president, manager, owner of the company, or senior partner. In larger companies division vice-presidents or district managers can be added to the list, along with the subject matter director, chief engineer, or vice-president–engineering, and perhaps the division or district engineer.

As a fundamental, unless the job is a bid to a city, county, state, or the federal government and will be awarded on the low figure, sales effort will be required. Even in those cases many states require prequalification of all contractors wishing to bid, and a number of states after prequalification "rate" the contractors according to ability and finance as to size of jobs they may bid and amount of volume they may have under contract for the state at any one time. In practically all other cases, sustained sales effort is required and must be made by top people in the company.

Employment of salesmen as such is usually futile. The decision on award of any sizable job is normally made by the president or chairman, whether it be a railroad, a power company, or an industrial concern, and he usually wants to settle the contract with the head or at least with a top executive of the contracting firm. Designating a salesman as "assistant to the president" or "assistant to the vice-president"—or any other high sounding but meaningless title—seldom helps to get the work. The salesman returns home without seeing the right man and some other contractor with better representation gets the job.

2-2. Contract Negotiation

Contracts are secured by negotiating with the owner, the chairman or president, or in some cases the chief engineer. The contractor who gets shunted off to the "purchasing agent" is seldom in much danger of getting any work.

The contractor, in determining the approach to the owner, must know what kind of proposal to present. Frequently a preliminary meeting with the owner or the top officer whom he has designated to receive proposals, analyze them, and recommend award, will pay off. The contractor, if he is acute and a good listener, should be able to come away from this meeting with a fair knowledge of the desired date of delivery of the completed project or readiness for constructive possession and operation. The contractor will also have found out the type of proposal the owner may wish presented to him, and how soon he wishes to receive it.

The owner, if he maintains a competent engineering department or has employed outside architect engineers, may issue a complete set of plans and specifications which will clearly state the terms of the proposal, in this case usually firm price, in lump sum or units. The contractor must comply, presenting the proposal at the time and in accordance with the terms set forth in the proposal form.

If the owner is one for whom the contractor has done work in the past, the contractor will know the owner's preferences and be in a better position to judge the type of proposal most likely to be acceptable.

The owner, if he is an old client and there is strong mutual trust, may well elect to have a straight cost-plus contract or even a year-to-year arrangement with an understood fee. Such an arrangement benefits both parties. In many cases the contractor furnishes design and testing engineers to supplement or replace the owner's engineers, even furnishing the owner with flow and production sheets and estimates for financing and budgetary purposes. This arrangement becomes a true package deal and, although the fees are relatively low, the continuity of work is attractive to the contractor, and the savings in overhead to the owner plus accurate foreknowledge of budget cost justify the practice.

The contractor may deduce that the prospective client will consider a target estimate type of contract, in which, as plans advance, an estimated cost is agreed upon plus an agreed fee. Overruns are shared in by the contractor until his fee is gone, and then bare costs are borne by the owner while savings are shared in by the contractor, until a maximum fee—perhaps 10 or 12 percent—is reached, after which savings accrue to the owner.

Frequently the contractor selling or bidding the job finds that the owner prefers a "guaranteed max" type of contract, particularly where similar or identical plants have been built for the same owner. This can be a remunerative type of contract if all goes well, but highly dangerous if troubles develop. Inasmuch as the contractor guarantees the top cost, labor troubles, strikes, and rapid inflation may well not only destroy his fee but result in a disastrous loss. Since he shares savings with the owner and the maximum fee is limited, he is in a sense betting against himself.

In many cases the contractor may be able to sell a job otherwise unavailable if he can contrive a "lease-back" arrangement. In order to accomplish this he must have sources of financing the plant or commercial building during construction and also for the time during the lease period when the lease has not yet paid for the building. In times of ample available money and relatively low interest rates, it is not too difficult to get the funds if the contractor owns the land and his credit is good. When interest rates are high and money tight, he must find a lender who has the funds and will take an equity position with him in the project. This may cut his profit in half but still make it possible to "sell" a satisfactory job.

2-3. Bidding Practices

Contrary to the practice in private industry, most government as well as state, county, and city projects are bid, and these, too, come under the prime function of top management. Usually such bids are opened at a designated time and in public.

Occasionally a government body will award to selected contractors, usually in joint venture, a contract for military construction outside the United States, or for Atomic Energy Commission or other "restricted" or "secret" work within the country, on "cost plus a fixed fee." The fee is very small and decreases with the size of the job percentage wise. This is not to the large national contractor "good work," and even with conversion of some military construction to "cost plus an incentive fee" where the contractor's performance is "rated" periodically and he may receive "points" (known to the trade as "Brownie points") for exceptional performance, these entitle him to only a small bonus even if he scores 95 out of a possible 100 percent. Thus such work remains a prestige effort rather than a highly profitable operation.

2-4. Liaison with Owners and Their Representatives

Whatever the work, the prime "selling" function of management still carries with it the necessary ability and organization to make accurate and rapid estimates of cost on all firm-price jobs and jobs where plans are complete. Even where plans are sketchy or nonexistent, the contractor must be prepared to make accurate "educated guesses" to help him sell target estimate, cost plus, and even guaranteed maximum contracts.

Selling the contract does not, however, stop with the signing of it. Top management must follow through by regular contacts with the owner and his staff people. The relationship created by the award of the project must be developed and improved. The contractor must at all times keep in mind that the owner, in selecting him, expects and will demand the personal attention of the contractor's management as assurance that every effort is being made to prosecute the work so it will be delivered on time and at the lowest possible cost.

To the owner the plant or facility he has contracted for is of the greatest possible importance. He must justify to himself and to his board the decision to invest in it and his selection of the contractor. If the work is delayed for any reason or the costs overrun, it is automatically the contractor's fault—unless the contractor has informed the owner *far enough in advance* of circumstances beyond his control which may lead to a delay or an overrun in cost or both.

This means that the contractor cannot and must not neglect the follow-through of continually "selling" himself and his organization to his clients. The traditional use of entertainment may sometimes aid in cementing a cordial relationship—*if* everything continues to go well on the job. Let the job go sour for any reason (even when obviously due to the fault of the owner's staff) and it is the contractor's neck, unless he has properly, confidentially, and *personally* alerted the owner far enough in advance about what to expect. Frequently such contact by the contractor with the owner will eliminate the trouble; more often the early warning to the owner will prevent severe criticism and pave the way for profitable work from the same source in the future.

The most practical approach is keeping the communication lines open and cordial at all times. The contractor or his chosen top officer can then communicate by mail or word of mouth directly with the owner or his delegated top officer on all matters of importance affecting the cost, time of completion, or quality of the work. The district or division manager, with the help of the contractor's chief engineer or his district engineer, will keep the lines open with the next echelon in the owner's organization on general contract matters. At the same time he must cover the project in the field frequently with his project manager, sitting in with the project manager and the owner's counterpart, his project manager or project chief engineer, on day-to-day matters as well as those that, by their nature, will require determination by higher authority later.

The owner, on the other hand, will be constantly informed, verbally and by written reports from his staff, how the job is going. This information may be and frequently is at variance with the contractor's, which will usually tend to explain that any difficulties that may arise are caused by circumstances beyond the contractor's control or by mistakes of the owner's staff.

The intelligent contractor must first insist that he has at

all times prompt current reports, verbally and in writing, of the true situation existing on the job and the earliest possible advance warning of trouble spots. He in turn must keep his district or project manager fully informed at all times of his dealings with the owner, and of the owner's position and attitude in the light of his own reports from his project people.

Frequently unpleasant and even disastrous situations have been averted because the contractor had received reports of problem situations on the job and discussed them fully and frankly with the owner *before* the owner was aware of them. It is imperative that the contractor impress on his district and field staff that all reports must be truthful, accurate, and unbiased. He or his top operating officer must, by personal visits to the job, verify their accuracy.

If the problems are clearly beyond the control of the project management, whether they be area-wide strikes, unpredictable or unseasonable crippling weather, failure of the owner's engineer to deliver timely and workable plans, or other unavoidable obstacles, the owner must be promptly and fully advised, with proven reasons and probable effects on the overall project. Conversely, should the project management be covering up its own mistakes and lying to protect itself, the only possible course is to replace the project management promptly and immediately and fully advise the owner of the action taken and the reasons. The owner may be seriously concerned and even angry— but he *will have* faith in the contractor's honesty and *will* believe him in the future.

From time to time a personal enmity may develop between the owner's representative and the project manager. This may be caused by lack of cooperation on the part of either one, but more frequently by both parties. The contractor, if he has sold himself well with the owner and is trusted, can by offering to transfer *his* project manager or other responsible employee, succeed in getting the owner to also move *his* representative or to so instruct him that the problem will be eliminated in the future. Thus, getting the work (and keeping it) is and always will be a continuing twenty-four hour a day job for the contractor and his top organization people.

2-5. Construction Financing

Financing the work until paid for is the second prime responsibility of top management. Seldom do current payments (estimates) from the owner, after withholding a percentage to guarantee completion (usually 10 percent), provide sufficient funds to pay the operating costs and leave enough margin to pay for equipment if purchased for cash, or on equipment purchase bank loans or possibly equipment dealers' time payments, plus the contractor's general overhead.

The contractor must keep his own capital in as highly liquid state as possible in order to cover his overhead, stand behind any losing jobs, and maintain compensating balances with his banks. This subject is covered fully in other chapters but should be discussed briefly here.

Let us assume a small contractor who has a cash working capital of $100,000 plus nonliquid assets in equipment, furniture and fixtures, and possibly some real estate and a small office building valued at another $100,000, all clear or having a value of that amount in excess of mortgage.

He has an opportunity to get a contract with a value of 1 million dollars which he believes will net him a profit after all costs but before taxes of $100,000, provided he can finance and bond the job. The construction time is one year.

He first goes to his bank, furnishes his latest financial statement (if it is not already on file), and establishes a line of credit or requests a specific credit for use on the particular job if secured (stating his requirements). In this latter case (preferred by many banks) a schedule of repayment is agreed upon and the money is available if the contractor is successful in his proposal. When the contractor is established and well known to the bank, he may already have a line of credit and may borrow within that line or allocate a portion or all of it for use on the specific project *after* he has been awarded the work.

The procedure of borrowing only on specific projects (covered in detail elsewhere) is very sound. The bank knows it will be kept currently advised of the progress and financial status of the job, and the contractor's cash is conserved for overhead, compensating balances on deposit in the bank or banks that loan on the job, and to maintain adequate reserves to pick up a loss if it occurs.

After arranging his financing, the contractor next goes to his bonding company. Presumably he already has on file his current financial statement and experience record. It may well be that he has a partner or backer who, for a participation in the action (30 to 50 percent of the profits), is willing to advance a considerable portion or all of the capital required above bank loans. If a full partner, he must also sign on the notes at the bank jointly and severally and likewise at the bonding company, and he must assume full joint and several obligations for all and any commitments his partner may make. In this case he probably acquires a full 50 percent interest in the partnership. Alternately, the backer may be willing to commit for only one job at a time, in which case the arrangement becomes a limited partnership or joint venture, and the joint-venture partner assumes responsibility for costs incurred on the one job only and not for other commitments the contractor may have made elsewhere. In this situation also the backer probably will demand and receive a full 50 percent of profits on the single job in which he has an interest. Finally the backer may be unwilling to risk more than the capital he furnishes for the particular job. In this circumstance he may advance all or part of the required funds, in return accepting an agreement of profit sharing plus perhaps notes of the contractor which will be subordinated, or follow after all obligations directly stemming from the one particular job, including obligations to the bank and the bonding company.

The bonding company, having been apprised of the entire circumstances, may then decide that the available money is sufficient to operate the job as well as allow funds in case of loss to pay lienable debts that may arise against the work.

It should be made clear here that the bank, in loaning the money, may advance it on unsecured notes or may demand that securities, equipment, or real estate be pledged to guarantee the loan. In case of failure or bankruptcy of the contractor, the bank may foreclose on the pledged collateral or, if there is no collateral, must take its place with all the other creditors after the lienable obligations have been paid.

The surety company, on the other hand, guarantees to the owner, public or private, that the job will be completed and that all lienable obligations will be paid. Lien laws vary considerably in different states, however. In general, labor, fuel, explosives supplies and parts, and materials used up in the work, embedded in or becoming part of the work, are considered lienable obligations against the property and thus charges against the job if not paid by completion. The usual practice in most states is to allow a period of thirty or perhaps sixty days for lien creditors to file against the work. It is these obligations that the bonding company assumes plus guaranteeing completion to the owner.

The contractor likewise, in most states, if not paid in full (excepting disputed claims) in a reasonable time after completion of the work may also file a lien against the project. This type of lien is *not* guaranteed by the bonding company.

The surety company, having satisfied itself that the contractor's available funds are sufficient to perform the proposed work, issues a bid bond to be presented with the contractor's bid proposal. This is a statement but not an outright guarantee that the surety company expects to write the performance and payment bond if the work is awarded to the contractor. It does, however, guarantee to pay to the owner a certain sum of money (usually stipulated by the owner in advance) in case the contractor is awarded the work and refuses to sign the contract and furnish the required performance and payment bond. Even the bid bond, however, in some states is not inflexible. In some states and frequently in federal bid proposals, if the contractor is able to reasonably prove a serious error in the typing or computation of his bid, he may be permitted to correct it to the proper figure (if still low) or be excused from penalty and the work awarded to others.

The surety charge for the bid bond is low. When the owner demands a bank guarantee or cashier's check (most common on foreign work), the bank issues a bid check, not expected to be cashed, also at a nominal charge.

Large and well-established contractors usually have credit lines or job credit limits set up in several major banks and often use different banks on a geographical basis for job loans in their particular area. This custom greatly simplifies the entire procedure, and the contractor's financial vice-president or treasurer may make the arrangements for financing a job when awarded with a single phone call or letter, followed by sending the bank signed notes for various amounts as needed. The same procedure is followed with bid and surety bonds. The surety office near the contractor's head office or the district office preparing the bid will have authority to issue bid bonds, and, if successful, surety bonds on any jobs bid within the contractor's normal limits.

On very large projects, domestic or foreign, the contractor's financial and bonding arrangements require special procedures. These large jobs are normally joint ventures. The contracting group may occasionally incorporate, as did Six Companies, Inc., on Hoover Dam. More frequently and recently the joint venture is used and a fictitious name chosen relating to the name or location of the project, such as "Yellowstone River Constructors."

The most common procedure is for a contractor wishing to sponsor the job to contact one or more contractors with whom he has had previous joint ventures and in whom he has confidence. It may be that one or more other contractors are also planning on sponsoring the particular project and are putting together a joint venture even though the bid may be several months off. Our contractor in this case may decide to accept a participation in the proposed venture under another's sponsorship. The procedures in any case are identical up to the point where the work is awarded and the organization, methods, plant, and equipment are decided.

Let us assume the job is in the hundred million dollar bracket. Our contractor has a 25 percent interest, is making a complete job estimate, and the sponsor has 40 percent. He calls his bonding company and assures himself that the bond will be granted and that the bonding company is satisfied with the financial stability and competence of the sponsor and all others in the proposed group.

He inquires as to the amount of capital the bonding company will require deposited in the joint-venture bank for operating funds. He also inquires whether the bonding company will permit each contractor to sign the bond for his percentage of the contract or whether they must all sign joint and several on the bond.

He is well aware that this technicality is not as important as it sounds, since in any case each partner must sign the *contract* joint and several. Still there is great advantage in keeping his liability to the bonding company as small as possible.

Finally the sponsor is advised of the terms of the bond as far as our contractor's bonding company is concerned. The sponsor, having checked with the other partners, advises his bonding company to contact all the others (there may be duplications) and be prepared to sign the bid bond at the proper time and place.

The partners each proceed with a comprehensive bid estimate, usually prepared in a form which can readily be compared with the sponsor's estimate and readily understood by the principals of the joint venture. The principals are men of experience, but usually they have not gone through the estimates prepared by the joint-venture partners' engineers and reconciled or noted differences. It is most important, therefore, even though the bid may be partially or wholly lump sum, that all quantities in both direct cost and total cost be expressed in such units as cubic yards, square yards, pounds or tons, etc., before the partners attempt to reconcile total cost and agree on contingency and profit. The failure of estimating engineers to furnish the principals with clear figures in units greatly hampers the principals in attempting to agree on proper pricing, and has often resulted in disasters because they did not understand the lump-sum figures prepared by their own engineers.

2-6. Accounting

Management must have the ability and (depending on the size of the company) the organization to keep proper accounts and current accurate costs, as well as to maintain adequate reserves for taxes, Social Security, and other employee benefits, and to make sound and timely purchases.

Proper accounts in the smaller companies, operating in a local area and performing two or three jobs at the same time, are relatively simple and may require only a time-keeper to cover the jobs and turn in the payrolls to the home-office chief clerk, who may have an assistant or bookkeeper and perhaps a stenographer. The owner or head of the company, when he is not out selling new work, spends his time on the jobs with his superintendents or foremen in charge; hence he is fully familiar with the job conditions and is acquainted with each workman and his ability. He thus is constantly aware of what is going on in the field as well as in the office and, as long as he does not attempt to expand too fast, can make a comfortable living in his chosen field.

2-7. Effects of Job Size

As the jobs become larger, more numerous, and more spread out, the contractor must of necessity organize well or he cannot survive. As he grows he must delegate au-

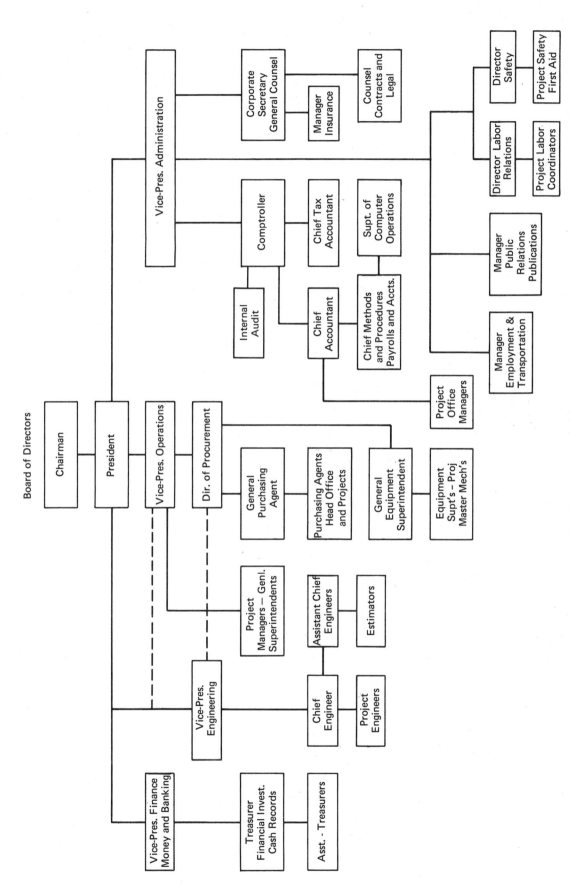

Figure 2-7-1 Organization chart: Direct reporting or vertical organization

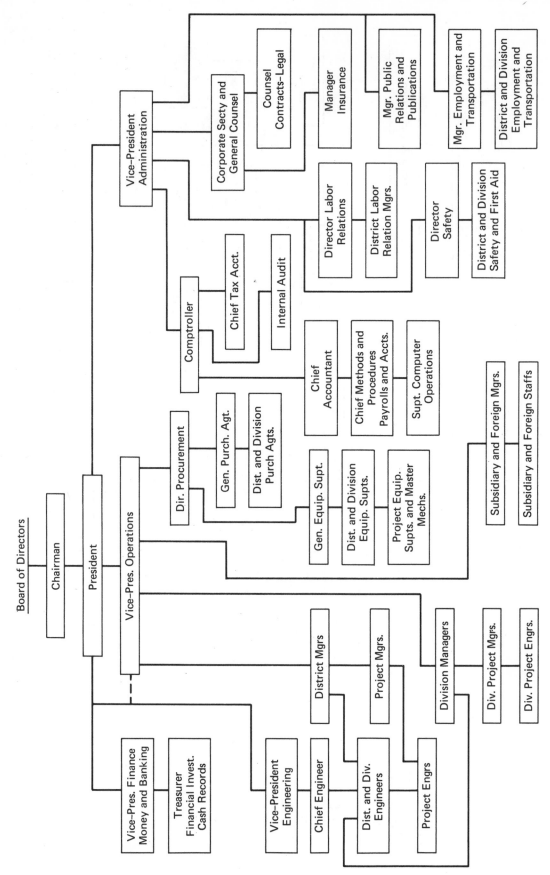

Figure 2-7-2 Organization chart: Indirect reporting or horizontal organization

thority more and more. The daily visit of a timekeeper is no longer adequate. Foremen become superintendents and superintendents become project managers. A project office manager becomes necessary and soon is business manager with a staff of bookkeepers, timekeepers, and stenographers who soon become secretaries. Major purchases, while still the function of the head office, must be supplemented for local purchases by a project purchasing agent. Safety is headed by a safety engineer who has a first-aid station. Labor relations come under a separate department and visits from the labor relations director may be augmented by a project labor relations representative. A master mechanic becomes mechanical superintendent with a staff, and if the project is remote and requires boarding and feeding all or part of the crew, there will be a steward, chef and a staff of cooks, waiters, and dishwaters ("pearl divers"), and janitors ("bull cooks") to take care of the camp and make the workmen's beds.

The head office, meanwhile, has mushroomed. The contractor now has a board of directors, a vice-president–engineering and chief engineer with a staff of assistants and estimators. The chief clerk is now financial vice-president and there is a corporate secretary and chief counsel as well as a treasurer and comptroller. Since there are now an executive vice-president–operations and an executive vice-president–administration, the functions are divided between them. Reporting by functions are district or division managers and their staffs, director of labor relations, director of employment, director of procurement, director of safety, as well as the manager of transportation, chief equipment superintendent, master mechanic, and their staffs.

Many payroll and accounting functions as well as engineering controls and engineering design techniques are now guided and processed by network equipment and procedures and operated by a new and different kind of technician (see Chapters 17 and 18).

Does our contractor, now president and chairman of the board, have anything to do? Indeed he does. He worries about money—and sells. He plugs holes in the organization as they occur—and sells. He corrects others' mistakes for which he alone is responsible—and sells. He considers new fields and new methods, explores new equipment—and sells. In his spare time he just worries—and sells! (See Figures 2-7-1 and 2-7-2.)

2-8. Job Supervision

Performing the work properly and on time is always the most important function of management. If the contractor cannot meet the schedule and control costs within his estimates, all his efforts and all his controls and all his organization will amount to nothing. He will run out of money and go broke or, at best, his reputation having preceded him, he will find it impossible to get new work at prices for which he can perform it.

The contractor, whether he be small, medium sized, or large, must have adequate job supervision. The project supervisor in smaller companies may be the contractor himself or a partner. As the organization grows and authority must be delegated, the project supervisor will become the project manager or, in cases of large and complex jobs, perhaps a vice-president, who in turn reports to a district or division manager (who is also a senior vice-president and perhaps a member of the board) or to the executive vice-president direct.

Whatever the type of organization or size of the concern, the project supervisor must be experienced and proficient in the particular field of work to be performed; he must have authority and competence to make sound decisions on the spot, subject to review by headquarters when necessary or in some cases advance approval by telephone. The competent project supervisor may have acquired his know-how in a number of different ways.

In earlier years a man became a foreman in his craft with only two qualifications: he had to be able to do more work than any man in the crew and he had to be able to whip any man in it. A carpenter, if he could learn to read plans, might be promoted to foreman, carpenter superintendent, and finally general superintendent. A horse-drawn scraper driver ("mule skinner") or later a caterpillar operator ("cat skinner") could rise from the ranks to grade foreman and grading superintendent or higher. In tunnels a driller, mucking machine operator, powderman, or miner could aspire to shift boss ("shifter"), assistant superintendent ("walker") or finally superintendent, and so on through various crafts and types of work.

These supervisors who came up through the ranks with little or no formal education had, however, one great advantage: they knew *how* to do the work and *how* to get it done. They were "ramrods" and drove their crews, but all had fixed preferences both in methods and equipment and stubbornly resisted any change in either. Only a few of them are still active and all are past middle age. They left performance records difficult to meet and got production from their men far greater than is possible to achieve with the highly paid, pampered craftsmen of today.

The modern superintendent and project manager is usually college bred and frequently holds degrees in engineering, business administration, and sometimes law, or, less frequently, in construction management from one of the great universities giving such courses. He is knowledgeable and has a broad understanding of the problems and their solutions inherent in the profession of his choice.

Today's project supervisor must have adequate engineering assistance, and he must have the ability to initiate and understand the making of engineering costs and progress charts and to know that they are being kept accurately and up to date, thus insuring current knowledge of the status of the job.

Outrageously high labor rates that are constantly climbing higher have long since passed the point of diminishing returns. The contractor and his project managers and all his staff must of sheer necessity exert every bit of ingenuity and exhaust every possible plan, using every feasible method and innovation, mechanical or otherwise, to cut down the number of craftsmen they employ. A piece of equipment or a tool, however expensive, becomes an economic necessity if it promises to replace two, three, or a dozen men. The tool hopefully will work; the men often won't.

The project manager or project vice-president must spend more and more of his time in his office, analyzing his engineering cost and progress reports and rescheduling his work to take advantage of every possible saving in labor. This causes the project supervisor to depend more and more on his staff, his general superintendent and the shift and crafts superintendents, his project engineer, his business manager, and the equipment superintendent.

2-9. Records and Reports

Adequate job accounting and constant cooperation between engineering and accounting in prompt and accurate costing of all items of work become of extreme urgency.

The basic index of the job status is the monthly cost and

revenue statement. This report is compiled by the accounting department from progress figures furnished by the engineers and from the monthly estimate for payment by the owner's engineers. It shows the amount earned during the previous month and the amount of cash due after withholding the retained percentage and any other deductions, such as materials furnished by the owner, etc. In some cases the amount earned is prepared and billed by the contractor's engineer and approved for payment by the owner's engineer. Sometimes a figure for "company estimate" may be added where preliminary plant or work has been started but has not advanced sufficiently for the owner's engineer to include quantities. Occasionally a figure for "overpayment on incomplete work" may appear if the owner's engineer has made advance partial payment on incomplete work not fully earned.

This "cost and revenue summary" is prepared as soon as labor and material costs and any other job expenditures can be ascertained. In any case it is furnished to the project supervisor, business manager, and project engineer with copies for the home-office executive heads and the comptroller's office, financial vice-president, engineering vice-president, and chief engineer. Every effort is made to have this in the hands of all concerned not later than the tenth of the month following date of estimate.

In the same period but usually a few days later an engineering report is also prepared by the project engineer's staff, with cooperation from the accounting department, and is sent out to all parties who receive the cost and revenue summary. This second report gives in detail the unit costs of each item of work performed in the past month and to date.

It is quite true that even the most carefully prepared cost and revenue summary, together with the engineering cost report, cannot state with complete accuracy the exact condition on the job *for that month*. However, these reports after two or three months do establish a reliable *trend*.

The president or chairman and others in the home office, as well as the district or division management and project executives, analyze the reports and carefully note the increase or decrease in home office *investment* in the project compared to the original control estimate and schedule of investment. They can then often come up with a suprisingly accurate forecast of what the job will make or lose at completion.

There *are* exceptions, of course, usually adverse. A strike in the area, for example, midway of the job, an unusually severe winter, refusal of the owner to pay for changes verbally approved by his representative and entered as additional "company estimate" in revenue, or improper allowance for early pay items unbalanced upward and corresponding reductions in later pay items, or many other reasons, all may throw the best forecast awry. However, in a large company operating a considerable number of projects simultaneously, the forecasts will tend to balance out fairly well, so that by mid-year, barring a catastrophe, a conservative estimate prepared by the controller or financial vice-president and carefully reviewed by top management will closely predict the fiscal or calendar year end results.

There are other reports (covered in detail in other chapters) which should be mentioned here. Daily bank balances, together with a record of borrowings, are prepared by the financial vice-president for review by the president and chairman. These, too, give management a quick and fairly reliable insight to trends.

There are a number of other reports (also discussed elsewhere in detail) that serve specific purposes both at the job site and the home office. Daily excavation yardage, daily tunnel footage, reports on daily yardage in rock fills or breakwaters, concrete pours, etc., all have their place, as do daily or weekly status reports on large building jobs. There is always a tendency in the field, as well as in district offices and headquarters, to initiate specific reports for purposes that appear necessary at the time, but they are not stopped when the need is past. This tends to increase the number of reports beyond good reason and build up whole series of reports, being prepared at considerable expense, that go the rounds and are initialed by somebody's secretary but read by nobody. A sound procedure is to have management review a list of all reports periodically and cancel all those that have become obsolete or redundant.

2-10. Labor Relations

The subject of labor relations, always sensitive, must be handled by the project supervisor and his labor-relations manager, as well as by the director of labor relations and his staff at the home office. The purpose is to maintain peace without giving in to unreasonable demands for featherbedding, fringe benefits, and even outright pay raises contrary to such union contracts as may be in force in the area at the time. The project office and the home office should maintain constant contact with each other so they are mutually completely informed and present a united front on policy as well as action.

It must be remembered that union officials and representatives outnumber the contractor's labor-relations people by at least twenty to one, and that they are spending full time, every day, to secure new concessions from the contractor without regard to their merit or justification. The subject of labor relations is covered in detail in Chapter 24 and will only be commented on in this chapter for background.

The contractor who resides in an area where the construction unions are relatively weak, may start out by operating "open shop" rather than signing a union contract. This may work very well up to the time when he begins to spread out and work regionally or nationally. At this time he will find himself dealing with the building trades or heavy construction unions and will be bound either by contracts which he negotiates with them or regional or local contracts signed by the appropriate Associated General Contractors (AGC) chapter or other contractors' association.

A considerable number of larger contractors working regionally or nationally, particularly in building and large industrial plant engineering and construction, have made a practice of signing international agreements direct with the building trades unions. These agreements vary, but essentially the contractor, in return for a "no strike" clause, gives up his bargaining rights and agrees in advance to pay and work under whatever wages and conditions prevail in the area or are arrived at by negotiation or strikes. These "International Agreements" are patently in favor of the unions and have contributed to a degree in the inflationary and successful demands of the construction workers' unions.

More contractors have elected to joint the AGC or various regional contractors' associations. Gradually the AGC local and regional chapters (but not the National Chapter) began electing from its members labor negotiating committees who would deal directly with the Heavy

Construction and Building Trades Unions in an effort to arrive at mutually fair contracts for a duration of one, two, or three years. Until the 1960's the practice was that tentative agreements were worked out and had to be approved by the local union members and, in some cases, by the individual contractors' association members and, in others, by the majority of the members. This practice worked quite well for some time and the rate of increase of wages and fringes was maintained fairly well within the growth of productivity as well as the overall inflationary period.

Through the 1960's and to the present time, the relation between the workmen's productive capacity and what he earned was lost.

A number of AGC chapters passed regulations that whatever came out of the labor committee's negotiations was binding on all the members. Hence the unions might approve a contract subject to ratification by a majority of its members, while the contractors were automatically bound by the offer of their labor committee. This meant that the unions could and did renege on offers of settlement and come back for more, while the contractors were put in the position of their committee's offer being the minimum settlement.

During the same period several things transpired. First, a number of local or regional AGC chapters split in two, one chapter being "Building" and the other "Heavy," covering the same general area. Concurrently a number of chapters having a majority of builders and "Building" members took over control of the labor negotiations.

In both these situations the "Building" oriented labor committee tended to give the unions anything they demanded away from the cities, in heavy construction territory, provided they were not too badly hurt by rate increases in the cities. The concessions consisted of such juicy fringes as travel time fifteen miles or more from home, double time for extra shifts and overtime, free board, portal-to-portal pay, and many others that the building people were not affected by.

At the same time the sub and specialty crafts were working out large benefits for themselves with the specialty and subcontractors. This was relatively simple, since the normal practice in a building job was to subcontract 70 to 90 percent of the work, and the custom among the subcontractors, plumbers, electricians, roofers, and many others was to "pool" their bids on any sizable job at their craft association office a day or so before the general contractor's bid was to go in. The craft association members who would normally be interested in the work would get together and decide whose turn it was to submit the low bid. The low bidder would compensate the other subs who had been invited to submit proposals for "failure to get the job," putting this sometimes substantial amount in his bid. Since in many areas there was virtually no competition as a result, the sub craft unions concerned had little trouble negotiating "sweetheart contracts" with the sub and specialty contractors' associations in their line of work, and they in turn simply added the cost of extra fringes and rates to the low bids, passing on the extra cost to the general contractors and hence to the owner.

In many states the practice was aided and abetted by legislatures passing laws that required that all subcontractors be named in advance in a bid. The excuse was that this would prevent the successful general contractor from "shopping around" to get lower bids after he had the job and thus increasing his margin of profit. This, of course, worked exactly as intended by both the union and subcontractors' lobbies. The bad part came later.

In a comparatively short time in the early 1960's the sub-craft unions had negotiated rates and fringes to fantastic heights. A certain number of the sub-craft members had to be employed in heavy construction—electricians, iron workers, plumbers, roofers, and others.

The heavy engineering general contractor was confronted with a dilemma. He found himself paying operating engineers (shovel runners, tractor operators, heavy-duty mechanics, etc.) from $3.50 to $4.75 per hour, working full eight-hour shifts, with time and a half on overtime or Saturday. If a camp was involved on the job, the operating engineer was paying $6 to $7 a day for board and room. A similar situation prevailed with carpenters, teamsters, laborers, and others in the heavy construction crafts. In the sub crafts, however—plumbers, electricians, boilermakers, steam fitters, etc.—it was a different story. The spoiled pets of phony "sweetheart contracts," they received $6 to $7 dollars or more an hour, travel time to the job, portal-to-portal pay, double time after forty hours, free board and room, and other special concessions.

In many cases sub and specialty craft unions refused to permit their members to work direct for general contractors not engaged in their line of work and not signatory to the craft association contract with the particular union. This forced the general contractor to "sub" at cost plus all the functions needed to be performed by a member of the sub or specialty union. In many cases the general contractor, if his work volume justified it, would buy a small sub or specialty contracting company or several of them, members of associations in their particular fields and signatory to the craft union contracts in their specialty. The obvious effect was that the trades unions working for general and heavy engineering contractors tailored their demands to equalize the differences between their contracts and those of the sub and specialized crafts and any differences in the highest rate contracts in certain urban areas.

The high increases in construction labor rates and fringes in the late 1960's and to the present can be largely accounted for by the whipsaw policy of construction craft unions, which has been successful in achieving 20, 30, and even 40 percent yearly increases in an industry already paying the highest labor rates in the country and in the world. It has contributed to a large degree to accelerated inflation, and to the troubles of contractors who in this period in many lines appear to have nearly priced themselves out of the market.

2-11. Types of Contractor Organizations

There are several types of organization commonly followed in the construction industry, ranging from the very simple to the highly complex.

Owner-Operator or Partnership This is the simplest type of organization. Many large and successful companies began as small operations managed by one or two men. The owner or senior partner was usually the top man in the office. He assumed the duties of work promoter and salesman; he negotiated with the banks and the bonding company, arranging credit and bond underwriting for a single job, or a line of credit at the bank and bonding limits with the bonding company. He alone, or perhaps with a single bookkeeper and a stenographer, acted as treasurer, bookkeeper, tax accountant, purchasing agent, expediter, employment agent, chief engineer and estimator, and finally, general worrier.

Today he must be not a jack of all trades but a master of several. The school of hard knocks, while still a must, is no longer enough. A university degree in engineering, business administration with some business law, or better yet a master's with construction management and several years with a large and possibly diversified contractor will have fitted him to start, on his own or with a partner, a successful contracting venture.

The man in the field, whether or not he is a partner in the venture, must be general superintendent, project engineer, job planner, mechanical or equipment superintendent, job organizer, and general pusher. He will have a few foremen, including a mechanic or mechanic foreman and two or three mechanics, a job clerk, and one or more timekeepers. He should have the same or nearly the same educational qualifications as his partner, plus several years in the field as foreman, assistant superintendent, and superintendent in the type of work which he and his partner intend to pursue.

The limit to which the venture can grow is the limit of energy and know-how of both the man in the office and the man in the field. The partnership, if successful, tends to branch out. New and competent superintendents are developed and will employ and develop their own organizations.

At this point the rate of expansion of the venture rests on the ability of the partners, particularly the field manager, to pick and develop discerning job superintendents and competent lieutenants, who in turn may in time become superintendents and project managers in their own right and develop their own organizations.

Meanwhile the partner or manager in the head office must keep up with the field in selecting and building an organization that can competently estimate, program, and provide sound and current cost accounting as the field operations spread out. He must also spend wisely, conserve and manage the assets frugally, and promote new work as rapidly as he can increase the bank credit and expand the bonding limits.

As the venture continues to be successful it will operate a greater number of jobs simultaneously. Depending on its original location and the nature of the work, it will spread out geographically and become a regional or national organization.

Corporations   At some point it will become a corporation. One of the partners will become president. A board of directors will come into being, perhaps an "inside" board of top key employees or perhaps there may be some "outside" members from the bank or the surety company, or a "backer" who has invested in the stock or made credits available. The man in the head office acquires a staff and may be chairman of the board or president or both. The man in the field becomes general manager or vice-president or both. A functional or vertical construction organization is developed. It may continue to grow and reach great size.

A successful company in building or industrial construction may branch out into other fields: highways, dams, tunnels, pipelines, or bridges. Conversely, a heavy engineering company in expanding may organize or acquire by purchase a building or industrial engineering complex of its own. In either case this leads to geographical expansion.

District Organizations   In any event, as long as the jobs report to the home office direct, whether it be to the presi-

dent, executive vice-president–Operations, or subject matter vice-president or director, the company remains a functional or vertical type organization.

As the company continues to expand and cover more geographical area it may be decided to have a number of branch offices in areas of considerable business potential, usually called district or division offices and headed up by top men, both nationally and abroad.

The top men may be called district managers, vice-presidents, or division vice-presidents, but they and their staffs will supplement, without replacing, home-office functions, such as estimating, accounting, purchasing, safety, labor relations, sales, and supervision of operations in the area. A considerable degree of regional autonomy will be granted the district and division managements. They will estimate and bid independently on all but the very largest projects in their terriory and those joint ventures where the home-office management usually sits in with the partners. On the basis of estimates by engineers of the company and the partners, they determine the final price. Even in these cases the district will probably have made at least a check estimate and expressed its thoughts to home-office officials and perhaps to the partners.

Major policy matters are dictated by the home-office management or referred to the home office for final decision. Engineering forms and principles are consistent with home-office practice. All monthly and, if requested, weekly and even daily production reports are sent to the home office. Frequently payrolls, payroll reports, and engineering reports and schedules are handled by data processing procedures, and all payroll checks, including deductions for withholding, Social Security, and company insurance are processed and paid from the home office by data processing methods. The president, executive vice-president and administrative vice-president maintain constant contact both with the district or division managers and, by frequent visits, with the major jobs.

The relative merits of the vertical or functional type of organization and the district or horizontal type are dependent on several factors.

The vertical type of organization has the clear advantage of direct control and direct reporting with the projects, wherever they may be located. The system works well as long as the company specializes within rather limited types of work or in a highly populated region of urban areas, and with a limited number of subject specialists, whether they be subject matter directors or vice-presidents reporting to the executive vice-president–operations.

The virtual elimination of expensive district offices in favor of local sales office in key spots, if the spots are covered by bona fide competent key officials, constitutes an undeniable economy in overhead.

A large company specializing in high-rise buildings in urban areas may develop sufficient work in reach of the home office in New York, Boston, or perhaps Philadelphia, Washington, Atlanta, Chicago, Los Angeles, or San Francisco to use up its capacity and possibly attain $100 million a year in volume or more.

A company specializing in engineering and construction of even a dozen types of industrial complexes can, as its subject matter specialists are sufficiently capable, achieve large volume and national scope operating from the home office. This is particularly true if its major clients are national or international industrial companies operating from home bases in the United States.

Construction organizations following a single line of work, such as tunnels or bridges, dredging, or pipelines, may and frequently do develop very efficient organizations in their fields and operate several large jobs at once. There is seldom sufficient consistent volume in any one area in these special types of work to warrant consideration of permanent branch or district offices, and these companies are able to operate rather modest home offices in locations of their choice regardless of where the work at the moment may be located.

Contractors operating in the real estate and housing development field, too, have no need of district offices. They choose the sites they wish to develop, regionally or nationally, and, subject to their ability to finance each venture, whether it be housing or commercial lease-back, a project office at the site suffices regardless of the location of the home office.

The case, however, for the horizontal or district type of organization cannot be ignored. It has basic advantages which must be considered when it can be properly applied. Larger national companies doing diversified work have in many cases reached their status by the district or horizontal route. There is great advantage to a diversified national contractor in maintaining permanent, well-managed district or division offices in key spots.

It is true that the office expense of maintaining an adequate permanent representation in several key spots—rental, salaries which may be to a degree duplicated in the home office, and other expenses—may appear on casual observation to result in an overall increase in the overhead burden. On closer study, however, such may not be the case.

First: The operation of a permanent district office in a key city, manned by a competent district manager or district vice-president and a staff, all of whom are or become local residents, establishes the company among possible clients, local labor unions, city officials, and others as a local company, and greatly assists in competing against local home-based contractors who may not have the resources or national reputation.

Second: Knowledge of the territory by competent resident executives is of great advantage. This familiarity with weather, soil and ground conditions, labor supply and efficiency, and union problems, plus acquaintance with local officials, local customs, and possible clients, makes for profitable jobs and greater volume.

Third: The additional overhead may on careful analysis be much less than would be supposed or even nonexistent. The district vice-president does replace a like position in the home office and reports direct to the operations vice-president or president. The district engineer reports to the chief engineer and eliminates an engineer of similar status at home. Likewise the district master mechanic replaces a mechanical superintendent at home. The district business manager performs certain functions which do not have to be repeated. The office rental, although expensive, reduces equivalent space which would have to be supplied at the home office. Finally, the district vice-president, being in the area, can cover his territory much more thoroughly and efficiently by car and can spend more time on his jobs than if he came from the home office. The result is closer association with the project managers and the work, more efficiency, and lower costs.

These are in essence the main advantages of competent district resident representation, although many other arguments can be presented.

In national diversified construction companies or those that are growing to attain that status, much thought should be given to a district or horizontal organization, and results as well as costs should be carefully considered. The overall results may be well worthwhile.

A concentration of expensive executives in the home office can be both costly and highly destructive. Instead of being out covering the work and trying to promote new jobs in their own territory, the executives responsible for areas, which in this type of company should be districts, tend to sit in meetings in the home office, arguing semantics, and *talking* about what they *should be out doing*.

3 BUILDING AN ORGANIZATION

J. B. BONNY

President and Chairman (Retired)
Morrison-Knudsen Company, Inc.
Boise, Idaho

HENRY C. BOSCHEN

Chairman (Retired)
Raymond International, Inc.
New York, New York

3-1. Personal Selection of Key Organization

In a young or smaller contracting entity, doing perhaps a few thousand to a million dollars a year in work, and likewise in a growing medium-sized company in the one to five or even ten million dollar a year class, the selection and assignment of key organization are strictly personal matters. The determination is made gradually and on the basis of obvious necessity to fill a spot. Selection may be made by the owner, one or both partners, or by the president with the help perhaps of his general superintendent, engineer, accountant, and office manager. Selection is made on the basis of acquaintanceship and belief in the capacity of the person to be employed, his loyalty, and responsibility as an individual. In any case selection of each key individual is made on the basis of the individual's performance, character, capacity, and personality in the positions that he has been serving in the organization. In making these selections, the owner or partners have to consider thoroughly how the individual will react in his new position of increased responsibility in the growing company.

Not all the selections thus made will work out. The wife's nephew may abscond with the payroll. The brilliant classmate in college may prove lazy, drunken, or incompetent. The highly touted executive lured away from a competitor may turn out to be a dud whom the competitor wanted to get rid of anyway, or he may be a natural "drifter" who will leave to accept the first attractive offer.

For any one of a hundred reasons, probably two out of three of the budding geniuses hired on the friendship or family relationship basis will quit, become incapacitated, die, or have to be fired. A number of others will prove to be fair employees, but capable of taking just so much responsibility and no more. A competent engineer may be able to estimate or follow costs or make reports under guidance, but be unable to train or lead subordinates. A good bookkeeper on a small job may be utterly helpless on a large one. An effective foreman or assistant superin-

tendent or mechanic foreman, who when faced with higher responsibility simply wilts, may live out his working life contented and happy that he has successfully avoided the trials of higher authority.

The management must not become discouraged or frustrated. There will be some of those selected, sometimes the least promising initially, who will develop and grow in capability and depth to fill the key spots and become top officers as the company spreads out and seizes its rightful place as one of the leaders in the profession.

Management will undoubtedly be tempted to look primarily outside of its own organization for the added talent needed to keep pace with growth. This temptation should be looked at with some skepticism. It is very easy to imagine glowing characteristics in an outsider whom you do not know very well. It comes as hard reality after employing such a person to find that one of the young men in your own company would have been a better selection.

As the company expands and becomes regional or national in scope, it may or may not diversify into a number of construction fields. The decision to diversify or not and the location of the head office will influence the decision on whether to retain the original vertical organization or adopt a horizontal or district and division type of organization, which in turn will have an effect on the selection and assignment of key personnel.

The selection and assignment of top organization becomes a matter, both in the home office and in the field, of what the job is for which the individual is employed, and what the scope and extent of his authority are to be, as well as what he is to be trained for in the future. The operation of the personnel department will not be discussed here since it is covered in detail elsewhere. Here an attempt will be made to define the principles of selecting and assigning or transferring those individuals who, by reason of their status and responsibilities, come under the direct purview of management.

3-2. Nucleus of Administrative Organization

In the head office the nucleus of the administrative organization has been formed. There is a president who is probably chairman also. There is an operating vice-president who may be called vice-president–operations or vice-president and general manager or general superintendent. There is an office manager who may also be secretary- treasurer, a purchasing agent, and, if the company is engaged in heavy construction, a general master mechanic. On the operating side there is a chief engineer who reports directly to the operating vice-president and president.

3-3. Advantage of Training and Advancing Key Personnel

The future growth and success of the company rests with key personnel, how they handle their jobs in an expanding business and, equally important, how they select their assistants, since in each general category one or more will become successor manager or head up a new category as one becomes necessary. As has been mentioned before, history has demonstrated time after time that the best way to grow is to raise and train your own key personnel. In this manner you are dealing with known quantities rather than bidding in the open market place for what frequently develop to be inferior personnel. It must be recognized, however, that in a growing organization a certain amount of outside employment is usually necessary to fill both key and lower echelon positions.

Let us say the office manager finds he cannot, himself, adequately fill the positions of secretary, treasurer, chief accountant, tax expert, employment manager, and controller. After lengthy discussion with the president ("They aren't in the budget. There's not enough room in the office. Can't you get by another year? Etc."), the office manager is given permission to fill these spots as he is able to transfer or find competent people and after discussion on each one with the president. This, when completed, will automatically make the office manager either business manager or vice-president–administration unless the wife has another nephew working in the field or somewhere, in which case *he* becomes vice-president–administration and the office manager settles for business manager (with a raise) or financial vice-president reporting to the president (with a larger raise depending on how hard he fights).

3-4. Preparing the Administrative Manual

At this time the office manager proceeds to prepare an administrative manual setting forth the duties, scope, and degree of authority for his own position and each of those he proposes to fill.

Let us assume that the first position in priority becomes the corporate secretary and that this position calls for a man with legal training who, in addition to keeping the corporate records, can cover contracts and contractual matters, subcontracts, and insurance and insurance records. He should also be able to handle law suits against the company or by the company against others, with the help as needed of outside or regularly retained attorneys.

The office manager, having arrived at a principle of selection for his key position, can commence his search with a definite specification for it in mind. Under these circumstances he soon finds a young attorney, living in and familiar with the area of the home office and not yet too deeply involved in his own legal business, who recognizes the possibilities in a growing company and joins the staff.

The office manager, soon to become vice-president–administration (the wife's nephew didn't dare tackle the job after all) carefully reviews his administrative manual again. The chief accountant's post calls for broad accounting experience with C.P.A. rating preferred, ample experience in field project accounting, and the ability to work with and guide field and home-office accountants, recommending sound assignments and promotions and recognizing shortcomings in his own or field accounting staffs. There is one man in the organization—office manager on the largest contract the company has. Can he be released for transfer? If so, who will take his place? There is one job just being completed. The job was not too large but the young office manager not only handled his work well but seemed to have excellent relations with the project manager, the project engineer, and others on the staff, and with the owner's engineer. His record showed two years of accounting in college and a C.P.A. license in his home state. The man who was the original choice for the larger job, although having an excellent record, had no C.P.A. rating. He was in good health and, having seniority, was past his middle fifties.

The office manager appealed to the president who thought the matter over carefully. "Your manual calls for almost the same specifications for treasurer, except that the treasurer must be able to work with the bankers and surety bond people. Other than that the job is not as arduous as that of chief accountant and Frank (the older accountant) has the age to gain the bankers' respect. Besides he could retire in ten years as financial vice-president with plenty of time to raise a replacement. Give the accountant's job to the kid. We'll make Frank treasurer."

3-5. Adding to the Manual—Labor Relations and Safety

The office manager, now vice-president–administration, pondered on his problems and added Labor Relations and Safety to his manual, as the president had been handling labor relations and there was no head-office director responsible for safety. When he had defined his requirements in the manual he had an idea. The safety director must understand safety rules and preach them to the project managers, the foreman, and the men. Even on jobs large enough to boast a combination safety and first-aid man and very large jobs with separate safety and first-aid setups or even a doctor, the safety director must not only get on the jobs and make sure safety rules are enforced but he must also talk to the union representatives and convince them that the job is safe.

The manual for the labor relations director called for an individual with broad knowledge of labor union rules and an understanding of the inner workings of various unions and their probable courses of action. The position demands a person of strong character to protect the company's interests in labor matters, who must also be acceptable and understanding to the labor groups with whom he will be dealing. This latter point is often overlooked. In spite of the general impression that exists, in many minds, of the difficulty of dealing with labor leaders, a great deal can be accomplished by cultivating their friendships and confidence, working with them, and establishing a reputation of being a fair dealer and straight shooter. The position could be filled with a dedicated man with a college degree or by an officer of the Building Trades or one of the unions.

The labor relations director and the safety director had one thing in common: they both had to deal frequently with the unions.

Perhaps until the company grew so large that it wouldn't be feasible, one man could hold both jobs. And so it happened. A college graduate with a degree in public relations and labor relations was working as safety and first-aid man on one of the jobs. After an interview and some checking he was transferred to the head office in charge of labor relations, public relations, and safety. Until greater growth made it mandatory to separate the jobs, he would do well.

3-6.   Specification for Employment Manager

Meanwhile a specification in the manual was written for the employment manager. The president, operating vice-president, and the project managers had been doing the hiring and firing, in the main successfully. Unfortunately the girl who had been keeping a running card record of foremen, timekeepers, mechanics, and job engineers' helpers was neither very thorough or very up to date. As a result, several people fired for cause from one job would show up on another and have to be fired all over again. What was needed was an employment manager to select and recommend as well as employ on order competent men in the grades just beneath management selection range (although naturally the president or operating vice-president would occasionally send out some old foreman or equipment operator without telling anyone and merely giving the "oldtimer" a slip to give the project manager).

In any event the now vice-president–administration had written a specification for the employment manager and put it in his manual. It showed he must be well acquainted with all the crafts, have a good memory for names and faces, and be able to keep orderly and functional records. Above all he must be able to keep the confidence of the top field people as well as the home office. He must furnish competent help when needed and still avoid hurting any applicant's feelings or making any applicant think he was neglected. As it turned out, a younger brother of the secretary and general counsel proved to have all the requisites, and since he did not report to the secretary he could be and was hired.

3-7.   Specification for Tax Expert and Controller

There remained on the VP–administration's list a tax expert and controller, and accordingly the manual was enlarged to include a specification for each.

The tax expert was easier to define than to find. The specification called for a man with a college degree in business administration, experience in accounting, and actual practice in tax matters. He must be "company minded" and ingenious in discovering and presenting all legitimate means of accomplishing tax avoidance, while conscientious and careful not to cross the line and commit tax evasion. Tax law today is so technical and complicated that the person who handles tax matters requires continuous schooling and study of the constantly changing laws and accounting practices. He really has to be a cross between a tax lawyer and accountant.

The controller also should have a business administration education and training. He must be highly competent in proper accounting methods, as well as budget prepara-

tion, and have a real ability to make sound forecasts based on trends developed from job cost and revenue summaries and engineer's unit cost reports. He must also have experience in tax matters and their effect on sound forecasts. One of the most difficult characteristics to encounter in a controller is the ability to develop the necessary accounting procedures and at the same time avoid an overly complicated mass of paper work. It is extremely easy to fall into the error of providing too elaborate a system which is costly and detracts from its value to the organization because of the pure volume of paper work involved.

The VP–administration realized that one man could not cover these two jobs indefinitely as the company continued to grow. He also realized that the president was becoming continually more concerned at the badly ruptured budget and would become even more so when he realized that a growing number of top assignments both in the head office and the field must be created and existing ones enlarged by adding more personnel. If the company growth slowed down and the percentage to cover home office overhead increased on volume, that increase from 3 to 4 or 5 percent would come directly out of profits—and hence from the hide of the VP–administration.

Again he searched for one man who could serve not only as tax expert but also as controller. In the person of a young man in the company's auditing firm he found the proper background and characteristics and after some pleading got the auditing firm to release him to the company.

3-8.   Purchasing and Equipment Management Functions

As the company continued to grow and spread, it became obvious that other departments both in the home office and in the field must be expanded and subdivided to take care of greater volume and the complexity of a greater variety of work. The purchasing agent was gradually going from frantic to erratic to numb. He had been trying to fill orders prepared by the chief engineer for permanent and embedded materials, listening to the president and the general master mechanic on purchases of construction equipment, parts lists, and tools, to the president and the VP–operations on fuel, lubricants, tires, explosives, wire rope and major supplies as well as cement and other items, and again to the president only on railroad routings. Finally he reached a point where his ulcers were battling with his high blood pressure and something had to be done.

The VP–administration, after conference with the president and VP–operations, was delegated to complete the functional manual, himself writing the specifications for the purchasing and equipment management functions for the president's approval, enlarging the manual to include the requirements for top field office personnel. Next, together with the chief engineer and the VP–operations, he was to prepare a manual for the chief engineer and his assistant chief engineers as well as project engineers. Finally, a manual was to be prepared covering requirements for project managers and general superintendents on projects of various sizes and types.

All this would take time and the pressing matter of purchasing and equipment management was tackled first. Certain major orders and major job purchase commitments had to be made with the president's approval and even to a degree with his handling. He could sometimes get a little extra discount on cement or explosives or perhaps on reinforcing bars or structural steel by talking to top officers of the vendor company. Occasionally a call from the president and a timely order for equipment or tires or automotive units might get a job awarded that might have been hanging in the balance. A vital decision on a large equipment order might hang on the stubborn prejudice of a valued project manager or the VP–operations' insistence on standardization of certain classes of equipment because of lower maintenance or lower parts inventories.

The manual, after due deliberation, was presented and approved. Purchasing, although an administrative function, must be controlled by operations. The breadth of company operations will have some influence on whether or not purchasing is controlled by administration or operations. If the company is largely dedicated to one type of activity, it often is desirable to place the purchasing function under the control of operations. On the other hand, if the company engages in a wide variety of activities, purchasing will probably be administered in a more satisfactory fashion if it is put under the control of the VP–administration.

3-9. Director of Procurement—Liasion Between Operations and Purchasing

The answer in the manual was a director of procurement. He must be over the general purchasing agent, who would continue to solicit quotations and order routine items. He would be over the general master mechanic, who would become general equipment superintendent and guide field policy and performance on equipment maintenance, preventive maintenance, and overhaul The general master mechanic would recommend replacements and advise the director of procurement of available units for transfer as determined by project management. He would in a sense replace the general master mechanic, who would become a traveling master mechanic reporting to him.

The director of procurement thus became, in the manual, an executive liaison between operations and purchasing of nontechnical material and equipment, and between enginering and operations and purchasing on technical and specification purchases. He had to be a man with technical knowledge of accounting and experience with equipment management and operations, as well as purchasing. In addition to all this, he had to have a flair for diplomacy, since the VP–operations, the chief engineer, and the VP–administration, to say nothing of the president, were by the nature of his job all his direct bosses.

The general purchasing agent could not fill this new job and didn't want it. Hopefully, with some of the pressure off, his ulcers might heal and his blood pressure subside. With a young purchasing agent to help him he might last quite a while.

Meanwhile a candidate for director of procurement appeared. A project office manager, previously promoted from warehouseman on a large job, he had knowledge of accounting and job purchasing and familiarity with equipment. He was transferred to the home office and the VP–administration heaved a sigh of relief. The president had been chiding him lately and he was tired.

3-10. Chief Engineer as a Key Operating Officer

The chief engineer as a key operating officer reported to the VP–operations and the president. His duties (listed elsewhere in detail) were mutitudinous and his responsibilities fearsome. He never had enough help and couldn't really trust them to finalize the estimate on a big job if he

had. He and his assistants were automatically to blame if they came in with an admittedly close estimate on a job, the company got it—and lost money. On the other hand, if they raised their estimates to safe levels and stopped getting new work and the volume fell off, the president would take great pains to explain that if they got no work they wouldn't need a chief engineer, or in fact anybody. Finally, if an estimate made with great thought and infinite care was successful and the job made more than usual, the project manager received great praise, a raise, or a project bonus, and nobody thought of praising, let alone raising, the chief engineer and his assistants.

The VP–administration and the chief engineer spent some time together writing a manual for the engineering department, including the project engineers. It wasn't easy.

The company was beginning to branch out geographically and to diversify, although it still retained its vertical or functional type of organization. It was decided to consider a number of fields not yet entered but almost certain to become of interest in the relatively near future. This circumstance did not immediately affect the home-office engineering department as such to any great degree, but it was advisable to prepare for this eventuality by setting up at least a skeleton organization on paper in addition to making such changes as were urgent and overdue.

As finally approved by the president these changes were made and covered in the engineering manual.

3-11. The Engineering Manual—Skeleton Organization

The chief engineer was made VP–engineering. He would continue to supervise in a general way all sizable estimates, and review and sit in on large contract bids, both company and joint venture. He would have executive authority over all engineering functions, as well as procedures and engineering reports from the field. He would be responsible for preparing the plan of job operations, and he would assign all project engineers and their immediate assistants, subject to concurrence of the project managers. He would report only to the VP–operations and the president.

A chief engineer was promoted from assistant. He would be primarily responsible for the preparation of all estimates and would consult with, and recommend major estimated bid costs to the VP–engineering. He would sit in with him and the VP–operations or president on the final determination of the company bid and later, if the project were bid as a joint venture, with all concerned.

A young engineer, previously designated office engineer, was promoted to assistant chief engineer. He received no other title for policy reasons. His duties were, however, outlined in the manual. In conjunction with the project manager, the project engineer, and the general counsel, he was to prepare and, if requested by management, to present such claims and change orders as could not properly be settled by project personnel on the spot. Management would determine the possible necessity of calling in outside consultants and attorneys depending on the circumstances.

The manual as prepared also covered fully the duties and scope of project engineering personnel. That subject is discussed in detail elsewhere.

On projects of any size, a cost engineer is usually provided. His responsibilities are to advise the job personnel and headquarters on the current costs and projections of all phases of the project.

At this point the company, expanding rapidly and successfully both by diversifying and geographical spread, is participating in many joint ventures with others both at home and abroad.

3-12. Key Operating People in a Vertical or Functional Organization

As it continues to remain a vertical or functional type organization it must add a number of key operating people to its head office staff.

These people become "subject matter" vice-presidents, managers, or directors. A few may be promoted from within the organization from those holding positions in fields in which the company is familiar and skilled. These will all be men with engineering or business training, experience in the field, and proven ability to select and train subordinates.

If the company started out in pile driving and manufacturing, for example, there will be a vice-president in charge of that department reporting to the vice-president–operations, now executive vice-president–operations. There may also be a vice-president in charge of underground construction, another on airports and military construction, etc., all of them with knowledge of the subject plus ability to pick and manage organization.

The project manager in these cases is picked for knowing the subject plus having an organization or having access to one that can do the job. On a very large job he may even be a vice-president himself but he still reports to the subject-matter vice-president or director.

If the company decides to branch out into high-rise buildings or housing developments, the question of financing may be involved. If so, the president, VP–administration, and treasurer work jointly to arrange finance during and after construction, using the estimate arrived at by the VP–engineering and his staff. If they are wise, they allow a fair percentage on top for contingencies.

In these cases the project manager, since most of the work will be subcontracted, may well be selected from the home-office engineering staff or be transferred from a project engineer's post in the field. He must have the ability to coordinate the subcontractors, as well as handle the field supervisors and an enginering staff that keeps track of costs, works with the project office manager, and prepares estimates for changes and charts progress. The engineering staff will also work with the accounting department in preparing codes of work items to furnish proper records and controls for management.

3-13. Dangers of Expanding in Unfamiliar Fields

Unless the company has key executives with sound background and broad knowledge of technical mine or mine and mill construction and operation, chemical, metallurgical, refinery, or manufacturing plants, as well as access to engineering organization for design and in the field, the company may well hesitate to enter the competition. The cost of developing plans for the first time is very high, and the building of such an organization from scratch almost prohibitive. However, in this area lies opportunity to design and construct, and the very lucrative design and construction management or turnkey contracts.

The secret of success in this mostly private-owner sector is the ability to make high-level sales and back them up with personable and competent key executives and an ample number of design and supervising engineers and inspectors.

3-14. Expansion by Acquisition—Becoming a Horizontal or Delegated Organization

Let us assume, therefore, that the company decides to purchase control of a medium-sized industrially oriented engineering and construction concern located in a city away from the home-office location. The sale is consummated by merger and it is decided to retain the identity and operating procedures of the acquired concern. The president of the acquired concern is near retirement age. Other executives, including the chief engineer, are much younger. The secretary-treasurer, however, is at retirement age and wants to retire. The problem is to introduce into the organization an executive vice-president who will learn quickly, adapt to the organization, and be ready to step into the president's spot when he retires. He must have had sufficient time with the owning company to remain loyal and yet not try to change the policies of the newly acquired organization overnight. In short, he must be one of the key people in the parent company being trained and pointed for successor management.

After much soul searching and many discussions, the decision was made to transfer a building project manager who was showing real promise and replace him with his assistant. The president of the parent company became chairman of the board of the acquired concern. A young and trusted project business manager was promoted to secretary-treasurer of the new subsidiary.

The now parent company had almost doubled its volume of work and had, without thinking about it, started to become a horizontal or delegated type of organization.

The next move was to acquire a Canadian company, this time by purchase. The organization, including the president, being active and willing to remain under new control was retained, and only the executive VP–operations of the parent company and the controller had new places to watch.

3-15. Formation of Districts

The decision to establish the first two districts on the West Coast, one north and one south, came at about the same time. The advantages appeared to outweigh the cost of additional offices. A large volume of work was being offered at each place and a district manager domiciled there would become more familiar with local conditions and become acquainted with local officials and heads of various companies likely to be letting work.

The problem of organizing the districts proved less difficult than was anticipated. It took fewer additional people than one would have thought. The district managers, with more authority than they had previously enjoyed as subject-matter directors in the home office and more responsibility, had new freedom, worked harder, and within a short time the competition between districts began to produce results. The small district staffs, consisting of district engineers promoted from the head office estimating department and district business managers transferred from completed projects, plus their secretaries, stenographers, and switchboard operators, made effective teams.

At this point, the great danger is that overhead costs will increase more rapidly than earnings. Extreme care and control must be exercised to keep the two in balance. The company will quickly establish norms on this relationship and will insist that each provide a center holding its overhead and operating costs to the agreed upon ratio of earnings produced within the center. Only in this hard-nosed attitude will the desired earnings result from the expansion.

3-16. Development of Overseas Cost-Plus Organization

Shortly before World War II the company in joint venture with others had been employed on a large cost-plus-a-fixed-fee contract for bases and other military construction in the Pacific. Up to the time that war action had forced termination of the contract, a large staff of competent supervisors had been developed. True, they were "cost plus" oriented and might not be the most economical operators, but many were available for foreign assignment and understood the problems of service of supply abroad. They would do well in areas where large postwar projects were being financed by U.S. government agencies, or where United States and Canadian companies were seizing the opportunity to open up power projects, pipelines, mining areas, oil develoments, etc.

Many of these areas were virtual vacuums. European contractors had not yet recovered sufficiently to be a threat, and local contractors had neither the skill nor the finances and could only be given small percentages in joint ventures as "window dressing" or subcontracts on noncritical portions of the work.

The company management realized it must work fast. Other United States companies sensing the opportunities were organizing to seize them. Some, who had been joint-venture partners in the past, would be tough competition.

A vice-president from an operating division was assigned to follow and head up the foreign prospects. The president and executive VP–operations would back him up and fly to any hot spot when needed. Appraisals of key talent from the war years were assembled. As soon as a sizable job was obtained in a foreign country, an office was established in the capital or principal city and a subsidiary was incorporated in that country. The subsidiary company would be headed by a president or vice-president reporting to the home office in the capacity of a district manager. He would be carefully chosen from among those heading up major projects in the Pacific before the war and having had foreign service since the war.

The requirements were tough and the living conditions not always the best, but the salaries were much higher and largely tax free. The specifications for the positions of district manager, foreign company head, and foreign project manager were almost identical. In each case the man should have a college degree, preferably civil engineering or business administration. He must be sober (and when drinking at a cocktail party, be able to conduct himself properly). He must have knowledge and experience in the fields in which the company is engaged in the country. He must be able to sell the company's services intelligently in any category where work is offered. The project manager must be able to handle his key men under difficult conditions and keep them from quitting or going on a drunk in sheer frustration. He must also be able to discuss their problems with the local labor, most of whom were grossly underpaid and obliged to build their own dirt-floored shacks and "rustle their own rabbits" or pay high prices in the company commissary.

The district manager or foreign company vice-president or president must be able to deal with high government officials, grafting customs officers, venal police, and equally venal union officials, and at all times remain cool and urbane. He must also be able to "smell" trouble on one or more jobs and talk project managers or superintendents out of quitting—or fire them if necessary.

Oddly enough the company management soon found out that knowledge of the local language was not a "must." The district manager and project manager could understand through an interpreter. Government officials could nearly always speak English, and felt superior (and therefore kindly) that the Gringo or Yanqui (depending on the location and country) was not as competent in languages as they were. In any event the Americans would soon pick up enough of the language to get by, and many excellent foreign district managers and project managers with some understanding of the local language would refrain from speaking it, thus gaining time to think of a critical answer while the interpreter was translating.

Foreign assignments soon became schools for key foreign personnel. District managers and the officers of foreign subsidiaries and foreign domiciled corporations were given quarters for themselves and their families or living allowances plus higher salaries partly or wholly tax free. By custom, taxes levied in their assigned countries were paid by the company. Thus, with cost of living at or near zero, with seniority gained prior to "going foreign," and with twenty years or less of foreign service, an executive could look forward to a comfortable retirement at home with a small fortune saved up.

Project managers, project engineers, and business managers likewise, with quarters furnished at the job and the same advantages as the district manager and his staff, could look forward to the same situation at the end of their foreign service, assuming work continued in the country or developed in a country where they could be transferred.

During the first half of the 1950's numerous opportunities both at home and abroad were being presented, opportunities which taxed the ingenuity and vision of key executives. Most of them carried traps for the unwary, and many contractors who failed to recognize the signs and understand the changing ground rules suffered severe losses and even disaster.

3-17. Updating Field Management at Home

Field management at home unless updated and upgraded became ineffective. Superintendents who had been accustomed to having entire organizations move with them from job to job discovered this could no longer be done. Unions were demanding that members of their locals be employed fully before contractors' skilled men, even though members of locals elsewhere, could be hired. This taxed the capabilities of former top producers among superintendents. They not only had to work with strangers but frequently with antagonistic and lazy strangers beholden only to their own local unions for their jobs but also well aware they could not be fired except for flagrant derelictions. This situation was compounded by various unions' successful demands that all foremen, including anyone who touched a piece of equipment or a tool even to demonstrate its use, must be a member of the local union. Management's representatives became at least assistant superintendents; even the master mechanic, upon becoming mechanical superintendent, was lost without his tools.

The project manager or superintendent who had been turning in excellent earnings by counting on his "pot gang" of foremen, skilled equipment operators, and craft foremen to pull him over the rough spots found that no longer adequate. Promoting the foremen to assistant superintendents failed to take care of the problem. The ex-foremen could no longer work with their men. They had to relay instructions through foremen, usually strangers, who being union members themselves lacked the "company loyalty" that had been the real backbone of every construction organization through the years.

Costs began to creep up, not violently but definitely up, 3, 4, or 5 percent a year, depending on the amount of hand work involved.

In heavy construction, equipment was beginning to make great strides, and for a time equipment efficiency was increasing in productive capacity faster than increases in its cost. It took no more labor to operate a large machine than a small one. In dragline work, shovel operation, truck performance, tractor and scraper output, tunnel drilling, loading, shooting and mucking, and even lining with concrete "guns" or pumps, the production was still going up faster than the impact of higher wages.

Superintendents and project managers, district managers, vice-presidents, and presidents or company owners had to change their thinking to survive. The operating cost and cost of ownership of equipment became more and more important. Careful scheduling and work analysis became not only the province of the chief engineer, the district engineer, and the project engineer, but of necessity had to be understood and followed by all the executive and operating personnel as well. Labor was an increasingly expensive commodity that must not be wasted. A good project manager or superintendent must have more than the ability to "push" a job. He must understand scheduling, sequence of operations, and how to reduce the number of man-hours of labor on a job and keep them at the lowest possible figure.

On high-rise and industrial buildings and even housing, architects and engineers were learning to simplify designs, raise prefabricated panels, and cut "gingerbread" to reduce hand labor. Project managers in their turn must, to survive, initiate new and faster methods.

3-18.   Military Programs Abroad

These problems required not only a change at home in the management approach to the staffing and controls of the "bread and butter" work, but widely affected the military construction programs abroad as well as in the United States. In Spain and in North Africa a large number of locals had to be trained by American operators and mechanics, as well as by clerical and accounting personnel and engineers.

Many of these locals soon became quite proficient at their assignments. Since their rates could be raised as they became more skilled and were put into a higher classification, there was little grumbling and low turnover. It soon became standard practice to have two or three and sometimes four classifications in the same job. Starting with the lowest or apprentice rate, the "learner" merely rode with the American instructor until he was able to operate the tractor, scraper, dragline, truck, or what not. Then he would be allowed to operate on his own (with a raise), but he would be watched. Finally after four to six months he might be designated tractor operator first class with another raise, or perhaps Monighan dragline operator, shovel operator, heavy truck driver, motor patrol, batch-mix plant operator, welder or mechanic first class, all dependent on his skill, willingness to work, and dependable attendance. Meanwhile office help and even engineering assistants in the field or in the office were going through similar training and receiving similar promotions.

This was likewise an education to project managers and superintendents, master mechanics, chief clerks, business managers, and project engineers. Their thinking had to be completely revised if they were to continue to succeed in foreign work.

3-19.   Civil Contracts Abroad—Developing Skills of Locals

Those supervisors who quickly learned that the locals in almost every country, no matter how unfamiliar with American equipment and American methods, could learn to operate complicated equipment and master complicated jobs, began to "phase out" high-priced instructors as locals were developed to handle their work. This paid big dividends, since the highest paid locals in the "developing" countries received only a small fraction of what their instructors had received. The possibility of promotion and advancement in the company, unheard of before in their history, plus the use of the word "please" when issuing an order—understood in every language—and fair and unbiased treatment paid extra and even more valuable dividends.

A "company loyalty" was built up among the locals, as well as a "company pride." It was possible in North Africa and other Moslem countries, for example, to tell how long the operator had worked by merely walking by his shovel, dragline, or "cat." At three months the operator wore shoes, at five he had acquired work gloves, at seven he sported a blue mechanic's shirt and pants, moderately clean, and by the end of a year the fez or burnoose had disappeared and on his close cropped head he flaunted at a jaunty angle a mechanic's cap complete with visor.

Meanwhile the percentage of Americans and Europeans employed had gradually gone down. About ninety-five percent of the employees were locals, with little or no loss of efficiency and the payroll cut by half or more.

Contractors who were fortunate enough to get additional contracts in an area found that they had acquired ready-made organizations. The previously employed locals would flock to the new job and being already trained needed no "instructors." Too, it proved that, within the limits of their familiarity, local subcontractors could perform concrete work, rubble and ashlar masonry, and brick work buildings or structures more economically and with quality equal or better than stateside. The district manager and project managers who learned these things turned in fine profits and excellent records on target estimate or cost plus a bonus for early completion of work.

A few illustrations of situations that could and did occur might give the reader an idea of the possibilities that always exist for the application of American contractor ingenuity and plain, old-fashioned "people psychology."

In Mexico and some other Latin American countries there developed a two- and finally a three-class system which was based not on race or citizenship but primarily on education, training, and capability. As more and more nationals of the country replaced their American instructors, the heavy equipment operators obtained higher rates for performance and reliability plus weekly bonuses for exceptional production. Many nationals became foremen, and a few became shift and craft superintendents and master mechanics, and in turn shared in the weekly production bonuses. Field rodmen and chainmen on survey crews became office engineers, cost engineers, and even assistant project engineers and project engineers. Timekeepers became bookkeepers, accountants, and business managers. The resident American staff shrank from perhaps 10 percent to 5, to 2, and finally perhaps to only a project manager and an assistant.

3-20. Feeding and Housing Abroad

Long before this occurred, however, the problem had become centered on feeding and housing. Those in the executive class were given housing for themselves and their families and the privilege of purchasing food and supplies at or near cost in the company commissary, or meals in the mess hall. The skilled craftsmen and operators were given barracks lodging two to four persons and having adjacent showers and toilets—typical bachelors' quarters. Finally the unskilled "obreros" would build thatched huts with mud floors and walls of adobe. They had the privilege of the mess hall, where clean, wholesome but native-style food was served at very low prices commensurate with their earnings.

Thus was born the "Gold Room." This was a separate, tastefully painted mess room with table cloths, silverware, and "china" plates and cups, operated like a "short order" house in the states, where the Americans and higher paid white- and blue-collar employees, at reasonable cost for their wages and salaries, could order ham or bacon and eggs with toast or hot cakes for breakfast, and steak or roast with salad and bread or rolls with gravy and ice cream or pie with coffee or tea for lunch or dinner. There was no restriction on who ate in the mess hall or the "Gold Room" except the price of the meal.

It was no surprise that local executive and administrative employees and even the skilled blue-collar workers preferred the "gringo" or "yanqui" style food, for many of them had been to the United States and some had been educated in American universities. The surprise was that the peon obreros soon came to save up for a week or two for a meal in the Gold Room for themselves and their families. Thus a feeling of class or race distinction was nonexistent and the goal of eating in the Gold Room became an incentive which bred many loyal and finally skilled employees.

3-21. Special Situations with Local Personnel Abroad

The degree to which the local nationals were able to learn special skills and adapt themselves to complicated situations was remarkable. On a contract on the island of Sumatra for an urea fertilizer plant, the design required an inordinate amount of overhead welding of stainless-steel pipe to extremely rigid specifications. A careful check showed that fourteen months after start of construction an increasing number of welders certified for this most difficult task would be required, reaching a peak requirement of sixty near the end of the job. Welders certified for overhead stainless-steel pipe welding designed for high pressure were almost nonexistent in the United States, and had they been available the cost of flying them to Sumatra, boarding and rooming them, and paying each one more than fifteen hundred a month until their return and arrival home would have been staggering.

The pipe and mechanical superintendent, who was himself certified, came up with a suggestion. Establish a welder's school for the natives. It would destroy some good stainless-steel pipe and might not work, but if it did the savings would be enormous.

The school was started and the superintendent spent two hours after work every day teaching his group. No one told them the extreme difficulty of the task they were to perform. As a result, with no inhibitions they learned quick-ly. Before their services were required sixty Indonesians qualified and received their specially engraved certificates of competence. The difficult welding job went off without a hitch. There were almost no failed welds when tested and a group of highly skilled artisans had been developed.

In Holland a welding problem of a different sort developed. A huge gas field was discovered and contracts called for the first section of a loop "big inch" line around the country. The problems were formidable. The water table was high, varying from inches to a few feet from the ground surface, which made it necessary for much of the excavation to be done under water by draglines; wheel or ladder trenching machines could only be used on the drier sections.

Dutch labor, both willing and efficient, was skilled in operating all equipment. There had been many ditches, canals, and trenches built and numerous concrete drains and water lines. Welded steel lines, however, were nonexistent and experienced Dutch welders nearly so. There was not sufficient time to train local welders and skilled men must be brought in from other countries.

Dutch regulations controlled not only minimum but also maximum rates for Dutch nationals. There was, however, no maximum established for foreign workers brought in under contract. It would be necessary to pay skilled welders from other countries at least twice the highest permissible rate for Dutch welders had sufficient numbers been available, and they would have to be given air fare to and from their homes and free board and room.

A vigorous recruiting program was instituted throughout Europe. A few qualified men were recruited and shipped in from England, Ireland, France, Italy, and a number from Germany. It looked as though the problem was solved.

The contractor, however, had failed to reckon with Dutch temperament and prejudices. No sooner had the German welders arrived than the project manager was visited by a committee from the Dutch construction unions. In typical Dutch fashion they got immediately to the point. They were not concerned with what the contractor elected to pay his foreign welders, with one exception. If the contractor persisted in paying imported Germans higher than the legal rate for Dutch employees, they would strike until the Germans were sent home.

The contractor had no alternative but to send the Germans home. This left a huge hole in the welding crew. To meet the tight schedule twenty more welders must be brought in from somewhere. The project manager who had worked on many foreign jobs remembered that in Pakistan two years before on a large cross-country gas line it had been necessary to bring in a welding crew from Lebanon. These men had been trained on the original "big inch" line in Saudi Arabia and were both competent and reliable. If they could be found and hired the problem would be solved. After several wires and two transatlantic phone calls they were found just completing a pipeline job in Libya for an American contractor who, glad to save their fare back to Beirut, arranged with them to fly to Amsterdam. This time the problem *was* solved, the contract was finished on time and with a profit. This led to several other contracts on the same project, equally satisfactory.

3-22. Personnel Problems in Combat Zones

When the military construction program was started in Viet Nam for the Navy Bureau of Yards and Docks (later Facilities Engineering Command), the problems, covered in detail elsewhere, were myriad. In addition to service of supply, accountability, equipment maintenance, warehousing, and the necessity of men working in a combat zone and frequently in danger from enemy action, the problem of wage scales again presented itself.

The Vietnamese government, attempting to slow inflation, fixed maximum wage scales for all local employees, and these were supposed to have been observed by all the services as well as the contractors, architect engineers, etc. These rates were absurdly low and had been tied back to rates in effect three years before. The workmen, male and female, were on the whole very efficient. Many who had been taught by the French were able to do carpenter work, plumbing, and electrical work, even to the point of winding motors, and they took to the larger American equipment readily. The women learned English quickly, soon becoming adept at clerical and bookkeeping tasks. Also, much hard work, even to sorting rock in the quarries, was performed by their country sisters.

To make matters even more complicated, a rate for United States employees and certain third-country nationals (Canadian, British, Australian, West German, Dutch, French, etc.) had to be established and approved by the OICC (Officer in Charge of Construction).

Finally, contracts had to be executed and approved for the employment of "other third-country nationals"—mainly Koreans and Filipinos. These were generally employed in clerical and semisupervisory positions above most Vietnamese but under the Americans and favored third-country nationals.

All Vietnamese employees had to be cleared in the area where they worked by the district Vietnamese general for "Security" and issued a card. Frequently the general or his subordinates would charge the Vietnamese as much as a month's wages for the "security clearance" card, wait three months, and then cancel all security clearances and force the employees to pay all over again to get reinstated. This procedure in certain areas by venal South Vietnamese officers caused delay to the work and sometimes shortage of workers. It was also very difficult to catch these venal officials, since any Vietnamese reporting the practice knew he would either be drafted immediately into the army or jailed as a "security risk."

As time went on the problem was aggravated by more and more inflation and the continued stubborn refusal of the Vietnamese officials to approve needed wage increases even when presented by the contractor and approved by the Navy. This led to a series of strikes which, since the contractor was sympathetic, resulted in some increases in wages for the locals who had performed hard work and worked long hours for very little, in the main loyally and uncomplainingly.

The cost of the work due to the amount of production put out by the Vietnamese employees was for the most part reasonable and in line with costs for similar work elsewhere, although performed under war conditions and subject to multiple confusions.

During the balance of the decade work in the United States remained at a high level. Opportunities were plentiful in every field from housing developments to high-rise buildings to chemical, food, manufacturing, and metallurgical plants. Highway work was booming and a great surge existed in power plants, both steam and hydro, as well as irrigation works, bridges, and pipelines for oil and gas transmission. Costs were still increasing at a safely predictable rate of 5 percent a year. A contractor with a good basic organization and a flair for picking and training men in key spots could branch out and diversify just as fast as

he could build sound organization. Smaller and medium-sized contractors banded together in joint ventures, took larger contracts, and themselves became large.

3-23. The Changing Foreign Picture

Concurrently the foreign picture was changing. The custom of starting with large loans or grants by the United States to various countries for construction projects had led a number of American contractors to turn their eyes toward foreign work. At first the work was usually contracted on a basis of design and construct on a cost-plus-a-fee basis, or construct on a fee basis or target estimate, with others performing the design and inspection. Certain agencies making loans or grants included in their agreements with foreign governments a "buy American" clause which effectively limited the competition to United States based companies and made the field most attractive. A number of American companies came to rely on their foreign work for 40, 50, and even 60 percent of their total volume. This situation, however, did not last.

European companies based in England, France, West Germany, Italy, Holland, and the Scandinavian countries, with the economy recovery in their areas, were able to grow in an expanding market. Their various governments began a policy of lending their nationals up to 80, 90, and even 100 percent of the face value of a contract obtained abroad. Insurance was put in effect guaranteeing payment by the foreign government of the amounts due. In one case the nationals of the country were insured against loss by overrun in costs up to the limit of the equivalent of $5 million.

Armed with these concessions the European contractors were able to invade what were previously United States preserves in Turkey, North Africa, Iran, Latin America, and even Pakistan and India. Moreover, they were able to offer to perform major works on firm unit and lump-sum prices where the previous rule had been cost plus.

The American contractors were in trouble abroad. They soon found it was impossible to take firm-price work in competition with "consortiums" of British, French, German, or Italian firms and make any money with cost-plus-a-fee organizations grown fat and happy from long feeding in lush pastures. United States oriented companies were forced to trim their foreign organizations seriously and retrain or replace key foreign staff as fast as cost-minded supervisors could be made available.

A number of companies who had formed or purchased control of French, British, or Italian construction and engineering companies and retained key executives native to the country were able to compete successfully both in the country of residence, and alone or with consortiums (joint ventures) of those nationals and secure preferential treatment abroad, permitting work to be bid, financed, and most hazards insured when contracts were bid or negotiated. Since income taxes were waived on work outside the home country and it was frequently possible to arrange waiving of income tax by the host country where the work was to be performed, the United States contractor had only to worry about when and how much tax he would eventually have to pay in the United States.

Generally, however, work abroad became less and less rewarding; the competition became keener and more difficult to meet, and the hazards multiplied. A number of countries in Africa, Asia, and Latin America, where apparently stable governments were overthrown by coup or revolution, became vociferously anti-American and proceeded to expropriate property and equipment of United States firms. In the case of some contractors, they refused to approve progress estimates and they impounded or seized construction equipment.

On those projects wholly or partially financed by U.S. government agencies, it was theoretically possible to secure insurance from a government agency (A.I.D.) covering riot, civil commotion, wars, expropriation, and inability to repatriate funds invested in the project. Unfortunately this almost never worked.

First: A.I.D. insisted that application for the insurance must be made prior to bidding the work and later withdrawn if the project was not awarded.

Second: A.I.D. would not make a firm commitment to provide the insurance without the prior approval of the "host" country where the work was to be performed. Frequently the "host" country would refuse to approve the issuance of the insurance and the contractor would be forced to proceed with the work if awarded, since in most cases he had been obliged to furnish a bid guarantee with his proposal. In other cases the approval of the insurance would be delayed for many months and, when finally granted, the project would be completed or nearly so and the contractor would be obligated to pay the full premium for practically no protection.

Third: The insurance, even if secured in the early stages of the work, was of marginal value. If the owner (the "host" country) decided not to pay progress estimates and the contractor suspended work in accordance with his rights under the contract, A.I.D. would refuse to pay the insurance on the grounds that the issue must be first decided in the courts of the "host" country. The contractor was then stymied, because the case would be arbitrarily decided against him regardless of the facts or of its merits. If the case was too open and shut in the contractor's favor, the trial would be interminably delayed or never come up at all.

Fourth: Finally, in spite of a law requiring that federal grants or loans be withheld in cases of expropriation or refusal to pay by any country, if the United States company being injured filed a complaint regarding the action taken by the "host" country," the contractor was still helpless. In practice this did not work either. Congress would continue to appropriate and the government agency involved would continue to grant to the "host" country large sums for various and sundry purposes, while the contractor was paying taxes (if he had anything left to pay taxes with) to subsidize the rascals who had robbed him in the first place, and could now feel free to rob the next American that came along.

In the meantime European construction companies, properly insured by their respective countries that will protect the interests of their own nationals, very realistically continued to get a larger and larger share in world construction markets, while American construction markets shrank to those solvent countries that could and would pay their bills and neither needed nor expected any United States aid.

3-24. The Changing Domestic Picture

Construction problems constantly change. The problems of the 1940's and 1950's are not the same as the problems of the 1960's, nor will the problems of the 1970's fail to change also. The one constant factor is that all of the problems concern people, and the contractor and his key executives must not only recognize the changes as they develop and change with the times, but they must learn to anticipate change and adjust to it as fast as it occurs, not after it has happened.

In the middle 1960's cost began to accelerate. From a predictable 5 percent a year, costs rose to 7 to 9 and 10 percent a year and are still rising at a faster rate each period. Labor has outdistanced costs, and demands for 15, 20, and 25 percent increases per year are the rule and not the exception. The construction industry, criticized and belabored for not "holding the line," has in fact "held the line" better than most industries—to its own disadvantage.

Housing costs have advanced to the point of driving prospective buyers to "packaged houses" and "mobile homes"; yet contractors, by adopting assembly-line methods, have succeeded in turning out houses which in price represent less than half the increase in unit labor costs where hand labor has to be used. The total increase, pricing housing nearly out of the market, is the cost of mortgage credit and the cost of labor.

Contractor ingenuity and planning have prevented the housing situation from being even more serious than it is. In high-rise buildings, factories, plants, and all major structures prices have gone up less by far than the hourly cost of labor. Architectural simplification and contractor ingenuity have made the difference.

Dams, canals, tunnels, highways, and pipelines cost 50 percent more on the average than they did in 1960, whereas labor costs have gone up 100 percent and labor efficiency (the willingness to work) has dropped 50 percent. Equipment manufacturers deserve much credit for producing faster equipment, bigger equipment—equipment that is much more productive and less subject to breakdown. But to the contractor must go the credit for the foresight to buy it and use it, as well as the ability to devise new methods, better communication and controls over construction scheduling and operations, engineering functions, and accounting and payroll procedures.

During the 1970's and beyond there will be new problems, new methods devised, and new and more perfect and accurate controls. There will also be new opportunities and new needs in housing, high-rise building, and processing and manufacturing plants. Transportation problems will have to be solved and environmental controls will become necessary and commonplace regardless of cost. Needs will increase with the population and will have to be satisfied.

No matter how crowded the field or how tough the competition, contractors who organize well, choose their staffs with understanding and good judgment, deal fairly, and look ahead will succeed.

There have always been some good construction jobs as well as some bad ones. There will always (as long as private enterprise continues to exist) be contracting opportunities. There will always be some contractors with the knowledge and wit to seize them. As Mark Twain was once reported to have said: "Let us thank God for the fools, but for them the rest of us could not possibly succeed."

4 BID STRATEGY

J. B. BONNY

President and Chairman (Retired)
Morrison-Knudsen Company, Inc.
Boise, Idaho

BID STRATEGY is the most difficult of all the functions of management to define. It is the least subject to the rules of logic, impossible of scientific engineering analysis, and yet basically so important that an average contractor with a fair knowledge of the business and a competent organization who lacks an adequate concept of the art is almost foredoomed to failure.

4-1. Basic Considerations in Bidding

Bid strategy, for the purpose of this Handbook, is considered to be the science (as poker is a science) of outguessing and outmaneuvering the competition in order to secure more profitable work. It therefore encompasses guidelines for preparing bids on lump-sum or unit price on building or heavy construction work.

In addition, bidding must cover the strategy of determining what work to bid on or, in cases of cost plus, C.P.F.F., target estimate, guaranteed max, or even turnkey projects, what to propose on. Also it must include how to approach the proposed structure itself so as to have the greatest possible chance of presenting a more attractive offer than the competition and what to offer to succeed in this without destroying profitabilities. The concept must include how to word or price the proposal, since each one is different from all others, in order to get the maximum advantage of changes in quantities, changes in design, and latent or obscure physical and subsurface conditions. There must be an accurate appraisal of the current competitive potential of each of the known or suspected bidders on the given project. A sound knowledge of the owner and the owner's engineer is imperative, since the price or terms of the proposal must be adjusted to take into consideration what their attitude and policy will be under any given set of circumstances that may be encountered during the construction period of the project.

Some labor areas are controlled by local unions so powerful and so corrupt that a stranger venturing into the area can expect nothing but trouble and frustration from the tight club consisting of the local contractors and union officers and business agents. In those areas, almost exclusively in the eastern half of the country, for an outsider to bid in a lump-sum or unit-price job is suicide, and even taking a cost-plus job may do damage to his reputation beyond repair.

These and other situations are discussed subsequently. The happenings recited all occurred; the actual locations and names have been disguised for obvious reasons. A most prominent and successful contractor once said: "We have lots of people who can and do make suggestions of what to get into and where to go, but almost no one who instinctively knows *where* to stay away from and *what* not to try."

In a very general way the topic of what projects to bid or propose on really becomes a dissertation on what projects to stay away from. Having passed the test of being a type of project that may be desirable, there may well be other circumstances that make it better left alone. Likewise, certain types of projects which should ordinarily be avoided for a dozen good reasons may, under certain conditions and circumstances, be the very ones to bid or propose on and may promise less competition and more profit than many others.

All contractors, from the local or regional house builder to the largest national or international diversified constructor, face from time to time a dearth of work to bid or propose on, with consequent increase of competition to the point where the low bidder is almost sure to lose his money and the others are spending theirs in order to prepare proposals that have no chance. Under these circumstances the low bidder, who is probably already in financial difficulty, may have bid in desperation just to keep going until he can "grab a good one." Also, in cases where there are ten or fifteen bidders, there is always one who fails to analyze exactly what he has to do. Occasionally, in these circumstances, the low bidder may have made a more careful analysis of the true conditions and come up with a more workable method.

An illustration of this method of outsmarting the opposition occurred on a large turnpike project in the eastern

34

part of the country. The contract provisions called for bids on an assumed quantity of rock, an assumed quantity of earth, and a very small quantity of removal of slide material by simply 'dozing any slide material out to widen the grade. The sponsor of the joint venture, who had recently completed a railroad job in the vicinity, pointed out that in his opinion the quantity of rock was overstated, that the dirt was understated, and that the cuts would slide and the slide removal material could run into several million yards instead of the stated few thousands. The other partner in the joint venture agreed and the excavation was priced identically for rock, earth, and slide removal and was low by nearly 20 percent. The results were as anticipated; the profits were generous and proved once again that there is no substitute for knowledge.

The contractor, regardless of size or geographical scope, must remember that even in times of shortage of work and maximum competition there are always some good jobs. The contractor who attempts to cover the field and submit proposals on everything that will be offered may, while his prices are being driven down by the competition, miss entirely the good jobs that by careful analysis would have yielded satisfactory profits.

Periods of high volume of work and rising construction prices do not lack their own pitfalls. The tendency, particularly among the large and growing contractors, to propose on everything and spread out too fast can be most dangerous. Overworked engineers trying to cover three or five or ten bids a week will make mistakes. The management tends to relax as the volume rises and fails to follow the practices of practical bid strategy, analysis, and follow-through on methods that can make a poor job a fair one or an average job a very good one. Hasty analysis and preparation of proposals by the engineers and inadequate analysis by the chief engineer, the VP–operations, and the president will deprive the company of the profits it should make and perhaps throw it into losses that in a prosperous era are the result of a lack of intelligent thought and competent bid strategy.

4-2.   When to Bid

The first requisite of successful contracting is an intelligent analysis of whether or not to submit a proposal on a given job. This must be determined by careful study of the contract conditions and meticulous examination of the site of the work by the estimating engineers, by the chief engineer, if advisable, and very possibly by the VP–operations and perhaps the president, unless one of them has had previous experience in the immediate area and has clearly in his mind any difficulties to look for and how it may be possible to avoid them. Even in such cases careless assumptions may spell disaster.

Let us take the case of a large national and international construction and engineering company specializing in designing and constructing large chemical, metallurgical, cement, food, and beverage plants, as well as tire, automobile, and equipment manufacturing facilities. The company is in turn a wholly owned subsidiary of a still larger diversified construction company domiciled in the West. The chairman of the owning company is also chairman of the subsidiary. The president of the subsidiary was trained by the owner company and has had several fairly successful years with the subsidiary as executive vice-president and finally president on retirement of the elderly former president who had been acquired when the parent company took over.

The subsidiary company had designed and successfully completed on a guaranteed max basis two large breweries in the Midwest and central South for a national brewing company and was about to start a third on the eastern Gulf coast. The plans for all these plants were nearly identical. There was no question raised as to the desirability of submitting a proposal on a fourth, this one in the New England states. The subsidiary president simply authorized the bid and went ahead with the estimate, taking previous costs, adding a small contingency and a small factor for published wage rates and a normal markup cf 10 percent for profit, and sent in the bid without bothering to consult with the chairman who in the past had suffered painful experiences on two previous jobs in the general vicinity.

When the guaranteed max bids were received, the subsidiary company had been undercut about 5 percent by another engineering construction firm, itself the subsidiary of a well-known and successful parent.

The president of the first firm called the executive vice-president of the brewery company and was informed that the low bidder, having had no previous experience, would not get the job; instead it would be awarded to the second bidder because of his previous and current experience in the field.

Pleased at getting the award, particularly when not low bidder, the president promptly called the chairman at the home office of the parent company and told him what a good job he was getting for 5 percent over the low bid. The chairman, pleased, was still dubious and inquired regarding the local conditions and labor and the accuracy of the investigation that preceded the bid. He was assured that it had been thorough and accurate, and besides there was 5 percent contingency and 10 percent markup before they could get hurt. The job was a cinch to make the estimate and they would share in savings.

The president, through carelessness and overconfidence, had let his company in for a disaster. He had not followed the basic rules nor had his engineering staff. No one had properly investigated the job; they had simply taken the plans on the previous almost identical work and added a factor for higher wage rates.

The simplest application of common sense and bid strategy would have added up to a different story. An investigation of the proposed plant location would have shown that the site was in anything but a good spot. In the summer, rain would turn the foundations into a sea of mud requiring the removal of 2 to 10 feet of topsoil and replacement with stable gravel or crushed rock. Certain areas would have to rest on pile foundations for which nothing had been provided in the estimate. Both these unconsidered items added up to more time and extra cost.

The climate in the proposed area was abominable. Summer rains caused much of the work, until it could be roofed over, to be suspended for days at a time. A short dry fall was immediately followed by a long cold and snowy winter; since most of the plant was still limited to foundations, work had to be suspended for several months. This meant that all the work would have to be accelerated when spring or summer weather permitted, again losing efficiency and aggravating the costs.

Had anyone bothered to inquire they would have discovered that labor was scarce and the quality was poor. Most of the available men were local farmers who, when the fall harvest started, simply disappeared.

The local labor unions were both arrogant and intransigent. Unable to supply men, not to mention competent men, they refused to permit bringing in skilled artisans from the locals in the larger New England cities and New York state. Unauthorized and illegal strikes were fomented and called frequently by job stewards, and days would be lost trying to reach local and international union officials to get the men back to work.

The owner's representative, as is typical when things are going badly, was highly critcal of the contractor's organization and sent almost daily reports to the owner condemning whatever was done. Computer analysis and scheduling of the job, in effect from the start, accomplished nothing, since men could not be secured to perform the urgent items of work as so scheduled.

Early and urgent appeals to the owner to permit a slower and more realistic schedule and for additional compensation to help make up for overruns in cost were largely ignored. The owner quite reasonably pointed out that he had already agreed to pay much more for the facility than the lowest bid just to avoid what was occurring. The job was behind schedule and the contractor must find the way to catch up.

In the end the job was completed and the extent of the disaster had to be admitted. The subsidiary company president was fired and the parent company had to take the loss.

The new president was advised by the parent company chairman that the old president had been fired for failing to use his head. If the job had been properly analyzed and sensibly bid, one of two things would have happened. Either the difference between the low and second bidder would have been so great that the low bidder would have refused to sign the contract and the second bidder would have a decent contract, or in the remote chance that the low bidder was awarded the work and signed the contract he would have been in such serious trouble that he would never bid on a brewery again and the second bidder would be a hero and continue to get future work on every brewery the owner built.

A sound understanding of the abilities of the owner's engineers, their general attitude and policies, and the general practices of a federal, state, or regional agency or public commission is a must.

Specifications and contract terms vary and all or nearly all contain "gray areas" of possible misunderstanding or variations of interpretation, as between the contractor and the engineer during the progress of the work. The simpler and less detailed the contract documents are, if written fairly and with intent to protect the rights of both parties to the contract, the fewer "gray areas" will exist.

The contractor who is deciding whether to bid or, if he decides to bid, what to add on or subtract from his bid "for the engineer" is exercising one of the most important elements of "bid strategy."

4-3. Influence of the Owner's Engineer

If the work to be bid on is complicated but within the capabilities of the contractor and the time is short for construction, the contractor must decide whether the engineer can be trusted to avoid engineer-caused delays and cooperate toward expediting the work to permit construction on time and within the bid cost.

If the contractor decides that the engineer *is* competent, *will* make prompt decisions, and *will* render fair rulings in "gray areas," then subject to his decision on his ability to beat the competition he will bid the job on a basis to win, and with minimum contingency money to cover only direct physical hazards. On the other hand, if the engineer

has a reputation for trying to "break" contractors and frequently ends up in court, the wise contractor will simply "pass up" this one and look elsewhere for work.

One of the areas where the engineer can't help the contractor a great deal but can injure him severely is subway, tunnel work, and underground powerhouses.

The problem frequently arises around the installation of steel "roof bolts," long steel bolts drilled into laminated or soft rock from 3 to 20 feet or more and grouted in before installing a steel plate against the roof with a large nut and perhaps wire mesh to keep the rock from falling until concreted. This method used with varying centers from 2 to 8 feet will, if placed immediately on opening up the face of the tunnel or perhaps the underground powerhouse heading, prevent the otherwise unstable rock from getting in motion and eventually crashing in large rock falls hazardous to men working below; at the same time it protects the safety of the entire roof until concrete lined.

In rock faces that show a tendency to "slack" or break down when exposed to air, a thin coat of gunite, cement sand, and water blown on the surface under air pressure will seal the roof and sides and prevent the slacking. Furthermore, should pressure develop in the roof, the gunite will show minute cracks prior to structural failure and permit additional roof bolting or rib supports to be installed before the rock actually falls.

Small quantities of roof bolts, mesh, and gunite are frequently included in the bid quantities and specifications, usually with the stipulation "as ordered in advance by the engineer" and with the further stipulation that "any quantity placed by the contractor for his own convenience will be for his own account," or perhaps a very small stipulated price, insufficient to pay for the bare materials let alone the labor, for "quantities placed by the contractor for safety protection during construction."

These items can run into considerable money—on a large job into one or more millions—and may spell the difference between profit and loss on the project.

In tunnels where the ground is so soft that it requires the installation of steel supporting ribs at regular intervals, the engineer may authorize permanent ribs at say 8-foot centers as a pay item. The ground may show signs of swelling and the contractor, knowing that quick action is required to prevent a cave-in, may be forced to install "jump sets" between the authorized spacing to provide necessary support.

In the case of the rock bolts and gunite and that of rib spacing, the engineer, if so inclined, will simply disappear at the time the decision must be made, claiming later that the work was performed "for the convenience of the contractor" and refuse to pay. This in spite of the fact that, if the contractor had not provided the additional protection, a disaster might have occurred for which the engineer would be responsible.

The writer is reminded of a case involving a large vehicular tunnel in a western city. The specifications called for an "A" section with lighter ribs and lagging and 18 inches of concrete in the better ground and a "B" section with closely spaced and heavier ribs and 2 feet of concrete in the least stable section. It became necessary early in the driving to install much heavier ribs and lagging to hold the ground. There were many houses on the tunnel line some 500 feet above, and a cave-in would have resulted in heavy property damage and probable loss of life.

The contractor claimed payment as "B" section at a much higher price for the length of the tunnel protected with the heavy supports. The city engineer refused, saying that the contractor had used the heavy supports for his own convenience and the fact that it was so well supported that the city could use the lighter 18-inch lining throughout made it "A" section, and paid at the lower price.

The contractor learned something about bid strategy at that time. If he felt that he *had* to bid on a job for hostile and sadistic engineers, he should have bid the same price for the "A" and the "B" sections. His bid would have been a million dollars higher but he would still have been low and would have ended up with a profit instead of a loss.

In one large eastern city there is an "authority" or "commission" commanding a large slice of all the available construction funds. The policy of this body has consistently been to try to break every contractor foolish enough to bid on its work. If the specifications are badly written, calling, for example, for concrete piles under a building or area to insufficient depth, and as a result the contractor following the letter of the specifications comes up with an unstable foundation, the authority promptly sues him to make him furnish and drive longer and heavier piles at his own expense.

The tale of the injustices and malfeasances of this particular body are too numerous to mention. Suffice it to say that their vicious tactics gradually became known. They received fewer and fewer bids on their work until finally came the day when they had none at all. The "authority" is currently supposed to be being "reorganized" and is begging by letter to contractors all over the country to get proposals submitted on its work. They *may* have reformed and some courageous contractor *may* succeed in making some money on their work, but certainly good bid strategy would be to wait and see.

The writer remembers a conversation some years ago with a general who was at that time Chief of the Army Corps of Engineers. The general, an upright and fair-minded man, was a great credit to the service and was sincerely looking for better army-contractor relations.

He asked a question: "Why is it that on an identical piece of work you will bid 10 percent lower to a railroad or power company than you will to the Corps?"

After some thought the answer was: The railroads and most power companies are only interested in getting the most for their money. Their specifications are simple and concise. If a contractor does his work well and on time, and gets hurt through some unforeseen circumstance not his own fault, they see him out. It is therefore possible to bid the work without contingency, and because there is little or no interference with the operation it is most economical."

"I see," said the general. "It makes sense. I'm going to point this out to my district engineers."

4-4. Influence of Competition

A shrewd analysis of the probable competition and its potentials in deciding whether to submit a proposal on a given project, and prior to bidding if the decision is affirmative, is most essential. A simple rather straightforward project in a familar area with average to good climate and an ample supply of competent labor, supervised by capable and reasonable engineers, may nevertheless offer little or no possibility for profit.

Let us assume that the job in question is a high masonry dam for a local public agency. Careful inquiry from material and equipment salesmen indicated there would prob-

ably be eight bids (some agencies publish a list of those procuring plans and even a list of probable bidders). All but two were joint ventures of national contractors.

The two bidding alone and separately were both capable of handling the job and each had the reputation of turning in very low bids if they particularly wanted a job to fit idle equipment or personnel. Of the two, Company A would be very hard to beat. It was just completing a sizable dam in a neighboring state. The equipment would fit this job and a whole organization was available.

The president of the second company thought it over. His company had no equipment available. It was tied up on another dam just in its early stages, as was its top dam organization. A subway tunnel in soft ground to be driven under air was up for bids two weeks later. It was the first section on a large new project. There were no sandhogs in the area and they would have to be trained. No one, however, had any advantage on the subway job and the competition would not exceed four or five or five contractors. It could be a contracting opportunity and just might go for a good price.

The president talked it over with the executive VP–operations and the VP–engineering. They finally agreed that they would concentrate on the subway, bid it with plenty of contingency and a good margin, and pass up the dam. The strategy paid off. The dam went "too cheap" to to Company "A." The bid on the subway was successful. The job was profitable and led to several others as the transit system went forward.

It is common practice among government agencies and municipal or state authorities, in letting contracts on large projects extending over a considerable mileage, to split the project into sections. This is also true for combination flood control and power dams, as well as straight power installations. In the case of the power and flood control dams and installations, the only purpose of splitting up the project seems to be to make the individual contracts sufficiently small to attract a larger number of bidders, thus increasing the competition and presumably reducing the price. It does not always work out this way, however, since frequently the contractor on the dam is the only one really interested when the powerhouse comes along. The subcontractors on the machinery installation, electrical installation, etc., tend to quote the contractor on the job their best figures, and since he has a lower overhead and no move-in cost, being already on the job, he frequently gets the powerhouse and perhaps the switchyard and even the transmission line at a higher price than it would have cost the agency or the district had the entire job been let as one contract in the first place.

This situation is reversed, however, when large aqueducts, for example, and irrigation or municipal water supply canals and tunnels are contracted in increments over a period of time as the money becomes available or the plans are finalized for each section.

The first contractor on the first section let is likely to have the best job on the entire line unless, as occasionally happens, the low bidder overestimates the competition and cuts his bid to at or near the break-even point. In this case the succeeding sections will be bid in progressively lower until, when the last few sections come up, there is no profit in them for anybody. This is, of course, bid strategy in reverse. It results from false pride in wanting a section of the big and well publicized project "no matter what" and is bound to be expensive.

This is reminiscent of a meeting between three joint-venture partners to finalize the bid on a large dam on the

Mexican-American border in the Southwest. One partner was very enthusiastic and wanted the prestige job badly. He had come in with a cost and bid figure several million below the other two partners. A long argument ensued and nothing was being resolved. Finally the sponsor, a large and very successful southern contractor, turned to the partner with the low figure and drawled: "Sam, falling in love with a job is like falling in love with a loose woman—it's bound to cost you a lot of money." That settled the matter. Another joint venture got the job at about the price Sam wanted to bid. Some three years later when the job was done, one of the partners in the "successful" joint venture admitted that it had lost "over three million dollars."

4-5. Overcaution

Not infrequently an excess of caution can cause a contractor to pass up a very good job. The contractor who had wisely chosen to pass up the dam in favor of the more complicated and less attractive subway work had been most successful. His first and two succeeding tunnels had made money, and an excellent crew of sandhogs, many of whom were hard core minority unemployables, had been trained and worked well. One of the later sections consisted of a section of tunnel and a station in filled ground running under a number of buildings and ending up under 30 feet of water. The president, who had been born and raised in the city, had read the history of how more than a hundred years before a number of ships had been abandoned and sunk in the mud by their crews who took off for the gold fields. The number was estimated at anything up to three hundred vessels buried under the later filled land.

The contractor with whom they had joint ventured on the other subway sections suggested that he would like to sponsor this last section in the filled ground. He pointed out that the specifications provided for daily pay and time extension, in case hulls, anchor chains, or other ships' debris were encountered. The president demurred, saying the authority could not pay enough to get rid of these old hulls if they were encountered. He suggested to his friend that he find another partner, as he preferred to have no part of that section.

Another partner was found and the job bid in and awarded. The job went smoothly and not so much as an old mast or anchor was encountered. Needless to say, the job was highly profitable, and the president of the company who had turned it down was very red in the face and had to admit to himself that he didn't know all there was to know about bid strategy.

4-6. Influence of Experience

The decision having been made to submit a proposal on a given contract and adequate study having been made to ascertain the fundamentals affecting cost, the final decision on bid price, target price, or guaranteed max hinges on who the competition is and how each one is likely to approach the particular job. Many contractors for one reason or another have certain types of work which they will favor and "hit harder" than others. It may be that they are more experienced and have better types of organizations for one kind of work over another or that they are "long" on equipment in certain fields and "short" in others. Whatever the reason, the tendency must be considered in evaluating the probable proposal each one of the competition

will turn in. Certain contractors are very hard to beat in their home areas, and others may favor certain areas because of better official connections, union relations, climate, or past experience.

Given experience and a good memory the proposed bidder can come up with a fair probability among, say, eight expected bidders as to which ones are not likely to be among the first three. He may decide at the last minute to waive submitting his proposal rather than look too high to the owner, or he may bid as low as he dares, in order to maintain his position for future invitations, hoping he does not really have to take the contract at his price.

One well-known and very successful western contractor, an artist at bid strategy, could not resist turning up at a bid letting at the last minute, too late for anyone to change his bid, wearing an old, stained, and motheaten brown hat. This was notice to the competition that he was going to be low bidder and *had* the job. Strangely enough the signal was almost always correct, and whenever he showed up in his "low bidder hat" most of the competition would give up and leave, figuring they had no chance.

4-7. Influence of Propaganda

Bidding propaganda between competitors sometimes takes the form of elaborate planning and skillful execution. When proposals are called for on major projects away from the population centers or bids are scheduled to be opened away from the site of the work, at engineer district or division headquarters, naval districts or perhaps state capitols, competing contractors tend to gather at one or more leading hotels to finalize their proposals and pick up any available gossip. Gathered there also are the various equipment dealers' representatives, material and supply salesmen, and interested surety bond and insurance people, as well as prospective subcontractors.

Opinions are expressed by all concerned. Interested contractors, if at all loquacious, tend to hold forth on the difficulties of the job, the dangers of the tough specifications, the shortage of good labor, the unpredictable weather, the vicious labor unions, the impossibly short schedule, the insufficient funds or any other factor surrounding the job. Such comments may lead the listener to believe that the speaker has been so disillusioned about the work that he will certainly bid a very high price and, due to the multitude of difficulties, may not submit a proposal at all.

Meanwhile the suppliers' representatives are circulating among the various contractors, joint venturers, and their engineers and picking up any and all crumbs of information to pass on to all the bidders or to their regular clients and particular friends.

Unbelievably, from all this chatter, exaggeration, and untruths a pattern may sometimes develop that gives some indication of how the competition actually feels about the job. A material salesman may report (particularly to the contractor who uses his price or perhaps gives him a chance to meet *his* competitors) having heard two engineers with two different contractors in a competing joint venture say something to each other like: 'We haven't a chance unless Jim (chief engineer of a third partner in the venture) gets off his high horse and gets realistic." This indicates lack of agreement among the estimators in *that* joint venture which may influence their price up and out of range. This does not necessarily change the answer, because several other bidders may be "pointing" at the particular contract; to be of any value, therefore, it is necessary to get some sort of fix on what each one is thinking.

By this time our contractor and his joint-venture partners are in general agreement as to where they should like their bid to be and also how low they would be willing to go. The engineers are in the process of writing up the proposal, which is going to be presented at 10 a.m. the following day. They have penciled in four or five major items which can be quickly erased and inked in at the last minute. This will take care of final material prices, seldom received until the last hour before the bid opening, as well as a hunch or change of opinion by the partners just before the last moment of inking, sealing, and delivery.

Let us say that the partners, after analyzing all reported conversation, decide that Company X is going to go after the job and will be the toughest competitor. The partners have confidence in the sponsor and will no doubt go along with the president of Company X in the final maneuvering.

One of the partners of our contractor has joint ventured with Company X in the past and thinks he may be able to get a reaction from its president. Our contractor and his partner have agreed on a total cost, without interest, surety bond, contingency, or markup, of $17 million. On the basis of a $20 million bid, the surety bond (1 percent of bid for the first two years) will be $200,000. A contingency of 5 percent of bid is agreed for labor increases and material and supply parts raises. This makes a million for contingency. Interest, although paid by each partner on his proportionate contribution to operating capital, is nevertheless a true cost. It is agreed that up to $5 million cash will be needed for an average of twelve months, and interest is pegged on the scale at $450,000. These items add a burden to cost of $1,650,000, making the new final total cost (subject to final changes in material quotes) $18,650,000. The partners did some quick scribbling. The sponsor scratched his head. "That's a very poor place to come out," he said. "I know it's mostly superstition but look how often it works. If we bid 10 percent markup, we are about $500,000 over the $20 million mark. If we bid 12 percent, we are $20,880,000. We have to go clear down to 7 percent to get under $20 million and then we're not quite $300,000 under the $20 million. Someone, if he gets down there, will squeeze out a little more to get under $19,500,000, and the job's no good at that figure."

After much discussion it was agreed that Bill, one of the partners who had previously joint ventured with Company X and had a good knowledge of their reactions, would try to find Jim, their president, and try to get an indication out of him. The others would await his return.

Two hours passed and the partners were getting tired and bored. Finally Bill returned shaking his head. "I found Jim in the bar; we talked quite a while. He didn't say much, but you know he has one eyelid that twitches a little when he is excited. When I said that anyone bidding under thirty million was crazy, his eye twitched. I have a hunch he'll bid over $25 million and under $30 million. Unless we find out something else in the morning, I suggest we get down under $25 million."

In the morning, nothing new having developed, the bid was settled at $23,500,000. Bill's hunch was right; Company X and partners were second low at $27 million. Bill's joint venture left $3,500,000 on the table, but since the engineers' estimate, announced at the bid opening, was $3 million under the low bid, they could not have bid much more and received award.

The custom of last-minute bid maneuvering does not always work. As often as not too much interpretation of the gossip results in raising the normal bid too high and losing the work, or in cutting too much off and getting a job with little or no profit or even with a loss. Proper analysis, however, of rumor, gossip, and statements from the competition frequently results in getting a better price if weighed with all the other factors and used judiciously.

4-8. Contract Modifications

In public works contracts as a rule the wording of the proposal cannot be changed without invalidating the bid by making it "unresponsive." Frequently in proposals to private companies the contractor may word his proposal as he sees fit and, in most cases, modify the proposal form of the owner by offering concessions or price reduction if unsatisfactory clauses are eliminated.

Let us take, for example, a case where the owner's engineer, to protect himself and the owner, has included a clause which in essence states: "Any errors or omissions in the plans or specifications or statements which do not properly describe the subsurface or latent conditions in connection with the project and which result in additional cost to the contractor shall be for the contractor's account."

Various versions of this paragraph are cropping up more and more in private, district, or commission contracts where private architect-engineers are employed. The contractor, on reading this or a similar paragraph, properly assumes that the engineer is attempting to pass on to the contractor by means of an unfair specification a responsibility that is by its very nature solely for the account of the engineer paid by the owner.

The contractor is immediately faced with a problem. Shall he refuse to bid and pass up what might otherwise be a very desirable contract? Shall he stipulate in his proposal that the said paragraph is null and void? There is, in fact, another way open to him. He can prepare and price his bid as though the objectionable paragraph did not exist. He can then add on to his base proposal an amount sufficient to protect him in any event, perhaps 15 to 20 percent. He then attached a stipulation to his bid offering to reduce the price to his original proposed figure if the offending paragraph is removed. The contractor has not by this stipulation made his bid nonresponsive, because he is offering to do the work in accordance with the terms of the plans and specifications. He has merely offered to cut his price substantially if certain wording to which he does not wish to agree has been removed. It is a rare owner who will insist on retaining a palpably unreasonable paragraph when he learns it will cost him a substantial sum to leave it in. If the contractor's deductive stipulation makes him low, the owner or his engineer will usually award him the contract.

On federal, state, and even county or city work, a stipulation that is nonresponsive is almost always thrown out and voids the bid, but there are certain circumstances where a deductive stipulation may be accepted. For example, if a contract for several different sections is opened for bids at the same time, there are significant alternatives open to the contractor. If the sum of all the sections makes a single contract too large for him to finance and equip, the contractor may stipulate that he will accept award of any one section and no more, thus improving his chances of getting a job if he is low on one or more sections.

The large contractor may, on the other hand, tie all the sections together and state that he will accept award of all sections bid upon or none. This procedure frequently permits the large contractor to get award of an entire project even though certain sections may have been priced below him. Finally, in bidding a project of several sections, the bidder may make deductions for award of certain combinations of sections such as: "If awarded all sections one through six, deduct $1 million from the final estimate and retained percentage" or "If sections one, two, and three are awarded, deduct $500,000 etc." By these stipulations the bidder has improved his chances of getting the combination of work he most desires but has not rendered his bid invalid. In each case he has quoted a base price in accordance with the plans and specifications and invitation to bid.

4-9. Unbalanced Bids

In most government and in many state and local contracts offered for bids on a unit price rather than lump-sum basis there is included a clause which in essence states: "If in the judgment of the engineer a bidder has unbalanced his proposal so as to cause his unit price to be unreasonably high on certain units of work and/or unreasonably low on certain other units, the said proposal may be declared nonresponsive and rejected."

Certain types of unbalancing are, however, practiced in normal bidding procedure, are a part of bid strategy, and unless carried to an obvious and flagrant degree are ignored by the engineer, who then loses his chance to question any unit price after award of the contract.

A few illustrations of the strategy of unbalancing will serve to show the reasons and the intended results. Unbalancing to help finance the job is most common.

The project is a large masonry dam with powerhouse. If each item is bid with only its own proportionate charge for move-in, plant, and equipment and overhead as well as profit, the mass concrete in the dam will carry the distributable costs and profit over most of the time of construction. This results in little or no early pay for move-in, plant and repair facilities, camp construction if any, and financing costs, bond and insurance premiums, and defers the time when withholding a percentage of the monthly pay estimates, usually set at 50 percent completion, is waived and estimates paid in full thereafter. The interest alone is significant, and the sooner the contractor or joint venture is relieved of a large investment in the work, the less interest has to be paid, which, in turn, reduces total cost and increases the margin of profit.

It behooves the bidder therefore, within the limits of what the engineer will accept, to increase the bid price of early construction items and deduct the increase from later items, in each case from items which it appears will not greatly increase or decrease. To be on sound ground, the contractor may compute the early expense items separately, charging them largely or all to the earlier work items. If the engineer questions the high price of early items compared to later ones, it can be shown that the procedure is merely one of writing off early costs at an accelerated rate rather than deferring them throughout the life of the job. This procedure is rarely questioned, since it may well result in a lower offer to the owner.

In general practice, unbalance to generate early money is discussed by items between the contractor's chief engineer and the president or VP–operations or the joint venturers' chief engineer and the principals at the final meeting after costs have been agreed upon and final markup discussed.

To return to our dam and powerhouse bid. Let us assume that an agreed total cost, including overhead on the project, has been decided at $34 million, and that a markup for contingency and profit of 15 percent brings the total agreed bid to $38,100,000, subject to any last-minute changes in material prices that will be deducted from or added to the mass concrete item and the powerhouse, which with the exception of the machinery installation is bid as a lump sum.

The principal early items consist of a diversion tunnel concrete lined, 2000 feet long and 30 feet in diameter, round and lined with 12 inches of concrete. It is bid by the foot complete, but unit prices are provided for a small amount of rock bolts "as ordered by the engineer" and likewise a small amount of steel rib supports. The agreed balanced bid price is $350 a foot for the tunnel, $1 per foot for the rock bolts, and 60 cents per pound for the steel supports.

The next early item is excavation for the dam foundation bottom and sides, 500,000 yards including the keyway trench up the steep sides. The normal bid price for this was agreed at $4.50 per yard.

The river was a hazardous one in a steep canyon with a spring snow runoff of up to 50,000 cubic feet per second—all the single diversion tunnel could handle with 150 feet of upstream cofferdam. It would require quite a bit of pumping—up to 2000 second feet—to keep the bottom dry until concreted. The diversion bid lump-sum balanced was $1,200,000. However, as is frequently the case, the specifications called for 40 percent of the bid price to be paid on this item when the water was diverted through the tunnel and the balance when the dam reached elevation 2255, well above flood danger. The cofferdam, although a fixed sum, had been rendered relatively unattractive for use in unbalancing for early money.

The mass concrete in the dam amounted to 1,500,000 cubic yards and worked out to bid at $15 per cubic yard, $22,500,000, not including separate unit prices for cooling pipe installed, reinforcing steel in certain critical spots including grout galleries, etc., and final grouting of both the concrete and surrounding rock.

The last two important late items were the powerhouse foundation and concrete bid agreed at $1,100,000 lump sum, and the gated overflow spillway excavation and concrete at $4 per yard for 50,000 yards of dirt and rock and $30 per yard for 30,000 yards of spillway concrete. Numerous minor quantities made up the total of the bid schedule in addition to a bid price for furnishing cement in bulk for the job and a lump sum for cooling concrete before and after placing to predetermined temperatures.

It is agreed that no items unbalanced for early money shall be bid at less than total cost plus 5 percent. Analysis of later items shows that between mass concrete, spillway, and powerhouse there exists approximately $2,100,000 available to add to early items for quick pay. A rough calculation in the meeting shows that raising the tunnel price to $650 per foot and the dam excavation price to $7.50 per cubic yard will take care of the early money, with corresponding decreases to cost plus 5 percent in the mass concrete and the powerhouse.

The next consideration becomes the possibility of unbalancing selected items up for those that may increase, deducting likewise from those that will remain the same or may decrease. This is an inherently dangerous practice and should only be attempted within reasonable limits. Only rarely can a substantial profit be acquired by unbalancing for quantity increases, and a substantial and intimate knowledge of the site and detailed and concentrated study of the specifications is a must.

To return to our dam. The basic geology was igneous. Nearly vertical side walls showed columnar basalt in spots, particularly above the dam crest where the base of an ancient lava flow began. Below, within the dam's water level, lay conglomerates, weathered to a degree tested competent but, by detailed examination, possibly subject to caves and underground channels. It might take a lot of grout to seal off the reservoir. Too, the small amount of rock bolting and steel supports in the tunnel might overrun considerably and it might be necessary to gunite all or part of the roof before concrete, although no quantity had been provided, and, if used, would have to be negotiated.

A check on the unbalancing for early money already agreed upon revealed that, if the dam excavation greatly increased, there was already an unbalance agreed to that might prove highly profitable. True, each additional yard within the dam's prism of rock would require nearly a yard of concrete at the low unbalanced price. However, above the dam's crest and under the columnar basalt was a different story. A moderate increase in the excavation below the crest would cause a much larger excavation of the basalt above. The dam excavation might double or more with a relatively slight increase in the concrete quantities. Thus a fool-proof unbalance had already been agreed upon.

There remained only the grout and tunnel bolting and supports to consider. A mere $100,000 would double the price at the estimated quantity of bolts and tunnel supports. Another hundred would raise the grout price as listed in the specification quantities 50 percent. A suggestion was made. "The present quote on cement is 30 cents a barrel too high. Let's bid the cement at bare cost and put the cut in to increase the grout, tunnel bolts, and supports. With any luck we can get a cut in the cement later and the unbalance will have cost nothing. At worst, the higher prices on grout, bolts, and supports will more than pay for themselves." It was so agreed, the bid was successful, and the job profitable.

There are occasions when unbalancing to take advantage of probable increase may take an odd turn and even work to the contractor's disadvantage. The practice in one branch of the government for a number of years was to include in many but not all job specifications a paragraph to the effect that "any stated quantities which overrun or underrun the estimated units by more than 25 percent may be the subject of a change order to arrive at a new unit price commensurate with the effect of such change upon the cost." This paragraph was assumed by all parties to be for the protection of the contractor, since a serious decrease in a significant item might well cause the contractor to lose a large part of the plant and equipment and other organization costs set up for the particular unit. Likewise, a large disproportionate increase in a unit might cause the contractor additional cost in special added plant and equipment and overhead as well as delaying other significant units with attendant acceleration and expense. Furthermore, if the contractor had inadvertently bid the particular unit too low and below his actual cost or unbalanced too far down in favor of another unit he had thought would increase, he could claim reason for an increase and thus come out whole.

The clause in question became, by usage, an accepted principle and, on occasion, widely varying quantities from the stated bid unit quantity were the subject of additive change orders, even though the particular paragraph was not included in the job specifications.

The writer is reminded, however, of at least one occasion where it did not work out that way. An experienced contractor in tunnel work bid on a large job in an area with which he was most familiar. The specifications carried a bid unit price for pumping water encountered in the tunnel. The unit price called for a quotation per thousand gallons for an insignificant quantity. The contractor, convinced that the quantity would overrun many times, bid $3 per thousand for pumping and, being the successful bidder, started work.

The water was encountered where the contractor expected to find it; more than 2000 gallons a minute flowed to a sump was pumped up a rather shallow shaft.

After some 50,000,000 had been pumped with no apparent relief in sight, the enginers became aware that the water pumping might cost more than the tunnel. The principle of excess quantity was invoked in reverse. The The contractor was asked to negotiate a new price under the "changes" clause and refused. The government resorted to the clause which stipulated that, in case a changed price could not be agreed upon, the contractor must continue the work, the government paying "force account" until the matter was settled. "Force account" consisted of bare field cost plus 15 percent for overhead and "profit." Since 15 percent did not quite cover the overhead, the profit was zero.

Some two years later after numerous meetings, conferences, and appeal board hearings, the matter was compromised at a figure which paid the contractor's lawyers and a very small profit. But worst of all, the basis was laid down, if not generally accepted, that the principle of overrun quantities could be interpreted for the owner as well as against him.

A more conventional approach to the problem of quantity changes, although not connected with any unbalance of prices, occurred in connection with a large dam built in the west for a government agency. The high masonry dam called for some 2 million yards of concrete in a gravity arch section, a type of structure so designed that the weight and strength of the concrete would stand safely without the additional stability of the curved arch.

Shortly after the excavation and the camp, shop, and roads were started, the government agency, having completed final checks and model tests, decided to decrease the amount of mass concrete by approximately one fourth by simply thinning up the entire section.

The agency notified the contractor to make preparations to negotiate a deductive change order covering the new plans. Both parties were surprised at the results.

The form costs remained the same for the narrower blocks and increased the cost per yard on the reduced yardage proportionally.

The pouring cost increased per yard because the capacity of the batching, mixing, and cableway plants was such that the new smaller blocks reduced the size of the pours and limited the full productive capacity of the plant.

The cleanup of each lift remained almost the same since the number of lifts and blocks had not decreased.

The aggregate plant cost did not change nor did the long belt line from the storage piles to the batcher. The saving per yard was only the operating cost on the eliminated yardage of concrete, about 30 cents per yard.

The cement saving was minimal, since the three sacks per yard first contemplated for the original section was raised to four sacks for the new section.

The cooling cost did not go down because the four sacks of cement per yard required more cooling.

The overhead and distributable costs did not decrease since with reduced concrete plant output the time remained the same.

The agency agreed that the contractor should not suffer a loss of profit due to a change for the government's benefit, so there were no savings there.

Savings in fuel and power were computed and came to only a few cents per yard.

When the change order was written after very careful analysis by both parties, the saving to the government was about 10 percent of the bid price of the eliminated concrete and the contractor came out about as he had expected to in the first place.

4-10. Contingency Allowances

Being able to recognize and provide for unusual hazards is one of the most difficult phases of bid strategy. It comes within the subject of contingency and varies with every job. The anticipated troubles may not happen and frequently do not. Unusual hazards almost never happen as forecasted, but something else completely unthought of usually does.

Many contractors contend that if you include 5 percent in every job for contingency, sooner or later you will use it all up, although it may be all on one job. Even more contend that no matter what you add for markup, half of it on the average will go to take care of unexpected contingencies. Be that as it may, an analysis of contractors' profits and a discussion of the number of annual failures leads one to believe that among other things, all too little attention is paid to unusual hazards. To attempt to list the things that can happen is hopeless—anything can happen and usually does! Sometimes, too, unusual circumstances prevent an occurrence forecast, provided for and expected, much to the relief of all concerned.

Let us take the case of the international contractor who, currently building a large jet airport near a city in Australia succeeded in negotiating on unit price a large railroad in the "outback." The climate along the 400 miles of the railroad was ferocious—desert heat and barren rock. It would be necessary to use movable camps along the right-of-way and, in the early stages, fly in most supplies to small fields along the line.

The principal difficulty with the airport near the large city was and had been labor. Not only was the labor largely unskilled but it was universally intransigent and in short supply.

Based on current experience and the need for more than two thousand men on the railroad it was concluded that it would be necessary to import practically all the labor. The company had long done work in Mexico. Railroad workers there were abundant, skilled, and would be attracted by a contract at Australian rates.

Australia discouraged the immigration or employment of Asiatics, but the Labor Department heads, after some discussion, agreed to the importation of Mexicans on contract provided that the Australian labor unions did not object.

Further discussion with the union resulted in an agreement that if the company advertised for the full complement of men in the newspapers and on the radio for two weeks without getting sufficient response, the union, if the company paid union dues for the Mexicans, would permit them to come in on contract for the railroad job.

Sure that they would be needed, the company's Mexico City office put out inquiries and within days received enough applications to ensure the quota. Meanwhile a ship had been located available for charter and able to carry two thousand Mexican nationals.

Then the results of the ads in the Australian papers began to appear. Australians applied by the thousands, three men for every job. They wouldn't work 10 miles from home, but the appeal of the frontier in the "outback" aroused their spirit of adventure.

The job was manned, the men worked well, and the vice-president in charge had learned something about the great continent of Australia as well as the unpredictability of unusual hazards.

4-11. Construction in Other Countries

As a company grows and becomes diversified, it tends to spread geographically. At this time the question of foreign work arises. The subject of the problems and desirability of foreign commitments is discussed elsewhere and only the subject of related bid strategy will be mentioned here.

In practice, work in Canada is considered only an extension of the problems to be encountered in the United States. Canadian contractors bid work freely in this country, often in joint venture with United States companies, and a like custom prevails in Canada.

For all practical purposes, a Canadian subsidiary or a branch of the company operating as a division in Canada becomes the same as a district in the United States and is headed by a Canadian or American manager reporting to the home office direct, and not through one of the two or more foreign vice-presidents covering various areas abroad.

Latin America as an area has become less and less attractive in recent years. Mexico, the most stable of all countries south of the border, has developed considerable capability among its own national contractors and with it a strong national pride, so that little if any opportunity exists for outsiders in the construction field.

The Central American countries, Guatemala, Salvador, Costa Rica, Nicaragua, and Panama, have all had a series of political regimes, each more anti-North American than the last, as have all the South American republics with the exception of Venezuela and Brazil, where opportunites have become more and more limited. The Caribbean Island countries outside of Cuba still have some United States financed work to offer, but again the field is limited.

Europe and Great Britain offer opportunities in certain fields, but the European contractors, many of whom are strong, well financed, and capable, are taking most of the market. The American contractor attempting to compete must not only know his business but have a real flair for bid strategy. He is dealing with experts.

In all other areas, with the exception of Australia and New Zealand, the French, West German, and Italian contractor is very tough competition. These firms have developed a bid strategy that up to now American firms have failed or refused to learn.

The European contractor approaches a bid as the start of a negotiation. He usually bids a figure that the American cannot meet. His method is simple. First he analyzes the plans and specifications to see what may have been left out. He then decides by careful inquiries what the owner expects to pay. With the help of his own government in guaranteeing loans, he offers liberal credit for most or all of the financing required. He then reduces his price to the point which he believes is below competition, but he adds stipulations to his proposal reducing the work he will do or otherwise cutting the cost. He does not expect these stipulations to stand; he only wants to be sure of coming up with what appears to be the low figure to get a chance to talk.

Since many of his stipulations are impossible of acceptance, by putting in the low bid he is, after the bids are open, able to determine how much to charge for the *removal* of each objectionable stipulation and still *remain low*.

The American contractor has, in the meantime, assumed that he must do what the specification says and has priced his work accordingly. Having very few or no stipulations to give up, he really has nothing with which to negotiate.

In Europe and the areas of high competition with European contractors, the exception being strictly United States financed work, it is often desirable to join forces with local contractors in what is commonly called a "consortium" rather than a joint venture. The principal difference is that the consortium frequently divides the work into sections where physically feasible. Each of the principal partners then takes responsibility for his portion of the work, the administrative functions being handled by a committee of the partners rather than a sponsor. In other cases where subdivision of the work is not practical there may, indeed, be a sponsor who heads up the operation chairing an operating committee and who usually by agreement is paid a management bonus in addition to his participation. Thus there may be eight or ten partners with small participations of 5 to 10 percent and one, the sponsor, who has slightly more, perhaps 15 to 20 percent plus an agreed upon management fee, frequently the first 3 to 5 percent of the total profit off the top. That this system works quite well testifies to the sincerity and spirit of cooperation in a multinational consortium.

In England and the former British colonies in the Commonwealth, the practice more resembles the American joint venture, except in those cases where, by agreement, a management fee or cut off the top for the sponsor is set up.

A good illustration of the formation of an international consortium is the case of a meeting to set the organization up and agree on percentages and sponsorship to bid on a "big inch" gas pipeline running from Holland across France, a strip of West Germany, Switzerland, and into Italy. The engineer for the owners, a group of oil companies, was American.

The proposed consortium consisted of a Dutch contractor 60 percent owned by a large American contractor, a French company that was French staffed but owned by the same American company, a West German contractor, a Swiss contractor, and two Italian contractors, one an independent and the other 50 percent owned by the American company and 50 percent owned by a large Italian industrial complex.

The Dutch company was represented by its American general manager, the French company by its French president, the Swiss by its president (director general), the West German by its president and an English-speaking German attorney. The independent Italian company was represented by its president and the Italian affiliate by its Italian president. The president of the American company was present at the meeting which, by agreement, was chaired by the Italian president of the company affiliated with the American company.

The Italian chairman spoke fluent French, German, and English as well as Italian. The head of the French company spoke English, and understood Italian, as well as French. The manager of the Dutch company spoke Dutch and German in addition to English. The Swiss spoke German and French. The German professed to speak only German and used his English-speaking attorney to interpret. The independent Italian professed only French in addition to his own language. The president of the American company, unbeknownst to any one but his French president, understood French as well as English.

After lengthy discussion it was agreed that each would address the meeting in his own language. The chairman would translate the German into Italian as well as French. The Frenchman would translate the French into English, the German attorney English to German and German to English. In this cumbersome roundabout way the conference dragged on. At last it was agreed that the Dutch pipeline company, who had been successfully working for the same American engineering company in Holland, would sponsor and organize the trenching, welding, and laying of the pipeline. The Swiss would drive the Swiss part of the tunnel through the Alps, and the independent Italian company the Italian part of the tunnel.

The sponsorship finally settled, the decision on percentages of participation was introduced. Each member of the consortium presented lengthy and involved arguments as to services he could render in his own country and why he should have a larger participation.

The German insisted on detailed translation of everything into German and vigorously questioned every suggestion.

The meeting, held in Milan, was in the Board Room of the Italian affiliated company's industrial half-owner, a luxurious and beautifully furnished penthouse office looking across the square on the city's principal and plushest hotel.

After some four hours grinding away, the terms of the consortium were settled and everyone heaved a sigh of relief. The Italian chairman gave a tired smile and said in English: "It's now nearly seven o'clock and I have to catch a nine o'clock plane back to Rome. Let's go over to the hotel, I'll buy everyone a drink." There was a chorus of "accepted" and the German spoke up in perfect English and said: "After that I shall also buy one or two."

The American president snorted at the German. "The drinks are on me; if I had known you spoke English we could have saved at least an hour and a half."

"True," said the German, "but I never said I didn't speak English, I merely brought my attorney along who also does. Besides, you never said you spoke French; I found out you did by watching your eyes."

They failed to get the job—another international consortium was low. But the American learned that a common language usually existed in any group and a few inquiries could save a lot of time.

Many countries have their quota of successful and competent business men. Most of them are fair-minded and good joint-venture partners; many are extremely helpful in their own countries and prove good friends as well as good business associates.

As an illustration, our American contractor, having done several pipeline jobs in Pakistan, had met three brothers who together had built up a large and successful commercial empire ranging from cotton mills to banks to insurance companies and had acquired control of a sizable building construction company, originally founded as a subsidiary of a well-known British contracting firm. The notable head of the government development organization for whom the American contractor had done pipeline work was persuaded to leave government service temporarily to head up the contracting firm and get it established.

A large canal job came up for bids. It required excavation of more than 100 million cubic yards of earth and would take at least three 20-yard Monighan draglines plus a complement of tractor and scraper equipment, batch plants, concrete mixers, and mobile truck units. To follow in a year was a huge dam and headworks that would be a most desirable job, currently being engineered by a well known British engineering firm in London.

A meeting was held with the three brothers and the head of their recently acquired construction company. An agreement was reached for a joint venture, 60 percent to the American firm and 40 to the Pakistani partners. The prices were to be finalized prior to the bidding in Lahore. The bids would be opened by the Pakistan Authority and their American canal consultants.

Of the five bids one was French, one was Italian, one was German, and two were American. This time American know-how and the big American draglines came out on top. The American-Pakistani joint venture was low and was awarded the work.

At the after bid meeting the older brother spoke up. "We will help all we can. I understand the rupees to operate the job are only allowed to be borrowed in Pakistan by special permission of the Finance Minister. We will secure that permission and our bank will lend the job all the lakhs of rupees you need at 5 percent. Meanwhile we will put up our share of the capital in rupees. Let us know how much you want and when you want it." Then, with a smile, "When you get the first big dragline going, I would like to ride on it. In fact we all would."

Unfortunately when the dam came up for bids, the contract was won by the American competitor who had been second on the canal. Thus, for the time being at least, ended a fine relationship with the completion of the canal, but the friendship and mutual respect remained.

The Middle East in general today is no place for the American contractor. With the exception of some United States financed work, gobbled up by German companies in Saudi Arabia, American presence in Egypt, Sudan, and most of the new black controlled states has disappeared or is fast disappearing. With the possible exception of Morocco, the North African area is lost.

Turkey, Iran, Afghanistan, Pakistan, and India continue to give the American contractor a chance if he can out-think his European counterpart.

As he goes farther East, however, the Japanese contractors have made more and more inroads in the areas thought to be British and American. The Japanese, moreover, are offering generous financing along European lines and competing with them in their own territory to the exclusion of the Americans, who in some places have been refused the right to even bid where the financing was European or Japanese and the United States had refused to participate.

Canada, Australia, and New Zealand remain areas where American contractors with or without local partners are given an equal chance. Indonesia, too, for the time being at least, still welcomes American know-how.

4-12. Playing Percentages

This is certainly an important factor in bid strategy and is most difficult to define. Almost any stated policy is subject to equally logical counterarguments.

"Go and get it back where you put it," has been used erroneously to mean that if you have a losing job in an area or for a particular political body or owner, you must know how and why you lost on it, and it is logical to use the hard won knowledge to get the loss back in the same area or from the same owner. This is nonsense and if followed is most likely to result in further loss and more bad experience. Certain areas, certain owners, and some agencies time after time trap the unwary.

Before moving into any new area or working for any strange owner or agency, time should be spent to find out what has happened to others in the past. The information may be difficult to find but it does exist. If others have had bad results and taken losses, there must be a reason. Don't assume you are smarter than the last hopeful who went broke: find out why if you can, but stay away until you do.

If you take a loss you can find out why, and you are foolish if you don't. The percentage says if you lost money on what looked like a good bid, you will lose it if you go back again for that or some other reason. Stay away and bid on something or somewhere where others' experiences and your own have been *good*. There are reasons for that, too.

A well-known contractor often said: "If you bid on an item which will probably increase, bid it low, and you will get lots of it." He did not of course mean what the statement at first hearing sounds like. He meant quite correctly that if an item was likely to cost less as the volume increased—as, for example, thicker base cover on a highway or a heavier layer of concrete pavement—it would be better not to raise the price to where the engineer couldn't afford to increase the quantity; rather hold to a fair price and gain from the increased volume.

The most reliable method of playing percentages is to watch the percentage of low bids, i.e., how many bids out of fifty or a hundred submitted turn out to be low bid. In times of plenty of work one low bid out of five is still too many; perhaps raising markup until it is one out of eight would be safer. In times of high competition and shortage of work one low bid out of eight would be too many; perhaps it should be one out of ten or twelve. The contractor who is "always the low bidder" *does* end up in the poorhouse.

5 CONSTRUCTION CONTRACTS

DARIO DE BENEDICTIS
Member, California Bar
San Francisco, California

ROBERT M. McLEOD
Member, California Bar
San Francisco, California

THIS CHAPTER is concerned with the various types of contracts commonly used in the construction industry. A simple but accurate definition of a contract is "a promise or a set of promises for the breach of which the law gives a remedy, or the performance of which the law in some way recognizes as a duty."[1] A construction contract is a contract under which one party promises to furnish services and materials to build a structure or to improve real property for another party who promises to pay for the work performed.

There are many ways in which contracts in general can be classified. This chapter discusses briefly the form of construction contracts and the types of contracts most frequently used in the construction industry, as well as the advantages and disadvantages of the various types of contracts and of certain contract clauses commonly found in construction contracts.

5-1. The Form of Construction Contracts

As is true of contracts generally, the form of construction contracts varies tremendously from contract to contract. Some are short and simply stated; others are long, detailed, and involved. Some cover small, inexpensive projects; others cover huge, costly projects of long duration.

However, as is also true of contracts generally, the repeated use of specific forms or of particular terms, conditions, or clauses by the parties to these various contracts has resulted in some standardization of contract forms and of specific clauses of such contracts.

In the area of private contracts, the American Institute of Architects, as far back as 1911, adopted standard forms of contracts which have been revised from time to time.[2] In addition, many private firms that are engaged in the construction industry, whether as owners or general contractors or subcontractors or suppliers, have developed their own standard forms of contracts, subcontracts, or purchase orders.

In the public works area, the public agencies involved, frequently pursuant to statutes or administrative regulations which specify or require the use of particular forms or clauses, have over the years developed standardized forms of contracts.[3] The most significant of these are construction contracts in which the United States is a party, because the United States has for many years contracted for a large volume of construction and, in doing so, has developed its own contract forms and provisions, which have been much publicized in standard government publications, such as the *Federal Register* and *Code of Federal Regulations*, and which have been interpreted in many court decisions over the years. More recently, private publishers, such as Commerce Clearing House, Inc., Federal Publications, Inc., Matthew Bender & Co., Inc., and others, offer a variety of publication services dealing with all phases of federal government contracting, including construction contracts.

These standard forms are not the work of any single man or any single agency. They have evolved from repeated use and periodic revisions. They have been subjected to the enactment of specific laws, or to court decisions which have affected their evolution. In addition, there have been negotiations between parties to such contracts, or a general review of existing forms. Frequently there are standing committees, in both private and public areas of construction, charged with the responsibility to review, revise, and update contract forms and then to publicize the results of their efforts.

A construction contract ordinarily consists of up to six clearly distinguishable parts, of which four identified below as parts 2, 3, 4, and 5, are the basic elements of most construction contracts, as follows:

The six basic elements

1. Bid form
2. Agreement form
3. General conditions or standard specifications
4. Special provisions
5. Plans
6. Addenda

Sometimes called "the specifications."

The first part includes the proposal or bid form,[4] the invitation for bids,[5] and the instructions to bidders.[6] Additionally, bidders are sometimes provided with prebid information or data, such as materials information pamphlets, soils reports, geological data, test or drilling reports, etc. All of this material may or may not become part of the contract, frequently depending on whether the document itself or the form of agreement subsequently entered into expressly includes or excludes such material. The bid form is the most important item in this first group from the contracting point of view, since it is the vehicle by which the offer of the contractor to enter into the contract is communicated. On public works projects, literal, formal, and absolutely strict adherence to bidding requirements is essential to avoid possible rejection or disqualification of the bid and loss of a potential contract.

Second is the contract form or agreement itself.[7] This document identifies the parties to the agreement, the date, the contract price, the basic commitment of the contracor to construct the described project in accordance with the referenced specifications to the contract, and the signature of the parties. Usually this form is quite brief, but it incorporates by reference the remaining parts of the contract.

The third part contains the general provisions,[8] sometimes called general conditions or standard specifications. It is here that specific clauses, applicable generally to all projects, are found. These clauses deal with various subjects, such as definitions, changes, payment, inspections, legal relationships, indemnity, and many others.

The fourth part consists of the special provisions or special conditions, also sometimes called technical conditions or technical specifications. These are specific clauses setting forth conditions or requirements peculiar to the particular project, and they supplement and sometimes modify the general conditions or standard specifications. The third and fourth parts, as herein described, are sometimes collectively referred to as "the specifications" to which the work to be performed under the contract must conform, and they are made a part of the contract.

The fifth part consists of the plans or drawings. They include drawings, profiles, cross sections, and the like, which are made a part of the contract. They show the location, character, dimensions, and details of the work to be performed.

The sixth part includes any addenda which supplement, modify, or amend any of the foregoing parts, which were available prior to the opening of bids on competitively bid work and prior to the execution of the contract, and which are made a part of the contract. These are to be distinguished from supplemental agreements or change orders which are negotiated and entered into *after* the contract has been entered into by the parties and are normally addressed to matters arising during actual performance of the work.

The format of the various parts of the contract and the inclusion or exclusion of various clauses and the terminology of specific clauses may vary among public agencies and with private parties, but most construction contracts, in more or less elaborate form, contain at least the four basic elements listed above. Generally speaking, if the contract documents are complete and detailed and are expressed in clear and unambiguous language, many problems that would otherwise arise during performance of the work would be avoided. In that sense, the form of a contract is most important.

5-2.   Oral vs. Written Contracts

A very elementary and obvious distinction, but one that is nonetheless quite important, is between oral and written contracts.

To begin with, some contracts are required by a special statute, commonly known as the Statute of Frauds, to be in writing to be enforceable. These include, among others, contracts to buy and sell real property, contracts to answer for the debts or obligations of a third party (an example of which is a faithful performance bond or a labor and material payment bond issued by a surety; to be binding they have to be in writing and signed by the surety), contracts to buy and sell tangible personal property in excess of a minimum dollar amount (usually $500, but the amount varies from state to state), and contracts that cannot be completed within one year (such as long-term leases).

However, except for contracts of the type just cited, an oral contract is just as valid and enforceable as a written contract. However, if it becomes necessary to sue to enforce an oral contract, there can be serious evidentiary and proof problems.

A construction contract, being one for services and materials, is ordinarily not required to be in writing to be enforceable. In fact, a California court recently held that an *oral* agreement for the construction of a project, in accordance with a set of elaborate plans and specifications, which were in writing, was a valid and enforceable contract.[9] However, because of the complexities usually involved even in the simplest construction contract, it is not surprising that most construction contracts are written, frequently with very elaborate and detailed terms and conditions.

There are other differences between oral contracts and written contracts. In some states, there is a difference in the period of time in which suit can be brought to recover damages for a breach.[10] Assignments and powers of attorney affecting contracts which are required by law to be in writing are also required to be in writing to be enforceable.

Oral amendments to written contracts present special problems and are a common source of litigation. Obviously, if all amendments or changes to a written contract were reduced to writing and signed by both parties to the original contract, these problems as to the form of the amendment would be avoided.

Some contracts contain a clause that specifies that amendments will be recognized only if in writing and signed by both parties. The binding effect of this type of clause may vary from state to state. From the contractor's point of view, it is important to observe such a clause and adhere strictly to its requirements. However, it may not be fatal if a change or amendment to the contract has not been put in writing, especially if the parties have actually performed according to the oral change or if one party has confirmed the change in a letter to the other party.

Special laws may control or affect this entire area, so that if a problem could or does arise, appropriate advice should be obtained.

5-3.   Private vs. Public Contracts

The classification of contracts can proceed along a number of different lines. We have already referred to differences between oral and written contracts. Another distinction is between public and private contracts. A public contract is one in which one or more of the parties involved is a governmental agency, i.e., the United States of America, a state, county or city, or any branch thereof. Although such contracts are in general said to be subject to the same rules of law that apply to private contracts, there are many differences. For example, in a few states,[11] a contractor having a contract with the state to build a public project is precluded by law under the doctrine of sovereign immunity from suing the state for breach of contract. Most states have waived their sovereign immunity in such cases, but special procedures or resorting to special courts may be necessary.

There are other significant differences between public and private construction contracts. The listing here of these differences is intended to be illustrative rather than exhaustive. At the outset, it should be noted that a contractor bidding on public works projects usually must take the contract as he finds it.[12] In fact, if he qualifies his bid so as to change any of the terms in the proposed contract, his bid would be declared nonresponsive.[13] However, this should not deter a contractor from seeking clarification of any ambiguities in the plans, terms, or specifications, if this is necessary to prepare his bid properly.

5-4. Competitive Bidding Requirements

Virtually all public construction contracts in the United States are subject to competitive bidding requirements. Since every citizen is presumed to know the law, a contractor who deals with a public agency does so at his peril. It follows, therefore, that a contractor should observe all of the requirements for competitive bidding. The failure to do so may make his bid nonresponsive or, if an award is made, may render the contract voidable or void and preclude the contractor from recovering his full compensation for work performed. Submitted bids that do not satisfy the competitive bidding requirements, except for such minor irregularities as can sometimes be waived, are nonresponsive and are subject to being rejected.

Once bids to enter into a public works contract have been opened, a bidder cannot withdraw or amend his bid for a stated period of time. There are exceptions in some jurisdictions if a clerical mistake has been made.[14] To obtain relief, a bidder must act promptly and he must demonstrate that the mistake was nonculpable and was not a mistake of judgment.[15] In some cases a lawsuit must be filed to obtain the necessary relief.[16]

On bids for federal contracts, a bid mistake may be corrected and the bid as corrected considered for award if there is clear and convincing evidence to establish both the existence of the mistake and the amount of the bid actually intended, unless the correction would result in displacing one or more bids. If a displacement would result, correction is allowed only when the existence of the mistake and what the amount of the bid actually intended can be ascertained from the invitation for bid and the bid documents alone, without external evidence or documentation.[17]

In any event, if there is a bid mistake, prompt action is absolutely necessary in order to obtain relief. In some jurisdictions it is a matter of public policy that a bidder cannot be relieved from any kind of mistake.[18] It has recently been ruled by a federal court that a disappointed bidder for a federal contract has standing to challenge the award of the contract to another bidder whose bid was alleged to be subject to some infirmity.[19]

5-5. Extra Work, Changes, and Changed Conditions

Another area where public contracts may differ from private contracts pertains to extra work, changes, and changed conditions. Federal contracts and contracts with some state and local governmental agencies (and some private contracts) contain a clause formerly referred to as a changed conditions clause and now commonly designated as a differing site conditions clause.[20] Such a clause generally provides that, if a contractor encounters subsurface or latent physical conditions at the site differing from those indicated in the plans or specifications, or if he encounters conditions differing from those ordinarily encountered, he should promptly call those conditions to the attention of the contracting officer. He may be entitled to an equitable adjustment in the contract price if he is then required to perform work that was more difficult and more costly as a result of encountering the changed conditions.

In the absence of a changed conditions clause, the risk of encountering conditions differing from those anticipated is generally with the contractor, unless there is some basis for claiming that the owner is liable, such as a misrepresentation in the plans and specifications of conditions to be encountered,[21] or a breach of warranty of the accuracy or correctness of the plans and specifications.[22]

With respect to changes and alterations in the work, public construction contracts usually contain an express clause governing these matters.[23] If the change falls within the scope of the changes clause, a contractor ordinarily has little difficulty so long as he carefully observes and follows the notice, pricing, and administrative requirements of the applicable changes clause of his contract.

However, a contractor may have a problem if the proposed change is beyond the scope of the work originally contemplated. On federal contracts, if a series of changes, or even a single change, is of such a magnitude as to be beyond the scope of the changes clause, the change is known as a "cardinal change."[24] This is treated as a breach of contract which may ultimately require submission of any claim for additional compensation or for additional time to the United States Court of Claims or a United States District Court. Disputes concerning ordinary changes (i.e., changes within the scope of the changes clause of the contract) are handled under the contract. In the latter type of cases, if a disagreement does develop between the government and the contractor, the controversy may be resolved administratively under the disputes clause of the contract and, if necessary by an appeal to the appropriate board of contract appeals.

On all public construction contracts, there is the further problem with respect to changes that are clearly and obviously outside the scope of the changes clause of the contract. In such instances a contractor may be ineligible to be paid for such work because it is subject to competitive bidding before the required work can be lawfully performed. This is true also of "extra" work, i.e., work beyond any originally contemplated by the contract. In such circumstances, a contractor may not be entitled to be paid for "extra" work or work outside the scope of the changes clause unless it is found that such work was exempt from competitive bidding requirements (e.g., if the amount involved is below the statutory minimum).[25]

Another difference between government contracts and private contracts pertains to the remedies for claims arising under the contract. Federal contracts all contain a disputes clause[26] which provides a method for the determination of disputed claims arising under the contract. In the first instance the contracting office will make his determination of the facts. If the contractor disagrees, he may elect, within the time limits specified in the contract, to appeal to the appropriate board of contract appeals, with the right of a limited review by the United States Court of Claims. Outright breaches of contract (which are considered to be claims that do not arise under the contract and hence not subject to the contract disputes clause) are required to be presented to the United States Court of Claims or to one of the United States District Courts. However, as a precautionary matter to exhaust all possible remedies, such disputes are usually first taken through the procedure applicable to claims arising under the disputes clause of the contract.

Remedies available to a contractor for breach of contract with a state will depend upon the state in which the work is performed. As noted above, some states still assert their sovereign immunity and prohibit any suit against the state,[27] but most states have waived their sovereign immunity and permit such suits, or at least afford some kind of remedy. The doctrine of sovereign immunity has no application to municipal corporations, such as cities and townships, or to certain public corporations, such as turnpike authorities, even though created by the state in which they are located. However, a written notice of claim is frequently required by such local agencies as a condition precedent to enforce such a claim. States frequently have similar requirements as well. Such requirements must be strictly observed to avoid denial of an otherwise valid claim.[28]

As to private contracts, the parties may, to a considerable extent,[29] specify remedies and procedures for enforcement of such remedies for breach of contract, such as clauses which provide for arbitration of disputes, liquidated damages, termination for default, and the like. It should be noted that any notice provisions or time schedules set forth in such clauses must be strictly observed as to time and manner of compliance in order to exercise such remedies.[30]

5-6. Fixed-Price Contracts

One of the most important types of construction contracts is the fixed-price or lump-sum contract. Under such a contract the contractor agrees to perform the entire work specified in the contract at a price agreed to and fixed at the time the contract is entered into. Frequently the price is determined by competitive bidding. As noted in a preceding section, public works contracts are, with few exceptions, required by law to be let on the basis of competitive bidding and must be for a fixed price. Of course, where competitive bidding is not required by law, it is also possible to arrive at a fixed price by negotiation between the parties to the contract without resort to competitive bidding.

A fixed-price contract is normally contrasted with cost-plus contracts, in which the price is determined by actual costs as they are incurred plus whatever additional payment in the way of a fee for profit the parties have agreed to. The cost-plus type of contract will be discussed more fully later in this chapter.

A variant of the fixed-price contract is the unit-price contract. Some contractors consider unit-price contracts as something different from lump-sum or fixed-price contracts and they therefore treat such contracts as a separate species. The significant characteristic of a unit-price contract is that, although quantities may vary (which means the total compensation paid to the contractor will vary), the price *per unit* is fixed at the time the contract is entered into and is intended to remain unchanged throughout performance of the work.

For example, a highway construction contract typically contains some items for which a lump-sum or fixed price for the entire item is required, but there are many items which are contracted for on a unit-price basis, such as roadway excavation per units of volume, aggregate base materials by weight or volume, concrete for pavement by volume, fencing or pipe by units of length, etc. Such a contract normally contains estimated or approximate quantities and a bidder is required to quote his unit price and is also asked to extend the price based upon the estimated quantity for each such item. The total prices for such items are added to the total of those items quoted on a lump-sum basis to arrive at a total bid price. This total bid price is then used as a basis for bid comparison and the invitation for bid and contract documents so specify.

The *actual* final contract price, i.e., the total amount ultimately paid to the successful bidder who enters into the contract, is not his total *bid* price but is determined by multiplying *actual* quantities, determined in the manner specified in the contract, with the agreed applicable unit price. In other words, payment is not based on the approximate or estimated quantities but instead on actual results, and the unit price remains fixed.

A problem peculiar to unit prices sometimes arises if there is a substantial overrun or underrun of quantities. A contractor makes his determination of the unit price, anticipating that the actual quantities will be reasonably close to the quantities listed in the invitation for bids. His price might well have varied if the quantities had been substantially different. If, then, the *actual* quantities do differ substantially, the contractor or the owner may feel he is entitled to a price adjustment.

A carefully drafted contract will contain a clause specifying what percentage increase or decrease shall trigger a price adjustment and will spell out the cutoff point for applying unit prices and the basis, e.g., force account or cost plus a specified markup, for pricing the balance. In the absence of such a clause it has been held that the contractor will be paid on the agreed unit-price basis for actual quantities even when the contractor had a high profit on each unit,[31] there being no other basis for a price adjustment. It is true, as noted below, that a contractor may have cause to adjust or modify the unit price on the same kind of grounds or by applying the same principles which apply to adjusting or modifying the price of fixed-price or lump-sum contracts.

Some fixed-price contracts contain a bonus provision which typically allows the contractor to receive stated bonus payments for completion of the work before the specified contract completion date. In most, if not all, such contracts there is a liquidated damages provision requiring the contractor to pay the owner stipulated amounts (usually at least equal to the amount of the bonus payments) if the work is completed after the contract completion date. However, many contracts contain a liquidated damages clause for failure to complete the work within the time stated in the contract, and the enforceability of such a clause does not depend on there being a provision for payment of a bonus for early completion.

It is not uncommon for owners to require the contractor to guarantee a maximum price beyond which the owner will not be required to pay. Most of such contracts provide that the guaranteed maximum price will be adjusted if the owner makes changes in the work or if certain other contingencies occur. However, occasionally the guaranteed maximum price is specified to be an absolute ceiling on the price to the owner and not subject to change. Contracts of this type should be evaluated most carefully before they are entered into.

A major characteristic of the fixed-price type of contract is the principle that a contractor agrees to perform the work at the agreed price no matter what it may actually cost him to do the work, and conversely the owner agrees to pay the price whether the contractor enjoys an unusually large profit or suffers a huge loss.

Another way to express it is to say that the contractor assumes the risk of all unforeseen conditions, such as increases in labor and material costs, unanticipated subsurface soil conditions, adverse weather, and the like. The owner, on the other hand, is able to anticipate that his building will be built or his project completed at a known price, determined before commencement of any work, regardless of the actual cost to the contractor.

There are circumstances where a contractor may be entitled to an adjustment in the price under a fixed-price contract. We shall consider some of these circumstances as illustrative, but not exhaustively so, of the kinds of problems that can arise and the effect that contract clauses covering those situations might have.

We shall use as examples construction contracts with the United States and specific contract clauses found in such contracts. These contract forms and contract clauses are in the public record,[32] and there is a history of their use over the years to which we can refer. It must be emphasized that every contract is governed first by its own terms and second by the law of the appropriate jurisdiction, whether it be federal or state. Hence, although basic principles of contract law apply generally, the specifics of each contract, of each set of facts, and of the law applicable to each situation, govern most and control the result.

Perhaps the most commonly recurring problem under fixed-price contracts arises in connection with changes or extra work. The underlying rule is that a contractor is entitled to perform the contract in accordance with the original plans and specifications without interference from the owner. Without a changes clause, an owner has no legal right unilaterally to change the work.[33] Since it usually happens that owners, for a variety of reasons, wish or are required to make changes in the work as it progresses, a well-drafted changes clause [34] is an important part of any construction contract. However, only changes within the scope of the changes clause will be governed by it.

If a dispute arises under the changes clause found in federal contracts, determination of such a dispute is governed by the disputes clause,[35] which gives the contracting officer the right to make a determination of the facts, subject to an appeal to the appropriate contract appeals board.

A change beyond the scope of the changes clause is called a "cardinal" change and is governed generally by the rules applicable to breaches of contract.[36] Hence, the dispute in those cases is determined by the United States Court of Claims or a United States District Court, depending on the amount involved, although proceeding through the contracting office and contract appeals board may be necessary to exhaust administrative remedies.

A common pitfall to contractors who have claims based on asserted changes is the failure of the contractor to give the notice in the time or manner required by the changes clause of the contract; another is to fail to get a *written*

change order (discussed in Section 5-2 of this chapter) or to get the order issued prior to doing the work. If for any reason no change order is forthcoming where one is required, the contractor is faced with the alternative either of not doing the changed work or proceeding with the work and notifying the contracting officer in writing that the work is being done under protest, and reserving the right to claim additional compensation and time.

A problem can also arise if the changes clause fails to spell out a precise pricing method applicable to changed work. In a recent California case,[37] it was ruled that when the contract (in this case a detailed purchase order) provided that the owner could require changes at a price to be agreed upon by the parties in the future, the validity of the contract depended on future events. If no agreement on price could be reached, the contractor would be required nonetheless to perform *minor* changes for a "reasonable" price, to be determined subsequently. If, however, the change requested was a *major* one, the contractor could not be compelled to perform the changed work if no price had been agreed upon following a good faith attempt to do so. In the latter situation, the court held that the contractor could walk off the job and rescind the contract if the owner insisted on a major change being performed without having reached an agreement as to price after a good faith attempt to do so.

Still another problem concerns deletions or omissions of work. In an early New York case, a contractor with the City of New York recovered his loss of profits from the City which had deleted merely 2½ percent of the work under a changes clause which failed to spell out that the City was authorized to delete *any* work.[38] Thereafter the City modified its changes clause, reserving the right "to omit any portion of the work without constituting grounds for any claim by the contractor. . . ." In a subsequent case, the City, acting pursuant to the revised changes clause, deleted substantial items of work from a subway contract, including the underpinning of a large building in the path of the subway. The City had decided it was more economical to purchase the building, raze it, and later resell the land, than to pay the contractor to underpin it. The contractor sued for loss of profits, relying on the earlier case but lost, the court pointing out the broad scope of the revised changes clause, quoted above, and noting that the deletion was consistent with the basic purpose of the entire contract, which was to construct a subway, and that the underpinning portion was incidental to the main purpose.[39]

However, even in the case of deletions, the specific changes clause in the particular contract at hand must be considered. For example, the City of Los Angeles let a contract to construct runways and related facilities for the Los Angeles International Airport. The contract contained a changes clause authorizing additions or deletions up to 25 percent and stipulated that the City could omit items found unnecessary to the project. Because the City had failed to make arrangements for moving a bypass road, it deleted a substantial portion of a runway (20 percent) and later re-let that work to someone else. The contractor sued and recovered his loss of profits on the deleted work, the court finding that the work was necessary (the City having subsequently contracted with another contractor to complete it) but was deleted solely because the City had been unable to furnish the job site as required by the contract.[40]

Extra work as distinguished from changed work is work completely outside the basic contract and therefore not governed by its terms. Instead, it is governed by a separate, independent contract, unless the parties agree otherwise, or unless the basic contract expressly provides that extra work, as such, is subject to the original contract.[41]

One potential problem, applicable to public works contracts, is the possible application of competitive bidding requirements to extra work. Generally speaking, the more obviously that the work is an "extra" and not merely a change, the more likely it is that courts will hold that the parties should have resorted to competitive bidding, with the result that compensation for any extra work actually performed may be denied to the contractor unless competitive bidding requirements were complied with.[42]

5-7. Unforeseen Difficulties, Differing Site Conditions Clause

When a contractor encounters unforeseen difficulties, such as unexpected adverse soil conditions, a high-water level, extremely bad weather, or the like, the general rule is that any additional construction costs incurred by the contractor must be borne by him, because this is the very risk he assumed in a fixed-price contract. This result is frequently reinforced by commonly used contract clauses which require preliminary site investigation by the contractor, or which disclaim responsibility by the owner, or which specify that payment of the agreed price is intended to cover all risks of unforeseen conditions, or which place responsibility for the work upon the contractor until the work has been accepted by the owner.

One exception is found where there has been a deliberate misrepresentation of the facts or concealment of facts which would have affected the price bid by the contractor. In such cases, general clauses disclaiming responsibility are of no avail to the owner.[43]

Another exception may exist if the contract contains a differing site conditions clause.[44] If the contractor encounters subsurface or latent physical conditions which (a) materially differ from those indicated in the contract or (b) are of an unusual nature differing from those ordinarily encountered, then the contractor must promptly notify the contracting officer. He in turn must investigate the conditions encountered and, if he finds that the contractor has encountered conditions of the type indicated, he will agree to an equitable adjustment in the contract price. If the parties disagree on whether the clause is properly being invoked or as to the amount of the equitable adjustment, such disputes arising under federal contracts are handled through the standard disputes clause.

The advantage of a changed conditions clause to the contractor is dual. If the conditions encountered differ from those indicated in the contract, he does not have to prove the added element of fraud or misrepresentation or willful concealment to be eligible for a price adjustment. If the conditions encountered are of an "unusual" nature rather than being different from the conditions indicated in the plans and specification, the clause may nonetheless provide a basis for a price adjustment. This is true even in a situation when, without such a clause, the risk of encountering such unusual conditions is otherwise entirely on the contractor. Thus the risk of encountering subsurface or latent physical conditions of an unusual nature differing from those ordinarily encountered is shifted from the contractor to the owner.

The advantage to the owner of including such a clause in his contract is that the contractor need not include a "contingency factor" in his bid price, thereby presumably resulting in a lower price to the owner for the work.

Where labor or material costs escalate during performance of the contract, the added costs are normally borne by the contractor. If an escalation clause is included, then, of course, that clause would govern according to its terms.

A responsibility for the work clause makes it clear that risks of damage to work while under construction and before final acceptance by the owner are assumed by the contractor, unless expressly excluded. It is essential, therefore, that careful consideration be given to this clause.

In summary, a lump-sum contract is intended to place the risk of all unanticipated or unforeseen additional costs on the contractor so that the owner can expect the project to be constructed at the fixed price agreed to by the parties at the time the contract is entered into. Problems arise when such additional costs are encountered. The owner will contend that such costs must be borne by the contractor because they fall within the scope of the risks assumed by the contractor. The contractor will attempt to find an exception to the general rule so as to pass the additional costs on to the owner. In all cases, the specific terms and conditions of the contract and the particular facts which apply to the situation at hand are of paramount importance and must be taken into consideration.

5-8. Cost-Plus Contracts

Frequently contrasted with fixed-price or lump-sum contracts are the cost-plus type of contracts. Under cost-plus contracts, the contractor agrees to perform the specified work at his cost, plus an additional amount, known as the fee, to cover the contractor's profit.

There are many variations of cost-plus contracts and we shall enumerate and describe a few:

(a) Cost- Plus- a- Fixed- Fee This contract contemplates reimbursement to the contractor of his actual costs, but his compensation by way of profit is fixed at an agreed amount at the time of contracting. The fee is frequently determined by a percentage of the originally estimated cost, but it will not vary if actual costs subsequently do vary.

(b) Cost- Plus- a- Percentage- of- Cost Here the contractor will be reimbursed for his actual costs and he will receive additional compensation measured by an agreed percentage of the actual costs. His fee, in these circumstances, will vary in proportion to the amount of the actual costs.

(c) Cost- Plus- an- Incentive- Fee Here the contractor receives reimbursement for actual costs plus compensation based on a special formula for sharing in actual costs over or under target costs.

Additional variations cover special situations, many of which are not common to the construction industry, such as *cost contracts* (with no fee, used for example when the contractor is a nonprofit corporation, or when the contractor is producing a pilot or model, anticipating award of a subsequent contract if the model turns out to be successful), *costs-with-guaranteed-maximums* (with provisions for sharing or not sharing of overruns, with maximums based on dollars, or on man-hours), etc.

A variation of both fixed-price and cost-plus type contracts are the so-called *"target"* contracts. There are many types of target contracts, but common to all is an agreement between the parties on estimated target cost, target profit or fee, ceiling price, or some combination of these or other targets.

In a *fixed-price target contract* it is common for the parties to negotiate a target cost, a target profit, and a ceiling price as well as a formula for determining the final profit and price. Upon completion of the work required under the contract, the final cost is determined and, if less than the target cost, the final profit will be increased according to the formula to an amount which is greater than the target profit. On the other hand, if the final cost exceeds the target cost, the contractor's final profit will be less than the target profit. If the final cost exceeds the ceiling price, the contractor will incur any out of pocket loss since he must pay all costs in excess of the ceiling price.

In a *cost-plus type target contract*, the parties negotiate a target cost and a target fee. Upon completion of the work, the contractor's fee is increased above the target fee if the actual cost is less than the target cost, and the fee is decreased if the actual cost exceeds the target cost, but the contractor will not incur a loss since he will be reimbursed by the owner for all costs incurred.

When the owner desires to start work before complete plans and estimates have been prepared, the contract may provide for initial work to be done on a cost-plus type basis with a target estimate of cost, to be followed by a subsequent agreement on a detailed estimate based on complete plans at which time the basis of payment may either continue on some form of cost-plus basis or the parties may enter into a fixed price contract based on the detailed estimate.

A *letter of intent or letter contract* is a preliminary contract under which the contractor may commence work and in which the owner and the contractor agree to negotiate a definitive contract of a specified type. Such letter contracts with the federal government specify the maximum costs and fee for which the government will be liable prior to the execution of the definitive contract. They also provide for the payments to be made to the contractor in the event the letter contract is terminated because a definitive contract cannot be agreed upon or for other reasons, such as the contractor's default or the convenience of the government.

Descriptive terms, such as *turnkey* or *grass roots*, are sometimes applied to construction or design and construction contracts. These terms are used when the contractor undertakes to design and build or to build a project from its very "grass roots" to final, ultimate completion as a completed plant or facility ready to be put into operation immediately and the contractor turns over a key to the entire, completed facility to the owner. Such terms would more accurately apply to the job or project rather than to the contract, since a turnkey project may be constructed or designed and constructed under the provisions of any of the types of contracts discussed in this chapter.

Turning back to the cost-plus contracts, it should be noted that there are a number of advantages and disadvantages. On the whole, this type of contract is to be preferred by a contractor, since it eliminates, or at least minimizes, many of the problems associated with fixed-price contracts. For example, the encountering of unforeseen difficulties, such as adverse soil conditions, inclement weather, escalation of wages or material prices, or changes in the work, are usually handled without any controversy or having to resort to claims procedures, since the actual cost of performing the work as changed as a result of these various conditions will be passed on to the owner, which is just what this type of contract contemplates.

Conversely, the owner does not favor a cost-plus contract, because ordinarily there is little or no incentive on the part of the contractor to reduce costs, especially on cost-plus-a-percentage-of-costs contracts.

Of course, there are problem areas, the most central of which is the proper definition of allowable or reimbursable costs. For federal contracts, the procurement regulations contain an elaborate and detailed description of allowable costs.[45] The definition of allowable costs should receive the careful attention of the draftsman of any cost-plus contract in order to avoid potential problems that might arise during the course of construction in ascertaining which costs are allowable and which are not. Sources of difficulties in this area include proper methods for determining and allocating home-office overhead to the particular project, compensation for a partner, owner or executive of the contractor, employment of a consultant, equipment usage (including such problems as allowability of payments on leases with options to buy, "rental rates" on contractor-owned equipment, allowability of costs of repairs, maintenance and tire replacement, standby time, premium for overtime, or use on more than one shift, etc.), charges for hand tools, and replacement of damaged or defective work.

With respect to public works projects, cost-plus contracts are permitted only under limited conditions. For one thing, cost-plus-a-percentage-of-cost contracts are absolutely prohibited by federal statute,[46] since they offer no inducement to contractors to reduce costs. Where procurement by means of cost-plus contracts is permitted for federal procurement, it proceeds on the basis of a negotiated procurement with such competitive bidding as may be feasible. There are some seventeen areas where formal advertising is not required and cost-plus contracts may be employed for federal procurement. Among these are contracts let because of emergencies, contracts under $2500, contracts for personal or professional services, contracts for work to be performed outside the United States, and experimental, research, and development contracts.

Normally, a construction contract with the United States of America to be performed within the United States of America must be a fixed-price contract awarded on the basis of formal advertising and competitive bidding. The exceptions generally involve either foreign work or emergency work.

5-9. Subcontracts

One special area of contracting is that of subcontracts. A subcontract is a contract between a general contractor or prime contractor and another contractor, frequently a specialty contractor, whereby the latter agrees to perform a portion of the work which the prime contractor has undertaken to perform under his contract with the owner. Generally to be classified as a subcontractor, he must perform work on the construction site,[47] but this is not always the test.[48]

A subcontract, then, is a contract and is subject to the law of contracts, but there are some special rules. For one thing, listing of subcontractors may be required of the prime contractor when submitting his bid to the owner, and on public work projects there may be statutory or contractual limitations on subcontracting in general, on changes in subcontractors, etc.[49] Moreover, the distinction between a subcontractor and a materialman or supplier may be of critical importance in determining rights under mechanics' lien laws or with respect to labor and material payment bonds, and related remedial provisions.[50]

Normally, a subcontractor may not sue the owner because he has no contract with the owner (i.e., there is no privity of contract). However, the prime contractor may sue the owner on behalf of a subcontractor and, when the owner is the federal government, the *Blair* doctrine[51] provides that the prime contractor may recover for the benefit of the subcontractor unless the prime contractor has no liability to the subcontractor. This latter eventuality is referred to in the next paragraph.

Another special rule applies to subcontractors on federal public works projects and is commonly referred to as the *Severin* doctrine.[52] Under this doctrine, if a subcontractor has waived any claim against or released the prime contractor from claims which are based on claims against the United States, the prime contractor cannot recover from the United States on any such claim.

Sometimes a prime contractor will use a purchase order

in circumstances where a subcontract might be more appropriate. Ordinarily, a purchase order is used to cover the purchase of materials from a supplier and it will contain provisions approriate to such transactions which usually are governed by the law of sales, which may differ in various respects from the law applicable to contracts generally. Subcontracts, on the other hand, cover situations where both labor and materials are to be furnished, and they usually contain provisions tailored for just that situation. It is important to use the proper type of contract, including special contract clauses, to fit the specific transaction covered.

Finally, it should be noted that a subcontract ordinarily does not create any contractual relationship between the subcontractor and the owner. The distinction in these relationships should be observed and preserved at all times.

REFERENCES

1. *Restatement of Contracts*, §1 (1932).
2. These forms are collected and published in book form by Little, Brown & Company. The most recent edition by William Stanley Parker and Faneuil Adams entitled "The A.I.A. Standard Contract Forms and the Law" was published in 1954. Individual copies of the various forms published by the A.I.A. (some as recently revised as 1970) may be purchased from the national headquarters of the A.I.A. in Washington, D.C., or from the offices of local chapters.
3. See, e.g., *Standard Form 23: Construction Contract,* 41 C.F.R. §1-16.901.23 for the form of the standard construction contract employed by agencies of the United States of America. Also see *The State Contract Act, California Government Code,* §§14250, *et seq.,* for statutory requirements for contracts with agencies of the State of California.
4. See *Standard Form 21: Bid Form (Construction Contract),* 41 C.F.R. §1-16.901.21.
5. See *Standard Form 20: Invitation for Bids (Construction Contracts),* 41 C.F.R. §1-16.901.20.
6. See *Standard Form 22: Instructions to Bidders (Construction Contracts),* 41 C.F.R. §1-16.901.22.
7. See reference 3.
8. See *Standard Form 23-A: General Provisions (Construction Contract),* 41 C.F.R. §1-16.901.23-A.
9. *Stephan v. Maloof,* 274 Cal. App. 2d 843, 79 Cal. Rptr. 461 (1970).
10. In California, the statute of limitations applicable to written contracts is 4 years, to oral contracts, 2 years. *California Code of Civil Procedure,* §§337, 339. However, the *Uniform Commercial Code,* with respect to the purchase and sale of Goods, has a 4-year statute of limitations on sales contracts without differentiating between oral and written contracts. *Uniform Commercial Code,* §2-725.
11. In a number of states, the doctrine of sovereign immunity persists, so that suit cannot be brought against the state. Included are Georgia, Maine, Maryland, New Jersey, Oklahoma, Texas, and Vermont. In some of these states, a legislative bill may be available as an informal remedy.
12. *Haggart Construction Co. v. State of Montana,* 149 Mont. 442, 427 P.2d 686 (1967).
13. *Armed Services Procurement Regulations,* 32 C.F.R. §2-404.2; *Federal Procurement Regulations,* 41 C.F.R. §2-404.2.
14. *M. F. Kemper Construction Co. v. City of Los Angeles,* 37 Cal. 2d 696, 235 P.2d 7 (1951).
15. *White v. Berenda Mesa Water* Dist., 7 Cal. App. 3rd 894, 87 Cal. Rptr. 338 (1970) (in which case relief was allowed to a bidder who had negligently underestimated the amount of hard rock to be excavated, which the court

ruled was a *mixed* mistake of fact *and* judgment). See generally Annot., 52 A.L.R. 2d 792 (1957).
16. *California Government Code,* §§14350 *et seq.*
17. *Armed Services Procurement Regulations,* 32 C.F.R. §§2-406.2, 2-406.3; *Federal Procurement Regulations,* 41 C.F.R. §§1-2.406.2, 1-2.406.3.
18. *A. J. Collela Inc. v. County of Allegheny,* 391 Pa. 102, 137 A. 2d 265 (1958).
19. *Scanwell Laboratories Inc. v. Shaffer,* 424 F.2d 859 (App. D.C. 1970).
20. The following is the current version of the differing site conditions clause:
    "Differing Site Conditions (1968 FEB)
    "(a) The Contractor shall promptly, and before such conditions are disturbed, notify the Contracting Officer in writing of: (1) subsurface or latent physical conditions at the site differing materially from those indicated in this contract, or (2) unknown physical conditions at the site, of an unusual nature, differing materially from those ordinarily encountered and generally recognized as inhering in work of the character provided for in this contract. The Contracting Officer shall promptly investigate the conditions, and if he finds that such conditions do materially so differ and cause an increase or decrease in the Contractor's cost of, or the time required for, performance of any part of the work under this contract, whether or not changed as a result of such conditions, an equitable adjustment shall be made and the contract modified in writing accordingly.
    "(b) No claim of the Contractor under this clause shall be allowed unless the Contractor has given the notice required in (a) above; provided, however, the time prescribed therefore may be extended by the Government.
    "(c) No claim by the Contractor for an equitable adjustment hereunder shall be allowed if asserted after final payment under this contract." *Armed Services Procurement Regulations,* 32 C.F.R. §7-602.4.
21. *Hollerbach v. United States,* 233 U. _. 164 (1913); *E. H. Morrill Co. v. State of California,* 65 Cal. 2d 787 56 Cal. Rptr. 412, 423 P.2d, 240 (1967); *Haggart Construction Co. v. State of Montana,* 149 Mont. 442, 427, P.2d. 686 (1967).
22. *United States v. Spearin,* 248 U. S. 132 (1918); *Souza & McCue Construction Co. v. Superior Court,* 57 Cal. 2d 508, 20 Cal. Rptr. 634, 370 P.2d 338 (1962).
23. There exists much confusion as to the meaning and usage of such terms as "changes," "alteration," "extra work," and "additional work," and the court decisions are not always consistent in the treatment of these terms or the work intended to be covered by them. See Annot., *Public*

Construction Contract—Extras, 1 A.L.R. 3rd 1273 (1965); Greenberg, *Problems Relating to Changes and Changed Conditions of Public Contracts*, 3 *A.B.A. Public Contract Law Journal*, 135 (1970).

The following is the current version of the changes clause:

"Changes (1968 FEB)

"(a) The Contracting Officer may, at any time, without notice to the sureties, by written order designated or indicated to be a change order, make any change in the work within the general scope of the contract, including but not limited to changes:

"(i) in the specifications (including drawings and designs);

"(ii) in the method or manner of performance of the work;

"(iii) in the Government-furnished facilities, equipment, material, services, or site; or

"(iv) directing acceleration in the performance of the work.

"(b) Any other written order or an oral order (which terms as used in this paragraph (b) shall include direction instruction, interpretation or determination) from the Contracting Officer, which causes any such change, shall be treated as a change order under this clause, *provided* that the Contractor gives the Contracting Officer written notice stating the date, circumstances, and source of the order and that the Contractor regards the order as a change order." *Armed Services Procurement Regulations*, 32 C.F.R. §7-602.3.

There is no standard form of "extra work" clause applicable to federal construction contracts. Some state contracts do contain extra work clauses. See reference 25 below. Federal supply contracts, on the other hand, are required to contain the following standard clause:

"Extras (1949 JUL)

"Except as otherwise provided in this contract, no payment for extras shall be made unless such extras and the price therefor have been authorized in writing by the Contracting Officer." *Armed Services Procurement Regulations*, 32 C.F.R. §7-103.3.

24. *Saddler v. United States*, 152 Ct. Cl. 557, 287 F.2d 411 (1961); *Peter Kiewit Sons' Co. v. Summit Construction Co.*, 422 F.2d 242 (C.A. 8th 1969) (rule applied to private parties, i.e., earthwork subcontract where substantial change in backfill requirements was held to be beyond the scope of the changes clause and therefore a breach of contract).

25. *Schneider v. United States*, 19 Ct. Cl. 547 (1884); *Borough Construction Co. v. City of New York*, 200 N.Y. 149, 93 N.E. 480 (1910). *City Street Improvement Co. v. Kroh*, 158 Cal. 308, 321, 110 Pac., 933, 939 (1910), in which the court defined extra work as "work not included in the contract" and stated that the public agency could lawfully authorize extra work only in accordance with the statutory authority to contract in the first place.

26. The current version of the disputes clause reads as follows:

"Disputes (1964 JUN)

"(a) Except as otherwise provided in this contract, any dispute concerning a question of fact arising under this contract which is not disposed of by agreement shall be decided by the Contracting Officer, who shall reduce his decision to writing and mail or otherwise furnish a copy thereof to the Contractor. The decision of the Contracting Officer shall be final and conclusive unless, within 30 days from the date of receipt of such copy, the Contractor mails or otherwise furnishes to the Contracting Officer a written appeal addressed to the head of the agency involved. The decision of the head of the agency or his duly authorized representative for the determination of such appeals shall be final and conclusive. This provision shall not be pleaded in any suit involving a question of fact arising under this contract as limiting judicial review of any such decision to cases where fraud by such official or his representative or board is alleged: *Provided, however*, that any such decision shall be final and conclusive unless

the same is fraudulent or capricious or arbitrary or so grossly erroneous as necessarily to imply bad faith or is not supported by substantial evidence. In connection with any appeal proceeding under this clause, the Contractor shall be afforded an opportunity to be heard and to offer evidence in support of this appeal. Pending final decision of a dispute hereunder, the Contractor shall proceed diligently with the performance of the contract and in accordance with the Contracting Officer's decision.

"(b) This 'Disputes' clause does not preclude consideration of questions of law in connection with decisions provided for in paragraph (a) above. Nothing in this contract, however, shall be construed as making final the decision of any administrative official, representative, or board on a question of law." *Armed Services Procurement Regulations*, 32 C.F.R. §7-602.6.

27. See reference 11.

28. See, for example, *Transbay Construction Co. v. City and County of San Francisco* (CCA 9th 1943), where a contractor was denied any right to recover damages for breach of contract by the owner city and county solely because the contractor was late in filing a notice of claim as required by the city charter.

29. The right to specify in advance remedies for breach of contract is somewhat limited. For example, caution must be exercised in drafting a liquidated damages clause for breach of contract to avoid having the clause declared void as imposing a penalty. *Cf. California Civil Code*, §§1670, 1671.

30. *Monson v. Fischer*, 118 Cal. App. 503, 519, 5 P.2d 628, 634 (1931).

31. *Perini Corp. v. United States*, 180 Ct. Cl. 768, 381 F.2d 402 (1967).

32. See generally Title 41, *United States Code, Public Contracts*; Title 41, *Code of Federal Regulations* (containing the Federal Procurement Regulations and procurement regulations of various federal agencies such as Federal Aviation Administration, Department of Agriculture, etc.); and Title 32, *Code of Federal Regulations* (containing the Armed Services Procurement Regulations). Also see reference 3.

33. See 4 *Corbin on Contracts* 815 (1951).

34. See reference 23.

35. See reference 26.

36. See reference 24.

37. *Coleman Engineering Co. v. North American Aviation, Inc.*, 65 Cal. 2d 396, 55 Cal. Reptr. 1, 420 P.2d 713 (1966).

38. *Litchfield Construction Co. v. City of New York*, 244 N.Y. 251, 155 N.E. 116 (1926).

39. *Del Balso Construction Corp. v. City of New York*, 278 N.Y. 154, 15 N.E. 2d 559 (1938).

40. *Hensler v. City of Los Angeles*, 124 Cal. App. 2d 71, 268 P.2d 12 (1954).

41. *Thos. J. Dyer Co. v. Bishop International Engineering Co.*, 303 F.2d 655 (C.A. 6th 1962). Also see reference 23 and 25.

42. See reference 25.

43. See reference 21.

44. See reference 20.

45. See Section 15, *Armed Services Procurement Regulations*, and parallel sections of the procurement regulations issued by other agencies of the Federal Government in 32 C.F.R. and 41 C.F.R.

46. *Armed Services Procurement Act*. §4(b), 10 U.S.C. §2306; *Federal Property and Administrative Services Act*, §304, 41 U.S.C. §254.

47. *Kingston Trust Co. v. State of New York*, 291 N.Y.S. 2d 208 (Sup. Ct. Spec. Term 1968); Annot., 141 A.L.R. 321 (1942).

48. In some jurisdictions, even if a material man installs the materials purchased, he is not deemed a subcontractor if the value of the installation services is slight compared to the cost of the materials; see 141 A.L.R. 321, 334.

Theisen v. County of Los Angeles, 54 Cal. 2d 170, 183, 5 Cal. Rptr. 161, 169, 352 P.2d 529, 537 (1960) (in which performance of work at the site was rejected as the test of who is a subcontractor as distinguished from a materialman in determining right to pursue mechanics' lien and related remedies; instead the court adopted as the test whether "in the course of performance of the prime contract he constructs a definite, substantial part of the work of improvement in accord with the plans and specifications of such contract." The court further noted that supplying off-the-shelf type items indicated that the supplier was a materialman, not a subcontractor.)

Also see *Comptroller General's Decision,* No. 13-166, 999 (August 25, 1969), 14 CCF ¶82, 983 (which held that listing of subcontractors did not cover suppliers of "off-the-shelf" items but did include those firms the bidder intends to use for the "manufacture, fabricating, installing or otherwise performing" work under the prime contract).

49. In California, the Subletting and Subcontracting Fair Prac-tices Act requires bidders on all public works contracts in excess of $5000 to list subcontractors for any portion of the work in excess of one half of 1 percent of the total bid, *California Government Code,* §4100 *et seq.;* except contracts for the construction of streets and highways, which are governed by contractual limitations.

On federal work, the Comptroller General, at the request of the General Services Administration, has permitted that agency to require listing of proposed subcontractors and authorized rejection of bids as nonresponsive which do not comply. 43 Comp. Gen. 206 (1963).

50. See references 47 and 48. Also see provisions of *Miller Act,* 40 U.S.C. §270a *et seq.* and *California Government Code,* §4200 *et seq.*

51. *United States v. Blair,* 321 U.S. 730 (1944).

52. *Severin v. United States,* 99 Ct. Cl. 435 (1943), *cert. denied,* 322 U.S. 733 (1944).

6 ARRANGING FOR FINANCING

EARDLEY W. GLASS

Financial Vice-President (Retired)
Morrison-Knudsen Company, Inc.
Boise, Idaho

ARRANGING for financing is the first and paramount necessity facing a contractor. From the very start as a small individual operator or a partner, through all the stages of growth to the largest national or worldwide diversified construction organization, the problem of assuring adequate funds to operate contract commitments, provide for overhead, pay wages and salaries, buy and maintain adequate equipment, purchase supplies and materials, and stand behind losses if they occur remains always present. As the company grows larger the problems become more complex, but the principles never change.

6-1. Initial Organization of a Construction Company

In the beginning there is one man or perhaps two partners who have saved or acquired sufficient funds to invest, as capital—enough to finance a small job or possibly two. They have selected a surety company and, based on their individual experience and reputation and available funds, have been assigned a bonding limit, the contract amount that the surety will grant them and bond at a given time.

Having arranged for a stated limit of bonding credit the partners, with no credit line for unsecured loans at the bank of their choice and no property or securities which can be pledged for a loan, use their capital to finance the first few small jobs. These jobs, if successfully completed, result in a profit, increase their capital, raise their bonding capacity, and make it possible to bid larger jobs if they can arrange bank credit.

6-2. Approach to the Bank

Let us say the surety company now will bond the partnership up to a $1 million or roughly ten times their available working capital. The job they wish to bid on will, however, be safe, require more than the $100,000 available, par-

ticularly since some of that is still tied up to complete payment on equipment previously purchased and in retained percentage not yet received on completed work. This reduces cash and the partners now go to the bank.

The partners will have with them the partnership agreement, a financial statement of assets and liabilities, a statement of earnings for the year and to date, and each partner will furnish his personal financial statement. (In the case of a single proprietorship, the owner will furnish the same information, including his personal assets and liabilities.)

Since the partners want to bid on a contract at or near their newly established bonding limit of one million dollars, they will make an estimate of cash required and cash flow to determine how much of a loan they will require and for how long, when the peak borrowing will be reached, and when the loan will be retired. The partners request a line of credit up to $100,000. This will give them ample operating funds in case they are successful in getting the job and ensure paying off the loan with reasonable reserves of their own capital to absorb any losses that may possibly occur.

Having considered the statements the banker will ask the following questions:

- What is the purpose of the loan, i.e., what job do they have in mind, what is its duration, and what are the principal hazards involved?
- What is the assurance that the loan will be repaid on schedule?
- What will be the source of payment?

If the questions are answered to the banker's satisfaction he will issue a line of credit, after first considering the credit reputation and general integrity of the partners. The line of credit will have a definite date on which it will expire, the interest rate that must be paid, and the notes written against this line of credit will have to be signed jointly and severally by both partners.

The partners now know the limits of the amount of bond they will be granted and also the amount of credit they can expect.

They bid on a job within their bonding and bank limit and get an award.

6-3. Importance of Bank Credit

A contracting organization cannot continue to grow and to prosper without access to larger and larger bank credits. Continued success, a favorable reputation for satisfactory performance, unquestioned integrity, and a record of keeping the bank informed on the current status of all contracts and any changes in the contractor's financial picture—all these add up to the only way to get and hold the confidence of the bank. The contractor must recognize at all times that he is renting money placed in the bank's hands by its depositors. The bank, in order to make a profit for its stockholders, must rent its depositors' money safely and to best advantage.

Let us assume that the partnership, having established its initial bonding limit and a credit limit at the bank, succeeds in getting away to a good start on a contract in the million dollar class. It has not had to use completely the $100,000 credit limit and, when the job is approximately half completed, the contractor has been able to maintain an average compensating balance of, say, 20 percent of the loan. He finds that after paying his payroll and bills he is able to pay back a portion of the loan ahead of the time agreed with the bank. This puts the contractor in the class of a preferred customer with the bank and lays the foundation for an even more generous credit line in the future.

Meanwhile an even larger and more promising job is advertised for bids before the completion of the first. The contractor determines that the equipment purchased for the first one will fit the newly advertised job which is in the $1,500,000 bracket. The schedule of the new job will permit the transfer of equipment in time as it becomes available from the operating work and thus will permit the contractor to place a very competitive bid, with a chance to get the job and still make a satisfactory profit.

The partners revise their financial statement to reflect the expected profit from the existing job and update the loan repayment schedule. This time they go to the banker first. They show him that the new job, if secured, will require a loan limit of $150,000; however, it will not exceed $100,000 at any time before the present loan is fully liquidated ahead of schedule. Thus they are asking for an increase in their line of credit from $100,000 to $150,000. Meanwhile their own capital will have increased to more than $150,000.

The banker gladly grants them the new credit line and they then proceed to their bonding company with a letter from the bank authorizing their new credit line and a forecast of earnings on the half completed job. The bonding company, under these improved circumstances, approves an increase to a million and a half bond on bidding the new work, counting the existing work as essentially completed when the new work must start.

The contractor, realizing the importance of maintaining the position of confidence and trust which has been created in the minds of the banker and the surety company by successful performance of work and a conscientious policy of frequent and truthful disclosure of current position, looks forward to further expansion of the business.

The partners have in the meantime been watching the performance of three of the younger foremen on the work and feel that two of them are ready to head up jobs of their own. They also think the estimator is ready for assignment as chief engineer, and the office man is ready to take over as secretary-treasurer.

Accordingly, after further conversation with both the bank and the surety company, they proceed to incorporate, one partner becoming president and chairman and the other vice-president and general manager. Shortly they have three jobs going, still financed by a somewhat larger but single line of credit.

6-4. Individual Project Financing

As a contractor expands and has several projects under construction at the same time, the procedure of having only a single line of credit begins to present problems to both the bank and the contractor. The original arrangement when operating a single contract stipulated not only a specific loan limit but also a terminal date when the loan must be paid off.

As the business grew and several jobs were underway at once, more and more credit was required as new work was taken. This resulted in constant borrowing within the line of credit. The amounts would vary within the line as jobs were completed and paid for and new ones taken, but they would never be paid off completely at any one time. This procedure resulted in frequent extensions of the notes as due dates arrived, and the loans went on and on without firm maturity dates, thus becoming in effect a capital loan.

Since this procedure is unsatisfactory to a growing construction concern and contrary to bank policy, the result was development of the individual project loan. Within the overall framework of a general loan limit each project, as it was acquired, became the subject of a separate project loan. A comprehensive study of the project by the contractor of cash flow, amount of maximum investment, and schedule of repayment is furnished the bank. The bank is assured that as funds are received from the job in excess of operating costs they will be used to liquidate the specific job loan. The bank is further assured that the loan will be completely liquidated on or before the scheduled completion of the contract and is advised that, if the project the bank is financing should develop a loss and there should not be sufficient funds from the job to pay the loan, the contractor will liquidate the loan out of his own funds on schedule or ahead of time.

This method of individual job financing is very flexible and, in addition to the loans being in the main self-liquidating, it opens the avenue of credit from a number of different banks and consequently larger and more adequate sources of construction funds.

6-5. Location of Banks

The contractor who is wise will establish credit and carry deposits with selected major banks in areas or regions where he has jobs or expects to develop them.

When a project develops in the geographical area of a major bank, usually in the largest city in the region, it is of real advantage to the contractor if he has already opened an account, established his credit by furnishing current financial statements, and discussed possible project loans in advance. The bank has had ample time to determine his reputation, capability, bonding capacity, and general credit. Banking and using project loans in the area or region of projects performed in no way disturbs the contractor's loan credit in other areas where he is working and other regional banks are used. Once the contractor establishes the practice of giving his business to a bank in the region of his jobs, he has successfully firmed up and greatly broadened his credit picture. He will also have the benefit of the bank's resources in determining the capability and credit of local and regional subcontractors and suppliers.

The contractor should always remember that if he borrows a *little more* than he really needs and keeps a *little higher* compensating balance on deposit in the bank he will save in the long run. He will get the lowest possible interest rates, and he will get the money he needs to borrow when times are rough and money is tight. The contractor who thus operates his deposit and loan policies becomes a preferred customer, whereas his penurious competitor who holds his deposits to the bare minimum soon finds his credit restricted and his interest rates higher.

6-6. Interest Rates

Interest rates charged the contractor are the product of several different factors and are a matter of negotiation with each individual bank. The bank considers the net worth of the contractor, the ratio of current assets to current liabilities, and frequently recognizes operating assets, the total of cash, securities and short-term receivables plus equipment in determining the loan. The bank weighs its experience with the individuals in the company with whom it has been dealing and reviews their history of paying loans on or before due. It also considers the profit record of the company as well as its practice of maintaining adequate balances. Finally, it weighs carefully the probability of the contractor continuing to get work in the bank's geographical territory.

The bank will also weigh its own availability of funds to loan and, depending on all these answers, will come up with an interest rate at or above the published "prime rate" which banks generally charge their most favored customers.

The contractor who "shops around" with several competing banks in an effort to get the very lowest interest rate is going out of his way to injure himself. There will come a day when he really needs a loan that he will not get.

6-7. How to Keep Good Bank Relationship

Developing and maintaining a sound and lasting bank relationship require both time and attention. The basic fundamentals of square dealing and paying off all notes on time are, of course, the first requisites, but it is wise to reinforce them. No man should withhold information from his doctor or his lawyer. The contractor should add to these his surety bondsman and his banker. The contact having been made, deposits placed, and a credit line established, the line of communication must be kept open.

An audited annual statement must be furnished each bank the contractor is doing business with (as well as his surety company). In addition, an annual and semiannual report of operations should be prepared and furnished them. These reports should give in considerable detail a description of each job, showing the location of the job, the type of work, the owner or client, the total amount of the contract, and the scheduled date of completion. There should also be a brief narrative as to the status of the project, the investment in the job, and the bank loan standing against it. There should be a frank, truthful analysis as to whether the job is showing a profit or loss and probable "cost to complete." Both the bank and the surety company are very receptive to these reports, and they accomplish a great deal in cementing a feeling of mutual confidence. The contractor need not fear any adverse reaction if the reports show one or two jobs in the losing column as long as the overall picture indicates a profit. Banks and surety companies are far from being fools. They know very well that there never was a contractor who always made money on every job. They will do much more to back a contractor whom they know is honest.

It is also very desirable to have the president and the vice-president–finance visit both the banks and the surety company from time to time. An excellent schedule is to arrive fairly soon after they have received the annual statement and the year-end report of operations. The general outlook should be discussed very freely and any new developments, favorable or unfavorable, should be pointed out. Major changes in the operation as they develop, new contracts of significant proportions, or an unforeseen loss that may affect the current situation should be brought to the banker's attention promptly, at the scheduled visit or by phone or letter at the time they occur.

6-8. Equipment Financing

Construction equipment, particularly to the heavy construction contractor, is, next to people, the most vital part of his ability to perform work. Good equipment is constantly being improved, and becoming more and more expensive as it becomes increasingly larger and more productive. To stay in competition the contractor must have his fleet constantly analyzed, not only to replace worn-out equipment as it becomes more costly to repair and maintain, but to replace otherwise satisfactory units as they become obsolete. This requires financing and lots of it, so the contractor can conserve his cash and remain sufficiently liquid to manage his business and have reserves to take advantage of contracting opportunities as they occur.

Equipment can be financed in several ways.

First: Most dealers from whom equipment is purchased generally have available financing arrangement. After the required down payment is made (usually from 10 to 25 percent of the purchase price), the balance can be paid off in monthly installments. A note and a conditional sales contract is given the dealer by the buyer. It may be for the full amount of the purchase with a credit showing the amount of down payment, or simply the balance due stating the monthly payments, which include principal and interest. The vendor retains ownership of the specific unit or units of equipment until the last payment is made.

The interest rate is usually quite high, perhaps 50 percent higher than the dealer can obtain from his bank, and sometimes, if the dealer is financing his own paper, the note and conditional sales contract may carry interest on the full purchase price until the entire purchase price is paid. This type of contract, charging as it does "interest on interest," is extremely expensive and should be avoided at all costs. If the purchaser has any reasonable credit rating he can always insist on the interest being charged "on a declining balance." The dealer frequently will decide to discount the note with his bank. In this case the contract will usually carry simple interest on a declining balance as payments are made. If, however, the dealer has named a higher rate of interest than the current "prime rate" charged by the bank, he may be able to discount the note for its face value, thus getting paid immediately for the full list price of the equipment including his commission, and may not have to pay the manufacturer the wholesale price of the equipment for thirty, sixty, or possibly ninety days. The banks, in turn, usually look with favor on discounting this "third-party paper," since they have two parties to look to for payment, the purchaser and, if he should fail, the vendor, who would be forced in this circumstance to make up the payments if the equipment, when recaptured, did not on resale yield enough to pay off the note.

Second: Various commercial credit corporations finance construction equipment. These credit corporations do a professional job but (depending on the contractor's financial and credit status) usually charge high interest rates, always higher than the normal bank interest rate. Should the contractor lack a good credit rating, they are inclined to charge full interest on the original loan until paid. Although interest is an expense and can be taken as an income tax deduction, it must be borne in mind that a deduction is of no use unless it can be charged off against a profit.

Furthermore, commercial credit companies will require that the equipment be collateralized to secure the loan. Thus a chattel mortgage must be given which is recorded with the county clerk in the county and state where the equipment is to be used. This can work a hardship or at best create a complication in case the contractor finishes his job before the equipment is paid for and wishes to ship it to another job in another county or state.

Third: A number of companies have been formed that buy equipment and lease it back to the user on a term basis, and even some banks have divisions or separate companies that make a practice of this. Under certain circumstances this arrangement may have merit, but it must be analyzed carefully.

It should be borne in mind that the leasing companies are essentially using free cash and/or a high credit rating to invest in large volumes of equipment (serving a number of lessees at once). Their gain stems from (a) volume and/or cash discounts direct from the manufacturer; (b) leasing the equipment at a monthly rate higher than bank interest and amortizing the equipment at a more rapid rate than its normal life from the full list price; (c) An offer by a lessee, having paid half or more of the purchase price, to buy the equipment for a price 10 to 15 percent above the remaining unamortized list price. Although the lease cannot provide this option for tax reasons, the practice of these lessors is to sell to the lessee for a reasonable markup above list price, crediting perhaps 90 percent of rentals paid, thus coming out with a handsome profit on the entire transaction.

There are certain circumstances under which a contractor may wish to consider the lease arrangement. For example, a contractor may have a fairly long-term job that will last as long as or perhaps a little longer than the equipment lease. Since the equipment rental is entirely a tax deductible expense, the contractor may charge the rental to the operating cost of the job. If he is fortunate and picks up another job that can use the equipment at or near the end of the lease period, he can purchase the equipment for 10 or 15 percent of its original list price and come up with quite a bargain going into the second job.

Another situation which may properly cause the contractor to consider an equipment lease arrangement comes about as follows: The contractor together with others in a joint venture has bid and been awarded a large contract running for a period of two to three years. It is understood that plant and equipment to be used throughout the life of the job will be purchased and owned by the joint venture. However, there is a large volume of excavation equipment needed for the first twelve to fourteen months of the job. The joint-venture agreement states that equipment which is needed only in the earlier phases shall be furnished by the partners in proportion to the participation each partner holds in the contract. Each partner shall furnish his part of the equipment at a predetermined rental scale, frequently at the Equipment Dealers' Manual rental rates at which rates equipment dealers will normally rent new or used equipment for short to medium terms. The contractor calculates that he can rent his share of the equipment from the equipment rental company at the same rates as the joint venure is paying him for the use of it. If it is rented by the joint venture for fourteen months and then returned, it will have generated in rental about 70 percent of its list price. The contractor at this point finds that he can purchase the equipment after the rental credit at about 40 percent of the list price. This makes a very good bargain, particularly if he has in the meantime been successful in securing another job to move it on to.

Finally, when the contractor has good bank credit and an adequate amount of available cash, he may deal directly with the bank. This is accomplished by placing a so-called "equipment note" with the bank. A number of items may be tabulated in the note which lists the equipment being financed and the purchase price of each item. Since the contractor is going to pay the manufacturer or dealer cash, he will be able to secure a cash discount of possibly 10 to 15 percent off the list price of the equipment. Generally speaking, the banks prefer to finance about 75 percent of the original cost. This leaves the contractor with 25 percent to finance out of his own pocket. The terms of the loan are flexible both as to interest and repayment dates, the length of time varying with the expected life of the equipment. Heavy excavating equipment may be financed up to four years, heavy trucks and tractors up to three years, lighter trucks, compressors, concrete mixers, pickups, etc., for two to two and one half years. Payments on the note are made monthly or quarterly. The bank is advised that should any of the equipment be sold or moved out of the United States the bank will be paid whatever equity remains unpaid.

This method has several advantages. The bank considers these equipment loans specific and not capital loans, since they are tied to specific items and are for a definite time; furthermore they tend to be self-liquidating as the equipment is used. As the experience of the banks developed, equipment loans tended to become preferred loans. Thus the equipment loans did not seriously restrict the contractor's ability to negotiate project loans. Finally as time went on the banks would extend equipment loans even on equipment destined to be used overseas when proper insurance was carried.

It should again be emphasized that there is no real substitute for bank credit when it is available to the contractor for operations. If a loan is requested that has a specific purpose, has a definite maturity date, and can thus be termed an operating loan, it is attractive to banks. If, in addition, the contractor does not neglect to maintain adequate compensating balances averaged on a one-year basis (usually 10 to twenty percent of all loans depending on whether money is scarce or plentiful), he will, as long as he is honest with the bank and reasonably successful, have available credit to supply money for all his legitimate operating needs.

6-9. Financing Other Than Banks

Most contractors, as they grow and expand geographically, need funds for general purposes and to supply additional capital. This so-called "capital financing" is not properly in the province of commercial banks, and the contractor must look to other sources for these funds.

Funds for capital purposes in the form of long-term financing are available from a number of sources. Insurance companies, pension funds, and investment bankers are always looking for good, long-term prospects (ten to twenty years and sometimes longer) in the form of notes, debentures, or preferred stock. In the case of loans in the form of notes or debentures, these institutions will make very thorough investigations into the company, its record of earnings and payment of dividends, its management and management policies, the age of its officers, possible successor management, and any other aspect of the company's picture and general trend of the industry that might affect adversely the positive assurance of the loan being repaid.

Generally all such long-term loans are made with definite conditions on maintaining certain balance sheet restrictions. These restrictions vary with different institutions and with their appraisal of the "quality" of the company borrowing the money, but almost always require that a minimum stated working capital be maintained and that there be a stated ratio of current assets to current liabilities and a limit to total debt. Normally dividends cannot be paid unless the earned surplus is in excess of a stipulated amount. These requirements are ordinarily not onerous if the company continues to be profitable and does not encounter losses that impair its financial structure. They are designed to minimize the risk of the lending institutions and in reality act to prevent the borrower from getting into injudicious fiscal and operating situations.

Since these loans are made without compensating balance restrictions, the contractor has all the funds for use for the purpose for which the loan was made. Repayment on a loan of this type varies, as does the interest rate. Payments are usually required on a quarterly basis but in some cases may be made semiannually on principal plus interest. The length of time varies, usually ten to twenty years. In certain cases when larger principal payments are bunched toward the end of the period or a substantially larger payment is made at the end, there may be a requirement that the borrower must accumulate and maintain a "sinking fund" to assure payment of the later amounts when due.

Sometimes loans are made with contingency or added interest based on future possible profits of the company, or perhaps the interest rate may vary up or down, stipulating perhaps 1 percent above the "prime rate." It may also happen, and frequently does, that the lender may want "a piece of the action" and will insist on an option to buy common stock at a predetermined price for a predetermined number of shares or percentage of the outstanding shares when the option is exercised. The lending agency may possibly stipulate that in acquiring the debentures it must also receive a stated number of convertible preferred shares which will pay a fixed dividend subject to earnings during the life of the loan or until converted at a pre-agreed price into common stock. Thus the lender, depending on the supply of money at the time the loan is made and the credit rating of the borrower, may by its loan requirements enjoy substantial benefits as the company continues to grow.

6-10. Public Ownership of Stock

At some point a contractor may wish to "go public" by selling an interest in the company to the public. This may be desirable for a number of reasons. The contractor, growing rapidly, may find himself in the situation where further borrowing of capital for expansion becomes difficult by reason of his relatively small capital structure in stock compared to his loans. It is true that he cannot deduct dividends which may be paid, as a cost to reduce income tax, but must in any case pay dividends from surplus accumulated or from current earnings after tax. At the same time the selling of common stock to the public does not carry with it the obligation to pay dividends and the company may be proscribed from doing so by previous indenture agreements covering long-term borrowing, at least until such time as earnings make dividends both permissible and desirable.

The procedures that must be followed in order to sell an interest in the company to the public are, to say the least, quite complicated. Registration with the Securities and Exchange Commission is mandatory and must be accompanied by financial statements, earnings records, and disclosure of any details of the business affecting the value of the stock.

The legal procedures necessary involve securing the services of outside public accountants, lawyers, and a recognized stockbroker to handle, for a fee or markup, the sale of the stock in the first instance and later to act as agent for those who wish to buy or sell stock. A bank or banks must be selected to act as transfer agent when stock is bought or sold or additional stock authorized to be issued for sale or distributed to existing stockholders in the form of a "stock split" or "stock dividend."

The probability is, if the company is fairly new, that its stock will be listed on the "over the counter" market, which is not a stock exchange as such but a published list of the price established by brokers who deal in the stock on behalf of their customers who wish to buy or sell the shares. At some later date the company may or may not wish to be listed on the New York Stock Exchange, the American Exchange, or some other. This will depend to a great degree on how much trading in the stock develops. If the float or percentage of the stock traded is relatively small in relation to the total number of shares or if the total number of shares issued is not large, the company may find the facilities of the "over the counter" market sufficient. A number of large companies whose shares tend to remain traded in small volume have refrained from being listed on a major exchange for many years. Others more national or international in scope whose stock is traded in considerable volume have become listed on a major exchange to broaden their market and enable the stock to be traded more freely in large blocks.

6-11. Joint-Venture Financing

A joint venture, as has been stated elsewhere, is a limited partnership of two or more contractors, perhaps up to six or seven, banded together for the purpose of bidding or negotiating a single job or project. If the proposal is successful, the work will be performed by the joint venture, usually under the direct management of one of the partners acting as sponsor.

The selection of the sponsor and determination of the percentage of participation of each of the partners is a frequent cause for disagreement and may cause one or more members to withdraw without agreement by reason of being offered too small a share, desiring to become sponsor, or even a shortage of funds or lack of bonding capacity. Occasionally the bid or proposal may be dropped and the job go to others because a joint-venture agreement cannot be reached and none of the proposed members has the individual credit and bonding capacity to go it alone. At other times, although a joint-venture agreement may have been reached, the differences of opinion on the price to be bid may be so wide as to preclude arriving at a figure that all can agree on for the bid. More frequently, however, agreement is reached either by each member voting for his bid figure, the totals then being averaged to produce a compromise to which all will agree, or by the sponsor, having the largest percentage and largest risk, dominating the decision and a price is agreed upon at or near his proposed figure.

Prior to the partners meeting to determine price, however, arrangements must be made for a bid bond, a performance and payment bond, and financing for the job if they are successful in obtaining the contract. This is an urgent necessity and, if not handled properly, may lead to serious problems and even disastrous loss to one or more of the partners if the project loses a large sum of money or some of the partners cannot produce their share.

It should be understood that in the usual practice each partner is required to sign the contract "joint and several," which means that in case of default each partner, regardless of his percentage, is obligated to complete the contract if any or all of the other partners default or become bankrupt.

Surety bonds likewise are written in most cases providing that the joint-venture partners sign the bond joint and several, and thus each one becomes liable to the surety company for the full amount of the bond penalty in case of default by the joint venture.

In some cases the joint venturers, having already signed joint and several on the contract and the bond, may decide to borrow funds from the bank in the name of the joint venture, signing the notes joint and several, resulting in each partner becoming fully responsible for the entire amount if the others are unable or refuse to pay their share. This practice is not recommended and can prove disastrous.

The safest and most businesslike procedure, which can usually be put in effect if the surety bond limits and bank credit of each of the partners is good, is to eliminate the joint and several obligations on the bond and have each partner sign the bond for his percentage "severally." By the same token, each partner should furnish his pro rata share of the joint-venture capital individually, which is then under the control of the sponsor and whomever each partner elects to represent him to supervise and approve deposits as well as withdrawals.

This procedure can be accomplished in the following manner: A request to the surety company is made before the bid to authorize each partner to sign the bond "severally." The surety company does not look on this procedure with favor, but when all the partners are considered good risks it will usually comply, with a stipulation that the partners provide a capital fund at the start of sufficient amount (frequently 10 percent or more of the face of the contract) to be deposited to the joint venture account and that no return of capital be made to the partners without prior approval of the surety. This deposit causes the partners to provide more early money than might be required if calls were simply made to the partners as funds were needed.

At the same time the funds not immediately needed can be invested so they draw interest and thus reduce the additional cost of the higher early investment. If this procedure is agreed upon before the bid and when the joint venture is formed, it acts to protect not only the surety company but each member of the joint venture.

When the surety will not consent to the bond being signed severally because of the relative weakness of one or more of the partners, the sponsor should insist on sufficient capital being put up at the start to assure that each partner has furnished sufficient funds so that his commitment for operating funds is protected. If one (or more) of the partners cannot meet this condition, his participation should be reduced to what he can handle or he should not be in the joint venture at all.

A large and financially sound contractor has often been tempted to take an interest in a joint venture which a smaller and less well-financed contractor is to sponsor because of special knowledge and perhaps equipment and organization in a particular type of work, although his finances are not adequate. In these cases, of course, the surety will insist on joint and several signing on the bond, relying on the large and financially sound partner, who may know very little about the work, to carry the load and bail out the job if anything goes wrong. To aggravate the situation, the weak sponsor will no doubt insist on joint and several notes to finance the venture, the bank also relying on the strong partner and going along. Several strong and successful contractors in their own fields have found out to their sorrow that "the grass is not greener over the fence" and such a commitment places them in the position where "you can't win, but you can lose."

Another type of financing has developed whereby a financial institution becomes a partner. A financing agency, insurance company, or perhaps a pension fund, wishing to have a "piece of the action" will put up all the money in return for, usually, 50 percent of the profits. A self-liquidating project, such as a high-rise building, shopping center, or perhaps an underground garage, lends itself most readily to this type of financing. The financing institution will agree to furnish all the cash involved, the interim construction loan, and the long-term mortgage money. Since the long-term mortgage money will not usually cover all the financing required, the balance or equity money will also be put up by the financial entity. When the project is completed and begins operations, the first available funds are used to return the financial partner's equity money, after which the contractor will have his 50 percent of the ownership and future profits. At this point the contractor or the financial partner or both may elect to sell their shares at a profit or continue to carry on the operation.

There are usually a number of such operations available to a contractor. They all require careful study to determine their economic feasibility, but when properly analyzed many have proved to be highly profitable with minimum risk to the contractor.

6-12. Financing Subcontractors

Frequently it becomes necessary or advisable for the prime contractor to assist in financing one or more of his subcontractors on a project. It may be that the subcontractor has good credit and can put up a bond for his portion of the work, but he may find it inconvenient or impossible because of other commitments or contemplated commitments to provide all the cash to operate the specific portion of the job he has offered to sub. In general practice, the prime contractor makes the subcontractor wait for his pay until he himself receives the money on the monthly or periodic estimate covering his own work, as well as the subcontractor's work. He also in most cases withholds the same percentage from the subcontractor's estimate as the owner withholds from the entire estimate, and does not pay the subcontractor the retained percentage until the work is completed and accepted by the owner—or at least the portion for which the subcontractor is responsible.

When the subcontractor is able to furnish a valid payment and performance bond, there is no basic objection to the prime contractor assisting the subcontractor in his financing. This is usually done by paying his payrolls and payroll charges for him, as well as purchasing his supplies and materials or at least paying for those the subcontractor has purchased. The prime contractor will bill the subcontractor for the cost of these services and deduct the charges from the amount owed under the contract. This procedure has some actual advantages in that the prime contractor is still protected under the bond, and he has the additional advantage of being certain that items which could become lienable against the work if not paid have actually been paid, and will not delay the final payment on the contract.

All too frequently a prime contractor may be persuaded to award a subcontract by reason of lower price, previous acquaintance, or some other reason, and the subcontractor cannot furnish bond. In this case the prime contractor has taken on full responsibility for this portion of the work as well as his own. He does not have a valid subcontract and should not consider the arrangement as such. There may be a valid reason, such as unusual skill on a technical portion of the work, past employment creating special confidence, or some other factor, which makes the prime contractor willing to go along, considering the price paid above whatever the final cost turns out to be as a bonus. He should, however, consider the arrangement for what it is—payment for supervision of a portion of the work. It is not a subcontract and should not be called one.

6-13. Financing Foreign Operations

When a contractor, having reached great size, has acquired national scope and experience and capability in various diversified lines of work, he may become involved in foreign work. This may come about in several ways, and the problems of financing the work vary greatly. They do not lend themselves to broad general definition but must be analyzed according to *where the work is, what kind of work is it, what is the duration,* and finally, *whom is it for?* Not, however, in that order.

Let us assume that a large national contractor with good credit and broad experience in joint venture with others is awarded a cost-plus-a-fee contract for the United States government for various military installations in the Pacific. The duration, as in all such assignments, is unknown. The fee is small, but the volume of work may prove very large. All direct costs are reimbursed except that the cost of interest to finance the work until reimbursed is not allowed and must be borne by the contractor. Inasmuch as reimbursement is paid twice a month, the money required is estimated to be $5 million. The joint-venture partners go to a large bank on the West Coast. Since the work is foreign, the bank offers to finance the job on an overdraft basis. The terms are simple. There will be a charge of 0.5 percent on $4 million for "earmarked" or availability of funds. The partners will deposit $1 million of their own funds, which will act as a compensating balance. The

bank will then supply up to $4 million additional on a daily overdraft basis at the current "prime rate." The $1 million put up by the partners, however, can be used up if necessary from time to time, since the agreement merely states that averaged over the year there shall be a balance in the account of at least $1 million.

The key to the cost of this financing is the speed of getting the list of payroll checks and vouchers for other expenditures to the government paymaster at job headquarters. By the nature of things, the voucher checks for payment of bills and paychecks paid employees come into the bank for payment in from two to three weeks. The government reimbursement check is handed to a messenger who goes out on *that night's plane* and the check is deposited to the joint-venture account for the next day's business.

It soon develops that the cost of financing the job, relying on the "float," becomes one of the more minor costs against the meager fee; other expenses that are not reimbursed, such as executive time and home-office expense of the partners, are much more serious. Since the fee is accumulated and paid without retained percentage, usually once a month, the partners' cash investment in the job is reduced in a few months to the point where the overdraft agreement with the bank can be cut in half and the average balances shown nearly eliminate further borrowing. The bank, however, is happy because the average balances have risen to the point where the joint venture has become a large depositor rather than a large borrower. The balance in the joint-venture account is credited to the sponsor and the other joint-venture partners as a substantial compensating balance against loans they may wish to make for other purposes.

Other military installations continue to be offered abroad, in the Pacific, in the Atlantic, in Greenland, Iran, Pakistan, and elsewhere. These after a time tend to be offered on a basis of firm unit-price bids in competition.

Our contractor in joint venture with others bids and receives award on a large and geographically widespread series of military cantonments in Iran. The job is in excess of $100 million and will require more than $15 million of construction equipment. Ten percent of the monthly estimates are withheld until the job is half completed, at which time, if the job is satisfactory and on schedule, no further withholding from the estimates is made.

The government, which has the option on foreign jobs, elects not to require a surety bond, the cost of which was stated in the bid and can be deducted from the overall price.

The amount of funds estimated was large. Fifteen million dollars for the equipment to be shipped to Iran plus move-in and operating money of $10 million, all of which had to be made available by the partners from their own funds or funds which they would borrow.

The partners approached a large East Coast bank with whom the sponsor had been dealing for some time. Much to their surprise the bank, which had been making equipment loans to the sponsor, offered "since this is a U.S. government job" to finance the equipment on the same basis as at home, the contractor to pay 25 percent and the bank to furnish the balance at the usual rates. The bank also offered to provide an $8 million overdraft loan, the partners to provide $2 million for the account.

The joint venture then had only to worry about raising $3,750,000 for the equipment and $2 million for operating money. Each partner had no problem furnishing his share and the job proceeded.

It was soon found that a considerable amount of the building work, which was largely brickwork, could be subcontracted to local contractors who were capable and could furnish guarantees of performance. These subcontracts reduced the amount of cash necessary to carry the work since the subcontractors were paid in the local currency, rials, with a 10 percent retained percentage as the prime contractor was paid. Local payrolls and expenditures and local purchases in rials were made by writing a dollar check on the American bank and depositing it for rials in the local bank in Teheran.

Since the local bank did not insist on waiting for collection and credited the dollar check to the joint-venture account when received, and since it took several days for the payroll checks to come in for collection as well as the payments on subcontractors' estimates, there was a slight but significant float time between the deposit time of the dollar check and the actual time when the rial checks came in for collection. Furthermore, the deposit in the local bank of the dollar check for the rial deposit did not reach the American bank as fast as the deposit of the U.S. government estimate check, which was flown over by courier as soon as received. Finally, the American payroll which, although covering only one tenth as many men as the Iranian payroll, was about equal in dollars, and the checks for American purchases, floated in two or three weeks after the monthly estimate, had been deposited in the U.S. bank, creating a favorable float and greatly reducing the amount of overdraft required.

After a few months the float acted to reduce the loan requirement in the United States bank from $8 million to $4 million, and in a year, when the job was half done and no further percentage was deducted, the estimates and float eliminated the further necessity of borrowing for operations, leaving only the remaining equipment notes to be paid off.

The contractor, meanwhile, had bought a well-known Canadian company that was on the whole doing quite well. There was no language barrier and business was conducted in a manner similar to that in the United States, with the difference only that branch banking was permitted, most of the major banks maintaining branches in every Province. The system of overdraft was not the rule, although it was possible to finance work without maintaining large compensating balances, at a slightly higher interest rate but at an overall interest cost of not much more than in the United States.

The Canadian operation soon led to work in Australia and New Zealand, sometimes with Canadian, American and Australian (or New Zealand) partners and sometimes alone. In these Dominion countries credit was good and, being in the sterling bloc, there was no problem of conversion of profits. There was, however, from time to time a fairly minor adjustment in the rate of exchange. It behooved the contractor working in these countries, therefore, to finance his work insofar as possible by borrowing the local currency without any major transfer of dollars, at least for the construction operations.

Certain other countries would come under the head of "safe" places to work—England, France, Holland, Belgium, and West Germany. If the contractor had special knowledge in a particular field and had picked a good construction partner in the country, he might find it possible to get work building pipelines or perhaps subways or tunnels as well as industrial construction for American firms building plants in those areas. The financing problems would be no more severe than at home, and bank loans could be obtained with the guarantee of a United States bank and, in some instances, on the credit of the American contractor alone.

Italy, Spain, and Portugal also developed numerous construction opportunities for the American contractor with a local or European partner. In Portugal, for example, the longest and largest suspension bridge in Europe was bid and secured by a combination of an American contractor, an American Steel and Bridge company, and a French banker who was able to provide all the funds, some $30 million (in escudos) above the commitment made by the U.S. Export-Import Bank and the World Bank. The job, as it was financed, provided for advances to the contractor, to be recaptured in later estimates, in sufficient amounts so that only a small sum of dollar financing for tugs and barges and United States concrete mixing and pouring equipment was necessary.

Certain countries in the Middle and Far East where American financing was involved, or where the U.S. Export-Import Bank, A.I.D., and perhaps the World Bank were making loans, offered opportunities for American contractors who could arrange with American banks for the necessary construction loans, again on an overdraft basis. Iran, Turkey, Afghanistan, Thailand, Taiwan, and even Indonesia offered good opportunities, and those projects that had been financed partly or wholly by United States agencies would not only permit the contractor to take out his profits but would make advances that greatly reduced his construction financing cost.

A number of large and rather attractive projects have been built in several countries in Africa. The financing has in general come from European sources, and the work has largely gone to European contractors, notably French, Italian, and West German.

Parts of Latin America and particularly most South American contractors, once a favored area for American investment and American contractors, have with some exceptions become too hazardous to invest in or perform construction work. The surge of nationalistic and anti-American feeling and the failure of the United States agencies to protect American nationals have made any investment or contract a form of Russian roulette. There seems to be no way in which a contractor can reasonably justify a contract investment in many of the countries of South America.

Perhaps an illustration will make the problem clear. An American contractor with ample South American experience has, in joint venture with others, bid and been awarded a large highway contract through a remote area of a South American country. The contract is financed by two U.S. government agencies. The work is engineered by a large and well-known American engineering firm. The contractor, nevertheless, has stipulated a sizeable advance payment sufficient to purchase all the necessary equipment and deliver it to the job, the advance to be recovered in increments as the job is built. The job progresses well except for certain arguments regarding quantities of slide material, which the contract specifications include as a part of the excavation.

Suddenly the president of the country is overthrown and driven to exile, with the government taken over by a leftist "Junta." The new government moves against a large American oil company operating in the country and expropriates or nationalizes its holdings. The contractor is continuing work, although the government has failed for several months to pay the portion of the monthly estimate it is obligated to pay in local currency. It has been approving the dollar portion of the estimate which is paid by the United States. Next the new government refuses to approve payment of the dollar portion and throws the American engineer who has approved the estimates off the job. One United States agency now refuses to make the dollar payments owed the contractor.

The outcome, of course, is that the contractor, having spent all the advance plus a lot of his own money, is forced to shut down the job. The government seizes his equipment and the United States agencies do nothing.

The answer is obviously that no matter what the contractor did he could not win. This sequence of events multiplied by a number of similar incidents in several other South American countries indicates why financing work in these areas is impossible.

There is one shining example of an area where the Americans apparently can *still* work—Brazil. If the present government stays in power and *if* it can better control inflation, it may be possible for United States contractors to finance and perform work in that country. *At least they can try*.

7 CONSTRUCTION FINANCING

GEORGE F. KERNAN

Vice-President (Retired)
Continental Illinois National Bank and Trust Company
Chicago, Illinois

THIS CHAPTER will deal primarily with the financing of contractors to support their own activities during major construction projects. These include the large, frequently long-term, usually multimillion dollar projects, such as dams, superhighways, bridges, and many other works for public bodies, as well as factories, power plants, and large commercial and sometimes residential structures for private ownership. Such projects are ultimately paid for with public funds, or by private ownership, funded by mortgages, debentures, or other long-term or "permanent" debt or equity instruments. It is not our intention to review such permanent financing, but rather the financing required by the contractor in his work as mentioned above, and also the substantial area of "interim" financing for project owners as contrasted with contractors that bridge the time gap between the beginning of construction and conclusion of the "permanent" financing.

7-1. Hazards of Construction Financing

Contractor financing has long been considered hazardous and challenging for many reasons. For one thing, any unsecured loans to a contractor literally stand "last in line" for payment in the event of trouble. Other creditor positions ranking ahead of such bank loans include taxes, wages and materials, both protected by lien rights, and finally the bonding company whose rights of subrogation in effect give it a pledge of receivables arising out of contracts covered by performance bonds.

Other characteristics of the construction business that make contractor financing hazardous can be grouped into three broad categories: (a) contingencies over which the contractor has no control, (b) contingencies within the purview of the contractor, and (c) contingencies of time and management.

Contingencies over which the contractor has no control include natural hazards, such as floods (involving extra costs for pumping out or containing water) and unseasonable cold (involving extra costs of protecting the structure from cold so that concrete can be poured, or carrying a job over because it could not be closed in before winter). This category also covers strikes affecting either suppliers and their ability to deliver materials on time, or strikes by the contractor's building trade mechanics, as well as the outright failure of suppliers to fabricate and deliver the materials at the time required to keep the project on schedule.

Encompassed in the second category are the risks inherent in the contractor's own field that he might not be able to anticipate or that he could overlook through inexperience. For example, soil and rock conditions are a fundamental determinant of cost in the laying of foundations, sewers, or anchors for bridges and dams. If water is present underground, expensive and time-consuming pumping and sheathing techniques may be required before any foundations can be poured. The contractor who has not anticipated such conditions, or who has not incorporated a "water clause" into the contract, must bear all such extra costs himself. Such oversights by the contractor can mean the difference between a healthy profit on a contract or a disastrous loss.

The third category usually involves two basic risks, time and poor management, which affect all types of contractors because of the inherent nature of their work. A contract usually involves a fairly long period, generally a year or more, between submission of the bid and completion. During this interim, prices of materials or the scale of wages paid to the building trades may rise substantially, or the productivity of labor may decrease. Many large, well-run construction companies were plagued by such problems in the severely inflationary period of the late 1960's. Of course, such problems are reflected in the company's profit statement.

Construction companies often engage in several projects

concurrently. Although this practice helps spread the risk of some loss over a wider base and improves revenues and cash flows, it makes the construction company susceptible to differing standards of management among the projects. If the company allots more attention and better managers to one project, perhaps because of its size or importance, labor working on other projects quickly senses this neglect and poor production per man may result. Lack of coordination or neglect can cause poor delivery schedules and necessitate expensive overtime shifts to bring the project up to schedule. Delegation of authority often does not work well in the construction industry. The complexity and interrelationship of the tasks involved, coupled with the critical time element, demand constant attention and coordination by management. A seemingly insignificant bottleneck in one area can quickly bring the whole project to a grinding and expensive halt.

7-2. Reasons Why Banks Extend Construction Financing

When one even briefly considers the dangers of the construction industry, he might be amazed to discover that they do raise the volume of funds required by this largest industry in the American economy that currently accounts for approximately $90 billion of our gross national product over and above the industry's own equity. There are numerous reasons why banks are willing to extend construction financing in spite of the hazards. Some construction loans are written at higher interest rates than general commercial lending, and this compensates at least in part for their greater risk.

Construction financing often develops relationships between the bank and the contractor which lead to other corporate and personal banking relationships. Such construction loans are instrumental in developing new business and industry within the bank's geographical market. This development manifests itself to the bank in terms of increased payrolls in the area, hence increased savings and new accounts. A bank invariably benefits from new business development in its area.

7-3. Steps a Bank Can Take to Lessen the Risk Factor

Given the inherent risks of construction loans and their substantial revenues for a participating bank, one asks what steps, if any, a bank can take to lessen the risk factor in lending to the contractor. Several protective evaluations greatly reduce the risk of a future write-off, and any bank can and should make these before commiting itself to a construction loan.

The most crucial evaluation to be made by a bank concerns the honesty of the potential borrower. This fact cannot be overemphasized. As a lender, the bank will depend almost entirely on the honesty of the contractor for most of the information on which it bases its credit judgment. The nature of a construction company makes it difficult for a bank to audit and verify much of the financial information provided. This subject will be discussed in greater detail shortly. Also, the bank must depend entirely on the contractor to complete the project according to specifications, to pay his suppliers, and to collect final payment in full so that he is able to retire the loan.

The contractor whom a bank has nurtured from modest beginnings is well known and presents no great problems. However, the situation is different in the case of the new customer or new contractor. Suppliers, architects, and owners for whom he has worked previously should be consulted. However, they can only verify his past actions. There is no way for the bank to be absolutely sure that the contractor will successfully execute the new project.

There is little that the bank can demand as collateral. Construction companies are traditionally established with modest capital. Some of the equipment they use may be designed for a particular project and have undeterminable salvage or resale value. Most important of all, their output has little or no value except in its completed form. If a construction company fails to complete a project according to specifications, for instance a $10 million dam, the prospective owner refuses to take possession and does not make final payment for the project.

The bank's construction loan becomes tied up in special equipment that often cannot be resold and a half-finished concrete monolith. An uncompleted dam has no sale value. The bank is forced either to write off its construction loan or arrange additional financing to complete the project. It would be foolhardy for a bank to extend a loan to a company whose management it did not trust implicitly.

The second evaluation which the lender must make concerns the capability and experience of the prospective borrower. The bank must consider whether the contractor has been previously successful in the type of work that it seeks to finance. Is the company new in the particular area of construction in question? This is an extremely important consideration, since the experience of many companies has shown that expertise and fine results in one area of construction do not guarantee similar results in other unrelated areas. In the 1950's and early 1960's, some successful contractors branched into paving in order to cash in on the massive federal highway building program, but found to their regret that they should have remained in building construction. The special unanticipated problems of terrain preparation devoured their profit margins. Other successful builders have encountered difficulties by branching into land development, promotion of apartment buildings and other new development projects.

The organization of the borrower must be analyzed as well. Is the company a one-man organization or does it have depth? Would the death of one individual, most likely the owner, paralyze the firm? Good record keeping and sound financial management are desirable and provide some protection to the lender, but they are expensive. Overhead can be excessive for the size of the company and unwarranted for the type of project in question, or it can be inadequate for the size and complexity of the projects undertaken.

7-4. Understanding a Contractor's Motivation

Another important aspect of the contractor's capability involves his motivation and understanding of the project. Is the contractor obsessed with the prestige of large projects that are proportionately less profitable than smaller ones? Does he understand the danger of taking on a project too large for his organization and capital? Both the borrower and lender must be more impressed by quality than quantity. Contractors' failures are sometimes caused by the firm's taking on too much work for its size. Some contractors obtain work by submitting unrealistically low bids for public projects, thus freezing out competitors, with the expectation of making their profit by extra payments obtained through negotiation with public officials.

A company saddled with excessive overhead might be

tempted to underbid on a project merely to utilize its expensive equipment and meet its overhead obligations. The well-run construction company knows when to avoid submitting the successful bid. Certain projects can offer minimal profit margins or none at all.

7-5.   Intricacies of a Construction Contract

The intricacies of the construction contract can mean the difference between success and disaster. If management has not protected itself against the many hazards of construction by means of contingency clauses in the contract, any potential profits are in jeopardy. If the payment schedules to subcontractors are too high, the contractor can precipitate a cash flow crisis. If no retainage is kept to assure execution according to specifications, the contractor is not protected against inferior work by the subcontractor that can endanger the project. If a contract is unsuitable to the company and cannot be negotiated, management should not make any bid. Unfortunately, not all construction companies realize this, and some have a tendency to submit bids on any and all contracts.

7-6.   Contractor's Financial Statements Vary Widely

Another protective evaluation to be made by the bank concerns the financial condition and responsibility of the borrower. There is no room for the inexperienced or careless loan officer in this regard. The loan officer must realize from the outset that financial statements and accounting practices of contractors vary widely and cannot be judged by the standards or techniques applied to manufacturing companies. Many of the accounting practices peculiar to contractors will be discussed later in the chapter.

In dealing with any contractor, the bank should stress maintenance of the following: (1) Audited statements prepared on a yearly basis by a C.P.A. who is familiar with construction industry accounting practices and the projects being financed: (2) interim statements filed as frequently as necessary, at least quarterly; (3) job progress reports which list by each major contract the work performed to date (labor, material and overhead), billings to date, a projection of costs necessary to complete the project, and anticipated profit.

7-7.   Contractor's Honesty Must Be Emphasized

The contractor's honesty must be emphasized again. These financial statements are important tools for the loan officer, but they are only as useful as they are accurate. Even the most knowledgeable loan officer must accept the accuracy of many of these statements on trust alone. There are few meaningful rules of thumb or minimal ratios useful in appraising a contractor's financial statements. Rather, the composition of his assets and liabilities and the status of major jobs are most important considerations. It is easily possible for a contractor to err in presenting the status of a job or its quality.

A contractor needing funds to take available discounts from suppliers would be deserving of bank credit if he had a sizable retainage built up on large projects nearing completion and substantial receivables for work in progress, despite the fact his bank loans were greater than his working capital or his debt-to-equity ratio was high. If, however, the contractor's receivables covered work completed but in dispute or work completed for a public body that had run out of money, or if the loan would be used to finance losses on an unprofitable job, he might not be de-

serving of bank credit despite an extremely low debt-to-equity ratio or a substantial working capital position.

The possible variations on these circumstances seem almost infinite when one considers all of the possible developments during the course of a large construction project scheduled over several years. It is no wonder that a rather colorful but accurate maxim in banking advises the construction loan officer to "live with the contractor like a wife" from the opening bid until the owner takes possession of the completed structure. The banker must realize that once a project is begun, he can exercise little control over it. Everything rides on the skill of the contractor.

7-8.   What Constitutes a Current Asset on a Contractor's Statement?

When the bank has satisfied itself as to the integrity of the company's management, it must examine the company's financial statements. One of the most important items considered in the case of a manufacturing concern is its working capital position. In a sense, this is also true for a construction company. This inquiry immediately raises the question of what constitutes a current asset in a contractor's statement. Unfortunately, this question has no easy answer. Cash offers no classification problem unless it is restricted for some reason or earmarked for a capital expenditure project, such as the purchase of equipment. Receivables are almost always given full value as a current asset; this includes not only current estimates for work done but also retainages which will not be paid until the end of the job. However, most jobs run for more than a year. One dam was under construction for more than six years. Is the retainage due on a six-year job a current asset?

Receivables in the form of advances to joint ventures offer another difficult problem of asset classification. If a joint venture bought some expensive equipment for a three-year job, there would be a real question as to whether the capital contributions or advances to the joint venture from the participating companies were current assets on the partners' statements. This problem underlines the importance of financial statements and job progress reports of joint ventures in the evaluation of capital contributions and advances.

In light of these classification problems, one must consider whether or not segmentation of assets into current and long-term categories serves any useful purpose for the contractor. The original basis for such classification stemmed from situations where there was a definite seasonal pattern requiring bank credit for peak needs, and in contemplation of a seasonal cleanup of bank borrowings. Probably less importance should be placed on working capital and more thought should be given to net worth, and even more to the contractor's ability to perform the contract and to generate an adequate cash flow properly to service his indebtedness of all classes. The real value of assets, such as equipment, receivables, and work in progress, can easily be erroneously determined, and thereby misstated or overvalued. They often have questionable salvage and their depreciation schedules may not be realistic, particularly as working conditions change.

7-9.   Contractor's Business Can Rarely Repay Loans by Conversion of Balance Sheet Assets

The fact is that a contractor's business rarely presents an opportunity to service and repay bank loans or other indebtedness out of the conversion of balance sheet assets in the conventional pattern of a manufacturer or a merchandiser. It is the successful conclusion of the job that returns the investment with a profit that keeps a contractor in good financial health and pays his debts. It has been said that when a contractor gets into financial trouble, the only items on his balance sheet worth 100 cents on the dollar are his debts!

7-10. Banker Must Have Detailed Knowledge of Cost System

Another problem arises when the banker tries to analyze work in progress on financial statements without a detailed knowledge of the cost system used. Even with well-prepared, audited figures, an entirely different picture of the contractor's financial standing can be obtained, depending upon the method used to prepare the report (i.e., percentage of completion method vs. completed contract method, or allocation of overhead to the job vs. allocation at fiscal year-end, to name a few methods). Every contractor, regardless of size, must have a good, workable cost system so that he knows where he stands on his work in progress at all times. This is becoming mandatory for any contractor if he is going to survive and progress under the conditions prevailing today.

Assuming that a contractor has an adequate cost system, his work in progress data can still be misleading, because the sale price is generally set first, by the contract, and the ultimate cost is determined only when the job is completed. The point easily overlooked is the effect that this basic fact has on the contractor's profit-and-loss statement and balance sheet. All profits on unfinished contracts must be estimated, and accuracy is largely dependent upon the type of job and the stage of completion. Although the unfinished construction contract may hold the key to the true financial position of a contractor, most audits provide only abbreviated schedules of work in progress.

Work in progress figures are also greatly influenced by the accounting method used. Although there are two recognized methods to be used in the preparation of contractor's statements—(1) completed contracts and (2) percentage of completion—it should not be overlooked that auditors tend to compute that costs incurred to date are in proportion to the billing, with the difference reflecting profit. This is often far from reality, because overbilling in some segments of the industry is a common and normal way of doing business. The banker must realize that when this practice is followed, the anticipated profit or over-billings appear in the current asset section, distorting the true working capital position and invalidating any ratios developed from these figures.

Some loan officers are of the opinion that contractors cannot overbill because their estimates must be approved by the architect, engineer, or some governmental body. However, this view overlooks the fact that the contractor can obtain a substantial amount of overbilling on a contract by unbalancing his bid through front-end loading or overestimating the percentage completed. Such overbilling may not be detected by normal examination of estimates.

7-11. Steps Taken to Regulate Financial Information to Lenders

Bankers and other lenders have taken steps to regulate the financial information they receive from contractors in an effort to overcome many of the inherent difficulties of contractors' statements. The Robert Morris Associates have led the vanguard in this area. This group of commercial bankers has done much to assist the development of meaningful financial data on this industry for use of its members. Figures 7-11-1 and 7-11-2 are two of the forms which they have developed out of their analysis of construction accounting.

The basic function of these forms is to clarify financial information through more complete disclosure and greater detail than would be obtained in traditional financial statements. Figure 7-11-1 contains a detailed balance sheet of company operations. This balance sheet has been carefully designed to segment "assets due from completed contracts" and "assets due from uncompleted contracts," since the latter asset category may never be fully realized.

Another worthwhile distinction is made between land and buildings used in the business and those not used in the business. Land and buildings in these categories may be subject to entirely different depreciation schedules because they have unrelated useful lives and salvage values. When the land and buildings are categorized in this way, the lender has a more accurate picture of the company's important noncash expenses, i.e., depreciation and future replacement costs.

Figure 7-11-1, contracts in progress, contributes important information regarding profit and loss to date computed on uncompleted contracts compared with estimated profit or loss on the contract balance of the projects in progress. This exhibit makes it more difficult for the contractor, perhaps, inadvertently, to realize a profit in one year on an uncompleted project which is later completed at a loss. If a project is completed at a loss, this loss ought to be reflected over the entire period in which the project was in progress. This is consistent with the accounting principle of a periodic matching of costs and revenues for any project. This form reduces the possibility of inadequate or incorrect accounting procedures that allow a company to show a profit for three years and a great loss for the fourth year on a four-year job. Unless some great, unforeseen costs arose in the fourth year, such an accounting would be incorrect and misleading because the company actually was losing money on the job from its start.

The various schedules on page 4 of Figure 7-11-1 are designed to indicate the unencumbered value of different assets and the maturity schedule of the encumbrances. This information enables the lender to compute the liquidity value of the company's assets and the amount of additional collateral the company can provide. In addition, the various maturity schedules provide a longer-range view of the company's liquidity position. If a company had a large amount of debt falling due at one time and most of its assets were already pledged as collateral, it might face the possibility of default, illiquidity, and possible bankruptcy unless it arranged some further cash infusion.

Figure 7-11-2 provides detailed information regarding the contracts in progress, with an emphasis on the relationship between the costs incurred and the revenues generated by each project. The casual observer might feel that these forms indicated an overemphasis on work in progress. However, in light of the small capitalization of many construction firms in comparison with manufacturing firms and the special illiquid nature of a construction company's assets, it is possible for a poorly managed company to get into trouble despite an increasing volume of business due to inadequate cash flow and a heavy debt structure. In the case of a bankruptcy, the institution which provided the construction financing might find itself having to write off

Carried in Stock by Cadwallader & Johnson, Chicago. Robert Morris Associates—Form No. C1111H

CONTRACTOR'S FINANCIAL STATEMENT
AND SUPPORTING INFORMATION

INDIVIDUAL ☐
PARTNERSHIP ☐
CORPORATION ☐

For the purpose of procuring and maintaining credit from time to time in any form whatsoever with the below named Bank, for claims and demands against the undersigned, the undersigned submits the following as being a true and accurate statement of its financial condition on the following date, and agrees that if any change occurs that materially reduces the means or ability of the undersigned to pay all claims or demands against it, the undersigned will immediately and without delay notify the said Bank, and unless the Bank is so notified it may continue to rely upon the statement herein given as a true and accurate statement of the financial condition of the undersigned as of the close of business on the date set forth in the following line.

Statement Date_____19____Submitted To_____

Submitted By_____

Date Statement Signed_____19____Signature_____Title_____

EXHIBIT A: BALANCE SHEET (OMIT CENTS IN ALL EXHIBITS AND SCHEDULES)		ASSETS
1. Cash (Excluding Items in Line 8)..SCH. (B)		$
2. Notes Receivable for Contracts Completed...................................SCH. (A)		
3. Accounts Receivable for Contracts Completed (Excl. Retainage)............SCH. (A)		
4. Amounts Due on Uncompleted Contracts:		
(a) Approved by Engineers or Architects.................................EXH. (C)		
(b) Subject to Approval of Engineers or Architects.....................EXH. (C)		
(c) Retainage...EXH. (C)		
5. Retainage Due on Contracts Completed.......................................		
6. Labor, Material and Direct Overhead Charged to Jobs (Not Included Above).....		
7. Inventory Not Applicable to Contracts in Progress...........................		
8. Deposits for Bids and Other Guarantees......................................		
9. Stocks, Bonds and Special Assessment Obligations Received in Payment of Contracts Completed or Uncompleted................SCH. (C)		
10. Other Security Investments...SCH. (C)		
11. Other Receivables..		
12. Land and Buildings—Not Used in the Business—*Less Reserve For Depreciation*_____SCH. (D)		
—Used in the Business—*Less Reserve For Depreciation*_____SCH. (D)		
13. Machinery, Equipment, Vehicles, etc. (Net)................................SCH. (E)		
14. Deferred Charges and Prepaid Expense.......................................		
15. Other Assets...		
16.		
TOTAL ASSETS		$
		LIABILITIES
17. Notes Payable to Banks—For Bid Checks....................................SCH. (B)		$
—Other...SCH. (B)		
18. Accounts and/or Notes Payable for Equipment..............................SCH. (E)		
19. Advances or Loans From Officers, Partners, Friends or Relatives............		
20. Accounts Payable for Material Purchases....................................		
21. Accrued Expenses Other Than Federal Income Taxes...........................		
22. Federal Income Taxes—Due $_____Accrued $_____		
23. Due to Sub-Contractors—On Uncompleted Contracts.........................EXH. (C)		
—On Completed Contracts...		
24. Mortgage on Real Estate (Amount Due in Next 12 Mos. $_____)....SCH. (D)		
25. Other Liabilities*...		
26.		
TOTAL Lines 17 Thru 26..........$		
27. Reserve for Unearned Income OR Profit or Loss to Date on Uncompleted Contracts....EXH. (C)		
28. Preferred Stock Outstanding (Par Value $_____).................		
29. Common Stock Outstanding (Par Value $_____)....................		
30. Earned Surplus (**Net Worth if Not a Corporation**)........................EXH. (B)		
31. Capital Surplus..		
TOTAL Lines 28 Thru 31..........$		
TOTAL LIABILITIES		$

*See SCH. (H) For Contingent Liabilities.

Page One

Figure 7-11-1

EXHIBIT B: PROFIT AND LOSS STATEMENT From_____19_____to_____19_____

1. Contracts Completed During Period .... $
2. Less—Direct Costs Including Material, Labor, Job Overhead and Sub-Contracts ....
3. Gross Profit on Contracts Completed ....
4. Other Income ....
5.   Total Income ....
6. Less—General and Administrative Expense ....
7.   Interest Paid $_____Other Expense $_____
8.   Provision for Federal Income Taxes ....
9. Net Profit or Loss ....
10. Add—Earned Surplus **(Net Worth if Not a Corporation)** at Beginning of Period ....
11. Adjustments (Net) $_____Dividends or Withdrawals $_____
12. Earned Surplus **(Net Worth if Not a Corporation)** at Close of Period .... $
13. Memo: Depreciation Included in Above Expenses .... $

EXHIBIT C: CONTRACTS IN PROGRESS { Items (1) thru (4) Based on Actual Figures. { Items (5) thru (8) Based on Estimated Figures.

    Note: Page Three contains a spread of the individual uncompleted contracts making up the grand total in the column to the right.

GRAND TOTALS ON UNCOMPLETED CONTRACTS TO DATE_____19_____

1. TOTAL CONTRACT PRICE .... $
    COSTS TO DATE:
    (A) Labor ....
    (B) Material ....
    (C) Direct Overhead Charged to Contract to Date ....
    (D) Paid to Sub-Contractors to Date ....
    (E) Owing to Sub-Contractors ....
    (F) ....
2.   Total Items (A) Thru (F) .... $
    ESTIMATES RECEIVED OR DUE:
    (G) Approved Draws Received ....
    (H) Due and Subject to Draw Pending Architect's or Engineer's Approval ....
    (I) Retainage on (G) and (H) ....
    (J) ....
3.   Total Items (G) Thru (J) .... $
4. **PROFIT OR LOSS TO DATE** (Total of Line 2 Minus Total of Line 3) .... $
5. CONTRACT BALANCE TO BE COMPLETED AFTER STATEMENT DATE .... $
    ESTIMATED COST TO COMPLETE CONTRACT:
    (K) Labor ....
    (L) Material ....
    (M) Direct Overhead ....
    (N) Let or to Be Let to Sub-Contractors ....
    (O) ....
6.   Total Items (K) Thru (O) .... $
7. ESTIMATED DRAWS ON CONTRACT BALANCE (Including Retainage) .... $
8. **ESTIMATED PROFIT OR LOSS ON CONTRACT BALANCE** (Total of Lines 6 Minus 7) ... $
9. COST OF EQUIPMENT ACQUIRED TO DATE .... $
10. ESTIMATE OF ADDITIONAL EQUIPMENT NEEDED TO COMPLETE CONTRACT .... $
11. EQUIPMENT RENTALS PAID TO DATE AND INCLUDED IN LINE (C) .... $
12. LOCATION OF JOB ....
13. NATURE OF WORK ....
14. CONTRACT COMPLETION DATE ....
15. ORIGINAL CONTRACT PRICE (BID) .... $
16. CONTRACT PRICE OF NEXT LOWEST BIDDER .... $
17. ADDITIONS OR EXTRAS TO DATE .... $
18. CREDITS OR OTHER DEDUCTIONS TO DATE .... $
19. INCREASE OR DECREASE IN ORIGINAL PROFIT ESTIMATE BASED ON EXPERIENCE TO DATE ... $
20. State Contract Provisions for Liquidated Damages, Penalties or Bonuses with Respect to Completion Within Specified Time.

Page Two

Figure 7–11–1 (Continued)

SCH. A. ACCOUNTS & NOTES RECEIVABLE FOR COMPLETED CONTRACTS (List 4 Largest—Group Others in Line 5)				
DUE FROM	Original Contract Price	Amount Due Excl. Unapproved Claims	Maturity Date	Claims Not Approved For Payment
1				
2				
3				
4				
5.				

EXHIBIT C: (Continued): (If you have more than six contracts in progress list the five largest contracts in Columns 1 to 5 and lump all the smaller ones in Column 6).

Col. FOR WHOM CONTRACT PERFORMED	NAME OF SURETY COMPANY	NAME OF PRINCIPAL SUB-CONTRACTOR
COL. (1)		
COL. (2)		
COL. (3)		
COL. (4)		
COL. (5)		
COL. (6)		

	COL. (1)	COL. (2)	COL. (3)	COL (4)	COL. (5)	COL. (6)
1						
A						
B						
C						
D						
E						
F						
2						
G						
H						
I						
J						
3						
4						
5						
K						
L						
M						
N						
O						
6						
7						
8						
9						
10						
11						
12						
13						
14						
15						
16						
17						
18						
19						
20						

Figure 7–11–1   (Continued)

SCH. B. CASH AND NOTES PAYABLE BANKS: (Cash Not in Deposit Column Below $_____).

NAME OF BANK	CASH ON DEPOSIT	AMOUNT OWING	PORTION COLLATERALIZED	MAXIMUM BANK LOANS DURING PAST 12 MONTHS

SCH. C. STOCKS, BONDS, ETC.:

DESCRIPTION	UNITS	BOOK VALUE	MARKET VALUE	AMOUNT DEPOSITED AS COLLATERAL OR IN ESCROW (STATE WHICH)

SCH. D. LAND AND BUILDINGS: (Use asterisk if not used in business). (Use hundreds of dollars).

LOCATION	IMPROVEMENTS	BOOK VALUE	ENCUMBRANCE Amount	ENCUMBRANCE Maturity	TITLE RECORDED IN NAME OF

SCH. E. MACHINERY, EQUIPMENT, VEHICLES, ETC.: (Use separate line for each general classification).

No. of Items	DESCRIPTION	AGE Note (1)	PURCHASE PRICE	PRESENT BOOK VALUE	ENCUMBRANCE Note (2)	DATE DUE (Final)

Note (1) If two or more items are lumped give sum of ages.   Note (2) Use asterisk if payable monthly.

SCH. F. Principal Contracts Completed Past 3 Years.   AMOUNT (Omit 000)
FOR WHOM PERFORMED

SCH. G. PRINCIPAL MATERIAL SUPPLIERS AT PRESENT:
NAME AND ADDRESS

FOR WHOM PERFORMED	AMOUNT (Omit 000)	NAME AND ADDRESS

SCH. H. CONTINGENT LIABILITIES:

	YES	NO
1. Have any notes or accounts receivable been sold or discounted?		
2. Are there any liens for labor or material filed on your work anywhere?		
3. Are there any disputes over payment of labor or materials on any contract?		
4. Are there any suits or claims or judgments pending or unsettled?		
5. Are you endorsor or guarantor on any contracts, accounts or notes of others?		
6. Do you have any financial interest in a joint venture with other contractors?		
7. Are there any other contingent liabilities?		

If any known contingent liabilities exist of a nature not set forth in the above questions one to seven inclusive, please attach supplemental schedule giving full particulars. The bank is correct in assuming no other contingent liabilities exist if no such schedule is attached.

Figure 7–11–1 (Continued)

NAME _____

SUBMITTED TO _____

CONTRACT INDICATE WITH WHOM, WHETHER LUMP SUM, FEE, ETC., PERCENTAGE OF RETENTION, AND BRIEF DESCRIPTION OF JOB	TOTAL CONTRACT AMOUNT, INCLUDING EXTRAS	AMOUNT BEING SUBCONTRACTED	PER CENT COMPLETED	ESTIMATED COMPLETION DATE
	$	$		

Robert Morris Associates, Form C-127
Carried in Stock by Cadwallader & Johnson, Inc., 225 W. Huron St., Chicago 10, Ill.

Figure 7–11–2

CONTRACTOR'S SUPPLEMENTAL STATEMENT

CONTRACTS IN PROCESS OR TO BE
CLOSED OUT IN CURRENT FISCAL YEAR

DATE _____

| COST OF WORK PERFORMED | TOTAL AMOUNT BILLED | RECEIVABLES | | ESTIMATED JOB PROFIT TO DATE | ANTICIPATED TOTAL PROFIT ON COMPLETION |
		AMOUNT BILLED AND NOW OWING, EXCLUSIVE OF RETAINER	AMOUNT OF RETAINER		
$	$	$	$	$	$

This interim form is submitted for the purpose of providing supplemental information in connection with establishing or maintaining credit with the above named bank. The information contained herein is true and correct to the best of my knowledge and belief.

_____
Signature

_____
Title

Figure 7–11–2   (Continued)

the loan completely because the partially completed project had no market value.

These forms and the explanation of their function should further strengthen the position which has already been made: that the lender should judge a company's credit worthiness only with reference to statements audited by a C.P.A. who knows the construction industry well. The audited statements should be backed up by interim statements, at least on a quarterly basis, and the lender should demand frequent and detailed job progress reports. Of course, as stated previously, the paramount consideration is still the integrity of the borrower.

The loan officer involved with construction financing should be familiar with the several types of accounting systems that can be used to draw up contracts. These various systems have different ramifications for the construction company and hence for the bank's loan. Each has certain advantages and disadvantages over the other. However, one may be preferable to another, depending on the characteristics of the particular company in question. The alert bank officer should be able to recognize the best system for his borrower and consider the ramifications of that system on the loan.

7-12. Completed-Contract Accounting vs. Percentage-of-Completion Methods

As mentioned earlier, there are two generally accepted methods of accounting: (1) the completed-contract method, which is preferable when lack of dependable estimates or inherent hazards cause forecasts to be doubtful; and (2) the percentage-of-completion method, which is preferable when estimates of costs to complete and the extent of progress toward completion are reasonably dependable.

The completed-contract method is relatively simple to follow. During the period of construction, both costs and billing are accumulated in the accounts, and no profit or loss is recorded until the contract is completed or substantially completed. Although it is a common accounting practice to treat general and administrative expenses as current expenses, it may be more appropriate to allocate some general and administrative expenses to contract costs when the completed-contract method is used. These costs then are accumulated rather than charged to the year's expense in order to produce a better matching of costs and revenues. This method is important, particularly in years when no contracts are completed.

The principal advantage of the completed-contract method is that of certainty. The profits and losses are finally determined on results rather than on periodic estimates which may involve unforeseen costs and possible losses. Its major disadvantage is that it does not reflect current performance when the contract extends through more than one accounting period, and it may result in irregular recognition of income. In cases of single proprietorships and partnerships, this may result in greater income tax liabilities because of the effect of graduated personal tax rates as compared to the flat corporate rates. This method does permit the maximum postponement of tax payment.

Some feel that the profit-and-loss statement has little meaning when it shows no profit for a job well along or when a substantial profit accrues to one year from a job which has been in progress for several years. If overhead expenses have been absorbed in the period when they occurred, the profit-and-loss statement has been distorted

further. Also, when there is a sizable loss under this system, the customer may be tempted to explain it away due to the method of bookkeeping. In such cases, it is wise for the loan officer to refer back to his spread sheets and see what has happened to the net worth of the company over the past five years.

The percentage-of-completion method is more difficult to follow because it requires an allocation of income during the period of construction. Income is allocated to the current accounting period in terms of a percentage of estimated total income indicated by an appropriate measure of progress. A frequently used measure is the relationship of incurred costs to date to estimated total costs, when estimated total costs to completion are based on the most recent information. Work performed is also used as a basis of income allocation.

When this method is used, it may be necessary to disregard certain recorded costs as a measure of performance in the early stages of some contracts to avoid misleading statements. This is necessary in those contracts where material costs account for the preponderance of the total contract. In such cases, it is possible for the contractor to have incurred a substantial cost without having performed any work on the contract. This example illustrates just one of the many problems which the contractor confronts as he tries to determine an appropriate income allocation.

The percentage-of-completion method has the advantage of recognizing income on a current basis rather than on an irregular basis dictated by contract completion dates. It also reflects the status of uncompleted contracts, which frequently represent the majority of the contractor's business in any accounting period. The major disadvantage is its necessary dependence on estimates of ultimate costs which are subject to uncertainties and frequently tend to be on the low side.

With regard to any given contract, the two methods of accounting are mutually exclusive. However, both methods have one procedure in common: the application of the long-established accounting practice of anticipating losses when current estimates of total contract costs indicate an ultimate loss. In brief, from an accounting standpoint, the entire loss accrues at the moment when current estimates of total contract costs indicate a loss. Under either method, the provision for the loss should represent the best judgment that can be made under the circumstances. The use of the completed-contract method does not, of itself, preclude the fact that a loss will become clearly apparent at some intermediate stage of completion.

These two accounting methods apply to guaranteed-price, lump-sum, or fixed-fee contracts. The lump-sum contract is one that stipulates the exact amount of money paid to the contractor by the owner when he takes possession of the project. A second and not uncommon type of long-term contract is the cost-plus contract. Such contracts usually contain many technicalities, including defined and excluded costs, guaranteed maximums, and penalty and bonus payments. The "plus" factor of the cost-plus contract may be expressed as a percentage of certain costs or of total costs or as a fixed amount. It is generally accepted procedure to accrue revenues and thereby recognize profits on the basis of partial performance when total profit can be estimated with reasonable accuracy and ultimate realization seems assured. When estimates are unreliable, the completed-contract method is preferable.

There might be circumstances in which a contractor could employ a combination of accounting methods, provided there is consistency in applying the same accounting treatment to the same set of conditions from one accounting period to another. The selection of an accounting method should be governed by a set of ground rules followed consistently. For example, a contractor might consistently use the percentage-of-completion method in accounting for major projects that require several years to complete, and use the completed-contract method for repair work that may require only a few weeks or months to complete. The contractor and banker must remember that the two accounting methods are mutually exclusive for one specific contract.

7-13. Supplementary Schedules, Cash Budget, and Equipment Projections

There are two other supplementary schedules which the loan officer should obtain from the potential borrower—the cash budget and the equipment projections. The cash budget represents the company's estimate of the cash required to run the jobs and meet the various overhead commitments. This forecast is best prepared on a monthly basis for the first year of the project and then perhaps on a quarterly basis beyond this point. This forecast should indicate the maximum amount of the financial requirement necessary to complete the job, when it will occur, and how soon after the project is completed the loan will be retired completely.

This budget forecast indicates what equipment purchases will be necessary before the project is finished. The lender should know how much equipment must be purchased to handle the job and what will be done with the equipment when the job is completed. All too frequently, this important information is unavailable. Yet such a schedule might reveal important changes in the balance sheet which will be caused by a job. For example, if the contractor has calculated a $100,000 profit on a job but has to purchase $500,000 of equipment and charge job expenses with an equipment rental of $200,000, he is going to have $200,000 less working capital at the end of the job. The question arises as to whether he can afford to tie up so much of his net worth in equipment. Cash budgets and equipment schedules help the lending officer anticipate such problems, which may otherwise be overlooked.

When the loan officer is reviewing the equipment projections, he must give careful consideration to the costs of preventive equipment maintenance. Much of the complex machinery used in large construction projects requires expensive maintenance at regular intervals if it is to remain operational for the life of a contract and into the future. Two items of cost must be considered: the out-of-pocket costs of maintaining the piece of equipment and the cost of time lost when the machine is not operating due to maintenance programs. If a giant earth-moving machine must be overhauled for a month, that lost month's work may represent thousands of dollars in delays and schedule readjustments. On-the-job visits are necessary if the loan officer wants to satisfy himself that the equipment is not being abused through lack of maintenance or misuse. The projected profit on a job may disappear if an expensive piece of equipment is destroyed and must be replaced before a job is finished.

7-14. Equipment Financing

Banks are a major source of the financing which enables construction companies to purchase equipment. There are few hard and fast rules relating to equipment financing.

Each company seems to develop its own pattern depending on its size, financial strength, and the kind of equipment it uses. Generally, the very large construction companies that have long-term debt in their capital structure have limits on what percentage of the equipment's value can be financed. Often, these limits allow 75 to 80 percent of the value to be financed. Some financially strong construction companies may offer unsecured notes to finance equipment.

Several procedures should always be followed by the bank in equipment financing to avoid needless confusion and reduce risk. The equipment being financed ought to be carefully identified, ideally by serial number. Each piece of equipment should be fully insured. A tractor might be driven off a cliff or ruined on a rock formation. The bank should insist that the finance agreement contain a "loss payable" clause to the lender in the event that any insured piece is lost. Appropriate amortization schedules for all equipment are mandatory. If a light pickup truck working on a rough construction site might be ruined within a year, it must be amortized in a year. However, a heavy-duty, $50,000 truck might have a useful life of three or more years. It could be amortized over a three-year period. Amortization schedules should be sufficiently detailed to account for the varying lives of different pieces of equipment. Equipment financing should be set up in a block of loans which is separate and distinct from working fund loans, since equipment is moved from job to job and under normal circumstances assists in producing gross earnings on more than one contract operation. Special equipment, which has been purchased for a specific job, should be charged off entirely on the job for which it was procured.

The ratio of monthly depreciation to the original cost of a particular piece of equipment changes in different circumstances. Because new equipment involves only limited maintenance or overhaul expenditures, the contractor can afford to take heavier write-offs in the early stages of ownership. This is especially true of standard equipment. As equipment becomes older and maintenance costs rise, the contractor is justified in making write-offs at a lower rate in order to maintain a consistent level of charges to his work. In the construction business, equipment investment turnover is a necessary and continuing process. A contractor cannot expect to remain in business unless his work is priced to give him an adequate margin for equipment write-off and replacement.

In recent years, some finance companies, seeking to utilize their rapidly accumulating funds, have begun to lease equipment to contractors. Leasing can be particularly advantageous to the smaller contractor who does not have the resources to purchase a piece of equipment costing several hundred thousand dollars. Generally, lease payments are tax deductible. The question of whether or not it is financially advantageous for a contractor to lease rather than purchase a particular piece of equipment can have different answers, depending on the contractor's tax situation, the length and specific terms of the lease, the cash flow aspects of the situation, and the alternative depreciation schedules and market values of the equipment if it were purchased.

7-15. Bid Deposit Loans

The bid deposit loan is a special kind of loan which should be separated from regular loans in the bank's portfolio. Many banks maintain a separate record of these in the liability ledger because they feel the credit risk is a great deal less in this type of lending to contractors. Bid loans fill a special need of the contractor. On public work, the contractor frequently has to make a bid deposit of between 5 and 10 percent of his bid price in the form of a cashier's check or a certified check when he submits a bid for a project. If the contractor is not the low bidder, his check is returned. If he is the low bidder, then he is required to furnish a performance and payment bond.

When the contractor is unable to furnish the bond, his bid check is forfeited and the job is awarded to the next lowest bidder, provided he can meet the terms of the bond. Although bid check loans are regarded as lower in risk than general contractor loans, the possibility of default still exists. It is wise for the lending officer to check with the surety company as to whether the surety bond would be available to the contractor in case he is the low bidder. Otherwise, the bank might be faced with the choice of losing its bid check loan or having reluctantly to finance a job that cannot be bonded. Here again, the factor transforming such loans from relatively secure instruments to highly speculative ones is the bank's knowledge of the contractor. There is no substitute for familiarity or understanding in these situations.

7-16.   The Matter of Bonding

In the matter of bonding, it should be pointed out that bonds are obtained by the owner or general contractor to ensure completion and payment for the project, but they do not protect the contractor's lender. In fact, they usually constitute a lien having precedence over the contractor's lender.

If it is at all possible, a single general contractor should be responsible for all work. His performance on the general contract can and usually should be guaranteed by a surety company. Sometimes a property owner or developer acts as his own general contractor either through the company which owns the property or through another body.

7-17.   Owners' "Interim" Loans and "Permanent" Financing

Up to this point, we have been discussing loans to contractors which are to finance the contractor's own operations and which are repaid by monies which the contractor receives from the owner. Because most private owners must in turn obtain what is known as "interim" financing in order to obtain funds necessary to make progress payments and, lastly, the final payment, to contractors, it is important for both contractor and his lending banker to know something of the requirements which the owner must satisfy in order to obtain these interim loans and, to some extent, the "permanent" financing as well.

Interim loans require certain documentation designed to protect the interim lender. We will describe them in detail later, but it can be briefly stated that the documentation is intended to ensure that the building or project will be completed and delivered lien free, so there will be no problem in obtaining the permanent financing (quite often a mortgage) which will pay off the interim loans. The contractor is required to furnish many of these documents and, therefore, a brief explanation of the procedures would appear ot be in order at this time.

Interim construction loans are customarily obtained by project owners from banks and such other institutions as real-estate investment trusts to provide funds during the construction of large privately owned apartment and com-

mercial projects. These are complex loans and are generally available only from financing institutions that have developed the necessary techniques or sophistication. The lender's management of the loan is primarily concentrated in three areas: (1) legal documentation, (2) project cost and financial strength of the borrower, and (3) architectural design and adequacy of construction.

The interim lender will generally require that permanent financing has been arranged before he considers a construction loan, and he will analyze the commitment for such permanent financing in considerable detail to assure himself that all conditions of the long-term lender can be reasonably made.

The loan will be governed by a construction loan agreement to be executed by the owner and construction lender. It will be secured by a mortgage on the property as well as other appropriate collateral documents, such as assignment of leases, chattel mortgages, etc. In many cases, the construction lender will plan to assign these documentations to the permanent lender upon completion of construction, in which case there will be an agreement between the permanent lender, the interim lender, and the borrower.

The loan agreement may also call for performance and material and labor payment bonds, hazard insurance, and a requirement that the contractor agree to perform for the construction lender in the event that the owner defaults and the construction lender elects to complete the project. Additionally, there will be appropriate provisions for title policies, surveys, project budgets, lien waivers, periodic inspections by a representative or agent of the lender, and appropriate guarantees by the borrower. Because of the complexity of this documentation, the lender will normally have an attorney prepare most of it, some of which items will also be submitted to the permanent lender for his prior approval.

The interim lender will require the owner to submit a detailed budget for all costs of the project when application is initially made for the loan, with updated budgets to be submitted at the time of each advance under the loan agreement. Additionally, the lender will want to review a copy of the general contractor's trade payment breakdown at the time of each advance to determine the status of these costs.

On major projects, the lender may engage an independent architect who will examine the plans and specifications to determine their adequacy. Review of plans and specifications will generally also be made by the permanent lender, whose comments will be submitted to the construction lender. During the course of construction, the interim lender's architect will periodically visit the project to determine conformity of construction plans and specifications, quality of workmanship, and reasonableness of sums requested by the contractor. It should be emphasized that the lender's architect is not duplicating the project architect's work but is making a more cursory review of the project and its progress to assure the construction lender that no major problems are in the making.

Construction loan funds will generally be disbursed once a month in connection with the contractor's payment requests. As previously indicated, the lender may require an updated project budget, contractor's trade breakdown, and a report from his inspecting architect. He will probably require that lien waivers and affidavits from the general contractors, subcontractors, and major suppliers be submitted to and approved by a title insurance company in states where mechanics' lien laws make such a practice prudent. In addition to the title company's insuring the lender against mechanic's liens in appropriate states, it will also certify as to whether there have been any changes in the conditions of title since their last report.

In negotiating for interim financing of a project for its owner, the following checklist may be helpful to the loan officer:

1. Acceptable permanent financing commitment
2. Project cost analysis
3. Appraisal of project on "when completed" basis
4. Disclosure of all owners plus financial statements of each
5. Requirement that principals endorse interim borrowings
6. Formal construction loan agreement
7. Attorney's fees, commitment, or standby fees and interest rate
8. Title guarantee policy showing merchantable title to include loan disbursement guarantee
9. Survey of entire property with location of improvements, i.e., buildings, easements, utilities, parking facilities, streets, building lines, etc., shown thereon
10. Permanent or short-term mortgage loan papers and other documentation as required under loan commitment
11. Adequate hazard insurance with mortgage clause attached in favor of lender, plus adequate public liability insurance
12. Sworn detailed construction statement setting forth all general and subcontracts and the amount of each
13. Construction contract and assignment thereof
14. Contingency reserve to cover possible increased construction costs, etc.
15. Completion or performance contract bond including labor and material coverage. A.I.A. form with dual obligee rider
16. The bank to act as payout agent, or some person or firm approved by bank, who has had considerable experience in construction payouts be employed at the expense of borrowers to represent the bank in approval of payout certificates who can certify that material and labor represented by the certificates has actually been used in the construction to date of payout
17. A special building account should be used for disbursements of all funds in connection with the construction
18. Payments to contractors subject to a holdback of not less than 10 percent
19. Owner's funds to be disbursed prior to bank's interim financing funds and evidenced by supporting waivers of lien, contractor's affidavit, and architect's certification. Interim construction funds to be disbursed upon certification by the architect, general contractor, owner, and approval of bank's inspecting engineer certifying:
    - Construction is proceeding in accordance with plans and specifications approved by permanent lender, completion within time required.
    - Undisbursed loan balance is sufficient to fully complete the project.

If the loan officer satisfies these points as he negotiates an interim construction loan with an owner, he will avoid many of the pitfalls that permit problems and tend to increase the riskiness of the project.

The loan officer must not be lulled into a false sense of security when making a construction loan to an owner who has a commitment for permanent financing. The permanent loan commitment is not "security" for the construction loan and it does not function as security. The availability of the permanent loan as a source of repayment lessens with every default of the borrower, whereas security becomes more available with every default. The second point, which is a logical deduction from the first, is that the best protection of the permanent loan is the careful administration of the interim loan by the lender.

Rarely are the most significant provisions of the permanent loan met at the time the initial interim construction loan is being closed. Often final plans and specifications have not been completed at this time. Little or no work has been done and key leases are still being negotiated. The construction lender can expect little more than a gradual satisfaction of conditions. He must assure himself that the remaining conditions can be satisfied in the normal course of operations. If the preliminary plans have been carefully evaluated and approved as feasible by the construction lender's architect, he should be able to proceed without the final plans and specifications.

If there is good intent by the prospective tenants to execute leases and the location of the project is sound, it is reasonable to expect that satisfactory leases will be signed. If detailed plans and specifications are not available, the construction lender and the borrower must recognize that quick approval by the permanent lender will not be coming. The construction lender must rely on the judgment of his architect regarding the soundness of the preliminary plans and the correctness of any changes that are made. The construction lender's counsel must pass judgment about leases and the terms of easements. The major point to be realized is that the construction lender cannot pass responsibilities off to the permanent lender. If he does so, he greatly increases the risk that the borrower will default and forfeit the permanent financing.

Another major consideration to be made at the time the construction financing contract is being drawn up is that of the actual yield of the loan. The yield can vary considerably based on fees negotiated, the interest rate charged, the term of the loan, and the average amount of funds used by the borrower. Various banks have different yield requirements which depend upon the nature of the bank's business, its liquidity position, competition from other institutions, internal management decisions founded on business projections, and the availability and desirability of alternative investments. The unusual risks involved in construction lending become significant factors in the determination of yield requirements by a bank.

Even though the credit of the developer, the contractor, and subcontractors is impeccable, a number of contingent circumstances might cause a loss or a delay resulting in a yield lower than was anticipated when the loan was originated. Excessive rain, fire, earthquake, cave-ins during excavation, or labor strikes can delay a construction job or bring it to a halt. Such delays jeopardize the value of a permanent loan commitment, which expires on a given date. A major tenant might fail to execute an important lease if the project is not delivered on time.

These inherent risks are well known in the construction industry. They justify additional compensation for the interim construction loan. This added compensation generally takes the form of loan fees which range from 1 to 4 percent. The fees charged perform a dual function—they compensate for added costs involved with interim loans which are expensive to process, properly service, and control; and they result in higher overall yields to compensate for the risks taken.

Careful consideration of the particular construction project, the contractor's reputation and past performance, and the opinion of qualified appraisers can result in a realistic determination of anticipated yields. By doing this, bank management can be more certain that the fees and interest charges in their portfolio of construction loans do, in fact, result in the yield returns they expect from this form of investment.

7-18.   Banks' Sensitivity to Supply and Demand of Money in the Economy

Banks are acutely sensitive to the supply and demand of money in the economy. In a period of broad economic expansion and rising expectations, the demands for funds by corporations rise substantially. Funds are needed for physical expansion, inventory expansion, new equipment, or major projects. However, in times of economic expansion, people seek more profitable alternative investments and the banks' supply of funds from deposits declines. The bank becomes caught in a vise of increased demand and smaller supply of funds. Of course, as the bank tries to meet all of its legitimate loan demands, the price of money must rise. This is translated into higher interest rates and more restrictive terms on loans.

The construction company is likely to undertake larger volumes of business in such good economic conditions. Since the construction company is often capital-poor and relies primarily on banks for the money needed to finance its projects, it is especially sensitive and vulnerable to changes in the price of money. In a period of tight money, the construction company may expect a number of adverse consequences. The interest cost of the loan will climb. The additional fees charged will also be larger. The bank may issue a shorter term loan and insist that a smaller percentage of the loan be outstanding in any one period. The myriad other terms of the loan agreement may be written more stringently. The bank's strategy is to discourage marginal projects by means of these more stringent terms. If the construction company is not a regular customer of the bank, its application for a construction loan may be refused completely.

In recent years, United States banks have borrowed Euro-dollars (American dollars held by Europeans) in an effort to increase their money supply and, hence, their capacity to lend. Although this form of borrowing has helped somewhat to alleviate the shortage of funds available to business, it has done so at an extremely high price. In the tight money crisis of 1969, banks were paying as much as 12 percent for Euro-dollars. These borrowings only increased the average cost of money to the corporation, since the banks had to cover their costs and operating expenses. Although the supply of Euro-dollars expands the money supply, it does so without lowering the cost of money to the borrower. Indeed, such borrowings increase the costs to the borrower. In situations of tight money, the good credit risk, undertaking a worthy project, will obtain financing. It is the marginal project that is priced into oblivion.

7-19.   Construction Loans in Foreign Countries

Participation in interim construction loans with foreigners is a special area of banking best left to those banks that have experience and exposure to such situations. All of the many complications encountered in domestic construction financing are multiplied in foreign dealings. The greatest difference between domestic and foreign construction lending is the loss of security. In dealing with United States projects, the bank need not give any consideration to the stability of the government. This stability is taken for granted. However, this is not the case in many other countries. Governments might be overthrown or contracts voided. The succeeding government does not necessarily guarantee the obligations of its predecessor. The currencies of many other nations are unstable. A devaluation could bankrupt a particular project.

There are some guidelines for a bank to follow in considering interim construction financing with foreign companies in foreign locations. The bank must insist that all financial transactions be made in United States dollars. The bank should require its client, the construction company, not to undertake any foreign work until the owner of the project makes a deposit of United States dollars in a United States bank that is sufficient to initiate a project, close it down, and fund a return trip for all of the United States employees on the job. There have been situations in which American firms have been stranded in foreign countries with no recourse because such provisions were not made.

There is another manner of financing overseas construction which lowers the risk to the construction company. It operates in this manner. The construction company establishes a relationship with a bank in the country where the project is located. That bank then approves an overdraft in the company's account which funds its operations. The construction company's United States bank guarantees the foreign bank that the overdraft will be paid when the project is completed. During the term of the overdraft, the construction company pays a set interest rate on it. This becomes a convenient working arrangement for the contractor. National banks in the United States are not empowered to enter into such guarantee agreements. However, some state-chartered banks can offer this service. The national bank could issue a performance letter of credit to the construction company in lieu of a guarantee.

The construction company contemplating a foreign project should thoroughly research all aspects of the job. Climate and terrain are especially important considerations. A project that could be finished within a year in a temperate climate might take two or three times as long to complete in a steaming jungle or frozen tundra. Large areas of the world are subject to rainy seasons which make earth preparation and unsheltered construction difficult, if not impossible, for several months of the year. Local work habits and social customs can present scheduling problems. Drinking water and minimal living conditions could be extremely expensive to maintain.

The question of maintenance becomes more complicated in an extremely hot or cold climate that is harmful to machinery. Replacement parts might have to be shipped thousands of miles by sea or air. In such situations, it is crucial that obsolescence and replacement schedules be accurate. Lack of proper attention to such matters can precipitate costly work stoppages and delays.

As was stated at the beginning of the chapter, construction banking is one of the most challenging of all bank lending activities. This chapter has touched on some of the major problems associated with it, but by no means has it covered the entire subject, for each project is in many ways unique and presents its own idiosyncratic problems. If the chapter contains a single maxim, it is that the nature of construction financing impels a bank to deal only with contractors whose integrity and competence are above question. The greatest safeguard for the well-being of any undertaking is the contractor himself. He alone controls the quality and pace of construction. The relationship between the bank and the contractor is one of trust. If that trust is poorly placed by the bank, no contract can effect a satisfactory remedy.

8 SURETY BONDS

DAN GORTON

Vice-President (Retired)
Fidelity and Deposit Company of Maryland
Baltimore, Maryland

BECAUSE SURETY BONDS usually are obtained through an insurance agent or broker and from an insurance company, the tendency is to classify bonds as insurance. This interpretation is a misconception.

8-1. How Surety Bonds Differ from Insurance

An insurance policy is a two-party instrument—the insured and the insurance company or insurer. The policy protects the insured against specified types of losses. The insured buys the policy for its protection against fire or accident or some other similar loss.

A surety bond, however, is a three-party instrument—the principal, who in a contract bond is the contractor; the obligee or owner, who is a party requiring the protection of the bond; and the surety. The surety guarantees that the principal will make whole any loss the obligee might sustain by reason of the principal's failure to carry out and perform all the conditions of the agreement entered into between the principal and the obligee. If the principal fails or is unable to fulfill his contractual obligations to the obligee, the surety then performs in accordance with the conditions of the bond. The surety can then seek recovery of its loss from its principal. Thus, it should be clear that when he furnishes a bond, a contractor is not providing insurance for himself, but affording protection to the owner. The surety is similar to the endorser of a note, who must make good the maker's failure to pay.

8-2. Purposes of Surety Bonds

Webster's dictionary defines the word "surety" as: "That which confirms or makes sure; a guarantee; one who makes a pledge in behalf of another and accepts certain accruing responsibilities." Webster defines "suretyship" as: "The obligation of a person to answer for the debt, default, or miscarriage of another."

For centuries the only surety bonds available were personal bonds signed by individuals who generally were friends of the principal. When signing such a bond, the personal surety actually pledged his personal assets as a guarantee that the principal would carry out the terms of whatever agreement was made between the principal and another party. The personal surety's liability under the bond usually was much broader in scope than he anticipated, and many a personal surety has suffered bankruptcy through signing what he probably believed to be an innocuous instrument required only as a matter of form.

In the latter part of the nineteenth century several corporations were formed for the sole purpose of furnishing surety bonds on a commercial basis. Although corporate suretyship was welcomed in the business world, its early development was slow. The demand for the various forms of bonds had not been fully established, and the surety companies were slow to fulfill whatever demand existed because they had to feel their way with extreme care and caution. They soon found that they could not underwrite the bond business in the same manner that insurance is handled.

Developments revealed that although some forms of bonds were of a nominal nature, many others were strict financial guarantees, and that bonds falling into both these categories nearly always contained a hidden or latent liability. This element of risk necessitated the surety's thorough scrutiny and analysis of the bond forms and agreements. The surety, in addition to making a careful investigation of the background, capabilities, and financial responsibility of the applicant, also had to ascertain, as far as humanly possible, that the applicant could and would fulfill all the obligations placed on it under the agreement or statute requiring the bond.

With the rapid growth of American business in the early part of the twentieth century, surety bonds became more and more necessary in order to bridge the uncertainties in various forms of contract agreements. Further, in certain large cities in which graft and corruption were rampant and more or less the "order of the day," contracts for public improvements were let to unqualified, inexperienced, and often unscrupulous contractors. Specifications not only as to material but also as to workmanship were disregarded, sometimes with tragic results and heavy loss of public funds. Unsatisfactory work had to be redone on some of these contracts. In a few instances buildings actually collapsed either before or shortly after completion, and material suppliers and even laborers were defrauded of their pay.

Public bodies were bound by law to award contracts to the lowest, responsible bidder. However, because in that era there was no hard and fast rule for determining the interpretation of the word "responsible," the awarding authorities often let the contract involved to the closest political crony. Even in localities in which the "spoils system" was not the custom, this question still was a problem. The answer was found in the use of surety bonds. The federal government, as well as many states, counties, and cities either changed or passed laws which included the requirement of performance bonds of contractors on all types of construction and supply contracts.

Surety companies, in their underwriting, were expected to weed out the unworthy and irresponsible bidders. If the contractor could pass the scrutiny and investigation of the surety underwriter and could furnish corporate surety bond on the contract involved, he would be considered a responsible bidder. Thus, the problem of determining the responsibility of the contractor was shifted from the shoulders of the public officials awarding the contract to those of the surety underwriters.

During the earlier years of the industry's development, most of the bond business was handled by a few companies, many of them specializing in surety to the near exclusion of other lines of insurance. Later a number of multiple-line insurance companies entered the surety field. There were no standards or guidelines to go by, and each company charged its own rates and set up its own underwriting formulas. This resulted in confusion and chaos. Reinsurance facilities were very limited. A company that was charging $15 per thousand on the contract price for its bonds was not interested in reinsuring a bond written by another company at a $12.50 rate. A conservative company was wary of reinsurance offered by a company known as a loose underwriter.

The United States government finally insisted on the creation of a statistical bureau to gather experience data from the various companies in business at that time. The data included premium income, expenses, losses, and other vital factors. The data procured enabled the companies to promulgate rates on the various types of bonds then being written by the industry. The premiums were not calculated as an insurance charge based on anticipated losses, but instead were a service fee correlated to the costs of underwriting and handling the prequalification procedures of the underwriter and the agent.

The government also adopted rules and regulations covering the qualifying of surety companies and prohibiting a company from assuming a single surety obligation for more than 10 percent of its capital and surplus. This necessitates surety companies' joining together as cosureties on large or jumbo-sized bonds. Such regulations are in the best interests of the companies and the public. The same limitation has been applied to banks on their loan limits for many years.

Some of the newcomers and other sureties that had been unable to overcome the keen competition of the more experienced and established sureties later undertook to provide suretyship for premium rates substantially lower than the established rates used by the experienced companies, with the result that a number of such companies failed—

but not all. The more successful of these rate-deviating companies tended to be more restrictive in their underwriting requirements and limited themselves to handling only certain classes of business.

The birth date of the very first surety bond ever written is lost in the maze of antiquity. It was not until near the close of the nineteenth century that the surety industry's path in world affairs slowly became more and more definitive. It emerged from a long period of comparative obscurity and uncertainty, and today performs one of the most vitally important and indispensable roles in our nation's economy.

Surety bonds are a requirement not only in the contracting field but also in many other facets of modern business. They are a requisite in our court procedures, in our banking structures, and in guaranteeing the faithful performance of public officials, administrators, and law enforcement agencies. The guarantee of a surety bond in the contracting field is all-important not only to the obligee but also to the material houses and to the suppliers of machinery, plumbing, electrical equipment, steel, cement, and other items vital to the construction of buildings, dams, bridges, highways, pipelines and other projects. Surety bonds enable all the above-named suppliers to sell their respective products to the contractor on a bonded job, safe and secure in the knowledge that if the contractor fails to pay such suppliers, the surety will perform pursuant to the conditions of the surety bond.

8-3. Underwriters

The home offices and branch offices of insurance companies and surety companies are staffed by diverse personnel, a certain proportion of whom are underwriters. There are various types of underwriters. Some handle surety bonds, others handle fire, automobile, life insurance, and other forms of coverages. An underwriter is one who, after making an analysis of the underwriting problems and following certain guidelines, makes the decision either to accept or decline the risk.

In nearly all lines of insurance, losses will occur with almost mathematical certainty because both the underwriting and the premiums are predicated on the law of averages. However, this should not be true of the surety bond business, which, as stated previously, is not insurance. Under the suretyship theory, the underwriter is underwriting against loss of any kind. It is his duty to try to foresee every possibility of loss and to plug all loopholes. His underwriting capabilities should increase in direct proportion to his length of service in the surety business; and his analyses of the losses occurring on bonds which he has underwritten should broaden the scope of his knowledge.

The situation of a surety can be compared to that of a bank. The surety extends bonding credit in reliance on the stability and indemnity of its principal, just as a bank extends credit on the ability of its borrower to repay the loan when due, or otherwise indemnify the bank against loss. Theoretically, neither should sustain any loss. However, this, as to the surety, applies only if the underwriter has performed his job correctly and carefully by making certain that the principal possesses the integrity, experience, ability, and equipment, as well as the necessary capital, not only to finance the work properly but also to absorb any possible loss it might sustain. It is the underwriter's duty to eliminate all unsound risks and unworthy contractors, just as banks will decline banking credit if any serious question arises as to the prospective borrower's ability to repay a loan when due. Because theories do not always prove out in practice, surety companies and banks do sustain losses.

8-4. Reason for Contractor Failures

There are many causes for contractor failure and consequent surety losses. One oldtime contractor, when asked what was the major cause of contractor failure, replied: "Running out of money." His boiled-down opinion undoubtedly is correct, but surety underwriters want to know why a contractor runs out of money. Surety statistics indicate that "overexpansion," i.e., taking on more work than a contractor's working capital can handle, probably is the major cause of failure in the building field.

Consistent low bidding, subcontract failure, and unforeseen labor troubles also are constant factors in the failure of both building and heavy engineering contractors. Unknown soil conditions have contributed to huge, unanticipated costs that have wrecked many concerns in the heavy construction field. Loss of key personnel, due either to death or retirement or the inroads made by competitors, has caused a great many failures.

Dishonesty has played a leading part in contract losses. In many instances contractors have diverted progress payments either for personal use or occasionally for gambling, instead of using the money to pay for labor and materials. In still other cases, contractors have spent anticipated profits before they were earned, only to discover that such profits were wiped out during the course of the work. In the mid-1950's a considerable number of contractors failed because they attempted to branch out into other fields, such as speculative building.

Of course, the surety underwriter cannot foresee all the contingencies; nevertheless, he must keep them in mind and strive to underwrite against the hazards. That is why underwriters, and the agents and brokers through whom they deal, often suggest and recommend that a contracting concern purchase substantial life insurance on its key officers or partners, as well as a fidelity blanket bond, on all its employees, and that the subcontractors on the job be bonded. Additional financial support of various types often can offset low bids or unforeseen conditions. All these requirements are a protection not only to the surety but also to the contractor.

8-5. Contractor-Underwriter Relations

Surety companies are always eager to obtain a good contractor account. It is to the surety's interests to keep that contractor in business. As the contractor prospers, his annual surety premiums increase year by year. It is not to the interests of either the contractor or the surety for the surety underwriter to grant excessive lines of credit, thereby very possibly leading a contractor to financial disaster. The underwriter must temper optimism with conservatism, and in this way the underwriter is a kind of balance—an interested but objective party in the whole process of successful construction contracting.

Sureties spend a great deal of money to train and develop these underwriters. In a recent survey of twelve leading companies it was determined that, in a five-year period through 1969, only 79 additional men out of more than 1100 hired still were on the payrolls. The aggregate cost of training all these people was more than $41 million, but the sureties retained less than one in ten of the trainees. In the first two years the average cost of training each man exceeded $30,000. The contractor rate of survival may be somewhat better than this, but the early stages of development may be just as trying of patience, ability, and attraction as the stress of becoming an underwriter. Although he may not always be right, an experienced underwriter represents a substantial corporate investment, and it only makes good sense for a contractor to take full advantage of that experience by consultation and close association with the underwriter—usually accomplished through his agent or broker.

The public in general has a misapprehension as to the amount of profits contractors make. Those not familiar with the contracting business are under the impression that contractors make 10 to 15 percent on all their work. The public does not realize the responsibility of contractors as to their subcontractors' payment of obligations, or for the costs of rectifying architectural errors, or for taking care of warranties that may be in force for a period of years. Also, very often contractors do not realize the extent and import of contract failures on the public pocketbook.

In 1945, according to Dun and Bradstreet, 92 contractors failed, leaving total liabilities of $3,600,000. In 1950 there were 912 failures—ten times the number in 1945. The liabilities increased eight times—to $25,600,000. Ten years later, in 1960, there were 2600 failures, leaving liabilities of $201 million—an increase of 800 percent. In 1968 the number of failures dropped to 2200, but the liabilities increased another 60 percent and totaled $323 million.

These figures represent actual bankruptcies. They do not take into consideration the large number of contractors who lost all or most of their working capital and who quit business without going into bankruptcy.

Naturally, the surety companies suffered heavy losses on many of these contractor failures. How much these losses totaled is not determinable, but it is known that over the past ten years the industry has had practically no profits. Bonds on federal contract work have resulted in a net loss to the surety industry.

Up to 1940 and the war period it was most unusual for a surety to sustain a loss of more than 10 or 15 percent of the contract price. That probably is one reason for the underwriting requirement of 10 to 15 percent ratio of working capital to total workload.

Of recent years, however, this percentage of loss to contract price has grown by leaps and bounds. It is not unusual today for a contractor and a surety to sustain a loss of 40 to 50 percent of the contract price; in fact, some outstanding cases have reached 100 percent of the contract price. In most of these cases the difference in the first few bids was not extreme, but unusual conditions and problems developed during the performance of the work. If surety bond rates were correlated to losses in certain segments of the business, it is clear that the bond costs would be considerably higher than they are. Consequently, the rate structure of the industry acts as a stabilizer, influencing companies to adhere to careful underwriting, qualifying only the more qualified contractors, if they intend to make a profit.

The economic pressures resulting from the costs of the Vietnam War, inflation, and other problems facing our country today create more and more problems for the underwriters. Contractors, large and small, must have bank credit. With the tight money situation that has existed in 1968, 1969, and 1970, contractors and their sureties have found themselves face to face with severe financing problems. They no longer can depend on their line of credit being available. Government edicts and shortage of bank money cause the picture to change almost daily. Contractors have found it necessary to bid only contracts that they can finance out of their own funds. Those with commitments from banks for financing jumbo, long-term contracts find themselves facing increases in interest rates of 3 to 5 percent over the figures they used when bidding their contracts. Contractors find it difficult to keep within estimates when interest rates, material, and labor costs constantly increase at an abnormal pace.

The underwriter must keep abreast of all these problems. Close contact with the contractor, his surety, and his agent or broker never was more important than it is today. Each contractor has only himself to worry about, but the man who is handling his surety account and the company that employs him have to worry about hundreds of accounts. The "thin" cases he approved because of his faith in the contractor come back to haunt him. It is the responsibility of the underwriter to keep his underwriting requirements keyed to the changing conditions, including those changes that may take place in the forseeable future.

8-6. Surety Bond Forms

A contract bond generally is in the form of a joint and several guarantee of principal and surety to pay a specified sum of money, known as the bond penalty; subject to the condition that if the principal performs the obligation or undertaking for which the bond is given, the bond is void. If, however, the principal fails to perform the obligation or undertaking, the bond shall be in full force and effect. A contract bond would be in substantially the following form:

"KNOW ALL MEN BY THESE PRESENTS: That X Y Z Construction Company, a corporation of the State of _____ _____, as principal, and Q R W Surety Company, a corporation of the State of _____, as surety, are held and firmly bound unto the State of _____ in the full and just sum of One Hundred Fifty Thousand and no /100 Dollars ($150,000.00), to the payment of which the said principal and the said surety hereby bind themselves, their heirs, executors, administrators, successors and assigns, jointly and severally, firmly by these presents.
"Signed, sealed and dated this _____ day of _____, 19___.
"WHEREAS, the said X Y Z Construction Company was, on the _____ day of _____, 19___, duly awarded a contract by the State of _____ for the construction of the Steel Bridge over the Water Wash on Highway _____ between A City and B City, _____ County, _____(state), all in accordance with the plans and specifications therefor.
"NOW, THEREFORE, the condition of the foregoing obligation is such that if the said principal shall well and truly and faithfully *construct said bridge and appurtenances in accordance with all the terms of said contract, and in strict accordance with the plans and specifications therefor, and shall further pay all labor and material obligations incurred* in the performance of said work, then this obligation shall be null and void; otherwise to be and remain in full force and effect.

> (s) X Y Z CONSTRUCTION COMPANY
> By _____
> President
> Q R W SURETY COMPANY
> By _____
> Attorney-in-Fact   "

The form of bond outlined above combines in one form

the guarantee of the performance of the contract and the payment of labor and material obligations.

The U. S. Corps of Engineers and other federal departments require separate bonds, one guaranteeing performance and the other the payment of obligations. Various state and political subdivisions have adopted somewhat similar regulations. Most public bodies stipulate and prepare their bonds, and in many instances federal and state statutes stipulate the form.

The bond penalties vary in almost every instance, but most often require 50 percent of the contract price. Inasmuch as there are so many variations to bond penalties, surety rates generally are tied to the contract price unless the bond penalty is of inordinately nominal amount. After all, the sureties are guaranteeing the performance of the entire contract and the payment of all the costs incurred in performing the job. The separate bonds, when required, are rated as one bond.

This bond is a plain and simple instrument, and to the layman it might not appear to be too hazardous an obligation. The teeth of the instrument are in the underlined portion hereinabove set forth, which ties all the conditions of the contract, plans, and specifications into the bond, as though fully written out in the text.

Contracts contain time limits, penalties, and other provisions that can cost the contractor money. Plans can be and often are vague and misleading, and either the architect or the engineer for the obligee is the authority who will interpret the design of or outline as to how the work shall be done. His interpretation may differ from that of the contractor, and this may result in a contract that is more costly to the contractor.

Specifications may include warranties of various kinds. Usually all construction work is guaranteed against faulty workmanship and materials for a period of at least one year. Roofing may carry a guarantee of from five to twenty years. Paving may be guaranteed for five years. Machinery, as well as heating and steam plants, may have to be guaranteed both as to workmanship and efficiency for a certain period of time.

When the contractor signs the contract, he assumes all these responsibilities; and when he signs the bond, along with the surety, he further guarantees compliance with the warranties. In this connection it is also the practice of our courts to read into the bond compliance with the laws of the state or of the public body awarding the contract. Therefore, the surety, in its underwriting, must be informed as to any unusual provisions in the contract or specifications which may increase the hazard both to it and to the contractor.

Sometimes an overzealous public attorney will insert wording in the bond form that goes beyond the statutory requirement. The additional liability created thereby is binding on both the principal and the surety. This can be costly to the surety and probably to the contractor. Years ago an aggressive attorney for the Highway Department of a western state added to the state form, after the words, "Pay all labor and material obligations," the following: "And all just debts, dues, and demands." The words last quoted were not required by the public works statute of the state involved. The courts held this additional wording to include the payment of bank loans, even though the contractor had used the loans to purchase equipment, automobiles, and other similar supplies, which, under the laws of that particular state, were not classed as "material." A change in the state's law finally eliminated this dangerous wording.

In another instance, language was inserted in the contract specifications and proposed contract which not only relieved the owner of "any and all liability of any kind or nature whatsoever" on the site where the work or improvement was to be performed but also "on any adjacent, contiguous and non-contiguous property." One of the prospective bidders obtained a cost quotation for the coverage necessary to protect him under such an unfair obligation, and the cost of the insurance would have nearly doubled the cost of the entire project, which was estimated at only $19,000.

Sureties, in concert with contractor organizations, are constantly working to make contract terms carry reasonable obligations and to outlaw some of the grossly unfair, if not unconscionable, conditions which owners and their attorneys or architects attempt to put into contracts.

It is axiomatic that terms and conditions which are not in the contractor's best interests are not in the best interests of his surety. The close identification of contractor and surety is apparent, and this serves to explain some of the requirements an underwriter may make. Such requirements are not merely self-serving but are of critical and mutual benefit.

8-7. Qualifying for Bond

An experienced contractor who has definite arrangements for credit with his surety company already has furnished the surety with periodical financial statements, references, personnel records, and other vital data. Therefore, when he subsequently bids a contract requiring a bond, the surety requires of him specific information only as to the size, hazards, and nature of the proposed contract and the status of his other work on hand.

On a new contractor account the surety underwriter must build or create a credit file containing complete information as to the contractor's integrity, experience, organization, equipment, and finances. The underwriter normally utilizes the services of Dun and Bradstreet and other trade-reporting agencies to aid in accumulating all data possible on the contractor's background, credit record, and general reliability. The surety must place *faith* in its contractor clients. If the contractor's credit record and background indicate he does not pay his bills but instead is being sued and attached constantly, a surety is not likely to be willing to guarantee that he will pay the bills incurred in the performance of a contract. If the contractor has a past record of failures, the surety cannot be certain that he will complete a difficult contract. The surety cannot afford to gamble on such a risk. When these circumstances prevail, the surety either will decline the risk or will extend limited credit only—until it has satisfied itself that the contractor is worthy of the trust and credit extended.

The contractor who does business as an individual, whether operating in his own name or as sole owner of a corporation, represents to the surety underwriter an additional hazard. Always present is the danger that something may happen to the "mainspring" of the company, and that unless there is a capable successor to carry on the work, severe losses may result not only to the contractor's estate but also to the surety.

Surety men often recommend that the contractor, along with his attorney or attorneys, consult with and seek the guidance of the trust department of his bank, to the end that a trust, supported by adequate business life insurance, be set up for the purpose of finishing the contractor's contracts in the event of his death. This protects both his estate and his surety. Any monies remaining after completion of the contracts involved will benefit the contractor's family. Life insurance on the officers or partners, payable to the company or partnership, has saved many concerns from a financial debacle when something has happened to take the key man out of the picture. The borrowing capacity under a life insurance policy also can assist in financing in times of trouble.

Equipment is necessary in all forms of construction work. The average building contractor does not need a large amount of equipment. He usually can get by with necessary trucks, hoists, and smaller items. As a rule, a building contractor rents a great deal of his equipment and sublets a certain portion of the work, such as the substructure, to excavating contractors. On the other hand, a contractor specializing in the heavy engineering field, such as roads, bridges, dams and other similar contracts, requires a large amount of equipment. It usually represents his heaviest investment.

The surety underwriter will need a list of the larger, important items, together with data as to the age and condition of every piece. The underwriter does not regard the equipment as working capital but instead as a necessity to perform the contract. The equipment also is an indicator as to the number and size of contracts the contractor is capable of handling without requiring additional monies for the purchase of equipment.

Too much equipment is so much "dead weight" for the contractor to carry. His equipment should not be out of proportion to his working capital requirements. In a surety underwriter's view, too many contractors are overimpressed with equipment efficiencies, and after acquiring more of it find themselves without work to pay for it. Taking jobs without profit just to support the equipment is, of course, playing with dynamite, and when inevitable losses occur on the job there is no margin for absorbing them.

8-8. Accounting and Financial Records

There was a day when a contractor could get by with the very minimum of an accounting system. Many kept their books "under their hats." This no longer is true. The complexities of modern business require the keeping of painstaking accounting records. Proper records of income tax and Social Security withholdings, as well as union dues retentions, present only a few of the contractor's accounting problems. He also must furnish the Internal Revenue Department precise records not only on all equipment and its depreciation, but also on all income and expenses.

With the small margin of profit available on construction work today, the contractor must have a cost-accounting system that will enable him to check his costs almost daily. He must be in a position to determine at all times how his work is progressing. Thirty days' delay in discovering that his costs are exceeding his estimates may be too late for the contractor to take critical, correctve measures, and profit may be completely wiped out.

Under some extenuating circumstances a few surety companies will accept an unaudited financial statement which either the contractor or his bookkeeper has prepared; but most surety underwriters will insist on an unqualified and verified statement of a C.P.A. This eliminates the time the underwriter is obliged to spend verifying and investigating the statement, and hastens the decision sought by the contractor. It also saves the underwriter from asking questions of the contractor and others, which may prove embarrassing.

A certified statement has numerous additonal advantages to the contractor himself. The thoroughness of processing an audit for unqualified certification often discloses faulty pricing methods, unrecognized costs of overhead expenditures, and other blind spots which may make the contractor more competitive in his bidding, or may make operating profitability more possible and probable. The additional cost of a certified audit, therefore, is more than offset by advantages in operations, pricing, and dealing with tax people, the banker, and the surety.

After the surety underwriter is satisfied as to the contractor's integrity, experience, organization, and accounting system, the final requisite to qualify the contractor for bond credit is his financial worth, and particularly his actual working capital. His working capital normally is the principal gauge which will determine the amount of surety credit an underwriter will allow. The surety underwriter prefers a C.P.A. statement, but, if one is not available, he will furnish financial statement forms for the contractor's use. He may request a small contractor to furnish interim statements and profit-and-loss sheets on a semiannual basis.

Periodically, the surety will review progress and profitability of other work on hand. If the contractor is making money, his bond credit will improve at more frequent intervals. If his financial statement indicates he is losing money, the surety is enabled to adjust its underwriting credit accordingly, and overextension of credit to the detriment of both contractor and surety may be avoided.

Before going into some of the surety underwriters' treatment of financial information it would serve good purpose to make some comments about the various accounting methods used in the contracting business, even though this subject will be more thoroughly treated in another section of this book. For all practical purposes, surety underwriters will reject a so-called "cash basis statement" for purposes of establishing a line of credit. The cash basis statement is simply the computation of collections against disbursements and does not bring into account the accrual of such items as receivables and payables, which comprise the essence of a financial condition.

So far as financial reporting is concerned, it must be appreciated that there is a unique distinction between accounting for contractors and accounting for many other types of businesses, such as one might find in either a retailing or manufacturing operation. One of the reasons for this is that inventory, which represents a large part of the investment, is of an entirely different character in construction. Inventory for the shoe store, for instance, is the proper accounting of goods carried on the shelves and in stock—something comparatively uncomplicated except for the details of quantities.

For the contractor, however, one must take inventory at the fiscal year of his equity and investment in uncompleted work, or what accountants call "contracts in process." This has no reference to materials in the contractor's warehouse or yard. A computation of contracts in process is something far more complicated than counting shoe boxes along the shelves. The contractor's equity in contracts in process is computed by determining the extent to which he has recovered or will recover both the costs expended in performing the work and the profit he has earned to date. Consequently, for any contractor who is in the process of completing a sizeable work program, the computation of net worth vitally depends upon how accurate or realistic is the estimate of his equity in uncompleted contracts. What might be acceptable as an inventory figure in a bare

balance sheet of a retail business is unsatisfactory for evaluation of a contractor's financial condition.

Another method sometimes used is called "simple accrual." This form also is unsatisfactory to an underwriter, and should be unsatisfactory to the contractor, because it could portray a distorted picture of the true situation. Neither the surety underwriter nor the contractor should fool himself with this kind of information. Simple accrual means that whenever billings exceed costs, they are indiscriminately lumped into net worth, with no attempt to break down what portion thereof is earned and what is unearned.

If the contractor has overbilled, perhaps in an unbalanced payment situation, the statement will obviously be overinflated, particularly if he is paying taxes on a completed-contract basis. There would be substantial, unreserved taxes which must be recognized and could well go undetected, grossly misleading both the contractor and the underwriter. In the reverse situation, where a contractor may have underbilled at the time the statement is made, then, of course, the simple accrual statement will reflect less working capital and net worth than that to which he is entitled.

There are two other methods, both of which are acceptable to a surety underwriter, that usually present a more complete and accurate statement of the true conditions. They are "completed contract" and "percentage of completion." The completed-contract method means that the accountant does not consider as profit or income any billings or costs accumulated on the contracts in the process of completion. He is deferring such profits or losses as may exist until the contracts are completed, at which time the entire revenues and expenditures are taken into the statement of earnings in that fiscal year in which the job is completed. Therefore, the balance sheet will show retained earnings comprised of net income accumulated only on jobs that have been completed, and the status of uncompleted contracts will be carried as a special item identified in one of several ways.

If the contractor has some jobs where billings exceed costs and others where the costs exceed billings, the accountant will either show these excesses separately, or he may simply net them out. The surety underwriter prefers a separate showing. The underwriter must determine if the costs shown are recoverable or if such costs result from a losing contract or contracts, thus becoming an unrecoverable loss. In such a situation, of course, a determination must be made as to what additional losses may be sustained before job completion is realized, and an arbitrary reserve should be included in the analysis of liabilites.

Every financial statement, in addition to a balance sheet, profit-and-loss statement, and supporting schedules, should include a job-by-job breakdown showing at least the contract price, the contractor's estimated costs when originally bid, total amount billed to date, total costs incurred to date, and revised estimated costs to complete.

The method most preferred by underwriters is the percentage of completion. In using this method the contractor makes a determination as to the profit he has earned on contracts completed and contracts uncompleted. Profits earned on uncompleted contracts are not shown as billings in excess of costs. To obtain this earned profit, it is necessary to ascertain what the final profit will be, as shown by costs incurred to date, and estimated costs to complete; the sum of which, taken against the total contract price, yields the gross profit.

The ratio of costs to date to total costs is the percentage

of the job that is completed. This percentage of the gross profit, or loss, is what the contractor has earned, or lost. If the billings in excess of costs figure is larger than the earned profit, that difference is referred to as unearned income. When this occurs, the accountant will show this figure in the balance sheet as "billings in excess of costs and earned income." When the job shows a profit earned, but costs exceed billings, it will be shown as "costs and earned profit in excess of billings," or as an asset. Accordingly, net income is comprised only of earned income, with any excess billings excluded and carried as a liability. Having such a statement, together with all the necessary schedules, will preclude many of the mechanical processes required in the time-consuming inquiries and analyses described as being necessary in the other methods.

An accountant performs only the services for which he has been compensated under his agreement. The quality or lack of quality of the contractor's books and internal procedures may be contributory to problems which give or deny the contractor the credit to which he may be entitled, or disastrously overextend credit on the basis of a statement which, in fact, does not warrant such credit. Items which are "netted out" should always be supported by exhibits or schedules. When overhead and general administrative expenses are allocated to uncompleted jobs, an explanation should be made. If receivables are doubtful or uncollectible, they should not be included in a netted-out process, but shown separately and explained.

Profit-and-loss statements that do not reconcile with net worth, or statements reflecting net worth changes from one fiscal year to the next, which result from factors not reflected in the profit-and-loss statement, should be completely explained.

There are, of course, a number of other practices that will cause delay and further explanations, and it is hoped by this brief introduction to the attitudes of the surety underwriter that the contractor may keep such problems in mind at the time he retains his accountant and establishes a policy and procedures.

8-9. Net Quick Worth

In arriving at the contractor's working capital (in surety parlance, the contractor's "net quick worth"), the surety underwriter allows the following as quick assets, or those immediately convertible to cash:

1. Cash in bank usable in the performance of the work.
2. Good and undisputed accounts receivable from completed contracts.
3. Earned estimates due on uncompleted contracts, including the retained percentages due on completion of the work.
4. Other valid and collectible accounts receivable.
5. Notes receivable if due within one year and if satisfactory evidence is presented indicating that the makers are able to pay when due. If secured, the surety underwriter requires full information as to the amount and nature of the collateral given as security.
6. Certified checks deposited with bids.
7. Corporate stocks of quality companies with reasonable stability, usually listed on either the New York Exchange, the American Stock Exchange, or on other recognized exchanges, receive credit in the discretion of the underwriter, who may discount them somewhat for market fluctuation. Some unlisted stocks are of the blue-chip variety, but more often than not such un-

listed stocks are not of a quality acceptable for substantial credit, due to their price unpredictability. Generally, the underwriter disallows stocks in family corporations, although exceptions may be made if the principal fully controls the family corporation and the financial condition is worthy of such consideration. If the corporation is in a financial position to loan money to the contractor and subordinate the debt, the underwriter may give credit for a reasonable percentage of the stocks' value. On federal, state, county, city, and other listed municipal bonds having a ready market, the underwriter will normally allow their current market value as a quick asset. On government "E" bonds the surety rarely allows credit, due to certain legal restrictions limiting immediate availability to creditors.

8. Allowable is the cost of materials on hand, purchased for use on contracts under way—if they are not included in any estimate of either the architect or engineer; or if they are paid for; or if the cost thereof is included in the financial statement as a liability. Materials stockpiled for possible future use may or may not receive credit as a quick asset, depending on the type and demand for them at a given time.
9. Cash surrender value only of life insurance, if payable to the corporation or the partnership, or if the contractor has a right to the cash surrender value.
10. The surety underwriter will consider miscellaneous investment assets on their merits, but usually he either discounts them heavily or disregards them entirely.

From the total of these current assets the surety underwriter must deduct the total current liabilities, or those obligations which are or may be due within one year, as follows:

1. All accounts payable, due material houses, subcontractors, or others.
2. Notes payable to banks or others, if due either on demand or within one year of the date of the statement. On long-term notes, that portion due within one year is charged.
3. Notes secured by chattel mortgage on equipment, if the mortgage provides that the entire balance falls due in the event of default on any payment, or if the mortgagee seems insecure.
4. Balance due on equipment purchased and falling due within one year from the date of the statement.
5. Accrued liabilities for other expenditures, including payroll, insurance, employee withholds, and the like.
6. All Social Security and income tax withholds due either federal or state governments.
7. Notes due officers, partners, or stockholders, unless either the indemnity of the holders of the note or a written subordination agreement is furnished the surety.

From the total of the quick assets, the surety underwriter deducts the total of the quick liabilities, in order to arrive at the contractor's net quick working capital. Job financing must also contemplate ability to absorb losses quickly, and this net quick working capital figure will be used by the underwriter to gauge the surety credit to which the contractor is entitled.

However, the underwriter's final decision may be influenced by the contractor's secondary, or fixed, assets. For instance, in normal times, substantial holdings of clear, income-bearing real estate could enable the contractor to raise money readily to aid in financing his work. On the long-term contracts exceeding one or two years, secondary assets take on greater importance to the underwriter, who looks upon them as dependable and predictable resources available for recoupment or refinancing when time is less urgent than in the immediate and more reliably foreseeable future.

Of material assistance to the underwriter is full information as to the amount of bank credit available to the contractor on the strength of his own assets and not loaned against estimates due or to become due from his contracts. From such data he is able to form some idea as to how much the contractor can borrow to carry him between payments and over the tight spots.

8-10. Indemnity

If a contractor is doing business either under his own name or under a firm name, the surety will have his indemnity when he signs the application. If the contractor is married, the surety may require his wife's idemnity, as well, particularly in community property states.

In the case of a partnership, each general partner is automatically and legally obligated for all the partnership debts. When the application is signed for the partnership by any one of the partners, all the partnership assets are pledged to the surety. A limited partner is obligated only to the extent of his investment in the partnership. Generally, the surety will require the signatures of the wives of the general partners. The surety will accept a limited partner's personal indemnity, if offered, but generally the fact that he is a limited partner means he will not risk more than his invested interest and that his personal and unlimited indemnity is not available.

A corporation usually is formed to limit the contractor's liability to the capital invested. Often a corporation also is formed primarily for some tax advantage. As a general rule, on small corporations a surety company will insist upon the personal indemnity of the owning contractor, as well as that of his wife. The surety may make additional bond credit available if the financial statement of the individual giving his indemnity justifies it.

The additional indemnity of outsiders, where there is legal consideration for it, also is acceptable to strengthen a borderline case. A surety does not like to rely upon indemnity, but if the indemnitor puts up a proper share of the working capital and agrees to leave the money at risk until the work is completed and the bond exonerated, the underwriter will allow full credit for the indemnitor's net quick worth, or at least for the secondary asset strength added to the whole picture.

On occasion the surety will suggest that a proposed additional indemnitor become a joint venturer in the bid, instead of an indemnitor. This usually is because the principal is going to require considerable financing and is inordinately light in fulfilling the underwriter's requirements. Another reason is that if the indemnitor should die, the probate court may not permit the administrator or executor of the deceased's estate to advance money to the contractor, and a critical source of money is at least temporarily shut off. The surety might be able to hold up the distribution of the indemnitor's estate and eventually recover any loss the surety might sustain; but in the meantime the contractor, without financing, might not be able to complete the work or liquidate the contract debts.

All the surety is trying to do, in suggesting that the proposed indemnitor become a coventurer or put up cash in addition to indemnity, is to see that the contract is properly and adequately financed. The surety is plugging a loophole, which, in many cases, has caused losses to both the contractor and the surety.

When arranging for a bond on a contract, the surety will request the contractor to prepare and sign an application for it. Some contractors, with established surety credit, sign general indemnity agreements, which apply to any and all bonds thereafter written. Both the application and the general agreement contain very strong indemnity provisions, under which the contractor agrees to protect the surety against any and all loss or expense of any kind which it may sustain by reason of having executed the bond. When a contractor defaults and the surety becomes involved, the surety must be able to move rapidly. The surety's representative must be free to negotiate with the owner (the obligee in the bond), with creditors, subcontractors and others. The indemnity agreement must and does provide for every contingency which the surety might face in seeing that the contract is completed and all obligations for which the surety is responsible under the bond are liquidated.

The general terms of a typical indemnity agreement are about as follows:

"The Contractor shall exonerate, indemnify, and keep indemnified the Surety from and against any and all liability for losses and/or expenses of whatsoever kind or nature (including, but not limited to, interest, court costs and counsel fees) and from and against any and all such losses and/or expenses which the Surety may sustain and incur: (1) By reason of having executed or procured the execution of the Bonds, (2) By reason of the failure of the Contractor to perform or comply with the covenants and conditions of this Agreement or (3) In enforcing any of the covenants and conditions of this Agreement. Payment by reason of the aforesaid causes shall be made to the Surety by the Contractor as soon as liability exists or is asserted against the Surety, whether or not the Surety shall have made any payment therefor. Such payment shall be equal to the amount of the reserve set by the Surety. In the event of any payment by the Surety the Contractor further agrees that in any accounting between the Surety and the Contractor, the Surety shall be entitled to charge for any and all disbursements made by it in good faith in and about the matters herein contemplated by this Agreement under the belief that it is or was liable for the sums and amounts so disbursed, or that it was necessary or expedient to make such disbursements, whether or not such liability, necessity or expediency existed; and that the vouchers or other evidence of any such payments made by the Surety shall be prima facia evidence of the fact and amount of the liability to the Surety.

"The Contractor does hereby assign, transfer and set over to the Surety, as Collateral, to secure the obligations in any and all of the paragraphs of this Agreement and any other indebtedness and liabilities of the Contractor to the Surety, whether heretofore or hereafter incurred, the assignment in the case of each contract to become effective as of the date of the bond covering such contract, but only in the event of (1) abandonment, forfeiture or breach of any contracts referred to in the Bonds or of any breach of any said Bonds; or (2) of any breach of the provisions of any of the paragraphs of this Agreement; or (3) of a default in discharging such other indebtedness or liabilities when due; or (4) of any assignment by the Contractor for the benefit of creditors, or of the appointment, or of any application for the appointment, of a receiver or trustee for the Contractor whether insolvent or not; or (5) of any proceeding which deprives the Contractor of the use of any of the machinery, equipment, plant, tools or material referred to in section (b) of this paragraph; or (6) of the Contractor's dying, absconding, disappearing, incompetency, being convicted of a felony, or imprisoned if the Contractor be an individual: (a) All the rights of the Contractor in, and growing in any manner out of, all contracts referred to in the Bonds, or in, or growing in any manner out of the Bonds; (b) All the rights, title and interest of the Contractor in and to all machinery, equipment, plant, tools and materials which are now, or may hereafter be, about or upon the site or sites of any and all of the contractual work referred to in the Bonds or elsewhere, including materials purchased for or chargeable to any and all contracts referred to be in the bonds, materials which may be in process of construction, in storage elsewhere, or in transportation to any and all of said sites; (c) All the rights, title and interest of the Contractor in and to all subcontracts let or to be let in connection with any and all contracts referred to in the Bonds, and in and to all surety bonds supporting such subcontracts; (d) All actions, causes of actions, claims and demands whatsoever which the Contractor may have or acquire against any subcontractor, laborer or materialman, or any person furnishing or agreeing to furnish or supply labor, material, supplies, machinery, tools or other equipment in connection with or on account of any and all contracts referred to in the Bonds; and against any surety or sureties of any subcontractor, laborer, or materialman; (e) Any and all percentages retained and any and all sums that may be due or hereafter become due on account of any and all contracts referred to in the Bonds and all other contracts whether bonded or not in which the Contractor has an interest."

The agreement outlined above is the result of the surety industry's combined claim experience over a period of more than seventy-five years. It has met the tests of our state and federal courts. Actually, however, it rarely is necessary to enforce the more stringent provisions contained therein. When a contractor is in involuntary default, he generally welcomes the help and advice of his surety. The surety is not going to spend any more money than is absolutely necessary in settling the claims under a bond. If a contractor is insolvent, the surety's chances of full reimbursement are quite doubtful. It always is to its best interests to keep the adjustment costs as low as possible.

8-11. Financial Aid from Sureties

In this connection it should be mentioned that the surety's function is to extend bonding credit—not banking credit. Except in rare instances, it does not make loans to, or otherwise finance, a contractor. One reason is that money advanced to a contractor does not reduce the amount of the bond. In those rare instances where the surety aids in financing, any money advanced is covered by collateral, in addition to full control by the surety of estimates due on the contract. When a contractor is in financial trouble, all his work on hand usually is involved. Some of the contracts may not be bonded, and the surety is not willing to advance financing for completion of the unbonded jobs.

The surety will give more help to a contractor who has proven himself trustworthy than to one whose honesty is questionable. However, the surety can participate and be of great assistance in the negotiating of the contractor's claims against the owner for additional compensation, for the remission of liquidated damages, and other problems of that kind. Sometimes the surety is able to persuade an owner to withdraw default notices. The surety may be successful

in getting public bodies to release to the contractor all or part of the retained percentages, in order to help the contractor in his financing or to speed up completion of the contract and the payment of material accounts.

The condition of a contractor's finances, the nature and status of the contract, the availability of secondary assets, the scope of his experience, and many other factors influence the surety's decisions in claim matters.

When a contractor is in financial trouble, his bank usually is involved. Although the interest of the surety and of the bank in the unpaid contract balances, including retained percentages, is frequently adverse, they have a certain common interest; and most sureties and banks find it best to cooperate in working out the contractor's problems.

In settling claims a surety is and must be extremely careful not to give up any of the contractor's rights to claims against either the owner or others. The surety cannot recover from the contractor monies it has wrongfully paid out. At the same time, any additional monies it can obtain from the owner, in the contractor's behalf, are going to reduce the surety's loss. A good, efficient, and experienced contract claim department is one of the best possible assets a surety company can have, and often it is a great benefit to the contractor.

8-12. Basis for Bond Credit

All other factors being satisfactory, the final gauge as to the amount of surety credit extended to a contractor is his ability to finance contracts and absorb losses. Generally speaking, his financing ability is based on his net quick worth, augmented in some cases by his ability to borrow against his secondary assets, such as real estate and equipment. Ordinary, unsecured bank loans are of assistance in carrying the contractor over tight spots, but they are not working capital or loss-absorbing monies. Although the loan is deposited in the bank, it is offset on the other side of the ledger as a quick liability.

The contractor's immediate loss-absorbing power consists of his net quick worth, plus the amount of money he is able to obtain through the liquidation of his fixed assets. Normally, the latter takes considerable time to accomplish.

The ratio of surety credit to net quick worth will vary greatly between one contractor and another, as well as between surety companies. Building contracts, for instance, often can be financed on a 5 percent basis, or less; i.e., 5 percent of the contract price. On such jobs, 75 to 80 percent of the contract usually is sublet. Subcontractors are paid as the contractor is paid. The contractor withholds the same percentage of retainage as is deducted by the owner, in making progress payments to subcontractors. Thus, if a building contractor can meet the early moving-in and mobilization expense, and has good credit, he can get by with very little capital outlay. If the job is bonded, credit is granted more freely by the suppliers of material. The severe test comes when the job has been completed, or substantially so.

Suppliers and subcontractors on public work usually have thirty days, after formal acceptance by the owner or the filing of completion notice with the public body, in which to file liens or notices of unpaid bills with the owner. On private work the suppliers of material and the subcontractors do not dare let the lien period expire. On public work the situation varies in different states and jurisdictions In some states the suppliers still would have a right against the bond, regardless of whether or not they had filed notice of unpaid bills, more commonly called a "stop notice." No lien or stop notice provisions exist in the laws governing federal contracts.

Again, it must be remembered that the reason for a surety's requirement of net quick working capital for a given work program is based upon quick loss-absorbing power; and this also embraces, among other things, anticipated financing ability to take care of cash flow interruptions arising from a multiplicity of causes, such as disputes, broke or defaulting suppliers or subcontractors, breakdowns in time schedules, and delays in release of retention money. Most contracts provide that the owner will withhold 10 percent of the contract price from the contractor until the expiration of some period of time, usually thirty days after the contract work is completed. This is the so-called "lien period" or, as some contractors lamentably call it, "the lean period," during which the contractor must meet obligations without benefit of cash inflow.

The effect on the contractor of the retention money privilege of the owner must be reckoned with at the time a job is bid. Although some of the impact is transferred to subcontractors, who by agreement receive their money when the prime contractor receives his, a number of jobs with bunched completion dates will put serious strain on his quick financial resources. Even if broke or defaulting subcontractors are bonded, the contractor will have to finance the inevitable time lag and delays in effecting recoveries from the sureties of the subcontractors. Under the most favorable circumstances, that portion of retention money allocable to profit may be urgently needed to finance losses or other setbacks on other jobs.

Consequently, financing is only one of the many areas of consideration in an underwriter's concept of net quick loss-absorbing power. If the underwriter considers 5 or 10 percent in net working capital as related to the price of a new contract, plus the uncomplete portion of other work on hand, to be a sound requirement for bond qualification, it is only incidentally related to the retention provisions of a contract. Through many years of experience, the underwriter knows that 5 to 10 percent in working capital, depending upon the quality of secondary or fixed assets, will usually see a contractor through most building contracts.

The situation changes when heavy construction work is involved. Most such work is performed by the general contractor. Very few subcontracts are involved. The work often is in isolated areas which require the setting-up of camps, repair shops, bunkhouses, and dining facilities. The moving-in of heavy equipment calls for a substantial outlay of working capital. Very few jobs provide for estimates or payment covering moving-in and setting-up expenses. The cost thereof usually has to be spread over the various bid items.

Heavy construction contracts often require the investment of from 20 to 35 percent, or more, of the contract price before the returns from the job will begin to reimburse the contractor for his early investment. Working capital of 10 percent of his contract work on hand is hardly adequate for either the contractor or his surety. On work of this type the contractor must have access to additional funds, as needed. Good secondary assets, such as real estate and equipment, usually justify a bank's granting loans. Such loans should be arranged for in advance and not left to chance at a later date.

If a building contractor has $25,000 of working capital, the so-called ten to one ratio would permit him to handle $250,000 of work, all other things being equal. Theoretically, the contractor could sustain a 10 percent loss and still pay off his creditors. However, if he has extended himself by taking on $250,000 of additional work, he would be able to absorb a loss of only 5 percent.

On building contracts today the average contractor cannot make 10 percent profit on a contract. He may have a job profit of 5 percent, or more, but after adding office and other internal expenses, he is lucky to come out with a net profit of 2 or 3 percent. On the other hand, if a subcontractor diverts funds and fails to pay his bills, the contractor easily can sustain a loss equal to or in excess of his estimated profit.

A few years ago a subcontractor submitted sub-bids to all the contractors who were bidding a group of buildings. Each building was a separate contract. In each case the subcontractor's bid was from 10 to 15 percent lower than those of his nearest competitors. In each case the low general contractor used the low subcontractor's bid. In every case the difference in the general contractor's low bid about equaled the difference in the particular sub-bids. When the subcontractor went broke, which in this case was inevitable, he caused the failure of two of the general contractors. Several other contractors, who had used this subcontractor on jobs which had been fully completed some months earlier, were shocked to have claims filed against them and their bondsmen for unpaid materials supplied this subcontractor. The claims were all for substantial amounts. A mistake in accepting or using a low sub-bid, as did these contractors, can easily wipe out the profit on a building contract, as well as a large share of the contractor's working capital.

Contractors who have been in business for a long period of time and have built up their working capital to substantial amounts nearly always also have accumulated substantial secondary assets and excellent bank credit. They also have developed a following of subcontractors on whom they have found they can depend. These subcontractors have confidence in the contractor and will see him through tight spots, without getting nervous and pushing the panic button if the contractor comes under pressure. The surety underwriter may be justified in stretching the surety credit for such a client. Peculiarly enough, most such contractors who have gained this enviable position rarely overextend themselves. Many of them are tougher underwriters as to their work loads than are their surety underwriters.

It is the newer, overambitious contractors who present most of the surety's problems. Some surety companies will not accept a contract account unless the contractor has at least $25,000 or $50,000 of working capital. In a way, this is not fair, because in order to get contracts today a young contractor must have surety credit. All of the large contractors started in a small way, and the newer contractors should have the same opportunity.

It is not out of order for a practical surety underwriter to reduce the credit ratio for the small contractor (provided he passes muster otherwise), as follows: A ratio of work to capital of four to one if the contractor's quick net worth is $10,000; of five to one on capital of $15,000; and of seven to one on capital of $25,000. Again, it must be understood that the ratio would vary, depending on the type of work and the terms of the contract. In any event, the small contractor cannot expect credit of from ten to twenty times his working capital, as might be extended to larger, more experienced contractors with strong secondary assets and a proven track record.

Surety companies definitely are opposed to granting fixed and unconditional credit lines. The contractor must have and be given a general idea of how much work he can undertake at any one time. However, the experienced underwriter never will grant the maximum line of credit. He nearly always will suggest a figure somewhat less than would be the maximum line. He usually will and should qualify his credit line as to the type of work, its duration, and the area involved. He also will provide that the credit line he is granting will be conditioned on no adverse changes having taken place in the contractor's financial picture. In addition, the underwriter will ask the contractor to consult with him when the contractor's work load exceeds the figure the underwriter has set as the maximum.

For the underwriter to give an unconditional line of credit is utter folly. One of the major causes for the failure of a contractor is his undertaking of work in fields in which he is not experienced. Sometimes a contractor will bid contracts in areas far from where he is domiciled and where different weather, ground, supply, and labor conditions prevail. This may result in his learning too late that what appeared to him to be comparatively high prices for the work involved in the new area actually were justified by physical, labor, and other conditions, all of which were unfamiliar to him. Surety companies also expect to be consulted on any work being bid in foreign countries.

If a contractor's bid is substantially lower than the bids of the next two or three bidders, the surety expects the contractor to review the bids with the underwriter. Often the difference can be justified easily—particularly on heavy construction work where experience, ingenuity, and special equipment may be involved.

It is more difficult to justify a substantially low bid on a building contract. There is a fixed amount of material involved in such a contract; competition is keen; labor rates are set; and ingenuity cannot account for any material difference in cost. The difference often is due to an irresponsible sub-bid accepted by the contractor. It is not probable that one contractor can build a building for $80,000, when half a dozen other contractors, of equal ability and responsibility, have bid from $100,000 to $105,000 for the same job. Surety credit lines generally provide that the contractor's bid must be in line with the competition.

Often the surety, in granting a line of credit, also will provide that no one contract shall exceed a certain percentage of the contractor's total work on hand. In other words, the surety may grant surety credit to accommodate a $1 million work load, provided no one job exceeds $500,000. The surety does not want the contractor to put all his eggs in one basket. Assume that the contractor has approximately $100,000 of working capital and undertakes a $1 million job. If he loses 10 percent on this job, his working capital is wiped out. On the other hand, if he takes two jobs of $500,000 each, he can lose 10 percent on one of the contracts without crippling himself. His loss also might be offset largely, if not entirely, by profits on the other contract.

Smaller contractors generally are imbued with the idea that there are fewer competitors to face if only they can step up into a higher or larger contract bracket. This may be true when multimillion dollar or jumbo contracts are concerned, but it is not necessarily true when a contractor is moving from a $250,000 to a $500,000 bracket, or from a $500,000 to $1 million and upward bracket. All the contractor is doing is moving up among more experienced and tougher competition, and he should make such a move in slower and sounder steps.

A younger and more ambitious contractor is inclined to push for more credit, but generally he does not realize that when he takes on a work load that reduces his ratio from ten to one to five to one, or less, he is not operating on his own capital, but instead on his surety's credit. When a surety underwriter furnishes bond in such an overloaded case, without secondary assets or indemnity to justify his action, he is doing his client a disservice that might result in the latter's failure. The surety company will remain in business, but the contractor may be finished.

This problem is not a matter of surety ratio or of how far the surety will go. Instead, it is a matter of how far the contractor should go. It is a matter of plain horse sense. If a contractor is making progress and already has accumulated a fair-sized working capital, why should he jeopardize all he has accumulated by taking on substantially more work than his capital really justifies? He would be stretching his finances to the breaking point. His financial picture could be compared to a rubber band, which, when stretched too far, snaps at its weakest point. An accident, a subcontractor failure, or some other disaster which normally would not wreck the contractor, may cause the financial rubber band to snap.

Admittedly, the construction business is extremely hazardous. Every job has its financially dangerous points and periods. Why should a contractor challenge fate by putting too much pressure on his finances? In order to get work today a contractor has to figure out his costs very accurately. From there on, he cannot add much of anything to his bid but instead has to find ways to deduct and deduct so as to reach a competitive figure. This means that as a rule he cannot add sufficient profit to justify his risking all he has accumulated over many years of hard work. One mistake, and he has to start all over again.

A surety underwriter can caution a contractor against jeopardizing all he has worked for, and even can refuse to write further bonds for him. However, if the contractor insists, his insurance agent or broker usually can find some other surety company that will take the gamble of writing the bond in order to gain a possible good account. If, however, two or three companies turn him down, in addition to the surety that has worked with him, that should be enough to persuade the contractor that it may be a serious mistake for him to seek out the bond of an unknowledgeable underwriter who may gamble with the contractor's, as well as the surety's, money.

Sooner or later every successful contractor has to make a choice as to the path to follow. Generally, up to this point he has taken an active part in pricing out and estimating his contract bids. He, personally, has supervised his work, even though he has superintendents nominally in charge of each job. With an excellent performance record behind him and substantial working capital, he often is invited to bid larger and more complicated work. He has to decide if he is going to continue his present pattern or become a "volume" operator. If he follows the latter path, he will have to increase his organization. He will find himself dependent largely on others to supervise the work. In other words, the man who has estimated the job will not be the man who runs the job. His overhead will increase and his margin of profit will decrease. His financing problems will be greater because he will be handling larger jobs, scattered over a larger area. When business is slack, he will have to dip into his working capital in order to keep his large organization intact. No longer will he be working for himself, but instead for the organization which he must maintain.

Some men have the organizing ability to expand and still keep in fairly close touch with all the details, but others have found it impossible to stand the pace. Any program of expansion should be planned very carefully so that the estimating, supervising, and financing are closely coordinated. A surety underwriter, when checking a contractor's financial reports and current mercantile reports, usually can spot any change in a contractor's operating procedure. If he detects any signs of deterioration, naturally his underwriting decisions will be affected.

The wise contractor, before taking on work beyond reasonable limits, will sit down with his surety underwriter, his insurance agent or broker, and his banker for a discussion of his plans. Both his surety and his bank will want to know his plans for finances. Should he need more money, his banker will express his opinions as to its availability. The surety underwriter then can determine whether or not he is safe in committing his company suretywise.

As stated previously, sometimes a borderline case can be strengthened by the personal indemnity of either the contractor or some other interested party. The bank undoubtedly will exact the same guarantee in connection with its loans.

The contractor should keep in mind that surety companies value their accounts. They want them to grow and flourish—not to die on the vine or fail because of over-ambition or inflated egos. Although some surety underwriters may be young and not too experienced, they are backed by trained home-office personnel, whom they must consult on any contract bond of size. If the local underwriter misses anything in underwriting, his superiors undoubtedly will notice the fact and call it to his attention. More often than not, the requirements or observations of the surety in the long pull are beneficial to the contractor.

A surety underwriter is not a contractor, nor is he an engineer. Having a working and talking knowledge of the contracting field is helpful to the underwriter in his underwriting, but too much knowledge of engineering, for instance, may be detrimental to an underwriter. Such knowledge could constrain his thinking and cause him to overlook old, sound underwriting principles by gambling on his engineering knowledge. The old saying, "A little knowledge can be a dangerous thing," aptly applies here.

The principles of sound surety underwriting have stood up through good times and bad, through depressions and wars. The contractors who have followed these same principles are the successful operators of today.

8-13. Bonding of Subcontractors

A surety underwriter is concerned with the contractor's methods of handling subcontractors. Most owners specify to bidders that they reserve the right to reject any or all bids, regardless of whether or not such rejected bid may be low. It would behoove general contractors to exercise that same discretion in the acceptance and use of subcontractor bids. Too often, because of the press of time and last-minute arrangements necessary just prior to bid-letting, the general contractor is sorely tempted to accept a subcontractor's, or even a supplier's, low bid, not knowing fully the performance record, trade payment record, and financial stability of such bidder. The general contractor, fearing that his competitors may use a real low subcontract bid, often will use it himself, but will add something to it to allow for any pre-award negotiations with other subcontractors or suppliers who he may have to substitute for the questionable subcontractor.

Of course, in states or situations where there exists a listing law, or similar requirement, under which substitution of subcontractors is prohibited or is permitted only under stringent circumstances and with the owner's consent, the risk of dealing with unknown subcontractors or suppliers is more acute.

Sometimes the general contractor is persuaded to use the unknown subcontractor's price merely on the latter's statement that he will furnish the general contractor a performance and payment bond to cover the subcontract, if the general contractor is awarded the job. Under such circumstances, the general contractor should require the subcontractor to obtain a letter from the surety company, confirming that it will furnish the subcontractor a bond on the subcontract.

It is most important for the general contractor to know the degree of influence the bonding of subcontractors has upon the surety underwriter who extends him his own line of credit.

The immediate impression is that the general contractor should not be charged, in the calculation of uncomplete work on hand against his surety line of credit, for that portion of a job which is subcontracted and under bond. There are some neophytes in the surety business who, for lack of experience, have proceeded upon this premise, contrary to the more realistic approach of the more seasoned underwriter. The fact is that an underwriter may be more inclined to go along with an unusual type of job or with an unusually low bid than he would be to simply add an arithmetical equivalent to a line of credit.

Consider the question of recovery by a general contractor on a defaulted subcontractor's bond.

First, there is the question of the default itself. There may be a good legal question arising in this respect, which can be resolved only through litigation.

Secondly, in any default situation there is the matter of credits due the subcontractor against the debits chargeable to him, and, in any number of such situations, such questions as this also are a matter of litigation.

Third, the situation may be even more complicated if the subcontractor is, in fact or in the opinion of his surety, in a financial position of solvency or of ability to hold his surety harmless. The surety would not prejudice its own rights of recovery of a loss by acting precipitously against the request of the subcontractor.

Fourth, the subcontractor, especially a questionable one who may have had considerable difficulty in obtaining bond in the first place, may have furnished the bond of a company only incidentally in the surety business or one that is remote from the scene and uses a fire or casualty adjuster to cope with the loss.

In short, there are many ways in which the general contractor may suffer time delays in effecting final recovery or settlement of a loss arising from a bonded, defaulting subcontractor, notwithstanding the assurance he has of some amount of ultimate recovery. In the meantime, the entire job may be delayed, other subcontractors may suffer, and the general contractor may be forced to advance sums of money in order to complete, rectify, or remedy the work of the defaulting contractor. Consequently, in the thinking of a surety underwriter, the fact that a subcontractor is bonded is definitely a plus for the general contractor; but it is not a device on which the contractor can rely in order to substantially increase his line of credit. The bonding of subcontractors is just good business sense.

Most contractors develop a roster of subcontractors whom they have checked out at one time or another with other contractors, with their bank, and with their agent or surety underwriter. In prequalifying subcontractors with the surety underwriter, the contractor may ascertain quickly the subcontractor's reputation for paying his bills, a rough estimate of his worth, and his line of credit with his surety.

Parenthetically, it should be stated that most sureties for the general contractor do not wish to become surety for the subcontractor on the same job. Obviously in situations of disputes, the surety may find itself in the position of "choosing sides," or in a conflict of interest situation. Thus, the opinion of the surety underwriter, who for legal reasons will not give his evaluation in the form of fixed advice, usually is an unbiased one.

In recent years some public bodies have set up job specifications to allow for separate and independent bids to them by mechanical, plumbing, and/or electrical contractors. One of the conditions of making an award to the general contractor is that he accept an assignment of such independently made low bids for the mechanical, plumbing, or electrical work—just as though the general contractor, himself, had used such subcontract bid in submitting his own bid. Obvious conflicts may arise in these situations, and the general contractor's surety underwriter should be advised well in advance of the bid date so that he may make such inquiries about the possible "assigned subcontractors" as he may consider necessary in the circumstances.

8-14. Bonding of Joint Ventures

When a contractor has an opportunity to bid a contract of a size or nature he does not feel he should undertake alone, the common practice today is to create a joint venture with another contractor, or several contractors. A joint venture is a temporary partnership between two or more contractors, formed for the purpose of performing a specific contract. The joint venture is capitalized by each of the contractor's putting up his pro rata share of a working fund. Normally, one of the joint ventures is the sponsor, and he supervises the work. He uses his own organization, augmented by any men from among his partners' organizations who will fit in key positions. The other members of the joint venture contribute capital and generally are called on for technical assistance or advice, as needed.

It is most important that a written joint-venture agreement be entered into, setting forth the relative interest of each coventurer in the contract. Profits and losses are shared on the basis of each of the contractors' percentage of interest. Also they are jointly and severally bound to complete the entire contract and to pay all obligations incurred in doing so. This applies as well to the so-called unit-type joint ventures, where one contractor agrees to do one part of the work and another contractor does still another part. If any of the joint venturers is unable to meet his share of the obligation, the owner will look to any of the remaining or surviving joint venturers to absorb his share. The joint venturers also are jointly and severally obligated to protect the sureties on the bond.

It therefore is essential that a contractor going into a joint venture make certain that each participant is able to provide his share of working capital and absorb his share of any loss that could occur. It must be remembered that whereas liquid financial positions may look excellent at the outset of the joint-venture work, each of the contractors may have other work on hand or other obligations which may negate that working capital and make it unavailable to the joint venture when it is needed most.

Each participant also should be knowledgeable on the type of work involved, and some of the joint venturers

should be capable of handling the contract if something should happen to the sponsoring contractor.

In addition to providing for contributions to the venture's capital and sharing in profits and losses, the agreement should provide for the type of accounting system and audits desired in the circumstances. It also should provide for a substantial blanket fidelity bond on the officers and employees. A large fidelity loss can be as costly as a catastrophe on the work itself.

The agreement also should provide for handling and revising the various participants' interests in the event that one or more of the contractors become insolvent, die, or for some other reason are unable to meet the obligations imposed by the agreement.

Joint-venture agreements should be very carefully drawn by an experienced contract attorney. A tax attorney also should be consulted, especially on joint ventures involving large-sized or foreign contracts.

Joint ventures created by, or consisting of, capable and well-financed contractors have been remarkably successful. However, joint ventures created by two or more financially weak or undercapitalized contractors, merely to get enough money together to start a contract larger than any of them has heretofore handled, generally have not been successful and often have ended in disputes and sometimes heavy losses. Everyone is happy when a contract is making money, but when reverses take place and money is being lost, love flies out the window.

Joint ventures also are created to handle a contract consisting of two different kinds of work. An illustration is a highway contract which incorporates not only a substantial amount of road work but also one or more sizable bridges. Sometimes a road builder and a bridge builder will create a joint venture on the basis that each agrees to handle his share of the work. Each contractor will estimate his portion of the contract as if it were a separate job. Each contractor will add whatever profit and contingency allowance he feels to be proper. The bids will be combined and submitted to the owner as one proposal. Again, although the contractors can make any deal they want as to sharing the work and profits and losses, both the owner and the surety will consider them as one entity, and will look to both contractors to perform the entire job and to pay all the obligations incurred. The surety normally will insist on the joint and several indemnity of all the joint venturers.

This type of joint venture can develop into a debacle if one of the members of the joint venture fails and is unable to finish his portion or pay his bills, or both. The remaining joint venturer has to assume the cost of completion and the payment of the obligations. Caution should prevail when a contractor is approached on a joint venture of this kind.

On extremely large contracts with substantial bonds, surety companies will sometimes agree to accept what is called "joint and several but limited dollar indemnity" from the joint venturers. The sureties may agree to limiting each venturer's indemnity liability to an amount equal to the individual contractor's proportionate share of the aggregate of the performance and payment bonds.

For instance, on a $10 million contract the performance and payment bonds might be 50 percent, or $5 million each—a total of $10 million. If a contractor has a 10 percent interest in the contract, his indemnity to surety could be limited to $1 million. It is conceivable that this might be of some value to a contractor at some time, but it is very unlikely and so far has never shown itself to be a significant advantage because he is jointly and severally liable, without limitation, to the owner or obligee in the bond for the full completion of the contract and the payment of all contract obligations. The contractor's liability to surety is secondary to his liability to the owner.

If the bond penalty is comparatively small in relation to contract price, as it often is in foreign countries, the sureties may agree to joint and several but limited indemnity; but often the limitation is double or triple each of the contractor's proportionate share of the bond penalty. The surety may have to pay out the full penalty of the bond, plus heavy expenses, and the double or triple indemnity will take care of the excess loss over the bond penalty. It will also, in some cases, make up one or more of the other joint-venturers' shares of the surety loss, in the event such joint venturers are unable to respond financially for their shares.

Each joint venture must be carefully analyzed by a surety underwriter. No two are alike. He must check the financial picture of each participating contractor, as well as study the other work each has on hand.

Sometimes the sponsoring surety is faced with a problem created by the surety for a financially weak member of the joint venture. This surety may agree to go along for his client's share of the bond, even though his client may already be overextended. In doing this, he is relying on the financial responsibility of the other joint venturers, knowing full well that all the contractors will have to go broke before the owner or obligee makes demand on the bond. His client is "riding in on his partner's backs." It is true that the overall or combined financial picture may be very favorable, but in justice to all the other participants, his client's portion either should be reduced or some other contractor substituted in his place. The sponsoring surety has no alternative in such a situation but to inform the sponsoring contractor of the problem. Then, if the sponsoring contractor and his other joint venturers wish to assume the risk, it is their choice.

Actually, the joint venturers should furnish each other with up-to-date financial statements and the status of their work on hand. Experienced contractors can quickly detect a contractor who is overextended and might have difficulty meeting financial demands. Some surety underwriters, as well as agents and brokers, lack the courage to tell a client that he should not participate in such ventures until his work load is reduced to a point where he is entitled to participate. They do not realize that they have a responsibility to the other venturers and sureties, as well as to their own client. They appear to be preoccupied with the fear of offending their contractor account.

The surety companies have no hard and fast rule as to the percentage of capital fund to be set up by the joint venturers. Circumstances as to the financial worth of the participants, amount of other work on hand, the effect of mobilization costs, payment schedules, and similar conditions will vary greatly. The financial requirements of the job will also have a substantial effect. If a contract is going to require very heavy early financing, the contractors must prepare in advance to meet the demands of the job.

In some cases a working fund of 5 to 10 percent of the contract will be established before the bond is executed or contracts are signed. This is almost always a preliminary investment which will enable the sponsor to set up the equipment, camp, and other starting essentials. Within ninety days or six months an equal amount, or even more, may be needed, and the contractors must be prepared to meet the call. If all the joint venturers are financially able to meet the estimated financial contributions, the sureties

do not lay down any fixed rule as to the amount of the capital fund to be established at that time.

If, however, some of the joint-venture participants are light financially and have other work which will also need additional financing, the sureties may insist that a substantial capital fund be established in the beginning. This is to make sure the money is available when needed. Such a requirement is a protection to everyone, including the sureties. It is possible in such cases that the sureties will exact an agreement from the contractors that no money will be returned to the participants without obtaining the consent or approval of the sureties to any redistribution.

There have been times during economic depressions when the sureties have insisted on the capital funds being deposited in trust accounts, subject to the signatures of the selected sureties. When this is done the sureties give the contractors or joint venture a "working" fund on a revolving basis. All estimates received from the job are assigned to the sureties and deposited in the trust account. The contractors' revolving account is reimbursed monthly or semimonthly, as the case may require, by the amount of expenditures made in performing the work. The trust fund would prevent the working capital from being attached and being diverted for purposes other than performing the contract bonded by the sureties. This procedure is rarely, if ever, followed today; but during the depression years it enabled contractors to get bonds on hazardous jobs, and also enabled surety companies to get enough surety companies to participate to handle the sizable bonds required.

The surety industry has always spread its risks by way of reinsurance or cosuretyship. In doing so it protects itself against catastrophic losses. Consequently, it favors joint ventures by contractors to accomplish the same result. Not all joint ventures on major jobs have made money, but very few, if any, joint ventures have failed or caused losses to their sureties. Such losses as have occurred were in cases when two or three small concerns tried to make one concern large enough to handle a job that no one of them could have handled alone.

The key to successful joint ventures is the proper and cautious selection of the joint-venture partners, making sure that each one has the ability, the money, and the fortitude to see the contract through to completion.

The fact that two, three, or more of the participants will independently estimate the cost of the contract reduces the chances of errors and omissions in the bid and helps to assure a profit.

The dangers of joint ventures are underfinancing or overextension of the participants. Taking on work of a type in which the sponsor and his associates have not had experience, and going into foreign countries or areas where labor, political problems, and physical conditions can upset all calculations, are additional and frequent causes of serious trouble.

On jumbo risks, such as large dams, tunnels, and other heavy construction contracts, the sureties are faced with handling performance and payment bonds of 50 percent, or more, on contracts in amounts ranging from $100 million to $350 million, and higher. Federal and state laws prohibit a surety from assuming a single obligation in excess of more than 10 percent of its capital and surplus on any one bond; hence, such jumbo risks often tax the ability of the industry to provide the required bonds. A joint venture consisting of several contractors usually undertakes these types of contracts. The sureties of the various joint-venture contractors jointly underwrite and sign the bonds as cosureties. Each surety may call in other sureties to assume the excess bonds over its own underwriting limits. Although each surety is expected to take care of its contractor's share of the bond, the sureties are jointly and severally liable for the full bond penalty, limited only to the stated dollar amount for which each company signed.

8-15.   Selection of a Bonding Company

The proper selection of a bonding company is of the utmost importance to any contractor. It is a matter that should not be left to chance, but one that should be studied and investigated carefully. It also is a matter that should not be left to others to decide. A contractor's entire future may depend on the proper selection of a bonding company. There are companies that specialize in the bonding field and do a national business through branch offices, which are well equipped with trained surety men. Some of these companies write very little business in the other lines common to the casualty-surety field. Other companies are casualty and fire insurance underwriters, with well-equipped surety departments. Numerous companies include surety bonds as one of the many lines of business which they handle; however, their volume of surety business does not justify the maintenance of fully manned, experienced surety departments. These companies are and may be substantial and capable fire and casualty insurance writers, but to them the surety field is a sideline, intended only to round out their service to the public.

In obtaining a line of surety bond credit, the contractor should be concerned primarily about the quality of service he receives from his surety; the quality and depth of expertise of its counsel in the many complex matters in which he may require advice; and the quality of a surety's commitment and help to him in difficult situations. Premium rates in and unto themselves should not be the sole determining factor in the selection of a surety.

The cost of a performance and payment bond usually represents less than 0.75 percent of the total contract price on any substantial job. Therefore, the difference in rates charged by surety companies reflects a very minute savings to the owner, who ultimately pays for the bond. The savings in most cases are only a very small fraction of the total cost of the job. Bear in mind the great multiplicity of ways in which the contractor may vary his bid with respect to subcontractors and suppliers selected, overhead costs, labor costs, and methods and procedures—all far more substantial and significant items than the comparatively less-important cost of the bond.

The government annually publishes a list of surety companies that have qualified to write surety business. The list sets the amount of maximum liability each company can sign for and accept on any one bond. The amount represents 10 percent of the capital and surplus of each company, as established by the Secretary of the Treasury.

Some of the fire and casualty companies have tremendously large capital and surpluses and consequently quite large underwriting limits, but this does not necessarily assure the contractor of the kind of surety expertise he may require. The 1969 list of qualified surety writers contains 248 companies, including about a dozen foreign companies. It also includes a number of so-called "professional reinsurers," which do not write business directly but which, as reinsurers or coinsurers, solicit and accept participation from the direct-writing companies.

In 1967 the total surety bond premiums of all the companies totaled about $322 million. Approximately two thirds of this volume represented contract bond premiums.

This included all forms of contract, subcontract, and supply contract bonds. It also included subdivision bonds and other forms of bonds common to the construction business. This means that in 1967 the total premium income of the surety industry, arising out of the contract field, amounted to approximately $215 million. The 1968 contract bond premiums totaled approximately $228 million; and it is estimated that the 1970 contract bond premiums will exceed $240 million.

In 1967 the ten leading surety bond writers wrote contract premiums totaling approximately $110 million—more than one half of the total contract premiums written by the entire industry. This concentration of the business is not due to any difficulty in entering the competition for bonds. Nor does it mean that all the other companies are any less qualified to handle the business. It does point up that certain companies are aggressively and seriously in the business of surety bonds, have expensively trained and developed underwriting experts in large numbers, and maintain nationwide facilities capable of giving contractors the benefit of services wherever they may be needed.

Most companies today do write insurance lines needed by contractors, and some of these companies ask for the contractor's surety business in order to sweeten their casualty premiums. This is true particularly when a contractor's compensation and public liability experience is bad. In many instances a contractor has had his bond business switched from company to company, in an effort to obtain more favorable casualty insurance rates, or, when casualty loss experience is particularly bad, to get such insurance at any price. This sometimes means that the contractor falls into the hands of a company with poorly staffed surety underwriters and inadequate service facilities to handle his surety account. The advantages of experienced underwriting and technical processing may be lost to the contractor in such circumstances. A contractor may be well-advised not to permit his bond business to be used as a leverage in this manner, but instead to select a company that is qualified, through experience, knowledge and organization, to handle his bond business properly. It is far cheaper, in the long run, to put into practice "safety programs" on all his contracts, thereby reducing his bad experience and making his casualty business, in itself, attractive to an insurance company.

Bids of contractors more than once have been found unresponsive and have been thrown out due to a technical imperfection in the bond prepared by inexperienced and careless persons.

A contractor often finds he needs his bid bond at the last minute, just before the deadline for filing his bid. He cannot afford to do business with a company branch or a general agent who must submit most bond matters to some far-distant office for approval. Many companies that are prominent in the fire-casualty field of insurance, but are small writers in the bond field, will have only one person in an office handling bonds. If he is out of the office or on vacation or ill, the surety department practically ceases business until he returns.

Therefore, in selecting a bonding company a contractor is well-advised to consider picking one of the leaders in the surety field—one with a trained organization, and particularly one that also has maintained a low loss ratio in surety over the years. A low loss ratio in surety may connote careful underwriting, but, most importantly, it indicates a stable of successful contractors who probably have obtained considerable good advice along the route to their success. This information is available to insurance agents and brokers. It also is obtainable through Best's Insurance Reports, to which most large banks subscribe.

In practice, most contractors make their choice of sureties through the advice of their insurance agent, who normally represents not one but perhaps several companies which write bonds. These agents are under no obligation to recommend any particular company; and the professional ones, in doing the best job for their contractor-clients, are careful to urge review of the character, underwriting philosophy, and facilities of the surety with whom they deal. Choice of a rate-deviating company is not the principal consideration of the more sophisticated and experienced surety producers.

Over the years there have been many instances when a multiple-line insurance company has tried to find a place in the sun for its surety department. Departing from sound underwriting principles, the overeager newcomer may begin by extending unreasonably large credit lines to contractors —often doubling, or more, what would be a fair and sound credit extension. It may also increase the commission rates to agents and brokers. Some shortsighted contractors and insurance agents and brokers, taking advantage of the line of least resistance, switch accounts to such a company. For a year, or even a little longer, the company may seem to be succeeding in picking up accounts and making money. Then the losses begin to pile up, and in the next three or four years the contractors, one by one, disappear from the scene, either broke or badly crippled. The company discharges the surety underwriter responsible and either closes up or severely curtails its surety department. In some cases such activities have cost the insurance company involved millions of dollars. The company deserves such losses; but, on the other hand, its activities have ended the careers of many capable contractors, who, had they been properly advised to consider more carefully the possible consequences of free-swinging suretyship, might have reached high pinnacles in the contracting business.

It is suggested that a contractor not only should ascertain the history of the insurance company with which he is doing business, but also should check its underwriting record over a period of several years. If the company is a comparative newcomer, either he or his agent should inquire into the credentials of the particular underwriter and his service facilities, and form a judgment regarding the desirability of a relationship on which he can depend for years to come.

The relationship of the contractor to the surety is a very personal one. Each should have full confidence in the other. In frequent discussions of contracts coming up for bid, the contractor and the underwriter gain an understanding of each other's thinking, which helps immeasurably in instilling confidence in each other. A contractor makes a mistake if he does not have direct contact with the man who may be a deciding factor in his future success.

Although there are a number of agents and brokers who, because of their experience and proven ability, enjoy authority to execute bonds for accounts which have established lines of surety credit, most of the more successful and professional agents and brokers are anxious for their contractor-clients to meet the underwriters. To leave all bond negotiations in the hands of a third party, no matter how capable the agent or broker may be, is unwise, to say the least. Although there are loan brokers who put the parties together and arrange loans, how far would a sound banker go in extending credit if he never had met the borrower? A contractor ultimately makes his own arrange-

ments for loans, and the banker's action is governed not only by what he sees in the financial statement before him but also by what he sees in the man before him. A surety underwriter is another form of banker—extending credit instead of cash. His confidence in the individual contractor goes a long way in his underwriting decisions. Denial of direct consultation betwen the contractor and underwriter may well deprive the contractor of the full line of credit to which he is entitled, and surely deprives him of the expertise that could be most helpful to his profitable growth.

8-16. Selection of Agents or Brokers

Included in any discussion of the selection of a surety company there also should be a discussion, to some extent, of the selection of an insurance agent or broker.

An insurance agent or broker handles many different forms of insurance coverages, and he is thoroughly knowledgeable and well-qualified to discuss many with authority. It would be humanly impossible for him to be an authority on every line, because the field is so broad, complex, and all-embracive. The average insurance agent or broker daily handles automobile, fire, public liability, and compensation insurance. With many producers, surety bonds are a comparatively minute part of the total premium volume, and some agents rarely have contact with them. As a result, the average insurance agent and/or broker probably least understands the surety field. He has to depend largely on the surety underwriter for guidance, and in such circumstances the wisdom of the producer and of his contractor meeting directly with the underwriter is amply manifest.

The insurance problem in handling a contractor's account probably is the most exacting and intricate in the insurance field. The contractor would be well-advised to select an agent or broker who has broad experience in the contract field. The contractor does not need an "order taker," but instead an authority on the subject. The contractor should take part in making the decision as to the insurance companies and surety company selected. Bear in mind that the carrier of other insurance quite often is not, for the reasons previously given, the same as the company writing the contractor's bonds. Too often an agent or broker will place business with a company that is offering excess commissions or other inducements in order to get business. A first-rate company does not have to go to such lengths. It will attract business because of its service and knowledge. The same applies in the surety bond field.

The contractor should make a special inquiry into the reasons for an agent's not wanting the casualty manager or surety manager to meet his contractor-client. Some agents are concerned that perhaps the underwriter may someday become an agent himself; and, having little confidence in their own ability, these agents put up self-protecting barriers to the underwriter. Naturally, the casualty or surety manager, upon never meeting the contractor, may wonder what is being concealed or withheld and what representations or misrepresentations are being made. The insurance agent or broker who handles a contract account in a straightforward manner with the surety is a great aid in building the confidence necessary to create a happy, lasting, and successful relationship.

There are relatively few insurance agents or brokers in the country who have any appreciable *discretionary* underwriting authority. A great many agents hold a power of attorney authorizing them to sign bonds in behalf of a surety company as an "attorney-in-fact." Except in rare cases, such authority is limited in dollar amount, and the agent has a letter of instructions requiring that, before executing any bonds, he first must submit the proposition either to the company branch office or home office for approval. In those rare cases where an agent is given an unlimited power of attorney, it usually is subject to the same conditions. The issuance of such a power is done to save time and to enable the agent to sign the bond locally, after either writing, wiring, or telephoning the company for authority.

There are a few agents who do have some discretionary authority in reasonable amounts, but it usually is exercised only after the account has been qualified and accepted by the surety and a line of credit established.

Home offices give their surety managers in the field varying amounts of discretionary authority, depending on their experience and capability. A surety manager in a branch office may commit his company, but usually, and quite correctly, he is held strictly accountable for his underwriting. It is important for a contractor not to accept from an insurance agent, or others, an oral commitment agreeing to write bonds unless he knows the person has the authority from his company to make such a commitment and will follow through on it. It always is better to have any such understanding in writing. An agent or broker rarely has any such binding authority.

There have been many instances when a contractor has been low bidder on a job, having acted under the impression that his bond was committed or assured, when, in fact, the company actually had no knowledge of the bid.

Even when the surety issues a bid bond which takes the place of a certified check, the surety usually reserves the right to decline to execute the final bond if the contractor's bid is out of line.

The foregoing statements are not intended to criticize or disparage those many thousands of fine, capable insurance agents or brokers who do an excellent job of looking after their contractor-clients' interests in every respect. However, there are agents who are and always will be "order takers," who go through the motions of getting the answers but who never quite understand what it is all about. There are, unfortunately, similar people acting as underwriters in the surety business. Needless to say, their surety careers usually are fairly brief, although sometimes expensive for their employers.

8-17.   Rules for Successful Contracting

Successful contractors are not created in a day, or by the successful completion of one or two contracts. The really successful ones are molded by years of experience gained through combatting the elements, economic and labor problems, wars, and depressions. Such contractors have had to fight their way back to the top, sometimes more than once, after suffering severe reverses. Fire tempers steel. Adversity tempers good contractors.

Summarized below are the simple rules which, if followed, will help a contractor prepare to meet the adverse problems he is bound to face at various times in his career. These rules will, in fact, assist you, as the contractor, in avoiding situations that could lead to disaster:

1. Have a C. P. A. set up an efficient cost-accounting system that will keep you abreast of your contract costs and enable you to stop losses before they ruin the job and possibly you, the contractor.
2. Keep your financial condition liquid so that you can meet emergencies without the necessity of calling on others for financial assistance. You should not tie up too much money in equipment. Enough is enough. Anything more is a weight around your neck. Cash is a good "ace in the hole."
3. Invest your monies in assets that, in time of need, can be liquidated quickly. Mortgaged real estate or non-income-bearing real estate cannot always be liquidated without loss.
4. Purchase substantial business life insurance on all your key personnel. This is particularly advisable on individual operators, solely owned corporations, and partnerships.
5. Keep your volume of work well within your financing and loss-absorbing power so that you will not be entirely wiped out by a contract loss and unable to start anew. "Make haste slowly but surely."
6. Keep your organization compact and intact. Thus, you will know each man's capabilities after he has been with you for years. You cannot always rely or depend on strangers. When business is poor, you can "carry" a tight organization. In good times you can always expand by hiring temporary help.
7. Pay your subcontractors and materialmen promptly and discount your bills if possible. In addition to saving you 1 or 2 percent, you will build up your credit, and, when you need it, your suppliers and subs will give you a "break."
8. Stay out of disputes and lawsuits whenever possible. If you can compromise or arbitrate a dispute, it is far cheaper for you than going to court. Credit reporting agencies always note the number of lawsuits, attachments, and judgments recorded against contractors; and material suppliers, surety companies, and banks always receive copies for their credit files.
9. Consult with your lawyer before you get into trouble —not afterward. More and more frequently architects and engineers are using the contract terms as a means of foisting onto the contractor responsibility for things which the owner, architects, and engineers should rightfully assme.
10. Last, but far from the least, recognize your surety, your insurance agent or broker, and your banker as integral parts of your organization. They should never be too busy to consult with and advise you. They all play an important part in your operation and success. They all want you to succeed. You are an asset to all three of them, and they can be a real asset to you if you will let them assist you.

9 EQUIPMENT MAINTENANCE AND REPAIR

WILLIAM S. LAMBIE, JR.
Vice-President, Parts and Service
Caterpillar Tractor Co.
Peoria, Illinois

FIELD MAINTENANCE

Field maintenance consists of regularly scheduled lubrication, fueling, inspection and adjustments, preventive repair, and system analysis performed in the field.

This work, when properly planned, performed, and recorded, constitutes a "preventive maintenance program" as outlined later.

9-1. Lubrication

Equipment manufacturers normally provide detailed lubrication and maintenance instructions with machines or through the dealer. These instructions are based on actual field tests and extensive research, and they should be followed closely.

Correct lubricants are essential to equipment maintenance. Suppliers must be instructed to provide the lubricants specified by the equipment maufacturer.

Adherence to recommended lubrication change periods is equally important. Greasing and oil changes must be performed within recommended time intervals (Figures 9-1-1 and 9-1-2). Failure to do so will increase operating costs.

Use only clean lubricants. Contaminants in lubricants, or dirt which has entered the system during oil changes, often account for early breakdown of engines or power trains.

Only a complete job of lubrication assures component service life. A grease job cannot prevent a machine breakdown if one fitting is regularly forgotten.

9-2. Fueling

Correct grade and quality of fuel is vital to machine performance and engine life. Fuel suppliers must comply with the manufacturer's recommendations.

Fuel should be protected during storage, transportation, and vehicle refueling. Improper fuel storage, transport, and delivery methods can contaminate the best fuel and cause engine damage resulting in costly repairs and downtime. Fueling should never take place in extremely dusty air. Tank caps, fueling nozzles, and tank necks must be wiped clean.

Whenever possible, fuel tanks should be filled (Figure 9-2-1) after work stoppage to prevent water condensation in empty tanks at night (Figure 9-2-2). Draining the condensate from fuel tanks is a regular part of the fueling operation. It should be frequently drained from storage tanks also.

Fuel spillage should be held to a minimum—it is both wasteful and a safety hazard. Most construction machinery can be equipped with a "Fast-Fill Fuel System." This fueling method forces fuel under pressure into the tank, without need to open the tank cap. Thus spillage and contamination from dirt are avoided. In addition, such a system reduces fueling time by one half or more.

9-3. Inspection and Adjustments

Regular inspection of machines is an important part of field maintenance (Figure 9-3-1). This is not the "walk-around" check the operator makes before starting to work, although this, too, is important. Regular inspection means periodic investigation of the condition of all components: engine, transmission, drive train, hydraulics, structural members, and wearing components, such as undercarriages, tires, and cutting edges (Figure 9-3-2).

To ensure thorough inspection of all vital parts, detailed inspection report sheets should be used for each machine listing specific points of inspection. Generalities will **not** suffice. For example: "undercarriage" is not adequate. Individual important components such as track rollers,

Figure 9–0–1 Over-the-highway type trucks with bottom dump beds

Figure 9–0–2 Track-laying tractor with "U" dozer moving dirt

Figure 9–1–1 Serviceman checking crankcase oil.

links, track shoes, etc., should be listed (Figures 9-3-3 and 9-3-4).

Maintenance instructions provided by the equipment manufacturer should be used as a reference in preparing the inspection report.

Regular adjustments of brakes, clutches, linkages, steering, etc., are often performed by inspection or lubrication personnel as part of the maintenance work done at scheduled service intervals (Figure 9-3-5). More serious findings are reported to the equipment manager or service shop foreman for action.

9-4. Preventive Repair

Minor repairs that can be performed in the field are part of field maintenance. Included are welding cracked metal parts (Figure 9-4-1) and loose wear plates, or replacing lost bolts, straps, or clips. Also included would be replacing a broken track shoe, shimming a ball joint, or replacing a brake shoe.

The need for a minor repair is usually noticed during operation or inspection of the machine. A minor fault or malfunction should be repaired at the first opportune time. Such tasks can frequently be done at the end of a work shift. No preventive repair, however, should be patchwork or designed to "just get by." All repairs, minor or major, should last for the life of the component (Figure 9-4-2).

9-5. Systems Analysis

Today manufacturers provide a growing number of tools, test devices, and procedures to analyze transmissions, hydraulic systems, and engines. System condition can be determined without need for disassembly. Use of such equipment eliminates guesswork (Figures 9-5-1 and 9-5-2) The chances for unexpected failure are reduced. Component replacement or exchange can be conveniently scheduled.

Current test devices analyze radiator air flow, coolant temperatures, engine manifold pressures, hydraulic pump flow, transmission oil pressures, hydraulic control leakage, and many other aspects of system performance (Figures 9-5-3 and 9-5-4).

Testing personnel must be specially trained to get reli-

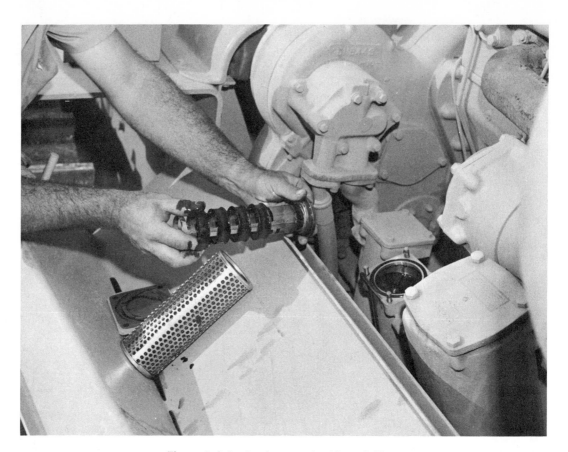

Figure 9–1–2 Serviceman checking oil filter

Figure 9–2–1 Serviceman refueling motor scraper

Figure 9–2–2 Serviceman refueling rubber-tired type dozer

Figure 9–3–1 One of servicing personnel inspecting undercarriage of track-laying tractor

Figure 9–3–2 Serviceman checking slack in tread linkage assembly of track-laying type tractor

able results. For this reason, system analysis is usually performed by a specialist in the field. Equipment service contractors or dealer service personnel can provide this specialized service.

9-6. Time and Cost Records

Field maintenance must be backed by a record system that includes maintenance and repair history and cost data (Figure 9-6-1). The history file provides information for making replacement or repair decisions. It also allows management to evaluate the effectiveness of field maintenance. The machine owner who maintains accurate machine records has the data needed to evaluate equipment performance and operating cost and to plan equipment replacement. Several record-keeping systems and preventive maintenance programs are discussed later.

9-7. Organization for Field Maintenance

For an equipment maintenance organization, success or failure depends upon the skill of personnel. High-caliber supervision and a well-trained work force are needed. Beyond this is the need for a team commitment to service—a sense of urgency in accomplishing the work and pride in doing it well. To build this type of organization is the real challenge to management.

Personnel requirements for equipment maintenance organizations will vary with the scope of the job. With a large staff, individual jobs can be well defined. Work specialization will increase proportionately with size of project and equipment fleet. In smaller operations the owner, operator, or a single mechanic may perform all maintenance functions.

Basic considerations in establishing an equipment maintenance organization are size of the operation, type of equipment, and type of work. Usually, the number of maintenance personnel and supervisors needed depends on these factors (Figure 9-7-1).

It is not the purpose of this section to outline in detail every job assignment but, rather, to convey the importance of equipment maintenance management. No matter how small or large the organization, to obtain satisfactory results, the responsibility for each maintenance function must be defined and assigned.

MAINTENANCE AND REPAIR RESPONSIBILITIES

Type of Work	Responsibility
Equipment maintenance	Mechanic, operator, driver, or equipment dealer
Minor adjustments	Mechanic, operator, driver, or equipment dealer
Inspection report	Equipment dealer
Service work	Equipment dealer

In a small operation, the machine owner assembles a team of versatile employees who have overlapping capabilities and responsibilities. The accountant or bookkeeper, whether full time or part time, is important. With his assistance, costs can be analyzed and activities geared to improve profit.

Owners of small operations cannot usually afford to tie up capital in efficient service tools and shop equipment, and on adequate replacement parts stock. The owner should rely on the equipment dealer or other outside source for major repair work.

Figure 9-7-2 shows a possible structure for a larger operation. In this type of organization there must be a clear definition of functions and responsibilities between foremen.

Figure 9–3–3 Serviceman measuring blade linkage slack on motor patrol

If the machine owner decides to have service work performed by an outside source, or if he utilizes dealer component exchange service, he can reduce his mechanical staff. Repair of smaller equipment, such as compressors, light trucks, pumps, etc., may require a few shop personnel.

The equipment manager's office staff is charged with maintenance of complete equipment records. This staff should be under direct control of the equipment manager.

9-8.   Equipment Manager Responsibilities

The equipment manager should be accountable for all equipment maintenance and repair. Experience in equipment maintenance and repair is important to an equipment manager. Equally important, however, is his ability to instruct, lead, and inspire people. Additionally, the manager must be able to evaluate facts and make good management decisions.

Equipment Manager Records   For individual machines, the equipment manager should keep records that show:

- Hours worked per day, hours worked per month, and total hours worked per year
- Idle time in hours and the total number of hours lost in downtime. These should be identified as "Downtime—Idle" (which is time lost not due to machine failure) or "Downtime—Actual Repair Time"
- Preventive field maintenance
- Repairs and failures
- Labor cost
- Cost of parts, lubricants and fuel

Analyzing records of failures and repairs will reveal the need for changes in the preventive maintenance program, correction of operator habits, or purchase of different equipment.

Analysis of the cost and extent of field maintenance will help management to evaluate and reduce owning and operating costs. The machine application should be described along with the operating conditions. Severe operating conditions will account for increased cost of maintenance and repair.

Figure 9–3–4   Serviceman providing a general check of undercarriage of track-laying tractor

Regular evaluation of the machine by the equipment manager, job superintendent, and operator will establish the machine's overall acceptability. Factual performance data are the best bridge across a possible communication gap between the equipment manager and machine operators, as well as between the equipment manager and job superintendent.

Repair-or-Replacement Decisions   Good machine records and cost accounting facilitate better management decisions. For example, if a large crawler tractor has accumulated high maintenance expense, one might come to the easy decision to replace it rather than repair it again. However, if records show that the working conditions were severe, not much might be gained by replacing it. With proper machine records, a great deal of guesswork is eliminated and the probability of reaching the best decision is increased.

Assistance in Equipment Selection   Equipment cost records can be valuable in selecting new machines. See Chapter 10.

Maintenance Program Coordination   Coordination of maintenance programs and construction operations is a most important responsibility. With cooperation between operating personnel and mechanics, and the support of equipment dealer service personnel, the equipment manager can develop and implement a successful preventive maintenance program. Personal experience in the field will help him to improve and tailor service procedures to minimize equipment downtime and maintenance cost.

Personnel Training   Training service and maintenance personnel is another task for the equipment management. Wherever specialization is feasible, individual job skill training will increase efficiency of personnel. Equipment dealers will assist equipment managers in personnel training, and provide literature and training aids.

Equipment Manager Authority   Decisions affecting the life and operating cost of machinery are not always controlled by the equipment manager. The authority of the equipment manager may or may not include, for example, the right to shut down a machine for repairs. However, it

Figure 9–3–5 Operator checking brakes and operating characteristics of front end loader

Figure 9–4–1 Welder repairing cracked dozer blade with mobile equipment

Figure 9-4-2 Completed weld repair on dozer blade

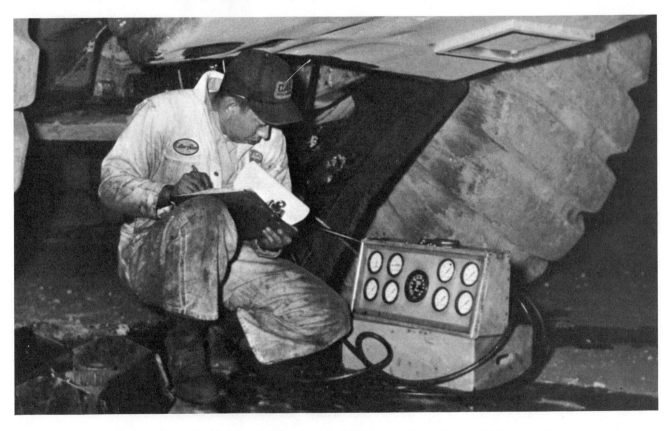

Figure 9-5-1 Serviceman checking power shift transmission with special instrument

Figure 9–5–2 Serviceman using special testing equipment for testing engines and other components

is always his responsibility to notify management of the possible consequences of abusing a machine, or running it when repairs are obviously required.

9-9. Lubrication and Fueling Personnel

Because lubrication and fueling are everyday activities, they are logically specialized areas (Figures 9-9-1 and 9-9-2). Further separation of lubrication responsibilities from fueling responsibilities will depend on the number of machines serviced. Not every machine owner will have so many machines that he can use a full-time lubrication specialist, but one person should carry this responsibility.

The areas of responsibility are:

- Fueling of machines
- Periodic lubrication
- Cleaning mud and debris from critical areas
- Recording fuel usage by machine
- Reporting service meter hours
- Reporting observed service needs

A strict definition of authority and responsibility is needed to make the system work. Local union contracts and practices vary from area to area. These differences must be considered in preparing job descriptions.

9-10. Inspection Personnel

In a large organization, maintenance inspection personnel would handle only inspections and adjustments (Figures 9-10-1, 9-10-2, and 9-10-3). For example, if power loss on a wheel tractor-scraper was reported, the inspection personnel would determine the cause and make the adjustments needed. If a turbocharger was near failure, inspection personnel would report this to the equipment manager. Mechanical personnel would exchange the unit. Here, too, job descriptions may vary due to crew size and other factors.

The areas of responsibility are as follows:

- Regular inspecting and adjusting
- Scheduling machines for maintenance checks at most opportune times

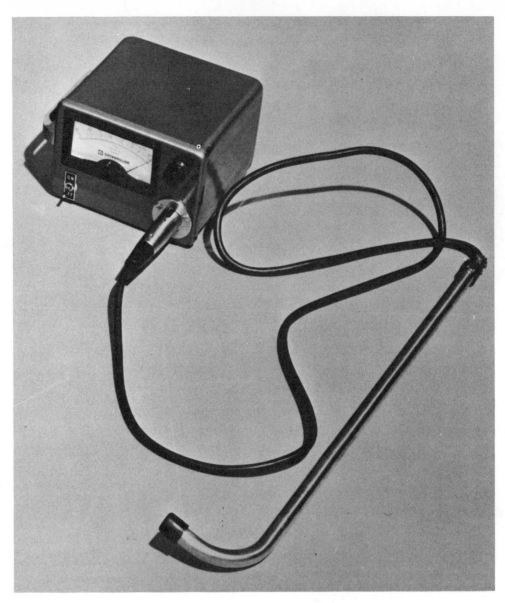

Figure 9–5–3   Air flow meter for checking radiator cores

- Advising the equipment manager of apparent service needs
- Keeping records of their work on machines

9-11. Contributing Personnel

Even though it is not his primary assignment, the machine operator is an integral part of an equipment maintenance organization. His daily operator's report will most likely contain remarks on machine performance and early indications of possible failures. Since every operator wants "his" machine to perform as well as possible, he rarely fails to report needed maintenance or repair work.

In the framework of this discussion, the operator, except in very small organizations, has only a reporting function in regard to machine maintenance. However, his reports are of vital importance. They are the first step toward prevention of repairs and costly downtime.

In a similar way, the operations foreman is in the position to detect and report early indications of maintenance

or service needs, such as reduction of machine production, slow machine, etc.

9-12. Field Maintenance Equipment

Requirements for field maintenance equipment will vary greatly with the size and location of the site, the type of heavy machinery used, and the work schedule of the operation.

On small jobs, a contractor may use a single truck for fuel, lubricants, tools, filters, water, and all other supplies. The expense of on-site service facilities (Figures 9-12-1 and 9-12-2) is warranted only when the project is of sufficient size and/or duration, or is too remotely located to be serviced from a home base.

Whenever on-site service facilities are planned, they must be located a safe distance from other structures— away from the job traffic to avoid dust and congestion, yet easily accessible. Such stations may include fuel storage

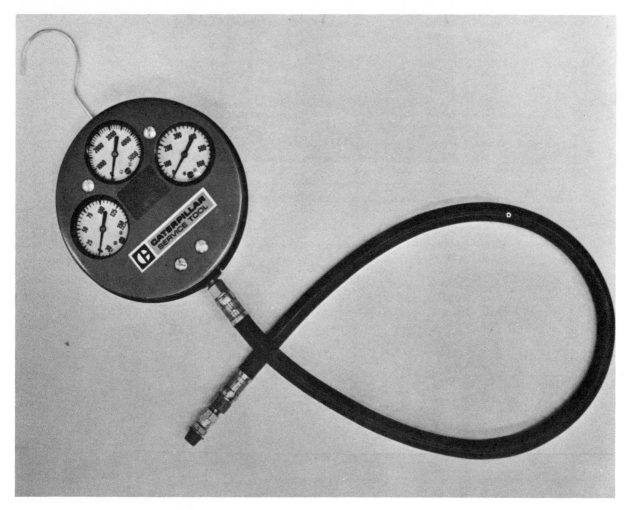

Figure 9–5–4   Multipressure indicators for checking hydraulic controls and transmissions

tanks, wash racks, grease pits, tire change equipment, and welding facilities. Such a maintenance facility may reduce the need for such mobile equipment as fueling and lubrication trucks.

On large earth-moving jobs, and especially on around-the-clock operations, mobile maintenance equipment is the best solution.

Fueling Trucks   These will be necessary if equipment is operating so far from the base camp that it is impractical to bring it in for refueling. The fueling truck (Figures 9-12-3 and 9-12-4) should be equipped with high capacity fuel pumps to reduce filling time. This is particularly advantageous if earth-moving machines are equipped with fast-fill systems (dry break). To aid night refueling, an engine-driven generator and flood lights are needed. The size and type of the fuel truck will be dictated by the number and type of machines to be serviced, their distance from the base of operations and from each other, and the terrain over which the truck must operate.

Lubrication Trucks   These are practical in the same circumstances as those requiring fueling trucks. On small jobs or where small equipment is used, it is common to have one truck equipped to provide both fuel and lubrication service (Figure 9-12-5). On large jobs with bigger equipment involved, it is usually practical to use separate trucks.

The lubrication truck should be equipped to provide all oils and greases and other expendable items, such as filter elements, required for routine maintenance of the equipment fleet (Figure 9-12-6). Lubricants required in large quantities, such as engine and hydraulic oil and cooling water or antifreeze solutions, should be carried in built-in tanks of suitable size.

Dispensing equipment may be mounted on the drums of gear oil, chassis grease, and other lubricants which are normally dispensed in small quantities. Individual air-powered pumps are normally used to dispense these lubricants. Proper pumps are needed to provide high-volume, low-pressure flow for low-viscosity oils, and the low-volume, high-pressure delivery to dispense heavy greases. Meters for recording consumption can be attached to the pumps. They permit keeping detailed maintenance records on each machine serviced.

The truck's air compressor provides power for the delivery pumps. It may also be used to inflate tires, run impact wrenches, and blow dirt off machines. For night operations, an electric generator can be installed to power floodlights. An air- or electric-powered, small capacity, high-pressure pump, attached to a 50- or 75-gallon reservoir of water and detergent, will provide spot cleaning

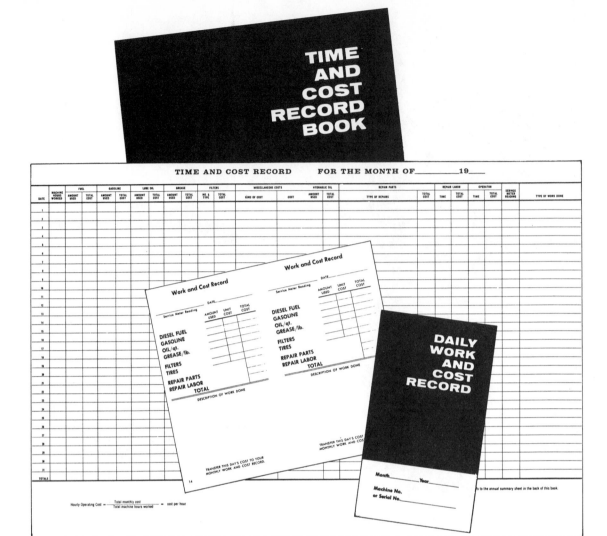

Figure 9–6–1   Equipment time and cost records

for lubrication or for inspection of highly stressed areas such as scraper goosenecks. The truck should have storage space for engine and hydraulic oil filters, air cleaner elements, and similar expandable parts.

9-13.   Systems Analysis and Testing Equipment

Regular analysis of all major machines systems and components tells the owner the condition of his machinery. Frequently, the same test equipment is used for troubleshooting, for determining the cause of a failure, and for operational checks after reconditioning. Therefore, the test equipment is utilized in field maintenance as well as in the service shop.

Some system analysis tests are simple; some are complex and technical; sometimes a detailed procedure requires only a few special tools; but all can help to determine machine condition or find the cause of a malfunction without the need for removal or disassembly of the component.

Most testing equipment is developed by equipment manufacturers for use on their products. For this reason, the equipment dealer makes frequent use of it and has developed know-how in system or component testing. By con-

tracting machine checkups out to the dealer specialist, the machine owner can minimize his tool investment and get reliable results.

Some examples of system analysis procedure and equipment follow.

Pressure Tester for Hydraulic Tranmissions   Test boxes are connected to pressure taps, pump line and tachometer drives. Accurate pressure gauges reveal individual clutch pressures, shift points, and converter charging pressures. Actual or impending failures can be located, and the need for adjustments or repairs determined.

Hydraulic Flow Meter   This test equipment is used to determine the operating condition of a hydraulic implement system (Figure 9-13-1). By measuring the oil flow at certain speeds and varying pressures, the condition of pumps, control valves, check valves, and cylinders of various circuits can be checked.

Engine Test Equipment   Equipment of this type is diversified and sophisticated. With the use of appropriate testers, rack settings, fuel consumption, smoke emission, low and high idle speeds, torque characteristics, injec-

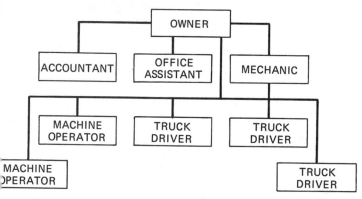

Figure 9–7–1 Organization chart: 1- to 20-man contracting operation

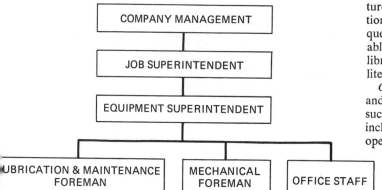

Figure 9–7–2 Organization chart: 20- to 100-man contracting operation

tion timing, manifold pressures, turbocharger performance and other engine data can be determined and analyzed.

Cooling System Testers These test groups determine air and water flows, coolant temperatures, flow speeds, and thermostat condition.

Undercarriage Inspection Tools To check the alignment and wear of track-type undercarriage parts, most equipment manufacturers provide special tools and procedures for their products. When used by experienced personnel, critcal wear parts can be monitored, their service life closely predicted, and replacement needs scheduled.

Other Test Equipment Included are analyzers for electrical systems, tire wear gauges, oil sample testers, battery hydrometers, and many others.

9-14. Maintenance Reference Material
All manufacturers of construction machinery provide literature which contains necessary information for the operation, maintenance, and service of the machine. Too frequently, this vital information is lost, stolen, or not available to personnel who need it. A well-organized reference library is the best insurance against this. Replacement literature can be obtained through the equipment dealer.

Operating instructions cover proper starting, stopping, and operating procedures. In addition, all prestart checks such as fuel, oil, water, and "walk-around" inspection are included. These are important, because proper machine operation is the first step of preventive maintenance.

Figure 9–9–1 A 15 cubic yard loader being serviced on the job by a lubrication and fueling crew

Figure 9–9–2 A 15 cubic yard loader being serviced by a field maintenance crew

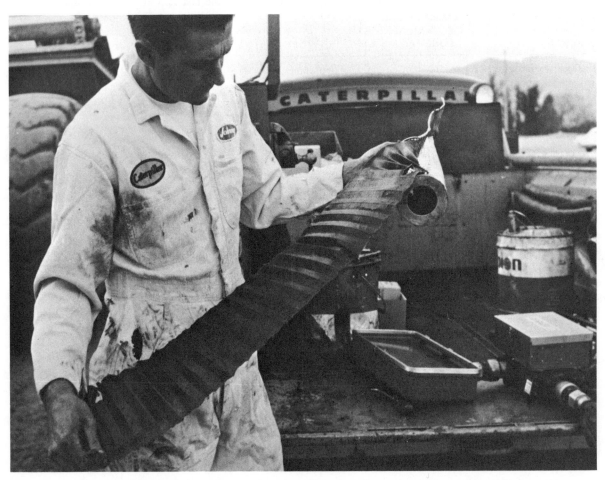

Figure 9–10–1 Serviceman examining filter elements removed from equipment

Figure 9–10–2 Serviceman in the control seat recording his findings after inspection of a piece of equipment

Figure 9–10–3 Servicemen examining the hub of drive wheels for bottom dump off highway truck

Figure 9–12–1   On-site maintenance base set up in place

Figure 9–12–2   Bins and repair parts in on-site warehouse

Figure 9–12–3   Typical job fueling truck

Figure 9–12–4   Motor scraper spread lined up and being refueled by fuel truck

Figure 9–12–5   Combination fueling and grease truck with lubrication hoses used to service parked equipment

Figure 9–12–6   Typical lubrication or grease truck for on-site lubrication

Figure 9–13–1 Hydraulic pumps, valves, and cylinders being checked in the field with a hydraulic oil flow meter

Lubrication and maintenance instructions provide the field maintenance personnel with detailed information concerning all scheduled service recommended for the machine In addition, the types of lubricants, oils, fuel, coolants and filters are specified.

Warning and instruction plates are usually mounted on the machine, in clear view. They serve to identify various tanks, valves, pressure taps, points of danger, and access covers. On properly maintained machines, such plates are kept clean and are replaced when lost or damaged.

Operator's and Maintenance Report Sheets Whenever possible, equipment maintenance should not be left to the memory of the maintenance personnel. Detailed service check lists used at each service interval assure the most satisfactory results. Examples of such report forms are illustrated and discussed subsequently.

Safety Instructions Unless properly performed, heavy equipment maintenance can be a dangerous business. Safety instructions found in reference literature and on warning plates must be closely observed. The field maintenance personnel should be thoroughly briefed on each new piece of equipment before they start to service a machine. The equipment dealer is the best source for safety procedures.

PREVENTIVE MAINTENANCE

Preventive maintenance is the term applied to all phases of field maintenance when they are properly planned, properly performed, and properly recorded. Only with a well-organized program is a machine owner able to monitor the condition of his equipment, to predict production based on machine availability, and to schedule repairs and overhauls in advance. Analysis of accurately compiled maintenance records supplies him virtually all information needed for performance evaluation, repair or replacement decisions, and cost-of-operation reviews.

Previously, this chapter discussed the actual maintenance work as well as the organization, personnel, and equipment needed to maintain heavy equipment in the field. Subsequently, the methods of controlling, recording and analyzing the cost and effectiveness of field maintenance will be discussed. Several preventive maintenance programs are included.

The days of the hit or miss "grease-monkey" are history. Machine owners now recognize that preventive maintenance is more economical than machine repair. Under an idealized preventive maintenance program, remedial or emergency repairs in the field would be reduced to zero.

CATERPILLAR

- PLANNED
- PREVENTIVE
- MAINTENANCE

MACHINE HISTORY FOLDER

NAME _____

ADDRESS _____

UNIT NO. _____ SERIAL NO. _____

PURCHASED FROM _____ DATE _____

ADDRESS _____ WARRANTY _____

ORIGINAL ENGINE	ENGINE REPLACEMENT	TRANSMISSION	FINAL DRIVE
Make			
Model			
Serial			

CAPACITIES

Cooling System _____ gal.   Hydraulic System _____ gal.   Tandem Drives _____ gal.

Engine Crankcase _____ qts.   Oil Clutch _____ qts.   Gear Housings _____ qts.

Transmission _____ gal.   Differential _____ qts.

Torque Converter _____ qts.   Final Drives _____ gal.

CABLE SPECIFICATIONS

Hoist: Length _____ Dia. and Type _____

Hoist: Length _____ Dia. and Type _____

Hoist: Length _____ Dia. and Type _____

TIRES

Size _____ Front _____ Back _____ Pressure _____ Front _____ Back

G. E. T. PART NUMBERS

CUTTING EDGES	BITS	BUCKETS	RIPPERS
Center _____	End _____	Tip _____	Tip _____
	Router _____	Adapter _____	Protector _____
Ends _____	Overlay _____		

FILTER PART NUMBERS

Crankcase _____ Steering Clutch _____ Hydraulic System _____

Fuel _____ Torque Divider _____ Air Cleaners _____

Transmission _____ Final Drive _____

PE017058-00

Caterpillar, Cat and ▨ are Trademarks of Caterpillar Tractor Co.

PRINTED IN U.S.A.

MAINTENANCE AND REPAIR COST RECORD

FOR (Month)	SERVICE METER READING (When Performed)	DESCRIPTION	TOTAL COST

MONTHLY MACHINE AVAILABILITY

Date	Hours Worked	Hours Available for Work	Possible Working Hours	Standby Hours	Wait-for-Repair Hours	Availability %	Utilization %
Jan							
Feb							
Mar							
Apr							
May							
June							
July							
Aug							
Sept							
Oct							
Nov							
Dec							

NOTE: Hours available includes time to perform preventive maintenance.

AVAILABILITY = Hours Worked
_____ Possible working hours
minus Wait-for-repair hours
minus Standby hours

UTILIZATION = Hours Available
_____ Hours Worked

Multiply results by 100 to get % figure

Figure 9–16–1

CATERPILLAR
● Planned
● Preventive
● Maintenance

MECHANIC'S DAILY MAINTENANCE REPORT

Wheel Loader

(√) = O.K.   (X) = Adjustment Made

(R) = Repair Needed and Not Completed

In the Case of Fuel, Oil or Lubricants, Enter Amount Added

LIST AT BOTTOM OF SHEET AN EXPLANATION OF REPAIRS MADE

OR REASON THEY WERE NOT MADE

NOTE: Clean All Caps of Dirt Before Filling Tanks, Radiators, Engine, Transmission, Differential, Reservoirs

NOTE: Wipe All Lube Fittings Before Lubing — Keep Lube Nozzle Clean

NOTE: Do Not Use Rags, Cans, Wooden Plugs or Any Other Than Correct Tank Cap

NOTE: Use Only Caterpillar Recommended Lubricants

Date _____ Unit No. _____

Serial No. _____ Service Meter Reading _____

1. Check Oil Level of Transmission and Final Drives and Look for Leaks (Amount Added _____) ☐
2. Check Oil Level of Hydraulic System and Look for Leaks (Amount Added _____) ☐
3. Check Oil Level of Starting Engine Crankcase (Amount Added _____) ☐
4. Check Oil Level of Diesel Engine Crankcase (Amount Added _____) ☐
5. Fill Fuel Tank (Amount Added _____) ☐
6. Check Cooling System Hoses, Coolant Level and Look for Leaks, Radiator Trash ☐
7. Check Battery Terminals and Ammeter (Charge Rate) ☐
8. Check Condition of Turbocharger ☐
9. Drain Air System Reservoirs ☐
10. Lubricate Steering Hydraulic Cylinder Bearings ☐
11. Lubricate Bucket Control Cylinders and Linkage Bearings ☐
12. Lubricate Bucket Positioner Bearings ☐
13. Lubricate Bucket Positioner Telescoping Tube ☐
14. Lubricate Lift Cylinder Trunnion and Piston Rod Bearings and Lift Arm Bearings ☐
15. Lubricate Steering/Frame Upper Pivot Bearings ☐
16. Make Visual Inspection of Equipment ☐
17. Make Necessary Repairs ☐

After First 10 Hours of Service:

Change Starting Engine Crankcase Filter Element ☐

REMARKS _____

Mechanic: _____

Oiler: _____

Figure 9-16-2

Regular inspections, adjustments, and preventive repairs would detect problems and prevent failures. All repair work could be scheduled at the most opportune time.

9-15. Preventive Maintenance Programs

There is no single preventive maintenance program that fits every job, and there is no single way to do the job. New ideas, equipment and methods are introduced continually. The goal, however, remains the same: Elimination of premature equipment failure by means of complete field maintenance.

Preventive maintenance programs of any scope require teamwork and a master plan. However, each equipment application in the field will require modifications of this plan to match the working conditions, size of operation, and type of equipment. Every phase of a preventive maintenance program must be carefully planned, necessary reporting forms prepared, and each assignment clearly defined. Personnel must be trained on the equipment and ready to perform their duties.

The following preventive maintenance plans are selected examples which range from simple cost and time bookkeeping forms to computerized equipment control systems.

9-16. Plan I—Planned Preventive Maintenance Packet

This packet consists of a complete set of reporting and recording forms and is available from Caterpillar dealers (Form PEO17057). Based on practical experience in the field, the packet helps to organize the needed information on several forms that permit detailed recording of trends, costs, and other pertinent data. Although the packet is not an actual procedure for a preventive maintenance program, it will assist in planning and implementing programs.

The packet consists of the following forms:

Machine History Folder (Figure 9-16-1) The front cover provides space for owner and machine identification, capacities, specifications, and for numbers of frequently needed filter elements and machine parts. On the back cover, space is provided for listing major repairs and for computing machine availability. The folder is used to store the forms listed as follows.

CATERPILLAR
- Planned
  - Preventive
    - Maintenance

MECHANIC'S 50-HOUR MAINTENANCE REPORT

Wheel Loader

(√) = O.K. (X) = Adjustment Made

(R) = Repair Needed and Not Completed

LIST AT BOTTOM OF SHEET AN EXPLANATION OF REPAIRS MADE

OR REASON THEY WERE NOT MADE

NOTE: Clean All Caps of Dirt Before Filling Tanks, Radiators, Engine, Transmission, Differential, Reservoirs

NOTE: Wipe All Lube Fittings Before Lubing — Keep Lube Nozzle Clean

NOTE: Do Not Use Rags, Cans, Wooden Plugs or Any Other Than Correct Tank Cap

NOTE: Use Only Caterpillar Recommended Lubricants

Date _____ Unit No. _____

Serial No. _____ Service Meter Reading _____

1. Drain Fuel Filter Element Housing ☐
2. Clean Air Induction System Precleaner ☐
3. Check Electrolyte Level of Battery ☐
4. Check Condition and Pressure of Tires ☐
5. Check Adjustment of Cooling System Fan Belts ☐
6. Lubricate Drive Shaft Spline ☐
7. Lubricate Rear Axle Trunnion Bearings ☐
8. Lubricate Steering Rod Sockets, Steering Link Rear and Front Sockets, Steering Intermediate Bellcrank, Steering Booster Rear and Front Sockets and Control Valve/Rod Socket (if applicable) ☐

After First 50 Hours of Service:

1. Change Transmission Filter Element and Wash Screens ☐

REMARKS _____

Inspection By: _____

Master Mechanic: _____

Equipment Superintendent: _____

Figure 9–16–3

Mechanic's Daily Maintenance Report (Figure 9-16-2)

Mechanic's Periodic Maintenance Report (50, 100, 500, etc., hours) (Figure 9-16-3) A series of forms is available for each major type of machine, such as track-type tractors, wheel loaders, or motor graders. Each sheet specifies all points to be lubricated and inspected. There are separate sheets for each recommended service interval, such as 50, 100, 500 hours, etc.

Operator's Daily Report (Figure 9-16-4) Directed to the shop foreman and management, this report is filled out daily by the operator. It provides space for listing operating deficiencies.

The Service Work Order (Figure 9-16-5) On this form, specific work requests are to be listed. Parts and service time needed are listed in the boxed areas.

Equipment Condition Report (Figure 9-16-6) Used for a general machine inspection, this form notifies the management of the machine's general condition.

Equipment Repair Report (Figure 9-16-7) This report accompanies any condition report which calls for needed service or repair work. The form provides a simple, organized means of listing the repairs and the parts needed.

Equipment Efficiency Report (Figure 9-16-8) This form is used to complete the overall performance of each piece of equipment. It makes possible comparison of individual machines with others in the fleet.

The forms contained in this "Planned Preventive Maintenance Packet" are widely and successfully used by small-to-medium-sized contractors. All information needed for a workable preventive maintenance program can be recorded on them leaving the means of evaluations to the individual contractor.

9-17. Plan II—The Tag and Chart Program

All preventive maintenance programs have the same goal, yet approaches differ. Whereas Plan I stresses a complete reporting and recording system, Plan II emphasizes timely performance of needed maintenace.

The Tag and Chart system is well accepted by operators and maintenance people who have used it. It is especially helpful if operators are inexperienced. This program can easily be combined with Plan I. Charts are based on the forms provided in the "Planned Preventive Maintenance

CATERPILLAR

- Planned
- Preventive
- Maintenance

OPERATOR'S DAILY REPORT

Date _____ Unit No. _____ Serial No. _____

Before operating equipment, make the following checks:

Radiator _____ ☐

Oil Levels _____ ☐

Visual Inspection _____ ☐

Service Meter Reading beginning of shift _____

Service Meter Reading at end of shift _____

Standby Hours _____

Wait-For-Repair Hours _____

Total Hours Operated _____

OBSERVATIONS _____

_____

_____

Operator's Signature: _____

Figure 9–16–4

Packet," and tags remind the operator when to call in the maintenance crew.

To implement this maintenance program, three items are needed for each machine (Figure 9-17-1).

- Periodic Service Record Cards
- Machine I.D. Tags
- Hourly Service Reminder Tags

Large 6″ × 16″ plastic laminated cards have printed on them the service intervals for the respective machine, followed by spaces for entering the date of the service work done, and the signature of the serviceman performing it. Stored in the machine, this service record is readily accessible.

The "Machine I.D. Tag" is attached to the key ring of the disconnect key. Also attached to this key ring is the respective "Hourly Service Reminder Tag."

Plan II works like this: Whenever the hours recorded

on the service meter approach the hours on the "Hourly Service Reminder Tag" (within 4 to 5 hours), the operator gives the key ring, along with the "Periodic Service Record" card, to the equipment foreman. Maintenance personnel perform the specified work for that particular service meter reading (1125 hours, for example) and enter the date of the service work on the textured space provided on the "Periodic Service Record," along with their signature. Then, they remove the 1125-hour tag from the key ring and attach the 1250-hour tag. If additional work is required, a supplemental work order is filled out and placed in the machine history folder.

9-18. Plan III—The On-Board Computer by Bissett-Berman

Human error in equipment maintenance accounts for a substantial share of machine breakdowns. To avoid this problem, electronic equipment is used to an increasing ex-

CATERPILLAR Planned Preventive Maintenance

SERVICE WORK ORDER

WORK ORDER NO. _____ AUTHORIZED BY _____ DATE _____

Unit No. _____ | Serial No. _____ | Service Meter Reading _____

INSTRUCTIONS: _____

MATERIAL

QUAN.	PART NO.

OUTSIDE EXPENSE				DESCRIPTION OF WORK COMPLETED	DATE
QUAN.	PART NO.	DESCRIPTION	PRICE		

Total Outside Expense

INTERNAL EXPENSE

Total Internal Exp.

TOTAL LABOR _____
TOTAL PARTS _____
TOTAL OUTSIDE EXPENSE _____
TOTAL _____

TOTAL

Figure 9–16–5

CATERPILLAR
- Planned
  - Preventive
    - Maintenance

EQUIPMENT CONDITION REPORT

Date _____   Unit No. _____   Serial No. _____   Service Meter
Reading _____

REPAIRS NEEDED: _____

_____

_____

_____

_____

_____

_____

_____

_____

_____

_____

_____

_____

_____

_____

_____

RECOMMENDATION _____

_____

_____

_____

_____

_____

_____

Inspection By: _____

Approved By: _____

      Master Mechanic   Equipment Superintendent

Figure 9–16–6

tent. Whereas Plan II uses reminder tags to alert the operator to initiate periodic maintenance activities, an electronic timing device is utilized in Plan III for the same purpose (Figures 9-18-1 and 9-18-2).

The computer is mounted either on the outside of the machine or inside the cab (Figures 9-18-3 and 9-18-4). Buzzers, flashing lights, bells, or even sirens serve as an alarm to attract the operator's attention. Flashing lights, mounted on machine cabs, can signal service trucks while the machine is completing its work cycle.

The on-board computer is primarily a timing device (Figure 9-18-5) which, at a prescribed number of operating hours, warns the operator that the machine needs service. The serviceman lifts the cover on the computer and removes the top punch card which lists all service work required at that time. The same information has been

● Planned
　● Preventive
　　● Maintenance

EQUIPMENT REPAIR REPORT

Date _____ Unit No. _____ Serial No. _____

Repairs Made at _____ by _____

Date Listed on Equipment Condition Report _____Service Meter Reading _____

Date Started _____ Date Completed _____ Man Hours to Complete _____

NEW PARTS INSTALLED

Item	Make	Part Number	Source	Cost

PARTS RECONDITIONED

REMARKS (Special Attention to Job Superintendent) _____

_____

_____

_____

_____

_____ Signed: _____

NOTE: A copy of Equipment Condition Report on which this repair was originally listed should always accompany this form.

Figure 9-16-7

  • Planned
  • Preventive
  • Maintenance

Job No. _____

Date _____

EQUIPMENT EFFICIENCY REPORT

For the Month of _____ , 19 _____

Unit Number	Serial Number	Service Meter Reading	(A) Possible Working Hours	(B) Hours Available for Work	(C) Hours Worked	(D) Standby Hours	(E) Wait-For-Repair Hours	(F) Repair Hours	(G) Availability % $\frac{C}{A \cdot D \cdot E}$	(H) Utilization % $\frac{C}{B}$	Total Cost for Month of _____

Equipment Superintendent _____   Job Superintendent _____   Home Office _____

Figure 9–16–8

PLAN II

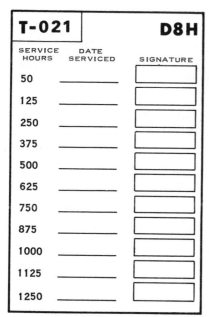

T-021		D8H
SERVICE HOURS	DATE SERVICED	SIGNATURE
50	_____	
125	_____	
250	_____	
375	_____	
500	_____	
625	_____	
750	_____	
875	_____	
1000	_____	
1125	_____	
1250	_____	

PERIODIC SERVICE RECORD

T-021

MACHINE
I. D. TAG

1125 HOURS

HOURLY SERVICE
REMINDER TAG

Figure 9–17–1 Periodic service record and typical tags for hourly service reminder time

Sentry/100 SERVICE COMPUTER DATA CARD

EQUIP ....JOB............ CAT 657 SCRP   ENG HRS............   SERVICE INTERVAL....

★ ★ Change ★ ★

....Diesel Engine Oil & Filter
....Scrp Hyd Sys Filter Element
....Transmission Oil Filter

★ ★ Lubricate ★ ★

....Belt Adjusting Pulley Brackets
....Brake Camshaft & Anchor Pins
....Fan, Hub & Adj Pulley Bearing
....Instrument Panel Bracket
....Steering Column Upper Brg
....Water Pump Bearings

★ ★ Check ★ ★

....Battery Level & Terminals
....Differential Oil Level
....Hydraulic Retarder Oil Level
....Planetary Oil Level
....Steering Gear Shft Lwr Brg

★ ★ Wash ★ ★

....Diesel Eng Crankcase Breather
....Differential Breather
....Hyd Retarder Breather

BISSETT BERMAN
2941 NEBRASKA AVE. • P.O. BOX 655 SANTA MONICA, CALIF. 90406

DATE SERVICE:_____   BY:_____   ☐ NOTE EXCEPTIONS ON BACK

© THE BISSETT-BERMAN CORPORATION 1967   IBM L36334

Figure 9–18–1 Service computer data cards for motor patrol

"punched" into the card. After completing the service work, the serviceman signs the card and sends it to the main office, where it is "read and recorded" onto a computer tape. If additional work is required, there is space on the card to indicate this. When received by data processing at the main office, this additional information is punched into the card before it is read and recorded.

This system is the most advanced in its field. The on-board computer is rugged and built of solid-state components. An office computer system can also be utilized for recording other pertinent information, such as accidents or unscheduled maintenance. Such data, however, have to be punched separately, since they cannot be recorded on the on-board computer. The final tape in the office will contain all historical information for fast recovery.

9-19. Plan IV—Total Computer Control System; Equipment Failure and Repair Data Retrieval System

This system utilizes accurate information recorded on prepared forms by the field operations foreman and by the shop or maintenance foreman. Processing the forms by computer makes results quickly available for easy analysis.

One of the outstanding features of this system is that it eliminates the human tendency to account for only the major machine failures. A substantial amount of downtime may result from minor failures. By charging each and every failure with its downtime, evaluation of maintenance program and equipment quality is easier and more thorough.

Sentry/100 SERVICE COMPUTER DATA CARD

EQUIP ....JOB............ CAT 14 MTR GR   ENG HRS............   SERVICE INTERVAL....

★ ★ Change ★ ★

....Diesel Engine Oil & Filter
....Hyd Filter Element
....Transmission Oil Filter

★ ★ Lubricate ★ ★

....Fan, Hub Adj Pulley Bearing
....Front Wheel Bearing
....Hyd Pump Shaft Brg

★ ★ Check ★ ★

....Battery Level & Terminals
....Blade Lift Control Hsgs Oil
....Brake Master Cylinder Fluid
....Circle Center Shift Control Hsg
,...Circle Reverse Control Hsg Oil
....Front Wheel Lean Control Hsg
....Power Control Hsg Oil
....Pwr Control Shft Wrm & Gear Hsg
....Steering Gear Hsg Oil
....Tandem Drive Hsg Oil
....Transfer Gear Hsg Oil
....Transmission Oil Level

★ ★ Wash ★ ★

....Diesel Eng Crankcase Breather
....Tandem Drive Hsg Breather
....Trans Hsg Breathers

BISSETT BERMAN
2941 NEBRASKA AVE. • P.O. BOX 655 SANTA MONICA, CALIF. 90406

DATE SERVICE:_____   BY:_____   ☐ NOTE EXCEPTIONS ON BACK

© THE BISSETT-BERMAN CORPORATION 1967   IBM L36334

Figure 9–18–2 Service computer data card for scraper

Figure 9–18–3 "On-board computer" installation of equipment on inside of cab

The information input from the field is recorded on three form sheets: the Operating Foreman's Daily Machine Record, the Down Equipment Status Report, and the Repair Maintenance Record.

The *Operating Foreman's Daily Machine Record* (Figure 9-19-1) records each machine's activity for each day. The time record is divided into two major parts, "Available Hours" and "Down Hours." Within each part, Idle Time may be recorded. Total Idle Time must be broken down into "Idle—Available" and "Idle—Repair." "Available Hours" is defined as the total hours the machine is available for use. Under available hours, specific activities which affect service life can be coded, thereby permitting more consistent evaluation of machine performance.

Each machine must be accounted for 24 hours per day, or the computer will report an error. The computer will also report the failure of listing a machine for a respective day.

For example, a tractor used only for pushloading broke down at 11:30 a.m. The machine sat idle for an hour before repair work was started and first-shift serviceman worked on it 1.5 hours. The second shift performed 8 hours of repair work. There was no third shift. On the next day, first-shift servicemen finished the repair and the machine was returned to work. The machine was operated for 8 hours on the second shift and was idle on the third

shift. This sample form (Figure 9-19-1) has been consolidated to show the two-days of activity for this machine only. Normally, the activities of all machines on one day would be shown here—arranged by equipment category, date, and shift.

The Operating Foreman's Daily Machine Record can be shortened by printing on it only those codes that apply to the types of equipment that will be listed on that particular sheet. Separate forms may be used for loaders, tractors, hauling units, or other categories of equipment. Refer to the Operating Vehicle Activity Code (Figure 9-19-2) for complete listing.

The *Down Equipment Status Record* (Figure 9-19-3) is used by the shop or maintenance foremen who assigns work order numbers. In the previous example, a failure occurred during the first shift. Because no mechanic was available, 1 hour was lost. One-half hour was charged to cleaning, and 1 hour to repair. The second shift reported 8 hours of repair, etc.

On the *Repair-Maintenance Record* (Figure 9-19-4), each failure is assigned to a particular machine area, and its cause is identified. The type of repair, how the repair was done, and where, are all recorded. For accuracy, a separate record sheet is made out for each coded machine area that is repaired. For example, the drive shaft and the differential were damaged and subsequently repaired

Figure 9–18–4   "On-board computer" installation of equipment on outside of cab

at the same time. Since these two repairs fell into separate machine area codes, one record sheet for each was made out, each giving all pertinent information.

The *machine area codes* are broken down into seven major categories (Figure 9-19-5)

100—Engine
200—Power Train
300—Undercarriage
400—Miscellaneous
500—Hydraulics
600—Linkage
700—Ground Engaging Tools

Computer Output   Any computer output, of course, is only as accurate and informative as the input data. The following information is most commonly deemed useful:
Machine Status Report
    Breakdown of Hours of Work in Each Application
    Breakdown of Machine Utilization
    Hours Worked
    Hours Repaired
    Hours of Preventive Maintenance
Repair-Maintenance Record
    Breakdown by Machine Type
    Consumables Usage
    Fuel and Oil Consumption by Machine
    Although it may appear complicated, this system is rela-

tively simple to implement. It is very versatile. Information on failure rates, mean time of failure, mean time of repair, and other valuable data can be easily obtained through this system.

PREVENTIVE MAINTENANCE SOURCES

In the past, many equipment users, no matter how large the operation, felt they could implement preventive maintenance with their own personnel. Unfortunately, planning, managing and control of the preventive maintenance program was frequently left to the experience and memory of one or two men. As a result, preventive maintenance practices were less effective than anticipated.

Today, more complex and productive construction equipment requires a well-organized preventive maintenance program. Complete records are especially important. With costs for each machine recorded, guesswork and personal opinion have given away to factual cost analysis.

More equipment owners are realizing that preventive maintenance is much more economical than machine repair. Yet preventive maintenance is an investment that must be managed and controlled. Hiring a preventive maintenance specialist is a practical way to accomplish this economically. If a man capable of being a specialist is not employed by the construction company, his services can be purchased.

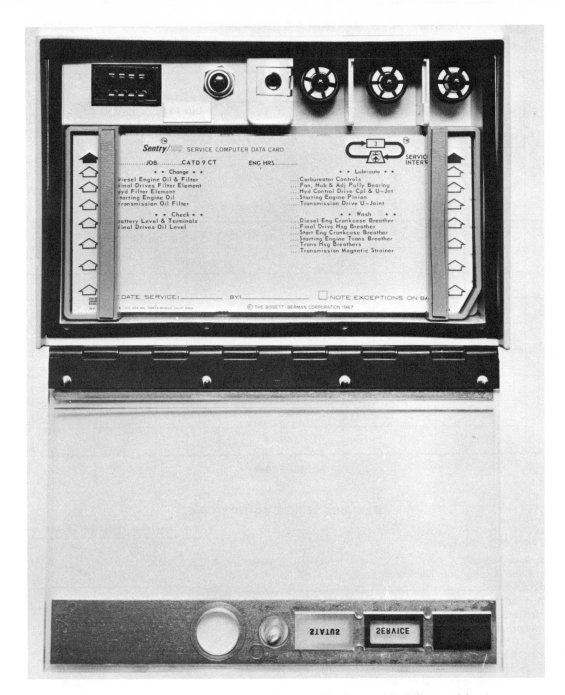

Figure 9–18–5 Close-up of "on-board computer" with computer data card in place

Purchased maintenance service appeared first in the tire industry. Construction equipment users purchased tire service contracts because they recognized the economy of competent tire maintenance and their own lack of knowledge in that particular field. A similar trend has developed in equipment maintenance. Today's equipment user finds it difficult to be competent in all phases of maintenance. It is frequently to his advantage to rely on the outside specialist.

9-20. Machine Owner

Smaller contractors, owning ten or fewer machines and operating within a 50-mile radius of their home facilities, may prefer to handle preventive maintenance with their own personnel. A fleet this size does not permit extensive job specialization of maintenance personnel nor permit diversity of maintenance equipment. A small crew can familiarize itself with the fundamentals necessary for field maintenance.

System analysis, however, is the exception. This should be left to a specialist, such as the equipment dealer. Since smaller owners must rely on outside sources for most repair and overhaul work, it is logical for them to seek outside sources for such specialized work as system analysis. For these owners an inspection report contract with the eqiupment dealer is often the best program.

Companies operating larger equipment fleets or working on jobs far away from home facilities should carefully

OPERATING FOREMAN'S DAILY MACHINE RECORD

TRACK-TYPE TRACTORS

1ST SHIFT FOREMAN _____ 2ND SHIFT FOREMAN _____ 3RD SHIFT FOREMAN _____

DATE	EQUIPMENT I.D. NO.	SHIFT	AVAILABLE HOURS BY ACTIVITY							DOWN HOURS BY ACTIVITY				FAIL TIME		ENG. ADDITIONS		
			BULLDOZING SOIL (71)	BULLDOZING SOIL PLUS 100' (73)	PIONEER (81)	PUSH LOAD (83)	NON-ENDURANCE (91)	OTHER	IDLE – AVAILABLE	REPAIR	SERVICE	NOT SCHEDULED	IDLE-DOWN	FIRST SHIFT	SECOND SHIFT	FUEL / SHIFT	OIL / SHIFT	REMARKS
1/30	TT-1082	1			5.5				1	.5	1		11:30		55	2		
		2							8									
		3											8					
1/31	TT-1082	1			3.5				4	.5								
		2			8													
		3					8											
		1																
		2																
		3																

Figure 9–19–1

OPERATING VEHICLE ACTIVITY CODE

Machines Code No. Applies to

Code No.	Description	Loaders	Tractors	Haulers	Carry Loader
11	Shot Rock Loading	X			
12	Shot Rock Loading into Trucks				X
21	Bank Loading	X			
22	Bank Loading into Trucks				X
31	Stockpile Loading	X			
32	Stockpile Loading into Trucks				X
41	Load + Carry + 100 Ft	X			
42	Load + Carry – 500 – 1000 Ft Haul				X
43	Self Loaded – 500 – 2000 Ft Haul			X	
44	Push or Top Loaded – 500 – 2000 Ft Haul			X	
52	Load + Carry – 1000 – 1500 Ft Haul				X
53	Self-Loaded – 2000 – 5000 Ft Haul			X	
54	Push or Top Loaded – 2000 – 5000 Ft Haul			X	
62	Load + Carry – 1500 + Ft Haul				X
63	Self Loaded – 5000 + Ft Haul			X	
64	Push or Top Loaded – 500 Ft Haul			X	
71	Bulldozing Soil – less than 100 Ft		X		
72	Stripping – Load + Carry – 1 Bucket				X
73	Bulldozing Soil – more than 100 Ft		X		
81	Pioneering		X		
82	Stripping – Load + Carry – 2 Buckets				X
83	Pushloading Scrapers		X		
91	Nonendurance Operation – Special Test	X	X	X	X
92	Specific Cycle – Engrg. Requested	X	X	X	
93	Shows + Engrg. Inspections	X	X	X	

Figure 9–19–2

DOWN EQUIPMENT STATUS RECORD

FOREMAN _____

DATE	EQUIP. NO.	SHIFT	WORK ORDER NO.	CLOCK TIME AT START OF REPAIR		HOURS SPENT ON:					
						ACTIVE REPAIR	WAITING ON PARTS	NO SHOP	NO MECHANIC	CLEANING	NO DIRECTION
1/30	TT-1082	I	34	12:30		1			1	.5	
		II	34			8					
1/31	TT-1082	I	34			4				.5	

Figure 9–19–3

analyze the cost of preventive maintenance. By purchasing a service contract they can eliminate the need for specialized equipment and facilities, avoid hiring and training extra personnel, and reduce management's burden of responsibility.

Big operations, possibly in remote regions, will require even more in-depth planning and maintenance expense evaluation. Tight time limits that require nonstop work schedules could turn field maintenance operations into pit-stop type service stations. For them, total service facilities must be provided. The type and amount of equipment used will be an important factor in planning.

Whatever the plan, it must be based on facts and figures. Past experience, knowledgeable personnel, and accurate bookkeeping are essential for success in preventive maintenance.

9-21. Maintenance Contracts

Service contracts, which have become increasingly popular, are of two basic categories:

(a) the inspection report contract, and (b) the full maintenance contract.

The *inspection report* contract covers the "regular inspection" and "system analysis" portion of field maintenance. Under this contract, the machine owner periodically receives a written report on each machine, stating its condition and maintenance or repair requirements. The report is based on periodic system analysis and inspection of machine components. With this information, the machine owner can decide what repair work will be done, and when. He may arrange for the work to be done in his own shop by his own personnel, or he may contract it to an outside source.

The contracted inspector, whether he be the equipment dealer or another contracted agent, could also be hired to make adjustments and/or minor repairs. Depending on the locale, perhaps an independent service shop could be hired to do certain repair work. The plan should be based on the cost and availability of service, as well as on service quality and/or time required.

The other type of service contract is the *full maintenance contract*. The degree of coverage by this type of contract varies with machine owner's needs and/or the dealer's or other maintenance contractor's capabilities (Figures 9-21-1 and 9-21-2). A broad total maintenance contract might cover fueling, lubrication, and all other phases of preventive maintenance, as well as remedial repair and overhaul (Figure 9-21-3). Some full maintenance contracts are billed on a cost-per-hour basis; others, on a maximum cost.

Contracted equipment service is becoming more popular because of its benefits, a primary one being cost. Machine owners who keep accurate records on maintenance and repair work are able rationally to weigh the advantages of

MACHINE AREA WORKED ON (CIRCLE ONE)	REPAIR – MAINTENANCE RECORD					

MACHINE AREA WORKED ON (CIRCLE ONE)
100. ENGINE
101. BASIC ENGINE
102. CYLINDER HEAD & VALVES
103. TURBO – AIR INDUCTION SYSTEM
104. FUEL SYSTEM
105. COOLING SYSTEM
106. ELECTRICAL SYSTEM
107. LUBE SYSTEM
108. STARTING ENGINE/MOTOR
200. POWER TRAIN
201. CLUTCH
202. TORQUE DIVIDER
203. TORQUE CONVERTER
204. AUXILIARY TRANSMISSION
205. MAIN TRANSMISSION
206. TRANSMISSION CONTROLS
207. DRIVE LINE – U–JOINT
208. DIFFERENTIAL – BEVEL GEAR
209. STEERING CLUTCH
210. STEERING CONTROLS
211. BRAKE SYSTEM
212. FINAL DRIVE
300. UNDERCARRIAGE
301. TRACKS – ROLLERS – IDLERS
302. TIRES
303. SUSPENSION
400. MISCELLANEOUS
401. MAIN FRAME
402. GUARDS – FENDERS
403. OPERATOR STATION
404. AIR CONTROL SYSTEM
405. HYDRAULIC CONTROL SYSTEM
500. ATTACHMENTS
501. WINCH – CABLE CONTROL UNIT
502. BUCKET – BOWL – BODY
503. RIPPER – SCARIFIER

DATE DOWN	EQUIP. NO.	WORK ORDER NO.	SMR	MAN HOURS TO REPAIR	MECHANICS INITIALS
1 / 30	TT – 1082	34			

WORK PERFORMED:

LIST PART WHICH YOU THINK CAUSED FAILURE:	PART NO.	DESCRIPTION

FAILURE CAUSE (CIRCLE ONE)	REPAIR – MAINTENANCE TYPE (CIRCLE ONE)	REPAIR SHOP CODE (CIRCLE ONE)
01. NORMAL WEAR	01. SCHEDULED REPAIR – SHOP	01. OWNER'S SHOP OR JOB
02. EXCESSIVE WEAR	02. SCHEDULED REPAIR – ON JOB	02. CATERPILLAR DEALER
03. BROKEN	03. UNSCHEDULED REPAIR – SHOP	03. I–H DEALER
04. CRACKED	04. UNSCHEDULED REPAIR – ON JOB	04. AC DEALER
05. CORRODED	05. PREVENTATIVE MAINTEN- ANCE	05. EUC. DEALER
06. BENT	06. INSPECTION	06.
07. SEIZED		07.
08. OVER HEATED		08.
09. LEAKS	REPAIR – MAINTENANCE ACTION (CIRCLE ONE)	09.
10. MALFUNCTION		10.
11. OVER CHARGING	01. REPAIR	
12. UNDER CHARGING	02. REBUILD	
13. HYD. PRESS. LOW	03. INSTALL	
14. HYD. PRESS. HIGH	04. ADJUST	
15. VIBRATION	05. CLEANING	
16. FOREIGN MATERIAL	06. PAINTING	
17. LACK OF POWER	07. OIL CHANGE	
18. WARPED	08.	
19. SLIPPING		
20. EXCESSIVE CLEAR- ANCE		
21. TOO HARD		

Figure 9–19–4

service contracts. The same records will help them to determine the extent of such contracts.

In considering a service contract, the equipment owner must gain a clear understanding of each party's obligations. He should insist that the service contractor explicitly describe the work to be done.

To the machine owner, the greatest benefit of a full maintenance contract is its known cost per hour of machine operation. It eliminates one variable cost factor and thus reduces risk. Knowing the cost of maintenance before submitting a bid is as important as knowing the cost of equipment and materials. Further discussion on service contracts and total cost bidding may be found subsequently under Equipment Maintenance Trends.

REPAIR AND OVERHAUL

This discussion of repair and overhaul deals with problems facing a machine owner after an equipment failure, or after normal service life of a machine has been achieved. The individual sections in this discussions cover types of repair and overhaul work, the organization and facilities needed, and the available sources of service work. Included are methods for calculating cost of repair and overhaul work

done in the owner's shop. The owner may thus compare cost with outside service sources. Parts inventory and costs are discussed.

9-22. Remedial Repair

Remedial repair is work needed to restore a component which has failed prematurely due to improper maintenance, lack of inspection and adjustment, or machine abuse. Remedial repairs often are needed immediately.

Proper equipment maintenance and inspection and regular system analysis can hold remedial repair to a minimum. The hydraulic cylinder wiper seal is a good example. Seal-wear is unavoidable. Therefore, as one part of preventive maintenance, the wear progress should be monitored and replacements made when necessary. If this is not done, the hydraulic pump, control valve, and cylinder will soon need repair because contaminants will have entered the hydraulic system. In this case, repair of the pump, valve, and cylinders constitutes remedial repair.

Essentially, remedial repair has two objectives: First is the rapid return of the equipment to operational condition; second is to determine the reason for failure and make correction to avoid a repeat failure. A typical example is a worn hydraulic pump due to dirt. Replacing the pump

MACHINE AREA FAILURE CODE LIST

Description	Track-Type Tractors and Loaders	Wheel Loaders		Hauling Units	
		Front	Rear	Front	Rear
Engine Area					
Basic Engine	101	101		101	111
Cyl. Head and Valves	102	102		102	112
Turbo-Air Induction System	103	103		103	113
Fuel System	104	104		104	114
Cooling System	105	105		105	115
Electrical System	106	106		106	116
Lube System	107	107		107	117
Air or Gas Start System	108	108		108	118
Power Train Area					
Drive Line—U-Joint	205	205		205	215
Differential—Bevel Gear	206	206		206	216
Steering Clutch	207				
Steering Controls	208				
Brake System	209	209	219	209	219
Final Drive	210	210		210	220
Drop Box				221	231
Tires		222	232	222	232
Track Links, Pins, Shoes	223				
Rollers	224				
Idlers	225				
Track Roller Frame	226				
Transmission Area					
Main Transmission	305	305		305	315
Controls	306	306		306	316
Pumps	307	307		307	317
Converter	308	308		308	318
Torque Divider				309[a]	
Miscellaneous					
Sheet Metal	403	403		403	
General Maint. (125 hr. serv., etc.)	404	404		404	
Retarder				405	
Fuel Tank	407	407		407	417
Fiberglass				408[a]	
Pump Drive Group				409[a]	
Hydraulic System					
Hydraulic Controls and Valves	501	501		501	
Hydraulic Implement Pump	502	502		502	
Hydraulic Lines	503	503		503	
Hydraulic Hoses	504	504		504	
O-Rings	505	505		505	
Steering System		506		506	
Steering Pump	507	507		507	
Loading Mechanism Pump				508	
Lift, Tilt, Bowl Cylinders	509	509		509	
Cushion Hitch System				510	
Vehicle Structure					
Vehicle Structure	601	601		601	
Differential Case and Covers				602	
Control System & Indicators	603	603		603	
Air System				604	
Air Lines				605	
Sub Assemblies				606	
Suspension System (Bar, Struts, etc.)	607			607	
Loader Frame	608	608			
Attachments, Tools, Etc.					
Bucket, Blade, Body, Bowl	701	701		701	
Teeth	702	702		702	
Cutting Edges	703	703		703	

Figure 9-19-5 (Cont.)

MACHINE AREA FAILURE CODE LIST *(cont.)*

Description	Track-Type Tractors and Loaders	Wheel Loaders		Hauling Units	
		Front	Rear	Front	Rear
Loader or Dozer Linkage	704	704			
Ripper Linkage	705				
Kickout System	706	706			
Ejector				708	
Push Frame				709	
Loading Mechanism (Hoe, Elevator)				710	
Hitch and Draft Frame				711	
Cushion Hitch				712	
Apron				713	
Pivot or Linkage Pins	714	714		714	

[a]Trucks only.

Figure 9–19–5   (Continued)

without finding and correcting the dirt will result in an early repeat failure. Troubleshooting information available from the equipment manufacturer will help determine the initial cause and contain instructions for correcting the problem.

Generally, remedial repairs correct premature failures resulting from error or oversight in the preventive maintenance program. This underscores the importance of a well-planned and well-executed maintenance.

9-23. Component Overhaul

An overhaul consists of that repair work necessary to restore a machine component to productivity after its normal wear life. In overhaul, individual components and parts are inspected, measured, and tested. Decisions are made to reuse, repair and reuse, or replace the parts. The goal is restoration that will assure additional trouble-free performance.

By system analysis and an effective preventive maintenance program, the need for component overhauls is determined prior to a breakdown. Overhaul work can than be scheduled at an opportune time.

In overhaul, a careful inspection of all parts may reveal impending service needs on other components and systems. Each overhauled component must be tested in operation before it is installed in the machine.

9-24. Component Exchange

Component exchange is a method of rapidly returning the machine to productivity. The principle is simple. The worn

Figure 9–21–1 Maintenance contractor performing regular machine maintenance in field

Figure 9–21–2 Service contractor's facility at job site

Figure 9–21–3 Track-laying tractor being repaired in the shop

Figure 9–24–1 Truck-mounted cranes lifting transmission from tractor in field

Figure 9–24–2 Exchange parts and components in storage at equipment dealer's parts facility

Figure 9–24–3 Exchange components in packaged form at equipment dealer's parts facility

or failed component is removed and directly replaced by a unit that was previously rebuilt and tested (Figures 9-24-1, 9-24-2, and 9-24-3).

The failed component, in turn, is sent to the service shop and carefully rebuilt, tested, and stored for future exchange needs. Two advantages are obvious: downtime is reduced substantially, and repairs can be made at a time that is convenient.

Today, large machine owners and most equipment dealers maintain component exchange inventory which includes major components such as engines and transmissions, and accessory groups such as generators, starters, and pumps.

The price of exchange components is usually based on the actual time and parts needed to restore a failed component to "like new" condition. Since paying overtime can be avoided, it normally costs less than doing the repair work at the time of the failure. In addition, the reduction of equipment downtime makes component exchange the major trend today in equipment repair.

On large earth-moving projects, some users replace engines at regular intervals to assure continuous machine availability. They replace drive trains and transmissions by the same method. Theoretically, through component exchange, the life of a machine could be infinite.

9-25. Machine Overhaul

Sometimes a complete machine overhaul is required. Even then, not every part will have to be replaced. Each system of the machine should be tested, its condition diagnosed, and remaining service life estimated. When a complete machine overhaul is in process, all wearing parts that will not last for an acceptable time period should be replaced.

Thus, parts with some remaining life will be replaced.

Overhaul restores the machine to an acceptable level of productivity and availability, and assures it additional normal and predictable wear life.

9-26. Service Facilities

Most machine owners have some type of equipment service facilities. Normally these are permanent structures located near the home offices. A small contractor may use his off-season machine storage building for limited service work during the work season. For large repair and overhaul jobs, he must rely on dealer or independent service shops.

Owners with larger equipment spreads may need more complete service faciilties, full-time shop personnel, parts inventory, and office space. The size of these facilities will depend on the type and amount of equipment service performed.

Service facilities on the job site are prevalent on large projects and in remote locations. The duration of the project is critical to the decision to provide such facilities.

When on-site service facilities are contemplated, the owner should consider combining them with field maintenance stations or other on-site facilities. Use of mobile field service equipment, such as service trucks, truck-mounted cranes, and welding equipment, can reduce the space needed for on-site service shops and make such combinations possible.

9-27. On-Site Service Facilities

The first step in determining the requirements for job-site facilities is the identification of the project. On-site facilities

Figure 9–27–1   One bay of open-canopy type on-site service facility with tractor being overhauled

Figure 9–27–2   General arrangement of open-canopy type on-site service facility, parking space, etc.

Figure 9–27–3 Mobile service vehicles (crane) in use at on-site service facility

should be considered when the project fits into one of the following categories: (a) Long-term job; or (b) job in a remote area. Facility needs vary with each construction project. Nevertheless, a number of basic requirements are valid for all on-site facilities:

Site Selecting the site for the service shop is an integral part of the total operational plan of the job. A centrally located facility will serve the project efficiently until the job is finished. Only after locating cut-and-fill areas and plotting the haul roads can an intelligent site selection for service facilities be made.

Accessibility The shop should be accessible to both on and off highway equipment. To assure continued shop availability, all-weather roads should be provided.

Obstructions Overhead cables and underground pipelines must not infringe on free movement of heavy equipment close to the facility.

Utilities With the exception of electric power, utilities are usually not available for on-site facilities. However, a sufficient water supply is essential to operate wash racks and sanitary facilities.

Space Requirements For job-site service facilities, space requirements are controlled by such factors as:

- Number, types and sizes of machines (shovels, trucks, tractors, loaders, scrapers, etc.)
- Type, size, and duration of construction project

- Extent of service work expected to be done in the on-site facility
- Availability of parts and exchange components
- Proximity of equipment dealer service facilities
- Climatic conditions at the site
- Operations schedule (one shift or around-the-clock)

Housing of on-site service facilities will vary upon need. Two types of housing are open-canopy type and enclosed structures.

Open-canopy type structures may be used in warm-weather regions, with mobile homes and/or truck trailers for office, locker room, and tool and parts storage (Figures 9-27-1 and 9-27-2). The structural components may be prefabricated and used again when moved to another job site. Since overhead cranes are rarely installed in temporary shops, mobile cranes are required (Figures 9-27-3 and 9-27-4).

Enclosed structures of prefabricated components (preferably of the drive-through bay design that adds flexibility to the utilization of floor space) are better suited for larger and long-lasting projects. Such structures should contain separate office space, washrooms, tool storage, and parts storage areas. The minimum service door size is generally 20 × 20 feet. Overhead cranes or jib cranes are desirable. As an alternative, mobile cranes may be used inside and around the shop building. A 10-ton lift capacity will handle most shop requirements, and a lift height of 24 feet above floor level will accommodate all but the largest machinery.

If field maintenance stations are contemplated, they should be located in proximity to the service facilities. This

Figure 9–27–4　On-site service shop with overhead doors and crane

will facilitate making minor repairs during fueling operations and thus improve utilization of service and maintenance personnel.

9-28. Machine Owner's Permanent Service Facility

Under some conditions, permanent service facilities at the construction company's home base may be a worthwhile investment. Not only should these be matched to present equipment needs, but they should have a flexible layout with provision for future expansion.

Site Selection　Efficiency and cost of a service facility are greatly affected by location. The place must be accessible for employees and suppliers. Proper transportation connections are indispensable.

The lot must accommodate all activities related to the operation. Its size should be at least ten times the floor area of the building in order to accommodate employee, visitor, and equipment parking, open-air storage, and traffic access to all entrances.

The land should be level and dry. The subsoil must support heavy loads. Natural hazards, such as flooding and possible earth slides limit the usefulness of the facility. Obstructions like overhead cables, underground pipelines, ditches, and gulleys should be avoided.

Local building codes, zoning regulations, and future highway plans should be investigated before the site is selected.

Space for Service　The size of operation, the amount and type of machinery, and the service work to be done determine how much space is needed for service. A minimum requirement would include a two-bay service shop with adjoining office, parts, tools, and sanitary facility space. The bigger the operation, the more diversified the structure must be.

Office space, washrooms, lunch areas, locker rooms, tool crib, and parts storage must be matched in size to the repair shop and its staff. If the service facility is located in proximity to general offices, the communication systems, janitorial facilities, and lunch rooms can be combined.

For the service shop, an outside door of a minimum of 20 × 20 feet is needed. Overhead traveling cranes with lift capacity of 10 to 15 tons are desirable. Their lift height above floor level should be 24 feet or more. An all-weather steam and/or water cleaning bay should be provided within or adjacent to the service facility.

The type of service work performed may require specialization areas within the service facility. These may include welding and metal fabrication, painting and sandblasting booths; track rebuild, electrical engine, and transmission shops; and tire repair and hydraulics repair areas, to name a few.

Even though a large investment in specialized tools and repair equipment is usually not economical for a fleet owner, lack of facilities at dealer shops or other qualified service sources in his area could make specialization necessary.

Evaluation of dealers' service and parts support capabilities can determine what is necessary for either on-site or permanent facilities. On-site facilities could be completely unnecessary. Outside service sources should be thoroughly studied before planning any service facility.

9-29. Contacting Job-Site Suppliers

The equipment manager should arrange to meet with local suppliers to provide materials, fuels, and other needs. Once the job is in progress, the equipment manager should keep these communication lines open. One earth-moving contractor makes it a practice to call each supplier once a week to relay any problems, give any complaints, and get the same from the suppliers. Understanding each other's needs makes work easier for all parties.

Oil Company Contacts Early contact with local oil and fuel suppliers will be most valuable. Working conditions, climatic considerations, and fuel and oil specifications for the equipment should be discussed in detail with the selected supplier.

Tire Supplier Contacts Choice of a tire supplier is important, since tire costs may be a major expense item. Competent tire dealers will analyze job conditions and assist in selecting proper tire equipment. A tire maintenance contract should be considered on the basis of achieving minimum operating cost (Figures 9-29-1 and 9-29-2).

Equipment Dealer Contacts When a construction company plans to bring equipment into an area, it is wise to notify local dealers so that they can arrange suitable parts backup. Early meetings between the equipment manager with the dealer's parts manager will help coordinate parts orders for the particular type of machinery used on the job. Critical items can be ordered and stocked in advance. If the dealer is close by and deliveries are made regularly, there is no need for a large parts inventory at the job site.

The possibility of a dealer's on-site parts facility should be considered where the work is in a remote location.

This is also the logical time to discuss equipment service needs and to inspect the dealer's service facilities.

9-30. Parts and Supplies Inventory

Occasionally equipment users and dealers combine efforts to locate a suitable temporary facility for the parts and service needs of a large construction job. This will reduce the equipment owner's need for parts inventory and/or service facilities, and perhaps even his personnel requirements.

Unless the shop facility is located at an extremely remote site, or if regular deliveries from parts dealers cannot be depended upon, the parts inventory of a machine owner should be held to a minimum. Usually a 10-day supply will be adequate. Maintenance supplies like filters can be purchased in bulk and stored either at the job site or kept at the home base and delivered to the site as needed. Lubricants, bulk cable, hose, and expendables can be handled the same way. Excessive fuel storage should be avoided for safety and economy.

Stock of general usage items such as hardware, wear parts, bulbs, and breathers should be closely matched to foreseeable need. Parts improvements by the manufacturer can make inventory items obsolete. But the main reason for holding the parts stock to a minimum, of course, is that inventory is an investment in nonproductive items. See Section 9-43, Parts Inventory Cost.

The parts storage area should be just as carefully planned as other facilities. Large items like track, buckets, etc., can be stored in an open storage area surrounded by a fence for theft and safety control.

Virtually all other parts must be stored in a protected area, away from dust, rain, moisture, and fumes. Proper storage shelves, bins, and racks are necessary. A book-

Figure 9-29-1 Service contractor's tire servicing truck with crane

Figure 9–29–2 Service contractor working on on-site tire change

keeping system that identifies parts in stock and their location is also important. A part that cannot be found quickly is a poor investment.

9-31. Safety

Parts storage and service facilities, regardless of type, harbor many safety hazards. Building layout and arrangement of activities within and around the building will influence the accident rate. In initial facility planning, physical separation of welding, fuel storage areas, battery charging, painting work, electrical cables, and overhead lifting devices should be carefully considered. Tire service should be separated from other activities by a room partition. Exhaust fumes must be ventilated. Fire extinguishers and fire escape doors are essential. Storage racks with adequate capacity should be provided.

9-32. Service Shop Management

The principal task of the shop manager is to build a skilled service team, well trained and able to handle economically any repair work assigned. On all construction projects with on-site service facilities, the shop manager or master mechanic should report to the project equipment manager (Figures 9-32-1 through 9-32-4).

The shop manager of a centralized service facility should report to the general manager of the equipment owner's home base, since the centralized service facility shop manager will receive work orders from various project equipment managers. The general manager will be in a better position to establish priorities.

A shop manager, even more than his mechanics, must be constantly alert for equipment design changes, new servicing procedures, tooling changes, and improved recording systems.

The responsibilities of a shop manager are:

● Supervision and training of shop personnel
● Coordination and control of all work done
● Time and cost control of shop work
● Parts, supplies, and equipment control
● Counseling and reporting to management on service matters.

Hiring, training, and retaining good mechanics has become increasingly difficult. Training unskilled people is a good investment only if they stay with the organization. Congenial atmosphere, attractive facilities, and financial incentives bring the best results.

Training personnel for specialized repair on new equipment is necessary, and it can be used as a job program incentive. Most equipment dealers are willing to assist the machine owner's shop manager in all aspects of training. Acquisition of special tools and service literature for new machinery is also the responsibility of the shop manager.

The shop manager must be able to analyze the cost and time involved for every service work request he receives. His analysis will aid the equipment manager in scheduling field operations. The equipment manager will also rely on the shop manager's report of machine condition after repair.

Time and cost control in the service shop is important. Several shop cost accounting systems are included in the Preventive Maintenance section of this chapter. Further

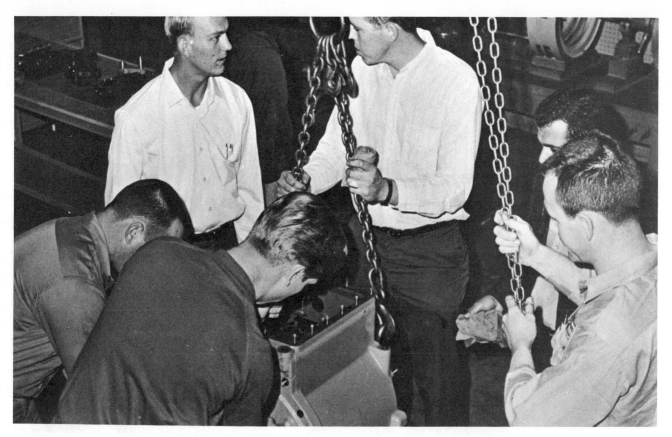

Figure 9–32–1   Process of training service personnel regarding new products and components (engine part on chain blocks)

Figure 9–32–2   On-machine training to diagnose failure causes

Figure 9–32–3 Mechanics performing minor repairs in the field to avoid transporting the machine to the shop

Figure 9–32–4 Sheet metal repair (fender) being made in the field in off-shift periods

Figure 9-33-1 Servicemen on field call evaluating repair problem and doing minor repair

details on actual shop cost are listed in Sections 9-38 and 9-39.

The shop manager controls all parts, tools, and supplies used in the service shop. Stock of critical parts must be monitored, ordering schedules revised when needed, and decisions made on tool replacement or repair.

Repair-or-replace decisions by the maangement will be strongly affected by the service record of an earth-moving machine, and by the shop manager's evaluation of its condition.

9-33. Shop and Field Service Personnel

The responsibilities of the mechanical personnel are:

- To handle preventive and remedial repairs
- To make basic failure diagnosis
- To keep records of machine repair time and parts usage

A heavy equipment mechanic must be able, when called in the field or assigned to a machine in the shop, to determine the scope of the work required to repair the machine. His analysis of the machine failure, especially when he is on a field call, will help the shop manager decide what to do. Two-way radio equipment is frequently used for transmitting initial field reports to the shop manager or equipment superintendent.

Based on the serviceman's field report, the shop manager can give the equipment manager a rough estimate of downtime. He can send test equipment, needed parts, or additional help to the field. Sometimes, an exchange component and a crane truck, with additional service personnel and tools, can perform the repair on location (Figures 9-33-1 and 9-33-2). If the machine must be transported to the shop, those arrangements can be initiated.

Once a machine is brought to the shop, the findings of the field report verified, and a repair method determined, the shop mechanic must be able to perform the work properly. He must also record parts, supplies, and time used repairing the machine. He should record the machine downtime as "Downtime—Idle" and "Downtime—Actual Repair."

9-34. Other Service Personnel

The size of the service facility will determine the type of personnel needed as well as total number of people employed. The larger the shop, the more defined the job assignments will be.

In large service facilities, a parts and tool foreman is in charge of the parts and supplies inventory. He monitors the stock and handles ordering and receiving for shop and field. He normally maintains a complete parts catalog file, keeps records on all materials and costs, and dispenses parts and tools to the mechanics.

Figure 9–33–2 Large earth-moving equipment being released from servicing

The drivers of supply trucks, low-boys, and towing equipment should also be included in the discussion of the service shop personnel. Their job assignments must be flexible enough to permit full utilization of their time. On smaller jobs, a truck driver might take over fueling assignments, operate a crane truck, or pick up and deliver parts.

Shop helpers and janitors are under the direct supervision of the shop manager. They assist the skilled mechanics in their work. Cleanup in and around the shop increases job efficiency and safety. Also, cost of washing incoming equipment, unloading supplies, etc., is reasonable if these jobs are done by the lower paid, unskilled workers.

Bookkeeping personnel and equipment will relieve the shop manager of lengthy record keeping and cost accounting. In centralized facilities, this assignment may be allocated to company personnel in other departments, but records must be at the shop manager's disposal at all times.

9-35. Field Service Vehicles

A number of field service vehicles may be required. All types will not be needed on one particular job and, in some cases, the functions of two or more vehicles may be combined. Exact requirements for a particular job depend on the equipment to be serviced and its location with respect to service facilities.

Field Service Trucks In some cases, a 1/2- or 3/4-ton pickup truck, equipped with hand tools and selected special tools, will be adequate for field troubleshooting and minor repairs (Figures 9-35-1 and 9-35-2). If there is no shop on the job site, or if some equipment is located too far away from the home base service shop, better equipped service trucks will be required. A 1- or 1½-ton truck, with

a field service body and truck-mounted crane may be needed under these conditions. A 4000-pound capacity crane will lift smaller engines and transmissions and is the largest crane that can be used on this size of truck. The crane can save time when removing components such as cylinder heads, hydraulic pumps, and hydraulic cylinders from larger machines.

The equipment carried on this truck will depend upon the machines to be serviced and conditions at the job site— a large assortment of hand tools and common pullers, along with the special tools needed for the individual machines in the field. Oxyacetylene cutting and welding equipment, and electric generators to power drills, impact wrenches and lights could be needed. Commonly used parts such as nuts, bolts, washers, etc., must be included.

Field Welding Truck A field welding truck is necessary at sites far away from the repair shop. It normally is a 3/4- or 1-ton truck equipped with an engine driven arc welder. A 500-amp welder permits use of the larger electrodes and, if an engine-driven, 5-horsepower air compresser is installed, carbon-arc cutting equipment can be used too. If the carbon-arc equipment is not used, oxyacetylene cutting equipment should be installed (Figures 9-35-3 through 9-35-6).

Truck-Mounted Cranes A truck-mounted crane will be required only if major component exchanges are made in the field. The truck should be big enough to carry the largest exchange component, and the crane should have sufficient lifting capacity to remove and install it.

Although each job must be considered individually, a 6-ton capacity crane is of sufficient size to handle the track roller frame from the largest track-type tractors, and it

Figure 9–35–1 Well-equipped field service truck needed to perform repairs on location

has sufficient reach and capacity to install engines and transmissions in most pieces of earth-moving equipment. If the on-site service shop is not equipped with overhead cranes, these mobile cranes will be used for shop work.

Other Field Service Vehicles Depending on the size of operation and location of the service facility, highly mobile radio-equipped vehicles may be used to reach a piece of equipment quickly. The radio-equipped cars, pickup trucks,

Figure 9–35–2 Another view of the same well-equipped field service truck

Figure 9–35–3   One type of field welding truck with gasoline driven electric welder, etc.

or station wagons, should carry troubleshooting aids and some essential tools and supplies. On smaller projects, this could be the shop manager's car or master mechanic's pickup truck.

For every service shop, trucks for material and parts pickup and delivery are needed. They can be equipped with cranes, winches, and hydraulic lift jacks. Number and size vary with the job and shop needs.

Figure 9–35–4   Another type of field welding truck and mounted equipment

Figure 9–35–5 Fuel storage tanks on-site

Figure 9–35–6 Open-air type tire storage on job site

Figure 9–36–1 Welder's area in service shop facility

Tractors with low-boy trailers may be a part of the equipment fleet, or part of the operational equipment group. Regardless of their classification, they are indispensable for transporting machinery to and from the service facility. The cost of equipment transportation in case of repair or overhaul work is part of the service expense.

The load capacity of equipment transport vehicles must be sufficient to transport the heaviest track-type machine. Rubber-tired machines may be towed by loading one axle and letting the other axle trail. Legal hauling restrictions may be an important consideration.

9-36. Service Shop Equipment

All tools and equipment needed in a heavy equipment shop are not listed here. To perform all kinds of repair and overhaul on construction machinery, a substantial investment in tools, test benches, and machine tools is required. Wash racks, compressors, tire-mounting equipment, mobile cranes, welding and cutting equipment, lifting equipment, painting facilities, and test equipment system analyzers are some of the essentials. Some special tools for each type and brand of machinery may be needed (Figure 9-36-1).

Before he invests too heavily in shop facilities, the machine owner must analyze his service needs and compare the cost of doing his own work to prices charged by outside sources.

9-37. Service Sources and Cost

When planning repair or overhaul work on heavy equipment, three sources—machine owner's service shop, independent service shop, and equipment dealer service shop—should be considered.

In finding the most economical and effective source for work, past experiences of the equipment user will be important. In addition, he must weigh such factors as size of the service operation, available exchange parts groups, current work load in the shop, and mechanic's familiarity with the equipment, when determining where to place an overhaul job.

9-38. Machine Owner's Service Shop

An overhaul or repair performed in the machine owner's shop has several advantages. For one, proximity. Mechanics can start the work immediately. Another is the year-round utilization of skilled mechanical help which the owner needs during the construction season. Some owners feel that they can perform the overhaul cheaper than having it done outside.

There are disadvantages, too. One is the problem of training mechanics and keeping them up to date on new products and service procedures. The equipment owner's shop personnel, if not aware of the latest shop methods in

SHOP COST ANALYSIS

Servicemen Salaries: Average Hourly Wage paid Servicemen $_____

OVERHEAD COSTS

*Direct Overhead Expense

Management Salaries: Salaries and bonuses of master mechanics and shop foremen $_____

Other Shop Personnel: Cost of auxiliary shop workers (mechanics' helpers, the fellow on the steam cleaner, etc.) $_____

Employee Benefits: Life & Health Insurance, pensions, state unemployment, FICA $_____

Allowed Time: Wages paid for vacations, holidays and sick leave $_____

Truck Expense: Operating expense, including insurance and taxes of trucks and other transportation equipment engaged in repair work $_____

Shop Re-work and Warranty: Value of parts and labor invested in doing a repair job over $_____

Repair and Maintenance of Shop Equipment: Cost to repair and maintain welding equipment, air compressor and steam cleaner $_____

Small Tools and Supplies: Uniforms, gasoline, oil, welding rods, drills, grinders, etc. Expenses of maintaining tool cribs should be included $_____

Lost and Clean-up Time: All labor expense for non-productive time $_____

Service Training: Wages paid and the expenses of training meetings $_____

Equipment Depreciation: Depreciation on transportation and other shop equipment $_____

Miscellaneous Direct Expense: All other expense of a direct nature $_____

**Allocated Overhead Expense

Administrative: Allocated % x top management salaries and bonuses. (Equipment superintendent) $_____

Office: Allocated % x office salaries, supplies and postage $_____

Occupancy: Allocated % x real estate, insurance, taxes, utilities, repairs, maintenance, rent, depreciation and amortization of leasehold improvements $_____

Other Allocated Expense: Allocated % x insurance and taxes (other than real estate), telephone, janitors, watchmen and other miscellaneous allocated expense $_____

Total Overhead $_____

$$\frac{\text{Total Overhead}}{\text{***Total Service Hours}} = \text{Hourly Overhead Rate}$$

Hourly Rate of Servicemen $_____

Hourly Overhead Rate   +$_____

Cost of Your Shop Per Hour $_____

*Direct Expenses: Costs incurred directly by the Maintenance and Repair Department

***Total Service Hours: Number of Man Hours charged to repair jobs during the accounting period

**Allocated Expenses: Expenses incurred by the entire organization and whose services the Maintenance and Repair Dept. uses. e.g. Payroll office. These expenses are allocated to the Maintenance and Repair Department in proportion to the usage of the services. The usage is the allocated percentage figure.

PEG07066

Figure 9–38–1 Shop cost analysis

repair and overhaul, can miss profitable shortcuts and critical service points.

The shop cost analysis from (Figure 9-38-1) lists the time and expense factors that affect total shop cost per hour.

9-39. Calculating Hourly Cost of Owner-Performed Service

Machine owners frequently say that it is senseless to pay $8.50 an hour for an equipment dealer shop mechanic to do the work his own man can do for $3.50 an hour. Nevertheless, the higher rate may be a bargain when total actual costs are analyzed. Proper cost accounting can tell the machine owner where to have repair work done, when to buy new equipment, and when to change maintenance practices (see Figure 9-39-1).

Comprehensive cost analysis establishes the real hourly repair costs in a machine owner's service shop. The results permit the owner to compare rates charged by equipment dealers and independent service shops.

First, total hours worked by all the repair and maintenance personnel are determined. In most cases, this figure is available from payroll records. The analysis must be based on a period of at least six or nine months. The total hours worked, however, do not represent the total productive time. Nonproductive time must be deducted. Every day, some time is lost in cleanup work, travel, and waiting.

SAMPLE SHOP COST ANALYSIS

Service Labor Cost

1. Total hours worked by all servicemen (from payroll records)		10,000 hr	
2. Hours of lost and cleanup time (12% of total hours worked)	1,200 hr		
3. Hours of rework time (6% of work done over)	600 hr		
4. Hours of training time	180 hr		
5. Total unproductive time (add lines 2, 3, 4)		1,980 hr	
6. Actual service hours (line 1 minus line 5)			8,020 hr
7. Total wages and bonuses paid to servicemen		$42,500	
8. Hourly rate of servicemen per service hour (line 7 ÷ line 6)			$ 5.30

Direct Overhead Expense

9. Total shop management salaries and bonuses (for time spent in managerial activity)	$ 3,800		
10. Other unskilled labor expense (not already included)	$ 6,400		
11. Total employee benefits (insurance, FICA)	$ 1,750		
12. Truck expense and other transportation expense	$ 2,185		
13. Repair and maintenance of shop equipment (noncapitalized expenditures)	$ 890		
14. Tools and supplies consumed (includes tool rentals)	$ 6,445		
15. Equipment depreciation (trucks and all shop equipment, but not fixtures, building, or office)	$ 3,600		
16. Miscellaneous direct expenses	$ 50		
17. Total direct overhead expense (add lines 9-16)		$25,120	
18. Hourly direct overhead rate (line 17 ÷ line 6)			$ 3.13

Allocated Overhead Expense

19. Administrative overhead expense not charged to *nonrevenue* operations	$		
20. Office (allocated % × office salaries, supplies, postage, and depreciation on office equipment)	$ 755		
21. Occupancy (allocated % × real estate, insurance, taxes, utilities, repairs, rent, depreciation, and amortization of leasehold improvements)	$11,400		
22. Other allocated expense (allocated % × telephone, janitors, watchmen, and miscellaneous)	$ 1,000		
23. Total allocated (add lines 19-22)		$13,155	
24. Hourly allocated overhead rate (line 23 ÷ line 6)			$ 1.64
25. Hourly overhead rate (line 18 plus line 24) line 18 was $3.13			$ 4.77
26. Total shop cost per service hour (line 8 plus line 25) line 8 was $5.30			$10.07

Opportunity Cost

27. Total investment in shop facilities, equipment, and parts inventory (exclude small tools)		$75,000	
28. Opportunity cost (line 27 × current bank interest rate)		$ 6,000	
29. Opportunity cost per service hour (line 28 ÷ line 6)			$ 0.75
30. Sum of shop cost per hour and opportunity cost per hour (add lines 26 and 29)			$10.82

Figure 9–39–1

If servicemen travel long distances to reach machines, or inexperienced mechanics wait for assistance, the total productive hours worked are reduced by as much as 30 to 35 percent. The actual rate of *lost time* can be determined by a time study analysis or can be estimated from experience. Assume that five repair and maintenance mechanics, including the shop foreman, worked 10,000 hours last year. A conservative estimate of lost time for the year is 12 percent. Thus, 1200 hours were spent on nonproductive chores.

When work isn't done correctly the first time and must be done over, it affects the hourly repair shop rate. *Redo work* losses vary with the mechanic's skill, the complexity of the job, and the type of machine. Redo work rates as high as 20 percent are not uncommon.

Redo rates can be determined by work sampling. All work done in the shop for a given period is checked and the rate determined. The redo rate can also be estimated by the shop foreman or equipment superintendent. In the sample analysis, 6 percent, or 600 hours, was spent on redo work.

Training is the third element that affects the total hours worked. Although training is a valuable and necessary part of any shop operation, men in training are not usually doing productive work. If a formal training program exists, training time can be determined accurately. The next best alternative is to estimate training hours. In the sample analysis, 180 hours per year is estimated.

Lost time, redo time, and training time are added to determine the nonproductive time per year. This total is then subtracted from the total hours worked to find the number of productive hours worked. To arrive at an hourly cost, determine total wages paid for the year. In the sample analysis, total maintenance and repair payroll is $42,500, including overtime and bonuses. Divide total wages by productive hours to establish the wage rate per productive hour. In this example, the $5.30 rate is well above the base scale of $3.50. The wage rate per productive hour represents the true wage rate, but there are other costs that must be added.

Direct overhead expenses are also involved in maintenance and repair. These are the costs that are directly assigned by the accounting section to maintenance and repair activities.

One direct shop overhead expense is made up of supervisory salaries and bonuses. For example, one third of the shop manager's salary, $3800, is added. This represents the foreman's time spent supervising mechanics and not doing work himself. In addition, $6400 is added for salaries of cleanup personnel, truck drivers, mechanic's helpers, and other people who contribute to the maintenance and repair cost.

Total employee benefits of $1750 must be added. These include insurance, pensions and such other benefits as U.S. Social Security.

Truck and transportation expenses totaling $2185 are included. This covers gasoline, oil, tires, repair, and license as well as shipping charges for shop equipment and supplies.

Another part of direct overhead expense is repair and maintenance of shop equipment amounting to $890. Tools and supplies, including tool rentals and tool crib expenses of $6445, must be added at this point.

Next, $3600 depreciation expense for trucks and shop equipment is included. And finally, $50 for miscellaneous direct expense should be added.

The total direct overhead expense is $25,120. Actual costs for each of these items should be available in the accounting records.

By dividing the total direct overhead expense ($25,120) by the total productive hours worked (8020 hours), hourly direct overhead expense amounts to $3.13.

Allocated overhead expenses are still to be added. Allocated expenses are items that cannot be directly identified with any specific activity. These expenses are divided and charged to all company activities. Normally, administrative overhead expense is not charged to *nonrevenue* operations. However, if it is felt this expense is a valid allocation to the owner-repair expense picture, then the same administrative expense allocation should be applied to the dealer or outside repair cost to achieve comparable costing. Since these costs are identical, they cancel each other and have been eliminated from this discussion.

Allocated overhead expenses applicable to this shop cost analysis are: Allocated office salaries, supplies, postage, and depreciation of office equipment for a total of $755. Occupancy representing the allocated percentage of real-estate depreciation and repair, insurance, taxes, and utilities amounts to $11,400. Finally, other allocated expenses for telephone, janitorial service, and miscellaneous items amount to $1000. The total of allocated expense is $13,155. The hourly rate of allocated expense is determined by dividing total allocated expense by productive hours. In this case, the figure is $1.64 per hour. The total overhead rate per hour is now $4.77.

By adding the hourly productive wage rate of $5.30 and the total overhead rate of $4.77, the cost-per-productive shop hour climbs to $10.07.

There is one other repair and maintenance cost called "Opportunity Cost." It cannot be precisely calculated, but it should be considered.

By letting an equipment dealer do most of the machine repairs and assigning his own mechanics to preventive maintenance, the equipment owner can eliminate his investment in repair equipment, parts inventory, and shop facilities. If this money were invested at the productive end of his business, it would earn interest. In this case, let's assume an investment of $75,000 and 8 percent. The $75,000 investment could earn $6000. On an hourly basis, this equals $0.75 per hour. Added to the $10.07 rate, the hourly repair and maintenance cost increases to $10.82.

This example illustrates the difference between base pay rate of $3.50 an hour and an actual cost rate of $10.82 per hour. All too often in the construction industry, the actual cost of repair work done in the owner's service facilities is not properly calculated. Without this information, alternative approaches to maintenance and repair work cannot be intelligently evaluated.

9-40. Independent Service Shop

In an independent shop, the hourly rate for service may be lower. But the independent shop must purchase parts and make a profit on resale. This places a middleman between the user and the source.

When repairs are made in an independent shop, the equipment owner should make a special request that replacement parts from the equipment manufacturer be obtained and used. He should avoid situations where work and parts will not be guaranteed.

Independent shops are just that—independent. Therefore, their services and equipment will vary. In many cases, a low price per hour may reflect a small investment in equipment. Also, the independent shop may have difficulty keeping servicemen up to date on new machines and methods. In some cases these two factors combined can cause slower repair jobs, and the lower hourly rate will be no bargain.

Acquaintance with an independent shop and its reputation is probably the most reliable means of judging what the equipment owner can get before he decides to have service work performed there.

9-41. Equipment Dealer Service Shop

Equipment dealers are informed of the latest service developments and their mechanics are up to date on their particular brand of machinery. This gives the equipment owner assurance of maximum benefit for his repair dollar.

The dealer has replacement parts with the latest im-

provements at his disposal. He has special tools and proper facilities to perform the overhaul work effectively and efficiently (Figures 9-41-1 and 9-41-2).

9-42. Calculating Service Costs Performed by Equipment Dealer

Calculating the expense of repair and overhaul work performed by an equipment dealer is generally a simple task. In most cases, only hourly cost and mileage have to be considered.

The hourly rate charged by the equipment dealer, in many cases, is lower than the true hourly cost of doing the job in the owner's shop. In addition, the dealer normally completes the work in a shorter time due to special tools and know-how. Likewise, the dealer's service work is warranted.

Figure 9–41–1 Fuel injector test bench

Figure 9–41–2 Hydraulic transmission test stand

Considering these facts, more and more equipment users, backed by actual service cost records, take their service work to the specialized shops and reduce their operating cost.

9-43. Parts Inventory Cost

In an ideal situation, the dealer parts department would be located next to the job site and, when a part was needed, a man could walk over and get it (Figure 9-43-1). Since this is seldom the case, some parts must be stocked so that equipment maintenance is not delayed. How much to stock is hard to calculate because many factors affect part usage. The average figure is 1.4 × the weekly average usage (10 days), but job conditions and type of operation must be considered. The dealer's ability to deliver parts ordered is of vital concern when calculating inventory protection (Figure 9-43-2).

It is wise to monitor any parts shortage that results in extended downtime. When it occurs, reasons for delay must be investigated and immediate remedies made. If parts

requirements change, the dealer must be notified at once so that he can adjust his parts inventory.

Calculating Parts Inventory Cost Along with all its advantages, the equipment owner must consider costs of keeping a parts inventory. Keeping a "full stock" of required parts will tie up available capital. To illustrate the expense of a large parts inventory, a short study of the inventory "hidden" costs follows.

[*Depreciation and Obsolescence Costs*] Machine improvements can result in obsolescence of present parts in stock. Price reductions after purchase and deterioration of materials in stock constitute losses that are not unusual. These costs can run slightly above 4 percent of the inventory cost.

[*Financing Costs*] At present interest rates, this cost amounts to 12 percent.

[*Handling Costs*] Costs incurred in shipping and receiving, trucking, and cost of labor and parts department help

Figure 9–43–1 Parts inventory at job-site facility

Figure 9–43–2 Parts inventory of the equipment dealer

will vary with the size of operation. However, 6 percent of the cost of inventory is a realistic figure.

[*Storage Area Cost*]   Directly related to the size of the storage area, this cost should run about 3 percent of inventory costs.

[*Taxes*]   Taxes vary, but for this example, 1.5 percent will be used.

[*Insurance*]   The cost of inventory and real-estate protection and industrial compensation is approximately 1.5 percent.

[*Total Hidden Inventory Costs—28 percent of Inventory Cost*]   Based on this cost factor, a $50,000 inventory will cost an additional $14,000 annually. This figure is very realistic and it underscores the importance of keeping the inventory to a bare minimum.

Only the items that prevent downtime and that enable the owner's personnel to perform proper maintenance should be kept in stock. Reliance on total maintenance contracts can further reduce inventory.

EQUIPMENT MAINTENANCE TRENDS

9-44.   Efforts to Shorten Service Time

With the ever-increasing cost of labor, efforts by equipment manufacturers and users have been concentrated on reducing the hours required for the upkeep of machinery (Figures 9-44-1 through 9-44-4).

One advancement in this direction is the widespread use of sealed bearings. All too often, bearing life has been shortened because of too little or too much grease. When the number of grease fittings to be serviced is reduced, it is easier to remember the remaining fittings.

In carrying this advantage one step further, on some equipment all grease fittings are put at one location. This makes lubrication faster, and fittings are not overlooked.

Another means to shorten service time is the fast-fill fuel system. This reduces fueling time and prevents fuel contamination and spillage. On some 24-hour operations, the fueling, greasing, oil, and filter change takes less than five minutes! Fueling and lubrication become a pit stop operation.

In large operations, there is more to the pit stop system than fast-fill fueling. The machines are equipped with a "service center" that centralizes all the drain/intake outlets for crankcase, transmission, hydraulic system, differential, water, and fuel at one point. The oil, for instance, is "sucked out" and then, without change of couplings, a metered amount of fresh oil is pumped in. Systems like this can be sound investments if operations make full use of them.

9-45.   Total Maintenance Contracts

Another successful solution to the overall equipment maintenance and repair problem is to get out of the service business altogether and contract for all service. Such agreements, called total maintenance contracts, may be made with the equipment dealer. He performs *all* service work on equipment at an agreed price per unit. This price can be based on time periods, service meter readings, or production.

Under such agreements, the user knows exactly what his expenditures for maintenance are, thus permitting him to bid jobs more accurately. Total maintenance contracts leave all decisions concerning service to the service contractor and allow the equipment user to spend his time and effort on production. Total maintenance contracts reduce the number of personnel on the job, may reduce operation costs, and frequently increase product availability.

Figure 9-44-1   Use of automatic wash rack on truck before entering maintenance facility

Figure 9–44–2   Complete maintenance pit stop facility and 120-ton truck therein

Figure 9–44–3   Dual pit stop facility for large earth-moving equipment

Figure 9-44-4 Injecting grease in grease fittings of prime mover all located in one place to facilitate servicing

9-46. Total Cost Contract

The total maintenance contract concept can be broadened to the point of fixing total equipment cost. This is a total cost contract under which the equipment dealer guarantees the owning costs of a machine for a specific time period. This approach may include provisions for the dealer to buy the machine back for a predetermined price when the contract ends.

If an equipment owner is working in one particular area and machine application remains relatively constant, the dealer might wish to extend this service to him. To ensure the lowest operating cost and to guarantee satisfactory performance, the dealer must retain a control of equipment maintenance.

9-47. Guaranteed Machine Availability

From total maintenance contracts and total cost contracts, it is only one step further to guaranteed machine availability.

Such contracts are usually tied into equipment purchases and offered by the equipment dealer to users performing relatively constant operations. Relying upon the quality of his product, the dealer guarantees the user a certain percentage of "available time for use" on the machine. The percentage will vary with the type of work performed and the machine category. Under favorable working conditions, this "guaranteed availability" might be 90 percent or higher. All equipment maintenance under such contracts is controlled and performed by the dealer.

9-48. Critical Item Monitoring

In the past, time of repair was dictated by the time of machine failure. With an adequate method of evaluating machine system performance and overall condition, it is possible to locate a wearing part, to monitor its gradual wear, and to correct or replace it before major trouble arises.

Many times, major hydraulic components have been replaced when the culprit was only a sticking relief valve. Today, with the hydraulic testing units used by equipment dealers, such mistakes are rare. Test equipment is also used on engines and power shift transmissions. When used by experienced personnel, testing equipment can diagnose situations with astonishing ease, pinpoint particular service needs, and save costs in parts and repair time.

First indications of coming trouble can be determined by oil analysis. This is a popular method of equipment condition control. By isolating certain contaminants in the oil—for example, the sudden increase in bronze content—it is possible to track down trouble. With oil analysis, a slightly cracked block or head, a faulty air cleaner, or a failure to change oil can be detected. By scheduled sample analysis, it is possible to monitor the condition and gradual wear of the components without ever disassembling the machine.

Monitoring of machine systems for signs of wear or impending failure is a developing science.

10 METHODS OF DECIDING OVERHAUL OR REPLACEMENT

WILLIAM S. LAMBIE, JR.

Vice-President, Parts and Service
Caterpillar Tractor Co.
Peoria, Illinois

EDITOR'S NOTE: *The matter of determining when to overhaul or replace equipment is a very complicated subject. In this chapter the author has presented a system for evaluating depreciation cost, resale value, maintenance cost, and other factors involved.*

The editor knows of no construction company that is using, or has used, this system. (It is, he understands, an adaptation of systems used in manufacturing and other industries.) However, an understanding of the principles involved should be helpful to contractor management.

A LARGE PORTION of a construction company's capital is often invested in its equipment spread, whether it owns one machine or several hundred. This important investment can be managed instinctively, and some successful contractors have used this approach. Generally, however, management by instinct does not produce good results. Sound management practices are needed to ensure the best return on equipment investment.

10-1. Equipment Management

An equipment fleet can be managed to provide minimum operating, maintenance, and repair costs, and maximum productivity and return on investment. This chapter discusses how to determine costs, and how to use cost factors in determining the optimum time to replace equipment. Guidelines are provided for use in selecting new machines.

To manage properly an equipment spread, an owner should:

1. Inspect every major piece of equipment regularly to determine its actual condition.

2. Keep complete records for each piece of equipment, covering maintenance and repair costs, downtime, and performance.
3. Use this information to:
   (a) Develop a maintenance program.
   (b) Highlight repair costs, investigate causes, and take corrective action.
   (c) Compare actual machine performance with that promised by the manufacturer and that used in estimating jobs.
   (d) Develop a long-range, economically sound replacement program.

Good cost and performance records on equipment can make the difference between profit and loss, i.e., between good management decisions and poor ones. But the records must be available to decision makers, and they must be used.

10-2. Determining Equipment Owning and Operating Costs

Total cost of owning and operating a machine can be used for determining the optimum time for equipment replacement, or for any other purpose in which a knowledge of actual cost of equipment operation is required. Determination of this total cost involves six factors:

1. Annual net loss in machine value
2. Investment, insurance, and taxes
3. Maintenance and repair costs
4. Other operating costs
5. Downtime
6. Productivity differential

The following sections will discuss these cost factors and how they are computed. For illustrative purposes, a hypothetical machine costing $100,000 and averaging 2000 working hours per year will be used. The charts shown do not purport to be valid cost estimates but are intended simply as examples.

10-3. Annual Net Loss in Machine Value

Annual net loss in machine value represents market value loss minus any tax effect.

Loss in market value of a piece of construction equipment is the difference in price it would bring on the open market from the beginning to the end of the year. This difference reflects not only the physical condition of the machine, but obsolescence of the model. Used equipment market conditions are also a factor. Generally, equipment dealers can provide assistance in estimating used machine prices.

At the end of the first year, the hypothetical machine costing $100,000 might be worth $80,000, i.e., it could be sold after one year for $80,000 on the open market. Loss in market value for the machine during the first year would be, then, $20,000. Loss in market value for subsequent years is obtained by subtracting the price the machine would bring at the end of each year from its market price at the start of that year.

In this discussion, depreciation is used to describe that portion of the purchase price arbitrarily allocated to each year's operation for accounting and tax purposes. Do not confuse loss in market value with depreciation. As stated previously, loss in market value is the difference in actual price of the machine on the open market from beginning to end of the year. Depreciation is an accounting and tax term, and it represents an arbitrary allocation of purchase price to each year's operation.

Three commonly used methods of establishing depreciation are: straight line, sum of year's digits, and declining balance. (Consult Chapter 28 on Taxes.)

In many countries the profits of an enterprise are shared with the government through taxes. Any losses suffered in the business result in a reduction of profits and, therefore, taxes. In effect then, the government "shares" in the loss. This "sharing" is called tax gain or tax shield.

For this example, assume that the machine owner is a United States company paying a corporate tax of 50 percent on profits. The owner is allowed a $25,000 depreciation during the first year of ownership of his $100,000 machine. Tax gain, or tax shield, amounts to 50 percent of the allowable depreciation, or $12,500. But if the machine were sold for $80,000 at the end of the year, there would be a taxable gain of $5000 which would result in $2500 additional tax. This, in effect, reduces the tax shield to $10,000.

Earlier it was noted that the machine worth $100,000 at the beginning of the year would bring $80,000 on the open market at year end. Loss in market value was $20,000. The $20,000 loss in market value was partially offset by a tax shield of $10,000. Thus, the net loss in machine value was $10,000.

Assuming 2000 working hours, hourly cost of net loss in machine value was $10,000 ÷ 2000, or $5 during the first year.

Table 10-3-1 provides a complete example of annual net value loss calculation. In this example, loss in market value is partially offset by a tax shield based upon the declining balance method of calculating depreciation. This method normally offers the largest tax shield in the early years under United States law, but it is not allowed in some countries. An eight-year life and a $20,000 salvage value have been assumed. Tax effect of depreciation should be calculated according to the life and method used on the owner's tax return.

10-4. Investment, Insurance, and Taxes

The second factor in determining total equipment owning and operating cost is investment, insurance, and tax costs. Whether the machine is bought with cash or under an installment arrangement, interest should be applied to the entire purchase price, since any funds paid could be invested at interest if they were not used for this equipment.

TABLE 10-3-1. HOURLY COST OF ANNUAL NET LOSS IN MACHINE VALUE

	Years from Present							
	1	2	3	4	5	6	7	8
Acquisition cost	$100,000	$100,000	$100,000	$100,000	$100,000	$100,000	$100,000	$100,000
Market value at End of year	80,000	65,000	55,000	45,000	35,000	30,000	25,000	20,000
Loss in market value	20,000	35,000	45,000	55,000	65,000	70,000	75,000	80,000
Depreciation, Dbl. declining	25,000	43,750	57,813	68,360	76,270	80,000	80,000	80,000
Gain or loss on disposition	5,000	8,750	12,813	13,360	11,270	10,000	5,000	0
Tax shield (50% line 4—line 5)	10,000	17,500	22,500	27,500	32,500	35,000	37,500	40,000
Total net value loss after tax	10,000	17,500	22,500	27,500	32,500	35,000	37,500	40,000
Current year cost after tax	10,000	7,500	5,000	5,000	5,000	2,500	2,500	2,500
Current hourly cost after tax	$5.00	$3.75	$2.50	$2.50	$2.50	$1.25	$1.25	$1.25

TABLE 10-4-1. INVESTMENT COST

	Years from Present							
	1	*2*	*3*	*4*	*5*	*6*	*7*	*8*
Value at start of year	$100,000	$80,000	$65,000	$55,000	$45,000	$35,000	$30,000	$25,000
Value at end of year	80,000	65,000	55,000	45,000	35,000	30,000	25,000	20,000
Average annual investment	90,000	72,500	60,000	50,000	40,000	32,500	27,500	22,500
Investment cost at 15%	13,500	10,875	9,000	7,500	6,000	4,875	4,125	3,375
Investment cost after tax (50%)	6,750	5,437	4,500	3,750	3,000	2,438	4,063	1,688
Investment cost per hour	$3.38	$2.72	$2.25	$1.88	$1.50	$1.22	$1.03	$0.84

Other actual finance, insurance, and tax charges should be included as follows:

	Percent
Interest	9
Service charge on loan (average per year)	2
Property taxes	3
Insurance	1
	—
Total rate for investment, insurance, and taxes	15

The average yearly investment in a machine is determined by adding its market value at the end of the year to its value at the start of the year, and dividing the figure by two. Investment cost is computed by multiplying the total cost rate by the average annual investment.

$$A = R \cdot \frac{S + E}{2}$$

Where A = annual investment cost,
R = total cost rate for interest, taxes and insurance,
S = market value at start of year,
E = market value at end of year.

For example, assume that the combined annual cost of interest, insurance, and taxes equals 15 percent of the average yearly investment. If a machine was worth $100,-000 at the beginning of the year and $80,000 at the end of the year, the average yearly investment would be $90,000. Multiplied by 15 percent, this yields an investment cost for the first year of $13,500. The same procedure is used for computing investment costs for subsequent years.

Investment cost may be deductible from profits and therefore "shared" with the government. Assume, again,

that a United States company pays a corporate tax of 50 percent. The investment cost of $13,500 calculated above would be reduced to $6750 for this company.

Table 10-4-1 provides a complete example of investment cost calculation.

10-5. Maintenance and Repair Costs

A major operating expense for earth-moving equipment is maintenance and repair. Over a five- to eight-year period, an amount equal to 100 percent of the purchase price could be spent. Under severe conditions, this amount could be reached in three or four years.

Cost of equipment maintenance and repair is affected by machine design, operating conditions, operator skill, efficiency of the servicing provided, quality of service, and many other factors. Extreme cost variation is to be expected, and "average" figures are very dangerous. An equipment owner should use his own records of maintenance and repair costs. They are the only reliable guide to sound equipment management decisions.

Table 10-5-1 provides an example of maintenance and repair costs that might be estimated for the $100,000 machine under consideration. Assume that these data are based upon actual owner records of maintenace and repair cost.

The maintenance and repair cost figures for the machine described in Table 10-5-1 increase rather irregularly due to major overhauls or other extensive repairs, in the year in which they occur. Be sure to include the full cost of repairs made by owner employees, as well as outside sources. Include an allocation of overhead expense.

Annual maintenance and repair expense reduces profit. It is, therefore, subject to the same 50 percent reduction or tax shield used in Table 10-3-1 and Table 10-4-1.

TABLE 10-5-1. MAINTENANCE AND REPAIR COSTS

	Years from Present							
	1	*2*	*3*	*4*	*5*	*6*	*7*	*8*
Annual maintenance and repair cost	$3,500	$4,000	$6,300	$7,500	$24,500	$17,500	$14,000	$12,000
Annual cost after tax (50%)	1,750	2,000	3,150	3,750	12,250	8,750	7,000	6,000
Cost per hour	$0.88	$1.00	$1.58	$1.88	$6.13	$4.38	$3.50	$3.00

TABLE 10-6-1.   OTHER OPERATING COSTS

	Years from Present							
	1	2	3	4	5	6	7	8
Operator's wages	$12,000	$12,000	$12,000	$12,000	$12,000	$12,000	$12,000	$12,000
Fuel	6,800	6,800	6,800	6,800	6,800	6,800	6,800	6,800
Other consumables	625	625	625	625	625	625	625	625
Annual cost	19,425	19,425	19,425	19,425	19,425	19,425	19,425	19,425
Annual cost after tax (50%)	9,713	9,713	9,713	9,713	9,713	9,713	9,713	9,713
Cost per hour	$4.86	$4.86	$4.86	$4.86	$4.86	$4.86	$4.86	$4.86

10-6.   Other Operating Costs

Other operating costs include operator's wages, fuel lubricants, and other consumables.

"Operator's wages" (see Table 10-6-1) should include the total salary and benefits package. The hourly operator's wage may be $3 per hour; however, cost of benefits provided could increase the figure to as much as twice that amount, or $6 per hour. If the machine works 2000 hours per year, the annual cost for the operator would be $12,000.

"Fuel" cost can be determined on the basis of actual experience or from data furnished by the manufacturer. For this example an annual cost of $6800 is used.

Cost for engine oil, hydraulic oil, and grease has been labeled "Other consumables" in the example. Again, actual experience is the best guide to these costs. An annual cost of $625 is used in the example.

As with the other expense items, any applicable tax credit should be applied to other operating costs. In the example, a 50 percent tax is assumed.

10-7.   Other Cost Categories

The cost categories covered up to this time are tangible costs. The next two cost categories are less tangible, but no less important. They are the cost of downtime and the cost due to increased productivity offered by a new model.

Downtime   The trend today is toward faster, more productive—and more expensive—machines. As productivity increases, the cost of lost time due to machine breakdown also rises.

Downtime is the time lost due to a machine breakdown which results in its not being available for work. Downtime should not be considered if productivity of the machine is not actually required. The equipment must have been scheduled to operate, and conditions such that it could have been operated.

Although it is desirable to have 100 percent machine availability, usage generally results in increased downtime as a machine grows older. This loss of machine availability can vary greatly with the make, age, and especially the service provided. Actual records should be used in computing downtime expense.

In Table 10-7-1 $25 per hour is used to represent the cost of renting a replacement machine during the time the owner's machine is not available.

Productivity Differential   Productivity differential is the cost of retaining a machine when an improved, more productive model is available. It is the extra cost of doing the job with the existing machine. It is not part of equipment cost except for comparison with a new machine and has the effect of adjusting the comparison of cost per hour to cost per unit of work accomplished.

The productivity advantage offered by a new machine may be estimated using specification data and other information. Whenever practicable, estimates should be verified through observed performance. An on-site demonstration can be requested of the equipment dealer, or machine performance can be observed at some other site.

Productivity differential cost may either be added as a cost of the present equipment or subtracted from the cost of the new machine. In this example, it will be considered an additive cost. As with downtime cost, the cost of productivity differential should not be considered if that extra productivity is not actually usable.

Table 10-7-2 assumes that a significantly improved piece of equipment becomes available during the fifth year an owner is using a current model. The new machine offers 30 percent greater productivity. A rental cost of $25 per hour has been used again.

TABLE 10-7-1.   DOWNTIME COST

	Years from Present							
	1	2	3	4	5	6	7	8
Availability, %	97	94	91	88	85	82	79	76
Hours of downtime	60	120	180	240	300	360	420	480
Annual cost in dollars at $25 per hour	1,500	3,000	4,500	6,000	7,500	9,000	10,500	12,000
Annual cost in dollars after tax (50%)	750	1,500	2,250	3,000	3,750	4,500	5,250	6,000
Downtime cost in dollars per hour	0.38	0.75	1.13	1.50	1.88	2.25	2.63	3.00

TABLE 10-7-2. PRODUCTIVITY DIFFERENTIAL COST

	Years from Present							
	1	2	3	4	5	6	7	8
Productivity diff. factor, %	—	—	—	—	30	30	30	30
Extra hours required to equal new machine	—	—	—	—	600	600	600	600
Annual cost in dollars at $25 per hr	—	—	—	—	15,000	15,000	15,000	15,000
Annual cost in dollars after tax (50%)	—	—	—	—	7,500	7,500	7,500	7,500
Production diff. dollar cost per hour	—	—	—	—	3.75	3.75	3.75	3.75

TABLE 10-7-3. SUMMARY OF COSTS PER HOUR

	Years from Present							
	1	2	3	4	5	6	7	8
Net machine value loss	$ 5.00	$ 3.75	$ 2.50	$ 2.50	$ 2.50	$ 1.25	$ 1.25	$ 1.25
Investment, insurance, and taxes	3.38	2.72	2.25	1.88	1.50	1.22	1.03	.84
Maintenance and repair	0.88	1.00	1.58	1.88	6.13	4.38	3.50	3.00
Other operating costs	4.86	4.86	4.86	4.86	4.86	4.86	4.86	4.86
Downtime costs	0.38	0.75	1.13	1.50	1.88	2.25	2.63	3.00
Productivity differential cost	—	—	—	—	3.75	3.75	3.75	3.75
Total equipment cost per hour	$14.00	$13.08	$12.32	$12.62	$20.62	$17.71	$17.02	$16.70

Summary The separate cost factors affecting construction equipment have each been discussed and example cost data shown in the previous tables. The cost data are summarized in Table 10-7-3. Factual data compiled in such a form as this will allow the equipment manager to make effective, timely decisions.

All of the foregoing tables are in terms of constant dollars; there has been no attempt to build inflation into the future years. Price increases due to inflation may be considered, or may be ignored, depending upon the intended usage of the equipment cost data. When the data are to be used for comparative purposes, constant dollar value may be assumed.

Equipment managers should make cost estimates and prepare the cost schedules to reflect actual cost and usage experience.

Note that the points of change in the "maintenance and repair" and the "productivity differential costs" are critical. Table 10-7-3 contains the assumption that a new model with 30 percent greater productivity will become available in the fifth year. The estimated timing of these events is critical.

10-8. Using Cost Data for Equipment Replacement Decisions

The method of determining costs as described in the preceding section can provide a valuable tool for planning an equipment replacement program. However, many unpredictable occurrences may affect such a plan. Managers should reestimate costs periodically for each machine, considering extreme usage, repair costs, a large amount of idle time, or other unexpected conditions. Actual cost experience should always be reflected.

A new estimate should be made before deciding whether or not a specific unit should be replaced. In this way the corrected estimate will not only be up to date, it will be converted to current dollars.

Table 10-8-1 illustrates actual vs. estimated figures for the example $100,000 machine through the fourth year of ownership. The estimates for Year 4 through Year 8 have been examined to ensure their current validity, and revised as necessary. In Table 10-8-1, the former estimate is shown in parentheses (see Table 10-7-3).

Note that in Year 3 a new model was introduced with a 10 percent increase in productivity. This was not anticipated when the current unit was purchased. But it is now estimated that no further increase in productivity will be available in the next four years.

After preparing this revision (Table 10-8-1) reflecting cost of the machine presently owned, a similar analysis should be prepared for the new machine under consideration. Because of its improved productivity, the new model costs $115,000. After preparing tables for the new machine, assume that a summary gives the values in Table 10-8-2.

The economical time to replace equipment is when next year's cost of the *old unit* will exceed next year's cost of the *new unit*. In this case, the old unit will cost $17.65 per hour next year (Year 5, Table 10-8-1), but the new unit will cost only $16.50 per hour (Year 1, Table 10-8-2). The machine should, therefore, be replaced. On the other hand, if the major overhaul could be put off a year, reducing the repair and downtime cost by $2.00, it would be more economical to keep the old machine for another year.

TABLE 10-8-1. REVISED SUMMARY OF COSTS PER HOUR

	Actual Cost, in Dollars				Reestimation, in Dollars			
	1	2	3	4	5	6	7	8
Machine value loss	5.00	3.75	2.50	2.00	2.00	1.75	1.25	1.25
	(5.00)	(3.75)	(2.50)	(2.50)	(2.50)	(1.25)	(1.25)	(1.25)
Investment cost	3.38	2.80	2.40	2.10	2.00	1.50	1.25	1.00
	(3.38)	(2.72)	(2.25)	(1.88)	(1.50)	(1.22)	(1.03)	(.84)
Maintenance and repair	0.95	1.10	1.90	2.00	5.25	4.75	4.25	4.50
	(0.88)	(1.00)	(1.58)	(1.88)	(6.13)	(4.38)	(3.50)	(3.00)
Other operating costs	4.85	4.94	5.02	5.14	5.15	5.15	5.15	5.15
	(4.86)	(4.86)	(4.86)	(4.86)	(4.86)	(4.86)	(4.86)	(4.86)
Downtime	0.40	0.75	1.35	1.60	2.00	2.30	2.75	3.50
	(0.38)	(0.75)	(1.13)	(1.50)	(1.88)	(2.25)	(2.63)	(3.00)
Productivity differential	0	0	1.25	1.25	1.25	1.25	1.25	1.25
	(0)	(0)	(0)	(0)	(3.75)	(3.75)	(3.75)	(3.75)
Total	14.58	13.34	14.42	14.09	17.65	16.70	15.90	16.65
	(14.50)	(13.08)	(12.32)	(12.62)	(20.62)	(17.71)	(17.02)	(16.70)

TABLE 10-8-2. SUMMARY OF COSTS PER HOUR FOR NEW MACHINE UNDER CONSIDERATION

	Years from Present							
	1	2	3	4	5	6	7	8
Machine value loss	$ 5.75	$ 4.25	$ 3.00	$ 3.00	$ 2.75	$ 2.75	$ 2.75	$ 2.75
Investment cost	3.85	3.10	2.60	2.15	1.75	1.40	1.20	1.00
Maintenance and repair	1.10	1.25	1.75	2.15	7.00	5.00	4.00	3.50
Other operating costs	5.25	5.25	5.25	5.25	5.25	5.25	5.25	5.25
Downtime	0.55	1.00	1.25	1.50	2.30	2.40	3.00	3.50
Productivity differential	—	—	—	—	1.00	1.00	1.00	1.00
Total	$16.50	$14.85	$13.85	$14.05	$20.05	$17.80	$17.20	$17.00

All costs have been reduced in these examples to costs per hour because many machine users are accustomed to thinking in those terms. The tables and comparisons could have been made in terms of annual costs with the same validity. If this seems more meaningful, omit the last line on the first six tables and prepare Tables 10-7-3, 10-8-1, and 10-8-2, using annual costs. All figures will then be 2000 times larger but will have the same relationship to each other.

10-9. Selecting Construction Equipment

Lowest overall cost is the prime consideration in deciding when to replace a piece of construction equipment. Once the decision has been made to buy a new machine, there are generally four factors to consider:

- Machine productivity
- Product features and attachments
- Dealer support
- Price

Machine Productivity In some types of operation, the production requirement is a known quantity. The best size of equipment can be chosen to deliver that production at the lowest cost. More frequently, determination of machine size is not this simple.

Past experience is a major factor in determining the size machine that will deliver the best productivity for the job. The buyer's own experience can be supplemented, upon request, by that of his equipment dealer. Today, many

dealers are equipped to provide data and counseling on machine capability under various operating conditions.

Distinguish primary usage from secondary usage. Suppose that a track-type tractor is under consideration. Is it being purchased primarily for bulldozing, land clearing, ripping, or push-loading scrapers? Primary usage is an important consideration in determining size and in choosing machine attachments. Efficient productivity in the primary usage should be given priority. Some compromise may be demanded in secondary usage.

Operating conditions have a substantial effect on machine productivity. Wheel tractor-scrapers, for example, are more productive on a flat, straight, well-maintained haul road than they are on an uphill, winding, rough road. They can be loaded faster with some materials than others. Some materials are difficult to eject and spread on the fill, whereas others dump clean and fast. All conditions to be encountered in the work cycle must be taken into account in assessing machine productivity.

Conditions under which construction equipment operates are seldom static and predictable. There is no substitute for experience and judgment in defining them. "Experience" is having a good understanding of the area in which machines will work. Topography, soil types, and climate are major factors. "Judgment" is the ability to evaluate possible combinations of operating conditions in practical terms. What will the machine encounter, most of the time, during its productive life? Local equipment dealers can be of real help in evaluating local job conditions and the effect upon a specific machine's productivity.

When considering larger pieces of equipment, transportability between work sites becomes vital to a final choice. Determine the legal restrictions to movement over public highways. Are special permits obtainable? Can the machine be partially disassembled or otherwise adapted to legal transport?

Operator wage can be a factor to consider in size selection. As the size and productivity of a machine increases, the operator's wage becomes a smaller percentage of the cost of moving a cubic yard of material. However, a machine that is too large can be an economic burden and an application handicap. Conversely, a machine that is too small will probably generate abnormal maintenance and repair costs. Very likely it will face applications that actually constitute machine abuse. The objective is to determine the right size for the work to be done.

Product Features and Attachments

Product Features and Attachments   Construction equipment is available with a wide variety of features and attachments that offer greater productivity, broader applications, versatility, increased operating safety, and improved operator convenience. These features should be evaluated with complete objectivity. Will they increase the productivity of the machine? Will they enable the machine to work in an area or do a job that it normally would not be able to do? Are they necessary to ensure the safety of operators and other personnel?

Machine productivity can be increased in many ways. Choosing the right machine with the right attachments is one of them. Automatic bucket controls, special-purpose buckets, and optional counterweights may increase wheel loader production, for example.

Production can also be increased by decreasing downtime. The ability to diagnose malfunctions and correct them quickly, or the ability to spot potential trouble during routine maintenance and correct it before expensive overhaul is necessary, also will increase productivity.

Adding special attachments, such as a retarder for speed control on grades, a scarifier, or special application cutting edges, can enable a machine to do work that ordinarily it might not be able to do economically.

Safety features deserve particular attention. The safety of operators and other personnel is a machine owner's responsibility that cannot be overemphasized. The availability of special cabs, warning lights and indicators, windshield wipers, defrosters, and seat belts should all be examined and evaluated.

Selection of tires for wheel-type machines, and track for track-type units, is frequently a critical process. Special tires are available today for high-speed hauling applications, for abrasive underfoot conditions, or for work where flotation is a problem. Track options are available for track-type tractors to meet a variety of special working conditions. Technical counsel should be sought when unusual conditions are expected. Working together, the machine user and equipment suppliers can achieve the most economical answer to a specific tire or track selection problem.

Experience, local regulations and practices, and factual cost and production records should enable the prospective purchaser to make a sound choice of machine features and attachments.

Dealer Support   From the moment of purchase to the final resale, the equipment dealer plays an important part in determining whether a particular machine can become an efficient part of an economical earth-moving system.

The availability of parts, service facilities, and qualified personnel must all be considered. No machine can work forever. When it does require service, will there be trained personnel available with the proper tools and equipment to diagnose and correct the problem? Are parts available or must a machine sit idle waiting for them? Can major assemblies be quickly and easily changed, and are replacement assemblies available on an exchange basis?

Dealer support can take many forms. Agreements are offered that provide regular machine inspection, service, guaranteed availability, and predetermined prices when the machine is returned for trade. Long- and short-term lease arrangements can be negotiated to guarantee availability, allow for purchase, or guarantee a total cost per machine hour.

Price   The final factor to consider in machine selection is price. Purchase price should be just one part of the overall selection process; resale price should also be considered, along with maintenance and repair costs. Although a machine costs less initially, it may be more expensive during its working life. A low initial price is a worthwhile consideration only when coupled with satisfactory performance and dealer parts and service support.

All of the factors must be considered and weighed to arrive at the best decision. The total cost of owning and operating a machine, and not the machine price, should be the decision maker in equipment selection.

11 CHARGES FOR USE OF EQUIPMENT

MARK A. ROBINSON
Director of Procurement
Morrison-Knudsen Company, Inc.
Boise, Idaho

ONE of the largest single factors of construction costs is the investment in and the operation of equipment. Proper management of this important asset is essential for any successful construction company.

Funds to purchase equipment are generated from the sale of common or preferred stock, debentures, bank loans, or equipment notes. At times, however, it may be more desirable to lease equipment (see Section 11-3).

Once a piece of equipment has been acquired, it is important to utilize it, because idle equipment cannot earn any profit. In order that it may be used, it is essential to maintain its condition through preventive maintenance; that subject was covered in Chapter 9.

11-1. Charges Affecting Ownership Costs

A clear distinction should be made between ownership and operating costs. There are six items that make up annual equipment expense: Interest on investment, taxes, insurance, depreciation, storage and overhead, and overhaul or major repairs. Not included would be normal operational costs, such as operator's wages, fuel, lubricants, minor repairs, supplies, loading, erecting, dismantling, and transporting.

Interest on Investments In order that a contractor know his true equipment expense, he should consider his average cost of obtaining money to purchase equipment. Therefore, interest expense should be calculated on normal bank loans, both long term and short term, equipment notes, and any other form of borrowing. Many times money is borrowed and a portion is left on deposit or is required by the bank as a compensating balance. This must be considered in determining true interest expense.

Taxes A fixed equipment expense is the personal property tax that is assessed normally by the county assessor in which the equipment is located. This tax varies by state and by county within the states. The costs are normally absorbed as a part of home-office equipment expense.

Insurance Equipment should be covered from the standpoint of operational liability for damage to persons or property as well as loss or damage to the equipment itself. This is discussed in detail in Chapter 29. Premium expense for this coverage is definitely a part of equipment expense.

Depreciation Although there are many variations of equipment depreciation allowed by the Internal Revenue Service, the three most common methods are straight line, declining balance, and sum of the digits.

The straight-line method is the most common and the simplest to compute. The estimated salvage value is deducted from the original cost and this balance is amortized over the estimated life of the unit. For example, assume that a tractor cost $50,000, had a useful life of five years, and was estimated to have a salvage value of $5,000. To calculate depreciation: $50,000 − $5000 = $45,000 × 20% = $9000 depreciation per year.

The *declining-balance method* allows for greater depreciation in the earlier years of the life of the asset. Normally a rate equal to twice that of the straight-line method is used, but the depreciation is subtracted each year from the cost, so that it applies to a declining balance. Assuming the same facts as in the straight-line example, the first year depreciation on the $50,000 tractor would be 40% × $45,000 or $18,000 the first year, 40% × $27,000 or $10,800 the second, 40% × $16,200 or $6480 the third, etc., until the unit is down to its estimated salvage.

The *sum-of-the-digits method* also allows for greater depreciation in the earlier years of the life of the asset. A different fraction is applied each year to the cost less salvage value of the asset. The bottom number of the fraction

TABLE 11-1-1. COMPARISON OF EQUIPMENT DEPRECIATION METHODS

Assume: Cost $50,000
Salvage 10% or $5000
Useful life   5 years

Years	Cost To Be Depreciated	Depreciation Rate	Amount	Balance
Straight Line				
1	$45,000	20%	$9,000	$36,000
2	45,000	20%	9,000	27,000
3	45,000	20%	9,000	18,000
4	45,000	20%	9,000	9,000
5	45,000	20%	9,000	0
Declining Balance				
1	$45,000	40%	$18,000	$27,000
2	27,000	40%	10,800	16,200
3	16,200	40%	6,480	9,720
4	9,720	40%	3,888	5,832
5	5,832	40%	2,333	3,499
Sum of Digits				
1	$45,000	5/15	$15,000	$30,000
2	45,000	4/15	12,000	18,000
3	45,000	3/15	9,000	9,000
4	45,000	2/15	6,000	3,000
5	45,000	1/15	3,000	0

or denominator remains constant and is the sum of the digits representing the years of estimated useful life of the asset. In the previous example of five-year life, the denominator would be 15, which is the sum of $1 + 2 + 3 + 4 + 5 = 15$. The numerator or top number of the fraction is the remaining life of the asset. On a five-year life, for the first year the fraction would be $5/15 \times \$45,000$ or $15,000 depreciation. The second year would be $4/15 \times \$45,000$ or 12,000.

A comparison of the three methods is shown in Table 11-1-1. It is important for all contractors to establish a depreciation method with the Internal Revenue Service.

Storage and Overhead Most contractors maintain some sort of a yard where equipment and supplies are stored between jobs. The cost of operating this yard and the overhead related thereto is a proper equipment expense.

Overhaul or Major Repairs A part of ownership cost is the cost of those major overhauls that are usually accomplished between jobs and are done at the contractor's home shop or by dealers. These costs are separate from running repairs, which are usually minor in nature. With the increasing component exchange program that is being offered by many dealers, there will be fewer major overhauls. Most components can be readily removed and replaced in the field. With planned replacement of major units, an overhaul as such would normally not be necessary.

11-2. Equipment Records

It is imperative that accurate and complete records be maintained on major equipment. This is not only necessary from the accounting standpoint, but from the control position.

Original Information At the time purchase is approved by management, a purchase order should be written giving model numbers and a full description of the item. The vendor at the time of invoicing should show this complete description, including options called for as well as serial and engine numbers. Payment should not be made unless this information is furnished. Likewise, an instruction manual and parts book should be furnished at this time.

Master Card A master card should be made at the time of payment. This card gives a complete description of the equipment, including model, serial number, engine number, date of purchase, cost, weight, original freight, and a description of any additions or changes. This card can also serve as a master location file. A representative sample is shown in Figure 11-2-1.

Depending on the sophistication of the control, a copy of the equipment card should follow the equipment and be retained in the job office. If the company is operating on a divisional or district basis, a copy should also be retained there.

Preventive Maintenance File At this same time a preventive maintenance folder should be made up and forwarded to the field with the equipment.

Sub-Ledger A sub-ledger is necessary in the home office to record the investment depreciation and investment tax credit. This sub-ledger is kept in balance at all times with the equipment asset account and the depreciation reserve. A computer listing is the most common.

11-3. Lease or Buy?

There has been a definite trend in recent years toward leasing of equipment. The question of lease or buy has been and is continually being appraised in the construction industry, which perhaps leases more than other industries.

The dictionary describes a lease as "A contract by which one party (Lessor) gives another party (Lessee) the use and possession of property for a specific time for fixed payments." This differs from other forms of acquisition inasmuch as the ownership is vested in the lessor, but the lessee has possession and full use of the asset. Leasing has

Form No. ....

EQUIPMENT CONTROL CARD

				Code No. ....
Type	Size	Make		Company No. ....
Model	Serial #	Eng. No.		_____
Wt.	Year	New	Extras Added	
P.O. #	Vo. #	Used		
Vendor		Date Purch.		_____

	Date	Reference	Amount					
Initial cost				Transferred to				
Sales tax				Contract	Dist.	Date	Ref.	
Freight								
Additions								
Transferred to contract	Dist.	Date	Ref.					

Figure 11–2–1 Equipment control card

also been described as a means of making a profit with the capital of others. Although this later statement depends on circumstances, leasing is a definite way to acquire medium and long-range capital.

In construction, short-term leasing is quite common primarily because a special machine is required for a limited time. Other reasons for leasing include:

- Use of newest most productive machine.
- Easy way to pay.
- Payment fully deductible for tax purposes as opposed to depreciation.
- Try before buy. The particular machine may be an entirely new model and the contractor may wish to experiment with it before committting capital.
- Conserve cash from the standpoint of no down payment as with normal equipment financing.
- Obsolescence not the responsibility of the lessee.
- Minimize maintenance by use of newer machine.
- Manufacturer may wish to control distribution of his machines and therefore will only lease.

A well-financed company should generally stay away from most leases except short-term requirements.

1. Many leases end in ultimate acquisition, either from a start of a lease on a rental purchase basis or lease with purchase at fair market value. This is the most expensive form of acquiring equipment.
2. Particularly in construction, it will be determined at job level that a certain piece of equipment will be re-

quired for a limited time and a lease will be executed. Job conditions may change or a superintendent may decide that he must have this particular item beyond the lease period. If this is not carefully policed, a contractor will pay for the equipment more than once, including financing charges.
3. Depending on the nature of the lease involved, the question of the Investment Tax Credit is sometimes unresolved. There is no question on a direct ownership.

11-4. Types of Leases

In addition to the short-term lease that is in a category by itself, basically there are two types of leases: the *financial* and the *operational*. In addition to these there are leases with a purchase option. Also, particularly in the automotive field, *open-* and *closed-end leases* are offered.

Financial Lease This type is strictly as described and normally is for the useful life of the asset. The lessor recovers the cost of the asset, plus interest or carrying charges, over the term of the lease.

Operational Lease This type is for a period less than the normal life of the asset and, therefore, the commitment is for less than the purchase price of the asset. In this type of arrangement the lessor will perform some sort of service in addition to furnishing the asset, such as routine maintenance. Also, there are other factors regarding utilization of the asset that would change the lease payments, such as a base price, plus a premium of so much per yard over a minimum established quota. In business other than construction the premium could be calculated on a percentage of sales. Charges for this type lease will include the carrying or financial, the ownership risk as applied to obsolescence, and a charge for additional services performed.

Closed-End Leases   In this type of lease the lessor furnishes the equipment for a specified length of time, normally less than the normal life, and he takes it back accepting normal wear and tear. The lessee is expected to repair any damage before returning the unit.

Open-End Leases   In this type of lease the lessee returns the unit after a specified period of time and the lessor sells the unit on the open market. The lessor has maintained a depreciation reserve during its life. Sale prices exceeding the book value are credited to the lessee, and sale prices less than book value are charged to the lessee.

11-5.   Terms and Conditions of Leases

Anyone entering into a lease agreement should fully understand all the terms and conditions, because conditions vary, as do many lease forms and practices. Some of the points to consider follow:

Length of Lease   The starting time should be defined as well as the end. Also, if the lease terminates prior to the ending time given, what, if any, is the penalty for the shorter period?

Hours of Operation   First, how are these determined— by operators' hours or engine hour meter readings? What are the agreed operating hours per month? Many firms leasing equipment use 176 hours as a base for single shift, but some rental firms use 200 hours per month. What is the premium for second- and third-shift operation? Normal practice is 50 percent additional rental or lease payments for each additional shift beyond the first shift.

When Are Lease Payments Due?   Normal practice is that these are paid in advance on a monthly basis.

Is a Purchase Option a Part of the Agreement?   If so, are the purchase price, percentage of rentals to apply, option date, interest charge, and purchase discount clearly spelled out? By including the purchase option the deal is not a true lease, but should be considered, particularly if the usage time is indefinite.

Who Bears the Expense of Maintenance and Major Repairs?   Normally, maintenance and operating repairs are borne by the lessee, with the lessor assuming the responsibility of such major repairs as engine or transmission failures.

Taxes and Insurance Responsibility   These should be clearly spelled out. Normally, but not always, they are the responsibility of the lessee.

Transportation Costs Must Be Defined   Normally, the lessee pays both ways, but sometimes the lessor will assume return freight on the assumption that another lease can be arranged. If erection of machines is involved, the responsibility for the expense must be clarified.

Condition of Equipment upon Return to Lessor   This condition must be defined. Some rental or leasing firms have tire or track clauses in their agreements that the equipment must be returned with a certain percentage of tread on tires or a percentage of track wear remaining. Some tire clauses charge the lessee for every thirty-second inch of wear beyond new condition.

11-6.   Internal Charges for Equipment Usage

Any construction company, regardless of size, should establish a firm basis for charging usage of equipment on their contract work. Costs as set forth in Section 11-1 are the basis of these charges. Older companies with good cost records would use their costs as a basis. Those without historical records or with new types of equipment can start with a percentage of the rates as set forth in the *Associated Equipment Distributors Compilation of National Averaged Rental Rates,* hereafter referred to as AED, or the Associated General Contractors of America *Contractors Equipment Ownership Expense,* referred to as AGC rates. Other possible basis of charges would be a percentage of what is known as Blue Book rentals, which is published by the National Research and Appraisal Company.

Normally, a construction company would set up an equipment pool, either on an overall basis or divisional or departmental basis. This would be administered by either the procurement director or the equipment superintendent. To the pool is charged the expenses enumerated in Section 11-1 and credited to this fund are the intercompany rentals.

Rentals are normally based on 176 hours or 200 hours per month, with a 50 percent additional charge for second and third shifts. Rentals should be charged upon arrival of the equipment at the job site and cease when the equipment is shipped from the job or is declared idle and available for transfer. As some project managers or superintendents have a tendency to keep equipment on their job beyond the time it is needed, it is well to enforce this rule.

An annual review must be made of the equipment by class to see that the rental rates being charged cover the costs that are incurred. The intent of an equipment pool covering straight contract work is to break even. There is no point to accumulate excessive profits in the equipment pool at the expense of the jobs, nor is there any point in recording losses in the pool which benefit the jobs. Correct equipment costs should be charged so that job costs are accurate. These costs then can be used for future estimates.

11-7.   Associated General Contractors' Ownership Expense

Basis for Normal Usage   The AGC schedule of rates is expressed as a percentage of original cost, including original freight from factory to job site and including original erection. In periods of fluctuating equipment prices, this would seem to be more accurate than other methods.

This percentage or expense per working month is based on the following six items: Depreciation, major repairs or overhaul, interest on investment, storage, incidentals and equipment overhaul, insurance, and taxes.

As a part of this calculation is the average number of working months per year based on average climatic conditions, normal business conditions, and normal seasonal fluctuations in the construction industry.

A typical calculation of the AGC manual for tractors, crawlers, diesel engine, power shift, and drawbar horsepower 165 to 400 is:

Depreciation, %	20
Overhaul, major repairs, paint, %	15
Interest, taxes, storage, insurance, %	11
Total ownership expense, %	46
Average use, months per year	7
Expense per working month = 46 ÷ 7 = 6.6%	

Let us assume that a contractor through experience has found that 70 percent of AGC will support his equipment pool for internal operations. To determine a proper rate on a new tractor costing $70,000, including freight, we apply the above rate on experience percentage. $70,000 × 6.6% × 70% = $3234 internal rental per month.

Shorter Working Seasons The length of the working season can materially effect internal rates. An example would be a contractor working in the continental United States as well as in Alaska. An adjustment, therefore, must be made to take into consideration the shorter working season and higher operating costs encountered. The premium for northern operations would normally be in the 25 to 30 percent range, which means the rate shown above would be as follows: Alaska rate, $3234 + 30% = $4204.

Types of Jobs Another possible consideration that at least should be watched is the difference between severe and ordinary jobs. A contractor working on a rock job will experience more costly equipment repairs than one working on a dirt job. Major overhauls will be more frequent and replacement sooner. Depending upon the size of the equipment pool and type of jobs underway during a particular period, it may be necessary to adjust rates to compensate for this situation. An example would be:

Job A—Severe conditions, rock	
Tractor internal rate	$3234
Adjustment + 10%	323
Rate to be charged	$3557
Job B—Mild conditions, common	
Tractor internal rate	$3234
Adjustment − 10%	323
Rate to be charged	$2911

New and Used Rates Another possible arrangement that can be made in setting internal rates is differentiating between new and used equipment. It is a logical fact that new equipment is less costly to operate than used equipment. In an equipment pool operation, equipment operating costs are absorbed by the job. It is, therefore, common practice to charge a higher rental for a new piece of equipment for a limited period, such as the first twelve or eighteen months of its life. By charging more in the early life of the equipment, the pool is kept in better balance, particularly where the sum-of-the-digits or declining-balance method of depreciation is used. The pool must absorb much higher depreciation during this period.

Management may wish to collect new rentals for twelve or eighteen months and at the end of this period reduce these rates 20 to 40 percent. The previous example of a tractor that carried a normal internal rate of $3234 would carry say a 70 percent rate of $2264 for its life beyond eighteen months.

Job Owned In certain circumstances it may be desirable for an individual job to own its own equipment.

This would apply to a long-term continuing contract, such as a mining operation. In this case, depreciation would be taken on the home-office books to coincide with the other company-owned units. This would be charged to this job along with any other costs relating to the equipment. The job would absorb its own overhaul.

11-8. External Charges for Equipment Usage

Joint Ventures There are times when a construction company will form a partnership or joint venture with another firm or firms to do a certain job. Provision should be made to establish rental rates on equipment to provide a profit. This, of course, depends on the participating interest of the partners and must be adjusted accordingly

It is standard procedure for a joint venture to own its own equipment on most jobs of any duration. Equipment charges to work items are handled on the basis of actual costs, as discussed previously. Partners furnishing additional equipment, usually for short periods of time, would set their rates in the area of reasonable third party rates which will allow some profit.

Rates for Extra Work and Claims On any contract there are usually extra work items that are authorized and paid for by the client during the normal progress of the job. Some agencies, such as the Corps of Engineers and some state highway departments, publish and incorporate into the contract schedules of equipment usage rates. These are normally calculated on a cost basis with the understanding that profit and overhead are allowed separately. If these are a part of the contract, the contractor is bound to use them. In the absence of these, the rates should be negotiated on the basis of AED rates or AGC rates with a percentage for profit.

Rates for Cost-Plus, Fixed-Fee, or Target-Estimate Contracts Frequently a construction company has an opportunity to submit a proposal on cost-plus or fee work. Sometimes this is done by submitting a target estimate. Provision must be made for charges to the client for equipment usage. In this situation, in determining rates the length of the contract is an important consideration. Other factors such as the amount of equipment, the general availability of equipment, and general market conditions must be considered. In arriving at an understanding on rates, the number of shifts contemplated must be discussed, as well as the rates for additional shifts. Likewise, it is well to provide for a standby rate in case of winter shutdown or unseasonable weather.

In cost-plus work, it is normal that all operating repairs be included as costs. There is always a question on whether major repairs are included in rental rates or are to be treated as reimbursable costs. This should be clarified at the outset. If major repairs can be included in costs, there will be less chance of disagreements. This should be arranged at the expense of reducing overall rates, if necessary.

11-9. Plant Equipment

Distinguishing Between Types Depreciation and charges for mobile construction equipment should be handled as set forth in Sections 11-6 through 11-8. Specialized plant items, which are necessary on larger construction projects, such as dams and tunnels are normally handled differently, both as regards depreciation and rental or usage. The salvage value is much less, which means that the cost of these two items must be written off during the progress of the work.

Method of Depreciation   Plant items can be depreciated on a time basis as is usually done on mobile equipment, but it is usually desirable to write plant off on a unit basis. In the case of tunnels, a footage basis is customary. For concrete dam on the concrete portion of the plant, a yardage basis would be used. The same would be true on an earth-fill dam for items applicable to the dam fill.

[*Dam Plant Items*]   A large plant is normally required for the construction of concrete dams, starting with the plant for concrete aggregate, the concrete mixing plant, placing units such as cableways, whirlies and trestle, temporary buildings, and air, water, and power lines.

[*Tunnel Plant Items*]   For underground work, plant items normally consist of temporary buildings, including office, shop, warehouse, and compressor house. Also required are air, water, ventilation, and power lines. Tunnel jumbos with drills have little salvage and, therefore, must be written off as plant. Compressors, muckers, cars, and pumps have a longer life and will normally be depreciated in a normal manner.

11-10.   Transferring Equipment Between Jobs

Any medium-sized or large construction company will have more than one project underway at the same time. Each job is a separate profit center and, therefore, accountable for all costs related to its operation, including equipment costs. For a company to utilize its equipment to the fullest, it is, therefore, necessary to transfer it between jobs.

Preinspection   The company equipment superintendent or master mechanic should arrange to inspect all equipment prior to transfer. In this way the last using job can be instructed to make any required repairs that were the results of their operations. In some instances the equipment superintendent may authorize shipment prior to the repairs being made due to the time factor. Having inspected the equipment, he is in a position to authorize the receiving job to make the necessary repairs and charge back these repairs to the last using job.

Accounting for Transfers   Any equipment pool can be broken down into divisions, districts, or areas. Each division accumulates costs and revenue in the form of rentals to jobs. As equipment is transferred between divisions, it is important to record the cost, less depreciation or book value, in the proper division so that current depreciation is charged to the correct account.

A transfer card or form of some kind should be made out the day the equipment is shipped. This should list the item by company number, show serial number, and list any attachments. A form in triplicate is desirable, the original going to the home office with a copy to the receiving job and a copy retained by the shipping job.

Cost and Load Out   A simple rule to follow in handling load-out costs is to have the shipping job absorb load-out on direct transfers between jobs. When the central shops ship, the pool owning the equipment should be charged. Sometimes a job is finished and there is no place to ship the equipment. This happens frequently in the case of large draglines. This cost also should be charged to the last using pool account providing the job books are closed.

Transportation Expense   The receiving job should pay all incoming freight on equipment, with one exception—freight on new equipment. The Internal Revenue Service normally requires that first freight be capitalized with the cost of the machine.

If equipment is not shipped to another job, the last using job would normally pay for the return of equipment to a district or home-office yard.

In the case of rail transportation, it is important to investigate the possibility of "storage in transit" possibilities, particularly in regard to the home office or yard. "Storage in transit" is a term used when commodities, including equipment, are shipped from one point to another point for storage and after storage are shipped on to a third point. Railroads normally offer special rates under these circumstances that result in substantial savings. For example, let us assume that a contractor headquartered in Salt Lake City, Utah, has a job on the West Coast in the vicinity of Sacramento, California. He completes the California job and has no place at present to use a power shovel weighing 200,000 pounds. He decides to ship this machine to his yard in Salt Lake for storage. The bill of lading is marked "For Storage in Transit" and he pays the normal rate of $1.30 per hundred weight. Later this contractor obtains work in North Carolina where he can use this machine. Normal freight costs would be:

Freight—Sacramento to Salt Lake City	
200,000 lbs. @ $1.30	$ 2,600
Freight—Salt Lake City to Columbia, North Carolina	
200,000 lbs. @ $3.90	7,800
Total	$10,400

By having the bill of lading marked "For Storage in Transit," the transcontinental rate of $3.90 cwt., would apply as follows:

Freight—Sacramento to Salt Lake City	
200,000 lbs. @ $1.30	$2,600
Freight—Sacramento to North Carolina	
200,000 lbs. @ $3.90 − $7,800	
Less Sacramento to Salt Lake previously paid—2,600	5,200
Total	$7,800
Savings—$2,600	

Arbitration of Disputes   Good practice dictates that working equipment should be carefully checked over before being shipped to a job. Most project managers or superintendents will always make available their poorest unit for transfer. As mentioned previously, wherever possible, arrangements should be made for a pre-inspection prior to transfer. In instances where this is not possible and a failure occurs, the company equipment superintendent is the logical one to allocate the costs incurred. A full report would be requested from both job master mechanics, as well as a copy of the preventive maintenance record. From this the equipment superintendent makes his decision. If his decision is not satisfactory to either or both parties, the matter is referred to the operations manager or next in line of authority.

11-11. Operating Cost vs. Overhauls— Determining Replacement

To manage an equipment fleet properly, it is necessary that accurate records be kept of all operating costs. This not only provides management information for bidding purposes, but indirectly assists in determining replacement time. Supplemental records should be maintained at the job level to record downtime, availability, and performance.

Normal operating repairs may be broken down in the following categories:

1. Repair parts
2. Oil and grease
3. Fuel or power
4. Tires and tubes
5. Cable and teeth
6. Outside repairs
7. Shop costs
8. Rental or depreciation
9. Operating labor
10. Repair labor

Items 1 through 7 are usually totaled for a supply subtotal, and 9 and 10 are combined to give a labor subtotal. Day-to-day maintenance items are considered a part of operating costs, whereas a complete check and reconditioning of major components would be classed as overhaul chargeable to the equipment pool. These costs are provided for when the rental rates are established. This would normally occur at the conclusion of one job and prior to the start of another. This major repair can be accomplished by: Last using job, dealer facility, contractors central or district shop, and receiving job.

Management should carefully analyze all related costs, including labor rates, source of parts, and transportation expense before determining the place of overhaul.

11-12. Disposition of Equipment

It has been said many times that the timing on the disposition of equipment is as important as the original decision to purchase. Disposition time would be determined by a combination of equipment condition, current market conditions, and availability of new more efficient machines.

At the time management agrees to dispose of a piece or a fleet of equipment, the equipment superintendent or master mechanic should be instructed to steam clean, make any obvious repairs, and paint. Major or costly repairs at this point are to be avoided, as the cost of this is not recovered in the sale price.

General guidelines with regard to sale price should be established at this time with whoever is handling the sale and management. The following points must be considered: Age, condition, whether or not current model, book value, and replacement cost.

Once the decision has been made to dispose of a piece or a fleet of equipment, three alternatives are open: trade to dealers, sale to third parties, or auction sales.

11-13. Trade to Dealers

A dealer handling a specific line of machinery is normally in a better position to merchandise used items of the same make. Although the Internal Revenue method of recording trades is not always advantageous to the contractor, trading in of like items is often the logical way to dispose of used and to procure new equipment.

Contractors operating in the northern climates, where the working season is eight months or less, many times turn in old equipment to dealers in the fall against delivery of new equipment in the spring or at the start of their next contract. Although this allows the dealer to operate on the contractor's money in the amount of the true value of the equipment until replacement, the dealer will usually overallow sufficient to offset this. The dealer will overhaul this equipment during the winter so that it will be ready at the start of the next construction season for his customers who prefer to buy good used equipment instead of new.

Should a contractor secure new work out of the operating area of the dealer that made the trade, he must look to other dealers to service the new equipment that he purchases. Many manufacturers make provision for this by allowing the servicing dealer a fixed percentage of the profits on the sale. This is withheld from the selling dealer.

It is well to secure bids on trades from more than one dealer. Further, the dealers over-allowance which would be given as a discount on an outright purchase must be considered. In fact, an over-allowance on any trade must be adjusted so that some discount is taken on the new purchase. If this were not done, a contractor's equipment assets will eventually end up being overstated.

11-14. Sale to Third Parties

Equipment that can be disposed of to outsiders or even dealers for cash is most desirable. This puts cash in the bank promptly and leaves the contractor in a better position to replace when and where he wishes. He also is in a better position to select the proper make, type, and size.

Used-equipment dealers who work on a straight commission are usually helpful in handling dispositions. In such transactions, prices must be set so that dealers are protected for their commissions. It is well to insist that they advise of any quotations they make or, in other words, that they record their customer. In this way the seller can protect these dealers. Likewise, the contractor will make casual sales at published prices that include dealer commissions.

In selling equipment outright it is necessary to advertise. This can be done in local newspapers, contractor publications, trade journals, or company published mailing lists or bulletins. Most larger construction companies prepare periodic sales bulletins, usually every six months.

11-15. Auction Sales

For fast liquidation of larger equipment fleets, an auction is normally the best procedure to follow provided that certain fundamentals are followed. An auction is usually justified when a company is curtailing operations or has hit a slack period that has resulted in a large accumulation of used equipment. Frequently, at the conclusion of large joint-venture projects, it is desirable for the partners to liquidate the assets and distribute cash. This is much faster than routine sales and stops selling and yard expense in short order.

Types of auctions include:

Straight Commission There are many reputable auctioneers available with a variety of terms. The most common is the auctioneer who sells for a straight commission. The cost of preparing the equipment for the sale is borne by the owner. Advertising and all sale expense is for the account of the auctioneer. Collections are also made by the auctioneer.

Guaranteed Minimum   Here the auctioneer guarantees a minimum total sale proceeds. Anything over this minimum is usually divided between the owner and the auctioneer. The cost of preparing the equipment is handled in various ways, but normally it is the owner's responsibility.

Outright Sale   Lastly, there are some auctioneer selling organizations that will purchase large lots of equipment outright, prepare it for sale at their expense, and then hold an auction. This type of selling organization will normally combine the equipment of more than one owner.

11-16. Auction Preparation and Planning

To have a successful sale it is important to have a good variety of items and some better than average major items, such as several good current crawler tractors with scrapers and graders. An auction that offers only older obsolete units will not attract good buyers. The latter type can be sold, however, with the better units.

Surplus materials, minor equipment, and small tools, if lotted properly, will normally sell well and bring their fair value.

As mentioned previously in this section, it is recommended that all sale items be cleaned and painted. For an auction all items should be repaired so that they are in running condition. Prospective buyers are invited to inspect the sale items a few days before the auction date. Buyers interested in the larger units will usually start the equipment so that they can test the controls and operation. An hour before auction time on sale day all engines on all equipment are started.

It is desirable to select a sale site close to the location of the major items to be sold. Any site selected should be within a reasonable distance from a commercial airport and to overnight accommodations. Likewise, the site should be graveled or paved and have good drainage. Temporary sanitary facilities are necessary, as is commercial transportation, including rail and truck.

It is desirable from the owner's viewpoint to hold an auction as soon as possible once the decision has been made to hold one. Sufficient time should be allowed for preparation, and consideration should be given to weather, time of year, holidays, and day of week. In the northern climates, late fall, winter, or early spring auctions are not desirable. Auctions at the beginning or near the end of normal construction season are usually more successful. Auctions held from Tuesday through Thursday are more successful than those held at the beginning or end of the week. Buyers do not like to travel over weekends. Days close to holidays are to be avoided.

12 PURCHASING, EXPEDITING, TRAFFIC, AND TRANSPORTATION

JOHN LANGBERG

Procurement Manager (Retired)
International Engineering Company, Inc.
San Francisco, California

A GOOD PURCHASING department, properly utilized, will contribute substantially to the profit-making potential of the contractor's organization. Close coordination of the purchasing activity with management, engineering, estimating, operating, and other divisions or subdivisions of the company will exploit purchasing's potential for profit to the fullest.

12-1. Coordination with Management

Purchasing keeps management informed of important changes in price trends, delivery patterns, product and labor availability, general business conditions, new or radically improved equipment, materials or techniques, sales and development opportunities, or other information generated in its day-to-day operations which could affect company policy or activity.

Through such indicators as price changes, lead times, interest or lack of interest in bidding, firm pricing or price in effect at time of shipment, conversations with manufacturing company executives, and the like, the purchasing department can sometimes spot a changing situation a number of weeks before it becomes general knowledge.

12-2. Coordination with Engineering

Purchasing personnel work closely with engineering personnel at the headquarters, regional, and project levels to determine specific contract requirements to be purchased and subcontracts to be placed. They will normally review such awards with engineering prior to placement to assure completeness of bills of material and adequacy of specifications and technical write-up.

Purchasing furnishes the engineering department with information, catalog and specification data relating to new or changed design equipment, and new or improved materials, techniques, or procedures of which it may not be aware.

The purchasing department, when requested, contacts the logical manufacturing or supply sources for pricing information or data required by engineering personnel. It is preferable for purchasing personnel, rather than engineering, to make such inquiries and requests of vendors for several reasons. It saves engineering time for the initial contact and in the inevitable sales follow-up which, in addition to being time consuming, is usually unnecessary at this point. Another good reason for restricting such contacts with vendors to the purchasing department is the greater assurance that the source to be contacted will be determined with subsequent pricing in mind should the inquiry ever result in purchase. For example, an inquiry to a domestic dealer will sometimes adversely affect a manufacturer's direct pricing on a purchase for export. On a domestic purchase, it could involve two commissions—one for the dealer originally contacted and another for the dealer representing the manufacturer at the point of delivery.

12-3. Coordination with Estimating Group

Competent purchasing personnel, working closely with the estimating group, will develop better and more complete prebid coverage, thereby contributing substantially to the success of the overall bid effort. The direct participation of the purchasing department in estimating provides a more meaningful relationship with manufacturers, suppliers, and

175

subcontractors, both before the bid and after award, assuring more efficient placement of subsequent purchase orders and subcontracts and greater potential for cost reduction and profit.

Since prebid pricing for materials and work can and often does vary drastically, the intensive market coverage that the Purchasing Department can generate is necessary to assure the estimating group of prices and subcontract proposals as low as the one being received by competitive bidding groups. Further, a good purchasing department will have the resources and know-how to locate suppliers and subcontractors untapped by or not quoting the other bidders and who can offer the needed competitive edge for a tight bid.

Through its knowledge of marketing, price structure, discounts, sales practices, and business trends, purchasing personnel can help the estimators evaluate the pricing under consideration and can project, with some assurance, the extent of price reductions, if any, likely to be effected subsequent to award.

Purchasing will also review material costs included in subcontract proposals and determine areas for cost reduction. Sometimes these savings can be considerable, particularly with medium to smaller subcontractors who may not be aware of the most direct supply channels and lowest cost sources.

Working with company bidding groups at bid time during close out, purchasing people are best able to finalize sub and material quotes and contacts before and after bid.

12-4. Coordination with Operating Departments

Purchasing coordinates with or works in the contractor's operating department at every level—headquarters, regional or divisional, and project. Its major responsibility is to obtain the needed materials, services, and equipment in the quality, quantity, and time required, at the lowest ultimate cost.

It will usually be found that centralized headquarters purchasing combined with local job-site purchasing and expediting under headquarters control will provide maximum savings and trouble-free procurement. The contractor, operating on a national or international basis, cannot possibly staff each job with procurement specialists for the many varied categories of materials, equipment, subcontracts, and services that must be provided. Assuming that the contractor could afford to do so, which he cannot, there are not a sufficient number of such specialists available to man each construction job properly. Thus, the procurement specialists must of necessity be concentrated in company headquarters and will normally, in complete cooperation with divisional and project management and engineering, place the major orders for materials going into the construction (permanent materials), the major subcontracts, and all items of high dollar value. In addition, the central or headquarters buying staff will normally contract for major operating supplies and materials, providing for job-site release as needed, by authorized project personnel. Contract type items will usually include such items as tires; fuels, oils, and lubricants; portland cement; aggregates and similar items that can be purchased to best advantage by the more experienced buyers available to the home-office purchasing department.

Major equipment, defined by a number of contractors as any individual item of equipment valued at $1000 or more, is best purchased on basis of joint review and decision of project, headquarters, and purchasing management. This is of particular importance for the basic equipment on which the job will depend for its profitability—earth-moving equipment, hauling equipment, loading equipment, etc. Almost invariably, major equipment purchases are effected at the central headquarters purchasing department by staff personnel knowledgeable in the types of equipment under procurement and with the specification guidance of experienced equipment specialists usually found in the contractor's equipment division.

Projects, particularly those that are large, will be staffed with a project or field purchasing agent to handle the purchase of miscellaneous operating supplies, spare parts, and other items not feasible for centralized headquarters procurement. The project purchasing agent will also be responsible for the issuance, on a timely basis, of delivery orders or releases against purchase orders or contracts placed for the project by the central headquarters purchasing department. The project purchasing agent also has the responsibility for expediting deliveries and performance under all purchases and contracts for the project, regardless of whether centrally ordered or job ordered.

On very large projects involving many individual purchases of complex or engineered items and numerous subcontracts for various phases of the work, it will be advantageous to staff the project engineering section with a materials engineer. The materials engineer will make certain that the necessary purchase orders and subcontracts are placed to cover all items of materials and work required by project plans and specifications and will coordinate the overall expediting activity to assure subcontractor and materials availability on hand precisely when needed.

For domestic operations the percentage of centralized buying at the contractor's headquarters as compared to the percentage of decentralized purchasing at job site will vary considerably, even for similar projects. It will depend on a number of factors, such as size and type of job; its proximity to the contractor's home office; complexity, variety, quantities, and dollar value of materials required; number of major subcontracts to be placed; strength and caliber of project or divisional management; owner's wishes; and local relationships.

12-5. Purchasing Responsibilities and Procedures for Overseas Operations

Overseas procedures and operations cannot be as flexible as those for domestic jobs. Materials originating in the United States will obviously be purchased in the United States, either by the contractor's headquarters purchasing department or sometimes, in the case of very large projects of lengthy duration, from special stateside offices set up for the project to service its engineering, purchasing, shipping, accounting, personnel, and other requirements involving stateside coordination.

The project purchasing department at the overseas location will purchase and arrange for deliveries of products and materials available in that country and also neighboring countries, if import regulations and/or governmental attitudes or decrees do not restrict or discourage such import.

Purchases made for the project in other overseas countries distant from either the job or the United States will best be handled or controlled by the project. Often the larger contractor will have an area headquarters in or relatively close to Germany, England, Italy, Belgium, France, Holland, Japan, and other major overseas producers and exporters. Such headquarters can, of course, be of great help in the actual purchase and shipping of needed materials from that area.

However, if there are a great number of purchases of high dollar value to be effected in a given area over a considerable time span, it will pay to set up a separate purchasing office in the area, staffed with the required specialists and linguists. Impressive savings can be realized through direct contacts with the manufacturing companies involved and by having an office on the scene to handle the shipping, export, and other formalities. The contractor can, of course, handle such overseas purchases from the United States through either import and export organizations or through the stateside offices or sales agents of leading manufacturers, many of whom do have offices in the United States. All too often, however, their pricing is geared to domestic United States pricing and marketing and does not reflect the greater savings and more expeditious handling to be realized from direct overseas purchasing and shipping effected by the contractor's own forces at the point of purchase. The number of sources available to the contractor's buying office overseas will be much greater than those available to the contractor in the United States or at job site. These will include the smaller, but equally competent manufacturing companies, difficult to communicate with in other than their own language, but often able to quote better prices with equal or better delivery and quality.

Regardless of where the orders for overseas projects are placed, the control remains at the project. It furnishes the direction for purchase and the requirements to be met.

12-6.   Organization

The purchasing department should be organized and staffed to suit the type, volume, and complexity of the buying, estimating, expediting, shipping, reporting, and associated activity expected of it.

The size and makeup of the contractor's headquarters purchasing group will depend on whether the contractor prefers centralized purchasing, decentralized purchasing, or a combination of both; whether he buys for domestic or overseas projects or both; and whether he is involved heavily in estimating (Figure 12-6-1).

If purchasing is completely centralized, the volume of items to be procured and what comprises that volume will determine the size and composition of the department. Medium to large companies will, for their volume, require a top-flight purchasing executive to organize and control the overall departmental activity. The chief purchasing executive will usually bear the title of procurement manager, procurement director, or vice-president–purchasing. For our purposes we will, in this chapter, refer to the chief purchasing executive as procurement manager.

Centralized purchasing in volume will also have need of a capable purchasing agent to allocate, supervise, and control the day-to-day work load with the support of assistant purchasing agents, traffic personnel, buyers, expediters, and clerical people in numbers and of suitable qualifications to do an effective buying job.

More often than not, however, the medium- or large-size contractor's purchasing reflects a combination of centralized and decentralized buying with centralized control. This will require the same basic organization as fully centralized staffing but with a lesser number of assistants, buyers, and supporting personnel.

If the headquarters purchasing department is also responsible for purchasing and shipping to overseas projects, a separate section is advisable, headed by an overseas purchasing agent, experienced in export buying, documentation, and exporting, who reports directly to the procurement manager. Supporting personnel will depend on the work load but will normally include assistant purchasing agents, buyers, and expediting, traffic, and clerical personnel.

For the completely decentralized purchasing operation, a top procurement manager would certainly be needed to coordinate properly the purchasing activity on a worldwide basis. His effectiveness would depend not only on his abilities, but to a large extent on the quality of each of the contractor's decentralized buying organizations; their willing-

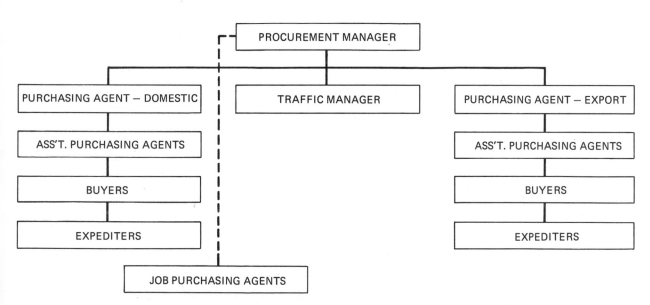

Figure 12–6–1   Typical organization chart for purchasing department, with responsibility for domestic and export buying, and staff supervisory responsibility for such purchasing as is effected at projects. The number of assistant purchasing agents, buyers, and expeditors will vary considerably, depending upon the type, volume, and complexity of the buying, estimating, shipping, and related activities expected of it

ness to cooperate; the number, regularity, and thoroughness of audits; and management support.

Decentralized purchasing does not lend itself to the close corporate controls possible with centralized procurement, nor does it take advantage of the considerable savings possible through centralized purchasing with its buyer specialization, consolidation of various job requirements for improved pricing, and access to better sources.

12-7.   Field or Job Purchasing Organization

If the contractor's purchasing is completely centralized at corporate headquarters, the project will have no need for a job purchasing agent. Such emergency needs or expediting requirements as may develop can usually be handled by the office or business manager in addition to his normal duties.

When, as more often happens, the contractor's project purchasing procedures involve centralized purchasing of major items at corporate headquarters and decentralized purchasing of the balance at job site, a job purchasing agent is advisable. If the work load justifies, he may require a clerk-typist to assist him.

If project purchasing is completely decentralized and is to be accomplished by the job purchasing agent, he will, depending on the volume, complexity, and variety of materials to be purchased, require assistance from the engineering section for the purchase of permanent materials or will need to add qualified buyers to his staff. On those projects where the contractor is doing most or all of the work with his own forces, purchases of permanent materials will total millions of dollars on large jobs and involve a multitude of complex architectural, mechanical, structural, and electrical materials and equipment. This will, for maximum dollar savings, require specialized buying personnel experienced in this type of procurement, knowledgeable in the materials and sources, and capable of working closely with the project engineer or members of his staff to ensure purchase of materials and equipment to required specifications for timely deliveries at the lowest ultimate cost.

12-8.   Personnel

In view of the many millions of dollars spent annually by the purchasing department of a medium- to large-volume contractor, it is in the contractor's interest to select his top purchasing executive with particular care. The contractor in effect gives his procurement executive a blank signed check to spend the company's funds as he sees fit. If that trust is misplaced or if the purchasing executive is not equal to the task, it can be very costly to the contractor before he realizes this to be the case. Honesty, ability, education, and experience are essential to successful, profit-making purchasing.

The outstanding procurement executive is likely to be an engineering or business administration graduate. He will be knowledgeable in construction materials, equipment, and marketing. He will be effective in departmental administration, in the selection and training of qualified personnel, and in supervising and directing their work. He will have the strong desire, judgment, and know-how to purchase effectively at lowest costs and the ability to infuse this desire in his subordinates.

The purchasing agent, buyers, and other key purchasing personnel should also have strong educational backgrounds, preferably in engineering or business administration, and the abilities and experience to perform their responsibilities,

shown below in the job description for the principal purchasing classifications.

Job Description—Procurement Manager   Directly responsible to management for the procurement of equipment, materials, and supplies for all company activities, also on behalf of various clients on a worldwide basis. Directs the activities of purchasing agents, assistant purchasing agents, buyers, expediters, traffic and supporting personnel to ensure that all properly requisitioned items are obtained and shipped at the lowest possible cost to meet required specifications and delivery schedules, and that Purchase Orders are issued in a timely and efficient manner to reputable vendors. Ensures that the procurement activities are conducted in compliance with sound business practice and contractual requirements and in conformance with all instructions, requirements, and regulations of clients and/or regulatory or financing agencies, such as the World Bank, United States Agency for International Development (U.S.A.I.D.), Inter-American Development Bank, etc.

Job Description—Purchasing Agent   Responsible to the procurement manager or project manager for the operation of the purchasing department in accordance with established policies and procedures. Ensures by personal actions and/or through the delegation of authority that requisitioned items are obtained at the lowest possible prices to meet required specifications and delivery schedules and that purchase orders are issued in a timely and efficient manner to reputable vendors. Screens requisitions and assigns them to appropriate assistant purchasing agents or buyers for action. Responsible for the allover maintenance of files and records of requisitions and the preparation and documentation of purchase orders. Assumes the responsibilities of the procurement manager in his absence.

Job Description—Traffic Manager   Directly responsible to the procurement manager for the routing and the dispatch by rail, truck, air, or sea of materials and equipment to stateside and overseas locations. Handles rate negotiations with carriers and shipping conferences. Maintains information relating to classifications, rates, and shipping schedules. Checks carrier billings for accuracy and takes all needed action for recovery if indicated. Traces and expedites shipments.

Job Description—Assistant Purchasing Agent   Assists the purchasing agent in the placing of purchase orders for properly requisitioned items at the lowest possible prices to meet required specifications and delivery schedules. Receives assigned requisitions from the procurement manager or the purchasing agent and supervises their disposition by the buyers or, subject to review and approval of the purchasing agent or procurement manager, enters into negotiation with vendors and personally places orders. Ensures that the files and records of the department are maintained and that the preparation of purchase orders is accomplished in accordance with established policies and procedures. Ensures that price quotations are obtained, sources of supply are determined, and purchases are negotiated in an expeditious and efficient manner.

Job Description—Buyer   Responsible for handling of purchase requisitions assigned to him by the procurement manager, purchasing agent, or assistant purchasing agent. Reviews requisitions, issues invitations for bids, interviews vendors to obtain information relative to product, price, ability of vendor to produce product, and delivery date. Reviews proposals from vendors and, subject to review and approval of the procurement manager or purchasing agent, negotiates the contract with acceptable bidder, keeping in mind the necessity to obtain items at the lowest possible price to meet required specifications and delivery schedules. Initiates preparation of purchase orders. Supervises records, as appropriate, pertaining to items purchased, costs, delivery, and product performance.

Job Description—Administrative Assistant (Procurement) Directly responsible to the procurement manager. Aids him in a staff capacity by coordinating departmental operations as directed; records control and special studies; handles routine correspondence for the procurement manager; is responsible for translations; coordinates collection and preparation of reports. Responsible for the work flow through the department's clerical sections—reproduction, typing, distribution, and filing. Supervises clerical personnel.

Job Description—Chief Expediter Directly responsible to the procurement manager. Ensures that supplies, materials, and equipment are shipped by vendor on promised shipping date. With regard to export orders, makes sure that the material arrives at the port of embarkation as scheduled. Supervises personnel within the expediting section charged with the timely and economical delivery of material and supplies to the ports for overseas shipments. Responsible for the control and flow of documents within his section and from his to other sections that ensure that delivery is accomplished in accordance with established policies and procedures.

Job Description—Expediter Under supervision of chief expediter, ensures that the timely and economical delivery of material and supplies assigned to him for expediting action is accomplished through established processes and procedures.

12-9. Purchasing Function

The purchasing function in a well-diversified contracting company of medium to large size will generally include the following:

1. Receive properly approved and authorized purchase requisitions or direction for ordering materials, equipment, and services.
2. Receive, or determine from available plans and specifications, listing of various items of materials and work for which prices and subcontract quotations are needed for bidding in process.
3. Request quotations and subcontract proposals.
4. Receive and evaluate quotations and subcontract bids.
5. Coordinate with management, engineering, operating and legal divisions.
6. Determine priorities for purchases and deliveries.
7. Issue purchase orders.
8. Assign subcontract awards.
9. Expedite manufacture or performance.
10. Determine rates for truck, rail, air, barge and ocean transportation.
11. If feasible, negotiate freight reductions.
12. Determine and set routings for all type of transportation, within contractual scope.
13. Check receipt of materials.

14. Check vendor invoices and carrier freight bills for accuracy. Take any action necessary for correction or recovery.
15. Arrange for issuance and transmittal of all required export documentation and authorizations for export and import.
16. Maintain source listings for materials and subcontractors either manually or by computer or both.
17. Maintain current catalog files for construction materials, equipment, and services; buyers' guides; and other reference data of value.
18. Maintain purchase order, subcontract, and correspondence files.
19. Maintain commodity and pricing records.
20. Maintain equipment inventory and equipment availability records.
21. Sell obsolete or excess equipment.
22. Issue periodic shipping and purchasing reports.

How well some of the more important of these functions are performed can drastically affect the contractor's profit picture. Therefore, it is well to review these in greater detail.

12-10. Purchase Requisitions

The authority to purchase is established on receipt of a purchase requisition or request to purchase, signed by an individual with management authority to approve such releases for purchase. Requests for purchase carrying the proper authority are frequently made verbally or by cable or telex, but should be confirmed for record keeping and control.

To ensure purchase of the requisitioned items in the quality and quantity desired, the descriptions of product must be accurate and complete and either include detailed specifications or adequate reference to recognized government and nongovernment specifications or standards, such as those of the American Society of Testing Materials (A.S.T.M. Specifications), American Standards Association (A.S.A. Specification), U.S. Government Federal Specifications, and others. In any event, the requisition must clearly state what is wanted to ensure purchase of the proper product and to eliminate any possibility of confusion on the part of potential bidders. The requisition and its attachments, if any, should incorporate all information needed by the bidders to calculate costs, such as required delivery schedules, type of packing, delivery point or points, and method of transportation, routing, etc. Much of this information will be standardized for a given project and can be given to the bidders by the purchasing department in its "Request for Quotation" (Figure 12-10-1). The initiating office should include any nonstandard information or direction with the purchase requisition.

The purchase requisition is verified for completeness and its receipt is acknowledged by returning a signed copy of the covering transmittal sheet to the sender. A copy of the requisition is furnished the procurement manager for any comments or direction he may wish to give the purchasing agent, who will assign the purchase action to an appropriate buyer or assistant purchasing agent. Before the buyer receives the purchase requisition, it is logged for purposes of control; then sufficient copies are run off for bidding and for distribution to engineering, expediting, or other personnel that will be involved. The buyer is given the purchase requisition in a purchase request folder which

A.B.C. CONSTRUCTION COMPANY

DIRECT ALL CORRESPONDENCE TO_____

_____

REQUEST FOR QUOTATION

TO_____   DATE_____

_____   NUMBER_____.

_____   SHEET_____OF_____

                                                     REFERENCE ALL REPLIES_____

GENTLEMEN:

PLEASE SUBMIT A QUOTATION ON THE FOLLOWING:

☐ ITEMS SHOWN BELOW   ☐ ITEMS ON ATTACHED MATERIAL LIST

☐ ITEMS SHOWN BELOW AS PER PLANS AND SPECIFICATIONS REFERENCED

_____
TEAR THIS LINE WHEN USING CONTINUATION SHEET

ENCLOSURES

( ) PLANS   YOUR QUOTATION MUST REACH
                                       THIS OFFICE NOT LATER THAN_____

( ) SPECIFICATIONS

( ) MATERIAL LISTS   A.B.C. CONSTRUCTION COMPANY

( ) PURCHASE CONDITIONS   BY_____

Figure 12-10-1

follows the purchase action from its inception to completion, with a copy of each new document—bid list, bid invitation, prebid correspondence, and records of pertinent conversations, quotations, and other documentation—inserted in the folder as the purchase action continues to finalization.

12-11. Source Selection

The most important preliminary to a successful purchase is the determination and preparation of the list of potential bidders—the companies that are to be sent the "Request for Quotation." It is vital that the "List of Bidders" (Figure

12-11-1) be selected with the utmost care and with all required research. Not only is it important to have the right sources listed, it is equally important that the wrong sources not be solicited. Reasons for listing the right sources are obvious—if these are not contacted, there will be no quotations or purchase from such sources. But why the concern over inclusion of so-called "wrong" sources on the list of bidders? The "wrong" sources will often be brokers or secondary sources for requirements of sufficient magnitude or for type of project that would ordinarily permit purchase direct from prime manufacturing sources.

Inclusion of other than direct sources on bid lists for such requirements will often discourage or prevent the direct sources from quoting. Secondary sources have been known to indicate or imply to the prime source that a special relationship exists between it and the buyer that would cause the buyer to prefer to deal through their company—personal friendship, entertainment, the desire of a company executive that they receive the order, and other influences that seem remote, but do happen. Or the prime source may not wish to get involved in a dispute over the matter of price protection, particularly if it should become convinced that the dealer or broker can obtain the business even at the necessarily higher price. The entire price structure can and often will change when the "wrong" sources are contacted for pricing on any given inquiry.

To select the right bid sources to receive invitations, the buyer will normally draw on the following:

1. His experience
2. Past purchases of similar items or services
3. Source listings of manufacturers and subcontractors—manual or computerized
4. Dealer listings, if requisition readily recognizable of no interest to prime sources
5. Buyers' guides and catalog files of United States and

LIST OF BIDDERS AND RECORD OF RECEIPT

REQ'N. No._____

BUYER _____

DATE ASSIGNED_____

INVITATION DATE_____

BID DUE DATE _____

BID DUE TIME _____

	TIME REC'D	DATE REC'D	NO BID OR LATE BID	DATE OPENED	TIME OPENED
CO.					
ADD.					
CITY					
ATTN: PHONE					
CO.					
ADD.					
CITY					
ATTN: PHONE					
CO.					
ADD.					
CITY					
ATTN: PHONE					
CO.					
ADD.					
CITY					
ATTN: PHONE					

APPROVED_____

Figure 12–11–1

International Manufacturers, subcontractors, etc. Among the good ones are *Thomas Register of U.S. Manufacturers, McCrae's Blue Book, Sub-Contractors Register.* For foreign production are such guides as *Germany Speaks, Guide of the Federation of British Industries, Canadian Trade Index,* and similar buying guides available for France, Italy, Belgium, Holland, Switzerland, Austria, Sweden, and Japan, the principal overseas producing countries open to United States contractors for construction materials and equipment.

6. Also, in connection with foreign production, the commercial attaché at the embassy or in the major consulates of the particular country can frequently provide valuable source information. The business counselor at the U.S. Embassy in that specific country can also provide similar information

7. Products of stateside production purchased in the United States for export to overseas projects will frequently be available to better advantage through the international divisions of major manufacturers. In addition, such divisions are better geared to handle export packing, export marking, and documentation

The number of sources to be contacted will depend on the type of product to be purchased, requisition size, number of producers, availability of product in the quantity and time needed, point or points of delivery, extent of price and delivery variance to be anticipated, and other similar competitive factors. In the case of items to be fabricated to plans and specifications—for example, "miscellaneous metals"—prices will be found to vary considerably, depending on such things as plant location, facilities, labor, work load, backlog, fabricator's profit picture; in other words, on how well the work and timing fit the fabricator's facilities and schedules and how anxious he is to obtain the order. Bid invitations for specially fabricated items or for products that vary in price should be sent to fabricators or producers on a nationwide basis rather than to a given area, where there may not be as many variables.

The buyer will prepare the proposed bid list for the review and approval of the purchasing agent. If anticipated dollar value is unusually large or should the procurement manager have requested to see the proposed list, the purchasing agent will pass it on to him for review and approval prior to bid solicitation.

12-12. Soliciting Quotations

Bids are solicited, from the companies listed on the approved bid list, in writing, by telephone, or verbally, depending on circumstances. Normally, if time permits, formal, written bid requests to the logical sources, clearly detailing the requirement, will avoid misunderstandings and will result in more responsive bidding and the lowest pricing.

12-13. Written Bid Procedures

The written request for quotation can be issued by means of a form request, accompanied by a copy of the requisition and all applicable information the bidder will require to estimate his costs properly and submit his bid without need for any further clarification. Such information would include pertinent specifications and drawings; closing time and date for receipt of bids; the contemplated packing, marking, and shipping instructions; anticipated F.O.B. points or, in the case of export orders, F.A.S., C.&F., or

C.I.F., as may be desired; delivery schedules; invoicing; and other documentation required.

When bids are solicited on behalf of a client, government or private, it is a good practice to instruct bidders to submit their quotations in preprinted envelopes furnished by the contractor's purchasing department with the bid request, identified on the outside as a "Sealed Bid in response to Inquiry #_____." Bids received in these envelopes are delivered unopened by the mailroom to the procurement manager for safekeeping until the date and hour set for opening bids.

Procurement for the contractor's own account need not be as formal as that for a client but, if the quantities and dollar values are high, it will save money and assure better buying practices if the formal system of bidding is utilized. Bidders receiving formal bid requests are naturally more inclined to believe that a considerable number of requests have been issued. They will, if seriously interested in the business, quote better pricing under the circumstance than for an informal request, which often leaves the impression of little or no competition. Formal requests do, in fact, permit a wide bid solicitation at very little extra cost as compared with a limited bid solicitation—merely the added cost of paper and postage. Purchasing control is simplified with sealed bids, opened and time stamped in the office of either the procurement manager or purchasing agent, at the designated time and date. Under this method, the contractor can be reasonably sure there has been no disclosure of competitive pricing or other information which could adversely affect either the company or the bidders.

12-14. Telephone or Verbal Bid Procedures

When the procurement manager or purchasing agent considers it impractical to solicit sealed bids, usually for reasons of urgency, he will direct the buyer to solicit bids by telephone from the companies listed on the approved bid list. For telephone bidding, the list of bidders will normally be limited to three or four bidders, depending on the dollar value involved and availability of product. Care should be taken to list the three or four most likely sources to produce the desired result—purchase at lowest ultimate cost. Obviously this would include sources that have been consistently low on previous inquiries and have performed satisfactorily. Particular care must be taken with telephone inquiries to be certain the bidder is given complete, accurate details of the requirement and that he clearly understands all the details. Telephone or verbal quotations are recorded by the buyer on a bid tabulation schedule (Figure 12-14-1) which reflects the name of the bidder, individual who quoted, the date and time of quotation, price, delivery, and other pertinent information. The bidder is requested to confirm his quotation promptly in writing.

12-15. Addenda to Requests for Proposal

It will frequently be necessary to modify a Request for Quotation subsequent to issuance but before its due date. This is normally accomplished through issuance of numbered addenda—Addendum No. 1, Addendum No. 2, etc. These are issued simultaneously to all bidders with a request that the bidder return one signed copy of the addendum with his bid to evidence its receipt and that the changes have been taken into account in the bidder's proposal.

Addenda will be issued for a variety of reasons, including, among others:

A.B.C CONSTRUCTION COMPANY

TABULATION OF BIDS

Project _____   Requisition No. _____

Location _____   Page _____ of _____

Line Item No.	Quantity	Unit	Description

OPENING DATE _____ 19 _____

AWARD TO BIDDER NO. _____
Approved: _____
Procurement Manager

DATE _____

1 - LOWEST PRICE.
2 - EARLY DELIVERY.
3 - ONLY AVAILABLE SOURCES.
4 - BETTER QUALITY.

5 - FEATURES OF MERIT.
6 - REQUIRED DESIGN.
7 - NEGOTIATED.
8 - PROPRIETARY ARTICLE FROM MANUFACTURER OR AUTHORIZED DEALER.

1		2		3		4		5	
UNIT		UNIT		UNIT		UNIT		UNIT	

Figure 12–14–1

- Extension of bid due date
- Changes or amplification of description or specifications
- Quantity changes or deletions
- Changed delivery requirements

12-16. Quotations for Contractor's Bidding Program

To be competitive, it is important that the contractor obtain quotations from as many as possible of the suppliers and subcontractors preparing quotations for the estimate and, hopefully, the lowest prices, all factors considered. To ensure thorough coverage, the following actions should be taken:

A listing of the various items of materials and work for which prices and subcontract proposals are needed should be given to the procurement manager by the engineering department or estimating group as soon after receipt of the plans and specifications as possible.

The procurement manager or purchasing staff member assigned by him will, if computer facilities and computerized source listings are available to the company, coordinate with the data processing department for computerized pricing solicitation covering the required materials and subcontract work. When properly programmed, the computer source listings will include the names, addresses, and telephone numbers of the leading manufacturers or suppliers of construction materials and the major subcontractors for any phase of the work the contractor may wish to sub.

The listings will normally be programmed to index the sources separately, by commodity and work classifications, for those suppliers and subcontractors operating on a national or international scale and separately for the smaller supply sources and subcontractors operating on a sectional basis. The contractor can, with minimum effort and in minimum time, carefully pinpoint his bid solicitation to the specific types of materials and subwork involved, on a regional or national basis, or both.

When computer facilities are not available, purchasing personnel coordinating with the estimating group can and should effect the price solicitation manually. This can be equally effective, but will, of course, take more time for preparation of mailing lists, various needed insertions in form letter requests for pricing, and other details. In a manner similar to the computerized request, the manual solicitation will indicate the job under consideration, that the contractor plans to bid it, that he would welcome their best proposal for applicable work and/or materials, the timing for price submissions, and the name and location of individuals to contact within the contractor's organization.

As soon as the list of plan holders is available, the individual handling the solicitations for pricing will screen the listing for manufacturers, suppliers, and subcontractors of the materials and work that will be involved. A form letter or telephone call requesting pricing should be directed to those firms not previously contacted by computerized or manual price solicitation.

The computerized bid solicitation and letter to plan holders will normally provide at least minimum coverage by alerting the larger firms and the subcontractors and suppliers holding plans that the contractor intends to bid. To provide additional or more specialized coverage, the purchasing group should utilize whatever additional listings, guides, or information as may be available to help locate the additional bidders. Useful sources would include subcontract registers in the region, associated general contractors listings of specialty contractors and material suppliers, union listings of subcontractors signed up with its local for work in the area, *Thomas Register of U.S. Manufacturers,* and similar guides and trade groups.

The additional sources developed through such means should be contacted promptly to determine whether they know of the job; indicate the items or work that should interest them; advise where plans and specifications can be seen or obtained and the deposit required, if any; give the bid date; and give the location for submission of their subpricing or materials quotes at a date early enough to permit evaluation and any needed clarification. If takeoffs have been made of various items and materials for which pricing is required, transmittal of such takeoffs or bills of material to specified and other logical sources with a cover letter requesting pricing and attaching copies of the applicable specification paragraphs or sections will expedite and normally assure timely quotations. Care should be taken, however, to make certain that the takeoffs or other data supplied do not divulge confidential information or data not readily available to the other prime contractors bidding the work.

The individual responsible for solicitation and price coverage should keep listings of all the firms solicited for pricing. These lists should be divided into the various product and subcontract groupings according to bid items or specification sections, whichever lends itself better to the bid closeout. The listings should include the computer printout of firms contacted; also, names, addresses, and phone numbers of those firms solicited for pricing by mail or telephone.

As returns are received, records should be made on these listings including response and other information that will determine coverage to be expected. For those product or subcontract groupings with poor response, further solicitation and follow-up is obviously indicated, probably by telephone to determine the reason for poor response and to best decide how and where to obtain responsive pricing for the item or work.

12-17. Evaluation of Quotations

For ready comparison, all quotations or subcontract proposals should be tabulated by the buyer on a bid tabulation schedule, including the following information, as a minimum, for each of the bids received:

1. Name of bidder
2. Unit prices and extensions for items quoted
3. Applicable discounts
4. F.O.B. points
5. Adequacy of packing and extra charge for required packing, if any
6. Delivery costs or charges to destination
7. Payment terms; cash discount, if any
8. Firm price or details of escalation
9. Delivery promise
10. Bond or other guarantees, if applicable
11. Exceptions or qualifications to specifications
12. Difference in cost of installation as compared with specification or other bidders
13. Any other factors bearing on award, such as bidders' financial condition, reserve production capacity, etc.

The buyer will, either personally or through the purchasing agent or procurement manager, depending on circumstances, coordinate with engineering, operating, legal, traffic, or other sections for their review as may be needed and for any assistance required to complete an effective evaluation of each proposal and to make his recommendation for award. Pertinent comments or evaluations by the other sections involved will normally be confirmed and made part of the purchase order or subcontract file.

The bid tabulation schedule should include the buyer's recommendation for award and show his basis for such recommendation. The purchasing agent will review and approve all bid tabulations prior to awards. All major or unusual awards will also be reviewed by the procurement manager before authorizations are given to vendors.

This procedure should be basically the same at either the headquarters or project levels, except that the job purchasing agent will normally report to the project manager.

12-18. Award and Issuance of Purchase Orders

The award will normally be placed with the bidder whose pricing, quality, service, delivery, product installation cost, and other factors, taken as a whole, will reflect the lowest ultimate cost to the contractor and/or the client.

Purchase orders issued by a central headquarters purchasing department for a specific project will be based on purchase requests, plans and specifications, bills of materials, or takeoffs furnished by the project to the central headquarters for purchase. The procurement manager or purchasing agent at the corporate headquarters will coordinate closely with the project manager and the project engineer to make sure that all required items are requisitioned and that proposed purchases for major items meet with their approval and will do the job effectively at the lowest overall cost. The procurement manager will also clear major purchases with corporate management, normally as a matter of company policy, but most importantly to keep management fully informed concerning any purchasing that can have an important bearing on profit.

Purchase orders issued at the project will be based on purchase requests of operating supervisors, approved by

PURCHASE ORDER

A.B.C. CONSTRUCTION COMPANY

_____

ADDRESS OF ISSUING OFFICE

ISSUED TO		PURCHASE ORDER NO. _____
		REQN. NO. _____
SHIP TO		PROJECT _____
		(THESE NUMBERS MUST APPEAR ON ALL INVOICES, CORRESPONDENCE AND SHIPPING MEMORANDA)
SHIP VIA	F.O.B. POINT	DATE _____

DELIVERY	TERMS	SELLER AGREES TO MANUFACTURE, SELL, AND DELIVER SUPPLIES OR SERVICES SPECIFIED HEREIN, SUBJECT TO THE TERMS AND CONDITIONS ON THE FACE AND REVERSE SIDE HEREOF.

ITEM No.	QUANTITY	DESCRIPTION OF MATERIAL	UNIT PRICE	AMOUNT	COST DIST.

APPROVED

_____

By _____
Procurement Manager

VENDOR'S COPY

Figure 12–18–1

PURCHASE REQUISITION NUMBER_____

BILL OF MATERIAL

Location _____

Page 1 of _____

Delivery Required Jobsite _____

Code No. _____

Line Item No.	Quantity	Unit	Description

Prepared by _____ Date _____. Approved by _____ Date _____

Figure 12–18–2

the project manager, or by personnel designated by him to approve such requests.

Purchase orders for materials or equipment to be exported will be issued on the basis of numerically controlled requisitions from the overseas project headquarters, properly approved and accompanied by all details and authorizations required to purchase the necessary items and to arrange the overseas shipment. Urgent or emergency purchases are often directed by cable or overseas telex message, but always confirmed by the required purchase requisition and supporting documentation. Close control is required to ensure receipt of all requisitions and to time purchases and shipments in order to avoid costly delays and work stoppages overseas due to unavailability of required product in the proper quantity and to the correct specification.

Purchase orders will be issued on a form, designed to contractor's requirements to provide the seller with the following minimum required information:

1. Name and address of purchaser
2. Name, address of seller; also name and telephone number of contact for expediting purposes
3. Purchase order date
4. Purchase order number; often coded to designate the applicable project and numerical sequence of orders for that project
5. Consignee, shipping instructions, and routing
6. Destination
7. F.O.B. point
8. Terms
9. Description and quantities of the products ordered, including or referring to all applicable specifications and/or drawings needed to assure complete understanding on the part of both the purchaser and the seller and effect a binding contract.
10. Unit and total pricing and discounts.
11. Delivery or shipping schedule.
12. Drawings and/or technical data required for review and approval.
13. Requirements for inspection and/or certifications.
14. Invoicing requirements.
15. General terms and conditions, which will include protective clauses to cover failure to perform, termination, price changes, warranty, patent infringement, taxes, governmental requirements, and other contingencies.
16. Applicable special conditions.
17. Provision for vendor's acknowledgment.
18. Authorized signature.

Purchase orders for materials to be exported will, in addition, provide necessary instructions and/or information for export packing, marking, and moving the ordered items from point of origin to the appropriate port of embarkation or, if purchased on an overseas delivered basis, to the overseas destination. These would ordinarily include:

19. Export packing instructions
20. Export markings
21. Individual, office, or forwarding agent to contact for shipping instructions to designated port
22. Import licence information—number, date, validity, allowable tolerances, information required for amendment, etc.

The original signed purchase order is mailed to the vendor accompanied by an acknowledgment copy to be signed and returned to the purchaser by the seller as evidence of receipt and acceptance of the purchase order. Often the vendor will return his own form acknowledgment of the purchase order. Any points of important difference between those stipulated in the purchase order and the vendor's acknowledgment should be reconciled promptly, in writing.

Copies of domestic purchase orders will be distributed to the project manager, project engineer, office manager, and job purchasing agent for purposes of information, materials control, expediting, accounting, receiving, and payment. Export purchase order copies will be distributed to the overseas project headquarters in reproducible form or in required number of copies, as well as to stateside departments or offices involved, expediting, accounting, forwarding agent, and export packers, if applicable.

12-19. Expediting

Making the purchase is one phase of the procurement action, a most important one, but obviously the action is not complete until the items reach the job in proper quantity, quality, and condition for use or installation. Delays can result in extra costs far in excess of any savings the buyer may have effected in the buying stage of the procurement.

A good buyer will have determined, prior to the purchase order placement, that the seller's delivery projection, as stated in the purchase order, was realistic and also would meet the Contractor's job requirement. Systematic follow-up nevertheless required to ensure deliveries as promised, particularly delivery of specially fabricated items and those requiring a long lead time in production.

At the project level, the follow-up or expediting is generally a function of the job purchasing agent. The materials engineer, on jobs so staffed, will coordinate the expediting activity in connection with subcontracts and permanent materials.

Centralized headquarters purchasing departments and the separate purchasing offices of large overseas projects will usually be staffed with expediters within the purchasing department to expedite purchases, prepare status reports, and verify shipments.

Procedures for Expediting Purchases The expediter makes initial contact with the vendor soon after the purchase order issuance to determine the following:

1. The purchase order has been received by vendor and is in progress.
2. Factory order, shop order, or work order reference number.
3. Name, address, and phone number of customer service representative or other individual to contact for progress reports and order servicing.
4. Confirmation that no problems are anticipated and that shipment or delivery will be as stated in the purchase order.
5. Factory scheduled dates for completion of each major phase of the purchase order progress—engineering, production, packing, shipping, and transportation.

The expediter will continue periodic contacts with the vendor or factory by mail, telephone, telex, telegram, or personal visit, as needed, to ensure maintenance of factory schedule and deliveries as required by purchase order terms and conditions. Should he at any time find a delay in the schedule, the expediter will press for corrective action with the vendor, factory, supplier, or subsupplier involved in the delay. A good expediter will have the ability, knowledge, tact, and aggressiveness to recommend actions or sources to eliminate or alleviate whatever problems are delaying the order.

If, however, the expediter cannot obtain the required improvement in schedules, he will review the problem with the chief expediter or with the original buyer and request his assistance. The buyer is, of course, better able to put needed pressure on the vendor. In the first place, it was the buyer who worked out the original schedule with the vendor on what was understood to be a realistic basis. Also, and most important, the vendor is looking to the buyer for additional business in the future. In fact, many purchasing departments are so staffed that the individual buyers do expedite all of their own purchases. This works out well if volume of purchases is not overly heavy or if the buyer is given sufficient assistance to handle both buying and expediting.

12-20.   Status Reports

The expediting section of most medium-to large-size purchasing departments is usually responsible for the issuance of status reports. These include:

Purchase Order Status Change Reports   Issued for individual purchase orders each time there is basic change from previously reported status (Figure 12-20-1). Prepared by the expediter to record and advise of delay or improvement in schedule, readiness for shipment, shipments, receipt at terminal, and other pertinent information relating to completion of the order by the vendor and its delivery to job site. Distribution of the purchase order status change report

ABC CONSTRUCTION COMPANY

REPORT OF CHANGES

TO: Foreign Business Manager

Reporting Date:

FROM: _____
(issuing office)

SUBJECT: _____

Requisition & P.O. No. _____ Originator's No. _____

Items: _____

Last Known Status			Status Information Required
	Date Request to Purchase Sent to:		
	Date P.O. Assigned		
	Date P.O. Issued		
	Shop Drawings Received		
	Shop Drawings Approved		
	Shop Drawings forwarded to _____		
	Inspection Scheduled		
	Promised Delivery Date		
	Revised Delivery Date		
	En Route to Port		
	En Route to Export Packers		
	Received at Export Packers		
	Packing Lists Received by Forwarder		
	Received at Port		
	ETD from Port		
	Shipped from Port		
	Name of Port		
	Name of Ship		
	ETA _____		

REMARKS:

_____
Procurement Manager

Figure 12-20-1

is made to the project headquarters for information and scheduling and to the traffic section and forwarding agent, if involved, for information and any required action. Copy of the report is retained in the expediting file to record status of the order as of the date of the report.

Procurement Status Reports for Individual Projects Issued on a scheduled basis, usually once or twice monthly, to reflect status of all requisitioned line items open for purchase for a given project and status of all purchase order line items undelivered at job site (Figure 12-20-2). *Special Status Reports* Issued whenever considered advisable by the procurement manager or requested by the project or management to provide earlier or more comprehensive information relating to specific purchases or shipments.

12-21. Verifying Shipments

For domestic shipments, the purchase order instructions will state the name and location of the project or other point of desired delivery. Materials or equipment covered by the purchase order will be received and verified for conformance to specifications and purchase order requirement at destination.

Export Shipment Packed at Contractor-Designated Export Packers For export shipments, the purchase order instructions will, in those instances not advisable or advantageous for the vendor to export pack, direct the vendor to ship to a contractor designated export packing firm for packing. Materials, equipment, etc., will be received and verified by the export packer against the vendor's packing list and the purchase order line items for compliance with order requirements. The contractor's expediting section will also receive a copy of the packing list from the vendor and immediately check it against the purchase order. Any discrepancies noted by the export packing firm in its physical check or by the expediting section in its check of the vendor's packing list will be reconciled or corrected prior to export packing.

An export packing list will be prepared by the export packer after packing, indicating the contents, export markings, net weight, gross weight, dimensions and cube of each box, container, or piece number. The export packing list will be distributed to the expediting section for information and status reporting purposes and to either the contractor's traffic section or to the forwarding agent, if a forwarder is being utilized, for issuance of pier delivery instructions and arrangements for overseas shipment.

Export Shipments Packed by Vendor When the vendor does the export packing and preparation, the purchase order instructions will direct the vendor to contact the contractor's traffic section, or forwarding agent if appropriate, for shipping and pier delivery instructions. Such requests are to be accompanied by vendor's packing list reflecting number of packages, pieces, boxes, etc., and contents, export markings, net weight, gross weight, and dimensions and cube of each. The vendor will simultaneously furnish the contractor's expediting section with a copy of the packing list. The expediter will check the packing list against the purchase order for accuracy of supply and to record and report the impending shipment to the overseas headquarters. Should the expediter note any discrepancy, he will immediately coordinate with the vendor and the traffic section or forwarding agent for corrective action prior to the issuance of shipping or pier delivery instructions.

12-22. Traffic and Transportation

The well-staffed purchasing department will include a traffic manager, responsible to the procurement manager, for the efficient routing and dispatch of all shipments via any means of transportation, tracing of shipments, rate negotiations, proper carrier billings, and other traffic and transportation functions.

The extent of traffic section staffing will depend on scope and volume of its activity. If it is to handle the movement of a great number of purchase orders to various projects, ports of embarkation, and overseas destinations, the traffic manager will obviously require supporting personnel. On the other hand, if the various projects are responsible for their own cargo movements and a forwarding agent is used to coordinate and arrange the shipments to overseas points, a much smaller staff will be needed, possibly the traffic manager only, with a secretary or assistance from the stenographic pool.

For very large projects involving a continuous flow of shipments to be exported in great numbers, it may be more economical and will certainly provide maximum control to handle the forwarding and shipping arrangements with the contractor's own forces. However, such projects are relatively few and far between. For the normal-sized project, it will prove advantageous to the contractor to utilize a reputable, experienced export forwarding agent to handle the multitude of details involved in arranging for overseas shipments and in the preparation, handling, and distribution of the needed formalities and documentation. Because the forwarding agent's costs are distributed over the greater consolidated volume of work for a number of clients, his costs for forwarding work are understandably lower per shipment. In addition, as a licensed forwarding agent, the forwarder receives a brokerage from the steamship line for booking cargo, writing bills of lading, and other steamship line related services.

In most instances this brokerage amounts to $1\frac{1}{4}$ percent of the ocean freight, which can be a considerable sum when large tonnages are handled. This brokerage is not available to the contractor if he handles his own shipping arrangements. It can readily be seen, therefore, that the forwarding agent can perform this service profitably, when combined fee and brokerage are considered, at lesser cost to the contractor.

Proper classification and routing of shipments will result in lowest cost and speediest delivery possible for the selected type of transportation. The traffic section will maintain classification, tariff and routing schedules for rail, truck, air, and sea transportation to enable it to select the most direct routing and to classify the shipment to carry the lowest legitimate rate.

Directly or indirectly, the purchaser pays the freight. If the purchase order is placed on an F.O.B. shipping point basis with freight for the purchaser's account, the routing and charges for freight can be readily controlled and will be auditable. Purchases F.O.B. destination or F.A.S. vessel include the freight charges and can only be controlled if the extent of freight included in the price is determined with the vendor prior to placing the order. Most bidders are willing to furnish this information when requested since they normally include in their pricing what they consider to be the actual freight cost.

It is surprising how often the bidder unknowingly includes far more than is actually involved for freight either through improper classification—domestic rate when export rate is applicable—miscalculation, or other reason. The

PROCUREMENT STATUS REPORT

DATED _____   CONTRACT NO. _____   LOCATION _____   PAGE _____ OF _____

REQ'N NO.	DATE REQ'N REC'D	DESCRIPTION	NO. LINE ITEMS	DATE P.O. ISSUED	P.O. NO.	VENDOR	TOTAL VALUE	DATE PROMISED DELIVERY	DATE VENDOR SHIPPED	DATE REC'D AT PORT OR PACKERS	VESSEL AND DATE SAILED	ESTIMATED TIME OF ARRIVAL OR REMARKS

Figure 12–20–2

difference can be very considerable and, obviously, if the bidder is willing to break out his freight charges, he is willing to adjust to the proper rate when brought to his attention.

Determination of the actual freight rate used by the bidder or vendor in calculating a delivered price is also important should it be possible to effect a reduction in the existing freight rate. This is often possible if the movement involves substantial tonnages of bulk materials—for example, portland cement—from mills to places which have not previously had such movements in volume. The existing rail rates, established for infrequent movements of relatively few carloads, are often understandably high and subject to downward revision if approached on a proper competitive basis. Any such reductions in the existing rates should reduce the lowest delivered pricing by the same extent.

Similar rate reductions are possible in truck and air freight rates if the existing rates are such that substantial tonnages and competitive aspects with other types of carriers or with one's own equipment justifies request for rate reduction and the approval of the federal, state, or other regulatory body involved. Project ocean freight rates, substantially lower than existing tariff rates, are also obtainable from shipping conferences, associations of steamship lines plying the same route. Project rates are normally granted where heavy tonnages are contemplated or to combat competition of nonmember lines.

The traffic section reviews freight billings to determine whether proper rates have been applied by the carriers. Where overcharges are discovered, the traffic section will take the required action with the carrier for recovery. Such overcharges usually stem from improper classification, application of domestic rates for export shipments, etc., and often involve substantial differences.

The traffic section is also responsible, when the need arises, for tracing shipments en route to ports or other destination and in the hands of carriers.

12-23. Equipment Inventory and Availability

Inventory and availability of major equipment is normally maintained at the central headquarters purchasing depart-ment. The control records will be kept either manually in the purchasing department or, more often at larger companies, in the electronic data processing section. The inventory will include, in addition to make, model, and year of the equipment, such information as date and cost of purchase, depreciation, and current location.

A more complete description of each piece of equipment is kept on individual record cards, giving such additional information as the purchase order number under which it was acquired; modifications or additions, if any, to the equipment and dates of such changes; serial numbers of all major components; and jobs or locations to which assigned by inclusive dates from acquisition to the present.

The equipment is allocated to the contractor's various operating divisions according to their individual needs. Each division controls and is responsible for the equipment assigned to it, within the contractor's established policies for usage rates, depreciation, overhaul and repairs, and time worked. The procurement manager, with consent of the divisional manager to which it is assigned and when authorized by management, transfers equipment from one division to another as needed, when the equipment is not in current use and no immediate need is anticipated.

12-24. Other Functions

Other functions of contractor's purchasing departments will vary considerably, depending on the strength and ability of its purchasing management, availability of suitable personnel, corporate policy, and other factors. Often purchasing department responsibilities will include such related and nonrelated responsibilities as:

- Selling of used and obsolete equipment
- Salvage and reclaiming operations
- Operation of maintenance and repair shops for automotive and construction equipment
- Operation of heavy machine shops
- Operation of electrical equipment repair and fabrication facilities
- Aircraft maintenance and operations
- Mobile and Aircraft radio maintenance and repairs

13 FUNCTIONS AND ORGANIZATION OF CONTRACTOR'S ENGINEERING SECTION

J. P. FREIN

Vice-President and Director (Retired)
Morrison-Knudsen Company, Inc.
Boise, Idaho

So MANY FACTORS are involved that no single rule for organizing the contractor's engineering section and determining its functions can be applied to all companies. Among the factors affecting these critical decisions are each company's scope and kind of contracting, its size, its degree of specialization, and how widely it operates. These factors will determine whether, at the least, the company owner performs engineering functions himself, or, at the most, an engineering section with diverse talents assumes broad functions and responsibilities. Some contractors, unable to justify economically large sections with diverse talents, employ outside specialists for functions that their own sections cannot perform. Almost all contractors find outside consultants necessary and desirable to deal with specialized or complex engineering.

The guidelines in this chapter apply to general contracting companies of medium to large size and reasonably diversified in the work they perform. It includes, but is not limited to large buildings, bridges, roads and streets, dams, tunnels, canals, heavy excavation, quarrying, aggregate processing, and steel erection. The companies subcontract specialty work, but under some conditions might do it themselves.

The guidelines are based both on experience under varied circumstances and on observation of the organization and operation of many contractors. Such observation is made possible by the numerous joint ventures existing today. In these enterprises, two or more companies jointly bid for construction contracts and, when successful, perform the work. A high degree of understanding and coordination is necessary for the partners to reach agreement, both on bid amounts and how work is handled. Although the joint-venture partners on one project may be individual competitors on another, repeated joint-venture associations result in some similarity in the way that the companies operate.

13-1. Basic Purpose of Engineering Section

The contractor's engineering section is an indispensable tool in helping top construction management in analyzing construction and engineering problems and in providing it with the necessary facts for accurate ultimate decisions. Engineers are suited by their training and background to study and evaluate the numerous questions that arise, and to keep current on developments related to them. Because of their close association with management, engineers often succeed to top management positions or form an inseparable part of management.

The contractor's engineering personnel are also confronted by the numerous incidental and collateral issues that apply to the usual business functions about which they must be knowledgeable. In addition, almost every construction job is physically unique within itself (except possibly for general classification), presenting new and different problems to be solved. Construction engineering, done successfully, is a very demanding line of endeavor requiring more strenuous effort on the part of the individual than most other means of livelihood. Where widespread operations are involved, extensive traveling can be necessary, as well as the requirement that the construction engineer move himself and his family from one job location to another to meet the ever-changing work sites. On the plus side, however, is the existence of a minimum amount of routine, the absorbing nature of the work itself, the unlimited possibility of accomplishment and the satisfaction which accompanies it, and the possibility of advancement to offical capacities in the industry which permit more permanent residence and community activities.

13-2.   Levels of Engineering Functions

The contractor's engneering functions can be adapted at several different levels, as follows:

1. The project level, where responsibility generally is limited to a single project.
2. The district, division, or area level, which has responsibility for several projects
3. The main-office level, which has responsibility for all company functions down to the project level
4. The main-office level of the contractor with localized operations or few projects; responsibility would be directly for the projects.

Management at the various levels supervises the engineering staffs at those levels in the following manner:

(a) Project engineers report to the project managers
(b) District, division, or area engineers report to district, division or area managers
(c) A chief engineer or director of engineering, supervising the company's entire engineering function is responsible to the company president or to someone he designates

13-3.   Contractor's General Engineering Organization

Without attempting to give a final answer on how all contracting companies should organize their engineering sections, but to show the possible levels of engineering functions and their relationship to management at various levels, Figures 13-3-1, 13-3-2, and 13-3-3 are offered as typical for a small company, a medium-size company, and a large company. For the sake of clarity, the figures do not show other departments of the contractor's organization. They are intended only to show possible lines of authority and technical assistance that might exist between management and its engineering personnel. As stated previously, each situation must be considered in the light of such factors as specialization, extent of diversification and area of operations, applying to each company.

To clarify Figures 13-3-1, 13-3-2, and 13-3-3, a brief outline of the relative authority and responsibilities of the principal engineering personnel of a contractor's organization follows. The descriptions of their classifications apply to a large diversified company with a number of districts, divisions, and subsidiaries generally operating as separate entities, such as shown in Figure 13-3-3. It would be necessary to scale them down to fit smaller companies.

Vice-President–Engineering   The vice-president–engineering reports to the executive vice-president–operations and, through him, to the president. He develops or approves company-wide engineering procedures, supervises their execution, and coordinates the estimating of major projects. He also assists and advises the executive vice-president–operations on engineering matters.

ORGANIZATION CHART –

COORDINATION ENGINEERING WITH MANAGEMENT TYPICAL FOR SMALL ORGANIZATION

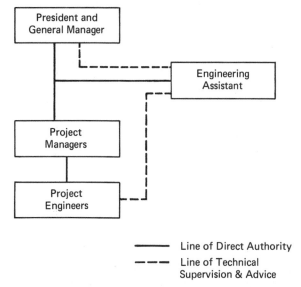

Figure 13–3–1   Organization chart: Coordination engineering with management, typical for small organization

Chief Engineer   The chief engineer reports to the vice-president–engineering and, in his absence, assumes his duties. He is in charge of the estimating of major projects, the development of construction methods, the accumulation of experience cost data, and the functions of the home-office engineering department. He assists and advises the districts, projects, and subsidiary companies in engineering matters. In cooperation with district and project managers, he selects, trains, and assigns key district and project engineering personnel.

Assistant Chief Engineer   The assistant chief engineer reports to and assists the chief engineer in the general performance of his duties. He acts for the chief engineer in his absence.

The assistant chief engineer also assists the chief engineer in matters pertaining to change orders, claims, and contract settlements. He provides technical advice and assistance to the district and project personnel in such matters during the course of contract performance. He is responsible for the final preparation of all claims remaining after the physical completion of the work, and serves as technical advisor to attorneys employed for the presentation of appeals to the highest administrative authorities of the clients and in connection with litigation.

Chief Estimating Engineer   The chief estimating engineer is assigned to the home office and reports to the chief engineer and/or the assistant chief engineer. He functions in the preparation and coordination of estimates being made by the home office and performs special delegated assignments.

District or Division Engineer   The district or division engineer reports to the district or division manager. He supervises the district engineering department, including estimating, plant design, preparation of control estimates

ORGANIZATION CHART —

COORDINATION ENGINEERING WITH MANAGEMENT TYPICAL FOR MEDIUM-SIZE ORGANIZATION

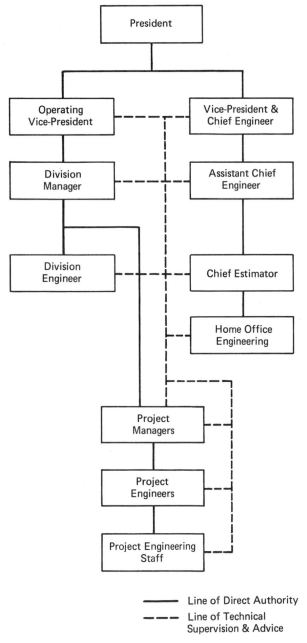

——— Line of Direct Authority

‒ ‒ ‒ Line of Technical Supervision & Advice

Figure 13–3–2 Organization chart: Coordination engineering with management, typical for medium-sized organization

and development of construction methods. He has advisory control over and lends technical assistance and advice to project engineers. He is under the advisory control of and receives technical advice and assistance from the vice-president–engineering, as well as from the chief engineer and the assistant chief engineer. The district or division engineer performs the functions of the engineering department for the district or division to which he is assigned.

Project Engineer The project engineer reports to the project manager. He is in charge of all engineering matters

including recommendation of construction methods; engineering layouts; plant design; cost distribution of labor, equipment, and materials; requisitioning of permanent materials; estimating change orders; claims for extra work; payment estimates; and subcontract estimates. He is under the advisory control of and receives technical advice and assistance from the district or division engineer as well as directly from the vice-president–engineering and chief engineer. The project engineer performs the function of the engineering department for the project to which he is assigned.

13-4. Functions and Duties of the Contractor's Engineering Staff

The engineering staffs at various levels would be expected to exercise their judgment in basic engineering matters, direct the company's engineering functions and advise management on the many phases of the business for which it would be responsible.

As mentioned previously, there can be no single rule fixing the limits of the activities of the contractor's engineering section. General practice in the contracting field is to delegate the following functions to the engineering personnel:

1. Maintaining a continuing market survey of prospective work being offered and advertised in trade papers and other published information; consolidating the information and reporting it periodically to the management as a basis for making decisions to bid.
2. Reporting bid results to interested company officers.
3. Planning, scheduling, and programming construction operations to be performed.
4. Estimating the cost of prospective work and preparing proposals for the same under the general direction of the company management.
5. Arranging for bid bonds or financial guarantees for proposals in compliance with tender documents.
6. Coordinating with joint-venture partners concerning compliance with tender documents, reconciliation of prebid estimates, and finalizing of the tender.
7. Soliciting, receiving, tabulating, and analyzing material prices and subcontract proposals for use in the preparation of bids.
8. Reviewing permanent material and subcontract proposals for projects entering the construction stages and and advising on commitments to be made concerning them.
9. Drafting subcontract conditions for those portions pertaining to such matters as technical aspects, construction details, and time of performance.
10. Making cost and method studies to determine the relative economy of alternate methods of construction.
11. Preparing budget-control estimates for jobs awarded.
12. Making detailed designs of construction plant facilities to be used on the work, or arranging for, and supervising the designs.
13. Developing new and better methods of performing construction operations pertaining to unique situations to conditions peculiar to given projects.
14. Assisting in the maintenance of contractural relations, and the interpretation of contracts, plans, and specifications relating to contractural obligations.
15. Supervising and performing engineering layout of work.
16. Maintaining suitable references and specifications for

**COORDINATION ENGINEERING WITH MANAGEMENT
TYPICAL FOR LARGE ORGANIZATION**

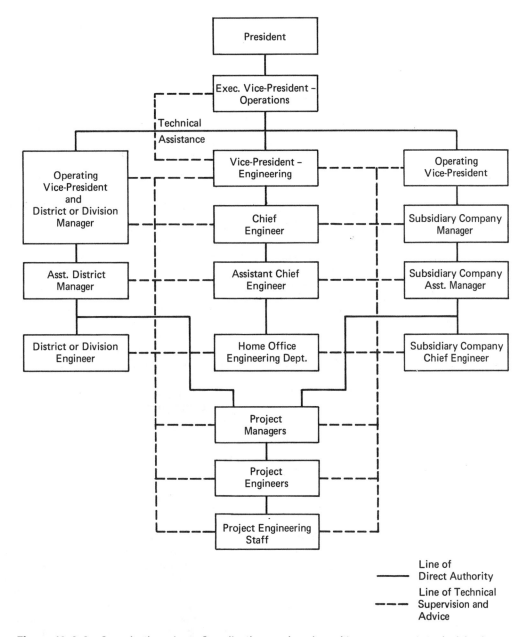

Figure 13–3–3 Organization chart: Coordination engineering with management, typical for large organization

construction machinery, and for permanent materials required for the work.

17. Requisitioning, scheduling, expediting, and control of permanent materials and supervising a running inventory of them.

18. Keeping a complete, current record of work being performed under the contracts and of contractural transactions.

19. Preparing or checking payment estimates for work performed under the contracts.

20. Supervising inspection of contract work, and helping to exercise quality control.

21. Accumulating data for and preparing statements covering extra work beyond that specified in the original contract.

22. Keeping accurate current records of work being performed by subcontractors and preparing subcontract payment estimates.

23. Accumulating data of performance experience for construction plant and equipment on the work.

24. Directing or performing the coding of costs for cost-accounting purposes.

25. Preparing periodic reports of costs and construction activities on current work.

26. Preparing requests or claims covering questions relating to payment for work performed beyond contract specifications.
27. Assisting in the settlement of contracts.
28. Preparing final reports for completed work.
29. Performing special assignments delegated by management.
30. Performing other engineering functions that may be necessary.

This list does not mean that all contracting firms should necessarily charge their engineering sections with every function but some provision must be made for them in the organizational setup. Management often cannot devote time to details, although it might perform some of the above functions.

Many of the functions obviously require legal, accounting, purchasing, insurance, or other expert knowledge. Engineering personnel should be able to obtain the advice of competent specialists—either within or outside the company. This relieves management of the burdens of compiling the record which it presumably would review later in any case. With engineering personnel doing the ground work, the overall progress of the matters is expedited instead of being bogged down awaiting separate action by the numerous people involved. This course of action also saves the expense of a permanent staff of experts in many fields. The question of what should be referred to specialists is a matter of judgment. However, since the matters are related to engineering functions and the personnel handling them have both experience and knowledge of the matters, they should be free to reach conclusions.

Following are detailed descriptions of the engineering duties and functions listed above:

13-5. Market Surveys of Prospective Work Being Offered

Reports of prospective construction work come from many sources, but primarily are found in news items and advertisements in trade papers, special bulletins, professional construction report papers, or other published data. They also may result from personal contacts.

A report or bulletin containing prospective work indicated by these sources should be prepared for all those concerned. It should be issued or revised periodically to reflect changes in the construction market and in the plans of the contractor's individual interest. In a large organization with many projects, the bulletin should be issued weekly, with the prospective work divided into the four following categories:

● Bids to be submitted
● Decision to bid pending
● Work to be advertised in the near future
● Projects for long range planning

Projects will advance to the more current categories as time passes or may be dropped for one reason or another. Projects will be dropped from the reports after their bid dates are passed, of course, or they might be dropped earlier if it is decided not to bid them.

For greatest convenience, the prospective reports should include the following data, arranged in separate columns:

Description of Project
    (a) Owner's name

    (b) Project name or brief description
    (c) Location
Bid Date and Location
    (a) The bid date—firm or approximate
    (b) Time that bids are to be received
    (c) City where bids are to be received
Estimate
    (a) Name of company division or individual to handle bid
    (b) Approximate value of proposed contract

If all the above information is not available for the initial reports, it should be added as it is obtained.

Prospective work reports will assist in preplanning the use of estimating personnel, the time of management, and the scheduling of company owned equipment. Figure 13-5-1 is an example of a prospective work report.

13-6. Reporting Bid Results to Company Officers

Since the engineering personnel normally presents the company's tender, it logically should report the results to the company's officers and to supervisors whose activities might be affected by the bid. This can be accomplished by a very brief report, prepared as soon as possible after the results are known. The report should include the names and the amounts of the low three or four bidders and the high bidder, and the amount of the company's offer if not among that number. Explanatory notes can be added if necessary or desirable for a better description of the company's competitive position.

13-7. Planning, Scheduling, and Programming Construction Operations

Planning of construction operations must be initiated when the prebid estimate is prepared. It involves the development of a concept for performing the various items of work necessary to execute the contract efficiently and on time. From a practical standpoint, planning is an inseparable part of estimating since an unpriced plan is an unknown quantity. The construction manager normally relies heavily on his engineer in the evaluation of alternative methods, and between them develop or create a plan on which the contractor's proposal is based.

The subjects of cost and method studies and estimating are discussed in Chapter 14, and therefore the discussion here emphasizes the scheduling and programming of construction operations only.

With respect to the work in progress, the contractor's engineering staff, under the direction of and with the cooperation of the project management, should prepare detailed programs for performance of the work. Sometimes the contract conditions establish the form, the method, and the detail required for project programming. The contractor must, in his own interest, and whether or not it is provided for in the contract, maintain comprehensive programming throughout the work. Such programming is the primary basis for control, efficiency, and good management of the operations. It might decide the contractor's success or failure.

The kind of work and its complexity will usually dictate the extent to which programming is necessary, the method to accomplish it, and the form in which it is expressed. It is imperative, however, that all programming be directed by someone who thoroughly understands the requirements of the contract and specifications, the construction designs

JOHN DOE CONSTRUCTION CO.

Sept. 2, 1971

WEEKLY PROPOSED WORK BULLETIN

Bids To Be Submitted

Bid Date & Place	Description	Approx. Scope and Company Handling
Sept. 10, 1971, 2 p.m., San Francisco, Calif.	Highway Department Mendocino County 5 miles of freeway	Western Division $10,000,000.00
Sept. 15, 1971, 11 a.m., Roanoke, Va.	Electric and Power Co. General Construction of Main Dam Dikes I, II, and III, Canals B&C and lining Canal A for Stevens Power Station, Virginia	Southern Division $9,500,000.00

Decision To Bid Pending

Oct. 2, 1971, 10 a.m., Omaha, Nebr.	Engineers, Omaha District Underground Storage facilities for Military Complex, Wyoming	Central Division $15,000,000.00
	Stevens Company On, Mass.	Eastern Division $5,500,000.00

Work To Be Advertised In Near Future

Dec. 1971	Data Processing Mfgrs. Office and Computer Manufacturing Plant Macon, Georgia	Bldg. Division $3,000,000.00

Figure 13–5–1

they call for, the specific method intended for accomplishing the work, the amount of production obtainable under the prevailing circumstances, the quantities of work to be performed, and the time necessary to accomplish it.

All important to successful programming of construction work is assured adherence to the plan of operations on which it is predicted and the sequences called for in the plan insofar as they can be accomplished. Late deliveries beyond the contractor's control, labor stoppages, equipment breakdowns, and severe weather conditions can disrupt the plan, requiring reprogramming of the work. But only unavoidable circumstances should be tolerated as legitimate reasons for deviations unless the original plan proves to be faulty.

Every possible element should be considered in the original programming. Sufficient allowances should be made for seasonal effects and weather, uncertainties of labor supply, delayed deliveries of equipment and materials, time for erection of necessary construction plant, time for approval or issuance of drawings by the owner's architect or engineer, and for any other circumstances that affect the work and that can be measured by facts or historic record. When sub-trades or subcontractors are involved, they should be consulted in advance concerning their ability to conform to the program. It is a mistake in developing a construction program to be too theoretical in assigning rates of production and other factors which approach the ultimate possible achievement, and to fail to recognize actual experience obtainable in the field. A practical construction manager or engineer will develop a general feel for what might be accomplished efficiency under known conditions and set goals in the programming which can be improved upon if circumstances ultimately permit.

As is generally recognized, the prospective construction project is "built on paper" in the prebid estimate of cost. An important part of the estimate is the proposed construction program which permits the pricing of the intended operations. The successful bidder will take time after the bid to re-evaluate the job based upon more exact information that may be available then. Vendors and subcontractors are usually more cooperative in working out details with the successful bidder, since he is the one who can write the order. Since the bidder at this stage becomes the contractor, he has additional incentive to develop the most efficient construction plan in order to make a satisfactory profit.

During the course of the contract work, accurate and up-to-date records of progress should be kept currently. They should show in some graphic or diagrammatic way how actual progress compares with original progress estimates. Project management thus can see at a glance if the work is on schedule and that the allowable contract time will be met, if it is behind schedule and subject to late completion and financial penalties normally specified in the contract, or if early completion with savings in costs is possible. Early completion almost invariably results in savings due to improved production and in lesser overhead costs over the shorter construction period.

Alert construction managers and engineers constantly strive to complete work ahead of schedule, provided it does not result in excessive overall costs. Astute manager's observe operations for the possibility of changing methods, and for increasing or decreasing the work force or the number of pieces of equipment in use to produce ultimate cost savings. No such changes should be made unless an intelligent, impartial study indicates a cost saving—or if no cost saving, a saving in time without an increase in cost of the actual field operations.

Periodic reprogramming is essential for controlling future activities. Many construction contracts require programming at monthly or quarterly intervals, with the frequency depending on the kind and complexity of the work involved. In this connection, it is important that all interested parties be kept informed of the current program, that subcontractors agree to all changes, that material and equipment deliveries be rescheduled and reordered accordingly, and that the owner-client and his engineer be aware of program changes and give official approval where it is required, advisable, or indicated.

The construction industry has possibly been one of the last segments of American business to adopt scientific controls for better planning and coordination of operations. This might be because of the nature of the construction, in which organizations are formed and disbanded for each job, and because unique problems exist at practically every construction site. Construction projects are of shorter duration than manufacturing, mining, and other continuing activities, and thus are less adaptable to minute control. However, competent construction managers and engineers of the past have relied heavily upon good planning and programming to the extent that it could be applied.

With the advent of the Critical Path Method (CPM), PERT, and other scientific control systems with applications to the computer, a finer degree of programming has become more feasible. If the activity is computerized, it permits rapid reprogramming, which reflects the effect on subsequent operations of the changes that are made in earlier ones and reestablishes the critical sequences throughout the contract execution. The CPM is discussed elsewhere in this volume and is mentioned here only to recognize its availability and proficiency when applied to the art of construction programming under warranted conditions.

Care should be taken in the selection of a competent supervisor for CPM programming and application. An effort should be made to verify the accuracy of computer printout sheets and to scan results of the CPM programming for obvious errors or inconsistencies which can result from mistakes in the input information or the basic computer program. To borrow a phrase from one authority in describing the application of computerized CPM, "Garbage in, Garbage out." The fact that information was developed by computer is no assurance of absolute accuracy.

Remarkable as computers are in the use of "memory banks" and rapid computation abilities, the most important element in successful computerized CPM application is that the input be based on a complete knowledge of the construction problems, the methods selected for their solution, reasonably accurate production possibilities, other factors affecting the work, and their proper inclusion in the development of the CPM program. The construction manager and his engineer must supply this element and follow through with a full understanding of the programming results.

Preliminary and basic scheduling of the work is mandatory to establish a concept of methods and operations for its performance within the allowable or a reasonable period of time, regardless of the system of programming ultimately chosen. An ordinary bar graph serves admirably for this function and can be refined and modified to reflect the development of more information growing out of the selection of methods and crews for doing the work and the productivity factors assigned to the various operations. The graph, although limited in reflecting greatly refined programming without the excessive use of engineering manpower, can satisfactorily serve throughout the life of the contract work to show the general status of the important work items or subdivisions of the work. It can be made in sufficient detail to meet the requirements of monthly or final reports intended for the principals of the client-owner and the contractor who have only limited time to gain an understanding of the project's status.

An example of a bar graph schedule or program is shown in Figure 13-7-1 at the end of the chapter.

13-8. Estimating the Cost of Prospective Work and Preparing Bid Proposals

This function is usually reserved for the contractor's headquarters office or a district or division office, because it requires careful supervison by top managemeint. Analyzing and pricing prospective work is one of the most crucial elements in whether the company succeeds or fails. Its direction should be entrusted only to a highly qualified and reliable individual with an engineering background. Top management, although perhaps unable to analyze the detailed pricing, should at all times be aware of its general operation and final proposals.

To assist management in its direction of finalizing proposals, the engineering personnel should have converted the various parts of the estimates into costs per units of work and provide previous basic cost experience in forms that can be related to the problem at hand.

For general contractors handling widely diversified work, estimating requires many talents based on diversified experience. Estimating engineers with field background are the most sought after for this assignment. Qualified people who serve usually as project managers, and who are cost conscious based upon their own experience in analyzing operations on their previous work, are particularly valuable to assist or to actually do some of the estimating. However, this function requires engineering direction because of the care necessary to understand and consider all of the things that apply to a given project—those things covered by the tender documents or normally inherent in the particular kind of work involved.

The subject of estimating is so extensive and complex that an entire chapter (Chapter 14) is devoted to it, with the comments here serving only to identify it with high-level engineering duties.

13-9. Arranging for Bid Guarantees

Tender documents generally provide that some guarantee be given by the bidder and submitted with his proposal as evidence of good faith and to ensure that the bidder will accept the contract for his price if he is low. The amount and form of such guarantees are normally specified. In the United States a bidder's bond is usually required. Cash deposits are sometimes acceptable. In foreign countries the obligation is sometimes satisfied by depositing negotiable government bonds, usually issued by a foreign government to which the proposal is presented.

Establishing the contractor's bonding capacity and the handling of company funds are usually the functions of the corporate secretary and treasurer. Ultimately, authorization for the issuance of guarantees must come from such officers. Because it is the function of the contractor's engineering staff to evaluate all conditions of the tender documents, thus ensuring that all conditions are met in order not to prejudice the bid, the engineering personnel normally initiate the request for the issuance of the bid guarantee documents to any appropriate company officers or directly to the bonding company if a bid bond is acceptable. Indemnity companies normally want to be informed about some particulars that might measure the risk involved, and the contractor's engineer is the logical one to convey such information to the bonding company.

It is important that this matter be handled far enough in advance to permit the transfer of funds, the arrangement for coinsurers by the bonding company, and the completion of the required formalities.

13-10. Preconstruction Liaison with Joint-Venture Partners

Joint ventures, or limited partnerships, for the execution of a given construction contract have been a practice for some years. The feasibility of joint ventures on large contracts has been clearly demonstrated.

As practiced in the United States, the principals of companies interested in a prospective construction job hold preliminary conversations within a circle of firms enjoying mutual respect. Agreement to participate in a joint venture to bid the job grows out of these contacts, and the group appoints one of its members as a sponsor to be responsible for the management of the contract activities and operations. Other members contribute bonding capacity, job financing, and other facilities in proportion to their percentage of the venture. For the purpose of establishing an agreed contract price, each company prepares cost estimates of the work.

Joint-venture members sometimes perform part of the work under a subcontract to the venture when it involves a specialty at which the member is proficient. When the physical conditions permit and there is no question of relative economy, the work might be divided among the members, with each performing a portion under the coordination of one of the members who handles the overall contractural matters. This arrangement is similar to the European-style consortium, except that a consortium is a loose assembly of specialists, each of whom performs his specialty and provides his own indemnification to the owner-client directly. Between them, consortium members perform a complete job with one partner coordinating the individual efforts of the members.

The contractor's engineering people, although not participating in organization of the joint venture, normally handle the details, coordinate the prebid formalities required by the tender documents, and establish a basis for comparing ideas of cost of the work leading to an agreed amount of the tender. The following comment might be more along the lines of what is required for an American-style joint venture. However, it can apply to any of the arrangements involving a combination of companies joined together to execute a contract. This describes the role of the managing sponsor and the things he must do to insure that the joint venture is prepared to submit a wholly compliant tender.

The sponsor's engineer normally lists from the tender documents all the required preliminary formalities, which can include:

- Formalization of the right of the member partners to do business in the particular city, state, county or political subdivision
- Compliance with local licensing laws
- Authorization of the boards of directors of the member corporations to bid the proposed work
- Submission of financial statements to the prospective owner-client
- Statements listing the officers of member companies

- Posting of the necessary indemnification, circulating bid bond papers for execution by the members surety, etc.

All such matters complied with in advance by the various members or placed in the hands of the sponsor for submission with the tender are usually coordinated by the sponsor's engineer, but they could be handled by others in the sponsor's organization.

The matter of coordinating the pricing is more universally accepted as an engineering function. It can be a difficult problem because of the variety of systems used by the people involved for estimating the cost of construction work. To compare only the ultimate total estimated cost is unsatisfactory, since it is virtually impossible to reconcile wide differences and agree on total cost. A better knowledge of the various components of the estimate is necessary.

It is almost impossible to regulate or prescribe the estimating system to be used by all joint-venture members. An estimator should not be required to deviate from his company's estimating procedure because he usually finds himself on unfamiliar ground and fails to recognize cost factors that might otherwise be meaningful and useful as checks. It is mistake to establish inflexible construction methods in advance of estimating because unique concepts of how to accomplish certain contract requirements and their possible resulting economy might be lost to the joint-venture group.

Agreement between the partners on the cost of the proposed work can be reached in different ways. However, years of effort to develop a satisfactory system have led to the following formula that seems to meet the requirements reasonably well as might be applied to large hydroelectric plant construction. It can be modified for application to other types of construction.

The first step is for the sponsor's engineer to accumulate and distribute to the partners the basic cost elements which would necessarily enter into all estimates and on which individual judgment is not involved. These would normally include the following:

- Climatological data, from local weather bureaus
- Transportation facilities serving the sites, applicable freight rates, and tariffs
- Commercial power available and applicable rates
- Fuel and lubrication oil quotations usually available in advance
- Applicable quoted tax rates
- Workmen's compensation rates
- Property and liability insurance rates and rates for other insurance
- Prevailing wage rates for labor, including working conditions, fringe benefits, etc.

These basic data would be applied in all estimates, even though some quotations would be tentative and would have to be adjusted when better information was obtainable. In addition, the sponsor's engineer should establish tentative prices for permanent materials and subcontract services. It should be understood that, although such prices would be used in all estimates, they would be superseded by the latest firm quotations obtainable, and the agreed cost adjusted accordingly prior to the bid preparation.

A tentative price for concrete aggregates is often provided in advance for all to use in their estimates and then adjusted to a quoted price by outsiders or to an agreed cost of aggregates processed by the joint venture. All joint-venture members would be expected to make their own individual investigations relative to the foregoing, to obtain independent quotations for permanent materials and sub-contracts, and to come to the joint-venture prebid meetings prepared to offer such data for consideration and use in adjusting the otherwise agreed cost.

The purpose of this procedure is to establish a degree of uniformity. It also eliminates the necessity of comparing every detailed quotation and material price in every partner's estimate at the prebid meeting for reconciling ideas of cost, without sacrificing accuracy or limiting the exercise of judgment by the partners.

Time also can be saved at the prebid joint-venture meetings by exchanging takeoff information in advance. The sponsor's engineer should arrange for his staff to make an early, complete quantity survey of the job. It should include such subsidiary things as quantities of excavation, forming and finishing areas for concrete and weights of permanent metal parts, and it should be listed on a statement form, with columns reserved for entering similar information developed by the other partners. An exchange of this information between all interested parties by mail will indicate large differencs in computations and allow for a resurvey of the parts in question, allowing such questions to be resolved in advance of getting together to discuss costs.

It is often desirable for the joint-venture members to offer items of plant and equipment which they feel might be applicable to the work and which can be made available. This can be accomplished by circulating lists of items offered, with prices and comments on location and conditions. This information should be released early enough to permit the various estimators to consider the offerings in their pricing.

The foregoing steps are preliminary to establishing a format for listing estimating results that can be extracted from any and all of the estimates prepared by the partners regardless of the estimating system used. To pursue possible differences in ideas of costs in the various parts of the estimates, a number of different comparisons are usually found desirable. The sponsor's engineer should outline these in writing to the other companies far enough in advance of the prebid meetings to permit each member to come to the meeting with his information already extracted from his estimate and posted to summary-comparison sheets provided by the sponsor. Following are some recommended comparisons.

General Comparison—This to cover all costs for the proposed job broken down as follows:
Direct Costs (for costs of work operations)
1. Labor (including payroll taxes)
2. Permanent materials
3. Specific plant
4. Equipment charge (rental, depreciation, or use value)
5. Supplies and expense
6. Subcontracts
7. *Total—Direct Costs*
Indirect Costs (general expense, overhead, etc.)
Total—All Costs

As a guide to the partners, the element of cost to be included in the foregoing headings should be described so that when the estimating results of each company are set down side by side on a columnar comparison sheet, differences in the general main elements of costs are apparent. Then it will be desirable to examine the particulars in the breakdown shown by the general comparison.

There often can be a misunderstanding about what should be considered as *Direct* vs. *Indirect Cost*. Some contractors are inclined to consider more things as distributable or indirect costs, and a listing of elements to be included for each will prove helpful. The following comparison is suggested:

Indirect Costs
Salaries
    Management and Supervision (to general foreman level)
    Office and administrative
    Engineering
    Warehouse and purchasing
    Labor relations
    Safety and first aid
    Watchmen and guards
       Total Salaries
Plus Other
    Automotive expense (rental, operation and licenses) serving the INDIRECT SALARY group
    Miscellaneous operating costs of office, warehouse, engineering and overhead functions
    General Plant (Buildings, utilities and general roads)
    Camp construction and operation
    Insurance (other than payroll insurance)
    Taxes (other than taxes on payroll)
       Subtotal—other than *Salaries*
    Total Indirect Cost (except bond)

Except for items listed under *Indirect Costs,* all other costs would be considered as *Direct Costs* charged directly to work items or specific construction operations.

An analysis of total labor is often found to be helpful, broken down as follows:

    Direct Labor—charged directly to operations
    Specific plant installation labor (prorated to direct costs)
    General plant construction labor
    Camp construction and operation labor
Total Labor (excluding indirect cost salaries shown under indirect costs)

A comparison of direct cost for groups of similar or related items will sometimes show where much of the difference in evaluation occurs. The categories for separate comparison might logically be chosen by summarizing numbered work items into:

(a)   Diversion and care of the river
(b)   Excavation—all items for all classes
(c)   Embankment and fills
(d)   Cement and pozzolan (cost delivered)
(e)   Concrete items
(f)   Reinforcing steel
(g)   Miscellaneous iron and steel
(h)   Hydraulic gates and machinery
(i)   Penstocks
(j)   Architectural features
(k)   Elevators
(l)   Electrical
(m)   Foundation drilling and grouting
(n)   Instrumentation

Sufficiently large differences of costs shown by the above comparison can be pursued down to the individual work items and operations if desired. Large differences often occur in concrete estimates, and subdivisions of costs in this connection for individual concrete items or groups of concrete items can be compared against the following breakdown:

- Concrete plant and equipment
- Plant and equipment operation
- Concrete forms
- Finishing
- Treating construction joints
- Curing and cleanup
- Admixture
- Cement wastage
- Point and patch
- Specials

The cost of concrete aggregates often is a substantial matter for large heavy construction jobs. If the aggregates are to be made by the contractor on the job from natural deposits (quarries, or sand and gravel pits) a comparison of the breakdown of the principal elements of cost therefore is often found to be necessary to understand wide divergence of opinion about cost. These would possibly involve:

- Pit or quarry operation
- Aggregate plant costs
- Plant operation costs
- Transportation—aggregate plant to mixing plant
- Sand plant and sand plant operation

Because the estimates by the various joint-venture partners will be predicated upon different concepts of how to perform the work, each estimate will be based upon a different list of plant and equipment. A comparison of the different categories of plant and equipment cost might help to explain large differences in total cost.

For the purpose of such a comparison, it is necessary to define what constitutes "plant and equipment" costs. For example, equipment can be defined as articles having a purchase value of more than $500 (items of less value being small tools), with moving or functioning parts and performing an operating function. The cost of equipment would be considered the first cost, or purchase price, plus sales and use taxes. Plant costs include freight and move-in, the cost of structural members and nonmoving parts supporting mounted articles of equipment, and all costs of constructing and erecting built-in-place construction plant, including applicable sales and use taxes. Total plant and equipment costs therefore consist of the sum of such costs less an amount calculated for salvage at the conclusion of the contract work. When such is practicable, rentals for plant and equipment items can be substituted for the amount resulting from the computation of first cost less salvage.

To help localize differences in plant and equipment costs, a comparative listing of total plant and equipment costs set down side-by-side for each estimate on a columnar sheet is suggested, broken down into categories or kinds of plants. An example follows:

PLANT AND EQUIPMENT SUMMARY

1. Open-cut excavation embankment plant and equipment
2. Underground plant and equipment
3. Concrete batching and mixing plant and equipment
4. Cement and pozzolan handling plant and equipment
5. Concrete transporting and placing plant and equipment
6. Compressed air plant and equipment
7. Job water supply plant and equipment
8. Electrical supply and distribution
9. Heating plant
10. Job roads and bridges

The total plant and equipment cost for the job will be a sum of the above and should be shown for each estimate being compared. As additional information for each estimate being compared, there should be set down side-by-side the total first, or purchase cost of all mobile or installed equipment included in the plant and equipment category, and the amount calculated as total salvage at the end of the job for such items, with the resulting amount charged in the estimate therefore against the operations on the job. This will point up the reasonableness of the salvage factor used by the various parties.

Prebid joint-venture meetings to resolve the question of the bid amount and other conditions normally are set only several days before the tender date. Two separate group sessions are involved: one for the engineers and estimators of the various member partners to reconcile estimates, and another for the principals of the bid partners to review the reconciliations of cost and to approve the bid amount.

If the engineers and estimators arrive at an "agreed total cost" for the project and the principals approve it, the total amount of the tender can be added up to include the following:

TOTAL AGREED COST
LABOR ESCALATION
MATERIALS ESCALATION
CONTINGENCIES
BOND
_____
equals   TOTAL ALL COSTS
+ MARK UP (PROFIT FACTOR)
equals   TOTAL AMOUNT OF THE BID

13-11. Soliciting, Receiving, Tabulating, and Analyzing Material Prices and Subcontract Proposals

Since the contractor's engineering personnel who prepare prebid estimates know the specific requirements for the proposed work because of their understanding of the tender documents, they should initiate all inquiries on prices for permanent materials and equipment entering the work, and for subcontract proposals for special items of construction or for services.

Contacts and negotiations with vendors can be handled through competent purchasing department personnel. They can make preliminary summary analyses of quotations and conditions applying thereto, passing them on to the estimating group. The latter should review the purchasing department's findings to ensure compliance with the contract specifications and to verify that proposed delivery dates will meet the proposed construction program.

At the prebid stages, direct communication between prospective subcontractors and the estimating staff is important, since almost invariably a detailed mutual study of the pertinent parts of the tender documents is necessary, together with an understanding of how the intended subcontract operations are meant to be coordinated with other activities on the job. Prospective subcontractors should be instructed to submit written proposals in advance of the bid date, quoting in compliance with the tender documents (or specifically outlining alternatives), stating the times of their intended performance, and giving any special conditions upon which their pricing is based.

Material and subcontract prices should be tabulated and summarized with respect to the work items or subsection of the tender documents to which they apply so that they can be readily compared. Quite generally, vendors and prospective subcontractors delay giving their quotations until only hours before bid time. Unless the shortage of time prevents considering such proposals, they must be evaluated for all conditions and incorporated in the overall proposal for the project. Extreme care is necessary in evaluating quotations under such circumstances.

When it is known that a quotation will arrive late, vendors and subcontractors should be asked to submit in advance a copy of their quotations giving all conditions, but leaving the pricing blank to be filled in later. The conditions of the proposal can then be evaluated in advance and will not have to be considered under the stress of finalizing the general proposal.

Other conditions of the general proposal having been resolved earlier, it can be adjusted to reflect the best permanent materials and subcontract prices at hand.

13-12.   Reviewing Permanent Materials and Subcontract Proposals

As a matter of good faith, it is important to give the formal orders for performance to the vendors and subcontractors whose pricing was used in the overall bid. In some areas, local practices protect the vendors and subcontractors by providing a so-called "depository" where copies of their proposals are left in advance of bid openings. It then is mandatory for general contractors to award the order of performance to those whose prices are used. Under other conditions, general contractors are required to state in their proposals the names of subcontractors who will perform various parts of the work. In such cases, it is necessary to arrive at complete understandings with subcontractors in advance of the bid. This is often difficult to accomplish.

Nonetheless, when the successful bidder is ready to start construction, he must have complete knowledge of material sources, prices, and delivery dates, and must also be assured of complete understanding with subcontractors. The same engineering-estimating personnel that initiated inquiries and evaluated proposals at prebid stages should review all permanent materials and subcontract proposals, making sure that they conform to the contract requirements, that they fit the more refined program that now is possible, and that they will produce the lowest ultimate economy to the venture.

Conferences with the interested parties are often necessary during the review to promote mutual understanding. Almost universally, under the stress of working on short time prior to the bid, some points are overlooked and their clarification becomes essential. If the construction program is later modified but final orders have not been written, vendors and subcontractors should be given an opportunity to consider such changes and resubmit their proposals. Only those indicated as logical contenders by the original quotations need to be consulted.

The engineering-estimating group participating in a review will accomplish the best results if the purchasing department people who will write the orders are on hand and the task is a joint effort. As each matter is settled, a draft of the purchase order should be prepared and approved by everyone before its formal release.

Perhaps not necessary, but often appreciated, is the practice by which the general contractor notifies unsuccessful vendors and subcontractors that orders have been given to others, and expresses gratitude for their cooperation in quoting. This would more nearly fit the function of a purchasing department except where the inquiries are handled directly by the engineering-estimating personnel. It is a valuable asset to have a circle of friends among the subtrades.

13-13.   Drafting Technical and Special Conditions of Subcontracts

There are a number of standard construction subcontract forms, such as those compiled by the Associated General Contractors of America and the American Institute of Architects. Some of the larger contracting firms have their own standard subcontract forms. All of these set forth good practice in the normal relationship between the parties, and necessarily provide that all of the conditions and requirements of the general contract will apply to the subcontract.

However, even with the names, dates, and places filled in the proper blanks, there remains the description of special conditions that will govern the transaction and a comprehensive statement of the work to be done or the services to be rendered under the subcontract. These could be of a highly technical nature, but in any case they should be stated to tie the affected work to the exact sections of the specifications or drawings involved and to the general contractor's plan of operations.

Although the composition and the wording might be referred to a legal counsel, the basic information should come from the engineering staff with the background study of the particular job and its requirements in the immediate area of the subcontract work in question. Normally a draft by the appropriate engineering employee is sufficient, but it should be reviewed by someone with a legal background, and then submitted for preliminary agreement by the other party, the prospective subcontractor.

13-14.   Making Cost and Method Studies of Alternative Methods of Construction

In the highly competitive construction market and with the ever increasing cost of machinery, labor, and materials, the successful contractor must use the most economical methods available to accomplish the work. The contractor's engineer is usually responsible for keeping up with new developments in construction machinery which promise greater efficiency and lower cost.

Construction probably presents more opportunities to exercise ingenuity in the performance of the work than is found in other industries. Some of the work is similar but, particularly in the heavy construction field, almost every job is unique. Every job, therefore, requires the contractor to choose one of a number of possible solutions, or devise something new and different to better meet the particular situation. If nothing else, the arrangement of known components to accomplish a job is often important in influencing the ultimate result. Alternative location of haul roads, for instance, where different grades and distances are involved, can have a substantial effect in cycle time of the hauling units that make thousands of trips over the same route.

Throughout the years, great strides have been made in the advancement of heavy excavating and earth- and rock-handling equipment. Increased speeds and load capacities resulted in mechanical and physical problems that were costly to the manufacturers and their contractor-clients alike. But the mechanical experts have done a good job in the improvement of their products. Before the contractor invests heavily in newly developed machinery, his engineer must study and research its feasibility. He also must compare the estimated cost of the results to be expected from the new machinery with the known results from equipment with a substantial experience background.

To the extent that time permits, cost and methods studies should be applied when the prebid estimate for a job is prepared so that the apparently cheapest, feasible method of construction can be selected as the basis for the proposal. After a successful tender has been placed and more time is available, a complete review of the methods and cost of accomplishing the work should be concluded before investments in plant and equipment are made, and before personnel and facilities are moved to thte site. In addition to assuring at this stage that the proper plant equipment and methods are being chosen, the postbid period also may present the opportunity to reconsider the matter of subcontracting parts of the work. It may be desirable to subcontract some of the work the general contractor had allocated to his own forces.

When the work is in progress, and additional facts are available, it is wise to reevaluate the effectiveness of the methods being used. Whoever does so should be alert to possible changes that will produce greater efficiency and lower the ultimate cost. It is seldom desirable at this stage to change the basic concept of plant and equipment, but looking ahead with respect to such subsidiary operations as methods of concrete forming, the use of new devices for compaction of incidental fills, and better blast hole spacing for rock excavations can result in substantial savings.

If the actual cost of the work is exceeding the estimated cost, a definite cost study should be made to determine the reasons for the difference. One might be that field supervisors are failing to obtain maximum efficiency from the work force. Time studies or observation of the workmen can result in improved efficiency. Human nature is such that a sincere individual will try to improve his work when he knows it is being observed.

Top management of the company or joint venture must decide finally concerning the adaptation of the basic construction schemes to be used on the jobs and the investment involved. The prospective project manager should be fully informed, participate if possible in the methods studies, and be in a position to provide his recommendations or alternatives.

13-15. Preparing Budget Control Estimates for Jobs Awarded

When it is known that a successful bid has been presented and the contract award is imminent, the low bidder must prepare himself for doing the work and reassure himself that the work will result in a profit. The contractor's engineering personnel are usually assigned to reestimating the work on a more careful basis and establishing a measure for cost control for use throughout the life of the contract, thus establishing a budget or standard for everyone to follow.

In preparing the budget-control estimate, the detailed pricing of all expected expenditures for purchases, labor, plant, equipment, and services should be reviewed and set down in order to provide an absolute and authentic reference. The methods of performing the work also should be

reviewed, as mentioned above, and the most promising from the standpoint of time and cost selected by management and incorporated in the budget estimate.

The budget-control estimate should be subdivided precisely according to the headings established in the tender documents, and the results determined should be directly comparable to the payment items called for in the contract. However, substantially more detail is required to fix the cost of the separate operations contributing to the cost of the work items.

The form of the control estimate should follow a standardized format familiar to all interested parties, and from which an understanding can be obtained with minimum study.

Cost-accounting code numbers should be assigned to the various subdivisions of the budget-control estimate. This will establish the cost-accounting key to be employed for the work, and the actual cost resulting on the work will be directly comparable to the amounts figured in the budget-control estimate.

The reader is referred to Chapters 14 and 15 on the subject of estimating. They suggest a format for preparing estimates that establishes the necessary standardization for producing a direct relationship between the prebid estimate, the budget-control estimate, and the reported actual cost. However, regardless of the detailed format of estimating adopted by the contractor, it is important to standardize a format and to have the actual cost of the work reported so that it is directly comparable to that shown in the budget-control estimate.

13-16. Detailed Design of Construction Plant Facilities

As soon as possible after the presentation of a successful bid, and when the concept of construction is decided, and probably simultaneously with the preparation of the budget-control estimate, the contractor's engineering staff should proceed with the layout of the stationary plant and equipment and other built-in facilities. Depending upon the kind and scope of the work involved, this can vary widely from minimum requirements for building projects to maximum for heavy construction jobs. Although each construction job must be considered separately and only general rules in this regard can be suggested, the following examples with respect to heavy construction are offered.

Temporary haul roads on large excavation jobs employing large-capacity dump trucks are an important part of the construction plant. For the most economical results, these must be laid out to account for differences in the elevations of the points of loading and dumping and still provide a minimum haul distance with grades suitable to the capabilities of the equipment without sacrificing speed. Proper surfacing and maintenance of the roads are important to give low rolling resistance, again affecting speed and gradability. Necessary bridges and culverts along the temporary haul roads will require special engineering design, both from the standpoint of the areas beneath for passing stream flows or runoff water and from the standpoint of the most economical structural design that can be employed.

For a large concerte dam, the amount and complexity of the plant and equipment design involved approach the maximum. Following a sequence suggested by the flow of materials toward final placement into the permanent work, the following deserves attention. The sources of concrete aggregates would have to be analyzed for chemical composition, soundness, reactive characteristics, etc., and, in the case of water-deposited materials, for their natural gradation. Studies would then be necessary for establishing the various steps and processes required as the result of crushing, screening, rod-mill grinding, hydraulic separation, and the like to produce suitable aggregates that could be blended into the desired product. Once the processing devices, the make, size, mechanical function, capacity, etc., have been selected, the desired physical arrangement to utilize gravity to the maximum should be determined. Then the conveyors and handling systems plus the necessary structural supports for all parts should be added. Processing water delivery and power supply and distribution are important incidentals. The best means of conveying the finished aggregate product to the place of use (the batching and mixing plant) is the next problem to resolve.

Batching and mixing plants are generally manufactured products available on the market, but they have various components with different capacities, mechanical and electronic features, etc., that affect the operation of the plant and the production of the diversely specified mixtures of concrete. Selection of the right plant having been made, the location, arrangement, addition of utilities (compressed air, water and electric power), and the coordination with other operational features are necessary, plus the design of foundations and incidental structural features for the batching and mixing plant, the cement and pozzolan plant facilities, and aggregate stockpiling arrangement.

Modern-day mass concrete is usually specified on the basis of a maximum temperature when placed in the forms. To accomplish this, it is necessary to provide a substantial plant for precooling the ingredients, or to cool the concrete after placement by pumping cold water through embedded pipe. Specialists in these processes are recommended, but even after selection of the mechanical devices forming a part of cooling plants, their arrangement, and coordination with other plant features, the structures to support and house them still must be designed by the construction engineer.

A variety of methods can be used to deliver the concrete from the batching and mixing plant to the site of placement. They include trucking or industrial railroad for hauling to the point of pickup by crawler cranes, high gantry cranes, tower cranes or possibly overhead cableways, etc. Presumably all of these methods would have been considered, unless the general character of the structure or the surrounding terrain rule out alternatives, the cost of each having been analyzed in some detail in the prebid estimate. The plant design might also involve haul roads, bridges, and railroad trackage for the delivery system, plus high trestles for placing cranes, or, in the case of overhead cableways, the span, the design of tower runways, the design of the cableway towers structurally, and the layout of electric lighting and utilities.

Most important in the accomplishment of the plant design features is to be constantly aware of the cost of the facilities being developed in order to stay within or below the prebid estimate, or the budget-control estimate if one has been prepared, inasmuch as the contract price, which would already have been established, is not adjustable for overruns in plant cost.

13-17. Developing New and Better Methods of Construction

With competition being a key factor in the construction industry, the contractor must find new and more economi-

cal ways of performing work in order to survive. Research and development programs seldom exist in construction, although such activities are carried on as a side line by the contractor's engineering force since that group is obliged to work continuously with construction plant and equipment matters, production, and the resulting costs attributable to such facilities.

Manufacturers of stock items of construction equipment such as dump trucks, power shovels, rock drills, cableways, and the like carry the burden of changes in machinery design. They are also confronted with providing tools more useful to the contractor in the performance of his work and offering products on the market that are more salable than those manufactured by others.

It is generally possible to work directly with manufacturers and their engineers on specific, large-scale needs and suggested modifications and developments if they are reasonable and marketable. This means of bringing forth new ideas should not be overlooked, but close contact with construction equipment people can pay dividends inasmuch as it affords an opportunity to learn about the things still on the drawing boards, or then in the testing stages of prototypes for machines developed by others. Early knowledge about these things could well permit their application in the field and the cost advantages attributable to them before the competition becomes aware of their existence.

For specialized applications, some decided advantages have been realized by the construction industry borrowing devices and ideas from other industries. Examples are rigging devices and supplies from the loggers, large hydraulic monitors from placer miners, and excavating wheels from the strip miners. No rule can be advanced for utilizing this avenue except to be generally interested in what is being accomplished in other fields and alert to the possibility of borrowing techniques for construction.

Continuous processes present the greatest possibility of large production and substantial savings. An ideal example is a wheel-type excavator digging from a bank or stockpile and loading the output on a conveyor system which carries the material in a continuous flow to its final position or temporary surge pile not requiring intermittent handling. The occasions for applying such a principle are rare because of the costly setup involved, but they do come along now and then. The excavating wheel is far from being a universally applied tool. The materials being dug must be ideally suited, with no solid rock or large tightly embedded stones occurring in the excavation pit.

The use of the continuous miner or mole for underground excavation work offers the same general facility if the excavated material is carried away from the mining machine fast enough to avoid piling up along the line. Continuous miners have been successfully used for the production of various kinds of ore and for tunnel excavations. But they also have been applied with catastrophic results to underground conditions not suited to their use. Many makes of boring machines have been and will be developed with ever-increasing thrust pressures against the face of excavations with different means for scoring and cutting the materials to be excavated. Although the inventors may envision that their devices are capable of rapid advance for the conditions at hand, experience has still proved that ground conditions most favorable to the use of continuous mining machines involve stiff clays, massive homogeneous formations of chalk, shale, lightly cemented granular materials, and some of the sedimentary rock formations. Harder natural materials have been bored with continuous miners, but sustained production has been sometimes difficult to maintain and cutter costs have been unreasonably high.

Therefore, the first element for the development of new methods for performing construction operations is an understanding of the problem at hand, a good appraisal of the effectiveness of the devices being considered, and as always necessary, a satisfactory estimate of the production possibilities and the resulting costs. Inventive pride and excessive enthusiasm in the development of new ideas has led to some unhappy results in the past. Therefore, this particular function, in which the management has a substantial responsibility, should be tempered with sound judgment and a reasonable degree of caution.

13-18. Maintenance of Contractual Relations

Management at all levels has the immediate responsibility of maintaining satisfactory contractual relations with the owner-client and his representatives. This generally is done at the level of the counterpart of the contractor's manager, but the contractor's engineering staff is the tool by which management keeps abreast of many contractual matters.

The owner's engineer represents the client. He administers the contract and interprets its provisions. His counterpart is the contractor's engineer, who must be in a position to agree or disagree with the conclusion of the owner's engineer. Neither the owner's nor the contractor's engineer is logically considered to have the final word in the legal interpretation of all contractual terms, and seldom does either have any legal background or training. Nonetheless, they must exercise the usual prerogatives of contracting parties the world over, develop and have a good understanding of their contractual rights, and direct their efforts toward compliance with the contractual terms.

Rarely is it economical or possible—or even psychologically advisable—for either party to have a legal expert on hand at all times at the job site. Since questions of word interpretation invariably arise, it is advisable for the contractor's engineer to solicit and receive expert legal advice. However, legal control alone, with a disregard for firsthand knowledge of the technical aspects or the personalities involved, can prove disastrous. Thus the contractor's management, assisted by its engineering personnel, must deal to some extent in the realm of legal matters. In the case of the contractor's project engineer, virtually his entire function under the direction of the project manager is to make sure that the contractor complies with the contract and is properly compensated.

Contractual questions too numerous to mention can arise during the life of a contract. They can include whether the work being required by the owner's engineer is within the scope of the contract, if measurements of work accomplished are in accordance with the prescribed methods, if payments are being made under the proper work items, if the contractor's work is in accordance with the specifications, and many other questions that originate with the contractor's engineer in the process of performing his functions.

As such questions arise, the contractor's engineer is obliged to discuss them with the owner's engineer in the hope of resolving them there and then. If they cannot be resolved, they should be documented and referred to the management for action.

The project engineer is often called upon to draft letters about contractual matters to the client's representative and to keep the management up to date on associated developments. The project engineer's office should be required to maintain a complete file of all exchanges of correspondence supported by the technical part of the record that deals with the engineering functions.

Unfortunately, it is not always possible for the owner's engineer and the contractor's engineer to resolve honest differences of opinion. Such situations call for adroit handling by the contractor's engineer, and a serious effort should be made to avoid any personal conflict. If the parties can face up to the right of each to his own convictions, usually neither is offended by an appeal to higher authority.

The contractor's engineer, however, should not in any way compromise the contractor's rights or position. He should keep the record complete at all times and see that proper and timely appeals can be made by the management if it so elects.

Discussion of important contractural matters should be reserved for the contractor's project manager, chief engineer, division vice-president, or operating vice-president.

13-19. Supervising and Performing Engineering Layout in the Field

Construction contracts vary in their specifications for field surveying for controlling the location, lines, and grades of the work to be performed, and for making measurements of the completed work for payment purposes.

Basic control lines for the location of the work, and controlling benchmarks to establish elevation usually are provided by the owner or client on the basis of previously made surveys. Beyond this point, the contract might fix further field survey work as the responsibility of the client-owner or the contractor. Some contracts require that the contractor lay out the work and that the owner check it. Often it is expedient for the contractor to provide partial layout work even though such is the normal function of the owner.

Assuming that under the contract the contractor is responsible for laying out the work, the contractor's project engineer must make a study of surveying requirements from time to time and organize accordingly. Considerable care and continuous supervision of this function are necessary because they are basic elements controlling the progress of the work and a substantial element of cost.

For smaller jobs, the project manager or the project engineer might organize and direct the survey crews. For larger jobs, it sometimes has been necessary to provide a field engineer. Under the project engineer's supervision, he keeps in constant touch with the status of the various construction operations, even checking as to whether the many active construction crews have sufficient line and grade marks or stakes to avoid delays or costly unproductive standby time.

Planning ahead and establishing the necessary reference points, off or adjacent to the site, that would permit quick transfer for line and grade must be done. Devising survey patterns that permit repeated use of the same reference points for setting line and grade at a number of different places on the job is advisable. Completing computations of measurements in advance of their use in the field by the survey crews has always been good practice and part of good planning as well.

An important function of the survey crews, when required, is field measurements of the quantities of work performed for the purpose of computing partial payments monthly and for establishing the final quantities of work performed under the contract. Cross sections or topography for earthwork starting with the original undisturbed surface of the ground, plus successive changes in the area during the progress of the work, might be necessary. Linear or volumetric measurements of other features of the work are usually required throughout the job.

Another important survey function might result from the requirement in some contracts that the contractor prepare and submit "as built" drawings, minutely descriptive of the work put into place. They become necessary when the plans, specifications, and preconstruction drawings are not sufficiently complete to serve as permanent records for the owner-client. Care must be exercised in making a good field record of all location work, usually descriptive of incidental features as well as the major parts of the work. In other words, the "as built" record is a complete design prepared after the fact.

Survey field notes should be carefully organized and supervised. They should be promptly reduced to the elevations or measurement for which they were taken, so that any questions about them can be resolved before the site conditions are changed. Field notes can be reduced at least in part by the field survey crews during inactive or bad-weather periods; but office personnel, with the use of electronic computer devices, can more often handle this function economically. Constant communication between the engineering office and the field survey crews should be maintained, and every effort should be made to promptly incorporate the field work into the permanent job record. Delay in carrying out this detail can raise serious questions about payments under the contract and whether certain features of the work have been properly installed.

The time of survey crews should be scheduled in detail before starting the work, and certainly in more detail during the progress of the work to meet the day-to-day requirements. When repetitious survey operations and continuing construction features are involved, greater efficiency can be realized by assigning them to the same survey party rather than to different parties.

The accuracy of surveys is vastly important. Work put in place to improper lines and grades can be disastrous. Make a practice of checking back to the original references in the field, making closures of the field work or, as a minimum, provide some observation to check accuracy. These precautions normally take little time and need not affect the productivity of the survey or construction crews

In many places in the United States, survey crews are members of the technical engineers union. Their employment is subject to wages, hours, fringes, and working conditions covered by contracts negotiated with employers in the area. Be aware of all the particulars involved because the number of men, the programming of their services, and the costs involved are materially affected. These factors differ from place to place. Be sure to obtain copies of current labor contracts before beginning work, and keep current on renegotiated conditions applying from time to time.

Some construction contracts provide that the owner or client will supply to the contractor all field layout work for construction purposes. Others provide that the owner will perform partial layout work and that the contractor complete it. In the first case, constant communication and almost minute coordination are required. Prevailing specification provisions usually state in detail the kind of notices to be given relative to starting and finishing of specific construction operations, number of shifts to be worked, etc. In fact, when the owner-client assumes the responsibilities of the field layout work, the contractor must assist him in the programming, scheduling of survey crews, and

their day-to-day assignment as though the contractor were supplying such services. When the contractor supplies part of the field layout services, the same is true to a limited degree, with the contractor organizing his own crews for the part reserved to him.

There are a number of time- and labor-saving survey devices and services available which deserve careful consideration and possible application when their accuracy and cost are satisfactory. The application of laser-beam devices for the projection of line and grade has revolutionized field surveying under certain circumstances. Instruments permitting the surveyor to read a linear distance between two points by setting up on one end and sighting in on the other without traversing the distance has obvious merits.

Photogrammetry is performing an ever-increasing part in field surveys, particularly in preliminary location work. Services are available whereby aerial surveys, reduced to ground topography, can form the basis of electronically computed quantities reflected by the difference in the topography. Thus the amounts of materials excavated from large areas, such as borrow pits or strip mine properties, can be very quickly determined without ground surveys except for the establishment of reference points and markers. Its application would be limited to large operations. Firms offering such services at a contract price should be consulted when a possible application presents itself.

13-20. References and Specifications for Construction Machinery and Permanent Materials

In every construction organization, large or small, an individual or group must undertake the responsibility of maintaining satisfactory specification references for the tools used to perform the work and the items that enter the completed work. The master files covering these items might be divided between several departments in the contractor's organization, or may be more logically assigned to an individual.

With respect to the construction machinery, the mechanical department must have immediately available descriptive information and parts lists covering the items of plant and equipment which it is obliged to maintain. Its people usually are less conscious of the importance of a paper record. On the other hand, the contractor's engineering personnel, designing the job plant facilities around the various pieces of equipment and needing the detailed specifications and operating characteristics of them for the purpose of estimating costs, must have ready access to such information. During the construction operations, the job engineer must make cost and method studies, redesign plant features, and advise on repairs.

The items to be incorporated into the permanent work are usually specified in detail in the contract documents, but on occasion their design and plan of installation are submitted as a condition of the tender by the contractor at the time of bidding the job. Sometimes these features of the work can be quite extensive, involving hydraulic turbines, generators, hydraulic gates and valves, and electrical installations. Although the purchasing department assists in the negotiations with subcontractors and vendors and in finalizing commitments with respect to the technical requirements, the basic information must come from the technical parts of the contract documents, subject to interpretation of the engineering personnel.

The engineering personnel is also charged throughout the job with the following through on matters of contractual compliance and deliveries, as well as the coordination of the output of several possible manufacturers supplying parts of the same general installation. Thus the engineering department needs the records for repeated reference throughout the life of the contract work.

The contractor's engineering staff usually keeps the master record of all applicable specifications and descriptive information. This is logical because that staff is intimately involved in the origination and use of descriptive and specification information pertaining to construction plant and equipment and to materials, machinery, etc., permanently entering the work, and because engineering personnel normally appreciate the need for such information. This practice has been found to produce the desired results.

13-21. Advise About Requisitioning, Scheduling, Expediting, and Inventorying Permanent Materials

Having all the materials and parts entering the completed project available at the right time is just as critical to the continued construction progress as the coordination of other activities on the job, or maybe more so when specially fabricated items are involved.

It is assumed that purchase orders and subcontract commitments for the important items will have been made in the early stages of organizing the job as previously suggested, based upon agreed conditions and a bracket of delivery dates that fit the programming then contemplated. Commitments for the less important and readily available items can be deferred. Large-volume purchases for portland cement, pozzolan, etc., are usually firmed up early subject to delivery schedules to be provided periodically during the job.

Planning is necessary to ensure the proper flow of the items that become component parts of the permanent work. Such items must be scheduled for delivery to fit the approved construction program, and rescheduled for changes made from time to time. Approvals by the owner-client or his engineer are often required of shop drawings and technical information pertaining to manufactured or specially fabricated items before shop work is started. Samples and specimens of materials must occasionally be submitted and approved. Shop inspections must be arranged when required.

A careful record of the status of these matters must be kept current under the direction of people familiar with the technical requirements of the contract and the detailed programming of the work as a whole. They must also appreciate the specific configuration and function of the parts involved.

Also necessary are current inventories of such common items as cement, masonry units, and others subject to limited storage facilities at the site or deterioration. Someone familiar with the job's short-term minute requirements should undertake this responsibility and requisition additional supplies so that stocks are adequate for continued operations.

The various steps necessary to assure on-time delivery of materials and devices entering the permanent work need to be followed through currently and expedited to avoid any delay along the line. It is important to know when parts will be shipped so that everything is available for timely assembly on the job. Vendors should provide shipping manifests and bills of lading as promptly as possible. Shipping dates and estimated dates of delivery should be set up to allow careful inspection of the shipment upon arrival and repair possible damage during transit, as well as to permit partial assembly of parts at the job site when this is desirable.

Arrangements and plans for temporary job-site storage should be made well before shipments are received. Some articles such as electrical devices may require special protection. All parts should be carefully identified on delivery and stored properly. Any shortages should be determined immediately and replacements ordered. Care should be taken that parts for permanent installations are not diverted to other purposes.

The purchasing, traffic, and warehousing departments of the usual contractor's organization must perform many of the functions mentioned above, but obviously they cannot supply the technical knowledge necessary for the successful acquisition and delivery of materials entering the permanent construction. It has often proven desirable, particularly in the heavy construction field, for the engineering personnel to be in constant consultation with other departments working with those functions.

13-22. Records of Work Performed and Contractual Transactions

The contractor's project engineer is generally regarded as having the responsibility for preparing and keeping job records that might affect the contractual position of the company. The detail and general character of such records, and the expenditures that might be justified in this connection, depend upon the kind and complexity of the contract work involved. Building work, for instance, would be relatively simple compared with most fixed-price heavy construction.

Some contracts provide that the owner's engineer will prepare the final and interim payments under the contract. Still others fix the function of preparing payment estimates upon the contractor, and it is intended that the accuracy of such estimates would be reviewed by the owner's engineer. These situations have a decided influence upon the character of records to be kept.

An indispensable part of the records for any job is a comprehensive daily diary logging the important prevailing happenings at the site of the work. This would include the weather condition and its severity, comment on each major construction operation in progress, subcontracted work as well, showing some measure of accomplishment such as quantities realized, linear advance, truck loads handled, and the like. It should record all unusual incidents, the people involved, the time of day, and a short descriptive narrative. Any important conversations that occur, commencement and completion of major construction operations or parts of the work should also be included.

A comprehensive daily diary is unsurpassed as evidence for proof of job facts at a later date. It is seldom evident at the time the diary is being written what situations it might be called upon to support after the work is done and the organization is disbanded, so a decided effort is necessary to prudently record the things about which questions might arise. The daily diary notes could well serve currently for supplying information for daily, weekly, or monthly reports to the main office if required.

Another conclusive record of current construction progress is a satisfactory photographic album. It need not be extensive or expensive, but the photographs should be dated, carefully filed and preserved and have brief descriptions. Pictures taken periodically from the same location and at the same angle are generally desirable to show the advancement of the work in progress. Photographs should be taken of unusual or special events. A camera comparable to the usual family type, if kept handy with a supply of unexposed film, will usually suffice, and it can be operated by a knowledgeable member of the engineering staff without special photographic ambitions. Small-frame prints are adequate for the album record. They can be enlarged for special purposes as needed.

The project engineer should keep a current file of all contract drawings and specifications for ready reference. Although there will no doubt be other such files for use by the construction forces, the one kept by the project engineer should be the key copy, and it should be complete and up to date at all times, including the latest changed and supplemental drawings and specifications. It should constitute the basis for transactions with the owner, the owner's engineer, subcontractors, and suppliers.

It is good practice for the project engineer to be assigned the duties of receiving all contract drawings and specifications, date stamping all copies for receipt, disbursing them to the field forces for use in the construction and otherwise, and keeping at least one file copy of every superseded drawing or specification. The latter record can have no substitute when it comes to preparing statements for changes in the work and extra work.

Even though part of the above-described duties tends to be of a clerical rather than an engineering nature, the importance of keeping minute control of this part of the contractural relationship, understanding the effect thereof and immediately dispensing the information involved, needs no explanation. The project engineer is in the best position to serve this responsibility, possibly assisted by his staff.

Correspondence files covering the exchange of expressions between the contractor and the client-owner relative to things that affect the contract should be maintained by the project engineer's office whether or not the particular matters originate with the project engineer. Since the project engineer is usually involved with drafting letters to the client-owner or his engineer, he must be totally informed about all contractual matters in case they would have a bearing on the matters that he is obliged to handle. The same general situation prevails with respect to correspondence pertaining to subcontractors, fabricators, and major suppliers when the exchange relates to work to be done or materials to be supplied.

As mentioned earlier, it is a usual duty of the project engineer to keep a close control of the status and delivery of permanent materials and devices incorporated in the permanent work.

Of extreme importance to the contractor is the matter of compiling currently an engineering record of work accomplished under the contract. In the case of either lump-sum or unit-price contracts, such records usually constitute the basis for partial and final payments for work performed, because they measure the status of the job.

For fixed-price contracts the amount of total payments to the contractor is based upon the total quantities computed and compiled for each of the items appearing on the bid schedule. It therefore is incumbent upon the project engineer to organize and conduct the necessary field work to obtain the required basic information, and to make such computations and compile such records as will accurately reflect the amount of work actually performed. The kinds of computations involved will be quite variable: cross sections for earth and rock excavations and embankments, volumetric computations for concrete and masonry,

shipping weights for some of the manufactured and fabricated metal items, computed weights of reinforcing steel (length by unit weights), computed square or linear measure, etc.

It is important to keep the record of computed quantities as up to date as possible in order that check measurements or remeasurements in the field can be made before site conditions are changed. Thus, more accurate monthly or interim estimates can be made during the progress of the work. This practice also is of considerable advantage from the standpoint of accumulating currently the record required for establishing the final payment to be made and avoids long delays at the conclusion of the work while the final computations are being made.

In making quantity computations, due consideration must be taken of the criteria set up in the contract for making such computations, or usual good and accepted engineering practice must be followed to avoid the question as to their acceptability or accuracy. The results of computations should be spot checked or reviewed to make sure that no serious error exists, such checks being possible by comparison with truck loads, batch count, etc.

A careful record should be kept of the number of batches of different concrete composition produced each day, and the part of the work into which they were placed. In the case of large batch plants, this can be accomplished by the printer on the batch recorder, with the tapes providing a permanent record. They should be saved until all questions on the job have been answered.

If the owner's engineer is responsible for making all the field surveys and preparing the payment estimates under the contract, and if the particulars are made available as a matter of practice to the contractor, there is some justification for utilizing the owner's records without the contractor duplicating all of the same efforts. Check computations by the project engineer will be necessary in many instances, however, and the contractor must keep a reasonably complete record of quantities measured, computed, and checked to cover the work performed.

The prevailing conditions for each job will dictate the specific records to be kept.

13-23. Preparing or Checking Payment Estimates for Work Performed

As mentioned earlier, some lump-sum or fixed unit-price contracts provide that the owner's engineer prepare the payment estimates, and others require that that function be performed by the contractor.

When Payment Estimates Are Prepared by Owner's Engineer   The contractor should establish an arrangement with the client-owner or his engineer for access to the record and detailed computations supporting payments that are being made under the contract. All particulars can be discussed and understood, and he can obtain photographic prints of documents of particular interest.

The contractor's project engineer should review all monthly estimates promptly upon receipt and be in a position to make a timely record of any difference of opinion that might exist. Partial interim payments are sometimes made on approximated quantities based only upon rough calculations rather than refined calculations. This is normal practice provided that refined calculations will have been made for as much of the previously completed work as could be reasonably expected. As soon as computations are represented as final for a given part of the work, their accuracy should be verified by the project engineer and all particulars pertaining thereto recorded and accumulated, for later incorporation into the proposed final payment under the contract.

For checking payment estimates by the owner's engineer, the contractor's project engineer will utilize the record prepared in his own office, as mentioned earlier. Such record will serve as background study preparatory to reviewing the monthly payment estimates by the owner's engineer, as a basis of conversation and compromise with respect to items for which questions exist, or the basic record upon which formal protest or claims may be founded for items upon which there continue to be areas of disagreement.

Checking payment estimates by the owner's engineer necessarily involves interpretations of the contract, the drawings, and the specifications. The payment estimates themselves will reflect interpretations by the owner's engineer, and the contractor's project engineer must necessarily verify them, calling upon legal opinions if necessary. The contract documents should set forth specific work items appearing on the bid schedule which forms the basis of the payment estimates.

It is important to verify whether payments are properly classified according to the contract, as well as whether computations are properly made and in accordance with prescribed criteria or good practice. Erroneous classification for payment purposes can result in the application of the wrong payment prices and can present a vastly different financial picture for the contractor. If ambiguities exist in the documents, or if the contractor's project engineer and the owner's engineer interpret the wording differently, a written record should be filed promptly with the owner or his engineer documenting the subject and asking for a review and adjustment of the matter.

Statements covering extra work or work on change orders should be reviewed in light of the wording of such orders and the accumulated costs, quantities, prices, or otherwise that formed the basis of agreement for compensation to the contractor. When such extra work or change orders are to be based upon time and material for labor, equipment, rentals, etc., the original billing will have originated with the contractor based upon time and material accounting which most likely would have been verified in advance by the owner's engineer.

Subcontractors are not parties to the general contract, and owners are prone to avoid discussions with them about contract matters in order not to become involved in any possible disputes between the contractor and his subcontractors. Presumably the subcontracts will contain the provisions of the general contract, and the owner and his engineer will consider the matter only in light of the general contract. To the extent that subcontract quantities and related general contract conditions are concerned, the contractor's project engineer has the responsibility to follow through with verification of payments, interpretations of the documents, records, protests, etc., as though the work had not been subcontracted.

Other matters involved with payments under the contract, such as penalties for late completion, bonuses due for timely performances, percentages temporarily retained, and anything else that affects the revenue under the contract, should be reviewed and verified and, to the extent justified, become matters of agreement or disagreement.

If serious shortages of credits for work performed exists with respect to the interim payments, these can result in large financial burdens to the contractor in that his investment in the job is increased and are probably the source of additional interest charges not actually justified. Although the contractor's project engineer should be in a position to evaluate the amount of such shortages from his records for the purpose of monthly cost and revenue statements, every effort should be made to influence the owner's engineer to keep payments as current as possible. Unnecessary extra costs resulting from this situation should be recoverable based upon usual contract conditions.

The function of review of the owner's payment estimates by the contractor's project engineer is a matter requiring great tact. It should be approached on the basis of a broad understanding of the position of the owner's engineer and the prevailing circumstances. It should not be regarded as a policing action by either party, but as a mutual effort to apply properly the terms of the contract. However, it should not be overlooked that bias or honest differences of opinion can creep into the situation. The contractor's engineer has the responsibility to report to his immediate superior the results of this activity, as well as any subjects which in his judgment require formal handling as contractual matters and are subject to the giving of timely notice or the filing of written protests and appeals.

When Payment Estimates Are Prepared by the Contractor The situation in this case reverses the position of the contractor's project engineer and owner's engineer from the standpoint of the responsibilities of providing the necessary support for the details summarized on the payment estimates. The contract conditions on this premise usually provide that under this arrangement the contractor will submit to the owner or his engineer, on a given date each month a statement of all work performed and/or earnings realized under the contract through the last day of the preceding month. They also usually provide that the owner will review the statement and, on or before another given time, make the interim payment based upon the monthly estimate prepared by the contractor, less previous payments and less any amounts with which the owner disagrees, specified amounts to be retained under the contract being withheld.

It is presumed that the contractor has followed the criteria for the preparation of the payment estimates set out by the contract, that the representatives of the owner's engineer have currently observed the basic field work and the computations prepared by the contractor, and that the owner's people will have been well informed of the particulars supporting the earnings reflected by the payment estimates before such estimates are actually prepared. Thus the time necessary for approval of the statements should be a minimum.

Total cooperation to the extent contemplated by the contract should be the keynote of the attitude of the contractor's project engineer in performing the function of preparing payment estimates. Prompt execution of proper directives by the owner's engineer, providing him with details and particulars, etc., can facilitate and expedite his approvals and result in prompt payments to the contractor.

The reversal of position resulting from the contractor rather than the owner preparing the payment estimates does not change the status of the contractor with respect to all of his contractual rights. The contractor's project engineer should be alert to the matter of reporting to the management all points of disagreement so that timely protests can be filed if that course is followed.

Final Contract Payment The final contract payment estimate must be a summary of all quantities of work performed and all earnings realized under the contract, including all extra work, changed work, or work covered by any amendments or extensions of the original contract for which final settlement is intended. It is required that all items listed on the final estimate be based upon refined final computations and records of account, except that the parties to the contract can still disagree concerning particulars. The contractor is free to reserve the right to pursue any claims at the time of final settlement of the contract, in which case the final settlement would specifically except them.

The computations and records upon which final settlement is made should be preserved for future reference for a reasonable length of time. In case claims remain to be arbitrated or referred to the courts for judgment, the job record will be indispensable. Otherwise there is always the possibility of claims being made by the owner or third parties predicated upon alleged latent subsurface conditions which develop, and the record will be needed for defense.

13-24. Job Inspection and Quality Control

Contractors of integrity will follow a policy of compliance with the contract with respect to the quality of the work and the other contractural conditions. Under some conditions the contract may require that the contractor provide inspection and perform tests to verify that the work is satisfactory, all subject utilimately to observations and approval of the owner's representative. If so, the contractor's engineering staff usually directs the program and compiles the necessary records, with the construction forces providing the necessary labor, equipment, and supervision. The existence of such arrangements should be written into subcontracts.

When the owner's representative furnishes inspection for quality control and the contractor does the sampling, testing, etc., the contractor's engineer should satisfy himself that the inspection and tests required do not exceed or conflict with the contractor's contractural obligations. Sometimes those observing the work—whether the owner's representative, the contractor's engineer or the construction forces—inadvertently disregard the specifications and strive for excellence in the work beyond the stated tolerances or values. Such a situation generally results in unnecessary costs and possible delays. In addition, there is the ever-present possibility of differences in interpretation of the written requirements. When the contract gives the owner's engineer great latitude of decision, unusual or unreasonable inspection might be enforced contrary to the intent of the specifications.

It sometimes happens that conditions are stated in the contract which are not truly representative, or that final results are specified which actually are unattainable, and that the contractor is directed to satisfy these conditions. An example would be when borrow pits are specifically designated for use without processing, and the specifications provide for an embankment material to be of a specific gradation. However, testing shows a definite gap in certain sizes that cannot be corrected by normal selective borrowing. Importation from other outside sources of the missing size fractions and blending on the fill could accomplish the specified end result, but such was not contemplated by the contract.

Or take the case of plastic fill materials going into an impervious-type embankment from designated pits in which the character of materials is variable. The specifications provide for the fill to be placed to a given percentage of optimum compaction. Recognized tests were employed. Borrow and fill operations proceeded with usual types of excavating and transporting equipment, and the materials were laid down to the specified thickness of layers and subjected to the required compactive effort. In situ tests from the embankments showed that some parts of the fills measured up to the specified compaction and others did not. Additional compactive effort or removing and replacing the fills only partially produced the desired results. The work was seriously delayed and the average production upon which the bid was based could not be accomplished because of the disrupted operations and the extra cost of removing and replacing some of the fill materials without compensation.

Upon scrutinizing the inspection procedures, it developed that compaction tests from the embankment were gauged against a series of laboratory tests of borrow-pit material made earlier and averaged to produce a curve to be used as a criteria for all embankment tests. The curve created by the original laboratory tests of the borrow materials was never representative of all the materials in the borrow pits and erroneously created compaction criteria that could not be met by the materials of less than near maximum stability, although such materials were designated as suitable by the contract. Compaction criteria for each kind of material occurring in the designated borrow pits was guaranteed by the contract. When the laboratory and embankment tests for compaction were coordinated, contractor's problems were basically solved.

The foregoing examples from actual experience pinpoint only a very small area of possible difficulties resulting from erroneous inspection, or inspection made without full recognition of the attendant facts. Such situations can creep into most every phase of supply and performance of the work, whether pertaining to the usual things the contractor has to do, or the highly complicated special parts and machinery involved in unique installations.

A general practitioner engineer, such as the contractor's project engineer, could not be expectd to be well versed in all of the specialties with which he comes in contact, but he should be alert to all difficulties in the performance of the work growing out of meeting the prevailing inspection requirements, and the wording of the contract specifications. If such difficulties persist, and the reason or the necessary remedial action is not apparent, they should be taken up with the management, and consultation with specialists is recommended.

A change in manufacturing or construction procedures, or correction of inspection criteria and requirements, could well result in clearing up the difficulties at an early stage and avoid costly delays. They might disclose that test results or inspection requirements are impossible of accomplishment, in which case the contract should be changed. Every effort should be made to keep the owner's representative informed of any observations or investigation of inspection requirements and to make them a matter of joint interest with both sides facing up to the facts, having in mind at all times the quality control necessary to comply with the contractor's contractural obligations.

Some contracts provide that the owner's representative be provided with certified test results of given parts and, in some cases, with sample specimens of the product involved. The required tests might be specified to conform to such acknowledged standards as A.S.T.M. and Federal Government Standards, or they might be set forth specifically in the contract documents in point. Orders to suppliers, fabricators, manufacturers, or subcontractors should comply with the general contract. Dates for submitting samples and test results should be established in order to permit manufacture, fabrication, and installation to proceed in coordination with the general construction program.

The contractor's engineer should undertake the responsibility of seeing that these conditions are met on time, that the test results and samples are conveyed to the owner, that the owner's approval is obtained and a complete record of all incidents relating thereto be made in writing, and that a complete file of the same be maintained at least until after final settlement of the contract. For better coordination of this particular detail, the contractor's engineer normally should handle the correspondence with third parties involved because of his more complete knowledge of the technical requirements.

When excavations reveal that subsurface conditions differ materially from those indicated by boring logs, test data, etc., and from conditions generally prevailing in such circumstances, the contract usually states that contractor is to notify the owner immediately. The contractor shall not disturb the site until the owner has had an opportunity to study the matter and advise the contractor as to how to proceed. If the owner finds unexpected subsurface conditions and if the cost of performing the work is increased as a result, the contract price is adjusted. The construction forces may or may not know what to expect with respect to subsurface conditions, and the owner's engineer may overlook or choose not to make an issue of them if they are different. Although job management may already have made its observations in the field, the contractor's engineer should periodically inspect all operations and be on the alert for differing subsurface conditions, making a complete record of any such matters and the developments.

Another case emphasizing the importance of the contractor's engineer making periodic inspections and recording his observations in the daily diary is evident when the job is completed under the careful continuous supervision of the owner's engineer. The job is accepted, final payments under the contract are made, and releases are given by the parties. Then, after several years, the structure settles or shifts, or some other defect occurs. The owner files suit against the contractor, alleging he failed to construct the work according to the specifications. If the contractor has made his own inspection and has supporting diary notes and photographs, he can refute the charges made in the suit. Naturally it is impossible to record all things that might require substantiation later, and it is not economically feasible to compile the vast files that would be necessary to do so. But a general approach to the problem and periodic inspections by the contractor's engineer are justified.

13-25.   Records of Costs and Extra Work

Although it is the function of the field forces to report the cost of extra work, and the function of the accounting department to summarize and make book entries, it takes someone more conversant with the terms of the contract than they to distinguish between the work specified as part of the contract work and that which is outside its scope. On the normal medium-to-large construction job, this responsibility is undertaken by the contractor's project engineer.

As discussed elsewhere, it is assumed that the contractor's project engineer will supervise the assignment of cost codes to the various charges against the work and, in that

function, along with his own field inspection, will be in a position to observe the kind and purpose of the various charges being made. The project engineer also will receive copies of all correspondence and revised drawings from the owner which in themselves constitute orders to perform specific items of work. Such orders may classify some of the work as extra, or may categorically state that any and all work called for is not to be classed as extra work, and that payment therefore is to be made at the unit or lump-sum prices stated in the contract.

Regardless of the source of the charges or statements to the contrary by the owner in transmitting drawings or directing the work, the contractor's project engineer, together with the project management, should determine whether the work involved falls within or without the scope of the contract as written. If there is a question in this regard, assistance from higher authority or legal counsel should be solicited. If it is determined that any work called for is outside the terms of the contract, notice should be promptly served upon the owner, and the issuance of an extra order providing for additional payments under the contract should be requested.

Some contracts state that no work should be done on extra work until the extra-work order has been issued by the owner and the terms of payments have been established. This is a worthwhile rule to follow where it can be applied, but unfortunately circumstances sometimes make its application impracticable. Examples include work that is initiated by the contractor's foreman upon the direction of the owner's inspector before any determination is made as to its status; when a conflict of opinion exists as to whether the work is extra and failure to perform it promptly would disrupt other operations under the contract; or when delays in negotiations of price and terms of payment are occurring, and the work in question must proceed as a logical part of the program.

Proceeding with the alleged extra work without all of the terms and conditions being worked out is a calculated risk, but a management decision must be made when deferring it would have far-reaching effects, delay the overall project, or increase the costs of performing the total contract work. In the development stages of the extra-work order or while attempts are being made to negotiate it, a complete written record should be kept and ample written notices directed by the contractor to the owner to protect his right to claim extra payment whether the extra-work order is written or not. The contractor's project engineer should assume the responsibility of making such a record.

Most heavy construction contracts set forth a plan for handling of extra-work orders and provide that the price of work can be compensated in one of several ways; i.e., at negotiated lump-sum prices, at negotiated unit prices, in which the quantities of work must be measured and certified to by the owner; or at actual necessary costs for labor, materials, equipment rentals and operation, and incidentals, plus an allowance for overhead, supervision, use of tools, and profit.

Field costs of the extra work would in any case be reported and compiled in the usual way according to the contractor's system of accounting. However, when the cost of the extra work is to be compensated on a cost basis or when the terms of the extra work order are not yet established or are in dispute, special handling and additional care are warranted. The contractor's project engineer should then undertake the supervision of the accumulation and reporting of the charges involved, with the actual work of making the reports and entering them in the books being functions of the field forces and accounting departments, respectively.

With respect to such cases, and when a firm price has not yet been negotiated, all labor involved and equipment used should be reported in detail on shift time cards, giving the actual time of each man and of each machine employed for each shift. This information should be summarized by the shift foremen on a special form or on a standard sheet with a carbon copy. The matter should be discussed openly with the owner's shift inspector familiar and the crews and machines employed, and the inspector should be asked to sign the summary to indicate its correctness, regardless of whether or not there still exists a question between the parties as to whether the work involved is to be compensated as extra work. The inspector should be given a copy of the summary for his records.

A careful record of all outside purchases and disbursements from the contractor's warehouses made expressly for the work involved should be recorded in the proper form and cost coded accordingly. At the time of preparing a statement or invoice with respect to the work in question, copies of the supporting records of time and material should be attached and sent to the owner. Work under approved extra-work orders should be invoiced monthly. Matters in question should be reserved for the proper time of presentation. The contractor's project engineer should be currently aware of all such activities involved, including the record being made, and should pass on the legitimacy of the charges and their completeness. The charges might even be cost coded by him or under his direction.

The underlying principle that the contractor's engineering people check all payments under the contract, if such is adopted, should apply to payments for extra work, as well as the originally contemplated contract work.

13-26. Records of Subcontract Performance and Payment Estimates

Subcontractors, as agents for the prime or general contractor, perform part of prime contract work. As mentioned elsewhere, the subcontractor is regarded as having no official positon with the owner. The contractor, therefore, must handle with the owner all questions relating to the subcontracted work covered under the general contract, including payments. It has been suggested that the contractor's project engineer keep all the necessary records and verify all payments made to the contractor by the owner. With such a background, the contractor's project engineer would logically be the one to keep track of all earnings by the subcontractor and to prepare the statements covering interim and final payments due the subcontractor.

Work performed under subcontract can be compensated in a number of ways as established in the subcontract agreements. Usually, but not necessarily, it is paid for in the same manner as specified in the general contract, and the work items in the subcontract are related directly to the items of the same description listed in the general contract, with all provisions pertaining thereto to apply to and become part of the subcontract conditions. If such is the case, payments to the general contractor can be the direct measure of what should be paid to the subcontractor in terms of quantities of work performed multiplied by the subcontract prices. Assuming firm unit-price contracts and subcontracts, the contractors' project engineer, upon checking payments made to the general contractor by the owner, will have immediately available the quantities of work per-

formed by the subcontractor that should be listed in the subcontractor's payment estimate. This would be the ideal situation, but unfortunately there often are many deviations and complications, and the project engineer's records must reflect them completely.

There are many reasons why special and complete records of the subcontractor's activities and earnings should be kept up to date and open to review by the subcontractor, among which are the following. The subcontract conditions might specify pay lines different from those called for in the general contract. The conditions might cover only a distinguishable part of a work item called for in the general contract. They might be compensated on a different unit of measure. They might involve change orders and extra-work orders not directly related to the conditions of the general contract. They might also provide for penalties and bonuses not stated in the prime contract. Coordination between the project manager and project engineer should be maintained at all times, and a written record and exchange of correspondence made of all verbal arrangements between the parties that might tend to modify or change the written word of the subcontract.

Monthly or interim payments under the subcontracts are necessarily approximate, as in the case of the prime contract, because minutely accurate field measurements of all work performed are not justified repeatedly, and approximate measurements are the rule. Also, some work items involving only office computations or summaries of weights can be approximate for interim payments, but reasonably full use of all the data of record should be employed. Lump-sum items not subject to analysis in terms of quantities of work-item operations must be considered in the interim payments on the basis of overall percentage of competion.

Relating the partial work done to the original bid estimate in terms of cost of the various parts of the item is a good gauge, unless the general contract provides a formula for making interim payments. As the work progresses, however, the contractor's records and computations should be finalized to the extent possible for those portions of the work completed from time to time so as to permit the preparation of more accurate interim payment estimates and to minimize the time for computations necessary for the preparation of the final estimate.

Interim payments would be expected to state all earnings to date for work performed up to that time, plus any other considerations provided in the subcontract but less the amount of the retained percentage contemplated in the subcontract to ensure completion of the work and less the total amount of all previous payments made to the subcontractor. The balance is the amount due for such interim payments.

Disputes between the contractor and subcontractor concerning payments, when and if they arise, and if the subcontract provisions incorporate all of the conditions of the general contract, should be resolved upon the same concept as the question involved is settled between the contractor and the owner, provided that the contractor will have presented to the owner for his consideration the complete position of the subcontractor. As in the case of all contracts, there might be unresolved questions at the time that the work has been concluded, and these would have to be considered at the time of final settlement.

When the work has been completed, and allowing a reasonable time thereafter, or within the time specified in the subcontract, the contractor's engineer shall complete all of the computations involving the work performed or other considerations of payments under the subcontract, and shall prepare a final estimate or statement of earnings covering the complete subcontract revenue due the subcontractor, less interim or periodic payments previously made. The statement must be submitted to the subcontractor for his review and approval. Subject to any possible changes or corrections agreed between the parties, it should form the basis of the final settlement with the subcontractor, except that a form of a release should be drafted by the contractor and submitted to the subcontractor for his approval and signature.

If the subcontractor still believes that additional considerations are due under the subcontract, he can sign the release subject to his reserving specific claims which should be described in some detail and written into the release form. Such claims can be pursued through arbitration, court action, or otherwise, as the case may dictate. When the release is obtained from the subcontractor, the contractor would normally turn back all money retained under the subcontract to ensure completion of the work. If, however, there are serious disagreements at the time final settlement is due, and such disagreements cannot be resolved by amicable negotiations, legal advice should be solicited before all payments are made under the subcontract so that the correct contractural position is maintained for the contractor.

The form of records to be kept of all subcontract earnings should, of course, follow good practice for the particular kind of work involved; i.e., cross sections for excavation, driller's logs for drilling, shipping papers, and freight bills for weight determination, and field measurements of various kinds. When payments under the general contract govern, statements of quantities prepared by the owner are advisable. The form of monthly or interim payment statements and the final payment estimate covering subcontract earnings should preferably follow the form of the payment estimates issued by the owner under the prime contract, so that all subject matter can be directly related when such is possible. Otherwise subcontract payment estimates should be in some acceptable format, but the same format should be used for all statements for ease of comparison.

13-27. Accumulating Performance and Production Data

The stock in trade of any contractor, or one of its principal elements, is his know-how relative to the performance of the work. Experience and the knowledge of how to evaluate it is an indispensable part of developing such know-how. Knowing the production capabilities of the crews and facilities at hand is the first step before attempts are made to improve them, and this information is available by continuous study of construction operations under way. It is, of course, directly related to the costs of the work being performed. The contractor's project engineer is the logical one to conduct such studies for management's use in analyzing the accomplishments on the current job and for accumulating basic data for estimating future work.

How the performance and production data are accumulated is a matter for determination to fit prevailing circumstances. A usual source of some of the basic information is from daily field reports, time cards, plant and machine operation reports, and automatic recording devices mounted on the machinery being used. Something more than just the time of use is needed to point up production. The cooperation of the foreman making the report also is

needed to give the kind of different work handled and the time, temporary downtime for maintenance and repairs, prevailing conditions affecting production such as weather, ground conditions, etc. They also should provide such measures of production as truck loads handled, number of batches produced, number of piles driven, and number of precast concrete members produced or set. The quantities of work on daily reports probably are only approximate. For accurate studies, computed quantities that become available as time goes on should be substituted.

At the time of starting any given operation, the management and its engineering personnel should agree on the kind of information regarding time, performance, and production that is needed, and arrangements should then be made for the field forces to report it accordingly. The accounting personnel also should be advised so that they can participate in its compilation. Reasonable care should be taken to avoid asking for so much information that the reports become burdensome, leading to lack of interest by the field personnel.

Field observations and summaries of quantities and the like from engineering records will be necessary in any case. Data taken from the daily reports should be summarized by the accounting staff and made available to the project manager daily if desired. They should also be made available to the project engineer at suitable times for performance and production studies. Monthly periods might be sufficient and convenient for inclusion in monthly project reports, although data on more critical operations might be desired at shorter intervals.

With respect to most all items of plant and equipment, an important performance characteristic is the "availability factor" of the unit or the facility; i.e., the percentage of the total possible working time that it is capable of withstanding the wear and tear to which it is being subjected, or, in other words, a measure of its mechanical integrity. The availability factor can be expected to vary for the same mechanical unit under different kinds of service and operating conditions. For instance, tunneling equipment should suffer less downtime for repairs in softer than in harder ground tunnel. For open excavations, equipment handling rock would be expected to be out of service more often than when it is handling common excavation. Wet conditions are harder on equipment than dry conditions, etc.

To provide the basic information, daily plant and equipment reports should state the following for all major units, and perhaps for some lesser units on which it is desired to build experience records.

(a) Maximum total time possible, in hours
(b) Idle time, in hours
(c) Time for field servicing, in hours
(d) Time down for repair, in hours
(e) Time working, in hours

The "availability factor" could be computed from the above as:

$$\frac{(e)}{(a) - ((b) + (c) + (d))}$$

"Servicing" time is to be construed as the time that the mechanical unit is taken out of the operations for greasing, oiling, and refueling. For some types of plant and equipment, these functions can be accomplished while the unit is in operation. They might be done on an "off shift," i.e., on a third shift when only two shifts a day are scheduled for work; or at the shift change time, i.e., the interval between the time one crew shift ends and the next shift begins, as in the case of seven and one-half hours working time in an eight-hour period. The lunch break for the crews also might be used for servicing. This can be accomplished by scheduling the working time of the servicing crews differently from the operating crews. If servicing time can be handled without withdrawing the mechanical unit from the operations during possible productive time, no consideration need be given it in the above formula for computing the availability factor.

Daily plant and equipment field reports should, of course, show the operation on which the mechanical unit is used and its location, such as for a power shovel: "Loading out rock excavation from lower part of spillway to trucks hauling to spoil." This can even be abbreviated or covered partially by cost coding. A measure of accomplishment should also be reported, such as: "Trucks loaded . . . for the shift." Note should be made of any unusual condition or occurrence. The accounting and engineering people receiving these field reports can generally fill in facts from their own knowledge or observations, such as the size of power shovel, trucks, etc.

Where a number of like units in an operating fleet are used, such as 100-ton bottom-dump trucks, 35 cubic yard end-dump trucks, 250 horsepower tractor-bull dozers, 24 cubic yard motor scrapers, all of the same size, make, and model, they can be reported summarily and included in the records accordingly.

Performance and production data, directly related to the cost of the work, that have been found to be most useful for comparison with the budget and control estimate for the job and for use by the estimating engineers are listed on the facing page with respect to variety of construction operations. These factors are based upon a preconceived system of estimating. However, they are basic to most ways of evaluating construction work. They can be stated for individual shifts or days, or as averages for any given period of hours, including the total job period. It will be desirable to modify this list as job conditions indicate, and as changes in construction methods are instituted.

To facilitate listing the information the following abbreviations are used: cubic yards — cu yds; hour = hr; linear feet = Li. ft; pounds = lbs.

13-28.   Cost Coding Charges for Cost Accounting

The subject of coordination of estimating and the cost-accounting breakdown, and how the engineering and accounting personnel can accomplish it, are described in Chapter 16. It is mentioned here to give it recognition as a suggested function of the contractor's engineering section. The description in this connection has to do with its handling and administration during the course of the contract after the cost code has been established and the cost books have been set up.

It is not always apparent to the uninitiated as to which particular operation or work account the costs of certain work should be charged. Since it should conform to the basic estimate for the job prepared by the engineering department (including the budget-control estimate) the engineering people should resolve such questions unless the management takes an active part in that detail.

The manner in which this control is exercised will be subject to the accounting department in compiling the costs in the cost accounts. Basically, however, charges against the various cost accounts originate on daily labor

time cards or reports, daily machine reports for plant and equipment, warehouse disbursement or credit tickets, invoices and freight bills coded directly as received, and such other original records of expenditures or liabilities accruing as the result of the performance of the work. Assuming such original sources of charges to the costs, some arrangement must be made to apply the proper cost codes to them.

The cost accountant can be trained to recognize the application involved in most cases, and when there are questions about any of the charges the contractor's engineer should be consulted. In the case of major items of construction involved, some of them can be highly repetitive and prevail throughout the job. These can be readily recognized by the cost accountant, and the cost coding therefore is assigned by him without consultation. For more complicated construction jobs, technical matters will invariably present questions requiring the opinion of engineering personnel familiar with such technical aspects and the basis upon which the cost coding key was predicated. Then too, if special time and methods studies are initiated during the job, they would properly be identified by someone knowing of their existence.

On smaller jobs, it may be advisable for the project engineer to perform the actual cost coding on the original sources of the charges. On larger construction jobs, the cost coding might be done by the cost accountant and edited by the project engineer before it is made a permanent part of the cost books, thus conserving the project engineer's time. On still larger jobs, when the expense is justified and there is a substantial volume of cost work, a special appointee, namely a "cost engineer," is advisable. His functions would include performing the cost coding, coordinating with the accounting department in the assembly of cost information, conducting special cost studies, compiling production and performance data, and making periodical cost reports based upon cost book information compiled by the accounting department.

13-29.   Reports of Costs and Construction Activities

Some plan of periodic reports to help the project management locate and correct operational difficulties shown by excessive costs, and for conveying the job status to off-job principals and officers, is generally justified. This function can be readily overdone or underdone, and it must be tailored to fit the situation at hand and the overall company requirements and desires.

The management should decide at the start of the job what reports should be distributed. Some of the larger contracting companies have established policy on this matter and possibly have stated it in manuals drafted to control the conduct of their personnel. Unless properly controlled, the matter of reporting can become troublesome and costly. If various persons on or off the job have the right to modify the contents of the reports, such reports can become voluminous and burdensome and can contain details of limited interest.

Active items of work should be covered in detail to the extent desired during the time when such work is being performed. When the work has been completed and is no longer of primary current interest, future coverage should be in summary or abbreviated form, or, in the case of daily or weekly reports, dropped entirely. This will tend to minimize the volume of the periodic reports and reduce the difficulty of preparing and reading them. To facilitate

	Rock	Common
Open Excavations (loading and hauling)		
15 cu yd power shovel		
6 cu yd power shovel		
3½ cu yd power shovel		
plus other size rigs	Cu yds/hr	Cu yds/hr
5 cu yd draglines		
3 cu yd draglines		
etc.		
100 ton bottom dump trucks		
35 cu yd end dump trucks	Cu yds/hr	Cu yds/hr
18 cu yd end dump trucks	Trips/hr	Trips/hr
10 cu yd end dump trucks		
etc.		
Open Excavation (dragline casting)		
18 cu yd dragline		
12 cu yd dragline		
6 cu yd dragline	Cu yds/hr	Cu yds/hr
3½ cu yd dragline		
etc.		

Rock Excavation (drilling and blasting)

Rotary drill—9″ blast hole	L. ft hole/hr drilled
Rotary drill—6″ blast hole	Hole spacing/ft
Percussion drill—4″ blast hole	L. ft drilling/cu yd
Percussion drill—3″ blast hole	No. of drill bits used
Percussion drill—2″ blast hole	Bit life—L. ft/bit
etc.	
Explosives used	
Ammonia nitrate	Total quantity used
Gelatin dynamite	Quantity per cu yd
Ammonia dynamite	Summarize different blasting
Blasting caps	patterns separately
Prima cord	
etc.	

Hard Rock Tunneling

Excavation	
Tunnel advance	Average l. ft/day (3 shifts)
Drilling heading	No. of holes drilled in face
Drilling heading	No. of bits used
Drilling heading	Bit life—L. ft drilling/bit
Drilling heading	Quantity of drill steel, lbs/cu yd
Powder consumption	
Kinds used	Pounds total and lbs/cu yd
Blasting caps—by	
various delay interval	Total No. and caps/cu yd

Round Cycle

Move in and set up	.................hr ....min
Drilling time	.................hr ....min
Move out	.................hr ....min
Blast and clear fumes	.................hr ....min
Move in to muck	.................hr ....min
Mucking time	.................hr ....min
Move out	.................hr ....min
Total round cycle	.................hr ....min

the reading of the periodic reports by the management and other interested parties, and for ease of comparison of the contents of one report with another, the periodic reports should be made to follow the same general format, with the contents changed to fit the circumstances existing at the particular time involved.

The distribution of periodic reports should be held to a minimum because the cost and progress information contained therein is basically the confidential stock in trade of the company paying for its assembly and preparation. Care should be exercised not to disclose to outsiders the possible cost advantages or disadvantages growing out of new or highly desirable methods employed on the work, the controls in force, etc. It is also important not to disclose any information that might be damaging to the position of the contracting company. Periodic reports should be filed and made available only to those duly authorized to see them. Much of what is mentioned with respect to the periodic reports should apply emphatically to the final report.

The contractor's project engineer is usually responsible for the assembly of cost and progress information entering the reports, and for the composition of the reports themselves. The compilation of the record supporting such cost and progress information is necessarily a joint effort of the job enginering and accounting staffs to accumulate a true and accurate picture of the status of the job.

There are several possible kinds of periodic reports, daily reports, weekly reports, and monthly reports. Daily reports should be made available to the project manager the day following the one being reported. Weekly reports should be placed in the hands of the project manager by the middle of the week following that being reported. If desired by off-job management, it should be posted at the time of delivery to the project manager.

Monthly reports usually await the closing of the books after the conclusion of the month's business to reflect properly the cost and progress record. But they should be delivered to the project manager and posted to the off-job management not later than the 13th of the month following that being reported. The accounting department usually prepares separately a cost and revenue statement for each month. Although summarizing the financial position of the entire job, and not presenting the cost accounting results, this must be coordinated with and delivered to the management at the same time or before the monthly cost and progress reports.

13-30. Daily Reports

The usual purpose of daily reports is to provide the project manager with current information as to the general status of the job, but particularly with respect to principal job operations, and the progress being made or maintained. Daily reports should be initiated only upon the direction of the project manager and designed to suit the particular operations in which he is interested. They ordinarily incorporate the following types of information:

1. The total number of men on the payroll
2. The number of truck loads handled for the principal excavation items and the resulting computed approximate quantities of work accomplished
3. The batch count quantities of concrete of the different kinds placed
4. The total length in linear feet of tunnel or shaft driven, preferably for each shift
5. A report of the total hours worked by the major items of plant and equipment, and any delays in the operations and the cause, including mechanical failures
6. A summary of unit and total costs for the principal items of work as taken from the daily labor distribution record, providing the possibility of studying the cost trend from day to day
7. Other items of importance as may be directed.

Because daily reports are usually designed around a specific need to follow more closely selected operations during the construction period and are subject to change from time to time as that need changes, no suggested sample daily report form is furnished. It is recommended that it be uniform from day to day and easy to read. Although a loose-leaf diary is not desirable from the standpoint of possibly losing pages, a loose-leaf page can be composed to serve both the purpose of a daily diary and a daily report.

13-31. Weekly Reports

Weekly reports are normally an enlargement of the daily reports and summarize the week's activities for the benefit of the project manager and off-job management and others closely following the progress of the job. They are of particular advantage when a project is experiencing operational or managerial difficulties and remedial measures are being tried. Examples of weekly reports are shown at the end of this chapter as Figures 13-31-1 and 13-31-2. Basically these convey physical progress accomplished for the week, approximate quantities and revenue for bid schedule work items realized, and a commentary on unusual things that happened during the period.

13-32. Monthly Reports

The contents of monthly reports should be selected to suit the needs of the job as determined by the management. Unless company policy is already established, the format and basic information to be reported should be agreed upon at the start of the work and carried through the life of the contract as a standard, modifying only the text to reflect developments from month to month. Basically the monthly report is an interim cost and progress report, which can be of considerable assistance in conveying to the off-job management and officers of the company a reasonably complete picture of how the job is progressing both physically and financially, as well as helping the job management and engineering personnel to analyze previous accomplishments and to project forward in the process of making future plans.

Monthly reports for highly specialized work will necessitate special means of reporting progress and costs. This would be true of tunnel work, underground chamber excavations, and large dragline excavations, such as for canals and mechanical installations. Nonetheless, reporting for such work would be expected to follow in some measure the general trend of the original bid estimate and to emphasize the elements of progress and costs used in the budget-control estimate. To facilitate job control and understanding, lump-sum contracts should be reported according to the breakdown created by the subdivision of the specifications, as well as that followed in the budget-control estimate.

The following is intended to be typical of large dam construction based upon a firm unit-price contract and can be considered only as a suggested guide based upon past practice. It can be adapted as the occasion requires. Compared to the average construction contract, it might be representative of a more complex situation than involved for a given job to be reported, in which case a simplification of the report would be justified. Conversely, additional detail might be advisable to convey the desired information as requested by management.

A description of the various component parts of the suggested form of monthly report follows, and the forms indicated by figure number as mentioned for each part are *included at the end of this chapter* to illustrate the ideas intended.

(a) General Progress Report (Narrative), Figure 13-32-1. This would provide a general background of conditions affecting the progress of the work, to describe happenings of particular interest, to summarize the amount of work accomplished for the principal work items in progess, and to explain parts of the remaining sections of the monthly report which are not readily apparent as the result of the graphs, figures, etc., presented.

(b) General Job Status Report, Figure 13-32-2 The purpose of this section is to provide in its simplest form the magnitude of the contract work involved, how it has been affected by extra work, contract change orders and contract advances provided by the owner, the volume of work already accomplished and still to be completed, the latter being expressed in terms of contract value, and percentages related to the total of the work involved. The latter also is stated in terms of days and percentages of the total, the total allowable contract time, and contract time elapsed, so that the time remaining can be readily computed.

A comparison of the percentage of the contract value completed to date and the percentage of time lapsed to date tells immediately the general picture of whether the operations should be expedited. The contract values stated also are often useful for considering not only the situation of the particular job involved, but also the overall company picture. The basic information for this particular part of the report is normally available from the files of the contractor's project engineer covering payments from the owner or estimates of work completed and time used to date, but the accounting personnel should at least verify that the values employed check with those entered into the job books.

(c) Graphic Progress Curve, Figure 13-32-3 Suggested in this instance is a simple two-line graph, the time being shown as the ordinates, and the percentages of completion as the abscissas. Time would be divided into monthly periods for the total length of the contract, and the percentage of completion related to the contract value completed to date vs. the total contract value or, in the case of lump-sum contracts or lump-sum items the values would have to be related to estimated values completed to date.

Shown thereon would be a line representing the construction progress estimated for the entire contract time at the inception of the work, and plotted against it month by month would be the actual progress accomplished to the end of each succeeding month. The resulting picture would show the past performance related to what was expected according to the contractor's original evaluation of the contract work, whether or not progress at the same rate would result in completion within the allowable contract time, or whether expediting of the work by working more hours, adding equipment and crews, etc., might be justified to meet the prescribed completion date.

For some contracts there are interim completion dates for certain sections of the work. Sometimes it is desirable to add curves illustrative of the situation pertaining to these, or possibly illustrative of certain dominant work items or physically separated sections of the work that are of particular interest to the recipients of the monthly reports. After the original drafting of the form of the graphs, little work is involved in adding the monthly plotting and having it reproduced for inclusion in the monthly report.

(d) Stage Development Chart, Figure 13-32-4 What is intended in this connection is some kind of a pictorial way of illustrating progress and completed work by stages with respect to the whole. It should convey at a glance the physical status of the work; the portions already done, those remaining, their geometric arrangement, indicated size, elevation, etc., when such factors are important. Such a diagram adapted to a tunnel, a road, or a canal could be depicted as a linear outline, with the stationing or mile posts shown for the full length of the project.

A more complex structure could be shown in outline— one such outline for each of the various phases of the work to be accomplished—structural steel framing, concrete encasement, masonry, plumbing, electrical, interior finish, etc. Areas representing the completed work could be cross-hatched or shaded, except that a differentiation in marking for that portion representing the work for the month being reported could be made, possibly shown in color, lighter shading, or the like.

To illustrate the point, a suggested sample is included at the end of this chapter which shows a profile—an elevation of a concrete dam with the foundation outline plotted as found after excavation, monolith joint lines occurring transversely, and the individual vertical lifts of concrete with top of pour elevations given, the location of penstocks and openings through the dam shown in phantom, etc. Previous concrete placement is shaded except that monoliths poured during the month being reported are cross-hatched.

The profile illustrates the following: essentially all of the foundation is covered, eliminating possible delays due to excavation and foundation preparation; the low blocks near the base of the dam are reaching a safer elevation above the normal water surface with the river diverted through the diversion tunnel; unprecedentedly high flows in the river if large enough to do so could overtop the low blocks with a minimum possible delay for cleanup and repair of damage; and relatively few lifts of concrete remain where the work might be held up awaiting penstock pipe installation or resulting in slow concrete placement around the obstructions created thereby.

(e) Monthly Cost Statement Summary, Figure 13-32-5 This would serve as an interim evaluation of the then accruing costs of the various parts of the work. It is particularly adaptable to firm unit-price contracts, in which case the cost of all of the various work items appearing on the bid schedule would be reported separately. If greater detail is desired, the costs of the various components of a bid schedule item can be presented also. In the case of lump-sum contracts, the cost of the various sections of the work as divided in the contract specifications can be shown separately, but more important is the arrangement permitting direct comparison with the original prebid estimate and the budget-control estimate. This would certainly apply to firm unit-price contracts also.

In the early stages of the contract work, or until representative operations are developed in the field for any given work item or subdivision of the work, the reported monthly costs are likely to be inaccurate and misleading inasmuch as they might reflect only the period for the so-called "learning curve." Then too, there can be a lag in the accounting when all the charges may not have been posted to the cost books, and at the early stages of the operations the percentage distortion of the unit costs can be very great. There comes a time, however, when the operations in the field have developed a regular routine pattern contemplated in the job planning and when the accounts have been brought up to date. The current periodic costs should then be comparable to the budget. This presupposes that the initial purchase cost or a fair charge for the facilities or equipment, their cost of move-in and erection on the site, and other costs, have been entered in the cost books on an equitable basis of amortization similar to that used in the budget-control estimate.

The monthly cost statement is necessarily produced as a joint effort of both engineering and accounting personnel, the former supplying the breakdown of work items and the subdivisions desired and required by management, the quantities of work accomplished as reflected in the engineering records, and the coordination with the budget-control estimate. The latter would supply the monetary values as compiled in the cost accounts and broken down to fit the cost coding key. When completed, the monthly cost statement is useful to management and engineering personnel in making future projections of cost.

Referring to the column headings in Figure 13-32-5, the subdivisions are in conformity with the breakdown envisioned in the suggested estimating system intended for general application to construction work, including its application to the budget-control estimate. Quantities of work performed are stated by work items, in total for the contract (marked "Original"), and for the quantities completed to the end of the month or period being reported (marked "To Date"). The "Estimated Cost to Date" column should carry the "Total Unit Cost" developed in the control estimate, and the extension thereof against the "To Date" quantities should reflect the expected total, including the so-called direct cost and indirect cost (general expense and overhead), of particular items of work, or of the total work performed up to that time when summarized.

The column marked "Actual Cost" broken down further to "Unit [cost] This Month" and "Unit [cost] to Date," and "Total [cost] to Date," as listed for each work item or division thereof are posted with the money values summarized in the cost books, the totals of which must at all times include the direct and the indirect cost and must balance with the general accounts of the enterprise kept by the accounting department. These figures become the basis for evaluating the accuracy of the estimating and the economic position of the job.

The column marked "Revenue to Date" is simply a statement of earnings under the contract and reflects the amount of payments made by the owner to the contractor adjusted for any shortages or overages which are properly justified to overcome approximations included in the interim payment estimates. "Gains or (Loss)" shown in the next column is the direct result of a comparison between the "Actual Cost," and the "Revenue to Date," and the total difference so reflected should check with the profit or loss at any particular time appearing on the corresponding statement of general accounts for the project. The profit or loss for individual bid schedule items or subdivisions of work covered by the Monthly Cost Statement provide the "red flags" that something is wrong when cost exceeds the revenue, and becomes the basis for initiating an investigation to determine "why."

The other columns on the monthly cost statement provide the basic sources of the charges entering the cost of the work and are stated simply as follows:

"Labor and Insurance"
"Permanent Materials"
"Specific Plant"
"Equipment Rental"
"Supplies"
"Subcontracts"
"Total Direct Cost"
"General Expense and Overhead"

The first six categories of cost entered here constitute the so-called "Direct Cost," and when applicable portions of "General Expense" and "General Plant" are added, the sum constitutes the "Total Cost" and is the same as the "Actual Cost" listed in the third main column from the left.

This same form is recommended for use in the preparation of the "Final Report" except that no "This Month" figures will appear and the "To Date" values should be marked "Final."

Considerable time can be saved in the preparation of the monthly cost statement by having the "Description," the "Original" quantities, the unit revenue, and all of the "Unit Estimated" figures typed on reproducible forms which can be used for the addition of the current information. Work items on which no work has occurred to date should be left blank except as noted, and completed items which do not change can be filled in permanently.

(f) Monthly Cost Statement, General Expense, Figure 13-32-6 Ex. 9 Although the monthly cost statement is designed to include a distribution of the indirect costs against the various work items or subdivisions of work, it is important for management to keep informed of the trends of the cost with respect to the various elements that make up such indirect costs. Two general categories are involved—namely, "General Expense" and "General Plant," which together constitute the total overhead costs for the job. Total overhead is normally distributed to the various work items or prorated against them based upon the relationship that the "Direct cost" for the item bears to the "total direct cost" for the whole job or, in other words, on a straight-line proration basis.

Referring specifically to the "general expense" portion of the "indirect costs," this basically embraces all of the overhead costs, not involved with the "built-in" facilities, which are of a general nature and serve all of the field operations, or cannot be conveniently written against the direct cost for given work items. What might be included under this category is to some extent a matter of choice, but good practice indicates that whenever it is possible and correctly applicable, charges should be directed to the specific field operation involved and the indirect cost kept to a minimum.

Figure 13-32-6, a Monthly Cost Statement, "General Expense" at the end of this chapter, is designed for use in monthly reports. It provides a practical breakdown into categories that again are coordinated with the estimating system described elsewhere. It sets down the total estimated costs for each subdivision of the breakdown and provides columns for the "Actual Cost" and the resulting "overRun" or "underRun" of cost for each such subdivision both on a "Current Month" and a "Job to Date" basis. A column is provided also for the "Balance Unamortized" for recording amounts held in reserve for later distribution against the work. Reserved costs should be held to a minimum, but some consideration of them is justified for initial organizational costs, advance bond and insurance premiums, and payment of taxes, supplies on hand, etc.

The "Monthly Cost Statement, of General Expense" is compiled from the records of the accounting department but is included as an inseparable part of the monthly report assembled by the engineering personnel.

(g) Monthly Cost Statement, General Plant, Figure 13-32-7

As mentioned in the preceding subsection, the handling and reporting of general plant is very much the same as that for "General Expense" and constitutes an inseparable part of the total job overhead. Basically, according to this concept, it embraces all of the temporary job buildings, general utilities, and general roads and yards. It also includes the cost of constructing camp facilities when they are required. A complete listing of the items involved in "General Plant" as suggested here is given in Figure 13-32-7 at the end of this chapter. The proposed form for reporting "General Plant" costs monthly is arranged very much as the form for reporting "General Expense," i.e., columns are provided to show the total "Estimated Job Costs," "Actual Costs" vs. the applicable "overrun" or "underrun" for each subdivision breakdown shown on the form.

The statement of "General Plant" is also compiled from the records of the accounting department and is necessarily an inseparable part of the monthly report assembled by the engineering department.

(h) Specific Plant Operations Report, Figure 13-32-8

This report for monthly preparation is suggested for large and major built-in construction plants as differentiated from mobile equipment, which should be otherwise accounted for. The term "Specific Plant" should be construed to mean plant facilities that serve specific field operations and whose initial cost and the cost of their operations can be charged as "Direct Cost" or as field cost, and not as part of the distributable "General Expense" or "General Plant" serving all operations alike.

Into this category would fall rock scalping plants, aggregate crushing and processing facilities, conveyor transport systems, batching and mixing plants, cement and pozzolan storage and handling plants, cableways, gantry crane trestles, air compressor plants, concrete cooling facilities, and the like. Such of these items that incur continued operating costs, that are given special treatment in the budget-control estimate, and whose interim operating cost is substantial, and a matter for interest and scrutiny for the purpose of job control should be reported on some such periodic statement. This form would not be adaptable to small operations. The quantities used in its preparation are a matter of engineering records, and the monetary values come from the cost accounts.

(i) Hourly Equipment Operating Costs, Figure 13-32-9

This report for mobile equipment shown at the end of the chapter is suggested as a necessary part of the monthly report inasmuch as the information being reported is a basic element in all estimating. It is usually a substantial part of the overall cost and is most useful in evaluating the mechanical integrity of the equipment provided, the effectiveness of preventive maintenance and repair programs in force, and availability factor under the attending circumstances. It pinpoints excessive idle time, which is usually a managerial matter. It is a subject of continual ref-

erence by the estimating engineers investigating new work.

For the sake of studying cost trends, the proposed form is set up for showing the hourly time and costs for the period being reported as well as the "To Date" data. Like articles of equipment should be reported in summary regardless of the number of pieces unless the relative severity of service is such as to justify obtaining separate values for each situation. The basic information would be shown on daily equipment reports and assembled in the cost records by the accounting department. The engineering department should be constantly aware of the development of costs in this category, assist the management in its evaluation, and see that it is made a part of the monthly reports.

(j) Special Plant Summary Report, Figure 13-32-10

Where necessary and desired, this report at the end of the chapter is intended to summarize the in place cost of specific built-in construction plant items. Its usefulness to the project management for job control is limited to the period starting in the early stages of the plant work and ending with the completion of its erection in the field. Its effectiveness on the cost of the work items is permanent inasmuch as any overruns in the specific plant cost, as a matter of its amortization, will increase the cost of such work items for which it is used over and above that shown in the budget-control estimate. Underruns in cost of specific plant would, of course, have the opposite effect.

Final plant design details are necessarily fully developed only after the time of bidding the job, approximate design concepts having served for the prebid estimating period. The budget-control estimate should reflect refinement in plant designs but, invariably, things develop in the ultimate acquisition and erection phases which must be followed carefully.

When and if questions arise in that phase involving changes in design of the plant, the management should have its disposal all pertinent cost information. The special plant summary report is recommended for that purpose and to substantiate the ultimate specific plant costs amortized against the work items. When all plant charges are reported thereon, this report becomes a useful reference for the estimating engineers figuring new work.

It is recommended that it be included in the monthly reports starting at a time when the costs reported thereon become meaningful with respect to distinguishable parts of the plant cost, and that it be discontinued as part of the monthly reports when the plant facilities are completed.

Typical specific plant items normally provided for a large concrete dam or facilities of similar character are suitable for reporting in this fashion and would include, but are not limited to, the following:

Aggregate processing plant
Overhead cableway
Concrete delivery railroad track
Concrete delivery railroad rolling stock
Miscellaneous concrete placing plant
Concrete mixing plant
Cement handling plant
Pozzolan handling plant
Concrete cooling plants
Aggregate stockpiling facilities
Compressed air plant
Job water plant
Walkways and scaffolding

(k) Report on Concrete Form Costs, Figure 13-32-11

Usually on the larger heavy construction jobs there are certain important subsidiary parts of the work that need further amplification than is provided on the other cost reports. One such matter is forms for concrete work Concrete forms vary in complexity and cost, and a proper understanding of the subject takes a special study of the various categories involved according to the same schedule of pricing used in the budget-control estimate. This is a subdivision of the work involving substantial amounts of money and it deserves particular attention at all stages, but principally early in the job when the forming concepts are being developed if those contemplated in the budget-control estimate are not used. Figure 13-32-11 at the end of the chapter is a particularly valuable tool for job management, and its inclusion in the regular monthly report is desirable where the conditions apply. The basic quantities involved for the various categories are a matter of engineering records. The money values to fit such categories would come from cost-accounting records coded in advance to fit the same.

(l) Other subjects as directed by the management

13-33. Final Project Reports

(a) Purpose and General Form   The final project report should represent a reasonably complete record of the job experience and should be compiled in a good standard form for ease of future reference. The amount of detail would depend upon the size and kind of job, how diversified it is, and a number of other things. Each case should be determined independently.

For diversified construction operations a library of suitable final project reports can go a long way toward taking the guesswork out of estimating, thereby reducing the financial hazards. Experience in one's field of endeavor is an invaluable asset, but individual recollection of particulars is not complete, and failure to record an experience in all probability would result in losing of some of its value.

Assuming a medium-to-large, diversified heavy construction project, some basic rules to follow for the preparation of the final project report are offered below. The contractor's project engineer is usually delagated the responsibility of compiling the final project report. The report in general should be descriptive enough to convey to an individual, not familiar with the physical makeup of the work or methods employed, an understanding of the costs and operations sufficiently clear to permit its being related to similar work elsewhere. Sufficient information should be given to permit the possibility of factoring the then prevailing economic structure to other known economic conditions.

The necessary basic information for the report should be available from the project engineering or other job records. To minimize the difficulties of writing the final report, its intended contents should be outlined early in the life of the job and assembly of the necessary information should proceed throughout the construction period. The completing of plant facilities needed for the performance of the work or completion of physically separate parts of the work or items of the payment schedule will permit writing up those features well in advance of when the data are required for assembly, with only minor changes or additions needed to complete it.

All reported information should be stated in a clear, concise manner. Care should be taken to leave sufficient margins to avoid obscuring parts of the text in the binding. Reports should be compiled on sheets 8½ x 11 inches, legibly typed and bound in suitable thicknesses for convenient filing. Double sheets or longer should still be bound in on the 11-inch side, and folded accordion style to the overall size of 8½ x 11 inches. Sheets should be composed with the written material progressing across the 8-inch width of page rather than across the 11-inch dimension, wherever such is possible. Hand lettering should be neat.

The following offers suggested sections of the report as typical examples only, which should produce a reasonably complete reference, with the understanding that the author make those changes believed necessary.

[*Table of Contents*]   The Table of Contents should list the various sections and subjects treated in the report, identified by page numbers. The descriptions should be brief and to the point, but should be sufficiently complete to convey an understanding of the general character of information contained in the suggested subsections outlined below.

[*General Description of Contract Work, Etc.*]   The author of the final project report should provide sufficient general descriptive information to acquaint the reader with such facts as location, the general geographical conditions of the site, surrounding populations, source of labor, remoteness, or convenience of such public facilities as transportation, housing, and schools; whether the job operated a camp; etc. There should be a general description of the principal contract work, the concept and methods for its general accomplishment. The narrative descriptions should be supplemented with reduced scale drawings showing the specific vicinity mapping, and several general drawings illustrating the structure or work to be done. The principal construction features should be described, along with an outline of construction methods and concept for their handling. Particular construction problems should be mentioned in the narrative.

The narrative should briefly outline the construction plant for the project, and reduced scale drawings might be appropriate. Add some comment about its efficiency, what might be done to improve it for future use, etc.

Something should be included to describe the owner's organization, such as names and titles of the various people, the overall organizational setup, how well it functioned, etc.

[*Weather Summary*]   There should be a section devoted to the weather conditions that prevailed during the project, quoting governmental weather reports and attaching them. Unusual weather should be described in detail to develop how it compares relative to the weather that might have been expected based upon the historical record. Meteorological data should cover total precipitation, snow fall, wind, and temperatures quoted as daily and monthly minimums and means, and storm severity.

[*Hydrology*]   In case of a dam, bridge, or other structure subjected to flowing streams or tidal currents, the general circumstances should be described and a tabular record of the official historical extremes prevailing during the job should be included.

[*Statement of Original Bids*]   Assuming that a firm-price competitive bid job is involved, a complete abstract of bids received by the owner should be included, together with comment on any negotiations that might have taken place, details of award, and related matters of particular interest. If the job was other than a firm-price competitive bid job, describe the details of negotiations.

[*Contractor's Organization*] Some treatment should be given to the contractor's organization, giving names, titles, functions, and the period that the individuals were on the job. An organization chart might be helpful.

[*Economic Data*] A description of basic economic conditions is often desirable, including the range of delivered prices for major construction supplies such as explosives, fuel oils and lubricants, power, lumber, wire nails, and some treatment of the escalation of prices during the contract period.

[*Labor Condition and Rates*] To make the reported costs meaningful, a summary on labor conditions and costs should be provided. It should include union agreements and their duration, wage rates, working conditions and fringes, taxes and employee benefits, relative productivity of the laboring people provided, availability of laboring people, their skills, competition in the labor market, labor disputes, strikes, and slow downs.

[*Safety Program and Accident Prevention*] A description of programs in force is advisable, together with a statement of the accident record accomplished. It should be compared with national and local average records for the particular class of work. Safety programs are normally a contract requirement for the larger projects, and the basic necessary information can come from the records of their directors.

[*Construction Progress Reports*] A section of the final report should describe the job progress for the life of the contract work. The form of such reports will in a large measure depend upon the character of the work involved. Some narrative description is usually justified, and one or more types of progress charts might be found desirable to illustrate better the overall program and possibly some of the major features and construction operations. The following forms might be considered:

1. A graphic progress curve covering the overall project progress under the contract based upon time and earnings, with the contract time broken down into months represented by the ordinates, and the percentage of completion being represented by the abscissas. On such a graph would be the originally estimated progress shown in the budget-control estimate, with the actual progress superimposed. This chart would be in the same form as suggested for the graphic progress chart for the monthly reports, except that the information would be carried to the conclusion of the work. A representative sample partially completed can be seen in Figure 13-32-3 at the end of this chapter.

2. Stage development charts, similar to the one described for the monthly report and illustrated in Figure 13-32-4 at the end of this chapter, are what is intended in this instance, except that these would show the accomplishments for the full contract time broken down into job periods defined by working seasons, river diversion stages, or other desirable sections. This is particularly desirable for mass concrete structures such as dams, but the principle can be adapted to other forms of construction.

3. A bar graph—construction program similar to the one shown in Figure 13-7-1 at the end of this chapter is often desirable except that both the estimated and actual progress can be shown, the former as an open bar and the latter as a shaded bar, both being drawn within the same horizontal lines on the chart dividing the individual work items. This provides a picture of relative progress for the various items.

4. Charts illustrating the major and controlling construction operations, such as mass excavations, mass concrete and tunnel advance, are sometimes desirable. These can take one of a number of forms. A simple solution is the use of a background divided into small squares with the divisions of the abscissa measuring time individually for one month, and the division of the ordinate measuring a given number of work units accomplished, such as cubic yards and linear feet. Progress for each month throughout the contract could be plotted by vertical lines—side by side—within the monthly period indicated.

[*General Cost Statement-Summary*] The form of this report would be the same as that illustrated by Figure 13-32-5, "Monthly Cost Statement Summary," except that the reported information would reflect the final bid-item cost experience to the completion of the job. The same would be true of any components of the bid items shown on this form, or, in the case of lump-sum contracts, the subdivisions or breakdowns of the total as elected by the contractor.

It is contemplated that at the time this report is prepared, all costs will have been recorded, including all amounts previously reserved for plant move-in and erection, dismantling, or for any other reason, and that the rentals, or first cost of plant and equipment less salvage values as agreed by the management, are also written against the job. As in the case of the "Monthly Cost Statement Summary," the final "General Cost Statement Summary," should reflect a comparison between the budget-control estimate unit costs vs. the actual costs, the revenue received vs. the actual costs, and the amount of profit or loss, for each item appearing on the summary.

As described for the monthly cost summary, the "General Cost Statement-Summary" would be a joint effort of both engineering and accounting personnel but included in the final project report prepared by the contractor's project engineers.

[*Cost Statement, General Expense*] The form of this report would be the same as that illustrated in Figure 13-32-6, "Monthly Cost Statement, General Expense," except that the reported information would reflect total cost to completion, and there would be no "Current Month" costs reported or "Balance Unamortized." Other comment concerning source of information, its assembly, etc., for the monthly reporting would apply.

[*Cost Statement, General Plant*] As in the case of the "Cost Statement, General Expense," this is to be the finalized version of the information reported, including the full intended write-off of the purchase cost of plant items, and the full cost of its move-in and erection and dismantling. The same principles described for the "Cost Statement, General Expense" will apply. Basic form of the report is shown in Figure 13-32-7, except that no interim costs or unamortized balance are to be shown.

[*Specific Plant Operations Report*] This would follow

the same pattern as the monthly specific plant operations report illustrated in Figure 13-32-8, except that all costs would be final to completion of the job and there would be no information shown for the period.

[*Hourly Equipment Operating Costs*] Except that all costs reported would be "To Date" and final for the contract, and the "This Month" cost would be omitted, this report would be similar in all respects to the one of the same title described for the monthly reporting. It is illustrated in Figure 13-32-9 at the end of this chapter.

[*Specific Plant Summary Report*] Referring to the report of the same name described for the monthly reporting function, it is seen that as the plant facilities are completed, the same form as shown in Figure 13-32-10 should reflect the final specific plant cost. For the final version of the specific plant cost report, a review of the subject should be conducted by the engineering and accounting people involved in its preparation to verify that the actual costs shown correspond with those in the job cost books. The final report should provide the actual costs broken down to correspond with the cost of the various specific plant items. These are further broken down to the first cost, etc., shown in the budget-control estimate, with a comparison of "control" and "actual" costs in the same detail.

[*Concrete Form Cost Report*] As mentioned with respect to the monthly report under this title, the matter of cost of forms for concrete is one of the largest subsidiary items for which detailed cost information is desired. The final report should treat the subject adequately if the project being reported involves that kind of work. Form cost, being a high labor content operation, needs close scrutinizing and careful analysis to avoid runaway cost trends. Man-hour production (square feet of forms per man-hour) is the most meaningful cost measure available, assuming the same class of forms and the application of the same methods. The labor costs per square foot of forms is convertible into man-hours by the application of an average total wage rate inclusive of fringe costs, tax, insurance, etc.

Concrete form costs are necessarily priced in the estimating process (in the budget-control estimate) based upon the quantity of each class of concrete forms, a preconceived plan for their fabrication, method of handling, and the erection costs, principally labor. The concrete form cost report should provide information directly comparable.

Therefore, the "Concrete Form Cost Report" would need to be constituted to fit the particular project being reported, although some will bear a similarity one to another, depending upon the class of work involved. Figure 13-32-11 at the end of this chapter is intended only to convey a general idea in that connection. The final report would not show experience in the "This Month" columns, but all figures would be for the complete job, "To Date."

Although concrete form costs are suggested for special treatment under this heading, it is pointed out that other similar subsidiary items of cost of particular importance may be justified. The management or the contractor's engineering personnel may see fit to arrange for its treatment.

[*Photographic Record*] No final report for a large construction job would be complete without a photographic record, the contents and extent of which would be subject to individual judgment of the kind and complexity of the work involved. Pictures very often answer questions that no one thinks to ask, and they often have a decided bearing on subjects in point on later occasions.

The kind of pictures suggested for inclusion in the "Photographic Record" in the final report would be a minimum assortment of job progress snapshots taken successively from the same vantage points at several locations—these to illustrate stages of actual construction, pictures of the important items of plant and equipment provided for the work, closeup views of major construction operations or those of an unusual or unique nature, important construction features called for on the plans and not otherwise included in the progress photographs taken. They also should include pictures of foundation details, particularly if there are difficulties or any questions related thereto, photographs of particular details involved in points of disagreement with the owner, or other things of obvious importance.

The master copy of the final report retained for official use need be the only one that would contain the more complete photographic record. Other copies could be accompanied by an abridged version, and include only subjects of particular interest. To be properly effective, job photographs must be numbered for positive identification and marked with the date they were taken. Attached or accompanying each must be written a description giving the location from which it was taken and what it purports to show or illustrate.

[*Claims Pending at Contract Settlement*] It is assumed that the final settlement of the contract with the owner will have occurred by the time of completing the "Final Project Report." Presumably the costs shown in the final report will reflect only the charges in the cost books, and give no monetary value to pending claims or possible future recoveries. To be complete, therefore, the record should provide an outline of the matters still contemplated for further arbitration, civil suit, or appeal to contract review boards authorized to hear such matters. Substantial delays are usual for the handling of contract claims by any of the means mentioned above, and it is often ill-advised to hold the record open for their conclusion. It is assumed that a separate procedure will be instituted for the pursuit of claims.

The outline of the claims pending at the time of contract settlement referred to above need only give a brief description of the questions in point, the amounts of money involved, and the particular work items that would be affected by its recovery, decisions received from the administrative officers of the client on the subject actions contemplated by the contractor, and the status of the matters at the time of their reporting in the "Final Project Report."

13-34. Contract Adjustments, Change Orders, and Claims

As previously discussed, an essential function of the contractor's engineering section is to be alert to requirements beyond the conditions of this contract. It is responsible for securing contract adjustments for any extra costs that may be justified. There are many occasions for requesting contract changes. They include changed conditions, unspecified requirements by the owner's engineer, changes in design, changes in the technical specifications, or anything else that alters the contractor's contractual position. The best means of dealing with the situation is to negotiate change orders with the owner's representative.

If this is unsuccessful, a record should be prepared to support possible future claims. The protests and appeals required by the contract should be filed in a timely manner. At the time of final settlement, the contractor should reserve the right to pursue the claims by such means as arbitration, appeals board hearings, or the courts.

It is not always easy to recognize situations that justify requests for contract adjustments. They must have a legal and moral basis. They must be supported by a reasonable interpretation of the contract conditions to be valid. Unfortunately differences of opinion sometimes cloud the issues.

In deciding whether to seek contract adjustments, the rule of reason should be the first test. The contractor's contractual position should be protected from the outset, and he should carefully comply with the provisions governing timely protests and appeals.

The project manager or project enginer probably would initiate the questions. Any matters of consequence should be discussed immediately with the contractor's district or division managers and enginers. Competent legal advice should be obtained in the early stages, and a decision made to pursue or abandon the issue.

If a decision is made to pursue the issue, the project engineer should build a suitable record to support the case. He should carefully assist the project manager in making the protests and appeals specified in the contract.

Typical of how the contractor's position. can be foreclosed might be seen from Article 6, "Disputes," as covered by U. S. Government form of Contract 23 in use for so many years. It provides:

"(a) Except as otherwise provided in the contract, any dispute concerning a question of fact arising under this contract which is not disposed of by agreement shall be decided by the Contracting Officer who shall reduce his decision to writing and mail or otherwise furnish a copy thereof to the contractor. The decision of the contracting officer shall be final and conclusive unless within 30 days from the date of receipt of such copy, the contractor mails or otherwise furnishes to the Contracting Officer a written appeal addressed to the head of the agency involved. The decision of the head of the agency or his duly authorized representative for the determination of such appeals shall be final and conclusive. This provision shall not be pleaded in any suit involving a question of fact arising under this contract as limiting review of any such decision to cases where fraud by such official or his representative or board is alleged; Provided however, that any such decision shall be final and conclusive unless the same is fraudulent, or capricious or arbitrary or so grossly erroneous as necessarily to imply bad faith or is not supported by substantial evidence. In connection with any appeal proceeding under this clause, the Contractor shall be afforded an opportunity to be heard and to offer evidence in support of his appeal. Pending final decision of a dispute hereunder, the contractor shall proceed diligently with the performance of the contract and in accordance with the Contracting Officer's decision.

(b) The Disputes clause does not preclude consideration of questions of law in connection with decisions provided for in paragraph (a) above. Nothing in this contract, however, shall be construed as making final the decision of any administrative official, representative, or board on a question of law."

Thus it is seen that protests and appeals should be handled in a timely manner. The timing and sequence of appeals, and avenues open to review of decisions by contract and appeals boards are different for individual contracts. The project engineer should be aware of the specific requirements and inform the project manager about their application and the impending expiration of protests and appeals dates. If specific procedures in this regard are not spelled out in the contract, the handling of questions should follow good and proven practice that legal counsel might suggest.

Ordinarily, these matters go through the following stages:

(a) Questions pertaining to contractual matters or payments under the contract are discussed with the authorized representative· of the client in an effort to resolve them. If an unfavorable verbal decision is received the next step is taken as follows:

(b) A written protest substantiating the contractor's position with appended descriptive information and supporting data is filed with the proper officer of the client. If an unfavorable written reply is given by such officer, then the contractor:

(c) Makes a further appeal to the successive lines of authority or to the principal administrative officer of the client. Upon receipt of a further denial, and if the contractor feels that his case is of sufficient financial consequence and sufficiently meritorious to warrant, he:

(d) Proceeds to prepare a claim, enlarging upon the development of its basic premise, the supporting information—computations, copies of pertinent correspondence, diagrams, drawings, photographs, or whatever else is necessary, together with a statement of the contractor's legal position. The claim, when prepared, can then be held for later presentation for review in arbitration, by the owner's contract appeal board (if any) or for filing with the courts.

It is not always readily apparent what the full effect of changes ordered by the owner-client or his engineer might be in terms of added cost or time of performance. Extensive study might be required to determine the effect.

Until the study is completed, it might be impractical or impossible to comply with the requirement for filing protests and appeals or to state the possible financial consequences.

In such cases, the owner-client should be advised in writing immediately. He should be asked for a time extension long enough to permit careful evaluation. These steps make a record that protects the contractor's position.

Time is a big element usually convertible into money values. Most contracts are written on the premise that time is of the essence, and failure of the contractor to complete the work within a given time sets into operation a provision for penalties against the contractor at a given rate per day for each day the construction time exceeds the given contract time. On the other side of the coin, delays to the contractor's operations in one or more subdivisions of the work can upset the balance, sequence or critical path of the various phases.

These can, in turn, delay the ultimate completion of the contract, can cause loss of economies inherent in the contractor's original plan of coordination of operations, can prolong the work resulting in all or any part of it having to be done in unseasonable weather at greater cost, or into periods where prices have risen due to inflationary trends, and many other things. Almost certainly longer construction periods, because they extend the time that the normal fixed cost is operating, will increase the contractor's overhead expense. Very important, therefore, is the necessity of watching the contract time and, when submitting requests to the owner for contract adjustment, to mention time as well as money values involved, asking extension in the contract time when justified.

The wording of specific construction contracts and the responsibilities assumed by the contractor will largely determine what can be considered as a basis for requesting adjustments under the contract. Contract time can be adjusted under some contracts for inclement weather, strikes, embargos, catastrophes, acts of God, acts of the enemy, civil disorders, etc.

Other contracts foreclose any possibility of adjusting the contract time for any cause, although its legality for specific things should be reviewed by legal counsel. Some contracts provide that the contractor assume all the hazards of the subsurface ground conditions. Others state that latent, unforeseen, and unusual subsurface conditions differing materially from those indicated in the contract or from those ordinarily encountered or inhering in work of the character in question, form the basis of a contract adjustment.

Some contracts permit time extensions, but not for monetary relief. Contracts may also include provisions for reimbursing the contractor for escalation of prices of labor and materials due to increases during the contract period. Therefore, consult the contract or obtain expert legal advice about what might rightly form a basis for contract adjustments if there is any question.

Management should decide how far to pursue claims. Sometimes a long period of time—possibly years—are involved, which can be very expensive. The attention of legal, management, and engineering personnel is diverted from other important matters to one whose outcome is uncertain. Arbitration, although sometimes speedier than than normal court processes, still is not fast. There is no assurance of recovery by either means. It is imperative, therefore, that all the facts and documentation in the case be accurately and completely prepared by the engineering staff, so that the company officers will thoroughly understand it before deciding what action to take.

13-35.   Assisting in Contract Settlements

The atmosphere usually surrounding the settlement of a contract reflects the association during the construction of the contractor and the client. It therefore benefits both parties to have enjoyed congenial relationships from the outset. This atmosphere can be greatly influenced by the conduct of the contractor's project engineer in the performance of his duties and dealings with the engineer for the client or owner.

In all probability the majority of the matters to be discussed or considered at the time of contract settlements will have to do with questions of payments, extra work, change orders, prospective claims, and the like which the contractor's engineer had under consideration at an earlier date with the client's engineer, and for which he had prepared the supporting data. They will be composed largely of the subjects of previous protests and appeals, denied or left in abeyance for final determination, and the occasion of the meetings at the time of final settlement provides an opportunity of review and, possibly, compromise with respect to those things that have a defensible basis.

Present at the time will probably be the principals of both the client-owner and the contractor who have the authority to make the necessary decisions for the parties. But presentation of the subjects will in all probability be in accordance with the information and supporting data compiled by the contractor's engineer and approved in advance by the management. Therefore, the contractor's engineer will necessarily be a participant in the settlement assisting the management.

Considerable time and effort should be spent in preparing for contract settlement meetings. All the subjects to be presented should be reviewed carefully in advance and an informal agenda prepared. Since it is mostly an appeal by the contractor, he will be expected to conduct the meeting. Since the occasion may be the last opportunity to present additional information for consideration by an administrative officer of the client-owner, it is good policy to offer additional data or arguments which might open up the possibility of negotiated settlement. Thus, in the preparation of the data to be submitted at the contract settlement meetings, the contractor's engineer should attempt to develop more attractive ways of presenting the subject matter, as well as provide new information for consideration.

As an outgrowth of contract settlement meetings, the client-owner usually requests a signed document from the contractor—basically a release—acknowledging receipt of all payment due under the contract. The contractor is free at that time to state any exceptions with respect to specific matters reserved for possible further actions in arbitration, for review by the owner's contract appeal boards or through the courts. The contractor's engineer, based upon his detailed knowledge of the subject matter, should either draft the wording of the statement of exceptions or check such statement to be sure that the items as listed can be supported by the record.

13-36.   Special Assignments

Occasions arise from time to time when the management might make assignments to the contractor's engineering personnel of special functions beyond their usual duties. These could be quite varied, but several examples are noted below:

(a) Act as consultants to design engineering firms, owners, public works agencies, or others who need advice concerning the construction phases of their work. The contractor might collect a fee for such services rendered.

(b) Make special investigations with respect to large prospective work programs or construction market possibilities requiring substantial expenditures for expenses and the necessity of being away from usual duties.

(c) Engage in research and development activities requiring a substantial amount of time and money not a part of the contractor's usual activities. These could be in connection with the design and construction of specialized construction machinery or devices.

(d) Participate in the activities of industry or engineering organizations requiring absences from one's usual duties, but which the management feels could produce company benefits in the way of good will or public awareness.

(e) Make public appearances for civic groups and the like.

(f) Write technical data for publication to promote enhancement of the company's image.

Activities of this kind must be limited and conducted only with the knowledge and consent of the management.

13-37. Annual Review and Reports to Management

Alert businesses annually review the effectiveness of their operations and accomplishments in order to adjust or correct policies or procedures.

In a smaller company the personal contacts are more frequent, the extent of activities is less, the individuals are generally familiar with one another's thinking, and are conversant in some detail with the accomplishments of the company. Under such circumstances, the need for formal review of activities and reports is not as great as with larger companies. Few of us, however, do better as time goes on unless we measure our past accomplishments and set new goals for the future. Even with a small company, the members of the engineering staff should engage in some such program, whether or not it is required.

Larger companies, in which less personal contact exists, find it necessary to keep an up-to-date evaluation of their people. In determining future policies with respect to business activities, they also find it necessary to establish a formal system of year-end study and reports for the advice and guidance of the management.

The reports are sometimes addressed to the board of directors of corporations for consideration at the time of their annual meetings, and summarize the accomplishments for the period of the previous year. As applying to the engineering activities outlined herein, subjects of particular importance would include a review of bids and proposals submitted, changes in economic conditions affecting costs of the work, the forecast of volume of business to be offered in the coming year, and review of contract settlements and claims pending.

An annual review of the company's bidding activities should indicate the extent of competition in the industry and how competitive the company has been. In addition, the review might disclose certain types of work that should be avoided. The analysis should cover the bidding activities of each division separately and then should be summarized for the company as a whole.

To be meaningful, the statistics should show the following:

1. Number of company bids or proposals submitted
2. Number of jobs awarded to company
3. Percentage of awards by number of jobs
4. Value of jobs bid based upon the low bids
5. Value of jobs bid based upon the company bids
6. Percentage by value company bids were higher than the low bids
7. Average number of bids per job
8. Company average bid position by number of bids
9. Number of times company bid was the highest bid
10. Value of work awarded to company during the year
11. A listing of jobs awarded to company giving brief description, title, date of bid, and bid amount of each
12. Comment on the construction market analyzed.

A report on the changes in economic conditions affecting costs of the work during the previous year can have a particular influence on future policy and bidding on new work. Although it is assumed that future estimating will be on the basis of quoted prices for labor, equipment, material, and supplies, the report should give a picture of inflationary trends and the amount of escalation which must be written into all proposals unless reimbursement is guaranteed by the contract. It would, of course, be impossible to summarize specific prices for comparison here, but information compiled by the U.S. Departments of Labor and Commerce, by *Engineering News-Record,* and other sources will, with a minimum amount of research, provide the basic information needed to make the presentation suggested. Only the principal elements of construction costs need to be treated on an average nationwide basis. For these, the current price and the percentage rise in prices during the twelve preceding months should be given. The elements suggested are:

(a) Construction labor, including costs of all fringe benefits and labor contract conditions, for common labor and skilled crafts separately
(b) Construction equipment
(c) Construction equipment repair costs
(d) Portland cement
(e) Lumber—S4S pine, per M. ft bm
(f) Lumber—S4S fir, per M ft bm
Plyform 5/8″, per sq ft
Plyform ¾″ per sq ft
(g) Structural steel, per ton
(h) Portland cement, per bbl
(i) Common brick, per M
(j) Ready-Mix Concrete, per cu yd

Forecasting future price rises is particularly speculative. But to the extent that such might be indicated by the available statistics, continuing labor contracts, publications, and forecasts of reliable agencies or authorities, a presentation of what might be expected along these lines should be presented as part of the report.

Published statistics compiled by the U.S. Departments of Labor and Commerce, by *Engineering News-Record,* and by other such sources provide a listing annually of the estimated amount of construction work by categories performed during the previous twelve months, the amount under such categories forecast for the following twelve-month period, and the resulting percentage change for each. This information is further summarized under the headings of heavy construction and building construction. An accumulation of such data, plus opinions by reliable sources on the subject, can provide management with a desirable basis for studying the prospective construction market. Because the contractor's engineering personnel follow the subject generally, it is suggested that they compile the report—forecast of volume of business to be offered in the coming year—following the general format of the statistical data mentioned above.

Some procedure is used by almost every business firm for periodically analyzing its commitments and the status of outstanding accounts. Since construction contracts tend to involve large amounts of money and extend over long periods of time, and because uncompleted contracts have a substantial bearing upon the company's future bonding capacity, their status is important to the financial condition of the firm and to future policy.

When there are unsettled questions relative to contract conditions, or claims growing out of disagreements between the parties, contract settlements can be delayed for long periods of time and large amounts of money can be held in abeyance at substantial costs for interest. The cost of pursuing claims also can be substantial.

Smaller contracting firms with few contract commitments

have little difficulty keeping in mind the particulars of such matters. But for larger, more diversified companies, a summary or review of contract settlements and pending claims is desirable for study annually by the management, and by the board of directors of corporations at their principal annual meeting.

Assuming that the contractor's engineering personnel will have been in minute contact with all contractural matters, will have undertaken the preparation of claims growing out of the contract work, and will have participated in the contract settlements, as suggested elsewhere, it is logical that they would be in the best position to prepare the proposed review of contract settlements and pending claims. The suggested report need not be lengthy. But it should give a brief summary of contract settlements, listing each one during the previous year and providing particulars that would seem to be of interest. Pending claims should be listed, giving the current status, description of the premise involved, the amount of money in question and an evaluation of their merits. Claims settled during the previous year should also be reported, giving the particulars of interest.

13-38.   Other Engineering Functions

Because the construction industry is so diversified and a general contractor could be involved in a great variety of work, it would be virtually impossible to state in an abstract way all the conditions that might require attention from the contractor's engineering staff. This treatise is not intended to limit the contractor engineer's participation in the company's activities, but rather to promote maximum cooperation with the management. The contractor's engineer should call management's attention to any loose ends properly fitting in his sphere of influence. With management's approval, he should assume the related responsibilities.

13-39.   Employment, Training, and Reassignment of Engineering Personnel

Most large construction companies have programs for the employment, training, and advancement of engineering personnel. The engineering field is the principal source of future company management, and it is important to select applicants who appear to have the greatest potential.

Newly graduated engineers should be evaluated by their courses of study, scholastic ratings, and extracurricular activities. These should receive preferred attention in the evaluations, but personality traits, industriousness, apparent ability to get along with others, and willingness to work are equally important. An applicant's willingness to travel and live with his family in temporary quarters is a prerequisite with most heavy construction contractors operating on a national or international scale.

Company personnel officers can make initial contacts and investigate engineering applicants. But an engineering officer of the company should conduct the final personal interview and make the decision on employment.

When experienced engineers are considered for hire, it is usually because they have special talents needed in the company. They should be evaluated on the same basis as newly graduated engineers, but special attention should be given to their experience and references. Since experienced engineers should be expected to perform a special service, they should be carefully interviewed by an engineering officer familiar with the service.

The training and development of engineering personnel are an important factor in accomplishing the current work load and in building an organization for the future operation of the company. The subject deserves the establishment of some procedures, however simple or complex, and these should have management approval. Following are suggestions based upon experience and proven results by relatively large contracting organizations. The underlying principles could be applied to smaller firms, however.

Inexperienced engineers and engineering trainees should first be placed in positions where they can learn about the company—its policies, its operations, its procedures for handling the various engineering functions—and become acquainted with the people with whom they will work. A desirable place to start is in the analysis of prospective work and in the estimating functions. Quantity surveys, layouts of construction plant, selection of construction methods, observing work pricing by experienced estimators, and computations and assembly of the prebid estimates furnish a good background area for numerous engineering functions.

This phase of the training begins to open up the subjects of contract documents (plans, specifications, and contract provisions), the basic concepts of the varied and numerous structures offered for bid, the preplanning necessary to develop a concept of construction to be used on the jobs, construction organization, and the method of pricing based upon an historic background. In other words, it offers the best perspective of the company and the various ramifications of its business and capabilities. This phase could be termed a prejob-site assignment, with the activities best suited for performance in the contractor's principal engineering office or in a district or division office where the same functions are carried on. Such prejob-site assignment period should be limited to about two years. It could be less, depending upon the individual's tendencies and abilities and the company's need for engineering personnel in the field.

The next step should be a job-site assignment to one of the many engineering functions described elsewhere. It should be in a secondary capacity at first, with responsibilities added as rapidly as possible. Engineers who have had this type of experience generally agree that such basic training gave them a better understanding of the processes, and helped them to be more useful and productive than if they had been given a field assignment at the outset.

A personnel file should be kept (normally by the personnel department) on each engineering employee. It would include a current history of each individual—his assignments, the dates, salary rates, etc. More important, however, is that the responsible engineering department officer and the engineering supervisor in charge be continuously aware of the performance of the engineering personnel under their direction. They should make a record of each employee's accomplishments that would indicate his ability to handle greater responsibilities and his tendencies toward different phases of the work. The engineer officer in charge should arrange to work personally with his subordinates and become acquainted with them to promote better relationships and cooperative effort in the organization. This will also help him to study the personalities involved and to initiate earned promotions.

Because the construction business is more susceptible to change than almost any other, the occasions for reassignment of personnel are quite frequent. It is necessary that the responsible parties have a good record and understanding of the talents available for the numerous situations that present themselves as new construction jobs are being staffed and organized.

Moving engineering personnel from job to job or to other situations in the district or main offices requires frequent attention. The usual problem for the employer is finding the right man with sufficient capabilities to fill the vacancy. Management is continuously looking for replacements in its ranks and logically draws upon engineering personnel with an understanding of the problems involved.

The general concept suggested for a large contracting firm is that a district or division control the reassignment of engineering personnel within the district or division. The shifting of engineering personnel between districts or divisions should, however, be cleared through the home-office engineering department to avoid assignments that are inadvisable due to ability and adaptability.

Because of its wide contact throughout the company, the office of the chief engineer should have at least a general, and preferably an intimate knowledge of the abilities of individuals in the various company engineering staffs. It would expect to receive numerous inquiries for engineering assistance from widely separated sections of the company. It also should operate as a clearing house for reassignments in many instances and is in a position to render a service in that connection.

The district or division engineers should be encouraged to employ and develop engineering personnel serving their needs. When a shortage of work results in a surplus of engineering personnel, the district or division engineer should circulate advance written notice of individuals scheduled for release, giving their names, position previously held, and date of availability. Copies of such notices should be sent to the office of the chief engineer and to the home-office personnel department.

The chief engineer, with the agreement of the vice-president in charge of engineering, should make assignment and reassignments of district engineers, assistant chief engineers, and other engineering personnel attached to the main-office engineering staff.

The suggested procedures outlined above would be greatly simplified for a smaller contracting firm. However, the basic principles of organizing and replenishing the engineering staffs would be substantially the same.

13-40. Suggested Project Engineering Organization (Large Company Concept)

Because of the wide variation of job requirements, there can be no standard project engineering organization. The uncomplicated contracts might be served adequately by one man, the project engineer. As the number of functions is increased and as the individual functions demand more time and attention, it will be necessary to increase the project engineering staff.

The matter of project engineering organization is directed by management, subject to technical advice and supervision of the engineering officers of the company. It should be organized to produce the greatest possible efficiency at a minimum cost. The district or division engineer has the responsibility of advising and instructing the project engineer in the formation and functions of the project engineering staff. Consideration must be given to facilitating and supporting the operations on the job to protect properly the contractual position of the company.

To help determine how to properly organize the project engineering staff, an outline organization chart is shown in Figure 13-40-1. It illustrates a large complex job.

Simplification, by combining duties for the various members of the staff, would be appropriate in most instances.

13-41. Index of Engineering Responsibilities

Table 13-41-1 is intended only for quick reference to engineering responsibilities more fully described in the text. The charts are predicated upon a large contracting firm. They can be modified to fit different concepts of organization or simplified for smaller companies by combining functions and eliminating certain of the engineering positions mentioned. The listing of engineering duties on the charts should be helpful in organizing a contractor's engineering staff and could serve as a check list, even though some functions shown were assigned to other than engineering personnel.

13-42. Procedures for Cost-Plus and Special Types of Contracts

Basically, cost-plus and special types of contracts should be governed by the same engineering procedures as described for firm-price contracts. It is, however, recognized that special treatment with respect to certain particulars would be necessary to comply with the wishes of the owners.

Unless the owner specifies, in detail, the mechanics of the accounting to be employed or establishes a cost classification totally incompatible with usual company practice, company practice should be employed to the fullest extent possible, with the idea of regrouping cost totals on the statement to be provided to the owner. If this procedure is followed, the information accumulated would be directly comparable to that developed by the company elsewhere and, therefore, would be more usable. Standard company forms should be used to the extent that they can be applied.

Functions and duties of engineering personnel should be as necessary to suit the occasion, but they should be in conformity with the usual company practice insofar as the same is applicable.

Cost information included in periodic reports, if it is not in conformity with company procedures, should be defined by notes to avoid misinterpretation as to how inclusive it might be. Progress and production data should be kept and reported in the usual forms outlined hereinbefore.

District and project engineering relationships as referenced here should apply the same as for fixed-price contracts. Particular attention shall be provided to ensure satisfactory compliance with the owner's wishes in the furtherance of the construction operations and the protection of the company's contractural position.

13-43. Assembly of Brochures and Special Information

The office of the chief engineer should maintain a detailed file of information on the company's activities and current standing. It should describe the company's background in general construction, specialized construction, engineering accomplishments, organizational makeup, historical records of officers and supervisors, and photographs.

These data should be accumulated as a source for the preparation of brochures and for supplementing proposals in order to present the capabilities of the company. The information should be in such form that it can be selected to suit different occasions or to emphasize experience and qualifications along specific lines of endeavor.

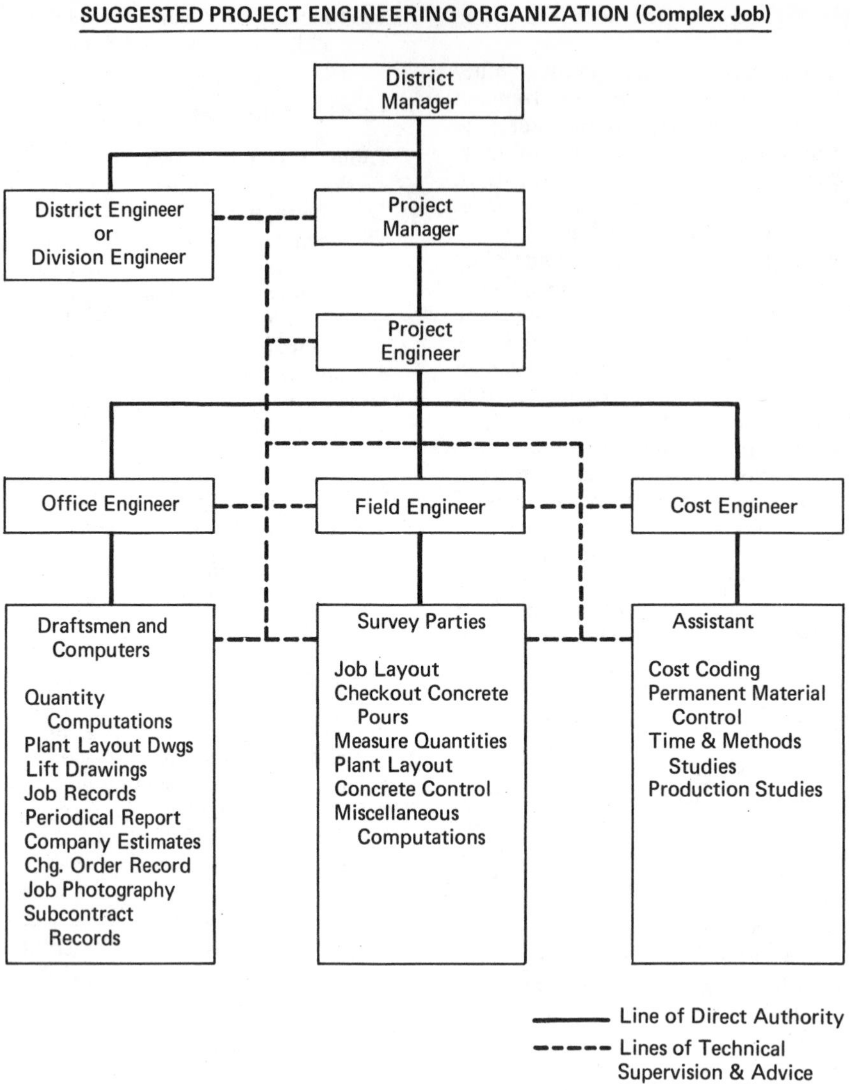

SUGGESTED PROJECT ENGINEERING ORGANIZATION (Complex Job)

Figure 13–40–1 Suggested project engineering organization (complex job)

It should be an engineering function to prepare brochures and assemble information, either at district level or home-office level. The various districts, divisions, and subsidaries should be encouraged to prepare, insofar as they can, through their principal engineering officers, brochure data and special information necessary for their requirements. However, the office of the chief engineer should cooperate by supplying specialized information, photographs, etc., for preparing brochure data and supplementing proposals. In addition, the office of the chief engineer should obtain from other departments the necessary accounting and fiscal information.

13-44. Evaluation of Contract Documents

Contract documents are often very complex. They generally consist of the following parts:

1. Invitation or solicitation to bidders
2. Instructions to bidders
3. General conditions of contract with sample copies of bid and performance bonds and sample copy of contract form to be signed by the successful bidder.

4. Special conditions of contract—usually dealing with specific circumstances applying to the particular project or site
5. Technical and payment specifications applying
6. Construction drawings illustrating the structures to be built or the work to be done
7. Addenda or modifications to the foregoing documents issued before the appointed date for presentation of proposals
8. Other data issued by the owner in advance of the receipt of bids and specifically classified by the owner as being part of the contract or prebid documents

Sometimes the owner or his engineer will issue supplementary information which he regards as an aid to the bidders, the accuracy of which he does not guarantee and which he specifically classifies as not forming a part of the contract documents. This could include meteorological information, hydrological data, quotations for services by third parties, and the like. It might also include subsurface explorations and logs of the findings of drilling contractors, geologists, etc., the interpretation of which must be made by the contractor at his own risk. This does not mean to

TABLE 13-41-1. ENGINEERING DUTIES AND ENGINEERING OFFICERS RESPONSIBILITIES—LARGE ORGANIZATION

KEY

Performs specific functions 1	Collaborates in function 4
Assists in function 2	Overall Supervision and Advice 5
Supervises function 3	Administrative responsibility 6

List of Engineering Duties	Vice-Pres.–Engineering	Chief Engineer	Asst. Chief Engineer	Dist./Div. Engineer	Project Engineer
General Functions					
Assist and advise vice-president–Operations and vice-president–Administration	1	2			
Develop or approve engineering procedures	1	2			
Coordinate estimating for major projects	1	2			
Direct estimating for major projects	3	1			
Develop construction methods	3	1	1	1	1
Accumulate experience and cost data	3	1	1	1	1
Supervise the home-office engineering department	3	1	2		
Assist district, division, project, and subsidiary companies	3	1	2		
Select, train, and assign key district personnel	3	1	2		
Select, train, and assign key project personnel	3	3		1	
Direct preparation of claims	3	3	1		
Direct preparation of change order requests	3	3	1		
Direct contract settlements	3	3	1		
Technical advisor to attorney regarding claims	3	3	1		
Direct district or division engineering department	3	3	3	1	
Direct estimating activities at district level	3	3		1	
Technical supervision of project engineering	3	3	3	1	
Preparing Proposals for New Work					
Solicit, tabulate, and analyze quotations	6	5	5	1	
Determine applicable wage rates and working condtions	6	5	5	1	
Determine applicable taxes and assessments	6	5	5	1	
Determine applicable insurance coverage and rates	6	5	5	1	
Ascertain availability of company equipment	6	5	5	1	
Provide advanced notices to surety companies	6	5	5	1	
Obtain bid bonds or bid guarantees	6	5	5	1	
Assemble and submit supplemental data with bid	6	5	5	1	
Obtain signatures of officers on bid and bond forms	6	5	5	1	
Check all bid documents for compliance	6	5	5	1	
Coordinate joint venture estimating activities	1[a]	2[a]		1[b]	
Arrange time and place of joint-venture prebid meetings	1[a]	2[a]		1[b]	
Notification to joint-venture partners and company officers	1[a]	2[a]		1[b]	
Preside at estimators joint-venture prebid meetings	1[a]	2[a]		1[b]	
Establish wage rates and working conditions for estimating	6	5		1[c]	
Setup suggested cost breakdown for estimating comparison	6	5		1[c]	
Establish tentative material and subcontract prices for estimating	6	3		1[c]	
Postbid Analysis of Jobs Awarded					
Review and revise original bid estimates	3	1[a]	2	1[b]	2
Study alternate methods and estimated cost of work	3	1[a]	2	1[b]	2
Resubmit revised bid summaries	3	1[a]	2	1[b]	2
Conduct postbid analysis meeting	1[a]	2[a]		1[b]	
Budget-control Estimate					
Prepare budget-control estimates	6	3	3	1	2
Coordinate with district and project management	6	3	3	1	2
Project Functions					
Planning, scheduling, and programming operations	6	5	5	4	1
Direct plant design	6	5	5	1	2
Prepare construction plant layouts	6	5	5	4	1
Compile and preserve details pertaining to plant	6	5	5	3	1
Perform or direct time and methods studies	6	5	5	3	1

(cont.)

TABLE 13-41-1 *(cont.)*

List of Engineering Duties	Vice-Pres.–Engineering	Chief Engineer	Asst. Chief Engineer	Dist./Div. Engineer	Project Engineer
Interpret contracts, plans, and specifications	6	5	5	4	1
Review all revised plans for changes and extras	6	5	5	3	1
Maintain complete file of all drawings issued	6	5	5	3	1
Maintain complete correspondence files with owner	6	5	5	3	1
Draft letters of protest regarding changes for project manager	6	5	5	3	1
Accumulation of job performance data	6	5	5	3	1
Prepare classification of cost accounts	6	5	5	3	1
Assign cost codes to records of job expenditures	6	5	5	3	1
Assemble cost and progress information for reports	6	5	5	3	1
Prepare daily reports, if required	6	5	5	3	1
Prepare weekly reports, if required	6	5	5	3	1
Prepare monthly cost and progress reports	6	5	5	4	1
Maintain progress charts	6	5	5	3	1
Analysis of permanent material and subcontract prices	6	5	5	4	1
Permanent materials control	6	5	5	3	1
Keep project diary	6	5	5	3	1
Prepare and preserve job engineering records	6	5	5	3	1
Check or prepare contract payment estimates	6	5	5	3	1
Draft letters to owners regarding payment estimates	6	5	5	3	1
Identify requirements of owner outside contract	6	5	5	4	1
Draft letters regarding extra work and changed work	6	5	5	3	1
Compute quantities of contract work performed	6	5	5	3	1
Compute quantities of extra work performed	6	5	5	3	1
Compile complete records regarding claim matters	6	5	5	4	1
Drafting letters to owners regarding claims	6	5	5	4	1
Provide job surveys and layout work	6	5	5	3	1
Maintain records of subcontract activities	6	5	5	3	1
Compute quantities of subcontract work	6	5	5	3	1
Prepare subcontractor payment estimates	6	5	5	3	1
Check owner's final estimate	6	5	5	4	1
Notify vice-president–Engineering and chief engineer of approaching final estimate and contract settlement				1	2
Review of terms of proposed contract settlement	1	1	2	2	
Prepare final job report	6	5	5	4	1

[a]Major jobs.
[b]Ordinary jobs.
[c]All jobs.

say, however, that the owner or his engineer is free to provide misleading, erroneous, or fallacious information which will or can lead the bidder at his peril into the wrong conclusions. The prebid period being relatively short, there is usually not time for a bidder to make his own subsurface explorations. In any case, the cost would not be justified, so he would be expected to rely on the information laid before him.

Interpretation of contracts might involve purely legal questions, for which legal advice should be solicited. Regardless, however, the contractor's engineering personnel have to interpret contractual provisions by the rule of reason until expert legal advice is obtained. This should apply to the prebid stages during the preparation of the initial estimate and through the construction stages when many apparently new questions present themselves. Basically, the contractor's engineering personnel have the responsibility for finding questionable contract requirements and reporting them to the management for a decision on how to proceed. Some of the questions might involve calculated risks, which contractors are prone to accept, but management must make the decision.

Some helpful guidelines in recognizing and evaluating questionable contract conditions are enumerated below. They are predicated upon the theory that the owner or his engineer has the responsibility for being precise in presenting the conditions for which he is soliciting prices. Any ambiguities should be interpreted against the party who draws the contract. Unfortunately, some construction contracts contain totally unfair and unreasonable provisions. They are intended to be so by those who drafted them and are quite well understood by both parties in advance of receipt of proposals. Such conditions should serve as a red-flag warning of the possibility that the contractor should seek business elsewhere. It seems, however, that some bidders will be found to agree to such provisions, presumably on the possibility of negotiating themselves out of trouble if it occurs. This is hardly acceptable business practice.

It is the responsibility of the owner and his engineer to draft contract conditions which are equitable and understandable to the parties involved. They should completely disclose all known circumstances and produce competent and complete contract documents, including specifications, designs, and drawings—all in a timely manner.

All the questions concerning the contract documents that might arise cannot be spelled out, but for acceptability be sure to check that:

1. The language of the contract documents is understandable in every respect.
2. Questions about possible ambiguities are resolved with the owner's engineer in advance of the bid.
3. The duties and responsibility of the owner and his engineer, as well as the contractor, are well defined and equitable.
4. Unusual and excessive indemnification by the contractor is not required.
5. Excessive financial investment by the contractor is not required.
6. The specified contract time is sufficient for normal speed of production.
7. Unreasonable penalties or liquidated damages provisions are not included.
8. Equitable extensions of contract time are provided for delays beyond the contractor's control.
9. All rights-of-ways and easements are provided by the owner.
10. Extra costs of suspending the work at the direction of the owner or his engineer are compensable.
11. Provisions for paying for extra work or changed work are equitable.
12. Provision is made for adjusting the contract price for unforeseen, unknown, and unusual conditions not contemplated by the parties, such as subsurface conditions.
13. Escalation due to increases in costs of labor and/or materials can be made a matter of adjustments in payments under the contract.
14. A fair means of handling disputes about contract matters is provided to permit appeals to higher authority, arbitration, etc.
15. The technical specifications are practicable and understandable.
16. The construction designs and contract drawings are practicable and complete.
17. The schedule of bid items includes all work necessary to produce the completed works.
18. The contractor is not expected to absorb the cost of items omitted or overlooked that would be required to produce a completed job.
19. The owner has made all necessary arrangements with utility and transportation companies and public bodies for the necessary interruption of their services or use of their rights of way, public roads or waterways.
20. The funding or means of financing the project is adequate.
21. Advance payments can be obtained to cover plant and move-in costs and for the cost of materials when delivered to the jobsite.

Every effort should be made to keep the management informed and to consult with the management concerning possible actions to be taken or how to consider their effects.

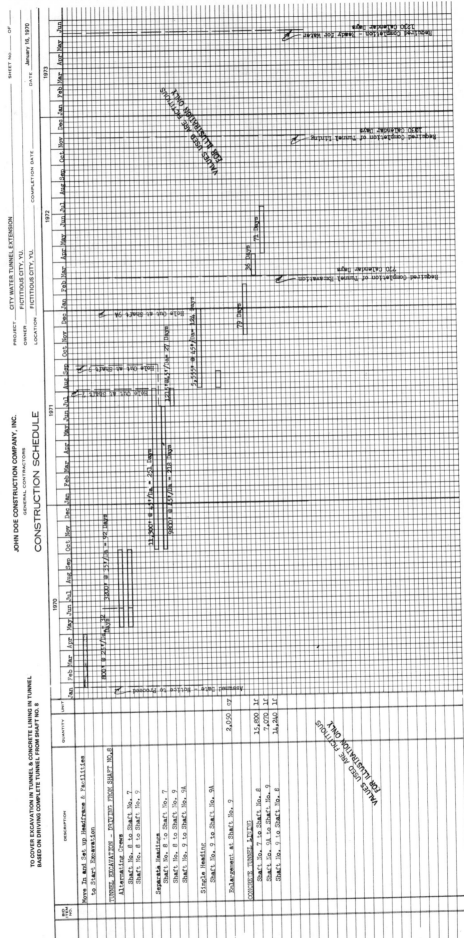

Figure 13-7-1   Bar graph—construction program

JOHN DOE CONSTRUCTION COMPANY, INC.

DOUBLE DUTY CANAL PROJECT SECTION 3

CONTRACT NO. 980

SPECIFICATION NO. 1901

WEEKLY REPORT NO. _____

FOR WEEK ENDING ___ Sept. 24, 1969

REVENUE

	ITEM	UNIT	UNIT PRICE	THIS WEEK QUANTITY	THIS WEEK AMOUNT	TO DATE QUANTITY	TO DATE AMOUNT
1	Canal Excavation, Common	C. Y.	.56			282,108	157,980.48
2	Canal Excavation, Rock	C. Y.	.56			208,082	116,525.92
3	Compact Embankment	C. Y.	.26			–	–
4	Excavation Wasteway Channel—Com.	C. Y.	.34			1,800	612.00
5	" " " —Rock	C. Y.	2.24				
6	" Structures—Com.	C. Y.	1.12	1,000	1,120.00	18,458	20,672.96
7	" " —Rock	C. Y.	3.90			156	608.40
8	" Siphon Barrel—Com.	C. Y.	1.58	5,000	7,900.00	259,371	409,806.18
9	" " —Rock	C. Y.	3.92			1,350	5,292.00
10	Trim Earth for Lining	S. Y.	1.12			–	–
11	Prepare Rock for Lining	S. Y.	1.12			–	–
12	Backfill About Structures	C. Y.	.39			1,494	582.66
13	Backfill About Siphon	C. Y.	.22	22,641	4,981.02	41,192	9,062.24
14	Compact Backfill Structures	C. Y.	2.50			2,860	7,150.00
15	Riprap	C. Y.	2.50				
16	Construct Reverse Filters	C. Y.	7.30			100	730.00
17	Construct Feeder Drains	L. F.	1.50			–	–
18	Construct Under Drains	L. F.	1.90			–	–
19	Concrete For Structures	C. Y.	42.00	57.5	2,415.00	1,578.10	66,280.20
20	Siphon Barrel	C. Y.	20.00	2,187.9	43,758.00	17,000.91	352,018.20
21	Structure Pave	C. Y.	20.00			–	–
22	Canal Lining	C. Y.	20.00			–	–
23	Place Reinforcing Steel	Lbs.	.029	346,753	10,055.84	2,871,985	83,287.57
24	Place Elastic Joint Filler	S. F.				–	–
25	Place Rubber Water Stop	L. F.	1.25			104	65.00
26	Install Clay Pipe Rev. Filter	L. F.				–	–
27	Install C. I. Pipe Rev. Filter	Lbs.	.10			1,000	100.00
28	Lay 12" Asbestos Cem. Pipe		1.90			–	–
29	Erect Steel Hwy. Bridge	Lbs.	.06			–	–
30	Place Metal Water Stops	Lbs.	.09	7,734	696.00	72,408	6,516.72
31	Erect Timber Railings	M.F.B.M.	125.00			–	–
32	Install Radial Gates	Lbs.	.06			–	–
33	Install Gate Hoists and Controls	Lbs.	.06			–	–
34	Install Cylinder Gate, Frame	Lbs.	.09			3,000	270.00
35	Miscellaneous Metal Work	Lbs.	.28			1,100	308.00
36	Place Bit. Mt'l in Floats	Cwt.	2.25			–	–
37	Install Electrical Conduit	L. F.	1.00			–	–
38	Install Electrical Conductors	Lbs.	1.00			–	–
39	Install Electrical Apparatus	Lbs.	1.00			–	–
40	Barbed Wire Fence	Mi.	1000.00			.225	225.00
41	Combination Fence	Mi.	1000.00			–	–
42	Cattle Guards	Ea.	190.00			–	–
43	CO #1 (b) ½ Rein. Steel	lb.	.0399	7,762	309.70	47,141	1,880.93
44	(c) ⅝ " "	lb.	.03885	28,342	1,101.09	169,936	6,602.01
45	(d) Over ⅝ Rein. Steel	lb.	.0378	238,619	9,019.80	3,094,804	116,983.59
46							
47							
	TOTAL				81,356.45		1,363,560.06

VALUES USED ARE FICTITIOUS FOR ILLUSTRATION ONLY

Figure 13-31-1 Sample weekly report

To: Mr. John Doe
Location: Main Office
Subject: **PROGRESS REPORT FOR WEEK ENDING NOVEMBER 6, 1971, CONTRACT 1001**

Date: November 7, 1971
From: Olie Olson
Location: Fictitious City, Yu.

Tunnel Excavation:

This was one of our best weeks when you consider the amount of work done rather than progress. We had only two good driving days, and the rest was in steel support ground. We drove a total of 234 feet this week, an average of 39.0 LF per day. Included in this total is 108 feet of full steel support, which is more support than was set in many weeks when we were in support ground all week. This bad ground did not contain much water, so our pumping situation did not get any more serious. The grouting last week succeeded in cutting off 160 gpm of inflow. We had one short section on Wednesday where a run got started and went about 12 feet up into the back. We broke some 6 × 8 crown bars at first, but then changed to 12 × 12 crown bars and caught it up all right.[a]

Driving Progress

	North Heading		Job Total
Monday	7 Rd.	53 LF	
Tuesday	4	36	
Wednesday	2	11	VALUES USED ARE FICTITIOUS
Thursday	5	37	FOR ILLUSTRATION ONLY.
Friday	5	36	
Saturday	9	61	
Total for week	32 Rd.	234 LF	
Total to date	15,858 LF		31,948 LF
% Driven	76.89 %		88.23 %

Round Records

	Eight-Foot Rounds		Ten-Foot Rounds	
	Total	Ave.	Total	Ave.
No. of rounds	29		1	
No. of holes	921	31.8	34	34.0
Total drilling	7,368	254.0	340	340.0
Drilling time	1,460	50.3	60	60.0
Mucking time	1,710	59.0	90	90.0
Advance	212.5	7.32	10.5	10.5
Powder factor	3.95 E/c.y.			
No. of rounds	2			
No. of holes	57	28.5		
Total drilling	228	114.0		
Drilling time	90	45.0		
Mucking time	125	62.5		
Advance	11.0	5.5		

Concrete Plant:

Erection of the batch plant started on Tuesday, and was about 10% complete at the end of the week. Rain on two days and unloading of cars cut into erection time. The remainder of the batch plant and tunnel forms were unloaded on Friday. The heading has been cleaned up for 300 feet for initial setup of the tunnel forms starting Monday. The railroad continues work on the siding, and now has it about 35% complete.

Slow deliveries will probably delay the start of concrete operations until well into December. Blaw-Knox has shipped the first agitator cars now, but this is two weeks behind schedule. Arco Welding is three weeks behind in shipping chassis for the agitators, and do not expect to ship for two more weeks.

CL/AMP/md
Att.

/s/ Olie Olson
Olie Olson

(b) WEEKLY PROGRESS REPORT[a]

Week Ending November 6, 1971

CONTRACT NO. 1000—Page 1 of 1

Job: CITY WATER TUNNEL EXTENSION

Item No.[a]	Bid Quantity	Changes	Estimated Quantity	Unit	Quantity this Week			Quantity to Date		
					Day	Swing	Graveyard	Day	Swing	Graveyard
2	195,000		195,000	CY	499	452	406	54,691	57,516	55,997[b]
3	2,000	−1,200	800	CY		8	10	217	201	181[b]
4	37,210		37,210	LF	22	19	18	2,589	2,771	2,701[b]
5	250,000	+150,000	400,000	MFG	5,120			193,210		[b]
18	3,000,000	−2,300,000	700,000	LF	7,830	8,160	9,850	163,610	164,990	162,060[b]
19	1,000	−700	300	MFBM	3.6	4.0	4.2	65.5	63.1	63.9[b]
21	20,000		20,000	LF	70	105	90	1,930	1,525	1,620
44	43		43	Mo	1			20.8		[b]
45	43		43	Mo	1			20.8		[b]
47	43		43	Mo	1			62.4		[b]

VALUES USED ARE FICTITIOUS
FOR ILLUSTRATION ONLY.

[a]Report only items worked during week.
[b]Adjusted to agree with October estimate.

Figure 13-31-2. Sample weekly reports

GENERAL PROGRESS REPORT
EVER GREEN DAM PROJECT
JANUARY, 1971

WEATHER

The weather for the first three weeks of January was unusually mild with temperatures up to 50 degrees during the day time. During the last of January the weather returned to normal and one inch of snow fell on the evening of January 28,1971

The river flow in the Arkansas began to increase this month, and 80,000 CFS was reported on January 31,1971

CONCRETE

For the month of January 11,645.25 cubic yards of concrete was placed in the Powerhouse, Intake and Erection Bay Sump.

The first section of the Intake steel roof form of the south bay of Unit 3 was set; and concrete was placed in the first pour of Intake roof this month. Concrete was also placed in the lower throat of the draft tube of Unit one; the intermediate and trailing draft tube pier of Unit two; and the draft tube roof of Unit 3.

Total concrete placed to date is 417,178.05 cubic yards.

EXCAVATION AND EMBANKMENT

Work continued in the following excavation items: Approach Channel, Tailrace and Powerhouse.

Placing impervious, pervious, transition and riprap fill in the Right Embankment continued this month. Rock was hauled from the Powerhouse excavation to Rockfill other than Embankments. Riprap was placed in the Left Embankment this month also.

DRILLING & GROUTING

A. Seepage Cutoff (East side of the River)

To date 61 holes have been drilled in the "A" line and 14 holes in the "C" line. All the holes in both "A" & "C" lines have been annular grouted, and 61 have been grouted with injection grout in the "A" line. Five holes have been grouted with injection grout in the "C" line.

B. Foundation Drilling & Grouting (AX Holes)

In the Gallery of the Right Gravity Dam 31 holes have been drilled and grouted to date. Eight holes in the Spillway Gallery have been drilled and five of the eight holes have been grouted.

Figure 13–32–1 Monthly report—general (narrative) progress report

C.   Drilling Anchorage Holes, Future Intake Units.

     To Date 64 holes of the 78 holes have been drilled.

<u>PROGRESS OF MAJOR WORK ITEMS</u>

<u>Excavation</u>

<u>Item</u>	<u>Completed this Period</u>	
2.16 Powerhouse	43,000	CY
2.17 Tailrace	199,000	CY
2.18 Approach Channel	29,000	CY

<u>Embankment</u>

2.39 Impervious	2,700	CY
2.40 Pervious	25,000	CY
2.41 Transition	1,400	CY
2.47 Rock Fill Other Than Embankments	9,000	CY
2.48 Riprap	7,100	CY

<u>Concrete</u>

2.54 Powerhouse	11,645.25	CY

<u>Forms</u>

2.62 Straight	47,302.1	SF
2.63 Curve	10,356.3	SF

<u>Salient Events for January 1961</u>

Jan. 11	Carpenters set first section of the steel intake roof form in Unit three.
Jan. 28	Carpenters stripped Draft Tube roof form from center and north bay of Unit three.
Jan. 30	Placed resteel on Draft Tube roof of Unit one.
Jan. 31	Placed concrete in first pour of the Draft Tube roof of Unit one.

-2-

Figure 13–32–1   (Continued)

GENERAL JOB STATUS REPORT

JOB _____ **EVER GREEN DAM** _____   CONTRACT NO. _____ **1005** _____   DATE _____ **June 1, 1971** _____

CLIENT _____ **George VII Mining and Development Company, Ltd.** _____

Amount of Original Contract	$ 65,123,405
Approved Change Orders to Date	62,037
Anticipated Over-run or (Under-run) in Uncompleted Work	25,622
Actual Over-run or (Under-run) in Completed Work	
Rental Revenue from Contract Owner not included above	
Estimated Total Amount of Principal Contract	$65,211,164
Other Contract Work not Included in Principal Contract	
Total Estimated Contract Volume	$ 65,211,164

Contract Revenue to Date	$ 27,176,039	
Less: Contract Advances for Materials on Hand	(2,657,123)	
Contract Advances for Plant and Move-in	(1,038,416)	
Revenue Reduction for Uncompleted Work	(896,014)	
Other		
Total Amount of Work Completed to Date		$ 22,584,486
Uncompleted Contract Volume		$ 42,646,678

Percent Complete Based on Original Contract	34.68 %
Percent Complete Based on Total Estimated Contract	34.63 %
Time Allotted by Original Contract	1,400 Days
Extension of Contract Time	10 Days
Total Contract Time	1,410 Days
Contract Time Elapsed	434 Days
Percent of Original Contract Time Elapsed	31.00 %
Percent of Total Time, Including Extensions, Elapsed	30.78 %
Date Contract was Physically Completed—If Completed	
Expected Date of Physical Completion—If Not Complete	**January 1, 1974**

VALUES USED ARE FICTITIOUS FOR ILLUSTRATION ONLY

Figure 13–32–2   Monthly report—general job status report.

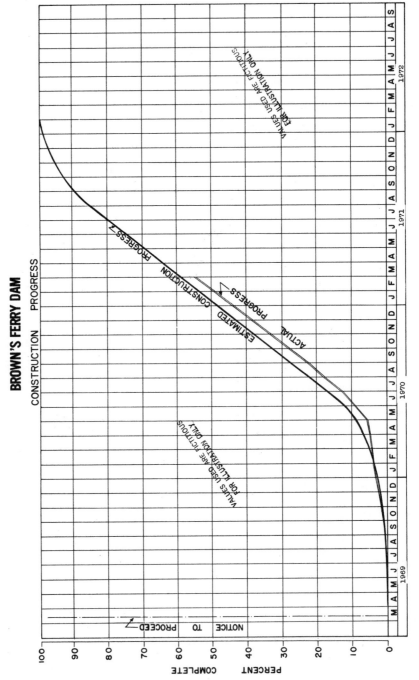

Figure 13-32-3 Monthly report—graphic progress curve

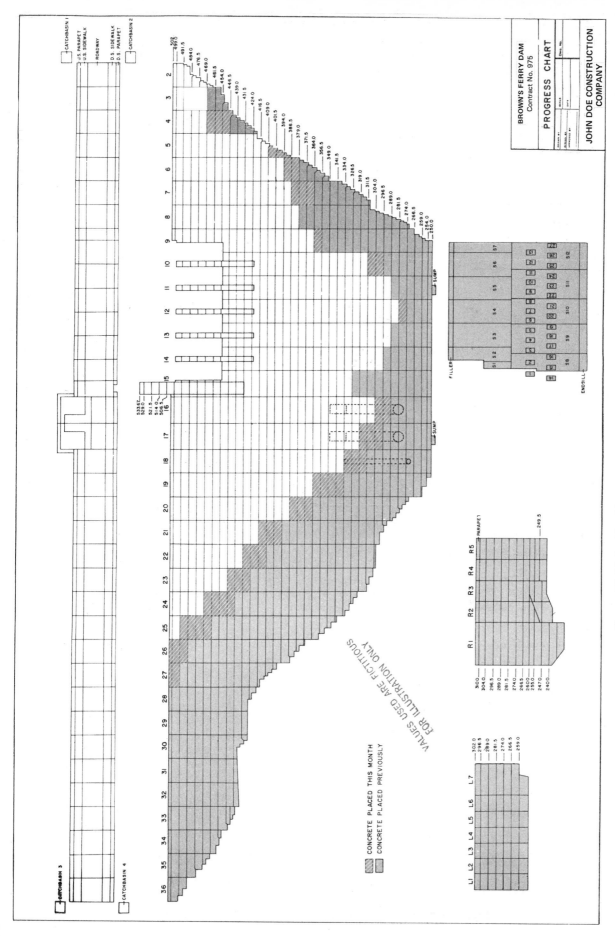

Figure 13-32-4 Monthly report—progress report in profile

JOHN DOE CONSTRUCTION COMPANY, INC.

MONTHLY COST STATEMENT

AT CLOSE OF BUSINESS   Feb. 28, 1971

PROJECT___ BROWN'S FERRY DAM   CONTRACT No.___ 975

CODE NO.	DESCRIPTION	UNIT	QUANTITIES ORIGINAL	QUANTITIES TO DATE	ESTIMATED COST TO DATE UNIT	ESTIMATED COST TO DATE AMOUNT	ACTUAL COST UNIT THIS MONTH	ACTUAL COST UNIT TO DATE	ACTUAL COST TOTAL TO DATE	REVENUE TO DATE UNIT BID	REVENUE TO DATE AMOUNT	GAIN OR (LOSS) AMOUNT
7004	Left Abutment Stripping	LS	100 %	100 %	74800	74,800			71,277 70	85,000.	85,000 00	13,722 30
7005	Preliminary Rock Cleanup	SQ.	600	590	9.50	5,604	7.71		4,546 95	10 00	5,904 00	1,357 05
7006	Close Line Drilling	SQ FT	30,000	28,376	.60	17,026	0.52		14,778 94	0.60	17,026 00	2,247 06
7007	Foundation Protection	SQ YD	5,000	5,444	5.75	31,303	2.69		14,667 51	6.00	32,665 00	17,997 49
7008a	Diamond Drill 1¼" Grout Holes	LF	15,000	8,676	1.92	16,658	1.854		16,085 03	2.00	17,352 00	1,266 97
7008b	Diamond Drill 1½" Grout Holes	LF	10,000		1.44					1.60		
7009	Moving Drill Onto Grout Holes	Ea.	800	319	25.00	4,975	20.00		6,380 00	25.00	7,975 00	1,595 00
7010a	Diamond Drill 3" Drain Holes	LF	10,000	4,410	5.92	26,107	5.78		25,510 98	6.00	26,460 00	949 02
7010b	Diamond Drill 3" Drain Holes	LF	7,000		4.42					4.60		
7011a	Core Drill 3" Exploratory Holes	LF	8,000	8,000	6.56	52,480	6.37		57,576 38	6.70	58,785 00	1,208 62
7011b	Core Drill 3" Exploratory Holes	LF	5,500	1,037	4.83	5,009				5.00		
7012	Connections to Grout Holes	Ea.	800	395	5.00	1,975	4.59		1,813 10	5.00	1,975 00	161 90
7013	Not Used											
7014	Not Used											
7015	Percussion Drilling	LF	21,500	17,023	1.29	21,960	0.46		7,887 90	1.50	25,534 00	17,646 10
7016a	Pressure Testing	Ea.	600	509	17.25	8,780	17.90		8,602 10	17.50	8,908 00	305 90
7016b	Pressure Testing	Ea.	200		11.00		11.00			11.00		
7017a	Furnish Cement Grouting	CF	4,500	4,400	1.90	8,550	1.85		8,465 90	2.00	9,183 00	717 10
7017b	Furnish Cement Grouting	CF	4,000	87	2.00	174				2.10		
7018a	Pressure Grouting	CF	4,500	4,162	3.70	15,399	3.82		15,885 23	4.00	16,648 00	762 77
7018b	Pressure Grouting	CF	3,500		3.85					4.00		
7019	Stilling Basin Anchors	LF	17,700	17,939	1.29	23,140	1.04		18,600 08	1.30	23,320 00	4,719 92
7020	Training Wall Anchors	LF	8,000	3,475	1.36	4,726	1.12		3,880 99	1.40	4,865 00	984 01
7021	12" RCCP	LF	100	36	4.60	166	6.26		225 34	4.80	173 00	(52 34)
7022	18" RCCP	LF	100		7.90					8.50		
7023	Uncompacted Backfill	CY	4,300	2,600	1.43	3,718	2.81		7,304 46	1.50	3,900 00	(3,404 46)
7024	Dumped Rock Backfill	CY	18,000	11,496	1.66	19,083	1.87		21,465 84	1.70	19,543 00	(1,922 84)
7025	Backing Material	CY	560	275	4.30	1,183	2.07		568 84	4.50	1,238 00	669 16
7026	Dumped Riprap	CY	1,000	525	2.24	1,176	3.89		2,041 77	2.30	1,208 00	(833 77)
7027	CONCRETE IN STILLING BASIN BAFFLES											
.01	Plant Operations and Amortization	CY	13,700	14,220				7.99	113,581 12			
.02	Place and Vibrate	CY	13,700	14,220				0.41	5,833 11			
.03	Point and Patch	SF	22,986	27,771				0.06	1,617 15			
.04	Treat Construction Joints	SF	6,201	10,614				0.06	665 23			
.05	Cure and Cleanup	CY	13,700	14,220				0.40	5,661 84			
.06	Finishing – Float	SF	56,483	61,434				0.17	10,512 01			
.21	Forms	SF	22,986	27,771				1.27	35,143 04			
.34	Aggregate Wastage	CY	13,700	14,220				0.06	828 63			
.35	Admixture	CY	13,700	14,220				0.03	422 14			
	Total	CY	13,700	14,220	11.65	165,663		12.25	174,264 27	13.30	189,126 00	14,861 73

VALUES USED ARE FICTITIOUS FOR ILLUSTRATION ONLY

JOHN DOE CONSTRUCTION COMPANY, INC.

MONTHLY COST STATEMENT

AT CLOSE OF BUSINESS   Feb. 28, 1971

PROJECT___ BROWN'S FERRY DAM   CONTRACT No.___ 975

CODE NO.	DESCRIPTION	UNIT	QUANTITIES ORIGINAL	QUANTITIES TO DATE	ESTIMATED COST TO DATE UNIT	ESTIMATED COST TO DATE AMOUNT	ACTUAL COST UNIT THIS MONTH	ACTUAL COST UNIT TO DATE	ACTUAL COST TOTAL TO DATE	REVENUE TO DATE UNIT BID	REVENUE TO DATE AMOUNT	GAIN OR (LOSS) AMOUNT
	Modifications											
7101	Mod. #1 Spillway Bridge	LS	100 %		330.26					363.30		
7102	Mod. #2 Pressure Recording Device Slots	LS	100 %	98	5,265.62	5,190			5,938.76	5879.60	5,762.00	(174.79)
7104	Mod. #4 Grout Holes In Erection Bay Area Addition 8-M-4-Diamond Drilling 1½" Grout Holes	LS LF	100 % 2,050	528	1.92	1,014	1.80		951.98	2.00	1,062.00	110.02
7105	Mod. #5 Gallery Truck Recess	LS	100 %	100 %	974.07	974			127.05	1071.48	1,071.00	943.95
7109	Mod. #9 Delete Everdur Conduit	LS	100 %	100 %						(1,002.00)	(1,002.00)	(1,002.00)
7111	Mod. #11 Install Telephone Cable											
7112	Mod. #12 24" Fishwater Line	LS	100 %	99.5%					8,433.21		13,065.00	4,531.79
	Totals					8,891,675			9,354,292.90		10,237,215.94	882,923.0
	Other Operations											6,767.5
	Totals To Date					8,891,675			9,354,292.90		10,237,215.94	889,690.6

VALUES USED ARE FICTITIOUS FOR ILLUSTRATION ONLY.

Figure 13-32-5 Monthly report—monthly cost statement summary

LABOR AND INSURANCE			PERMANENT MATERIALS			SPECIFIC PLANT			EQUIPMENT RENTAL			SUPPLIES			SUBCONTRACTS			TOTAL DIRECT COST			GEN. EX. & OVERHEAD		
UNIT EST.	UNIT COST	AMOUNT	UNIT EST.	UNIT COST	AMOUNT	UNIT EST.	UNIT COST	AMOUNT	UNIT EST.	UNIT COST	AMOUNT	UNIT EST.	UNIT COST	AMOUNT	UNIT EST.	UNIT COST	AMOUNT	UNIT EST.	UNIT COST	AMOUNT	UNIT EST.	UNIT COST	AMOUNT
26210		24,977 55							16478		16,727 45	22312		19,527 13				65000		61,345 26	9800		9,332 44
4.50	4.00	2,356 99							2.50	2.50	1,475 98	1.20	0.16	78 92					8.20	6.63	3,911 89	1.30		585 06
.35	0.16	4,623 60							.05	0.05	1,505 81	.12	0.22	6,379 45					.52	0.44	12,508 86	.08		2,270 08
2.00	0.38	2,084 22							1.00	1.00	5,444 30	2.00	0.56	3,055 76					5.00	1.94	10,584 28	.75		4,083 23
.10		8 41							.05	0.05	433 80	.05			1.467	1.55	13,447 80		.667	1.60	13,890 01	.253		2,195 02
.10									.05			.05			1.05			.25			.190		
															22.50	17.50	5,582 50	22.50	17.50	5,582 50	2.50		797 50
.10	0.01	65 28							.05	0.05	220 50	.05			4.95	4.95	21,829 50	5.15	5.01	22,115 28	.770		3,395 70
.10									.05			.05			5.643			5.843			.577		
.10	0.14	1,284 64							.05	0.05	465 31	.05	0.08	663 62	5.50	5.29	47,629 50	5.70	5.54	50,043 07	.860		7,533 31
.10									.05			.05			4.00			4.20			.630		
	0.10	40 10													5.00	4.49	1,773 00	5.00	4.59	1,813 10			
	0.17	2,907 18								0.01	207 90		0.10	1,706 82	1.116	0.01	104 00	1.116	0.29	4,925 90	.174		2,962 00
															15.00	14.65	7,456 85	15.00	14.65	7,456 85	2.25		1,145 25
															10.00			10.00			1.00		
.10	0.14	653 19	1.444	1.349	6,188 57				.02	0.02	91 74	.08	0.08	354 30				.644	1.59	7,287 80	.256		1,178 10
.10				1.50					.02			.08						.70			.300		
.10	0.17	710 60							.05	0.05	208 10	.05	0.03	111 53	3.00	3.07	12,774 00	3.20	3.32	13,804 23	.500		2,081 00
.10									.05			.05			3.143			3.343			.507		
.30	0.09	1,558 91	.694	.710	12,754 27				.03	0.03	538 16	.10	0.04	770 95				.124	0.87	15,622 29	.166		2,977 79
.30	0.11	389 31	.75	0.71	2,467 25				.03	0.03	104 25	.10	0.08	294 68				.18	0.94	3,255 49	.180		625 50
1.60	3.66	132 40	2.00	1.80	64 94				.10	0.10	3 60	.30	0.08	2 80				4.00	5.67	203 74	.600		21 60
2.26				4.00					.15			.46						6.87			1.03		
.86	0.97	2,515 91							.13	0.16	402 79	.25	1.50	3,891 76				.24	2.62	6,810 46	.19		494 00
1.00	0.58	6,719 05							.15	0.17	1,953 39	.29	0.89	10,264 28				.44	1.65	18,936 72	.22		2,529 12
1.00	0.35	96 02	2.30						.13	0.13	35 75	.29	1.01	277 57				.72	1.49	409 34	.58		159 50
1.33	1.06	557 23							.23	0.23	120 75	.39	2.31	1,211 54				.95	3.60	1,889 52	.29		152 25
0.94	0.75	10,610 64	2.40	2.92	41,572 88	1.556	1.81	25,738 20	.155	0.26	3,671 53	.81	0.72	10,231 27				5.861	6.46	91,824 52	1.53		21,756 60
0.33	0.39	5,604 64								0.02	228 47	.33	0.41	5,833 11				.33	0.41	5,833 11			
0.04	0.06	1,570 98										.005	0.01	46 17				.045	0.06	1,517 15			
0.30	0.05	491 62										.02	0.06	173 61				.32	0.06	665 23			
0.23	0.34	4,823 12										.03	0.03	838 72				.26	0.40	5,861 84			
0.14	0.14	8,456 81									93 50	.01	0.03	1,961 70				.15	0.17	10,512 01			
0.94	0.72	19,934 36									65 49	.65	0.55	15,143 19				1.59	1.27	35,143 04			
												.06	0.06	828 63				.06	0.06	828 63			
												.03	0.03	422 14				.03	0.03	422 14			
3.90	3.62	51,492 17	2.40	2.92	41,572 88	1.525	1.81	25,738 20	.155	0.26	3,830 52	2.15	2.10	29,873 90				10.12	10.72	152,507 67	1.53	1.53	21,756 60

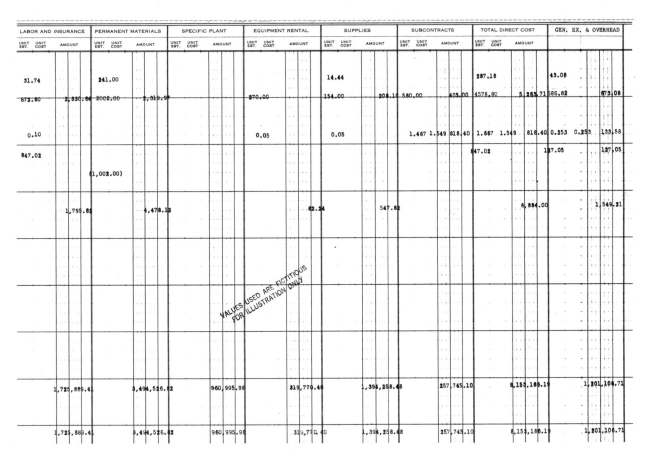

LABOR AND INSURANCE			PERMANENT MATERIALS			SPECIFIC PLANT			EQUIPMENT RENTAL			SUPPLIES			SUBCONTRACTS			TOTAL DIRECT COST			GEN. EX. & OVERHEAD		
UNIT EST.	UNIT COST	AMOUNT	UNIT EST.	UNIT COST	AMOUNT	UNIT EST.	UNIT COST	AMOUNT	UNIT EST.	UNIT COST	AMOUNT	UNIT EST.	UNIT COST	AMOUNT	UNIT EST.	UNIT COST	AMOUNT	UNIT EST.	UNIT COST	AMOUNT	UNIT EST.	UNIT COST	AMOUNT
31.74			241.00									14.44						287.18			43.08		
672.80		2,830.80	2002.00		2,319.90				270.00			154.00		208.10	580.00		405.00	4578.80		5,265.71	686.82		673.08
0.10									0.05			0.05			1.467	1.549	818.40	1.667	1.549	818.40	0.253	0.253	133.58
847.02																		847.02			127.05		127.05
			(1,002.00)																				
		1,795.82			4,476.12						62.24			547.82						6,884.00			1,549.21
		1,725,889.41			3,494,526.82			960,995.98			319,770.40			1,394,258.48			257,745.10			8,153,185.19			1,301,106.71
		1,725,889.41			3,494,526.82			960,995.98			319,770.40			1,394,258.48			257,745.10			8,153,185.19			1,301,106.71

Figure 13-32-5 (Continued)

MONTHLY COST STATEMENT
GENERAL EXPENSE

JOHN DOE CONSTRUCTION COMPANY, INC.

LOCATION EVERGREEN DAM
CONTRACT 100b   DATE June 1, 1971

VALUES USED ARE FICTITIOUS FOR ILLUSTRATION ONLY

ACCT NO.	ACCOUNT	ESTIMATED TOTAL EXPENSE	CURRENT MONTH			JOB-TO-DATE			BALANCE UNAMORTIZED
			ACTUAL COST	ESTIMATED COST	(OVER) UNDERRUN	ACTUAL COST	ESTIMATED COST	(OVER) UNDERRUN	
101	Management and Supervision Salaries	442,800	10,624	9,225	(1,399)	246,239	207,585	(38,654)	38,556
102	Office Salaries	369,200	6,428	7,692	1,264	142,786	173,081	30,295	22,357
103	Engineering Salaries	685,400	32,511	14,279	(18,292)	462,296	321,316	(140,980)	72,386
104	Warehouse and Purchasing Salaries	243,250	5,033	5,068	35	112,709	114,036	1,327	17,648
105	Labor Relations Salaries		864		(864)	17,088		(17,088)	2,676
106	Safety and First Aid Salaries	83,000	2,046	1,729	(317)	45,025	38,910	(6,115)	7,050
107	Watchman, Guard and Fire Guard Salaries	18,000	2,139	375	(1,764)	44,157	8,438	(35,719)	6,914
108	Retirement Plan Expense (J. V. jobs only)		662		(662)	22,380		(22,380)	3,505
109	General Automotive Expense	232,828	5,756	4,851	(905)	114,226	109,150	(5,076)	17,886
110	Manhaul		59		(59)	212		(212)	33
111	Automotive License Fees and Permits	10,000	89	208	119	10,449	4,688	(5,761)	1,636
112	Stationery and Office Supplies	19,200	415	400	(15)	20,543	9,001	(11,542)	3,217
113	Office Rent (Including Temporary)					750		(750)	117
114	Office Equipment—Rental or Depreciation	44,000	2,020	917	(1,103)	25,339	20,627	(4,712)	3,968
115	Office Expense—Other		5,305		(5,305)	10,977		(10,977)	1,719
116	Engineering Supplies and Expense	14,400	375	300	(75)	16,027	6,751	(9,276)	2,510
117	Engineering Equipment—Rental or Depr.		410		(410)	5,378		(5,378)	842
118	Photography and Reproduction	2,400	85	50	(35)	7,811	1,125	(6,686)	1,223
119	Outside Engineering Expense	3,000		-63	63	11,573	1,406	(10,167)	1,812
120	Consultants Expense								
121	Warehouse and Purchasing Expense	23,600	68	492	424	1,762	11,064	9,302	276
122	Labor Relations—Supplies and Expense					178		(178)	28
123	Safety and First Aid Supplies and Expense	41,200	1,104	858	(246)	29,719	19,315	(10,404)	4,653
124	Watchman, Guard and Fire Guard Expense	3,000	7	63	56	4,231	1,406	(2,825)	662
125	Telephone and Telegraph	24,000	1,946	500	(1,446)	44,094	11,251	(32,843)	6,905
126	Radio Communications								
127	Heat, Light, Power & Janitor Service	19,200	516	400	(116)	14,905	9,001	(5,904)	2,334
128	Building Maintenance—Owner Facilities	14,400		300	300	1,863	6,751	4,888	292
129	Legal Expense	5,000		104	104	27,721	2,344	(25,377)	4,341
130	Travel Expense—Job Connected	10,000		208	208	2,088	4,688	2,600	327
131	Travel Expense—Mobilization of Personnel	9,000		187	187	7,527	4,219	(3,308)	1,179
132	Entertainment, Donations and Contributions	4,000		83	83	5,745	1,875	(3,870)	900
133	Bank Charges	3,000		63	63		1,406	1,406	
134	Fire Protection	3,000		63	63		1,406	1,406	
135	Sanitary Facilities	8,000	2,122	167	(1,955)	9,849	3,750	(6,099)	1,542
136	Drinking Water and Ice	27,008	251	563	312	6,932	12,661	5,729	1,085
137	Misc. Job Road and Yard Maintenance	19,200		400	400		9,001	9,001	
138	Job Signs								
139	Winter Shut-Down	13,050							
140	Central Office General Expense			272	272		6,118	6,118	
153	District Office General Expense								
154	Gain or Loss—Commissary Operations								
155	Gain or Loss—Mess Hall Operations		1,666		(1,666)	24,883		(24,883)	3,896
156	Gain or Loss—Barracks Operations								
157	Gain or Loss—Residence Operations	50,000	9,162	1,042	(8,120)	148,333	23,440	(124,893)	23,226
158	Gain or Loss—Trailer Camp Operations		1,451		(1,451)	22,480		(22,480)	3,520
159	Meal Expense		493		(493)	7,956		(7,956)	1,246
160	Insurance—Other than Payroll	389,000	12,545	8,104	(4,441)	243,228	182,363	(60,865)	38,085
161	Taxes—Other than Payroll	655,617	20,397	13,659	(6,738)	296,406	307,353	10,947	46,411
162	Surety Bond Premium	105,000		2,188	2,188	71,036	49,224	(71,036)	11,123
163	Employees Bonus	5,000		104	104	57,300	2,344	(8,076)	8,973
164	Small Tool Depr. & Minor Equip.	13,000		2,600	2,600		6,094	2,344	
165	Cash Discounts Earned		(2,600)	271	271	(73,802)		73,802	(11,555)
166	Miscellaneous General Expense & Cleanup	13,000	271	63	63	3,073	6,094	3,021	
167	Miscellaneous Frgt.	3,000	63			1,406	1,406	1,406	481
168									
	TOTALS	3,614,753	124,009	75,311	(48,698)	2,273,472	1,694,594	(578,878)	355,985

Figure 13-32-6 Monthly report—monthly cost statement, general expense

JOHN DOE CONSTRUCTION COMPANY, INC.

MONTHLY COST STATEMENT
GENERAL PLANT

LOCATION EVER GREEN DAM
CONTRACT NO. 1005   DATE June 1, 1971

(VALUES USED ARE FICTITIOUS FOR ILLUSTRATION ONLY)

ACCT. NO.	GENERAL PLANT	ESTIMATED JOB TOTAL	CURRENT MONTH ACTUAL COST	CURRENT MONTH ESTIMATED COST	CURRENT MONTH (OVER) UNDERRUN	JOB TO DATE ACTUAL COST	JOB TO DATE ESTIMATED COST	JOB TO DATE (OVER) UNDERRUN	BALANCE UNAMORTIZED
201	TEMPORARY BUILDINGS								
01	Office	180,531				240,162	180,531	(59,631)	132,089
02	Warehouse	16,640				46,642	16,640	(30,002)	25,653
03	Equipment Repair Shop	14,167				43,646	14,167	(29,479)	24,005
04	Carpenter Shop	58,487				38,008	58,487	20,479	20,904
05	Electrical Repair Shop	20,725				20,113	20,725	612	11,062
06	Pipefitters Shop	3,830				9,858	3,830	(6,028)	5,422
07	Riggers Loft	3,744				7,502	3,744	(3,758)	4,126
08	Hospital or First Aid Bldg. or Trailer	3,920				8,651	3,920	(4,731)	4,758
09	Field Offices	5,824				12,708	5,824	(6,884)	6,989
10	Tool Sheds	10,000				17,884	10,000	(7,884)	9,836
11	Change House	4,000				1,989	4,000	2,011	1,095
12	Powder House	2,000					2,000	2,000	
13	Cap House								
14	Tire Repair Shop	3,920				7,227	3,920	(3,307)	3,975
15	Oil Storage	600					600	600	
16	Grease Rack	1,000					1,000	1,000	
17	Service Station	2,000					2,000	2,000	
18	Fuel Storage	2,800					2,800	2,800	
19	Inspector's Field Office								
20	Inspector's Concrete Testing Facilities								
21	Paint Shop								
22	Workmen's Shelters	7,224				17,477	7,224	(10,253)	9,612
23	Compressor House	1,650					1,650	1,650	
24	Erection Shop								
25									
29	General Yard and Site Grading and Surface	18,000				8,457	18,000	9,543	4,652
202	UTILITIES INSTALLATION (Except Camp)	23,500				82,501	23,500	(59,001)	45,382
01	Electrical Supply and Distribution					37,503		(37,503)	20,627
02	Telephone Lines and Job Telephone Systems	4,500				12,668	4,500	(8,168)	6,967
03	Water Supply Storage and Distribution	8,000				20,275	8,000	(12,275)	11,151
04	Sewer System and Treatment Facilities	8,000				12,055	8,000	(4,055)	6,637
05	Radio Communications	3,000					3,000	3,000	
203	TEMPORARY ROADS								
01	General and Access Roads and Bridges					259,949		(259,949)	143,012
204	CAMP BUILDINGS								
01	Barracks					11,446		(11,446)	6,295
02	Mess Hall								
03	Commissary								
04	Recreation Hall								
05	Bull Cook's Dormitory								
06	Trailer Court, Including Utilities					39,419		(39,419)	21,680
07	Trailer Court Wash House								
08	Residences					209,084		(209,084)	115,037
09	Site Grading and Surfacing, Except Trailer Ct.								
205	UTILITIES FOR CAMP, Other than Trailer Court								
01	Power Supply and Distribution								
02	Telephone System								
03	Water Supply and Distribution								
04	Sewer System and Treatment Facilities								
	TOTAL GENERAL PLANT					582,612	204,031	(378,581)	320,483

Figure 13-32-7   Monthly report—monthly cost statement, general plant

SPECIFIC PLANT OPERATIONS

Period Ending _____ March 1970

PLANT DESCRIPTION ___ 25 Ton Cableway (excluding Track Maintenance)

JOB ___ Brown's Ferry Dam ___ CONTRACT NO. ___ 975 ___ P/R Tax & Ins. ____ % Overtime ____ %

	AVAILABLE	OPERATING	EFFICIENCY	QUANTITY	PRODUCTION	
To Date:	5952 Hrs.	5246 Hrs.	88 %	479736 CY	98.0 CY/Hr.	
Period:	483 Hrs.	447 Hrs.	93 %	40144 CY	104.0 CY/Hr.	
Control:			90 %		____ CY/Hr.	

Utilization of Total Machine Time

Pouring	Yarding	Idle	Repair
4,442 Hr.	804 Hr.	422 Hr.	273 Hr.
386 Hrs.	61 Hr.	21 Hr.	14 Hr.

OPERATIONS:	PERIOD			TO DATE			CONTROL	
	COST	COST/HR	COST/CY	COST	COST/HR	COST/CY	COST/HR	COST/CY
1. Operating Labor (Incl. Tax & Ins.)	3,429	7.67	.0854	39,502	7.53	.0823	6.50	.0650
2. Repair of Service Labor (ditto)	3,688	8.25	.0919	41,181	7.85	.0858	7.10	.0710
3.								
4. Equipment Rental or Depreciation	54	0.12	.0013	630	.12	.0013	0.20	.0020
5. Fuel or Power	4,461	9.98	.1111	45,902	8.75	.0957	8.50	.0850
6. Oil and Grease (in above)								
7. Repair Parts & Outside Repairs	930	2.08	.0232	20,040	3.82	.0418	3.25	.0325
8. Operating Supplies	2,901	6.49	.0723	31,528	6.01	.0657	5.90	.0590
9. Wire Rope	12,069	27.00	.3001	35,410	6.75	.0738	6.82	.0682
10. Shop Costs								
TOTAL OPERATING	27,531	61.59	.6853	214,194	40.83	.4465	38.27	.3827
AMORTIZATION:								
Specific Plant	60,345	135.00	1.5032	708,210	135.00	1.4762	122.00	1.2200
Equipment								
TOTAL AMORTIZATION	60,345	135.00	1.5032	708,210	135.00	1.4762	122.00	1.2200
TOTAL COST	87,876	196.59	2.1876	922,404	175.83	1.9227	160.27	1.6027

REMARKS: Production in Cu. Yds./Hr. = Quantity in Cu. Yds. divided by operating hours less yarding time. Wire rope cost for the period is disproportionate because it includes large replacements. Operations served by this machine include all work associated with handling and placing the concrete, metal work, gates, etc for the main dam and appurtenances.

Figure 13-32-8 Monthly report—specific plant operations report

JOHN DOE CONSTRUCTION COMPANY, INC.
HOURLY EQUIPMENT OPERATING COSTS

MONTH_____ April 1970
CONTRACT 1001
LOCATION__ Oak Hollow Dam

Equipment Account Number	8382		8478		8487		8489		8527		8606		8607	
Type of Equipment (Show Model, Size, etc.)	Euclid Loader 1Q-BV		Vibrating Rollers Essick - VR72T		LeTourneau Tamper N 50-55		Sheepsfoot Peterson 4 Drum		Scrapers Euclid - TSS 24		Tractors Caterpillar D-8		Tractor Caterpillar D-9	
New Cost														
Horsepower														
Shipping Weight														
Fuel or Power Consumption per hour (To Date)	5.18				16.09				16.24		10.84		10.51	
Legend Code	(B)(C)(D)**		(C)**		(B)(D)**				(B)(C)(D)**		(D)*			
Number of Units	1		6		2		1		11		4		1	
	This Month	To Date	This Month	To Date	This Month	To Date	This Month	To Date	This Month	To Date	This Month	To Date	This Month	To Date
Operating Hours		3,159		23,668 1/2		4,588 1/2		2,535		81,194 1/2		17,944		3,552
Repair Hours														
Operating Efficiency % (Hours worked ÷ Possible Working Hours less hours avail. but not used)		58.67				84.74				71.46		84.61		52.05
LABOR (Inc. Taxes & Ins.)														
.01 Operating (Inc. Oiler)		3.070				4.535				3.904		4.448		4.675
.02 Repair & Service		11.353		2.549		6.382		2.588		3.699		1.780		8.633
Total Labor		14.423		2.549		10.917		2.588		7.603		6.228		13.308
RENTAL OR DEPRECIATION														
.04 Rental M-K New Rate														
M-K Used Rate														
M-K Special Rate														
Outside Rate														
.04 Depreciation—Monthly % Amount		1.627		.643		4.991		4.808		4.261		4.100		5.084
Total Rental or Depreciation (J)		1.627		.643		4.991		4.808		4.261		4.100		5.084
OPERATING SUPPLIES & EXPENSE														
.03 Repair Parts		15.385		4.630		8.589		1.812		6.024		2.797		8.637
.05 Fuel or Power		.681		.685		1.820		.071		1.910		1.240		1.242
.06 Oil and Grease		.578		.161		.307		.006		.780		.284		.613
.07 Cable and Teeth		1.594		.001						.279		.331		1.461
.08 Tires and Tubes						.028				3.668				
.09 Outside Repairs		.171		.036				.730		.016		.126		.244
.10 Shop Costs		2.986		.803		1.152				1.289		.478		2.279
Total Supplies & Expense		21.395		6.316		11.896		2.619		13.966		5.256		14.476
TOTAL HOURLY COST		37.445		9.508		27.804		10.015		25.830		15.584		32.868

LEGEND: (Note here conditions encountered in the work that affect the operating cost of the equipment.)

Please refer to the notes on page 1 of the Equipment Report.

.066 cyl. hd. cr.
25.764 Revised Total

VALUES USED ARE FICTITIOUS FOR ILLUSTRATION ONLY

H.D. Mechanic Rate Per Hr. $ 3.125
Mech. Helper Rate Per Hr. $ 2.55
Diesel Cost Per Gal. $ .12
Gasoline Cost Per Gal. $ .27
Elec. Power Per K.W.H. $
P.R. Tax & Insur, Percent 13.0 %
Overtime Worked Percent(F) 15.56 %

Figure 13-32-9 Monthly report—hourly equipment operating costs

JOHN DOE CONSTRUCTION COMPANY, INC.
SPECIAL PLANT REPORT

JOB_____ BY_____ DATE March 1, 1970

JOB | BROWN'S FERRY DAM
FOR | SPECIFIC PLANT ANALYSIS

BID DATE_____

VALUES USED ARE FICTITIOUS
FOR ILLUSTRATION ONLY

ITEM NO.	DESCRIPTION	QUANTITY	UNIT	LABOR (1)	INSTALLED EQUIP. (2)	INITIAL FREIGHT. (3)	EQUIP. DEPREC. (4)	SUPPLIES (5)	SUBCONTRACTS (6)	TOTAL COST	ITEM NO.
2191 .00	8 CY Cableway		Control	83,760	401,600	27,776	17,020	162,520		692,676	
			Actual	65,703	390,424	29,360	6,434	150,203	3,485	645,609	
.01	Tail Tower Track		Control	8,932	4,560	1,440	2,138	13,558		30,628	
			Actual	11,307	5,770	1,920	24,479	19,102		40,578	
.02	Concrete Delivery Truck		Control	2,700	2,920	640	500	7,450		13,210	
			Actual	3,410	5,026	1,675	840	4,598		15,549	
.03	Concrete Delivery R.R. Equip.		Control	840	7,261	525	260	1,660		10,546	
			Actual	2,129	30,579	2,309	166	4,498	52	40,113	
.04	Misc. Concrete Placing Equip.		Control	500	31,020	3,120	300	28,420		63,360	
			Actual	2,370	32,472	4,645	31	5,504		45,022	
.05	Mixing Plant		Control	43,792	93,730	6,550	4,918	16,388		165,478	
			Actual	34,900	164,524	11,757	1,614	40,070	860	253,825	
.06	Cement Handling		Control	8,488	15,350	3,650	1,327	2,407		31,222	
			Actual	13,198	25,902	3,079	1,355	16,721	250	60,505	
.07	Flyash Handling		Control	6,875	17,542	1,060	820	1,805		28,102	
			Actual	7,669	19,755	1,790	721	10,454	188	40,977	
.09	Cooling Plant		Control	60,100	151,147	6,260	2,800	183,050		403,357	
			Actual	79,326	182,870	7,619	2,487	159,399	10,911	459,298	
.10	Aggregate Stockpiling		Control	24,118	30,524	1,900	2,598	24,858		83,958	
			Actual	70,892	51,677	2,614	5,461	137,871		268,515	
.12	Bunkers and Washing Plant		Control	7,582	34,688	1,000	1,968	11,373		56,611	
			Actual	15,155	11,565	509	392	27,234		54,955	
.13	Air Plant		Control	7,932	31,318	2,522	1,938	7,758		51,968	
			Actual	14,226	88,032	5,239	503	17,660	78	125,638	
.14	Water Plant		Control	6,380	27,400	1,200	1,070	13,020		49,070	
			Actual	10,568	35,020	1,156	1,284	21,157	11,150	80,343	
	Flyash Plant at Siding		Control	Not Used Flyash Delivered F.o.b. Jobsite							
			Actual								
	Walkways, Scaffolds & Ladders		Control	25,000			0	10,000		35,000	
			Actual								
	Sub Total Less Aggregate Plant		Control	286,999	849,560	57,743	37,617	483,267		1,715,186	
			Actual	330,853	1,060,790	73,672	23,767	634,871	26,874	2,130,927	
.15	Aggregate Plant		Control	74,760	162,827	12,800	0	179,225		429,612	
			Actual	112,245	456,482	26,785	11,075	275,725		882,312	
	Total Specific Plant		Control	361,759	1,011,387	70,543	37,617	663,492	26,874	2,144,798	
			Actual	443,098	1,517,272	100,457	34,842	890,596		3,013,239	

VALUES USED ILLUSTRATION
FOR ILLUSTRATION ONLY

Figure 13-32-10   Monthly report—special plant report

JOHN DOE CONSTRUCTION COMPANY, INC.
SPECIAL PLANT REPORT

JOB
BY
DATE March 1, 1970

JOB BROWN'S FERRY DAM
FOR SPECIFIC PLANT ANALYSIS
BID DATE

VALUES USED ARE FICTITIOUS
FOR ILLUSTRATION ONLY

ITEM NO.	DESCRIPTION		QUANTITY	UNIT	LABOR (1)	INSTALLED EQUIP. (2)	INITIAL FREIGHT (3)	EQUIP. DEPREC. (4)	SUPPLIES (5)	SUBCONTRACTS (6)	TOTAL COST	ITEM NO.
2191	**8 CY CABLEWAY**											
.001	Excavation and Grading	Control	2,000	cy	3,000			1,000	1,000		5,000	
		Actual			3,921			960	4,519		9,400	
.002	Foundation Concrete	Control	600	cy	12,000			3,000	15,000		30,000	
		Actual	564	cy	6,000			526	14,733		21,259	
					10.63			0.93	26.13		37.69	
.003	Counterweight Concrete	Control	200	cy	3,000			1,000	5,000		9,000	
		Actual	211	cy	3,303			98	7,695		11,096	
					15.65			0.46	36.47		52.58	
.004	Structural Steel	Control	280	ton	2,760	62,750	12,250	920	920		79,600	
		Actual	286	ton	18,457	70,133	12,606	2,610	13,146	3,400	120,352	
					64.53	245.22	44.07	9.12	45.96	11.89	419.70	
.005	Machinery	Control	180	ton	28,000	296,674	13,326	5,600	5,600		349,200	
		Actual			6,743	204,932	10,785	662	9,875	85	233,080	
.006	Guys, Lines and Backstays	Control			27,000			4,500	79,000		110,500	
		Actual			12,390			1,189	70,743		86,265	
.007	Electrical Installation	Control			5,000	44,176	2,200	1,000	53,000		103,376	
		Actual			11,943	113,416	5,969	283	18,867		150,476	
.008	Air and Water System	Control			0			0	0		0	
		Actual			701			33	907		1,641	
.009	Enclosures and Stairs	Control	1,000	S.M	3,000			0	3,000		6,000	
		Actual			2,247			73	9,720		12,040	
	Total Cableway	Control			83,760	401,600	27,776	17,020	162,520		692,676	
		Actual			65,703	396,424	25,360	6,434	159,203	3,485	645,609	
2191	**TALL TOWER TRACK**											
.011	Excavation and Grading	Control	13,000	C.Y.	4,300			1,500	2,800		8,600	
		Actual	30,200	C.Y.	5,893			1,632	7,290		14,815	
					0.20	3.32	1.10	0.05	0.24		0.49	
.013	Ballast, Ties and Rail	Control	1,400	L.F.	4,632	4,560	440	638	10,758		22,028	
		Actual	1,740	L.F.	5,414	5,770	4,920	847	847		25,763	
					3.11		1.10	0.49	6.79		14.81	
	Total Tall Tower Track	Control			8,932	4,560	440	2,138	13,558		30,628	
		Actual			11,307	5,770	1,920	2,479	19,102		40,578	

Figure 13–32–10 (Continued)

JOHN DOE CONSTRUCTION COMPANY, INC.
SPECIAL PLANT REPORT

JOB _____ BY _____ DATE March 1, 1970

JOB BROWN'S FERRY DAM
FOR SPECIFIC PLANT ANALYSIS
BID DATE _____

VALUES USED ARE FICTITIOUS FOR ILLUSTRATION ONLY

ITEM NO.	DESCRIPTION		QUANTITY	UNIT	LABOR (1)	INSTALLED EQUIP. (2)	INITIAL FREIGHT (3)	EQUIP. DEPREC. (4)	SUPPLIES (5)	SUBCONTRACTS (6)	TOTAL COST
2191	**CONCRETE DELIVERY TRACK**										
.021	Excavation and Grading	Control	1,500	C.Y.	550			200	350		1,100
		Actual			983			531	669		2,183
.024	Ballast Ties and Rails	Control	800	L.F.	2,150	1,920	640	300	7,100		12,110
	(1F Track)	Actual	860	L.F.	2,427 (2.82)	5,026 (5.84)	1,675 (1.95)	309 (0.36)	3,929 (4.55)		13,366 (15.54)
	Total Concrete Delivery Track	Control			2,700	1,920	640	500	7,450		13,210
		Actual			3,410	5,025	1,675	840	4,598		15,549
2191	**DELIVERY RAILWAY EQUIPMENT**										
.031	25 Diesel Locomotive	Control	1	Each	240	7,261	525	160	160		8,346
		Actual	2	Each	856	30,679	2,309	46	1,576		35,466
.032	Delivery Cars	Control	1	Each	600			100	1,500	52	2,200
		Actual	2	Each	1,273	280		120	2,922		4,647
	Total Delivery Equipment	Control			840	7,261	525	260	1,660	52	10,546
		Actual			2,129	30,959	2,309	166	4,498		40,113
2191	**MISCELLANEOUS CONCRETE EQUIPMENT**										
.041	Concrete Buckets	Control			500	21,380	3,120	300	10,300		35,600
		Actual			734	31,087	4,645	15	2,697		39,182
.042	Vibrators and Others	Control			0	9,640		0	18,120		27,760
		Actual			5	4,385		3	1,434		2,827
					1,631			9	1,373		3,013
	Total Placing Equipment	Control				31,020	3,120	300	28,420		63,360
		Actual			2,370	32,472	4,645	31	5,504		45,022
2191	**MIXING PLANT**										
.051	Excavation and Grading	Control			2,000			1,000	1,000		4,000
		Actual			1,332			239	210		1,783
.052	All Concrete	Control	150	C.Y.	4,500			750	3,000		8,250
		Actual	64	C.Y. (40.08)	2,565			125 (1.95)	3,958 (61.84)		6,648 (103.84)
.053	Drainage	Control			815			1	66		882
		Actual									

Figure 13-32-10 (Continued)

JOHN DOE CONSTRUCTION COMPANY, INC.
SPECIAL PLANT REPORT

JOB ___ BROWN'S FERRY DAM
FOR ___ SPECIFIC PLANT ANALYSIS
PAGE 4 of 9 OF ___
SHEET ___ OF ___
EST NO ___
BID DATE ___

JOB ___ BY ___ DATE March 1, 1970

ITEM NO.	DESCRIPTION		QUANTITY	UNIT	LABOR (1)	INSTALLED EQUIP. (2)	INITIAL FREIGHT (3)	EQUIP. DEPREC. (4)	SUPPLIES (5)	SUBCONTRACT (6)	TOTAL COST	ITEM NO.
.054	Structural Steel	Control	114	ton	10,900	10,934	3,300	1,320	2,480		14,700	
		Actual			6,299			779		627	31,238	
		(unit)			55.25	95.94	28.95	6.83	81.57	5.50	274.04	
.055	Machinery Installation	Control			13,392	93,730	6,650	1,848	2,908		118,528	
		Actual			6,234	150,829	8,457	113	4,573		170,206	
.056	Electrical Installation	Control			7,000				1,000		8,000	
		Actual			7,013	2,861		49	10,196		20,119	
.057	Air and Water	Control			0						0	
		Actual			3,885			100	2,177		6,162	
.058	Enclosures and Stairs	Control			6,000				6,000		12,000	
		Actual			6,757			208	9,591	233	16,789	
	Total Mixing Plant	Control			43,792	93,730	6,650	4,918	16,388		165,478	
		Actual			34,900	164,624	11,757	1,614	40,070	860	253,825	
	CEMENT HANDLING AND STORAGE											
.061	Excavation and Grading	Control	100	CY	50				13		75	
		Actual			557			108	211		896	
.062	Concrete	Control	50	CY	750				500		1,375	
		Actual	72	CY	3,175			125 / 165	4,196		7,536	
		(unit)			44.10			2.29	58.28		104.67	
.063	Silo Erection	Control	48	ton	4,938	7,600	2,400	892	1,192		17,022	
		Actual	45	ton	4,739	10,290	2,257	701	6,361	250	24,598	
		(unit)			105.31	228.66	50.16	15.58	141.36	5.56	546.63	
.064	Elevator, Screw and Topper	Control	25	ton	2,500	7,750	1,250	300	600		12,400	
		Actual			4,206	15,612	822	371	5,711		26,752	
.065	Electrical Installation	Control			250	0		0	100		350	
		Actual			222			5	119		346	
.066	Air and Water	Control			0	0		0	0		0	
		Actual			253			2	81		336	
.067	Enclosures and Stairs	Control			0			0	0		0	
		Actual			46			3	12		61	
	Total Cement Handling	Control			8,488	15,350	3,650	1,327	2,407		31,222	
		Actual			13,198	25,902	3,079	1,355	16,721	250	60,505	

2191

VALUES USED ARE FICTITIOUS FOR ILLUSTRATION ONLY

Figure 13-32-10 (Continued)

JOHN DOE CONSTRUCTION COMPANY, INC.
SPECIAL PLANT REPORT

JOB _____ BY _____ DATE March 1, 1970

JOB BROWN'S FERRY DAM
FOR SPECIFIC PLANT ANALYSIS
BID DATE _____

VALUES USED ARE FICTITIOUS FOR ILLUSTRATION ONLY

ITEM NO.	DESCRIPTION	QUANTITY	UNIT		(1) LABOR	(2) INSTALLED EQUIP.	(3) INITIAL FREIGHT	(4) EQUIP. DEPREC.	(5) SUPPLIES	(6) SUBCONTRACT	TOTAL COST
2191	**FLY ASH HANDLING AND STORAGE – JOB**										
.071	Excavation and Grading			Control	50			10	15		75
				Actual	513			98	180		791
.072	Concrete	79	CY	Control	750			125	500		1,375
				Actual	2,681			94	3,526		6,301
				(unit)	*33.94*			*1.19*	*44.68*		*79.76*
.073	Silo Erection	27	ton	Control	2,700	7,902	700	360	540	150	12,202
				Actual	1,609	9,959	1,356	346	3,265		16,685
				(unit)	*59.52*	*119.07*		*22.81*	*20.93*	*5.56*	*617.89*
.074	Elevator, Screw and Hopper			Control	3,125	9,640	360	325	650		14,100
				Actual	2,654	9,796	434	175	3,521	98	16,618
.075	Electrical Installation			Control	250				100		350
				Actual	106	0		5	81		192
.076	Air and Water			Control	0			0	0		0
				Actual	106	0		3	182		291
.077	Enclosures and Stairs			Control	0			0	0		0
				Actual		0			99		99
	Total Flyash Handling			Control	6,875	17,542	1,060	820	1,805		28,102
				Actual	7,669	15,755	1,790	721	10,854	188	40,977
2191	**COOLING PLANT**										
.091	Excavation and Grading	1,500	CY	Control	1,500			300	450		2,250
				Actual	2,360			195	261		2,834
.092	Concrete	100	CY	Control	3,000			500	2,000		5,500
				Actual	8,638			137	7,504		16,679
				(unit)	*36.60*			*0.58*	*33.49*		*70.67*
.093	Structural Steel and Tanks	236		Control	2,000			0	18,000		20,000
				Actual	8,636	1,216		256	32,330	40	42,478
.094	Gates, Feeders and Chutes			Control	0	0		0	5,000		5,000
				Actual	952	7,956	372	33	955		10,268
.095	Machinery			Control	20,000	151,147	6,260	2,000	99,006		278,407
				Actual	8,228	182,870	7,619	145	33,861	173	232,876

Figure 13-32-10 (Continued)

JOHN DOE CONSTRUCTION COMPANY, INC.
SPECIAL PLANT REPORT

JOB BROWN'S FERRY DAM
FOR SPECIFIC PLANT ANALYSIS

BY _____ DATE March 1, 1970 _____ BID DATE _____

VALUES USED ARE FICTITIOUS FOR ILLUSTRATION ONLY

ITEM NO.	DESCRIPTION	QUANTITY	UNIT	LABOR (1)	INSTALLED EQUIP. (2)	INITIAL FREIGHT (3)	EQUIP. DEPREC. (4)	SUPPLIES (5)	SUBCONTRACT (6)	TOTAL COST
.096	Piping — Control			3,000	0		0	12,000		15,000
	Actual			32,027	2,201		1,110	32,012		67,350
.097	Electrical Installation — Control			7,000	0		0	9,000		16,000
	Actual			8,931	4,767	174	179	9,752		23,803
.098	Enclosures and Insulation — Control			23,600			0	25,600	10,094	49,200
	Actual			7,163			325	28,449		46,031
.099	Cooling Tower — Control			0			0	12,000	604	12,000
	Actual			2,391			109	13,895		16,999
	Total Cooling Plant — Control			60,100	151,147	6,260	2,800	183,050	10,911	403,357
	Actual			79,326	199,010	8,165	2,487	159,399		459,298
2191	AGGREGATE STOCKPILING SYSTEM									
.101	Excavation and Grading — Control	1,100	C.Y.	600	0		220	330		1,150
	Actual			6,411			2,450	7,376		16,237
.102	Concrete — Control	240	C.Y.	7,200	0		1,200	4,800		13,200
	Actual	455	C.Y.	30.69 / 13,966			0.73 / 331	39.43 / 11,941		0.85 / 32,238
.103	Dump Hoppers — Control	0	each	0			0			
	Actual	5		1,616	2,007		137	6,988		10,748
.104	Bin Dividers and Bulkheads — Control			600	0		0	840		1,440
	Actual			3,171			150	6,532		9,853
.105	Structural Steel (Bents, Trusses-Control Chutes)			2,438	0		322	9,292		12,052
	Actual			11,348			834	30,045		42,227
.106	Conveyor Machinery (Belting, Drives and Idlers) — Control			10,880	26,960	1,700	816	3,696		43,552
	Actual			21,648	44,439	2,339	1,005	33,750		103,181
.107	Electrical Installation — Control			1,000	0		0	1,000		9,218
	Actual			6,496			89	2,633		9,218
.108	Air and Water — Control			0	0		0			0
	Actual			1,213			54	1,468		2,735

VALUES USED ARE FICTITIOUS FOR ILLUSTRATION ONLY

Figure 13-32-10 (Continued)

JOHN DOE CONSTRUCTION COMPANY, INC.
SPECIAL PLANT REPORT

JOB ____ BY ____ DATE March 1, 1970

JOB | BROWN'S FERRY DAM

FOR SPECIFIC PLANT ANALYSIS

PAGE 7 of 9

SHEET ____ OF ____

EST NO. ____

BID DATE ____

ITEM NO.	DESCRIPTION		QUANTITY	UNIT	(1) LABOR	(2) INSTALLED EQUIP.	(3) INITIAL FREIGHT	(4) EQUIP. DEPREC.	(5) SUPPLIES	(6) SUBCONTRACT	TOTAL COST	ITEM NO.
109	Enclosures and Stairs	Control			1,000				4,000		5,000	
		Actual			423			2	4,541		4,966	
110	Re-laim Tunnel	Control	0	lf		0		0	0		0	
		Actual	324		6.47 / 20 97			31 5	65.15 / 21,107		72.60 / 23,519	
111	Gates and Feeder	Control	3	each	400	4,064	200	0	900		5,564	
		Actual	13	each	185.53 / 2,412	402.38 / 5,231	21.15 / 27 5	88	148.92 / 1,936		764.77 / 9,942	
112	Rock Ladder	Control	2	each	0	0		0	0		0	
		Actual			91			6	3,554		3,651	
	Total Stockpile System	Control			24,118	30,524	1,900	2,558	24,858		83,858	
		Actual			70,892	51,677	2,614	54 61	137,871		268,515	
2191	**RESCREEN AND WASHING PLANT**											
121	Excavation and Drainage	Control	0		50			10	15		75	
		Actual			186			24	89		299	
122	Concrete	Control	0	cy	200			50	300		550	
		Actual	25		851			0.48 / 12	23.12 / 57 8		57.64 / 1,441	
123	Structural Steel (Bents, Chutes, and Trusses)	Control	25	ton	2,332	2,928		308	9,208		14,776	
		Actual			3,653			155	19,073		22,881	
124	Screen Installation	Control			3,750	34,760	1,000	1,500	750		38,760	
		Actual			2,821	11,565	609	71	3,960		19,026	
125	Conveyor Machinery	Control			0	0		0	0		0	
		Actual			523			35	1,122		1,680	
126	Air and Water System	Control			500			0	500		1,000	
		Actual			1,830			64	945		2,839	
127	Electrical System	Control			750			100	600		1,450	
		Actual			5,291			31	1,467		6,789	
	Total Rescreen and Washing	Control			7,582	34,688	1,000	1,968	11,373		56,611	
		Actual			15,155	11,565	609	392	27,234		54,955	

Figure 13-32-10 (Continued)

252

JOHN DOE CONSTRUCTION COMPANY, INC.
SPECIAL PLANT REPORT

BY _____ DATE March 1, 1970

JOB _____ BROWN'S FERRY DAM
FOR SPECIFIC PLANT ANALYSIS

PAGE 8 of 9
SHEET ___ OF ___
BID DATE _____
EST. NO. _____

VALUES USED FOR ILLUSTRATION ONLY

ITEM NO.	DESCRIPTION		QUANTITY	UNIT	LABOR (1)	INSTALLED EQUIP. (2)	INITIAL FREIGHT (3)	EQUIP. DEPREC. (4)	SUPPLIES (5)	SUBCONTRACT (6)	TOTAL COST
2191	AIR PLANT										
.131	Excavation and Grading	Control			100			50	50		200
		Actual			320			47	138		505
.132	Concrete	Control	60	C.Y.	1,200			600	1,500		3,300
		Actual	18	C.Y.	1,143			16	735		1,694
			63.22					0.89	40.83		104.94
.133	Compressor Installation	Control	3,300	C.F.M.	4,112	28,618	1,722	1,008	2,928		38,388
		Actual			6,786	85,256	4,487	187	4,586	78	101,380
.134	Install Valves and Headers	Control	4,000	L.F.	2,520	3,200	800	280	3,280		10,080
		Actual			5,977	2,776	652	253	12,201		21,859
	Total Air Plant	Control			7,932	31,818	2,522	1,938	7,758		51,968
		Actual			14,226	88,032	5,139	503	17,660	78	125,638
2191	WATER PLANT										
.141	Wells	Control	0		0	0		0	0		0
		Actual			416			69	451	1,274	2,210
.142	Pumps & Chlorinator Install	Control	3	Ea.	1,500	17,900	600	500	500		21,000
		Actual			2,209	21,963	1,156	168	1,422		26,918
.143	Reservoirs and Tanks	Control	2	Ea.	1,180	9,500	600	220	1,420		12,920
		Actual	1	Ea.	556			170	786	9,771	11,283
.144	Main Headers	Control	7,000	L.F.	3,700	13,065	1,200	350	11,100		15,150
		Actual			7,407			877	18,498	105	39,952
	Total Water Plant	Control			6,380	27,400	1,200	1,070	13,020		49,070
		Actual			10,568	35,028	1,156	1,284	21,157	11,150	80,343

Figure 13-32-10 (Continued)

JOHN DOE CONSTRUCTION COMPANY, INC.
SPECIAL PLANT REPORT

JOB ___   BY ___   DATE March 1, 1970

JOB BROWN'S FERRY DAM
FOR ___   BID DATE ___

VALUES USED ARE FICTITIOUS FOR ILLUSTRATION ONLY

No sub-breakdown furnished on this plant. The value of salvageable equipment shown under quantity column.

ITEM NO.	DESCRIPTION	QUANTITY	UNIT	Cost of Salvageable Items	LABOR (1)	INSTALLED EQUIP. (2)	INITIAL FREIGHT (3)	EQUIP. DEPREC. (4)	SUPPLIES (5)	SUBCONTRACTS (6)	TOTAL COST
	AGGREGATE PLANT										
151	Excavation and Grading				9,801			2,417	10,929		23,147
152	Concrete	573	cy		14,753			616	23,466		38,835
								1.07	40.95		67.78
153	Structural Steel				22,081		2,190	2,316			30,700
154	Conveyor Machinery	27,810			11,378	95,155	7,162	609	49,214		163,518
155	Electrical System	4,635			14,648	36,598	747	400	14,572		66,965
156	Air Plan Installation	2,780			865		220	21	779		1,465
157	Water Plant (To Site)	12,980			5,947	13,867	1,020	1,582	7,772		30,211
158	Water at Plant Site	1,000			2,649		1,043	388	3,123		6,160
159	Reclaim Tunnels				1,749		431	152	14,230		16,131
160	Truck Loading Hoppers				4,369	5,719	1,825	305	24,610		35,434
161	Cribbing & Wall Primary Crusher				1,030		45	20	45		1,095
162	Primary Crusher Installation	23,175			3,382	80,712	4,247	297	2,596		94,434
163	Secondary and Fine Crushing	37,090			3,130	97,788	5,147	347	6,269		112,681
164	Screens and Feeders	6,500			1,464	46,317	2,438	134	4,946		55,299
165	Rod Mill	38,517			3,113	47,168	3,110	326	6,933		61,365
166	Sand Plant Machinery Installation	8,350			6,274	33,158	1,745	871	16,599		58,647
167	Enclosures and Walkways				5,103			126	4,543		9,972
168	Drainage and Rod Mill Waste				309		650	148	365		822
	Total Aggregate Plant	162,827			74,760	162,827	12,800	11,075	179,225		429,612
					112,245	456,482	24,785		275,725		882,312

Figure 13-32-10 (Continued)

JOHN DOE CONSTRUCTION COMPANY, INC.

CONCRETE FORM COSTS
BROWN'S FERRY DAM

Thru March 26, 1972

		FORMED AREA–S.F.		LABOR			SUPPLIES			TOTAL DIRECT COST		
		THIS MONTH	TO DATE	UNIT COST THIS MONTH	UNIT COST TO DATE	CONTROL	UNIT COST THIS MO.	UNIT COST TO DATE	CONTROL	UNIT COST THIS MONTH	UNIT COST TO DATE	CONTROL
32	CONC. — MASS											
.11	BUILT IN PLACE	4,711	81,301	1.26	.93	1.30	.83	.65	0.65	2.09	1.58	1.95
	REUSABLE WOOD FORMS											
.13	Shop Fabricate	1,505	22,119	.64	.33		.30	.48		.94	.81	
.14	Set & Strip	3,344	62,193	1.24	.92		.29	.40		1.53	1.32	
	TOTAL			1.36	1.04	0.42	.46	.57	0.25	1.82	1.61	0.67
	SPECIAL WOOD FORMS											
.15	Shop Fabricate	6,406	87,967	.39	.34		.59	.54		.98	.88	
.16	Set & Strip	6,946	93,114	1.32	.84		.52	.37		1.84	1.21	
	TOTAL			1.64	1.16	1.20	1.03	.88	0.54	2.67	2.04	1.74
	STEEL CANTILEVER PANELS											
.17	Purchase & Assemble	64,850	435,347				.29	.29		.29	.29	
.18	Set & Strip	64,850	435,347	.29	.27		.16	.16		.45	.43	
	TOTAL			.29	.27	0.42	.45	.45	0.25	.74	.72	0.67
	SPECIAL STEEL FORMS											
.19	Purchase & Assemble	2,628	13,405				.65	.65		.65	.65	
.20	Set & Strip	2,628	13,405	1.74	1.14		.25	.24		1.99	1.38	
	TOTAL			1.74	1.14	1.52	.90	.89	0.72	2.64	2.03	2.24
	TOTAL ACCOUNT 7032	82,479	685,350		.55	0.59		.56	0.32		1.11	0.91
	FORM RATIO Control (1.422)	1.48	1.42									

NOTE: Concrete items 27 and 31
100% Complete as of
November 30, 1970

Figure 13-32-11 Monthly report—concrete form costs

14 COST ESTIMATING FOR LUMP-SUM AND UNIT-PRICE CONTRACTS

J. P. FREIN

Vice-President and Director (Retired)
Morrison-Knudsen Company, Inc.
Boise, Idaho

This presentation does not purport to offer values of cost for pricing construction work. Its intent is to relate basic concepts and principles gained from study and experience in the development of estimating procedures which have been successfully applied, as well as to suggest useful formats.

Estimating construction costs, like other phases of the business, has undergone extensive changes both in refinement and the development of greater accuracy. The construction industry was possibly one of the last in the business community to follow the trend toward greater scientific development of its operations. In the past, it was envisioned that unit bid prices and costs for given kinds of work fell within a certain range for different areas. Estimating consisted largely of the estimator applying his experience to similar situations and assigning prices with little computation to back them up. Construction methods were somewhat standard. Wage rates and prices for materials and supplies fluctuated but little from time to time and place to place. Labor's productivity was assured. More work was done by hand and labor was plentiful. Competition in the construction industry was not particularly strong by present-day standards.

As time has passed, labor costs have risen and vary widely from area to area. Working rules tend to reduce productivity; and fringe benefits, which also vary from area to area, must be considered carefully in figuring the total labor bill. Competition increased, as has the number of contractors.

Mechanization to combat rising labor rates increased as time went on. It has continued to accelerate in what has become an ever-growing effort to cut costs to meet the rugged competition prevailing in the industry. Today's successful contractor is the one who can conceive of a better way to do the work, develop new methods of performance that can be relied upon, and exercise the greatest prudence in his purchasing and commitments for services. He is continuously aware of opportunities in the construction market, is selective in his bidding, and is knowledgeable and astute in his negotiations and dealings with his clients.

The present-day estimator must be ready to adjust to changing requirements of his profession. Estimating construction costs is a very demanding endeavor, one requiring the estimator's full attention and intense study to make sure that all conditions have been properly considered and priced. Although each problem requires individual consideration, estimators are prone to look back at their previous estimates for somewhat similar situations and to be influenced by them in working out the current problem. For that reason, a proper record at the time of preparing an estimate becomes important. It is second only to resolving new situations that present themselves.

Numerous books have been written about estimating, many of them offering values for the pricing of individual operations. But proper consideration of all elements of cost comes first, and it is the purpose of this presentation to offer a procedure and format for doing so. If it is not complete, it is because all situations cannot be seen in advance. However, it suggests a plan of systematic study that should be helpful in recognizing the important factors of estimating.

Obviously the prices at which new business is obtained —and therefore the procedures for figuring them—are of prime importance in the question of whether the business is a success or a failure. The management and control of the operations are equally important, but the management of the estimating must come first. Because the estimating is essentially a laborious detail, it is too often slighted by the top management that has not had the oportunity or the time to become informed about the contract conditions or all of the cost elements that must be considered. Someone with suitable ability and temperament must direct the estimating.

Estimating construction costs, especially those of a diversified project, consists of building the job on paper, including careful programming, selection of plant and equipment, evaluation of production possibilities, the formation of the construction crews, as well as assigning a price to all of the elements. It is often desirable for regular field personnel to assist in the estimating of prospective work, preferably the intended project manager. Field personnel, although knowledgeable about field operations and helpful in advising on possible day-to-day job problems, often are not temperamentally suited to the detailed considerations involved in the study needed for accurate pricing. They are subject to new job assignments, and their availability for estimating is sometimes uncertain. Therefore, highly competent estimating personnel and competent supervision are essential.

Since this chapter is devoted to the estimating of firm-price work, it presupposes that plans and specifications for the work are to be complete and sufficiently comprehensive to permit a detailed analysis of the work and that the specific requirements of the client will be spelled out in the tender documents. Where this condition is not met, it is assumed that a cost plus, target estimate-incentive fee, or other type contract, discussed elsewhere, will be applied.

14-1. Organization for Estimating Department

Many factors enter into the question of how an estimating department should be organized. They include the contractor's degree of specialization or, conversely, how diversified he might be, whether activities are localized or widespread, the size of the business, etc. The more specialized and localized the business is, the more that current and previous cost experience will apply, thus minimizing the problems of accurately estimating prospective new business. Under some situations, the contractor himself might adequately and efficiently serve the estimating function without detracting from his additional functions of management of the company.

As the size of the business increases, however, the estimating function is invariably delegated to others who become specialists in that line. Except for routine matters, a greater degree of knowledge and experience becomes necessary as the demand becomes more complex. Obviously there is no fixed rule for organizing an estimating department that applies to all contracting firms, except to do so on the most efficient and economical basis. The following comment is directed toward a medium-to-large size, well diversified, somewhat dispersed contracting organization. The suggested procedures can be scaled down to fit smaller companies.

Quantity Survey and Take-off Work   This is a subsidiary function to the estimating or actual pricing. For a firm-price, lump-sum contract, its accuracy is essential since it establishes the amount of work to be priced and for which payments will ultimately be received. Items omitted result in loss of revenue to the company. For firm unit-price contracts, it is equally important because it forms the basis for the amortization of plant and equipment, and it enters into the application of construction methods and many other facets of the planning. For unit-price contracts, the accuracy of the total units of work to be performed under the various items determines the scope and magnitude of the proposed contract. If quantities are not stated in the tender documents, they must be computed before the estimate and plans for pricing can be established. Even if the tender documents state quantities for a unit-price job, such quantities should be verified. They are generally regarded as being only approximate and inaccuracies in them would distort the programming, plant requirements, and other factors affecting the pricing.

Takeoff work requires a good understanding of the tender documents. That is especially true of the plans and specifications, which tend to become more technical as time goes by. Takeoff work should be done in such a way that the computations can be subsequently followed for verification. Computation sheets should be dated and bear the initials of the individual performing the work. They should include a brief description of the part involved, illustrated by dimensioned sketches, and should show the plans and specification references.

The takeoff work should be planned and performed to fit the plan for pricing the item in question. The estimator pricing the work should direct the takeoff work, if he does not actually perform it himself. His time can be better used if he delegates the takeoff to an assistant and directs it in some detail. Aside from computing or verifying the quantities of work under the various items, takeoff work also requires determining the amounts of the various kinds and classes of work, such as the different kinds of concrete forms involving different pricing, the computation of the amounts of temporary shoring, haul distances and grades for hauls by dump trucks, and numerous other things necessary to support the detailed estimate breakdowns.

Drafting and miscellaneous engineering computations and design are generally part of the takeoff work and are performed by the same individuals. The personnel for such functions must be capable and have the necessary technical training.

Takeoff departments provide a desirable place for inexperienced engineers to break into the construction business. They will become acquainted with the basic elements of what forms part of a complete construction proposal, studies of construction methods, and plant studies. These are essentials involved in carrying on a construction business, and such experience admirably fits beginners for more important functions later.

The Selection of Estimating Personnel   Individuals who price construction projects must be temperamentally suited, honest, and experienced in the work involved in the estimate. If the estimator has not had first-hand experience in the work, he should be a competent construction engineer capable of accurate analysis of construction problems as a whole, and of proper evaluation of the references available to him.

No amount of theory can provide full appreciation of job conditions, "tricks of the trade," alternative approaches to problems, the interferences that sometimes hinder the process of getting the work done, and the amount of leeway that should be allowed under different circumstances. Theoretical answers often are impractical and sometimes are overly optimistic.

Besides practical experience, the responsible estimator should be emotionally suited to intense study and analysis of the problems he will have to solve. Casual or superficial consideration leads to possible gross error unless the subject is a highly repetitive one and uncomplicated by nature.

Another quality the estimator must have is honesty in representing both his own ability to evaluate a given situation and confidence that he has studied the matter adequately. Slipshod estimating to save face is an inexcusable characteristic, and one that cannot be tolerated. Beware of the individual who provides quick but inaccurate answers.

Estimating personnel thus should be selected for their practical background and personal qualities. Those with project engineering backgrounds are practically suited for general construction estimating or for estimating specialized kinds of work in which they are experienced, such as underground work and cross-country pipelining. Specialists in mechanical and electrical fields also are necessary for evaluating their specialties or for the selection, designing, and pricing of construction plants involving their specialities.

Persons with backgrounds as project managers, and speciality supervisors, often are competent estimators because of their skills in planning and organizing construction operations. Although many in this class avoid the paper work necessary in the compilation and preparation of an estimate, they still can contribute their experience to the process.

Typical Estimating Organizations Estimating is generally recognized to be an engineering responsibility and therefore would come under the direction of the principal engineering officer of the company, presumably the chief engineer, in an overall capacity, or district or division engineers under the direction of the chief engineer for a company with widely dispersed operations.

It is presumed that the activities of an engineering office (head office or district office) are sufficient to support a regular estimating staff consisting of technical people to make the necessary quantity surveys and takeoff and engineering studies. This group would be supervised by an officer engineer who would assign the work and would coordinate it with the estimators responsible for the ultimate pricing. The office engineering group would look to the responsible estimator of each job to outline the form and detail in which the takeoff information should be prepared to fit the plan of pricing.

A group of estimators would be available for the detailed preparation of estimates as they are required, for specific parts of estimates according to the special talents of the estimator, or for the assembly and coordination of the various parts of estimates. Estimators cannot be proficient in all kinds of construction work, particularly in acknowledged specialties. The broader the estimator's background, knowledge, and experience, however, the greater his value to an estimating group. The principal engineering officer would be responsible for evaluating whether the estimators are capable of properly evaluating the specific problems involved. If questions arise on special problems beyond the capabilities of the estimators, he would be obliged to hire special talents or find other ways to answer them.

In the general contracting field, estimators tend to develop capabilities along specific kinds of work, such as buildings, heavy excavations, heavy concrete work, mechanical installations, plant and equipment design and pricing, tunnel and underground works, utility work, grading and paving, and cross-country pipeline work.

Some estimators develop special abilities in the evaluation of work overseas. Proper organization for estimating consists largely of selecting the right man for each job.

Smaller companies, and particularly those following specialized lines, will have less complicated estimating organizational problems. The owner of the business could do all the estimating, or he could employ several estimators of proven qualities to perform it. The larger and more diversified a company becomes, the greater is the problem of organizing the estimating functions.

14-2. Estimating Manual

A company with a number of estimators and division, district, or area offices should establish guidelines for preparing estimates and should determine their formats. An estimating manual is an essential means of control to tie the function into other company activities.

The estimating manual should do the following:

1. Serve as an indispensable check list and guide to estimating engineers making site investigations and accumulating basic data for prospective work.
2. Assist in the training of estimating engineers.
3. Minimize the effort necessary to supervise the preparation of estimates and to relate them to basic data from the files.
4. Set up a step-by-step procedure for estimating to establish a sequence and format that can be easily read and followed by other company estimators.
5. Direct the estimator in the proper breakdown of costs to parallel the established company accounting system, thus allowing a direct comparison of the costs reported later during construction.
6. Direct the project engineers and others on the work in the preparation and interpretation of company estimates and cost reports.
7. Assist in summarizing various parts of the estimates into their various elements for analysis and comparison with other estimates.
8. Assist in establishing the cost-accounting breakdown for new work for the benefit of all concerned.

Standardized estimating, even for one individual, is a decided advantage. A standard form facilitates references from one estimate to another and often eliminates the need for reconstructing a step-by-step analysis of similar problems. It helps estimators understand what is presented, know how the figures are constituted and what they include, and accept the figures with some degree of confidence. For the supervisor reviewing an estimate before preparing a bid, it can save valuable time that he might otherwise spend going through it detail by detail.

Printed forms establishing a format for the various parts of an estimate can help in controlling the estimating activities where a large number of people are involved. They also result in a substantial time saving, since it is not necessary to write out all of the details involved in the estimate. Printed forms should be approved and adopted by the proper authority, avoiding the possibility that others may draft different forms and that the effect of standardization is lost. If computers are used in the estimating, standardization is a necessity because of the preprogramming necessary for their adaptation.

In Section 13-4 entitled "Functions and Duties of the

Contractor's Engineering Staff" a plan is described for using the same basic arrangement for the prebid estimate, the budget-control estimate, and the reports of actual cost prepared by the job so that the information in all of them would be comparable. This emphasizes the need for estimate standardization.

Veteran estimators might object to standardization and the application of procedures contained in an estimating manual. They might do so on the theory that it is inconsistent with what they were accustomed to in the past and that they cannot do their best under such circumstances. This is justified only if the individual is the only one involved in all of the attending considerations, but coordination is necessary when others are participating.

Burdensome and unnecessary procedures are not suggested here simply for the sake of standardization. The desirable controls and the methods of applying them must be tailored to fit each situation. Simplicity—to the extent that it produces the desired control—should be the basic ingredient. It necessarily follows, however, that the larger, more dispersed and diversified a company is, the more that controls will be necessary and justifiable.

14-3. Building vs. Heavy Construction Estimating

Serious mistakes have been made in failing to differentiate between estimating for standard building construction and diversified heavy construction. To be sure, there are similarities between the operations required for both of them, but there are many dissimilar conditions that make the units of work for heavy construction cost more than for building construction. This is true with the possible exception of mass excavations and the like, or where traffic conditions seriously affect the work.

Exotic architectural features of buildings and foundation conditions need special considerations. However, most building construction details follow a somewhat accepted standard, using the same materials, methods, and labor crafts. Usually there is repetition of the same operations— site to site, story to story, floor to floor—and the greater the repetition the greater is the possibility of efficiency. The work is almost invariably situated close to the source of materials, services, and transportation. Factory representatives of the materials and facilities are more readily available to service their accounts and assist in general.

Labor is probably the biggest element of difference. Competent craftsmen, although needed for both building and heavy construction, usually find ready employment on buildings in cities. They are inclined to take advantage of the opportunity for a more permanent home, shorter travel time and distance, and good pay and working conditions.

Much building work is performed by local contractors, or outside contractors that operate locally. Over the years, they have developed a supervisory organization and have a labor following. Such connections have many advantages for workmen and result in better teamwork, organization, and labor productivity. There are many advantages to the contractor as well, including an intimate knowledge of the labor market, a firsthand association with labor officials, and the chance of better relations with the unions.

By contrast, heavy construction often is remotely situated. It is usually handled by contractors who do not normally operate in the area. The work is usually some distance from the source of materials, facilities, and labor. The contractor's association with local vendors, purveyors of services and supplies, and the labor market and its leaders, often begins with and ends with each project.

The shorter the job the less is the opportunity to develop the desirable associations and to organize greater efficiencies into the operations. The task of maintaining an efficient work force of adequate size is a substantial one.

Construction methods and machines used on building projects are standardized to a greater degree than for heavy construction. The machines are generally of smaller capacity and less complex. The work is more or less ordinary and highly repetitive. Experienced specialists and specialized devices are readily available.

Almost invariably, there is a great difference in the cost of concrete forms for building and for heavy construction. Usually the concrete sections are less massive for buildings, the concrete placing rates are slower, and the concrete goes into the forms with less impact than for heavy work. Lighter forms with simplified tying systems therefore are more readily adaptable, all contributing to lesser costs. Since some of the formed concrete surfaces for buildings are to be covered later by exterior materials, such as masonry, plaster and prebuilt panels, tolerances, surface finish, and exacting workmanship required for exposed heavy construction surfaces are not necessary. This permits less sophisticated form construction and less effort for their alignment. Sometimes it seems that the form specifications for heavy construction are unusually stringent, but such is the industry's practice. The same differences exist in other construction operations in the two fields.

Although most construction operations can be analyzed theoretically, the only true indicator of the difference between similar work in building and heavy construction is historical cost experience, including all of the elements. When no actual experience or historical cost background is available, a theoretical approach must be used, allowing for all of the factors entering into the particular operation relying upon estimated values.

14-4. Historical Cost Background for Estimating

Regardless of the kind, size, scope, or dispersal of a contractor's operation, successful pricing of prospective business naturally goes back to how well the estimator knows his business and the cost factors entering into it. The most reliable information on this subject can be found in the history of actual previous costs. It reflects all the elements involving management and business practices, the efficiencies —or lack of them—in the supervision and organization of the operation, labor productivity, etc. Needless to say, historic cost information must be adjusted for differences in wage rates, location, and other conditions affecting cost. But the basic historical cost record is the most reliable source to start from, and in the construction industry it constitutes a valuable part of the contractor's stock in trade.

To accumulate historical cost background, it is necessary to have had actual practical experience in the various lines of work in question, preferably by the construction company involved. If not the company, then certainly the individual or specialist handling the matter must have had the experience. Many things in construction are unique, or peculiar to a job or location, but some basic principles for which there is precedent will almost certainly apply. In the development of construction methods involving new and untried devices, knowledge and experience in the general field, including the cost, must be a consideration in the pricing of the prospective work.

Historical cost information must be reliable and should be understood in detail by the individual interpreting and using it. It is dangerous to assign bid prices for work based upon cost reports by others unless the particulars of their makeup are known. Even an estimator's own cost reports become questionable in time unless they are supplemented by sufficient description and documentation.

All too often, because of the need to be competitive, estimators and bidders will be influenced by prices for construction work appearing in published abstracts of bids. Keeping in step with economic ranges of the market is vastly important for continued survival, but accepting unsupported prices might lead to an accelerated business failure. Indemnity companies can give ample testimony to this. It does not follow that because one party feels that he can perform work for his price (and perhaps he can), another party can or will be able to do so also. A better reason must be found for undertaking the substantial financial exposure that goes with bidding a firm price job.

A number of books on construction costs purport to provide unit costs of the work and sometimes production rates and labor productivity. They doubtless are based upon research and possibly the author's experience for a particular set of circumstances, a particular location, and a particular degree of productivity of labor. All are variable elements, however, so their tabulated pricing can only lack accuracy. If the item being priced is unfamiliar and not of much consequence, such "cookbook" pricing will have little effect upon the overall financial outcome of the project. Coming up with a general price range is better than none at all.

Organization of a good, usable record of cost and production experience presents problems, especially when it will be used by a number of people. Project reports are necessarily the basic source of information emphasizing the importance of planning and directing the preparation of job reports—interim as well as final. It is often desirable to classify the information in project reports by similar or like operations for ready comparison of experience—one place to another—and to set the selected data down in a way that future experience can be added. Production rates and man-hour data are an important part of the record of cost data.

It would be burdensome to classify and record all or even a part of the historical background experienced in the diversified heavy construction field. Even if it were possible, numerous details would never be of further interest. Therefore, the data to record should be carefully selected on the basis of how frequently it might be referred to in the future, how important an operation might be in the pricing of prospective work, and how readily it can be classified. Although each case must be judged upon prevailing circumstances, the following are suggested as items that might be included in the historical record of the company:

1. Equipment hourly operating costs for each category or similar major units
2. Costs for major items of construction plant—first cost, erection and operation costs—such as crushing plants, aggregate processing plants, concrete batching, mixing, precooling and delivery systems, cableways, high-lift cranes, concrete placing trestles, etc.
3. Costs for concrete forms of various kinds by separate categories; for example, wood, steel, straight, and curved
4. Pile-driving costs for piling of different types
5. Erection costs for different kinds of steel
6. Machinery erection costs
7. Cross-country pipelining costs for different operations and conditions
8. Costs of blast hole drilling for major rock excavations

9. Costs for various kinds of explosives and methods of blasting
10. Tunnel excavation, including various operations and elements of cost involved
11. Tunnel concrete, including various systems of forming, conveying, and placing the concrete
12. Steel reinforcement, including various kinds of fabrication and erection problems for different types of structures.

The length of such a list depends on the needs of each company. A highly qualified estimator will develop a definite competence in using references provided to him and will compile additional ones based upon his own experience.

Cost and production references can be programmed into a computer by classifications and produced on printout sheets to serve a specific purpose. This subject is discussed in Chapter 19.

14-5. Initial Considerations in Planning an Estimate

Estimating construction work costs money and consumes the time of people—the top management as well as the engineering and the estimating personnel carrying most of the burden. Simple jobs and those conveniently located require less time and expense to estimate, and therefore require less critical scrutiny from the standpoint of deciding whether or not to bid. Basically, however, it is wasteful to spend any effort analyzing and estimating prospective work, large or small, promising little or no possibility of being secured on a bid basis or involving hazards out of proportion to the investment. The ultimate decision of whether to bid usually rests with top management. But analysis of the conditions leading to the decision is usually a function of the engineering and estimating groups. The following discussion is sufficiently broad to cover the analysis of a large proposed construction project, but only its pertinent parts need to be considered for each problem.

The first factor to consider is the proposed financing. This should include how firm the financing is, whether it is adequate, how readily available it is, and whether advance payments for such items as mobilization and aquisition of materials are provided without requiring a disproportionate investment by the contractor. The reputations of the prospective client and his engineer—whether they are fair in their dealings and in contract interpretations—should be considered next.

A major factor in the analysis is the availability of sufficient estimating talent, a matter for the company's top engineering officers to determine. Equally important is whether the company is properly situated geographically to undertake the work, whether it has able supervisors available, and whether crews, plant, equipment, or other facilities can be assigned to the work if the bid is successful. Another question is whether it is wise to utilize the necessary bonding capacity and financial resources for the work under consideration, or whether they should be held for other prospective work. The severity of the competition also is an important consideration; i.e., whether a local contractor or a competitor already moved in and set up at the site has overwhelming advantages. Local labor conditions should be considered as well.

Competent staff members should be assigned as early as possible to supervise and estimate the proposed work, so they can also help in the deliberations leading to a decision on whether or not to bid.

This decision will depend in part on a general preliminary review of the contract documents. The contract conditions should be considered on the basis of their fairness to both parties, financial implications and provisions, schedule of payments, possible penalties against the contractor, procedures for appeals, means of resolving disputes, arbitration, and indemnification provisions.

A study of the plans and specifications should be subject to later detailed study to determine scope, magnitude and complexity, the contract time available and its adequacy, the completeness of the designs, plans and specifications, and whether site explorations by the client and his engineer appear sufficiently extensive to disclose project conditions.

If, based on a study of these considerations, it is decided to bid on the project, the following procedure is suggested.

The persons who will supervise or help with the estimate should make a more thorough analysis of the tender documents. This will permit them to start preparing the estimate and develop a general, if tentative, plan for performing the work. They should keep notes of points about provisions of the documents that might require further investigation.

The takeoff or quantity survey should be started and directed by the estimator to fit a concept of breakdown and pricing for the various parts of the work. He should start gathering prices and other basic information needed to proceed with the estimate. Plans should be made to investigate the site. At this stage, it should be possible to arrange for sufficient help to cover all phases of the estimate so it will be completed well before the bid date.

14-6. Site Visits and Investigations

Careful investigation of the site by the principal estimators, following preliminary study of the tender documents, is advisable. Detailed reports by others than the estimators have been used for pricing when time and distance prevent site inspections. But firsthand observations will help to answer questions later. Underwater work or work where surface observations provide no clue to its nature should be investigated by some other means.

Contract plans and the list of questions and notes prepared earlier should be referred to during the site inspection. The investigators should accurately orient themselves in relation to the location of the proposed work. They should try to visualize the arrangement or movement of the plant and equipment based on preliminary ideas of work methods. In addition, they should locate haul roads and other construction facilities, keeping in mind the clearing involved, grades, clearances, and other construction problems.

The client's engineer should be asked to help interpret the preliminary layout survey marks, to show the site and geological features, and to answer questions. If he cannot answer all questions, he usually can suggest a source for the information. Photographs of site features and important details, particularly in color, are helpful during the estimating period.

Site inspections should be as simple or detailed as the circumstances require. No all-inclusive check list of what to look for during the first site visit is possible, but the following items are suggested for a large hydroelectric development or similar project.

Transportation Facilities   Railroads with detailed locations of side tracks or available unloading areas, as well as storage areas at siding locations and their adaptability to unloading and storage facilities for cement, pozzolan, stockpiled aggregates, and other materials. The possibility and economy of locating a railroad connection at the job site should be considered.

Airlines serving the area and the location of their terminals. Location of nearby landing strips for private aircraft and locations close to the job site where strips could be developed.

If water transportation might be used, dock sites and existing unloading facilities and their capacities should also be investigated. Truck lines and contract haulers available to serve the site, as well as the kinds of equipment they have and can make available, are important information.

Public Highways and Roads   Obtain maps showing existing public highways and roads that might serve the job. Investigate their surface conditions, bridge loading, or other limitations, and the advisability and possibility of improving them.

Public Utilities   Electric power sources, the characteristics of the supply, the quantity obtainable, the possibility of outages, the location of terminals or what arrangements can be made with the supplier to provide substations, the distances from the available source to points of distribution at the job site, and rates should be noted. Natural gas service, if available, might be considered. Existing domestic or potable water services should be investigated.

Job Water Supply   A job water supply system should be considered, including its location, quantity, and the quality and condition of the source.

Meteorological Data   All weather information available from local sources. Weather Bureau statistics should be obtained from government sources, but comment by local residents is sometimes helpful.

Geological Evidence and Observations   In addition to discussing the geology of the site with the client's engineer or his geologist, the site investigator should observe the surface exposures for clues to subsurface conditions that might affect the work and study groundwater data. He also should study the available cores and samples of subsurface materials produced in the site explorations and study exposures by test pits in order to determine the sources of additional geological information if it is necessary.

Natural Materials to Handle or Process   The site investigator should make a detailed check of the characteristics of the natural materials to be handled during the work. If excavation is required, he should investigate such things as the amount of stripping, the depths to which it should be carried, the stability of materials and their probable angle of repose, susceptibility to moisture, and location in relation to the work areas.

Other considerations are the haul routes, terrain, distances, locations, and character of disposal areas or embankment sites. If rock work is involved, the investigator should determine its general character and kind, its reaction to different methods of drilling and blasting, its massiveness of formation, the amount of fragmentation possible when it is blasted, and how readily controlled drilling and blasting could produce excavations to desired line and grade in view of seams and cleavages in the formation.

Great care is normally required in investigating the natural construction materials required for the work. If commercial sources might fill the need, local suppliers should be contacted for compliance with the specifications and price information, including the cost of handling and transportation and production and delivery rates. On larger heavy construction jobs, the general contractor usually produces concrete aggregates, base course materials, riprap, derrick stone, and similar products.

If that is the case, site investigations should locate material sources. It should be determined whether they are privately owned or will be provided free by the client or owner, the amount of royalties to be paid, the difficulties of developing the sites, depths, or usable materials and unusable overburden, haul distances to the point of use, the quality and characteristics of the materials, whether the finished product can be made to comply with the specifications, etc.

Local Conditions Affecting the Work   Investigations of the site should include a study of local traffic conditions and whether they might delay the work. Consideration also should be given to steps that might have to be taken for protection of improved private property adjacent to the work and for protection of the public.

Local Labor Conditions   The local labor market should be studied to determine the available supply of all classes of labor, its competence, and the effect of work practices on productivity. Another factor to be considered is whether other work in the area will compete for the available labor.

Housing for Construction Workers   A general survey should be made of available housing facilities in the area. It should include mobile home spaces or the possibility and cost of developing them. If construction camps would seem necessary, possible camp sites should be studied on the ground.

Envisioning General Arrangement of Construction Facilities   Many factors must be considered in the job planning, much of which is done by laying out the proposed concept and arrangement on topographic and other maps of the site. Observations should be made during the initial inspection trip to determine whether the estimator's preliminary ideas are compatible with the local terrain and developments. It may be decided that a complete modification of the tentative construction concept is necessary.

Local Sources of Supply   The site investigation should include a study of the extent to which local vendors, machine shops, etc., might serve the job's needs, the proximity of principal marketing areas, and the location of parts for equipment repairs.

*   *   *

Site investigators should make adequate notes and write reports. They are important if others will use the data, or if the estimate will not be completed for some time after the site visit.

A second site visit might be necessary to obtain additional information during preparation of the estimate. One person normally can handle this assignment.

14-7.   Compiling Basic Data for a Specific Estimate

Apart from the technical and judgment factors entering an estimate, there are many things—basically economic by nature—that must be explored before making an estimate. Many will require special inquiries to develop the necessary information. These generally consist of the following elements:

- Prevailing labor rates, fringe benefits, and working conditions, plus changes provided in current labor agreements or laws
- Payroll taxes and insurance, social benefits, etc.
- Taxes not computed on payroll amounts
- Insurance not computed on payroll amounts
- Firm quotations for the principal items of construction supplies and expendable materials
- Firm prices for permanent materials and installed items of equipment
- Subcontract quotations and conditions
- Available company owned plant and equipment
- Approved plant and equipment rental rates or amortization
- Local laws affecting the cost of the work

For contractors operating locally in repetitive work, the answers to many of the above elements will be apparent to the estimator. However, they can change from place to place and from time to time. For work in new and unfamiliar areas, care should be taken to make the information current and complete. Inquiries for such information should be made promptly after tender documents are received for any project on which a bid is contemplated.

14-8.   Labor Rates, Fringe Benefits, and Working Conditions

This element of cost is probably the most variable, and information is difficult to develop. Expert advice on it is important. The best approach is to obtain copies of current labor contracts containing the information. They can be supplied by local contractors who are parties to the agreements and by labor union officials. Owner-clients sometimes negotiate labor agreements applicable to the work before the call for bids.

In some foreign countries, wages, working conditions, and fringe benefits are provided by law. Sometimes these elements are a matter of local precedent, although this condition is rapidly disappearing with the continued enlightenment of laboring classes around the world.

The labor-relations departments maintained by some companies would be best equipped to develop labor rates and conditions relating to prospective work. Otherwise a knowledgeable individual should be assigned to investigate the labor situation. Because no substantial pricing can be done until the labor information is obtained, the chief estimator or engineering officer in charge should issue a memorandum to start it as early as possible.

The various labor crafts in construction usually, but not necessarily, negotiate their own wages and conditions. It is desirable to obtain the wage and working condition information along the lines provided in the labor agreements, but reduced to abstract form for convenient use. Labor rate and working conditions information should be made available to the estimators as soon as possible. Figure 14-8-1 suggested as a satisfactory format for this information. Separate listings for each craft should be made on sheets arranged as shown in the figure.

Sheet ....of ....

Date ....

BASIC LABOR DATA FOR ESTIMATE

Project ....

Location ....

Kind of Project ....

Craft and Classification	Hourly Wage Rate		
	Current to	Current to	Remarks

LABORERS

General

Common
Air tool Operator
etc.
etc.
etc.

Conditions

Foreman: Base rate plus $...
Health & welfare ....per hr
Shift arrangement ....hrs work
Travel time pay ....per day
Subsistence ....per day
Travel expense ....per day
Other

Foremen's rates are usually used as the base rate for each craft and classification, plus an additional rate per hour. Shift arrangement is generally stated as the number of hours worked per shift for a full shift's pay of 8 hours, such as 8–7½–7. This means the day shift works 8 hours, the swing shift 7½, and the graveyard shift 7 hours, with an allowance of a 30-minute lunch period in addition. Overtime is usually classified as time and a half or double time, and the condition for work on Saturdays, Sundays, and holidays should be shown under this heading. Travel time, subsistence and travel expense, if they apply, usually involve a per diem allowance. Any other considerations should also be described.

Following are crafts and classifications usually found on United States construction, which should be listed separately:

Laborers

General-common
　Air tool operator
　Dump man
　Pipe labor—metallic culvert
　　　　　nonmetallic
　Timber faller
Concrete laborer
　Mason or plaster tenderer
　Vibrator operator
　Form stripper
　Steel form handler
　Sand blaster
　Gunite—nozzleman
　　　　Gun tender
Rock work—high scaler
　Jackhammer operator

Wagon drill operator
Chuck tender
Diamond drill operator
Powderman
Powderman helper
Riprap placer
Underground (free air)
　Shifter
　Powderman
　Miner—mucker
　　　　driller
　　　　chuck tender
　　　　timberman
　Steel form handler
Paving
　Form setter
　Fine grader
　Form stripper
　Asphalt raker

Operating Engineers

Crawler tractor w/attachments over 100 hp
　　　　　　　　　　under 100 hp
Motor Scrapers_____cy. to_____cy.
Rubber-tired tractors w/attachments—earth mover
　　　　　　_____cy. to_____cy.
　　　　　Over _____cy.
　　Multiple engine
　Front end loader
　Overshot loader
　Hoist—single drum
　　　　multiple drum
　Power shovel, dragline & crane
　　Over 5 cu yd

Figure 14–8–1　List of basic rates for estimate

Under 5 cu yd
Motor grader
Smooth roller
  Finish work
  Other work
Elevating grader
Trenching machine
Asphalt hot plant
Crushing plant
Concrete batch plant
Mixer operator—stationary
Mixer operator—paving type
Concrete paving spreader
Concrete finishing machine
Air compressor—portable
            stationary
            two or more
Water pump—under 4 inches
          4 inches and over
Concrete or grout pump
Churn drill
Heavy duty mechanic or welder
Oiler
Fireman
Dinkey locomotive
Brakeman
Mucking machine operator
Cableway signalman

Teamsters

Pickup and flatrack_____ton to_____ton capacity
Large flatracks _____ton to_____ton capacity
Large flatracks _____ton to_____ton capacity
Large flatracks _____ton to_____ton capacity
        Over _____ton capacity
Dump trucks
  End dumps _____cy. to_____cy. capacity
        _____cy. to_____cy. capacity
        _____cy. to_____cy. capacity
        _____cy. to_____cy. capacity
        Over _____cy. capacity
  Bottom dumps_____cy. to_____cy. capacity

_____cy. to_____cy. capacity
_____cy. to_____cy. capacity
_____cy. to_____cy. capacity
    Over _____cy. capacity
Water truck _____to_____gal. capacity
    Over_____gal. capacity
Transit mix trucks _____to_____cy. capacity
    Over_____cy. capacity

Warehousemen
Warehouse clerks

Carpenters

Apprentice
Journeyman
Stationary saw operator

Millwrights

Apprentice
Journeyman

Pile Drivers

Apprentice
Journeyman

Boiler Makers

Apprentice
Journeyman

Plumbers and Steamfitters

Apprentice
Journeyman
Pipeline welder

Painters

Apprentice
Journeyman—brush
Journeyman—spray

Cement Masons

Apprentice
Journeyman
Troweling machine
  operator
Epoxy, mastic, etc.

Iron Workers

Apprentice
Journeyman—reinforcing
Journeyman—structural

Electricians

Apprentice
Journeyman

Surveyors

Party chief
Instrumentman
Rodman
Chainman

If the foregoing crafts and classifications of labor are listed on printed forms, additional space should be provided under each heading and on a blank page for writing in any additional classifications.

Figure 14–8–1   List of basic rates for estimate (Continued)

14-9.   Social Benefit Payroll Taxes and Insurance

This element is intended to cover taxes and insurance paid by the employer and based on the amount of employee earnings. As will be discussed later, their costs are generally figured as incidental to the labor cost for any given item and are considered as part of the direct cost.

In the U.S.A., they include:

1. Payroll taxes (social benefits)
    F.I.C.A.—for old age benefits
    Federal unemployment compensation insurance
    State unemployment compensation insurance
    These are changed by law from time to time and should be verified. In 1971, as an example, they were:
        F.I.C.A.—(O.A.B.)—4.8% on the first $7800 of earnings.
        Federal unemployment—0.4% on the first $3000 of earnings.
        State unemployment—variable for different states, but—2.7% + on the first $3600 of earnings.

2. Payroll Insurance
    Workmen's compensation
    Public liability and property damage
    Employer's liability and occupational disease.
    Rates for these vary with location and classes of work, etc. They are usually figured as a percentage of payroll cost, although workmen's compensation sometimes is quoted per man-hour. The company's insurance carrier should be consulted for rates, but when workmen's compensation insurance is subject to plans administered by state governments, state agencies should be asked for the provisions of the plans. Many insurance plans provide for reductions or increases based on a company's historical loss record, and this fact should be considered.

14-10.   Taxes Other Than Payroll

These taxes can be assessed for virtually any reason that local, state, and federal governments want to use for producing revenue. They must be studied before new work proposals are completed because unless they are included in the estimate, they are not otherwise reimbursed. Tax experts or public accountants usually can provide information about these taxes. However, it sometimes is necessary to consult local government officials or local experts about existing practices and interpretations of the law. Tax information should be gathered by the company's tax expert or financial people, at the request of the engineering officer supervising the estimate.

Taxes of this type include the following:

1. Sales and use taxes on the purchase of goods, materials and services.
2. Business tax on the contract value
3. Manufacturing tax on the value of things prefabricated off the construction site
4. So-called "turn-over-tax" on all expenditures, often to to be duplicated in the case of subcontracted work
5. Income tax, federal and state, on the earnings of the company
6. Property tax—mostly by local subdivisions, such as counties, on the value of property at the site
7. Road and highway taxes
8. Imposts and duties on the value of imports

14-11. Insurance Other Than Payroll

This includes all kinds of insurance other than workmen's compensation, public liability, employer's liability, and social benefit insurance described above. Tender documents often will include a provision for insurance coverage, the limits and the parties to be insured, principally the owner-client. Otherwise, the company can elect the kind of coverages and their limits. When the insurance coverage is determined, quotations should be obtained from qualified carriers.

The kinds of insurance in this category include the following:

1. Comprehensive liability
2. Coverage on contractor's plant and equipment
3. Builder's risk
4. Property while in transit, including marine coverage
5. Fidelity and depositor's foregery
6. Money and security
7. Business interruption and/or extra expense
8. Engineer's or architect's professional liability
9. Faulty workmanship
10. Debris removal
11. Fire suppression

14-12. Quotations for Expendable Construction Materials and Supplies

Delivered prices for expendable items and services to be used in quantity in the construction and do not become parts of the completed structure should be requested as soon as possible after the plans and specifications are received. This is necessary so that reasonably accurate values for them can be included at the start of the estimating. Purchasing agents are in a good position to develop this information. It would include such things as:

- Fuel oil
- Lubricating oils and greases
- Lumber and timber for concrete forms
- Wire and nails

- Patented form ties
- Explosives
- Blasting supplies
- Drill steel and bits
- Welding rods

Special-purpose items that are indicated as the estimate progresses

This category is not intended to include the cost of plant and equipment items and repair parts that are discussed elsewhere.

14-13. Quotations for Permanent Materials and Installed Items of Equipment

The items involved here are those that will become part of the completed work and are presumed to be shown on the plans or detailed in the specifications. Since their true values constitute an important part of the pricing of the work, firm quotations should be requested as soon as possible after the plans and specifications are received.

The estimating section of the engineering department should initiate a takeoff of the required items and make a list showing the bid schedule item number or project specification section references. The list should also describe the items, including brief specifications necessary to convey a general idea of them, and the quantities required.

Items of a similar nature should be arranged in groups so that lists of them can be sent to the appropriate vendors and suppliers. Following are typical groups into which items might be arranged for this purpose:

- Culvert and drain pipe
- Plumbing fixtures, piping and pumps
- Hydraulic equipment, gates, and valves
- Reinforcing steel
- Fabricated steel items and vessels
- Electrical materials and equipment
- Masonry units
- Cement
- Miscellaneous

The character of the proposed work will determine the different subdivisions. A typical listing under the various classifications is illustrated at the top of the next page to suggest an arrangement for presentation to vendors and suppliers. Purchasing agents can be helpful in preparing a mailing list of vendors and suppliers.

After the take-off of permanent materials and the lists are completed, the lists should be sent to prospective vendors or suppliers. They should be accompanied by a letter giving the name of the project, the name and address of the owner or awarding agency, the contract and specification number, the location of the project, public transportation facilities serving it, the date of receipt of bids by the owner, the date when quotations should be furnished, the time that delivery of materials is required, and any other information that might help the vendors and suppliers investigate the project.

Vendors quite generally delay the delivery of their quotations until the last minute, which makes it impossible to hold up all affected pricing until they arrive. It is often desirable to assign tentative estimated prices to permit overall estimate to go forward, and then to adjust the pricing at the time of finalizing the bid, using the best quotations received from the vendors.

GRUNION CREEK POWER HOUSE
Material List "E"
Hydraulic Equipment, Gates, Valves, Etc.

Item No.	Description	Quantity
20.	Intake valves and hydraulic devices for station service unit Consists of: 2 (6'6" × 8'0") complete gate valve assemblies, associated hydraulic equipment and piping. Project Specification Par. 8-04	1 job
21.	Diffusion chamber gates See Project Specification Par. 8-04 48" × 48" sluice gates C.I. 48. Each required with operating mechanisms	1 job
23.	Valves 6" and under See Project Specification Par. 8-07 and piping plans for various systems	5,500 lbs
24.	Valves 8" and over Same reference as for Item 23	60,700 lbs

14-14. Subcontract Quotations and Conditions

After a preliminary review of the tender documents and a firm decision to bid the job, management should help decide whether to subcontract parts of the work or perform them with the general contractor's own forces. Work the general contractor is not equipped to handle obviously should be subcontracted. Subcontractor's should be approved by management on the basis of their reputations of performance and practices.

The procedure for takeoff and listing of parts considered for subcontracting and the solicitation of subcontractors for proposals should follow the steps outlined for requesting quotations on permanent materials. Since subcontract matters involve construction progress, the limits of activities and adherence to detailed contract conditions, they should be controlled by an engineering officer familiar with the technical requirements and contract definitions. However, they still can be a joint effort of purchasing and engineering personnel.

14-15. Listing Available Company-Owned Plant and Equipment

A factor in weighing the advantages and disadvantages of competing for a job is the plant and equipment the contractor has available for it. Therefore, the estimators should be provided with a list of company-owned plant and equipment which can be used on the job, its location, when it will be available, its condition, etc. Their consideration of the availability of company-owned plant and equipment might develop considerable savings in construction time and money in estimating the new work, and would reduce the investment in it.

14-16. Schedule of Approved Plant and Equipment Rental Rates or Amortization

When the estimate planning has reached the point where concepts of construction methods and plant and equipment have been tentatively developed, guidelines for establishing rental rates for amortizing plant and equipment values should be determined so that this element of cost can be written into the estimate. This consideration should cover both new and used plant and equipment.

Rental rates are often predetermined and a detailed schedule of them might be incorporated in the policy procedures of the company. They might be related to the formulas or pricing of this factor as established by Associated General Contractors of America, or the Associated Equipment Dealers, both well known to United States contractors. Another alternative is that plant and equipment charges can be made against the work based on the first cost of the facilities as charged to the job, less a salvage value stated in terms of a percentage of the value.

Management should decide on the method to be used before estimating, unless there is a predetermined policy. Where estimating systems are used that do not involve careful distribution of plant and equipment cost, these cost factors can be summarily considered at the end, but that does not fit the concept of estimating presented here.

14-17. Abstract of Local Laws Affecting Cost of Work

When a contractor operates for some time in an area, the effects of local laws become well known and periodic changes publicized. But when a contractor moves into a new area, he should obtain advice on how local laws might affect his operations and therefore the cost of work. Local representatives of service companies, manufacturers, vendors, and suppliers can explain the effect of the laws on the handling and delivery of their own products or services, but some careful overall research is often desirable. Things that might be disclosed include the following:

- Limits on the hours of work
- Blasting limitations
- Procedures and limitations relative to the handling and storing of explosives
- Load limits on streets and roads
- Limitations for controlling regular traffic, and encroachment on public ways
- Zoning ordinances limiting the construction of temporary plant facilities
- Measures to be taken for the protection of the public
- Building codes that define the construction designs to be employed, means of supporting deep excavations, etc

14-18. Preliminary and Refined Construction Programs

The earliest appraisal of a proposed project must include consideration of the available time, coupled with seasonal effects, water stages, and many other factors. A preliminary construction program, considered along with the quantities of work to be performed, gives a measure of the magnitude of the necessary operations. It is a step in the selection of construction methods and the development of a general concept for operating the proposed project. Sometimes a tentative construction schedule in the form of a bar graph is included in the tender documents. If not, the estimators must make a study of time and construction sequences as early as possible.

As the estimate progresses, the various phases of construction are studied and ideas of method and production are crystallized. The program can then be refined as all the required sequences and times of performance become known. Thus the programming of the work may be an ever-changing process during the prebid considerations and is an inseparable part of preparing an estimate for construction work.

At the conclusion of the estimating, a refined construction program is produced that often is required to be attached to the bid documents. This refined program may be in the form of a bar graph or critical path network, depending on the wishes of those who compile it, or the tender documents may specify its form.

14-19. Principal Divisions of Estimated Costs

Subdivisions of estimated cost and the accounting breakdown must be coordinated so they can be directly related for evaluating the estimating accuracy, for better control of field operations, and for the general guidance of all concerned. This subject is covered in more detail elsewhere herein.

Contracting firms have different ideas about how to break down costs. All of them are presumed to produce the same general end result, but with varying degrees of accuracy. No system of cost estimating and cost control is perfect, but experience has proved that the concept described here provides a reliable and reasonably accurate solution without being unnecessarily burdensome.

Basically, the principle involved is to charge directly to the specific construction operations all elements of cost that can be handled that way and to prorate those that cannot be readily broken down, distributing such costs proportionately to the individual work items or subdivisions of direct cost. Unit-price contract conditions will establish the work items, since these normally constitute the items of the bid schedule. Lump-sum contracts will normally require that the estimator establish his own list of work items, but the same principle of estimating and cost control applies.

Thus, the estimate can be divided into two principal parts constituting "Direct Costs," or the cost of the actual field operations, and "Indirect Costs," including such elements as overhead, general expense and general plant, which are not normally considered as direct field operations. Both the "Direct Costs" and the "Indirect Costs" consist of charges from various sources and a further breakdown under each heading is suggested to correspond to the source of such charges. The following "Principal Divisions of Estimated Cost" will illustrate this point.

DIRECT COST (to be charged direct to work items)
1. Labor and insurance
2. Permanent materials
3. Specific plant
4. Equipment rental
5. Supplies
6. Subcontracts

_____

TOTAL DIRECT COST

INDIRECT COST (to be distributed to work items)
7. Labor and insurance
8. Permanent materials
9. Plant
10. Equipment rental
11. Supplies
12. Subcontracts

_____

TOTAL INDIRECT COST

_____

TOTAL ALL COSTS
(Sum of direct and indirect costs)

This arrangement is further illustrated by Figures 14-19-1 and 14-19-2 (direct and indirect cost—estimate summaries, respectively) at the end of the chapter and the particulars of which are described further in the following:

14-20. Direct Costs

The direct cost after detailing should be posted directly to the "Estimate Summary" sheet, Figure 14-19-1. The charges to be included under the several subdivisions corresponding to the column headings are as follows:

1. Labor and Insurance This element should consist of the basic wages, plus the cost of all fringes paid to workmen and foremen in the performance of field operations and the cost of tax, insurance, and social benefit charges described elsewhere. The basic information would normally come from foremen's reports. An exception would be that the labor and markup charges (fringes, tax, and insurance) for the construction of specific plant items should be charged to "Specific Plant" as explained later. Labor and markup charges applying to the "Indirect Cost" items should not be included in this category.

2. Permanent Materials Under this heading should be the purchase price of materials, parts or installed equipment that are built in or become a component part of the structure. Here, too, should be sales and use taxes, freight, marine insurance, wharfage charges, customs duty and other charges making up the delivered cost of the materials, parts or installed equipment at the nearest railhead if by rail or to the job site if by truck. The cost of operations performed by the contractor's own forces, including unloading rail cars, hauling to job site, storing, etc., is to be charged to the other appropriate subdivisions of direct cost. The cost of such products as concrete aggregates, riprap, quarry stone, etc., processed by the contractor from natural construction material sources, at or near the site, should not be charged as permanent materials. Instead they should be charged as costs applicable to field operations under other headings.

3. Specific Plant This subdivision of cost includes all costs involved in providing at the job site built-in construction plant facilities. They are not to be confused with mobile or movable construction equipment covered in the following category. Specific plant items are those used to provide a specific service with respect to definite construction operations. Their cost can be readily charged as direct cost to specific work items or groups of similar work items. Such facilities as rock or aggregate processing plants, batching and mixing plants, cement handling plants, cableways, whirley crane trestles, railway trackage in tunnels, and tunnel ventilating systems would fall into this subdivision.

Specific plant costs include the first cost (purchase cost, plus sales and use tax) and freight and handling of all parts shipped to the job to become part of the finished plant. They also include the cost of all excavations, foundations, handling and plant erection, and dismantling. Costs for such parts as installed stationary equipment (vibrating screens, crushers, mixers, batching equipment, and others for which a salvage value is anticipated at the end of the job) should be included. This is done on the basis of an agreed amortization established before starting the estimate or at rental rates also established beforehand but applying for the full estimated time.

4. Equipment Rental This cost covers the use of major mobile or movable equipment. Such equipment does not include that installed in specific plants and minor equipment with a purchase value of, say, $500 or less that would be charged to the work as supplies and that can be charged to the various work items on the basis of time required or other suitable measure. Equipment charges usually are established by rental rates that are known or established before starting the estimate.

Charges to the job based upon the first cost of equipment, less an agreed salvage, can be distributed on the basis of the number of work units to be handled. Preferably, however, they should be distributed on a rental rate related to the estimated total time the equipment is to be used on the job. (See comment under Section 14-7, "Compiling Basic Data for a Specific Estimate—schedule of approved plant and equipment rental rates or amortization.") The cost of freight, move-in, erection, and dismantling of mobile and movable equipment should be compiled separately and distributed to the items of work that the equipment serves. It should be listed in the column headed "Specific Plant." Equipment dismantling is estimated initially.

5. Supplies and Expense This subdivision of cost should include all expendable items to perform field operations. They are not component parts of the completed work and can be charged as direct cost to specific work items. Included as well in this subdivision are services used in the performance of field operations and tangible property. The cost of supplies is the full delivered cost, plus sales and use taxes, freight and shipping costs, and anything else necessary to accomplish delivery. Tangible property would be such items as fuel, lubricants and other petroleum products, repair parts for plant and equipment, electric power, cable, teeth, tires, lumber and timber for concrete forms, form ties, wire, nails, small tools, explosives, blasting supplies, drill steel, and bits. Expenditures related to the acquisition or construction of plant facilities should be charged in the "Supplies" column under "Specific Plant" cost and not directly to the work item cost.

6. Subcontracts Under this category should be charged all payments to subcontractors who have performed part of the contract work under formal subcontracts. They should not include the cost of services rendered to the subcontractor for his own account. Nor should they include operations performed by the general contractor, not provided for in the subcontract but necessary to complete parts of the work.

Only the work subcontracted should be considered here, and supplementary costs charged in the labor or supplies column as direct cost under the appropriate work item. It is customary and advisable to require subcontractors to furnish payment and performance bonds. The cost of the bond is a proper part of the subcontract cost, even if paid separately by the general contractor. Any part of the general contract work might be subcontracted unless limited by the general contract, but speciality work is particularly adaptable to this means of handling and is sometimes necessary.

14-21. Indirect Costs

It will be noted that column headings (Figure 14-19-2) provide for the same general classification of column headings for indirect cost as for direct cost (Figure 14-19-1). The basic description of charges to be entered under the various column headings for direct cost will apply generally to indirect cost, although the latter must be more fully detailed to be understood. Figure 14-19-2 provides a list of the elements of cost that should be included. They will be discussed in more detail, along with comments on detailing indirect costs. Indirect costs should be prorated to the work items or bid schedule items, based on the proportion that the total indirect cost bears to the total direct cost, and the total cost which is the sum of both, listed in the "Total Cost" column on the Estimate Summary, Figure 14-19-1.

14-22. Hourly Equipment Operating Costs

For most heavy construction jobs, the cost of owning and operating mobile or movable equipment is a very substantial part of the job's cost. Equipment operating costs fall basically under the heading of direct cost. To a very limited degree, however, they can apply to the indirect cost in the way of constructing and maintaining general plant facilities, etc. Establishing accurate equipment operating costs therefore is important. They should be set up and reviewed early in the estimating so that they can be applied in detailing the estimated construction costs from the outset.

Equipment operating costs vary from place to place, even for identical articles of equipment and depending largely on the kind of service they perform. But many other things are involved, such as competent repair facilities, the quality of preventive maintenance, and the quality of the operators. For operating conditions similar to those included in historical background cost records, it is possible to project future costs.

Experience records are the most reliable references available. Some equipment manufacturers with no practical cost experience for their own products are prone to greatly understate their estimated operating costs and to overstate production probabilities. A word of caution, therefore— do the best possible job in establishing the hourly equipment operating costs.

The work that equipment will perform should be carefully considered. For instance, for power shovels, draglines, backhoes, and front-end loaders, the handling of poorly shot rock is the most severe; well-fragmented rock much less severe; stony earth less severe. Good earth would be mild service and the loading of stockpiled materials (sand, gravel, coal, etc.), light service. Craning service involving long periods of inactivity and very little impact is the lightest service possible. Dump-truck operation would fit the same general picture when handling the various materials mentioned above. Equipment operating costs will vary widely and be reduced materially as the severity of the service lessens, other factors also being considered.

The ability of a piece of equipment and its parts to withstand the stresses of the work being performed is, of course, the first consideration in selecting equipment. Nonetheless, physical and mechanical failures due to overtaxing of the machinery are disclosed by the operating cost records. On major excavation hauling jobs, truck tire costs are an important factor. Some construction equipment is under-tired. Even experts have difficulty evaluating tire size and construction to fit given needs. Overheating and resulting disintegration is a common fault due to miscalculation of load, speed, grade, and curves.

Figure 14-22-1 at the end of this chapter is suggested as a form to set up the estimated equipment operating costs for use in detailing the various construction item costs. This format has proved quite satisfactory and covers basically all the elements of equipment operating cost that would have to be considered.

The kind, type, model, and capacity of the equipment should be written at the top of each column. Like items of equipment doing similar work should be considered in the same column, with the number not being mentioned. Abbreviations for descriptive information is usually necessary. Horsepower, fuel consumption, new cost and shipping weight should be stated. At the top of the page, basic price information covering supply items, mechanics' wages, tax, insurance, and overtime are listed as they apply to the equipment.

In computing the operating cost for the equipment, the operators' and oilers' hourly wages with the cost of fringes should be listed in the appropriate column. This is assuming that they would be charged only for the actual hours operated, and that operators' and oilers' time during any repairs would be absorbed in the repair labor cost. Repair and service labor costs are the product of the average mechanic's wage, plus fringes and a factor developed by experience that is stated in terms of mechanic's time for repair and service per hour of operation of a given machine performing specific work.

Rental or depreciation is simply the agreed or established charge for equipment usage. It is reduced to the cost per operating hour based on the time it is estimated that the equipment will operate, the number of shifts per day, the time out of service for repairs, weather conditions, seasonal effect, etc. This element of cost is subject to final adjustment in arriving at the total agreed cost for the job, but the above computation permits a reasonably accurate distribution to the work items.

The supplies and expense part of equipment operating cost are predicated on historical experience. The listing of the various subdivisions of cost under this heading is self-explanatory, with the possible exception of outside repairs. This would cover billing by commercial houses for repairs done off the job site or for special services (welding, metalizing, rebabbitting bearings, etc.) that cannot be done on the job. The item of shop costs is intended to cover small quantities of welding rod, wiping rags, waste,

small tools, and small shop supplies that are minor in nature and for which separate accounting to a specific repair job would be impossible.

The resulting total hourly cost for operating equipment is intended to cover all costs, except insurances and taxes that do not apply. These are included in the overall cost of the job as part of the general expense.

14-23. Summary of Equipment and Move-In Cost

Elements of equipment cost to be considered apart from rentals or depreciation are freight and other shipping costs and the move-in and assembly of the equipment at the job site. Comparable costs for specific built-in plant facilities are covered elsewhere under comments on detailing direct cost. The distribution of such costs for specific plant facilities is readily done. But since the mobile or movable equipment could serve numerous work items of different character, the distribution of the move-in and set-up costs might require special study and computation.

When the work items using the equipment have been analyzed for operational cost, and it is known what equipment will be required, a complete listing of the equipment (specific plant components excluded) should be made. Figure 14-23-1 at the end of this chapter is suggested as a suitable format for the listing.

A description of the equipment appears on the left-hand side of the page. Only one listing is necessary for a group of identical articles unless they come from different shipping points. Columns across the top of the page, to apply to the listing on the left, are headed as follows:

"Condition" (described "New" or "Used—O.K.")
"Number of Units Required" (this to apply to each article shown)
"Location of Equipment" (probably factory if new or other job site if company owned)
"Unit Weight" (this is the shipping weight)
"Total Weight" (as extended)
"Freight Rate, cwt." (quoted by the carrier)
"Total Freight Cost" (as extended)
"Unit Cost of Equipment" (purchase price if new; appraised value if used)
"Total Cost of Equipment New"
"Total Value—Used Equipment"

From the computation and information summary shown in the various columns, it is possible to make the determination at the bottom of the sheet. In other words:

1. The total freight bill involved
2. The cost to unload and haul, railhead to job site, based on the total weight involved
3. Cost of assembly and dismantling at job site for those articles needing assembly
4. The total move-in and setup cost.

The cost of shipping the equipment out can be assumed to be borne by the next job. These costs are still divided according to the three principal sources of cost, namely labor, equipment, and expense (supplies). They must be properly distributed to the work items served, based on the time the equipment is used, on the number of units of work to be accomplished per item (all of which should be known at that time), or on an estimated equitable basis.

The summary shown in Figure 14-23-1 will serve other purposes. The cost of new equipment purchased for the job will help analyze the financing needed for the venture. The value of both new and used equipment is needed to analyze the possible return on the overall investment. These subjects are discussed later with relation to the bid computation.

14-24. Selection of Construction Methods and Procedures

The selection of construction methods and procedures on which to base the estimate may be apparent; or it may seem impossible, depending on the problem and how expert the estimator and his company are in the type of work to be undertaken. The sites may make access difficult, as well as the use of known devices for performing the otherwise normal operations. Study and research will almost invariably produce a plan for solving the problem. But without such a solution, attempts to price the work would be fruitless. Special technical problems demand advice of specialists or that they perform or direct the work.

The most economical solution is the one that considers contract compliance, time of performance, productivity, and efficiency. All of the bidders are not necessarily on the same footing, except for contract compliance. Control of patented devices, existing ownership, immediate availability of construction facilities, the techniques of specially trained help and other factors all can operate to favor of one bidder over another. Any bidder, however, is limited by the most ideal circumstances that he can logically expect for his performance, and basic prudence requires that he stop there.

Consideration of construction methods and their application must start with the factors that the contractor's experience has shown to be satisfactory and workable. Additional consideration then can be given to the possible use of improved devices, methods, or procedures. But they must have been checked out to the satisfaction of the bidder's principals. The correct bracket of cost for an operation can be readily priced for a method employing the bidder's historical cost experience. Untried methods or procedures normally require some speculative pricing and thus must be carefully considered.

The basic rules for selecting construction methods and procedures in estimating should include the following:

1. Stay with known principles on which historical cost experience is available, except that new devices or procedures may be considered if they use known principles with proven background for which accurate costs can be estimated in correlation with other uses.
2. If you are not sure that it would be possible to perform work by a given method or procedure, don't attempt to price it.
3. Price the work on sure methods and procedures and adapt or develop cost-cutting possibilities during the performance of the contract.
4. Hire reliable experts to price and perform specialized or unfamiliar work.

14-25. Preliminary Design of Construction Facilities

To facilitate the estimating, particularly on large, heavy construction projects, it is often desirable and necessary to make preliminary layouts and designs of various construction facilities expected to be used.

For some projects, the location and access to job buildings and shops are important features to indicate on maps. This will show how well they will serve the various sections of the work. Principal roads and haul roads often have to be designed for location, line, and grade. Plan layouts and profiles might be required.

Temporary bridges must be sketched to show construction details and weight limits. Excavating and grading for plant sites, cableway runways, etc., require detailing. How the various craning and handling facilities would reach the different sections of the work often must be illustrated. Plant foundations and supporting structures should be sketched up sufficiently to permit preliminary designs.

Before the bid, preliminary designs and layout need only provide a systematic and reasonably accurate record of what was contemplated during the estimating, to permit analysis of the structural problems, and to facilitate the takeoff of quantities necessary to build the temporary facilities so that they can be accurately priced.

If the bid results in the award of a contract, the preliminary designs and layouts form a foundation on which to start detailing the items of construction plant. They also form part of the plant study.

14-26. Detailing Direct Construction Costs

Preamble Many construction operations might be discussed in this section. However, since that would not be practical, it presents principal guidelines that can be generally adapted and examples of major features involved in the heavy construction field.

The estimator should remember that all parts of the estimate are subject to review by a supervisor before the bid price is fixed. He also should keep in mind that he may be asked to explain all of the particulars for the preparation of a subsequent budget-control estimate if the bid is successful and that the original estimate assumes great importance in the event of litigation. The estimate also provides the basic plans for construction plant and equipment and plays a prominent part in the refinement of designs and aquisition of plant and equipment.

Therefore, important features should be covered in detail in the original prebid estimate and the estimator's intent made clear by the sketches, descriptions, and references. Historical cost background should be referenced in the estimate and used freely either as a basis for computations or to help verify the accuracy of results already determined. To be useful, historical cost background would have to be adjusted for differences in labor rates and conditions, materials, etc., to fit conditions of the job being estimated.

Minor work operations would require less detail and supporting data. But "off-the-cuff" pricing should be limited and based on the estimator's specific knowledge of a given bracket of cost for certain operations. Competent estimators will invariably develop, with experience, a feel for unit costs of operations. Although it might not be sufficiently reliable for the basic pricing of new work, it should be used by the estimator to verify his computed costs.

General Concept of Job Operations Before starting to detail direct construction operations, the estimator must read the tender documents. This will help him develop a general concept of construction for the major features of the project, the general relationship of the different phases of the work, the general accessibility of the site, the major plant facilities that might be used and their approximate location. Preparation of a general job layout plan, such as is illustrated in Figure 14-26-1 at the end of the chapter, will assist in developing the overall job concept. The estimator can refine it as part of the record of the process used to price the work.

The general concept would also have to be sufficiently clear to the estimator to permit him to direct the start of takeoff work and quantity surveys for a preliminary plan of operations. Enough preliminary planning of a possible overall construction program must have been done to determine the magnitude and approximate production rates needed to complete the work on time.

At this stage, the principal estimator is ready to begin the detailed analysis of the various parts of the job or to assign them for analysis to others experienced in the various types of work involved. For example, if the project is a large hydroelectric development, the rock and common excavation for the foundations, plus the embankment construction, might be assigned to one individual, the concrete work to another, the hydraulic equipment to a third, and so on. The estimator should also arrange for preparation of the estimated hourly equipment operating costs, as shown in Figure 14-22-1, covering the mobile or movable equipment that is expected to be used.

A tentative list of this equipment should be made, along with an estimate of their operating costs. This information should be made available to all individuals working on the estimate. They thus will be aware of facilities that might be interchanged between operations. Items on the tentative equipment list would be subject to change as detailed study of methods and operations might indicate, and additions or deletions to the equipment operating costs would have to be made to coincide.

With development of the various construction features in the estimate, the general concept should be refined. In addition, the programming and time schedule should be adjusted to conform with the methods, plant and equipment, and procedures finally selected.

Programming of the Work for Estimating   Scheduling of the construction operations is an inseparable part of the estimating function. It should be done accurately enough and in sufficient detail to ensure that the work can be well coordinated and can be done efficiently and on time. If possible, the construction time might be shortened compared with that allowed in the tender documents. Ultimate cost savings then might be realized as a result of reducing the time of such continuing obligations as overhead. This, of course, is desirable if project financing plans permit earlier payments under the contract.

The tentative program envisioned in the initial general concept of job operations will invariably be refined by the estimators as they detail the construction operations assigned to them. At the outset, however, the best appraisal of the time required to perform the work should be in writing to assist those concerned with the detailing. Changes should be made during the estimating, so that the final modified program or construction schedule can be used as the basis for the bid.

Prebid construction programming usually will be satisfied by reasonably accurate bar graphs. The estimator should be sufficiently familiar with construction to make time and construction sequence judgments. Such a construction program or schedule is given in Figure 14-26-2 at the end of the chapter.

For the more complicated parts of the job, such as in the case of a large concrete job, the construction program for estimating cannot be prepared without separate analysis of the limits imposed by river diversion and specification requirements for height of pour, time between pours, differences in the heights of adjacent pours, etc. Such a separate analysis is illustrated in the concrete pour schedule shown in Figure 14-26-3. Dates are assigned when concrete lifts can be made using the production capabilities of the plant and equipment to the best advantage.

If the tender is successful, the prebid construction program forms the basis for greater refinement of it to be incorporated in the budget control estimate, then for the project management to use in directing the work. Later, the critical path method of programming can help develop maximum efficiency. In CPM programming, allowance should be made for sufficient float time and possible contingent lost time. The programming of interim dates should permit accomplishment of the ultimate final job completion on a realistic basis.

Special Construction Plant Cost   Contractors on most large heavy construction projects find it necessary to provide some built-in plant facilities, since mobile equipment alone would not be sufficient to perform all of the functions. Essentially, special construction plants are temporary and immovable facilities that must be built in or affixed to the site to serve a function in performing the work. They are thus distinguished from movable equipment that performs an operating function and is salvageable for reuse elsewhere. Built-in construction plants include, but are not limited to the following:

Rock crushing or separation plants
Concrete aggregate processing plants
Line conveyor transport systems
Overhead tramways
Concrete batching and mixing plants
Cement and pozzalan handling and storage plants
Concrete cooling and refrigeration plants
Industrial job railroad facilities
Whirley cranes, runways, and trestles
Overhead cableways, including runways and towers
Craning platforms
Special haul roads, bridges, and ramps
Derricks
Job electrical transformation and distribution systems
Pumping and underwatering systems
Shaft hoisting facilities, including guides, skips, and cages
Shaft head frames
Tunnel drop hole systems for concrete
Tunnel enlargements for passing tracks and other temporary facilities
Any other over excation or excess concrete adjoining the specified permanent works which are solely to facilitate the construction
Telephone and communication systems for operational control
General plant-site grading

To make it easier to recognize plant facilities that must be priced as part of the estimate, drawings should be made to outline those that are contemplated and/or to detail them sufficiently to permit accurate pricing, i.e., to develop the size of the excavations, concrete quantities, size of structural members, etc., and to give overall dimensions of the facilities to make sure they will fit the areas allocated by the estimate in the overall job layout plan. Many parts of the special plant facilities are produced commercially, including supporting structures. They can be shown as provided by the manufacturer's specifications.

Figure 14-26-4 at the end of this chapter shows a concrete plant layout, including facilities for aggregate storage and handling, rinsing screen for recleaning coarse aggregates, concrete batching and mixing plant, and cement storage and handling plants. This layout would have to be supplemented by additional sketches to show the sizes and shapes of foundations required by the bearing capacities of local soils, detail of the conveyor bents, etc.

Some construction plant facilities are very complex and highly technical. They must be designed sufficiently ahead of the original estimating to ensure reasonably accurate pricing that will cover the ultimate cost of providing and constructing the facilities if the tender is successful. In the case of whirley crane trestles, the sizes of structural members, length of trestle, legs to fit the terrain, cross bracing, trestle deck structure, crane and railroad track layouts, etc., must be calculated or known, and sketched out for the record. Similarly, in adapting an overhead cableway system to a given site, the general layout to provide coverage of the proposed permanent structure should be made by the estimator or his assistants. They should determine the character and amount of runway excavations, the sizes and structural design of towers, backstay systems, etc., and then coordinate the whole with the manufactured mechanical and appurtenant parts.

Prebid estimating of such items should carefully consider the intricacies of the complex stresses involved and the arrangement, shape, and size of members. Grave responsibilities are involved that should be shared with the company's principal engineering officer. Background in these matters is indispensable and outside experts should be consulted whenever necessary. Historical design and cost information obtained from others in the company is helpful in shortcutting the selection of plants, their detailing for estimating, as well as in their design and construction if the tender is successful.

Figure 14-26-5 to be appended at end of chapter illustrates the pricing of concrete plant and equipment. It assumes that the estimator has determined a concept of performing the job, the plant and equipment facilities that will be needed, has programmed the work in some detail, and is ready for the final prebid pricing.

Sheet 1 is a summary of the more detailed costs of concrete plant and equipment items shown on the nine following sheets. For the purpose of the prebid estimate, it is assumed that the plant and equipment are sufficient to do the entire concrete phase of the job. It is also assumed that its cost will be allocated and distributed to the various concrete work items shown in the bid schedule, which is usually done on the basis of the average cost per cubic yard or cubic meter of concrete as will be shown later.

In Figure 14-26-5, the costs are assembled under the headings of labor and insurance, equipment, supplies, and total direct costs. These include all the costs of providing the items necessary to develop the finished facilities, including move-in and erection. In the equipment column are posted the cost (rental or use value) of mobile equipment forming part of the finished facility, plus the usual charge for the equipment used for the move-in and erection of plant facilities. As compiled, Figure 14-26-5 absorbs—above the line for totaling—the full value of the equipment forming part of the facilities. In a separate column is listed the first cost of salvageable items of equipment for which credit for salvage can be computed later.

Salvage values should be applied with management's approval, since matters of policy are involved. The supervising engineering officer should, however, recommend a practical salvage value, the probable condition at completion, used equipment market, etc. In lieu of equipment charges on this basis, equipment values to be charged against a job can be predetermined and included in the equipment column.

When the plant and equipment values can be summarized, as on sheet 1 of Figure 14-26-5, amounts totaled in the labor and supplies column should be transferred to the specific plant column. Separate unit values per cubic yard or cubic meter of concrete for specific plant and equipment should be determined and used to distribute the costs of plant and equipment to the work items. The reason for carrying part of the cost as specific plant instead of labor and supplies is to help distinguish between the setup cost and the operations cost.

If approved salvage values are developed at this stage for the salvageable items, the unit costs for specific plant and equipment can be adjusted prior to distribution to the work items. Otherwise this adjustment would have to be made when the estimate is reviewed and the agreed cost is established. When the agreed salvage values are determined, they should be less the cost of dismantling the plant and storing the salvageable items. Otherwise, the cost of dismantling and storing should be detailed and included as part of the values shown in Figure 14-26-5.

The pricing of the various items of plant and equipment shown on sheets 2 to 10 of the exhibit is self-explanatory. However, all the details of consulting manufacturers, designing the facilities and supporting structures, analyzing erection problems, etc., must precede its listing for pricing. The pricing itself, if not done on the basis of quoted values from the outside, should be done on the basis of adjusted historical cost records. Such things as erection should be computed on the basis of theoretical crews and estimated production rates.

As indicated above, special construction plants can be required for a variety of construction operations other than heavy concrete work. Among them are tunnel excavation, open excavation, mechanical installations, etc. However, the principles for determining their values, summarizing their costs, and distributing it to the proper work items are basically the same.

Some construction work will involve only the use of mobile equipment for the direct field operations. In such cases, only the move-in and erection costs need to be added as a specific plant charge to the product of the equipment operating cost and the time of usage.

Specific Plant and Equipment Operating Costs When the special construction plants to be considered in the estimate are determined, the next step is to establish the basic cost of operating them. Typical of this step are the computations on the four sheets of Figure 14-26-6. The basic prices for operation are developed as the detailed programming is being completed and the time of operation can be determined. The costs should be reduced to the per-hour amount of use or the per cubic yard or cubic meter quantity of materials handled, then should be used to compute the proper charges against the individual work items. Thus the application of the specific plant operating cost is handled in the same manner as the operating cost for mobile equipment.

Figure 14-26-7 shows the method for computing the operating time of plant and equipment for a large concrete job on the basis of the quantity of work and estimated production rates based upon historical experience or calculated output. This format permits the breakdown of a given work item into several different methods of handling as in the case of concrete placing.

Summary of Concrete Cost A format for summarizing the cost for a given bid item on a mass concrete job is shown in Figure 14-26-8. It utilizes the elements of cost and production discussed earlier and adds other elements of cost necessary to account for all items entering into the field cost or direct cost to produce the finished product, i.e., the finished concrete for the item. The cost of plant and equipment, its operation, forms, finishing, treating horizontal construction joints and other listed items hence are all joined on the first sheet of Figure 14-26-8 and reduced to a cost per cubic yard. The summary total and unit costs for the bid quantity are transferred to the direct cost section of the estimate summary, to which the costs of all work items are posted.

The concrete cost summary in Figure 14-26-8 does not include the cost of cement, since it would be priced as a separate item of the bid schedule. Otherwise its cost should be added to the summary of concrete cost shown. The shown cost of aggregates is based on a firm subcontract price delivered in the contractor's stock piles near the concrete mixing plant. This might be a contractor-performed processing item, for which an additional example is provided. The cost of concrete forms is usually one of the largest elements entering the summary for major heavy concrete jobs and it deserves special consideration and treatment because of its importance.

Form Cost for Concrete Items Few construction operations vary so widely in cost and character as forms for concrete, and few involve a larger labor factor than wood forms. Labor productivity has a marked influence on the cost of forms; therefore every effort should be made to engineer them in advance. This will minimize the number of ties and points of adjustment requiring special attention. Important, too, is the development of a forming system that can be used repeatedly and assembled readily.

Steel forms usually provide these features, but their cost and initial assembly is relatively expensive. Inasmuch as their feasibility is contingent on the number of times they can be used, so the saving resulting from their repeated erection and stripping can more than offset their higher initial cost. Some steel forms that are usable in a number of places can be regarded as salvageable.

Wood forms should be designed well ahead of time and drawings prepared for their assembly and erection. Since carpenters' conferences to determine each step are expensive, predesign is necessary for efficient supervision. The number of possible reuses or prefabricated wood forms will determine their design and their economy in comparison with built-in-place forms, which rank among the most expensive.

Patented forming systems can be adaptable and economical. Many of them are composed of sections that can be pre-assembled into gang panels for large flat area work and crane handling and for repeated use. Or they can be assembled in individual panels to form a larger part of irregular flat areas, with built-in-place wood forms filling in the remaining areas. Companies furnishing patented forms on a purchase or rental arrangement often will make a prebid study of how their product can be applied.

Supported forms should be studied on the basis of the supporting system being priced separately from the contact surface forms, including joists and members for any panelizing. Moveable and reusable supports should be considered, as well as wheel mounting for easier moving from place to place.

The kind of form ties employed, their spacing, and the difficulty of installation and adjustment for dimensioning also are important factors. They must be considered from the standpoint of strength and the loads they might carry. The concrete pouring rate is another all-important factor, since it affects the ties as well as the forms.

For several reasons, forms for building-type work are usually less expensive than for heavy construction work. Forms for buildings generally do not need to be of the same exacting dimensions as for heavy construction. Building concrete often is covered by veneers of such materials as tile, brick, plaster, and flooring, so that irregularities in the surfaces will not be seen.

Concrete placing in building forms, because of the pouring rate and means of handling, imposes less load than in heavy construction. Heavy construction usually requires greater production in the placing of concrete and employs equipment of larger capacity that pours bigger individual bucket-load volumes. This results in greater impact on the forms and higher hydrostatic pressures due to the greater heights to which the forms are subjected, requiring stronger and better tied forms.

The shape of the forms, the accessibility of the erection area, the methods of handling, and the height above the ground—all play a part. Curved forms are more difficult to fabricate and erect than flat forms. The greater the degree of curvature, the greater the degree of difficulty.

The easiest forms to analyze are those whose uses are highly repetitive. They include forms for tunnel lining and for covered conduits that are mounted on wheeled trailer carriages. The work to be formed for successive concrete pours remains the same for some distance, and it is possible to develop a cycle of use involving stripping the form, moving forward, resetting the form, and pouring. Powered devices for moving and jacking the forms into position are part of them, and the structural parts can be made strong enough to allow a minimum number of ties.

So many variables are involved in concrete forms and their pricing that it is difficult to determine their cost, except for the simplest kind. Figure 14-26-9 gives an idea of the many different forms that can be found on a single heavy construction job. To price forms, it is necessary to classify them in order to evaluate the different types. This classification must start with the takeoff and must be directed by the estimator so he can apply known information. Sketches of supports, etc., sometimes are advisable to analyze form costs.

Historical cost background is the best guide for accuracy in pricing forms. Nothing else will serve as well, except when innovations are possible and it appears advisable to try something new. Familiarity with a wide variety of situations over a long period of time will allow the estimator to compare the current problem with past experience. Previous costs can be adjusted to account for variations in the prices of labor and material—and even labor productivity—to update the cost for new pricing.

If such adjusted pricing of concrete forms is not used for the initial pricing, it should be used to verify the estimated costs developed by other means. This will determine the potential calculated risk. Since forms comprise the largest part of the cost of most concrete jobs and there is a high labor factor, the possibility of inaccurate pricing is greater than in virtually any other construction operation.

Aggregate Production Estimate Figure 14-26-10 is a sample estimate of the production of concrete aggregates; its values are fictitious and for illustration only. The example is based on production at a river-run deposit explored by test pits. Samples were tested for the percentage of sizes occurring therein. The test pit data are assumed to be accurate and such physical characteristics as durability, soundness, and nonreactiveness proved.

The first step in making the estimate is to compare the pit gradation with that required by the specifications to fit the average concrete requirement. This is intended to determine if the size brackets indicated by the pit analysis meet the specification sizing. The quantity that is out of sizing would be considered as waste, except that the oversizes might be reduced by crushing to usable materials of smaller size.

The approved construction program determines the rate of aggregate production required to meet the concrete schedule. This factor, together with the gradation study, permits the estimator to compute the capacities needed for separation, washing, crushing, sand processing, etc. His next task is to choose the processing equipment that will accomplish the required aggregate production. He should prepare a diagrammatic flow sheet to be made a part of the estimate.

At the same time he considers the proposed plant site, its terrain, and the location of stockpiles, the estimator can determine the structure needed to support the various processing units, the feed arrangement for the plant, and the conveying equipment. This plant layout also should be sketched to scale on paper and made a part of the estimate.

Now the estimator is ready to solicit the prices of materials and equipment. He should estimate the cost of acquiring, moving in, and erecting the processing plant complete with all the necessary utility services.

A study of the aggregate pit operations must be made to evaluate the stripping of unusable materials and its probable wastage. After the loading and hauling of the pit-run materials to the processing plant are studied for capacity and efficiency, the overall pit operation can be priced for the entire time it will be underway. Surge piles of raw materials sometimes make it possible economically to stockpile pit-run materials to be fed to the plant by a bulldozer on a part-time basis.

Other factors to be considered in pricing the plant's operational cost are the number of men required to run it and the amount of servicing and repair it may need. The cost of power, supplies, and replacement of wearing parts can be computed or drawn from previous experience.

Pricing transportation of the aggregate to the batching and mixing plant or elsewhere completes the task of computing the total aggregate cost, except for summarizing the cost of all the steps mentioned above.

Rock Excavation Estimate Figure 14-26-11 is a simplified version of a rock excavation estimate. It employs a short form of assembling the pertinent data, but the same principles are applied as in the foregoing examples of other estimates. Wage rates and working conditions, as well as prices for materials and supplies, should come from current labor contracts or quotations. Equipment operating rates would be derived as explained earlier. The method and selection of equipment, as well as production rates, drill hole spacing, powder factor, etc., would be decided by the estimator.

14-27. Detailing and Distribution of General Expense and Overhead Costs

General expense and overhead costs are basically the same as the indirect costs referred to above. This category of costs consists of expenditures that, determined by time of use, direct functioning with respect to items or groups of items of work, or otherwise, cannot be charged directly to field operations for given subdivisions of cost. Some estimating systems group large amounts of cost, such as equipment depreciation, repair shop operation, the operation the operation of bull gangs or labor gangs, warehouse disbursements, etc. They then distribute the summary of such costs against amounts of the field costs otherwise accumulated in the accounts for specific work items.

This distorts the individual work-item costs, since a part of the charges incurred for a class of work, such as excavation, is directed against concrete work and vice versa. Any scheme of distributing such accumulated charges thus must be arbitrary and inaccurate. It is recommended, therefore, that as many charges as possible be made directly against the field operations. The remainder should be assembled under the heading of general expense and overhead for proration to the work-item costs commensurately, based on the relationship that the total general expense and overhead bear to the total of the direct costs of the items.

Figure 14-19-2 suggests a list of general expense and overhead charges that almost invariably include all the indirect charges involved in a heavy construction job covering variable classes of work. Special financing costs and unusual contract conditions might require the addition of other items. The kind of work being considered, size of job, etc., will control the concept of overhead organization to be used. Items listed and described in the left-hand column that do not apply should be ignored and the cost columns, therefore, left blank. Figure 14-19-2 will serve as a check list to avoid oversights and omissions. Care should be taken, however, to avoid putting figures in the blank spaces just because they are there.

Presumably, the construction program and time schedule will have been basically completed and approved by management before detailing the general expense and overhead and that they will be used to establish the time elements applying to the elements of cost where time is a measure of the expense. By this time, too, all the basic estimating data should have been accumulated, and historical cost background information will be adaptable to Figure 14-19-2.

Overhead Labor, Tax, and Insurance A desirable subdivision of overhead and general expense labor at the job site, which should facilitate comparison with the historical background information in Figure 14-19-2, is as follows:

- Management and supervision salaries
- Office salaries
- Engineering salaries
- Warehouse and purchasing salaries and wages
- Labor relations salaries
- Safety and first-aid salaries
- Watchmen and fire guard salaries

Individual job classifications shown under each of the above headings in Figure 14-19-2 are provided. These can be changed to fit variations in terminology and to describe combination job functions. Classifications can be added for any speciality personnel. Payroll tax and insurance constituting the markup on labor costs should be added into the summary under each heading to make the labor cost complete. They already have been described elsewhere.

Management and supervision salaries included all top management and supervisory personnel at the job site or in satellite offices directed by the job management. Personnel described as general foremen and of lesser rank are charged directly to the field operations for the various work items.

Job office salaries include the project office manager and all personnel involved with accounting, handling and administration of funds, accumulation of job costs, and similar functions assigned by the project management or the main office.

Engineering salaries should include salaries for the project engineer and engineering personnel under his direction. Warehouse and purchasing salaries and wages include personnel involved in these functions including the receiving, handling, and storage of repair parts and general supply items. The cost of handling and storing permanent materials, such as cement, installed machinery and structural steel, should be charged directly to the appropriate work items, and not to warehousing, purchasing, etc.

Labor-relations salaries, unless such services are contracted, should cover the company's labor-relations manager at the job site and his employes. Safety and first-aid salaries include saftey and first-aid personnel at the job site. Watchmen and fire guard salaries cover that category of personnel.

Retirement Plan Expense This account should cover only the company's portion of contributory or wholly company-financed retirement plans. The plans normally cover only salaried or key company personnel. Labor unions usually require enforcement of pension funds tied to the conditions of the labor contracts for hourly people.

General Automotive Expense This account covers automotive equipment provided for general use on the job. The equipment would include that assigned to the overhead and general-expense personnel in the categories shown above. It also would include other automotive equipment serving the job generally and not charged directly against the field operations for given work items. In the first group would be sedans, station wagons, pickups, and the like; and in the second group would be such vehicles as ambulances and service trucks hauling miscellaneous freight and general supplies around the job.

Charges for the use of automotive equipment should be computed either on the basis of monthly rentals or on the residual value. This is the difference between the purchase value, or first cost, less a salvage value at the conclusion of the work. The operating cost of automotive equipment should be derived from historical cost records on similar equipment for the time required or should be estimated on the most reliable information available. A form like Figure 14-22-1 can be used to record the estimated operating costs.

Man Haul This subdivision of cost is for the hauling of workmen from the project gate or parking area to the work area. Such hauls are normally made by buses or flatrack trucks equipped with seats. The reason for doing so is to avoid congestion in the work areas and to avoid accidents on the job. Only large jobs where movement of the workmen on foot is not feasible or access is difficult would be involved in this proposed breakdown. As suggested for automotive expense, costs should be broken down into equipment rental and operating costs. This permits posting to the several separate columns (labor, equipment, supplies, and total) across the top of Figure 14-19-2.

Automotive License Fees and Permits Usually, "off-highway" vehicles that operate only in construction areas need not be licensed. If licenses are required, the cost should be charged as part of the operating cost. The fees and permits referred to here are for the general automotive and man-haul vehicles forming part of the overhead and general expense.

Miscellaneous Operating Expense The listing of charges under this category of overhead and general expense is included in Figure 14-19-2. Basically, these elements of expense are those serving the functions of the overhead personnel and items not specifically charged directly to the field operations. Since they appear to be self-explanatory, they need not be repeated here.

Winter shutdown expense is for those projects closed down during the winter season. The cost of winterizing concrete pours or provisions to permit winter operations will be charged to the direct cost of the specific work item involved. Charges under this heading are concerned with winterizing only the job buildings and general facilities serving the overhead and general expense categories. Winter shutdown expense should include the cost of demobilizing and remobilizing because of the seasonal effect on the job, as well as the cost of moving personnel in and out or holding them at the site on a standby basis.

Temporary Buildings This general bracket of overhead and general expense covers temporary job buildings, general storage yards, and similar facilities. The facilities in Figure 14-19-2 are included as a check list.

General Utilities Installation Only the utilities listed in Figure 14-19-2 and serving the job generally are charged under this heading. Electrical and water systems directly serving the field operation should be charged to direct cost.

Temporary Roads These are for entering the cost of general access roads and bridges only. Haul roads serving such major operations as excavation or concrete work should be charged to the direct costs for each operation.

Camp Construction Construction camps may not be required, but when they are, the matter deserves serious consideration. Modern-day construction workers are willing to drive long distances in order to live in established communities that better serve them and their families. Only precedent and an understanding of the situation can help the estimator in making a reasonable camp analysis, including the quality, kind, and number of facilities to provide and the occupancy that might be expected. The job-site location, its remoteness, and the living standards of the workers all will have a bearing on the matter.

Company-built trailer courts have been well received. Companies have been able to attract a better quality of workmen, foremen, and supervisors by helping them to finance trailer-house purchases. On very large jobs, companies usually provide temporary residences for a limited number of key people and their families, and barracks or camps with single and double rooms for single personnel. It is advisable and sometimes necessary to provide messing, commissary, and recreation facilities for single personnel.

In recent years, factory-built camp buildings have become available to contractors. They resemble mobile homes in construction except that they are designed to serve the needs of single personnel, with bedrooms, toilets and showers, kitchen and dining rooms and recreation rooms. The completely furnished units are delivered as modules that can be grouped according to job needs.

Figure 14-19-2, under the heading of "Camp Construction," suggests a format for computing construction camp needs and pricing the facilities. That is, "Temporary Camp Buildings" and "Utilities for Camp Facilities," must be priced to give "Total Camp Construction" costs. Since some revenues will come from charges to employees using the facilities, the cost of camp construction is not a total loss. However, it is not a self-liquidating venture, except in unusual cases.

Camp Operations   Camp operations entail a number of considerations, and Figure 14-19-2 outlines them so that none of the elements will be overlooked. The elements include current operating costs and revenues from:

- Single status living quarters
- Camp houses (residences)
- Trailer parks and appurtenant facilities
- Mess hall and dining facilities
- Company commissary or store

The cost of free meals for guests or visiting head office personnel also should be detailed, as well as operating schools for employees' children.

Insurance Other than Payroll   Basically, this is industrial and fidelity insurance. A partial list of the coverage available is included in Figure 14-19-2. Many contractors choose to provide builder's risk insurance only on specific kinds of structures. They theorize that tunnels and the like are a minimum risk and insuring them against damage is not warranted. Contract conditions often require the owner to be named among the insured. In other instances, insurance must be handled through local agents to be valid. Care should be taken to assure that insurance is underwritten by reliable companies.

Taxes Other than Payroll   Under this heading come taxes other than workmen's compensation, social benefit taxes and insurance, general public liability and property damage, etc., which are computed on the amount of the payroll costs, and are charged directly to the cost of the operation or the function involved. The cost of permits and vehicle licenses do not come under this heading, but are charged directly to the operation or the equipment operating costs which are further distributable to the field operations or to "Automotive License Fees," etc., as part of the overhead. Sales and use taxes are added to the cost of materials or services, which are chargeable directly. The principal taxes to be summarized under this heading are county or city property taxes, state or city business taxes, and such special taxes as manufacturing taxes in the State of Washington.

Interest on Investment (Cash Flow Analysis)   Amounts paid as interest for money to perform the contract must be included as part of the cost of the work and should be listed under this heading. The computation of the amount of interest to be accrued should come from a cash flow analysis prepared by the estimator. This is based upon the accumulating demand for funds as shown by the estimate or agreed costs approved by the management, less amounts derived from earned revenues as indicated by the agreed construction program.

Interest is to be computed for periods of the job (possibly on a quarterly basis) on the total amount of the funds borrowed during the time borrowings are increasing, as well as on the residual values after it is possible to start retiring the loans from contract revenues. The cash flow analysis is a vastly important part of the estimating function. An exception to this plan for handling interest charges is equipment purchase notes. In that case, it is desirable to consider the total cost of the notes as an obligation against equipment operations, subject to amortization against work performed.

Central Office Expense   The cost of maintaining and operating the company's head office and its district or division offices is a real cost of doing business. It is properly chargeable against the various enterprises in which the company is engaged, including its individual construction projects. If a company operates only one job at a time, the entire head-office expense is chargeable against that job. If the company operates numerous jobs, distribution of the head-office expense is a matter of policy. A number of contractors have determined the percentage of the cost of their work that is required to amortise their total off-job costs.

When an estimate is prepared, it is a simple matter to apply this percentage to the total. Although the precentage should not exceed 3½ percent, it depends on the kind of work being performed by the contractor and the relative cost of permanent materials and installed equipment required by the contracts and of labor and supplies. It also depends on the efficiency of the company's overhead operations. As a matter of actual and later accounting, adjustments can be made in the charges for central office costs against the jobs being handled, with such adjustments being small.

Head-office costs include office rentals and operation, salaries, and other costs involved with the top management, engineering, accounting, purchasing, travel expenses, central warehousing, and auditing.

Contract Indemnity   The owner usually requires each contractor to file with his bid a bond or some other form of indemnification to ensure that he will sign the contract and comply with its terms if he is the successful bidder. A bond, sometimes in a form the owner prescribes, often is sufficient. Bid bonds normally are obtained from the bidder's usual bonding company and are relatively inexpensive. However, the tender documents might require a letter of credit from a specified or approved bank in a given amount.

The cost of such letters of credit will be fixed by the bank, usually a low percentage per month for a specified number of months. Other tender documents require cash or local government bonds of equivalent value to be deposited with the owner until the contract is awarded. This involves the cost of interest. Regardless of the type of tender indemnity, its cost should be included in the project estimate and management should be advised of the amount, cost, and conditions in advance.

The tender documents usually also specify the indemnity required to cover actual performance of the work and complete payment for labor, materials, and other expenses. This insurance to the owner that the contractor will complete the work at his own cost and expense normally can be covered in the United States by a bond written by a surety company acceptable to the owner.

Under ordinary circumstances, the amount of the bond is 50 percent or less of the bid amount. The cost of such bonds approximates 1 to 1½ percent of the bid amount, depending on the time of exposure and the bid amount. For different contract amounts, the rates are bracketed and progressively decrease for the accumulated values. Quotations should be obtained from the primary surety for each job being bid, but the following sets out the general form of computation required of the estimator:

Estimated Value of Bid $...............
First $ (of contract value) @ $..............per M = $...............
Next $................ @ $.............per M = $...............
Next $................ @ $.............per M = $...............
Next $................ @ $.............per M = $...............
Additional[a] $ .... @ $.............per M = $...............
        Subtotal $...............

Term Premium[b]
.... Mo. @ ....% = $...............
Total Performance and Payment Bond Premium[b] $...............

[a]Contract value over given amount quoted by surety.
[b]A surcharge to be added to the bond premium computed above based upon a rate quoted by the surety if contract time exceeds a given time (say 24 months).

The owner may require the performance and payment indemnity to be in some other form, such as letters of credit from approved banks, cash deposited with the owner or in an account he controls. Whatever is required, the cost must be included in the bid estimate. Overseas, indemnification normally takes some other form than a surety bond, but the amount usually is less than in the United States. The corporation secretary of chief financial officer of the company may be needed to help develop the basic information for the computations involved.

Total General Expense and Overhead A summary of the subjects described under this subsection should constitute the indirect or distributable costs, except that any special elements of a similar nature that are peculiar to a job should be added. In addition to the detail provided in Figure 14-19-2, supporting computations, quotations, etc., needed to make the general expense and overhead summary complete and understandable should be attached or covered by references.

14-28. Consideration of Contingencies

Two basic elements of contingent costs should be considered in preparing a firm-price construction proposal. They are (1) physical damage, and (2) economic changes. There might be other possible contingent elements of cost, but they would have to be determined in the light of their specific situations.

The amount of contingent cost that might be necessary is generally a matter of speculation. It involves a calculated risk and therefore should be approved by management. The possibility of occurrences producing contingent costs is based on details related to the preparation of the estimate, however. For that reason, the estimator should set down

the particulars and develop estimated contingent costs for a decision by the management.

Estimates of contingent costs can vary widely because of the unknown factors they are intended to cover. Contingent costs should never be included in the basic estimate, since it is an evaluation of known things or those falling within reasonable expectations for progress of the work. They should be treated separately, only being added in the final stage of the proposal preparation, and after the estimate of normal costs has been approved.

In some situations, it is possible to buy insurance against the occurrence of contingent risks or the partial repair of damages if they do occur. The contingent costs thus are reduced, but the insurance premiums are generally high and their inclusion in the estimated costs is a matter for management to decide. This does not apply to normally insured risks, such as the protection of the public and third parties, fire and windstorm damage, and the like.

Physical hazards could involve a number of things. Typical are flooding from abnormally high river stages and the overtopping of temporary cofferdams and protective works that delay or damage the project. They require detailed technical considerations and studies of historical stream flow records to determine the possibility of such occurrences, their frequency and severity, and the possible amounts of resulting damages. If this kind of risk is ultimately insured, such studies are normally required by the insurance underwriters.

Economic risks are more prevalent. However they often can be analyzed within computable limits when contract periods are shorter and indicated economic trends can be applied. Since labor contracts covering wages and working conditions often extend for three years, the changes they will make are known. The cost of materials and supplies is subject to smaller changes than labor. Equipment and parts experience greater price changes than materials and supplies because more labor is involved in their manufacture and marketing.

It is more difficult to analyze contracts extending for long periods. Even the self-styled economists have failed to judge the amount or the severity of change. The demands that labor will make beyond existing contracts are often unpredictable and are sometimes extreme.

Economic changes affecting construction therefore fall into two classes. A reasonably accurate amount of escalation for short periods can usually be figured from known conditions, but the change that will take place during longer periods is highly speculative.

The average estimator should figure the costs of construction on known prices, but he should compute and summarize the elements of escalation and contingent economic risk for management review before they are added to the proposal.

14-29. Determination and Application of Markup

Markup is normally composed of contingencies and profit factors that must be added to the sum of the estimated direct and indirect costs. Profit factors, as well as contingencies, usually are set by management after a review of the estimated costs. Estimated costs might be regarded as minimal or conservative, and profit margins are generally influenced by the prevailing opinion.

When several people review an estimate and they differ on how adequate or conservative the estimate is, they should reach agreement on a combined contingency and profit factor. This is applied without differentiating between the separate parts to an agreed cost. However, it is management's perogative to make such a determination and the estimators should help explain the cost estimate and define the possible contingencies.

If cost factors for contingencies and profit are separate in unit-price bids, they should be combined for easier distribution and application. Of course, all elements—estimated cost, contingencies, and profit—are added together for lump-sum bids to arrive at the total.

14-30. Completion of Estimate Summary and Bid Prices

The "Estimate Summary," Figure 14-19-1, at the end of this chapter, is designed to summarize the direct cost and its principal subdivisions by items, to permit the addition of the prorata part of general expense and overhead which is applicable, and to add the markup to arrive at bid prices. This form is particularly adaptable to unit-price bids, but to lump-sum bids as well for recording the costs applicable to the individual parts making up the lump sum as broken down in the estimate.

Columns at the left-hand side of the sheet are provided for "Item No.," "Description," or name of the item, the "Quantity," and the "Unit" in which the quantity is expressed. The items should be listed in the same order as in the bid schedule or enumerated as subdivisions of the work for lump-sum bids. The direct costs as broken down into their principal subdivisions—"Labor and Insurance," "Permanent Materials," "Specific Plant," "Equipment Rental," "Supplies," and "Subcontracts"—are recorded in the appropriate columns from computation sheets on which the direct costs are detailed.

The "Total Direct Cost" is also recorded and the unit cost for each subdivision should be shown for each item, including the "Total Direct Unit Cost." The total of each principal subdivision and the total of all directs costs for items listed should be determined by adding up the columns mentioned above. Thus far, Figure 14-19-1 serves as a summary of direct costs.

The next column, "Total Cost," is to include both the direct and indirect (general expense and overhead) cost. Since the concept of estimating described here includes in the direct cost all expenditures that can be allocated to specific work items, the remaining, or indirect cost can be prorated to the work items on the basis that indirect cost bears to direct cost.

Thus, when the total amount of general expense and overhead detailed in Figure 14-19-2 has been determined, and the total direct cost has been added up in Figure 14-19-1, the former divided by the latter will give a factor. When added to the "Direct Cost," it will produce the "Total Cost." It is preferable to factor up the total direct unit cost, to extend the result against the quantity, and then total the total cost as extended, because the unit cost can be rounded out to the desired decimal place. Any difference between this result and the sum of the totals of direct and indirect costs can readily be absorbed in adjustments made in the larger of the work items listed.

Markup (contingencies and profit) can be added on the same principle of proration. In this instance, relate the total markup to the total cost, apply the proration factor to the total unit cost, and round out the unit bid to eliminate excessive decimal places. To produce prices that can be readily extended against the quantities, the amounts computed for large items should be adjusted to bring the total amount of the bid in line with the sum of the total cost and total markup.

The last column of Figure 14-19-1 is left unnamed for use in making last-minute adjustments in the unit and amount of the bid for any items. They sometimes are necessary because of changed quotations.

14-31. Special Considerations in Estimating Foreign Projects

The basic guidelines for estimating construction in the United States also apply to estimating foreign work, with the possible exception of the things decreed by law and some other considerations. Investigations of prospective overseas work are often inconclusive, possibly because local laws and precedents are not well defined, or possibly because the investigator did not meet the "right" people. Lines of authority and influence are less clearly established and language barriers, nationalistic feelings, etc., all present problems and possible misunderstandings.

On the other hand, some administrative officers in foreign countries are permitted wide latitude in handling their business. Many potential problems can be solved through cordial relationships with them. A local partner or prominent contact is usually indispensable in the matter of getting along, or even obtaining basic information on which to base a proposal. However, potential bidders should beware of influence peddlers with no actual influence.

Project Financing   The engineering officer in charge of an estimate, if not the estimator himself, must understand the terms of the project financing. The financing scheme usually is treated in the tender documents. Sometimes financing arrangements are solicited as part of the tender. The estimating staff seldom takes part in the development of project financing. But, as part of the engineering staff's responsibility to study the tender conditions, the estimating group must alert management to potential problems and must know of any costs that should be included as part of the work expense.

Project financing sometimes involves the use of many currencies, depending on the origin of the loans made to the owner. Some loans are in the form of supplier credits requiring that the value of the loan be spent only in the country where the loan originates. Loans from the Export-Import Bank (U.S.A.) normally require expenditures in the U.S. A.I.D. (Agency for International Development), a U.S. State Department organization, usually imposes no restrictions on its loans. Loans through the World Bank, to which the U.S. government is a heavy contributor, are relatively free of spending limitations. The European Investment Bank of Brussels and Luxembourg, a Common Market institution, protects its own participants by requiring purchases within Common Market areas.

The estimator evaluating a proposal involving such financing arrangements must adhere to the pricing dictated by the suppliers in the favored areas. He is not free to use lower prices that might be obtainable elsewhere.

Another possible element of cost related to project financing results when a broker-financier is employed to form a group of lenders that will loan money to the owner. Broker-financiers charge high fees that must be added to the bid and may make it uncompetitive. Some broker-financiers might demand participation in the venture, in addition to a high fee.

Local Laws and Taxes During investigations of prospective foreign work, it is virtually impossible to learn all the local laws that might affect it. But prudence demands thorough research of the subject. Every foreign country presents a new situation. A reliable partner operating in the local construction field probably can furnish a good summary, but care should be taken to determine whether the laws apply equally to foreigners and nationals.

Reliable local attorneys familiar with how the laws apply to construction work might provide all of the answers. Another usually reliable source of general information on the subject is the local office of a highly reputable firm of charter accountants (certified public accountants). It should be one that makes a practice of serving locally based foreign companies and knows how the local laws have been interpreted, particularly in cases involving expenditures and the cost of doing business. The accounting firm also should be familiar with local taxes.

Local laws might require advance approval for bringing in key personnel to supervise the work. Work permits for expatriates might be stringently limited and good for only short periods before renewal, which might be difficult to secure. The use of the foreign country's nationals usually is a condition stated in the tender documents.

Often the importation of certain items is prohibited by law to protect home industry or local tax revenues from the sale of certain products, such as cement, on which the government fixes prices and which are heavily taxed. Applicable taxes fixed by law sometimes can be determined, but the amount that will be paid is uncertain. This may have to be negotiated with the agency making the assesssment. Important taxes sometimes assessed overseas include:

1. Transaction or business taxes payable on the contract amount
2. Stamp taxes involving a percentage of the value of all purchases and sales. Stamps of the right denominations must be posted on the record of the transaction. These taxes can be compounding
3. Turnover taxes similar to transaction taxes, except that they are payable by subcontractors and general contractors on monthly payments under the contract and subcontract regardless of any duplication
4. Corporate income taxes that sometimes must be paid whether or not a profit is realized
5. Personal income taxes. The income of an expatriate often is classed in a high bracket because it is larger than that paid to nationals
6. Property taxes are sometimes very high and are applicable to the value of the contractor's plant and equipment
7. Local subdivisions of government sometimes assess special taxes
8. Sales taxes of various kinds
9. Taxes on labor for social benefits can be devious and quite high
10. Custom duties and importation tax laws are sometimes difficult to evaluate and classify
11. Taxes on assets (bank accounts)

An important matter that must be investigated by any prospective bidder are laws concerning the importation and repatriation of funds for financing the construction operations, as well as related banking practices. An example is the situation permitting the unlimited importation of hard currencies as capital investment, but with repatriation or export of investment capital limited to a relatively small percentage of the total each year in amounts related directly to the amount imported. If the local economy is subject to rampant inflation, the value of the initial investment capital might well shrink substantially, thus resulting in sizable loss. The possible cost of such situations should be considered in the proposal.

Availability and Productivity of Local Labor The availability and productivity of local labor are vital matters when preparing a proposal for work in any location. But the particulars to be evaluated can vary widely, depending on the economy of the country and the stage of its development. Highly developed countries with strong economies generally have competent common and skilled labor forces accustomed to using modern equipment. Their productivity is usually good, except where civil instability exists or where radical tendencies prevail. Labor scarcity is common in highly developed countries, and labor importation usually is restricted.

Labor productivity in a foreign country should be evaluated only after considerable observation and research. A local partner is important in this matter, which involves the physical fitness of the people, their native intelligence and adaptability, local labor practices and working conditions, salary and wage rates, fringes, and contingencies. A contractor who has worked in a foreign country for some time will develop his own history of experience in many of these elements, and pricing prospective work will be a matter of applying known information.

Otherwise, such questions as the size and makeup of crews, the amount of supervision needed, the number of supervisors that must be imported from the contractor's own country, and the advisability of training for local workers are matters of speculation. However, experience in his own country and in similar areas will be helpful to the contractor in answering them.

For instance, it will be necessary to import working foremen to service and maintain sophisticated equipment if local workers have no such experience. They can be assisted by local workers.

The estimator should determine the number of local workers by figuring how many are needed to produce the same amount of work as an imported technician. A supply of helpers or common laborers can accomplish the preliminary or incidental operations. Local labor rates in developing countries usually are low, and the penalty for increasing the size of crews with additional nationals is minimal.

The working foreman will provide the technical know-how and supervision. He can help develop the talents of the nationals so that they can assume greater responsibilities and the number of imported supervisors can be reduced.

Totally inexperienced local help can be rapidly trained in certain functions to be as productive as highly paid expatriates. These functions are the highly repetitive operations. Truck driving is one, although the accident rate might be expected to be high. Operation of power shovels, draglines, large power drills, and similar equipment usually can be entrusted to local people after brief training periods, and their productivity can be developed to a relatively high degree. Such equipment as cableways, where safety is a factor, should be operated by the most intelligent technicians—whether nationals or expatriates.

In foreign construction, the use of third-country nationals in highly technical jobs and as supervisors is common. The reason is to utilize available talents, intelligence, and experience in jobs not suited to local help, and economies can be realized. Third-country nationals very often fit well in intermediate supervisory capacities, such as subforemen and bull gang bosses, handling less-complicated operations.

Important in evaluating foreign labor productivity is an understanding of the standard of work and the finish permissible in the area. If the specifications do not establish the excellence required, accepted practice should be investigated. Tolerances permitted in concrete surfaces, for instance, can make a big difference in the amount of labor required to accomplish it. The same principle would apply to numerous other construction operations.

Suitability of Construction Methods Sometimes the newest and most efficient method of performing work in a highly developed country is not the most economical in a less-developed country. Therefore, the question of what method to use deserves consideration. In some areas, excavation is handled by donkey trains, with all materials being hand loaded. Head baskets for transporting excavated materials, plastic concrete, and other construction materials are still used in some areas. Drilling rock by double jack, although antiquated by our standards, is common in remote places, as is the production of crushed stone by hand-handled hammers and small iron rings for sizing individual pieces of rock.

These examples emphasize extremes, but wage rates, working conditions, local standards of living, and the temperament of the people make the difference. The question of which method to use—new or old—may be determined by the allowed time of work performance.

Temporary Construction Communities Required On large jobs in remote foreign areas, the contractor must provide a reasonably self-sufficient community meeting the standards of the employees. For United States citizens, and highly paid Europeans, the standards must be high, with spacious living quarters (air conditioned in most places); medical care, including hospital facilities under the direction of qualified physicians; schools, stores and commissaries; recreational facilities and community centers, places for church services, and airplane landing strips.

Third-country nationals are likely to require lesser refinements because they are accustomed to somewhat lower standards. In developing countries, nationals will be content with minimal or even substandard facilities similar to those with which they have become accustomed.

Construction community facilities on the same project thus might be designed and built to three separate standards. It is customary for the employer to make a nominal charge for construction camp facilities, depending on the standard. These charges usually are not sufficient to defray the costs involved.

The kind of construction work and its magnitude will determine the type of construction camp facilities to be provided. Whether the work is seasonal or year-round is an important consideration. If rapid linear progress is involved, mobile camps probably will be necessary. Unless long-time, year-round jobs are involved, workmen may be employed only on a single basis and no accommodations for their families will be needed. All of these matters must be considered and priced by the estimator, with management concurring or directing.

Fringes and Inducements Offered Expatriates Competent supervisors probably are more important on foreign construction than on domestic work, where the operations can be observed more closely by the head office. Highly capable construction supervisors and technicians usually are satisfied with employment at home, and only special conditions will attract them to foreign jobs. These conditions include greater responsibilities and opportunities, the stress of domestic relationships, and the urge to travel. But the greatest influence is the chance to make more money and save it for early retirement.

The fringes and inducements that are offered can vary widely. Expatriates usually are employed on a contract for 18 to 24 months. Their pay scales often are higher than domestic standards by 25 percent or more, round-trip transportation, for them and their families, is paid on completion of the employment contract, and they are paid while en route. Other benefits sometimes offered include vacation allowances, allowances for shipment of baggage and personal effects, and furnished houses at stipulated rates. Employees must be guaranteed that their combined income tax (foreign and domestic) will not exceed the usual domestic tax. Third-country nationals are signed up on employment contracts with lesser conditions, but they conform with practices in their areas of origin.

Breakdown of Imported vs. Local Costs For many reasons, it is advisable and necessary to have an accurate separation of the costs of the work involving local expenditures as distinguished from the costs of imported items. This breakdown will facilitate the estimating by controlling the amount of local currency involved, the application of taxes, insurance, and other markup factors. The same breakdown should be followed in the cost accounting for the work.

The "Estimate Summary" in Figure 14-19-1 at the end of this chapter, designed for use on domestic work, should be enlarged to include the following column headings for adaptation to foreign work:

Item No.	Local permanent materials
Description	Specific plant
Quantity	Equipment depreciation
Unit	Imported supplies
Imported labor, tax and insurance	Local supplies
Local labor, tax, and insurance	Subcontracts
Imported permanent materials	Total direct cost
	Total cost
	Agreed cost
	Blank column heading

The detail of the estimate, of course, would have to be made in such a way as to provide the necessary detail for posting to the estimate summary.

14-32. Special Formats of Estimates for Specific Kinds of Construction Work

The procedures and formats for estimating construction work described earlier has proved satisfactory for estimating unit-firm price work in the heavy construction field. The same is true for detailing target estimate and other type contract work to the extent that it is done for proposal purposes and where the general character of work approximates the usual heavy or highway types. The same basic principles apply to estimating specialty lines of construction (using the techniques of the trade involved), but the format for computing and summarizing it might be different to suit the convenience of the estimator.

For instance, on work of highly repetitive limited operations, such as cross-country pipelining or tunnel excavation or tunnel lining, trade practice is to set up and price spreads of plant and equipment needed for the job. Next is the complete crew to perform all work operations and then the necessary materials and supplies. The summary of these comprise the total cost of the field operations. It can be reduced to a cost per linear foot of pipeline coated, wrapped, laid in the trench, and backfilled; the cost per linear foot of tunnel driven and supported; the cost of tunnel lining per linear foot; or the cost per cubic yard or cubic meter of concrete lining placed.

In other words, where there is little or no diversification of construction operations and the desired end result is one figure for the whole job, more straight-line computation can be used. But the basic elements of cost must somehow be there.

The same would apply to lump-sum contracts, where only one total amount is necessary for bid purposes. The most important factor to be considered by the estimator is that he treat the separate component construction operations in such a way that will permit him to check their general accuracy against available cost background.

People specializing in a given line of work and with substantial experience in it will have developed certain estimating routines. It is sometimes hazardous to direct them along other lines that may disrupt their train of thought and result in error. This assumes that their figures are understandable and can be reduced to the basic elements of labor, equipment, and supplies that can be coordinated with an overall picture. This should not be construed to indicate acceptance of "off-the-cuff" pricing without back-up detail or of breakdowns that are adjusted to make the picture look right. In these days of rapidly changing economic conditions, it is virtually impossible to remain current with accurate "off-the-cuff" pricing. Even so, it does not provide the necessary record.

14-33. Typical Building Estimate—Short Format

An estimate for a project involving only building work can be prepared along the lines described. Large and complicated jobs are admirably suited to such treatment, with the estimator setting up the various work items in sufficient detail to permit analysis and pricing, and referencing them to the tender documents. For uncomplicated building jobs, particularly smaller ones, a simplified format of estimate can be used in the interest of saving time where extensive detail for developing the estimated unit costs is not necessary.

To begin with, proposals for building projects are generally called for on a lump-sum basis. Contracturally, the unit prices for the many detailed construction operations are not considered unless specifically mentioned in the tender documents or unless changes in the work being executed bring them into play at a later date.

Tender documents for building projects are usually presented in two parts. A typical table of contents of them might approximate the following:

Part I—General

Division A   Advertisement for Bids
Division B   Instructions to Bidders
Division C   Form of Proposal
Division D   Contract Form
Division E   Performance and Payment Bonds

Division F   General Conditions
Division G   Special Conditions

Part II—Technical Specifications
Section 1   General Requirements
Section 2   Site Work
  2A   Demolition
  2B   Site Grading
  2C   Earthwork
  2D   Paving
  2E   Landscaping
Section 3   Concrete Work
  3A   Steel Reinforcement
  3B   Concrete, General
Section 4   Masonry Work
  4A   Mortars and Grout
  4B   Masonry, General
Section 5   Metal Work, Structural, and Miscellaneous
  5A   Structural Steel Material and Fabrication
  5B   Structural Steel Erection
  5C   Steel Joists
  5D   Steel Roof Decks
Section 6   Carpentry
  6A   Carpentry—Framing, Etc.
  6B   Glued Laminated Beams
  6C   Finish Wood Floors
  6D   Insulation
Section 7   Roofing, Sheet Metal, Etc.
  7A   Built-up Roofing
  7B   Dampproofing
  7C   Sheet Metal
  7D   Caulking
Section 8   Doors, Windows, and Glass
  8A   Hollow Metalwork
  8B   Rolling Doors
  8C   Wood Doors
  8D   Steel Louvers
  8E   Aluminum Windows
  8F   Glass and Glazing
Section 9   Finishes
  9A   Lathing and Plastering
  9B   Gypsum Wallboard
  9C   Acoustic Treatment
  9D   Ceramic Tile
  9E   Resilient Floor Covering
Section 10   Specialties
  10A   Finish Hardware
  10B   Toilet, Shower, and Miscellaneous Metal Partitions
  10C   Toilet Accessories
  10D   Miscellaneous
Section 11   Mechanical
  11A   Plumbing
  11B   Plumbing Fixtures
  11C   Heating and Ventilating
  11D   Swimming Pool Filtration System
Section 12   Electrical
  12A   Electrical General
  12B   Material and Installation
  12C   Lighting Fixtures
  12D   Fire-Alarm Systems

Part I—General *(cont.)*

Part II—Technical Specifications *(cont.)*

Part II—Technical Specifications *(cont.)*

Part I, "General," of tender documents varies widely in its contents, one job to another. Its provisions might have a profound effect on the estimating involved, particularly with respect to the overhead costs. Part II, the "Technical Specifications," however, relates primarily to the construction operations required and states all the particulars to be complied with in the physical accomplishment of the work, including any incidental or collateral items necessary. Many of the subdivisions in Part II relate specifically to specialty work, which the general contractor is probably not qualified to perform.

This work would logically be subcontracted to speciality contractors. They would offer to undertake it subject to complete compliance with both Part I and Part II of the tender docdments, insofar as they apply to the work in question, with the subcontractor assuming the responsibilities of the general contractor.

Some 60 to 75 percent of the work in the building field usually is subcontracted. This tends to minimize the amount of the plant and equipment facilities and organization needed by the general contractor at the site. However, the general contractor must provide for all things necessary to coordinate the operations of the several subcontractors involved, for rendering any special services to them, for assuring that they are totally compliant with the tender documents, and for their financial responsibility.

It follows that the portion of the work not subcontracted will be undertaken by the general contractor with his own forces and facilities. He will be required to make his own detailed take-off quantities and to set up the pricing sheets to permit complete coverage of all items of cost. For the purpose of illustration, it is assumed that the general contractor will perform all of the demolition, excavation and grading, all of the concrete work including forms, and all of the carpentry items. He will detail the costs of these so that they can be summarized under the usual headings of:

- Labor and insurance
- Permanent materials
- Specific plant
- Equipment rental
- Supplies

The sum of these represents the direct cost of the general contractor's work items. With the further addition of all of the quoted subcontract items, the total direct cost is produced.

The nonspecialty building items performed by the general contractor are generally of the type performed before and may be directly related to historical man-hour production and cost records maintained by the contractor. For such items, it is often possible to make a direct adjustment by factoring up the historical data without extensive detail in the determination of the direct cost. If the work items involved are peculiar to a given site, are complicated by specific conditions, or have no precedent in the contractor's previous experience, they should be analyzed with respect to their various components in the estimate of the direct cost.

The indirect costs can be detailed on a standard estimating sheet and the description of the elements or components can be written in rather than using the long printed forms illustrated elsewhere. In setting down the descriptions of the elements to be priced, however, the estimator should use the company's check list of what should be included. He also should have a list and description of requirements provided in the tender documents and not specifically included in the technical sections of the specifications.

Such requirements might be stated in the invitation to bid, instruction to bidders, or under the general and special provisions. It is sometimes convenient to include under indirect costs general services to subcontractors or the cost of facilities for the use of a number of the parties concerned.

As an illustration only, there is included at the end of this chapter Figure 14-33-1 consisting of five sheets setting forth a "Typical Building Estimate—Short Format" as described above. Sheets 1 and 2 comprise the estimate summary, sheet 3 covers the indirect costs, and sheets 4 and 5 suggest the form of typical direct cost detail.

JOHN DOE CONSTRUCTION COMPANY, INC.
GENERAL CONTRACTORS

ESTIMATE SUMMARY

BIDS OPENED _____ DATE _____ AT HORSE SHOE RIVER, IOBIA (LOCATION)
COMPLETION TIME _____ BID BOND _____
JOB: ROTTEN ROCK DAM AND APPURTENANCES
FOR: IOBIA HYDROELECTRIC COMM.
EST. BY _____   DATE _____   SHEET 1 of 2   NO _____

-111.726-

VALUES USED ARE FICTITIOUS FOR ILLUSTRATION ONLY

Item No.	Description	Quantity	Unit	Labor & Ins. Unit Cost	Labor & Ins. Amount	Perm. Mat. Unit Cost	Perm. Mat. Amount	Spec. Plant Unit Cost	Spec. Plant Amount	Equip. Rental Unit Cost	Equip. Rental Amount	Supplies Unit Cost	Supplies Amount	Sub-Contracts Unit Cost	Sub-Contracts Amount	Total Direct Unit Cost	Total Direct Amount	Total Cost Unit Cost	Total Cost Amount	Proposed Bid Unit Cost	Proposed Bid Amount
1	Care & diversion of river	sum	Job	L.S.	289580											.0523	80523	.81	99371	L.S.	1004464.00
2	Common exc.	720,000	c.y.	.1319	94968					.0685	49320	.0996	71712			.30	216000	.331	238320	0.37	266400
3	Rock exc.	270,000	c.y.	1.175	317290					.376	101421	.679	183413			2.23	602206	2.489	672030	3.05	823500
4	Borrow exc.	2,650,000	c.y.	.1198	317470					.0464	122960	.1367	362255			.3029	802685	.334	885100	.37	980500
5	Stripping right abutment	sum	Job	L.S.	14381						3435		4938				22744	L.S.	25115.06	L.S.	28000.00
6	Preliminary rock clean-up	4,500	s.f.	10.00	45000					2.50	11250	2.50	11250			15.00	67500	16.564	74538	18.50	83250
7	Close line drilling	12,000	s.f.	1.00	12000					.05	600	.15	1800			1.20	14400	1.325	15900	1.50	18000
8	Core drilling 3" dia. holes	34,600	l.f.	.13	4498					.05	1730	.05	1730	3.50	121100	3.73	129058	4.119	142517.4	4.60	159160
9	Diamond drilling, 3" dia. drain holes	28,000	l.f.	.13	3640					.05	1400	.05	1400	3.50	98000	3.73	104440	4.119	115332	4.60	128800
10	Diamond drilling, 1-1/2" dia. holes	57,000	l.f.	.13	7410					.05	2850	.05	2850	2.25	128290	2.48	141360	2.739	156123	3.05	173850
11	Core drilling, 10" dia. holes in core.	200	l.f.	.50	100					.25	50	.25	50	25.00	5000	26.00	5200	28.710	5742	32.10	6420
12	Percussion drilling	50,000	l.f.	.50	25000					.10	5000	.20	10000			.80	40000	.883	44150	1.00	50000
13	Setting pipes grouting under embkt.	440	ea.											15.00	6600	15.00	6600	16.564	7288.16	18.50	8140
14	Moving drill onto grout holes	2,600	ea. move											7.50	19500	7.50	19500	8.292	21533.20	9.25	24050
15	Pressure testing	2,600	ea. test											15.00	39000	15.00	39000	16.564	43066.40	18.50	48100
16	Furn. cement for grouting	50,000	bag	.10	5000		60000			.02	1000	.03	1500			1.35	67500	1.491	74550	1.65	82500
17	Pressure grouting	50,000	c.f.	.15	7500					.05	2500	.05	2500	2.25	112500	2.50	125000	2.761	138050	3.05	152500
18	Compacted fill (impervious)	2,500,000	c.y.	.0445	111250					.0211	52750	.0351	87750			.1007	251750	.111	277500	.125	312500
19	Compacted fill (pervious)	53,000	c.y.	.0768	4069					.0268	1422	.0608	3223			.1644	8714	.181	9593	.20	10600
20	Backing and filter meter.	95,000 (75,000)	c.y.	1.3602	129219			.4902	46569	.3323	31569	1.2129	115525			3.3956	322582	3.682	349827	4.12	391400
21	Dumped riprap	95,000	c.y.	.64	41996					.292	14582	.541	27097			1.673	83675	1.847	92350	2.05	102500
22	Spalls	83,000	c.y.	.543	45047					.246	20365	.473	39228			1.262	104640	1.273	115619	1.55	128650
23	Rock fill	540,000	c.y.	.180	97069					.0785	42395	.158	85239			.4165	224703	.460	248400	.50	270000
24	Additional rolling for compaction	500	hr.	4.20	2100					5.15	2575	5.15	2575			12.85	6425	14.190	7095	15.90	7950
25	Conc. in stilling basin, baffles & end sill	28,500	c.y.	4.8255	137527			1.016	28967	.592	16872	2.4207	68990			8.8547	252356	9.7682	278374	11.00	313500
26	Conc. in roadway, sidewalks, & parapets	1,800	c.y.	27.8035	50046			1.016	1829	.593	1067	10.3077	18561			39.7082	71488	43.1686	77703	48.00	86400
27	Conc. in thin slabs, beams & columns	8,600	c.y.	16.0055	137647			1.016	8738	.592	5091	6.6477	57170			24.2622	208646	26.7822	230327	30.00	258000
28	Conc. in spillway pier	7,500	c.y.	14.8105	111079			1.016	7623	.592	4440	6.3967	47975			22.8152	171117	25.1842	188882	20.00	150000
29	Conc. in training walls	25,000	c.y.	5.0695	126738			1.016	25410	.592	14800	2.3587	58967			9.0322	225815	9.9642	249105	11.10	277500
30	Mass conc. in dam	1,160,000	c.y.	3.3506	3886646			1.016	1179024	.592	686720	1.7352	2012832			6.6947	7765277	7.3822	8563352	8.25	9570000
31	Conc., fndtn. in core trench	750	c.y.	12.1735	9130			1.016	-	2.890	2168	5.3070	3980			20.3705	15278	22.1792	16634	24.80	18600
32*	Portland cement	950,000	bbl.			4.20	3990000									4.20	3990000	4.638	4405100	5.18	4921000
33*	Blended cement																				
	(a) Portland Cement	712,500	bbl.																		
	(b) Natural cement	237,500	bbl.																		
	SUBTOTAL THIS SHEET				6053240		4050000		1298160		1345343		3689337		529990		16916030		18670172		20858234

* Delete inapplicable item

Figure 14-19-1 Estimate summary

JOHN DOE CONSTRUCTION COMPANY, INC.

GENERAL EXPENSE AND OVERHEAD

EST. BY ___   CK'D BY ___   JOB ___   SHEET 1 OF 8   DATE ___

	LABOR AND INSURANCE		MATERIALS		PLANT		EQUIPMENT RENTAL		SUPPLIES		SUB-CONTRACTS		TOTAL	
	UNIT COST	AMOUNT	UNIT COST	AMOUNT	UNIT COST	AMOUNT	UNIT COST	AMOUNT	UNIT COST	AMOUNT	UNIT COST	AMOUNT	UNIT COST	AMOUNT
OVERHEAD SALARIES:														
MANAGEMENT & SUPERVISION SALARIES														
Project Manager Mos.														
General Superintendent Mos.														
Asst. General Superintendent &/or Shift Superintendents Mos.														
Excavation Superintendent Mos.														
Concrete Superintendent Mos.														
Carpenter Superintendent Mos.														
Rigging Superintendent Mos.														
Tunnel Superintendent Mos.														
Quarry Superintendent Mos.														
Master Mechanic Mos.														
Electrical Superintendent Mos.														
Pipefitter Mos.														
P.R.T. & Ins. ___% on $ ___ Labor														
Total Management & Supervision Salaries														
OFFICE SALARIES														
Office Manager Mos.														
Chief Accountant Mos.														
Cost Accountant Mos.														
Accountant Mos.														
Voucher Clerk Mos.														
Paymaster Mos.														
Chief Timekeeper Mos.														
Timekeepers Mos.														
Payroll Clerk Mos.														
Secretaries & Stenographers Mos.														
Switchboard - Receptionist Mos.														
Machine Operator Mos.														
P.R.T. & Ins. ___% on $ ___ Labor														
Total Office Salaries														
SUB-TOTAL FORWARD														

Figure 14-19-2   Detail of general expense and overhead

JOHN DOE CONSTRUCTION COMPANY, INC.

GENERAL EXPENSE AND OVERHEAD (CONTINUED)   SHEET 2 OF 8

EST. BY _____   JOB _____   CK'D BY _____   DATE _____

	LABOR AND INSURANCE		MATERIALS		PLANT		EQUIPMENT RENTAL		SUPPLIES		SUB-CONTRACTS		TOTAL	
	UNIT COST	AMOUNT	UNIT COST	AMOUNT	UNIT COST	AMOUNT	UNIT COST	AMOUNT	UNIT COST	AMOUNT	UNIT COST	AMOUNT	UNIT COST	AMOUNT
Brought Forward														
ENGINEERING SALARIES														
Project Engineer — Mos.														
Office Engineer — Mos.														
Cost & Materials Engineer — Mos.														
Draftsman — Mos.														
Field Engineer — Mos.														
Party Chief — Mos.														
Instrumentman — Mos.														
Rodman — Mos.														
P.R.T. & Ins. ___ % on $ ___ Labor														
Total Engineering Salaries														
WAREHOUSE & PURCHASING SALARIES & WAGES														
Purchasing Agent — Mos.														
Chief Warehouseman — Mos.														
Receiving Clerk — Mos.														
Records Clerk — Mos.														
Warehousemen — Mos.														
P.R.T. & Ins. ___ % on $ ___ Labor														
Total Warehouse & Purchasing Salaries														
LABOR RELATIONS SALARIES														
Labor Coordinator — Mos.														
P.R.T. & Ins. ___ % on $ ___ Labor														
Total Labor Relations Salaries														
SAFETY & FIRST AID SALARIES														
Safety Engineer — Mos.														
First Aid Man — Mos.														
Nurse — Mos.														
P.R.T. & Ins. ___ % on $ ___ Labor														
Total Safety & First Aid Salaries														
SUB-TOTAL FORWARD														

Figure 14-19-2 (Continued)

JOHN DOE CONSTRUCTION COMPANY, INC.

GENERAL EXPENSE AND OVERHEAD (CONTINUED)

SHEET 3 OF 8

EST. BY _____   JOB _____

CK'D BY _____   DATE _____

	LABOR AND INSURANCE		MATERIALS		PLANT		EQUIPMENT RENTAL		SUPPLIES		SUB-CONTRACTS		TOTAL	
	UNIT COST	AMOUNT	UNIT COST	AMOUNT	UNIT COST	AMOUNT	UNIT COST	AMOUNT	UNIT COST	AMOUNT	UNIT COST	AMOUNT	UNIT COST	AMOUNT
Brought Forward														
WATCHMEN, GUARDS & FIRE GUARDS—SALARIES														
Chief Guard Mos.														
Guards Mos.														
Watchmen Mos.														
P.R.T. & Ins. ____ % on $ ____ Labor														
Total Watch, Guard & Fire Guard Salaries														
RETIREMENT PLAN EXPENSE														
____ % on $ ____ Salaries of Eligible Employees														
SUB-TOTAL OVERHEAD SALARIES														
AUTOMOTIVE EXPENSE:														
GENERAL AUTOMOTIVE EXPENSE														
Purchase Less Salvage, or Rental—Sedans Mos.														
—Pickups Mos.														
—Ambulances Mos.														
—Service Trucks Mos.														
Operation Cost—Sedans Mos.														
—Pickups Mos.														
—Ambulances Mos.														
—Service Trucks Mos.														
Total General Automotive Expense														
MANHAUL														
Purchase Less Salvage, or Rental—Bus Mos.														
—Flatracks Mos.														
Operation Cost—Bus														
—Flatracks														
Total Manhaul														
AUTOMOTIVE LICENSE FEES & PERMITS														
____ Units ____ Yrs. = ____ Yrs.														
SUB-TOTAL AUTOMOTIVE EXPENSE														
SUB-TOTAL FORWARD														

Figure 14-19-2 (Continued)

JOHN DOE CONSTRUCTION COMPANY, INC.

GENERAL EXPENSE AND OVERHEAD (CONTINUED)

SHEET 4 OF 8

EST. BY | JOB | CK'D BY | DATE

		LABOR AND INSURANCE		MATERIALS		PLANT		EQUIPMENT RENTAL		SUPPLIES		SUB-CONTRACTS		TOTAL	
		UNIT COST	AMOUNT	UNIT COST	AMOUNT	UNIT COST	AMOUNT	UNIT COST	AMOUNT	UNIT COST	AMOUNT	UNIT COST	AMOUNT	UNIT COST	AMOUNT
Brought Forward															
MISCELLANEOUS OPERATING EXPENSE:															
Stationery and Office Supplies	Mo.														
Office Rent (Incl. Temporary)	Mo.														
Office Equipment - Rental or Depreciation	Mo.														
Office Expense—Other	Mo.														
Engineering Supplies and Expense	Mo.														
Engineering Equip.—Rental or Depreciation	Mo.														
Photography and Reproduction	Mo.														
Outside Engineering Expense	Mo.														
Consultants Expense															
Warehouse and Purchasing Expense	Mo.														
Labor Relations—Supplies & Expense	Mo.														
Safety and First Aid—Supplies & Expense	Mo.														
Watchman, Guards & Fire Guards Expense	Mo.														
Telephone and Telegraph	Mo.														
Radio Communications	Mo.														
Heat, Light, Power & Janitor Service	Mo.														
Building Maintenance—Owners Facilities	Mo.														
Legal Expense	Mo.														
Travel Expense—Job Connected	Mo.														
Travel Expense—Move-in Personnel	Men														
Entertainment, Donations & Contributions	Mo.														
Bank Charges	Mo.														
Fire Protection	Mo.														
Sanitary Facilities	Mo.														
DrinkingWater and Ice	Mo.														
Misc. Job Road and Yards Maintenance	Mo.														
Job Signs	Mo.														
Winter Shutdown															
Winterize Building & Equipment															
Move Out & In of Personnel															
Standby Time for Personnel															
SUB-TOTAL MISCELLANEOUS OPERATING EXPENSE															
SUB-TOTAL FORWARD															

Figure 14-19-2 (Continued)

JOHN DOE CONSTRUCTION COMPANY, INC.

GENERAL EXPENSE AND OVERHEAD (CONTINUED)

SHEET 5 OF 8

EST. BY ___ JOB ___ CK'D BY ___ DATE ___

	LABOR AND INSURANCE		MATERIALS		PLANT		EQUIPMENT RENTAL		SUPPLIES		SUB-CONTRACTS		TOTAL	
	UNIT COST	AMOUNT	UNIT COST	AMOUNT	UNIT COST	AMOUNT	UNIT COST	AMOUNT	UNIT COST	AMOUNT	UNIT COST	AMOUNT	UNIT COST	AMOUNT
GENERAL PLANT:														
TEMPORARY BUILDINGS														
Office														
Warehouse .... Buildings														
Shelving														
Yard, Surfacing, Sills, Fencing														
Equipment Repair Shop & Equip. Inst.														
Carpenter Shop Building														
Cutting Deck														
Electrical Repair Shop														
Pipefitters Shop														
Riggers Loft														
Hospital or First Aid Building or Trailer														
Field Offices														
Tool Sheds														
Change House—(Tunnel & Shaft Work)														
Powder House														
Cap House														
Tire Repair Shop														
Oil Storage														
Grease Rack														
Service Station														
Fuel Storage														
Inspectors Field Office														
Inspectors Concrete Testing Facilities														
Paint Shop														
Workmen's Shelters														
General Yard and Site Grading & Surfacing														
Total Temporary Buildings														
UTILITIES INSTALLATION—(Excluding Camp)														
Electrical Supply & Distribution														
Telephone Lines and Job Telephone System														
Water Supply Storage & Distribution														
Sewer System and Treatment Facilities														
Radio Communications														
Total Utilities Installation														
TEMPORARY ROADS														
.01 General & Access Roads & Bridges														
SUB-TOTAL GENERAL PLANT														
SUB-TOTAL FORWARD														

Figure 14-19-2 (Continued)

JOHN DOE CONSTRUCTION COMPANY, INC.

GENERAL EXPENSE AND OVERHEAD (CONTINUED) SHEET 6 OF 8

EST. BY CK'D BY JOB DATE

	LABOR AND INSURANCE		MATERIALS		PLANT		EQUIPMENT RENTAL		SUPPLIES		SUB-CONTRACTS		TOTAL	
	UNIT COST	AMOUNT	UNIT COST	AMOUNT	UNIT COST	AMOUNT	UNIT COST	AMOUNT	UNIT COST	AMOUNT	UNIT COST	AMOUNT	UNIT COST	AMOUNT
CAMP CONSTRUCTION:														
TEMPORARY BUILDINGS—CAMP														
Total Direct Labor Cost $														
Ave. Dir. Labor Cost/Mo.: (Mo.) $ /Mo.														
Ave. Hourly Wage Rate /Hr.														
Ave. Monthly Wage														
Average Number of Men														
Peak Manpower Required + % =														
Non-Manual Personnel														
Total Manpower at Peak														
Est. % Find Own Accom. =														
Est. % Live in Barracks =														
Est. % Live in Trailers =														
Est. Residences for Supervision = Units														
Barracks Men @ SF/Man = S.F.														
Mess Hall for Men														
Commissary														
Recreation Hall														
Bull Cooks Dormitory Men @ SF EA S.F.														
Trailer Court—Including Utilities														
Trailer Court Wash House														
Residences for Supervisory Personnel														
Site Grading and Surfacing (Other Than Trailer Court)														
Total Temporary Buildings—Camp														
UTILITIES FOR CAMP FACILITIES—Other than Trailer Ct.														
Power Supply and Distribution														
Telephone—Camp Telephone System														
Water Supply and Distribution														
Sewer System and Treatment Facilities														
Total Utilities for Camp Facilities														
SUB-TOTAL CAMP CONSTRUCTION COST														
SUB-TOTAL FORWARD														

Figure 14-19-2 (Continued)

JOHN DOE CONSTRUCTION COMPANY, INC.

GENERAL EXPENSE AND OVERHEAD (CONTINUED)   SHEET 7 OF 8

EST. BY _____   JOB _____   CK'D BY _____   DATE _____

	LABOR AND INSURANCE		MATERIALS		PLANT		EQUIPMENT RENTAL		SUPPLIES		SUB-CONTRACTS		TOTAL	
	UNIT COST	AMOUNT	UNIT COST	AMOUNT	UNIT COST	AMOUNT	UNIT COST	AMOUNT	UNIT COST	AMOUNT	UNIT COST	AMOUNT	UNIT COST	AMOUNT
CAMP OPERATIONS:														
COMMISSARY OPERATIONS—Gain or (Loss)														
MESS HALL OPERATIONS—Gain or (Loss)														
___ Mo. Operation × ___ Men														
× ___ Meals/Days × 7 Days = ___ Meals														
Meal Cost														
Less Revenue														
Total Mess Hall Operations														
BARRACKS OPERATIONS														
___ Mo. Operation × ___ Men														
× 7 Days = ___ Man Days														
Operation Cost														
Less Revenue														
Total Barracks Operations														
RESIDENCE OPERATIONS														
___ Mo. Operation × ___ Residences														
= ___ Mos. of Occupancy														
Operation Cost														
Revenue														
Total Residences Operations														
TRAILER CAMP OPERATIONS														
___ Mos. Operation × ___ Average Spaces														
Occupied = ___ Trailer Mos.														
Cost of Operation														
Revenue														
Total Trailer Camp Operations														
FREE MEALS EXPENSE														
SUB-TOTAL CAMP OPERATIONS														
CAMP CONSTRUCTION (From Sheet No. 6)														
NET COST OF CAMP & OPERATIONS														
SUB-TOTAL FORWARD														

Figure 14-19-2 (Continued)

JOHN DOE CONSTRUCTION COMPANY, INC.

GENERAL EXPENSE AND OVERHEAD (CONTINUED)

ITEM NO. ___ SHEET 8 OF 8

JOB ___

EST. BY ___ CK'D BY ___ DATE ___

	LABOR AND INSURANCE		MATERIALS		PLANT		EQUIPMENT RENTAL		SUPPLIES		SUB-CONTRACTS		TOTAL	
	UNIT COST	AMOUNT	UNIT COST	AMOUNT	UNIT COST	AMOUNT	UNIT COST	AMOUNT	UNIT COST	AMOUNT	UNIT COST	AMOUNT	UNIT COST	AMOUNT
Brought Forward														
INSURANCE OTHER THAN APPLICABLE TO PAYROLL:														
Excess P. L. and P. D.														
Contractors Plant and Equipment														
General Plant Buildings $ ___														
Camp Buildings														
Mobile Equipment														
Concrete Plant & Equipment														
Aggregate Plant														
General Expense Equipment														
Total Plant & Equipment Value ___														
___ Yrs. Use @ ___ % = ___ % of Above Value														
Builders Risk $ ___ Bid @ ___ %														
Fidelity Insurance														
Total Insurance Other Than Applicable To Payroll														
TAXES OTHER THAN APPLICABLE TO PAYROLL:														
County Tax on Plant & Equipment														
___ Yrs. Use @ ___ % of Value $ ___														
State Business Tax														
City Business Tax														
Total Taxes Other Than Payroll														
INTEREST ON INVESTMENT														
CENTRAL OFFICE EXPENSE														
___ % on Cost of Work														
CONTRACT INDEMNITY														
Tender Guarantee														
Performance and Payment Bonds														
Other Types of Contract Indemnity														
TOTAL GENERAL EXPENSE AND OVERHEAD														

Figure 14-19-2 (Continued)

JOHN DOE CONSTRUCTION COMPANY, INC.

HOURLY EQUIPMENT OPERATING COSTS

JOB		CONT. NO.		SHEET	OF	PAGE
LABOR:	H.D. MECHANIC	PER HR.	SUPPLIES:	DIESEL FUEL	PER GAL.	DATE
	MECH. HELPER	PER HR.		GASOLINE	PER GAL.	
	*P.R. TAX & INS.	%		BUTANE	PER GAL.	
	*OVERTIME	%		ELECT. POWER	PER KWH	BY

TYPE OF EQUIPMENT

Model or Capacity

Horsepower
Fuel or Power Consumption per hr.
New Cost
Shipping Weight

OPERATION

LABOR COST
Operator
Oiler
Repair & Service
Total

RENTAL OR DEPRECIATION
New
Used
Cost Per Hour

SUPPLIES & EXPENSE
Fuel or Power
Oil & Grease
Repair Parts
Cable & Teeth
Tires & Tubes
Outside Repairs
Shop Costs
Total

TOTAL HOURLY COST

TYPE OF EQUIPMENT

Model or Capacity

Horsepower
Fuel or Power Consumption per hr.
New Cost
Shipping Weight

OPERATION

LABOR COST
Operator
Oiler
Repair & Service
Total

RENTAL OR DEPRECIATION
New
Used
Cost Per Hour

SUPPLIES & EXPENSE
Fuel or Power
Oil & Grease
Repair Parts
Cable & Teeth
Tires & Tubes
Outside Repairs
Shop Costs
Total

TOTAL HOURLY COST

*Note: Cost to be included under Labor Subdivisions to be Exclusive of payroll taxes and ins. and overtime, or actual percentage of same indicated in space at top of sheet.

Figure 14-22-1 Estimated hourly equipment operating costs

SUMMARY OF EQUIPMENT AND MOVE-IN
LEONARD'S LANDING DAM STAGE II

EXCAVATION EQUIPMENT

VALUES USED ARE FICTITIOUS FOR ILLUSTRATION ONLY

Item	Condition	No. Units Req.	Location of Equipment	Unit Weight	Total Weight	Freight Rate c.wt.	Total Freight Cost	Unit Cost of Equip. New	Total Cost of Equip. New	Used Cost at 75% of New
NW 800 Shovel & D.L.		1	Factory	154,500	154,500	2.00	3090	54,210	54,210	
4500 Manitowoc Shovel & D.L.		1	Factory	381,810	381,810	2.00	7636	197,357	197,357	
600 CFM Compressor		1	"	8,400	8,400	3.00	252	13,750	13,750	
5 KW Light Plants		11	"	2,500	27,500	2.00	550	1,150	12,650	
50 Ton Compactor		1	L.A.	24,000	24,000	3.50	840	13,000	13,000	
Spiked Roller		1	"	11,000	11,000	2.00	220	6,000	6,000	
D-8 Cat Bare		6	"	37,575	225,450	2.00	4510	16,398	98,388	
D-8 Dozer		4	"	5,600	22,400	2.00	448	1,823	7,213	
17 C.I. End Dump		6	"	40,800	244,800	2.50	612	27,018	162,108	
23 C.I. End Dump		8	Boise	64,000	512,000	3.00	15360	41,278	330,224	247,668
20 T. Truck Crane		1	Factory	57,000	57,000	2.00	1140	27,700	27,700	
Motor Patrol		2	"	23,975	47,950	2.00	960	14,493	28,986	
4500 Gal. Water Truck		2	Boise	33,000	33,000	3.00	990	24,025	24,025	
Pickups (Direct cost only)		2	Jobsite	2,000	4,000	--	--	1,700	3,400	
Grease Truck		1	"	5,000	5,000	--	--	7,000	7,000	
Fuel Truck		1	"	4,000	4,000	--	--	7,000	7,000	
Flat Rack		2	"	4,000	8,000	--	--	5,000	10,000	
Shale & Shalk saw (Move-in under item)		1	New					39,262	39,262	
Shale & Chalk (Drill & Shoot)		1	"	7,000	7,000	3.50	245	15,000	15,000	
Horiz. Drill) (Move-in under		1	"					3,621	3,621	
Vert. Drill) Item		1	"	83,225	83,225	2.00	1664	14,271	14,271	
Twin D-8		1	"	10,000	10,000	2.00	200	41,621	41,621	
Ripper		1	"	12,000	12,000	3.00	360	3,500	3,500	
400 Amp. Welder		3	"	4,000				1,700	5,100	
Miscellaneous	Lot		"	50,000	50,000	2.50	1250	15,000	15,000	
315 cfm Comp Move-in under item		1	"					9,000	9,000	
Group Pump		1	"					1,500	1,500	
					1,923,025		40,327		1,150,985	1,068,429

		LABOR	EQUIP	EXPENSE
Unload & haul to site	1,000 tons	3,500	1,000	1,500
Erect Shovels	268 tons	1,340	134	268
Freight to Site				40,327
Total Cost	48,069	4,840	1,134	42,095

DISTRIBUTION:

Item No.	TOTAL	LABOR	EQUIP	EXPENSE
116	10,094	1,016	238	8,840
117	1,923	194	45	1,684
118	28,842	2,904	681	25,257
120	1,442	145	34	1,263
121	5,758	581	136	5,051

Figure 14-23-1 Summary of equipment and move-in costs

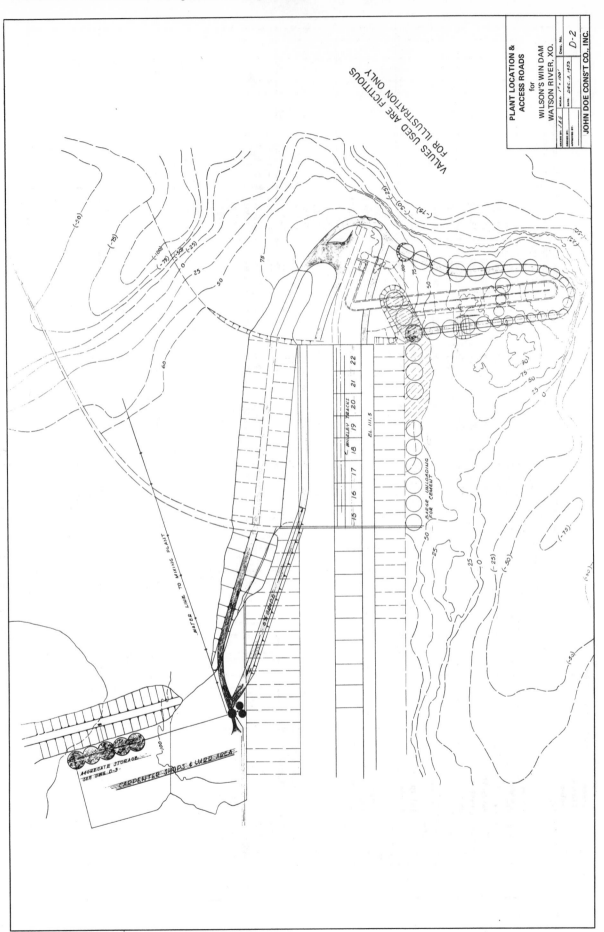

Figure 14-26-1 General job layout plan

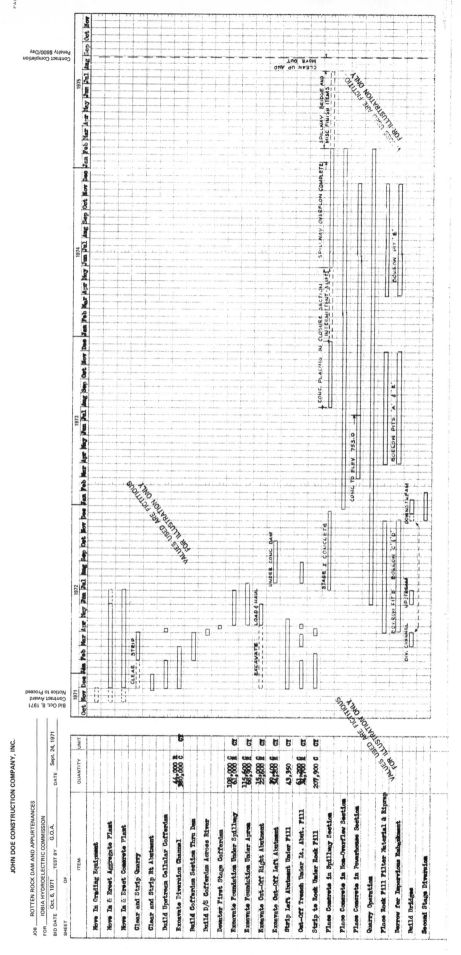

Figure 14-26-2 Construction program

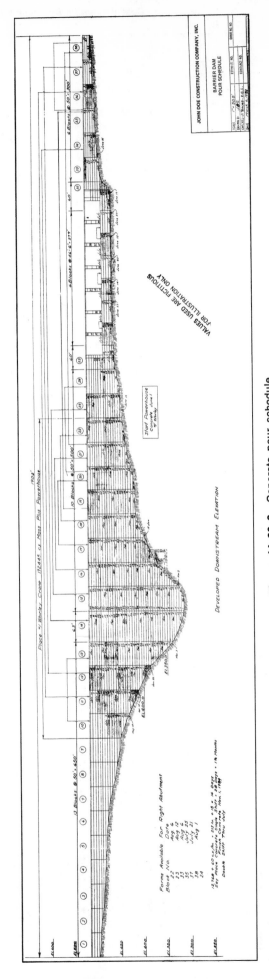

Figure 14-26-3 Concrete pour schedule

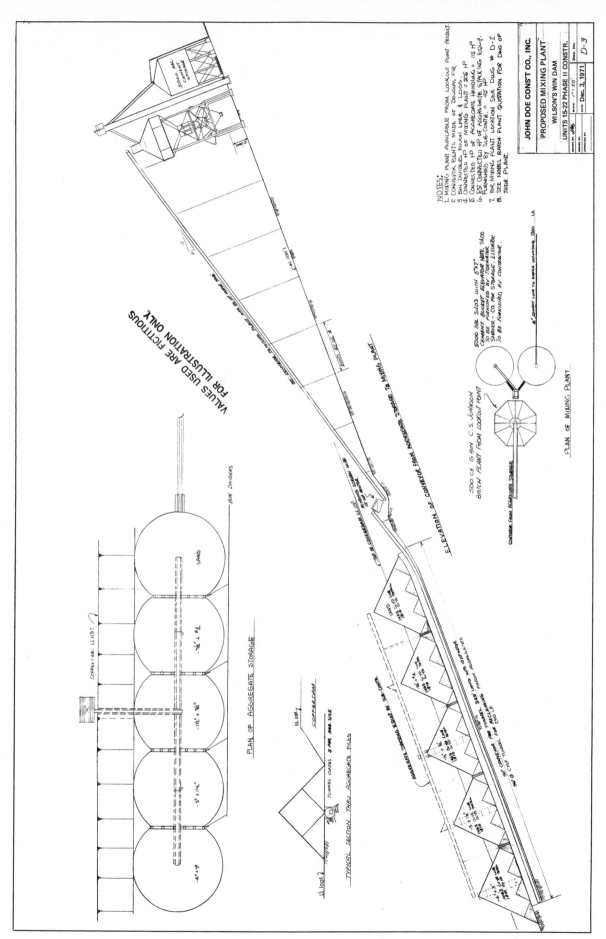

Figure 14-26-4 Construction plant layout—concrete plant

JOHN DOE CONSTRUCTION COMPANY, INC.

JOB: WILSON'S WIN DAM   ITEM NO.   PAGE 1 OF 10   SHEET

CONCRETE — GENERAL SUMMARY OF PLANT AND EQUIPMENT COST

BY   DATE 12-11-71

	LABOR AND INSURANCE		PERMANENT MATERIALS		SPECIFIC PLANT		EQUIPMENT		SUPPLIES		FIRST COST OF SALVAGEABLE ITEMS SUBCONTRACTS		TOTAL DIRECT COST		ESTIMATED COST
	UNIT COST	AMOUNT	UNIT COST	AMOUNT	UNIT COST	AMOUNT	UNIT COST	AMOUNT	UNIT COST	AMOUNT	UNIT COST	AMOUNT	UNIT COST	AMOUNT	
CONCRETE PLANT:															
Plant Site Grading and Roads		6,805						3,560		3,935				14,300	
Crawler Cranes (2 - 4500 Manitowocs)		2,528						327,116		19,758		326,800		349,402	
Whirley Crane		6,004						136,567		11,030		135,000		153,601	
Trestle & Track for Whirley		20,828						4,986		62,350		38,514		88,164	
Batching & Mixing Plant		12,990						109,537		19,402		107,407		141,929	
Aggregate Handling & Rinsing Plant		10,689						51,438		8,994		46,120		71,116	
Concrete Hauling Trucks & Haul Roads, etc.		---						91,000		4,130		91,000		95,130	
Cement Handling Plant		2,634						11,209		3,334		10,456		17,177	
Air Plant		4,650						21,763		18,143		34,773		44,556	
Water Plant		2,100						8,700		8,190		15,000		18,990	
Electrical Plant & Lighting System		15,750						19,950		18,028		21,730		53,725	
Miscellaneous Equipment										51,622		42,954		51,622	
Telephone & Signal System Allow		1,000								1,500		1,500		2,500	
Sand Drying Plant for Sand Blasting		1,000								2,000		2,000		3,000	
Walkways & Ladders		1,500								3,000				4,500	
Concrete Control Room & Temporary Field Office		1,305								2,625				3,930	
TOTAL COST OF CONCRETE PLANT		89,783						785,821		238,041		873,254	3.225	1,113,645	
Unit + Take-Off = 345,292 Cu. Yd.														3,225	
Transfer Labor and Supplies to Specific Plant	0														
Unit Cost for Distribution to Work Items							0.950	327,824	2.275	785,821	(1.00) (.345292)		3.225	1,113,645	
Salvage Recommended —															

VALUES USED ARE FICTITIOUS FOR ILLUSTRATION ONLY

Figure 14-26-5 Construction plant pricing

JOHN DOE CONSTRUCTION COMPANY, INC.

JOB: WILSON'S WIN DAM
CONCRETE – GENERAL PLANT COST
ITEM NO.
BY
DATE 12-14-71
SHEET 2 OF 10
PAGE

VALUES USED ARE FICTITIOUS FOR ILLUSTRATION ONLY

Description	Quantity	LABOR AND INSURANCE UNIT COST	AMOUNT	EQUIPMENT RENTAL UNIT COST	AMOUNT	SUPPLIES UNIT COST	AMOUNT	FIRST COST OF SALVAGEABLE ITEMS SUBCONTRACTS UNIT COST	AMOUNT	TOTAL DIRECT COST UNIT COST	AMOUNT
PLANT SITE GRADING & ROADS:											
Plant to be set up in an area already graded by a previous contract Say 3000 Cu. Yd. Excavation											
Common	1500 C.Y.	.20	300	.15	225	.15	225			.50	750
Rock	1500 C.Y.	.65	975	.25	375	.50	750			1.40	2100
Road Surfacing 7000 x 30 x 1 =	7800 C.Y.	.35	2730	.20	1560	.20	1560			.75	5850
Road Maintenance	28 Mos.	100.00	2800	50.00	1400	50.00	1400			200.00	5600
TOTAL			6805		3560		3935				14,300
CRAWLER CRANES:											
Concrete can be placed in all structures except for topping out of intake structure & fishway deck. (2) 4500 Manitowoc Cranes will be required for placing concrete in these structures.											
2 - 4500 Manitowoc w/live boom					326,800				326,800	1.50	326,800
Freight in	6320 cwt					1.50	9480			1.50	9,480
Unload & erect	316 Tons	8.00	2528	1.00	316	1.50	474			10.50	3,318
Sales Tax on 326,800.00							9804				9,804
TOTAL			2528		327,116		19,758		326,800		349,402
WHIRLEY CRANE:											
One whirley crane will be required on a trestle for the powerhouse structure & intake structure.											
1 Crane w/20 Ton Capacity & 125' Boom per- quote of Washington Iron Works					135,000				135,000		135,000
Freight on	2700 cwt					1.00	2700			1.00	2,700
Sales Tax on $135,000.00							4050				4,050
Unload & Haul	135 Tons	3.00	405	.50	68	1.50	202			5.00	675
Erect Crane	135 Tons	35.00	4725	10.00	1350	15.00	2025			60.00	8,100
Ballast Purchase	60 Tons					15.00	900			15.00	900
Ballast Install.	60 Tons	3.00	180	.50	30	1.50	90			5.00	300
Power Supply	688 L.F.	.50	344	.10	69	1.40	963			2.00	1,376
Wire Crane	1 Ea.		350		50		100				500
TOTAL			6004		136,567		11,030		135,000		153,601

Figure 14-26-5 (Continued)

JOHN DOE CONSTRUCTION COMPANY, INC.

JOB: WILSON'S WIN DAM — ITEM NO. — PAGE
CONCRETE – GENERAL PLANT COST — BY — SHEET 3 OF 10 — DATE 12-14-71

WHIRLEY TRESTLE: 32' O. C. Whirley Rails Figure
To Buy New Steel & Fabricate Trestle on Job

MATERIAL NEEDED:

Description	Quantity	Labor and Insurance Unit Cost	Labor and Insurance Amount	Equipment Rental Unit Cost	Equipment Rental Amount	Supplies Unit Cost	Supplies Amount	First Cost of Salvageable Items Subcontracts Unit Cost	First Cost of Salvageable Items Subcontracts Amount	Total Direct Cost Unit Cost	Total Direct Cost Amount
2408' – 36" WF x 300 # Beams =	722,400 #										
Stiffner Plate =	4,000 #										
Anchor Bolts =	1,000 #										
1204' – 105# Rail =	42,105 #										
34 Fr. Splice Bars =	1,870 #										
4 – Crane Stops =	1,000 #										
1204 – Rail Clips =	6,622 #										
21 Ea. Bracing 12" WF 40 x 32' =	26,880 #										
1204 – Bolts & Nuts =	1,050 #										
Connections @ 200#/GA =	8,800 #										
TOTAL	815,727 #										
Purchase Structural Sections	377 Tons					92.00	34,684		34,684	92.00	34,684
Purchase Used Rail & Accessories	26 Tons					80.00	2,080		2,080	80.00	2,080
Purchase New Anchor Bolts, Misc. Steel	5 Tons					350.00	1,750		1,750	350.00	1,750
Frt. on Structural Sections – Penna.	377 Tons					35.00	13,195			35.00	13,195
Frt. on Rail	26 Tons					30.00	780			30.00	780
Frt. on Anchor Bolts, etc. (Local)	5 Tons					30.00	150			30.00	150
Build Wood Cantilevered Walkway for Access to Whirley											
Purchase Structural Douglas Fir	10 M					100.00	1,000			100.00	1,000
Install	10 M	60.00	600	10.00	100	30.00	300			100.00	1,000
Misc. Rough Hdw.							100				100
Weld Beams Together – 1204 x 2 =	2408 Welding	2.50	6,020	.25	602	.50	1,204			3.25	7,826
Unload & Haul Steel to Site	408 Tons	3.50	1,428	.50	204	1.50	612			5.50	2,244
Erect Steel	408 Tons	30.00	12,240	10.00	4,080	10.00	4,080			50.00	20,400
Sales Tax on $38,514							1,155				1,155
Paint	18,000 Sq.Ft.	.03	540			.07	1,260			.10	1,800
TOTAL TRESTLE FOR WHIRLEY			20,828		4,986		62,350		38,514		88,164

VALUES USED ARE FICTITIOUS FOR ILLUSTRATION ONLY

Figure 14-26-5 (Continued)

JOHN DOE CONSTRUCTION COMPANY, INC.

JOB: WILSON'S WIN DAM   ITEM NO. _____   PAGE ____   SHEET 4 OF 10
CONCRETE – GENERAL PLANT COST   DATE 12-12-71   BY ____

BATCHING & MIXING PLANT:

Use Lookout Point's Plant w/Additional Mixer
List Plant quoted by Johnson for Wt. & Price

Item	Weight	Price
1 - 400 C.Y. 420 Ton Bin w/Cement Bin	136,000#	$27,647.00
4 - Aggregate Weigh Batchers	4,200#	4,224.00
1 - Cobble Batcher	1,350#	1,225.00
1 - Water Batcher	1,950#	2,142.00
1 - Admix. Batcher	750#	775.00
1 - Cem. Batcher w/Scale & Recorder	1,350#	1,441.00
1 - Recording Dial Scale	5,000#	11,220.00
1 - Set Test Weights	2,500#	966.00
1 - Dry Batch Collecting Cone	7,000#	1,680.00
1 - Swivel Chute	5,500#	4,317.00
3 - 2 C.Y. Tilting Mixers	49,500#	34,815.00
	7,000#	2,330.00
Sub-Total Lookout Point Plant	222,100#	92,802.00

CWT. Batch Hopper (New Cost)

VALUES USED ARE FICTITIOUS FOR ILLUSTRATION ONLY

Cost breakdown:

Description	Qty/Unit	Labor Unit Cost	Labor Amount	Equip. Rental Unit	Equip. Rental Amount	Supplies Unit Cost	Supplies Amount	First Cost of Salvageable Items / Sub-Contracts Amount	Total Direct Unit Cost	Total Direct Amount
Sub-Total Lookout Point Plant					92,802			92,802		92,802
Frt. on: Structural Steel 143,000# = 1450 cwt						1.60	2,288			2,288
Used Contr's. Equip. 79,100# = 791 cwt						2.85	2,254			2,254
NEW ADDITIONS:										
1 - Add. of New 2 C.Y. Tilting Mixer 16,500# $11,605.00					11,605			11,605		11,605
1 - 50 HP Compressor for Plant 2,000# 3,000.00					3,000			3,000		3,000
1 - Add. of Time Stamp to Mixer Record.					500					500
Frt. to Jobsite	185 cwt					3.50	648			648
Unload & Haul to Jobsite	120 Tons	3.50	420	1.00	120	1.50	180		6.00	720
Erection - Structural Steel	72 Tons	40.00	2,880	10.00	720	10.00	720		60.00	4,320
Erection - Mixers	34 Tons	25.00	850	5.00	170	5.00	170		35.00	1,190
Erection - Misc. Machinery	12 Tons	100.00	1,200	30.00	360	50.00	600		180.00	2,160
Foundations - Excavation, Common	10 C.Y.	4.00	40						4.00	40
Excavation, Rock	20 C.Y.	5.00	100	.50	10	1.00	20		6.50	130
Concrete Ftgs. & Slabs, etc.	50 C.Y.	15.00	750	5.00	250	25.00	1,250		45.00	2,250
Anchor Bolts	5,000 #		100				200			300
Reinforcing Steel		.03	150			.07	350		.10	500
Misc. Platforms Gaurds, etc.	L.S.		1,000				1,000			2,000
Painting			500				500			1,000
Enclose Plant & Insulate			2,000				3,000			5,000
Electrical Wiring			3,000				3,000			6,000
Sales Tax 3% on 107,407							3,222			3,222
TOTAL			12,990		109,557		19,402	107,407		141,929

VALUES USED ARE FICTITIOUS FOR ILLUSTRATION ONLY

Figure 14-26-5 (Continued)

JOHN DOE CONSTRUCTION COMPANY, INC.

JOB: WILSON'S WIN DAM
CONCRETE – GENERAL

ITEM NO.   BY   SHEET 5 OF 10   DATE 12-11-71   PAGE

Description	Quantity	Labor and Insurance Unit Cost	Labor and Insurance Amount	Permanent Materials Unit Cost	Permanent Materials Amount	Specific Plant Unit Cost	Specific Plant Amount	Equipment Rental Unit Cost	Equipment Rental Amount	Supplies Unit Cost	Supplies Amount	First Cost of Salvageable Items Subcontracts Unit Cost	First Cost of Salvageable Items Subcontracts Amount	Total Direct Cost Unit Cost	Total Direct Cost Amount	Estimated Cost
AGGREGATE HANDLING: (See Attached Schedule)																
Aggregate Sub-Contr. will furnish aggregate stacking																
Excavate Rock for 96" Ø Tunnel	785 C.Y.	1.50	1,178					.50	393	1.00	785			3.00	2,356	
Backfill	785 C.Y.	0.30	236					.15	118	.20	157			.65	511	
Misc. Concrete Ftgs.	10 C.Y.	15.00	150					5.00	50	25.00	250			45.00	450	
Mat'ls. & Equipment per Att. Schedule									49,350				46,120		49,350	
Frt. on Items not FOB Site											1,435				1,435	
Install 96" Ø CMP	290 L.F.	7.50	2,175					1.00	290	1.50	435			10.00	2,900	
Install Structural Steel	Tons		(Installed under separate items)													
Lumber	18 M	100.00	1,800					20.00	360	50.00	900			170.00	3,060	
Install Conveyor	670 L.F.	3.50	2,345					.25	168	.75	503			4.50	3,016	
Install Machinery	Tons		(Installed under separate items)													
Install Pump	1 Each		150						50		150				350	
Install Spray Piping			150						50		50				200	
Install Culvert Pipe 18"	100 L.F.	1.50	150					.10	10	.15	15			1.75	175	
Allowances for Drain Ditches			100						20		30				150	
Water Collection under Screen			100						25		100				225	
Transfer Chutes			50						10		40				100	
Electrical Connections	4 Each	75.00	300					20.00	80	40.00	160			135.00	540	
Tunnel Gates	15 Each	50.00	750					25.00	375	15.00	225			90.00	1,350	
Conveyor Walkway & Lighting	670 L.F.	1.50	1,005					.20	134	3.50	2,345			5.20	3,484	
Sales Tax on $46,120 (3%)											1,384				1,384	
Install Rinsing Screen			50						10		50				100	
			10,689						51,433		8,994		46,120		71,116	

Figure 14-26-5 (Continued)

JOHN DOE CONSTRUCTION COMPANY, INC.

WILSON'S WIN DAM   AGGREGATE HANDLING EQUIPMENT   ITEM NO. _____   BY _____   DATE 12-11-71   SHEET 5A OF 10   PAGE _____

DESCRIPTION	QUANTITY	LOCATION	UNIT WT.	TOTAL WT.	FRT. TO	UNIT COST	TOTAL COST		SALVAGABLE ITEMS
96" Ø CMP Tunnel	290 L.F.	Port.	290	66,700	1,167 JOBSITE	33.00	9,570		9,570
30" 5 Ply 32 oz. Belting	1,570 L.F.	Bol.	818	13,000	FOB Job	6.75	10,598		10,598
Troughing Rollers 30"	170 Ea.		80	13,600	FOB Job	30.00	5,100		5,100
Return Rollers 30"	70 Ea.		35	2,450	FOB Job	15.00	1,050		1,050
18"x32" Tail Pulley	2 Ea.		300	600	FOB Job	150.00	300		300
18" Dia. Screw Take-up	1 Ea.		120	120	FOB Job	75.00	75		75
10 HP Motorized Head	1 Ea.		265	265	FOB Job	957.00	957	10 HP	957
Tunnel Gates	15 Ea.	Mont.	1,100	16,500	FOB Job	350.00	5,250		5,250
Structural Lbr. for Conveyors	11 M					100.00	1,100		
Misc. Bolts	L.S	Port.		500	30	100.00	100		
Misc. Blocking Lbr.	2 M					70.00	140		
Bin Dividers	5 M					60.00	300		
Impact Rollers 30"	30 Ea.	Bol.	60	1,800	FOB Job	64.00	1,920		1,920
18" CMP Drains	100 L.F.	Port.			FOB Job	2.50	250		
Waterline to River 8"	1,200 L.F.	Beckley	8	9,600	168	1.20	1,440		
Pump - 500 Gal. per Min.	1 Ea.	Port.		4,000	FOB Job	2,200.00	2,200	40 HP	2,200
Spray Piping				1,000	50	500.00	500		
Gravity Take-up							1,500		1,500
Rinsing Screen			5,500	5,500	FOB Job	2,500.00	2,500	15 HP	2,500
50 HP Motorized Head Pulley			1,300	1,300	FOB Job	4,500.00	4,500	50 HP	4,500
				136,935	1,415.00		49,350	115 HP	46,120

VALUES USED ARE FICTITIOUS FOR ILLUSTRATION ONLY

Figure 14-26-5 (Continued)

JOHN DOE CONSTRUCTION COMPANY, INC.

JOB: WILSON'S WIN DAM   ITEM NO. ____   SHEET 6 OF 10   PAGE ____
CONCRETE – GENERAL PLANT COST
BY ____   DATE 12-14-71

Description		LABOR AND INSURANCE		PERMANENT MATERIALS		SPECIFIC PLANT		EQUIPMENT RENTAL		SUPPLIES		FIRST COST OF SALVAGEABLE ITEMS SUB-CONTRACTS		TOTAL DIRECT COST		ESTIMATED COST
		UNIT COST	AMOUNT	UNIT COST	AMOUNT	UNIT COST	AMOUNT	UNIT COST	AMOUNT	UNIT COST	AMOUNT	UNIT COST	AMOUNT	UNIT COST	AMOUNT	
CONCRETE HAULING TRUCKS & HAUL ROADS																
4 C.Y. Bucket @ 1 Bucket per Unit																
Load = 1.5 Min.																
Haul 2500' 4.1 Min.																
Unload 2.0 Min.																
7.6 Min. = 4 Trips/Hr. 16 C.Y./Hr.																
6 Units = 96 C.Y./Hr. Max.																
1 Spare																
Purchase 7 Euclid 10 Ton Flatracks FOB Factory	7 Ea.							13000.	91,000				91,000		91,000	
(Lookout Pt.) Frt. to Jobsite Drive										200.00	1,400				1,400	
Sales Tax 3% on $91,000.00											2,730				2,730	
Haul Roads figured under Plant Site & Grading																
Total Conc. Hauling Flatracks								13000.	91,000		4,130		91,000		95,130	

VALUES USED ARE FICTITIOUS FOR ILLUSTRATION ONLY

Figure 14-26-5 (Continued)

JOHN DOE CONSTRUCTION COMPANY, INC.

JOB: WILSON'S WIN DAM ITEM NO. _____ CONCRETE – GENERAL PLANT COST
BY _____ DATE 12-14-71 PAGE _____ SHEET 7 OF 10

VALUES USED ARE FICTITIOUS FOR ILLUSTRATION ONLY

Description	Qty	Labor Unit	Labor Amt	Equip. Rental Unit	Equip. Amt	Specific Plant Unit	Specific Plant Amt	Supplies Unit	Supplies Amt	Salvageable/Sub Amt	Total Dir. Unit	Total Dir. Amt
CEMENT HANDLING PLANT: Specifications call for 10,000 Bbl. storage at the site. As per Tidewater – Shaver Barge Co. quotation they will supply 10,000 Bbl. storage at the site if they deliver cement by water – contractor to supply screws under silos & bucket elevator.												
2-10" Dia. screw conveyors 5,200 #	4313.00											
1-14" x 8" bucket elevator 19,130 #	6143.00											
24,330 # $10,456.00										10456		10456
Frt. to jobsite	243 cwt										2.00	486
Unload & haul to jobsite	12½ tons	3.50	44	1.00	13						6.00	76
Erect	12½ tons	80.00	1000	30.00	375			20.00	625		160.00	2000
Electrical hookups	Allow.		600						400			1200
Foundations Excavate Rock	30 c.y.	5.00	150	.50	15			1.00	30		6.50	195
Concrete Trenches & Ftgs.	30 c.y.	15.00	450	5.00	150			25.00	750		45.00	1350
Anchor Bolts	Allow.		50						50			100
Re-Steel	3000 #	.03	90	.07	210						.10	300
Trench Covers	L.S.		100						200			300
Platforms & Ladders	L.S.		50						100			150
Painting	L.S.		100						150			250
3% sales tax on 10,456.00									314			314
TOTAL CEMENT PLANT			2634		11209				3334	10456		17177
AIR PLANT: Provide a max. of 1500 c.f.m. figure 3-Gardiner Denver water cooled skid mounted units w/125 HP synchronous motors												
3-WEJ units fob factory	3 Ea.			6745.00	20235					20235		20235
Elec. Starting equip.	3 sets			371.00	1113					1113		1113
Frt. to jobsite	280 cwt							2.50	700		2.50	700
Unload & set up compressors	3 Ea.	250.00	750	50.00	150			100.00	300		400.00	1200
Compressor building 20 x 30	600 sf	2.00	1200					2.00	1200		4.00	2400
Air Receiver	1 Ea.				25				1200	1200		1275
AIR PIPING: & DISTRIBUTION SYSTEM												
8" pipe	2500 lf	.30	750	.05	125			1.35	3375	3125	1.70	4250
6" pipe	2500 lf	.20	500	.03	75			1.01	2525	2100	1.24	3100
3" pipe	200 lf	.20	400	.02	40			.90	1800	1200	1.12	2240
ELECTRICAL HOOKUP	Allow		1000						1000			2000
Valves Fittings etc.									5000	5000		5000
3% sales tax on									1043			1043
			4650		21763				18143	34773		44556

Figure 14-26-5 (Continued)

JOHN DOE CONSTRUCTION COMPANY, INC.

JOB: WILSON'S WIN DAM ITEM NO. PAGE SHEET 8 OF 10
CONCRETE — GENERAL PLANT COST BY DATE 12-11-71

VALUES USED ARE FICTITIOUS FOR ILLUSTRATION ONLY

Description	Qty	Labor and Insurance Unit Cost	Amount	Permanent Materials Unit Cost	Amount	Specific Plant Unit Cost	Amount	Equipment Rental Unit Cost	Amount	Supplies Unit Cost	Amount	Salvageable Items Unit Cost	Amount	Total Direct Cost Unit Cost	Amount	Estimated Cost
WATER PLANT Figure to set up pumps on east end of cellular cofferdam. Provide 1500 gpm of water @ 50 # pressure on top of intake 225'																
Figure 3 – 750 gpm units (1 spare)	3 Ea.							2800.00	8400				8400		8400	
Starting equipment	3 sets							(Included above)								
Frt. to the Job																
Install Pumps	3 Ea.	150.00	450					50.00	150	150.00	450			350.00	1050	
Distribution Lines																
8" pipe	2000 lf	.30	600					.05	100	1.35	2700		2700	1.70	3400	
6" pipe	1000 lf	.20	200					.03	30	1.00	1000		1000	1.23	1230	
3" pipe	1000 lf	.10	100					.02	20	0.99	900		900	1.02	1020	
Valves, fittings, etc.	L.S.										2000		2000		2000	
Electrical Hookup	3 Ea.	150.00	450							1.50	450			300.00	900	
Shelter over pumps	L.S.		300								300			L.S.	600	
Sales tax on equip. 3% of $1300.00											390				390	
TOTAL WATER PLANT			2100						8700		8190		15000		18990	
MISCELLANEOUS EQUIPMENT																
4 c.y. buckets 1 rig & 2 man & 2 spare = 32000#	8 Ea.									1811.00	14488		14488		14488	
2 c.y. buckets 15000#	6 Ea.									1666.00	9996		9996		9996	
1 c.y. bucket 4000#	4 Ea.									900.00	2000		2000		2000	
Mass concrete vibrators 2000#	6 Ea.									700.00	4200		4200		4200	
Frequency changers 2000#	2 Ea.									2000.00	4000		4000		4000	
Wiggle tail vibrators 2000#	15 Ea.									350.00	5250		5250		5250	
Rigging Supplies											3000		3000		3000	
Small tools for concrete											5000		—		5000	
Frt. Allowance Say 80,000#										3.00/cwt	2400		—		2400	
Repairs on vibrators & buckets											(To be distributed to concrete items by the c.y.)					
Sales tax 3% on 42394											1288				1288	
TOTAL MISCELLANEOUS EQUIPMENT											51622		42394		51622	

Figure 14–26–5 (Continued)

JOHN DOE CONSTRUCTION COMPANY, INC.

JOB	WILSON'S WIN DAM	ITEM NO.		PAGE 76
	CONCRETE – GENERAL PLANT COST			SHEET 9 OF 10
		BY		DATE 12-13-71

Item		Labor and Insurance		Permanent Materials		Specific Plant		Equipment Rental		Supplies		First Cost of Salvageable Items Sub-Contracts		Total Direct Cost		Estimated Cost
		Unit Cost	Amount	Unit Cost	Amount	Unit Cost	Amount	Unit Cost	Amount	Unit Cost	Amount	Unit Cost	Amount	Unit Cost	Amount	
CONCRETE CONTROL ROOM & TEMPORARY FIELD OFFICE																
Control room as per spec's. 8' x 12' dust free & sound proof. Control room to be attached to mixing plant control platform	100 sf	3.00	300							3.00	300			6.00	600	
Portable temporary field office 10' x 12' =	120 sf	3.50	420							3.50	420			7.00	840	
Concrete testing room 10' x 14' =	140 sf	2.75	385							2.75	385			5.50	770	
Electrical wiring			100								200				300	
Telephone wire installation etc.			100								200				300	
Drinking Water	28 Mos. @									10.00	280				280	
Heating	28 Mos. @									30.00	840				840	
TOTAL			1305								2625				3930	
ELECTRICAL PLANT & LIGHTING SYSTEM																
Construct distribution line from substation near Haelys office to west end of powerhouse. Unit 15	4000 lf	1.50	6000					.10	400	.75	3000		3000	2.35	9400	
Transformers and substations:																
Main substation provided by PUD #1 Klickitat County for delivery of power at 7200/12470 volts.																
Secondary substation & stepdown transformers at powerhouse only	1500 KVA							1250.00	18750				18750		18750	
Install substation only			1000								500				1500	
Install transformers	1500 KVA	1.50	2250							3.75	5625			5.25	7875	
Miscellaneous Frt.											1000				1000	
Lighting System for Powerhouse Non-overflow & Fishladder		1.50	2000						500		5000				7500	
Miscellaneous Distribution Lines	3000 lf	1.50	4500					.10	300	.75	2250			2.35	7050	
Sales tax 3% on 21,750											653				653	
TOTAL COST LIGHTING PLANT			15750						19950		18028		21750		53278	

VALUES USED ARE FICTITIOUS FOR ILLUSTRATION ONLY

Figure 14-26-5 (Continued)

JOHN DOE CONSTRUCTION COMPANY, INC.

OPERATION	CONCRETE — GENERAL PLANT COST			ITEM NO.		SHEET 10 OF 10 PAGE
				JOB	WILSON'S WIN DAM	
QUANTITY	UNIT	EST. BY		CK'D BY		DATE 12-13-71

ELECTRIC POWER REQUIREMENTS	CONNECTED HORSEPOWER		
BATCHING PLANT:			
For Lookout Point correspondence	145 HP		
Add one 4 c.y. mixer	75 HP		
		220	
AGGREGATE HANDLING			
1 Rinsing screen	15 HP		
1 500 gal. per min. pump	40 HP		
2 conveyor head pulleys	55		
		110	
CEMENT HANDLING			
Bucket elevator	15 HP		
Screws	15 HP		
		30	
Air Plant		375	
Water Plant		180	
Whirley Crane Main Hoist	200 HP		
Slewer Motor	50 HP		
Travel	60 HP		
		310	
Job Lighting		167	
Shops & MK Co.		75	
Frequency changers		40	
Total Estimated HP		1507 HP	
Say		1500 HP	

VALUES USED ARE FICTITIOUS FOR ILLUSTRATION ONLY.

Figure 14–26–5   (Continued)

OPERATION CONCRETE PLANT OPERATION	ITEM NO.	SHEET 1 OF 4 PAGE
	JOB WILSON'S WIN DAM	
QUANTITY UNIT EST. BY	CK'D BY	DATE 12-14-71

	LABOR	EQUIP.	EXPENSE
BATCHING & MIXING PLANT			
LABOR:			
1-Foreman = 2.875			
1-Plant Operator = 2.75			
1-Oiler = 2.15			
1-Laborer = 2.10			
1-H.D. Mechanic = 2.65			
	12.53		
EXPENSE:			
Repair parts and expense			3.50
Electric Power:			
3 of 4 mixers operating @ 75 HP = 225 HP			
1 compressor = 50 HP			
1-bucket elevator = 15 HP			
1-screw = 10 HP			
300 HP			
= 225 KW			
@ 70% = 158 KW@ .0075			1.19
Lube, oil & grease			.20
Shop cost			150
Total cost per hour	12.53		5.39
Good hours =	7500.00		35,340

VALUES USED ARE FICTITIOUS FOR ILLUSTRATION ONLY.

Figure 14-26-6 Special plant operating cost

OPERATION	CONCRETE PLANT OPERATION		ITEM NO.	SHEET 2 OF 4 PAGE
			JOB	**WILSON'S WIN DAM**
QUANTITY	UNIT	EST. BY	CK'D BY	DATE 12-14-71

	LABOR	EQUIP.	EXPENSE
AGGREGATE HANDLING			
(Operate Rinsing Plant & Conveyors to Bins)			
LABOR:			
Foreman (in batch plant) = 0			
Conveyor operator oiler = 2.15			
Repair Labor = 0.50			
2.65	2.65		
EXPENSE			
Power = 110 HP @ .75 = 83 KW @ 70% = 58 KW			
58 KW @ .0075 = 0.44			
Lube, oil & grease = 0.20			
Pump parts = 0.15			
Screen parts & wire = 0.10			
Conveyor parts = 0.15			
Shop costs = 0.25			
1.29			1.29
Total Aggregate Handling /hr.	2.65		1.29
6,000 hours @ =	15900.00		7740.00
Unit ÷ 345292	0.046		0.022
WATER PLANT 18 Mos. operation			
2-750 gpm units ≠ one spare figure automatic			
system with one man attending 2 hrs/day = 18 x			
30 x 2 = 1080 hrs. Operation hrs. = 18 x 30 x 24 =			
12960 hrs.			
Labor			
Attendant Hd mech. 1080 hrs @ 2.65	2862		
Repair labor 12960 hrs @ .35	4536		
EXPENSE			
Power 2x60= 120 x .75 x .50 = 45 KW @ .0075= .34			
Lube .06			
Parts .20			
12960 hrs. @ .60			7776
TOTAL	7398		7776
Unit ÷ 345292 c.y.	0.021		0.023

VALUES USED ARE FICTITIOUS FOR ILLUSTRATION ONLY

Figure 14–26–6 (Continued)

OPERATION CONCRETE PLANT OPERATION		ITEM NO.	SHEET __3__ OF __4__ PAGE _____
		JOB **WILSON'S WIN DAM**	
QUANTITY UNIT EST. BY		CK'D BY	DATE 12-15-71

	LABOR	EQUIP.	EXPENSE
<u>AIR PLANT</u> 3 @ 562 cfm @ 125 HP ≈ 375 HP			
18 mos. 2 shifts x 22 days			
≈ 18 x 2 x 7.5 x 22 ≈ 5940 hrs. Say 6000 hrs			
1-Operator @ 2.45 = 2.45			
Repair Labor 3 @ .25 = .75			
6000 hrs. @ 3.20	19200		
Power 375 HP @ .75 = 281 KW @ 70% ≈ 197 KW			
197 KW @ .0075 ≈ 1.48			
Lube = .20			
Repair parts = .75			
Shop costs = .20			
6000 hrs. @ 2.63			15780
Total Cost 34980	19200		15780
Unit 345292 c.y. =	0.056		0.046
ELECTRIC PLANT			
Figure 18 mos. operation @ 22 days			
≈ 396 say 400 days @ 24 hrs. ≈ 9600 hrs.			
1-Chief electrican @ 800.00/mo. @ 18 mos. ≈	14400		
2-Elect – 4 – man @ 3.10 ≈ 6.20			
3-Electricans @ 2.85 ≈ 8.55			
Cost /hour 14.75			
Cost/day @ 8 hrs. ≈ 118.00			
400 days @ 118.00	47200		
Subsistence: 18 x 4 1/3 ≈ 78 weeks			
5 men 6 days ≈ 30 MD/Wk. @ 6.00 x 78	14040		
Travel time 18 men @ 20.00 x 2	720		
Travel expense 18 x 20	360		
1% Pension $76720	767		
EXPENSE			
Electric supplies 9600 hours @ 2.00			19200
Total Cost 96687	77487		19200
Unit – 345292 c.y. =	0.224		.056

Figure 14–26–6   (Continued)

		LABOR	EQUIP.	EXPENSE	

LIGHTING PLANT

    Figure 18 mos. @ 22 days = 400 days

    @ 12 hrs. = 4800 hrs.

    Allow 125 KW @ 4800 Hrs. = 600,000 KWH

        @ .0075 4500

    Light Bulbs   400 days @ 20.00 8000

                 TOTAL 12500

        Unit - 345292 c.y. 0.036

CEMENT HANDLING   477,560

    Sub-contract to ~~Whitaker~~ A. B. Jones

    ~~shovel~~ for cement in silos at batch

    plant. Cement bucket elevator operated

    with batch plant.

VALUES USED ARE FICTITIOUS FOR ILLUSTRATION ONLY

Figure 14-26-6 (Continued)

JOHN DOE CONSTRUCTION COMPANY, IN'

ESTIMATED PRODUCTION RATES AND OPERATING TIME
CONCRETE LOCATION & PLACING

JOB _____   DATE 11-20-71   BY _____

FOR _____   Manitowoc

JOB   WILSON'S WIN DAM

BID DATE 12-22-71

PAGE ____ SHEET ____ OF ____ EST NO ____

VALUES USED ARE FICTITIOUS FOR ILLUSTRATION ONLY

ITEM NO.	DESCRIPTION	QUANTITY	UNIT	TAKE-OFF Cubic Yards	PLACING RATE Cubic Yards/Hr.	PLACING CREW Hrs.	TRUCK CRANE	WHIRLEY Hrs.	PLACING CRANE Hrs.	CONCRETE TRUCKS No.	CONCRETE TRUCKS Hrs.	MIXING PLANT Hrs.	ITEM NO.
(17)	Concrete in Powerhouse												
(a.)	Intake concrete												
	Int. main piers under int. floor slabs			6942	30	231			231	2	462	70	
	Int. main piers El. 39.5 to 98.5			16246	30	542			542	2	1082	170	
	Int. main piers El. 98.5 to 178			6756	30	225		225		2	450	70	
	Int. main piers El. 128 to 185			5155	30	172		172		2	344	90	
	Int. floor slab 5' thick			14200	40	355			355	2	710	142	
	Int. floor slab temp. fill conc.			264	16	17			17	1	17	—	
	Int. intermediate piers El.39.5 to 98.50			18509	30	617			617	1	1234	200	
	Int. intermediate piers El.98.5 to 122.0			7001	30	233		233		2	466	70	
	Int. intermediate piers El.122 to 185.0			12770	30	406		406		2	816	180	
	Int. Area (A) El. 78.5 to 98.5			8133	50	163		163		3	489	70	
	Int. Area (B) El. 91 to 99.35			1020	20	51		51		2	51	10	
	Int. Area (C) El. 114 to El. 142.50			1904	36	53		53		2	106	15	
	Int. Area (D) El. 114 to 123			7674	40	192		192		2	384	60	
	D/S wall of int. El.98.5 to 142.50 4'-0"			4667	24	194		194		1	194	40	
	Supported slab El.111.5 int. gal. 2'-0"			945	24	39		39		1	39	10	
	Supported slab El.129.5 int. gal.			820	20	41		41		1	41	8	
	Supported slab El.142.5 int. gal.			827	24	41		41		1	41	6	
	D/S wall pilasters El.114 to 185			137	24	6		6		1	6	—	
	U/S dtv. wall bet. iceway & control El. 185			4779	Included in Per Item 24	207		207		1	207	45	
	D/S dtv. wall bet. gatewell & control 416			6368	24	265		265		1	265	90	
	Temp. blkhds. f/doorways D/S int. wall 134			587	20	29		29		1	29	5	
	Pre-cast Overflw.weir&trash sluice T.El.120.5			263	16	16		16		1	16	—	
	Int. deck supported over gatewells El. 185			586	16	17		17		1	17	—	
	Int. deck supported over ice&trash.wy.El.185			6519	16	37		37		1	37	5	
	Int. P-23				30	217		1	217	2	434	70	
	Crane bumpers on P-23 int. El. 185			2						—		—	
	Parapet wall on P-23 int. El. 185			4									
	Cant. deck on P-23 int. El. 185 2'-0"			27	10	3		3					
	Cant. beams on P-23 int. El. 185			7	10	1		1					
	Stub monolith of non-overflow E.unit22(Mass)			3243	50	65			65	3	195	5	
	Parapet wall on non-overflow E. unit 22			4				1					
	Gallery deck beams El. 142.5			5	8	5		5				5	
	Supported stairs curbs & risers int.			1	1			1					
	Toppling req'd. this contract			12	12			1					
	End wall of galleries												
	SUBTOTAL Concrete in Intake			136285	(30.7)	4444		2400	2044		8132	1318	

Figure 14-26-7   Estimated production rates and operating time

JOHN DOE CONSTRUCTION COMPANY, INC.

JOB WILSON'S WIN DAM
CONCRETE IN POWERHOUSE
ITEM NO. 17   SHEET 1 OF 4   PAGE   BY   DATE 11-23-71

		LABOR AND INSURANCE		SPECIFIC PLANT		SUPPLIES		SUB-CONTRACTS		TOTAL DIRECT COST		ESTIMATED COST
		UNIT COST	AMOUNT	UNIT COST	AMOUNT	UNIT COST	AMOUNT	UNIT COST	AMOUNT	UNIT COST	AMOUNT	
SUMMARY OF CONCRETE COST												
Plant Cost Without Salvage	246,675 C.Y.			3.225	795527					3.225	795527	
Plant Operation *per detail attached*	246,675 C.Y.	1.30	319843			0.60	147768			1.90	467611	
Plant Crews "	246,675 C.Y.	1.24	306864							1.24	306864	
Form Cost "	246,675 C.Y.	6.426	1585205			3.385	834955			9.811	2420160	
Finishing (unformed)	298,8745 F.	0.13	32198			0.02	6052			0.15	38250	
Treating Construction Joints "	246,675 C.Y.	0.83	204228			0.22	53732			1.05	257960	
Curing and clean-up "	246,675 C.Y.	0.60	148005			.15	37001			0.75	185006	
Admixture "	246,675 C.Y.					0.03	6290			0.03	6290	
Cement Wastage "	249,760 C.Y.					0.10	24975			0.10	24975	
Aggregate Wastage "	246,675 C.Y.	0.46	113220			0.05	11424			0.05	11424	
Point & Patch "	246,675 C.Y.		7555			0.08	18697			0.54	131917	
Handling Precast Slabs, etc. 521 units @ 14.50/043.50	246,675 C.Y. 0.03		1000			0.007	1823			0.04	9378	
Casting Yard L.S.							2000				3000	
Specials												
Oakum 21 CF 10.00/0/25.00			210				525			0.003	735	
Wedges Oak 4" x 3" x 3" 1200 EA .10/0/.25			120				300			0.002	420	
Expansion Joint ½" Thick 1388 SF .20/0/.20			278				278			0.002	556	
Tar Paper 22918 SF @ .04/0/.01			917				229			0.004	1146	
Vibrator Repair & Small Tools 246675		.02	4933			.02	4934			0.04	9867	
Subtotal		11.045	2724876	3.225	795527	4.665	1150983			18.935	4671086	
Cost of Aggregates 246675 C.I.								1.85	456349	1.85	456349	
Total Direct Cost on 246675 cu. yds. (Take-off quantity)		11.045	2724876	3.225	795527	4.666	1150983	1.85	456349	20.786	5127435	
Total Direct Cost on 250600 cu. yds.		11.045	2767877	3.225	808185	4.666	1169300	1.85	463610	20.786	5208972	

VALUES USED ARE FICTITIOUS FOR ILLUSTRATION ONLY

Figure 14-26-8   Summary of concrete costs

JOHN DOE CONSTRUCTION COMPANY, INC.

JOB   WILSON'S WIN DAM   ITEM NO. _____   PAGE _____

CONCRETE IN POWERHOUSE   SHEET 2 OF 4

Unit – Cu. Yd.   BY _____   DATE 11-23-71

	LABOR AND INSURANCE		PERMANENT MATERIALS		SPECIFIC PLANT		EQUIPMENT RENTAL		SUPPLIES		SUB-CONTRACTS		TOTAL DIRECT COST		ESTIMATED COST
	UNIT COST	AMOUNT	UNIT COST	AMOUNT	UNIT COST	AMOUNT	UNIT COST	AMOUNT	UNIT COST	AMOUNT	UNIT COST	AMOUNT	UNIT COST	AMOUNT	
PLANT COST															
General Plant 246675 C.Y. @					3.225	795527									
Special Plant C.Y. @															
PLANT COST Subtotal															
PLANT OPERATION															
Batching & Mixing Plant 4200 Hrs @	12.53	52626							5.39	22638				75264	
Wet Batch Trucks Bucket Hauling 14695 Hrs @	3.53	51873							2.25	33064				84937	
Placing Cranes — Whirleys 4674 Hrs @	6.45	30147							4.25	19865				50012	
Placing Cranes — Manitowoc 5039 Hrs @	8.75	44091							5.37	27059				71150	
Dinkies Hrs @															
Aggregate Handling & Rinsing C.I. @ .046 / 0 / .022															
Water Plant C.I. @ .021 / 0/ .023															
Air Plant C.I. @ .056 / 0 / .046															
Electric Plant C.I. @ .224 / 0 / .056															
Lighting Plant C.I. @ 0 / 0 / .036															
Conveyor Plant C.I. @ In agg. handling															
Cooling Plant C.Y. @ None Req'd															
Cement Handling C.I. @ None Req'd	85596								45142					130738	
246675 C.I. @ .347 / 0 / .183															
Shift Time & Overtime @ 10% on $26433		26433												26433	
Tax and Insurance @ 10% on $290766		29077												29077	
PLANT OPERATION Subtotal		319843								147768				467611	

VALUES USED ARE FICTITIOUS FOR ILLUSTRATION ONLY

Figure 14–26–8 (Continued)

JOHN DOE CONSTRUCTION COMPANY, INC.

JOB: WILSON'S WIN DAM   ITEM NO. ____   PAGE ____   SHEET 3 OF 4   DATE 11-23-71   BY ____

CONCRETE IN POWERHOUSE

VALUES USED ARE FICTITIOUS FOR ILLUSTRATION ONLY

Description	LABOR AND INSURANCE		PERMANENT MATERIALS		SPECIFIC PLANT		EQUIPMENT RENTAL		SUPPLIES		SUB-CONTRACTS		TOTAL DIRECT COST		ESTIMATED COST
	UNIT COST	AMOUNT	UNIT COST	AMOUNT	UNIT COST	AMOUNT	UNIT COST	AMOUNT	UNIT COST	AMOUNT	UNIT COST	AMOUNT	UNIT COST	AMOUNT	
PLACING CREWS															
1 Foreman @ 2.75 = 2.75															
4 Vibrator Men @ 2.25 = 9.00															
4 Laborers @ 2.10 = 8.40															
1 Signal Men @ 2.75 = 2.75															
1 Hookman @ 2.75 = 2.75															
DI for IW 8 1/3% on 5.50 = 0.46															
9713 Hrs. @26.11		253606													
Shift Time & Overtime @ 10% on $253606 =		25361													
Tax & Insurance on $278968															
PLACING CREW Subtotal		306864												306864	
TREATING CONSTRUCTION JOINTS															
775023 Sq. Ft. @ .20 ≠ 0 ≠ .05 (Horiz)		155005								38751					
214013 Sq. Ft. @ .23 ≠ ≠ .07 (Vert)		49223								14981					
CURING & CLEANUP															
2162568 Sq. Ft. @ ≠ 0 ≠		204228								53732				257960	
246675 Cu. Yds @ 60 ≠ 0 ≠ .15															
ADMIXTURE															
246681 Cu. Yds @ 1.5 bbls./C.Y. =		148005								37001				185006	
270021 Bbls Cement @															
7400 Gallons @ .85 gals./bb. =												6290		6290	
CEMENT WASTAGE															
270021 Bbls. of Cement @ 1.5 %															
5550 Bbls. of Wastage @ $4.50												24975		24975	
AGGREGATE WASTAGE															
246681 Cu. Yds. Concrete @ ____ tons/cu. yd															
Tons Total @ 2 ¢ $ Waste =															
6167 C.Y. Wastage @ 1.85 ≠									1.85	11424				11424	

Figure 14-26-8 (Continued)

JOHN DOE CONSTRUCTION COMPANY, INC.
WILSON'S WIN DAM
JOB CONCRETE IN POWERHOUSE

ITEM NO. 17   SHEET 4   PAGE 4 OF 4
DATE 11-23-71
BY

	LABOR AND INSURANCE		PERMANENT MATERIALS		SPECIFIC PLANT		EQUIPMENT RENTAL		SUPPLIES		SUB-CONTRACTS		TOTAL DIRECT COST		ESTIMATED COST
	UNIT COST	AMOUNT	UNIT COST	AMOUNT	UNIT COST	AMOUNT	UNIT COST	AMOUNT	UNIT COST	AMOUNT	UNIT COST	AMOUNT	UNIT COST	AMOUNT	
POINT & PATCH															
Exposed Sq. Ft. @ .065 / / / .010		94523								14542					
Unexposed Sq. Ft. @ .045 / / / .010		18697								4155					
Other Sq. Ft. @		113220								18697				131917	
SCREED & FINISH															
Troweled Finish 26480 Sq. Ft. @ .16 / 0 / .04		4237								1059					
Float Finish 174787 Sq. Ft. @ .11 / 0 / .02		19227								3496					
Screed Only 33568 Sq. Ft. @ .07 / 0 / .01		2350								336					
Roadway 58039 Sq. Ft. @ .11 / 0 / .02		6384								1161					
		32198								6052				98250	

VALUES USED ARE FICTITIOUS FOR ILLUSTRATION ONLY

Figure 14-26-8 (Continued)

JOHN DOE CONSTRUCTION COMPANY, INC.

WILSON'S WIN DAM — ITEM NO. 17 — SHEET 1 OF 6 — PAGE

CONCRETE IN POWERHOUSE — Unit – Cu. Yd. — DATE 11-23-71 — BY

VALUES USED ARE FICTITIOUS FOR ILLUSTRATION ONLY

FORMS / DESCRIPTION	QUANTITY	LABOR AND INSURANCE UNIT COST	AMOUNT	PERMANENT MATERIALS UNIT COST	AMOUNT	SPECIFIC PLANT UNIT COST	AMOUNT	EQUIPMENT RENTAL UNIT COST	AMOUNT	SUPPLIES UNIT COST	AMOUNT	SUB-CONTRACTS UNIT COST	AMOUNT	TOTAL DIRECT COST UNIT COST	AMOUNT	ESTIMATED COST
INTAKE FORMS:																
D/S Edge intake slab	2960 sf	1.50	4440							.45	1332			1.95	5772	
Temp. Conc. fill form edge & bulkheads	2960 sf	1.35	3996							.40	1184			1.75	5180	
Contraction Joints in Intake slab	4450 sf	1.50	6675							.45	2002			1.95	8677	
Trash rack blockout in intake slab	891 sf	1.50	1337							.75	668			2.25	2005	
Bulkhead gate blockout in intake slab	836 sf	1.50	1254							.75	627			2.25	1881	
Operating gate blockout in intake slab	2617 sf	1.50	3926							.75	1963			2.25	5889	
Interm. piers intake slab to intake roof																
Vertical Straight Pannels	125315 sf @	.65	81455							.40	50126			1.05	131,581	
Vertical curved D/S	17820 sf @	.65	11583							.40	7128			1.05	18711	
Vertical curved U/S	11808 sf	.65	7675							.40	4723			1.05	12398	
Blockout U/S nose	8320 sf	.65	5408							.40	3328			1.05	8736	
Interm. piers above roof																
Haunch sides	5003 sf	.647	3237							.408	2041			1.055	5278	
Haunch ends	3024 sf	.647	1956							.408	1234			1.055	3190	
Vertical Pannels	22049 sf	.471	10385							.251	5534			.722	15919	
Void under ice & trash vert. pannels ends	8178 sf	.814	6657							.494	4040			1.308	10697	
Vert. sides over trash vert. pannels	8282 sf @	.65	5383							.40	3313			1.05	8696	
Vert. sides over area "D" vert. pannels	8432 sf @	.65	5481							.40	3373			1.05	8854	
Edge opening for ice & trashway U/S wall	3994 sf	.65	2596							.40	1598			1.05	4194	
Vertical Construction Joints	616 sf	1.35	832							.45	277			1.80	1109	
Miscellaneous Edge Forms	896 sf	1.25	1120							.30	269			1.55	1389	
Edge openings for ice & trash roof supported																
main piers	1848 sf	1.25	2310							.60	1109			1.85	3419	
Main Piers																
Vertical Pannels to Int. Roof	90808 sf @	.65	59025							.40	36323			1.05	95348	
Curved Vert. U/S Intake Roof	5904 sf @	.65	3838							.40	2362			1.05	6200	
Vert. Pannel U/S nose sloping int. roof	6240 sf @	.65	4056							.40	2496			1.05	6552	
Blockout for trash guide U/S nose	10400 sf @	.65	6760							.40	4160			1.05	10920	
Haunch sides	4241 sf	.647	2744							.408	1730			1.055	4474	
Haunch ends	3312 sf	.647	2143							.408	1351			1.055	3494	
Vert. Pannels Between slots	11024 sf Gatewell	.471	5192							.251	2767			.722	7959	
Void Under ice & trash vert. Pannel ends	4713 sf	.814	3836							.494	2328			1.308	6164	
Vert. Pannels over ice & trash	4141 sf @	.65	2692							.40	1656			1.05	4348	
Vert. Pannels over area "D"	4216 sf @	.65	2740							.40	1686			1.05	4426	
Edge opening ice & trashway U/S wall	3423 sf	.65	2225							.40	1369			1.05	3594	
Vertical constr. joints U/S wall	5136 sf	1.35	6934							.45	2311			1.80	9245	
U/S wall vert. & sloping of void u/ice & Tra.way	11520 sf	.814	9377							.494	5691			1.308	15068	
Edge opening roof ice & trashway	1584 sf	1.25	1980							.60	950			1.85	2930	
MAIN PIER contraction joints	87588 sf	.50	43794							.67	58663			1.17	102,457	
SUBTOTAL	490,101 sf	.663	325042							.452	221712			1.115	546754	

Figure 14-26-9   Form cost for concrete items

JOHN DOE CONSTRUCTION COMPANY, INC.

JOB: WILSON'S WIN DAM
CONCRETE IN POWERHOUSE
Unit – Cu. Yd.

ITEM NO. 17   SHEET 2 OF 6   PAGE   DATE 11-23-71   BY

VALUES USED ARE FICTITIOUS FOR ILLUSTRATION ONLY

FORMS DESCRIPTION	QUANTITY	Labor and Insurance Unit Cost	Amount	Supplies Unit Cost	Amount	Specific Plant	Total Direct Cost Unit Cost	Amount
Total Brought Forward	490,101 sf		325042		221712			546754
Intake Forms Continued								
Supported roof under area "A"	16368 sf	1.356	22195	1.675	27416		3.031	49611
Supported roof under area "B"	4560 sf	1.356	6183	1.675	7638		3.031	13821
Supported roof under area "C"	11803 sf	1.356	16005	1.675	19770		3.031	35775
Supported roof under area "D"	21264 sf	1.356	28834	1.675	35617		3.031	64451
Area "B" sides	5040 sf	1.356	6834	1.675	8442		3.031	15276
Bulkhead slots vertical	61405 sf	.471	28922	.251	15413		.722	44335
Bulkhead supported sill	240 sf	1.25	300	.40	96		1.65	396
Operating gate slots vertical	67302 sf	.471	31699	.251	16893		.722	48592
Operating gate pass thru piers El.78.5-142.5	1801 sf	1.00	1801	.35	630		1.35	2431
Ext. D/S wall of intake gateways El.78.5-142.5	7198 sf	1.205	8677	.353	2167		1.558	10844
Int. D/S wall of int. gateways El.78.5-142.5	30272 sf	.452	13683	.367	11110		.819	24793
D/S wall of gate wells gal.si.El.78.5-185	50608 sf	.647	32738	.408	20645		1.055	53383
D/S wall o/gate wells gal. side sloping	3518 sf	.647	2276	.408	1435	See Detail	1.055	3711
D/S wall o/gate wells gatewell side vert.	43350 sf	.471	20418	.251	10881	See Detail	.722	31299
U/S wall o/gate wells El.90,35-185	39592 sf	.471	18695	.251	9963		.722	28658
D/S wall o/ice & trash sluiceway El.114-183	41935 sf	.705	29564	.391	16397		1.096	45961
U/S wall o/ice & trash sluice sloping st.	3202 sf	1.356	4342	1.675	5363		3.031	9705
U/S wall o/ice & trash sluice vertical	5840 sf	1.356	7919	1.675	9782		3.031	17701
Precast deck edges	5230 sf	1.25	6538	.35	1830		1.60	8368
Intake deck beams sides	2640 sf	1.75	4620	.75	1980		2.50	6600
Intake deck beams soffits	1440 sf	1.75	2520	.75	1080		2.50	3600
Intake deck supported	8770 sf	1.75	15348	.75	6577		2.50	21925
Intake deck supported edges	5856 sf	.75	10248	.75	4392		2.50	14640
Intake deck rail blockout	3540 sf	1.25	4425	.35	1239		1.60	5664
Trash sluice gate blockout	12759 sf	.60	7655	.25	3190		.85	10845
Parapet wall sides	270 sf	.75	203	.30	81		1.05	284
Parapet wall bulkheads	18 sf	.90	16	.30	6		1.25	22
Crane bumpers	85 sf	.90	77	.35	30		1.25	107
Supported deck gal. El.111.5	13144 sf	.90	11830	.75	9858		1.65	21688
Supported deck gal. El.129.5	15104 sf	.90	13594	.75	11328		1.65	24922
Supported deck gal. El.142.5	15136 sf	.90	13622	.75	11352		1.65	24974
Pilasters D/S gal. wall	412 sf	.75	309	.30	124		1.05	433
Door opening & pass thru D/S wall	4256 sf	.75	3192	.30	1277		1.05	4469
Edge openings in gal. slats	1249 sf	.75	937	.30	375		1.05	1312
Supported Stairways	496 sf	1.75	868	.75	372		2.50	1240
Supported Stairways edges & risers	394 sf	1.75	690	.75	295		2.50	985
Curbs Around Openings	126 sf	1.25	158	.35	44		1.60	202
SUBTOTAL	1,032,215 sf		720508		509924		1.92	1230432

Figure 14-26-9 (Continued)

JOHN DOE CONSTRUCTION COMPANY, INC.
WILSON'S WIN DAM
CONCRETE IN POWERHOUSE   Unit – Cu. Yd.

JOB   ITEM NO.   BY   DATE 11-23-71   SHEET 3 OF 6   PAGE

FORMS	DESCRIPTION	QUANTITY	LABOR AND INSURANCE Unit Cost	Amount	PERMANENT MATERIALS / SPECIFIC PLANT (note)	SUPPLIES Unit Cost	Amount	TOTAL DIRECT COST Unit Cost	Amount	ESTIMATED COST
	Total Brought Forward	1,032,215 sf		720508			501924		1230432	
	Intake Continued									
	Beam Sides in Gallerys	528 sf	.85	449		.35	185	1.20	634	
	Beam Soffits in Gallerys	528 sf	.85	449		.35	185	1.20	634	
	Pier – 23 Forms (Extra Cost Tie Contid – Forms	352 sf	1.25	440		.60	211	1.85	651	
	P-23 to rock (1989 sq. ft.)		.35	696		.35	696	.70	1392	
	Cantilevered Deck & Beam	439 sf	1.75	768		1.50	659	3.25	1427	
	Cantilevered Deck Beam Sides	208 sf	1.25	260	(U. S. Vertical on Dam)	.35	73	1.60	333	
	U/S Face of Stub Monolith	2395 sf	.55	1317		.25	599	.80	1916	
	Gallery in Stub Monolith	696 sf	1.25	870		.60	418	1.85	1288	
	East Face of P-23 Exposed	4233 sf	.50	2117	Contraction Joint Forms	.70	2963	1.20	5080	
	Contraction Joint Stub Monolith P-23	3495 sf	.55	1922	Contraction Joint Monolith Non Overflow	.25	874	.80	2796	
	D/S Wall of Gallery in Intake	562 sf	.75	422		.45	253	1.20	675	
	Vertical Construction Joints	280 sf	1.25	350		.40	112	1.65	462	
	Eastwall Gallery El. 129.5 to 142.5	322 sf	.65	209		.35	113	1.00	322	
	Cantilevered Eastwall Gallery El. 111.5 to 129.5	281 sf	.85	239	P. H. Contr. Joint Forms	.30	84	1.15	323	
	Cantilevered Eastwall Gallery El. 98.5 to 111.5	77 sf	.85	65	P. H. Contr. Joint Forms	.30	23	1.15	88	
	D/S Sloping Forms for Stub Monolith	825 sf	.70	578	D. S. Sloping Forms Non Overflow	.26	215	.96	793	
	D/S Vertical Forms for Stub Monolith	210 sf	.70	147	Same as Non Overflow	.26	55	.96	202	
	8' Wide – D/S Wall of Gate well P-23	529 sf	.65	344		.45	238	1.10	582	
	West Face Stub Monolith	205 sf	1.35	27		.45	9	1.80	36	
	Keys 2'-0" x 0'-3"	5296 lf	1.60	8474		.80	4237	2.40	12711	
	Keys 6" x 1"	2064 lf	.60	1238		.30	619	.90	1857	
	Sub-Total Intake Forms	1,047,667	.708	741440		.499	522560	1.206	1264000	
	FORM RATIO = 7.69 SF/c.y.									
	D/S of Main Unit Forms									
	D/S of Main Pier Sides	22836 sf	.90	20552		.25	5709	1.15	26261	
Center Pier 25572	Draft Tube Pier End D/S	1248 sf	.90	1123		.25	312	1.15	1435	
	Draft Tube Pier End U/S	1488 sf	1.165	1734		.438	651	1.63	2385	
	Draft Tube Pier Term Nose	1984 sf	.90	1786		.25	496	1.15	2282	
	Keys in Pier 1'-6" x 0'-3"	240 lf	1.25	300		.35	84	1.60	394	
	Draft Tube Supported Roof	31,008 sf	1.206	37,385		.352	10917	1.558	48302	
	Draft Tube Sides Straight	11,520 sf	.90	10,368		.25	2880	1.15	13248	
Plt. in place	Vert. Const. Joints D/S Edge 2408 sf									
Plt. in place	Vert. Const. Joints in Slab 14792 sf									
Plt. in place	Vert. Const. Joints Pier 6720 sf	23,920 sf	1.50	35,880		.45	10764	1.95	46644	
	Blockout in Draft Tube Piers	1,141 sf	1.25	1,426		.35	399	1.60	1825	
	Blockout in Draft Tube Slab	647 sf	1.50	971		.35	226	1.85	1197	
	Draft Tube Curved Walls	13,718 sf	1.075	14,748		.292	4006	1.3678	18754	
	SUBTOTAL	1,157,178 sf	.708	867,713			559004		1426717	

Figure 14–26–9 (Continued)

JOHN DOE CONSTRUCTION COMPANY, INC.
JOB
CONCRETE IN POWERHOUSE
WILSON'S WIN DAM
ITEM NO.   BY
Unit – Cu. Yd.
PAGE
SHEET 4 OF 6
DATE 11-23-71

VALUES USED ARE FICTITIOUS FOR ILLUSTRATION ONLY

FORMS DESCRIPTION	QUANTITY	LABOR AND INSURANCE UNIT COST	AMOUNT	PERMANENT MATERIALS	SUPPLIES UNIT COST	AMOUNT	TOTAL DIRECT COST UNIT COST	AMOUNT
Total Brought Forward	1,157,178 sf		867713			559004		1426717
Aux. Water conduit sloping U/S wall 1:1	8600 sf	.814	7000	Use Same Cost As Below Ice Sluiceway	.494	4248	1.308	11248
Aux. Water conduit vertical U/S wall	11868 sf	.814	9661	Use Same Cost As Below Ice Sluiceway	.494	5863	1.308	15524
Aux. Water conduit vertical D/S	15996 sf	.677	10829		.549	8782	1.226	19611
Aux. Water conduit supported roof	14448 sf	1.206	17424	Same as Roof or Draft Tube	.352	5086	1.558	22510
Aux. Water conduit D/S EX 7 Wall	18352 sf	.677	12424		.549	10075	1.226	22499
Main Piers Between Water Passage	5100 sf	.90	4590	Use Draft Tube CTR. Pier Cost	.25	1275	1.15	5865
U/S Wall of Water Passage Int.	8880 sf	.90	7992	Use Draft Tube CTR. Pier Cost	.25	2220	1.15	10212
D/S wall of water passage int.	12580 sf	.90	11322		.25	3145	1.15	14467
Edge Openings in wall between Aux. Pas.	2752 sf	3.00	8256		1.00	2752	4.00	11008
U/S wall of FT Channel	19680 sf	.90	17712		.25	4920	1.15	22632
D/S Wall of FT Channel	19680 sf	.90	17712		.25	4920	1.15	22632
Supported Deck El. 94 of Fish Cont. Gallery	5572 sf	1.206	6720		.352	1961	1.558	8681
U/S Int. wall of Fishway control gallery	10323 sf	.90	9295	Same as Roof of Draft Tube	.25	2582	1.15	11877
D/S Int. wall of Fishway control gallery	10328 sf	.90	9295		.25	2582	1.15	11877
D/S Ext. wall of Fishway control gallery	11704 sf	.90	10534		.25	2926	1.15	13460
Strut. Sides over FT Channel	2352 sf	1.00	2352		.35	823	1.35	3175
Strut. Soffits over FT Channel	1470 sf	1.50	2205		.50	735	2.00	2940
Col's. El. 94 to 106	6336 sf	1.75	11088		.60	3802	2.35	14890
Edge Door openings in F. C. Gallery U/S wall	270 sf	1.25	338		.35	95	1.60	433
Risers in openings in F. C. Gallery U/S	47 sf	.80	38		.35	16	1.15	54
Edge Escape hatch door gallery U/S wall	255 sf	1.25	319		.35	89	1.60	408
'Scape hatch supported slab	42 sf	2.00	84		1.00	42	3.00	126
Escape hatch walls	1059 sf	1.00	1059		.45	477	1.45	1536
Diffusion chamber gate guide struts, sides	284 sf	1.25	355		.35	99	1.60	454
Diffusion chamber gate guide struct. soffits	214 sf	1.50	321		.50	107	2.00	428
Stub walls between F7 & FC Channels	4393 sf	.65	2855		.30	1318	.95	4173
Supported deck over fishway control gallery	4180 sf	1.25	5225		.50	2090	1.75	7315
Edge forms in sum in Aux. water conduit	195 sf	1.00	195		.40	78	1.40	273
Draft tube end beam sides	4019 sf	.80	3215		.35	1407	1.15	4622
Draft tube end beam soffits	1523 sf	2.25	3427	Precast	1.00	1523	3.25	4950
Draft tube plug sides	3340 sf	.60	2004	Precast	.20	668	.80	2672
Draft tube plug soffits	714 sf	.20	143		.05	36	.25	179
Draft tube plug blockouts etc.	205 sf	2.50	513		1.50	308	4.00	821
Recesses for exhaust ducts	120 sf	1.25	150		.40	48	1.65	198
Blockout for gutter in fish control gallery	357 sf	2.00	714		.50	178	2.50	892
Diffusion chamber beam sides	941 sf	1.25	1176		.35	329	1.60	1505
Diffusion chamber beam soffits	315 sf	1.25	394		.35	110	1.60	504
Diffusion chamber D/S wall ext.	2782 sf	.90	2504		.25	696	1.15	3200
SUBTOTAL	1,368,450 sf		1069153			637415		1706568

Figure 14-26-9 (Continued)

JOHN DOE CONSTRUCTION COMPANY, INC.
WILSON'S WIN DAM

JOB — CONCRETE IN POWERHOUSE   Unit – Cu. Yd.

ITEM NO. — BY —   PAGE — SHEET 5 OF 6   DATE 11-23-71

FORMS / DESCRIPTION	QUANTITY	LABOR AND INSURANCE		PERMANENT MATERIALS		SPECIFIC PLANT		EQUIPMENT RENTAL		SUPPLIES		SUB-CONTRACTS		TOTAL DIRECT COST		ESTIMATED COST
		UNIT COST	AMOUNT	UNIT COST	AMOUNT	UNIT COST	AMOUNT	UNIT COST	AMOUNT	UNIT COST	AMOUNT	UNIT COST	AMOUNT	UNIT COST	AMOUNT	
Total Brought Forward	1,368,459 sf		1069153								67415					1706568	
Diffusion chamber D/S wall int.	4046 sf	.90	3641							.25	1012			1.15	4653	
Diffusion chamber walls transverse	8575 sf	1.25	10719							.35	3001			1.60	13720	
Diffusion chamber supported deck	2322 sf	1.35	3135							.50	1161			1.85	4296	
Diffusion chamber edge slate	773 sf	.80	618							.25	193			1.05	811	
Fishway deck beam sides	16524 sf	1.35	22307							.50	8262			1.85	30569	
Fishway deck beam soffits	6800 sf	2.25	15300							1.00	6800			3.25	22100	
Main Piers D/S intake to U/S Fishway I.S. dowels	8528 sf	2.85	17056							.50	4264			2.50	21320	
Edge openings in piers to U/S " I. S. dowels	3840 sf	1.25	4800							.35	1344			1.60	6144	
Contraction joints main pier	96571 sf	.85	82085							.20	19314			1.05	101399	
U/S wall o/fishway Cont. wall tie across	23923 sf	.90	39531	Use Wood Same as Conc. Cost in S. Case						.25	10981			1.15	50512	
8" blockout for temp fill in floor El.15.5%	1190 sf	1.50	1785	Use Same Cost as D.S. Edge Area "A" Intake						.50	595			2.00	2380	
blockout for valve pit	746 sf	1.00	746							.40	298			1.40	1044	
U/S wall of diffusion chamber	2909 sf	1.75	5091							.50	1455			2.25	6546	
Fishway deck cross beams sides	14951 sf	1.25	18689							.40	5980			1.65	24669	
Fishway deck cross beams soffits	4693 sf	1.25	5866							.50	2347			1.75	8213	
Cantilevered beams for light sides	592 sf	2.00	1184							1.00	592			3.00	1776	
Cantilevered beams for lights soffits	256 sf	2.00	572							1.00	256			3.00	768	
Fishway deck cantilevered over FT channel	976 sf	1.25	1220							.40	390			1.65	1610	
Fishway deck cantilevered over F.C. channel	6346 sf	1.25	7933							.50	3173			1.75	11106	
Fishway deck cantilevered over D/S of piers	2992 sf	1.50	4488							.75	2244			2.25	6732	
Edge openings in fishway decks	5256 sf	1.00	5256							.40	2102			1.40	7358	
Intermediate piers El. 597111.5	29968 sf	.80	23974							.70	20978			1.50	44952	
Blockout for guides	3158 sf	1.20	3790							.80	2526			2.00	6316	
Main D/S piers remainder	40248 sf	.80	32198							.70	28174			1.50	60372	
Main D/S piers formed with Aux. conduit	12904 sf	.677	8736							.549	7084			1.226	15820	
Fishway deck rail blockout	3192 sf	1.25	3990							.35	1117			1.60	5707	
Blockout sill for gates El. 59	488 sf	1.25	610							.35	171			1.60	781	
Misc. Edge in slab over aux. water cond.	555 sf	3.00	1665							1.00	555			4.00	2220	
Form in temp conc. wall fill																
Form in vert. constr. joints	516 sf	1.35	778							.35	202			1.70	980	
P-23 exposed end forms	2258 sf	1.25	2823							.45	1016			1.70	3839	
Beams cantilevered sides	370 sf	1.50	555							.50	185			2.00	740	
Beams cantilevered soffits	136 sf	2.25	306							.75	102			3.00	408	
Misc. vertical walls	7192 sf	1.35	9709							.45	3236			1.80	12945	
Sloping straight wall	1589 sf	1.30	2066							.50	794			1.80	2860	
Supported slab	594 sf	1.50	891							.50	297			2.00	1188	
Form up operating gate slots vert.	15200 sf	1.25	19000							.50	7600			1.75	26600	
Form up operating gate slots supported	398 sf	4.00	1592							2.00	796			6.00	2388	
SUBTOTAL	1,840,846 sf		1505247								808271					2313518	

VALUES USED ARE FICTITIOUS FOR ILLUSTRATION ONLY

Figure 14-26-9 (Continued)

JOHN DOE CONSTRUCTION COMPANY, INC.
WILSON'S WIN DAM
CONCRETE IN POWERHOUSE
Unit – Cu. Yd.

JOB ITEM NO. PAGE SHEET 6 OF 6 BY DATE 11-23-71

FORMS DESCRIPTION	QUANTITY	LABOR AND INSURANCE UNIT COST	AMOUNT	PERMANENT MATERIALS UNIT COST	AMOUNT	SPECIFIC PLANT UNIT COST	AMOUNT	EQUIPMENT RENTAL UNIT COST	AMOUNT	SUPPLIES UNIT COST	AMOUNT	SUB-CONTRACTS UNIT COST	AMOUNT	TOTAL DIRECT COST UNIT COST	AMOUNT	ESTIMATED COST
Total Brought Forward	1,840,846 sf		1,505,247								808,271				2,313,518	
Trashrack gate vert.	7779 sf	1.50	11669							.50	3889			2.00	15558	
Sluice gate vert.	6484 sf	1.50	9726							.50	3242			2.00	12968	
Bulkhead gate vert.	16655 sf	1.50	24983							.50	8327			2.00	33310	
Bulkhead gate supported	197 sf	2.00	394							1.00	197			3.00	591	
Draft tube gate vert.	8605 sf	1.50	12908							.50	4302			2.00	17210	
Draft tube gate supported	272 sf	4.00	1088							2.00	544			6.00	1632	
Precast edges for fishway stoplogs	5768 sf	.35	2019							.25	1442			.60	3461	
6" ∅ formed drain in intake joint	1152 lf	.35	403							.10	115			.45	518	
Sub Total Forms D/S of Intake	822027 sf															
Form Ratio = 7.52 sf/cu.yd.																
Total Forms For Item 17	1,869,694 sf															
Form Ratio 7.58 sf/cu.yd.																
West & East curved main piers Scroll case	4768 sf	1.075	5126							.292	1392			1.367	6518	
West & East straight main piers Scroll case	12936 sf	.90	11642							.25	3234			1.15	14876	
	1,869,694 sf	.848	1,585,205							.447	834,955			1.295	2,420,160	

VALUES USED ARE FICTITIOUS
FOR ILLUSTRATION ONLY

Figure 14-26-9 (Continued)

JOHN DOE CONSTRUCTION COMPANY, INC.

JOB   WILSON'S WIN DAM   ITEM NO.

BY   DATE   May 1971   SHEET   1 OF 13   PAGE

ANALYSIS OF TEST PITS – THIS CONTRACT

VALUES USED ARE FICTITIOUS FOR ILLUSTRATION ONLY

	–6"	–3"	–1½/3¼	–3/4/3/8	–3/8–¼	Sand ¼	–8/48	–8/16	–16/30	–30/50	–50/100	Pan
PIT NO.	6.7	18.3	28.7	8.6	1.8	17.4	5.4	3.4	6.3	39.7	27.3	17.9
		12.8	19.4	8.2	2.9	30.1	6.9	4.6	8.8	33.8	24.2	21.7
TOTALS	15.4	28.8	21.2	11.5	5.4	17.7	16.7	6.7	21.8	18.6	24.9	25.6
AVE.	46.5	80.6	69.3	28.3	9.1	65.2	29.0	11.7	91.9	76.4	66.2	
	15.5	26.9	23.1	9.4	3.4	23.7	9.7	4.9	7.3	30.6	25.5	22.0
OVERALL AVERAGE	14.5	25.1	21.6	11.9		20.2	9.7	7.3	7.3	30.6	25.5	22.0

Total C.Y. Concrete = 614,200 c.y.
% Req'd. for Conc. Mix

Pit Ave from above

Tons of Agg. required @ 1.75/c.y. 1,074,850 T.

Figure to use 80D shovel in pit @ 27.5 c.y./hr.

Loading production @ 1.5 ton/c.y. = 412 T.P.H. = Say 400 T.P.H.

TAKE 3300 Hrs. Operator – Products

3300 Hrs. @ 400 T.P.H. = 1,320,000 tons

3300 Hrs. Operation Balances on 4/$16 x $30 – Waste = 245150 tons = 18.5% waste

Figure 14–26–10 Aggregate production estimate

Sheet 2 of 13

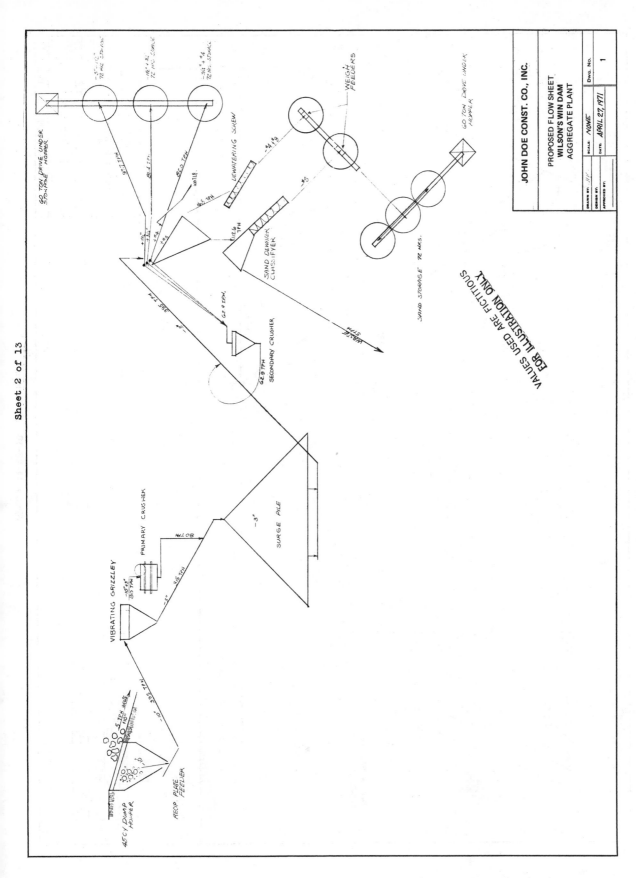

VALUES USED ARE FICTITIOUS
FOR ILLUSTRATION ONLY

JOHN DOE CONST. CO., INC.

PROPOSED FLOW SHEET
WILSON'S WIN DAM
AGGREGATE PLANT

Figure 14–26–10 (Continued)

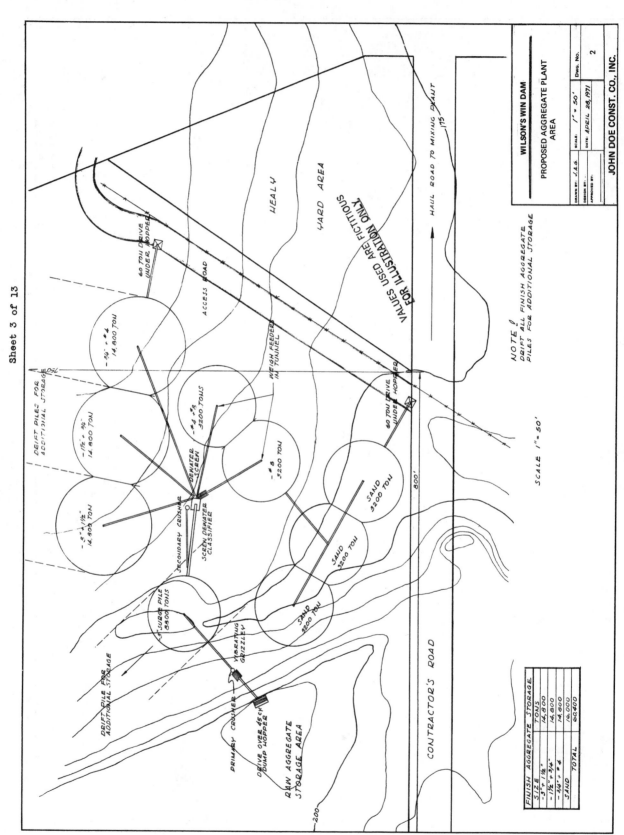

Figure 14-26-10   (Continued)

JOHN DOE CONSTRUCTION COMPANY, INC.

JOB WILSON'S WIN DAM
CONCRETE GENERAL AGGREGATE PROCESSING PLANT
ITEM NO. __ PAGE __ SHEET 4 OF 13 DATE May 1971 BY __

Description	Qty	Labor & Ins. Unit	Labor & Ins. Amount	Specific Plant Amount	Equip. Rental Unit	Equip. Rental Amount	Supplies Unit	Supplies Amount	First Cost Salvageable Amount	Total Direct Unit	Total Direct Amount
AGGREGATE PROCESSING PLANTS											
Equipment and Supplies per Detail Attached						171360		16843	(171360)		188203
Sales Tax on Equip & Supplies								5646			5646
Freight to Jobsite								8948			8948
Unload from Cars & Haul to Site	260 T.	3.50	910		1.00	260	1.50	390		6.00	1560
Electric Power Supply: 480 = Say 500 HP. 500 KVA											
Main Substation by P.U.D.											
Secondary Substation & Stepdown Transf.	500 KVA				12.50	6250			(6250)	12.50	6250
Install Substation			500					500			1000
Install Transformers	500 KVA	1.50	750				3.75	1875		5.25	2625
Misc. Freight								750			750
Transmission Line to Substation	800'	.75	600		.10	80	.75	600		1.60	1280
Misc. Distribution Lines at Plant	2500'	.50	1250		.10	250	.50	1250		1.10	2750
Install Plant Equipment:											
Machinery and Structural (Crushers)	28 Tons	25.00	700		5.00	140	5.00	140		35.00	980
Misc. Machinery	76 Tons	100.00	7600		30.00	2280	50.00	3800		180.00	13680
Excavate Loading Hopper in Rock	135 C.Y.	3.00	405		.50	68	1.50	202		5.00	675
Install 18 C.M.P. Drain Lines	700 L.Ft.	1.25	875		.20	140	.30	210		1.75	1225
Install Bin Dividers—Lumber	26 MBM	40.00	1040		5.00	130	5.00	130		50.00	1300
Structural Lumber	70 MBM	75.00	5250		10.00	700	15.00	1050		100.00	7000
Install Spray Piping	L.S.		150					50			200
Install Water Supply:											
8" Supply Line – on Ground	4250'	.25	1063		.08	340	.12	510		.45	1913
Valves and Fittings								500	(500)		500
Electrical Hookup			150			25		100			275
Install Pumps	2 Ea.	150.00	300		25.00	50	150.00	300		100.00	650
Plant Wiring											
Power Lines, Incl. above											
Elect. Connections	27 Ea.	75.00	2025				25.00	675		100.00	2700
Lighting System	Allow		1500					1500			3000
Chutes	7 Ea.	50.00	350		5.00	35	10.00	70		65.00	455
Concrete for Footings & Slabs	74 C.Y.	10.00	750		5.00	375	25.00	1875		40.00	3000
Special Rock Excav. for Foundations	50 C.Y.	5.00	250		1.50	75	3.0	150		9.50	475
Hand Ecav. for Conveyor Bents & Equip.	200 C.Y.	4.00	800								800
Excav. for Tunnel Conveyors – Uncl.	2100 C.Y.	.40	840		.25	525	.35	735		1.00	2100
Backfill for Tunnel	750 C.Y.	.30	225		.20	150	.20	150		.70	525
Install C.M.P. 96" ∅	885 l.f.	7.50	6638		1.00	885	1.50	1327		10.00	8850
Haul Road & Access Road Around Plant	Allow		500			250		250			1000
TOTAL DIRECT COST PLANT INSTALLED			35421 x	85947		184368		50526 x	(178110)		270315
TRANSFER TO PLANT						184368			(178110)		270315

VALUES USED ARE FICTITIOUS FOR ILLUSTRATION ONLY

Figure 14-26-10 (Continued)

JOHN DOE CONSTRUCTION COMPANY, INC.

JOB: WILSON'S WIN DAM ITEM NO. PAGE SHEET 5 OF 13 BY DATE May 1971

AGGREGATE PROCESSING EQUIPMENT

VALUES USED ARE FICTITIOUS FOR ILLUSTRATION ONLY

Quan.	Item	Condition	Location	Unit Weight	Total Weight	Frt. to Job	Unit Cost New or Used	Total Cost New or Used	Salvage	Motor HP Required
L.S.	Drive Over Beams	New	Portland	---	6000 #	30	---	720		500
1500 BM	12" x 12" Deck Planking	New	Portland	---	5000 #	-	---	135		200
2100 #	60# Rails to Form 10" Grizzley	New	Portland	---	2100 #	20	.30/#	63		
1 Ea.	45 c.y. Dump Hopper Steel	New	Portland	---	4000 #	20	600	600		
1 Ea.	400 T.P.H. Recip. Plate feeder with motor (48")	New	Boise	3500 #	3500 #	53	2000	2000		10
1 Ea.	Vibrating Grizzley with motor	New	Penn.	8000 #	8000 #	310	2800	2800		20
1 Ea.	10" x 36" Primary Crusher with motor	Used	Hungry Horse	15000 #	15000 #	#2.00 300	3750	3750		50 — Should use larger primary & eliminate most of waste.
70 l.f.	96" Ø C.M.P. #8 GA under surge	New	P Portland	260 #	18200 #	#.75 319	31.56	11362		
360 l.f.	96" Ø C.M.P. #8 GA under coarse agg.	New	Portland	260 #	93600 #	" 1638	31.56	11362		
270 l.f.	96" Ø C.M.P. #8 GA under blending sand	New	Portland	260 #	44200 #	" 774	31.56	5365		
285 l.f.	96" Ø C.M.P. #8 GA under finish sand	New	Portland	260 #	74100 #	1297	31.56	8995		
1 Ea.	Secondary Crushers #37 Kennedy with motor	New	Penn.	40000 #	40000 #	1800	25560	25560		60
4 Ea.	Dbl. 4 x 12 deck vibrating screens with motors	New	Penn.	5400 #	21600 #	1200	2800	11200		40
1 Ea.	Eagle sand dewatering screw 22" x 25' with motor	New	Penn.	7600 #	7600 #	115	1600	1600		10
2 Ea.	Eagle sand dewatering classifying screw 20"x22'w/mot.	New	Penn.	10900 #	10900 #	310	5800	5800		20
1 Ea.	Sand tank steel (Collecting chutes)	New	Job	3300 #	3300 #	---	500	500		
2 Ea.	60 ton drive under bins steel	New	Job	10000 #	20000 #	Ft. allow.	1800	3500		
82 Ea.	Troughing rolls 30" x 5" Dia.	New	Boise	80 #	6560 #	"	30.00	2260		
120 Ea.	Troughing rolls 24" x 5" Dia.	New	Boise	60 #	7200 #	"	28.50	3420		
18 Ea.	Troughing rolls 18" x 5" Dia.	New	Boise	45 #	18810 #	"	26.00	10868		
32 Ea.	Return rolls 30"	New	Boise	35 #	1120 #	"	15.00	480		
48 Ea.	Return rolls 24"	New	Boise	20 #	950 #	"	13.00	624		
168 Ea.	Return rolls 18"	New	Boise	15 #	2520 #	"	11.00	148		
692 l.f.	30" 5 ply 32 oz. belt	New	Boise	8.28 #	5730 #	"	6.75	4871		
919 l.f.	24" 5 ply 32 oz. belt	New	Boise	6.62 #	6084 #	"	4.04	3713		
3,495 l.f.	18" 4 ply 28 oz. belt	New	Boise	4.32 #	15098 #	"	3.22	11254		
3 Ea.	18" Dia. x 32" tail pulley	New	Boise	300 #	900 #	"	150.00	450		
1 Ea.	18" Dia. x 26" tail pulley	New	Boise	250 #	250 #	"	125.00	125		
9 Ea.	18" Dia. x 20" tail pulley	New	Boise	200 #	1800 #	"	100.00	900		
13 Ea.	18" Dia. screw takeups	New	Boise	100 #	1300 #	"	75.00/pr	975		
2 Ea.	30 HP Electric motor conveyor DE LM	New	Boise	750 #	1500 #	"	3000.00	6000		60
1 Ea.	20 HP electric motor conveyor B	New	Boise	500 #	500 #	"	2100.00	2100		20
2 Ea.	15 HP electric motor conveyor NO,QR	New	Boise	300 #	300 #	"	1500.00	3000		30
5 Ea.	10 HP electric motor conveyor F, G, H, I, K	New	Boise	265 #	1325 #	"	957.00	4785		50
2 Ea.	5 HP electric motor conveyor J, P,	New	Boise	200 #	400 #	"	650.00	1300		10
2 Ea.	Convexco weigh feeders	New	L.A. Montana	600 #	1200 #	100.00	1800.00	3600		10
16 Ea.	Tunnel gates	New	Job	1100 #	17600 #	---	350.00	5600		10
5 Ea.	Chutes to Conveyors	New	Job	---	500 #	---	100.00	500		
	Chute to sand tank	New	Job	---	100 #	---	50.00	50		
40 M	lumber for conveyors and towers						100.00	7000		
	TOTAL THIS PAGE				468857 #	8266		171135		390

VALUES USED ARE FICTITIOUS FOR ILLUSTRATION ONLY

Figure 14-26-10　(Continued)

JOHN DOE CONSTRUCTION COMPANY, INC.

JOB WILSON'S WIN DAM

AGGREGATE PROCESSING EQUIPMENT

ITEM NO.   PAGE

SHEET 6 OF 13

BY   DATE May 1971

VALUES USED ARE FICTITIOUS FOR ILLUSTRATION ONLY

Quan.	Item	Condition	Location	Unit Weight	Total Weight	Frt. to Job	Unit Cost New or Used	Total Cost New or Used	Salvage	Motor HP Req'd.
	Brought Forward				468857 #	8266		171135.00		
26 M	Bin Dividers 3480 sq. ft.	New	Job	-	-	-	75.00	1950.00		
L.S.	Spray piping etc.	New	Job	150 #	150 #	2	500	500.00		
3 Ea.	Impact rollers 30"	New	Boise	70 #	210 #	Frt. Allow	69.00	207.00		0
4 Ea.	Impact rollers 24"	New	Boise	60 #	-	Frt. Allow	64.00	-		
	Impact rollers 18"	New	Boise	50 #	200 #	Frt. Allow	59.00	236.00		
1 Ea.	Pump 750 GPM 4250 lf @ 140' head	New	Job	5000 #	5000 #	Frt. Allow	3750.00	3750.00		90
4250 lf	Water line from river 8"	New	Berkley	8 #	34000 #	680	1.20	5100.00		
700 lf	Misc. 18" C.M.P. for draining off waste	New	Job	-	-	-	2.25	1575.00		
1 Ea.	Chute to secondary crusher	New	Job		-	-	-	750.00		
L.S.	Nails spikes bolts rough HBW for LBR Str.	New	Job		-	-	-	3000.00		
	TOTAL EQUIPMENT				508417 #	8948		188203.00		480
	TOTAL SUPPLIES							171360.00		
	TOTAL							16843.00		
	(add sales tax 3%)									

Figure 14-26-10 (Continued)

JOHN DOE CONSTRUCTION COMPANY, INC.

JOB: WILSON'S WIN DAM   ITEM NO.   PAGE   SHEET 7 OF 13

CONCRETE GENERAL AGGREGATE PROCESSING   BY   DATE May 1971

		LABOR AND INSURANCE		PERMANENT MATERIALS		SPECIFIC PLANT		EQUIPMENT RENTAL		SUPPLIES		FIRST COST OF SALVAGEABLE ITEMS SUBCONTRACTS		TOTAL DIRECT COST		ESTIMATED COST
		UNIT COST	AMOUNT	UNIT COST	AMOUNT	UNIT COST	AMOUNT	UNIT COST	AMOUNT	UNIT COST	AMOUNT	UNIT COST	AMOUNT	UNIT COST	AMOUNT	
SUMMARY																
Clearing of pit area None Req'd.			-		-		-		-		-		-		-	
Stripping sand and gravel deposit	120,000 cy	.125	15037		-		-	.081	9695	.087	10463		-	.293	35195	
Aggregate pit oper. and haul to plant	880,000 cy	.158	139019		-		-	.146	128619	.163	143786		(310550)	.467	411124	
Construct & maintain haul roads to plant			12834		-		-		7442		7605		-		27881	
Processing plant operation	1,075,000 tons	.0757	81354		-		-	.0005	600	.0545	58653		-	.1307	140607	
Haul finished aggregate to B plant	1,075,000 tons	.0304	32714		-		-	.0299	32095	.0289	31115		-	.0892	95924	
Disposal of waste at processing plant	245,150 tons		7815		-		-		6881		6813				21509	
Fence around pit area					Not required on this contract										-	
Sales tax on misc. supplies												905				905	
SUB-TOTAL	1,075,000 tons	.268	288773					.172	185332	.241	259340			.681	733445	
UNIT /cu.y. Conc. ÷	614,200 cy	.470						.302		.422				1.194		
Processing plant erected	614,200 cy					.140	89947	.300	184368				(178110)	.440	270815	
Total cost of aggregate	614,200 cy (Concrete)	.470	288773			.140	89947	.602	369700	.422	259340			1.634	1003760	

VALUES USED ARE FICTITIOUS FOR ILLUSTRATION ONLY

SALVAGE on EQUIPMENT

Pit Equipment first cost - see detail $310,550.00

Less rentals charged in operations:

Pit stripping	3995	
Pit operation	128516	
Haul waste	6681	
Haul agg. to B. Plant	32095	171,287.00
Total not charged into work -	139,262.00 = 45% = 55% Write-off	

Salvageable items of processing plant
written into plant cost 100% = $178,110.00

VALUES USED ARE FICTITIOUS FOR ILLUSTRATION ONLY

Figure 14-26-10   (Continued)

JOHN DOE CONSTRUCTION COMPANY, INC.

JOB	WILSON'S WIN DAM	ITEM NO.		PAGE	
	CONCRETE GENERAL AGGREGATE PROCESSING			SHEET 8 OF 13	
		BY		DATE May 1971	

STRIPPING SAND AND GRAVEL DEPOSITS:

From separate detail, required to run plant 3300 hrs. @ 400 T.P.H. = 1,320,000 tons ÷ 1.5 = 880,000 c.y. For stripping required, figure to use ave. of all test holes in area. BU-A = 5.0', B-AT = 4.0', B-AO = 3.5', B-AZ = 4.0', B-AY = 4.0', B-BA = 2.0', -- Ave. = 3.8',
Ave. depth of raw material; B-AT = 25.0', B-AO = 50',
B-AZ = 27.0'

Figure ave. of 30' of useable aggregate = ave. depth of deposit of 34.0'

Area of pit $\frac{30'}{27} = 1.111$ c.y./s.f.   Required area =
880,000 / 1.112? = 792,000 sq. ft. $\frac{792,000 \times 4'}{27}$ = 117,500 c.y.
or say strip 120,000 c.y.

OPERATION: To eliminate move-in of cat and scraper spread figure to handle 2/3 with dozers and load and haul 1/3 w/80D and 25 c.y. Bott m dumps required for haul of material to aggregate plant.

120,000 c.y. ÷ 65 c.y./hr. w/dozer = 1845 hrs.
40,000 c.y. ÷ 275 c.y./hr. w/80D = 145 hrs.

VALUES USED IN ILLUSTRATION ONLY
FOR ILLUSTRATION ONLY

	LABOR AND INSURANCE AMOUNT	EQUIPMENT RENTAL UNIT COST	EQUIPMENT RENTAL AMOUNT	SUPPLIES UNIT COST	SUPPLIES AMOUNT	TOTAL DIRECT COST UNIT COST	TOTAL DIRECT COST AMOUNT
LABOR:							
Foreman 150 hrs. @ 3.10	458						
(Foreman in pit operation)							
Equip. operation:							
1-80D shovel 150 hrs. @ 7.35 / 6.98 / 5.10	1103		1047		765		
3-25 c.y. B. dumps 450 hrs. @ 4.55 / 6.55 / 6.35	2048		2948		2858		
1-D-8 dozer 1900 hrs. @ 4.25 / 3.00 / 3.60	8075		5700		6840		
(Rental 195 hrs./mo.)							
17% shift and overtime on $11,684 labor	1986						
10% payroll tax and ins. on $13,670 labor	1367						
TOTAL DIRECT COST 120,000 c.y.	15034	.081	9695	.087	10,463	.293	35,195

VALUES USED ARE FICTITIOUS
FOR ILLUSTRATION ONLY

Figure 14-26-10 (Continued)

JOHN DOE CONSTRUCTION COMPANY, INC.

JOB: WILSON'S WIN DAM   ITEM NO. ____   PAGE 9 OF 13   SHEET ____   DATE May 1971   BY ____

CONCRETE GENERAL AGGREGATE PROCESSING

AGGREGATE PIT OPERATION & HAUL TO AGGREGATE PROCESSING PLANT

Haul distance = 12,400' = 2.35 Mi. one way = 4.7 Mi. R. trip
Use 8CD shovel in Pit w/25c.y. B. dumps top speed =
27.1 M.P.H. - Figure ave. both ways = 18 M.P.H. Capacity
payload = 78,000# = 25 c.y. struck = 29.7 heaped w/3:1
slopes. Load: 78,000# ÷ 3000#/c.y. = 26 c.y. - Use 27 c.y.
per load   Cycle: Load 27/300 x 60 =   5.4 Min.
                   Haul 4.7 x 60   15.7 Min.
          Dump Turn & Acc. 18   1.3 Min.
          10% lost time on 22.4 min.   2.3 Min.
                                                24.7 Min.

60/24.7 = 2.43 Loads/hr. x 27/load = 65.5 c.y./unit hr.
275 ÷ 65.5 = 4.2 units = say 5 units
OPERATION 880,000 c.y. ÷ 275/hr. = 3200 hrs.
   Foreman 3200 hrs. @ 3.10

VALUES USED ARE FICTITIOUS

EQUIP. OPERATION
   1-8CD shovel 3200 hrs. @ 7.35 / 6.98 / 5.10
   5-25 c.y. B. dumps 16,000 hrs. @ 4.55 / 6.55 / 6.35
   Pickup 23 mo. @ 60.00 / 50.00

17% shift and overtime on $107,620 labor
10% P.R. Tax & Ins. on $125,915 labor
          Sub-Total Direct Cost 880,000 c.y.

EQUIPMENT & MOVE-IN:
Move-in: 25 c.y. B. dumps 6 @ 608 cwt. = 3648 cwt.
          8CD NW Comb. 1@ 1450 cwt. = 1450 cwt.
          Unload & haul to site 205 tons
Equipment
   6-25 c.y. B. dumps @ $42,440 Ea.= $254,640.00
   1- NW 8CD Comb. @ $54,210 Ea.= 54,210.00
   1-Pickup 1700   = 1,700.00
   3% Wa. State sales tax on $310,550.00
   (See summary for computed salvage value
   of pit equipment)

          TOTAL DIRECT COST 880,000 c.y.

Haul Cost computation (Haul Cost Only):

Item	Amount
Labor	72800
Equip.	104800
Supplies	101600
17%	12376
10%	8518
Total	300094
880,000 c.y. @ 2.35 Mi. = 2,068,500 c.y.-Mi.	2068500

.145/c.y. Mile   1.5 = .097/6 c.y.mi

Cost columns:

Description	Labor & Ins. Unit	Labor & Ins. Amount	Equip. Rental Unit	Equip. Rental Amount	Supplies Unit	Supplies Amount	−Salvageable Unit	−Salvageable Amount	Total Direct Unit	Total Direct Amount
Foreman 3200 hrs. @ 3.10		9920								
1-8CD shovel		23520		23336		16320				
5-25 c.y. B. dumps		72800		104800		101600				
Pickup		1380		1380		1150				
17% shift and overtime		18295								
10% P.R. Tax & Ins.		12592								
Sub-Total Direct Cost	.157	138,507	.146	128,516	.135	119,070			.438	386,093
Move-in 3648 cwt					3.37	12294				12294
Move-in 1450 cwt					2.00	2900				2900
Unload & haul 205 tons	2.50	512	.50	103	1.00	205	4.00			820
6-25 c.y. B. dumps								(254,640)		
1- NW 8CD Comb.								(54,210)		
3% sales tax						9317				9317
TOTAL DIRECT COST	.158	139019	.146	128619	.163	143786		(310,550)	.467	411424

Figure 14-26-10 (Continued)

JOHN DOE CONSTRUCTION COMPANY, INC.

JOB: WILSON'S WIN DAM
CONCRETE GENERAL AGGREGATE PROCESSING
ITEM NO.
PAGE
SHEET 10 OF 13
DATE May 1971
BY

VALUES USED ARE FICTITIOUS FOR ILLUSTRATION ONLY

Description	Labor and Insurance Unit Cost	Labor and Insurance Amount	Permanent Materials Unit Cost	Permanent Materials Amount	Specific Plant Unit Cost	Specific Plant Amount	Equipment Rental Unit Cost	Equipment Rental Amount	Supplies Unit Cost	Supplies Amount	Sub-Contracts Unit Cost	Sub-Contracts Amount	Total Direct Cost Unit Cost	Total Direct Cost Amount	Estimated Cost
CONSTRUCT AND MAINTAIN HAUL ROAD TO PROCESSING PLANT FROM PIT															
Construct Road:															
Pit to bridge access road = 3300'															
Previous road for spillway															
Estimate - 4300' = 77%															
Cost of construction: Refer to Previous															
detail w/ spillway estimate and take 75%		2850						1500		1465				5815	
Bridge access road to plant:															
From spillway job est. this road was proposed as bid item															
and later deleted.															
Quantities - Rock 5,000 c.y.															
Common 15,000 c.y.															
Figure that since Atkinson intends to use this same pit															
that cost of construction will be shared - say rock,2500cy	1.00	2500					.50	1250	.50	1250			2.00	5000	
Common,7500cy	.20	1500					.20	1500	.20	1500			.60	4500	
MAINTENANCE OF HAUL ROAD:															
Figure 1½ hrs. per 8 hr. shift with blade and water truck															
3200 hrs. ÷ 8 = 400 shifts x 1.5 = 600 hrs. Ea.															
(Other contractors will be required to stand part of															
maintenance cost)															
Blade 600 hrs.	4.00	2400													
4500 gal. tanker 600 hrs.	3.75	2250													
17% shift & O.T. on $4650.00 labor		790													
10% tax/hr. on $5440.00 labor		544													
TOTAL DIRECT COST		12834						7442		7605				27881	

Figure 14-26-10   (Continued)

JOHN DOE CONSTRUCTION COMPANY, INC.

JOB WILSON'S WIN DAM
CONCRETE GENERAL AGGREGATE PROCESSING

ITEM NO.
BY
DATE 12-10-71
PAGE
SHEET 11 OF 13

VALUES USED ARE FICTITIOUS FOR ILLUSTRATION ONLY

Description	Labor & Ins. Unit Cost	Labor & Ins. Amount	Equip. Rental Unit	Equip. Rental Amount	Supplies Unit Cost	Supplies Amount	Total Direct Unit Cost	Total Direct Amount
PROCESSING PLANT OPERATION: 3300 Hours								
LABOR:								
1- Plant Superintendent 23 Mo.	600.00	13800						13800
Tax & Ins. @ 8½%		1173						1173
Operation and repair labor								
1- Plant operator @ 2.75 = 2.75								
1- Oiler @ 2.15 = 2.15								
1- Dump grizzly @ 2.10 = 2.10								
1- Screening tower oper.@ 2.50 = 2.50								
1- H.D. Mechanic @ 2.65 = 2.65								
2- Bin filling laborers @ 2.10 = 4.20								
(2 shifts)								
7 men 3300 hrs. @ 16.35		53955						53955
10% overtime on $ 53,955 labor		5396						5396
10% P.R. tax &) $ 59,351 labor		5935						5935
(Ins. on								
EXPENSE:								
Elect. power 480 H.P. connected load @ .75 load factor					.0075	6683		6683
x .746 = 270 kw 270 kw x 3300 hrs = 891,000 kwh								
Lube, oil and grease 3,300 hrs.					2.50	8250		8250
Repair parts, screen cloth, etc. 1,075,000 tons					.04	43000		43000
DOZER WORK ON STORAGE PIERS:								
Spread piles for additional storage and clean up around								
plant - say 5% of total time = 165 hrs. - say 200 hrs.								
D-8 dozer - 200 hrs. @ 4.25 ≠ 3.00 ≠ 3.60		850		600		720		2170
17% shift & overtime on $850 labor	.0757	145						145
10% P.R. tax & Ins. on $995 labor	.132	100						100
TOTAL PLANT OPERATION 1,075,000 tons	.0757	81354	.0005	600	.0545	58653	.1307	140607
Per c.y. concrete 614,200 c.y.	.132		.001		.095		.228	

Figure 14-26-10 (Continued)

JOHN DOE CONSTRUCTION COMPANY, INC.

JOB WILSON'S WIN DAM
CONCRETE GENERAL AGGREGATE PROCESSING
ITEM NO.
BY
DATE May 1971
SHEET 12 OF 13
PAGE

HAUL AGGREGATES FROM PROCESSING PLANT TO BATCH PLANT:

Figure to load 25 c.y. bottom dumps from bins at aggregate plant. Two 45 c.y. capacity drive over hoppers provided at batch plant for unloading. Haul distance = 3800'. Because of traffic congestion figure ave. haul speed of 10 m.p.h.

Cycle: Load from bins at plant   1.0 min.
    Haul to plant $\frac{3800' \times 2 \times 60}{5280 \times 10}$ 7.9 min.

    Dump, turn & acc.   1.0 min.
    10% lost time on 9.9 min.   1.0 min.
    _____
    10.9 min.

$\frac{60}{10.9}$ = 5.5 loads x 40 tons = 220 tons/hr.

Two trucks will keep plant operating at full capacity. Figure that no additional hauling equip. will be needed as pit equip. can be used since concrete will be placed on second and third shifts and aggregate plant will operate on first shift.

Operation: 615,000 c.y. concrete @ 1.75 tons =
1,075,000 tons ÷ 220= 4900 hrs.
25 c.y. B. dump 4900 hrs. @ 4.55 ≠ 6.55 ≠ 6.35
25% lost time on operators ≥ 1225 hrs. @ 2.55

17% shift & overtime on $25,419 labor
10% payroll tax & ins. on $ labor

Total Direct Cost 1,075,000 tons

VALUES USED ARE FICTITIOUS FOR ILLUSTRATION ONLY

	LABOR AND INSURANCE		PERMANENT MATERIALS		SPECIFIC PLANT		EQUIPMENT RENTAL		SUPPLIES		SUB-CONTRACTS		TOTAL DIRECT COST		ESTIMATED COST
	UNIT COST	AMOUNT	UNIT COST	AMOUNT	UNIT COST	AMOUNT	UNIT COST	AMOUNT	UNIT COST	AMOUNT	UNIT COST	AMOUNT	UNIT COST	AMOUNT	
25 c.y. B. dump		22295						32095		31115				85505	
25% lost time		3124												3124	
17% shift & overtime		4321												4321	
10% payroll tax & ins.		2974												2974	
Total Direct Cost	.0304	32714					.0299	32095	.0289	31115			.0892	95924	

Figure 14-26-10 (Continued)

JOHN DOE CONSTRUCTION COMPANY, INC.

JOB WILSON'S WIN DAM   ITEM NO. _____   PAGE _____

CONCRETE GENERAL AGGREGATE PROCESSING   SHEET 13 OF 13

DATE May 1971   BY _____

	LABOR AND INSURANCE		PERMANENT MATERIALS		SPECIFIC PLANT		EQUIPMENT RENTAL		SUPPLIES		SUB-CONTRACTS		TOTAL DIRECT COST		ESTIMATED COST
	UNIT COST	AMOUNT	UNIT COST	AMOUNT	UNIT COST	AMOUNT	UNIT COST	AMOUNT	UNIT COST	AMOUNT	UNIT COST	AMOUNT	UNIT COST	AMOUNT	
DISPOSAL OF WASTE AT PLANT:															
From pit and processing plant analysis 16,500 tons of ≠ 10" material will have to be loaded and hauled away to waste, plus fine aggregate waste, making a total of 245,151 tons or 18.5%. Of this total 60,381 is — $100 material that will be sluiced away with wash water.															
Load & haul w/shovel – 16,500 ÷ 1.5 = 11,000 c.y.	.20	2200					.15	1650	.15	1650			.50	5500	
Load & haul w/bottom dumps															
245,151 – 16,500 – 60,381 = 168,270 tons	.0304	5115					.0299	5031	.0289	4863			.0892	15009	
CONSTRUCT WASTE WATER DITCH:															
Run ditch east parallel to R. R. tracks along road and dump into ravine at high fill previously set up for waste = 2800'		500						200		300				1000	
TOTAL DIRECT COST		7815						6881		6813				21509	

VALUES USED ARE FICTITIOUS FOR ILLUSTRATION ONLY

Figure 14-26-10 (Continued)

JOHN DOE CONSTRUCTION COMPANY, INC.

OPERATION	EXCAVATION FOR COFFERDAM	ITEM NO. 1	SHEET 1 OF 3 PAGE

JOB **WILSON'S WIN DAM**

QUANTITY	UNIT	EST. BY G.O.A.	CK'D BY	DATE 12/22/71

		LABOR	EQUIP.	EXPENSE

<u>Drill & Shoot Rock:</u> (Below El. 80 -)

Figure to drill this material from
the top of the previous fills,
casing the holes through the
gravel fills - from previous calculations,
Ave. casing Length = 12 Ft. -

Drill Patern = 5' x 5' or Say .9 C.Y.
per foot of hole =

Quantity = <u>2569 C.Y</u> = 2860 L. Ft. -
Say 3000' of Drill Hole -

105' Long x 75' Wide = 7875 S. F. Area

At 35 At 36 $\dfrac{2056 \times 27}{7875}$ - 7.04" Drill Hole -

3000' ÷ 7.04 - 427 Holes -

Casing- 427 Holes @ 12' = 5124' @ .65 3331

<u>Drill & Shoot:</u>

Thru Fill - 5124' @ 30'/Hr. = 171 Hrs.

" Rock - 3000' @ 20'/Hr. = <u>150 Hrs.</u>

                                          321 Hrs.

Operation -   321 ÷ 4 = 80 /Hr.

Foreman -	1 @ 2.60 =	2.60
Drillers -	4 @ 2.43 =	9.72
Helpers -	2 @ 2.35 =	4.70
Powderman -	1 @ 2.48 =	2.48
Powder Helpers -	1 @ 2.35 =	2.35
Comp. Oper. -	1 @ 2.43 =	2.43

85 Hrs. @ 24.28 2064

Equip. Oper.

Comp. (2)	160 Hrs. @ 1.50 / 2.01 / 2.75	240	322	440
W. Drills	320 Hrs. @ .40 / .38 / .65	128	122	208
Flatrack	80 Hrs. @ 2.73 / 1.00 / 1.60	218	80	128

Drill Steel - 2600 C.Y.@ ¼# = 650# @ .23 150

Bits - 3000' @ 400' = Say 8 @ 20.00 160

569 CY

6500

Powder --9070 @ 1# = 9100# @ .19 1729

Caps -- 430 Holes / 5% = 450 @ .25 113

| Sub-Total Forwarded 9433 | | 2650 | 524 | 6259 |

VALUES USED ARE FICTITIOUS FOR ILLUSTRATION ONLY

Figure 14-26-11 Rock excavation estimate

JOHN DOE CONSTRUCTION COMPANY, INC.

OPERATION   EXCAVATION FOR COFFERDAM   ITEM NO.   1   SHEET 2 OF 3 PAGE

JOB   WILSON'S WIN DAM

QUANTITY   UNIT   EST. BY   CK'D BY   DATE   12/22/71

	TOTAL	LABOR	EQUIP.	EXPENSE
Drill & Shoot Rock - Under Water				
Brought Forward	9433	2650	524	6259
Blasting Supplies -				
2569 C.Y. .04				103
Air Hose, Small Tools & Misc.				250
10% Overtime on $ 2650 Labor		265		
10% Tax & Ins.on $ 2915 Labor		292		
Sub-Total Direct Cost	10.343	3207	524	6612
Unit ÷ 2569 C.Y.	4.02	1.25	.20	2.57

VALUES USED ARE FICTITIOUS FOR ILLUSTRATION ONLY

Figure 14–26–11   (Continued)

JOHN DOE CONSTRUCTION COMPANY, INC.

OPERATION EXCAVATION FOR COFFERDAM	ITEM NO. 1	SHEET 3 OF 3 PAGE
	JOB WILSON'S WIN DAM	

QUANTITY	UNIT	EST. BY G.O.A.	CK'D BY	DATE 12/22/71

	LABOR	EQUIP.	EXPENSE
<u>LOAD MATERIAL IN TRUCKS & HAUL</u>			
<u>TO WASTE:</u>			
Use 4500 Manitowoc Dragline			
w/5 C.Y. Bucket to Remove All			
Material - Figure to Load Out			
75 C.Y. Solid Measure per Hour.			
Below Water - 2264 C.Y.			
Above Water - <u>2569 C.Y.</u>			
Total - 4833 C.Y.			
4833 ÷ 75 = 65 Hrs. Plus Moving			
= Say 70 Hrs.			
Fill Material @ 100 C.Y. Per Hr. -			
Over Blasted Material - 6500 C.Y.			
Fill Over South Channel - 1250 C.Y.			
Access Fills: 431			
<u>15,651</u>			
75% of 16,081 <u>12,000 C.Y.</u>			
19,750 C.Y.			
Say 20,000 C.Y. = 200 Hrs.			
Cycle: 17 C.Y. E-D @ 11 C.Y. Solid			
or 15.5 C.Y. Loose -			
Load - $\frac{11}{75}$ x 60 = 8.8 Min. -			
Haul - $\frac{1.14 \text{ Mi} \times 2 \times 60}{12.5 \text{ M.P.H.}}$ = 11.0 Min. -			
Dump, Turn & Acc. 2.0 Min. -			
12½% Lost Time on 21.8 = <u>2.7 Min.</u> -			
$\frac{60}{24.5}$ = 2.45 x 11 = 27 C.Y. 24.5 Min. -			
Per Hr. = 3 Trucks -			
Operation: 270 Hrs.			
Foreman 1 @ 3.30 = 3.30			
Dumpman 1 @ 2.20 = <u>2.20</u>			
270 Hrs. = 5.50	1485		
Sub-Total Forward	1485		

VALUES USED ARE FICTITIOUS FOR ILLUSTRATION ONLY

Figure 14–26–11 (Continued)

JOHN DOE CONSTRUCTION COMPANY, INC.

ESTIMATE SUMMARY

SHEET 1 OF 5 PAGE

JOB Pioneerville Town Hall

EST. BY _____ CK'D BY SUI DATE Dec. 2, 1970

DESCRIPTION	SPECIFICATION SECTION	LABOR AND INSURANCE		PERMANENT MATERIALS		SPECIFIC PLANT		EQUIPMENT RENTAL		SUPPLIES		SUB-CONTRACTS		TOTAL COST		NAME OF SUBCONTRACTOR (Identification)
		UNIT COST	AMOUNT	UNIT COST	AMOUNT	UNIT COST	AMOUNT	UNIT COST	AMOUNT	UNIT COST	AMOUNT	UNIT COST	AMOUNT	UNIT COST	AMOUNT	

Values used are fictitious for illustration only.

Direct Cost

Site Work
- Demolition — 8-A — — — — — — — — — — — — — — 300 —
- Site Grading — 2-B — 100 — — — — — 40 — 60 — — — 215 —
- Earthwork — 2-C — 826 — — — — — 163 — 244 — — — 915 —
- Paving — 2-D — 3503 — — — — — 526 — 576 — — — 10250 A.B. Shurtacka
- Landscaping — 2-E — — — — — — — — — — 10250 9100 9100 Plumers Inc.

Concrete Work
- Steel Reinforcement — 3-A — 10500 — 32740 — — — 1050 — 1472 — — — 45790 —
- Concrete — 3-B — 7620 — 37312 — — — 1125 — 16780 — — — 31837 —

Masonry Work
- Masonry in place — 4-A&B — — — — — — — — — — — 62100 62100 DiMauro & Rock Co.

Metal Work - Structural & Miscellaneous
- Structural Steel in place — 5A&B — — — — — — — — — — 85000 85000 Steel Structures Co.
- Steel Joists — 5C — — — — — — — — — — 7540 7540 Trusses Inc.
- Steel Roof Decks — 5D — — — — — — — — — — 2015 2015 Decks Co.

Carpentry
- Framing Etc. at Site — 6-A — 5200 — 10100 — — — 350 — 450 — — — 16100 —
- Glued Laminated Beams — 6-B — 4800 — 13000 — — — 200 — 400 — — — 9400 —
- Finish Wood Floors — 6-C — — — — — — — — — — 4000 4000 Wood Floor Co.
- Insulation — 6-D — — — — — — — — — — 2205 2205 T. Mahoney Co.

Roofing, Sheet Metal Etc.
- Built-up Roofing — 7-A — — — — — — — — — — 20100 20100 Toppers Inc.
- Damp-proofing — 7-B — — — — — — — — — — 1220 1220 "
- Sheet Metal — 7-C — — — — — — — — — — 2124 2124 "
- Caulking — 7-D — — — — — — — — — — 3500 3500 Caulkers Co.

Doors, Windows & Glass
- Hollow Metal Work — 8-A — — — — — — — — — — 6070 6070 Steel Doors Inc.
- Rolling Doors — 8-B — — — — — — — — — — 3200 3200 Steel Doors Inc.
- Wood Doors — 8-C — 600 — 1200 — — — 100 — 100 — — — 7200 —
- Steel Louvers — 8-D — — — — — — — — — — 600 600 Steel Specialists
- Aluminum Windows — 8-E — — — — — — — — — — 6010 6010 Window Windows
- Glass & Glazing — 8-F — — — — — — — — — — 1235 1235 N.Y. Glass Co.

Sub Totals-Carried Fwd — — 104049 98372 — — 35916 — 201100 — 233928 — 440025

Figure 14-33-1 Typical building estimate summary

JOHN DOE CONSTRUCTION COMPANY, INC.

ESTIMATE SUMMARY

SHEET 2 OF 5 PAGE

JOB Pioneerville Town Hall

EST. BY OAB.   CK'D BY   DATE Dec. 2, 1970

DESCRIPTION	SPECIFICATION SECTION	LABOR AND INSURANCE AMOUNT	PERMANENT MATERIALS AMOUNT	SPECIFIC PLANT AMOUNT	EQUIPMENT RENTAL AMOUNT	SUPPLIES AMOUNT	SUB-CONTRACTS AMOUNT	TOTAL COST AMOUNT	NAME OF SUBCONTRACTOR
SUB TOTALS - BROUGHT FWD.		104,844	90,372		3576	20,100	23,758	246,025	(Continuation)
FINISHES									
Lathing & Plastering	9-A						18,080	18,080	Murst Plastrs. Co.
Gypsum Wall Board	9-B						1214	1214	"
Acoustic Treatment	9-C						1302	1302	"
Ceramic Tile	9-D						2680	2680	Russell Tile Co.
Painting & Finishing	9-E						12,540	12,540	Paris Co.
Resilient Floor Covering	9-F						1200	1200	Floor Coverings
SPECIALTIES									
Finish Hardware *	10-A	1250	6010					7260	
Toilet, Shower & Misc. Metal Partitions	10-B						3000	3000	Bros. & Assoc. Inc.
Toilet Accessories	10-C						250	250	Accessories Co.
Miscellaneous	10-D						220	220	"
MECHANICAL									
Plumbing & Plumbing Fixtures	11-A&B						31,512	31,512	Union Plumbing
Heating and Ventilating	11-C						36,021	36,021	"
Swimming Pool Filtration System	11-D						12,300	12,300	"
ELECTRICAL									
Electrical Services, materials, & installations	12-A&B						22,601	22,601	West Const'n Co.
Lighting Fixtures	12-C						2100	2100	"
TOTAL DIRECT COSTS		105,649 9	104,382		8576	22,100	87,714	6,110,16	
TOTAL INDIRECT COSTS (For Detail Attached)		76,800		4000	6,100	105,00	30,000	187,600	
TOTAL ALL COSTS		182,099	104,382	4000	9576	32,600	40,779	79,996	
CONTINGENCIES		(To be added)	(To be added)						Tentative
PROFIT		(To be added)	(To be added)						Tentative
TOTAL BID		(To be added)						87,094	Tentative

* Mark items not subcontracted for which detailed estimates are attached

VALUES USED ARE FICTITIOUS FOR ILLUSTRATION ONLY

Figure 14-33-1   (Continued)

Figure 14-33-1 (Continued)

JOHN DOE CONSTRUCTION COMPANY, INC.

ESTIMATE DETAIL

SHEET 4 OF 5 PAGE

JOB   Pioneerville Town Hall

EST. BY OAB   CK'D BY SUI   DATE Dec. 2, 1970

		LABOR AND INSURANCE		PERMANENT MATERIALS		SPECIFIC PLANT		EQUIPMENT RENTAL		SUPPLIES		SUB-CONTRACTS		TOTAL DIRECT COST	
		UNIT COST	AMOUNT	UNIT COST	AMOUNT	UNIT COST	AMOUNT	UNIT COST	AMOUNT	UNIT COST	AMOUNT	UNIT COST	AMOUNT	UNIT COST	AMOUNT
DIRECT COST DETAIL – GENERAL CONTRACTOR			17,345												
ITEM No. 2-A DEMOLITION															
Remove Concrete Curb	60 L.F.	3.00	180							1.00	60	5.00	300		3.00
(from historical experience)															
ITEM No. 2-B SITE GRADING															
Area of Building Plot 91,180 S.F.															
Spread on site (loose) – 3227 CY															
Use Caterpillar Front End Loader @ Dozer															
160 CY/hr. Say Spread 2245															
Dozing Dress Say – 1100															
Front End Loader – Dozer 33 hrs. @ 10.00		10.00	330			5.00	165	8.00	264	23.00	252				
Per 105 @ O.T. on labor @ Plate			56									8.75			
			386												
ITEM No. 2-C EXCAVATION															
Basement Excavation – Machine															
disposal by haul & Dozing on site															
load w/ F.E. Loader 2800 CY @ 180 CY/H															
Foreman @ 7.00 – 4.00															
Dump Men @ 4.00 – 4.00															
F.E. Loader 75 hrs. @ 11.00 – 1		8.00	925						8.75	875	8.75	875			
Dozer Spread 16 hrs. @ 10.00 – 2.5		10.00	750						8.75	875	8.75	1500			
Hand Excavation – Footings etc.		1.00	160												
Back Fill – Machine 250 CY @ 6.00		0.25	90												
Back Fill – Hand Compacted 12 CY @ 6.00			100						6.65	380			1.25	314	
Dump truck haul to disposal			750						6.30	380			6.62	332	
operation (General Supply)													6.50	82	
Excavated from outside 150 hrs															
Total ITEM No. 2-C			3563							536			65.00	2252	2252
										2252	6345				

VALUES USED ARE FICTITIOUS FOR ILLUSTRATION ONLY

Figure 14-33-1 (Continued)

JOHN DOE CONSTRUCTION COMPANY, INC.

ESTIMATE DETAIL

SHEET 5 OF 5 PAGE

EST. BY OAB.   JOB Pioneerville Town Hall   CKD BY SUI   DATE Dec. 2, 1970

DIRECT COST DETAIL – GENERAL CONTRACTOR ITEMS (Cont'd)

CONCRETE WORK

Structural Steel Reinforcement – By quotation, detailed, listed

Cut, bent, tagged & delivered – 210,000#

Sales Tax – 2½% on above

Installation – Re-steel. Completed 210,000# @ 0.05

Ready Mix. Concrete – Delivered 2100 CY

Sales Tax on above @ 4%

Placing Concrete (Labor Factored)

In basement slab   800 CY

In footings   200 CY

In walls – 9" to 15"½   250 CY

In columns   130 CY

In precast floor slabs   700 CY

Roof slab   400 CY

Rough finishing – Slabs

Monolithic float finish

Curing & clean up – 312,000 ft² (Factored line items)

Forms for Concrete – Factored line items (supplemental detail attached)

For footings

For walls

For columns – Beams

For floor systems – Beams

Shoring for support slabs (Precast)

TOTAL CONCRETE WORK — ITEMS – 3 & 5 – B

Note:
The above is priced to include all of class concrete. Arena service for field operations.

CARPENTRY

Framing Fr. ——————— Item No. 6 H

Glued Laminated Beams – Item No. 6-5

DOORS – Wood Doors — Item No. 8-C

VALUES USED ARE FICTITIOUS FOR ILLUSTRATION ONLY.

Estimate should show detail of each operation.

Name of Supplier: R. Steel Inc. / Concrete Inc.

| LABOR AND INSURANCE | | PERMANENT MATERIALS | | SPECIFIC PLANT | | EQUIPMENT RENTAL | | SUPPLIES | | SUB-CONTRACTS | | TOTAL DIRECT COST | |
UNIT COST	AMOUNT	UNIT COST	AMOUNT	UNIT COST	AMOUNT	UNIT COST	AMOUNT	UNIT COST	AMOUNT	UNIT COST	AMOUNT	UNIT COST	AMOUNT

Figure 14-33-1   (Continued)

15 ESTIMATING OTHER THAN FIRM-PRICE CONTRACTS

J. P. FREIN

Vice-President and Director (Retired)
Morrison-Knudsen Company, Inc.
Boise, Idaho

ALTHOUGH the general rule in the construction industry is to offer work on a firm-price basis, there are conditions that warrant other arrangements. This chapter is devoted to the principles of pricing the contract work under such circumstances. It also discusses how the contractor is to be compensated.

Proposals other than the firm-price type generally are applied to proposed construction work not fully described by advance plans and specifications or perhaps only envisioned as a preliminary concept of what is required by the owner-client. Incomplete tender documents, however, are only one reason why an owner would award his work without competitive bids on a conditional basis. Some owners, for reasons of their own that might include the desire to avoid publicity of their project, prefer to negotiate in confidence with known constructors, sometimes on a variable-price basis.

There are circumstances where industrial organizations make a practice of negotiating their proposed construction with given contractors to obtain the superior technical advantages and talents of the contractor's organization. The urgency of saving time is possibly the strongest reason for not following the time-proved method of competitive bidding.

Proposals other than firm price expose the owner-client to some risks not common to fixed-price bids. Such transactions should not be entered into except between highly reputable firms and where there is a substantial amount of mutual confidence. Nonetheless, it is only prudent that all of the terms and conditions of the arrangement be completely documented in advance for the guidance and protection of both parties.

Before making a commitment for such a transaction, the owner often would like some degree of assurance that the project cost will not overrun an amount related to its economic feasibility. Inasmuch as there are usually uncertain-

ties and variables involved, the contractor must be able to evaluate any financial consequences so that he can properly advise the owner about the probable cost of the enterprise and avoid nonreimbursible expenditures on his own part.

Estimating construction operations, management, and administrative costs for all types of contracts should follow the patterns and principles described in Chapter 14 on "Cost Estimating Lump-Sum and Unit-Price Contracts." Admittedly, for other than firm-price contracts, this cannot always be done. But the basic principles of estimating should be compromised as little as possible in considering the less certain elements that present themselves.

History shows many variations of other than firm-price contracts. They formerly were more prevalent than they are today. Nonetheless, numerous examples of these conditional arrangements are in current use. Basically, they fall into the following general categories:

1. Cost plus a percentage fee
2. Cost plus a fixed fee
3. Cost plus award fee
4. Cost plus a fixed fee with guaranteed maximum
5. Target estimate with incentive fee and penalty
6. Turnkey proposals (variable conditions)

It is presumed that one of these would be selected to best fit the attending circumstances and the preferences of the parties.

15-1. Cost-Plus-a-Percentage-Fee Contracts

A typical example of this type of arrangement is seen at the end of this chapter as Appendix 15-A1 entitled "Form of Agreement—Straight Cost Plus Basis." The form of agreement and the conditions were taken from the manual of the Associated General Contractors of America. This is a very straightforward arrangement calling for complete reimbursement of all cost plus the agreed percentage fee for the contractor. It presupposes the availability of contract drawings and specifications prepared by third parties, provided that changes and modifications can be made in them without contract modification.

The AGC general conditions of contract for engineering construction can be adapted to fit particular needs. No mention is made of estimates of cost of the contract work to be prepared by the contractor and agreed upon by the owner, although this might well be assumed to be a necessity in any case.

Actually, a cost-plus-a-percentage-fee contract is usually applied to a situation where advance construction details are minimal. Sometimes only sketch plans and tentative specifications are available at the time of signing the contract for construction. Economic feasibility is usually involved and the parties to the contract, plus the third-party engineer-designer for the owner (unless it is a turnkey arrangement), jointly summarize the known facts, conditions, and technical requirements to be a part of the contractual considerations. Thus the concept of the work is developed for the benefit of all concerned. The owner almost invariably must know some bracket of probable cost, and the contractor is obliged to make the best estimate of what the contract work would cost on the basis of the available information.

The contractor, visualizing the physical proportions of the structures, type of construction, the various installations, problems inherent and typical of the character of work involved, proceeds to outline probable construction operations, quantities of work, etc., to be done. Usually there is some precedent for the problem in question, preferably one handled by the contractor. Adding his knowledge of construction management, labor requirements, productivity factors, etc., the contractor assigns probable cost factors for the accomplishment of the task at hand, drawing largely upon information from his historical cost records for similar construction operations performed elsewhere and filling in for the unusual things where necessary.

At this stage, little detailed estimating is done or is possible. But more care should be taken with respect to speciality items of large proportions, such as machinery to be furnished and installed. Some preliminary outside quotations should be solicited for unknown items. The entire matter should then be coordinated and assembled and a substantial contingency factor added.

This procedure is sometimes called conceptual estimating. It actually is the process of preparing a preliminary estimate using the best information then known by the parties. Its accuracy will depend largely upon the knowledge and ability of the contractor and how well his cost references and adaptations fit the proposed work. Such an estimate necessarily lacks accuracy. However, it is useful for indicating economic feasibility or for initiating financing programs for the project.

If the contract is awarded, the construction designs, details, and specifications are advanced by the engineer, and as more details become known, the contractor has the responsibility of re-estimating the various parts of the work in accordance with the detailed refined methods employed for fixed-price contracts. This procedure should be helpful to the engineer for instituting economies of design and to the contractor in the refinement of his plans for construction. It should also be helpful to the owner for keeping informed of the probable ultimate outcome, for ordering possible changes to effect savings, to add elaborations where

advisable, and to properly control the job financing throughout the construction period.

On the other hand, if the construction designs, specifications, and conditions are reasonably well known before the contract is signed, a detailed estimate should be made promptly by the contractor as part of his contractual function for the benefit of all concerned. Such an estimate should be refined in the manner normally applied to a fixed-price contract. It should be modified from time to time for changes in design, changes in scope or other factors appreciably effecting the costs under the contract. Thus, all interested parties can better execute their respective functions.

Since a cost-plus-a-percentage-fee arrangement contemplates complete reimbursement of all necessary costs incurred, plus a fee entirely commensurate with the amount of such cost, it is not necessary to consider the complications of adjusting the compensation to the contractor that are involved in other types of contracts. The contractor's know-how in handling construction problems is recognized in such arrangements, he is normally allowed considerable latitude in the handling of the operations, and his fee is in payment for his managerial performance.

15-2. Cost-Plus-a-Fixed-Fee Contracts

This type of contractual arrangement normally operates about the same as a cost plus a percentage fee agreement. However, the amount of the contractor's fee is stated as a fixed sum instead of varying with the amount of cost incurred. Reimbursement of all necessary costs of the work is a condition.

To apply the principles of this type of contract, there must be available beforehand sufficient preliminary designs and information of record for the contractor to prepare an approximate estimate of cost of all the work to be done under the contract. In addition, the conditions of contract must be explicit. The construction details at this stage are rarely complete. But they would be sufficient to indicate proportions, sizes, and character of construction that would permit preliminary pricing of the items enumerated by the owner for inclusion in the contract activities.

The contractor then proceeds to prepare an estimate using the principles described for the preliminary estimate for the cost plus a percentage fee contract. In other words, the pricing is based on cost factors known to the contractor for similar types of work or developed by preliminary inquiries or investigations. This process might also be classed as conceptual estimating, but it is recognized as possibly the best appraisal of contract amount that can be made at the time. After the estimate is reviewed by the owner, it is either approved by the parties or modified to suit mutual agreement.

When it is approved, the estimate then becomes the basis for figuring the contractor's fee which is determined by the application of a percentage agreeable to the parties. Regardless of how much the actual cost of the work might overrun or underrun the preliminary estimate, the contractor receives only the fixed fee agreed upon by the parties. But the overall scope of the work cannot be changed by the owner, nor can he introduce new contractual conditions without making an equitable adjustment in the contract. In other words, the owner must permit the further estimating of the additional work or the added cost of the wholly unexpected contractual conditions, with a proportionate adjustment in the contractor's fee.

Reestimating the cost of the project and bringing the financial forecasts up to date with the development of additional information and designs should be a requirement. It assists all parties concerned in the programming and direction of the work, the acquisition and delivery of the things needed for its performance and for financial control.

The cost-plus-a-fixed-fee contract arrangement is widely used by the U.S. government on wartime projects where time is vital and the details of construction are not well defined.

15-3. Cost-Plus-an-Award-Fee Contracts

The cost-plus-an-award-fee type of contract has been used by the U.S. government in recent years where conditions prevent fixed-price awards. As in the case of the cost-plus-fixed-fee type of contracts, the contractor must estimate the probable cost of the work from somewhat indefinite and incomplete facts and data to establish a tentative contract value at an early stage. Based upon such tentative contract value, if approved by the parties, the contractor's fee is computed in two parts. They are, (1) a base fee, or minimum fee, determined as a percentage of the agreed contract value and paid to the contractor in any case for a completed contract performance; and (2) as additional award fee that is graduated from nothing to another given percentage of tentative contract value for the degree of excellence of performance of the contractor. This degree of excellence is judged by a board of control composed of representatives of the owner and the contractor, with the weight of influence being in favor of the owner.

The evaluation of the contractor's performance to determine the amount of the award fee would be based on weighted factors such as:

1. Quality of work, 20 points max.
2. Control of the work and savings and reduction in cost, 40 points max.
3. Adherence to schedule and early completion, 40 points max.

Total weighting factor—100 points

The base fee is usually set at a relatively small percentage and the award fee at a larger percentage to provide incentive earnings for outstanding performance. Thus, if the base fee was 4 percent and the maximum award fee 6 percent, the contractor could earn 10 percent on the estimated cost of the work. But agreement of the members of the board of control could result in something less.

The amount of award fee established by the board of control may depend on the personality of the various members and the courage of their convictions in stating their opinions. Administrative officers of political subdivisions of governments often find it difficult to rule in favor of a contractor unless the record is clear and the contract provides unquestionable authority for their decisions. This factor is an important consideration where this type of contract is in question.

Other than the conditions dealing with how the fee is established and paid, the cost-plus-award-fee type of contract would normally operate like the cost-plus-a-fixed fee type of contract. The preliminary estimate of cost would be based on as close an approximation as feasible and without going into minute detail. Refinement of the estimate periodically, as additional and more specific information becomes available, is a necessity for the benefit of all concerned. Although the contract work would be subject to changes as time goes on, any changes in scope or in contract conditions would be the subject of a contract adjustment. It would include an adjustment in the amount of the contractor's fees, but not in the percentage amounts by which they are determined.

15-4. Cost Plus a Fee with Guaranteed Maximum

This variation of a cost-plus contract, although perhaps including variable contractural conditions with respect to details and based on reimbursing the contractor for the actual necessary costs of the work plus a fee, assures the owner that the proposed venture can cost him no more than a given agreed total amount. It also assures the owner that if the costs exceed that amount, any overruns of cost would reduce or nullify the contractor's fee and for even greater excesses of cost the contractor is obliged to pay with his own funds for the completion of the project.

In other words, the contract payment conditions decidedly favor the owner. He can, under the condition of the underrun of cost, derive the benefit of such underrun—or whatever proportionate part the contract might provide—and is protected against the cost exceeding an amount that he might agree to. The contractor, on the other hand, assumes the entire risk of a cost overrun and may be in a position to participate in underruns of cost. But he participates in only a part of such underruns to the extent that the contract might state.

Because the risk is greater, the initial pricing necessary for establishing the guaranteed maximum amount to be stated in this type of contract must be the very best. It should be, if such a measure were practical, even better than the estimating necessary for fixed-price work. Great care also must be taken to assure that the contract conditions and the specifications are very definitive and that there are ample means of adjusting contract payments for changes and unforeseen conditions.

For this type of contract, the contractor should make a very detailed estimate and justify it to the owner. He should see that the estimate or a variation is approved by the parties, as well as the amount of an estimated fee to be paid on a given schedule. The amount of the total estimated cost plus the total amount of the fee generally is used as the guaranteed maximum cost to be written into the contract, except for any contingent amounts agreed to by the parties and included therein.

It is difficult to understand why contractors accept a proposition of this kind, unless they possess certain natural advantages and could protect themselves against the ravages of inflation and other pertinent factors. However, the desire to obtain new work and overcome the competition seems to be a motivating influence out of all proportion to the exercise of good judgment.

Any way this type of contract is considered, it is the equivalent of a lump-sum bid, less the possibility of the contractor enjoying all the benefits of savings that he can effect in the execution of the contract. The estimating and pricing of the contract work should be as refined and complete as possible to make them, with the procedures described on pricing lump-sum and firm unit-price contracts being none too good for the purpose. If a contractor investigates prospective work offered on this basis and finds that all details are not well defined, it would be advisable for him to avoid it.

15-5. Target Estimate with Penalty and Incentive-Fee Contracts

This type of contract is particularly suited to projects for which only preliminary plans and specifications are available when it is offered for bids, but because of the urgency of early completion or for other causes it is desired to get the work underway promptly. Perhaps all of the initial subsurface explorations have not been completed in advance and unknowns exist in this area. A third-party engineer employed by the owner-client presumably will refine the plans and specifications after the contract award.

Conditions would be such that it would not be possible accurately to price the contract work in advance. If such were attempted, unreasonable amounts of contingencies would logically be added in the contractor's bid to produce a firm-price proposal. Nevertheless, sufficient reasonably firm data are available in the preliminary tender documents, along with the background and historical cost experience of the contractor, to make a reasonable approximation of the project cost. Hopefully, the owner-client has a background in the previous construction of somewhat similar facilities and appreciates the attending problems. If so, perhaps he will be reasonable about the particulars of the negotiations. Mutual respect and understanding between the parties again is the first essential with respect to the success of this type of contract.

Essential, too, is an understanding of the conditions of contracts to apply. They should be carefully drawn to cover the rights and responsibilities of the parties. This type of contract basically contemplates an initial estimate by the contractor, which should be as accurate as possible under the circumstances. The estimate should be detailed where it can be, rounded out by experience data where it cannot be detailed, and the contractor would be obliged to substantiate his figures.

Negotiations then proceed with the client-owner to reconcile ideas of cost of the project work. The initial estimate made by the contractor is then changed to add features or conditions desired by the owner and an attempt is made to agree on the value of the work to be done. When the parties agree on values, the cost estimate becomes the target estimate for purposes of the contract. This is recognized as approximate.

The target estimate costs are specifically defined to include definite services to be furnished by the contractor. Presumably these would be all of the services that would be provided by an integrated general contractor if the work was to be done on a firm-price basis. The scope of the proposed work is to be specifically stated to include definite structures involving certain areas, capacities, types of construction, etc.

Although all details of the construction cannot be stated, the description should be sufficient to bear a definite relation to the approximations made in the target estimate. As part of the target estimate, the equivalent of a bid schedule should be prepared, listing the items of work, the approximate quantities for each, and the unit prices forming the basis of the target estimate.

The contract conditions would be expected to provide for complete reimbursement to the contractor for all necessary costs. The kind and character of such costs should be enumerated or described. Expenditures for items exceeding a given amount, say $3000, might require prior approval of the owner. The contractor, however, should be permitted the exercise of his judgment in the choice of methods and management for which he would be hired in the first place.

The contractor would be required to keep books of account and cost accounts to fit a system and pattern agreeable to the owner, with periodic statements of the accounts and cost accounts provided to the owner. This should be in such form as to permit direct comparison to the target estimate.

The owner-client naturally would have reserved the right to make changes in the work, in its design, and even in its scope. As such changes are made, provided they do not constitute refinements in design or the like intended at the time of the initial negotiations, the target estimate should account for them. This detail, preferably, should await final completion of the work and settlement of the contract. Large overruns in quantities or materially different subsurface conditions that result in substantially greater costs than originally envisioned are usually considered to be occasions for modifying the target estimate.

As time goes on, and more things become known or apparent, re-estimates of cost should be made periodically by the contractor on a more refined basis. These should be provided to the owner-client as forecasts of ultimate cost, regardless of whether matters are involved that would ultimately constitute a basis for changing the target estimate. Such information is indispensable to the owner-client in exercising control, planning for the necessary financing, and evaluating the performance of the contractor.

The contractor should expect to be required to substantiate great differences between the target estimate and the actual cost of performance. This might be a condition requisite to his being able to continue as a party to the contract. Invariably, this type of contract also provides for termination of the contract for cause, with both parties having rights in that connection.

For his services, the contractor is to be paid a fee. It usually is stated in terms of a percentage of the total target estimate cost. This, in turn, would be stated in terms of a basic fee, which might be about 8 percent of the target estimate including any modifications thereto. However, if at the time of final settlement of the contract it is found that the actual costs exceed the target estimate costs, then the contractor is penalized at a given rate, say 1 percent, of the target estimate costs, for a stated amount that the actual cost overruns the target estimate cost.

If the contractor's right to proceed and complete the work is preserved, he would be entitled to a minimum fee, say 4 percent, of the target estimate cost, plus complete reimbursement of all necessary costs as defined in the contract. Should savings be realized and the actual necessary costs are less than the target estimate costs, the contractor would be rewarded for such underrun of cost by being paid a supplemental fee. It would be measured in terms of a given rate, say 1 percent of the target estimate costs, for a stated amount that the actual cost underruns the target estimate cost. However, there would be stated a maximum fee which the contractor could earn (basic fee plus supplemental fee), say 12 percent of the target estimate, including any modification thereto.

In lieu of this fee arrangement, provisions are sometimes made for a lower basic fee. There are no penalties for overrunning the target estimate costs, but a bonus for underrunning the target estimate costs by a division of the savings at a rate of, say 25 percent for the contractor and 75 percent for the owner, of any amounts by which the actual costs are less than the target estimate costs.

The target estimate with penalty and incentive fee type of contracts of the past have been known to operate quite satisfactorily. However, aside from understanding and mutual respect between the parties, proper estimating, pricing, and cost conrols have a lot to do with developing a satisfied owner-client. Careless estimating and pricing, as well as failure to exercise adequate cost controls and reporting of costs, render this type of contract completely unworkable.

In Appendix 15-A2 at the end of this chapter is a sample draft of a target estimate incentive fee type agreement based on actual experience and proved to be satisfactory. The detailed conditions stated in the draft give a good idea of how carefully the agreement must be drawn to avoid later confusion.

15-6. Turnkey Proposals and Contract Conditions

The term, "Turnkey Proposal," simply means a proposal for the design and construction of a given project. In other words, it means a complete, all-inclusive service whereby the contractor provides all of the engineering and construction functions needed by the owner-client under one contract. The detailed arrangements for compensating the contractor and the conditions of the contract can vary widely, except that, since the definitive pricing of the work must follow the engineering work, much of the initial pricing must be conditonal.

The turnkey type proposal is best suited to a situation when the initial basic information is minimal, there is only a preliminary undeveloped concept to go on, and the time for delivering the completed work is exceedingly short and of the greatest urgency. Normally, the owner-client would arrange for an engineering organization specializing along the lines of the character of work involved to initiate preliminary investigations, start engineering studies, and design the facility involved. When the designs are completed and the specifications and contract conditions are all prepared, the next step is advertising for bids or sending out invitations for bids to a select group of contractors. After proposals are submitted, the owner would study them and enter into negotiations with those presenting favorable bids. Next comes the award of the contract and, finally, its execution.

All of this takes a long time by normal processes. The turnkey type of contract tends to telescope all of the various steps into one, allowing construction of the facility to proceed along with the engineering and design work or only slightly behind the design of the various parts. The engineer and the contractor being one, it is possible to order construction materials as the various drawings come off the drawing boards, and design and construction of the various features can be coordinated and integrated.

There are many circumstances, including the time element, that justify employing the turnkey type contract concept. The repair of serious accidental damage to a facility in order to restore it to service and avoid the continuation of heavy financial loss is an example. The opportunity for the owner-client to avoid all the troublesome details of arranging for a new facility is another.

Perhaps the owner is not capable of handling such matters and wants to avoid the distraction from his regular business. The owner may be a manufacturer who wants to market a new product before his competitors learn of it. The matter of confidence is another important condition.

The selection of a contractor for this type of contract is vastly important. He should be highly qualified and experienced in the type of work, both as an engineer and constructor, and enjoy a good reputation for fair dealing and integrity. The owner must be satisfied that he can rely on his contractor with a minimum amount of supervision and have confidence in his advice.

Since the situation of the manufacturer discussed above occurs frequently, an example along those lines is provided below, starting with the initial negotiations between the owner-client and the contractor. The same general principles might be applied to virtually all turnkey construction.

The manufacturer has the idea for a certain product, as well as some background of chemical laboratory experimentation, that leads him to believe that a satisfactory product can be made with certain ingredients and by certain processes. He may have applied for a patent for the process. He has little or no idea how this can be implemented on a large manufacturing scale, the kind of processing facilities required, or the buildings in which to house them.

In addition, he is not familiar with how to go about solving such problems or with the matter of constructing the necessary factory. He knows that to avoid possible serious error and delay, the best thing is to engage a company that has handled such development work, so he starts negotiations. The parties proceed in somewhat the following manner:

The owner engages the contractor and they sign an agreement describing in detail the authority and responsibilities of the parties. Under this agreement, the contractor is to provide all technical investigations, engineering designs for mechanical and structural facilities, a complete construction service. He also will supply the know-how and ready application of his existing organization competent in estimating costs, purchasing, organizing and directing the field operations, employing consultants and specialists, and the handling of labor negotiations, legal matters, insurance, taxes, shipping and receiving items for use in the work, and other things required for the design and construction of the complete facility ready to operate.

The contractor would generally undertake the management of the enterprise for the owner. He would be reimbursed for all expenditures, but the owner would retain the right of approving matters affecting policy and also the larger commitments to be made by the contractor. The contractor would organize a detailed accounting system for the project and report periodically on the use of all funds advanced by the owner and the cost of the work.

Although these matters would cover the entire project, the owner would need assurance that things were not getting out of hand and that he would not be committed to a situation that later would prove to be impracticable. In other words, a check on the feasibility of the enterprise should be tested before an attempt was made to go all the way. A three-step provision written into the agreement would accomplish this.

First, the owner wants to know how long it would take and how much it would cost for the contractor-engineer to develop the initial designs to the point where a preliminary estimate of the possible total ultimate cost could be made. The owner employs the contractor-engineer to make such an estimate for a wholly reimbursible cost plus a small fee. This might be called the first order by the owner. A short time limit would be set for the contractor-engineer's reply.

The contractor-engineer then begins accumulating all the available facts from various sources. From the owner comes information about the product and its ingredients, the proposed process and the laboratory findings, the capacity of the plant, the packaging to be involved, the handling and crating for shipment, details of the site, and any other tangible facts.

From other sources, the contractor-engineer would develop the particulars of the machinery required and other matters to help him envision the construction problems and the cost. He proceeds with preliminary sketches and a description of the proposed facility, then prepares a report for the owner that includes an estimate of time and cost to develop the initial designs to the point where a definitive estimate would be possible. The report also would give an indication of feasibility of the enterprise.

The owner-client considers the information and decides whether or not to proceed with the project. If he decides to proceed, he issues what might be called his second order. It is an order to proceed with designs and engineering work necessary to make a definitive estimate of cost of the whole enterprise. To accomplish this, the owner-client budgets an amount of money in line with the estimate.

The contractor-engineer starts the engineering and designs, billing the owner for complete reimbursement of the necessary costs plus an agreed percentage fee. This cost and fee are considered chargeable against the budgeted funds. During the preparation of the designs covered by the owner's second order, the owner probably would direct the contractor-engineer to acquire the things needed and to begin site improvement work. This assures that general feasibility of the enterprise is indicated by the information being developed from time to time.

If the contractor-engineer prepares his initial definitive designs and engineering data within the time allowed, as well as within the amount of money budgeted, he submits his findings and technical data to the owner-client along with the preliminary definitive estimate of cost for the whole project. At this stage, designs and specifications would not be complete in all details and would have to be supplemented as the project proceeds.

If the contractor-engineer fails to accomplish the initial definitive engineering work within the time and money budgeted, the owner-client can cancel the contractor-engineer's right to proceed further with the project. As an alternative, the owner can consider the reasons for the failure and modify the terms governing the arrangement.

Assuming that the engineering data submitted by the contractor-engineer had been approved by the owner and the owner wants to make final arrangements for completion of the project, he issues a third order. This is the order to complete the engineering designs and proceed with the construction of the facility, all in accordance with a reimbursable cost-plus-a-fee basis. However, if the owner questions the feasibility of the project, or would have difficulty financing it, he would have a specific time to make a decision on what to do.

His alternative would be to cancel further activities on the project and settle with the contractor-engineer—a right he would have reserved. We assume, however, that he decides to proceed and that all the provisions in the original agreement for such an eventuality begin to operate. That is, it would have been set up for the contractor, when so ordered, to conclude all matters of design and construction according to prescribed terms and conditions applicable to a cost reimbursable basis plus an additional fee.

The amount of the fee and its conditions of payment could be quite variable. They could include a straight percentage fee based upon the cost of the contract work; a fixed fee agreed upon in the initial agreement, with the amount being a percentage of the definitive estimate submitted under the second order to the contractor; a fixed fee with penalties and incentives; or otherwise.

In any case, however, it is assumed that the fee might be on some kind of a sliding scale tied to actual cost and that interim payments of the same would be limited by periodic forecasts of the final outcome. This would require periodic estimates by the contractor of the ultimate cost of the enterprise, using information developed as time passes. The owner would want such estimates for his financial control and to keep abreast of performance with the contractor. They are indispensable to the contractor in his job planning and cost control throughout the period of construction. For ease of reference we will call these the third-order estimates.

If the procedure above is followed it is obvious that estimating plays an important part even if the information on which to base it is meager and of questionable reliability at first. However, as time goes on, the information becomes more reliable and more complete with the development of additional data and definitive designs. The estimates can be no more reliable than the basic assumptions and data used in their preparation—depending, of course, on the ability of the estimators preparing them.

For example, the estimate prepared according to the first order of the owner would take a lot of vision backed by the experience and knowledge of the contractor-engineer, and no theoretical alternative exists. The accuracy of the estimate might well be questionable, because the basic information is minimal and no procedures for estimating according to a system can be outlined. The estimate prepared according to the second order of the owner-client would be more reliable.

It is assumed that approximate quantities of work and its general character can be established for pricing based upon the historical cost background of the contractor-engineer and that the second-order estimate can be systematized to fit a given concept of estimating. However, the ultimate in refinement is not possible because all the details are not yet designed. Highly technical installations, special machinery and devices to be developed, etc., would be priced on the best appraisal permitted by similar experience.

Estimates prepared in accordance with the third order of the owner should be as refined as the engineering data would permit. They should follow a specific pattern and sequence so that they can be compared.

As mentioned earlier, the arrangement and conditions applying to turnkey type contracts can vary widely and can be made to suit individual situations and opinions. By sacrificing some degree of control by the owner and minimizing the possibility of his being able to readily identify the tendencies of the cost getting out of hand, turnkey type contracts can be greatly simplified. They can be made on a one-step basis with the contractor-engineer being given more of a free hand from the outset. Such an arrangement is illustrated in Appendex 15-A3 at the end of this chapter. It is offered for reference only.

_____ PROJECT

TABLE OF CONTENTS

FORM OF AGREEMENT

     Exhibit "A" - General Conditions of Contract,

           Plans and Specifications

     Exhibit "B" - Form of Employment Agreements

           (a) For Expatriate Personnel

           (b) For Assimilated Local Personnel

           (c)

     Exhibit "C" - Original Target Estimate

     Exhibit "D" - Schedule of Basic Prices used in
                    Original Target Estimate

APPENDIX 15-A1

**Form of Agreement—Straight Cost Plus Basis from The
Manual of The Associated General Contractors of America**

FORM OF AGREEMENT FOR ENGINEERING CONSTRUCTION
COST-PLUS BASIS

THIS AGREEMENT, made on the _____ day of _____, 19 _____,

by and between _____

party of the first part, hereinafter called the OWNER, and _____

_____

_____

party of the second part, hereinafter called the CONTRACTOR.

It is understood ENGINEER representing Owner shall be _____

_____

_____

WITNESSETH, That the Contractor and the Owner, for the considerations hereinafter named, agree as follows:

ARTICLE I — Scope of the Work

The Contractor hereby agrees to furnish all of the materials and all of the equipment and labor necessary, and to perform all of the work shown on the Drawings and described in the specification for the project entitled _____

_____

_____

all in accordance with the requirements and provisions of the following Documents which are hereby made a part of this Agreement:

(*a*) Drawings prepared for same by _____

numbered _____

and dated _____, 19 _____.

(*b*) Specifications consisting of:

1. "Standard General Specifications" issued by _____

_____, _____ Edition.

Cost-Plus—6

2. "Special Conditions" as prepared by ....

.... dated ....

3. The "General Conditions of Contract for Engineering Construction"—1966 Edition.

4. Addenda

   No. .... Date ....

ARTICLE II — Time of Completion

(*a*) The work to be completed under this Contract shall be commenced within .... calendar days after receipt of notice to proceed.

(*b*) The work shall be completed within .... calendar days after receipt of notice to proceed.

(*c*) Failure to complete the work within the number of calendar days stated in this Article, including extension granted thereto as determined by Section 19 of the General Conditions, shall entitle the Owner to deduct from the moneys due to the Contractor as "Liquidated Damages" an amount equal to $...................................... for each calendar day of delay in the completion of work.

(*d*) If the Contractor completes the work earlier than the date determined in accordance with Paragraph (*b*), and the Engineer shall so certify in writing, the Owner shall pay the Contractor an additional amount equal to $.................................... for each calendar day by which the time of completion so determined has been reduced.

ARTICLE III — The Contract Sum

(*a*) The Owner shall pay the Contractor for the performance of the work an amount equal to the actual "Cost of the Work" as defined below, plus ....% of all costs to reimburse the Contractor for indirect overhead and general supervision, plus ....% of all costs, including indirect overhead and general supervision, as commission or profit.

(*b*) The "Cost of the Work" shall be determined as the net sum of the following items:

   1. Job Office and all necessary temporary facilities such as buildings, use of land not furnished by the Owner, access roads and utilities. The costs of these items include construction, furnishings and equipment, maintenance during the period that they are needed, demolition and removal. Salvage values agreed on or received by the Contractor shall be credited to the Owner.

   2. All materials used on the work whether for temporary or permanent construction.

   3. All small tools and supplies; all fuel, lubricants, power, light, water and telephone service.

Cost-Plus—7

4. All plant and equipment at specified rental rates and terms of use. If the rental rates do not include an allowance for running repairs and repair parts needed for ordinary maintenance of the plant and equipment, then such items of cost are to be included in the Cost of the Work.

5. All transportation costs on equipment, materials and men.

6. All labor for the project and including the salaries of superintendents, foremen, engineers, inspectors, clerks and other employees while engaged on the work but excluding salaries of general supervisory employees or officers, who do not devote their full time to the work.

7. All payroll charges such as Social Security payments, unemployment insurance, workmen's compensation insurance premiums, pension and retirement allowances, and social insurance premiums, vacation and sick-leave allowances applicable to wages or salaries paid to employees for work done in connection with the contract.

8. All premiums on fire, public liability, property damage or other insurance coverage authorized or required by the Engineer or the Owner, or regularly paid by the Contractor in the conduct of his business.

9. All sales, use, excise, privilege, business, occupation, gross receipt and all other taxes paid by the Contractor in connection with the work, but excluding state income taxes based solely on net income derived from this contract and Federal income taxes.

10. All travel or other related expense of general supervisory employees for necessary visits to the job excluding expenses of such employees incurred at the Home Office of the Contractor.

11. All Subcontracts approved by the Engineer or Owner.

12. (Insert other costs proper for inclusion in this Contract.)

     a. --

     b. --

     c. --

13. Any other costs incurred by the Contractor as a direct result of executing the Order, subject to approval by the Engineer.

14. Credit to the Owner for the following items:

     a. Such discounts on invoices as may be obtainable provided that the Owner advances sufficient funds to pay the invoices within the discount period.

     b. The mutually agreed salvage value of materials, tools or equipment charged to the Owner and taken over by the Contractor for his use or sale at the completion of the work.

     c. Any rebates, refunds, returned deposits or other allowances properly credited to the Cost of the Work.

Cost-Plus—8

ARTICLE IV — Progress Payments

The Owner shall make payments on account of the Contract as follows:

As early as possible after the first day of each month the Contractor shall present to the Owner a statement of all costs incurred on account of the work involved in this Contract during the preceding month. This statement shall be accompanied by copies of supporting invoices and such copies of payrolls, or totalization and distribution of same, as may be required; lists of plant and equipment used, with rates for same, together with any other information necessary to allow the Engineer to verify the accuracy of the statement. To the total of the costs incurred there shall be added the full amount of the percentages earned as set up in Article III, and the full amount shall be paid to the Contractor within 10 days after the receipt of the statement.*

IN WITNESS WHEREOF the parties hereto have executed this Agreement, the day and year first above written.

-- OWNER

WITNESS:

---By: ---
                                                              Title

-- CONTRACTOR

WITNESS:

---By: ---
                                                              Title

---

* Here may be inserted a provision for a retained percentage if desired.

Cost-Plus—9

GENERAL CONDITIONS OF CONTRACT FOR ENGINEERING CONSTRUCTION
INDEX

SEC. 1—Definitions

(*a*) The Contract Documents shall consist of Advertisement for Bids or Notice to Contractors, Instructions to Bidders, Form of Bid or Proposal, the signed Agreement, the General and Special Conditions of Contract, the Drawings, and the Specifications, including all modifications thereof incorporated in any of the documents before the execution of Agreement.

(*b*) The Owner, the Contractor and the Engineer are those named as such in the Agreement. They are treated throughout the Contract Documents as if each were of singular number and masculine gender.

(*c*) Wherever in this Contract the word "Engineer" is used it shall be understood as referring to the Engineer of the Owner, acting personally or through assistants duly authorized in writing by the Engineer.

(*d*) Written notice shall be deemed to have been duly served if delivered in person to the individual or to a member of the firm or to an officer of the corporation for whom it is intended, or to an authorized representative of such individual, firm, or corporation, or if delivered at or sent by registered mail to the last business address known to him who gives the notice, with a copy sent to the central office of the Contractor.

(*e*) The term "Subcontractor" shall mean anyone (other than the Contractor) who furnishes at the site, under an Agreement with the Contractor, labor, or labor and materials, or labor and equipment, but shall not include any person who furnishes services of a personal nature.

(*f*) Work shall mean the furnishing of all labor, materials, equipment, and other incidentals necessary or convenient to the successful completion of the Contract and the carrying out of all the duties and obligations imposed by the Contract.

(*g*) Extra work shall mean such additional labor, materials, equipment, and other incidentals as are required to complete the Contract for the purpose for which it was intended but was not shown on the Drawings or called for in the Specifications, or is desired by the Owner in addition to that work called for in the Drawings and Specifications.

(*h*) Dispute shall mean lack of agreement between any parties that have any obligations, duties, or responsibilities under the terms of the Contract, Drawings, or Specifications.

SEC. 2—Execution and Correlation of Documents

The Contract Documents shall be signed in duplicate by the Owner and the Contractor.

The Contract Documents are complementary and what is called for by any one shall be as binding as if called for by all. In case of conflict between Drawings and Specifications, the Specifications shall govern. Materials or work described in words which so applied have a well-known technical or trade meaning shall be held to refer to such recognized standards.

15

SEC. 3.—Design, Drawings and Instructions

It is agreed that the Owner will be responsible for the adequacy of design and sufficiency of the Drawings and Specifications. The Owner, through the Engineer, or the Engineer as the Owner's representative, shall furnish Drawings and Specifications which adequately represent the requirements of the work to be performed under the Contract. All such Drawings and instructions shall be consistent with the Contract Documents and shall be true developments thereof. In the case of lump-sum Contracts, Drawings and Specifications which adequately represent the work to be done shall be furnished prior to the time of entering into the Contract. The Engineer may, during the life of the Contract, and in accordance with Section 18, issue additional instructions by means of Drawings or other media necessary to illustrate changes in the work.

SEC. 4—Copies of Drawings Furnished

Unless otherwise provided in the Contract Documents, the Engineer will furnish to the Contractor, free of charge, all copies of Drawings and Specifications reasonably necessary for the execution of the work.

SEC. 5—Order of Completion

The Contractor shall submit, at such times as may be reasonably requested by the Engineer, schedules which shall show the order in which the Contractor proposes to carry on the work, with dates at which the Contractor will start the several parts of the work, and estimated dates of completion of the several parts.

SEC. 6—Ownership of Drawings

All Drawings, Specifications and copies thereof furnished by the Engineer shall not be reused on other work, and, with the exception of the signed Contract, sets are to be returned to him on request, at the completion of the work.

SEC. 7—Familiarity with Work

The Owner shall make known to all prospective bidders, prior to the receipt of bids, all information that he may have as to subsurface conditions in the vicinity of the work, topographical maps, or other information that might assist the bidder in properly evaluating the amount and character of the work that might be required. Such information is given,

however, as being the best factual information available to the Owner. The Contractor, by careful examination, shall satisfy himself as to the nature and location of the work, the character of equipment and facilities needed preliminary to and during the prosecution of the work, the general and local conditions, and all other matters which can in any way affect the work under this Contract.

SEC. 8—Changed Conditions

The Contractor shall promptly, and before such conditions are disturbed, notify the Owner in writing of: (1) Subsurface or latent physical conditions at the site differing materially from those indicated in this Contract; or (2) previously unknown physical or other conditions at the site, of an unusual nature, differing materially from those ordinarily encountered and generally recognized as inherent in work of the character provided for in this Contract. The Engineer shall promptly investigate the conditions, and if he finds that such conditions do so materially differ and cause an increase or decrease in the cost of, or the time required for, performance of this Contract, an equitable adjustment shall be made and the Contract modified in writing accordingly. Any claim of the Contractor for adjustment hereunder shall not be allowed unless he has given notice as above required; provided that the Engineer may, if he determines the facts so justify, consider and adjust any such claims asserted before the date of final settlement of the Contract. If the parties fail to agree upon the adjustment to be made, the dispute shall be determined as provided in Section 39 hereof.

SEC. 9—Materials and Appliances

Unless otherwise stipulated, the Contractor shall provide and pay for all materials, labor, water, tools, equipment, light, power, transportation and other facilities necessary for the execution and completion of the work. Unless otherwise specified, all materials incorporated in the permanent work shall be new and both workmanship and materials shall be of good quality. The Contractor shall, if required, furnish satisfactory evidence as to the kind and quality of materials.

SEC. 10—Employees

The Contractor shall at all times enforce strict discipline and good order among his employees, and shall seek to avoid employing on the work any unfit person or anyone not skilled in the work assigned to him.

16

Adequate sanitary facilities shall be provided by the Contractor.

SEC. 11—Royalties and Patents

The Contractor shall pay all royalties and license fees. He shall defend all suits or claims for infringement of any patent rights and shall save the Owner harmless from loss on account thereof except that the Owner shall be responsible for all such loss when a particular process or the product of a particular manufacturer or manufacturers is specified, unless the Owner has notified the Contractor prior to the signing of the Contract that the particular process or product is patented or is believed to be patented.

SEC. 12—Surveys

Unless otherwise specified, the Owner shall furnish all land surveys and establish all base lines for locating the principal component parts of the work together with a suitable number of bench marks adjacent to the work. From the information provided by the Owner, the Contractor shall develop and make all detail surveys needed for construction such as slope stakes, batter boards, stakes for pile locations and other working points, lines and elevations.

The Contractor shall carefully preserve bench marks, reference points and stakes and, in case of willful or careless destruction, he shall be charged with the resulting expense and shall be responsible for any mistakes that may be caused by their unnecessary loss or disturbance.

SEC. 13—Permits, Licenses and Regulations

Permits and licenses of a temporary nature necessary for the prosecution of the work shall be secured and paid for by the Contractor. Permits, licenses and easements for permanent structures or permanent changes in existing facilities shall be secured and paid for by the Owner, unless otherwise specified. The Contractor shall give all notices and comply with all laws, ordinances, rules and regulations bearing on the conduct of the work as drawn and specified. If the Contractor observes that the Drawings and Specifications are at variance therewith, he shall promptly notify the Engineer in writing, and any necessary changes shall be adjusted as provided in the Contract for changes in the work.

SEC. 14—Protection of the Public and of Work and Property

The Contractor shall provide and maintain all necessary watchmen, barricades, warning lights and signs and take all necessary precautions for the protection and safety of the public. He shall continuously maintain adequate protection of all work from damage, and shall take all reasonable precautions to protect the Owner's property from injury or loss arising in connection with this Contract. He shall make good any damage, injury or loss to his work and to the property of the Owner resulting from lack of reasonable protective precautions, except such as may be due to errors in the Contract Documents, or caused by agents or employees of the Owner. He shall adequately protect adjacent private and public property, as provided by Law and the Contract Documents.

In an emergency affecting the safety of life, of the work, or of adjoining property, the Contractor is, without special instructions or authorization from the Engineer, hereby permitted to act at his discretion to prevent such threatened loss or injury. He shall also so act, without appeal, if so authorized or instructed by the Engineer.

Any compensation claimed by the Contractor on account of emergency work, shall be determined by agreement or by arbitration.

SEC. 15—Inspection of Work

The Owner shall provide sufficient competent personnel, working under the supervision of a qualified engineer, for the inspection of the work while such work is in progress to ascertain that the completed work will comply in all respects with the standards and requirements set forth in the Specifications. Notwithstanding such inspection, the Contractor will be held responsible for the acceptability of the finished work.

The Engineer and his representatives shall at all times have access to the work whenever it is in preparation or progress, and the Contractor shall provide proper facilities for such access, and for inspection.

If the Specifications, the Engineer's instructions, laws, ordinances, or any public authority require any work to be specially tested or approved, the Contractor shall give the Engineer timely notice of its readiness for inspection, and if the inspection is by an authority other than the Engineer, of the date fixed for such inspection. Inspections by the Engi-

17

neer shall be made promptly, and where practicable at the source of supply. If any work should be covered up without approval or consent of the Engineer, it must, if required by the Engineer, be uncovered for examination and properly restored at the Contractor's expense, unless the Engineer has unreasonably delayed inspection.

Re-examination of any work may be ordered by the Engineer, and, if so ordered, the work must be uncovered by the Contractor. If such work is found to be in accordance with the Contract Documents, the Owner shall pay the cost of re-examination and replacement. If such work is not in accordance with the Contract Documents, the Contractor shall pay such cost.

SEC. 16—Superintendence

The Contractor shall keep on his work, during its progress, a competent superintendent and any necessary assistants. The superintendent shall represent the Contractor, and all directions given to him shall be binding as if given to the Contractor. Important directions shall immediately be confirmed in writing to the Contractor. Other directions shall be so confirmed on written request in each case. The Contractor shall give efficient superintendence to the work, using his best skill and attention.

SEC. 17—Discrepancies

If the Contractor, in the course of the work, finds any discrepancy between the Drawings and the physical conditions of the locality, or any errors or omissions in Drawings or in the layout as given by survey points and instructions, he shall immediately inform the Engineer, in writing, and the Engineer shall promptly verify the same. Any work done after such discovery, until authorized, will be done at the Contractor's risk.

SEC. 18—Changes in the Work

The Owner may make changes in the Drawings and Specifications or scheduling of the Contract within the general scope at any time by a written order. If such changes add to or deduct from the Contractor's cost of the work, the Contract shall be adjusted accordingly. All such work shall be executed under the conditions of the original Contract except that any claim for extension of time caused thereby shall be adjusted at the time of ordering such change.

In giving instructions, the Engineer shall have authority to make minor changes in the work not involving extra cost, and not inconsistent with the purposes of the work, but otherwise, except in an emergency endangering life or property, no extra work or change shall be made unless in pursuance of a written order by the Engineer, and no claim for an addition to the Contract Sum shall be valid unless the additional work was so ordered.

The Contractor shall proceed with the work as changed and the value of any such extra work or change shall be determined as provided in the Agreement.

SEC. 19—Extension of Time

Extension of time stipulated in the Contract for completion of the work will be made when changes in the work occur, as provided in Section 18; when the work is suspended as provided in Section 23; and when the work of the Contractor is delayed on account of conditions which could not have been foreseen, or which were beyond the control of the Contractor, his Subcontractors or suppliers, and which were not the result of their fault or negligence. Extension of time for completion shall also be allowed for any delays in the progress of the work caused by any act (except as provided elsewhere in these General Conditions) or neglect of the Owner or of his employees or by other contractors employed by the Owner, or by any delay in the furnishing of Drawings and necessary information by the Engineer, or by any other cause which in the opinion of the Engineer entitled the Contractor to an extension of time, including but not restricted to, acts of the public enemy, acts of any government in either its sovereign or any applicable contractual capacity, acts of another contractor in the performance of a contract with the Owner, fires, floods, epidemics, quarantine restrictions, freight embargoes, unusually severe weather, or labor disputes.

The Contractor shall notify the Engineer promptly of any occurrence or conditions which in the Contractor's opinion entitle him to an extension of time. Such notice shall be in writing and shall be submitted in ample time to permit full investigation and evaluation of the Contractor's claim. The Engineer shall acknowledge receipt of the Contractor's notice within 5 days of its receipt. Failure to provide such notice shall constitute a waiver by the Contractor of any claim.

SEC. 20—Claims

If the Contractor claims that any instructions by Drawings or other media issued after the date of the

Contract involve extra cost under this Contract, he shall give the Engineer written notice thereof within _____ days after the receipt of such instructions, and in any event before proceeding to execute the work, except in emergency endangering life or property, and the procedure shall then be as provided for changes in the work. No such claim shall be valid unless so made.

SEC. 21—Deductions for Uncorrected Work

If the Engineer deems it inexpedient to correct work that has been damaged or that was not done in accordance with the Contract, an equitable deduction from the Contract price shall be made therefor, unless the Contractor elects to correct the work.

SEC. 22—Correction of Work Before Final Payment

The Contractor shall promptly remove from the premises all materials and work condemned by the Engineer as failing to meet Contract requirements, whether incorporated in the work or not. The Contractor shall promptly replace and re-execute his own work in accordance with the Contract and without expense to the Owner and shall bear the expense of making good all work of other contractors destroyed or damaged by such removal or replacement.

If the Contractor does not take action to remove such condemned materials and work within 10 days after written notice, the Owner may remove them and may store the material at the expense of the Contractor. If the Contractor does not pay the expense of such removal and storage within ten days' time thereafter, the Owner may, upon ten days' written notice, sell such materials at auction or at private sale and shall pay to the Contractor any net proceeds thereof, after deducting all the costs and expenses that should have been borne by the Contractor.

SEC. 23—Suspension of Work

The Owner may at any time suspend the work, or any part thereof, by giving _____ days' notice to the Contractor in writing. The work shall be resumed by the Contractor within ten (10) days after the date fixed in the written notice from the Owner to the Contractor so to do. The Owner shall reimburse the Contractor for expense incurred by the Contractor in connection with the work under this Contract as a result of such suspension.

If the work, or any part thereof, shall be stopped by notice in writing aforesaid, and if the Owner does not give notice in writing to the Contractor to resume work at a date within _____ days of the date fixed in the written notice to suspend, then the Contractor may abandon that portion of the work so suspended and he will be entitled to the estimates and payments for all work done on the portions so abandoned, if any, plus _____% of the value of the work so abandoned, to compensate for loss of overhead, plant expense, and anticipated profit.

SEC. 24—The Owner's Right to Terminate Contract

If the Contractor should be adjudged a bankrupt, or if he should make a general assignment for the benefit of his creditors, or if a receiver should be appointed as a result of his insolvency, or if he should be guilty of a substantial violation of the Contract, then the Owner, upon the certificate of the Engineer that sufficient cause exists to justify such action, may, without prejudice to any other right or remedy and after giving the Contractor and his Surety seven days' written notice, terminate the employment of the Contractor and take possession of the premises and of all materials, tools, equipment and other facilities installed on the work and paid for by the Owner, and finish the work by whatever method he may deem expedient. In such case the Contractor shall not be entitled to receive any further payment until the work is finished. If the unpaid balance of the Contract price shall exceed the expense of finishing the work, including compensation for additional managerial and administrative services, such excess shall be paid to the Contractor. If such expense shall exceed such unpaid balance, the Contractor shall pay the difference to the Owner. The expense incurred by the Owner as herein provided, and the damage incurred through the Contractor's default, shall be certified by the Engineer.

SEC. 25—Contractor's Right to Stop Work or Terminate Contract

If the work should be stopped under an order of any court, or other public authority, for a period of more than three months, through no act or fault of the Contractor or of anyone employed by him, or if the Engineer should fail to issue any estimate for payment within seven days after it is due, or if the Owner should fail to pay the Contractor within seven days of its maturity and presentation any sum certified by the Engineer or awarded by arbitrators, then the Contractor may, upon seven days' written

19

notice to the Owner and the Engineer, stop work or terminate this Contract and recover from the Owner payment for all work executed, plus any loss sustained upon any plant or materials plus reasonable profit and damages.

SEC. 26—Removal of Equipment

In the case of termination of this Contract before completion from any cause whatever, the Contractor, if notified to do so by the Owner, shall promptly remove any part or all of his equipment and supplies from the property of the Owner, failing which the Owner shall have the right to remove such equipment and supplies at the expense of the Contractor.

SEC. 27—Responsibility for Work

The Contractor assumes full responsibility for the work. Until its final acceptance, the Contractor shall be responsible for damage to or destruction of the work (except for any part covered by partial acceptance as set forth in Sec. 28). He agrees to make no claims against the Owner for damages to the work from any cause except negligence or willful acts of the Owner, acts of an Enemy, acts of war or as provided in Sec. 32.

SEC. 28—Partial Completion and Acceptance

If at any time prior to the issuance of the final certificate referred to in Section 42 hereinafter, any portion of the permanent construction has been satisfactorily completed, and if the Engineer determines that such portion of the permanent construction is not required for the operations of the Contractor but is needed by the Owner, the Engineer shall issue to the Contractor a certificate of partial completion, and thereupon or at any time thereafter the Owner may take over and use the portion of the permanent construction described in such certificate, and may exclude the Contractor therefrom.

The issuance of a certificate of partial completion shall not be construed to constitute an extension of the Contractor's time to complete the portion of the permanent construction to which it relates if he has failed to complete it in accordance with the terms of this Contract. The issuance of such a certificate shall not operate to release the Contractor or his sureties from any obligations under this Contract or the performance bond.

If such prior use increases the cost of or delays the work, the Contractor shall be entitled to extra compensation, or extension of time, or both, as the Engineer may determine, unless otherwise provided.

SEC. 29—Payments Withheld Prior to Final Acceptance of Work

The Owner, as a result of subsequently discovered evidence, may withhold or nullify the whole or part of any payment certificate to such extent as may be necessary to protect himself from loss caused by:

(a) Defective work not remedied.

(b) Claims filed or reasonable evidence indicating probable filing of claims by other parties against the Contractor.

(c) Failure of the Contractor to make payments properly to Subcontractors or for material or labor.

(d) Damage to another contractor.

When the above grounds are removed or the Contractor provides a Surety Bond satisfactory to the Owner which will protect the Owner in the amount withheld, payment shall be made for amounts withheld, because of them.

No moneys may be withheld under (b) and (c) above if a payment bond is included in the Contract.

SEC. 30—Contractor's Insurance

The Contractor shall secure and maintain such insurance policies as will protect himself, his Subcontractors, and unless otherwise specified, the Owner, from claims for bodily injuries, death or property damage which may arise from operations under this Contract whether such operations be by himself or by any Subcontractor or anyone employed by them directly or indirectly. The following insurance policies are required:

(a) Statutory Workmen's Compensation.

(b) Contractor's Public Liability and Property damage—

   Bodily Injury:
      each person _____ $_____
      each accident _____ $_____
   Property Damage:
      each accident _____ $_____
      aggregate _____ $_____

(c) Automobile Public Liability and Property Damage—

   Bodily Injury:
      each person _____ $_____
      each accident _____ $_____
   Property Damage:
      each accident _____ $_____

20

All policies shall be for not less than the amounts set forth above or as stated in the Special Conditions. Other forms of insurance shall also be provided if called for by the Special Conditions.

Certificates and/or copies of policy of such insurance shall be filed with the Engineer, and shall be subject to his approval as to adequacy of protection, within the requirements of the Specifications. Said certificates of insurance shall contain a 10 days' written notice of cancellation in favor of the Owner.

SEC. 31—Surety Bonds

The Owner shall have the right, prior to the signing of the Contract, to require the Contractor to furnish bond covering the faithful performance of the Contract and the payment of all obligations arising thereunder, in such form as the Owner may prescribe in the bidding documents and executed by one or more financially responsible sureties. If such bond is required prior to the receipt of bids, the premium shall be paid by the Contractor; if subsequent thereto, it shall be paid by the Owner. The Owner may require additional bond if the Contract is increased appreciably.

SEC. 32—Owner's Insurance

The Owner shall secure and maintain insurance to 100% of the insurable value thereof against fire, earthquake, flood, and such other perils as he may deem necessary and shall name the Contractor and Subcontractors as additional insured. Such insurance shall be upon the entire work in the Contract and any structures attached or adjacent thereto. He shall also secure and maintain such insurance as will protect him and his officers, agents, servants, and employees from liability to others for damages due to death, bodily injury, or property damage resulting from the performance of the work. The limits of such insurance shall be equal to the amounts stated in subparagraphs (b) and (c), of Section 30.

SEC. 33—Assignment

Neither party to the Contract shall assign the Contract or sublet it as a whole without the written consent of the other, nor shall the Contractor assign any moneys due to him or to become due to him hereunder, except to a bank or financial institution acceptable to the Owner.

SEC. 34—Rights of Various Interests

Whenever work being done by the Owner's or by other contractor's forces is contiguous to work covered by this Contract, the respective rights of the various interests involved shall be established by the Engineer, to secure the completion of the various portions of the work in general harmony.

SEC. 35—Separate Contracts

The Owner reserves the right to let other contracts in connection with this project. The Contractor shall afford other contractors reasonable opportunity for the introduction and storage of their materials and the execution of their work, and shall properly connect and coordinate his work with theirs.

If the proper execution or results of any part of the Contractor's work depends upon the work of any other contractor, the Contractor shall inspect and promptly report to the Engineer any defects in such work that render it unsuitable for such proper execution and results.

SEC. 36—Subcontracts

The Contractor shall, as soon as practicable after signing of the Contract, notify the Engineer in writing of the names of Subcontractors proposed for the work.

The Contractor agrees that he is as fully responsible to the Owner for the acts and omissions of his Subcontractors and of persons either directly or indirectly employed by them, as he is for the acts and omissions of persons directly employed by him.

Nothing contained in the Contract Documents shall create any contractual relation between any Subcontractor and the Owner.

SEC. 37—Engineer's Status

The Engineer shall perform technical inspection of the work. He has authority to stop the work whenever such stoppage may be necessary to insure the proper execution of the Contract. He shall also have authority to reject all work and materials which do not conform to the Contract and to decide questions which arise in the execution of the work.

SEC. 38—Engineer's Decisions

The Engineer shall, within a reasonable time after their presentation to him, make decisions in writing on all claims of the Owner or the Contractor and on all other matters relating to the execution and progress of the work or the interpretation of the Contract Documents.

SEC. 39—Arbitration

Any controversy or claim arising out of or relating to this Contract, or the breach thereof which cannot be resolved by mutual agreement, shall be settled by arbitration in accordance with the Rules of the American Arbitration Association, and judgment upon the award rendered by the Arbitrator(s) may be entered in any Court having jurisdiction thereof.

SEC. 40—Lands for Work

The Owner shall provide as indicated on Drawing No. _____ and not later than the date when needed by the Contractor the lands upon which the work under this Contract is to be done, rights-of-way for access to same, and such other lands which are designated on the Drawings for the use of the Contractor. Such lands and rights-of-way shall be adequate for the performance of the Contract. Any delay in the furnishing of these lands by the Owner shall be deemed proper cause for an equitable adjustment in both Contract price and time of completion.

The Contractor shall provide at his own expense and without liability to the Owner any additional land and access thereto that may be required for temporary construction facilities, or for storage of materials.

SEC. 41—Cleaning Up

The Contractor shall remove at his own expense from the Owner's property and from all public and private property all temporary structures, rubbish and waste materials resulting from his operations. This requirement shall not apply to property used for permanent disposal of rubbish or waste materials in accordance with permission of such disposal granted to the Contractor by the Owner thereof.

SEC. 42—Acceptance and Final Payment

(a) Upon receipt of written notice that the work is substantially completed or ready for final inspection and acceptance, the Engineer will promptly make such inspection, and when he finds the work acceptable under the Contract and the Contract fully performed or substantially completed he shall promptly issue a certificate, over his own signature, stating that the work required by this Contract has been completed or substantially completed and is accepted by him under the terms and conditions thereof, and the entire balance found to be due the Contractor, including the retained percentage, less a retention based on the Engineer's estimate of the fair value of the claims against the Contractor and the cost of completing the incomplete or unsatisfactory items of work with specified amounts for each incomplete or defective item of work, is due and payable. The date of substantial completion of a project or specified area of a project is the date when the construction is sufficiently completed in accordance with the Contract Documents as modified by any change orders agreed to by the parties so that the Owner can occupy the project or specified area of the project for the use for which it was intended.

(b) Before issuance of final payment, the Contractor, if required in the Special Conditions, shall certify in writing to the Engineer that all payrolls, material bills, and other indebtedness connected with the work have been paid, or otherwise satisfied, except that in case of disputed indebtedness or liens, if the Contract does not include a payment bond, the Contractor may submit in lieu of certification of payment a surety bond in the amount of the disputed indebtedness or liens, guaranteeing payment of all such disputed amounts, including all related costs and interest in connection with said disputed indebtedness or liens which the Owner may be compelled to pay upon adjudication.

(c) The making and acceptance of the final payment shall constitute a waiver of all claims by the Owner, other than those arising from unsettled liens, from faulty work appearing within the guarantee period provided in the Special Conditions, from the requirements of the Drawings and Specifications, or from manufacturer's guarantees. It shall also constitute a waiver of all claims by the Contractor, except those previously made and still unsettled.

(d) If after the work has been substantially completed, full completion thereof is materially delayed through no fault of the Contractor, and the Engineer so certifies, the Owner shall, upon certificate of the Engineer, and without terminating the Contract, make payment of the balance due for that portion of the work fully completed and accepted. Such payment shall be made under the terms and conditions governing final payment, except that it shall not constitute a waiver of claims.

(e) If the Owner fails to make payment as herein provided, there shall be added to each such payment daily interest at the rate of 6 per cent per annum commencing on the first day after said payment is due and continuing until the payment is delivered or mailed to the Contractor.

APPENDIX 15-A2

**Form of Agreement—Target Estimate Type Contract with
Provisions for Penalties & Incentive Fee for Performance**

FORM OF AGREEMENT

_____PROJECT

I N D E X

_____PROJECT

    (d) Specific Plant and Owner Equipment
    (e) Equipment Rental
    (f) Materials and Supplies
    (g) Traveling Expense
    (h) Transporting Personnel
    (i) Small Tools
    (j) Freight on Plant and Equipment
    (k) Taxes
    (l) First Aid and Safety
    (m) Any other costs, if approved in writing
    (n) Subcontracts
    (o) Permanent Materials
    (p) Insurance
    (q) Home Office Expense
    (r) Recruiting
    (s) Passports, medical examinations, shots, etc.
    (t) Legal and Auditing

3. Materials and Supplies, if not new, at agreed price
4. Discounts deducted to determine actual costs
5. Salvage

VII   Construction Plant and Equipment

1. Title in Owner on purchased equipment
2. Provision for renting

VIII   Target Estimates

A. Work prior to Target - Straight cost-plus _%
B. 1. When agreeable, convert to Target Estimate
   2. See Exh. C and D,_% Fee
   3. Adjustments, per Art. VIII, XIV, and XXI
   4. Original Estimated Quantities, Exh. C.
   5. Expect Changes
   6. At completion, adjust quantities and prices
   7. Adjust Original Target w/ revised quantities and prices
   8. This makes Final Adjusted Target Estimate

IX   Contractor's Fee
A. Cost plus _%
B. 1. Repeats Cost plus __% after conversion to Target
   2. When Final Adjusted Target Estimate determined
      (a) If less than Target, additional fee 25% savings, 12% max.
      (b) If over Target, reduce fee 25% of overrun, 4% min.
      (c) Major and Minor Equipment figured at 25% salvage. Adjust.
        1. If sales over 25%, excess credits cost
        2. If sales under 25%, do not increase cost
      (d) Inventories at end, allow 80% salvage
      (e) Customs considered excluded - otherwise add

_____PROJECT

X   Payment of Fee

        A. Contractor submits, Owner pays in 15 days
        B. 1. After Target, contractor accounts
           2. Monthly, Contractor submits costs and progress
           3. Monthly, Contractor submits fee
           4. If not over Target, Owner pays __%
           5. If over Target, __% less 25% of excess
           6. Upon completion, final fee, Owner pays in 60 days

XI   Manner and Times of Payment of Actual Costs
        1. Contractor submits by 15th
        2. Owner approves or rejects, paying approved part by 30th.
        3. Progress payment or approval still subject to further review
        4. Operating Fund of escudos and dollars to finance all costs
        5. Contractor forecasts cash, Owner responsible to furnish

XII   Time and Penalty Bonus
        - Penalty of __% per month from fee (still 4% min)
        - Bonus of __% per month to fee (no limit)

XIII   Accounting and Records
        1. Contractor keeps accounts and unit costs, subject to audit
        2. Contractor retains time cards, etc. w/ accounting records to Owner.
        3. Contractor prepares and submits monthly progress reports

XIV   Changes, Changed Conditions and Cost Variations - General

XV   Extra Work - General. Excluded from Target analysis

XVI   Workmen's Compensation and Compliance w/ Law; Insurance

        1. Contractor to comply with laws and regulations
        2. Contractor to carry PL and PD w/ Owner co-insured, w/ normal limits
        3. No Builders Risk or permanent structure insurance unless directed

XVII   Right of way, Easements, Permits and Regulations
        1. Owner to provide for permanent work
        2. Contractor to provide other, w/ Owner's help
        3. Contractor to issue and post all required notices

XVIII Liens and Litigation

        1. Hold harmless during construction
        2. Completed project to be free of liens
        3. Parties advise each other of claims and litigation
        4. If Contractor rides bills 30 days, Owner to pay amount

_____PROJECT

XIX   Taxes, Fines, or Penalties
        - Exemptions from import duties
            - Permanent materials
            - Machines for design or construction
            - Equipment, Tools, Vehicles and Implements - Export 6 mo.
                after completion
            - Above if worn out or expended need not be exported
            - Owner may declare a free zone to facilitate above
        - Exemptions from Taxes and Imposts
            - Taxes (except Income), Licenses, Contributions, Imposts, etc.
            - Charges for Social Benefits on 3rd Country Nationals
            - Owner to grant permits and licenses for personnel
            - Income Taxes on 3rd Country Nationals, and other taxes
            - Defines "associates"

XX   Title to Materials
        - Title passed to Owner through Contractor on payment or delivery

XXI   Termination or Suspension of Agreement; Default
        1. If Contractor refuses or neglects, upon 10 days Owner takes over
        2. If Contractor quit or go broke, Owner takes over
        3. Under 1 or 2, no further payments until final computation
        4. If suspended by Courts or Government, job reduced, then resume
        5. If Owner fails to pay, upon 10 days Contractor quits and gets paid

XXII   Subcontracts

        1. Require prior consent, w/ copies to Owner
        2. Contractor responsible for subcontractor
        3. Creates no contract between subcontractor and Owner

XXIII   Force Majeure

        - Time Extensions and Cost Adjustments resulting for
            - Delays by Owner (plant, equipment, funds, design approvals, and
                instructions
            - Delays by vendors selected by Owner
            - Strikes and Work Stoppages
            - War (declared or undeclared), riot, civil commotion, etc.
            - Epidemics
            - Storms causing over 2 weeks' shutdowns
        Contractor must notify Owner within 14 days

XXIV   Arbitration

        1. In event of dispute, submit to 3 arbitrators
        2. In 30 days, 2nd party countercharges and selects member
        3. If not, Swiss Judge selects 2nd arbitrator
        4. In 30 more days, 2 members or Swiss Judge select 3rd member
        5. Decisions by 2 of 3 members are binding

_____PROJECT

        6. 3 Arbitrators select meeting place
        7. If one dies, replace by same method
        8. Arbitrators decide who pays what part of expenses.

XXV   Law Applicable - Laws of Mozambique apply

XXVI   Successors and Assigns - General

XXVII   Final Acceptance - Contractor released upon final acceptance

XXVIII Suppression of Fires - General, excluded from "costs"

XXVIX   Independent Contractor - General

XXX   Not for Benefit of Third Parties - General

XXXI   Article Headings - To be ignored in interpretation

AGREEMENT

AGREEMENT, made this ___ day of _____, _____, between

_____ hereinafter called

the "Owner", and _____

_____

_____

_____

hereinafter called the "Contractor",

WITNESSETH:

WHEREAS, the Owner proposes to construct a _____

_____

_____

_____

_____

WHEREAS, the Contractor is engaged in the construction business,

is experienced in the construction of _____

_____ , and desires to perform construction work and services

incident to the construction above described for the Owner upon the terms

and conditions hereinafter set forth.

NOW, THEREFORE, The parties hereto agree as follows:

ARTICLE I.

Scope and Prosecution of Work

1. The Contractor shall, except as hereinafter otherwise provided, furnish the necessary materials, superintendence, labor, tools,
equipment, and transportation for construction of those portions and phases
of the _____ Project outlined in the plans and specifications which

are incorporated herein by reference as Exhibit "A".

2. Contractor at all times during the progress of the Project shall be and act subject to the general direction and control of Owner, insofar as results to be attained hereunder are concerned, but not as to the mode and manner of performing the work.

3. Contractor shall commence each stage of the Project immediately following authorization by the Owner, and shall prosecute the work continuously and diligently to completion, in accordance with a Construction Schedule to be agreed upon by the parties. Letter agreements confirming each Construction Schedule, signed by both parties, marked "Exhibit 'B'", shall be attached to and become a part of this Agreement.

4. It is expressly understood and agreed that the time limits set out in the Construction Schedules are of the essence; provided that adjustments of the time limits set out in each Construction Schedule may be made by the Engineer, as hereinafter defined, when delays have been caused by matters beyond the reasonable control of the parties; and provided, further, that each stage of the Project may be stopped or placed on a different Construction Schedule, in whole or in part, by written agreement of the parties.

5. Contractor shall exercise its best knowledge, skill, and experience in planning, purchasing materials, obtaining and furnishing skilled and competent labor and personnel, supplying tools and equipment, and performing all work incident to the full and satisfactory completion of the Project in accordance with the Construction Schedules and Owner's plans and specifications, and shall do everything possible on its part to establish and follow a fast, economical, and efficient construction program.

6. Contractor shall obtain the most favorable prices and discount privileges available, and shall pay all bills for purchases within the allowed best discount period.

7. The Contractor is authorized to purchase without prior approval any item or group of items required for the work which do not cost in excess of _____ F. O. B. plant or agency. For purchases in excess of _____ F. O. B. plant the Contractor will submit his requisition to the Owner for approval, and such approval shall be given within 10 days of submission of the requisition. If the Owner has not returned the requisition to the Contractor within 10 days of submission, then the requisition shall be considered approved without further action by the Owner. It is directly understood that the Contractor is responsible for the procurement of equipment, material, and supplies which he considers necessary for the proper and speedy execution of the work, and that therefore the Owner's approval of the Contractor's procurement requisitions will not be unreasonably withheld.

ARTICLE II.

Plans and Specifications

1. It is understood and agreed that the work to be done hereunder by the Contractor shall be performed in accordance with the General

Conditions shown in "Exhibit 'A'", attached hereto, and specifications, plans, drawings, schedules, and such other papers as shall be furnished by Owner, covering construction details of the Project, all of which shall be incorporated herein by reference and become a part of this Agreement.

2. Any conflict or contradiction between this contract and the exhibits, plans, drawings, or specifications thereof or between any of them shall be resolved and settled by Owner, subject only to review according to Article XXIV.

ARTICLE III.

Engineer and Chief Engineer

1. Wherever used in this Agreement, the word "Engineer" means a representative of the Chief Engineer of Owner, duly authorized to act for him, and wherever the words "Chief Engineer" are used, they mean the Chief Engineer of Owner in person.

2. Engineer shall have complete authority to interpret the plans and specifications, and to determine all questions of adequacy of workmanship and materials; to see that all work is performed in full accordance with the terms and requirements of this Agreement; and to suspend all or any part of the Project when necessary to insure the proper performance by Contractor with respect to technical compliance.

3. Engineer shall make inspections and tests of the work of Contractor, and if, in the opinion of the Engineer, any of the workmanship or materials is defective or unsatisfactory, or fails to conform to the plans or specifications, the Engineer may require Contractor to remove and replace said work, or otherwise remedy the deficiency.

ARTICLE IV.

Contractor: Project Management: Employees

1. Contractor shall constantly keep on the Project during its progress a competent Project Manager, and any information given to such Project Manager or any notices served upon him shall have the same force and effect as if given to or served upon Contractor. Contractor shall employ on the Project only persons who are qualified and competent to perform the work entrusted to them and who are neither disorderly, intemperate, nor otherwise unsatisfactory.

2. Owner shall have the right to require Contractor to remove from Owner's premises any person employed or furnished by Contractor in or about the site of the work or Owner's facilities if Owner in its sole discretion determines that the employment or presence of such person is detrimental to the performance of the work or to Owner's operation.

ARTICLE V.

Compensation

Owner shall pay Contractor, in the manner and at times hereinafter specified, a total remuneration consisting of Contractor's actual costs, determined in accordance with the provisions of Article VI hereof, plus a fee, determined in accordance with the provisions of Article IX hereof.

ARTICLE VI.

Contractor's Actual Costs

1. It is the intent of this Agreement that Contractor shall be paid by Owner for all actual costs, charges, and expenses incurred by Contractor in the performance of the work incident to the Project, within the general and specific instructions given Contractor by Owner.

2. Without limiting in any way the generality of the foregoing, but as an aid in determining the types of costs, charges, and expenses for which payment will be made, Contractor shall be paid for the following: -

(a) The cost to Contractor of salaries and wages and all other benefits paid to Contractor's Employees (excluding executive officers of Contractor) employed in the performance of the work incident to the Project. The Contractor will submit to the Owner and the Owner will approve the salary and wage schedules which will be used in employing personnel; it is understood and agreed that in view of the labor market conditions in _____ and the location of the Project, adjustments will probably be required in the salary and wage schedule to insure the proper staffing of the project, and approval for such adjustments shall not be unreasonably withheld. The Contractor as an allowed cost is authorized to pay bonuses to his employees in such amounts as do not in total exceed 10% of the aggregate payroll expenditures; all necessary expenses paid by Contractor to or for the benefit of such employees, and all costs and expenditures under or arising out of employment agreements entered into with such employees in the form which is attached to and made a part of this Contract.

(b) Contractor's expenditures for payroll taxes on salaries and wages, and applicable group insurance costs, retirement benefits, vacation allowances and sick leave benefits, for employees engaged on the Project in accordance with Contractor's established practices, and as approved by the Engineer; provided that such allowances and benefits provided by Contractor for employees not engaged on the Project exclusively shall be chargeable to actual costs only for the time the employee or employees are directly employed on the Project.

(c) The net cost to Contractor of establishing and maintaining a project construction office and such other temporary field, purchasing or staging offices as may be required for the Project, including office supplies, telephone and telegraph charges, and other office operating costs, and construction, maintenance and operation of warehouses, machine shops and other temporary facilities.

(d) The cost of major and specialized equipment procured for Owner by Contractor in accordance with the provisions of Section VII (1).

(e) Compensation for use of Contractor-owned equipment at rates specified in Article VII; and the cost of fuel, lubricants, equipment operating supplies and repair parts and services necessary to maintain Owner and Contractor-owned equipment in good operating condition.

(f) Materials and supplies purchased by Contractor for Contractor's use in the Project, including all actual necessary freight, express, and drayage paid by Contractor.

(g) Traveling expenses on Contractor's employees when traveling necessarily and solely in the interest of the Project.

(h) In accordance with Employment Agreements, the cost of transportation of personnel from points of residence or hire to the jobsite and return upon completion of employment, including expenses enroute and excess baggage charges when authorized by the Contractor.

(i) Small tools, office furniture and equipment, job apparatus, minor equipment, and similar items of property.

(j) Actual transportation expense of construction plant and construction equipment to the site of the Project, including loading and unloading costs.

(k) Excise and other taxes payable by Contractor as provided in Article XIX, directly applicable to the Project, except those based on or applicable to Contractor's Corporate Income or profit.

(1) First-aid and job safety expenses.

(m) Any costs of the Project, not provided for herein, which have been approved in writing by the Engineer.

(n) Subcontract work furnished by Contractor in accordance with terms of such subcontract, subject to the provisions of Article XXII hereof.

(o) Materials and supplies purchased by Contractor for physical incorporation in and as a part of the finished Project.

(p) Premiums for all necessary insurance coverages carried by Contractor applicable to the Project, including the cost of public liability and property damage insurance, and workmen's compensation and employee fidelity bonds, and the cost of all-risk insurance covering the permanent materials, work in progress, and Owner-owned equipment.

(q) Costs to cover expenses incurred in the Contractor's U. S. office, including the cost of procurement, expediting, shipping, accounting, personnel, salareies involved in recruitment in the United States, executive salaries, travel expenses outside of _____, general overhead, etc. For ease of accounting these costs are established as ___% of all other actual costs or of the adjusted Target Estimate.

(r) The cost of establishing and operating recruitment offices located elsewhere than in the United States.

(s) The cost of passports, medical examinations, inoculations for personnel recruited outside of Mozambique.

(t) Legal and such auditing fees as mutually agreed upon.

3. Materials and supplies furnished or purchased by Contractor in connection with the Project shall be charged at actual cost to Contractor, if new. If not new, they shall be charged at their fair and reasonable value as mutually agreed.

4. In determining the actual cost for purposes of this Agreement, there shall be deducted from gross cost all cash and trade discounts, refunds, allowances, and credits.

5. Upon completion of the Project, Contractor, at Owner's option, may purchase from Owner specialized and major equipment as defined by Article VII (1), stocks of any minor equipment or tools, as defined in subdivision (i) of Section 2 of this Article VI, and materials and supplies at salvage prices mutually agreed upon. All salvage value of, credits for, and returns from sale or other disposition of surplus materials, buildings, minor equipment, tools, and other property acquired by Contractor included in the cost of the Project, shall be credited against Contractor's actual cost.

ARTICLE VII.

Construction Plant and Equipment

1. Certain major items of specialized construction plant, machinery, and equipment will be required by construction of the Project, and the useful life of a substantial portion of major construction equipment will be largely expended in the performance of the construction work. The Contractor shall procure such construction equipment for the account of the Owner, and upon purchase thereof title shall pass to the Owner. Payments of the actual costs of equipment purchased will be made from funds advanced by Owner in accordance with the provisions of Section XI of the Agreement. Contractor, upon purchase of construction equipment, as provided in this Article VII (1), will deliver, or cause to be delivered to Owner, appropriate bills of sale or other evidence of ownership vesting legal title of such construction equipment in Owner.

2. Such construction equipment as is readily available from Contractor's existing fleet may be furnished on a rental basis. Compensation for the use of equipment so furnished shall be at the rates set forth on schedules for each stage of the Project, as agreed upon by the parties, which schedules shall set forth the basic monthly rental rate, the new price of the equipment if purchased new, or the used price agreed upon by the parties, and the estimated months the equipment is to be used. The rental rates set forth shall be based at one hundred per cent (100%) of the monthly rental rates published in the latest edition of Compilation of Rental Rates for Construction Equipment prepared by Associated Equipment Distributors, 30 East Cedar Street, Chicago 11, Illinois. The monthly rental rates so established shall apply for two hundred (200) hours per month or less. Compensation per hour for any use in excess of two hundred (200) hours per month shall be computed at one four-hundredth (1/400) of the monthly rate applicable to the first two hundred (200) hours per month or less.

ARTICLE VIII.

Target Estimates

A. The parties contemplate and agree that Contractor should commence operations hereunder as soon as possible and prior to completion of detailed plans and specifications and engineering studies. It is agreed, therefore, that the work shall be commenced and that the Contractor shall be paid its actual costs plus a fee of _____per cent (_%) of such actual costs until engineering and investigations have proceeded to the point that definite projections of costs can be made and estimates can be arrived at as to the anticipated cost of the several parts of the project. At such time, the parties may agree to convert this Contract from a straight cost-plus fee Contract to an Incentive Fee Target Estimate Contract by the terms of which the Contractor will be reimbursed for its actual costs, but its fee shall be increased for savings in costs effected by the Contractor or shall be decreased in the event the Contractor's costs overrun the target estimate of costs for performing the work.

B. 1. At such time as the parties agree to convert the Contract to an Incentive Fee Target Estimate form of agreement, Owner and Contractor shall agree on a target estimate to be known as the Original Target Estimate for the cost of the Project.

2. The Original Target Estimate for the Project shall be detailed in "Exhibit 'C'", and "Exhibit 'C'" shall show Contractor's estimated costs (Article VI). "Exhibit 'C'" also shall show Contractor's fee (Article IX), computed at_____per cent (_%) of estimated costs.

3. The Original Target Estimate shall not be adjusted except as provided in this Article VIII and Articles XII, XIV, and XXI.

4. The quantities and unit prices used in the preparation of the Original Target Estimate shall be shown in "Exhibit 'C'".

5. During the period of construction, it is anticipated that changes in the original plans, specifications, and time schedules will be made, that changed conditions may be encountered, and that the costs of labor, materials, equipment, and supplies may vary from those on which the Original Target Estimate is based, resulting in changes in the quantities of work or in the unit costs of performing the work or both.

6. At the completion of the Project, the quantities used in the preparation of the Original Target Estimate will be adjusted to reflect the final quantities, and the unit prices used in the preparation of the Original Target Estimate will be adjusted to reflect changes, changed conditions, and cost variations.

7. After the final quantities are determined as outlined in Section 6 of this Article VIII, the Original Target Estimate will be adjusted by applying the unit prices in "Exhibit 'C'", as may be revised as outlined in Section 6, to the final quantities. These unit prices, as may be revised, apply equally for increases and decreases in the quantities itemized.

8. The Original Target Estimate adjusted as hereinabove provided becomes the Final Adjusted Target Estimate.

ARTICLE IX.

Contractor's Fee

A. Contractor, in addition to the payment of its actual costs

(Article VI), shall be paid a fee in U. S. A. Dollars computed at _____ per cent (_%) of such actual costs.

B. 1. At such time as the Contract is converted to an Incentive Fee Target Estimate form of agreement, Contractor, in addition to the payment of its actual costs (Article VI), shall also be paid a fee in U. S. A. Dollars computed at _____ per cent (_%) of the Final Adjusted Target Estimate.

2. As soon as possible following completion of the Project, the total amount less fee paid by Owner to Contractor, shall be determined, and the Final Adjusted Target Estimate shall be made in accordance with the provisions of Article VIII.

(a) If the total of such payments adjusted as provided in paragraphs (c) and (d) less fee is less than the Final Adjusted Target Estimate, Contractor shall be paid an additional fee amounting to twenty-five per cent (25%) of the underrun difference, provided that the Contractor's fee, as so increased, shall not be more than _____ per cent (_%) of the Final Adjusted Target Estimate.

(b) If the total of such payments adjusted as provided in paragraphs (c) and (d) less fee is more than the Final Adjusted Target Estimate, Contractor's fee shall be reduced by twenty-five per cent (25%) of the amount of the overrun difference, provided that the Contractor's Fee, as so reduced, shall not be less than four per cent (4%) of the Final Adjusted Target Estimate.

(c) Although the cost of major and minor equipment and plant will be reimbursed to the Contractor, these items have been assigned an assumed salvage value of _____ per cent (_%), and only _____ per cent (_%) of their cost is included in the Target Estimate prices, and _____ per cent (_%) of the cost is the basic amount which will be included in the costs when they are to be compared to the Target Estimate in determining the amount of fee which is payable. If at the conclusion of the work, the Owner retains possession of the equipment, then there shall be used as a cost for comparison with the Target the original book value of the equipment less an assumed salvage value of _____ per cent (_%). If the Owner directs the Contractor to dispose of the equipment, then the Contractor will effect disposition, and the cost of the work will be adjusted as follows:

1. If the proceeds of the sales exceed _____ per cent (_%) of original book value, then the excess shall be treated as a credit to cost.

2. If the proceeds of the sales are less than _____ per cent (_%) of the original book value, then the original assumed cost of _____ per cent (_%) shall remain the cost of comparison with the Target Estimate.

(d) The Contractor, in the course of normal construction operations, will complete the work, having residual inventories of new and unused construction materials and supplies, permanent materials, spare parts, office engineering supplies, camp supplies, mess supplies, medical and safety supplies, etc. These inventories are the property of the Owner, and will

be turned over to the Owner. For the purpose of comparing costs to the
Target, the cost of the inventories turned over shall be treated as a
credit to the cost of the work to the extent of eighty per cent (80%) of
delivered cost.

(e) Customs Duties on equipment have been eliminated from the Target
Estimate. If Customs Duties must be paid on imports, then any such payments
will be allowed as a cost, and the total of such payments will be allowed as
a cost outside the Target Estimate.

ARTICLE X.

Payment of Fee

A. Contractor shall maintain careful accounting of its actual costs
as the work progresses, and at the end of each month shall submit to Owner
a statement of actual costs incurred in the performance of work to the
date of such statement and fee computed at _____ per cent (_%) of the
total actual costs less the amount of Contractor's fee previously paid by
Owner. Owner shall examine such statement of Contractor's fee and shall pay
to Contractor within fifteen (15) days the amount of such fee due Contractor, less payments previously made to Contractor.

B. 1. At such time as this Contract is converted to an Incentive
Fee Target Estimate form of agreement, Contractor shall establish a cost-
keeping procedure for the purpose of determining, as the work progresses,
whether or not and to what extent the costs have exceeded or are less than
the costs computed in accordance with the unit prices set forth in
"Exhibit 'C'".

2. At the end of each month, Contractor shall furnish to Owner a
statement of the total amounts expended during the month, together with a
statement showing a determination of what the cost of the work represented
by such expenditures would have been by application of the unit prices
shown in the Original Target Estimate.

3. Contractor shall submit to Owner a monthly statement showing
the total fee due Contractor for the work performed by it and the amount of
Contractor's fee previously paid by Owner.

4. If the actual costs for all work performed hereunder, from the
beginning of work on the Project through the period shown on said statement,
are not in excess of the determination of which such cost would have been
by application of the unit prices shown in the Original Target Estimate, the
fee shall be computed at eight per cent (8%) of the total actual costs.
Owner shall examine such statement of Contractor's fee and shall pay to
Contractor, within fifteen (15) days, the amount of such fee due Contractor,
less payments previously made to Contractor.

5. If, however, such actual costs shall have exceeded the determination of what such costs would have been by application of the unit
prices shown on the Original Target Estimate costs for such work, the fee
shall be computed at _____ per cent (_%) of such determined costs less

twenty-five per cent (25%) of the amount of the excess of actual costs over such determined costs, but not less than four per cent (4%) of actual costs.

6. Following the completion of the Project, the total fee due Contractor shall be determined as provided in Article IX (B-2). Owner shall pay to Contractor, within sixty (60) days after such determination, the amount of the total fee, less payments of fee previously paid, less any time penalty as provided in Article XII. If total fee due is less than the total of fee payments already paid Contractor, Contractor shall immediately refund to Owner the difference between the total payments made and the total fee.

ARTICLE XI.

Manner and Times of Payment of Actual Costs

1. On or before the fifteenth (15) day of the month next succeeding the month in which work was commenced, and on or before the fifteenth (15) day of each month thereafter, Contractor shall submit to Owner an invoice in triplicate, including such invoices and other supporting data as Owner may require, setting forth for the preceding calendar month the billing for items of actual cost incurred by Contractor, as provided in Article VI.

2. The invoices referred to in the preceding paragraph shall be checked by Owner, and said invoices in such amount as shall be approved by Owner shall be paid to Contractor not later than fifteen (15) days after receipt by Owner of said invoices. Items not approved and not paid shall be referred to the Chief Engineer for report and recommendation of Owner and Contractor for settlement.

3. Payment by Owner of monthly invoices submitted by Contractor shall not constitute final settlement thereof, and all invoices shall be subject to further checking or auditing by Owner within a reasonable length of time.

4. It is the intent of this contract that the funds required for expenditures required for the completion of the Project will be supplied by the Owner, and that the Contractor, at no time, will be required to finance expenditures with its own funds. To accomplish this objective, a Revolving Bank Account will be established by the Contractor at the _____.

The Owner will make an initial deposit in this account of _____. All expenditures made by the Contractor for costs incurred in connection with the Project shall be made from this account. All payments received from the Owner on billings under the provisions of this Article XI (2) and other receipts which are proper credits to the cost of the work will be deposited into the Revolving Account to replenish it to its initial balance. Any balance remaining in the account at the completion of the Project shall be returned to the Owner. This Revolving Account shall be considered a Trust Account, and the Contractor will furnish the Owner an advance payment bond to insure proper accounting for and return to the Owner of any unexpended balance. If the Owner provides any dollar advances, such funds will be deposited in a Dollar Revolving Account in the Contractor's name at _____, and such funds will be handled in the same manner as described above.

5. Within fifteen (15) days after execution of the Contract, the Contractor will supply the Owner with a forecast, by month, of its anticipated expenditures during the life of the project. On the tenth (10) day of each month thereafter the Contractor will supply a revised monthly forecast to the Owner. The Contractor does not guarantee the forecast, and the Owner will remain responsible for supplying funds as required to complete the work.

ARTICLE XII.

Escalation

Attached to this Contract as "Exhibit 'D'" is a tabulation of basic materials, fuels, power, and freight, together with the price used for each material in computing the Target Estimate prices. If the actual prices paid by the Contractor for materials and fuels listed in "Exhibit 'D'" vary from the prices shown in said exhibit, then there shall be added to or subtracted from the Final Target Estimate an amount to compensate for such variations. The salary and wage rates and conditions of employment submitted by the Contractor pursuant to the provisions of Article VI (2) and approved by the Owner are certified by the Contractor to be the salary and wage rates used in computing the Target Estimate prices. If the Contractor, with the Owner's approval as provided by Article VI (2), finds it necessary to vary the salary and wage rates or conditions of employment previously approved, then there shall be added to or subtracted from the Final Target Estimate an amount to compensate for such variations. The Contractor's fee shall be increased or decreased by virtue of any adjustments in the Final Target Estimate which arise from the application of this escalation article.

ARTICLE XIII.

Accounting and Records

1. Contractor shall maintain a system of accounts and accounting records satisfactory to Owner, together with supporting papers for all work performed under this Agreement and included in monthly statements submitted by Contractor to Owner, as provided in Article XI (1) of this Agreement. Contractor shall also keep such cost records as may be necessary to provide for control of field operations. Owner shall have the right to examine Contractor's original records of the cost of all work performed under this Agreement, and may audit the same during the progress of construction and upon completion of the Project.

2. Contractor shall retain all time cards, material records, and other accounting records for all work performed by Contractor hereunder, and Owner shall have the right to audit the same at all reasonable times. Upon completion of the Project, all such accounting records shall be forwarded to Owner for its permanent files.

3. Contractor shall prepare and submit to Owner comprehensive monthly progress reports of work performed by Contractor, including pictorial record of work progression.

ARTICLE XIV.

Changes, Changed Conditions, and Cost Variation

Changes in plans, specifications, and schedules may be made by Owner when found necessary or desirable. Changed conditions, that is conditions materially differing from those anticipated by the parties, may be encountered. Costs of labor, materials, equipment, and supplies may vary from those on which the Original Target Estimate is based. To the extent that any such causes shall affect the Contractor's actual costs, a modification and revision of the Target Estimates shall be mutually agreed upon and set forth in writing.

The technical analysis of the hydrology of the Zambezi River indicates the probability that the foundation work areas will be inundated or overtopped for four months per year until the concrete arch dam has been constructed to sufficient height to provide adequate head of water to force the floods through the diversion works. In the event that floods of sufficient magnitude occur at such periods as to prevent continued access to the work sites for eight consecutive months in any given calendar year, then the effect of such floods shall be considered as a changed condition, and the Target Estimate shall be adjusted accordingly.

ARTICLE XV.

Extra Work

From time to time, Owner may request Contractor to perform work outside the plans and specifications and not included in the Target Estimate. Contractor agrees to perform such extra work as may be directed by Owner, and Owner shall pay to Contractor the actual cost thereof (direct and indirect costs), as determined in accordance with Article VI, plus a fee of _____ per cent (_%) thereof, but neither the costs nor the fee paid for extra work shall be included in "actual costs" and "fee" under the provisions of Articles VIII, IX, and X, and the Target Estimates shall not be revised to reflect such extra work.

ARTICLE XVI.

Workmen's Compensation and Compliance with Law;
Insurance

1. Contractor shall comply with all provisions and requirements of the Workmen's Compensation and the Occupational Disease Disability Laws of the Province of Mozambique and other political subdivision, and with all applicable Federal, Provincial, or local laws and legal requirements relating to employment, and to any and all work under this Agreement and Contractor's performance thereof.

2. Contractor shall carry public liability and property damage insurance covering its liability for injuries or damage to the persons or property of third parties, and the policies therefor shall be endorsed to provide protection to the Owner against claims of third parties arising out

of the operations of Contractor hereunder. The limits of this coverage shall not be less than those ordinarily carried by Contractor on its other operations.

3. Contractor shall carry as a cost of the work insurance against loss or damage to the contract work, and to materials, equipment, and supplies which have been furnished by Owner, or purchased by Contractor and chargeable as actual costs under the provisions of Article VI hereof. The cost of replacing, restoring, or repairing loss or damage not covered by insurance or within the deductible limits thereof shall be borne by Owner, and such cost shall not be deemed to be a part of the cost of the Project.

ARTICLE XVII.

Right of Way, Easements, Permits, and Regulations

1. Owner will obtain all permits, licenses, easements, or other rights necessary for the use and occupancy of the lands of governmental subdivisions or agencies, or private property, necessary for the construction and permanent use and operation of the Project.

2. Whenever it is necessary, in performing work hereunder, to occupy, use, obstruct, or reconstruct any highway, highway bridge, public and private lands or public place, or to do anything in connection with public or private property, Contractor shall produre the necessary permits, easements, and licenses or other rights therefor (except those specified in Section 1 of this Article), and in doing such work shall be governed by and comply with all applicable laws, rules, and regulations of the Federal, Provincial, County, or other public authority having control of such highways, bridges, public or private lands, public places or public and private property, and with the terms and conditions in all permits, easements, and licenses issued by any such authority. Owner will cooperate with and assist Contractor in obtaining such permits, licenses, and easements necessary for the Project.

3. Contractor shall issue and post all notices required by all laws, orders, rules, and regulations relating to the Project.

ARTICLE XVIII.

Liens and Litigation

1. Contractor shall not suffer or permit any laborer's, materialmen's, or other liens to arise or exist upon or against the Project or other property of Owner, or against Owner, by reason of Contractor's operations under or performance of this Agreement.

2. Contractor further agrees that the completed Project, when turned over to Owner, shall be free and clear of any and all claims, liens, encumbrances, or charges of every description arising by reason of Contractor's operations under or performance of this Agreement.

3. Each of the parties hereto shall give the other written notice

of any claim or litigation which may affect the Project or the Owner or its property in any way arising as a result of or in connection with any of the work performed under this Agreement.

4. If, during the progress of the Project, Contractor shall allow any indebtedness to accrue to manufacturers, suppliers, subcontractors, or others, and shall fail to pay or discharge the same within thirty (30) days after any demand following the due date thereof, Owner may pay such bills directly, and such payments shall be considered a payment to Contractor under this Agreement.

ARTICLE XIX.

Taxes, Fines, or Penalties

Exemption from Import Duties:

The Owner (Government) will grant to the Contractor, its subcontractors and its associates exemption from import duties for all materials to be embodied in the work.

The Owner will grant to the Contractor, its subcontractors and its associates exemption from import duties on the machines to be utilized in the design or construction of the work.

The Owner will grant to the Contractor, its subcontractors and its associates authorization for the temporary importation with exemption of duties on equipment, tools, vehicles, and necessary implements for the performance of the work and preliminary works, which will be reexported within six (6) months following the date of the final acceptance of the work.

The equipment, vehicles, tools, and implements which are used up on the job or become worn out during the performance of the work need not be reexported if a report on their uselessness will be prepared in due time, signed by representatives of the Owner, Customs authorities, and the Contractor, its subcontractors and its associates.

The Owner may declare a free zone and set off the area to be occupied by the work and marginal work areas in order to facilitate the application of the exemption of duties referred to in this article on the equipment, vehicles, tools, implements, and materials which do not leave these areas.

Exemptions from Taxes and Imposts:

The Owner will grant to the Contractor, its subcontractors and its associates exemption from the payment of taxes, licenses, contributions, and imposts and others of a similar nature. This exemption does not apply to income taxes. The Owner will grant exemption of payment of all charges for social benefits imposed by Portuguese Legislation on the work done by the foreign personnel employed on the work, the Contractor, its subcontractors and its associates being responsible for the payment of these charges under the terms of the legislation in effect, relative to all Portuguese personnel employed on the work.

The Owner, through the competent authorities, will grant entry and exit visas in and out of Mozambique, and residence permits, working permits, as well as every necessary license for the foreign personnel which may be employed on the work by means of the Contractor, its subcontractors and its associates' request duly justified.

The Owner will grant the exemption of payment of taxes or income tax to the foreign personnel employed on the work under the terms of official authorization.

For the purpose of interpreting the foregoing, the term "associates" means all firms, Portuguese or foreign, which will take part in the preparation of design phases, serve engineering functions, or will perform parts of the work pursuant to contracts and agreements with the Contractor and its subcontractors and its associates or with the Owner.

ARTICLE XX.

Title to Materials

Title to all materials, articles, and supplies which Contractor may purchase for the Project, and the cost of which is reimbursable to Contractor hereunder, shall pass to Owner through Contractor, and such title shall vest in Owner at the time payment is made therefor by Contractor or upon delivery to Contractor, whichever of said events shall first occur. Such transfer of title shall be evidenced by appropriate instruments of transfer. This provision for transfer of title to Owner shall not (a) relieve Contractor of any of its duties or obligations under this Agreement or of any responsibility or liability for the safe delivery and safeguarding, custody, or warehousing of the materials or equipment, or (b) constitute any waiver of Owner's rights to absolute fulfillment by Contractor of all of the terms hereof; provided, however, that the cost of replacing, restoring, or repairing loss or damage to such materials, articles, equipment, supplies, and property shall be reimbursable to Contractor, and such cost will not be deemed to be a part of the Cost of the Project.

ARTICLE XXI.

Termination or Suspension of Agreement; Default

1. Should Contractor at any time cease or suspend the Project, or any part thereof, or should Contractor be adjudged a bankrupt or make a general assignment for the benefit of creditors, or should a receiver be appointed for Contractor, Owner, itself, shall have the right to provide labor and materials and proceed with, or to contract with or hire others to proceed with, and to finish the Project, and Owner shall have the right, at its option, to take over and assume any or all contracts made or orders placed by Contractor for materials, equipment, or supplies, and take over the complete plant of the Contractor.

2. In the event of discontinuance of all work on the Project under the provisions of Section 1 above, Contractor shall not be entitled to receive any further payment under this Agreement until the Project shall be

wholly finished, at which time Contractor shall be paid whatever net balance
is found to be due to Contractor for payments of actual costs, plus Contractor's fee computed as provided in Article IX for the work actually
performed.

3. If the Project is halted by reason of court action or litigation,
or by governmental act or authority, Owner may, at its option, terminate
this Agreement in its entirety; or Owner, in any event, may require Contractor to place its operations on a standby basis, or may direct Contractor
as to the amount, order, or rate of progress of work to be performed hereunder, and Contractor, upon such requirement or direction, shall make every
reasonable effort to reduce the actual costs to a minimum during such
period, and Contractor shall be paid in accordance with this Agreement for
the work performed by it during such period, and shall resume construction
immediately upon notification by Owner, and upon such resumption an adjustment of the Target Estimate to reflect cost changes during such period
shall be agreed upon by the parties in writing.

4. In the event Owner fails to make the payments to the Contractor
as herein provided, or otherwise fails to comply with the terms and provisions of this Agreement, and fails to remedy such conditions after
ten (10) days' written notice, Contractor shall have the right to terminate
this Agreement, or if due to Force Majeure or otherwise, when the Contractor is not in default of his obligations, Owner elects to terminate the
Agreement, then under such termination, Contractor shall be paid within
thirty (30) days, whatever net balance is found to be due to Contractor for
payments of actual costs plus Contractor's Fee, computed as provided in
Article IX for the work actually performed and an additional termination
fee equal to ten per cent (10%) of the base fee on that portion of the work
not performed at the time of termination.

ARTICLE XXII.

Subcontracts

1. No subcontracts shall be made in connection with this Agreement
without the advance written consent of Engineer, but such consent shall in
no way modify or affect any of the terms and provisions of this Agreement.
All subcontracts shall be in writing, and two copies thereof shall be
furnished to the Engineer.

2. Contractor shall be responsible to Owner (a) for all work performed by, and for the acts, omissions, or negligence of subcontractors
and of all employees or agents of subcontractors, and (b) for the compliance
by each subcontractor with the requirements of all of the applicable provisions of this Agreement, and of all applicable laws, rules, and regulations, to the same extent that Contractor would be responsible if Contractor
were doing the work directly.

3. No subcontract, and nothing contained herein or in any subcontract, shall be construed to create any contractual relationship
between any subcontractor and Owner, or to make Owner in any way liable
or responsible to any subcontractor or its employees.

ARTICLE XXIII.

Force Majeure

Extensions of time and adjustments in the Target Estimate to compensate for added costs will be granted for all causes beyond the Contractor's reasonable control, including but not limited to:

Delays by the Owner in supplying plant, equipment, funds, design approvals, and instructions.

Delays due to breach of contract delivery dates on the part of suppliers which have been selected by the Owner, or on the part of public transport operators.

Strikes or work stoppages.

War, hostilities (whether war be declared or not), invasion, act of foreign enemies, rebellion, revolution, insurrection or military or usurped power, civil war or riot, commotion or disorder.

Use or occupation by the Owner of any portion of the works prior to final acceptance of the entire project, or a cause solely due to the design of the works (when not designed by the Contractor), or any such operation of the forces of nature as reasonable foresight and ability on the part of the Contractor could not foresee or reasonably provide against.

Epidemics.

Storms causing a total of more than two weeks' shutdown of the work during the life of the Project.

Contractor must notify the Owner within a reasonable time of any delays for which the Contractor requests an extension of time. This notification shall be no later than fourteen (14) days after the occurrence.

ARTICLE XXIV.

Arbitration

1. In the event of any failure to agree upon any matter to be settled by mutual agreement, or in the event of any dispute or difference as to any matter of fact arising between the parties hereto out of or relating to this Agreement, or any stipulation herein (including a dispute arising upon the cancellation, termination, or completion of this agreement as to the amounts properly payable to the respective parties), which cannot be settled between the parties themselves, the parties hereto shall submit the matter in dispute or in question to arbitration by three arbitrators, one of the arbitrators to be chosen by each party hereto, and the third arbitrator by the two thus chosen. The party desiring such arbitration shall give to the other party written notice of its desire, specifying the questions to be arbitrated and naming the arbitrator chosen by it.

2. Within a reasonable time thereafter, not exceeding thirty (30) calendar days, the other party shall give in like manner like written notice, specifying any additional questions to be arbitrated and naming the arbitrator to act for it.

3. In the event either party shall fail to appoint an arbitrator within thirty (30) days after receipt of the written notice from the other party of its desire to arbitrate, the arbitrator for the party shall be appointed by the presiding judge of the highest court of Switzerland upon application thereto by the party filing notice for arbitration.

4. The two arbitrators thus chosen shall then select the third. In the event that the two arbitrators chosen by or for both parties shall fail within thirty (30) calendar days to select the third arbitrator, the third arbitrator, upon written request of either party hereto, shall be appointed by the presiding judge of the highest court of Switzerland.

5. The decision of any two of three arbitrators thus chosen, when reduced to writing and signed by them shall be final, conclusive, and binding upon both parties hereto.

6. The arbitrators so appointed shall determine where the arbitration shall be held. In making such determination they shall take into consideration the place where they can best obtain facts in order to enable them to arrive at a fair and sound judgment.

7. In the event any arbitrator so appointed shall die or resign or be unable for any reason to perform his duties, another arbitrator shall be appointed in lieu of such arbitrator, such appointment to be made in the same manner by the same party who appointed such arbitrator who has died, resigned, or been unable to act.

8. The arbitrators so appointed shall determine which party shall assume the expense of such arbitration or the proportion of such expense which each party shall bear; and the arbitration expenses so allocated shall be paid by the party or parties by which the same are directed to be paid.

ARTICLE XXV.

Law Applicable

The Contractor and _____ (unless otherwise agreed or specified) accept the laws of the Province of _____ as the proper law of the Contract.

ARTICLE XXVI.

Successors and Assigns

All the covenants and Agreements of Contractor herein contained shall be binding upon the successors and assigns of Contractor, and shall inure to the benefit of the successors and assigns of Owner; provided, however, that this Agreement shall not be assigned or transferred in whole or in part except with the advance written consent of Owner, nor shall the same be transferred by or through any action, either legal or equitable, or any court proceeding of any kind or character whatsoever.

ARTICLE XXVII.

Final Acceptance

Upon completion of all work to be performed by Contractor, and acceptance thereof by Engineer, Owner shall notify Contractor in writing that the Project has been satisfactorily completed, and such notice shall release Contractor from all obligations hereunder, except provisions of Article XVIII (2), and excepting claims in respect to which written notices shall have been previously given by Owner to Contractor.

ARTICLE XXVIII.

Suppression of Fires

To the extent not compensated by insurance, Contractor shall be reimbursed by Owner for any costs incurred in the suppression of fires during the progress of work hereunder, and such cost shall not be deemed to be a part of the cost of the work.

ARTICLE XXIX.

Independent Contractor

Contractor, in performing the services under this Agreement, shall act as an independent contractor and not as an agent or employee of Owner.

ARTICLE XXX.

Not for Benefit of Third Parties

The Agreement and each and every provision thereof is for the exclusive benefit of the parties hereto and not for the benefit of any third party.

ARTICLE XXXI.

Article Headings

All Article headings are inserted for convenience only and shall not affect any construction or interpretation of this Agreement.

IN WITNESS WHEREOF, the parties hereto have executed this Agreement at _____ on the day and year first above written.

Attest:

_____   _____

                                        By:_____

Attest:

                                        _____

_____   By:_____

APPENDIX 15-A3

**Form of Agreement—Turnkey Type of Contract with Base Fee
& Award Fee Based upon Kind of Performance of the Contract**

A G R E E M E N T

THIS AGREEMENT is made this _____ day of _____,

1969, between _____

_____

_____, (hereinafter called the

Owner) and _____

_____

_____

_____, (herein-

after called the "Contractor").

WHEREAS Owner proposes to _____

_____

_____ and,

WHEREAS Contractor is engaged in the design and construction

business and desires to design and construct the project upon the terms and

conditions hereinafter set forth,

NOW, THEREFORE, the parties hereto agree as follows:

ARTICLE I

SCOPE OF WORK

1. The Owner agrees to employ Contractor to do the following:

(a) Prepare preliminary studies, sketch plans and outline specifications, including preliminary estimates of cost covering the project
which is more fully described as follows:

Plant Site - _____

_____

_____

_____

_____

Description of the Work _____

_____

_____

_____

_____

_____

_____

_____

_____

_____

_____

_____

_____

_____

_____

_____

_____

_____

_____

_____

———————————————————————————————

———————————————————————————————

———————————————————————————————

———————————————————————————————

———————————————————————————————

———————————————————————————————

———————————————————————————————

———————————————————————————————

———————————————————————————————

———————————————————————————————

(b)   Develop the approved plan into final checked construction drawings and specifications, and prepare and present a detailed estimate of the cost of the project.

(c)   Erect, construct and equip the project in accordance with plans, specifications, general conditions and estimates, as approved by the Owner. Plans, specifications, general conditions and estimates, when so approved, are to become a part of this Agreement.

Contractor agrees to act as engineer, builder and advisor to the Owner in all matters concerning the work, and agrees to furnish qualified personnel to accomplish the planning, estimating, designing, drafting, purchasing, accounting, erecting and construction of the project.

Contractor agrees that, subject to prior approval by Owner, all contracts and purchase orders for necessary materials, supplies, equipment and services shall be made by Contractor in its name, title to materials, supplies and equipment to pass to Owner upon delivery at site of work, except equipment brought to site on rental basis; that all personnel hired by Contractor shall be employees of Contractor and not of Owner; that Contractor will observe and comply with all applicable local, state and federal laws and regulations; and that Contractor will maintain workmen's compensation and employers' liability insurance, comprehensive liability insurance, contractors' protective public liability insurance, automobile liability insurance, and other insurance required by Owner, in limits satisfactory to the Owner, evidence of such insurance shall be promptly transmitted to the Owner.

Contractor shall save harmless and protect the Owner against all laborers', materialmen's and mechanics' liens upon the building or premises on which the work is located arising out of labor or material furnished under this contract, provided Owner shall have paid amounts due Contractor.

2.   Contractor at all times during the progress of the work shall
be and act subject to the general direction and control of Owner insofar
as results to be attained hereunder are concerned, but not as to the mode
and manner of performing the work.

ARTICLE II

Owner's Responsibilities

The Owner will:

1.   Provide full information as to its requirements for the work.

2.   Assist Contractor by placing at its disposal all available information pertinent to the site of the work including previous reports and
any other data relative to design and construction of the work.

3.   Guarantee access to and make all provisions for Contractor to
enter upon public and private lands as required to perform its work under
this Agreement.

4.   Examine all studies, reports, sketches, estimates, specifications, drawings, proposals and other documents presented by Contractor and
shall render in writing decisions pertaining thereto within a reasonable
time so as not to delay the work of Contractor.

5.   Designate in writing a person to act as Owner's representative
with respect to the work to be performed under this Agreement; and such
person shall have complete authority to transmit instructions, receive information, interpret and define Owner's policies and decisions with respect
to materials, equipment elements and systems pertinent to the work covered
by this Agreement.

6.   Obtain approval of all governmental authorities having jurisdiction over the work and such approvals and consents from such other
individuals or bodies as may be necessary for the completion of the work,
with the exception of construction permits and licenses.

ARTICLE III

Contractor's Responsibilities

Contractor will:

1.   Designate a Project Design Engineer to act as Contractor's
representative during all design phases, and he shall have complete
authority to receive instructions and notices.

2.   Consult with Owner to determine the requirements of the work.

3.   Prepare engineering and architectural studies and report on
the work in sufficient detail to indicate clearly the problems involved and

the alternate solutions available to Owner, to include schematic layouts and sketches and a general cost estimate for the work and to set forth recommendations.

After agreement to proceed with the final design, Contractor will:

1. Prepare detailed construction drawings and specifications for the work.

2. Furnish to Owner engineering data for and assist in the preparation of the required documents so that Owner may secure approval of such governmental authorities as have jurisdiction over design criteria applicable to the work.

3. Advise Owner of any adjustment of the cost estimate for the work caused by changes in scope, design requirements or construction costs and furnish a revised cost estimate for the work based on the completed drawings and specifications.

4. Advise Owner of the proposed construction schedule.

After agreement to proceed with the construction phase, Contractor will:

1. Commence each stage of the work immediately following authorization by Owner, and shall prosecute the work continuously and diligently to completion, in accordance with a construction schedule agreed upon by the parties. It is expressly understood and agreed that the time limits set out in the construction schedules are of the essence, provided that adjustments of the time limits set out in the construction schedule may be made by Owner, as hereinafter defined, when delays have been caused by matters beyond the reasonable control of the parties; and provided, further, that each stage of the work may be stopped or placed on a different construction schedule, in whole or in part, by written agreement of the parties.

2. Exercise its best knowledge, skill and experience in planning, purchasing materials, obtaining and furnishing skilled and competent labor and personnel, supplying tools and equipment, and performing all work incident to the full and satisfactory completion of the work in accordance with the construction schedule and approved plans and specifications, and shall do everything possible on its part to establish and follow a fast, economical and efficient construction program.

3. Obtain the most favorable prices and discount privileges available, and shall pay all bills for purchases within the allowed best discount period.

4. Constantly keep at the work during its progress, a competent project manager, and any information given to such project manager or any notices served upon him shall have the same force and effect as if given to or served upon Contractor.

5. Employ on the project only persons who are qualified and competent to perform the work entrusted to them.

6. Secure all construction permits and licenses and all governmental and public utility charges and inspections necessary for the prosecution of the work.

7. Give all notices and comply with all laws, ordinances, rules and regulations applicable to the work.

8. Take all necessary precautions for the safety of, and provide the necessary protection to prevent damage, injury or loss to:

(a) all employees on the work and other persons who may be affected thereby,

(b) all the work and all materials or equipment to be incorporated therein, whether in storage on or off the site, and,

(c) other property at the site or adjacent thereto, including trees, shrubs, lawns, walks, pavements, roadways, structures and utilities not designated for removal, relocation or replacement in the course of construction.

ARTICLE IV

Compensation

Owner agrees to pay, and Contractor agrees to accept as full compensation, satisfaction and discharge for all work done, materials furnished, costs and expenses incurred and damages sustained and for each and every matter, thing or act performed, furnished or suffered in the performance of this Agreement, the following:

A. All actual costs, charges and expenses incurred by Contractor in the performance of the work, which shall include, but not be limited to:

1. ENGINEERING: The actual cost to Contractor as approved by Owner's representative for preparation of designs, plans, specifications, estimates and material lists; for checking shop details and for shop inspections; for architectural and engineering inspections of the work, more specifically, but not necessarily exclusively the following:

(a) Salaries and wages of engineering personnel, estimators, and stenographers and clerks for their time devoted to the work. (Salaries and wages to include a proportionate share of vacation, holiday and sick leave allowance.)

    (1) Fringe benefits in accordance with company practices, labor agreements or required by law or regulation.

    (2) Taxes on employees' earnings.

(3)   Workmen's compensation and other insurance premiums measured by payroll costs.

(b) Seventy-five per cent (75%) of salary and wage charges under item 1(a) above for engineering and main office overhead expenses.

(c) Traveling and living expenses for engineering personnel and estimators while away from their home office on this work.

(d) The cost of blueprints, photostats, other reproductions and models.

(e) Any sales, use or gross receipts taxes payable on purchases for the work or on receipts under this contract.

(f) The cost of toll telephone and telegraph service.

(g) Insurance premiums for coverage normally carried by company, the cost of which is measured by company receipts.

(h) The cost of soil bearing and other tests required for foundation design, or other purposes.

(i) The cost of surveys and outside consultants, if required.

(j) The cost of temporary drafting rooms and engineering offices including their equipment, maintenance and operation at the site, or elsewhere, except at Contractor's main and sub-offices.

(k) The cost of computer center time directly devoted to the engineering work, based on Contractor's hourly rate for each piece of equipment used.

(1) Any other costs not described above which are proper charges to the work and approved by the Owner's representative.

(m) Fee - In addition to the reimbursable costs set forth above, Contractor shall be paid by Owner an amount equal to _____ per cent (%) of the aggregate amount of the items set forth in Section 1(a) through (1), as a fee for engineering services.

2.   CONSTRUCTION: The actual cost to Contractor of constructing the work for wages, salaries, expenditures and expenses in connection therewith, as approved by Owner's representative, more specifically but not necessarily exclusively the following:

(a) The cost to Contractor of salaries and wages and all other benefits, except bonuses, paid to Contractor's employees (excluding executive officers and other employees performing generally a company-wide function at Contractor's home office); all necessary expenses paid by Contractor to or for the benefit of such employees, and all costs and expenditures under or arising out of employment agreements entered into with such employees. Cost of labor paid on an hourly basis shall not exceed the rates prevailing in the area, including rates determined by bargaining agreement, if any, to which Contractor is bound.

(b) Contractor's expenditures for payroll taxes on salaries and wages, and applicable group insurance costs, retirement benefits, vacation allowances and sick leave benefits, for employees engaged on the work in accordance with Contractor's established practices; provided that such allowances and benefits provided by Contractor for employees not engaged on the work exclusively shall be chargeable to actual costs only for the time the employee or employees are directly employed on the work.

(c) The cost to Contractor of establishing and maintaining a project construction office and such other temporary field offices as may be required for the project, including office supplies, telephone and telegraph charges, and other office operating costs, and construction, maintenance and operation of warehouses, machine shops and other temporary facilities.

(d) Compensation for use and/or availability at the work site of Contractor-owned equipment at rates specified in the current edition of Contractor's Equipment Red Book, and the cost of fuel, lubricants, equipment operating supplies and minor repair parts necessary to maintain said equipment in good operating condition (major repair parts will be paid for by Contractor).

(e) Consumable materials and supplies purchased by Contractor for use in the work, including all actual necessary freight, express and drayage paid by Contractor.

(f) Traveling expenses of Contractor's employees when traveling necessarily and solely in the interest of the project.

(g) Moving expenses to the job and return of salaried personnel assigned to the work.

(h) Small tools, office furniture and equipment, job apparatus, minor equipment and similar items of property, where the current value or original cost to Contractor of each item is $1,000.00 or less.

(i) Actual transportation expense of construction plant and construction equipment to the site of the work and return, including loading and unloading costs, and assembly and disassembly costs.

(j) Excise and other taxes payable by Contractor directly applicable to the work, except those based on income.

(k) First-aid and job safety expenses.

(l) Payment to subcontractors for work performed.

(m) Materials and supplies purchased by Contractor for physical incorporation in and as a part of the finished work, including all actual necessary freight, express and drayage.

(n) The cost of renting equipment from others.

(o) Premiums for all necessary insurance coverages carried by Contractor applicable to the work including the cost of general and automobile liability insurance, workmen's compensation insurance, employer's liability insurance, builder's risk insurance, contractor's equipment insurance, etc.

(p) Royalties, custom duties, licenses, permits, testing, and inspection fees incurred because of the work.

(q) Cost of repairing damages or payment for liabilities therefor, and for personal injuries, or of resetting work destroyed or repaired or replacement of materials, supplies and equipment as a result of accidents occurring during or because of the prosecution of the work (if not recoverable under insurance policies).

B. Home office costs amounting to three per cent (   ) of all actual costs determined in A2 above only.

C. Base fee amounting to four per cent (   ) of the sum of the costs of A2 and B above.

D. Award fee of up to eight per cent (   ) of the sum of costs of A (less A 1(m) and B above based on a Performance Evaluation Grade assigned every three months by Owner for Contractor's performance.

Owner will grade Contractor's total performance every three months (December 1, March 1, June 1 and September 1 of each year) for all work. Owner will have the right to request and receive such information from Contractor as may be reasonably required.

In arriving at the grade for each category, Owner will multiply the assigned rating in terms of percentage times the various weighting factors. The Performance Evaluation Grade shall be the sum of the grades for each category and shall be used to determine percentage of award fee earned by Contractor by entering the award fee payment schedule which is attached as Exhibit "A" (see p. 405).

CATEGORIES TO BE EVALUATED	Weighting Factor
(1) Quality of Work	20
(2) Early Completion	40
(3) Control and Reduction of Costs	40
Total Weighting Factor	100

In arriving at the assigned score and ultimate grade for each of the above categories, Owner will consider each of the above categories from the following standpoints:

(1) QUALITY

     Workmanship
     Skills demonstrated
     Completeness
     Technical Competence

(2) EARLY COMPLETION

     Meet or beat milestone or other schedule dates
     Anticipate schedule problems caused by changes
     Ability to recover in changing program
     Planning

(3) COSTS

     Cost consciousness
     Accomplishment within estimated costs
     Minimization of waste
     Control and cost accounting
     Efficiency

As soon as possible after the end of each three-month period, Owner's grading of Contractor's performance will be made known to Contractor. Should Contractor take any exceptions to Owner's grading, such exceptions must be made known to Owner in writing, within ten days after receipt of the grading. After receipt of any exceptions, Owner will make a final determination with respect to the final performance evaluation. Grade will be final and will not be subject to the "Arbitration" clause of this Agreement.

ARTICLE V

Manner and Times of Payment

1. On or before the fifteenth (15) day of the month next succeeding the month in which work was commenced, and on or before the fifteenth (15) day of each month thereafter, Contractor shall submit to Owner an invoice in triplicate, including such invoices and other supporting data as Owner may require, setting forth for the preceding calendar month the billing for items of actual cost incurred by Contractor, as provided in Article IV. In addition, each monthly invoice shall include Contractor's home office costs fee, base fee and award fee when determined.

2. The invoices referred to in the preceding paragraph shall be checked by Owner and said invoices in such amount as shall be approved by Owner shall be paid to Contractor not later than fifteen (15) days after receipt by Owner of said invoices together with applicable fees. Contractor shall be notified of the items not approved and the reasons thereof.

3. Payment by Owner of monthly invoices shall not constitute final settlement thereof, and all invoices shall be subject to further

checking or auditing by Owner within a reasonable length of time.

ARTICLE VI

Accounting and Records

1. Contractor shall maintain a system of accounts and accounting records in accordance with company policy, together with supporting papers for all work performed under this Agreement and including in monthly statements submitted by Contractor to Owner. Contractor shall also keep such cost records as may be necessary to provide for control of field operations. Owner shall have the right to examine Contractor's original records of the cost of all work performed under this Agreement, and may audit the same during the progress of construction and upon completion of the work.

2. Contractor shall retain all time cards, material records, and other accounting records for all work performed by Contractor hereunder, and Owner shall have the right to audit the same at all reasonable times. Upon completion of the work, all such accounting records shall be forwarded to Owner for its permanent files.

3. Contractor shall prepare and submit to Owner comprehensive monthly progress reports of work performed by Contractor, including pictorial record of work progression.

ARTICLE VII

Changes in Plans and Specifications

Changes in plans or specifications may be made by Owner when found necessary or desirable.

ARTICLE VIII

Authorized Purchases

For purchases in excess of $1,000.00, Contractor will submit its requisition to Owner for approval, and such approval shall be given within ten days of the submission. If Owner does not return the requisition to Contractor within ten days of submission, then it shall be considered approved without further action by Owner. It is understood that Contractor is responsible for the proper and speedy execution of the work, and that therefore Owner's approval of Contractor's procurement requisitions will not be unreasonably withheld.

ARTICLE IX

Subcontracts

No subcontracts shall be made in connection with this Agreement without the advance written consent of Owner, but such consent shall in no way modify or affect any of the terms and provisions of this Agreement. All subcontracts shall be in writing, and two copies thereof shall be furnished to Owner.

2. Contractor shall be responsible to Owner (a) for all work performed by, and for the acts, omissions or negligence of subcontractors and of all employees or agents of subcontractors, and (b) for the compliance by each subcontractor with the requirements of all of the applicable provisions of this Agreement, and of all applicable laws, rules and regulations, to the same extent that Contractor would be responsible if it were doing the work directly.

3. No subcontract, and nothing contained herein or in any subcontract, shall be construed to create any contractual relationship between any subcontractor and Owner or to make Owner in any way liable or responsible to any subcontractor or its employees.

ARTICLE X

Liens and Litigation

1. Contractor shall not suffer or permit any laborer's, materialmen's or other liens to arise or exist upon or against the work or other property of Owner or against Owner by reason of Contractor's operations under or performance of this Agreement.

2. If, during the progress of the work, Contractor shall allow any indebtedness to accrue to manufacturers, suppliers, subcontractors or others, and shall fail to pay or discharge the same within thirty (30) days after any demand, following the due date thereof, Owner may pay such bills directly, and such payment shall be considered a payment to Contractor under this Agreement.

ARTICLE XI

Taxes, Fines or Penalties

Contractor shall pay promptly all social security and unemployment compensation taxes, sales, use and other excise taxes, and all other taxes, fees, charges, or other costs due and payable by Contractor to the United States or to any state or political subdivision thereof in connection with any and all work performed, labor done and materials provided under this Agreement; and Contractor shall protect and defend, and hold Owner harmless from any cost or liability on account of any fines or other penalties, that may be assessed resulting from any acts of negligence, omission, wrongdoing or violation of law on the part of Contractor.

ARTICLE XII

Ownership of Documents

All documents, including original drawings, estimates, specifica-

tions, field notes and data are and remain the property of Contractor as
instruments of service. Owner shall be supplied a set of reproducible
record prints of drawings and copies of other documents but shall use them
solely in connection with the work and not for the purpose of making subsequent extensions or enlargements thereto and will not sell, publish or
display them publicly. Reuse for extensions of the work or for new
projects, shall require written permission of Contractor and shall entitle
it to further compensation at a rate to be agreed upon by Owner and Contractor.

ARTICLE XIII

Termination or Suspension of Agreement; Default

1. Should Contractor at any time, in the opinion of Owner fail,
refuse or neglect to provide proper or sufficient equipment, materials,
or workmen, or to make satisfactory progress with the work, Owner shall,
at any time, upon ten (10) days' written notice to Contractor, have the
right to discontinue any or all work, or any part thereof, and/or to
terminate Contractor's services in connection therewith. In the event of
any such termination of Contractor's work, Owner may proceed with or complete the work itself or through others, but this shall not relieve
Contractor of its responsibility hereunder for full satisfactory completion of all work not so terminated, or of any liability for damages for
breach of contract.

2. Should Contractor at any time cease or suspend the work or
any part thereof, or should Contractor be adjudged a bankrupt or make a
general assignment for the benefit of creditors, or should a receiver be
appointed for Contractor, Owner, itself, shall have the right to provide
labor and materials and proceed with, or to contract with or hire others
to proceed with, and to finish the work, and Owner shall have the right,
at its options, to take over and assume any or all contracts made or
orders placed by Contractor for materials, equipment and supplies.

3. In the event of discontinuance of all work on the project
under the provisions above, Contractor shall not be entitled to receive
any further payment under this Agreement until the work shall be wholly
finished, at which time Contractor shall be paid whatever net balance is
found to be due to Contractor for payments of actual costs, plus Contractor's base fee for the work actually performed.

4. If the work is halted by reason of court action or litigation, or by governmental act or authority, Owner may, at its option,
terminate this Agreement in its entirety; or Owner, in any event, may
require Contractor to place its operations on a standby basis, or may
direct Contractor as to the amount, order or rate of progress of work to
be performed hereunder, and Contractor, upon such requirement or direction,
shall make every reasonable effort to reduce the actual costs to a minimum
during such period, and Contractor shall be paid in accordance with this
Agreement for the work performed by it during such period, and shall resume construction immediately upon notification by Owner.

ARTICLE XIV

Guarantees and Correction of Work
After Substantial Completion

Contractor warrants and guarantees that all work, materials and
equipment will be of good quality and free from faults or defects. Upon
receipt of written instructions from Owner, it will correct all faults
and deficiencies in the work which appear within one year after substantial completion. Owner will give prompt written notice of observed
defects. The warranties and guarantees provided in this Article shall be
in addition to and not a limitation of any other remedies provided by this
Agreement or by law. If Contractor after notice, fails to proceed promptly
to comply with the terms of this guarantee, Owner may have the work corrected and Contractor shall be liable for all expenses incurred.

ARTICLE XV

Arbitration

Arbitration of all questions in dispute under this Agreement shall
be at the choice of either party and shall be in accordance with either
the provisions, then in effect, of the Standard Form of Arbitration Procedure of the American Institute of Architects or the rules, then in
effect, of the American Arbitration Association, as such party shall designate. This Agreement shall be specifically enforceable under the
prevailing arbitration law and judgment upon the award rendered may be
entered in the court of the forum, state, or federal, having jurisdiction.
The decision of the arbiters shall be a condition precedent to the right
of any legal action. If the applicable statute of limitations would bar
the institution of any legal or equitable proceedings based on a claim or
dispute under this Agreement, neither party shall have the right to seek
arbitration of such claim or dispute under this paragraph.

ARTICLE XVI

Successors and Assigns

All the covenants and agreements herein contained shall be binding upon, and inure to the benefit of the successors and assigns of the
parties; provided, however, that this Agreement shall not be assigned or
transferred in whole or in part except with the advance written consent
of the other party, nor shall the same be transferred by or through any
action, either legal or equitable, or any court proceeding of any kind or
character whatsoever.

ARTICLE XVII

Final Acceptance

Upon completion of all work to be performed by Contractor and acceptance thereof, Owner shall notify Contractor in writing that the work has been satisfactorily completed, and such notice shall release Contractor from all obligations hereunder, excepting claims in respect to which written notices shall have been previously given by Owner to Contractor.

ARTICLE XVIII

Force Majeure

Neither party shall be considered in default of performance of its obligations hereunder, or any of them, to the extent that performance of such obligations, or any of them, is delayed by

Force Majeure. Force Majeure shall include, but not be limited to, hostilities, revolution, civil commotion, strike, epidemic, fire, flood, wind, explosion or embargo, or any law, proclamation, regulation or ordinance of any government or governmental agency having or claiming to have jurisdiction in the premises, or any act of God, or any cause, whether of the same or different nature, existing or future, which is beyond the control of the parties hereto.

ARTICLE XIX

Independent Contractor

Contractor, in performing the services under this Agreement, shall act as an independent contractor and not as an agent or employee of the Owner.

ARTICLE XX

Not For Benefit of Third Parties

This Agreement and each and every provision thereof is for the exclusive benefit of the parties hereto and not for the benefit of any third party.

ARTICLE XXI

Interpretation

This Agreement shall be interpreted in accordance with laws of the State of Arizona.

ARTICLE XXII

Article Headings

All article headings are inserted for convenience only and shall not affect any construction or interpretation of this Agreement.

IN WITNESS WHEREOF, the parties hereto have executed this Agreement at _____, on the day and year first above written.

ATTEST:

_____   By:_____

ATTEST:

_____   By:_____

ATTEST:

_____   By:_____

ATTEST:

_____   By:_____

EXHIBIT "A"

PAYMENT SCHEDULE

16 COST CONTROLS, RELATION, AND COORDINATION WITH ENGINEERING AND ACCOUNTING

J. P. FREIN

Vice-President and Director (Retired)
Morrison-Knudsen Company, Inc.
Boise, Idaho

THIS CHAPTER explains the relative positions and the respective functions of the contractor's engineering department and his accounting department with respect to cost controls related to the work. The principal duties in this regard are outlined below.

16-1. Prebid Estimating and Bidding

The engineering department analyzes prospective work and estimates its cost in preparation for making the proposal. To do so, the engineering department must become familiar with all particulars of the contract and the technical specifications, as well as understand the physical aspects of the work to be done under the contract. It must develop with management a concept of how to do the work, what equipment and plant to use, etc.

After the concept is determined, the department initiates inquiries for quotations on all items needed, accumulates all the pertinent basic information, and then makes a detailed cost analysis of all construction operations necessary to complete the job, developing them step by step.

At this stage, another function of the engineering department is to coordinate with any joint-venture partners to see that all prebid requirements are met and to reconcile the prebid cost estimates of all the partners. The engineering department also is required to participate with management in finalizing the bid prices for all items concerned and oversee the filing of the proposal.

To this point, the accounting department has not taken any part in the preparations for bidding the job. If the proposal is unsuccessful and does not result in an award, the matter is dropped and the accounting department is aware only of the amount of money spent to analyze and bid the job.

16-2. Preparing the Record to Start a New Job

However, if the proposal is successful and a contract is obtained, it is necessary to prepare for proper accounting during execution of the work. Since the engineering department alone has the facts concerning the job to this point, both it and the accounting department jointly outline the information that the record should show.

Use of the record, while it is being accumulated and after it is summarized, is fourfold:

1. To verify the original bid estimate prepared by the engineering department and to localize conditions of profit and loss. This, incidentally, proves the feasibility of the construction method employed and the quality of the estimating performed.
2. As a tool for management in directing and supervising the work, spotting cost overruns, showing where special attention is needed, replacing personnel, changing of method, modifying operating procedures, etc.
3. To record historical cost information and to preserve it for use in bidding future work.
4. To keep the off-job top management and central office personnel informed about the progress of the work and any other particulars.

Preparing for the record at this stage consists primarily of adopting a cost-code key to ensure the accumulation and recording of the information serving the above purposes. The accounting department takes the position that it will conduct its functions in this connection to fashion a tool that will be most usable for all concerned. Relying on the knowledge and background of the engineering department from its detailed prebid study of the project, it attempts to follow the details of the prebid estimate as closely as possible under the guideline of company accounting practices.

At this time, the two departments agree on how each will function in the accumulation of information on the job and how things would be coordinated in contributing to future jointly prepared reports. Although this aspect normally would be a matter of established company procedure, it should be reviewed in light of the peculiarities and makeup of each job.

16-3. Selection of Accounting System

The accounting department is responsible for selecting or designing the accounting system to be used in recording the costs incurred, and for determining how they should be distributed to the individual subaccounts needed to provide the monetary information so that it can be used by management and the engineering department. Usually, the company will have developed certain accounting practices and principles and an adaptation of these is all that is required.

16-4. Accumulation of Physical and Other Data on the Job

For proper accounting, both on an interim and a final basis, it is necessary to keep accurate records of the amount of work performed and to which contract items it applied. The engineering department makes the necessary field measurements (cross sections, etc.), performs current field observations, or otherwise determines the amount of work executed under the contract. It also maintains complete records of this information.

The information is always available to the accounting department, but it is necessarily summarized monthly (or more often if required by management) for purposes of making reports and for checking the earnings under the contract. Thus, the engineering department accumulates all of the contract data pertaining to the performance of the job.

Maintenance of inventories, receipts, and disbursements of items handled through the warehouse, handling and payment of invoices, payrolls, and other transactions involving expenditures are the functions of the accounting department.

16-5. Applying the Cost Coding in the Field

Applying the cost coding to assure that expenditures are correctly charged against the proper work item requires a complete knowledge of the contract and specifications. On more complex jobs the engineering department—due to its background of detail estimating, programming, planning work and keeping current records of work in progress, and participation in preparation of the cost coding key—is in the best position to indicate how the payroll and other expenditures should be charged. A cost engineer is employed on larger jobs to coordinate with the accounting

department in the exchange of information and applying the cost coding.

On less complex jobs, where relatively few cost breakdowns are involved and where there is less question with respect to contractural or technical matters, it is often practicable for the accounting department to assign the codes for cost accounting.

16-6. Maintaining the Cost Accounts

It is a natural function of the accounting department to perform all the necessary entries in the cost accounts to keep them current, to summarize them periodically according to individual subaccounts, and to account for all expenditures. In addition, the department keeps the cost accounts in balance with the general books of account, which it must maintain for an overall fiscal picture of the job.

16-7. Computing or Verifying Earnings Under the Contract

Since the engineering department keeps complete current records of all work performed under the contract, knows all the contract terms and the specifications, and coordinates job activities with the owner's engineer, it is logical that the contractor's engineer verify earnings under the contract. The owner's engineer usually prepares the interim and final estimates of contract earnings, although the contractor's engineer sometimes prepares the statements of earnings for verification by the owner. If the owner's engineer prepares the payment estimates, the contractor's engineer checks and verifies them.

Interim payments often are only approximate and do not necessarily convey an accurate picture of the true earnings. Although the cash receipts under the contract are handled by the accounting department, the engineering department must provide the true amount of earnings, assuming that these are different from the actual payments under the contract. Data supplied by the engineering department in this regard help the accounting department make a true fiscal analysis of the job at the various accounting periods.

16-8. Preparing Interim and Final Cost Reports

Preparing of all cost reports is a joint effort of the accounting and engineering departments. The accounting department supplies the monetary values broken down according to the detailed cost accounts, and the engineering department supplies accurate quantities of work performed, production rates, haul distances, etc., as well as the earnings under each contract item. Man-hour data and equipment hours (worked, down for repairs, and idle) come from the accounting records, together with the amount of consumable supplies such as fuel, power, etc., that receive special attention in the reports.

Coordination of all the data contributed by the accounting and engineering departments provides detailed total and unit costs, overrun or underrun of the estimated costs, profit or loss on the individual contract items and their sub-parts, operating costs of plant and equipment, etc. All of these are made part of the cost reports to the extent that they can be used by those who receive the reports.

Interim and final reports also contain statistical data desired by the management. This might be contributed by either the accounting or engineering departments. The task of putting together the periodic reports is shared by those contributing information. Although the use of computers and business machines for this purpose is becoming prevalent for large jobs, it requires direction and an understanding of the subject to get the job done.

16-9.   Systems of Assigning Cost Account Codes

There is no fixed rule for determining what to call any of the numerous cost accounts that might be used on a job. The reason for assigning cost codes is to give each account a short name—usually a numeral—that it can be called by for the convenience of all those working with them. Contractors often develop their own systems for establishing cost codes—dividing the cost accounts into two general categories: recurring items and nonrecurring items.

For the former, the same general nomenclature is used from job to job, supplementing it only when necessary to deal with the peculiarities of an individual job. The nonrecurring items generally require reassignment for each job, although some similarity in the sequence of the nomenclature might be applied.

The following examples illustrate the subject. They are not intended to represent any particular system of coding in use; nor are they complete.

Recurring Items   As the term implies, these are items whose character could make them applicable in some degree to most jobs; parts of them are applicable to every job. In the cost coding, the subaccounts can be supplemented for individual job peculiarities.

With respect to "Plant Facilities," the cost coding should be arranged to cover the first in-place cost of the facilities. For instance, the basic code for a batching and mixing plant, might be Acct. No. 301-07. Parts of the first cost can be divided further into its components, such as:

Acct. 301-07 (a) for installation labor
    301-07 (d) for equipment rental and installed
                     equipment
    301-07 (e) for supplies
    301-07 (f) for subcontracts

The operating cost of the same plant facility then can be designated as follows:

    301-07-1 for operating labor, tax, and insurance
    301-07-2 for repair and service labor
    301-07-4 for equipment rental or depreciation
    301-07-5 for cost of fuel or power
    301-07-6 for oil and grease
    301-07-7 for repair parts
    301-07-8 for operating supplies
    301-07-9 for cable and wire rope
    301-07-10 for miscellaneous shop cost

A possible system of basic coding for recurring items is as follows:

General Expense and Overhead

Acct. No.	Account
101	Management and supervisory salaries
102	Office salaries
103	Engineering salaries
104	Warehouse and purchasing salaries

General Expense and Overhead (cont.)

Acct. No.	Account
105	Labor-relations salaries
106	Safety and first-aid salaries
107	Watchmen, guard, and fire guard salaries
108	Retirement plan expense
109	General automotive expense
110	Man haul
112	Stationery and office supplies
113	Office rent (including temporary)
114	Office equipment—rental and depreciation
115	Office expense—other
116	Engineering supplies and expense
117	Engineering equipment—rental or depreciation
118	Photography and reproduction
119	Outside engineering expense
120	Consultant's expense
121	Warehouse and purchasing expense
122	Labor relations—supplies and expense
123	Safety and first-aid supplies and expense
124	Watchmen, guards, and fire guards expense
125	Telephone and telegraph
126	Radio communications
127	Heat, light, power, and janitor service
128	Building maintenance—owner facilities
129	Legal expense
130	Travel expense—job connected
131	Travel expense—mobilization of personnel
132	Entertainment, donations, and contributions
133	Bank charges
134	Fire protection
135	Sanitary facilities
136	Drinking water and ice
137	Misc. job road and yard maintenance
138	Job signs
139	Winter shutdown
140	Central office general expense
141	Group insurance expense
142	Vacation expense—key employees
143	Data processing expense
144	Project control costs (CPM)
145	Gain or loss—camp service operations
146	Meal expense
147	Insurance—other than payroll
148	Taxes—other than payroll
149	Surety bond premium
150	Employee bonuses
151	Cash discounts earned
152	Miscellaneous general expense

General Plant

Acct. No.	Temporary Buildings
201-01	Office
201-02	Warehouse
201-03	Equipment repair shop
201-04	Carpenter shop
201-05	Electrical repair shop
201-06	Pipefitters shop
201-07	Riggers loft
201-08	Hospital or other
201-09	Field offices
201-10	Tool sheds
201-11	Change house
201-12	Powder house

General Plant *(cont.)*

Acct.
No.　　　　　　　*TEMPORARY BUILDINGS*

201-13　Cap house
201-14　Tire repair shop
201-15　Oil storage
201-16　Grease rack
201-17　Service station
201-18　Fuel storage
201-19　Inspectors' field office
201-20　Inspectors' concrete testing facilities
201-21　Paint shop
201-22　Workmen's shelters
201-23　General yard and site grading surface

202 UTILITIES INSTALLATION (except camp)
202-01　Electrical supply and distribution
202-02　Telephone lines and job telephone systems
202-03　Water supply storage and distribution
202-04　Sewer system and treatment facilities
202-05　Radio communications

203 TEMPORARY ROADS
203-01　General and access roads and bridges

204 CAMP BUILDINGS
204-01　Barracks
204-02　Mess hall
204-03　Commissary
204-04　Recreation hall
204-05　Bull cook's dormitory
204-06　Trailer court, including utilities
204-07　Trailer court wash house
204-08　Residences
204-09　Site grading and surfacing, except trailer court

205 UTILITIES FOR CAMP, other than trailer court
205-01　Power supply and distribution
205-02　Telephone system
205-03　Water supply and distribution
205-04　Sewer system and treatment facilities

Specific Plant

Acct.
No.　　　　　　　*Account*

301 SPECIFIC PLANT
301-01　Air production plant
301-02　Aggregate crushing and separation plant
301-03　Aggregate washing plant
301-04　Asphalt plant
301-05　Cableways
301-06　Cement storage
301-07　Concrete batching and mixing plant
301-08　Concrete placing plant
301-09　Derricks and cranes
301-10　Quarry
301-11　Refrigeration plant—aggregate cooling
301-12　Unwatering system
301-13　Water distribution system
301-14　Electrical generating plant

302 UNLOADING, ASSEMBLING, AND ERECTION
302-01　Shovels
302-02　Draglines
302-03　Tractors and scrapers
302-04　Euclids and trucks

Specific Plant *(cont.)*

Acct.
No.　　　　　　　*Account*

303 EQUIPMENT MOBILIZATION COST
303-01　Freight in on previously owned equipment
303-02　Local drayage and hauling
303-03　Sales tax on purchase of mobile equipment

304 DISMANTLE AND MOVE OUT
304-01　Specific production plants
304-02　Construction equipment

305 TUNNEL PLANT
305-01　Jumbo and drilling equipment
305-02　Railroad system
305-03　Air ventilation system
305-04　Water distribution system

306 FORMS FOR CONCRETE

Nonrecurring Items　Nonrecurring items consist mostly of the individual bid items or schedule-of-work items set up by the contract. They are almost always different for each job. Generally, they are designated by a numeral or symbol by the owner in his contract documents. Some owners try to use similar designations for somewhat similar work items for subsequent jobs, possibly to fit into their own accounting schemes. Government agencies dictate the breakdown of construction cost for some public utility companies to follow, suggesting account numbers to be used.

In any case, it is prudent for the contractor to employ the owner's designation as part of the coding of the individual work items. This avoids confusion and facilitates the owner's understanding of the accounts in the exchange of cost information and correspondence.

Typical of an owner's listing of work items is the following summary copied from the bid schedule included in the tender documents along with invitations to contractors to bid. In this instance, the items are designated by simple consecutive numerals, but they could be identified otherwise. The bid-item number becomes the basic code designation for the account to cover the work item. If the work items are numerous and there is a possibility of conflict with numbers given to recurring items, the letters W. I. prefixed to the owner's designation will usually suffice to avoid the conflict.

A breakdown of construction operations under the individual work item is customary. The breakdown is intended to correspond with the various steps of developing the item cost in the original bid estimate. These can be designated by suffixing to the basic code an alphabetical character. For instance, if the basic work item is, "Borrow Excavation," Item No. 4, as designated by the tender documents, a further breakdown could be shown as follows:

Item and
Code No.　　　　　　*Description*
　　4　　Borrow excavation
(or W.I. 4)
　　4 (a)　Clearing borrow pits
　　4 (b)　Stripping borrow pits
　　4 (c)　Prewetting borrow pits
　　4 (d)　Developing access for borrow pits
　　4 (e)　Load and haul materials from borrow pits
　　4 (f)　Road maintenance for borrow haul
　　4 (g)　Dressing exhausted borrow-pit areas

In addition to the above, the principal elements or subdivisions of cost should be further designated by still another suffix to cover each element where it applies. Thus, subitem No. 4(e) above might be divided into:

4 (e) 1—Labor and insurance
4 (e) 2—Permanent materials (not applicable in this instance)
4 (e) 3—Specific plant (not applicable in this instance)
4 (e) 4—Equipment rental
4 (e) 5—Supplies
4 (e) 6—Subcontracts

Typical Listing of Work Items and Accounts

*Item
and
Acct.
No.*   *Work-Item Account*

1. Diversion and care of river
2. Common excavation
3. Rock excavation
4. Borrow excavation
5. Stripping right abutment
6. Preliminary rock cleanup
7. Close drilling
8. Core drilling 3″ diam. holes
9. Diamond drilling 3″ diam. drain holes
10. Diamond drilling 2½″ diam. holes
11. Core drilling 10″ diam. holes in concrete
12. Percussion drilling
13. Setting pipes for grouting under embankment
14. Moving drill onto grout holes
15. Pressure testing
16. Furnish cement for grouting
17. Pressure grouting
18. Compacted fill (impervious)
19. Compacted fill (pervious)
20. Backing and filter materials
21. Dumped riprap
22. Spalls
23. Rock fill
24. Additional rolling for compaction
25. Concrete in stilling basin baffles
26. Concrete in roadway, sidewalks, and parapets
27. Concrete in thin slabs, beams, and columns
28. Concrete in spillway piers
29. Concrete in training walls
30. Mass concrete in dam
31. Concrete foundations in core trench
32. Portland cement

Typical Listing of Work Items and Accounts *(cont.)*

*Item
and
Acct.
No.*   *Work-Item Account*

16-10. Basic Arrangement of Coordination

The foregoing outline of how the engineering and accounting departments can coordinate cost controls is a matter of proved practice. However, neither this nor any other system can be made to work without the support and direction of management and its continual demand for excellence along these lines. Accurate control of cost information and its proper and effective use is one of the basic techniques for successful contracting.

Good relationships and cooperation between the chief engineering officer and the chief accounting officer to establish a definite routine is necessity. Such routine should be made a part of both the engineering and accounting manuals—if such are provided. In any case, the engineering and accounting personnel involved must understand their respective responsibilities to reach the common goal.

17 NETWORKING TECHNIQUES FOR PROJECT PLANNING, SCHEDULING, AND CONTROL

JOHN FONDAHL

Professor of Civil Engineering
Stanford University
Stanford, California

THE NETWORKING techniques have been available since the late 1950's. They appear to offer a tremendous potential for improving the planning and control of project-type work. They have been presented in countless management seminars and technical articles as well as many textbooks. They have been widely used in the construction industry and are frequently required by contract specifications. In spite of all this, it seems accurate to state that results have not lived up to expectations. Further improvements and more education are required for a truly effective level of application.

In the limited space of this chapter, the attempt will be made to describe the basic concepts and procedures of the networking techniques in plain terms. Their advantages and their shortcomings will be discussed. A number of improvements will be suggested and promising areas for further development will be indicated. The more sophisticated mathematical aspects and the elaborate details of many existing system packages will be omitted in favor of greater attention to fundamental matters.

Construction contracts require the performance of work having a definite starting point and end point; such work is referred to as a project. These projects are of sizable magnitude and duration; involve important commitments of men, material, equipment, and money; can be accomplished by numerous combinations of methods, sequencing, and intensity of resource application; and are obtained under competitive conditions that make sound decisions a necessity for profitable results. Under such conditions, project planning, scheduling, and control are of major importance.

In order to plan a project it is necessary to divide the job to be done into its component activities. Practically any construction project is sufficiently complex that the breakdown and various relationships to be considered should be recorded rather than carried in the memory of the planner and conveyed orally to the other parties involved. Therefore, some type of "paper model" of the project is usually developed in the formulation of a plan, to communicate results of the plan to others, and to serve as a basis for evaluating progress and controlling the work. During the past half century the bar (Gantt) chart has fulfilled this role.

The bar chart shows the project breakdown and the scheduling of each component activity plotted on a time scale. Using the scheduling information available from the bar chart and details concerning costs, crews, equipment, and material, that have been developed in the process of preparing the cost estimate for the project, other possible planning and control aids include:

1. Percentage Completion curve (S-curve) showing forecasted and actual cumulative percent completion plotted on a time scale and used primarily as a control tool.
2. Labor schedules developed for separate classifications of labor and showing the number of men required each time period.
3. Equipment schedules developed for separate types and sizes of equipment and showing the number of units required each time period.
4. Material schedules for important items of material or installed equipment indicating deadlines for such steps as preparation of shop drawings, approval of drawings or samples, beginning of fabrication, date of shipment, and date needed at job site.
5. Financial schedules indicating on a time scale the cumulative income received to date, the cumulative expenses to date, and the differences between the two to

show the amount of financing required during each time period.

In many cases only the bar chart has been used. If the other curves and schedules are also prepared, their reliability is dependent on scheduling data from the bar chart.

The bar chart has certain shortcomings that limit its usefulness for effective project planning and control. Since the networking techniques were created to overcome these weaknesses, it is worthwhile to discuss a few of the more more important ones. These include:

1. *Failure to require a detailed analysis and breakdown of activities.* The division of the job into its component parts can be very gross and the bars representing these components can be freely overlapped timewise. Although simplicity has its advantages, it permits a relatively sloppy preparation of a project plan which includes little information of value for control purpose. For example, the bar chart for a concrete dam project showed a single bar representing earth excavation that extended across the entire project duration. It was accurate since earthwork operations were in progress every month of the job span. However, it required little thought and provided minimum information. The networking techniques require a level of breakdown such that every activity must be completed before a following one commences. In those cases where an activity can start when another is partially complete, the earlier one must be broken down further. This requirement forces the planner to make a more detailed analysis which generally results in a better plan and schedule. In the case of the previous example, the earthwork activity would have had to be broken down into many additional activities representing excavation for various purposes such as abutments, Stage I dam foundation; Stage II, powerhouse substructure, tailrace, core trench, canal intake, etc. Considerably more thought would be required in order to do this and much more information for control purposes would be made available.
2. *Omission of many time-consuming activities.* Items such as the procurement of important materials, preparation, and approval of drawings and designs, and obtaining permits are seldom shown on a bar chart. The networking techniques require these and all other time-consuming activities to be normally included in the project breakdown, and estimates of their durations must be shown.
3. *Failure to communicate complete details of the project plan.* In addition to the fact that many activities are not shown or not broken down into sufficient detail, sequential relationships are not shown. Sometimes it may be deducted that one activity is dependent on performance of another one. However, that same activity may require the completion of several others which either are not shown or their relationships are not obvious. A person reviewing the project plan cannot tell whether the planner has taken into account all the restrictions and relationships that should have been considered. Networking techniques require that for each activity all of the necessary preceding activities be included on the plan and that the relationships be shown. The reviewer is able intelligently to judge the validity of the plan or detect omissions in it.
4. *Failure to indicate adequately the consequences of scheduling deviations.* The bar chart is frequently used as a control device where bars representing actual performance are plotted adjacent to the bars representing forecasted performance. When actual performance lags behind that forecasted, it becomes apparent that the activity involved is behind schedule. However, the effects of this slippage on project performance are not clearly indicated. They may range from a matter of no importance to one of vital importance. The networking techniques provide data for each activity that clearly indicate the effect of any slippage on project performance.
5. *Failure to provide a suitable project model for updating purposes.* When activity performance varies from its forecasted schedule, other activities "downstream" are frequently affected. It becomes necessary to update their scheduling if good control is to be maintained. The bar chart does not serve as a satisfactory model for such updating because it does not clearly indicate the relationships between activities. The network diagram serves as an ideal framework for project updating.

The major advantages of the bar chart relate to its simplicity. From the standpoint of the user, the bar chart is so simple that almost anyone can understand its meaning. Moreover, it is a well-established method of schedule presentation for supervision and field personnel at all levels. One possible solution to overcome the greater complexity of the network diagram plus separate tabulation of network scheduling data involves the use of a bar chart format for presentation to field personnel, as will be discussed later. From the standpoint of the planner, preparation of the bar chart is sufficiently simple that it will be made "in-house" and at low cost. There are considerable advantages in the preparation of a plan and schedule by a contractor's own staff. They are more likely to be realistic and more likely to be followed than if made by an outsider.

Unfortunately, network preparation and processesing have often been subcontracted to the lowest bidder who offers to satisfy contract scheduling specifications requirements. Under these conditions there is little possibility of effective application. One solution here is for better education of contractors' staffs in network planning and processing. Another related solution is the use of simpler networking techniques that do exist and that are within the in-house capabilities to perform. A third solution might consist of simpler specifications that do not require excessive data and that permit maximum freedom in the choice of networking methods. In the matter of cost, networking is more expensive since it requires more effort and involves more data. The extra cost will be justified only if the results do, in fact, make better planning, scheduling, and control possible and if they are applied.

17-1. Historical Background

A brief review of the historical development of the networking techniques will provide a better understanding of some of their characteristics.

Practical applications began in 1958 with the introduction of the Critical Path Method (CPM) and the Program Evaluation Research Task (PERT), later renamed Program Evaluation and Review Technique. CPM was the product of private enterprise, having been developed by a team organized by the Du Pont Company and concerned with improved planning of complicated industrial work. PERT was developed by a group under direction of the

Special Projects Office, U.S. Navy, to provide better control of the complex Polaris Missile Program.

Both CPM and PERT were introduced at approximately the same time and, despite their separate origins, they were very similar. Both required that a project be broken down into component activities which could be presented in the form of a network diagram showing their sequential relationships to one another. Both used the "arrow diagram" where network lines are arrows that represent activities and network nodes are events that represent the points in time at which activities may be commenced or are completed. Both required time estimates for each activity which were used in very routine-type calculations to determine project duration and scheduling data for each activity. These calculations were used to determine which activities must be kept on schedule if the calculated project duration was to be realized ("critical" activities) and which activities had extra time ("float" time in CPM and "slack" time in PERT) available for their performance.

Originally there were some differences between CPM and PERT. Although both used the same network diagram, CPM was an activity-oriented system and descriptive labels were applied to the arrows in the diagram. PERT was an event-oriented system and descriptive labels were applied to the nodes in the diagram. CPM was a deterministic system in which only a single-time estimate of each activity's duration was required. The type of work that PERT was developed to control involved considerable research and design efforts, and there was a reluctance to furnish single-time estimates. Therefore PERT allowed multiple-time estimates. Calculations involved a weighting of "optimistic," "pessimistic," and "most likely" estimates of the duration of each activity to obtain a single, expected duration. Relative uncertainty of performance times was reflected in statistical output that allowed prediction of the probability of project completion, or the achievement of any other intermediate project event, by a specified date.

The differences between CPM and PERT have largely disappeared over the years since their introduction. U.S. government agencies, for example, have seemed to favor the designation PERT but often proceed to describe a system that requires only single-time estimates for each activity and network diagrams having activity labeling. Therefore, it is now necessary to examine contract specifications or other detailed requirements rather than to depend on the title used to designate a networking system. In this chapter the more general phrase "networking techniques" has been used instead of CPM or PERT or other acronyms.

A third independent effort directed toward improved construction project planning and control was undertaken at Stanford University in 1958 under sponsorship of the U.S. Navy Bureau of Yards and Docks (now Facilities Engineering Command). Results of this research emphasized simple, noncomputer methods and a different type of network diagram. [1,2,3] Previous work at Stanford had been directed toward the application of industrial engineering techniques to construction. One such technique is the use of a flow diagram in which job operations are represented by circles, and lines are used to correct the circles and symbols representing other types of steps in job performance. When the 1958 research recognized the need for a project model, and having no knowledge of the arrow diagramming approach, it used what it called a *circle-and-connecting line diagram,* in which network nodes represented activities rather than events. Further work at Stanford included the development of two of the earliest computer programs, in 1962 and 1963, designed specifically for circle-and-connecting line diagramming. Later authors referred to this type of diagramming as *activity-on-node* diagramming,[4] simply *circle diagramming,*[5] or as *block* diagramming in Europe. When IBM introduced their Project Control System (PCS) in 1964, they incorporated the ability to accept data based on either arrow diagramming or *precedence* diagramming. This latter name has become a standard designation for this diagramming method, which is becoming increasingly popular and has important advantages over arrow diagramming.

17-2. Network Diagramming

The first step and the most important one in the application of networking techniques is the development of the network diagram. This diagram is a model of the plan of execution of the project. It shows all component activities required to accomplish the project and their sequential relationships. It is important because all the data calculated and all the reports produced, no matter how sophisticated a system or how large a computer may be used, are only as valid as this model on which they are based. This makes it essential that those who are most able and familiar with the work to be done participate actively in the development of the diagram. The mechanics of diagram construction can be very simple, but good judgment is required to obtain a model that is a reasonably realistic representation of the actual job. Any diagram is a compromise between complete accuracy with excessive detail and oversimplification with insufficient detail.

Some of the principal functions of the diagram were indicated at the beginning of this chapter in discussing the means of overcoming the weaknesses of the bar chart approach. They include: assisting the planners in keeping track of important relationships and forcing them to develop a detailed plan; conveying complete details of their planning to others who should review it; and serving as a working model for calculating scheduling data, other planning and control methods, updating, testing of alternative strategies, replanning and taking corrective action when needed.

The general procedure for diagram development starts with the selection of some component activity of the project and its representation on paper by an appropriate symbol. Next, the question is asked: "What activities must be completed before this activity can commence?" Symbols for these activities are placed on the drawing in such a manner that they will be recognized as directly preceding the original activity. The question is also asked: "What activities may commence when this activity is completed?" Also symbols for these activities are entered in such a manner that they will be recognized as directly following the original activity. This process is then repeated for each new activity entered on the diagram until the plan is complete. If activities are related and overlap one another, further breakdown is necessary. Precedence diagramming permits the overlapping of activities by the use of *lag factors,* as will be discussed later. In this case the processing method, usually a computer program, must take into account the overlaps even though the breakdown is not shown on the diagram. There is some difference of opinion concerning the best starting point for diagram development. The best advice is to roll out a sheet of paper and start somewhere; beginning, end, or middle of the project are all acceptable.

17-3.   Arrow Diagramming

In this method, the symbol used to represent an activity is an arrow. The length and direction of the arrow generally have no importance except that the flow of work on a network diagram is usually from left to right. Occasionally networks are drawn to a time scale, in which case the length of the arrow is significant. If one activity directly follows another, the arrow tail of the following activity should be located at the arrowhead of the preceding one. The terminals of each arrow become junctions, or nodes, in the diagram and represent *events*. An event is a point in time at which all activities whose arrowheads end at that node are completed and all activities whose arrow tails begin at that node may be started.

Figure 17-3-1a shows a single activity represented by an arrow. A descriptive label of the activity has been entered along the arrow. This is the more common labeling system, and the resulting diagram is characteristic of the CPM arrow diagram, also referred to as an *activity-on-arrow* diagram. An alternate system is shown in Figure 17-3-1b

where the terminal points receive the descriptive labeling. The resulting diagram is characteristic of the original PERT arrow diagram, also referred to as an *event-on-node* diagram. It is not commonly used for construction planning and even where specifications refer to PERT networks they now usually expect, or allow, activity labeling. However, in some cases it is desirable to show and label the very important, or "milestone," events even where most of the diagram is activity-labeled.

Figure 17-3-1c shows the beginning of network development following the question: "What activities must be completed before forms can be erected for foundation B." For computer processing the nodes are given labels, usually numeric. The beginning node of an activity is referred to as the *i*-node and the end node as the *j*-node. In *i-j* notation, the activity of Figure 17-3-1a or 17-3-1b is activity 15-16. Since a computer can compare numbers and make decisions on the basis of whether they are the same or not, this labeling allows the computer to determine sequencing relationships. For example, if the computer was required to determine those activities directly preceding activity 15-

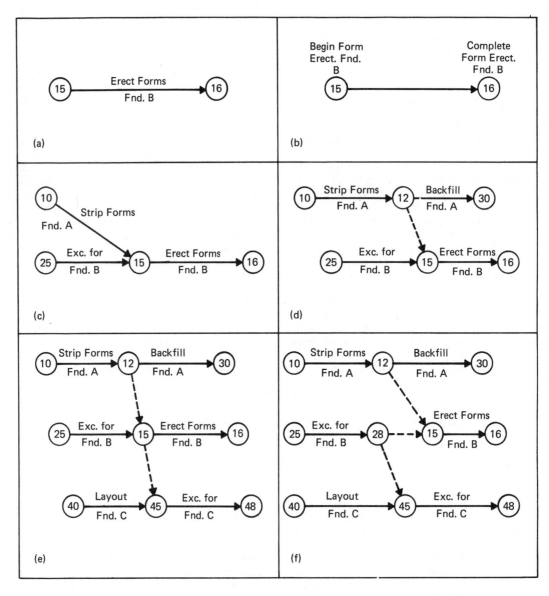

Figure 17-3-1 Arrow diagram development and use of dummy activities

16, it would simply match its *i*-label, 15, with the *j*-labels of all other activities in the network and select activities 10-15 and 25-15 as predecessors. Early computer programs required all *j*-labels of activities to be greater than their *i*-label. Now most programs permit random numbering of nodes and some allow the use of alphabetic and other characters as well.

An artificial device known as a *dummy* activity is necessary in arrow diagramming. It is an activity that has no physical significance and that is given a zero-time duration, but which is essential for showing correct relationships between activities. For example, in reference to Figure 17-3-1c, assume that the planner recognized that after the forms were stripped from foundation A, backfill may be placed there. If the arrow tail for this new activity is placed at node 15, it also indicates that backfill for foundation A cannot be commenced until excavation for foundation B is completed. Assuming this restriction is not valid, the diagramming solution shown in Figure 17-3-1d, whch introduces a dummy activity 12-15, allows the correct relationships to be presented. There are other types of situations in which dummies may be used, but the matter of developing the correct logic in the diagram is the most important application.

Considerable care is required in the use of dummies in order to avoid mistakes that sometimes are difficult to recognize. Suppose that in further development of the diagram of Figure 17-3-1d it is noted that excavation for foundation C may be started after excavation for foundation B is completed and also after layout work for foundation C is completed. This requires another dummy activity, since the head of the arrow for the layout activity cannot properly be terminated at node 15.

Figure 17-3-1e shows a solution that appears to satisfy the sequencing restrictions but is incorrect. The mistake results from the use of two dummies in series which establish an unintentional relationship. As shown, excavation for foundation C requires the occurrence of event 45, which requires the occurrence of event 15, which requires

the occurrence of event 12, which requires the completion of stripping forms for foundation A.

When time estimates are made and scheduling computation made, as discussed later, the beginning date for excavation of foundation C may be controlled by the completion date for stripping forms of foundation A. Actually there should be no relationship between these two activities. However, the computer has no way to recognize such a mistake and, even if it were to cause incorrect scheduling of many following activities as well as causing the calculated project duration to be wrong, no error messages would be printed by the computer. A correct solution, utilizing still another dummy, is shown in Figure 17-3-1f.

17-4. Precedence Diagramming

In this method the nodes of the network represent the activities, and the lines between nodes show sequential relationships. If it is desired to show important events on the diagram, they may also be represented as nodes by considering them to be activities of zero-time duration. A different node symbol may be used to distinguish events from activities, such as a square instead of a circle. Arrowheads should be placed on sequence lines in preliminary rough diagrams where occasionally preceding activities may be plotted to the right of following ones. They are often omitted in diagrams where the flow of work is consistently from left to right.

A precedence diagram of the network segment shown in the arrow diagram of Figure 17-3-1c would appear as shown in Figure 17-4-1a. The addition of the new activities discussed with relation to Figures 17-3-1d and 17-3-1f would result in the precedence diagramming segment shown in Figure 17-4-1b. No dummy activities are required. In a sense, each sequence line is a dummy activity; however, they are considered simply as indicating sequential relationships and no special thought is required to determine when or how to use them properly.

The use of *lag factors* (sometimes called lead-time fac-

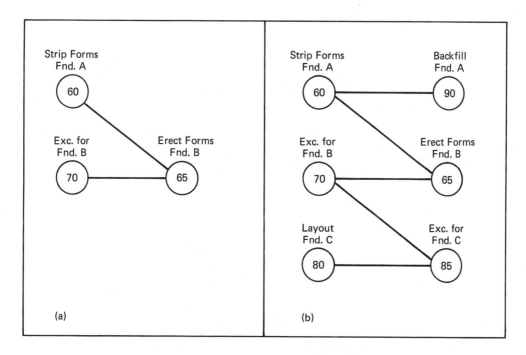

Figure 17-4-1 Procedure diagram development

tors) for overlapping activity duration has become a popular feature generally associated with precedence diagramming. Since each sequential relationship between activities is shown by a separate line, it is simple to indicate a lag factor adjacent to a sequence line on the diagram and to take this factor into account during processing. This reduces the amount of activity breakdown that is required and simplifies the diagram. The most common lag relationship is a start-to-start factor, indicating that the later activity must start at least a certain time period after the start of the earlier activiy. Some procedures and computer programs also allow start-to-finish, finish-to-start, and finish-to-finish lag relationships. The lag period is usually expressed in time periods, but may alternatively be expressed in percentage completion in some procedures.

Careless application of lag factors can be the source of serious errors. However, skillfully used with proper calculation procedures they can be very effective. Some types of construction, such as pipeline work and roadway construction, involve series of overlapping activities that are very cumbersome to break down to the extent ordinarily demanded by the network approach. By simultaneously using both start-to-start and finish-to-finish lag factors between successive activities in the series and using variable activity durations subject to minimum limits for the intermediate activities in the series, it may be possible to simplify the diagramming of such work. The calculation procedure, or computer program, must be properly designed for this. In any case, where lag factors are used in connection with computer processing, it is essential for the planner to understand the rules built into the computer program.

17-5. Comparison of Arrow and Precedence Diagramming

The decision concerning which diagramming system to use is of extreme importance and may determine the success or failure of networking applications. Major points of comparison between the systems include the following:

1. Simplicity This is the most important point of comparison since it has the greatest effect on the success of the system chosen. As already stated, all data calculated and reports produced will be valid only to the extent that the diagram represents a realistic plan that will actually be followed. Two skills are necessary to prepare the diagram. One involves a thorough knowledge of how to accomplish the job, how it may be broken down into appropriate components, and what the relationships between these components are. This is the more important skill and preferably should be furnished by a team of key personnel, including the estimator, the project manager, and representatives of field supervision and speciality contractors. The second skill is the ability to follow networking rules and to show the activities and relationships properly on paper in the form of a diagram.

Using precedence diagramming, the skill required for diagram construction is minor, and there is really no obstacle to the project team developing the diagram themselves. Activity nodes are placed wherever convenient and logical on the paper, labeled, and sequence lines drawn to nodes representing preceding or following activities. The principal rule that the team must understand (regardless of which type of diagramming is used) is that there should be sufficient breakdown of activities to satisfy the requirement that no activity starts until the activities preceding it are completed. The mechanics of precedence diagram construction give little opportunity for mistake, other than pure carelessness in drawing sequence lines between the intended preceding and following activities. One exception would occur if lag factors are used, because some expertness is necessary at least in reviewing the correctness of the final product.

Using arrow diagramming the degree of skill demanded for diagram construction is greater and the probability of error, generally in the proper use of dummy activities, is higher. Diagramming construction takes longer and is more frustrating, due to the requirement to locate arrowheads and tails at common nodes when there are multiple preceding or following activities and to erase and relocate activities when it becomes necessary to insert dummies in the course of diagram development. These problems reduce the effectiveness of the planning team.

All too frequently the task of diagram construction is turned over to the outside expert, and the network plan is no longer the product of the job team. This can be fatal to the success of the networking technique. If the job team develops the network themselves, agree that it is workable, and preferably put their names directly on the drawing, their objective is to prove that it will work. If the network is prepared by an outsider, job-level management may not consider itself responsible for the plan and may adopt the objective of proving that it will not work.

For truly effective use of networking techniques, diagram sketching should be a commonplace method of analysis. Subnetworks, as discussed later, should be developed, and individual activities can often be planned more effectly by expanding them into small, detailed networks. The simpler precedence diagramming approach encourages such practices.

2. Revisions Revisions to a precedence diagram are simpler. Consider the case shown in Figure 17-3-1d. When the backfill activity was added, the form stripping activity had to be erased and relocated before the necessary dummy arrow could be inserted. In a complete diagram, other changes might be required at the arrowhead of the backfill activity. Any change involving the use of dummies introduces the hazard of the type of error discussed in connection with Figure 17-3-1e and so requires some special degree of diagramming skill.

In the case of the corresponding precedence diagram of Figure 17-4-1, the change was accomplished much easier. A convenient blank space was found for the new activity node and label, and sequence lines were drawn to preceding and following activities. No relocation of existing activities was required. No special skill in handling dummies is involved.

Ease of revision is highly important in encouraging network updating by job personnel.

3. Numbering System With greater attention being devoted to integrated management systems, there are more opportunities for using the activity breakdown and numbering scheme for additional data purposes. The dual numbering system of the arrow notation is very awkward for such purposes, especially when the number of digits or characters needed may be in the range of 6 to 12. Even more important, however, is the matter that revisions in arrow diagrams often require activity numbers to be changed. For example, consider Figures 17-3-1c and 17-3-1d. With the insertion of the new activity it was necessary to change the designation of the strip forms activity from 10–15 to 10–12.

Precedence diagramming allows a single label and a permanent one. When the same revision was incorporated in the precedence diagram of Figure 17-4-1b, for example, their was no need to change the label of the strip forms activity.

4. Computer Applications   An early objection to precedence diagramming was that no computer programs were available for processing them. This is no longer the case.

Programs based on precedence diagramming can be at least as efficient in computer running time and probably more so. This is somewhat academic, since the time required for basic calculations is usually negligible and cost differences would be insignificant. However, charges for computer processing of networks are often based on the number of activities processed. Fewer activities are involved in precedence networks for two reasons. Dummy activities are not required and the lag factor feature of precedence diagramming programs may eliminate additional activity breakdown.

Sometimes a program designed for arrow diagramming is used to process a precedence diagram. This requires treating every sequence line as a dummy activity. If charges are based on an activity count, they will of course be more costly for the precedence diagram. However, the reason is that the program was not designed for the job it is doing.

5. Lag Factors   The use of lag factors can reduce the activity breakdown required, thereby simplifying diagrams and possibly reducing processing costs. Precedence diagramming lends itself to the use of lag factors since there is a separate sequence line between every pair of preceding and following activities.

6. Representation of Events   It is sometimes desirable or required by specifications to represent an important event on the diagram and to print out data concerning it. The nodes on an arrow diagram represent most events. However, to obtain printout information (in the more common activity-oriented system), an additional arrow and descriptive label is usually required as input data. The same is true for a precedence diagram. Any event can be represented in the same manner as an activity, giving it a zero-time duration. Therefore either system is capable of representing events.

7. Time Sealing   A network diagram may be plotted to a time scale. This provides representation of the calendar scheduling of each activity and clearly indicates those concurrent activities in progress on any given date. However, it requires considerable effort and cost to develop and then has a short period of validity. As activities deviate from their expected schedule, which probably will occur frequently, the scheduling of other activities is changed and the time-scaled diagram needs to be plotted again.

Since an arrow has length, it can be plotted in such a manner that its horizontal projection shows its duration and scheduling dates on a horizontal time scale. Time scaling of a precedence diagram is not quite so obvious, since activities are plotted as nodes, which are really points having no dimensions. However, if each activity node is plotted above its start date on the time scale, the result is essentially the same as a time-scaled arrow diagram. Either type of diagram can provide the same information and can be read with equal ease.

8. Standard Practice   The arrow diagram was used by the developers of both CPM and PERT. Several consulting firms that first began to provide workshops and services in these methods did an efficient job of education in the use of arrow diagramming. U.S. government agencies, which usually establish a pattern for other public agencies, have generally encouraged the use of arrow diagramming. Contract specifications often require furnishing of an *arrow diagram* rather than using the more general term, *network diagram*. The Associated General Contractors of America in their own publication[6] on CPM only describe arrow diagramming.

On the other hand, precedence diagramming has been gaining in popularity for several years[7] as project management discovers its advantages.

It already prevails in some other countries which have been able to survey American practice and select its best features or who have developed their own improvements. Many progressive U.S. contractors prefer and use precedence diagramming for all work where they are allowed freedom of choice. A strong case can be made for the superiority of precedence diagramming[8] that should eventually change the current standard practice in the United States favoring arrow diagramming.

17-6. Diagramming Trends

Diagramming practice offers one of the greatest challenges for increasing the effectiveness of networking techniques. Current practice is relatively satisfactory for the orignal planning effort. The mere act of developing the complete diagram, even if it were not put to further use, is valuable in assisting and forcing the planning team to do a very thorough job. Current diagramming practice is also relatively satisfactory to the owner who often specifies it. He has never before received such a complete presentation of the contractor's plan for his purposes of evaluation, control, and future claims negotiations. However, current practice is far from satisfactory in realizing the dynamic potentialities of network applications. Once developed, the network diagram is too big, takes too much effort, and costs too much to change freely. Although it is a suitable model for updating scheduling changes (unless it is time scaled), it discourages updating, experimentation, and replanning that involve other than trivial logic changes.

Several areas of improvement have already developed. One of the most promising is the use of subnetworks and the related abilities to concentrate on small portions of the project plan and to greatly condense the master network. This topic will be discussed later. Another area involves the construction of diagrams by computer controlled plotting equipment. A number of programs are available. Some plot all activities in rows of horizontal chains above a time scale. Where activities have more than one following activity, vertical connections between duplicate nodes at different levels may have to be supplied by the user. Other programs plot diagonal lines, where necessary, between nodes and may or may not be time scaled. To date these programs have not been widely used. Good diagram construction is still more of an art than a science. It is not difficult to develop a plotting program that succeeds in showing all activities of a project plan and their correct relationships. But for effective presentation, activities should be arranged in logical groupings and in such a manner as to minimize the confusion of many crossing lines at poor intersection angles. A program capable of substituting for good judgment and being economically attractive may yet be developed because the need exists.

Other possibilities are emerging as new hardware is developed and brought within reasonable cost. With a remote terminal, timesharing installation allowing interactive communication between the user in the field office and a large computer center, there are many opportunities for new systems. It is conceivable that network relationships might be given directly to the computer and become available to the user in response to specific queries at the office console. It is possible that visual displays, such as the cathode-ray tube, will allow temporary presentation of desired diagram segments for observation by the planner as needed. Many other conjectures could be made, but it can be stated with certainty that the development and use of the model of the project plan still offer opportunities for considerable improvement.

17-7.   Scheduling Calculations

After a satisfactory diagram of the project plan has been completed, the next major step is to supply activity time estimates and calculate scheduling data. As in the case of diagram construction, these time estimates affect all calculated data and the validity of all reports produced. Therefore it is essential that they be made by able personnel thoroughly acquainted with the work to be done.

Time estimates may be made in terms of any consistent and appropriate time units, such as weeks, calendar days, working days, hours, or minutes. Working days are most common for construction planning, although estimates in weeks and tenths of weeks are used in some systems.

There is no unique time estimate for the performance of most construction operations. Depending on method, crew size, overtime work, and other variables, an activity may be accomplished in different lengths of time. A choice is made of the mode of performance that appears to be a reasonable compromise between speed and cost for the overall conditions of the project involved. A single-time estimate is developed. (Even in the multiple-time-estimate PERT approach, all the time estimates for an activity are based on a single mode of performance and then are weighed by formula to furnish a single-time estimate.) If the resulting project duration is not satisfactory, replanning may include changes in activity durations. A formal technique for achieving the best time-cost balance for each activity will be discussed later under the topic of "time-cost trade-offs."

Six items of scheduling data are generally calculated for each activity. These include earliest start time, earliest finish time, latest start time, latest finish time, total float, and free float. Sometimes the term "slack" is used instead of "float." Once the diagram has been constructed and time estimates have been made, the scheduling calculations are routine and simple. The mathematics required involves adding or subtracting two numbers at a time. The procedures will be described utilizing the simple precedence diagram of Figure 17-8-1.

17-8.   Forward Pass Calculations

Starting at the beginning of the network, the earliest start and finish time for every activity is calculated. The calculations are based on the activity durations and the requirement that each activity may commence only after those preceding it have been completed. The earliest finish time

calculated for the final project activity indicates the project duration.

Referring to Figure 17-8-1a and starting with activity A, the earliest start time for that activity would be the beginning of day 1. (An alternate procedure starts with zero and interprets each number to represent the end of the corresponding day instead of its beginning.) Since the duration of activity A is five working days, its earliest finish time is $5 + 1 = 6$, the beginning of the sixth working day, or the end of the fifth working day. Since both activities B and C may start when activity A is complete, their earliest start times are the beginning of day six. Calculations are continued according to this procedure until values for all activities have been determined. Note that in cases where an activity, such as activities E or F, is preceded by more than one activity, the earliest start time is controlled by that preceding activity with the later completion date.

The calculations provide the earliest start and finish dates for each activity and the project duration. In the example, the project duration is 40 working days, since the final activity may be completed at the beginning of the 41st day which is the same as the end of the 40th day.

17-9.   Backward Pass Calculations

In Figure 17-8-1b, starting at the end of the project, the latest finish time for the final activity F is set equal to the project duration that was determined to be possible by the forward pass calculations. The activity duration is then subtracted to obtain the latest start time. Since preceding activities must be completed before following ones can start, the latest finish time for both activities D and E are set equal to the latest start time of activity F. The remaining latest finish and start times are calculated in a similar manner proceeding toward the beginning of the network diagram. They represent the latest possible start and finish dates for each activity that would not change the previously calculated project duration. Note that in the case of activities having more than one following activity, such as activities A and B, the latest finish time is controlled by the following activity having the earliest latest start date.

17-10.   Total Float and Critical Path Calculations

The difference between the latest finish time and the earliest finish time (or between latest start time and earliest start time) of an activity indicates the extra time available for scheduling or performing that activity. It is called *total float*. Figure 17-8-1c shows the calculations for the total float of each activity.

If the total float is zero, the activity must be held on schedule, since no extra time is available for its performance. Such activities are termed *critical activities*. There will exist at least one chain of critical activities from the beginning to the end of the network diagram. In the example it consists of activities A-B-D-F. Each such chain is referred to as a *critical path*. The sum of the activity durations along a critical path is equal to project duration. If any critical activity is delayed, project completion will be delayed by the same amount (unless the performance of other critical activities is accelerated or other changes in the project plan are made).

17-11.   Free Float and Interfering Float Calculations

Total float indicates how long the earliest finish time of an activity may be delayed before project duration is affected.

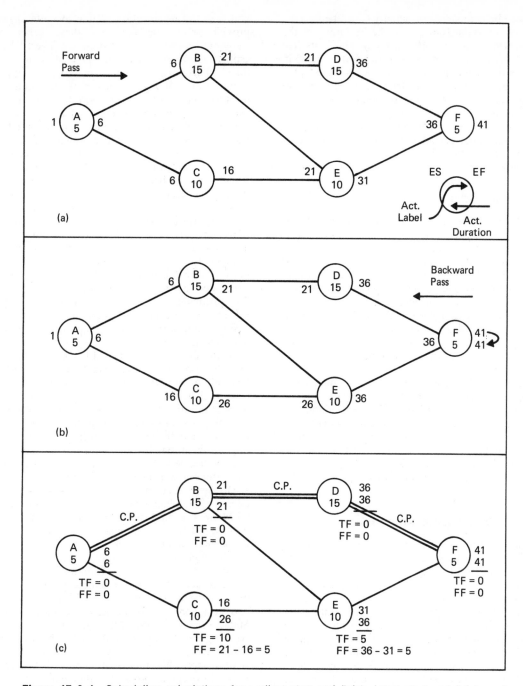

Figure 17–8–1 Scheduling calculations for earliest start and finish, latest start and finish, and total and free float times for activities in a precedence network

Another useful and more restrictive measure of extra time is termed *free float*. It indicates the length of time that the earliest finish date of an activity may be delayed without delaying any other activities as well as not delaying project completion. It therefore cannot exceed total float and is generally less.

The free float of an activity may be calculated by subtracting its earliest finish time from the earliest start time of its following activity. If there are more than one following activities, the one whose earliest start date occurs first is the controlling one. In Figure 17-8-1c, the free floats of the critical activities were set equal to zero simply by recalling that free float cannot exceed total float. The free float

of activities C and E were calculated in accordance with the rule given.

Another quantity called *interfering float* is not usually tabulated in scheduling reports, but it offers a useful concept for control purposes and is often shown in a modified form of bar chart. The interfering float of an activity is the difference between its total float and its free float.

Activity C in Figure 17-8-1 offers a good example for summarizing the meanings of the three types of float that have been defined. The completion of activity C can be delayed from the beginning of the 16th day to the beginning of the 26th day without delaying project completion. Its total float is therefore 10 days. This total float is composed of two periods. The completion of activity C can be delayed from the beginning of the 16th day to the beginning of the 21st day without delaying any other activity. This is its free float period of five days. If activity C is delayed to finish in the period after the beginning of the 21st day and up to the beginning of the 26th day, it will delay other activities, in this case only activity E, but will not delay project completion. This is its interfering float period of five days. Note that any delay of an activity into the interfering float portion of its total float period not only uses up its own float time but uses up the float time of at least one, and often many, other activities.

17-12. Calendar Dating

Having calculated earliest and latest start and finish times in numerical terms (e.g., the beginning of the 32nd day), it is generally preferable to convert these numbers to corresponding calander dates. It is a simple matter to construct a conversion table. Working day numbers can be listed. Then starting with the calendar date of project commencement opposite working day number 1, corresponding calendar dates can be listed opposite each working day number, omitting nonwork dates, such as Saturdays, Sundays, and holidays. Using this conversion table, calendar dates can be substituted for each of the numbers calculated in the forward and backward pass procedures.

When using computer processing, a deck of data cards may serve as the conversion table. It is possible to maintain several such decks, each for use under different conditions. For example, separate decks would represent five-day work weeks, six-day work weeks, working days for locations having different holidays, etc. By using the proper deck as part of the input data and specifying the date of project beginning, all reports can be printed directly in terms of calendar-dated results.

17-13. PERT Probabilistic Approach

Historically the PERT approach was based on multiple-time estimates for each activity. As already noted, systems now referred to as PERT often require only single-time estimates. However, for purposes of this section, the multiple-time-estimate procedure will be referred to as the PERT method and the single-time-estimate procedure as the CPM method. Only a brief, summary discussion of a few aspects of the probabilistic approach will be given here.

The PERT approach has seldom been popular in construction because the data requirements are considered excessive and the advantages appear marginal or nonexistent. Nonetheless, there are situations where estimators are unwilling to provide single-time estimates as a basis for scheduling computations. One example occurs when networking techniques are applied to the engineering phases of a project as well as the construction phases. Engineers and architects are probably justifiably reluctant to state a single-time estimate for their design work. Another case may occur in negotiated construction work involving poorly defined end products. Turnkey type contractors are often faced with both of these situations. In such cases, the opportunity to give a range of activity time estimates may offer the only means of obtaining data that will permit development of a schedule.

The conventional PERT approach required three time estimates for each activity, all based on the same method and manner of performance. They included an estimate of the most probable duration, the most optimistic duration, and the most pessimistic duration. An attempt was made to provide precise definitions of these terms. A simplified weighing formula was then used to convert these three time estimates into a single one referred to as the "expected time." This formula, based on statistical approximations, required multiplying the most probable estimate by 4, adding the results to the other two estimates, and dividing the sum by 6 to obtain the expected activity duration. Then conventional scheduling computations, similar to those already discussed, could be performed.

The multiple-time estimates for each activity define a probability distribution curve for its duration. This curve is usually skewed because the pessimistic estimate at one extreme is generally further removed from the most probable estimate than the optimistic estimate at the other extreme. This causes the "expected time" used in scheduling calculations generally to be greater than the most probable activity duration. If it is assumed that an estimator's single-time estimate for CPM calculations would be the same as his most probable estimate for PERT calculations, it can be concluded that the CPM method will result in project duration estimates that are less than obtained by the PERT method. Some argue that when an estimator provides only a single-time estimate, he will protect himself by making it on the high side.

It can also be demonstrated by statistical approximations that the project duration calculated by the PERT method has only a 50 percent chance of being satisfied. It is possible to calculate the probability of satisfying any other project duration. If a project duration were obtained by summing the most probable durations of the activities on the critical path, instead of using the calculated expected times, the probability of completion within this length of time would be well below 50 percent and might commonly be in the range of one chance in four to one chance in ten.

Do these concepts contain a message of interest to the construction planner? Assume that a contractor uses the CPM approach. Assume that his estimates are made by typical, confident, optimistic construction personnel who make no attempt to "protect" themselves by estimating on the high side but, rather, develop the most probable estimates for each activity duration. Statistically speaking, the contractor has only a remote probability of achieving his forecasted completion date, at least using the methods and costs on which his personnel based their estimates. Stated somewhat differently, the common approach to networking results in schedules with built-in incentives of which the contractor may be unaware, but which may require more than the usual amount of "blood, sweat, and tears" in order to prove that they actually can be met.

One major turnkey contractor has applied a different statistical approach to project duration estimates. This method uses the network diagram as a simulation model. Multiple-time estimates are made for each activity and define a probability distribution curve for the duration of that activity. A computer program, using random number selection methods, determines a duration for each activity and uses them in forward-pass calculations to determine a project duration. This procedure is repeated a number of times, say 1000, and results in a probability distribution for project duration. The procedure overcomes one of the weaknesses of the PERT approach because it recognizes that different paths of activities are capable of determining project duration, and it allows this to happen in the various calculation iterations. Results provide the expected project duration or the probability of achieving other completion dates. The principal disadvantage, as compared to the

PERT method, is that it is less efficient computationally and, therefore, more expensive to apply.

17-14. Communication and Control

Having developed a project plan and a schedule that are acceptable, the next step is to communicate this information to those parties who need it for job control or who have some other valid interest in project performance. The most common means of presentation is to show the project plan in the form of a network diagram and to show all scheduling data in a separate tabulation, often a computer printout. Figure 17-14-1 shows a small, simplified precedence diagram for a pier project. Figure 17-14-2 shows the master schedule for the same project. The advantages of this approach are that it simplifies updating and permits effective use of computer processing. Most updating involves scheduling changes. By keeping the diagram free of scheduling information, it requires updating less frequently. It is relatively simple to update scheduling information separately, especially by computer, but diagram revision generally requires more effort and expense. Of course, the advantage is lost if the diagram is time scaled. One compromise is to time scale the diagram initially, thereby locating activity symbols in correct positions relative to the original scheduling relationships, but to eliminate the time scale when schedule updating becomes necessary.

A second means of presentation, shown in Figure 17-4-3 is to show all information directly on the network diagram. The forward and backward pass calculations have been made on the diagram and results indicated above and below date boxes on each activity node. Then a calendar conversion table is constructed, and the numbers are converted to dates for insertion inside the boxes. The conversion table might be included on the drawing. The advantage of this approach is that all planning and scheduling data and results are available on a single sheet. Earliest and latest start and finish dates are shown for each activity. Total and free float for any activity can easily be obtained by subtracting the appropriate numbers given adjacent to date boxes. The critical path can be colored or shown in some other distinguishing manner. The disadvantage of the approach is that any planning or scheduling variations may affect a number of activities, requiring extensive revision of the diagram.

A third means of presentation is to show all information in the familiar bar chart form. An ordinary bar chart is constructed first, listing all activity descriptions and labels and plotting each bar length, equal to activity duration, extending from the earliest start date to the earliest finish date on the time scale. Then an extension is added to the bar of each noncritical activity to show the total float period. The end of the extended bar represents the latest finish date. If desired, a mark may be added along the

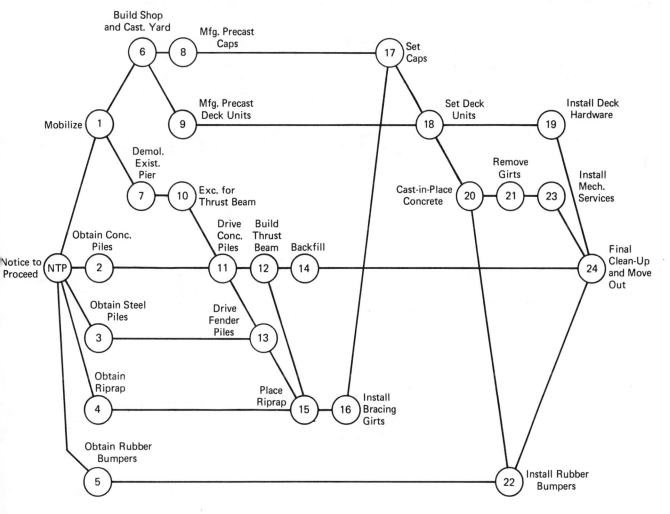

Figure 17-14-1 Precedence diagram for a pier construction project

STANFORD CONSTRUCTION INSTITUTE   6/21/71

SAMPLE PIER CONSTRUCTION PROJECT   THIS REPORT IS SORTED ACCORDING TO
                                                                        EARLY START ORDER

ACTIVITY LABEL	DUR.	CR	ACTIVITY DESCRIPTION	EARLIEST START	EARLIEST FINISH	LATEST START	LATEST FINISH	LATEST START	LATEST FINISH	FREE FLOAT	TOTAL FLOAT
NTP	0		*NOTICE TO PROCEED	JUN 28/71	JUN 28/71	JUN 28/71	JUN 28/71	JUN 28/71	JUN 28/71	0	0
1	10		*MOBILIZE	JUN 28/71	JUL 12/71	JUN 28/71	JUL 12/71	JUN 28/71	JUL 12/71	0	0
2	20		OBTAIN CONCRETE PILES	JUN 28/71	JUL 26/71	JUL 6/71	AUG 3/71	JUL 6/71	AUG 3/71	6	6
3	10		OBTAIN STEEL PILES	JUN 28/71	JUL 12/71	AUG 17/71	AUG 31/71	SEP 1/71	SEP 17/71	36	47
4	20		OBTAIN RIPRAP	JUN 28/71	JUL 26/71	AUG 23/71	SEP 22/71	AUG 23/71	SEP 22/71	40	40
5	60		OBTAIN RUBBER BUMPERS	JUN 28/71	SEP 22/71	OCT 19/71	JAN 17/72	OCT 19/71	JAN 17/72	79	79
6	3		BUILD SHOP AND CASTING YARD	JUL 12/71	JUL 15/71	JUL 12/71	JUL 15/71	SEP 20/71	SEP 23/71	0	48
7	15		*DEMOLISH EXISTING PIER	JUL 12/71	AUG 2/71	JUL 12/71	AUG 2/71	JUL 12/71	AUG 2/71	0	0
8	32		MANUFACTURE PRECAST CAPS	JUL 15/71	AUG 30/71	SEP 23/71	NOV 8/71	SEP 23/71	NOV 8/71	48	48
9	32		MANUFACTURE PRECAST DECK UNITS	JUL 15/71	AUG 30/71	OCT 7/71	NOV 22/71	OCT 7/71	NOV 22/71	58	58
10	1		*EXCAVATE FOR THRUST BEAM	AUG 2/71	AUG 3/71	AUG 2/71	AUG 3/71	AUG 2/71	AUG 3/71	0	0
11	20		*DRIVE CONCRETE PILES	AUG 3/71	AUG 31/71	AUG 3/71	AUG 31/71	AUG 3/71	AUG 31/71	0	0
12	14		*BUILD THRUST BEAM	AUG 31/71	SEP 22/71	AUG 31/71	SEP 22/71	AUG 31/71	SEP 22/71	0	0
13	3		DRIVE FENDER PILES	AUG 31/71	SEP 3/71	SEP 17/71	SEP 22/71	SEP 17/71	SEP 22/71	11	11
14	1		BACKFILL	SEP 22/71	SEP 23/71	FEB 4/72	FEB 7/72	FEB 4/72	FEB 7/72	93	93
15	10		*PLACE RIPRAP	SEP 22/71	OCT 6/71	SEP 22/71	OCT 6/71	SEP 22/71	OCT 6/71	0	0
16	23		*INSTALL BRACING GIRTS	OCT 6/71	NOV 8/71	OCT 6/71	NOV 8/71	OCT 6/71	NOV 8/71	0	0
17	10		*SET CAPS	NOV 8/71	NOV 22/71	NOV 8/71	NOV 22/71	NOV 8/71	NOV 22/71	0	0
18	16		*SET DECK UNITS	NOV 22/71	DEC 16/71	NOV 22/71	DEC 16/71	NOV 22/71	DEC 16/71	0	0
19	3		INSTALL DECK HARDWARE	DEC 16/71	DEC 21/71	FEB 2/72	FEB 7/72	FEB 2/72	FEB 7/72	32	32
20	20		*CAST-IN-PLACE CONCRETE	DEC 16/71	JAN 17/72	DEC 16/71	JAN 17/72	DEC 16/71	JAN 17/72	0	0
21	7		REMOVE GIRTS	JAN 17/72	JAN 26/72	JAN 17/72	JAN 26/72	JAN 20/72	JAN 31/72	0	3
22	15		*INSTALL RUBBER BUMPERS	JAN 17/72	FEB 7/72	JAN 17/72	FEB 7/72	JAN 17/72	FEB 7/72	0	0
23	5		INSTALL MECHANICAL SERVICES	JAN 26/72	FEB 2/72	JAN 31/72	FEB 7/72	JAN 31/72	FEB 7/72	3	3
24	2		*FINAL CLEAN-UP AND MOVE OUT	FEB 7/72	FEB 9/72	FEB 7/72	FEB 9/72	FEB 7/72	FEB 9/72	0	0

PROJECT DURATION IS 156 WORKING DAYS

Figure 17–14–2 A computer printout sheet showing calendar-dated master schedule. Activities on this report are listed chronologically in terms of early start dates. Additional listing can be requested according to alternative sorting priorities

bar to indicate the latest start date. Next, the total float bar may be subdivided to show the free float period and the interfering float period. The bar diagram now contains all the scheduling information including earliest and latest start and finish dates as well as total, free, and interfering float periods for all activities. If desired, sequential information may also be included. Labels of preceding activities may be shown to the left of each bar and labels of following activities to the right.

For use as a progress reporting and control device,

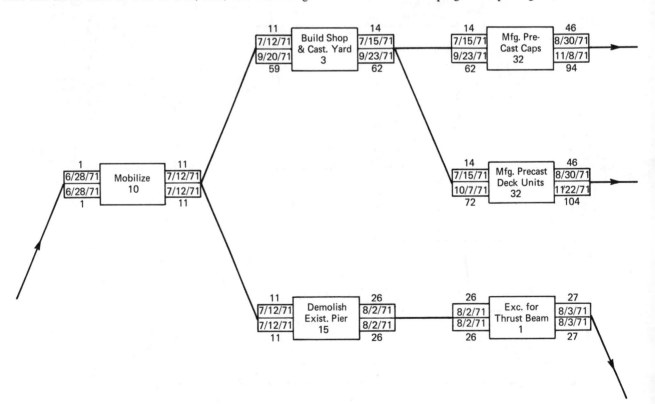

Figure 17–14–3 Partial precedence diagram including scheduling data

space can be reserved below each bar for plotting actual performance times. Then, for example, if completion of an activity is delayed into its interfering float period, it is possible to observe the labels of the following activities, locate these activity bars further down the diagram, determine which activities are affected and the seriousness of the delay, and continue tracing the effects down the chart if necessary.

A partial bar chart for the small pier project is shown in Figure 17-14-4. The principal advantage of this approach is that it allows networking information to be conveyed in a form that is relatively familiar and easy to interpret by project personnel who are already acquainted with the bar chart form of presentation. It has the disadvantage cited for the preceding method that any planning or scheduling variations may require extensive modifications of the chart. However, bar charts of this type can be produced relatively simply by computer processing, so this disadvantage is less serious.

The application and advantages of network scheduling data for project control purposes should be quite apparent. All critical activities are identified and can receive appropriate attention. Near-critical activities are also identified and can receive similar attention. For other activities, the amount of extra time available is quantitatively known. When scheduling variations occur for any activity, their

significance can be clearly determined. If a critical activity is delayed or a noncritical activity is delayed beyond its latest finish date, project duration has been extended by the same amount of time. In the latter case, a different critical path has been formed and may be determined. If a noncritical activity is delayed but is completed during its free float period, the matter is of no consequence since neither project duration nor any other activity is affected. If a noncritical activity is delayed but is completed during its interfering float period, the matter may or may not be of importance. It requires further investigation since it is known that one or more other activities have been delayed.

Determination of those activities delayed allows prompt notification to be given or action to be taken and consequences to be evaluated. Although, in this case, all activities affected have float time, from a practical viewpoint the delays sometimes can be as serious as the delay of a critical activity. For example, an activity to be performed by a subcontractor has five days of float time and is delayed only three days. However, due to other commitments, the delayed subcontractor is unable to return again for two weeks. The matter is one of considerable importance, since project duration will be extended and a new critical path created. These facts would only become known in advance if it were recognized that the original delay involved interfering float time and if the effects were investigated.

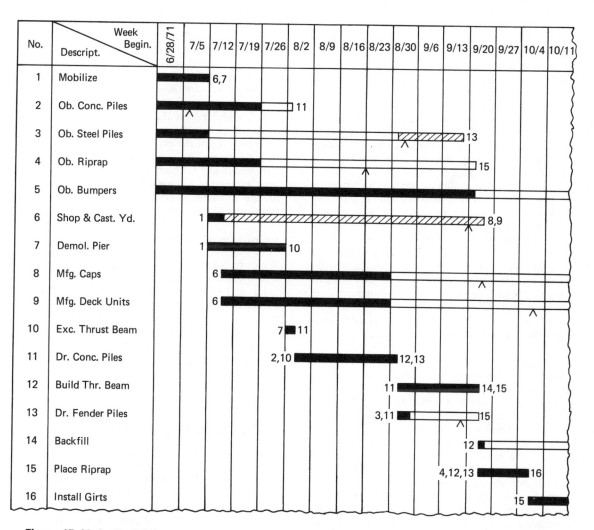

Figure 17–14–4 Partial bar chart, including scheduling data, float periods, and precedence relationships

The networking techniques provide a much more effective basis for project control than has been possible by other methods. However, the information available must be used, and it must be kept updated.

17-15. Updating

One of the major failures in the application of networking techniques has been the failure to utilize the dynamic potential of these procedures. The network diagram serves as a model of the project plan that permits changes to be incorporated as they occur and scheduling data to be updated to reflect these changes. All too often, however, only the original plan and scheduling data are ever produced. They continue to cover the office wall long after they are obsolete and bear little resemblance to the current progress of the job.

Many changes occur throughout the course of most construction projects. The most common are probably the unintentional variations between forecasted and actual duration of individual activities. Other common changes involve revisions in activity planning and estimated duration, postponement of activities, addition or deletion of activities, overlapping of acvtivities, and changes in sequencing of activities. Any single change that occurs may affect the scheduling of numerous other activities and may change the identity of the activities forming the critical path.

Updating frequency is often established by specification requirements. It is rarely less than monthly, often more infrequent, and sometimes poorly enforced. It is the contention here that updating should be more frequent than the strictest of these requirements and that there should be no necessity of enforcement since it is in the contractor's interest for proper control of his work. Except for activity duration or rescheduling changes confined to free float periods, updating should occur when changes are recognized. This could be twice a day or once a week, but seldom as rarely as once a month.

Current methods seem to make this concept for updating impractical, if not ridiculous. Schedule updating is generally accomplished by a complete recomputation of all data. The most rapid method for doing this is computer processing. Turnaround time for such processing is probably at least a day. Often a substantial minimum cost is involved regardless of how small the network might be. Even without these obstacles, the mere paperhandling requirements for all the various reports might be overwhelming.

However, an effective compromise that will permit very frequent updating seems possible and worthy of consideration. The proposed procedure is as follows:

1. Follow the normal procedure of developing an overall network diagram, complete scheduling data, and supplementary reports and schedules that are of value.
2. Update this network diagram and data monthly to get completely revised schedules and reports. Use computer processing methods if justified.
3. Using the "dateline cutoff" method of subnetworking, presented in the next section, prepare a subnetwork covering the span of the first month and overlapping at least a week into the second month.
4. Whenever changes occur during the month, incorporate them into the subnet and manually make forward pass computations only. This is practical since, regardless of the size of the project network, the subnet should

be relatively small and only forward-pass computations are involved.
5. These computations will permit determining those activities in the immediate future affected by each change, the revised date of their start times, and the effect on project duration. If project duration is affected, the change may result in different, or additional, critical activities. The critical path can be easily identified by "backtracking" (matching start times of activities with finish times of preceding activities) from the final node of the revised subnet. The essential information for good job control has now been updated. Additional information regarding revised float times could be obtained by the added effort of making backward pass calculations in the subnet, but this is not essential.
6. At the time of the complete monthly updating, use the results obtained to construct the subnet for the next month ahead, again overlapping at least a week into the following month. Continue this procedure month by month until project completion.

This procedure recognizes the importance of complete project planning and scheduling at the beginning of the job, and its subsequent revision periodically through the job. It recognizes the importance of continual updating for truly effective control. It recognizes the waste in effort and cost and the impracticality of completely updating all information for the entire span of the project every time a change occurs. For example, it is of negligible value for day-to-day control purposes to reschedule an activity that commences a year in the future since it will probably need to be rescheduled many more times in the interim and it will be updated at the end of each month. Finally, this procedure of determining the important effects of each change as it occurs, rather than lumping it with a number of other changes at the time of monthly updating, allows more intelligent claims negotiations.

In the case of the project office having a computer terminal, updating could be accomplished as changes were recognized and without the necessity of manual computation. The use of a simplified program, subnets, and printout limited to essential control information would make such frequent updating much more economically attractive.

17-16. Subnetworks

It can be advantageous to divide a large network diagram into smaller subnets. A small network is easier to comprehend and permits simpler methods of calculations and analysis. Subdivision may allow certain portions of the project to be isolated on the basis of location, responsibility, or calendar period. In some cases, subdivision may be necessary in order not to exceed the capacity limitations of a particular processing method or equipment. Contract specifications may require submittal of a master network diagram having a maximum number of activities and supported by detailed subnetworks.

17-17. Terminal Interfaces

The simplest type of subnetwork formation is to isolate and remove a portion of a larger network, replacing it by a single equivalent activity. In order to be able to do this, it is necessary that the portion to be removed only be connected to the remainder of the diagram at its terminal points. For example, in Figure 17-17-1, the portion of the network from activity D to activity V, inclusive, could be

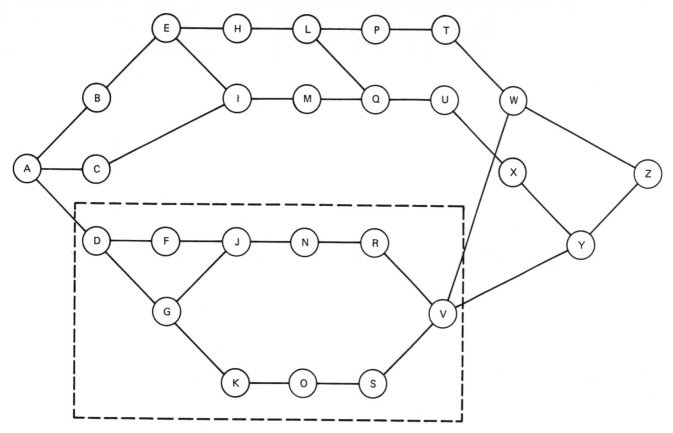

Figure 17-17-1 The portion of this network from activity D to activity V has terminal interfaces only and may be removed as a subnetwork and replaced by a single equivalent activity

replaced in the main network by a single activity whose duration was set equal to the duration of the subnetwork.

Unfortunately, there are limited opportunities to reduce the size of an existing network in this manner because there are generally one or more intermediate interfaces between the rest of the network and any grouping of related activities that one desires to treat as a subnet. However, approached from a different viewpoint, this method of subnetworking can be very important. A master project network can be constructed having a minimum degree of breakdown and showing little detail for separate operations. Then individual activities from this master network can be expanded into separate subnetworks for further analysis. For example, one activity on a master diagram might carry the descriptive label "Install Mechanical Work in Area 3." The specialty contractor who does this work might elect to develop a complete network to plan its performance.

17-18. Intermediate Interfaces

A second approach to subnetwork formation can be applied when there are intermediate references as well as terminal ones. Suppose, for example, that it is desired to isolate the same portion of the network of Figure 17-17-1 as before but that there are intermediate ties as shown in Figure 17-18-1 between activities C and F and between activities N and Q. The subnetwork cannot be replaced by a single activity, but it can be replaced by a condensed network. The condensed network consists of the interface activities, both terminal and intermediate, connected to one another by artificial activities having a duration equal to the longest path between them. If no path exists between

a pair of interface activities or if the longest path goes through another interface activity, no connection is required. Important "milestone" events may also be included in each subnetwork and treated in the same manner as interface activities and, of course, interface activities may be defined as events whenever this is preferable. Figure 17-18-2 shows the master network with both the isolated portion of the original network and the remainder portion replaced by condensed subnets. More impressive condensation might be accomplished in typical situations where the number of activities in the original network is much greater in relation to the number of interfaces.

In this approach, detailed planning is accomplished in work on the subnet level. Then each subnet is condensed and integrated into a master network. Time calculations are performed for the master network to obtain scheduling data for each interface activity and milestone event. With proper selection of activities and events, the integrated master network may serve directly as a top management control device. Scheduling data obtained for the master network is transferred back to the detailed networks. Time calculations are performed for the detailed networks based on this scheduling data for the interface and milestone activities and events. This method was presented by the U.S. Air Force a number of years ago,[9] not for the purpose of subdividing a single network, but to control a large program consisting of a number of contracts having some interface relationships with one another. Each contractor would develop his own network, including interfaces. Combinations of the separate networks into a master one for the entire program was very desirable but was impractical due to the excessive size of the resulting single network.

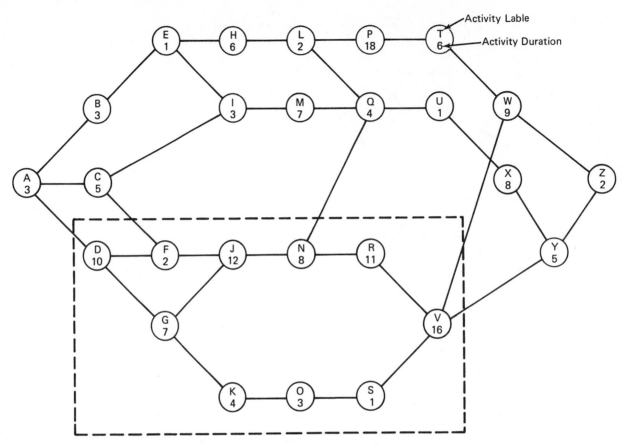

Figure 17–18–1 Same as Figure 17–17–1 except that there are also intermediate interfaces between activities C and F and between activities N and Q. A different approach to subnetwork utilization is required

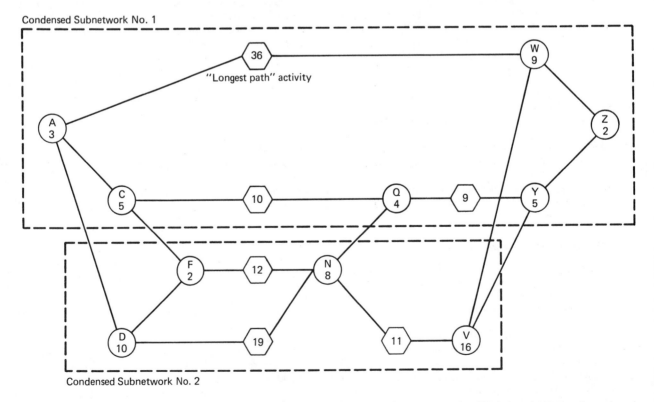

Figure 17–18–2 The network of Figure 17–18–1 has been divided into two subnetworks. Each has been condensed and then recombined into the master network in order to calculate or update scheduling data for interface activities

Therefore, the condensation and integration system was devised.

Procedures and computer programs have been developed for network condensation. These and other methods of condensation are features of major computer program packages.[10]

The practical limitation of this method is that the number of interfaces and milestones within a subnet must be very limited if any substantial condensation is to be achieved. It is also a somewhat cumbersome procedure, since any updating or replanning on the subnet level requires a new condensation of the subnet involved, a new calculation of the revised master network, and, finally, new scheduling calculations for all the subnetworks whose interface activities have altered scheduling.

17-19. Dateline Cutoff Method

A third type of subnetworking has been proposed in a research report of the Construction Institute at Stanford University.[3] It seems to have definite advantages for project control purposes, as discussed in the Section 17-15 on Updating, and also for detailed resource allocation analysis to be discussed later.

In this method, a subnetwork is developed by isolating a portion of the master network between specified cutoff dates. The first subnet would extend from the beginning of the master diagram up to and including all activities starting by a selected date. The remainder of the network, no matter how large, is replaced by an event representing project completion and artificial "tie" activities between this final event and each interface activity of the subnet.

An interface activity is any activity in the subnet that was connected to an activity in the remainder of the network which has been cut off. The duration of each tie activity is equal to the duration of the longest path between the nodes that it connects and can easily be determined from backward pass calculations already made for the master network. This subnet may be expanded in greater detail within the period that it covers.

One of the principal objectives of this type of subnetworking is to encourage very refined planning, updating, and control over a period in the immediate future where such action is justified, and leaving overall planning and updating of activities in the more distant future to the periodic recalculations of the entire master network. As soon as the project progressess well into the period covered by the first subnet, a second subnet is prepared for the next time period. This subnet commences with all activities in progress at the beginning of its time period, dropping all completed activities, and it extends to a second dateline cutoff. This process is repeated through successive time periods until the final subnet extends to project completion.

As an indication of the size reduction of master networks, suppose a building project involved 1000 activities and extended over a two-year period. On an average there would be approximately 40 new activities each month. Assuming a maximum of twice the average, the largest subnet for monthly time periods might contain 80 activities. To this must be added the number of necessary tie activities, which depends on the number of interfaces created by the cutoff date selected. Updating calculations on networks of this size range are possible in the field office by

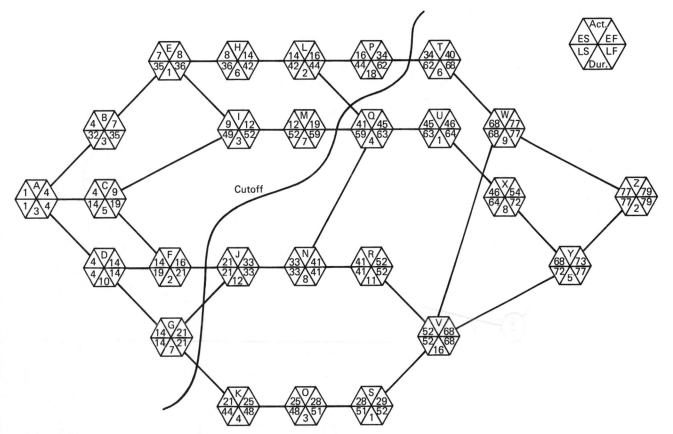

Figure 17–19–1 Same as Figure 17–18–1 except that forward and backward pass calculations have been shown. A subnetwork for the first 20 days of project duration is to be developed using the dateline cutoff method

manual methods. More detailed resource leveling studies also become practical.

Referring to the diagram of Figure 17-19-1 a dateline cutoff for the first subnet is chosen for the 20th working day. Figure 17-19-2 shows the resulting subnet. The durations of the necessary tie activities are calculated by subtracting the "effective late finish" time of the interface activity from the project completion date. For cases where the interface activity is only connected to activities outside the subnet, as in the case of activities F, G, M, and P, effective late finish time is the same as the late finish time calculated by the normal backward pass calculations. If the interface activity is followed by activities both within the subnet and outside the subnet, its effective late finish is the latest finish time that would be calculated by backward pass calculations that ignored its connections to the following activities within the subnet. This is the case with activity L where its effective late finish time is determined by the latest start time of activity Q rather than activity P.

The subnet will function in the same manner as if the remainder of the master network were still present. Revisions may be made and forward and backward pass calculations performed. The data for each activity within the subnet, as well as the project completion date, will be correctly updated. All changes in float times and identity of critical activities will also be obtained. Updating of data for activities following the subnet will not be accomplished until the subnet including these activities is constructed or until the master network is updated. If at any time changes are made in activities following the

current subnet, the master network must be updated in order to recalculate tie activity durations and to revise the project completion date.

17-20. Computer Applications

Computer processing can provide effective assistance for many networking applications. It may be a practical necessity to meet some specification requirements calling for a wide variety of reports and different arrangements of data. There are some procedures, such as certain resource leveling methods, that require so much data handling that they would be impossible to perform otherwise. However, the role of the computer in networking techniques has sometimes been overstated or misunderstood.

Quite often computer processing has been used simply as a means of satisfying the specification requirements with little attention paid to its potential contributions to good planning and control. This is probably especially true when a company employs an external data processing firm, often on a low bid basis, with a minimum of cooperation from, and guidance to, the project personnel. A better understanding of the advantages of the computer, the characteristics of good programming, and the types of reports possible should lead to improvements in computer utilization.

The usual reasons given for computer processing involve several common ideas that are largely misconceptions, such as: networking depends on difficult mathematical procedures, results will be obtained very rapidly because the computer is so fast, and errors will be eliminated

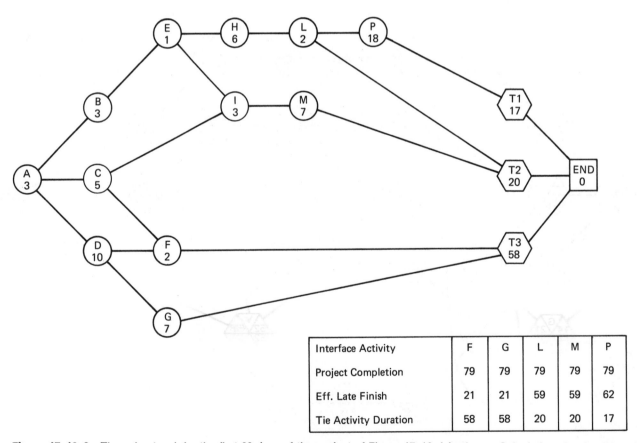

Interface Activity	F	G	L	M	P
Project Completion	79	79	79	79	79
Eff. Late Finish	21	21	59	59	62
Tie Activity Duration	58	58	20	20	17

Figure 17–19–2 The subnetwork for the first 20 days of the project of Figure 17–19–1 is shown. Calculations for durations of the tie activities, based on the dateline cutoff method, are also shown

because the computer doesn't make mistakes. The mathematics involved is seldom more difficult than adding or subtracting two numbers at a time. There are a large number of such calculations, and they involve no judgment. These are factors favoring computer processing rather than the difficulty of the procedures.

The idea that the computer is fast, although true once the data have been read in, can be misleading. All relationships from the diagram, all time estimates, and other input data must be entered on coding sheets and then punched on cards. Where processing is accomplished at an outside computer center, results may not be available until the following day. The output of an initial run is seldom error free and may require reprocessing one or more times. Therefore, although the computer may only require a few seconds to make all calculations involved, the complete turnaround time for obtaining valid results may be three to five days. If the calculations were performed in the office by an ordinary technician using noncomputer methods, they would probably take less than half a day even for a moderately large project network. It is also argued that the human will make mistakes and the computer will not. Again, the latter is true once correct data have been read into the computer. However, the computer processing does not guarantee error-free results, because the human element is still present in the preparation of input data. There are ample opportunities to omit data or make incorrect entries on the coding sheets or to transpose or mispunch data on the keypunch. Most such errors will not result in the generation of error messages but, rather, will be acceptable to the computer and will cause reports to be produced that appear to be valid.

What are the principal advantages of the computer in the application of networking techniques? Probably the most important is the ability to arrange calculated data in different orderings and formats and to produce many different reports for various purposes and levels of management. Once the basic calculations have been made and some additional data for calendar dating, resource requirements, and cost information have been provided, there are practically unlimited possibilities. Moreover, the computer can produce additional reports with little additional time or cost above that required for generation of just a master schedule. It certainly can do so more economically than the lowest paid office employee. Although this is a strong point of the computer if utilized intelligently, the creation of a multitude of voluminous reports can be self-defeating unless they are carefully designed to meet specific requirements and selectivity distributed to those who need the data in each report that they receive.

Another advantage of the computer is that updating is simplified. After an input deck of punched cards has been successfully used to produce valid reports, it is relatively easy to make changes in the few cards for activities needing revised data and then to make a new computer processing run. Since minor changes are usually involved, a successful printout of the updated reports will generally be obtained on each processing run. However, it is still not a casual matter to make an updating run. If the work is not done in-house, there is usually a substantial minimum fee even for processing small networks. Due to the effort and cost, there is some reluctance to make new runs until required by specifications or until a major change has occurred. This is the reason that it has been proposed to form small subnetworks allowing frequent manual updating in the field office for close control of the job.

There are some networking procedures where computer processing is essential. These are cases that involve so many calculations or so much data handling that manual calculations are simply not feasible. The principal examples are certain resource allocation methods. However, there are generally alternative methods in which computers are less essential and where the benefits gained by the ability to apply judgment control throughout the procedure may outweigh the advantage of rapid data manipulation.

The greatest advantages of computer processing for networking applications are just beginning to emerge. The utilization of remote consoles in project offices connected to large central computers on a timesharing basis will permit interactive procedures between man and machine. These will take better advantage of the computer's ability to store information and calculate and retrieve rapidly and of man's ability to make judgment decisions and changes during the intermediate steps of reaching a solution. The hardware is available for such applications and is becoming economically practical. The software is being developed.

A few of the more common and useful types of computer reports will be discussed. The examples shown in the figures were selected from a program developed at Stanford University in 1965. There are many program packages available from computer manufacturing companies and from data processing firms. Each have their own particular characteristics and variations.

17-21. Master Schedule

The principal report produced by most networking programs is one that lists all activities by label and description and gives their earliest and latest start and finish dates, their total and free float periods, and indicates the critical activities. Figure 17-14-2 illustrates such a report. The activities in this report are arranged chronologically in order of earliest start dates. Other arrangements may be requested, such as listings by card input order (allowing the user to group activities prior to processing in any order desired), latest finish order, total float order (criticality) etc. Data listing is generally based on more than one level of ordering, e.g., according to criticality and, for activities having the same total float, according to earliest start dates.

The master schedule may be presented in undated form with all scheduling data in terms of working day numbers, for example. More commonly, it is presented in calendar dated format. This requires calendar conversion information and the start date for the project to be included in the input data. Sometimes a supplementary report consisting of a calendar conversion table is also printed out. For each month it might list the calendar dates, the corresponding calendar day and working day numbers, and an explanation of the reason for each nonwork day. Programs generally permit a scheduled date to be specified for the start or finish of any activity. This date will override the computed date and be substituted for it in forward or backward pass calculations. When a scheduled latest finish date is specified that occurs earlier than the calculated earliest finish date, then negative float will result. In such cases, the critical path becomes the path having negative float of the greatest magnitude. The maximum negative float indicates that the project will be completed behind schedule by that number of days if it is performed as planned.

The program used to generate the master report of Figure 17-14-2 allows considerable freedom in labeling of activities. Numeric, alphabetic, or any other characters found on the keyboard of the keypunch machine may be used in the label. Early programs allowed only numeric labeling and often required that later activities or events have higher values than earlier ones. Completely random labeling allows much ingenuity to be exercised in selection of the labeling system. Portions of labels may indicate the nature of the activity, responsibility for activity performance, area or floor level, location of the activity on the job, or relationships with other coding systems used for cost control or estimating purposes. Sometimes, where large diagrams have been developed, the label has been used as a map grid indicator to locate the activity or event symbol on the network print.

A calendar-dated master report would represent the standard output obtained in a normal processing run. Other reports, including those discussed below, would generally be optional and could be called for when desired.

17-22.   Subschedules or Responsibility Schedules

The program used in the previous example allows each activity to be coded so that it will appear on one or more additional schedules that only include activities with the same code number. This allows preparation of special schedules similar to the master schedule but only containing those activities of interest to a specific recipient. For example, all electrical items may be shown on a report for the electrical subcontractor or all activities involving earthwork items may be included in a report for the excavation superintendent. A separate schedule can be furnished to the manager of each group within one's own organization (superintendents, foremen, purchasing, engineering, etc.) and to each outside organization (subcontractors, continuing suppliers, owner representatives, etc.) which includes only activities for which they have primary responsibility. By eliminating all other data, the reports are greatly reduced in size and are more likely to be used.

The subschedule activities may be arranged in any of the various orderings that are used in the master schedule. Some project managers prefer to omit certain activity data from the subschedules. For example, the lower echelon manager who receives the report needs to know the earliest, or scheduled, start and finish dates for his activities, and he is also entitled to know the amount of free float, since he can utilize this time without affecting others on the project. However, if he utilizes interfering float time he does cause delays to one or more other activities. Usually he is not in a position to judge the importance of the consequences of these delays. Therefore, the subschedule may omit latest start and finish dates and total float data. This will encourage the lower echelon manager to discuss any proposed changes that he would like to make that extend the completion of any of his activities beyond the free float period.

In addition to responsibility-type reports, exception-type reports may also be desired for different levels or purposes of project management. The project manager may, for example, desire a report containing only those activities that are critical or near-critical, or a report including only those activities whose scheduling has changed since a preceding report date. There are many other possibilities. If the ability to produce such reports is included in the design of a program package, a negligable additional expense is required to obtain them.

Some of the more sophisticated computer systems allow the user to design additional reports and to modify standard reports. The formats for these new or modified reports may be developed without the necessity of an experienced programmer making internal changes in the basic program structure.

17-23.   Bar Chart by Computer

A bar chart may be generated by the computer as an alternative and more familiar means of presentation of scheduling data. Opposite activity descriptions and below a calendar time scale, each bar may be composed of segments of different symbols representing the duration of a critical activity, the duration of a noncritical activity, the span of free float time, and the span of interfering float time. The bar chart may be of any length and height that is necessary. However, excessively large bar charts have little value except for wall papering. Therefore, a program may provide the opportunity to specify initial and final dates for the span of time covered by a bar chart. This reduces the height as well as the length, since only those activities in progress during the specified time span will be included.

The program previously cited has the ability to produce separate bar charts corresponding to the contents of each subschedule. Therefore, the electrical subcontractor, for example, may receive a bar chart showing the scheduling of only the electrical activities. These subschedule bar charts may also be limited, if desired, to designated time spans.

17-24.   Resource Schedules

The selected program is able to produce resource schedules for various labor classifications or equipment types. The information is presented in digital form showing, for each day, the number of resource units needed, the number available, and the difference between these two figures which would represent idle resoures or, if negative, the amount by which requirements exceed availability. This same information is also presented graphically in histogram form to offer a more rapid means of resource schedule evaluation. Figure 17-24-1 shows an example of a single resource schedule for a small project. The dotted line on the chart represents resource availability. Problem areas are obvious. They include the peaks which represent periods in which resource requirements are greater than the number of units available and valleys during which there are periods of idle resources.

Different symbols are used on the chart to represent resource units required by critical activities and those required by noncritical activities. If resource units required by critical activities exceed resource availablity, there is no possibility of rescheduling activities without extending project duration (unless changes are made in resource requirements or in the project plan) (Figure 17-24-2). However, if the excessive requirements result only from noncritical activities, there exists a possibility of rescheduling which may or may not solve the problem without extending project duration.

In order to obtain resource schedules, additional input must be provided. Each resource type may be assigned a number. Then data must be coded and cards punched for each activity requiring resources by giving the number of the resource type and the number of units required. The sample program permits up to 999 resource types, but not more than six resource types may be used by a single

```
STANFORD CONSTRUCTION INSTITUTE                      7/12/71

SAMPLE SMALL BUILDING PROJECT

RESOURCE POOL OF    4    CARPENTERS              RESOURCE NO.    1

REQUIRED      DIFFERENCE      0         10        20        90        100   DAY
---------------------------   .---------.---------.---------.---------.
I    0            4           I   .                                          1
I    0            4           I   .                                          2
I                            S
I                            S
I    0            4           I   .                                          3
I    0            4           I   .                                          4
I    16          -12          IXXXXXXXXXXXXXXXX                              5
I                            H
I    20          -16          I****XXXXXXXXXXXXXXX                           6
I                            S
I                            S
I    9           -5           I*XXXXXXXX                                     7
I    5           -1           I*XXXX                                         8
I    0            4           I   .                                          9
I    0            4           I   .                                          10
I    0            4           I   .                                          11
I                            S
I                            S
I    0            4           I   .                                          12
I    0            4           I   .                                          13
I    6           -2           I****XX                                        14
I    1            3           I*  .                                          15
I    0            4           I   .                                          16
I                            S
I                            S
I    2            2           IXX .                                          17
I    2            2           IXX .                                          18
I    0            4           I   .                                          19
I    2            2           IXX .                                          20
I    2            2           IXX .                                          21
I                            S
I                            S
I    8           -4           I****XXXX                                      22
I    4            0           I****                                          23
I                            H
I    2            2           I** .                                          24
I    0            4           I   .                                          25
I                            S
I                            S
I    0            4           I   .                                          26
I    0            4           I   .                                          27
I                            H
I    0            4           I   .                                          28
---------------------------   .---------.---------.---------.---------.
REQUIRED      DIFFERENCE      0         10        20        90        100   DAY
```

Figure 17–24–1 A computer printout sheet of daily resource requirements for one resource is shown in both digital and chart formats. The differences between a stated resource availability limit and actual requirements are also shown. Similar charts would be produced for other resources involving different labor classifications, equipment types, etc.

activity. A separate resource schedule is printed for each resource type.

This feature of the program develops resource schedules but does not level resources. However, a trial-and-error procedure may be followed to achieve some degree of resource leveling. By joint examination of resource schedules and the bar chart which are plotted to the same horizontal time scale, it is possible to determine which activities are responsible for resource problems and whether those activities can be rescheduled during float periods, preferably free float which doesn't cause additional downstream changes, to improve the project plan. Finally, using the ability to designate scheduled start times for activities that will override computed earliest start times, a new set of resource schedules can be obtained. This procedure may be repeated as necessary. More effec-

```
STANFORD CONSTRUCTION INSTITUTE                              7/12/71

SAMPLE SMALL BUILDING PROJECT

RESOURCE POOL OF    4    CARPENTERS                RESOURCE NO.    1

REQUIRED    DIFFERENCE    0        10      ?      90     100   DAY
---------------------     .--------.---------.---------.---------.
I     0          4        I   .                                      1
I     0          4        I   .                                      2
I                         S
I                         S
I     0          4        I   .                                      3
I     0          4        I   .                                      4
I     4          0        IXXXX                                      5
I                         H
I     4          0        I****                                      6
I                         S
I                         S
I     3          1        I*XX.                                      7
I     3          1        I*XX.                                      8
I     4          0        IXXXX                                      9
I     4          0        IXXXX                                     10
I     4          0        IXXXX                                     11
I                         S
I                         S
I     4          0        IXXXX                                     12
I     4          0        IXXXX                                     13
I     4          0        I****                                     14
I     3          1        I*XX.                                     15
I     4          0        IXXXX                                     16
I                         S
I                         S
I     2          2        IXX .                                     17
I     4          0        IXXXX                                     18
I     2          2        IXX .                                     19
I     4          0        IXXXX                                     20
I     4          0        I XXXX                                    21
I                         S
I                         S
I     4          0        I****                                     22
I     4          0        I****                                     23
I                         H
I     4          0        I**XX                                     24
I     2          2        IXX .                                     25
I                         S
I                         S
I     4          0        I****                                     26
I     0          4        I   .                                     27
I                         H
I     0          4        I   .                                     28
---------------------     .--------.---------.---------.---------.
REQUIRED    DIFFERENCE    0        10       2(     90     100   DAY
```

Figure 17–24–2 Computer printout sheet showing a resource schedule for the same project as shown in Figure 17–24–1 after rescheduling of activities to satisfy the resource availability limit. In this case leveling did not require extension of project duration, but in many cases it will

tive approaches to resource leveling will be discussed later in this chapter.

17-25. Monitor Report

Ordinarily when a computer processing run is made, the output is in the form of printed reports. However, the sample program also permits the master schedule data to be furnished in condensed form in other output forms such as punched cards. When the next updating run is made with a revised master deck of input cards, the condensed data from the preceding run (or any other earlier run) may be included in the input data. The result is illustrated by the report shown in Figure 17-25-1. Updated scheduling data are provided for each activity and beneath each data item is shown the number of days change, plus or minus, between the current report and the earlier report to which it is compared. Boxes are also provided for field updating. During the following time period on the project, actual or re-estimated start dates and activity durations

STANFORD CONSTRUCTION INSTITUTE

SAMPLE PIER CONSTRUCTION PROJECT
PROJECT PROGRESS REPORT - DATED 8/ 2/71   COMPARED TO PROJECT REPORT - DATED 6/21/71

ACTIVITY LABEL	DESCRIPTION	STATUS	DURATION (REVISED)	ACTUAL	START (REVISED)	ACTUAL	TOT.FLOAT (REVISED)	FREE FLT. (REVISED)	REMARKS
NTP	NOTICE TO PROCEED	COMPLT	0 (0)		JUN 28/71 (0)	9			
1	MOBILIZE	COMPLT	10 (0)		JUN 28/71 (0)	9			
2	OBTAIN CONCRETE PILES	CURRNT	25 (5)		JUN 28/71 (0)		0 (-6)	0 (-6)	
EWO-1	PERFORM PILE TEST	CURRNT NEW ACTIVITY	5 *8 days*		AUG 2/71		0	0	*Pile modifications required*
3	OBTAIN STEEL PILES	CURRNT	10 (0)		JUN 28/71 (0)		51 (4)	40 (4)	
4	OBTAIN RIPRAP	CURRNT	20 (0)		JUN 28/71 (0)		44 (4)	44 (4)	
6	BUILD SHOP AND CASTING YARD	CURRNT	3 (0)		JUL 15/71 (3)		49 (1)	0 (0)	
7	DEMOLISH EXISTING PIER	CURRNT	15 (0)		JUL 12/71 (0)		4 (4)	0 (0)	
8	MANUFACTURE PRECAST CAPS	CURRNT	32 (0)		JUL 20/71 (3)		49 (1)	49 (1)	
9	MANUFACTURE PRECAST DECK UNITS	CURRNT	32 (0)		JUL 20/71 (3)		59 (1)	59 (1)	
10	EXCAVATE FOR THRUST BEAM	CURRNT	1 (0)		AUG 2/71 (0)	*8/5/71*	4 (4)	4 (4)	*Weather delay*
11	DRIVE CONCRETE PILES	FUTURE	20 (0)	*30 days*	AUG 9/71 (4)		0 (0)	0 (0)	*Re-estimate based on EWO-1*
12	BUILD THRUST BEAM	FUTURE	14 (0)		SEP 7/71 (4)		0 (0)	0 (0)	
13	DRIVE FENDER PILES	FUTURE	3 (0)		SEP 7/71 (4)		11 (0)	11 (0)	
14	BACKFILL	FUTURE	1 (0)		SEP 28/71 (4)		90 (-3)	90 (-3)	
15	PLACE RIPRAP	FUTURE	10 (0)		SEP 28/71 (4)		0 (0)	0 (0)	
16	INSTALL BRACING GIRTS	FUTURE	23 (0)		OCT 12/71 (4)		0 (0)	0 (0)	
17	SET CAPS	FUTURE	10 (0)		NOV 12/71 (4)		0 (0)	0 (0)	
18	SET DECK UNITS	FUTURE	16 (0)		NOV 30/71 (4)		0 (0)	0 (0)	
19	INSTALL DECK HARDWARE	FUTURE	3 (0)		DEC 22/71 (4)		29 (-3)	29 (-3)	
20	CAST-IN-PLACE CONCRETE	FUTURE	20 (0)		DEC 22/71 (4)		0 (0)	0 (0)	
21	REMOVE GIRTS	FUTURE	7 (0)		JAN 21/72 (4)		0 (-3)	0 (0)	
23	INSTALL MECHANICAL SERVICES	FUTURE	5 (0)		FEB 1/72 (4)		0 (-3)	0 (-3)	
24	FINAL CLEAN-UP AND MOVE OUT	FUTURE	2 (0)		FEB 8/72 (1)		0 (0)	0 (0)	

PROJECT DURATION IS 157 WORKING DAYS - (REVISED   1 DAYS)

Figure 17-25-1 A computer monitor report comparing an undated project schedule to an earlier schedule indicating activity status and showing the magnitude of data change. The boxes are used by field personnel to enter data changes for the next scheduled updating

are indicated in these boxes. Then this report is sent in by the field office at the end of the time period to serve as a basis for revising the master input deck for the next updating run.

17-26. Progress Estimates—Cash Flow

Many specifications require that the contractor allocate the total contract amount to the various activities in the project and, based on the scheduling of these activities, calculate progress payments for each month of the project duration. Generally this must be updated each month to obtain a correct progress estimate for the current month and revised forecasts of all future payments.

Such a report is a simple addition to the computer system output but does require additional input of the dollar amounts or contract percentages assigned to each activity. This report shows the revenue earned each month. Some contractors also take the additional step of providing cost data associated with each activity and developing forecasts of monthly costs as well as monthly earnings. Then a subtraction of cumulative costs at the end of each month from cumulative earnings less retention will develop a forecast of financial requirements of the project over its life.

17-27. Other Reports by Computer

Comments on computer-produced network diagrams have been included in the earlier discussion of diagramming techniques. Many other computer-generated reports are possible. Probably the most important of those omitted from the above discussion are ones involving cost control. There are many ways in which costs can be associated with the work breakdown structure and that budgeted costs and actual costs can be developed and compared. The subject of cost control, however, is outside the scope of this chapter.

17-28. Computer Program Evaluation

Almost any person with a single course in computer programming can write a program that will provide the required calculated data. However, a number of factors distinguish a good program from a mediocre one. One is program efficiency. A good program requires less computer time for processing. It also utilizes the capacity of the computer fully. For example, it permits processing a larger network, faster, on a given piece of equipment than is possible with a poor program.

Of even more importance are the formats of the output reports. Many programs provide the desired information, but often in an unattractive jumble of data that are difficult to read. A good programmer studies the need of the user and gives considerable attention to arrangement of reports and to selectivity in data to be presented. Sometimes graphical presentation, as in the resource schedule example, is worth the extra effort.

Finally, a good program has a high degree of diagnostic ability in editing the input data and informing the user of errors present. In most cases, where the error may be critical, processing is terminated and error messages printed out. One type of error that occurs in complicated networks is the "loop," where a preceding activity in a chain of connected activities is later used as a following activity in the same chain. Most programs recognize the loop condition and terminate processing. The better programs also locate the loop and print out its components, since this may save many hours of searching.

17-29. Resource Allocation—Single-Project Leveling

A project schedule based on earliest start dates is generally a very uneconomical schedule and often a completely unrealistic one. If resource schedules were developed for the various labor classifications and equipment types required by all the activities in progress during each time period, some might have peaks that exceed availability limits and probably all would have undesirable variations that represent extra costs. In the absence of any formal procedures, project management will apply common sense to reschedule activities for the purpose of leveling resource requirements. Superintendents, for example, seldom have the men and equipment necessary to work on every activity that could be performed at a given time. They make daily decisions concerning which activities are less important and can be postponed until resource units are available.

Networking techniques offer data and a project model that should make it possible to do a better job of resource leveling. The knowledge of the identity of critical activities and the amounts of total and free floats of noncritical activities allows more intelligent decisions for rescheduling activities. The ability to generate resource schedules based on any given project schedule allows problems to be recognized and decisions to be made in advance rather than at the job site at the beginning of each work day.

Since any rescheduling of one activity will often affect the scheduling of other downstream activities and may change a number of different resource schedules, each change usually involves extensive new calculations and data updating. For this reason, computer processing has been considered important for formal procedures, and many different programs have been developed. There has been little success in achieving an "optimum" solution for the problem except for networks of very limited size. This goal has been somewhat academic in any case, since the mathematical definition of an optimum solution ignores many characteristics of the solution and many of the measures, besides rescheduling of activities, that are possible to improve a solution. However, numerous available procedures produce feasible solutions that will satisfy resource limitations and, in some cases, will achieve substantial leveling of resource requirements as well.

A simple procedure will be briefly outlined as a basis for discussing and understanding some of the features of resource allocation methods.

1. Set availability limits for each resource type. These may be definite cutoffs that cannot be exceeded, or there may be a range of units above the normal limits that involve a premium cost. By progressively lowering the limits set, the procedure being described can be used for resource leveling purposes also.
2. Commencing at the beginning of the project, consider all activities that are available for scheduling. Schedule the activity having the highest priority. If this causes new activities to become available for scheduling, add them to the list of available activities. Then select the activity with the next highest priority, scheduling it as early as possible without exceeding resource limitations. Whether an activity may or may not be interrupted can be specified in advance and will often determine the earliest date at which it can be scheduled.
3. The principal variations in different procedures involves the basis used for determining the priorities of the activities available for scheduling. There must be sufficient levels of priority rules established that a decision is reached in every situation. For example, primary priority might be given to the activity available for scheduling which has the lowest latest start date (on the basis that if an activity is scheduled later than its latest start date, it will delay project completion). If two or more activities have the same latest start date, secondary priority might be given to the activity whose updated earliest start date occurs latest (this would be equivalent to selecting the most nearly critical of the activities). If two or more activities are still eligible, a third degree of priority might be established arbitrarily based on the order of activity listing. There are many other possible bases for establishing priorities, such as activity duration (schedule longer activities first since they are more difficult to schedule later), number of resource units required (schedule activities with high requirements first since they offer greater difficulties later) etc.

4. Continue the above procedure until all activities are scheduled. Project duration may increase during the process since the start times of some activities may have to be delayed beyond their latest start dates. Should project duration change, latest start times of activities still to be scheduled will also change. However, the originally computed latest start dates may still serve as the basis for determining priorities, since all latest start dates of activities available for scheduling will be incremented by the same amount and their relative order will, therefore, be unchanged.

The simple procedure outlined above can easily be programmed since it follows definite rules. It is also sufficiently simple that it can be applied manually. Resource requirements of each activity and project resource availability need not be constant over their respective time spans.

Some procedures involve so many calculations or data manipulations that computer processing is essential. Varying degrees of sophistication exist in different programs. Some allow multiple combinations of activity duration and number of resource units utilized. Some consider penalties for interrupting activities and for delaying project completion. Some procedures do not alter the original project duration but offer no guarantee that resource requirements will be reduced below availability limits. Some programs recognize that no single set of priority rules consistently produce the best solution and allow the user to vary the rules for any processing run. Resource allocation programs are available from the major computer manufacturing companies and from various software firms.

Figure 17-24-2 represents the same resource schedule shown in Figure 17-24-1 after the foregoing procedure was applied. In this example, progress duration was not extended even though it would have been had it been necessary to do so in order to meet resource limitations. Other resource schedules were also involved and were concurrently modified to satisfy their allowable limits.

17-30. Multi-Project Leveling

Contractors are often confronted with the problem of assigning a limited number of resource units to several projects which are in progress concurrently. If networking procedures are being used for each of these projects, the diagrams could be combined into a single master plan, and resource allocation could be accomplished in the same manner as for a single project. The principal complexity would appear to be the matter of the size of the resulting single network.

However, there are other considerations that increase the difficulty of this problem. One involves the respective priorities of the various projects. Generally delays are more serious for some projects than for others. The consequences of delays may be set forth in liquidated damage clauses in the contracts or may be related to more intangible considerations. In any case, the priorities of the separate contracts should be considered when resource scheduling procedures require the completion of any of the projects to be delayed.

Another complexity involves the degree of mobility of certain resource types. Some resources may readily be shifted from project to project. Others may involve substantial move-in and move-out expenses, and transfers between projects may only be justified for an extended period of use having some minimum duration. Satisfactory resource allocation procedures for multi-project scheduling should recognize all of these additional complications.

17-31. Evaluation of Resource Allocation Procedures

One extreme of resource allocation procedures consists of a very sophisticated program utilizing the tremendous storage capacity and speed of a large computer and based on the application of networking techniques. The other extreme consists of a foreman, without even a bar chart schedule, making spot decisions as problems and needs arise during the course of each workday. The effectiveness of the results produced by the latter approach probably do not compare as unfavorably with those of the first approach as might be imagined. The foreman, assuming he has good judgment and an inclination to do his job well, can use his ingenuity in many ways that cannot be programmed into even the most elaborate computer algorithm. He can revise methods and crew sizes, share certain resource units between concurrent activities, borrow partial crews, and take dozens of other measures that are possible under the particular circumstances.

An ideal resource allocation procedure would combine the strong points of both approaches. It would take advantage of the power of the computer to store and manipulate information and to make rapid computations for resource schedules. It would take advantage of the networking techniques to provide a working model of the project plan, identification of critical activities, scheduling and float data, and dynamic updating as rescheduling was performed. But it would also take advantage of the tremendous ability of human judgment to modify methods and plans in order to overcome obstacles throughout the step-by-step process of developing a final solution. Such a procedure has become feasible with today's hardware. Remote consoles in the contractor's field office communicating with large, timesharing central computers permit the development of interactive procedures that can combine the strong points described above. Active research is in progress to develop such systems and to reduce the costs involved.

Resource allocation programs are already available from computer manufacturers and software companies that produce workable schedules satisfying resource restrictions. In general, they involve a "blackbox" type of approach where data are submitted, batch-processed with limited or no ability for the program to make data modifications during intermediate steps, and a solution is printed out without providing explanation of the specific obstacles encountered or the steps taken to overcome them.

Future programs will provide procedures in which the planner and manager have much greater control in guiding the development of schedules that achieve effective resource utilization.

17-32. Time-Cost Trade-offs

Time-cost trade-off procedures, potentially, are among the most powerful planning tools offered by the networking system. They were among the earliest applications proposed by the developers of the CPM. Computer programs for these procedures have been available for many years. The time-cost approach introduces cost data in its attempt to provide the most economical project plan and schedule. Since this is a major goal of construction management, it should follow that these procedures are widely used. They are not. The methods developed, although mathematically sound, have practical shortcomings that have not been overcome. However, it is worthwhile to define the problem and discuss the objectives because their importance ensures that eventually they will be applied more successfully.

17-33. The Problem and Objectives

In previous discussion it has been assumed that a method for performing each activity in a project has been selected and that either single or multiple-time estimates have been made based on that method. Actually there are many methods available and choices involved in the performance of most activities. These may affect both the time and cost of activity performance. For a given completion date, there is some combination of activity durations and costs

that gives the most economical schedule. The time-cost trade-off procedure attempts to develop this schedule. In a broader sense, there is some scheduling combination that gives the lowest possible total project cost and a corresponding completion date. The time-cost trade-off procedures may also be applied to determine this scheduling solution. These concepts will be described in greater detail.

Considering only direct costs, such as those for labor, equipment, and materials, there is some mode of performance for an activity that results in least cost. There is also a corresponding activity duration. These have generally been designated as the "normal" time and cost. There are a number of ways in which most activities can be expedited and these usually increase the direct costs. For example, crews can work overtime and extra shifts, larger crews working less efficiently can be employed,

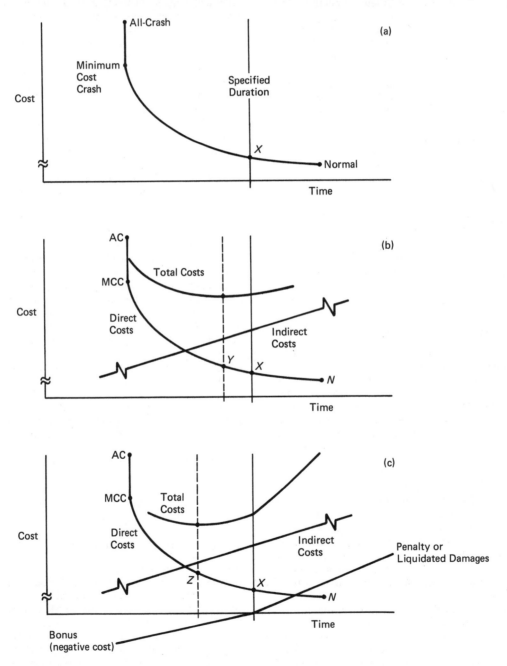

Figure 17-33-1   Project time-cost relationships

more expensive methods and equipment can be used, etc. The fastest conceivable method for performance of an activity results in a "crash" time and cost. If every activity in a project were performed at its normal time and cost, there would be a corresponding project "normal" schedule as represented in Figure 17-33-1a. The time coordinate of this point could be determined by making forward pass calculations using normal durations for each activity. The cost coordinate could be determined by adding the normal costs of all activities.

If every activity in the project were performed at its crash time and cost, coordinates of an "all-crash" project schedule, also shown in Figure 17-33-1a, could be determined in a similar manner. An all-crash performance would be a very uneconomical solution even where a crash effort was justified, because many activities never become critical. Money spent to expedite these activities is wasted. There exists a "minimum-cost crash" schedule, as shown in Figure 17-33-1a, which only includes expediting those activities that affect project duration. Generally, however, we are interested in schedules that fall between normal and crash extremes.

If the most economical schedule, based on direct costs only, for intermediate project durations were developed, a curve of the general shape shown in Figure 17-33-1a would result. This shape reflects the fact that as we progress from normal project duration toward crash project duration, it costs more and more to gain each time reduction. Suppose a project completion date was specified as indicated in Figure 17-33-1a. The schedule corresponding to point X would represent the most economical scheduling solution for completion at this date. This schedule could be developed by starting at a known point, the normal schedule, and using the time-cost trade-off procedures to modify this schedule and advance along the curve to point X.

A more comprehensive approach to the problem recognizes that the most important objective is not to determine the best schedule for an arbitrarily assigned completion date but, rather, to determine the schedule that gives lowest total cost. This requires a consideration of all other time-dependent costs that have not already been included. These costs are mostly of the "indirect" type that tend to increase with project duration. They include such items as

salaries, maintenance of offices and shops, utilities, insurance, interest, etc. The costs are generally not difficult to estimate and can be represented by a curve as shown in Figure 17-33-1b. The magnitude of these costs is not important, but the cost change per time unit (slope of the curve) over the range of project durations under consideration should be estimated. Shortening project duration tends to increase the direct costs and to decrease the indirect costs. As successive steps are taken to shorten project duration in the most economical manner, a solution is reached where any further reduction in project duration will cause a greater increase in direct costs than the corresponding saving in indirect costs. This point establishes the schedule for the minimum-cost coordinates of the total cost curve. Point Y of Figure 17-33-1b represents a better solution than point X.

Costs associated with variations in project duration from the owner's viewpoint are also an important cost element. In fixed-price contracts, these are often represented by a liquidated damage clause or a bonus-penalty clause. Figure 17-33-1c shows the addition of a bonus-penalty curve and the resultant shift of the optimum solution to a still faster schedule represented by point Z.

17-34. Procedures

Of the three curves shown in Figure 17-33-1c that are combined to produce the total cost curve, the direct cost curve is the one that is most difficult to develop and is the key to solving this problem. Only a sufficient portion of the curve to reach point Z is necessary, and cost slopes rather than total magnitudes of costs are all that are required.

The conventional computer solution to the problem employs a mathematical technique from operations research theory. It requires starting with data corresponding to a known point on the project curve. This is accomplished by using the data for the "normal" schedule. Data are also required for the cost slopes of each activity in the project. Figure 17-34-1a shows a simplified cost curve for a single activity determined by estimating time and cost coordinates for normal performance, point N, and crash performance, point C. The common approximation for this curve assumes that it consists of a straight, continuous line between

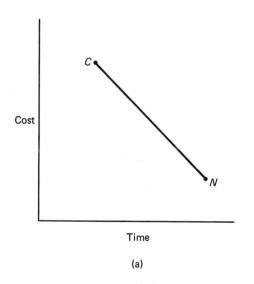

(a)

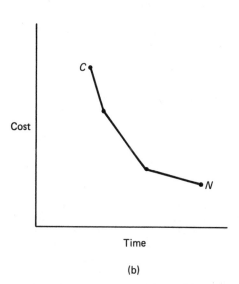

(b)

Figure 17–34–1 Activity time-cost relationships

these two extremes. This implies that activity performance can be expedited from normal duration to any other duration down to crash duration at a constant rate of cost increase. This is only a convenient approximation, since the actual activity time-cost relationship is probably neither continuous nor linear.

Mathematical procedures may allow closer approximations by permitting segmented activity time-cost curves of the type shown in Figure 17-34-1b or by allowing statements that a curve is noncontinuous between certain data points. However, it is seldom practical to take advantage of these closer approximations because an excessive amount of input data would be required. Given a minimum of the normal and crash time and cost coordinates for every activity, the computer procedure can produce the set of schedules that determines the project time-cost curve between normal and crash duration limits.

Although the foregoing solution is mathematically satisfactory, it has seldom been used for construction planning. The following practical obstacles may account for this:

1. Too many data are required. Since a batch processing method is involved, a complete set of input data is necessary. It consists of at least two separate time-cost estimates for every activity on the project. The differences between the normal and crash modes of performance are generally so great that these requirements are equivalent to a very detailed estimate of two completely different projects.
2. Some of the data requirements are unrealistic. For example, an estimator cannot be expected to provide a crash cost estimate for an individual activity when the existing conditions for the performance of other activities are unknown. The cost to crash a particular activity may be quite different if the balance of the project is already on a three-shift, seven-day week, than it would be if other activities are at the normal end of their time-cost curves. However, the procedure does require a crash cost amount for each activity, presuming that its performance can be isolated for estimating purposes.
3. Other interrelationships between activities are ignored. Many relationships actually exist between activities in addition to the sequential ones represented in the network diagram. Measures taken to expedite one activity may affect the duration, cost, and cost slope of another activity. Decisions to use a different method, pay the move-in cost for another piece of equipment, or work a six-day week generally produce results that cannot be limited to a single activity. These interrelationships should be considered but are not.
4. Activity time-cost curve approximations do not give the best results. Because of the excessive data requirements, the two-point curve of Figure 17-34-1a is generally adopted. Once an activity is selected for expediting based on a single cost slope, it is usually expedited all the way to crash duration. The curve type of Figure 17-34-1b is a more realistic one in most cases. It indicates that it is often uneconomical to expedite an activity all the way to its crash limit. Better results would be achieved by expediting several different activities over limited periods corresponding to the flatter segments of their time-cost curves. However, these opportunities are overlooked since the type of approximation of Figure 17-34-1a is usually considered necessary.

A less efficient but more practical method for determining the most favorable project plan and schedule is to start with an initial schedule and make successive modifications and updating runs. This is a trial-and-error procedure. Each time a change is made with the intent of changing the project duration and decreasing total costs, a recalculation of scheduling data is necessary to show whether the result was achieved and, if so, to provide a new set of data as the basis for the next proposed change.

A method developed in research efforts at Stanford University has overcome the four shortcomings discussed for the first procedure while providing a more systematic and efficient method than the second procedure. However, as this approach was developed, it became increasingly obvious that no procedure which attempts to separate the time-cost trade-off problem from the resource allocation problem will be truly effective.

17-35. Combined Resource Allocation and Time-Cost Trade-offs

Although resource allocation procedures may be applied independently to satisfy resource limitations and level resource requirements, it is of dubious value to attempt to apply time-cost trade-off techniques independently. The determination of the most economical project plan and schedule requires that resource scheduling and time-cost trade-offs be considered jointly and that all variable cost factors be included in the problem. Reasons for this viewpoint include the following:

1. Some times-cost trade-off possibilities involve resource scheduling changes. For example, the addition of an extra unit of a resource type permits a new set of resource schedules to be developed that may result in a reduced project duration. The cost of acquiring the extra unit divided by the reduction in duration provides a cost slope for this possible change.
2. The availability of idle resources offers opportunities for time-cost trade-offs at reduced cost slopes. The extra resource units may be used to shorten some activity durations with little or no increase in cost.
3. Time-cost trade-offs made without investigating effects on resource schedules may either be impractical or may cause variations in resource schedules that affect costs and should be considered as part of the cost slopes in order to reach correct conclusions.
4. Scarce resource types often result in a project schedule that does not have a critical path. Project duration is partly determined by activities that have been postponed due to lack of sufficient resource units. Therefore the conventional techniques for time-cost trade-offs, which depend on expediting critical activities, become inoperative. New approaches must be developed for these situations.

Current practice of treating resource allocation and time-cost trade-offs separately (in those rare cases where attempts are made to apply these techniques) and the respective procedures followed do not achieve a minimum total cost plan and schedule. The common objective of resource allocation procedures is to meet resource availability limits with the least increase in project duration. This does not necessarily represent a least-cost schedule. For example, it may be advantageous to extend project duration in order to reduce costs by an improved leveling of resource requirements below resource availability limits. The objective of time-cost trade-off procedures is to achieve minimum total costs, but as long as they ignore costs associated with resource variations, they fail to accompuish this objective.

A common objective for the joint application of both techniques must be to achieve a feasible schedule at minimum total cost. Total cost must include direct costs associated with activity performance, indirect costs associated with project duration, completion date costs reflected in bonus-penalty or liquidated damage clauses (in a broader sense, user costs, savings, or additional revenue that are shared with the contractor), and costs associated with resource variations and activity interruptions. The latter include idle time costs, move-in and move-out costs, learning curve and break-in costs, and costs dependent on the maximum level of resource requirements. At this time a well-designed, integrated, systematic procedure for minimum total cost planning is not generally available. Several research efforts are in progress to achieve such a system.

17-36. Contractor-Owner Relationships— Specification Requirements

Contract documents have customarily required that the contractor furnish data concerning his plan and schedule to the owner. In the past this requirement was often satisfied by a bar chart; more recently network diagrams, schedules, and other reports have been required. There are valid reasons for requirements of this type. The owner needs a basis for judging the contractor's progress and for determining whether the project is proceeding according to an accepted plan. The owner needs a basis for forecasting his obligations, such as the amounts of progres payments and the timing of owner-furnished materials and services.

Sometimes various actions or omissions of the owner and of others may result in claims for time extensions and equitable money adjustments. The effects of these actions or omissions on the project plan and schedule generally play an important role in settlement of such claims. Networking techniques were used very early after their introduction to substantiate major claims. Often the diagrams and data were developed "after the fact." It has become apparent that both contracting parties, for their proper protection, need a project plan and schedule that is established at the beginning of the job and is kept updated throughout the job.

Contract specifications should allow as much flexibility to the contractor as possible in providing an acceptable plan and schedule. For example, a requirement to furnish a "network diagram" rather than an "arrow diagram" allows the contractor to choose the particular type of diagramming that he can use most effectively. Requirements for excessive detail may cause the contractor to subcontract schedule preparation to an outside data processing firm rather than conducting a more beneficial in-house effort. If the owner feels the necessity for having data listed in a half dozen or more different orderings, he might well do this himself from a basic schedule furnished by the contractor.

There is some tendency for the owner to assume the planning and scheduling functions for the project and even include a network diagram and schedule in the prebid contract documents. This is an intrusion into the scope of the contractor's primary functions, and the implications should be carefully considered. Other requirements that should often be evaluated more carefully are the frequency (or infrequency) of updating, the desirability of time-scaled diagrams, and the usefulness of bar charts showing every activity for the entire project duration.

17-37. Claims

The application of networking techniques to claims settlements appears quite obvious, but it is often oversimplified. It is assumed that if a change is made, the network diagram can be modified accordingly and new scheduling calculations made. These will show the effect on project duration and indicate any time extension that should be granted. If only float time is reduced without any change in project duration, the contractor is not entitled to any adjustment. The scheduling data used as a basis for comparison for such determinations is generally assumed to be the earliest start schedule with its associated total and free float periods. This does not lead to fair conclusions.

The earliest start schedule is probably a very uneconomic one, if not completely impractical. It should only be a starting point for developing a more realistic schedule based on a consideration of resource limitations and a minimizing of resource variations with their associated idle time periods and work interruptions. For example, a resource schedule based on early start dates was shown in Figure 17-24-1. This schedule is a very poor one and probably is totally unacceptable. The resource schedule shown in Figure 17-24-2 is a much better one and was achieved by rescheduling activities within their float periods. A considerable amount of the float time was consumed in reaching this improved schedule. Moreover, in spite of the fact that there is still float time remaining for many activities, practically any delay of any activity would require project duration to be extended if resource requirements were held at existing levels.

This example illustrates several points that are ofen overlooked. First, if the owner had made changes in the original schedule that had only consumed float time, it would not have been correct to state that the contractor was unaffected costwise. The reduction in float time would have made it impossible for the contractor to achieve the more economical resource-leveled schedule within the same project duration. Second, if the owner delayed practically any activity in the project as rescheduled after resource leveling, even though the activity had float time, the result would probably be equivalent to delaying a critical activity. Either project duration or resource requirements would have to be increased, and both actions would probably result in additional costs. Third, there are definite advantages for the contractor to proceed beyond the basic scheduling computations and use the more advanced techniques, such as resource allocation, to present a realistic schedule. It will be simpler to demonstrate the adverse effects of the changes on a well-planned schedule that has already been modified to achieve economical performance.

The networking techniques can be used to demonstrate other matters of cost importance. For example, they might be used to illustrate loss of opportunities for learning curve effects, results due to changes in an activity's duration caused by rescheduling into a different season, results of increased activity duration due to inefficiencies associated with prolonged overtime work, consequences of delays associated with approval of drawings, testing, and inspection, etc. If a well-designed procedure for minimum total cost planning becomes available, as discussed at the end of the preceding section, it will have important applications in the determination of equitable cost adjustments in the field of contract claims.

17-38. Implementation

Based on the methods and recommendations of this chapter, a brief outline of procedures for networking implementation is proposed as follows:

1. Draw a network diagram as the job plan is developed. Avoid excessive detail. Work as a team composed of the key people involved in the project.
2. Estimate activity durations based on methods and crews that appear appropriate for the degree of time pressure existing.
3. Calculate project duration and determine critical activities. If the network is of moderate size this may be accomplished by manual forward pass calculations and by "backtracking" from the final activity to locate the critical path. Alternatively, the data from steps 1 and 2 can be coded, punched, and a computer run can be made.
4. If the project duration is not satisfactory, replan the work to change the duration of selected critical activities or to make sequencing changes. Update the diagram and time estimates accordingly and repeat steps 3 and 4 until a satisfactory plan and schedule are obtained.
5. Identify key resources and provide resource data input for all activities using these resources. Develop resource schedules. Determine if any major problem areas exist such as excessive resource requirements or unfavorable variations in those requirements.
6. Reschedule, if necessary, to eliminate any major problems in the requirements for key resources. This may be accomplished by trial-and-error replanning and re-

scheduling. Preferably, however, a resource allocation procedure would be used.

7. Having achieved an acceptable solution for the master plan, develop subreports to provide scheduling information to each lower echelon manager and subcontractor regarding activities for which he is responsible. Update the master plan and schedule and the key resource schedules monthly. After any necessary revisions to eliminate new problems that may have developed, also update subreports and distribute the new schedules.
8. Develop a subnetwork for the first period of the project using the "dateline cutoff" method. Extend the span of this subnet at least a week beyond the time for the monthly updating of the master schedule. Expand the activity breakdown of the subnet and develop input data for additional resource types.
9. Use the subnet for detailed planning and resource leveling of the work in the immediate future. Attempt to eliminate idle resources and expedite activity performance to achieve a more economical plan. Use the subnet to update data as soon as changes are recognized and to achieve close control of the project. Issue weekly memos to lower-echelon managers and subcontractors who have activities in progress during that week. These should provide updated start dates, details on methods and resource utilization, and current activity duration estimates.
10. Develop a new subnet immediately following each monthly update of the master schedule. Each subnet should extend beyond the date of the next monthly update. It will provide day-to-day project control, and permit continued detailed replanning of the project to meet problems that arise, to take advantage of opportunities that occur and to continually seek to improve the existing plan.

REFERENCES

1. Fondahl, John W., A Non-Computer Approach to the Critical Path Method for the Construction Industry, *Technical Report No. 9,* Department of Civil Engineering, Stanford University, 1961.
2. Fondahl, John W., Can Contractors' Own Personnel Apply CPM Without Computers? *The Contractor,* November and December, 1961.
3. Fondahl, John W., Methods for Extending the Range of NonComputer Critical Path Applications, *Technical Report No. 47,* Department of Civil Engineering, Stanford University, 1964.
4. Moder, Joseph J., and Cecil R. Phillips, *Project Management with CPM and PERT,* 2nd Ed., Van Nostrand Reinhold Co., New York, 1970.
5. Shaffer, L. R., J. B. Ritter, and W. L. Meyer: *The Critical-*

Path Method, McGraw-Hill Book Company, New York, 1965.
6. "CPM in Construction," The Associated General Contractors of America, Washington, D. C., 1965.
7. "Contractors Shift from Arrow to Precedence Diagrams for CPM," *Engineering News-Record,* May 6, 1965.
8. Fondahl, John W., Let's scrap the arrow diagram, *Western Construction,* August, 1968.
9. "USAF PERT Cost System Description Manual, Preliminary Draft," Appendices A and B, AFSC PERT Control Board, Hq. AFSC (SCCS), Washington, D.C., 1963.
10. *Project Management System/360 Application Description Manual,* 2nd Ed., International Business Machines Corporation, 1968.

18 USE OF COMPUTERS IN CONTRACTOR'S ENGINEERING ORGANIZATION

JOHN W. LEONARD

Chief Engineer
Morrison-Knudsen Company, Inc.
Boise, Idaho

THIS CHAPTER is devoted to the discussion of the use of computers in a contractor's engineering organization. One of the main functions of the contractor's engineering staff is to estimate the cost of the work. Thus, if a computer is to be used to any great degree, it has to be applied to some extent in this field. In the following we examine the computer capabilities to find its best qualities and how they can be applied to facilitate one of the principal tasks of the engineering staff—that of estimating.

18-1. Advantages of Computers

The most apparent benefits from the computer are:

1. It can be used on many of the repetitious, rather burdensome, and tiresome types of work. It can do them accurately.
2. It can do this and other types of work, in less time than it can be done manually.
3. It can update historical unit costs unemotionally. With the extremely rapid inflation that has been occurring in the past several years, contractors' historical unit costs rapidly become outdated; and relying on such costs can lead to disastrous results. Estimators have a habit of getting prices fixed in their minds—$20 a yard for concrete, $1 to drill, load, and shoot rock, etc.; and these are being outdated at a rate of 10 percent a year. The computer can update these costs and has the advantage of doing it unemotionally. It does not try to reduce them because they look too high.

4. The computer is an excellent tool for making comparisons. Again, it does this unemotionally and will make the same decision every time; thus, one method can be honestly compared to another method. The computer will never try to rationalize and possibly price down one method in favor of another because of pride of authorship or inventiveness with respect to its own concept of how to do the work.
5. With the proper additional add-on equipment supplementary to the basic computer, it can serve as a library. Contractor's engineering staffs accumulate historical records of productions, drill bit life (footage), the constitution of operating crews, cost of erection of various types of plant and equipment, and other similar data. These can be stored in a computer for retrieval, and might well be an easier way to disseminate such information among the estimators in the office as opposed to making a series of volumes for each estimator containing historical information.
6. It can also do things more accurately because it can carry on a series of calculations over a long period of time without mental fatigue and come up with a proper answer—for example, calculating the acceleration of a large haul unit and its round-trip cycle time.

The use of the computer in these fields can provide sufficient utility so that it can pay its way; and then additional uses can be made for some of the other functions of the engineering staff, such as design of cellular steel sheet-pile cofferdams, or for land cofferdams consisting of

Z-pile and walers. It can be used to assist the engineering staff in solicitation of vendors and to provide an up-to-date inventory of company-owned construction equipment with expected dates of availability. It can be used by the engineering department to assist in selection of equipment for a new project based upon its adaptability. It can also be employed for other uses, but it must pay its own way principally in facilitating the estimating functions, except insofar as it serves other departmental needs.

18-2.   Alternative Computer Services Available

Patently, there are economic limitations as to what can be spent on a computer; and, at the present time, there are two alternatives for making use of what is presently available within practical economic limits. These are: (1) the contractor's own computer facility (usually under lease), and (2) the facilities of a service bureau.

The Contractor's Own Computer Facility   Concerning the contractor's owning or leasing a computer, there are two types generally available. The first is a relatively high-cost, high-speed computer manufactured principally for the commerical market and oriented principally to book-keeping.

The second type is a lower cost, low-speed computer oriented principally for engineering. Both of these machines are oriented to punched card entry. Some of the readers will, no doubt, realize from this description the reference to the IBM machines. We have chosen to discuss these machines because IBM is the largest manufacturer of computers. Note that both of these machines are fundamentally oriented around a punched card entry and will, therefore,

usually require another format to be filled out by the engineer. The punched card will be punched from that format.

Facilities of a Service Bureau   The second alternative mentioned above, the service bureau, can be almost anything. It can have available the largest computers. It can provide for use of portions of larger computers. It can share the use of these computers with others, and communications with it can be made by a terminal, which is nothing but a typewriter located in the user's office. The printing out by such a facility would be about twice the speed of a fast typist, and about an average good reading speed.

Each of the above has its own disadvantages and advantages.

For the high-speed, high-cost computer referred to, the speed would be about 1000 lines per minute, and a lease cost would be between $5000 and $10,000 a month. This type of computer is shown in Figure 18-2-1.

This machine is very fast; to illustrate its speed of 1000 lines per minute, it could write roughly the equivalent of ten double-spaced typewritten pages in about a minute.

This size of computer probably would not be a high-cost item to an engineering department because of its adaptability to the needs of the other departments of the contractor's organization. They would share the burden of cost and help make effective use of the available time.

The average estimator would use this computer possibly in the order of half an hour per month. Therefore, the other uses—bookkeeping, payroll, cost accounting, inventory, etc.—could well utilize the balance of the time.

Another consideration is that the machine will not

Figure 18-2-1   Typical computer installation

Figure 18–2–2 Typical telecommunicator instrument for contact with service bureau computer

always be immediately available to the estimator, and he might have to wait for two to three hours. The type of computer that the contractor normally would use on commercial things does not always have a great deal of memory bank capacity (this can be thought of possibly in terms of "brain"), since it was designed in its simplest form for calculating payrolls and the like and not for making a series of long involved engineering calculations. Such a machine, therefore, has a limited memory or brain. Another disadvantage, as set forth above, is the card entry feature.

The second type of computer that the contractor might own or lease—the so-called engineering type—is slower and prints out at about 10 to 30 percent of the speed of the commercial computer. It, too, has a generally limited memory brain; but it costs less per unit of brain than the commercial type; therefore, it has a lower first cost or lease cost. Also, because of its low speed and lack of memory, discs, and ancillary equipment to handle large volumes of papers, it is less expensive to operate. A much smaller operating crew is required. This machine, then, can be thought of possibly for smaller firms to be used as a combination facility for payrolls and engineering.

The service bureau can have either of the above available on an hourly rental basis. The machines can be located in a separate building or even in a different town,

generally in any city of 100,000 to 150,000 population. If a service bureau is used, the data have to be taken to it and returned. Data are often transported back and forth to the service bureau by bus. Alternatively, however, access to the service bureau is by the "typewriter" terminal, shown in Figure 18-2-2.

The "typewriter" terminal has a great advantage in that it requires no card entry. An estimator can operate it at an acceptable rate of speed even at the "hunt and peck" level. Data can be transmitted no faster than the individual can type. Larger memory bank or brain facilities are available in this type of facility. However, if a time-sharing system is employed, operational problems will result if an effort is made to use too much of this computer's brain. This type of terminal, then, would appear to be something in which a small construction company or a district office of a medium-size or large contractor might be interested. Quite possibly, the best approach would be to use both the high-speed commercial machine for a portion of the work and the "typewriter" terminal for the balance.

18-3. Computer Languages

The seeming mystique of computers is in the language of the programs, which is the way of entering amounts and data into the computer in a form that it has been designed to accept. The computer has no capability to think. It is, however, a sophisticated device designed around the mechanics of following certain procedures for given input data; therefore, it requires the "computer language" to talk to it. Many of the language's developments today have been with respect to commercial application, but there have been some developments for engineering or other uses.

The language developed for commercial application is called COBOL. COBOL stands for Common Business Oriented Language. This was designed for commercial data processing and was developed by a committee of both computer manufacturers and users. Thus, the language means the same to all the people in the industry; and it also means the same to the computer. It was oriented to talk to the computer on business things. COBOL is meant to direct the machine to "Write stock record," "Move purchase record to monthly statement," and other such things.

There is a corresponding language called FORTRAN. This is for engineering or mathematical calculations. FORTRAN stands for FORmula TRANslator. It was originally developed for scientific problems and is, therefore, a language that talks in mathematical terms. It is being adapted to some extent to the commercial field for book-keeping uses. However, it lacks the advantages for commercial use that COBOL has. COBOL, on the other hand, is awkward to use in programming long mathematical computations.

With the development of the IBM 360 computer came Program Language One, or PL/1. This was developed by IBM and several user groups and was an attempt to close the gap between the commercial and scientific users. It was developed to be employed in both fields and has the advantage of FORTRAN in calculations, and the advantage of COBEL in writing out reports, estimates, etc.

Thus, with the language PL/1, the programmer can go either way. He can program the data recovery. He can program the long mental time calculations. He can do the updating type of work, and he can do the design type of computations. PL/1 is an easy program to use and most programmers, if not familiar with it, can pick it up rapidly. PL/1 has two additional features that seem desirable for the contractor's use. First, in using his own computer, generally at some stage in the game, he will find problems of memory (brain) limitations. These can be solved by so-called overlays, wherein the machine is programmed to do part of the problem and then, when the answers come forward, to go back and do the next part of the problem. The answers from the first portion become the input for the second portion. PL/1 does this well. It has procedures that permit it to write larger-size programs on smaller computers. Second, a portion of PL/1 can be used on the "typewriter" terminal, enabling a programmer to develop programs over the terminal to a computer in a service bureau.

18-4. Computer Programs

The command to the computer is the program. The program makes the data supplied to the computer meaningful. It tells the computer what to do with the data, such as what calculations to make, what steps to take if "A" exceeds "B" or if the answer to "A" times "B" is greater than the sum of "A" and "B," and then it prints out the answer. With effort, almost any problem can be solved by a custom-made program adapted to the computer.

In order to illustrate the steps necessary to develop a program, an example will be given of a program that compiles the gross wage rates to be used in a construction estimate.

The compilation of the hourly gross labor rate with the appropriate fringe benefits, payroll taxes and insurance, and workmen's compensation rates can be a long and laborious task. The computer program can compute this quite quickly once the base labor rates and the benefit schedule is entered. Depending upon the state, many different base rates and weekly wage limitations are used in calculating the gross labor rate. These differences are programmed into the computer so that the appropriate calculation can be made according to that state's criteria.

18-5. Sequences in Computer Application

The basic steps for computer application are:

- Develop procedure
- Make flow chart
- Write program
- Testing and debugging operation
- Documentation

Procedure The procedure that is to be programmed must be thoroughly understood. All information should be collected about the use of the procedure, and all of the necessary information about the input data must be obtained.

In the wage rate example program, all the different benefits and taxes must be thoroughly understood. Invariably, labor agreements differ from one place to another. Payroll taxes and workmen's compensation rates are calculated differently in different states. The program must provide a way of handling these differences.

Take, for instance, pension—it can be an add-on amount, a percentage of the base rate, or a percentage of the gross pay, which affects the procedure for figuring the gross pay that a laborer is to receive. To be included are travel expenses, subsistence, etc., but not taxes or fringe benefits that are paid to a labor union organization.

Inasmuch as this program had to be flexible, it was necessary to establish the input data along the lines of telling the program how to calculate each fringe benefit and add-on feature. Provisions were introduced for calculation of each addition to the basic pay scale by the proper method such as: (1) percent of the base rate, (2) percent of gross pay (as described above), (3) add-on amount in dollars per hour, (4) add-on amount for days worked, and (5) an add-on amount per week. Each rate was given a key to tell the program how to do the calculations.

The overtime or premium time was to be calculated on a percentage of the base rates. Therefore, the straight-time rates could be used in all crew calculations; then, depending upon the amount of overtime worked by the crew, the appropriate percentage for such overtime could be added on at the end.

Flow Chart The procedure for a computer function can be written in a text, but it cannot be thoroughly understood as to its logic, potential complications, or the intricacies of its program without the use of a flow chart.

The flow chart is a graphical representation of the logic and data flow of a program. A flow chart takes the sequences and interrelationships from hard-to-understand

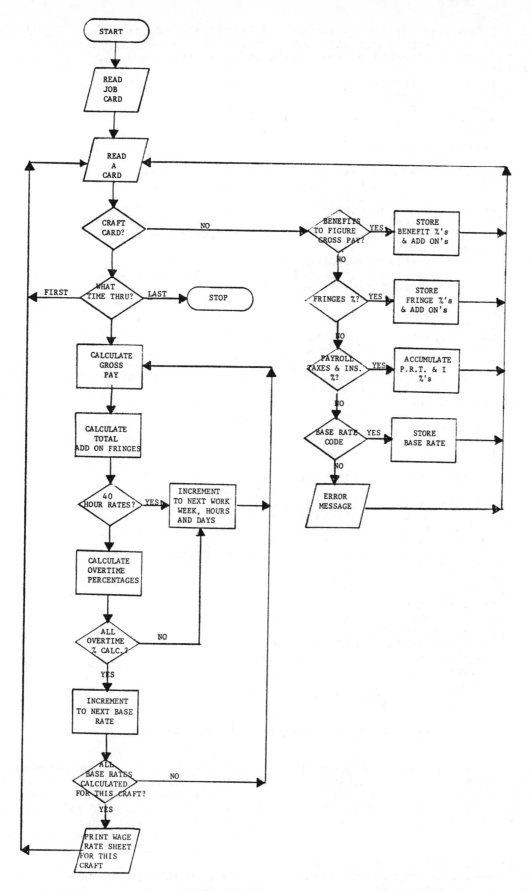

Figure 18–5–1 Flow chart

```
E020.. PROCEDURE OPTIONS%MAIN¤,.    /* LABOR RATE PROGRAM  */
DECLARE SYSIN FILE INPUT ENVIRONMENT%F%80¤ MEDIUM%SYSIPT,2501¤¤,.
DECLARE PRINT FILE OUTPUT ENV%F%132¤ MEDIUM%SYSLST,1403¤¤,.
DECLARE
        X           CHAR%1¤,
        D1          CHAR%6¤,
        DATE1       CHAR%8¤,
        PROJECT     CHAR%40¤,
        DE%5¤       CHAR%78¤,
        DA%5¤       CHAR%78¤,
        CARD        CHAR%80¤,
        A1          CHAR%20¤ INIT %@PERCENT OF BASE-RATE@¤,
        A2          CHAR%20¤ INIT %@PERCENT OF GROSS-PAY@¤,
        A3          CHAR%19¤ INIT %@ADD ON AMOUNT   $/HR@¤,
        A4          CHAR%20¤ INIT %@ADD ON AMOUNT   $/DAY@¤,
        A5          CHAR%21¤ INIT %@ADD ON AMOUNT   $/WEEK@¤,
        FRGROSS%10¤,FRADD%10¤,PRT%10¤,KEY1%10¤,KEY2%10¤,KEY3%10¤,
        WORKKEY%4¤,
        BASE%55¤,SHIFTIME%6¤,WORKCOMP%4¤ ,PERCENT%6¤,
        %FRDESC%10¤,FRADDESC%10¤,PRTDESC%10¤,CRAFT,EFFDATE,CLASS%55¤,DESC,
                COMPDESC%4¤, TYPE¤  CHAR%32¤,
        HD%15¤,
        LIMITS%54,4¤,
        ISW,M,N,.
ON ENDFILE%SYSIN¤ GO TO Y,.
ON ENDPAGE %PRINT¤ GO TO H7,.
D1#DATE,.
DATE1#SUBSTR%D1,3,2¤ CAT @-@ CAT SUBSTR%D1,5,2¤ CAT @-@ CAT
      SUBSTR%D1,1,2¤,.
CALL INTIAL,.
PRTDESC #@ @,.
PRT#0,.
ISW #   0,.
IND#0,.
S#1,.
GET FILE %SYSIN¤ EDIT %LIMITS,CARD¤
            %9%12%F%2¤,F%4¤¤,X%8¤¤,A%80¤¤,.
A..  GET FILE%SYSIN¤ EDIT%CARD¤%A%80¤¤,.
GET STRING %CARD¤ EDIT %K¤%F%2¤¤,.
IF K # 1 THEN DO,.
     PROJECT # SUBSTR%CARD,3,40¤,.
GET STRING %CARD¤ EDIT %KSTATE¤ %X%36¤,F%2¤¤,.
IF KSTATE#0 THEN KSTATE#54,.
     GO TO A,.
     END,.
IF K#4 THEN DO,.
     IF INDEX # 5 THEN GO TO A,.
     INDEX#INDEX+1,.
     DE%INDEX¤#SUBSTR%CARD,3,78¤,.
     GO TO A,.
     END,.
IF K # 5 THEN DO,.
     IF IND # 5 THEN GO TO A,.
     IND#IND+1,.
```

Figure 18–5–2 Typical written program—computer printout

texts and puts them down graphically so that they can be seen at a glance. The flow chart is a blueprint of the program to the programmer. It outlines the problems, logic, and solutions so that they can be programmed step by step. Figure 18-5-1 shows a flow chart for the wage rate program.

Program Once the flow chart has been made, with the decision points laid out and the exact data flow through the program outlined, the program is written.

A typical example of a written program, that applies to the computation of gross labor rates given above, is shown in Figure 18-5-2. Although it obviously is not presented in a form for conveying ideas through the normal process of being read, it is in a form to tell the computer how to proceed with the calculations employing variable data to be put into it later. It is in the basic language of the computer, and is included here only to illustrate how the computer is commanded to perform.

Testing and Debugging Operation The first test of a program should be with data as simple as possible. Attempts to test with complicated data at the beginning may result in difficulties. As the application of the program is proven

```
                    DA%IND¤#SUBSTR%CARD,3,78¤,.
                    GO TO A,.
                    END,.
            IF K # 7 THEN DO,.
                    GET STRING%CARD¤ EDIT %HD¤   %X%2¤,15F%5¤¤,.
                    GO TO A,.
                    END,.
            IF K # 9  THEN DO,.
                    PUT FILE %PRINT¤ PAGE EDIT %LIMITS¤ %SKIP,2%F%2¤,F%4¤¤¤,.
                    GO TO A,.
                    END,.
            GET STRING%CARD¤ EDIT%K,DESC,RATE,KEY,X¤%R%FF¤¤,.
            IF K#2 AND ISW#1 THEN GO TO H,.
            ISW#1,.
    START..
            IF K # 50 THEN GO TO G,.
            IF K GT 39 THEN GO TO E,.
            IF K GT 29 THEN GO TO D,.
            IF K GT 19 THEN GO TO C,.
            IF K GT  9 THEN GO TO B,.
            IF K  #  2 THEN CRAFT # DESC,.
    ELSE IF K  #  3 THEN EFFDATE#DESC,.
    ELSE IF K # 6 THEN S # 1 + RATE/100,.
    ELSE PUT FILE %PRINT¤ PAGE EDIT
            %@ERROR IN CONTROL CARDS.  CARD # @,K,@ IGNORED FOR @,CRAFT¤
            %A,F%2¤,A,A¤,.
            GO TO A,.
    B..     FRDESC%K-9¤#DESC,.
            FRGROSS%K-9¤#RATE,.
            KEY1%K-9¤#KEY,.
            GO TO A,.
    C..     FRADDESC%K-19¤#DESC,.
            FRADD%K-19¤   #RATE,.
            KEY2%K-19¤    #KEY ,.
            GO TO A,.
    D..     PRTDESC%K-29¤# DESC,.
            PRT%K-29¤    # RATE,.
            KEY3%K-29¤   # KEY,.
            GO TO A,.
    E..  IF K#40 THEN STHRS#RATE,.
    ELSE IF K#41 THEN OT   #RATE,.
    ELSE IF K#43 THEN GET STRING%DESC¤EDIT%SHIFTIME¤%6F%5¤¤,.
    ELSE IF K#44 THEN DO,.
                IF M # 4 THEN GO TO E1,.
                M#M+1,.
                WORKCOMP%M¤#RATE,. COMPDESC%M¤#DESC,.
                WORKKEY%M¤#KEY,.
        E1.. END,.
    ELSE PUT FILE %PRINT¤ PAGE EDIT
            %@ERROR IN CONTROL CARDS.  CARD # @,K,@ IGNORED FOR @,CRAFT¤
            %A,F%2¤,A,A¤,.
            GO TO A,.
    G..     N#N+1,.
            IF N GT 55 THEN DO,.
```

Figure 18–5–2 *(cont.)*

on simple matters, those presenting complications can follow.

For each module, about four test runs are required, with approximately four additional runs on the entire program. Testing should progress until every instruction conveyed by the program has been executed successfully. Real live data should be used in final testing. Error routines should be thoroughly checked out with data that contains every conceivable error that the program might encounter.

Documentation　The documentation for the programs is a written statement of the various things that are involved in the problem at hand and are especially important. The estimator usually has a limited knowledge of the computer, and to use the programs effectively he must understand the procedure behind the program and the input employed.

The documentation must describe how the program solves the problem. This writeup must include assumptions, equations, and exceptions used in the program. A section on how to use the program should be included for the use of the estimator.

The card layouts are a must in order for the keypunch operator to be able to punch the cards.

Sample punched cards for the wage rate program are

```
          IF N#56 THEN PUT FILE %PRINT□ PAGE EDIT
                 %@# CLASSIFICATIONS GREATER THEN 55... IGNORED FOR @,
                 CRAFT□%A,A,A□,.
             GO TO A,.
             END,.
       CLASS%N□#DESC,.
       BASE%N□ #RATE,.
       GO TO A,.
 H.. DO J # 1 TO 6,.
      IF SHIFTIME%J□ # 0 THEN SHIFTIME%J□ # 8,.
      END,.
      DO III # 1 TO M,.
      GO TO H8,.
 HO5..IF IND GT 0 THEN PUT FILE %PRINT□ SKIP,.
      DO J # 1 TO 5,.
          IF J GT IND THEN GO TO I1,.
          PUT FILE %PRINT□ EDIT%DA%J□□    %R%F1□□,.
      I1.. END,.
      IF INDEX GT 0 AND IND # 0 THEN PUT FILE %PRINT□ SKIP,.
      DO J # 1 TO 5,.
          IF J GT INDEX THEN GO TO I2,.
          PUT FILE %PRINT□ EDIT %DE%J□□    %R%F1□□,.
      I2.. END,.
      PUT FILE %PRINT□ SKIP%2□ EDIT
                      %@FRINGE BENEFITS TO FIGURE GROSS PAY@,@ @□%R%F□□,.
      DO J# 1 TO 10,.
      IF FRGROSS%J□#0 THEN GO TO H1,.
      IF KEY1%J□#1 THEN TYPE#A1,.
      IF KEY1%J□#2 THEN TYPE#A2,.
      IF KEY1%J□#3 THEN TYPE#A3,.
      IF KEY1%J□#4 THEN TYPE#A4,.
      IF KEY1%J□#5 THEN TYPE#A5,.
      PUT FILE %PRINT□ SKIP EDIT
                  %FRDESC%J□,FRGROSS%J□,TYPE□%R%FFF□□,.
 H1.. END,.
      PUT FILE %PRINT□ SKIP%2□ EDIT
                  %@ADD ON FRINGES@,@ @□%R%F□□,.
      DO J # 1 TO 10,.
      IF FRADD%J□#0 THEN GO TO H2,.
      IF KEY2%J□#1 THEN TYPE#A1,.
      IF KEY2%J□#2 THEN TYPE#A2,.
      IF KEY2%J□#3 THEN TYPE#A3,.
      IF KEY2%J□#4 THEN TYPE#A4,.
      IF KEY2%J□#5 THEN TYPE#A5,.
      PUT FILE %PRINT□ SKIP EDIT
                  %FRADDESC%J□,FRADD%J□,TYPE□%R%FFF□□,.
 H2.. END,.
      TYPE#A2,.
      PUT FILE %PRINT□ SKIP%2□ EDIT
                  %@PAYROLL TAXES AND INSURANCE@,@ @□%R%F□□,.
      ADDON#0,.
      DO J# 1 TO 10,.
      IF PRT%J□#0 THEN GO TO H3,.
      ADDON#ADDON+PRT%J□,.
```

Figure 18-5-2 *(cont.)*

presented in Figure 18-5-3 as an illustration, together by a statement of the basic data including basic wages and an example of a typical input form giving the additive things such as fringes, taxes and insurance, etc., to be considered in the computations. (See Figures 18-5-4 and 18-5-5.)

The code describes the type of card:

20-30	Fringe benefits
44	Workmen's compensation rate
50	Base wage rates

The key tells the program what type of rate and how it is to be applied:

1 Percent base rate
2 Percent gross pay
3 Dollar add-on amount per hour
4 Dollar add-on amount per day
5 Dollar add-on amount per week

Figure 18-5-6 is a sample printout of the compiled wage rates for one craft. This printout gives the gross wage rate to be used in estimating. This gross wage rate includes all

```
              PUT FILE %PRINT¤ SKIP EDIT
                        %PRTDESC%J¤,PRT%J¤,TYPE¤%R%FFF¤¤,.
  H3.. END,.
        IF WORKKEY%III¤ NE 3 THEN DO,.
        ADDON#ADDON+WORKCOMP%III¤,.
        PUT FILE %PRINT¤ SKIP EDIT
                     %@WORKMANS COMP RATE@,WORKCOMP%III¤,TYPE¤%R%FFF¤¤,.
        END,.
        PUT FILE %PRINT¤ SKIP EDIT
                        %@TOTAL PAYROLL TAXES AND INSURANCE@,ADDON,TYPE¤
           %A%34¤,F%6,3¤,X%1¤,A¤,.
        IF WORKKEY%III¤#3 THEN DO,.
        TYPE#A3,.
        PUT FILE %PRINT¤ SKIP EDIT
                        %@WORKMANS COMP RATE@,WORKCOMP%III¤,TYPE¤%R%FFF¤¤,.
        END,.
        IF S NE 1 THEN PUT FILE %PRINT¤ SKIP%2¤ EDIT
                %@WAGE RATES ARE ESCALATED@,%S-1¤*100,@PERCENT@¤
                %R%FFF¤¤,.
        PUT FILE %PRINT¤ SKIP%2¤ EDIT
                        %@SHIFTS@,SHIFTIME,@OVERTIME RATE@,OT,
           @STRAIGHT TIME HOURS@,STHRS¤ %A%7¤,F%6,2¤,F%9,2¤,F%5,2¤,
           F%9,2¤,2F%5,2¤,X%3¤,A%15¤,F%4,2¤,X%3¤,A,X%2¤,F%6,2¤¤,.
        GO TO H10,.
  H4.. IF N GT 55 THEN N#55,.
        DO II # 1 TO N,.
        DO I#  1 TO 7,.
        IF I#1 THEN DO,. DAYS#HD%1¤ ,. HRS#HD%2¤ ,. END,.
        IF I#2 THEN DO,. PERCENT%1¤#%16/%SHIFTIME%2¤+SHIFTIME%3¤¤-1¤
                *100,.        GO TO CONTINUE,.      END,.
        IF I#3 THEN DO,. PERCENT%2¤#%24/%SHIFTIME%4¤+SHIFTIME%5¤+
                     SHIFTIME%6¤¤-1¤*100,. GO TO CONTINUE,.  END,.
        IF I#4 THEN DO,. DAYS#HD%3¤ ,. HRS#HD%4¤ ,. END,.
        IF I#5 THEN DO,. DAYS#HD%5¤ ,. HRS#HD%6¤ ,. END,.
        IF I#6 THEN DO,. DAYS#HD% 7¤,. HRS#HD% 8¤,. END,.
        IF I#7 THEN DO,. DAYS#HD% 9¤,. HRS#HD%10¤,. END,.


        BASEWK#BASE%II¤*STHRS,.
        OTWK#%HRS-STHRS¤*OT*BASE%II¤,.
        PENALTYOT#BASEWK+OTWK-HRS*BASE%II¤,.
        CALL GROSSFRINGE,.
        GROSSPERWEEK#BASEWK+OTWK+GROSSFRINGES,.
        CALL FRINGE,.
        CALL PRTINS,.
        GROSSPERWEEK#GROSSPERWEEK+PRTI+FRINGES,.
        GROSSRATE#GROSSPERWEEK/HRS,.
        IF I#1 THEN
            DO,.
            FRINGERATE#GROSSRATE-BASE%II¤,.
            RATE40#GROSSRATE,.
            GO TO CONTINUE,.
            END,.
        PERCENT%I-1¤#%GROSSRATE/RATE40-1¤*100,.
```

Figure 18–5–2 *(cont.)*

fringe benefits, payroll taxes and insurance, and the workmen's compensation rate.

18-6. Alternative Sources for Development of Computer Programs

Computer programs can be created or obtained from six possible sources.

- Firms specializing in program development
- Contractor semi-oriented firms
- MIT, ICES, and other package programs
- Moonlighters
- Manufacturers
- Contractor's own people

Firms Specializing in Program Development The first is through the so-called software firms. These are firms who specialize in creating programs. Some of them are part of computer companies. Most of them are separate firms dealing in these services. They are called software firms because they furnish the paper-type requirements for the computer as opposed to the hardware equipment.

```
CONTINUE.. END,.


H6.. PUT FILE %PRINT□ EDIT
             %CLASS%II□,BASE%II□*S,FRINGERATE*S,RATE40*S,PERCENT□
             %SKIP,A%32□,2F%6,3□,F%7,3□,2F%8,2□,F%11,1□,3F%15,1□□,.
     END,.
     END,.
     CALL INTIAL,.
     CRAFT#DESC,.
     GO TO A,.

H7.. KSW#1,.
     GO TO H9,.
H8.. KSW#0,.
H9.. PUT FILE %PRINT□ PAGE EDIT
                %@WAGE RATES PER HOUR FOR-@,PROJECT,EFFDATE,@DATE @,
          DATE1□%X%9□,A,A%32□,X%2□,A%32□,X%3□,A,A□,.
     PUT FILE %PRINT□ SKIP%2□ EDIT
                   %@CRAFT - @ ,CRAFT,@TYPE OF WORK - @,COMPDESC%III□□
          %X%9□,A,X%3□,A,A,A□,.
     PUT FILE %PRINT□ EDIT%@    RUN #@,SUBSTR%PROJECT,37,4□,
          @    STATE @,KSTATE□   %A,A,A,F%2□□,.
     IF KSW#0 THEN GO TO H05,.
     GO TO H11,.
H10..KSW#0,.
H11..PUT FILE %PRINT□ SKIP%2□ EDIT
     %@CLASSIFICATION@,@-@,HD%2□,@ HOUR WEEK-   PERCENT CHANGE   @,
     @-PERCENT CHANGE FOR OVERTIME FROM @,HD%2□,@BASE   FRINGE TOTAL@,
     @RATE    RATE   RATE    2 SHFTS 3 SHFTS  @,HD%4□,@ HOUR WEEK    @,
     HD%6□,@ HOUR WEEK   @,HD% 8□,@ HOUR WEEK   @,HD%10□,@ HOUR WEEK@□
     %X%3□,A,X%19□,A,F%2□,A,A,F%2□,A,SKIP,X%54□,A,SKIP,X%33□,A,SKIP,
     X%33□,A,4%F%2□,A%13□□□,.
     IF KSW#0 THEN GO TO H4,.
     GO TO H6,.


GROSSFRINGE.. PROCEDURE,.
     GROSSFRINGES#0,.
     DO J# 1 TO 10,.
     ADDON#0,.
     IF FRGROSS%J□#0 THEN GO TO P,.
     IF KEY1%J□#1 THEN ADDON#%   FRGROSS%J□/100□*BASE%II□*HRS,.
     IF KEY1%J□#2 THEN ADDON#%FRGROSS%J□/100□*%BASEWK+OTWK□,.          EO
     IF KEY1%J□#3 THEN ADDON#FRGROSS%J□*HRS,.
     IF KEY1%J□#4 THEN ADDON#FRGROSS%J□*DAYS,.
     IF KEY1%J□#5 THEN ADDON#FRGROSS%J□,.
     GROSSFRINGES#GROSSFRINGES+ADDON,.
P.. END,.
END,.


FRINGE.. PROCEDURE,.
     FRINGES#0,.
     DO J# 1 TO 10,.
```

Figure 18–5–2 *(cont.)*

The software firms, in general, appear to be principally building-oriented because that seems to be a more repetitious field. They also like to think and talk in terms of entire computer systems. Quite probably the market for the software firms in the construction industry is just not a market fruitful enough for them at this time. They are also not knowledgeable in construction—particularly in the heavier construction field. Therefore much of the work has to be done by the contractor's staff.

Contractor Semi-Oriented Firms Another category is some of the consulting firms who are semi-oriented to contractors. Several of these have developed estimating programs which they may use or which may be used by contractors.

This type of program has been developed and used by some of these firms for developing estimates in the feasibility stage of projects, in some cases for engineers' estimates; and, in many cases, they have been used as check estimates for the nonsponsoring partner in joint ventures. This type of program uses the computer for simple mathematics such as multiplications, etc., and for the manipulation of data so as to accumulate man-hours by crafts. This tends to use the computer as a calculator-

```
                ADDON#0,.
                IF FRADD%J¤#0 THEN GO TO Q,.
                IF KEY2%J¤#1 THEN ADDON#% FRADD%J¤/100¤*BASE%II¤*HRS,.
                IF KEY2%J¤#2 THEN ADDON#% FRADD%J¤/100¤*GROSSPERWEEK,.
                IF KEY2%J¤#3 THEN ADDON#FRADD%J¤*HRS,.
                IF KEY2%J¤#4 THEN ADDON#FRADD%J¤*DAYS,.
                IF KEY2%J¤#5 THEN ADDON#FRADD%J¤,.
                FRINGES#FRINGES+ADDON,.
        Q..     END,.
        END,.


        PRTINS.. PROCEDURE,.
                PRTI#0,.
                DO J # 1 TO 10,.
                IF PRT%J¤ # 0 THEN GO TO R,.
                IF J # 2 THEN DO,.
                    DEDUCTOT#LIMITS%KSTATE,3¤*PENALTYOT,.
                    IF GROSSPERWEEK-DEDUCTOT LE LIMITS%KSTATE,4¤
                        OR LIMITS%KSTATE,4¤ #0
                        THEN PRTI # PRTI + PRT%J¤/100*%GROSSPERWEEK-DEDUCTOT¤,.
                        ELSE PRTI # PRTI + PRT%J¤/100*LIMITS%KSTATE,4¤,.
                    GO TO R,.
                    END,.
                PRTI#PRTI + PRT%J¤/100*GROSSPERWEEK,.
        R..     END,.
                IF WORKKEY%III¤ # 3 THEN PRTI # PRTI+WORKCOMP%III¤*HRS,.
                    ELSE DO,.
                    DEDUCTOT # LIMITS%KSTATE,1¤*PENALTYOT,.
                    IF GROSSPERWEEK-DEDUCTOT LE LIMITS%KSTATE,2¤
                        OR LIMITS%KSTATE,2¤ # 0
                        THEN PRTI # PRTI + WORKCOMP%III¤/100*
                            %GROSSPERWEEK-DEDUCTOT¤,.
                        ELSE PRTI # PRTI + WORKCOMP%III¤/100*LIMITS%KSTATE,2¤,.
                    END,.
        END PRTINS,.


        INTIAL.. PROCEDURE,.
                CRAFT    #à à,.
                EFFDATE  #à à,.
                FRDESC   #à à,.
                FRADDESC #à à,.
                CLASS    #à à,.
                INDEX#0,.
                M#0,.
                FRADD#  0,.
                FRGROSS#0,.
                N#0,.
                STHRS#40,.
                OT#1.5,.
                SHIFTIME#8,.
        END,.


        F..    FORMAT%A,SKIP,A¤,.
        F1..   FORMAT%SKIP,X%52¤,A%78¤¤,.

        FF..   FORMAT%F%2¤,A%32¤,F%8¤,X%1¤,F%1¤,A%36¤¤,.
        FFF..  FORMAT%X%2¤,A%32¤,F%6,3¤,X%1¤,A¤,.


        Y..    PUT FILE %PRINT¤ PAGE EDIT%àEND-OF-JOBàu%A¤,.
               CLOSE FILE %PRINT¤,.
        END..    /*  END OF PROGRAM   */
```

Figure 18–5–2 *(cont.)*

typewriter. This type of program will generally end up in output somewhat as follows:

An example printout of a calculator-typewriter is given in Figure 18-6-1 wherein an excavation job is illustrated. The column cost code on the left-hand side refers to the stored cost of the item listed. The hours are entered for the unit; referring to the code, it then calculates and totals up the separate columns. This system can be expanded to where each piece of equipment has assigned production rates for loading, operating, etc.

An example printout of a calculator-typewriter applicable to building construction is given in Figure 18-6-2.

These programs make use of operating cost banks, labor cost banks, and material cost banks. These are standard

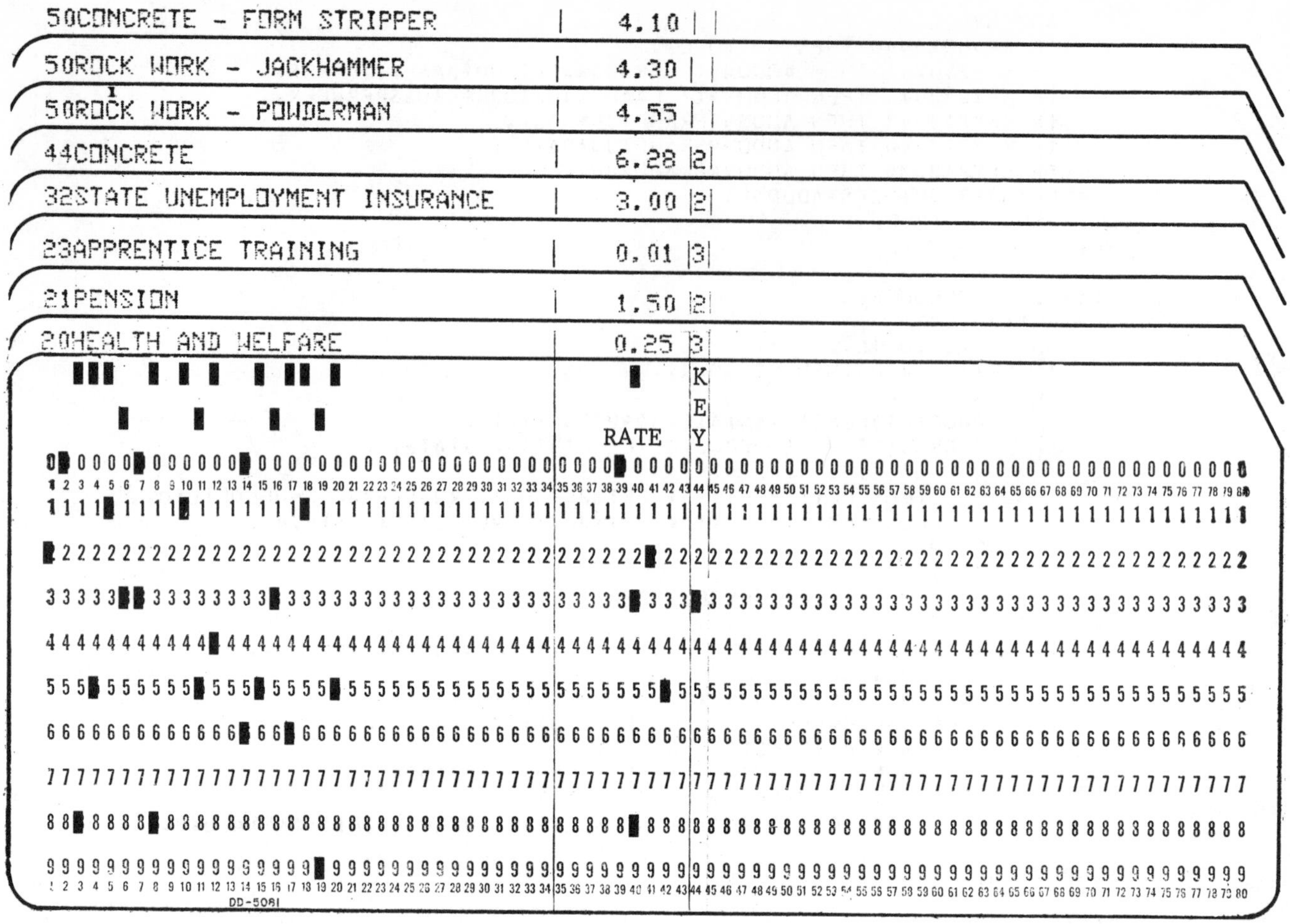

```
50CONCRETE - FORM STRIPPER          |  4.10 | |

50ROCK WORK - JACKHAMMER             |  4.30 | |

50ROCK WORK - POWDERMAN              |  4.55 | |

44CONCRETE                           |  6.28 |2|

32STATE UNEMPLOYMENT INSURANCE       |  3.00 |2|

23APPRENTICE TRAINING                |  0.01 |3|

21PENSION                            |  1.50 |2|

20HEALTH AND WELFARE                 |  0.25 |3|
```

Figure 18–5–3   Sample punch cards

operating costs that they will use throughout the estimate, such as wage rates and standard material prices. Small tools are often accounted for by a percentage allowance on labor.

These types of estimating systems have several disadvantages. They tend to have the estimator crew up everything, which requires more work and makes the system less flexible. It is very difficult to handle something that doesn't run full time with the crewed up spread that has been input to the computer and cards—for example, if the hand compaction spread only runs 2000 hours out of the 8000 hours total required. For concrete operations, with various crews on finishing, pouring, curing, and cleanup, the cost sometimes gets to be overstated, because it is difficult to take credit for the different uses of the crews from pour to cleanup, etc.

With the card entry and the separate crew, it becomes difficult to vary the production rates. Often it is desired to use one rate for a portion of the job, another rate because loading equipment will control, and another rate because of lack of haul units.

This concept makes it difficult to use "allow" numbers—"allow" $3000 for additional culverts or "allow" $300 for road barricade. This becomes a hardship and is too much of a struggle with small items.

The built-in factors that apply to labor with the operating cost banks, etc., might be too high for a portion of the operation because the production rates are very low or because the particular use in point is extremely light;

and, once these are in the computer bank, it becomes a major effort to change them. The estimator cannot understand all the programs, etc.; and he becomes baffled.

Another example of a problem is for an operation such as ripping—if it is especially difficult, there has to be added some downtime on the operators which would not show up in the operating cost bank. Standby equipment has to be added as a separate item to cover the cost of that which otherwise would not get into the estimate and is something that tends to get overlooked because of the inflexible nature of the computer.

If it is desired to have the entire estimate printed out, a lot of typing is involved. Not only does someone have to punch cards but another person has to read the estimators' writing (not often an easy chore) and arrange for it to be typewritten. The format that is required before a punched card can be punched is usually somewhat foreign to what the estimators are accustomed to because it has to be in a form that can be translated by the keypunch operator. This takes time. The estimate has to be put on the input sheets, punched by the keypunch operator, sent to the computer, held for computer time, and printed out. Oftentimes an estimator will be held up in his work because he cannot continue until he gets the answer back from the computer on work he has already done. When it does come back, he has cooled off on the subject and is not as alert when resuming work as he was earlier. Also, he is likely to believe that what is printed out on that green and white or blue and white paper is authentic.

LABORERS	WAGE RATES					
CLASSIFICATIONS	FROM TO	FROM TO	FROM TO	FROM TO	FROM TO	FROM TO
GENERAL — COMMON	4.10					
AIR TOOL OPERATOR	4.15					
BARKO AND TAMPER	4.20					
DUMPMAN	4.10					
FLAGMAN OR WATCHMAN						
PIPE LAYER — METALLIC CULVERT						
NON-METALLIC						
HYDRAULIC MONITOR						
CONCRETE — CONCRETE LABOR	4.10					
MASON TENDER	4.15					
VIBRATOR OPERATOR	4.20					
HEAVY VIBRATOR OPERATOR						
FORM STRIPPER	4.10					
SAND BLASTER	4.15					
GUNITE — NOZZLEMAN						
Multi-Tamper	4.20					
ROCK WORK — HIGHSCALER						
JACKHAMMER	4.30					
WAGON DRILL						
AIR TRACK DRILL						
DIAMOND DRILL						
CHUCKTENDER						
POWDERMAN	4.55					
POWDERMAN HELPER						
PAVING — FORM SETTER						
ASPHALT LABORER						
ASPHALT RAKER						
CONCRETE SAW						
HEADERBOARD MAN						
FOREMAN						
FOREMAN RATIO						
TUNNEL WORKERS —						
FREE AIR — WALKER						
SHIFTER						
LEADMAN						
SAFETY MINER						
MINER						
DRILLER — JACKLEG OR JACKHAMMER						
WAGON OR AIR TRACK						
DIAMOND						
POWDERMAN — HEADING						
PRIMER HOUSE						
FORM SETTER						
MOTORMAN OR MUCKER OPERATOR						
NOZZLEMAN — GUNITE						
SLICKLINE						

Figure 18–5–4 Basic labor rate listing

01 Project: .... State Code .... Run # .... Date ....

WAGE RATE PROGRAM—INPUT FORM

KEY: 1—% Base; 2—% Gross; 3—$/Hour; 4—$/Day; 5—$/Week

	Code	Alphabetic Description		Rate		Key
	CC 1-2	3	34	35	42	44
CRAFT	02	Laborers				
EFFECTIVE DATE F/WAGE RATE	03	From to				
NOTE (For Additional	04					
Information)	04					
	04					
	06					
FRINGES TO FIGURE GROSS PAY						
	10	Travel		10		4
	11	Subsistence		50		5
	12	Paid Vacation		5		1
	13	Paid Holidays				
	14					
	15					
	16					
ADD ON FRINGES	20	Health and Welfare		0.25		3
	21	Pension		1.5		2
	22	Savings				
	23	Apprentice Training		0.01		3
	24	Unemployment Fund				
	25	Industry Fund		0.05		3
	26					
	27					
	28					
PAYROLL TAXES and INSURANCE						
% of Gross Pay	30	F.I.C.A.		4.80		2
	31	Bodily Injury and Property Damage		2.5		2
	32	State Unemployment Insurance		3.0		2
	33	Federal Unemployment Insurance		0.40		2
	34	Excess Employer's Liability		0.04		2
	35					
	40	Straight Time, hours		40		
	41	Overtime Factor		1.5		

		1 Shift	2 Shifts		3 Shifts		
	43	8	8	7.5	8	7.5	7

		Type of Work		Rate		
WORKMAN'S COMPENSATION RATE	44	Concrete		6.28		2
Use Only as Many	44					
As Are Needed	44					

Figure 18–5–5 Wage rate program—input form

Having the entire estimate printed out by the computer has several advantages. The completed estimate is very easy to follow. It can be reviewed quickly by the responsible officers of the contractor, and it can be quickly referred to in prebid joint venture comparison meetings when making detailed comparisons and breakdowns. It has the value of consistency; and, if it is agreed that something must be changed, it is easy to make the change. For example, if one part of the operating cost seems to be in error, that change can be made in the entire estimate quickly and accurately. This, of course, is also advantageous when the estimate work is reviewed by the chief estimator, chief engineer, etc. The changes can be made by changing just a few cards and running out a whole new revised estimate.

One big advantage of the system of complete computer

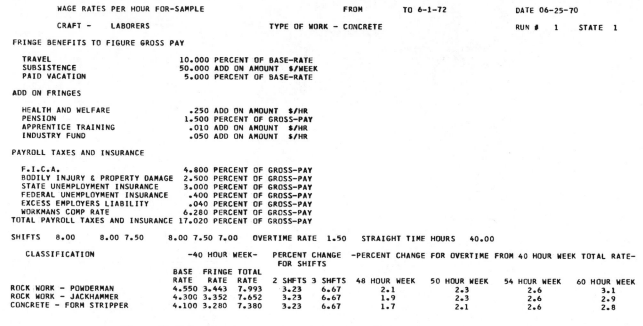

WAGE RATES PER HOUR FOR-SAMPLE   FROM   TO 6-1-72   DATE 06-25-70

   CRAFT -   LABORERS   TYPE OF WORK - CONCRETE   RUN #   1   STATE   1

FRINGE BENEFITS TO FIGURE GROSS PAY

     TRAVEL   10.000 PERCENT OF BASE-RATE
     SUBSISTENCE   50.000 ADD ON AMOUNT $/WEEK
     PAID VACATION   5.000 PERCENT OF BASE-RATE

ADD ON FRINGES

     HEALTH AND WELFARE   .250 ADD ON AMOUNT $/HR
     PENSION   1.500 PERCENT OF GROSS-PAY
     APPRENTICE TRAINING   .010 ADD ON AMOUNT $/HR
     INDUSTRY FUND   .050 ADD ON AMOUNT $/HR

PAYROLL TAXES AND INSURANCE

     F.I.C.A.   4.800 PERCENT OF GROSS-PAY
     BODILY INJURY & PROPERTY DAMAGE 2.500 PERCENT OF GROSS-PAY
     STATE UNEMPLOYMENT INSURANCE   3.000 PERCENT OF GROSS-PAY
     FEDERAL UNEMPLOYMENT INSURANCE   .400 PERCENT OF GROSS-PAY
     EXCESS EMPLOYERS LIABILITY   .040 PERCENT OF GROSS-PAY
     WORKMANS COMP RATE   6.280 PERCENT OF GROSS-PAY
   TOTAL PAYROLL TAXES AND INSURANCE 17.020 PERCENT OF GROSS-PAY

SHIFTS   8.00   8.00 7.50   8.00 7.50 7.00   OVERTIME RATE 1.50   STRAIGHT TIME HOURS 40.00

   CLASSIFICATION   -40 HOUR WEEK-   PERCENT CHANGE -PERCENT CHANGE FOR OVERTIME FROM 40 HOUR WEEK TOTAL RATE-
                                                 FOR SHIFTS
                               BASE FRINGE TOTAL
                               RATE   RATE RATE 2 SHFTS 3 SHFTS 48 HOUR WEEK 50 HOUR WEEK 54 HOUR WEEK 60 HOUR WEEK
ROCK WORK - POWDERMAN   4.550 3.443 7.993   3.23   6.67   2.1   2.3   2.6   3.1
ROCK WORK - JACKHAMMER   4.300 3.352 7.652   3.23   6.67   1.9   2.3   2.6   2.9
CONCRETE - FORM STRIPPER   4.100 3.280 7.380   3.23   6.67   1.7   2.1   2.6   2.8

Figure 18-5-6 Sample computer printout—compiled wage rates for one craft

EXHIBIT 2-2.1

MASS EXCAVATION

COST CODE		UNIT	LABOR	EQUIPMENT	SUPPLIES
0613	EXCAVATE AND HAUL	CY			
0010	FOREMAN - EXCAVATION	HR			
0101	PICKUP - EXCAVATION	HR			
0110	SHOVEL	HR			
0115	FRONT END LOADER	HR			
0130	SCRAPER	HR			
0135	TRUCK	HR			
0012	MEN	HR			
	SPREAD	CY			
0160	CAT 12 HAUL ROAD	HR			
0165	WATER TRUCK	HR			
0164	CAT GRADER	HR			
0163	CAT GRADER	HR			
0013	DUMP MAN	HR			
	COMPACT	CY			
0011	FOREMAN DUMP	HR			
0101	PICKUP FOREMAN	HR			
0125	COMPACTOR	HR			
0126	ROLLER	HR			
0164	CAT GRADER	HR			
0127	SHEEP'S FOOT	HR			
0128	PNEUMATIC TON	HR			
0100	CAT DOZERS	HR			
0014	GRADE MAN	HR			

Figure 18-6-1 Example of a printout of a calculator-typewriter illustrating an excavation job

estimating is that it tends to introduce the use of preprinted sheets for various operations. Figure 18-6-3 is a typical preprinted sheet that outlines the various elements of cost that must be considered in estimating the cost of providing material handling conveyors, and how the costs should be broken down.

However, there are definite space limitations in the computer. It can only handle 132 spaces in output. If the foregoing sheet were to be printed out by the computer after doing the required calculations, it would require two pages (which would need to be pasted or taped together) and would be shown in Figure 18-6-4.

Considering the foregoing, it would appear that the disadvantages of complete computerization of an estimate outweigh the advantages.

Package Programs Another alternative open to the computer user is employing so-called packaged or library programs. This is the type of programming that has been developed by a university such as the Massachusetts Institute of Technology, by the Integrated Civil Engineering System, or by computer manufacturers who have established programming libraries over a period of time and made them available to their customers with or without charge. Another source of these package programs is the various groups of computer users that have banded together to establish a library for joint use.

The ready-made programs or packaged or library programs are not a very fruitful field for the contractor. It is a disappointing field. Manufacturers have naturally developed programs for the most obvious market, which is the engineering design firms. Even these are not being too widely used by the engineering firms. Even though the charges for these programs are nominal, most engineering firms still develop their own programs; and the library-type package program will represent only 10 to 30 percent

ESTIMATE REPORT NUMBER FOUR

PAGE 9

RD CODE	TITLE/NOTES	DESCRIPTION	UNIT	QUANTITY	LABOR AMOUNT	MATRL AMOUNT	------MAN DAYS BY CRAFT-------					TOTAL
							1	2	3	4	5	
2	BUILDING FOUNDATION											
1210		FOUNDATION PREPARATION	SF	500.	200.	0.	5.1	0.0	0.0	0.0	0.0	5.1
2020		FORM FOOTINGS	SF	500.	400.	50.	4.1	0.0	9.0	0.0	0.0	13.1
2180		ANCHOR BOLTS	LB	400.	400.	30.	3.5	0.0	3.5	0.0	0.0	7.0
2510		CONCRETE	CY	40.	200.	1500.	3.0	1.5	0.5	0.5	0.0	5.5
2720		FINISH & CURE	SF	1200.	80.	20.	1.0	0.0	0.0	2.0	0.0	3.0
4630		BACKFILL	CY	100.	200.	350.	6.0	0.0	0.0	0.0	0.0	6.0
					1480.	1950.	22.7	1.5	13.0	2.5	0.0	39.7

THE FOLLOWING MATERIALS WERE USED IN THIS WORK

MATERIAL		QUANTITY	AMOUNT
SHIPLAP 1"	SF	1000.	100.
GRAVEL	CY	100.	350.
2 X 6 LUMBER	BM	400.	40.
4 X 4 LUMBER	BM	150.	20.
FORM OIL	GAL	20.	10.
CONCRETE	CY	40.	1500.
NAILS, BOLTS, ETC.	LB	25.	5.

RD CODE	TITLE/NOTES	DESCRIPTION	UNIT	QUANTITY	LABOR AMOUNT	MATRL AMOUNT	1	2	3	4	5	TOTAL
4	ACCESS FTG											
1050		EXCAVATION	CY	50.	200.	0.	5.0	0.0	0.0	0.0	0.0	5.0
2030		FORMS	SF	30.	750.	50.	9.2	0.0	9.2	0.0	0.0	18.4
2510		CONCRETE	CY	20.	25.	300.	0.5	0.1	0.1	0.0	0.0	0.7
2710		FINISH & CURE	SF	30.	0.	0.	0.0	0.0	0.0	0.0	0.0	0.0
4600		BACKFILL	CY	0.	0.	0.	0.0	0.0	0.0	0.0	0.0	0.0
					975.	350.	14.7	0.1	9.3	0.0	0.0	24.1

THE FOLLOWING MATERIALS WERE USED IN THIS WORK

MATERIAL		QUANTITY	AMOUNT
2 X 4 LUMBER	BM	15.	3.
FORM OIL	GAL	0.	0.
CONCRETE	CY	20.	300.
NAILS, BOLTS, ETC.	LB	1.	1.

RD CODE	TITLE/NOTES	DESCRIPTION	UNIT	QUANTITY	LABOR AMOUNT	MATRL AMOUNT	1	2	3	4	5	TOTAL
5	SUMP											
1060		EXCAVATION	CY	200.	300.	0.	4.4	0.0	0.0	0.0	0.0	4.4
1250		FOUNDATION PREPARATION	SF	400.	50.	0.	0.4	0.0	0.0	0.0	0.0	0.4
2080		FORM BEAMS	SF	250.	4500.	200.	25.0	0.0	56.7	0.0	0.0	81.7
2090		FORM WALLS	SF	250.	1300.	0.	10.3	0.0	10.7	0.0	0.0	21.0

Figure 18–6–2   Example of a printout of a calculator-typewriter application to building construction

of their total programs. Many of the engineering firms do not use them at all.

An engineering firm can make good use of the computer for selecting the most desirable road or canal location by comparing one alignment against four or five others. This sort of thing is the computer's finest moment. All of the original ground data must have been put in the computer. The templates consisting of road widths and side slopes are put in also, and six different alignments can be run to determine quantities and feasibility. Three of them might be eliminated quickly. One of them might probably be objectionable for other reasons. The engineer is, thus, left with two alignments. He can select the best portions of each and could then come up with one final optimum alignment.

The contractor normally does not have this problem to solve because the engineer would already have selected the alignment. The contractor, in computing quantities for a road or canal, would have to go through all this again in order to establish original ground data for only one run—the final alignment—not six alignments.

There are machines available for taking data from aerial photographs (with requisite surface controls) and developing original ground data which can be given to the computer along with template—road widths, side slopes, etc. This can be all done without any manual work using a special aerial photograph machine for making the input to the computer. The machine described above is presently known as a digitizer. Another special machine can take information off of contour drawings, but it is probably not economically feasible for a contractor to own these devices. Possibly some enterprising service bureau will provide a service to contractors for accomplishing these functions in the next few years.

Another example of package programming is one pre-

pared by IBM for instruction purposes and to demonstrate what can apply to takeoff work. The example is set forth in Figure 18-6-5. The first portion has to do with takeoff for forms and concrete—"Q" standing for quantities and "H," "L," and "W" being dimensions. The use of a code tells the computer how to calculate, the calculations being different for the square area of forms than for the volume of concrete in cubic yards. There is input into the computer the desired unit costs under the headings of "Labor," "Materials," and "Equipment," that apply to the various elements entering into cost computations for concrete and forms. These constitute the data bank for computations. The answer then combines the takeoff and the desired unit prices to give a summary for the job in the form as shown in Figure 18-6-5.

Some time ago one of the leading contractors attempted through the combined AGC/ASCE Committee to establish the practice of having the owner's engineer supply the contractors with original ground data in the form of punched cards. As of this writing, very little significant progress has been made along this line.

In many cases the original ground surface that applies to the contractor's work is the surface after some excavation has been done. For example, the foundation for an earthfill dam, or the ground surface upon which the fill is placed, is only determined after the cutoff trench to varying depths in rock has been completed and after the remainder of the foundation area has been excavated. Cross sections of this excavation have to then be plotted by hand, and it is just as quick to planimeter them to determine the areas involved as it is to wait to put them on punched cards and queue up for the computer output.

Moonlighters   Another source can best be described as moonlighters. These generally are engineers working for

CONVENTIONAL CHANNEL STRINGER STYLE

	Unit	LABOR AND INSURANCE		PERMANENT MATERIALS		SPECIFIC PLANT		EQUIPMENT RENTAL		SUPPLIES		SUB-CONTRACTS		TOTAL DIRECT COST		COST PER LF OF CONVEYOR	SALVAGE TAKEN
		UNIT COST	AMOUNT	UNIT COST	AMOUNT	UNIT COST	AMOUNT	UNIT COST	AMOUNT	UNIT COST	AMOUNT	UNIT COST	AMOUNT	UNIT COST	AMOUNT		
Purchase	LF																
Local Freight	ST																
Unload and Handle																	
Small Tool Allowance																	
Sales Tax																	
DRIVE STATION																	
Structure Excavation	CY																
Concrete	CY																
Structural Steel Support	LB																
Machinery Install	LB																
Electrical	HP																
Shelter	EA																
CONVEYOR LINE INSTALLATION																	
Concrete	CY																
Steel Fabricated Structures	LB																
Wood Ties	LF																
Machinery Installation (Idlers)	LB																
Install Belt - Per L.F. of Conveyor	LF																
Lighting	LF																
Electrical Control Wire - L.F. of Conveyor	LF																
Air Distribution Lines	LF																
Dismantle and Store	LF																
Clear and Grubb	AC																
Excavation Common	CY																
Excavation Rock	CY																
Culverts	LF																
Road Surfacing	LF																
Road Crossing	LF																
Substation																	
Transmission Line (or Parkway Cable)																	

Figure 18-6-3 Typical preprinted sheet showing elements of cost for material handling conveyor

457

```
OPERATION    CONVEYOR INSTALLATION                              ITEM          SHEET    OF        PAGE
                                                                JOB
QUANTITY                                              EST. BY-  CKD. BY-       DATE

DESCRIPTION                          QUANTITY  UNIT   LABOR  AND    PERMANENT       SPECIFIC        EQUIPMENT
                                                     INSURANCE      MATERIALS        PLANT           RENTAL
                                                     UNIT          UNIT            UNIT            UNIT
                                                     COST   AMOUNT COST   AMOUNT   COST   AMOUNT   COST   AMOUNT

CONVENTIONAL CHANNEL STRINGER STYLE
    PURCHASE                            LF
    LOCAL FREIGH
    UNLOAD AND HANDLE                   ST
    SMALL TOOL ALLOWANCE
    SALES TAX

DRIVE STATION
    STRUCTURE EXCAVATION                CY
    CONCRETE                            CY
    STRUCTURAL STEEL SUPPORT            LB
    MACHINERY INSTALL                   LB
    ELECTRICAL                          HP
    SHELTER                             EA

CONVEYOR LINE INSTALLATION
    CONCRETE                            CY
    STEEL FABRICATED STRUCTURES         LB
    WOOD TIES                           LF
    MACHINERY INSTALLATION (IDLERS)     LB
    INSTALL BELT - PER L.F. OF CONVEYOR LF
    LIGHTING                            LF
    ELECTRICAL CONTROL WIRE - L.F. OF CONVEYOR  LF
    AIR DISTRIBUTION LINES              LF
    DISMANTLE AND STORE                 LF
    CLEAR AND GRUBB                     AC
    EXCAVATION COMMON                   CY
    EXCAVATION ROCK                     CY
    CULVERTS                            LF
    ROAD SURFACING                      LF
    ROAD CROSSING                       LF
    SUBSTATION
    TRANSMISSION LINE (OR PARKWAY CABLE)
```

Figure 18–6–4 Computer printout sheet showing elements of cost for material handling conveyor

colleges or highway departments who have become interested in programming and do it as an avocation.

This could be a good way to start developing programs, but it is not very satisfactory for either party on a permanent basis. The programmer is at a disadvantage because, as the program develops, the user might decide to do something different; problems occur. The programmer is not sure just what is really required and what he can count on.

Manufacturers of Construction Equipment Many of the manufacturers of heavy, automotive-type equipment have developed programs from which one can get the performance of haul units such as end dump trucks, bottom dump trucks, or scrapers. The input into the computer is to include the length of the haul road, rolling resistance, grades, and speed limits if required on stretches of the haul road. The performance of the unit in question is usually described in the published standard performance charts prepared by the manufacturer. A factor, then, has to be applied to convert from the computer theoretical optimum basis to what could be expected under operation conditions. This factor is usually in the range of 50 to 70 percent of optimum. It allows for the normal lost time on the job. It recognizes that truck drivers are gregarious people and love to bunch up. It recognizes that the driver does not return full throttle empty and other such factors. Adjusted for such things, the manufacturer's computer programs can be applied.

Programs Developed by User's Personnel The ultimate method is for a contractor to develop his own programs with his own people, his own programmers, and his own engineers and estimators. This seems to be the best alternative—to have one's own programmer. However, a selection of this type should follow some general rules. The first one is to select as a programmer someone who has not done much commercial programming. He should like

mathematics and, generally, be the type of individual who is willing to experiment and explore new ground. He will have to be an instructor in some part in getting the estimators to use the programs, and some of the programs he may have to run himself as they may be too complicated for anyone else to set up since they may cover many variables.

18-7. Contractor-Developed System

Programming is both a science and an art. It is particularly an art in knowing what is possible; and, above all, it is an art in knowing when to quit. This is the problem in making a programmer out of an engineer. He will generally get carried away and never know when to quit.

If one starts out developing his own programs, he should expect to find that about 30 percent of those he has tried to develop for the first time don't work. The usual reason is that it is too difficult to develop and isn't worth the effort. The second reason is it won't sell—you can't get people to use it even after you have it. The third is there is not sufficient demand for it. You may think you have a need for it, but situations and times change; and it may no longer be required.

In the contractor's engineering organization, it appears to be best for this contractor to use his own computer, his own terminal facilities, and his own programmer, supplemented by use of service bureau facilities as necessary. However, the question is—where does one draw the line?

First, set some limitations as to what you are going to try to do with the computer and how much time you can save by it. Several people in the field of computer utilization use a six-to-one ratio—i.e., one hour of computer time must save six man-hours in order to be economical. This will take care of the cost of developing the programs, writing off the programs that turn out to be useless, etc. It will take care of the machine time, the machine oper-

SUB-CONTRACTS		TOTAL DIRECT COST		COST PER LIN FT OF CONVEYOR	SALVAGE TAKEN
UNIT COST	AMOUNT	UNIT COST	AMOUNT		

Figure 18–6–4 *(cont.)*

ators, and leave you with a slight benefit. If you accept, as a proper approach, the six-to-one ratio, it would seem to eliminate the use of the computer as a typewriter-calculator as described in Section 18-6.

The person in charge of this for the contractor needs a couple of books of instructions to bring him up to date. One book that is recommended is a computer dictionary. This is a very valuable tool.

If one looks at the process of preparing an estimate on a major construction job, the system seems to break down as best illustrated on the schedule sheet in Figure 18-7-1, where we have shown days and the operations of a typical estimate. We have marked the bars by legend for three types: first, that which the computer can help on (solid line); second, that which the computer cannot help on (broken line); and third, that on which it can help or does help in part (slashed line).

Concerning the parts for which data processing will not help, obviously the computer cannot receive the drawings and specifications and assign them to the various estimators, nor can it review the plans and specs or visit the site, and it probably would not be much good on the original quick scheduling and the adjusting of a bar graph. It can do little in the estimators' meetings or in the principals' meetings, and it cannot be used in the adjusting of subcontractors' and materials prices, or the cost prior to the bid.

It can, in part, select equipment, compare alternate haul units, and match them with a loading unit. It can provide the operating cost; and, accordingly, it can help on deciding the basic approach. It certainly can help in making

CAT	Q	H	L	W	AMT	$ AMT	10-22-69
3522	1				72.000		
3111	2	2.600	1.800	3.700	18.720		
3112	2	2.600	1.800	3.700	18.720		
3111	2	2.600	3.800	3.700	39.520		
3112	2	2.600	3.800	3.700	39.520		
3111	4	1.900	3.800	5.800	57.760		
3112	4	1.900	3.800	5.800	57.760		
3113	4	6.200	5.300	8.900	704.320		
3113	1	7.200	5.100	1.300	92.160		
3310	2	4.300	5.200	2.700	4.472		
3310	1	8.600	4.200	5.100	6.823		
3321	2	1.000	2.500	4.600	.852		
3350	3	1.600	4.200	5.800	73.080		
3310	4	1.500	9.000	9.000	18.000		
3101	4	1.500	9.000	9.000	216.000		
3310	5	1.800	11.000	11.000	40.333		
3101	5	1.800	11.000	11.000	396.000		
3310	1	1.400	7.000	7.000	2.541		
3101	1	1.400	7.000	7.000	39.200		
3310	5	1.300	6.800	6.700	10.968		

CAT	DESCRIPTION	UM	LABOR	MATERIAL	EQUIPMENT
3101	COLUMN FTG. FORMS	SF	.500	.150	.300
3102	WALL FTG. FORMS	SF	.500	.150	.110
3103	COLUMN FORMS	SF	.750	.150	.600
3104	WALL FORMS	SF	.750	.150	.220
3105	GRADE BEAM BOTTOM FORMS	SF	.500	.150	.900
3106	STRUCTURAL SLAB FORMS	SF	.800	.150	.240
3107	P. T. SLAB FORMS	SF	.800	.150	.055
3108	BEAM BOTTOM FORMS	SF	.750	.150	.470
3109	BEAM SIDE FORMS	SF	.750	.150	.420
3110	CURB FORMS	SF	.600	.150	1.250
3111	BEAM FP BOTTOM FORMS	SF	.600	.150	.210
3112	BEAM FP SIDE FORMS	SF	.600	.150	.350
3113	COLUMN FP FORMS	SF	.600	.150	.520
3310	COLUMN FOOTINGS	CY	12.000	14.500	8.000

CAT	DESCRIPTION	QUANTITY	OUTS	NET QUAN	UM	LABOR UNIT COST	LABOR COST	MAT UNIT COST	MAT COST	EQUIP UNIT COST	EQUIP COST	TOTAL COST
3101	COLUMN FTG. FORMS	1341.	0.	1341.	SF	.50	670.75	.15	201.22	.30	402.45	1274.42
3102	WALL FTG. FORMS	4487.	0.	4487.	SF	.50	2243.68	.15	673.10	.11	493.61	3410.39
3103	COLUMN FORMS	360.	0.	360.	SF	.75	270.00	.15	54.00	.60	216.00	540.00
3104	WALL FORMS	6860.	0.	6860.	SF	.75	5145.00	.15	1029.00	.22	1509.20	7683.20
3105	GRADE BEAM BOTTOM FORMS	13118.	0.	13118.	SF	.50	6559.00	.15	1967.701	.90	11806.20	20332.90

Figure 18–6–5 Computer printout sheet showing pricing and extensions for concrete form costs

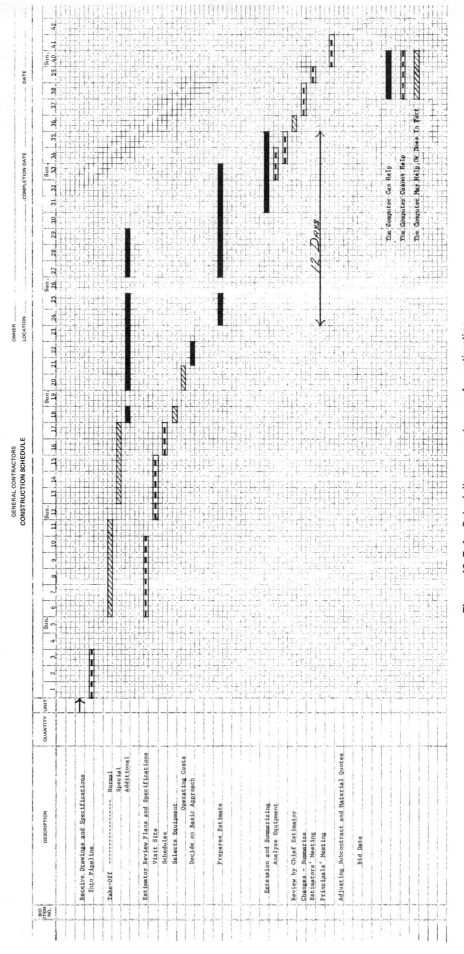

Figure 18-7-1 Scheduling computer use for estimating

	Forms	Estimator Himself	Computer	Updated Calculator	Clerical	Computer Information	Other	Later Maybe Computer
Read plans and specifications		XX						
Visit site		XX						
Thinking		XX						
Get available equipment						XX		
Deciding		XX						
Discussing with ace and plus		XX						
Routing		XX						
Operating cost		XX	XX					
Factoring labor			XX					
Titling estimate sheets							XX	
Printing estimate sheets	XX		XX					
Calculations				XX				
Directing takeoff		XX						
Item summary sheets		XX						XX
Prepare for review		XX						
Review changes		XX	XX					
Review estimate summary					XX			
Joint venture comparison sheets					XX			

Figure 18–7–2 Chart of how estimating functions are handled

some of the changes after review. It can do some of the takeoff. It very definitely can work in the preparation of the estimate and the extensions and summarizings. Figure 18-7-2 shows what the estimator does, what the computer does, and what the computer can supply information for.

In the early stages of the estimate, the engineering staff can develop such things as wage rates, equipment operating costs, cost of moving pieces of equipment to the site, and unit prices of concrete for curing. These can be done while the estimator is making his initial visit to the site, and time can be saved thereby. The above preliminary cost factors that require more printout, such as the operating cost and wage rates, are best done on one's own computer. The elements that do not require much printout, such as concrete unit prices and form costs, can be handled through the terminal. The design of cofferdams can be handled on the contractor's own computer facility; and, if one's own computer is tied up, the terminal can be used as time becomes more critical in the later stages of the estimate. More detailed descriptions and application of these programs are set out subsequently.

One of the simplest functions of the computer is the updating of historical production records expressed as man-hours. The estimator can generally fill out a format which has 80 spaces on it that correlate with the punched card. An example is shown in Figure 18-7-3. The first column would be the cost code—W106. The next column would be the description—conduit, for example. The next column would be size—6″, and so on. The field labor rate would be put in, usually expressed as dollars per man-hour to convert the stored man-hours to actual dollars. Several factors are left for updating stored material cost and stored labor cost. The printout will give the estimator what he should be using for this particular job. His input has been only the wage rates and whether he considers it a more-difficult-than-average job.

Logically, the question of what the cost would be to the contractor for the data processing described above is important, i.e., including the use of the computer. Roughly estimated it will be $35,000 per year. Thus, it is relatively inexpensive. As additional computer programs are devel-

oped for new applications, the average cost will probably go up because one is prone to develop the programs contributing the greatest economy first, and those providing lesser economy at later dates.

To show typical charges for time-sharing services by outside bureaus, sample forms of agreements are given in Figure 18-7-4.

Service bureau firms generally charge for storage—i.e., the cost of storing programs, the use of the terminal ($2 to $3 per hour), and additional charges based on how much of the memory bank of the computer the customer is using. This is broken down sometimes in "page" and CPU times. Twenty dollars an hour might be a reasonable overall cost of the terminal.

A typical charge for use of the contractor's own computer facility is shown in Figure 18-7-5. The charges are for hours of use priced at set rates. The computer is the basic item; but the additional equipment must be charged out—the transmitters, sorters, etc. The keypunch is billed separately along with programs or analysis done by the company's programming staff which is oriented to and works with its computer principally on commercial things. Some of the programs that the engineering and estimating departments use can be developed largely by the firm's own programmers. The nonbilling portion refers to the amounts of time and cost that were not allocated to any of the jobs run during the month and have to be distributed to all cost and kept track of to be sure of proper utilization.

18-8. Personnel Requirements

One term used throughout the computer world is a systems analyst. This person is defined as being one level above a programmer. He has the role of examining the administrative (and engineering) functions of a business, what procedures are being used or what problems need to be solved, and to recommend to the client firm how these things can best be done, and, of course, how they can best be done by computer. It is difficult for a person outside of the engineering-construction business to appreciate fully the multiplicity of possible conditions to be met and

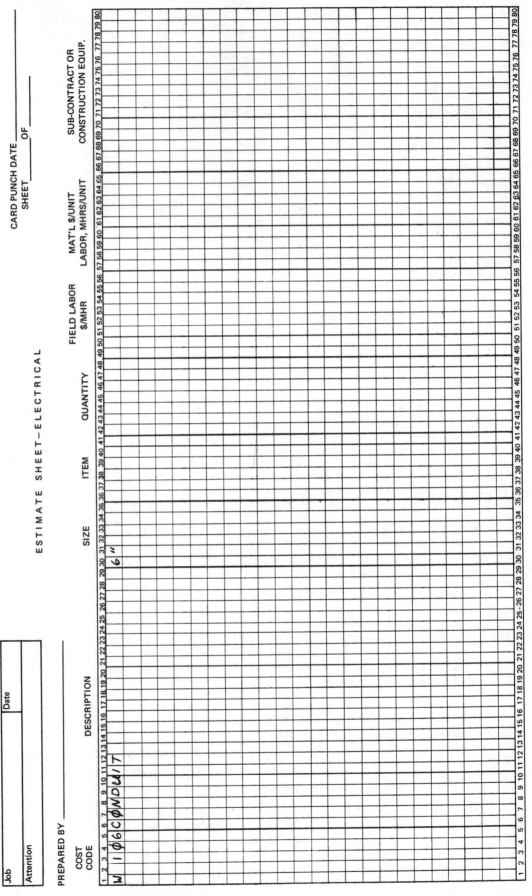

Figure 18–7–3　Preprinted sheet for estimating electrical work

CUSTOMER NAME & ADDRESS   TERMINAL LOCATION if Different

• •
• •
• •
•

DATE	OFFICE		SALESMAN		

Device Type	Quan.	Serial Number	Monthly Rental	Monthly Maint.	Purchase Price

REACT SCHEDULE OF RATES

Pages	Core Equiv.	Rate/Term Minute	Terminal Hour Extended Rate	CPU Rate Minute	Terminal Connect/Chg./Hour
1	4K	$0.025	$ 1.50	$4.00	$2.00
2	8K	0.05	3.00	4.00	2.00
3	12K	0.075	4.50	4.00	2.00
4	16K	0.10	6.00	4.00	2.00
5	20K	0.125	7.50	4.00	2.00
6	24K	0.15	9.00	4.00	2.00
7	28K	0.175	10.50	4.00	2.00
8	32K	0.20	12.00	4.00	2.00

REACT BACKGROUND PROGRAM RATES[a]

Level	Cost/CPU Hour
Priority 1	$240.00
Priority 2	$180.00
Priority 3	$150.00

REACT MONTHLY RANDOM ACCESS STORAGE COSTS (2314)

First 100K Storage—No Charge
Each Additional 100K—$12.00/Month

NIGHT-TIME BATCH RATE

Less than 10 hours/month	$250/Hour
10–25 hours/month	$225/Hour
Over 25 hours/month	$200/Hour
Night Time 1/0 Only Rate	$ 60/Hour

Six months guaranteed usage of over 20 Hours/Month—$180/Hour

[On back of agreement]

The charges stated are those currently in effect. All charges are subject to change upon 30 days written notice. This agreement is subject to the terms and conditions on the reverse side hereof and customer acknowledges he has read and herewith agrees to same.

CUSTOMER

By .... By ....

[a]Usage in excess of $1000/month will receive a 10% discount on the excess if paid as agreed.

Figure 18–7–4 Typical timesharing service agreements

those which are peculiar to a given situation. Therefore, many of the programs involved must necessarily be developed by the contractor's estimating department and a programmer. This arrangement will work fairly well. It is quite obvious that one must know what programs can do and which are the most important.

It is necessary to work with actual applications to demonstrate the feasibility of employing the computer, or it can readily become a laughing stock; and the lack of progress can turn people against its adoption. Then, as the programmer makes progress in adapting the use of it, he will tend to find other programs that will prove practicable.

ACCOUNT NO.-01.5118.06 ENGR. - H. O.   * * * N O N - C H A R G E A B L E * * *

ACTIVITY	COMPUTER	TRANSMITTER	REPRODUCER	COLLATOR	SORTER	CONVERTER	KEY-PUNCH	VERIFIER	AODO-X	PROG-ANAL	** TOTAL **
5-06	.09										.09
7-06	.01										.01
8-06	1.43										1.43
TOTAL ***	1.53										1.53
AMT. DUE	131.81										131.81

Figure 18-7-5 Computer printout—typical charge for use of the contractor's own computer facility

He will develop a better feel of what will or will not provide logical subjects. When his programming load drops off in the later stages of the game, he should become a rather excellent systems analyst as well as a programmer—a combination of both.

The personnel assigned exclusively to data processing seems to be about one person for every eight in the average engineering-estimating office. It is necessary to have the systems-analyst-programmer and one person who can run the keypunch and the typewriter terminal for communicating with computer service bureaus from time to time. Most of that person's time will be on the keypunch, because the estimators can use the terminal without help part of the time if they also can do typing. Probably half of the employee's time would be oriented to the computer.

18-9. Problems to be Anticipated

One of the problems of developing a computer program for use by the contractor's engineering organization is that it is difficult to coordinate it to the contractor's existing estimating system. Each contractor most generally has developed a system of estimating with which he is familiar. This system has been developed over a period of time, and every interested party in the company is usually familiar with it. Its use makes it easy for the senior estimator, chief engineer, to review cost estimates, a very important job.

If the computerization of estimating is attempted, the old estimating system goes out the window, and the format and procedures of estimating give way to how the computer can best handle the function involved and what sort of printout results. It makes no difference what has been done during the last number of years. This is an extremely difficult hurdle for the contractor's estimating group. The older estimators, who are sometimes the more qualified, resist the use of the computer and resist the change they have to make to accept a new discipline. The middle group sometimes becomes confused trying to change from one system to the other. Possibly they have not been thoroughly indoctrinated in the basic system, and this new form creates problems that tend to confuse the clarity of the estimate. The younger, newer estimator might take to it willingly; but he is not sufficiently experienced, or sometimes not sufficiently qualified to recognize computer errors that may be due to an incorrect format for the punch-card operator, the punch card itself, or perhaps a program in which this particular problem was not anticipated.

The contractor cannot shut down operations for six months while he converts to computer estimating. Several large contractors who have tried to adopt a full computer system of estimating in the heavy construction field have not found it to be a success principally for the many reasons outlined above, and have abandoned it. Other contractors in the building field have found that only a portion of their personnel use it.

Perhaps one can say that using the computer is like politics—it is the art of the possible. Use it on what it can do best, what can be most accepted, and what can be fitted into the contractor's existing estimating system. The adoption of computer estimating has to be done over a period of time and has to be done in a manner that can be used as it is injected into the estimating system.

18-10. Program for Adoption of the Computer

Once it is decided to use the computer, there are probably several steps that can be taken early in the game to help sell it to the personnel required to use it. The computer manufacturers can be approached to offer a one-day or two-day briefing and general course in explaining the computer—discussing computer languages, how the computer works, defining such things as "binary," "memory," etc. These manufacturers' representatives are experienced at this and they present the material so that it can be readily understood. Most engineers find it extremely interesting. Such a briefing should be held at an early stage.

Several months' use of a typewriter terminal for communicating with service bureaus by the programmer is helpful in programming, and it does make a nice demonstration tool to help the sale of the computer. It is interesting to see how one can type in things to a machine a long distance away and receive an answer that is responsive and useful. Even if one abandons the service bureaus and decides to use his own computer facility exclusively, a terminal is helpful in this indoctrinating and selling stage.

Considering the economic feasibility of providing a computer facility, the present stage of development of the computer (third generation), and generally adopting the six-to-one ratio previously described, programs can be developed for eight prominent areas, as follows:

- Updating historical cost information
- Design and long mental time calculations
- Choice of alternates
- Library
- Clerical
- Takeoff
- Accuracy
- Other

18-11. Updating Historical Cost Information for Reference Purposes

These programs take previously stored costs and factor them to the conditions of the job being estimated. Historical cost records for such elements as might apply to the job being estimated are drawn from the stored cost records and factored up for escalation, inflation, wage

```
ACCOUNT NO.-01.5118.06 ENGR. - H. O.              * * * C H A R G E A B L E * * *

 ACTIVITY   COMPUTER TRANSMITTER REPRODUCER   COLLATOR   SORTER   CONVERTER  KEY-PUNCH  VERIFIER   ADDO-X  PROG-ANAL ** TOTAL **

  C-06        2.30                                                             .10                                      2.40

  1-06        5.54                                                                                                      5.54

 TOTAL ***    8.24                                                             .10                                      8.34

 AMT. DUE   588.00                                                             .72                                    588.72
```

Figure 18-7-5 *(cont.)*

rate increases, new equipment prices and overtime, so that they might be applicable to the current job. These stored costs are actual experienced costs and costs from previous estimates where a considerable amount of time has gone into the study and development of them; when factored up they provide suitable references in the evaluation of costs to apply to the job in question.

Programs have been developed in the following areas:

(a) Evaluating previously reported equipment operating costs
(b) Establishing estimated equipment operating costs for inclusion in the estimate
(c) Move-in and erection costs
(d) Concrete form cost breakdown
(e) Concrete unit costs—such as curing and cleanup, construction joint treatment, point and patch, and finishing

(a) Evaluating Previously Reported Equipment Operating Costs   Costs are reported on major jobs that have a lot of hours on the major pieces of equipment to give a running accumulation of operating costs. These costs are then factored for inflation difference between the years, and a weighted average total is arrived at for the article of equipment of a given description. These costs are then studied to arrive at a base cost to be used in the estimating of a new job and which is to be included in the program.

These equipment costs are used along with previously estimated costs in the Estimated Hourly Equipment Oper-

ating Cost Program. The elements considered in this evaluation are given in Figure 18-11-1.

(b) Establishing Estimated Hourly Equipment Operating Cost for Inclusion in the Estimate   This part of the program takes previously estimated costs and actual costs and applies the appropriate gross labor rates for mechanics and operators, a factor for escalation, quoted fuel prices for the job, and the hours to be worked on the job to come up with an estimated cost to be used for the job being estimated. For pieces of large equipment such as a 35-ton truck, cost background from various sources are printed out for different kinds of service. The estimator must decide which kind of service applies to the job he is estimating and select the cost background that would most nearly apply. Such cost background includes the total labor (operator and mechanic repair cost), total rental, and total supplies. Figure 18-11-2 lists the input for this part of the program.

Figure 18-11-3 is a sample listing of the computer output for the hourly equipment operating costs taken from the computer. Note that the pickup is listed as a monthly cost. This is because the costs of automotive vehicles, such as the pickup, are reported on a monthly basis.

(c) Move-In and Erection Costs   This part of the program furnishes delivered cost for many items of new equipment, including purchase costs—new, plus freight from point of manufacture to job. Also included is erection cost per unit based on a fixed multiple of hourly gross

```
JOB    QUESTION MARK JOB                          DATE   04-13-70          COMPUT

JOINT VENTURE


              SHIFTS
              NUMBER OF DAYS                 6.00
              HRS PER SHIFT                  8.00
              SHIFTS PER DAY                 2.00
              NO. HOLIDAYS/YR.               4.00
              DAYS LOST DUE TO WEATHER/YR.  12.00

              HRS PER MONTH               394.67


              FUEL COSTS AND FACTORS
              DIESEL                     $   .150/GAL
              GASOLINE                   $   .260/GAL
              ELECTRICITY                $   .020/KW
              GROSS MECH WAGE RATE       $ 6.000
              PERCENT INCREASE FOR EXTRA O.T.  3.000%
              ESCALATION  FROM AUGUST 1968     7.000%
```

Figure 18-11-1   Computer printout—basic information of equipment operating cost

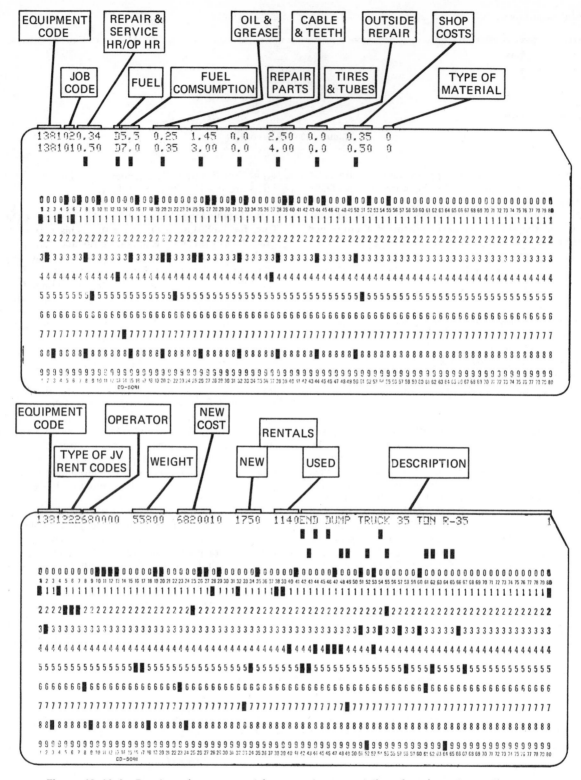

Figure 18–11–2   Punch card arrangement for computer computation of equipment operating cost

mechanic labor cost. For freight cost purposes, the point of origin of all units can be considered to be one of five major industrial centers in the United States. Equipment freight rates are required from each area as the appropriate freight from a specific city in that area to the job site or nearest railhead.

In the event that a secondary haul is required to reach the job site (i.e., rail plus truck or rail plus barge), the secondary freight rate from point of transfer to job site should be included.

In order to determine the erection cost per unit, it is necessary to include the proper gross mechanic wage rate for the project.

This part of the program considers that every item of

	LABOR R&S HR/HR	OPER	OILER	LABOR R&S	LABOR TOTAL	EQUIP RENT		FUEL GAL/HR	FUEL COST $/HR	OIL & GR	PARTS	CABLE & TEETH	TIRES & TUBES	OUT SIDE REP	SHOP COST	TOTAL SUPPLIES	TOTAL HOURLY COST
1041 PICKUP 1/2T 2WD MONTHLY COST				3600		$2500											
USE	7.80	.00		.00	48.20	48.20	100.00 G	190.00	49.40	6.00	28.90	.00	16.48	9.60	9.50	119.88	268.08
1381 END DUMP TRUCK 35 TON R-35				55800		$68200											
GRAND COULEE	.41	5.91	.00	2.53	8.45	6.82	D	6.60	.99	.30	1.67	.00	3.42	.00	.42	6.80	22.07
LIBBEY DAM	.60	5.91	.00	3.71	9.62	6.82	D	7.20	1.08	.30	1.73	.00	2.73	.00	.48	6.32	22.76
KETTLE	.60	5.91	.00	3.71	9.62	6.82	D	7.20	1.08	.30	2.08	.00	4.45	.00	.48	8.39	24.83
SEVILLE CANAL	.72	5.91	.00	4.45	10.36	6.82	D	8.40	1.26	.42	3.45	.00	3.77	.00	.71	9.60	26.78
CARTERS DAM	.72	5.91	.00	4.45	10.36	6.82	D	13.00	1.95	.42	3.74	.00	4.12	.00	.70	10.93	28.12
CASTAIC	.48	5.91	.00	2.97	8.88	6.82	D	7.20	1.08	.30	1.73	.00	3.42	.00	.60	7.14	22.84
DWORSHAK DAM	.55	5.91	.00	3.40	9.31	6.82	D	7.20	1.08	.42	2.76	.00	3.76	.00	.60	8.62	24.75
2884 DRILL - AIR TRACK - CM150-475				7500		$22500											
USE	.20	6.15	.00	1.24	7.38	4.50		.00	.00	.09	1.90	.00	.00	.45	.26	2.70	14.59
4121 CRANE TRUCK 70T 670TC				122095		$116800											
USE	.49	7.51	5.52	3.03	16.06	11.68	D	3.70	.55	.19	1.63	.73	.56	.00	.72	4.38	32.12

Figure 18-11-3 Typical computer printout sheet of previous equipment operating costs taken from storage bank

equipment for the job is new and furnished from the factory. Possible savings by use of used equipment is considered later in the preparation of the estimate.

(d) Concrete Form Cost Breakdown The best costs for use in estimates are from reported costs and from other estimates on which considerable time has been spent analyzing, correcting, and screening costs.

This part of the program adjusts such applicable costs by factoring them up for the job being estimated. Wage rates, escalation, price of lumber, increase for bolts and hardware are input to come up with the new costs to be used in the estimate.

Wage rates are based on fringes, payroll taxes and insurance, and overtime, but they do not include shift time, which is calculated from another program. There is no sentiment and sympathy induced by the computer, and it is in sentiment and sympathy that errors are made in pricing forms for concrete.

(e) Concrete Unit Costs. Historical costs are stored in the computer for the operations of point and patch, finishing, curing, cleanup, and construction joint treatment. The labor portion of these costs is updated by this part of the program with the applicable gross wage rate for finishers and air tool operators, etc. The cost of supplies involved is increased for inflation to arrive at a cost to be used for the subject job.

18-12. Time-Consuming and Lengthy Computations

Jobs that require extensive and time-consuming calculations provide the best subject for computerization. Programs that would fall in this category would include:

(a) The design of cellular cofferdams
(b) The design of land-type cofferdams
(c) Computing cableway cycles
(d) Tunnel rating for flood
(e) Flood storage computation
(f) Tunnel rating for diversion
(g) Diversion closure

(a) The Design of Cellular Cofferdams This provides a good example of when a timesharing terminal employing a service bureau can be used to good advantage. The purpose of this kind of program is to design a cellular-type cofferdam that will pass the presently in-vogue design criteria of various agencies that generally have to approve a contractor-designed cofferdam. The criteria to evaluate

are overturning, horizontal and vertical shear, Cummings effect, interlock stress, and the "tin can" effect—this is the separation of the sheets from the fill which pull away like a tin can from its contents.

The program will establish whether or not a cellular-type cofferdam can be used. The answer might be beyond the geometric limitations of the project. This has always been a question with respect to cellular cofferdams, as the economic limit of diameter of a cellular cofferdam is approximately 60 to 70 feet. Beyond that, it is necessary to employ higher grade and more costly sheetpiling or more expensive cloverleaf structures.

Given the conditions at the site as to foundations, water surface, and water-level variations, the program resolves the cofferdam diameter needed and resisting berm height, if any. Factors of safety in the order of 1.25 are acceptable in this computation. If the cofferdam does not pass this criteria, the program will suggest a new berm height for the diameter selected. If that berm height is not practical, the diameter, of course, must be changed. This is where the timesharing environment gives the estimator the opportunity to try a given diameter and berm height; and, if this is not acceptable, he has instantaneous response to change whichever one he wants, depending on the conditions of the job and geometry of the cofferdam.

In this first example, the message by the computer (1) "RESULTANT NOT IN MIDDLE THIRD" means the cofferdam is unstable in the overturning conditions. The computer program then suggests increasing the berm height 3 feet.

Figures 18-12-1 and 18-12-2 are copies of the presentation of the problem to the computer which already is in the position of the program involved, and Figure 18-12-3 is the printout of its reply giving the various values involved, which point up stability and desirable design characteristics.

However, the estimator might decide to increase the height 8 feet and reduce the diameter 5 feet. This puts the cofferdam in the stable condition and also reduces the interlock tension on the sheets, and the computer printout is given in Figures 18-12-4.

(b) The Design of Land-Type Cofferdams Many construction estimates require designs of land-type cofferdams. These are not like a cellular cofferdam design where the design is on a somewhat empirical basis. The purpose of this program is to establish the amount of steel sheetpiling that will be required for the cofferdam and the number of rows of walers and struts necessary. It will also help to select the most economical cofferdam based upon different sizes of sheets and/or variable strength of the sheets

```
cofferdam site 3      04-09-70

MAX PRESSURE                               2368.845 LBS./S.F.
INTERLOCK TENSION                         15510.902 LBS/INCH        ***
WATER FORCE                              140056.800 LBS
BERM FORCE                                  324.000 LBS
DRY WEIGHT                                  141.671 KIPS
SUBMERGED WEIGHT                            140.862 KIPS
TOTAL CELL WEIGHT                           282.533 KIPS
SUBMERGED MOMENT ARM                         21.988 FEET
TOTAL CELL FILL MOMENT                      570.416 KIP FT.
WATER OVERTURNING MOMENT                   3127.935 KIP FT.
ALLUVIUM OVERTURNING MOMENT                    .648 KIP FT.
TOTAL OVERTURNING MOMENT                   3698.999 KIP FT.
BERM RESISTING MOMENT                       466.667 KIP FT.
VERT BERM RESISTING MOMENT                  566.685 KIP FT.
VERT ALUV RESISTING MOMENT                    2.623 KIP FT.
TOTAL RESISTING MOMENT                     1035.975 KIP FT.
NET OVERTURNING MOMENT                     2663.024 KIP FT.
RESULTANT FORCE                             261.630 KIPS
RESULTANT ARM                                10.179 FEET
RESULTANT NOT IN MIDDLE THIRD
BERM WIDTH TO DEVELOP PASSIVE               100.000 FEET
PASSIVE BERM FORCE                           70.000 KIPS
INTERLOCK RESISTANCE TO SHEAR             16581.915 LBS/S.F.
BASE FORCE MAX                               10.333 KIPS/S.F.
BASE FORCE MIN                                -.638 KIPS/S.F.
CHANGED TO ZERO
POINT OF MAXIMUM SHEAR                        28.468 FEET
MAXIMUM SHEAR FORCE                       61802.232 LBS/S.F.
CUMMINGS                                   2312.633 KIP FT.
SLIDING                                             FS1 =   1.50
VERTICAL SHEAR POINT OF MAX SHEAR                   FS2 =   2.00
VERTICAL SHEAR POINT OF MIN RESISTANCE             FS3 =   3.64
CUMMINGS   HORIZONTAL SHEAR                         FS4 =   1.17
INCREASE BERM HEIGHT TO           23 FEET
TIN CAN EFFECT                                     FS5 =   1.29
DIAMETER =   65.24     BERM HEIGHT    20
CELLD
60
BERMHT
28
```

Figure 18–12–1 Computer printout sheet for design of a steel sheet piling cofferdam—Example No. 1

```
xeq
CELLULAR COFFERDAM DESIGN PROGRAM     04-09-70

ENTER INPUT DATA
ENTER JOB OR COFFERDAM NAME
cofferdam site 3
HIELEV
88
WAELEV
85
LOWAT
40
ALELEV
24
LOELEV
18
CELLD
65
BERMHT
20
```

Figure 18–12–2 Input to computer modifying basic data for design of steel sheet piling cofferdam given in Example No. 1

(high tensile versus regular) and possibly may reduce the number of walers and struts. The only way the larger sections of sheetpiling will prove economical or that high-strength sheets will be advantageous is to eliminate a complete row of walers and struts.

The surrounding soil is classified as sand, clay, and/or water and can be considered in layers so that these can be intermixed. The program allows for five different soil layers using pressures determined by soil test information.

This program works well on a timesharing setup with a service bureau.

(c) Computing Cableway Cycles This program will give the hours that a given cableway will take to pour a concrete dam from a given pickup point and tower location. This can be used, then, to pick a more economical pickup point location and the most economical travel and hoist speed of the cableway.

This program has to be a custom program for each job because the geometry of concrete dams is so different, one from another, that this has to be programmed in each time.

In setting up this program, the input to the computer

```
cofferdam site 3      04-09-70

MAX PRESSURE                              2416.185 LBS./S.F.
INTERLOCK TENSION                        14883.662 LBS/INCH        ***
WATER FORCE                             140056.800 LBS
BERM FORCE                                 324.000 LBS
DRY WEIGHT                                 110.845 KIPS
SUBMERGED WEIGHT                           143.269 KIPS
TOTAL CELL WEIGHT                          254.115 KIPS
SUBMERGED MOMENT ARM                        21.544 FEET
TOTAL CELL FILL MOMENT                     414.259 KIP FT.
WATER OVERTURNING MOMENT                  3127.935 KIP FT.
ALLUVIUM OVERTURNING MOMENT                  .648 KIP FT.
TOTAL OVERTURNING MOMENT                  3542.843 KIP FT.
BERM RESISTING MOMENT                     1280.533 KIP FT.
VERT BERM RESISTING MOMENT               1034.557 KIP FT.
VERT ALUV RESISTING MOMENT                   2.443 KIP FT.
TOTAL RESISTING MOMENT                    2317.533 KIP FT.
NET OVERTURNING MOMENT                    1225.310 KIP FT.
RESULTANT FORCE                            213.052 KIPS
RESULTANT ARM                               5.751 FEET
RESULTANT IN MIDDLE THIRD
BERM WIDTH TO DEVELOP PASSIVE              140.000 FEET
PASSIVE BERM FORCE                         137.200 KIPS
INTERLOCK RESISTANCE TO SHEAR           16913.295 LBS/S.F.
BASE FORCE MAX                              7.147 KIPS/S.F.
BASE FORCE MIN                              1.329 KIPS/S.F.
POINT OF MAXIMUM SHEAR                      37.189 FEET
MAXIMUM SHEAR FORCE                      55993.514 LBS/S.F.
CUMMINGS                                  2030.359 KIP FT.
SLIDING                                  FS1 =    1.88
VERTICAL SHEAR POINT OF MAX SHEAR        FS2 =    2.44
VERTICAL SHEAR POINT OF MIN RESISTANCE   FS3 =    8.13
CUMMINGS  HORIZONTAL SHEAR               FS4 =    1.33
TIN CAN EFFECT                           FS5 =    1.58
DIAMETER =  60.48    BERM HEIGHT    28
CELLD
999
**   201. XEQ "STOP".
```

Figure 18–12–3 Computer printout sheet giving results for design of steel sheet piling cofferdam, Example No. 1 modified

must consider coordinates of the fixed tower, radii to the movable tower, and the radii to the pickup points which establish the layout of the cableway as it applies to the site. Bucket capacity, fixed time, maximum pouring rate, and horizontal and vertical velocity give the program the characteristics of the cableway mechanism.

The geometry of the dam is programmed in so that the time and distance to pour each lift can be calculated and the hours are accumulated, together with the total volume of the dam and by separate concrete lifts.

(d) Tunnel Rating for Floods   With the river rating curve (flow versus elevation), tunnel dimensions and hydraulic loss coefficient, a tunnel can be rated as to the amount of flow of flood proportions the tunnel can handle. This rating is input to the flood storage program, and computations can be conveniently handled by the computer.

(e) Flood Storage Computation   To come up with the cofferdam height required to pass a design flood through a diversion tunnel, the upstream elevation of the pool will be printed out versus time for a given duration of flood. This is a practical computer application.

(f) Tunnel Rating for Diversion   The natural river tailwater curves are input along with the tunnel hydraulic characteristics. These must include flows of less magnitude than those necessary to accomplish closure for diversion to flows greater than those necessary to accomplish closure for diversion. With the above input the computer compiles the losses and headwater elevation for the diversion flows—usually an open-channel condition, and is a suitable computer problem.

(g) Diversion Closure   With the maximum allowable river velocity during closure, the computer will print the equilibrium conditions of flow for tunnel and river before start of closure and the equilibrium flow conditions with river closed to maximum velocity and the cross-sectional area left to be closed (Figure 18-12-5).

18-13.   Computer Assistance in the Choice of Alternatives

This type of programming gives the estimator a chance to run two or more solutions to a given problem. This gives him a chance to pick the more economical solution. Programs of this type are listed as follows:

coffer dam site 3   03-30-70

```
MAX PRESSURE                              2416.185 LBS./S.F.
INTERLOCK TENSION                        14883.662 LBS/INCH
WATER FORCE                             140056.800 LBS
BERM FORCE                                 324.000 LBS
DRY WEIGHT                                 110.845 KIPS
SUBMERGED WEIGHT                           143.269 KIPS
TOTAL CELL WEIGHT                          254.115 KIPS
SUBMERGED MOMENT ARM                        21.544 FEET
TOTAL CELL FILL MOMENT                     414.259 KIP FT.
WATER OVERTURNING MOMENT                  3127.935 KIP FT.
ALLUVIUM OVERTURNING MOMENT                  .648 KIP FT.
TOTAL OVERTURNING MOMENT                  3542.843 KIP FT.
BERM RESISTING MOMENT                     1280.533 KIP FT.
VERT BERM RESISTING MOMENT               1034.557 KIP FT.
VERT ALUV RESISTING MOMENT                   2.443 KIP FT.
TOTAL RESISTING MOMENT                    2317.533 KIP FT.
NET OVERTURNING MOMENT                    1225.310 KIP FT.
RESULTANT FORCE                            213.052 KIPS
RESULTANT ARM                               5.751 FEET
 RESULTANT IN MIDDLE THIRD
BERM WIDTH TO DEVELOP PASSIVE              140.000 FEET
PASSIVE BERM FORCE                         137.200 KIPS
INTERLOCK RESISTANCE TO SHEAR            16913.295 LBS/S.F.
BASE FORCE MAX                               7.147 KIPS/S.F.
BASE FORCE MIN                               1.329 KIPS/S.F.
POINT OF MAXIMUM SHEAR                      37.189 FEET
MAXIMUM SHEAR FORCE                      55993.514 LBS/S.F.
CUMMINGS                                  2030.359 KIP FT.
SLIDING                                       FS1 =   1.88
VERTICAL SHEAR POINT OF MAX SHEAR             FS2 =   2.44
VERTICAL SHEAR POINT OF MIN RESISTANCE        FS3 =   8.13
CUMMINGS  HORIZONTAL SHEAR                    FS4 =   1.33
TIN CAN EFFECT                                FS5 =   1.58
DIAMETER =  60.48      BERM HEIGHT   28
CELLD
65
BERMHT
20
```

Figure 18–12–4 Computer printout sheet of results for design of steel sheet piling cofferdam, Example No. 1 further modified

Figure 18–12–5 End dumping rock portion of cofferdam in lifts to produce river diversion

(a) Optimization of loading unit to hauling units
(b) Drill spacing
(c) Tunnel calculation program
(d) Aggregate pit analysis
(e) Cableway cycles

(a) Optimization of Loading Unit to Hauling Units Size of bucket of loading unit, size of truck, production rates, new costs of each, and hourly equipment operating costs of each are input to this program. The trucks are first matched to the loading unit; then different combinations are tried in order to come up with the most economical spread. The total production is output along with the total hourly cost and purchase cost of the spread. This is particularly adaptable to the use of the computer.

(b) Drill Hole Spacing for Tunnel Work With the tunnel dimensions and the spacing of the perimeter presplit and patterned holes, the number of holes can be calculated and plotted by use of the computer. The different possible patterns can then be input into the tunnel program which calculates the drilling and blasting data—i.e., the depths of drilling, amount of drilling required, etc.

Figure 18-13-1 is a computer printout of blast hole location and spacing for a circular tunnel with an excavation radius of 13 feet with three burn holes. Figure 18-13-2 is a computer printout of the same tunnel using only one burn hole.

(c) Tunnel Calculation Program Along with the drill spacing, the drilling speed, crew time, and advance per round are input to this program. The drilling and blasting data are calculated for different drill patterns to give the most economical setup. These two examples show how the drill pattern is changed to change the drilling and blasting data.

Figure 18-13-3 is a printout by the computer of values of the different elements to be applied in the cost estimate based on eleven drills.

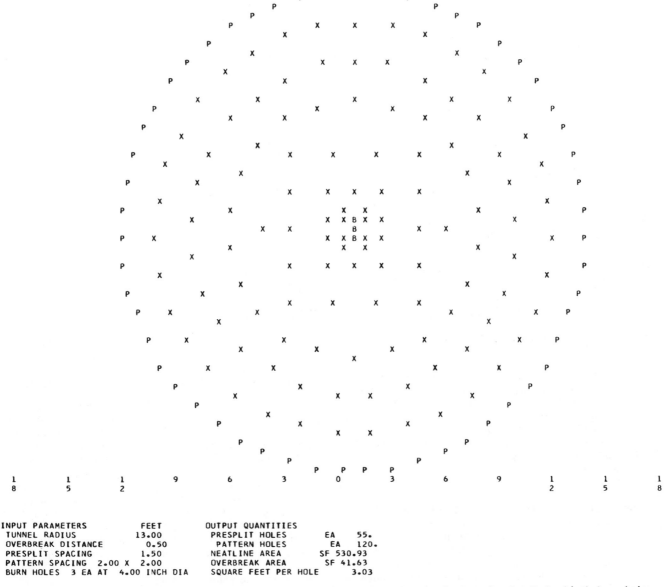

INPUT PARAMETERS	FEET	OUTPUT QUANTITIES		
TUNNEL RADIUS	13.00	PRESPLIT HOLES	EA	55.
OVERBREAK DISTANCE	0.50	PATTERN HOLES	EA	120.
PRESPLIT SPACING	1.50	NEATLINE AREA	SF	530.93
PATTERN SPACING	2.00 X 2.00	OVERBREAK AREA	SF	41.63
BURN HOLES	3 EA AT 4.00 INCH DIA	SQUARE FEET PER HOLE		3.03

Figure 18–13–1 Computer printout sheets showing plotting of drill pattern for circular tunnel, using 3- to 4-inch burn holes

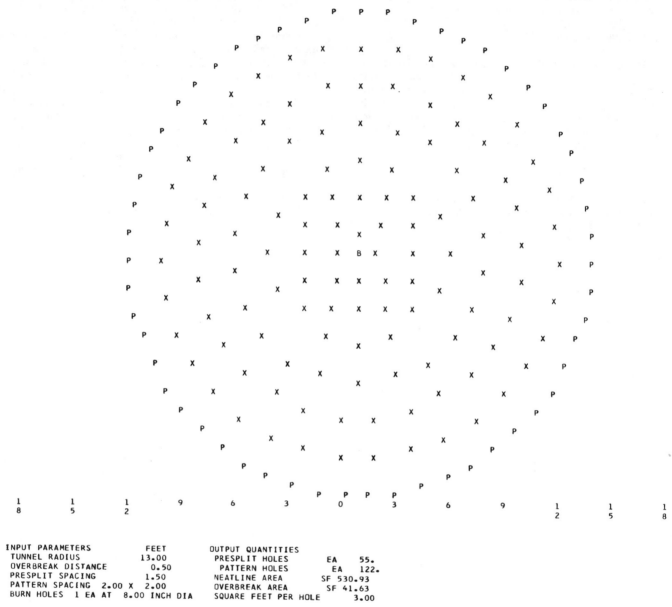

INPUT PARAMETERS	FEET	OUTPUT QUANTITIES		
TUNNEL RADIUS	13.00	PRESPLIT HOLES	EA	55.
OVERBREAK DISTANCE	0.50	PATTERN HOLES	EA	122.
PRESPLIT SPACING	1.50	NEATLINE AREA	SF	530.93
PATTERN SPACING 2.00 X 2.00		OVERBREAK AREA	SF	41.63
BURN HOLES 1 EA AT 8.00 INCH DIA		SQUARE FEET PER HOLE		3.00

Figure 18–13–2 Computer printout sheet showing plotting of drill pattern for circular tunnel, using only one 4-inch burn hole

(d) Aggregate Pit Analysis In designing a crushing and screening plant for an aggregate pit, many alternate setups of screens and crushers can be tried. With the mechanical analysis of the pit, this program can select the appropriate screening and crusher setups to come up with the specified mix of aggregates. This program gives the estimator the opportunity to try different blends and inputs of gravel to determine the most economical setup with the least wastage and crushing.

(e) Cableway Cycles as They Affect Acquisition of Equipment This program has been described earlier, but it also can serve for trying alternate layouts and cableway designs and speeds to compare newer type cableways with company-owned older models to see if the new cableways can increase the production significantly enough to justify buying a new cableway.

18-14. Storage of Historical Experience Data

Much historical experience (both cost and production) data can be stored on the computer for retrieval. The storing of information in the memory banks of the computer makes for easy updating and printing of the new information correlated with the old information. Printed books containing such information can be kept; however, they become outdated rapidly and are difficult to reprint. New information can be added readily to the computer-stored data and recalled for the individual estimators' particular use when desired, quite expeditiously.

Historical experience data pertaining to the following listed operations can be stored and reported:

- Aggregate plant operating cost
- Cableway performance information

```
        DRILLING AND BLASTING              ITEM NO.    SHEET 1 OF 1 PAGE    1

            TUNNEL AND SHAFT EXCAVATION      JOB      COMPUTER TEST TUNNEL

     QUANTITY      3000. UNIT    LF   EST. BY              DATE5-13-70

TYPE OF ROUND  BURN

SIZE AND SHAPE OF TUNNEL
CIRCULAR 13 FOOT RADIUS

TYPE OF SECTION
UNSUPPORTED

DRILLING DATA
   EXCAVATION PAY SECTION AREA           SF        531.00
   NUMBER OF DRIFTER HOLES @ 1.750 DIA   EA        175
   SQUARE FEET PER HOLE                SF/HOLE      3.034
   LF OF DRILLING PER PAY CY            LF/CY        9.91
   NO. OF LARGE DIA. BURNHOLES @ 4.000 DIA  EA        3
   LENGTH OF ROUND DRILLED               LF         10.00
   TOTAL DRILLING PER ROUND - DRIFTER    LF       1750.00
   DRILLING RATE - DRIFTERS 1 3/4      LF/HR        70.00
   TOTAL DRILL TIME PER ROUND - DRIFTER  HR         25.00
   NUMBER OF DRILLS                      EA     10 OF  1 BURN
   NUMBER OF HOLES PER DRILL - ACTUAL 17.5   USE 18.
   CREW TIME PER ROUND - DRILL           HR          2.57
   CREW TIME PER ROUND - DRILL           MIN       154.

   LENGTH OF TUNNEL                      LF       3000.00
   ADVANCE PER ROUND                     LF          9.00
   TOTAL NUMBER OF ROUNDS                EA        334
   TOTAL DRIFTER DRILLING - ALL ROUNDS   LF     584500.0
   TOTAL BURNHOLE DRILLING - ALL ROUNDS  LF      10020.0
   TOTAL DRIFTER DRILL HOURS - ALL ROUNDS  HR    8349.98
   TOTAL BURN, HOURS ALL ROUNDS @  20.00 LF/HR    501.0

   ESTIMATED BIT LIFE        LF OF DRILLING        350.
   TOTAL NUMBER OF BITS REQUIRED 1.750 DIA  EA    1670
   LARGE DIA. BURNHOLE BIT LIFE      LF OF DR     1200.
   TOTAL LARGE DIA. BITS REQUIRED 4.000 DIA EA       9
   DRILL STEEL REQUIRED @  0.048LB/LF     LB     28056.00
   BURNHOLE STEEL REQUIRED @  0.300LB/LF  LB      3006.00

BLASTING DATA

   EXCAV. PAY QUANT. PER LF SOLID MEASURE   CY       19.670
   EXCAVATION PAY QUANTITY                CY      59009.93
   POWDER FACTOR                        LB/CY        4.5
   TOTAL POWDER REQUIRED                  LB    265544.6

   TOTAL NUMBER OF HOLES - ALL ROUNDS     EA     58450.
   TOTAL CAPS REQUIRED      5. EXTRA/ROUND  EA    60120

USED INPUT POWDER FACTOR

SUBROUTINE NO   1
```

Figure 18–13–3   Computer printout sheet giving extended quantities and time for drilling and blasting for tunnel excavation

- Concrete plant operations
- Conveyor installations
- Conveyor operations
- Electrical plant information
- Fixed time production data

- Foreign equipment factors
- Metal and mechanical work
- Miscellaneous metal work
- Piping and mechanical work
- Production records—estimated and actual

Haul units—end dump trucks
Loading units—shovels up to 15 CY
Scrapers—self-loading
- Sheet pile cofferdam installation
- Drilling and blasting—production and cost
- Miscellaneous operating cost
- Piping work
- Excavation and fill
- Excavators—Erection costs and production
- Camp and project buildings
- Operating cost
- Concrete form cost summary
- Installation of hydraulic turbines, governors, and generators
- Electrical, powerhouse, building specialties, and miscellaneous
- Concrete plant installation and operation costs
- Aggregate production costs

18-15.   Performance of Clerical Functions

Programs for such functions relieve the estimator of long laborious clerical duties. Extensions and preparation of summary sheets are good examples. If the estimator decides to change a quantity or a unit price, a new extension may be quickly run out and new summary sheets printed with a minimum amount of hand corrections. Programs of this category include:
(a) Wage rate program, (b) Extension of concrete form costs, and (c) Extension of concrete item costs.

(a) Wage Rate Program   The wage rate program for computer is fully described in the foregoing. It is mentioned here only to emphasize that the routine extensions which would otherwise be time consuming can be readily handled by the computer to good advantage.

(b) Extension of Concrete Form Costs   This program takes, from the form unit cost program, the unit costs the estimator has chosen to use and extends these against the square footage of forms used for each bid item. This, then, is totaled and a unit price calculated for the total square footage of forms used in a given bid item. The unit price is then input to the concrete extension program to be totaled and distributed.

(c) Extension of Concrete Item Costs   This program relates to the extensions and summary sheets for a concrete job. After the estimator has evaluated the unit costs from the form cost program, the concrete unit cost program, etc., he selects the unit cost he wants to use. He establishes concrete placing rates, haul rates, etc. The applicable quantities are taken from the Takeoff Department. This program uses as input to the computer the dollar amounts for plant operations, placing crews, form costs, curing and cleanup, placing rates, etc., and distributes the cost to the various concrete work items on a cubic yard basis. Once this has been done, final corrections can easily be made and a new distribution and summary sheet can be printed out with these corrections as desired. Figure 18-15-1 depicts a large mass concrete job in process.

18-16.   Analyzing Performance of Off-Highway Trucking

One of the most difficult things to compute accurately is often the most significant item in a project—transporta-

Figure 18–15–1 Typical high trestle and gantry arrangement for concrete placing plant on high concrete dam

tion of earth. This is becoming more difficult to compute because of the extreme size of the transporting units presently employed. Many of these are in the 150-ton gross weight category (four or five times heavier than a maximum-size highway truck). It is extremely difficult to calculate properly the momentum and inertia for these; and this, of course, will effect how far or how fast they might roll up a slope before starting to slow down. Also, with the new torque converter drives, it is difficult to calculate the acceleration time of such units from starting or for any significant grade change.

Almost all of the major equipment manufacturers of transporting units have developed computer programs to calculate the performance of their products under varying conditions of grade, etc. These programs are generally obtainable from the manufacturers:

(a) Haul Cycle Program   The individual unit's capabilities are installed in the computer employing the published performance data, generally to include grade ability, speed, rim pull, etc. The fuels, net and gross, and weights are also part of the required data. The retarder performance data are also used. Many times speed limitations will have to be considered because the estimator knows that the road has many curves which will limit speed. The estimator still has to supply the fixed time, the allowance for dump, waiting, loading, spotting, etc., and assign the

Figure 18–16–1 Typical off-highway hauling equipment susceptible to analysis for a given haul by use of computer

load per unit—i.e., the amount of payload. The computer works out the total time cycle based on the input data mentioned above.

(b) Fuel Consumption It is possible with additional information from the manufacturers' to calculate fuel consumption concurrent with haul production, also employing the computer.

Figure 18-16-1 shows the kind of equipment susceptible to analysis as described above, but the principles of analysis would be the same for end dump haul trucks, motor scrapers, or other transporting units.

Figure 18-16-2 is a computer printout showing the development of a time cycle and fuel consumption demand for a 100-ton capacity, 1000 horsepower rear dump truck for given haul conditions.

18-17. Other Practical Computer Uses

This is a catch-all category, a portion of which is manipulating data in the company's computer for use in the estimate. This portion would include:

(a) Soliciting subcontract and material prices
(b) Analyzing availability of company-owned equipment
(c) Analyzing overhead personnel requirements
(d) Equipment performance comparisons
(e) Analyzing haul road conditions
(f) Analyzing river closure problems
(g) Sinking of a caisson

(a) Soliciting Subcontract and Material Prices This is a mailing list type of thing which is customary with many commercial firms engaged in soliciting the public for sales purposes or otherwise. Vendors can be classified by regions of the country and by type of work, and letters are then written to them calling their attention to a given job to be bid and requesting their quotations. The computer in this instance serves the function of selecting from its memory banks the subcontractors and vendors most likely to fit a given situation.

(b) Analyzing Availability of Company-Owned Equipment This is another type of control of inventories and forecast of inventories wherein the contractor forecasts the major pieces of equipment to become available. This, of course, can be used for estimating the work and should be available at an early stage so that the estimate can be calculated about the contractor's own available equipment as much as practical. The computer printout would give availability information.

(c) Analyzing Overhead Personnel Requirements This, again, is a manipulation of existing data, in the personnel files wherein each active contract of the contractor would report the number of people on each project with their classifications. This can be used by reference to estimate the overhead requirements for work of a similar nature and scope or as a general guide on most any project. The computer printout would show in this instance the number of people employed on a given project in overhead classifications.

(d) Equipment Performance Comparisons The engineering department is often called upon to assist in the organization of field operations and particularly in the initial startup stages. In this connection, the computer can be used to great advantage—for instance, in the selection of equipment by comparing Brand X to Brand Y. An excerpt from a comparison using two equipment companies' computers on a major job may be of interest here:

We attempted to put Brand X and Brand Y on the same basis giving them the same haul distances, rolling resistance, fixed times, dump speeds, and job traffic conditions. Of course, they were careful not to do what we asked them so it would be difficult to make a clear cut comparison. They wanted

```
PROGRAM E0010010 HAUL CYCLE ESTIMATING BY VEHICLE SIMULATION                                                03/10/70

EUCLID RX-105 REAR DUMP GM 12V-149T 1000 HORSEPOWER 100 TON LOAD

CALC. VELOCITY-MPH.    0.0      3.00     5.00     6.20     6.21     9.00     11.00    11.01    12.50    15.00    15.01    17.70
CALC. VELOCITY-MPH.   19.50    19.51    21.40    25.50    25.51    29.00    35.00    35.10     0.0      0.0      0.0      0.0
CALC. VELOCITY-MPH.    0.0      0.0      0.0      0.0      0.0      0.0      0.0      0.0      0.0      0.0      0.0      0.0

CALC. RIMPULL-LB...  104000.   70000.   54000.   49500.   39500.   29500.   27000.   24000.   21600.   19500.   17000.   16000.
CALC. RIMPULL-LB...   14500.   13600.   12600.   11000.   10100.    9700.    8500.    3000.       0.       0.       0.       0.
CALC. RIMPULL-LB...       0.       0.       0.       0.       0.       0.       0.       0.       0.       0.       0.       0.

GROSS ENGINE HP....       0.     684.     880.    1000.     799.     865.     968.     861.     880.     953.     831.     923.
GROSS ENGINE HP....     921.     864.     879.     914.     839.     916.     969.     343.       0.       0.       0.       0.
GROSS ENGINE HP....       0.       0.       0.       0.       0.       0.       0.       0.       0.       0.       0.       0.

CALC. RET. VEL.MPH.    0.0      5.00     6.50     8.60    12.00    15.50    19.00    24.00    30.00    37.40     0.0      0.0
CALC. RET. RP-LB...  -70000.  -70000.  -50000.  -40000.  -30000.  -24000.  -20000.  -16000.  -43000.  -10400.       0.       0.

PERF. CHART RES....    0.0      CHART WT....      0.      RET. CHART RES....    0.0      CHART WT....      0.
SPEED RATIO...   1.000

FUEL CURVE HP......       0.     100.     290.     500.     665.     730.     780.     835.     877.     920.     960.    1000.
FUEL CURVE GAL./HR.   10.0     13.5     25.0     32.0     37.8     40.9     43.3     46.0     48.2     50.7     53.1     55.7

WEIGHT EMPTY.......  133300.   ON DRIVE AXLE...  133300.
WEIGHT LOADED......  333300.   ON DRIVE AXLE...  333300.


HAUL OB-1 T  OVERBURDEN - TENTATIVE ELK R. NO. 2
```

...FROM START...			...ON ELEMENT...			...FOR LINE...		...MPH..		ACCEL.	ENG.			ROLL	GRADE	SPEED	END	RE
TIME	DIST.	GAL.	TIME	DIST.	GAL.	TIME	DIST.	RATE	START	END	FT/MIN.2	HP.	GRADE	RES.	LENGTH	LIMIT	SPEED	CO
													LOADED 0.0	4.00	200.	22.00	22.00	
0.01	1.	0.0	0.01	1.	0.0	0.010	1.4	26.7	0.0	3.0	25595.	342.						
0.02	5.	0.0	0.02	5.	0.0	0.010	3.7	43.4	3.0	5.0	16909.	782.						
0.03	9.	0.0	0.03	9.	0.0	0.008	3.9	51.9	5.0	6.2	13348.	940.						
0.03	9.	0.0	0.03	9.	0.0	0.000	0.0	49.5	6.2	6.2	10829.	900.						
0.06	31.	0.0	0.06	31.	0.0	0.033	22.3	45.9	6.2	9.0	7355.	832.						
0.10	61.	0.1	0.10	61.	0.1	0.034	29.9	50.5	9.0	11.0	5183.	917.						
0.10	61.	0.1	0.10	61.	0.1	0.000	0.2	50.4	11.0	11.0	4228.	915.						
0.14	103.	0.1	0.14	103.	0.1	0.040	41.2	47.9	11.0	12.5	3290.	871.						
0.22	200.	0.2	0.22	200.	0.2	0.081	97.4	50.5	12.5	14.8	2508.	917.						
													LOADED-12.00	2.00	1600.	27.00	22.00	
0.22	201.	0.2	0.00	1.	0.0	0.001	1.2	50.5	14.8	15.0	18720.	917.						
0.22	201.	0.2	0.00	1.	0.0	0.000	0.1	49.1	15.0	15.0	17921.	892.						
0.23	221.	0.2	0.01	21.	0.0	0.014	19.7	48.2	15.0	17.7	17313.	877.						
0.24	236.	0.2	0.02	36.	0.0	0.009	15.4	50.8	17.7	19.5	16878.	922.						
0.24	236.	0.2	0.02	36.	0.0	0.000	0.1	49.1	19.5	19.5	16461.	893.						
0.25	255.	0.2	0.03	55.	0.0	0.010	18.6	47.9	19.5	21.4	16131.	872.						
0.27	302.	0.2	0.06	102.	0.0	0.023	47.5	49.4	21.4	25.5	15680.	897.						
0.27	303.	0.2	0.06	103.	0.0	0.000	0.1	48.2	25.5	25.5	15245.	877.						
0.28	323.	0.2	0.07	123.	0.1	0.009	20.2	48.3	25.5	27.0	15020.	878.						
0.87	1705.	0.3	0.65	1505.	0.2	0.582	1382.4	10.0	27.0	27.0	0.	-2400.						
0.91	1800.	0.3	0.69	1600.	0.2	0.044	94.9	10.0	27.0	22.0	-10000.	0.						
													LOADED 0.0	4.00	200.	22.00	0.0	
0.92	1813.	0.3	0.01	13.	0.0	0.007	12.8	48.4	22.0	22.0	-182.	879.						
1.11	2000.	0.4	0.20	200.	0.0	0.193	187.2	10.0	22.0	0.0	-10000.	0.						
													EMPTY 0.0	4.00	200.	22.00	22.00	
1.11	2000.	0.4	0.00	0.	0.0	0.004	0.5	26.7	0.0	3.0	70946.	342.						

Figure 18–16–2 Computer printout sheet showing development of time cycle and fuel consumption for given off-highway haul

room to maneuver. As nature abhors a vacuum, salesmen abhor facts.

This problem vanishes using your own computer.

(e) Analyzing Haul Road Conditions Once the basic spread of equipment has been selected, the major haul roads can be adjusted and readjusted to come up with the most optimum haul road condition (curves, grades, etc.) for the selected piece of equipment based upon computer calculated time cycles.

(f) Analyzing River Closure Problems The lengthy calculations for the height of cofferdams required for a final river closure operation considering, particularly, flow through an opening partition, ponding during closure, etc., can be performed admirably by the computer, and is a substantial timesaver.

(g) Buoyancy Calculations to Control the Sinking of a Bridge Foundation Caisson The contractor's engineering department or some consulting engineering firm can be of service to the project during construction on some of the more complicated work, such as closure and sinking of a bridge foundation caisson by use of computer to make buoyancy calculations for differing attending conditions.

In the case of an air-floated caisson, this requires control of the depth of digging, the number of air domes on the dredge wells at any one time, the pressure in each dome, the height of the steel sheet walls, the height of the concrete walls, and the depth of fill concrete. Such control will maintain the stability of the caisson with reference to metacentric height, and penetration in the river or bay bottom, and generally it proves to be extremely comforting to the contractor, the engineer, and the owner. Figure 18-17-1 depicts the sinking of such a bridge foundation caisson.

..FROM START...			...ON ELEMENT...			...FOR LINE...		...MPH..		ACCEL.	ENG.		ROLL	GRADE	SPEED	END	RET.
TIME	DIST.	GAL.	TIME	DIST.	GAL.	TIME	DIST.	RATE	START	END	FT/MIN.2	HP.	GRADE	RES.	LENGTH	LIMIT	SPEED	CODE
.12	2002.	0.4	0.01	2.	0.0	0.004	1.3	43.4	3.0	5.0	49228.	782.						
.12	2003.	0.4	0.01	3.	0.0	0.003	1.3	51.9	5.0	6.2	40324.	940.						
.12	2003.	0.4	0.01	3.	0.0	0.000	0.0	49.5	6.2	6.2	34026.	900.						
.13	2010.	0.4	0.02	10.	0.0	0.010	6.5	45.9	6.2	9.0	25339.	832.						
.14	2017.	0.4	0.03	17.	0.0	0.009	7.8	50.5	9.0	11.0	19909.	917.						
.14	2017.	0.4	0.03	17.	0.0	0.000	0.0	50.4	11.0	11.0	17520.	915.						
.15	2026.	0.4	0.04	26.	0.0	0.009	8.9	47.9	11.0	12.5	15175.	871.						
.16	2046.	0.4	0.05	46.	0.0	0.017	20.1	50.5	12.5	15.0	13220.	917.						
.16	2047.	0.4	0.05	47.	0.0	0.000	0.1	49.1	15.0	15.0	11222.	892.						
.19	2082.	0.4	0.08	82.	0.1	0.024	35.1	48.2	15.0	17.7	9702.	877.						
.21	2112.	0.4	0.10	112.	0.1	0.018	30.1	50.8	17.7	19.5	8616.	922.						
.21	2112.	0.4	0.10	112.	0.1	0.000	0.2	49.1	19.5	19.5	7573.	893.						
.23	2156.	0.5	0.12	156.	0.1	0.025	44.4	47.9	19.5	21.4	6748.	872.						
.24	2174.	0.5	0.13	174.	0.1	0.009	17.9	49.4	21.4	22.0	5619.	897.						
.25	2200.	0.5	0.14	200.	0.1	0.013	25.7	29.5	22.0	22.0	0.	426.						
													EMPTY	12.00	2.00	1600.	27.00	22.00	1.
.30	2283.	0.5	0.05	83.	0.0	0.045	82.8	47.9	22.0	19.5	-4832.	872.						
.30	2283.	0.5	0.05	83.	0.0	0.000	0.4	49.1	19.5	19.5	-4007.	893.						
.35	2371.	0.6	0.10	171.	0.1	0.053	87.5	50.8	19.5	17.7	-2964.	922.						
.48	2552.	0.7	0.23	352.	0.2	0.126	181.4	48.2	17.7	15.0	-1878.	877.						
.48	2553.	0.7	0.23	353.	0.2	0.001	1.1	51.2	15.0	15.0	-722.	929.						
.42	3800.	1.5	1.17	1600.	1.0	0.944	1246.9	50.3	15.0	15.0	0.	912.						
													EMPTY	0.0	4.00	200.	22.00	0.0	1.
.42	3800.	1.5	0.00	0.	0.0	0.000	0.1	49.1	15.0	15.0	11222.	892.						
.45	3835.	1.5	0.02	35.	0.0	0.024	35.1	48.2	15.0	17.7	9702.	877.						
.46	3859.	1.5	0.04	59.	0.0	0.014	23.4	50.8	17.7	19.1	8616.	922.						
.63	4000.	1.5	0.21	200.	0.1	0.168	141.4	10.0	19.1	0.0	-10000.	0.						
		2.3			0.9	5.000		10.0										

LONS OF FUEL BURNED / 60 MIN. HOUR...   18.47

VEL TIME.   2.631   FIXED TIME...   5.000   CYCLE TIME...   7.631   TRIPS / 60.0 MIN. HR...   7.86

Figure 18–16–2 *(cont.)*

Figure 18–17–1 Typical air-floated bridge caisson for deep pier foundation construction for which buoyancy computations by computer are practical

19 COMPUTER CAPABILITIES IN CONSTRUCTION MANAGEMENT

W. F. MEYER
IBM Marketing Manager
New York, New York

EDITORS' NOTE: *Chapter 19 is not intended to conflict with reference to computers and network methods explained in other chapters. Those chapters do not attempt to explain what computers are capable of but in each case set forth when the use of computer techniques is economical and recommended. This chapter sets forth what the computer can be made to accomplish without demonstrating its economy in the various circumstances stated.*

DATA PROCESSING systems and techniques in the construction industry have been firmly established over the last few years. Present data processing users are found in all segments of the industry, and the areas of application are broad. The environment today is one of expanding application, extending from the mechanization of such functions as payroll, cost, and scheduling to the mechanization of certain parts of the estimating operation. Data processing systems are being used for the planning and scheduling of projects and for the control of project costs through well-functioning and responsive cost control systems. The construction industry environment is one that has fostered the recogition of data processing as a tool of management.

The prediction and control of costs constitute the main stimulant in the growing use of data processing systems and techniques. It is the prospect of improving these capabilities of prediction and control that has resulted in the growing investment in data processing systems within the construction industry. In the competitive environment of the industry, these capabilities become as important as the advantages gained through the recognition and use of new construction technology, and the contractor who does not utilize this new construction technology is at a decided competitive disadvantage.

19-1. Project Cycle

The life of a construction project can be viewed as a cycle. The link from the end of one project to the start of another is reflected by using the information and experience gained in the first project to estimate the new one. In addition, management has the need to extract data from various parts of the cycle in order to exercise control. These relationships are depicted in Figure 19-1-1.

Estimating Estimating, as a department, has the prime responsibility for preparation of the estimate of costs. The cost estimate is based upon factors such as:

- Historical cost performance records or standards
- Construction methods involved
- Knowledge of labor and equipment required, and anticipated productivity

After determining what resources are necessary to construct the project, estimating then relates specific costs to the project. Using the information developed by estimating, management has the responsibility of establishing the bid prices for the job. In addition to the estimate of costs, such factors as the following are used in determining bid prices.

- Who else is bidding?
- How much profit is required to justify the risk?
- Is new equipment needed?
- Are resources such as equipment and supervisory personnel available?
- Anticipated weather conditions

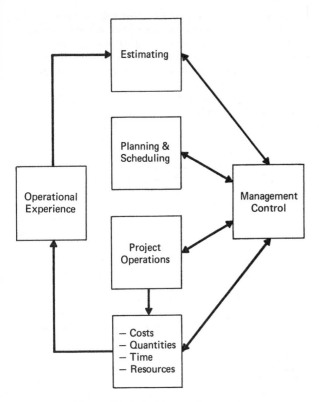

Figure 19-1-1 The project cycle

Project Control Operating System, will be described at the end of this chapter.

Experience has shown that few contractors have attempted to mechanize all aspects of project control at one time. Instead, most centralized systems have resulted from initial installation of a major area followed by a continuing development of procedures covering other major areas. Most contractors have begun application of data processing principles to their work by first mechanizing the phase which represents the most significant savings. Some contractors have felt that cost control should be the first step; others have stressed the project scheduling area. Still other companies have placed major emphasis on the estimating phase of operations. Whichever area was chosen, the experience gained in the first step proved invaluable to subsequent mechanization of other areas of project control.

As a preliminary step to augmenting any of these areas of project control the contractor must develop a method of coding which will be compatible with all systems. This is the *most* important part of the entire system, a code for the standard designation of each element involved in a job. Certain criteria can be set in the specifying of a good code structure.

- It must be flexible so that the same structure will apply to all jobs, although some of the coding may vary from job to job.
- It must be easily modified and expanded.
- It must provide for future applications.
- It must be compatible with an automated type of system.

A coding system should cover labor, material, equipment, subcontracting, and overhead. The common information base must include labor-type coding, job numbers, work performance elements, resources applied, and the degree of their interrelationship. The estimator first uses these codes to break the job down into separate categories. The coding should be compatible with activity numbers for scheduling and account numbers for maintaining cost control. A number of codes developed by industry associations can form the basis for codes developed by individual contractors to match their own needs.

Planning and Scheduling Planning and scheduling functions begin when notification is received of the successful submission of a bid. A plan is developed which demonstrates the sequence of work required for construction of the project. The plan is then implemented into a schedule. Dollars are budgeted for the anticipated costs of construction, and a pay schedule is negotiated with the contracting party or owner.

An overall operating schedule and budget evolve which, if realistic, should result in the successful completion of the project in the desired time and at the anticipated rate of profit. Again, management plays an important role in evaluating the results of the planning and scheduling functions.

Project Operations The project is turned over to a project manager or superintendent who has the direct responsibility of finishing the job on time and at a profit. During the life of the project, progress and costs undergo constant monitoring and are related to the estimate and schedule, thereby providing a basis for decisions affecting the course of the project. This same information, if acquired accurately and in sufficient detail, provides a source of operational experience that can aid in the estimating phase of future projects.

In the discussion that follows, the application of data processing to the three major areas of estimating, planning and scheduling, and project cost control will be explored in some detail. It should be kept in mind that these applications are fundamental "building blocks" or modules of the contractor's overall operating system. Each can be developed independently of the other, but when integrated they can form a total data processing control system. Finally, a conceptual approach to the integration of the data processing functions of a construction company, the

ESTIMATING

Few functions within the sphere of operations of the construction industry are more vital to success than is the estimating function. The financial success of a contractor depends upon his organization's ability to (1) estimate accurately the costs of proposed projects, and (2) perform the work at or below the estimate.

The prime objectives in the estimating functions are as follows:

- Determine the method of construction.
- Takeoff quantities of work.
- Apply costs to quantities.
- Classify and summarize estimated costs.
- Submit bid.

19-2. Data Processing Applications in Estimating

In recent years, a marked increase has become apparent in the number of contractors who are doing estimating work with data processing equipment. The range of application is large. Some contractors are performing only the printing of the estimate with data processing equipment, whereas other companies have been able to mechanize the bulk of the estimating function.

Several factors have resulted in this growing mechanization of the estimating function. First, companies using data processing systems in cost-capturing functions are accumulating data for estimating which can be used automatically. Proper coding of cost data allows much of the information to have application on subsequent bids. Some contractors have discovered that a great deal of the work load for standard items of work can be implemented on the data processing systems, thereby allowing estimators to concentrate on nonstandard work. Mechanization relieves the estimator of the clerical functions of looking up standards and performing calculations, and allows more time to be spent on the important aspect of determining the method of construction.

The increased detail reflected in bids is another factor that influences mechanization of the estimating functions. This increased detail is sometimes attributable to the requirements of the bid specifications, but more often it is dictated by the internal requirements of the company. For contractors who have mechanized the cost-capturing function, it is imperative that the job budget be set up and costs captured at the corresponding level of detail. The budget is prepared as a result of the estimate of costs.

The compelling requirement for accuracy is another significant factor. The volume of calculations involved in preparing an estimate is enormous. Quantity calculations, rate applications, extensions, summarizations, and cross-footings must be performed. Since an error in calculations could be disastrous, checks and cross-checks are made. The inherent accuracy of data processing systems significantly reduces the chances for error.

The methods and approaches adopted by contractors have encompassed most estimating functions. Some contractors have been able to develop fully automated estimating systems, whereas others have mechanized only selected functions.

For purposes of this discussion, estimating functions have been broken down into the following five data processing applications:

1. Quantity takeoff
2. Detail takeoff
3. Pricing
4. Printing
5. Evaluation of changes and alternatives

Each of the processing applications will be discussed individually in the presentation that follows.

19-3. Quantity Takeoff

The automatic performance of all estimating functions except the most simple quantity takeoff calculations requires an electronic computing system. A further requirement is that all of the construction functions be stated in terms of the mathematical formulas that describe each operation. The formulas are described to the computer by lists of instructions called *programs*. Each program is assigned a unique program identification number corresponding to the contractor's function or cost code.

From drawings and specifications, the estimator prepares the input to the data processing system. For example, an estimator can describe a footing in terms of the dimensions and characteristics of the structure. A form is filled out showing the function number for concrete footings of a certain type, the dimensions, and the number of identical footings to be found in the proposed project. These data are punched into a card and are processed through the data processing system.

Based upon the function number, the computer locates the correct program for footings and then performs calculations on the dimensions as dictated by the program. The results are multiplied by the number of identical footings found in the plans.

The output from the program states the volume of the structure in cubic yards of concrete required. Based upon the characteristics of the footing, the total pounds of reinforcing steel are computed, as well as the total square feet of forms and the total cubic yards of excavation. Each output record carries a code, which is later used in applying standards.

The method described above is now being used successfully. The system requires a coding scheme to describe functionally the construction operations. The coding system must be sufficiently detailed to ensure that the output is meaningful. In the footing example, the resultant cubic yards of concrete would be meaningless for costing purposes unless the method of placing (direct, buggy, etc.) is described.

Another requirement of a system that is to perform quantity calculations is that performance standards be meticulously accumulated. Many contractors who have not yet mechanized this function are presently engaged in capturing performance records in sufficient detail to be used for quantity calculations at some future time.

19-4. Detail Takeoff

The term "detail takeoff" is used to describe the operational step which breaks down the items of work from the quantity survey into the detail of the cost elements which apply. In a mechanized system, the detail takeoff consists of an operation whereby the cost or function codes are matched against a file which represents the contractor's catalog of standards. The catalog of standards is a file of cards, magnetic tape, or magnetic disk records developed through the cost-capturing function of prior projects.

The composition of the standards file differs considerably between contractors. Some organizations capture costs in terms of total costs and quantities set. Others capture the costs at a much more detailed level. Labor costs, for example, are reported and maintained by labor craft code within cost code.

The following example demonstrates how a mechanized system of detail takeoff operates. A record with a cost or function code and quantity, which apply to erection of a certain type of form, is processed against the standards file. The standards file reveals that erection of a square foot of the form requires 0.2 hour of laborer's time, 0.3 hour of carpenter's time, and 0.1 hour of foremen's time. This information is extended by the total square feet to determine the total hours by craft. Records are created which will be "input" to the pricing operation.

Another method of detail takeoff is becoming increasingly popular and is providing tangible advantages to contractors. This method relieves the estimator of the pricing, classifying, and printing of the estimate. The system further provides information on manpower, equipment, and material requirements to the planning department, if the bid becomes successful. This system is used primarily by contractors who are relating the estimate to the critical path method. A detailed discussion of the relationship of estimating to critical path is included in the section titled "Planning and Scheduling."

19-5. Pricing

Pricing, or costing, for the purpose of discussion, is defined as the application of rates to previously determined items of cost. The pricing function is often the first estimating operation to be automated by a contractor. Mechanization of pricing relieves the estimator of the responsibility of looking up rates and prices and performing the related calculations. The cards or records that are processed in the pricing procedure are the input to the printing procedure.

The pricing function can be classified into two major types. The first type involves application of standard unit costs. The second type involves application of prices and rates to the details of the elements of cost for each item of work.

The input to the standard unit cost method of pricing is furnished by cards punched with the following information:

● Project number
● Cost code
● Estimated quantity
● Efficiency factor

The efficiency factor permits the estimator to increase or decrease the standard unit cost based upon his judgment and knowledge of the particular work item being costed. For example, an estimator may wish to increase the historically developed unit cost by 10 percent for some item because of adverse working condtions on the proposed project. To accomplish this objective, he enters an efficiency factor of 110 percent.

The input records are matched against the standard cost file or cost code number. The unit cost obtained for each item is multiplied by the estimated quantity. The result is further multiplied by the efficiency factor. Records are created which are used subsequently in the printing procedure.

An input into the system which prices the details of the elements of costs is provided by cards which have been keypunched or automatically created during the detail takeoff process. The function of pricing details involves two steps. The first step is to look up the correct price or rate, and the second step is to multiply the quantity by the rate.

The procedure used in pricing depends upon the type of data processing system employed by the contractor. In punched-card installations, it is necessary to break apart the elements of cost by type, and to price each group individually. For example, labor cards must be broken out of the other cost cards and matched against a master labor rate file in craft-code sequence. The correct craft rate is obtained and then multiplied by the number of hours.

Contractors having electronic computing systems with auxiliary, random access storage price all elements of costs without breaking up the input file. The capability exists to have "on-line" the necessary rate tables for labor and equipment and the prices for standard materials. The data processing system looks up the proper rates or prices in the tables and multiplies the quantity by the rate. The result is reflected in input records. The records are then ready to be used in the printing of the estimate.

19-6. Printing

The term "printing" as used in this discussion includes the classification, summarization, and preparation of all printed reports for use in estimating. Several factors have influenced the mechanization of the printing function. The main factor has been the saving of the estimator's time in performing the necessary classification and printing of the multiplicity of detailed and summarized reports.

The inputs to the printing operation are cards or magnetic records created during earlier operations, or cards punched with the precomputed data for the printing function. The procedure used on data processing machines involves first sorting the records into the desired sequences. The next step is to print the desired reports. On data processing systems with adequate storage, the sort operation is sometimes eliminated, and the information for the reports is accumulated in the storage devices.

The output of the printing operation takes the form and sequence dictated by the individual contractor and the type of data processing system employed. A detailed estimate, in which each cost code is broken down by the individual elements of cost which apply, is depicted in Figure 19-6-1. Each cost code reflects a total estimated

GENERAL CONTRACTORS COMPANY

ESTIMATE DETAIL

Date: 03-04-71   Job 101   Page 11

Cost Code	Description	Quantity	Unit	Labor	Materials	Equipment	Subbid	Overhead	Total Estimate	Unit Cost
120761	Backfill first stage	6000	CYD	800 00	300 00	825 00			1925 00	321
290372	Set concrete	350	CYD	3000 00	7500 00	350 00			10850 00	31 000
708295	Erect office shack	1	LS	200 00	50 00	75 00	22 00		347 00	347 000
906702	Performance & payment bond	1	LS					465 00	465 00	465 000
	Total estimate			75000 00	24000 00	42000 00	6000 00	5000 00	152000 00	

Figure 19-6-1 Detail estimate broken out by elements of cost

GENERAL CONTRACTORS COMPANY

BID ITEM SUMMARIZATION

Job 201   Description: NORTH BRIDGE   Date: 03-04-71   Page 1

Bid Item	Description	Quantity	Unit	Total Cost	Unit Cost
1	Clear and grub	Job	LS	97500 00	
2	Remove structure	Job	LS	900 00	
2	Excavation including haul	600000 00	CYD	300000 00	5000
4	Backfill-foundation	12000 00	CYD	36000 00	3 0000
5	Reinforcing steel	2000000 00	LB	250000 00	01250
6	Concrete class A	100000 00	CYD	8000000 00	80 0000
7	Structural steel	120000	LB	48000 00	4000
	Total estimate			8732400 00	

Figure 19–6–2   Bid-item summary

cost and a calculated unit cost. This same type of report is seen as a summary report where the estimate has been made in more detail, i.e., labor craft within cost code.

A report prepared to summarize the estimated costs is illustrated in Figure 19-6-2. The report reflects a sequence which corresponds to the bid-item number assigned in the specification by the prospective client. From this report, management obtains the total cost and unit cost for each item which must be bid separately.

The bid-item summary and the example of the detailed estimate report are typical of the types of estimating reports prepared on data processing systems. Many other reports not shown here are printed to meet the requirements of construction management. The ability of data processing systems to classify, summarize, and print reports accurately and rapidly has resulted in many contractors mechanizing all or parts of the data processing potential of their estimating responsibility.

19-7.   Evaluation of Changes and Alternatives

Evaluation of changes and alternatives represents a composite of all previously mentioned estimating applications. The function is broken out separately here in order to stress the importance of data processing as a tool to aid construction management to answer quickly the "what if" questions of estimating.

Data processing permits the user in the construction industry to make minor or major changes to estimates and to be aware of the effect of the changes in a relatively short time. The only manual processing is the preparation of records which reflect the proposed changes. These records are then inserted in the estimate file and the old records pulled out. From then on, the process follows the automated procedure.

New estimate reports are created that reflect the changes. The reports are then submitted to management for approval or further change. The process continues until an acceptable estimate is prepared.

Contractors that have mechanized estimating functions are further provided the facility to evaluate quickly the effect of alternative proposals. Alternative proposals could represent different methods of construction, more than one type of equipment, or more than one type of material. The ability to evaluate quickly alternative solutions to the bid affords management the assurance that the pro-

posed bid represents the best package to submit to the prospective client.

PLANNING AND SCHEDULING

19-8.   Responsibilities of Planning and Scheduling

The responsibilities of planning and scheduling in the construction industry cover the allocation of time and dollars to produce acceptable budgets and schedules for construction projects. The planning phase involves budgeting of time and money, and the scheduling phase converts decisions made during the planning cycle into a schedule to be implemented by the field force.

The construction industry has been a leader in the use of data processing systems to assist in the planning and scheduling functions. The development of PERT (Program Evaluation and Review Technique), CPM (Critical Path Method), and other management science techniques has caused many contractors to re-evaluate their present methods. In many cases the use of PERT or CPM is stipulated by the client as part of the bid specifications. This is especially true of contracts administered by governmental agencies, such as the Department of Defense and the Corps of Engineers.

The traditional method for job planning and scheduling has been the Gantt, or bar chart, in which the job's major activities are listed vertically and bars representing time durations are drawn to the right along a horizontal time scale. Planning and scheduling the job are done simultaneously by deciding when each activity should begin and what its duration will be.

Gantt charts as project control tools have grave limitations:
1. If the resulting job finish date is not satisfactory, the activities which require replanning are not readily identifiable.
2. The relationships between these broadly defined job activities cannot be defined clearly because of insufficient detail.
3. It is difficult to identify any job planning and how schedules were established.

In 1953, responding to the U.S. Navy's desire for a more sophisticated planning system, the Line-of-Balance (LOB) method was developed. LOB uses a Gantt chart, but applies the 'level of activity intensity" concept to account for the inevitable variation in labor productivity. That is, if the time duration of the job is half over, for example, the labor has not necessarily completed half of its work.

The LOB method gave project management a better tool for anticipating deadline trouble. It also had some ability for resources planning, to intensify behind schedule activity. But there was still no detail of interdependence of activities, and it is as useless as a Gantt chart for replanning.

Then came the era of PERT. There were successive versions developed for defense research projects, finally leading to PERT/COST, in which the PERT concept was extended to include cost reporting. At about the same time, the management control technique known as the CPM was being developed in private industry.

PERT, CPM networks, and the newer technique of precedence diagramming all involve breaking a project down into detailed activities, all done in logical sequences. When all the activities are recorded and interrelationships accounted for, the result is an activity network that is a model of the entire project.

By adding up the time required by all the activities on each of the alternate paths through the network from beginning to end, the lengthiest time path through the network is found. This path determines the time required for the total project and is therefore called the critical path. All other paths have slack, a manageable resource in all projects.

Both PERT and CPM are primarily concerned with the completion of one activity before the next one can begin. Both methods assume that any given activity cannot begin until the preceding activity (or activities) on which it depends has been concluded.

Precedence diagramming, on the other hand, allows activities to begin before completion of their predecessors. It offers significant advantages in planning and constructing networks, and then manipulating activities for construction projects.

PERT and CPM project networks are both based on the arrow diagram. The section of an arrow diagram in Figure 19-8-1 shows that activities B and C cannot start until activity A has been completed; activity D cannot begin until both activities B and C are finished. Each end of the arrow has an event number, establishing a unique identification code for the activity between the end events or nodes.

The dummy activity does not really exist; it is simply a necessary convention in an arrow diagram so that activities B and C do not have the same code designation. The section of precedence network in Figure 19-8-2 corresponds to the arrow diagram in Figure 19-8-1. In the precedence method, the activities are shown in the blocks at ends

of relationship lines. The unique code number refers to the activity itself, not to its absolute position in the network. Therefore, in this example, these is no need for a dummy activity for activities B and C.

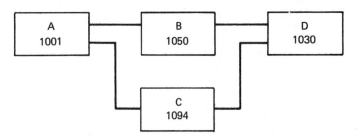

Figure 19–8–2 Precedence network

Any construction project can be represented either by an arrow diagram or a precedence network. A precedence network, however, is far less complex and requires much less time to prepare and fewer program instructions for a computer. Because event numbers are completely foreign to construction company operations, standard activities cannot be given standard codes. Thus, the coding for a critical path network cannot be directly used in an electronic data processing (EDP) system for tying together estimating, cost control, and other project management functions.

The activities in a precedence network, on the other hand, can be identified by codes that are part of the contractor's overall coding system. Changes in the activity sequence due to rescheduling a project do not affect the activity codes. They remain the same—meaningful beyond that project alone. This is particularly important. For example, cost data can be directly related to individaul activities, so that the precedence network can be useful long after a job is completed.

The following discussion is concerned with how data processing systems are now being used in the construction industry to facilitate project planning and scheduling. The presentation is divided into five major areas as follows:

- Planning for budget and schedule preparation
- Scheduling of project
- Output reports
- Simulation of alternatives
- Updating and revision of project schedules

19-9. Planning for Budget and Schedule Preparation

The importance of the planning phase of operations cannot be overemphasized. The success of the cost control and scheduling systems is dependent in large part upon the effort expended during the planning operation. The results of the work performed during the planning phase is represented by the project operations budget and the input to the scheduling system.

The basic function of planning is to develop and provide a plan of operations that is feasible in terms of time and costs. To develop this plan effectively requires such information as knowledge of construction methods, availability of resources, and anticipated productivity levels. Since this same information was needed at the time the estimate was prepared, the estimate is used frequently during the planning cycle.

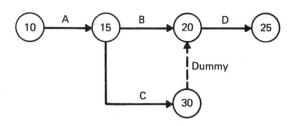

Figure 19–8–1 Arrow diagram and precedence network example

Budget Preparation   When the estimate of costs has been made at the same level at which costs will be captured, the estimate becomes the budget. However, where estimates have been made at other than cost code level, it is necessary for planning to break down the estimate into the cost codes which apply.

The construction budget is represented in data processing systems by a file of records containing cost code, estimated total cost, estimated quantity, unit of measure, and estimated unit cost. In most cases, the cost amounts are further subdivided into the various elements of cost (labor, materials, etc.) which apply. Information from the budget records is used to compare actual cost and quantity with data reported from the project site. From this comparison, management is informed of how the job is progressing costwise. Deviation of actual cost from estimated cost are pointed out for management review.

Schedule Preparation   As is the case in budgeting, planning obtains a great amount of information for scheduling from the estimate. Again, depending upon the level of the estimate, most of the data required to schedule the project may already be in a form that can be processed by machine.

To provide the necessary data to a scheduling system, planning must perform the following operations:

1. Break down the overall project into the items of work which should be independently scheduled. This operation may already have been performed during the estimating cycle.
2. Construct a network that displays graphically the sequence logic of the project. That is to say, items of work are placed on the diagram in the proper sequence which defines the interrelationship of each activity to its predecessor and successor activities.
3. Assign a unique code number to each activity, which conforms to the numbering conventions of the particular scheduling system employed.
4. Assign a duration, or the estimated time required to accomplish the function of the activity, to each item.
5. Provide a list of the activities to data processing for conversion into punched cards.

Planning must also furnish the data processing section with the instructions required to process the network data. These instructions relate to the scheduling program and are concerned with the preparation of a project calendar and selection of the desired data processing options.

Scheduling of Project   The availability of electronics computers has given impetus to the widespread use of CPM for construction project scheduling. CPM schedule calculations can, of course, be performed manually. However, manual network calculations for all but the simplest projects generally prove prohibitive because of the time involved. Not only is it time consuming to generate the initial schedule, but also the updating and rescheduling phase during the life of the project becomes extremely laborious. The use of PERT/CPM techniques in project planning and scheduling is simplified when computer programs are permitted to do all the paperwork and calculations except for original entry of the raw data. The computer program directs the computer to calculate a network schedule for the project and then produces, on request, the various reports needed for project planning and control.

A recently introduced computer program, called the Project Control System (PCS), can accept input data either for PERT/CPM or precedence networks. (PERT/CPM networks are converted to precedence networks inside the computer.) The PCS for the IBM 1130 computing system, for example, is able to process a network with as many as 2000 activities and 4500 precedence relationships. The equivalent program for the more powerful System/360 can handle 5000 activities with 12,500 precedence relationships. The data flow required for computerized scheduling of a network is shown in Figure 19-9-1.

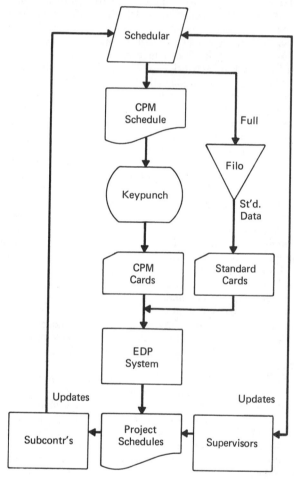

Figure 19-9-1   Data flow in computerized scheduling

The printed schedules are sent to the network planners for review. In nearly all cases, the calculated project completion date will not be adequate. The network planners will revise the network by any of a number of methods and will record the revisions on a prescribed form. Cards will be punched to record the modification data, sorted into the prescribed sequence, and then read into the data processing system. The original activity records will be located and updated to reflect the changes. The calculations will be performed and the revised schedules printed out and sent to the network planners for evaluation.

The initial generation of a project schedule is usually an iterative process before a calculated project completion date is deemed satisfactory. Once a desired completion date is obtained, the network-records file (cards, magnetic tape, or magnetic disk) is held for subsequent use in the recording of progress and change data reported from the job site.

The reports produced by the PCS program may cover the entire project network or any portion of it. Among the major reports are:

- Network listings: each activity, its description, duration, dates, and float
- Bar graphs: activity duration ploted against calendar time
- Milestone listings: for each milestone, activity number, description, dates, and float
- Resource reports: source utilization as a function of time for one or more resource codes
- Work status and progress reports: informing supervisors of work scheduled to be done in a given period of time. Data reported for an activity includes description, remaining duration, percent completion, start and finish dates, and preceding work items
- Cost reports: one report lists estimated and actual costs by activity, together with its description, start and finish dates and current status as indicated by remaining duration, percent completion, and float or slack, allowable. The other report lists estimated and actual cost by activity within a calendar month on a current and cumulative basis

In addition to input data assocated with a specific project, the contractor is able to introduce standard punched-card data on activities and costs that are part of almost every job and are in its coding system. An important element of scheduling, of course, is the updating of the project scheduled as reports are received from subcontractors and field supervisors.

A larger contractor in the Southwest, who helped develop and test the PCS program, uses precedence networks for very effective project planning and scheduling. The network may involve thousands of activity blocks and relationships. Typical reports generated by the computer for this firm's management are the activity sequence, bar chart schedule, and material requirements plot showing cumulative barrels of cement required each week for a project.

19-10. Simulation of Alternatives

Project scheduling, using the CPM on a data processing system, provides construction management with a tool to evaluate readily the effects of proposed alternatives. A construction company which uses the system can, in effect, simulate suggested changes in network sequence and, in short order, be aware of the results of the changes.

The ability to answer the "what if" questions of network planners is valuable in both the initial scheduling of a project and the subsequent updating and revising of schedules during the life of the project.

Simulation of alternatives is particularly valuable in initial project scheduling because of one basic reason: The probability is infinitesimal that a satisfactory project completion date is obtained from the first scheduling run. More often than not, initial project results in several runs before a desired completion date is attained.

Proposed changes to the sequence logic of a project network take several forms. Operation durations can be reduced by determining that additional resources are to be committed to a project, or certain operations will be scheduled on an overtime basis, or certain operations will be scheduled on a multiple-shift basis. In evaluating the results of proposed changes, management must consider not only the individual operations, but also the effects on the overall project.

A decision to commit additional equipment to a project because of a conflict in requirements at some period, for example, should be reviewed in terms of the implications on the network operations preceding and following the operations which cause the conflict. From a standpoint of economics, move-in and move-out charges on equipment generally preclude additonal equipment being assigned to the project for short periods.

Other "what if" questions that can be answered readily through the CPM are concerned with manpower and materials. For example, increasing the number of men in a particular craft can have the effect of not only decreasing operation durations, but also of permitting currently planned serial operations to be performed concurrently. An example of a proposed network change prompted by a decision concerning material can be found in the alternative of having materials prefabricated, rather than fabricated at the job site.

The important feature of rapid simulation of alternatives during the initial scheduling of a project is emphasized because construction projects tend to be considerably different from previous projects performed by the same contractor. As a result of these variations, the trial-and-error method of determining a required (or desired) completion date is often necessary.

Once the job is underway, the ability to determine the effects of proposed changes takes on new importance. Physical factors at the job site and proposed change orders are two items that can result in changes to existing network logic. The capability of quickly simulating the effect of proposed changes to a project schedule can and does provide a contractor with a tool to use in negotiations with the owner. Some contractors have found CPM analysis to be an effective sales tool.

19-11. Updating and Revision of Project Schedules

After the project is underway, the schedule must be constantly reviewed and analyzed in view of accomplishment and modification. Home-office and project management must remain informed of job status throughout the life of the project.

Updating Cycle The reporting cycle for updating networks varies considerably from one contractor to another. Some organizations update the network only under special circumstances, such as a long delay due to inclement weather conditions. Most contractors, however, have adopted a uniform period for progress reporting. Generally, the period is monthly, biweekly, or weekly. Often the reporting cycle is determined by the requirements of the owner. One government organization responsible for coordination of many construction contracts requires a biweekly reporting cycle.

19-12. Job-Site Reporting

The responsibility for job-site reporting is delegated to the project manager or superintendent. Depending upon the size of the job, the direct responsibility may be further delegated to the project engineer or the field office manager.

Progress reporting from the job site ordinarily involves

several different methods of determining what has been accomplished. Visual checking is one method that is used by job-site personnel. Another technique involves the recapitulation of quantities set in place for certain operations. Usually, several methods are used on a single project, and the reporting project bears the responsibility of reducing the progress data to a common factor used in the CPM program, such as remaining time or percent complete.

Methods of Reporting Progress   The two traditional methods of recording progress to computer-processed CPM networks are: (1) to report the percent complete, and (2) to report the remaining time required to complete operations with progress for the previous period. As an example, the IBM Project Control System allows a contractor to record progress to activities (within the same project) in any of the following methods:

- Percent complete this period
- Percent complete to date
- Remaining duration (the time required to complete an activity)
- Quantity set in place this period
- Quantity set in place to date

The latter two methods relating to quantities set are used only when an estimated work-item quantity has been previously set up in an activity record.

Provision for a full range of reporting options is particularly valuable to an organization that uses an integrated system to control progress and costs. An excavation work item, for example, lends itself well to a quantity type of progress reporting because quantity must be reported to determine costs per cubic yard. On the other hand, a percent complete or remaining time option is more appropriate for an operation involving the installation of a boiler.

Methods of Recording Progress   The requirement for accuracy in reporting progress data from the job site has resulted in major emphasis being placed on the method of recording the data. Much effort has been expended on the part of contractors to design systems where progress recording requires minimum effort on the part of field personnel.

Methods now in use range from forms completely filled out by job-site personnel to computer-generated cards and lists that require minimum effort. To facilitate job-site reporting further and to ensure validity and timeliness of the data, data communication units are being used by some construction companies for progress recording.

Regardless of the method utilized to record progress data, two basic questions must be answered by the field personnel at periodic intervals: "Which activities have been worked on this period?" "How much of each operation has been accomplished?" A good technique is to use the previous reporting periods report as input for the next period by crossing out information on the report and recording current status directly above it.

The use of a preprinted report significantly reduces the clerical time required by field personnel to record progress data. The accuracy of the data is also enhanced because the chances for transposition errors are reduced. Other methods involving the use of prepunched, preprinted cards are making significant contributions to the successful implementation of network scheduling. Since the value of management information is directly related to the validity and timeliness of the source data, maximum effort should

be expended to reduce the clerical requirements from the job site.

19-13.   Exception Reporting

In recent years, a trend to the management-by-exception principle has been noted. The principle addresses itself to a system of reporting where exceptional situations are specifically pointed out to management and normal situations are passed over.

The technique of management by exception is particularly appropriate to analysis of project schedules in the construction industry. The present trend in project scheduling is to reduce the size of reports through various means. One method is to print only items in progress or which are likely to begin within a specified period designated by the individual company. The theory behind the technique is that management has no need to concern itself, each reporting cycle, with activities not scheduled to begin for a long time.

Examples of reports which are appropriate to a parameter date type of reporting include an early-start-sequence list and an early-start-sequence bar chart. Both reports provide a convenient method of condensing an overall project network into the probable starting schedule for the forthcoming period.

Another method of exception reporting is to point out only the specific items that should be called to the attention of management for possible action. The IBM Project Control System, as an example, has the capability of informing management of the status of operations that should be closely reviewed. Called project action reports, this information is arranged in such a way as to pinpoint unusual situations, thereby eliminating the need for management to search through lengthy schedules to discover problem operations. Examples of project action reports are found in the following list.

- Critical items that should have been started, but were not started
- Critical items that should have been completed, but were not completed
- Critical items scheduled to be started next period
- Critical items scheduled to be completed next period

With this information, management can readily determine which operations are causing the schedule to slip and which operations must be started and completed in the next period to prevent further schedule slippage. When a project is proceeding according to plan, few operations well be listed on the reports. When the schedule is not being met, however, the action reports furnish management with a capsule look at the items that require attention and possible action.

PROJECT COST CONTROL

19-14.   Cost Control System

Cost control in the construction industry encompasses the gathering of costs data by some predetermined classification, the relating of actual costs to estimated costs, and the reporting to management of current cost status. An effective cost control system has two main prerequisites. First, the information must be accurate and, second, the information must be timely. The inherent speed and accuracy of data processing systems represent the compelling reason behind the adoption of mechanized cost control systems within the construction industry.

Objectives of Cost Control The two major objectives of cost control in the construction industry are:

1. To provide management with timely information concerning the relationship of actual costs to estimated costs.
2. Through the accumulation of cost information, to provide current performance information to be used in future estimates.

As a result of the comparison of estimated costs to actual costs incurred, management is informed of over-budget items. If the information is available soon enough, management decisions can be made to bring costs back in line for the remained of the project.

An efficient cost-keeping function can point out to management not only which costs are high, but also why the costs are over the budget. For example, when estimates of labor costs have been made in detail by craft, analysis of costs incurred can point out which crafts are performing unsatisfactorily compared to the estimate. Effective cost control can inform management of high costs because of excessive downtime and repair charges for a certain type of equipment. Inordinate costs for materials used in certain operations can be pointed out to management through cost analysis.

In the ensuing discussion, mechanized cost control systems are presented which were developed by contractors knowledgeable in construction cost characteristics. As was the case with "estimating," the elements of cost are considered to be labor, equipment, materials, subcontractor, and overhead.

Labor is probably the most important and certainly the most variable element of cost control. The basic input information for labor cost control is the time sheet or card. Although the exact format may vary, all time records contain the basic information. For each employee name, job number, and work week recorded at the top, the foreman jots down the R (regular), O (overtime), and D (doubletime) hours for each type of work (cost code) that the employee does each day or week.

Although this cost control system requires that the foreman or supervisor record this information manually, it is entered only once on this time sheet and yet is used in many ways. After this information is punched onto cards and entered into the computer, a wide variety of accounting functions can be performed in the system—payroll checks; workmen's compensation; federal, state and local taxes; union dues; and welfare fund reporting.

In addition, with a coding system common to the estimating and the planning and scheduling functions, the basic labor cost data can be used in other ways by management.

Since the time card is the nucleus of a payroll and labor cost system, the need for accuracy in labor reporting is of paramount importance. A good example of a technique used extensively is the utilization of IBM cards which have been prepunched and preprinted with standard information. For example, a set of cards prepunched and preprinted with employee badge number, employee name, and craft code is sent to the job site. Each recording of hours requires that the field office manager enter only the cost code and the number of hours.

The recording of the hours worked and the distribution can be by several methods. Keypunching can be eliminated by use of either the mark sense or port-a-punch principle. Another technique involves the recording of the data on a card, and in a subsequent operation, the card-punch operator reads the data and punches it into the same card.

Whichever approach is adopted, the primary objectives are to reduce the volume and complexity of clerical effort at the job site and reduce the volume of card punching. Both objectives are directed toward reduction of effort and increased accuracy in labor cost reporting.

Construction payrolls are often processed at the job site because of the distance from the home office and the short time allowed for payroll processing. The development of the data communication system has provided the industry with an effective tool to bridge the communications gap between the job and the home office.

Regardless of whether the payroll is processed at the home office or at the job site, the costs of labor must be properly distributed to the cost codes that apply. The labor cost data captured is necessary for cost control on the project as well as for historical performance standards.

A medium-sized Midwestern contractor, for example, considers his weekly labor report the key cost control tool for accurately evaluating labor cost performance in the past, present, and future. The weekly labor report is ready for the home-office and project managers no later than Tuesday afternoon. It summarizes the job cost situation current to the previous Sunday afternoon. The quantity of work performed and total labor are shown for as many as 300 labor classifications. Costs are included for the current period, previous period, and cumulative to date—actual versus estimate.

The report also projects a final cost for the job and for each labor classification code, comparing the projected final cost against the estimate.

A larger Southwestern contractor divides his labor cost information into several reports, one of which is the man-hours-to-date summary shown in the computer printout in Figure 19-14-1. The project manager is concerned only with the accumulated hours by carpenters, iron workers, finishers, etc.

To be effective, however, cost reports must be analyzed in a practical manner. Effective cost control dictates that the information in cost reports be evaluated in terms of the construction characteristics of the operation. As an example the setup costs involved in some operations are extremely high. Therefore, a projected loss based on an abnormally high unit cost during the beginning phases of work can be alarmingly high. As the operation progresses, however, projected profit or loss on the estimate takes on a more realistic nature. At that time, management can analyze the reason for the deviation and point out corrective measures to improve the situation.

19-15. Labor Cost Performance Records

The main prerequisite of a system which accumulates historical performance records to provide information to estimating is a comprehensive cost-coding system. The importance of the coding system cannot be overstressed. To be effective, the coding system must be distinguished by two characteristics. First, the code must be sufficiently comprehensive to encompass the myriad of different operations found in construction projects. Second, the code must be simple enough to be practically administered by field personnel oriented to construcion work, not paperwork.

JOB

DATE: 8/15/71

JOB COST STATEMENT

LABOR DISTRIBUTION THRU ....

MATERIAL INCLUDED THRU ....

PR TAXES & INSURANCE THRU ....

LABOR FRINGE BENEFITS THRU ....

Cost Code	Description	Unit	Quantities			As Sought Estimate		Used for Projection		Cost to Date	Projected Final Cost	Overrun or Savings	Engrs. Adjust
			Original	Adjusted	To Date	Original	Adjusted thru Co - - - -	Estimated Unit Price	Unit Price to Date				
0250104 ***	ASPHALT PAVING ALLO					99,676	99,676			.00	99,676		
	TOTAL OF SECTION 02						551,553			437,372.22	555,962	4,409	
0300002	CONC. MATERIALS					588,646	589,944			359,037.49	589,944		
0301101	P/CAPS FTGS GR DMS	SF	31379	32053	30053	18,964	19,858	.620	.515	15,464.69	16,507	3,351	
0301111	FOUND WALLS & PIERS	SF	3699	3699	3699	2,427	2,427	.656	.507	1,873.74	1,875	552	
0301121	COREWALL FORMS	SF	164435	165703	119268	101,769	102,510	.619	.489	58,366.85	81,029	21,481	
0301201	COLUMN FORMS	SF	74133	74240	55587	41,514	41,574	.560	.485	26,971.42	36,006	5,568	
0301251	RETAIN WALLS TILTED	SF	4023	4023	3627	3,862	3,862	.960	.931	3,377.03	3,745	117	
0301261	PARA WALL FORMS	SF	3735	3735		3,287	3,287	.880		.00	3,287		
0301271	CURB FORMS	SF	1965	1965	1700	1,790	1,790	.911	1.268	2,155.67	2,692	702	
0301301	SLAB FORMS	SF	544035	544035	399630	149,442	151,228	.278	.260	103,856.15	141,449	9,779	
0301351	BEAM FORMS	SF	141565	141869	100805	112,263	112,475	.793	.727	73,327.96	103,139	9,336	
0301651	STAIR FORMS	LS			15	4,673	4,673			2,690.90	4,673		
0301701	MISC. FORMS	LS			1	1,665	1,818			1,037.69	1,818		

Figure 19-14-1. Job cost statement for the structure project.

Offhand, these characteristics seem to be in direct opposition; however, further analysis of the project concept of industry reveals little conflict.

Once the project budget has been determined and set up at the cost-code level, the project personnel are concerned only with the cost codes applicable to the particular job. The job budget may involve only a small portion of the total number of possible cost classifications maintained at the home office. Thus, field reporting of costs is made only to preassigned cost classifications, simplifying the determination as to the particular codes which apply. The use of preprinted cards or forms for field reporting, as mentioned previously, further contributes to the simplification of data classification and accumulation.

In general, two viewpoints prevail among data processing users in construction which cover the use of captured labor performance standards. The first involves the use of the standards in an automated estimating system. This method involves a technique of automatically applying the labor standards to the estimated quantity of each cost item in a bid. The second method involves the retrieval of past labor costs to be used for information purposes by estimating.

The level at which the labor cost data are captured is also a variable factor. The degree of detail ranges from capturing labor unit hours by craft-within-cost-code at one end, to the determination of unit labor costs by cost code at the other end. Within the two extremes, many variations in level of detail can be found. Generally, the finest level of detail is found in the installations of contractors who have mechanized estimating functions to the greatest degree.

Captured labor cost information is maintained in several different forms. Punched cards, magnetic tape or disk, and printed reports all serve as a medium which provides estimating with information. The final job cost statement, printed when the job is closed to the estimate, functions as current cost information to estimating when bidding similar projects. This report is considered the end product of historical cost information by other contractors. Other construction companies, however, use the data captured from project to update the historical cost catalog. These data are maintained in one of the forms that can be processed by machine and can be quickly retrieved for informational use or for automatic application in subsequent estimates.

From a data processing standpoint, maintenance of performance records represents a valuable by-product of the cost control system. To control costs effectively at the job site, cost data must be classified in such a manner as to be compared to the project budget. The same information required for cost control is essential to historical costs maintenance. The updating of past cost records with current data provides management with the assurance that estimates reflect the effect of actual operations. Thus, maintenance of performance records serves not as an adjunct, but as an integral part of labor cost control in the construction industry.

19-16. Equipment Costs

Cost keeping for construction equipment, particularly in heavy construction, constitutes an important application area for data processing systems. In recent years, tremendous increases in productivity of construction equipment have been noted. The technological improvements have been accompanied by increases in the cost of the equipment, and, as a result, ownership of equipment constitutes a substantial part of the assets of many contractors. The need for effective control of equipment costs parallels the need for control of labor costs.

Several divergent practices are found in classification of equipment costs in the construction industry. One method incorporates the charging of all equipment costs to a single cost code. Another practice entails charging equipment costs to special classifications of cost codes pertaining only to equipment. Still another method can be found whereby equipment charges are considered one of the various elements of cost which apply to specific cost codes. In the latter method, for example, charges for a crane used in hoisting concrete would be directly charged as equipment costs to the concrete pouring operation.

Reporting of Equipment Costs The source document for a comprehensive system of equipment costing and record keeping will closely resemble a time card used for labor costing. From the equipment time card, the input to the data processing system is prepared. Records are created which update job costs as well as equipment records.

Data required from the project for use in establishing equipment charges are dependent upon the particular method employed. In general, however, the following information is needed and recorded by the foreman:

- Project number
- Equipment number
- Distribution code
- Distributed hours

The project number is the code assigned to identify the job for which the equipment is being used. An equipment number is a unique number assigned to each unit of equipment. The distribution code identifies the classification to which the distributed hours should be charged. This latter code includes not only the cost codes, but also special-purpose codes which identify the hours as downtime for repairs, idle time, downtime for weather, holidays, and any other categories desired. Distributed hours represent equipment time to be charged to each distribution code.

Equipment Cost Data Processing Cycle Following a preliminary audit of the equipment time cards which certifies that all available hours have been distributed, IBM cards are prepared which contain the project number, equipment number, distribution code, and distributed hours. The cards are processed against an equipment rate file that contains the ownership and operating rates. The correct rates are located and multiplied by the distributed hours. Wherever the distribution code is a job cost classification, the hours are multiplied by both the ownership rate and the operating rate. Otherwise, the hours are extended by the ownership rate only. Records are created with the equipment cost amounts to be charged to the particular projects. Other records are prepared which will be used to update the equipment records.

The equipment cost records are processed against a file that contains budgeted and actual costs for each cost classification. The actual costs are updated with the current period equipment costs, and an equipment cost analysis report may be printed.

The analysis report points out the significant aspects of the equipment situation to home-office and project management. A comprehensive equipment cost system includes provisions to show deviations from the budget, abnormally high percentages of downtime for repair, and unusual percentages of idle time. From the report, management can determine how the job is proceeding as to equipment costs, and the reasons for budget deviation. For example, exceedingly high proportions of available time charged as idle or standby time can point out a need to review the equipment pool committed to the project. Perhaps the pool can be reduced and excess equipment resource diverted to some other project where economical utilization can be realized. Conversely, a low percentage of idle time may point out a need for more equipment on the job site to prevent costly delays which could occur because of equipment not being available.

19-17. Equipment Records

Because of the large investment in equipment, many construction companies have emphasized development of systems to maintain equipment records on their data processing machines. The equipment record-keeping application includes such functions as recording of depreciation, scheduling of preventive maintenance, recording of maintenance charges and fuel consumption, and analysis of rates. Entries to system are provided by purchase orders for new equipment, equipment cost/time cards, and shop-work orders for maintenance. With this information, the data processing system can prepare various reports necessary for equipment management. Examples include: comparative analysis of productive time and repair time, evaluation of operating rates, and analysis of repair cost. From these types of information, management can determine when a piece of equipment justifies replacement because of prohibitive downtime and repair costs, when operating rates should be changed, and the relative efficiency of similar equipment available from more than one source.

19-18. Material Costs

The term "materials" as used in this discussion is construed to cover not only those materials, such as structural steel, which become part of the structure, but also supplies, such as dynamite, which are expended during the life of a job. No distinction is made since costing for both types can generally be handled in the same fashion.

Material costs are usually entered into the job costing procedure through the medium of the accounts payable operation. From the invoice of the supplier, the data concerning the material classification and price are obtained. When accounts payable are processed on a centralized basis, the invoice is matched to a material receiving slip from the job site. The invoice amount is distributed to the particular cost codes that apply.

One method of charging material costs to the job records the charge at the time the invoice is received and approved. Another method records the cost to the job at the time the cash disbursement is made. The distribution of material costs may be made to cost codes at the time of entry, or the amount may first increase the job inventory. In the latter case, the material costs are charged to the specific job cost codes as the materials are used.

Regardless of the system used by the individual contractor, data processing systems provide construction management with a tool to combine the related function of paying for the material and controlling its use. Maintenance of project inventories can be performed and tied directly into the material requirements aspect of the plan-

ning and scheduling system. Comparative analysis of actual material costs to budgeted costs is performed, and deviations are pointed out for management information. An extension of the material costing activity and accounts payable application provide information of on-site material inventory credit to management for progress billing purposes.

A further extension of the system provides management with the information required to analyze material purchases and answer the following questions.

- What did we buy?
- Who was the supplier?
- Where was it used?
- What did it cost?

19-19. Subcontractor Costs

A significant part of the total cost for many construction projects is represented by charges for work which is subcontracted. This cost element is particularly important in certain divisions of construction, such as building and industrial. From a cost control standpoint, subcontractor costs do not assume the importance of labor and equipment costs, since the subcontractor is normally awarded the contract on a fixed, lump-sum basis. However, an extremely important application for data processing systems exists in subcontractor accounting because of the importance of good relations between the general contractor and his subcontractors.

Reflecting subcontractor charges in job costing is performed in much the same manner as charging material costs to the job. When the progress bill of the subcontractor is received and approved by the project manager, the document enters the accounts payable system. The dollar amounts are coded to the cost classification set up in the job budget, and cards are punched to record the liability and the job cost. The cards for the distribution of the expense are channeled to the job costing procedure.

The cards which represent the liability are placed in the accounts payable file. Periodically, when checks are written, a list of outstanding obligations is printed and forwarded to management for approval. Approved invoices are selected from the file, and retention is calculated on the amount billed. Retention records are created and held for eventual payment when the job is closed.

Desired listings of selected subcontractor billings and payments are run at any time. Other transactions, such as change orders or advances, are reflected in the subcontractor file and provide access to a complete history of transactions involving a subcontractor on a job. This information is particularly helpful in reconciling discrepancies involving finances between the general contractor and a subcontractor.

Subcontractor cost distribution records are processed against the job cost file to update the actual costs and to be reflected in the monthly financial statement for the project. Management is provided information that shows the remaining balance of subcontractor cost classifications. Tied into the project schedule, this information is used for cash flow projections.

19-20. Overhead Costs

The fifth and last element of job cost is overhead. Sometimes known as general expense or burden, the term "overhead" as used in this discussion means all costs which are not directly chargeable to items of work within the project.

Overhead costs are normally classified into two categories: direct and indirect. Direct overhead includes such items as the salary of a watchman and field office expense. Indirect overhead charges would reflect costs for items such as proration of home-office expense.

A variety of methods are employed to charge overhead expenses to projects in the construction industry. Most contractors, however, set up accounts within the framework of a project cost code system to be used for direct overhead costs. Since many overhead costs can be controlled, comparative analysis of budgeted-to-actual expenses is an integral part of any job costing system.

Charges for overhead expenses enter the job cost system through transactions in payroll and accounts payable. Costs for supervisory personnel, for example, are entered in the same way as costs for direct labor. The major difference is that supervisory time is normally charged to a single cost account, and not distributed to specific operations as is done with direct labor costs.

From accounts payable processing, records are created to record overhead cost items such as costs for light and power. Again, these costs are ordinarily applied to specific overhead accounts, and not to the actual operations.

The data processing required for charging overhead costs in the job costing procedure parallels the material and subcontractor systems. Actual costs are updated for each classification and are reflected in weekly or monthly financial reports. Items with costs which exceed the budget and which are based upon the percent complete of the project are pointed out for management evaluation.

In addition to facilitating the control of overhead expenses during the life of the project, data processing systems provide an efficient method of capturing the actual cost data. This information is then made available to estimating to be used in preparing future estimates.

THE PROJECT CONTROL OPERATING SYSTEM

19-21. Concepts of the Project Control Operating System

This movement, rapidly becoming apparent in the construction industry, strives to reduce the related data from the estimating, planning and scheduling, and cost control operations of a construction company to one common level. Many contractors have already succeeded in developing integrated systems which encompass all of the data processing required to control construction projects. Other companies are presently expending time and effort in application development work pointed at a centralized method of control utilizing data processing systems.

The Project Control Operating System is a conceptual approach to the integration of the data processing functions of a construction company. The system does not precisely parallel any presently installed in a construction company; yet, it is representative of procedures and methods used by many construction companies. The technique and flow presented in the Project Control Operating System is a result of a composite of ideas stimulated by data processing users in all types of construction.

Not all contractors have adopted the same approach in the utilization of data processing systems for management information and control of construction projects. Vast differences in size of companies and in management philosophies preclude a blanket approach that is applicable in all respects for the entire industry. Therefore, the system presented is not to be considered as the only way to utilize data processing systems profitably. Instead, the discussion is directed toward the objective of stimulating the thoughts of creative construction people who can then modify the approach and tailor a system best suited to their individual company needs.

The modular concept is the central point of the Project Control Operating System. That is to say, major segments of the approach can and are being performed independent of the rest of the system. Recognizing the diverse nature of construction companies and projects, the system is presented in a modular fashion designed to permit a step at a time development of an overall comprehensive scheme of project control.

A major factor influencing the degree of mechanization is the type of data processing systems employed. Construction industry users presently use equipment ranging from minimum punched-card configurations to large EDP systems. Some parts of the system, particularly in the project scheduling area, are not feasible for other than electronic systems. To circumvent this obstacle, contractors can either apply the results of manual scheduling or purchase time on an outside system to perform the scheduling function.

The medium used in the system to capture the data from the estimating, planning and scheduling, and cost operations is represented by the critical path activity. Development of new techniques, such as the precedence or sequence method of CPM scheduling, has provided the impetus to designate the CPM activity or work item as a logical vehicle to serve as the control point of construction data processing operations. However, it is necessary to point out that most of the concept advanced in the Project Control Operating System is equally adaptable to systems where another level of control is utilized. A detailed exposition and network diagram of the system is given in the Appendix at the end of this chapter.

DATA COMMUNICATIONS

19-22. Development of Data Communications

The construction industry has always had a special communications problem. The job sites at which a particular contractor is active are remote from the home office, widely distributed, and often difficult to reach. A communication system that depends on mail, telephone, and messenger has considerable built-in delay. Even though an effective information handling procedure may have been developed and a computer does most of the clerical work, a manual time sheet is likely to be mailed or hand-carried to a central computer facility.

The delay is compounded when a project manager at the site tries to request up-to-date activity or cost information that is at least partly based on time sheets. The report he receives—even if requested by telephone—is not as timely as it could be. Worse—there may be much useful information he does not even bother to request since he knows it will be out of date when he finally gets it.

The need for better communications between construction sites and the home office has led to the development of data communication systems linking the sites and the home office. In such systems, communications terminals are located at the job site to permit direct communication between the foreman or project supervisor and the computer.

One type of terminal, already being used in data networks by many construction companies, provides one-way transmission of punched-card and numerical information from the field to the data processing center.

In using this data transmission terminal, the foreman is equipped with a stack of prepunched machine-readable cards. Each contains identifying data for the men reporting to the foreman. At the end of the day, the foreman enters all labor records on the terminal by placing the punched card in the unit and pressing the numerical keys that record the cost code number and number of hours for each task performed. He need not identify the employee, since the information has been prepunched into the card.

With a network of such terminals at all job sites the contractor can receive complete labor time records at the end of each workday. The unit that receives the card data at the home office need not have an operator. The timekeeper or foreman at the site simply dials the telephone number associated with the unit which answers automatically, ready to receive the information.

The same type of terminal can be used to transmit data on labor, materials, equipment, and progress of the job. This type of data terminal has the advantage of being easy to use and low in cost. But it has the disadvantage of being able to transmit in only one direction.

Two-way terminals also in use today, can be placed at job sites. The terminal is capable of input and output with punched cards or typewriters. Progress and exception reporting, work schedules and schedule changes, and many other information exchanges between the home office and site can pass over telephone lines. A supervisor on site can request an activity schedule or cost report from the trailer office terminal, for example, and immediately have it printed out at his elbow.

A step further—and not far in the future—will be a system of visual display stations at construction sites. Each display station consists of a cathode-ray tube display screen and a keyboard. Reports for project planning, scheduling, and control can be displayed on the screen by keying the proper request code. The project manager might request graphs, curves, and even sections of drawings—anything, as a matter of fact, that has been stored in the computer data files, that he has a right to see.

In addition, he can use the keyboard to change or update the data displayed on the screen, simultaneously updating the data stored in the computer file. Terminals at the home office or at other need-to-know locations could then have immediate access to the same updated data.

One way in which a display terminal can be extremely useful to project management on site is in time-cost balancing. Both time and cost for a project are established when the planning and scheduling have been done. However, the project manager on site often faces unanticipated events that require a change in schedule—and the change must produce the best results for the additional cost. Evaluating numerous possible alternatives, each with a different cost trade-off, is a difficult task when attempted by trial-and-error methods.

At a display terminal, however, the manager can present the appropriate alternatives directly to the computer and request an immediate report on the results achieved with each. He is then able to make an immediate decision among the alternatives, knowing that the one selected will be the best in terms of both time and cost.

SUMMARY

The past few years have brought about a tremendous increase in competition in the construction industry. This trend has resulted in three fundamental requirements to ensure profitable performance. The first concerns the necessity of estimate preparation based upon facts, not conjecture. The second is a need to produce practical budgets and schedules. The third requirement is to capture cost data quickly and accurately.

Many contractors have turned to data processing systems as a tool to satisfy the aforementioned requirements. To this end, the systems are used extensively in many areas concerned with project control.

In the area of job estimating, mechanized systems are providing construction management with a vehicle to assist the estimator in bid preparation from the quantity survey through the final printing of the estimate. Costs of bid preparation are significantly reduced but, more importantly, the trained estimator is permitted more time to concentrate on the methods of construction. More estimates can be prepared, thereby increasing the possibility of successful bids. Alternative methods can be quickly evaluated, thereby providing management the assurance that the best method is proposed. Lastly, the inherent accuracy of data processing systems ensures the validity of the bid.

Planning for allocation of time and costs to produce job budgets and schedules is particularly applicable on mechanized systems. Coupled with today's management-science techniques, the systems provide to management the ability to analyze various methods and evaluate project schedules to determine the most efficient job cycle. New approaches tied to the CPM permit labor and equipment to be scheduled in the most efficient manner, thereby reducing requirements and costs. Dynamic updating and schedule revision during the life of the project permit management to have fingertip knowledge and control of job progress.

Cost data captured from field operations serve a twofold purpose. First, management is informed of the cost status of the work in progress; and second, the same data are used to update cost experience for estimating information. Both functions are basic in project control, and both require that the information be timely and accurate.

Today, data processing systems are used in all three major areas of project control estimating, planning and scheduling, and cost control. Integration of these functions, coupled with home-office operations, provides construction management with an information system which ties together all aspects of construction control.

APPENDIX 19-A   PROJECT CONTROL OPERATING SYSTEM

The flow of the data within the system is presented in the form of a precedence sequence network (see Figure 19-A-1). Each of the operations or work items has been numbered, and a brief discussion of each item is found in the narrative. The work-item numbers have significance only in that specific items are identified in the discussion. Throughout the presentation, the relationships of the major functions to each other are emphasized. In addition, the modularity feature is stressed. The following presentation is directed to the component parts of the Project Control Operating System as depicted in Figure 19-A-1.

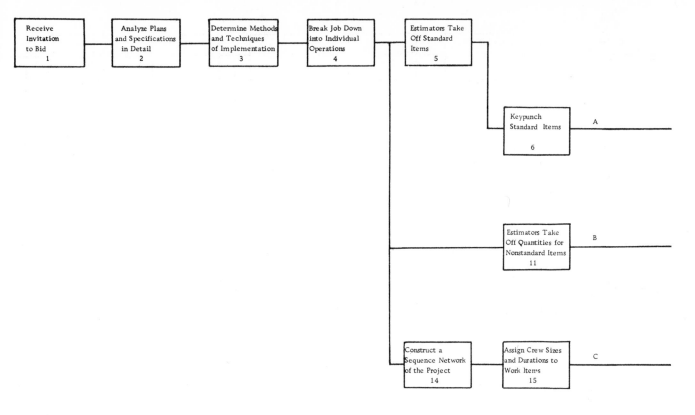

NETWORK NARRATIVE

Work Item	Description	Work Item	Description
1	Within each project, the overall system is initiated when an invitation to bid or negotiate is received.		operation has not yet been automated, the detail produced in work-items 5 to 22 may not be prepared until the job is awarded.
2	The plans and specifications are analyzed in detail by the estimators and planners.	5	When the estimating application is installed on the data processing system, the estimators take off standard items from the plans and specifications and fill in the prescribed forms with data
3	The methods and techniques required to implement the construction project are determined.		such as operation identification code, dimensions, and special handling instructions. The takeoff follows the scheme of the operations breakdown performed in work-item 4.
4	The method of takeoff is determined, and the job is broken down into the individual operations to be estimated. At this point, the modularity of the system is first emphasized. Utilizing the entire concept, the operations decided upon will constitute individually schedulable items of work and will conform as well to functional codes which identify each item to the estimating operations. Deviation from the system as presented, however, permits the functional breakdown of items to follow any scheme best suited to estimating. If the estimating	6	Cards are keypunched with the information from the takeoff forms describing each of the standard operations to be processed in the automated estimating procedure.
		7	The cards are processed through the data processing system, and the various estimating calculations are performed. Operation quantities are first determined and the cost experience file

Figure 19–A–1 Precedence sequence network for Project Control Operating System

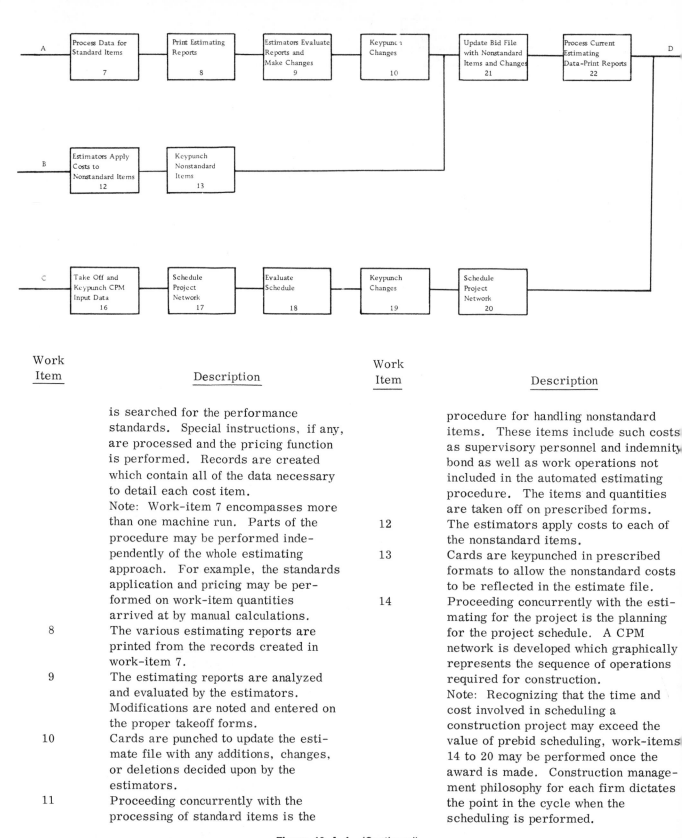

Work Item	Description
	is searched for the performance standards. Special instructions, if any, are processed and the pricing function is performed. Records are created which contain all of the data necessary to detail each cost item.
	Note: Work-item 7 encompasses more than one machine run. Parts of the procedure may be performed independently of the whole estimating approach. For example, the standards application and pricing may be performed on work-item quantities arrived at by manual calculations.
8	The various estimating reports are printed from the records created in work-item 7.
9	The estimating reports are analyzed and evaluated by the estimators. Modifications are noted and entered on the proper takeoff forms.
10	Cards are punched to update the estimate file with any additions, changes, or deletions decided upon by the estimators.
11	Proceeding concurrently with the processing of standard items is the

Work Item	Description
	procedure for handling nonstandard items. These items include such costs as supervisory personnel and indemnity bond as well as work operations not included in the automated estimating procedure. The items and quantities are taken off on prescribed forms.
12	The estimators apply costs to each of the nonstandard items.
13	Cards are keypunched in prescribed formats to allow the nonstandard costs to be reflected in the estimate file.
14	Proceeding concurrently with the estimating for the project is the planning for the project schedule. A CPM network is developed which graphically represents the sequence of operations required for construction.
	Note: Recognizing that the time and cost involved in scheduling a construction project may exceed the value of prebid scheduling, work-items 14 to 20 may be performed once the award is made. Construction management philosophy for each firm dictates the point in the cycle when the scheduling is performed.

Figure 19–A–1   (Continued)

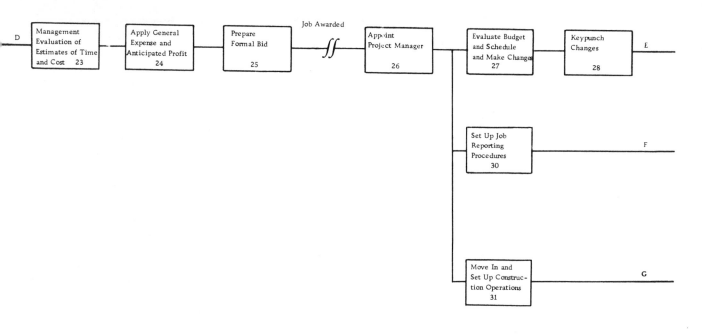

Work Item	Description	Work Item	Description
15	Crew sizes and durations are applied to each work item. Crew size implies the makeup of labor and equipment resources required to perform each operation.		planners are keypunched into cards for processing.
16	Network takeoff forms are filled out by the planners and the data are key-punched into cards.	20	The schedule file is updated, and the CPM calculations are performed for each proposed group of changes. The schedule information is presented in the sequence and format requested by planning.
17	The project network is scheduled on the data processing system. If network errors are found, they are pointed out and corrected. The network is processed again until all error situations have been cleared. The schedule is printed in the desired sequences.	21	Proceeding with the development of the project schedule is the revision of the estimate file to reflect changes and nonstandard items.
18	The schedule is evaluated and any changes necessary are reflected on the diagram by the planners. Alternatives may be proposed and evaluated in work-items 18 to 20. The process is an iterative one, continuing until a desired project completion date is achieved. Network changes involving reduced or increased costs must be reflected in the estimate of costs.	22	The updated estimate file is processed, and the desired estimating reports are printed.
		23	Management evaluation of the estimate of costs and the project schedule takes place. At this point, modifications to the estimate and/or the schedule may be made. If such is the case, the changes are reflected in the estimate and/or schedule file and the corresponding reports are reprinted.
19	Network changes prompted by the	24	After the estimate of costs and project schedule are approved, management determines the amount of general expense and profit to add to the costs.

Figure 19–A–1   (Continued)

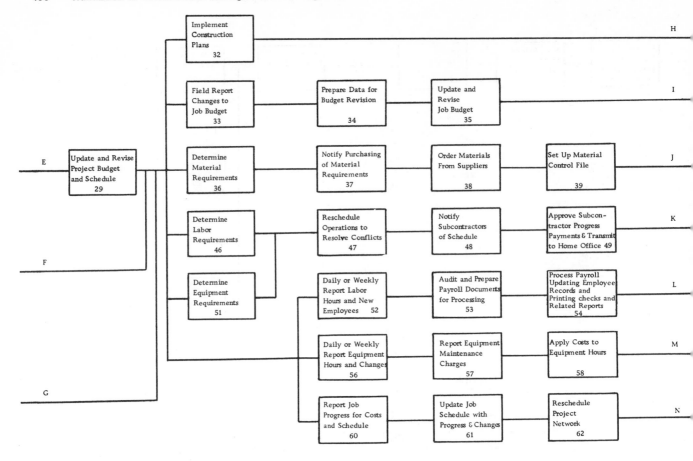

Work Item	Description	Work Item	Description
25	The formal bid documents are prepared and submitted to the owner. Note: Application of general expense and profit and subsequent preparation of the formal bid documents are not considered to be performed on data processing systems.	28	which cost data will be captured. Modifications to the job budget or schedule are keypunched into cards for processing.
26	Following the award of a construction contract, the implementation phase begins. The first step is for construction management to appoint a project manager whose responsibility is to coordinate all aspects of the project.	29	The project budget and schedule files are updated and revised to reflect any changes. The new budget and schedule is printed and the files are maintained for subsequent reporting of progress and costs.
27	The project manager evaluates the estimate and schedule, and proposes any changes he deems advisable. Approved modifications are recorded on prescribed input forms. Note: At this point the budget and schedule are set up, if not done previously. The job budget represents the detail of the estimate of costs set up and maintained at the same level in	30	The job reporting procedures are set up and forwarded to the project manager and field office manager. Included in this operation is the preparation of prepunched and preprinted cards or lists to be used by the field for reporting progress, costs, and change data. Special provisions peculiar to the project at hand are spelled out and included in the field reporting instructions.
		31	Depending upon the requirements of the owner, the construction operation may or may not begin before the final

Figure 19–A–1   (Continued)

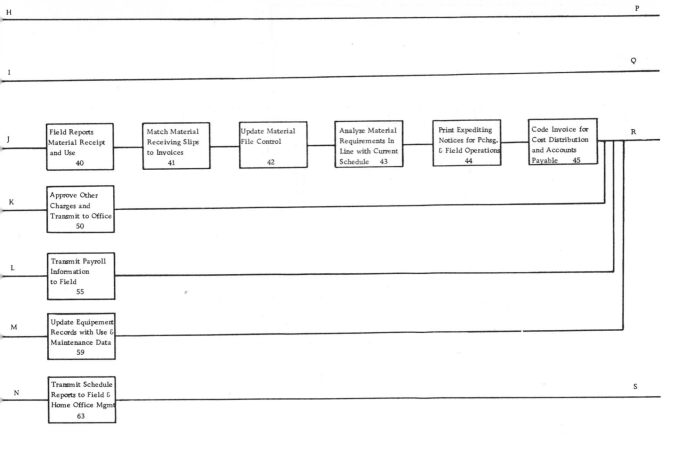

Figure 19–A–1 (Continued)

Work Item	Description	Work Item	Description
	budget and schedule is determined. In most cases the job is already underway before all planning and scheduling is completed.		Prescribed forms are filled out and cards are punched.
32	The construction phase is shown as a single work item and serves as a predecessor to work-item 74, Close Job to Estimate.	35	The job budget file is updated and revised to reflect the approved changes.
	Note: The arrangement of work items and work-item numbers in the network has significance only insofar as they apply in individual paths of the network. As an example, work-item 51, Determine Equipment Requirements, will probably take place before work-item 33 which covers the field reporting of changes to the job budget.	36	Material requirements for each work item were determined during the estimating or planning phase. Through the medium of the CPM work item, required dates for material can be determined by obtaining the anticipated start date for each operation.
		37	Reports are printed which show by date the amount of each type of material required. These reports are sent to the purchasing department where arrangements are made to obtain the material as needed.
33	As the job progresses, any modifications applicable to the budget because of change orders or changes in methods are reported by the field and transmitted to the home office.	38	The purchase orders are prepared manually or on the system. If prepared on the data processing system, records are created in work-item 37 which contain the necessary information to describe each item of material.
34	Following approval of budget changes, the data are prepared for processing.		Purchasing determines the supplier

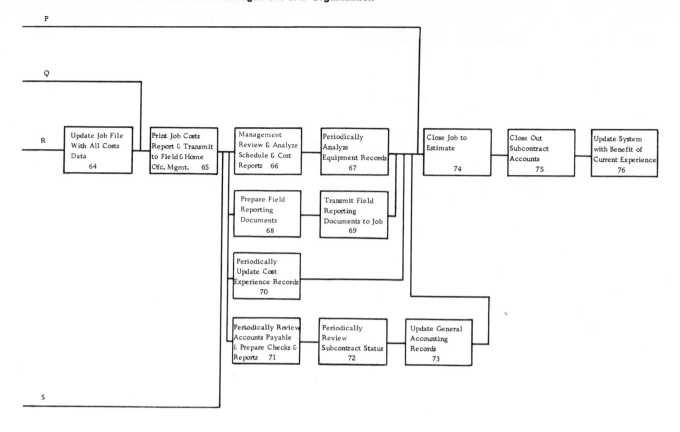

Work Item	Description	Work Item	Description
	and notifies the project. In addition, purchasing coordinates factors such as multiproject ordering or ordering in quantities which result in favorable discounts.		delivery commitments are in line with requirements. Continual review is necessary because of changes in the schedule which are prompted by job conditions.
39	A material control file is set up and maintained. Included in each record is a required date generated through the scheduling procedure and a promised date received by purchasing from the supplier.	44	Changes in required delivery dates necessitated by schedule changes are printed out and called to the attention of purchasing and project management. Material requiring expedited delivery receives major emphasis.
40	During the life of the project the field reports material receipts and issues on a periodic basis.	45	The invoices from the suppliers are coded, and cards are punched and placed in the accounts payable file. Material costs are distributed through coding either the receiving slip or the invoice with the charge codes and amounts.
41	Material receiving slips are matched to the supplier's invoice to control payments for materials. Discrepancies between billed items and received items are reconciled before the invoices are approved for payment.		
42	Cards are punched with the data concerning material receipts, use, and changes in status of on-order materials.		Note: Material control represents a most significant aspect of the Project Control Operating System. Few contractors presently utilize their data processing systems for material control because of the complexity of the operation. The major obstacle to date has been the requirement that
43	Periodically, the material control file is analyzed together with the updated project schedule to ensure that		

Figure 19–A–1 (Continued)

Work Item	Description	Work Item	Description
	material requirements be set up at the CPM activity level, thereby increasing the volume of processing required. Some recent government contracts have required that material requirements be carried at the CPM level, and contractors awarded the jobs have realized significant benefits from integrating material control into the project schedule.	50	All invoices received by the field are certified and approved for payment and transmitted to the home office.
46	Labor requirements for the job at hand are determined. This function can be performed in a variety of ways. One method is to first estimate labor requirements at the level of the CPM operation. With this technique, a tabulation or plot of labor requirements by craft or crew for each week can be quickly made. The project manager can use the report to smooth the labor curve and schedule operations accordingly. Another method is to use an electronic computer to schedule the job against an available labor pool. With this method, the system analyzes the labor requirements for each operation and schedules the start of operations based upon the availability of labor resources. Ordinarily, the criterion of priority in manpower scheduling is based upon the least amount of total slack. The same techniques are used in work-item 51, where equipment requirements are determined.	51	Equipment requirements are determined in the same manner as for labor requirements discussed under work-item 46.
		52	During the life of the project, the field is responsible for reporting labor hours for payroll and cost distribution purposes. Daily or weekly timecards may be used as the reporting document, and the data may be transmitted to the home office in a variety of ways. The use of IBM Tele-processing systems for two-way payroll communication permits complete centralization of payroll processing. The field is also responsible for reporting terminations and new hires, and filling out the required forms such as the W-4 for a new employee.
		53	The payroll documents sent from the field are audited and prepared for data processing by conversion into punched cards. Manual processing for this work item is virtually eliminated with the use of Tele-processing systems.
		54	All payroll processing takes place including calculations, updating all affected records and printing of payroll checks, registers, and allied reports. The records affecting job costing are maintained, or new records are created.
47	After labor and equipment requirements have been determined, conflicts are resolved by applying schedule dates to the network. As explained in work-item 46, schedule dates can be generated either by the project manager or through the use of a data processing system. In any event, the network changes are made and the project is rescheduled.	55	The payroll checks, registers, and other necessary supporting reports are transmitted back to the jobsite through Tele-processing systems or some other medium.
		56	Using a timecard system similar to labor costs, equipment hours are recorded and classified for distribution of equipment costs and updating of equipment records.
48	The subcontractors are notified of their schedule requirements. This item is repeated during the life of the project as often as major schedule changes dictate a requirement.	57	Maintenance charges, an integral part of an equipment recordkeeping system, are reported and charged to the specific items of equipment. A comprehensive system would also require fuel and lubrication used to permit analysis of costs of operation. The maintenance data is converted into punched cards for processing.
49	The field approves requests from sub-contractors for progress payments and transmits the documents back to the home office.	58	The home office, upon receiving the equipment timecards, converts the data

Figure 19–A–1 (Continued)

Work Item	Description	Work Item	Description
	to punched cards and applies the rates to the reported hours. Records are created which will charge the dollar amounts to the operations which utilize the equipment as well as update the equipment records with utilization information.		the reports may cover only one element of cost. A combination of reports is probably the most useful. A labor or labor and equipment report on a weekly basis, prepared concurrently with payroll and equipment processing, serves effectively as a cost control tool for management. Budget deviations are pointed out to management. Paramount among the objectives of the labor cost reports is that the reports timely and accurately portray the cost status of the project. This is done in order that management may be informed of adverse situations in sufficient time to propose remedial action.
59	Equipment records are updated with all of the data pertaining to maintenance charges incurred and the distribution of hours. An effective system requires a major breakdown of productive time and nonproductive time. Within the category of nonproductive time, further breakdowns are advisable which would segregate the hours by downtime for repairs, idle time, standing by, etc.		
60	Progress and charge data are reported periodically by the field. In order to evaluate labor costs to progress, the reporting period should be weekly. Job progress should be reported in the medium which best describes each operation. Quantities set are appropriate for most operations, while percent complete is indicative of progress on lump-sum work items.	66	Management reviews and analyzes cost reports and schedule reports. The facility of management exception reporting permits management to evaluate progress and costs through a detail analysis of only those items which require management attention.
61	Progress and change data are converted into cards and processed against the job schedule file. The file is updated and prepared for processing.	67	Whenever requested, the equipment records are analyzed. Special reports are printed which point out a number of factors necessary for effective equipment management. From these reports, management can determine when an item of equipment should be scrapped because of excessive maintenance costs and excessive downtime for repair. Another type of report can indicate when purchase option equipment should be captured. Operating rate analysis based upon current performance standards and costs can be performed, and items of equipment possibly requiring rate revision are pointed out. Another type of equipment analysis is a comparison of cost and productivity of similar equipment available from more than one manufacturer. Preventive maintenance scheduling is another integral factor in a comprehensive equipment and recordkeeping system. Equipment specification and maintenance of depreciation reserves are also performed.
62	The revised job schedule file is processed and the schedule calculations are performed. A new schedule is computed based upon the current status of the project. The updated project schedule is printed in the format and sequences dictated by construction management.		
63	The schedule reports are transmitted to home office and field management for evaluation and review.		
64	The file which contains the job budget and all costs to date is updated with the cost data for the preceding period. The data emanated from labor and equipment time recording, material control and costing, subcontractor payments, and accounts payable.		
65	Once all cost data have been reported and the job costs file updated, the cost reports are printed. The cost reports may be of the comprehensive type covering all elements of job costs, or	68	Field reporting documents to be used for progress and cost reporting from the jobsite are prepared on the data

Figure 19-A-1 (Continued)

Work Item	Description
	processing system. The type of document prepared depends upon the communication system used between the home office and project. Generally, the field reporting documents take the form of cards or lists which allow cost and progress reporting with minimum clerical effort on the part of field personnel.
69	The field reporting documents are transmitted to the field. The documents cover a period sufficiently long to encompass the overlap of field reporting cycles.
70	Cost experience records maintained for estimating purposes are updated periodically with the costs data captured from the project operations. The cost experience file is maintained at the level dictated by management policy. Ideally, the file would contain cost experience at the level of man-hours required for each craft to produce a single unit applying to each cost code. Other systems, however, might consider total labor costs or hours to produce a unit. Still other systems could capture only costs data, ignoring hours.
71	The accounts payable file containing invoices from suppliers and subcontractors is reviewed. Reports on items currently due are prepared and sent to management for approval of payment. Approved items are pulled from the file, and checks and remittance statements prepared. Retainage is calculated where appropriate, and corresponding records created and filed. Check registers are printed and check

Work Item	Description
	reconciliation records created.
72	The subcontractor file is pulled and reports are printed which show the current status of each subcontractor account. Maintenance of this file provides quick access to all transactions within a given account. Contract awards, progress bills, payments, retainage and changes in the original contract are all reflected in the file.
73	The general ledger of the company is periodically updated with the data concerning each project. This operation includes entries on the income side as well as on the expense side. After the records are updated, the required reports are printed.
74	At the end of the project the job is closed to the estimate. A report is printed which shows the original estimated costs and the actual costs incurred on each operation. Other reports reflect estimated and actual costs at a higher level. The reports are held for estimating and for management information.
75	The subcontractor accounts for the job are closed out, and all open invoices and retainage are paid.
76	As a result of the experience gained from the project, the system itself may be revised to incorporate desired new features. Just as every project updates the construction experience of a contractor, so too does the experience captured in the administrative functions of costing and scheduling point out areas where the approach may be improved.

Figure 19–A–1 (Continued)

THE PROJECT CONTROL OPERATING SYSTEM
SUMMARY

The concepts advanced in the foregoing presentation are directed toward a centralized system of application of data processing principles to control of construction projects. In evaluating the system, several significant advantages of the concept become readily apparent.

The first major advantage is that all control is centered at one level. Estimating, scheduling, and cost control are performed using the CPM work item as the gathering system for all related data. The ability is present to analyze job progress directly with job costs, thereby providing management information to control both time and costs. Resource requirements for labor, equipment, and materials are readily determined. Not only is construction management informed of the resources required, but also when the resources are needed.

Another major advantage of the system is found in the standardization of processing afforded the user. Exception processing is reduced to a minimum. As an example, cost items such as job utilities and salaries of supervisory personnel should certainly not be considered schedulable items. Yet these types of cost items can still be part of the overall system and classified as one or more work items. A simple code affixed to a nonschedulable work item identifies it as a work item to be ignored during the project scheduling process.

Economy of operation, particularly at the jobsite, is another advantage of the system. A centralized system of payroll processing can save many hours of time compared to payrolls produced at the jobsite. The recording required for payroll also provides entry to the job costing application.

Development of an integrated system of project control should be uppermost in the minds of contractors in converting applications to IBM data processing systems. A carefully thought-out installation plan for applications which result in a centralized system should be developed and maintained in much the same manner as a plan for project construction.

Figure 19–A–1   (Continued)

20 OFFICE ADMINISTRATION: HEADQUARTERS AND FIELD

DONALD G. PERRY

Partner
Lybrand, Ross Bros. and Montgomery
San Francisco, California

ADMINISTRATION of construction operations is carried on both at the home office and in the field. The home-office or headquarters-office administration is centered on controlling operations overall, whereas field-office administration is concerned solely with the immediate project at hand.

Within these parameters office administration can be concerned with a number of departments and even locations. Depending upon the size of the organization, there may be district offices that resemble both field and headquarters in that they may administer small projects in their geographical territory and supply administrative and staff support services to projects in the area.

20-1. Headquarters vs. District and Project Administration

Even the smallest company must differentiate between headquarters and project administration since the two functions are independent. Project administration is geared to provide administrative services for the specific project involved. The costs and services are directly related to the project and are not performed for the benefit of the company as a whole or for the benefit of other projects. These administrative services are accounting, purchasing, warehousing, personnel, and payroll.

The same services may well be performed at headquarters. However, these are generally for the benefit of the company as a whole rather than being oriented toward a specific project. Although certain headquarters administrative functions may be directed at specific projects from

time to time, such as personnel, they are there to provide support for all projects and are not limited to only the one, as in the case of field administration.

20-2. Field Administration

Field administration is oriented toward the successful completion of a single construction project. Being project oriented the administrative functions are geared to the project from inception to completion. Therefore, such things as accounting records and cost projections are on a total project basis and do not concern themselves with annual or other periodic closings, as may be true in the home office.

The size of the field administrative staff depends not only on the volume under the contract but also on the type of contract being performed. Construction management, for example, may require only one or two administrative people at the job site. If the contract also includes the construction itself, however, the size of the administrative staff would be much larger.

Field administration would include business management, purchasing, warehousing, and payroll. The chief administrative man on a field project would be the administrative manager, business manager, or office manager, depending on the size of the job. This individual reports directly to the project manager while remaining under advisory control of the home-office administrative staff. In those companies large enough to have district offices, the home-office control may be exercised through the district-office administrative manager rather than directly from the home office.

20-3. Functions of a Job Administrative Manager

Under a job administrative manager comes, first, general administration and office services: second, warehousing and procurement; and third, accounting and payroll. Administration of construction camp and messhall, where applicable, would also be under the job administrative manager. Although his direct line of authority does come from the project manager, the administrative manager has responsibility for maintaining free and open communication with the district and home office.

The size of the administrative staff on construction projects will vary with the volume of the job. Table 20-3-1 shows how one company has estimated the requirements for administrative staff on a job of normal complexity. This estimate does not contemplate client audit, as in the case of a cost reimbursement job where the administrative staff would be larger.

The project administrative manager has the responsibility of providing office space; arranging for post office, telephone, telegraph, and other utility service; notifying all concerned of these contacts; and arranging for the staffing of the administrative assignments.

Before work on the project commences, the job commitment ledger and general ledger with its subsidiaries must be started. Each job will have its own general ledger under the control of the project administrator. Although most established companies will have a standard chart of accounts, individual work order codes will have to be assigned on a job-by-job basis. These codes should be assigned jointly by the project administrator and the project or other engineer who is most familiar with the individual bid items.

Often temporary buildings are available from a central company supply. These may be trailers or other more permanent structures which can be erected quickly at a given job site. The home or district office can also supply office furniture, file cabinets, partitions, shelving, and other office equipment. On some large projects EDP equipment is maintained at the job site in order to process job costs, particularly labor, expeditiously.

20-4. Establishing a Chart of Accounts

With the establishment of a chart of accounts the accounting records can be opened. In addition to the general ledger, a number of subsidiary records are kept. These subsidiary ledgers can relate to employees' receivables, miscellaneous receivables, warehouse inventories, shop operations, plant operations, general plant erection, specific plant erection, general expense, major equipment, minor equipment, accounts payable, construction cost ledgers, equipment maintenance, and operation. Books of original entry may comprise cash receipts book, cash disbursements book, voucher register, payroll register, and general journal.

In addition, there may be other journals or registers required in individual company situations. Some companies combine the above functions in that job records may well be on a cash basis and the voucher register and cash disbursement record may be combined. Also, there

TABLE 20-3-1.   ADMINISTRATIVE STAFF

Man Count	100	200	400	400 w/Mach.	600	600 w/Mach.	800 w/Mach.	1000 w/Mach.
Office and Accounting								
Office manager	1	1	1	1	1	1	1	1
Chief accountant					1	1	1	1
Accountant			1	1	1		1	1
Accounting clerks	1	1	1	1	1	1	1	1
Voucher clerks			1		1	1	1	1
Machine operators				1		1	1	2
Stenographers	1	1	1	1	1	1	1	1
Equipment clerk			1	1	1	1	1	1
Paymaster		1	1	1	1	1	1	1
Timekeeper and payroll clerk			2	2	4	4	4	4
Typists			1		1			
	3	5	10	9	13	12	13	14
Warehouse and Purchasing								
Purchasing agent					1	1	1	1
Stenographer			1	1	1	1	1	1
Chief warehouseman					1	1	1	1
Warehouseman	1	1			1	1	1	1
Comb. purchasing agent and warehouseman			1	1				
Warehouse clerk		1	1	1	1	1	1	1
	1	2	3	3	5	5	5	5
Shop								
Shop clerk			1	1	1	1	1	1
	0	0	1	1	1	1	1	1
Totals	4	7	14	13	19	18	19	20

are cases where cash receipts come into the district or home office without passing through a job bank account. In the latter case, cash may be transferred by journal entry from the home office. These records may be kept by hand or, if volume is sufficient, a machine posting of one sort or another may be used. In some cases, as previouly mentioned, EDP equipment may be installed at the project site itself. In other cases, a company may have a centralized data processing department which will be connected to the project by a data transmission system utilizing telephone or telegraph lines. Even payroll can be centralized if regular labor distribution printouts are transmitted to the project at least two to three times a week. Payroll checks may either be prepared in the head office and sent or the printer at the project office can be programmed to prepare payroll checks.

20-5. Accounting Reports to the Home Office

Accounting reports should be submitted to the home office at least monthly. These should show the status of the job at the end of the month and as a minimum would consist of a balance sheet and a job summary of cost and revenue from inception to date. Other information which should be contained in the monthly statements include contract

volume and uncompleted work showing changes after the previous month-end, with such detail supporting schedules as the company may feel is necessary. Generally, these statements can be sent very soon after the month-end, particularly when the job accounting records are kept on the cash basis. Examples of standard reporting formats are given in Figures 20-5-1 through 20-5-5. Figure 20-5-1, Figure 20-5-2, and 20-5-3 show a set of forms for a general contractor in heavy construction.

Forms used by a consulting engineer on construction, management, and full construction contracts are shown in Figures 20-5-4 and 20-5-5.

20-6. Other Functions of Field Administration

Another function of field administration is purchasing. Purchased construction materials are a significant part of any contract. Although major items of equipment to be installed in a project, such as a turbine or generator for a power plant, may well be purchased centrally, the local purchases are significant. Often contract owners wish to utilize the purchasing abilities of their contractor but wish to keep control over the volume of purchase orders written and over the disbursement of funds. In these cases it is customary for the contract owner to approve the

SUMMARY OF CONTRACT VOLUME AND UNCOMPLETED WORK

Location .... Date .... Contract No. ....

Client's Name ....
Type of Contract: Unit Price or Lump Sum ☐ Cost-Plus-Fixed Fee ☐ Bid ☐ Negotiated ☐

Amount of Original Contract ....
Approved Change Orders to Date ....
Anticipated Over-run or (Under-run) in Uncompleted Work ....
Actual Over-run or (Under-run) in Completed Work ....
Rental Revenue from Contract Owner not included above ....
   Estimated Total Amount of Principal Contract ....
Order Contract Work not Included in Principal Contract ....
   Total Estimated Contract Volume ....
Contract Revenue to Date ....
Less: Contract Advances for Materials on Hand ....
   Contract Advances for Plant and Move-in ....
   Revenue Reduction for Uncompleted Work ....
   Other ....
     Total Amount of Work Completed to Date ....
Uncompleted Contract Volume ....
Percent Complete Based on Original Contract .... %
Percent Complete Based on Total Estimated Contract .... %
Time Allotted by Original Contract .... Days
Extension of Contract Time .... Days
   Total Contract Time .... Days
Contract Time Elapsed .... Days
Percent of Original Contract Time Elapsed .... %
Percent of Total Time, Including Extensions, Elapsed .... %
Date Contract was Physically Completed—If Completed ....
Expected Date of Physical Completion—If Not Complete ....

Figure 20—5—1 Contract volume report

SUMMARY OF COST AND REVENUE

(Lump Sum or Unit Price Contracts)

Location ....Contract No. ....

At Close of Business ....

Name of Client ....

Nature of Work ....

	Job to Date	Current Year to Date	Current Month
REVENUE AS SHOWN BY JOB BOOKS:			
Per Engineer's Est. No. ....Dated			
Per Company Est. No. ....Dated			
....			
....			
Other Revenue			
TOTAL REVENUE			
COST OF WORK DONE:			
Labor (including all payroll taxes & insurance)			
Permanent Materials			
Specific Plant			
Equipment Rental or Depreciation			
Supplies			
Sub-Contracts			
TOTAL DIRECT COSTS			
OVERHEAD:			
General Plant ....$			
Less: deferred portion ....$			
General Expense ....$			
Less: deferred portion ....$			
APPROXIMATE COST OF WORK DONE			
OPERATING PROFIT			
OTHER OPERATIONS			
NET PROFIT PER JOB BOOKS			

Estimated amount of contract, including change orders ....$............

Amount of Main office investment in Contract (4001) ....$............

Date set for completion ....

Per cent of work done ....Per cent of time elapsed ....

	Payrolls	Company Owned Equipment Depre- ciation	Company Owned Equipment Rental to Home Office
This month			
This year			
To date totals			

....

Business Manager   Project Manager

Figure 20–5–2   Cost and revenue report

purchase orders himself before they are issued. This type of procedure, of course, is utilized only on jobs where costs are being reimbursed to the contractor.

Whether or not these costs are included in reimbursable cost and revenue depends upon the degree of control which the contractor has over the purchases. If he is acting for the contract owner only in name and is not acting on his own account, these costs should be excluded. When he has a responsibility for the purchasing function or when purchase costs enter into computation of the fee, these costs may be included in the contractor's revenue and costs.

Commitments for purchases of material and equipment for a project are of vital interest to the engineer who is forecasting total costs. A record should be kept of all material required, and it should be carefully documented as it passes through the commitment ledger into firm commitment and then into accounts payable and finally into

BALANCE SHEET

At Close of Business ....Contract No. ....

	Detail	Detail	Whole Dollars

ASSETS

CASH:
Revolving Funds ....
    —Bank ....
    —Bank ....
    —Bank ....

ACCOUNTS RECEIVABLE:
Engineer's Estimate ....
Retained Percentage ....
Company's Estimate (reverse each mo.) ....
Employee's Amounts Receivable ....
Miscellaneous Receivables ....
Fees Receivable ....
Reimbursable Costs ....
Reimburseable Costs Unbilled ....
Billable Costs Accrued ....

INVENTORIES:
Warehouse Inventories ....

SHOP AND PLANT OPERATIONS:
Project Shops Operations ....
Credits for Shops Operations ....
Production Plants Operations ....
Credits for Production Plants Operations ....
Project Camp Service Operations ....

DEFERRED CHGS. & OTHER ASSETS:
General Plant ....
Credits for Amortization—Gen. Plant ....
Specific Plant ....
Credits for Amortization—Specific Plant ....
General Expense ....
Credits for Amortization—Gen. Exp. ....
Suspense Details ....
Refundable Deposits ....
Returnable Containers ....
Equipment Rental Suspense ....
Contract Advances to Subcontractors ....
Deferred Work Costs ....
Prepaid Insurance ....

EQUIPMENT—JOB OWNED:
Major Mobile and Construction Equip.—Cost ....
Reserve for Depreciation of Major Equip. ....
Major Equip. Installed in Plant—Cost ....
Reserve for Deprec. of Installed Major Equip. ......
Minor Equipment—Cost ....
Reserve for Depreciation of Minor Equip. ....
Minor Equipment Installed in Plant—Cost ....
Reserve for Input of Installed Minor Equip. ....
Small Tools and Expendable Equipment ....
Sm. Tools and Expend. Equip. Amortized ....

TOTAL ASSETS

Figure 20–5–3   Balance sheet, format for the engineer

cash disbursements. The engineer has as much need for this information on lump-sum and unit-price jobs as he does on cost-recovery jobs.

Warehousing as well as purchasing is both a headquarters and a field function. In the field the warehouseman keeps account of expendable tools and general supplies, such as welding rod and pipe fittings which do not go directly into the project itself. These items should be controlled in and out of the warehouse. Small tools should be signed out daily and returned at the end of each shift. Because these are attractive to persons accustomed to working with them, these tools, if not controlled, may be used to supply many a basement workshop. Access to the warehouse, therefore, should be limited to the warehousing personnel and should be under tight security on all shifts.

Payroll represents one of the largest cost elements on many construction projects. Control of labor hours and

BALANCE SHEET

At Close of Business ....Contract No. ....

LIABILITIES INVESTMENT AND OPERATIONS

	Detail	Whole Dollars	Whole Dollars

LIABILITIES:
Accounts Payable ....
Social Security and Withholding Taxes ....
Accrued Taxes and Insurance ....
Accrued Other Expenses ....
Reserves ....
Deferred Credits ....

   TOTAL LIABILITIES ....

INVESTMENT ....

   TOTAL LIABILITIES AND
   INVESTMENT ....

OPERATIONS:
Revenue from Operations:
   Engineer's Estimate ....
   Company Estimate ....
   Revenue Reduction for Uncompl. Work ....
   Fee from Client ....
   Revenue from Reimbursable Costs ....
   Equipment Rental Revenue ....
   Other Contract Revenue (Net) ....

      Total Contract Revenue ....

Cost of Contract Operations:
   Construction Costs—Work Accounts ....
   Reimbursable Costs ....
   Cost Relating to Equipment Revenue ....
   Non-Reimbursable Costs ....

      Total Contract Costs ....

   TOTAL GAIN OR LOSS FROM
   CONTRACT OPERATIONS ....

OTHER OPERATIONS:
   Penny Elimination ....
   Equipment Rental Revenue
      other than Contract Owner ....
   Gain or Loss from Sale of Equipment ....
   Miscellaneous Gains or Losses ....

      TOTAL OTHER OPERATIONS ....

      TOTAL PROFIT OR (LOSS) ....

TOTAL LIABILITIES INVESTMENT & OPERATIONS

Figure 20–5–3 *(cont.)*

efficient utilization of the various crafts is therefore of vital importance to a successful project. For this reason, prompt reporting of payroll earned and of hours worked plus the control to ensure that all persons said to be on the payroll are actually working, becomes essential. This is the function of the paymaster. Although time cards and time clocks are used on some projects, in many cases the control is by badge number. Each employee will be assigned a number which will be on a brass tag kept at the main gate of the project. From time to time each day the paymaster will account for all tags that are not left at the gate by having each man show him his own tag which he in turn checks off on his master list. On the basis of the paymaster's report, the labor distribution and the payroll are made up.

Employees are hired locally for the construction work force. This may be done through the union directly or the project may have its own personnel people to screen

FIELD FINANCIAL STATEMENT

Location ....   Job No. ....

As at ....19.........   Division ....

ACCOUNT	AMOUNT DR. OR (CR.)
OPERATING ACCOUNT	
PAYROLL ACCOUNT	
PETTY CASH	
RECEIVABLES—CLIENT BILLED	
RECEIVABLES—CLIENT RETAINED	
UNBILLED CONTRACT COSTS—CLIENT	
UNBILLED COSTS—INTEROFFICE	
UNBILLED CONTRACT COSTS—BILLING DEFERRED	
UNBILLED REVENUE OR WORK IN PROGRESS	
ACCOUNTS RECEIVABLE—EMPLOYEES	
ACCOUNTS RECEIVABLE—OTHER	
PREPAYMENTS—REFUNDABLE DEPOSITS	
TOTAL (1)	

INTEROFFICE ACCOUNT	
ACCOUNTS PAYABLE	
UNCLAIMED WAGES	
WITHHOLDING TAX—FEDERAL	
WITHHOLDING TAX—STATE/CITY	
FICA TAX—EMPLOYEES	
GROUP INSURANCE—EMPLOYEES	
COMPANY THRIFT—EMPLOYEE CONTRIBUTIONS	
STATE AND VOLUNTARY (CWS) DISABILITY INSURANCE—EMPLOYEES	
PERSONAL ACCIDENT INSURANCE—EMPLOYEES	
SALES AND USE TAXES	
FICA TAX—EMPLOYER'S	
FUI TAX—EMPLOYER'S	
SUI TAX—EMPLOYER'S	
ACCRUED PAYROLL	
EMPLOYEE ALLOWANCES—NON MANUAL	
WELFARE AND BENEFIT FUNDS—MANUAL	
ACCRUED INSURANCE PREMIUMS	
ADVANCES AND BILLINGS—CLIENTS	
SOCIAL AND RETIREMENT BENEFITS RECOVERY ACCRUAL	
SUBTOTAL	
GROSS MARGIN FROM PAGE 2	
TOTAL (MUST AGREE WITH (1) ABOVE)	

NOTES:

Figure 20–5–4 Balance sheet

STATEMENT OF REVENUE AND COSTS

Location ....

As at .... 19........

DESCRIPTION	Total All Jobs	Job No.	Job No.	Job No.
CONTRACT REVENUE:				
REIMBURSABLE COSTS—COST PLUS				
LUMP SUM OR UNIT PRICE				
EARNED FEE—COST PLUS				
TOTAL CONTRACT REVENUE				
CONTRACT COSTS:				
REIMBURSABLE				
FIELD COSTS				
EST. AND ENGINEERING—HOME AND BRANCH OFFICE				
PROC. AND OTHER SERVICES—HOME AND BRANCH OFFICE				
TOTAL REIMBURSABLE CONTRACT COSTS				
NON-REIMBURSABLE COSTS				
FIELD COSTS				
EST. AND ENGINEERING—HOME AND BRANCH OFFICE				
PROC. AND OTHER SERVICES—HOME AND BRANCH OFFICE				
TOTAL NON-REIMBURSABLE COSTS				
LUMP SUM OR UNIT PRICE COSTS				
FIELD COSTS				
EST. AND ENGINEERING—HOME AND BRANCH OFFICE				
PROC. AND OTHER SERVICES—HOME AND BRANCH OFFICE				
TOTAL LUMP SUM OR UNIT PRICE COSTS				
TOTAL CONTRACT COSTS				
GROSS MARGIN				
AGENCY COSTS INCLUDED ABOVE				
CLIENT FURNISHED MATERIALS INCLUDED ABOVE				

Figure 20–5–5

applicants. The personnel department should maintain employee records and should be involved with employment practices, union relations on a day-to-day basis, and the administration of local wage and hour requirements.

20-7. Field Administration in Joint-Venture Contracts

When a large contract is undertaken it may be performed as a joint venture rather than by a single company. Joint ventures are entered into to provide adequate financial strength for bonding purposes.

Whereas one contractor will be the key or sponsor, the others participate in the earnings or losses according to a prearranged formula, generally on a percentage basis. The sponsor usually receives the largest percentage, although not necessarily as much as 50 percent. Joint ventures, being under the control of the sponsor, utilize his accounting methods. Each joint venture is considered as a separate job and the accounting is similar to job financial accounting.

When a company is sponsor of a joint venture, it has a responsibility to provide financial data and information as required to the members of the venture. This may mean audited financial statements on an annual basis, and will mean regular reports and tax return information for each member of the venture.

20-8. Headquarters Administrative Function

The headquarters administrative function not only promulgates uniform requirements but also receives the product of the field reporting system and processes it for operating management's use. The uniform chart of accounts, reporting forms, and administrative procedures to be followed in the field are all produced originally from the headquarters administrative office. Much of the administration found in the field is duplicated at the home and district level for the company as a whole.

20-9. Administrative Services at District Level

At the district level the greatest blending of home-office and field-administrative functions is found. As mentioned previously the district office maintains administrative control over smaller jobs in the area. In addition, the district office supplies staff support to jobs under its control. This is in addition to any engineering or construction support which may be rendered.

In some cases district offices are formulated around a particular type of expertise. In large companies, therefore, one would find pipeline or refinery operations in a district office near oil-field installations. In most cases, however, the district office is organized on a geographical basis. Although there may be representatives of the different construction capabilities present in the office, generally these offices are to service a particular area.

Headquarters administrative functions relate primarily to the staff. Although some construction jobs may be administered out of the home office, they should be considered separately from the actual headquarters administration. The construction job accounting and administration done at the headquarters office can be looked on as merely another district office which is located physically at company headquarters. Headquarters administration includes primarily policy making, accounting, and procurement. Policy-making functions have been covered in other chapters in this handbook.

20-10. Accounting as a Headquarters Staff Function

Accounting is headed by the chief financial officer—generally the comptroller. The comptrollership function entails all aspects of accounting and may include the issuance of directives and accounting policy to the field, gathering and assembling data from field and district office locations, maintaining the corporate accounting records, preparing financial reports for the use of management, providing support to field and district office accounting personnel, designing a program of internal audit, maintaining relations and planning with the outside independent auditors, preparing special reports as may be required for bonding purposes, bid prequalification, and others, and maintaining relations with the public on financial matters.

Accounting policies should be formalized and directives issued to ensure that uniformity of reporting and treatment of financial matters is achieved. A uniform chart of accounts with names and some description should be available for all accounting locations. The design and issuance of this chart of accounts are the responsibility of the headquarters comptroller. The chart should be supplemented with an accounting manual containing further details on the accounting system in use.

The chart of accounts can be used for a multiplicity of purposes in that it can designate location, type of account, i.e., asset/liability or operations, and can fix responsibility as well. The accounting manual should stipulate the method of accounting to be followed; e.g., percentage of completion or completed contract method. It should also lay out the format for reporting to the headquarters office. The timing of the preparation and submission of financial information should also be spelled out in the accounting manual. Special requirements for special reports, such as Social Security taxes, withholdings for employee benefits, anticipated cash needs, and similar matters

should be covered in the accounting manual. Special year-end requirements should also be covered.

20-11. Gathering Data for Financial Reports for Top Management

The gathering and assembling of data from field and district offices form the basis for preparation of financial reports for top management. Often a single job may incur charges and receive revenue from more than one location, i.e., the district or home office may have incurred costs relating to a job. These will have to be added to the basic field statements as received in order to arrive at the total picture for the job. These occur under a variety of circumstances.

First, the headquarters or district procurement office may be handling some material or equipment purchases for the job which may not have been billed to the job at a given statement date. The home or district office may be supplying services to the job on a fee or fixed-price basis. In this case the charges to the job should be replaced by the actual cost to the district or home office. Another example is on cost-plus jobs, in which the company may not wish the contract owner to see nonreimbursable costs. These could be incurred at the job and then transferred by a charge to the district or home office. In each of these cases all costs and revenues accruing to a particular job should be assembled in one place, namely, the headquarters office. Costs and expenses not relating to contracts should also be gathered and summarized at the home office. These would come primarily from the district offices, since all costs incurred at a job site would normally be chargeable to that job. Table 20-11-1 illustrates this gathering of job cost and profit.

TABLE 20-11-1. GATHERING JOB COST AND PROFIT[a]

	Revenue	Cost	Margin
Job site	$100,000	$90,000	$10,000
District nonreimbursable		10,000	(10,000)
District purchasing	10,000	8,000	2,000
Home office	10,000	5,000	5,000
Combined	$100,000	$93,000	$ 7,000

[a]Revenue to district and home office is included in job-site cost and therefore is excluded in both cases from the combined revenue and cost.

20-12. Corporate Accounting Records

The corporate accounting records comprise those accounts that relate to corporate-wide matters and not to individual jobs. These would be corporate treasury cash accounts, the capital or shareholders, equity accounts, corporate long-term debt, income taxes, and may include a number of other items. In many companies accounts receivable are maintained at the headquarters office for control purposes. Equipment may be purchased on a corporate basis and then charged to the jobs as it is used on a rental basis. Investments in joint ventures or other corporate entities would be maintained in the headquarters accounts. In addition, accrued liabilities for items such as the company's share of employee benefits and other accruals not relating to jobs would be carried in the headquarters accounts.

Job accounting is sometimes done on the basis of the contractual documents; i.e., although costs are recorded as incurred, revenue may be recorded or billed in a way that does not adequately reflect the revenue actually earned on the job. In this case the adjustment from job books to financial reporting basis would be recorded in the headquarters accounts.

Preparation of the financial report used by management is a vital function of the comptroller's department. Without adequate information on which to base decisions, management cannot function effectively.

20-13. Emphasis on Job Results

Reports should encompass all operations of the company with the emphasis, of course, on job results. To be able to assess job profitability and performance, management needs to know not only the past history on each job, namely earnings or gross margin from inception to date, but also should be informed of changes in forecast total cost, changes in revenue, either agreed to or contemplated, as well as any changes in the scheduled job completion. These are necessary in planning future operations and commitments of personnel and resources. Without adequate information, management is unable to decide where its efforts are most needed. A suggested format for reporting the results of operations on a job gross margin level with forecast information as well is given in Table 20-13-1.

In addition to reports on individual jobs or summaries, management needs financial reports on overhead and general administrative costs. Without knowing where the company stands in these areas, management cannot successfully plan future operations. Another key report is that of work backlog by type of job and by length of time to work off the backlog of work in hand.

20-14. Prequalification and Other Reports

In addition to preparation of financial reports for the use of management there are other reports that must be prepared by the comptroller's department. Prequalification reports are generally required by most public jurisdictions that ask for bids from contractors. The purpose of these reports is to provide some assurance to the agency that the contractor is financially able to handle the volume which he may be asked to bid upon. These are prepared for each jurisdiction. This would involve state, county, city, highway departments, public works departments, in fact, any agency that may require services of a contractor. These are generally similar in content and format although, unfortunately, not uniform. The report generates a tangible net worth which is used to measure financial stability. In many cases these reports have to be audited by an independent public accountant.

20-15. Comptroller Support to Field and District Personnel

The headquarters comptroller must provide support to the field and district office accounting personnel. In addition to the policy directives mentioned before, the comptroller's office should be prepared to assist in providing additional personnel where needed and staffing new field offices as contracts are opened. He must also keep field people occupied between jobs.

20-16. Internal and External Audit Services

The comptroller's office is responsible for three areas of audit. These are internal audit, i.e., the company's own staff; external audit, i.e., the company's independent public accountants, and audits of individual jobs, i.e., the contract owners. Internal audit serves two primary functions: (1) to ensure adherence to the accounting policies and procedures promulgated by the headquarters office; and (2) to provide assurance that adequate controls are being followed at all company locations to prevent defalcation or similar irregularities. The internal audit staff should report to one of the higher executives in the company. Although functioning with the comptroller and policing his activities, it is preferable for the internal audit staff to speak with an independent and authoritative voice. The internal audit staff should preferably speak with the authority of the president. This is seldom possible, desirable though it may be.

The internal auditors should have a regular program designed to cover all company locations, including headquarters, district office, and job sites. The schedule should be planned well in advance and should include some flexibility for special assignments which may arise from time to time. The internal auditors should have a definite program for each level of examination in order to ensure that all areas are covered. This should include detailed procedural tests to ensure adherence to company policy, as well as control reviews for the protection of company assets.

In addition to audits of the financial records, the internal auditors should be conducting operational reviews to improve profitability as well. They should also be utilized in special studies, such as acquisition audits. They should be capable of auditing the company's EDP procedures, if any.

TABLE 20-13-1. MANAGEMENT REPORT
(000 omitted)

Job	% Complete	Month		Year to Date		Contract to Date		Forecast Total Contract	
		Revenue	Margin	Revenue	Margin	Revenue	Margin	Revenue	Margin
Highway	90	50	3	340	10	1400	58	1600	65
Office Building[a]	75	15	—	60	—	100	(20)	150	(20)
Apartment[b]	10	5	—	8	—	8	—	95	5

[a]Total loss on the office building was accrued in the prior year.
[b]The apartment has not progressed far enough to record income.

The comptroller's office is the main liaison with the company's independent public accountants. Whereas most bid prequalification reports require the opinion of an independent public accountant, this is only one reason for a company to engage one. An examination by independent public accountants provides assurance to management and to stockholders that the company's financial statements are prepared in accordance with generally accepted accounting principles and are a fair presentation of financial position and the results of operations.

The outside auditors do not undertake an examination in sufficient detail to provide assurance that there has been no fraud or other irregularities, since such an examination would not be appropriate for them to undertake. It would be prohibitively expensive and falls more into the domain of the internal audit staff. The outside auditors also do not provide assurance of compliance with all company policies and procedures, since they may well not visit every location and will not make a sufficiently in-depth review to cover the subject fully.

When a company has entered cost reimbursement type contracts it can expect to be subject to audit by contract owners. The comptroller should make provision for this audit by keeping sufficient detailed records and documents for the use of the owners' auditors and providing facilities and assistance to expedite their visit.

Maintaining relations with the public and stockholders on financial matters can also be the task of the comptroller's department. In a large company this may be divorced from the comptroller and given to the chief financial officer. Publicly held and traded companies must report to their stockholders and to the Securities and Exchange Commission (S.E.C.) quarterly with a more extensive annual report. These reports are required of all publicly traded companies and are not unique to the construction industry.

20-17. Reports to the Securities Exchange Commission

The annual report to the S.E.C. has been expanded. One requirement is that if the company takes up equity in undistributed earnings of any other "person," it must file for each such person the financial statements that would be required if it were registered with the S.E.C. This means that financial statements on joint ventures have to be filed. This may be changed. Any company experiencing hardship should consider discussing the problem with the S.E.C.

20-18. Relations with Banks and Financial Institutions

Relations with banks and financial institutions also falls under the chief financial officer. Since most contractors must borrow money in order to finance their investment in contracts and joint ventures, they must maintain close relationships with the banking community. Banks generally require audited statements and may require them in greater detail than would ordinarily be prepared. Maintaining good relations with the banking community, therefore, is a vital part of the financial departments.

Headquarters office purchasing performs a staff function to the purchasing offices in the field. They evaluate suppliers and notify field purchasing of acceptable suppliers and of price lists which have been negotiated for standard items. The headquarters office purchasing function will also handle any large or unusual procurement which may be necessary on a given job.

20-19. Headquarters Office Purchasing

The headquarters office purchasing people should review the quality of the work being performed in the field. This can be done by review of field purchase orders which can be sent to the headquarters office or through field office visits. This review should include ensuring that purchase orders are issued for all items procured except those under blanket order, that these purchase orders are issued in advance, and that they are properly priced to protect the company.

20-20. Tax Record Department

The tax record department has the responsibility for maintaining tax files covering federal, state, and foreign income taxes, state property taxes, Social Security taxes, and other taxes which may be imposed on the company. The tax department may be part of the comptroller's office or it may be separate. In a well run company a tax department is responsible for planning and for seeing that information returns are filed and payments made on a timely basis.

In a large company tax planning is a very important function. Knowledge of the alternative methods available under the various tax laws is the key to running a successful tax department. This would include state and local taxes, as well as Federal and foreign.

20-21. Equipment Department

The equipment department at the head office has responsibility for maintaining records of all equipment owned by the company as to its location and condition. Properly managed, an equipment operation can be highly profitable. By charging jobs the current rental for equipment and having an efficient maintenance operation, a company can realize a good rate of return on its investment in construction equipment.

The equipment department should not only know where all equipment is, it should prepare monthly billings for equipment rental to jobs and it should be responsible for determining when equipment should be traded in or sold.

20-22. Data Processing Department

The data processing department can be used by both engineering and administrative departments. In those companies where engineering and design is important, these fuctions may take precedent in both the configuration of equipment and its utilization. In this case it should be of a more scientific type, but it should still be able to perform business fuctions. Prime time may well be set aside for scientific use only, leaving the second or third shift available for the administrative function. The comments that follow relate to administrative function and not necessarily to engineering, although they may apply.

Electronic data processing equipment can be a useful tool to the administrator of a construction company. Coupled with its well-known ability to summarize and calculate rapidly, and its increasing storage facilities, EDP equipment, as now developed, is capable of making many decisions formerly made by management. These include determining cost overruns on individual segments of a job, editing time cards for overtime reported which has not been authorized, and reporting invalid job or charge numbers for direct or indirect labor, materials, or overhead accounts. Although there may be a tendency at first to get excessive detail from EDP equipment at the current status of development, exception reporting is more productive.

With the advent of exception reporting and similar decision making capabilities, controls over EDP equipment have to be stringently adhered to. These include control over data input to the equipment, controls over the processing of data, and controls over output. Input controls include transaction count known as batch controls and the use of totals of certain control numbers, such as employee badge numbers, to get a predetermined total. Editing such controls by people independent of computer operations and the use of key verifying as well as key punch ensure good data input.

Processing controls include the use of tape labels to ensure that the proper files are being processed, the preparation of clear and complete operating instructions covering all operating situations that can be anticipated, preventing operators from having access to details of the programs, and obtaining printout of all halts or deviations from normal operations with an explanation of the reason. In addition to the controls within the machine room there should be adequate documentation and flow charts of computer programs so that any changes can be easily made as required, and so that there will be a record as to the capabilities of the programs.

The headquarters data processing department can handle not only the headquarters accounting and business functions, but can also handle district and field office accounting through telephone tie lines. This would require maintaining capability in the field to input data to the system and to have printout capability in order to receive informaion back. This could be strictly a key punch and card or paper tape reader for input and low-speed offline printer for output. But this relatively inexpensive installation, by giving the job access to the capabilities of the headquarters computer, can give the job access to data processing power otherwise not feasible.

A data processing department offering services of both a business and scientific nature needs to establish an operating schedule suitable to its users. When field locations are accessing the computer, either the input has to be stored offline until it can be processed or other applications must be postponed. Depending upon the desired turn-around time the computer can accept field data as they become available. The demands of regular users, such as the payroll and accounting functions, should be met at the most feasible time before the unscheduled demand jobs are processed. As mentioned previously, this may mean running business programs during a second or third shift. The regular recurring users, however, should be served and the scheduled demands should be met, if practicable, since in many cases timely output from the computer is necessary for proper functioning of another department.

20-23. Warehousing Records

Warehousing, as well as equipment, can be a profitable operation. Similarly, charges to jobs for warehouse services coupled with an efficient warehouse operation can yield a good return.

Warehousing records should include detail of what is on hand, the description of the item, its location, the reorder point, and the reorder quantity. These records may be hand-posted card records maintained at the warehouse, similar records maintained in the headquarters office, or EDP records maintained on the EDP equipment. Provisions should be made for regular audit of the warehouse, including cycle counting the stock on hand. Access to the warehouse should be limited to authorized personnel only.

21 CORPORATE AND COST ACCOUNTING

FRANCIS DURAND

Partner
Ernst and Ernst
New Orleans, Louisiana

THE BASIC functions of accounting systems and procedures are to gather, process, and report on financial information. The end products are reports which may deal with the financial position of the company as of a specific date (balance sheet or statement of financial position), the results of operations during a specified period (statement of income), and cost data related to specific operations or projects (contract costs). In this chapter we will deal with the systems and procedures involved in producing the varied type of accounting reports related to the operations of a construction contractor.

21-1. Financial Statements

Financial statements, although they may be termed the end product of accounting systems and procedures, are the most appropriate starting point for this discussion. To produce the financial statements of a construction contractor, specific information must be available. The source data utilized to produce this information are also used to produce other reports. In fact, a policy must be adopted which provides that any information presented by supplemental type reports should be readily reconcilable to data presented in the financial statements. Hence, as we progress with the discussion of the financial statements, we will cover systems and procedures that are related to accomplishing other objectives.

The two principal financial statements are the balance sheet and the statement of income. These two statements are common to all business organizations; however, certain facets of a construction company's operations, not common to other types of organizations, affect these two statements. These facets involve the treatment of transactions related to long-term contracts.

21-2. Accounting Method Applicable to Long-Term Contracts

The term "long-term contracts" can cover many different types of contracts—lump sum, cost plus, etc. We are principally concerned here with those types of contracts in which the contractor is obligated to perform specific services for a fixed amount, and the time required to complete the contract extends over a substantial chronological period.

There are two principal methods of accounting for income and costs related to long-term contracts. They are referred to as the "completed-contract" and the "percentage-of-completion" methods. Both methods are recognized as acceptable by the Internal Revenue Service for determining taxable income (Regulations 1.451.3 under the 1954 Internal Revenue Code). *Accounting Research Bulletin No. 45,* promulgated by the Committee on Accounting Procedure of the American Institute of Certified Public Accountants (AICPA) in October 1955, states:

"The committee believes that in general when estimates of costs to complete and extent of progress toward completion of long-term contracts are reasonably dependable, the percentage-of-completion method is preferable. When lack of dependable estimates or inherent hazards cause forecasts to be doubtful, the completed-contract method is preferable. Disclosure of the method followed should be made." (Par. 15)

A review of the procedures employed in applying the two methods will indicate the reason for this statement.

Under the percentage-of-completion method of accounting for long-term contracts, income is recognized as work progresses on a contract. An acceptable method of determining this income is to relate costs incurred to date to total estimated costs to complete the contract. Hence, as of a specific date, recognizable contract income would be determined by the following formula:

$$\frac{\text{Costs incurred to date}}{\text{Total estimated costs}} \times \text{contract price} = \text{contract income to date}$$

Where the percentage-of-completion method is used, current assets on the balance sheet may include costs plus recognized income not billed as of the balance sheet date. Correspondingly, current liabilities may include billings in excess of costs incurred and income recognized to date. For statement presentation purposes, this method considers that a specified portion of the contract price has been earned; to the extent that this portion has not been billed to the owner, it therefore represents a current asset. Also, to the extent that amounts billed to the owner exceed that portion of the contract price which has been earned, a liability has been incurred for such excess.

The completed-contract method recognizes income when the contract is completed or substantially completed (when remaining costs are not significant in amount). Under this method an excess of accumulated costs over related billings should be classified in the balance sheet as a current asset, and an excess of accumulated billings over related costs should be classified as a liability, in most cases as a current liability.

Under either the percentage-of-completion or the completed-contract methods, provision should be made for expected losses. "If there is a close relationship between profitable and unprofitable contracts, such as in the case of contracts which are parts of the same project, the group may be treated as a unit in determining the necessity for a provision for losses." (Par. 11—ARB 45)

The principal advantages of the percentage-of-completion method are that income is recognized currently as work progresses under the contract, rather than irregularly as contracts are completed, and the reflections of the status of the uncompleted contracts through the current estimates of cost to complete or of progress toward completion. The principal disadvantage of this method is its dependency upon estimates of future costs which are subject to the uncertainties frequently inherent in long-term contracts.

The completed-contract method has the advantage of recognizing income on the basis of results as finally determined, rather than on the basis of estimates. Its principal disadvantage is its inability to reflect current performance when the work under a contract extends into more than one accounting period. This method may result in irregular recognition of income.

It is not at all uncommon for construction contractors to adopt the percentage-of-completion method of accounting for financial statement purposes and report income for federal income tax purposes under the completed-contract method of accounting. When this occurs, it gives rise to what is known as a "timing difference." "Accounting for Income Taxes" (Opinion No. 11, December 1967, of the Accounting Principles Board of the AICPA) covers "timing differences." When timing differences occur, the tax effect should be measured by the differences between income taxes computed with and without inclusion of the transaction creating the difference between taxable income and pretax accounting income. The difference between the two computations gives rise to "deferred tax" amounts, reflecting the tax effects which will reverse in future periods.

21-3. Other Factors

Allocation of equipment ownership and operating costs to projects (specific contracts) is, in the author's opinion, a procedure that must be followed to have proper accounting for long-term contracts. Various procedures may be employed to allocate such costs and they are discussed in a later section of this chapter which covers policies and procedures applicable to accounting for equipment costs.

Normally, general and administrative expenses are not allocated to contracts-in-progress. Such expenses are usually treated as "period costs," i.e., deducted in the year incurred. When the percentage-of-completion method of accounting is used, income is recognized periodically, offsetting the deduction of general and administrative expenses. This may not be the case, however, if the completed-contract method of accounting is employed. Hence, *Accounting Research Bulletin No. 45* states:

"When the completed-contract method is used, it may be appropriate to allocate general and administrative expenses to contract costs rather than periodic income. This may result in a better matching of costs and revenues than would result from treating such expenses as period costs, particularly in years when no contracts were completed." (Par. 10)

The contractor's election to defer, through allocation to contracts-in-progress, the deduction of general and administrative expenses may be particularly appropriate when the contractor is reporting income on the basis of the completed-contract method and he is engaged in only one or very few projects, thus resulting in no income being reportable during specific periods. In any event, the election to allocate should be consistently followed. *Accounting Research Bulletin No. 45* cautions its interpreters as follows:

"In any case there should be no excessive deferral of overhead costs, such as might occur if total overhead were assigned to abnormally few or abnormally small contracts in process." (Par. 10)

In 1965 the AICPA published a bulletin entitled *Audits of Construction Contractors*. This bulletin, with reference to the citation noted above, states:

"Here the committee apparently had in mind that, when construction volume was at a low point, only a reasonable amount of overhead should be assigned to contracts in process. The remaining general and administrative or overhead costs should then be shown as period expenses even though net losses were thereby produced."

Whether the contractor does, or does not, allocate overhead costs to contracts-in-progress for financial statement purposes, these costs *must* not be overlooked if proper financial management policies are to be pursued. An excessive amount of overhead costs in relation to the volume of projects being undertaken is just as costly as are excessive contract costs in relation to the contract price. Hence, appropriate consideration of these costs is essential in arriving at contract prices.

21-4. Illustrative Financial Statements

To illustrate some of the points brought out in the preceding discussion, Figures 21-4-1 and 21-4-2 are examples

CONSTRUCTION COMPANY, INC.

BALANCE SHEET

December 31, 19....

	Method of Accounting	
	Percentage of Completion	Completed Contract
ASSETS		
Current Assets		
Cash	$ 850,000	$ 850,000
Due on contracts, including retained percentages of $2,800,000	11,500,000	11,500,000
Costs and estimated earnings in excess of billings on contracts-in-progress	900,000	—
Costs in excess of billings on contracts-in-progress	—	100,000
Inventories of materials and supplies, at lower of cost or market	600,000	600,000
Prepaid expenses	350,000	350,000
TOTAL CURRENT ASSETS	14,200,000	13,400,000
Property, Plant, and Equipment—on the basis of cost		
Land	150,000	150,000
Buildings	400,000	400,000
Equipment	3,550,000	3,550,000
Allowances for depreciation	(2,000,000)	(2,000,000)
	2,100,000	2,100,000
	$16,300,000	$15,500,000
LIABILITIES AND SHAREHOLDERS' EQUITY		
Current Liabilities		
Notes payable to bank	$ 2,000,000	$ 2,000,000
Accounts payable	1,500,000	1,500,000
Accrued expenses	1,200,000	1,500,000
Billings in excess of costs and estimated earnings on contracts-in-progress	700,000	—
Billings in excess of costs on contracts-in-progress	—	3,700,000
Federal and State taxes on income	300,000	300,000
	5,700,000	9,000,000
Deferred Federal and State Taxes on Income	1,750,000	(300,000)
Shareholders' Equity		
Common stock, par value $1 a share: authorized, issued and outstanding, 500,000 shares	500,000	500,000
Retained earnings	8,350,000	6,300,000
	8,850,000	6,800,000
	$16,300,000	$15,500,000

Figure 21-4-1

CONSTRUCTION COMPANY, INC.

STATEMENT OF INCOME

Year ended December 31, 19....

	Method of Accounting	
	Percentage of Completion	Completed Contract
Contract revenues	$41,000,000	$44,000,000
Contract costs	37,700,000	41,400,000
	3,300,000	2,600,000
Overhead expenses	1,200,000	1,200,000
	2,100,000	1,400,000
Other income (expense) net	100,000	100,000
	2,200,000	1,500,000
Federal and state income taxes:		
Due currently	750,000	750,000
Deferred	350,000	—
	1,100,000	750,000
NET INCOME	$ 1,100,000	$ 750,000

Figure 21-4-2

CONSTRUCTION COMPANY, INC.

STATUS OF CONTRACTS IN PROGRESS

December 31, 19....

Estimated Profit (Loss)	Contract Revenue			Contract Costs		
	Cumulative	Prior Year	Current Year	Cumulative	Prior Year	Current Year
$ 800,000	$10,000,000	$ 8,500,000	$ 1,500,000	$ 9,200,000	$ 7,800,000	$ 1,400,000
500,000	8,000,000	7,900,000	100,000	7,500,000	7,300,000	200,000
(200,000)	11,000,000	7,000,000	4,000,000	11,200,000	7,100,000	4,100,000
1,500,000	15,000,000	10,000,000	5,000,000	13,500,000	9,000,000	4,500,000
2,600,000	44,000,000	33,400,000	10,600,000	41,400,000	31,200,000	10,200,000
2,000,000	16,000,000	10,000,000	6,000,000	14,400,000	9,100,000	5,300,000
(400,000)	7,000,000	2,000,000	5,000,000	7,200,000	2,100,000	5,100,000
1,500,000	10,800,000	4,000,000	6,800,000	9,900,000	3,800,000	6,100,000
4,000,000	9,600,000	—	9,600,000	8,000,000	—	8,000,000
(200,000)	3,000,000	—	3,000,000	3,100,000	—	3,100,000
6,900,000	46,400,000	16,000,000	30,400,000	42,600,000	15,000,000	27,600,000
$9,500,000	$90,400,000	$49,400,000	$41,000,000	$84,000,000	$46,200,000	$37,800,000

Losses		Contract Billings	Percentage of Completion		Completed Contract	
Provisions	Ending		Billings in Excess of Revenue	Revenue in Excess of Billings	Billings in Excess of Costs	Costs in Excess of Billings
$ —	$ —	$10,000,000				
—	—	8,000,000				
(100,000)	—	11,000,000				
—	—	15,000,000				
(100,000)	—	44,000,000				
—	—	16,200,000	$ 200,000	$ —	$ 1,800,000	$ —
(100,000)	200,000	7,500,000	500,000	—	300,000	—
—	—	10,300,000	—	500,000	400,000	—
—	—	9,200,000	—	400,000	1,200,000	—
100,000	100,000	3,000,000	—	—	—	100,000
—	300,000	46,200,000	700,000	900,000	3,700,000	100,000
$ (100,000)	$ 300,000	$90,200,000	$ 700,000	$ 900,000	$ 3,700,000	$ 100,000

Figure 21-4-3.

of balance sheets and statements of income based on the completed-contract and the percentage-of-completion methods of accounting for long-term contracts. These financial statements are in summary form and are based on the following factors pertinent to our discussion:

1. The status of contracts-in-progress, together with other data shown in Figure 21-4-3.
2. A tax rate of 50 percent applicable to the current year and to the prior years during which deferred income taxes were accrued.
3. The assumption that there are no transactions giving rise to deferred income taxes other than those related to the accounting for long-term contracts.

The financial statements given in Figures 21-4-1 and 21-4-2 would have to be expanded to comply with gen-

erally accepted accounting principles applicable to disclosure requirements. Certain data would have to be added, either in the body of the financial statements or in their notes. For example, the amount of depreciation included in costs and expenses should be indicated in either of these places. Such disclosures are not pertinent to this discussion and therefore have been omitted.

The balance sheet in Figure 21-4-1 indicates that retained earnings of Construction Company, Inc., at December 31, 1972, under the percentage-of-completion method of accounting amount to $8,350,000, or $2,050,000 greater than the comparable figure under the completed-contract method of accounting. These sums can be reconciled by reference to Figure 21-4-3, which presents the status of contracts-in-progress at December 31, 1972, as follows:

Corporate and Cost Accounting 519

CONSTRUCTION COMPANY, INC.

STATUS OF CONTRACTS IN PROGRESS

December 31, 19....

		Total Estimated Costs		
Contract No.	Contract Price	Actual to Date	Estimated to Complete	Total
Completed contracts:				
2501	$ 10,000,000	$ 9,200,000	$ —	$ 9,200,000
2525	8,000,000	7,500,000	—	7,500,000
2528	11,000,000	11,200,000	—	11,200,000
2531	15,000,000	13,500,000	—	13,500,000
TOTAL COMPLETED CONTRACTS	44,000,000	41,400,000	—	41,400,000
Contracts in progress:				
2520	20,000,000	14,400,000	3,600,000	18,000,000
2524	14,000,000	7,200,000	7,200,000	14,400,000
2529	18,000,000	9,900,000	6,600,000	16,500,000
2536	24,000,000	8,000,000	12,000,000	20,000,000
2541	6,000,000	3,100,000	3,100,000	6,200,000
TOTAL CONTRACTS-IN-PROGRESS	82,000,000	42,600,000	32,500,000	75,100,000
TOTALS	$126,000,000	$84,000,000	$32,500,000	$116,500,000

	Contract Income (Loss)			Provision for
Contract No.	Cumulative	Prior Year	Current Year	Beginning
Completed contracts:				
2501	$ 800,000	$ 700,000	$ 100,000	$ —
2525	500,000	600,000	(100,000)	—
2528	(200,000)	(100,000)	(100,000)	100,000
2531	1,500,000	1,000,000	500,000	—
TOTAL COMPLETED CONTRACTS	2,600,000	2,200,000	400,000	100,000
Contracts in progress:				
2520	1,600,000	900,000	700,000	—
2524	(200,000)	(100,000)	(100,000)	300,000
2529	900,000	200,000	700,000	—
2536	1,600,000	—	1,600,000	—
2541	(100,000)	—	(100,000)	—
TOTAL CONTRACTS-IN-PROGRESS	3,800,000	1,000,000	2,800,000	300,000
TOTALS	$ 6,400,000	$ 3,200,000	$ 3,200,000	$ 400,000

Figure 21–4–3 *(cont.)*

Retained earnings—completed contract method of accounting (Figure 21-4-1)		$ 6,300,000		
Add:				
Contract income on contracts-in-progress, cumulative (Figure 21-4-3)		3,800,000		
Difference in provisions for future losses:				
Completed-contract method:				
Contract No. 2524	$ 400,000			
Contract No. 2541	200,000			
	600,000			
Percentage-of-completion method:				
Ending balance of "provision for losses" (Figure 21-4-3)	300,000	300,000		
		10,400,000		

Deduct:

Taxes on contract income on contracts-in-progress at 50%	1,900,000		
Taxes on difference in provision for losses at 50%	150,000	2,050,000	
Retained earnings—percentage-of-completion method of accounting (Figure 21-4-1)		$ 8,350,000	

The amounts appearing Figure 21-4-1 which represent "costs and estimated earnings in excess of billings on contracts-in-progress" ($9,000,000), "costs in excess of billings on contracts-in-progress" ($100,000,), "billings in excess of costs and estimated earnings on contracts-in-progress" ($700,000), and "billings in excess of costs on contracts-in-progress" ($3,700,000), can be traced to Figure 21-4-3. Reference to this illustration will provide information as to how the sums are determined. Under the percentage-of-completion method, the two figures under discussion are determined by comparing the columns headed "Contract Revenue—Cumulative" and "Contract Billings." Under the completed-contract method, the column headed "Contract Billings" is compared to the column headed "Contract Costs—Cumulative." These comparisons are made on an individual contract basis.

Construction Company, Inc.'s balance sheet indicates that income taxes have been deferred. These amounts have been computed as follows:

Percentage-of-Completion Method

Contract income (Cumulative) on contracts-in-progress (Figure 21-4-3)	$3,800,000
Less provision for losses—ending balance (Figure 21-4-3)	300,000
	$3,500,000
Deferred income taxes at 50%	$1,750,000

Completed-Contract Method

Balance of provision for losses account (not separately stated in Figure 21-4-3):	
Contract No. 2524	$ 400,000
Contract No. 2541	200,000
	$ 600,000
Deferred income taxes (debit) at 50%	$ 300,000

The provision for future losses on contracts-in-progress would not be deductible for income tax purposes until the contracts were complete. The provision for income tax benefits to be realized when these sums are deducted is presented in Figure 21-4-1 as a reduction of the deferred income tax liability (percentage-of-completion method) for illustrative purposes. In actual practice, it would be more appropriate to present the sum representing deferred income tax benefits on the asset side of the balance sheet.

In Figure 21-4-1 the provision for future losses on contracts-in-progress has been included among "accrued expenses." Note that "accrued expenses" under the completed-contract method exceeds the comparable sum under the percentage-of-completion method by $300,000, the difference in the provision for losses under the two methods.

The sums related to contract revenues and contract costs, which are included on the statement of income (Figure 21-4-2), are derived from the data presented by the schedule—"Status of Contracts in Progress" (Figure 21-4-3). Contract revenues on the statement of income, under the percentage-of-completion method, are represented by the total of the column headed "Contract Revenue—Current Year" ($41,000,000) in Figure 21-4-3. Contract revenues under the completed-contract method are represented by the subtotal on completed contracts under the column headed "Contract Revenue—Cumulative" ($44,000,000). Contract costs on the statement of income are derived from the following data presented in Figure 21-4-3:

Percentage-of-Completion Method

Total of column headed "Contract Costs—Current Year" (Figure 21-4-3)	$37,800,000
Add (deduct) changes in provisions for losses:	
Contract No. 2528	(100,000)
Contract No. 2524	(100,000)
Contract No. 2541	100,000
Contract costs—Figure 21-4-2	$37,700,000

Completed-Contract Method

Subtotal of completed contracts under column headed "Contract Costs—Cumulative" (Figure 21-4-3)	$41,400,000
Add (deduct) changes in provision for losses:	
Contract No. 2528	(200,000)
Contract No. 2541	200,000
Contract costs—Figure 21-4-2	$41,400,000

In the course of reviewing the data on contract costs presented in Figure 21-4-3, it should be noted that the sums representing contract costs do not include any amounts related to a provision for future losses. Further, it is assumed that, when an estimate has been made that a loss will be incurred on contracts-in-progress at the beginning of the year, a provision for that loss has been provided in prior years in an amount equal to the sum shown in Figure 21-4-2 as the estimated loss. Hence, under either method of accounting, contract costs shown in Figure 21-4-3 have to be adjusted for losses that have been accrued in prior years, but incurred or realized during the current year. As an example, on Contract No. 2528, the following transactions could have occurred under the provision for losses:

	Method of Accounting	
	Percentage of Completion	Completed Contract
Provision for losses set up based on estimates as of the end of the year		
First year	$150,000	$150,000
Second year	50,000	50,000
	200,000	200,000
Deduct losses included in income for year (see note below):		
First year	50,000	
Second year	50,000	
	100,000	
Balance of provisions at beginning of year ended Dec. 31, 1972	100,000	200,000
Deduct losses included in income for year (see note below) ended Dec. 31, 1972	100,000	200,000
Balance of Provision at Dec. 31, 1972	$ 0	$ 0

Note: The term "included in income for year" is intended to represent the loss which would have been deducted from income without any provision for future losses.

The deduction for federal and state income taxes on the statement of income (Figure 21-4-2) is equivalent to

50 percent of the pretax income. The computation of the amount due currently, and the amount deferred, is based on employing the following procedures:

1. Construction Company, Inc. reports its taxable income under the completed-contract method; so we start with this method. Financial statement pre-tax income is $1,500,000. The only difference between this income figure and taxable income are transactions involving the provisions for future losses. During the year ended December 31, 1972, under the completed-contract method $200,000 was deleted from the provision for future losses due to the completion of Contract No. 2528, and $200,000 was added to the provision for future losses due to the estimate of losses to be incurred on Contract No. 2541. Hence, under this method, financial statement pre-tax income and taxable income are the same, $1,500,000, and operations for the year did not give rise to any changes in deferred taxes.
2. Under the percentage-of-completion method, taxes currently due are 50% of $1,500,000, taxable income for the year (see Step 1). Deferred taxes are computed as follows:

Contract income for current year on contracts-in-progress (Figure 21-4-3)	$2,800,000
Less Contract income—prior year on contracts completed during the year (Figure 21-4-3)	2,200,000
	600,000
Add changes in provisions for losses (Figure 21-4-3)	100,000
	$ 700,000
Deferred taxes at 50%	$ 350,000

Construction Company, Inc.'s statement of income is in summary form and, therefore, a simple reading of the statement might give rise to a question as to how certain overhead costs were treated. The term "overhead expenses" may cover only general and administrative expenses. Such things as construction overhead and equipment overhead may be included under contract costs. Correspondingly, costs allocated to specific contracts may or may not include equipment costs or construction overhead. Schedules accompanying the statement of income should furnish the data necessary to determine what treatment was accorded the costs under discussion.

21-5. Accounting Systems and Procedures

It is beyond the scope of this discussion to cover in minute detail all the aspects of contractors' accounting systems and the various procedures that are utilized to accomplish the prime objective of gathering, processing, and reporting on financial information. Instead of a detailed discussion of such systems and procedures, we will discuss broadly the following functions:

1. Headquarters office
2. Project office accounting—contract cost accounting
3. Internal auditing—control procedures
4. Accounting for equipment

21-6. Headquarters Office

Headquarters office accounting personnel are normally responsible for the preparation of financial statements and accompanying schedules. This primary responsibility entails responsibility for many varied functions, all of which are related to accounting systems and procedures.

The design, implementation, and maintenance of an effective accounting system is not a simple matter. One important rule must be followed: The accounting system and its related functions must be only as complex as are required to furnish the information management needs to effectively operate the business. It is possible to break this rule in two ways: One, by having a system so complex that it produces information of less value than the cost of producing such information; and, two, by using a system so simple that, although the cost of maintaining the system is very nominal, management is not receiving the information required to operate the business effectively.

Generally speaking the accounting system should be designed to furnish the following basic data:

1. The information required to prepare financial statements in accordance with generally accepted accounting principles. (The information referred to here relates to the reporting of financial data, such as classification of assets and liabilities within the current and noncurrent categories, reporting of income and costs by proper accounting periods.)
2. The information required to properly report contract income and contract costs. (This function is discussed in more detail in a later part of this chapter; however, it is mentioned here to stress the importance of proper coordination of project cost accounting and the information flowing through the general ledger accounts from which the financial statements are prepared.)
3. The information required to properly process equipment ownership and operating costs. (Comparable to the comments above related to cost accounting on contracts, it is important that procedures utilized to collect and process cost data on equipment ownership and operation be coordinated with those related to processing the data for presentation in the financial statements.)
4. The information required to properly file federal and state income tax returns, employment tax returns, and any other type of report required by government agencies.

Long-range planning is an effective management tool, and its use is becoming more and more essential to assure financial stability and maintenance of an acceptable growth pattern. If Mr. John Doe, who is the chief executive officer of a growing business organization, had a schedule of appointments in various cities on various dates, he would not embark on this trip without first planning an itinerary which would enable him successfully to keep his appointments. Mr. Doe would, in his planning, cover the mode of transportation, arrival and departure times, lodging, etc. As his trip progressed, he would adjust his plans as required for unanticipated delays, cancelled appointments, etc. As chief executive officer of a growing business organization, Mr. Doe should similarly plan the "itinerary" for the growth of his business. A 50 percent increase in equity through retention of profits during the ensuing five-year period might be called a "plan," but in essence it is only a goal. The "plan" should entail the steps needed to accomplish the goal and designate the parties responsible for successful completion of the steps involved.

Finance and accounting personnel located in the headquarters office should be assigned the responsibility of accumulating specific data related to design and implementation of the long-range plan. This responsibility may be broken down into several categories. First, the finance and accounting departments should make available to other departments pertinent information on the submission of data by the other departments for forecasting purposes (prior period costs and similar data). Second, the finance and accounting departments should accumulate data furnished by all the departments and complete, or recast, these data to furnish management with meaningful reports. Third, the accounting and finance departments should be in the position to furnish management with information related to operations in the past period, or past periods, for comparative purposes.

It is imperative that headquarters office accounting personnel extend their complete cooperation to the personnel of other departments so that these other department personnel have access to all the information usable in preparing forecast data. It is not within the realm of accepted practice, however, for accounting personnel to take it upon themselves to furnish source data for forecasting purposes which is dependent upon operations that are the responsibility of other departments. For example, it is not the function of the accounting department to compute estimated future costs on contracts in progress.

An integral part of any long- or short-range plan is the inclusion of procedures designed to furnish an evaluation of the progress toward attainment of the planned goal. These procedures are normally incorporated in the design of the accounting system. In the planning stage, planned operations are converted into forecasted dollars of income and costs, by chronological periods. Subsequently, actual dollars of income and costs must be accumulated in categories allowing a comparison of actual results of operation with those which had been forecasted in sufficient detail to provide management with the ability to take appropriate action to stop adverse trends.

Proper planning entails the assignment of responsibilities for successful completion of specific phases of the overall plan to specific executive and supervisory personnel. The accounting system must be designed so as to provide the information necessary to furnish management a comparison of planned income and costs with those resulting from actual operations, on a responsibility basis—"responsibility reporting." As an example, if equipment costs, ownership, and operating are the responsibility of a vice-president in charge of equipment, reports to management should compare actual costs and forecasted costs on equipment in such a manner that they are not intermingled with costs which are not the responsibility of the vice-president in charge of equipment. Likewise, within the category of equipment costs, as much detail should be furnished to the vice-president responsible for such costs as he needs to fulfill his responsibilities effectively. The responsibilty for the operation of central repair facilities may be assigned to a superintendent under the vice-president; if so, the actual results of operating this facility should be compared with the forecasted results.

Proper cash forecasting is essential to efficient operations of a contractor, due to the fact that fund requirements for equipment acquisition, working capital, etc., must be coordinated with cash flow from the payment of estimates, collection of retainage on contracts, etc. Therefore, cash forecasts on an annual basis may not suffice, and a breakdown of the forecast into shorter chronological periods is normally required. Procedures employed in cash forecasting should include periodic review of forecasts and appropriate revision based on factors known at the time of the review.

The coordination of cost data and the processing of that data for financial statements should be examined. Source data, which are utilized both for determining cost and for making entries into the general ledger from which the financial statements are prepared, basically consist of such things as vendor invoices, payrolls, and material transfers. The procedures relating to the processing of these data should give consideration to the fact that source data may be processed (1) for contract cost reports at the job-site office and (2) for financial statements at the headquarters office. Hence, at any given date, data recorded for determining contract costs on contracts-in-progress must also be recorded for preparing financial statements. Where the contractor's procedures involve the maintaining of the contract cost ledger in the headquarters office, the reconciliation of contract costs in this ledger to contract costs recorded in the general ledger should be no problem. However, when the contract cost ledger is maintained on the job site, it is necessary to institute procedures providing for timely forwarding of data to the job office and the headquarters office.

Disbursements related to the payment of payrolls and vendor invoices may be initiated at the job or project office or at the headquarters office. When such disbursements are initiated at the job office, the company's procedure on the subsequent handling of these disbursements should provide for a review or audit of the appropriate supporting data.

21-7. Project Office Accounting—Contract Cost Accounting

The prime objective of project office accounting is the determination of contract cost. Related duties may involve the preparation of payrolls, processing of invoices for payment, payment of invoices and issuing of payroll checks, etc. Whether the disbursements are initiated by the project office or by the headquarters office, one of the prime duties of project office accounting personnel is to process payroll data and vendor invoices for payment. This processing includes indicating the information necessary to record the data in job-cost ledgers and in the general ledger.

Project office accounting personnel are also normally responsible for the preparation of reports on contract costs, comparisons of actual costs estimates, explanation of variances, etc. It should be clearly understood that the duties of the project accounting personnel related to the explanation of variances (original estimates versus actual cost) should be limited to the accumulation of data furnished by engineering personnel or the project manager, and to explanations related to accounting errors, such as the failure to include costs in the proper period.

Job costs should be accumulated in a manner enabling the preparation of reports which summarize for each contract the following data:

1. Total estimated price, including change orders
2. Actual cost to date
3. Estimated cost to complete
4. Total of items two plus three (total revised cost)
5. Original estimated cost adjusted for change orders
6. Estimated project profit or loss:
   (a) Original
   (b) Adjusted to date

(c) Adjusted to date as of last report (or end of prior fiscal year)
7. Project price collected to date:
  (a) This report
  (b) Last report
8. Project profit or loss to date:
  (a) This report
  (b) As of the end of the last fiscal year

When the contractor is utilizing the percentage-of-completion method of accounting for financial statements, the difference between item 8(a) and item 8(b) should represent the year-to-date contract profit or loss included in the financial statements.

The contract cost ledger should accumulate the data necessary to furnish the information on costs referred to above, insofar as possible, on a line item (or bid item) basis. The identification and designation of the line items would be furnished by the original estimate. In other words, it is necessary to correlate the accumulation of cost data on contracts with the details of cost used in preparing the original estimates.

Procedures related to the processing of vendors' invoices, payroll data, and other source data should include the appropriate coding of this data for entry into the contract cost ledger. Since the status of contracts-in-progress is determined by considering both actual cost to date and estimated cost to complete, the proper coding of source data by line item is an important factor. Improper coding of costs by line item may result in the determination of unit costs (based on actual costs to date) which are incorrect and the projection of estimated costs to complete based on such inaccurate unit costs.

The project cost ledger may be maintained in any one of many various forms. Ledgers maintained on a manual basis will vary in form from those involved in a system that utilizes EDP. Also, cost reports that are the output of a system involving the use of EDP may vary in form from reports that are prepared manually from manually maintained cost ledgers. In either event, good accounting procedures would call for the furnishing of somewhat comparable information in the cost report.

Figure 21-27-1 presents a simple form of a job cost report. It would be possible to condense the data presented in Figure 21-27-1 by eliminating the data related to the breakdown of cost on line items between labor, materials, equipment, subcontract, and other costs. The remaining information, however, is basic insofar as furnishing management with proper data on contract cost. It would also be possible to expand upon the job cost report to furnish additional data. Other data that might be furnished consists of:

1. A breakdown of total cost charged to the job between direct cost and indirect or overhead cost
2. A breakdown of the job status to date between the data related to the current month and the data related to the prior periods
3. An accumulation of man-hours and equipment-hours and related cost per hour, on a line item basis
4. In certain instances, a breakdown of the contract price to the appropriate line items involved
5. An indication, on a line item basis, of the percentage of completion
6. An indication of the percentage of actual to estimated cost
7. An identification of cost with changes in the original contract price through change orders

Besides the job cost report, the contractor's procedures may provide for the furnishing of various other reports from the job or project office. As an example, the procedures may involve the rendering of daily cost summaries which would indicate the cost for the specific date covered, the accumulated costs to date, the total budgeted, and the balance remaining to be spent. Correspondingly, the procedures may call for weekly cost statements covering items comparable to those referred to on the daily cost statements. It is possible that the contractor's procedures may call for daily cost data and weekly cost data referred to above, but limited to labor only.

Project or job office accounting personnel also would normally have duties related to the accumulation of equipment costs. These duties principally relate to the reporting of data rather than the accumulation of data. Accounting for equipment costs are covered by a subsequent section of this chapter.

21-8. Internal Auditing—Control Procedures

It is not uncommon for internal auditors to be considered, by uninformed parties, as "figure checkers." Naturally, in performing his assigned duties, the internal auditor will be required to employ certain clerical procedures which may give rise to this misconception. A well-designed and fully implemented internal audit program, however, is intended to furnish management with the assurance that established company policy is being followed and that appropriate procedures are being followed to safeguard the company's assets. In addition, the internal audit function should call to management's attention adverse operational trends noted in the course of a review of financial data, a need for a specific procedure to be adopted or changed, suggestions for changes in company policy dealing with financial management, etc.

The internal audit function may be performed to a great extent by personnel who need not be designated, by title, as internal auditors. This is particularly true in the case of smaller and middle-size companies. The auditing function, performed either by an internal staff or by outside public accountants, consists basically of the performance of tests to determine whether proper policies and procedures exist and are being followed. The extent to which these tests are performed depends upon the amount of internal control that is built into the company's policies and procedures. For example, if company policy provides that a receiving report is to be prepared on all invoices covering materials, the auditor will examine a sufficient number of invoices to see that this procedure is being followed. If this test indicates a substantial number of exceptions, the test may be expanded.

A detailed discussion of all the tests which an internal auditor may perform in the course of making an examination is beyond the scope of this chapter. There are certain control procedures, however, that are deemed worthy of discussion. Management should be furnished with a summary giving the basic information related to contract performance discussed earlier in this chapter. Also, management should be furnished with an exception report covering individual line items where actual costs vary from estimated costs in excess of specified percentage or dollar amounts. It should be management's prerogative to establish the percentage or dollar amounts which would represent acceptable variances.

Operating personnel (project superintendents or other responsible individuals) should be advised of the line items

JOB COST REPORT

Name ....

Date Started ....

Job No. ....

Location ....

Cost Code	Description	Job Status to Date			Estimates to Complete			Revised Estimate			Original Estimate			Variances	
		Units	Actual Cost	Unit Cost	Units	Cost	Unit Cost	Units	Cost	Unit Cost	Units	Cost	Units	Cost
0010-	Excavation—earth-work cu yd													
-01	Labor													
-02	Materials													
-03	Equipment													
-04	Subcontracts													
-05	Other													
	Total													

TOTAL COSTS:

-01	Labor													
-02	Materials													
-03	Equipment													
-04	Subcontracts													
-05	Other													

Original Contract Price

Additions on Change Orders

Profit (Loss)

Profit (Loss) To Date

Profit (Loss) Prior Fiscal Year

Profit (Loss) Current Fiscal Year

Figure 21-7-1

involved on the exception report furnished to management. Procedures should be adopted which would call for these individuals to submit comments to management, such as the reasons for variances. These comments should be made part of a permanent file. Any further inquiry or action should then be instigated by management.

The exception report should basically include the following data:

1. Contract identification
2. Original estimated profit or loss
3. Revised estimated profit or loss
4. Excessive line item variances:
   (a) Line item identification
   (b) Original estimated cost
   (c) Actual cost to date
   (d) Estimated cost to complete
   (e) Total of (c) and (d)
   (f) Variance between (b) and (e)
   (g) Acceptable variance
   (h) Excessive variance, (f) minus (g)

A substantial amount of internal auditing can be accomplished by the use of cost comparisons. As an example, if actual unit costs on a line item to date have amounted to $1 per unit, and the job cost report indicates that estimated costs to complete are based on a cost per unit of 50¢, an inquiry should be instigated. Within specific categories, a comparison of current expenses with corresponding expenses for the preceding year, or with a predetermined forecast, may indicate a necessity for further investigation to determine the reason for variances.

21-9. Accounting for Equipment Costs

The fundamental objectives of maintaining adequate equipment records are to furnish data on proper cost distribution to projects and the accumulation of historical cost data which can serve as a basis for (a) estimating costs, (b) making repair or replace decisions, and (c) comparing costs on different makes of comparable equipment. One of the first decisions should be on the types of equipment covered by the record-keeping system. Major units of equipment (tugboats, barges, cranes, dozers, etc.) should certainly be covered. At some point, however, the cost of obtaining data on smaller units could exceed the value of the data obtained.

Equipment costs are normally charged to project costs on the basis of predetermined unit rates or unit costs. The determination of these unit rates or unit costs is based on data collected on a unit basis related to:

(a) Measurement of usage—hours, days, weeks, months, etc.
(b) Cost related to such usage

Cost may be segregated into two categories, those incurred as a result of having the equipment available for use (ownership) and those incurred as a result of operating the equipment (operating). Ownership costs would cover such items as depreciation, rental, taxes, licenses, insurance, etc. Maintenance labor, outside repairs, parts, tires, fuel, and similar type costs would be considered operating costs. The extent to which costs are segregated should be determined only on the basis of a review of the data required by management.

Ownership costs should be converted to basic daily, weekly, or monthly rates. This conversion can be accomplished as follows:

Cost of unit	$70,000
Estimated life	6 years
Estimated salvage value	$10,000
Estimated usage while assigned to jobs	9 out of 12 months
Cost of taxes, licenses and insurance	$ 6,200 per year
Assigned rates:	
Annual costs involved:	
Depreciation	$10,000
Other	6,200
	$16,200
Costs per day:	
Estimated usage 75% of 360 days	270 days
Per day ($16,200 ÷ 270)	$ 60

Initially, operating costs such as maintenance, etc., are estimated for a chronological period, together with the number of hours the equipment is to be used during such period, and the rate established as follows:

Estimated costs per year	$ 8,000
Estimated hours of use per year	800
Cost per hour	$ 10

After a sufficient period of time has elapsed, rates should be based on historical costs and historical utilization.

Vendor invoices, labor distribution reports, journal entries, cash disbursements, etc., furnish the source data from which costs are accumulated. Note again that accumulated costs by operating units should be reconcilable to general ledger control accounts. To accomplish this it may be necessary to segregate costs related to equipment not covered by the system in separate accounts.

Certain decisions will have to be made in connection with the accumulation of costs related to the operation of equipment:

1. Whether labor costs related to operators of rolling stock should be charged to project costs or to operating costs on the equipment. The author believes such costs should be charged to the project rather than to the equipment.
2. A decision must be reached as to whether the project or the equipment should be charged for fuel costs related to operation of the equipment. One procedure is that which involves charging line item cost for operating costs on the equipment at a rate which includes a predetermined factor per hour for fuel costs with an offsetting credit (equal to the fuel cost factor) to job overhead. Under this procedure actual fuel costs would then be charged to job overhead. If it is assumed that the hourly rate for fuel is reasonable, actual fuel costs (adjusted for inventories) should approximate the credit arising in overhead for hours of usage. Any substantial variance could indicate a breakdown in reporting procedures, unreasonable rates, or the improper recording of costs.
3. Equipment repairs can be classified into two different

types, those done at repair facilities and those done at the project site. Cost applicable to the maintenance of equipment repair facilities and labor applicable thereto should be distributed to the units on the basis of job orders. These costs then enter into the determination of the hourly rate covering operating costs on the equipment. Equipment repaired at the job site should be charged to the appropriate equipment unit costs. It is possible to have an exception to this procedure by charging the cost of repairing wrecked equipment in excess of any insurance recovery to project costs.

The determination of unit costs of equipment per hour, day, week, etc., requires the accumulation of hours, days, weeks, etc., by units. The accumulation of assigned days, i.e., the days that units are assigned to projects, can result from "checking-in" and "checking-out" procedures. This results in the accumulation of a number of days which a unit is assigned to a project and is no longer available for any other project. Hours of operation can be accumulated from equipment time cards. Project accountants could furnish the project foreman time cards for each unit of equipment. These cards would be turned in weekly (along with labor time cards) indicating daily hours of usage.

Ownership cost may be charged to the project on the basis of ownership daily rates and included in the job or project overhead. It is also possible to convert 50 percent of ownership costs to an hourly operating cost and then charging same to line item project costs, on the assumption that ownership costs are related to actual usage as well as to availability of equipment.

Operating costs are normally charged to the project on the basis of operating hourly rates. These costs are assigned to project line item costs.

The accounting system should minimally produce the following types of data in reports furnished to management:

1. Daily (or ownership) and hourly (operating) costs by unit.
2. A comparison of predetermined rates based on estimates, or based on historical costs and utilization during prior periods, with the costs referred to in item 1.
3. An explanation of variances, e.g., low utilization may be producing high ownership costs per day, or unanticipated maintenance costs may result in higher operating costs per hour.
4. Overabsorption or underabsorption of costs related to the maintenance of central repair facilities and personnel. A substantial underabsorption (based on rates which are reasonable) could indicate a potential problem area.

21-10. Evaluating Cost Trends and Cost of Future Work

The project or job cost ledgers can be designed to furnish data very valuable in evaluating cost trends and in determining estimated cost of future similar work. The job cost reports may be so designed that cost data for the current month includes the presentation of percentages of actual to estimated cost. These percentages for the current month can be compared with comparable percentages for the previous month or months. Assuming the data are available in the cost ledger, these comparisons can be made for labor, equipment, material, subcontract, and other costs. When the percentage of actual cost to estimated cost increases on a continuing basis from month to month, a serious problem may be indicated.

The job cost ledgers can furnish estimators with historical data of invaluable assistance in estimating the cost for bidding on future work. As an example, appropriately designed cost ledgers can furnish basic data related to specific line items, such as man-hours involved, equipment hours involved, or units of material. In estimating the cost of comparable future work, the estimator may utilize this historical data and arrive at the estimate of the future work by application of such things as current prices and rates.

21-11. Utilization of Computer Techniques

Electronic data processing is utilized by contractors of all sizes. Contractors whose operations involve, for accounting purposes, a large volume or number of transactions will find it more advantageous to utilize computer techniques than will contractors whose operations do not involve as large a volume of transactions. The decision to use computer techniques should be based on an evaluation of information which can be furnished by the application of computer techniques compared to the cost of applying these techniques.

It is possible to maintain all accounting and cost ledgers on EDP. However, the most commonly used computer technique is that related to the handling of payrolls. It is not necessary that the contractor own and operate his own equipment to take advantage of the techniques available for the processing of payrolls by EDP. Commercial banks and other organizations offer the required services for a negotiated fee. In connection with the processing of payrolls, the utilization of EDP should furnish the information required to:

1. Maintain appropriate payroll records
2. Distribute payroll costs to contracts in progress by line item.
3. Accumulate, where appropriate, man-hours by line item cost.

The application of computer techniques in the area of contract cost and equipment cost may be considered essential when the number of transactions involved is voluminous. In these cases, the use of these techniques is required in order to render reports on a timely basis.

In the course of evaluating the feasibility of adopting EDP, due consideration should be given to the potential growth of the company's operations. Such consideration should cover not only growth in the number or volume of transactions to be processed, but should also cover growth of management's requirements for information. Where a company's operations are not widespread and the number of contracts involved is not substantial, top management may stay abreast of the company's operations simply by being in close contact with the jobs in process. When, however, the number of jobs in process covers a wide geographical area and the number of contracts is substantial, management will be unable to keep abreast of the company's operations by being in close contact with the individual jobs in process; thus, management's need for information which can be produced by accounting and cost records becomes more acute. Therefore, prior to reaching a decision as to whether the adoption of EDP processing or computer techniques is advantageous, appropriate consideration must be given to the potential growth in the company's future operations.

21-12. Conclusion

At the beginning of this chapter, it was stated that the basic functions of accounting systems and procedures are to gather, process, and report on financial information. Throughout the chapter continual reference was made to the end product of the systems and procedures—the reports. The discussion covered basically two types of reports: those that are distributed to management and other parties (bankers, creditors, stockholders, etc.), the financial statements; and those that are prepared basically for the benefit of management, such as the report on contracts-in-progress. The functions of the accounting department and the end product produced by this department point to the fact that the accouning department is a service department performing a staff function. Hence, it is of extreme importance that accounting department personnel recognize that cooperation with personnel of other departments is essential to the efficient performance of their duties.

22 PAYROLL PROCEDURES

DONALD G. PERRY

Partner
Lybrand, Ross Bros. and Montgomery
San Francisco, California

OF ALL the costs involved in a construction project, payroll is the one of greatest concern to the contractor. Materials, equipment, components, and subcontractors can be fairly definitely priced prior to submission of the bid. Labor as charged through the payroll is the one element most difficult to control. The payroll system, therefore, is one of the key elements to a contractor's management of his construction.

22-1. Payroll System

A payroll system includes not only the payment of wages but also involves the accumulation of time and the distribution of both payroll and payroll-related costs. Through a well-run and efficient payroll system a contractor can tell very quickly whether or not he has problems in his labor productivity.

22-2. Timekeeping

Timekeeping or the recording of hours worked can be done in a variety of ways. Each person on the payroll can punch a time clock on and off the job. Another method is to have timekeepers, who log the workers on and off the project. Generally, when this is done there is a check from time to time during the day to ascertain that all the men are still on the project.

A refinement of the timekeeping function is the distribution of labor costs. On large projects the work done by the laborers will generally be chargeable by their labor craft. However, there may well be instances in which the contractor will want a more detailed breakdown of where the time is being charged. For example, he will want to know costs of doing extra work, costs related to unknown conditions, or costs resulting from change orders to the original contract. These time records may either be kept by the timekeeper or may be noted by the men on their own time cards.

22-3. Accumulation of Payroll Costs

The accumulation of payroll costs may be either a very simple system of accumulating man-hours of work or it may involve the detail breakdown by dollars as well. On most projects the original forecasts and estimates were set up on the basis of labor hours. Therefore, the control is generally on the same basis. With a manually operated system the time is reported to the office. It will be accumulated in the same categories as the hours were put into the original estimate. By maintaining cumulative totals of hours and comparing hours expended to original estimate in relation to the physical completion of the work, the contractor will know how productive his labor force is. Figure 22-3-1 is an example of a manual accumulation of labor hours.

22-4. Use of Electronic Data Processing (EDP)

On larger contracts the use of EDP equipment may become feasible. There are a variety of ways in which this may be utilized. First, the job site may have its own EDP equipment on hand. Second, a service bureau in the area may be used to prepare EDP records. The third use of EDP would be to have the data processed at the contractor's central office, using one of the various types of data transmission equipment available. In any case, fairly immediate output would be needed and the labor distribution should be available on a fairly frequent basis. Use of EDP equipment should enable production of these records with dollar costs as well as labor hours. An example of a EDP produced report is given in Figure 22-4-1:

22-5. Payment of Salaries and Wages

In addition to the distribution of costs, the payroll department also handles payment of salaries and wages. This may be by check or cash, and entails not only computation of gross pay but also the various deductions for taxes and

#	DATE	W-1 HM. FRAMES 700	W-3 WOOD DOORS & HDWE 1700	W-10 BLOCKING 3317	W-24 ROUGH FRAMING 2262	W-25 EXT. TRIM (NO DOORS) 1350	W-26 INT. TRIM (NO DOORS) 2483	W-32 WARDROBES 100		
1	1/28/									
2										
3	3/25/59	52 18								
4		52 18								
5	4/1/59	21 67								
6		73 75								
7	4/8/59	120 01								
8		193 76								
9	4/22	12 95								
10		206 71								
11	5/6			35 32						
12		206 71		35 32						
13	5/13			224 54	120 67					
14		206 71		259 86	120 67					
15	5/20	8 83			230 63					
16		215 54		259 86	351 30					
17	5/27	85 91		23 94	155 61					
18		301 45		283 80	506 91					
19	6/3	72 67		265 04						
20		374 12		548 84	506 91					
21	6/10	32 77		303 24						
22		406 89		852 08	506 91					
23	6/17	47 88		151 62	63 84		399 00			
24		454 77		1003 70	570 75		399 00			
25	6/24/59	63 84		383 04			319 20	47 88		
26		518.61		1386 74	570 75		718 20	47 88		
27	7/3/59			15 96	47 88	191 52	31 78	31 92		
28		518.61		1402 70	618 63	191 52	749 98	79 80		
29	7/10/59			89 30		536 07	72 28			
30		518.61		1492 00	618 63	727 59	822 36	79 80		
31	7/17/59			26 49		221 15	372 94			
32		518.61		1518 49	618 63	948 74	1195 20	79 80		
33	7/24/59			67 24		163 00	502 23			
34		518.61		1585 73	618 63	1111 74	1697 43	79 80		
35	8/3/59	35 32		35 32		206 12	216 48			
36		553.93		1621 05	618 63	1317 86	1913 91	79 80		
37	8/7/59		94 09				164 23			
38		553.93	94 09	1621 05	618 63	1317 86	2078 14	79 80		
39	8/14/59		93 34				40 11	93 59		

Figure 22–3–1 Analysis of labor cost—manual

fringe benefits. A statement showing the computation should be included with each payment.

22-6. Preparation of the Payroll

Preparation of the payroll for payment involves accumulating time charged since the last pay period, computation of the straight and premium time pay, and computation of the various deductions to be made. These deductions are either of a statutory or a contributory nature. Statutory deductions comprise those amounts which must be withheld by law, including federal and state income taxes, Social Security taxes, and other required deductions. Contributory type of deductions may or may not be voluntary, depending upon various circumstances, such as union contracts and the like. These include hospitalization, retirement, union dues, and credit union deductions.

ANALYSIS OF LABOR COST

Estimate Quantity	Estimate Unit Cost	Estimate Cost	Code / Description	Actual Qty in Place	Actual Labor Cost	Unit Cost	% Complete	To Date Projected Labor Cost	Difference	Carpenter Actual	Carpenter Estimate	Mason Actual	Mason Estimate	Laborer Actual	Laborer Estimate	Days Est.	Days Actual	Days Diff.
		8,000.00	8A MILLWORK INSTALLATION-JOB		2,122.05				8,000.00	519.00				216.00				
		11,200.00	8B MILLWORK FABRICATION-YARD		839.40				11,200.00			4.00		3.00				
16,000.00	.40	6,400.00	19 ERECT MISC. IRON & STEEL	7,750.00	1,505.50	.19	.48	3,040.00	3,360.00			280.00		92.00				
		4,200.00	21 FLASHING						2,100.00									
		1,400.00	30 HARDWARE						1,400.00									
		1,090.00	31 HOLLOW METAL		2,918.70					536.00								
		2,600.00	44 INSULATION						2,600.00									
		1,020.00	76 MISC. SPECIALTIES						1,020.00									
			62 TOILET ACCESSORIES															
		10,000.00	102 LAYOUT		503.10					89.00				28.00				
		3,000.00	103 TEMPORARY HEAT		8,520.13				3,000.00					107.50				
		23,900.00	107 FORKLIFT OR TOWER		16,359.40				23,900.00	1,797.00				2,001.50				
		9,800.00	112 SUPERVISION		16,267.37				1,500.00	24.00								
		6,200.00	115 FINAL CLEANUP		297.70				6,200.00	56.00				66.00				
		500.00	120 PUNCH LIST							4.00								
			121 YARD		3,058.99									744.00				
100.00	5.00		201 CEMENT BLOCKS	20.00	57.60	2.88	.20	288.00	212.00			8.00	16.00					
560.00	19.15	10,725.00	306 HAND EXCAVATION		9,748.80	19.15								2,275.00				
620.00	3.00	1,860.00	403 CONCRETE-WALLS & GR BEAMS	334.00	934.40	2.79	.53	1,729.80	130.20					235.50	496.00			
286.00	3.00	858.00	405 CONCRETE-SLABS ON GROUND	41.00	124.60	3.03	.14	866.58	8.58-					33.00	229.00			
2,154.00	3.01	6,491.00	406 CONCRETE-SLABS & BEAMS	1,988.00	5,744.13	2.88	.92	6,203.52	287.48			37.50		1,406.50	1,731.00			
551.00	6.31	3,478.00	408 CONCRETE-COLUMNS		2,283.30	4.39	.94	2,418.89	1,059.11					164.50	927.00			
75.00	25.00	1,875.00	412 CONCRETE-STAIR FILL & STEPS	85.50	622.58	7.28	1.14	546.00	1,329.00	72.50		1.00		80.50	500.00			
125,606.00	.10	13,159.00	417 FINISHING	111,300.00	10,202.81	.09	.88	11,304.54	1,854.46			1,521.00	2,309.00					
122,600.00	.13	16,340.00	419 RUBBING	89,000.00	12,620.67	.14	.72	17,164.00	824.00-			1,822.50	2,867.00					
128,100.00	.02	2,562.00	420 PROTECTION & CURING	18,800.00	517.34	.02	.14	2,562.00						740.50	683.00			
21.00	5.71	120.00	423 CONCRETE-MATS, ETC.	36.00	230.20	6.39	1.71	134.19	14.19-					47.00 / 62.00	32.00			
527,000.00	.17	90,800.00	601 FACE BRICK	497,800.00	77,882.64	.15	.94	79,050.00	11,750.00			9,511.00	10,667.00	6,559.50	8,000.00			
29,200.00	.54	15,768.00	603 MASONRY BACKUPS 4"	19,840.00	12,145.11	.61	.67	17,812.00	2,044.00-			1,125.00	1,756.00	1,539.00	1,536.00			
13,640.00	.54	7,451.00	604 MASONRY BACKUPS 6"	17,650.00	10,041.58	.56	1.29	7,638.40	187.40-			1,143.00	830.00	922.50	726.00			
63,080.00	.63	40,236.00	607 MASONRY PARTITIONS 4"	52,950.00	44,631.90	.84	.83	52,987.20	12,751.20-			5,286.50	4,258.00	4,009.50	4,258.00			
12,445.00	.87	10,940.00	608 MASONRY PARTITIONS 6" & 8"	7,240.00	3,980.73	.54	.58	6,720.30	4,219.70			564.00	1,158.00	244.50	1,158.00			
3,400.00	.35	1,203.00	611 FIRE BRICK	58.00	45.00	.74	.78		1,203.00			8.00	141.00		106.00			
74.00	4.50	333.00	614 FLUE LINING		45.00		1.00	54.76	278.24			8.00	37.00		32.00			
69,750.00	.22	15,345.00	615 SCAFFOLDING	69,750.00	24,651.63	.31	1.00	21,622.50	6,277.50-	3,092.50			2,312.00	1,702.00	578.00			
1,434,444.00	93.01	512,097.00	TOTALS	1,100,459.75	411,814.50	43.68	23.41	371,832.80	140,264.20	25,437.50	28,320.00	21,327.50	26,350.00	37,891.00	32,913.00			

Figure 22-4-1   Analysis of labor cost—EDP

22-7. Cumulative Records and Type of Payment

Payroll records include the accumulation of the current payroll and the amount to be paid to each employee, as well as cumulative records by employee and by type of payment. This means accumulating taxes and other deductions withheld both by employee and by type of withholding. Federal taxes are required to be paid currently. Others may be paid quarterly. Other items withheld from employees will have to be remitted as required. A company-operated benefit plan may well have requirements different from union operated plans, although in all cases these funds should be remitted very promptly.

22-8. Withheld Amounts and Employer Contributions

In addition to the withheld amounts there normally are contributions required from the employer. These may be either a direct matching of the amounts withheld or may be computed on a different basis. Determining these amounts will depend upon the type of employee benefit plan involved. Although mention has been made of company and union plans, there are also employee benefit plans which are operated by employer groups rather than individual companies. These arise generally when there is an industry-wide bargaining unit that will administer these welfare funds for the industry as a whole.

Payroll costs, therefore, include not only the total payroll but also the cost of the employer portion of the fringe benefit package and the matching of the items withheld. Other elements of cost may be noncontributory fringe benefit plans which are often negotiated under union contracts whereby union members receive medical or retirement benefits without contributing directly themselves. In addition there are certain costs which are based on payroll, such as workmen's compensation and liability insurance.

22-9. Company Expense and Allocation of Overhead

The company expense may be built into the payroll charge for labor distribution or may be applied separately to total construction costs. Through the use of EDP equipment, the labor distribution can easily include such costs as an added factor. In a less sophisticated system these costs may be included as an element of overhead. The overhead, of course, can be applied to the direct labor costs and allocated to the different sections of the construction contract.

22-10. Methods of Computing Payroll and Labor Distribution

There are various methods of computing payroll as well as labor distribution. One very simple method is to have all payroll factors shown on the time card and have the payroll prepared directly from this source. In this example the elements making up net payroll can be shown on the back of the time card to serve as a reminder to the payroll clerk that all items have been considered.

22-11. Payroll Journal, Earnings Statement, and Employees' Earnings Record

In addition to the payroll journal there are other records which are products of the payroll system. The first of these is the earnings statement. This is given to the employee with his payment and shows the gross pay less deductions and the amount of payment he is receiving. This enables the employee to see that the payroll has been computed correctly.

Also required is an employee earnings record. This accumulates the total payroll history for each employee, including all of his deductions in order to prepare Social Security and withholding statements. Since Social Security earnings must be reported quarterly, these employee earn-

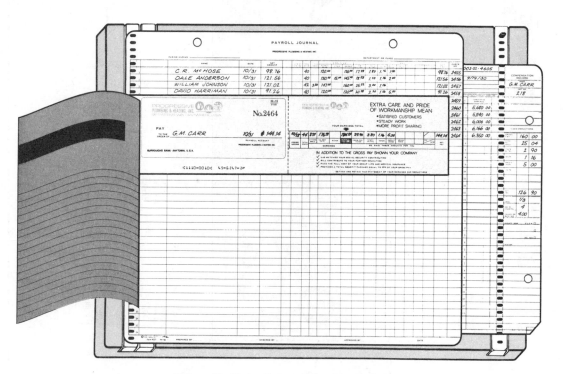

Figure 22-12-1 One write payroll system

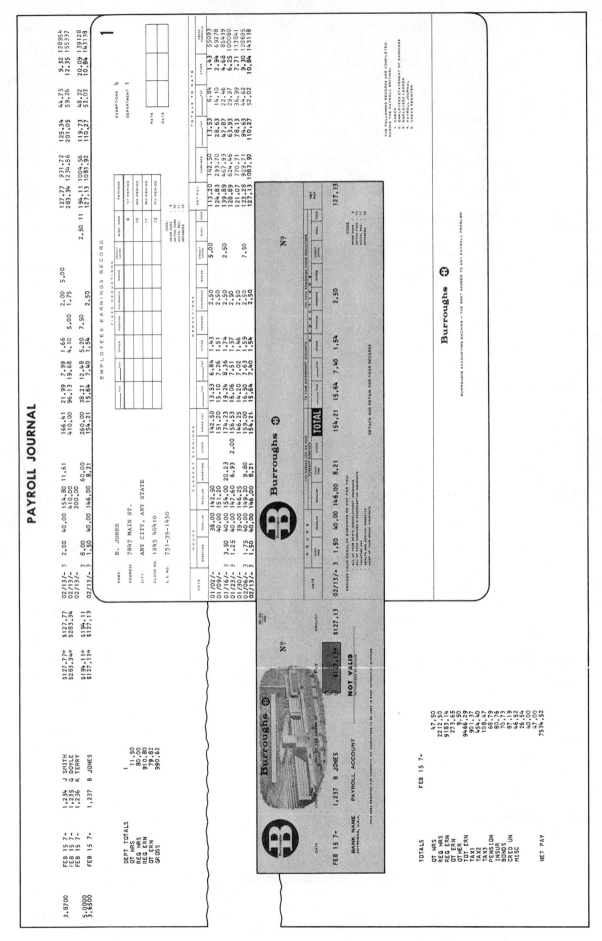

Figure 22-13-1 Mechanized payroll system

ings records are generally accumulated on a quarterly basis.

Since there are in essence three separate records which must be produced from the payroll system, i.e., earnings statement, employee earnings record, and payroll journal, various means have been devised to prepare these simultaneously.

22-12. The Manual System

The first of these is on a manual system. An illustration of one of these is given in Figure 22-12-1.

Under this system, the earnings statement has a carbon backing which reproduces on the payroll journal. Behind the payroll journal the earnings record card is inserted under a second carbon. By a single writing, therefore, the earnings statement, payroll journal, and earnings record are produced simultaneously. Through use of the employees name or employee number the earnings record can be controlled in that the same identification should appear on each line of the card. By only having to write the information once, errors in transcription have been eliminated. There still, of course, would have to be a close review and check on the original entry. This can be done through review and approval of the payroll journal. The final check is maintained through the signature on the payroll check itself.

22-13. Mechanized Payroll Preparation

The next level of sophistication in payroll preparation is mechanization. Under this method, essentially the same information is produced as in the pegboard manual system. However, by use of a bookkeeping machine with various memory banks, a number of the clerical functions can be replaced, primarily in the areas of addition and subtraction. Most bookkeeping machines can add across as well as in vertical columns, thereby accumulating gross pay and withholdings, and computing net pay and accumulating it. Through simultaneous printing the earnings statement, payroll register, and earnings record can be produced with one typing. Such a machine is shown in Figure 22-13-1.

The next level of sophistication arises through use of EDP equipment. Here the degree of input required can vary tremendously.

22-14. Methods of Applying Data Processing to Payroll Preparation

Under a rather simple system the employee name and hours can be entered with a prepunched card giving rate and withholding information. Under this system, any changes in the control information of rate and withholding have to be made in the card itself. The current week's payroll information and the master deck have to be collated in order to function properly. Loss of a card or having cards out of sequence can cause problems under this method. Under a more sophisticated method the employee master tape will have all the employee data on file available for use in computing the payroll. Under such a system a master tape will have employment information including earnings history, retirement plan vesting, and other similar information and will mantain the earnings record, as well as currently effective hourly rates and withholding information. Under this system the input data would be limited to employee identification by name or number and the pay period time worked. With such a system the distribution can be made by also giving the assignments on which the employee worked.

22-15. Use of a Bank Computer System

There is one other method that should be described, although it does not have wide application to construction projects. This is the use of a bank computer system to prepare payrolls. The reason that this has limited application in construction is that it is most functional where there is little variation from period to period in the hours worked or the employees on the payroll. With the rate of turnover experienced on construction projects and the variation which can exist in the hours worked, generally this system has limited application in the construction field. In small constructon, however, this can be of value to the contractor. Under this system, the bank keeps payroll records, prepares the tax returns, and generally takes the clerical burden away from the contractor. The contractor, furthermore, does not have responsibility for the payroll bank account or keeping track of outstanding checks and for preparing the annual withholding statements. In addition, it can be attractive to the employee in that the bank will deposit his earnings directly to his own bank account without his having to make out a deposit slip or go to the bank. The cost of this system being on an individual check basis, generally is excessive where the volume of employees working on a construction project is large. Therefore, its use is limited since, with large volume, a contractor can afford to have his own payroll department. In addition, there may be a delay in getting payroll distribution or, in many cases, distribution is not available from this system since it is primarily to relieve the burden of payroll disbursement and is not involved in distribution of cost which, of course, is of prime importance to the contractor.

Payroll is one of the most vital costs which the contractor has to control, and to do so he must have current and rapid payroll information available. The payroll system therefore is one of greatest importance to the contractor.

23 EMPLOYMENT PRACTICES AND RECORDS

R. E. BERNARD

Vice-President
Kaiser Engineers
Oakland, California

A CONSTRUCTION project is an organization of specialists, headed by the project manager. Included in the organization are various field engineers, scheduling and cost control analysts, estimators, quality control and material control experts, labor and craft supervisors, skilled craftsmen, and many others, all with a broad range of experience within their individual specialties and skillfully organized to work within the framework of a tightly scheduled project contract. It is a business of temporary employment and temporary assignments. Skilled craftsmen, hired by the hour and working under union agreement, complete their job specialty as needed during the construction phase of the project and are then laid off, to be rehired again by another company for another project. Even the smaller semipermanent staff of engineers and construction management personnel is on temporary assignment until the completion of the project, whereupon they must be reassigned to a new project.

Success and growth of a construction company in such a competitive business requires a diversity of project experience, involving more and different job specialties and taking their project teams on various job assignments throughout the country and throughout the world.

For the personnel department of such a company, this diversification of skills and constant state of flux requires special procedures and practices to meet the changing (almost day to day) personnel needs of the various projects. Proper operation requires a coordinated and harmonious effort with project management to determine their personnel needs. In addition, a flexible but systemized procedure is needed for finding, placing, and replacing needed personnel quickly to the close scheduling of the contract.

This chapter deals with personnel procedures and practices as they aid the planning, progress, and completion of a construction project. This involves manpower planning, determining manpower availability within and outside the company, selecting and assigning personnel to a specific job, and maintaining up-to-date personnel records for project and future use. Aspects of foreign projects are considered, showing differences in procedures when they exist. In addition, the chapter covers general procedures for both large and small companies (or large and small projects). In the case of the small company, greater responsibility for personnel actions is placed in the hands of the project manager, rather than the personnel department, for greater efficiency and economy of project operations.

This chapter is not meant to give detailed "how to" procedures for interviewing, hiring, orientation, upgrading, transfer, layoff, and dismissal, but does discuss these actions as seen from a management viewpoint, thus describing how the personnel department fits into the overall functioning of the company.

Closely allied to the personnel department functions is the personnel development activity which can be visualized as the process of meeting future company manpower needs through planned development of individuals within the company as opposed to recruitment and hiring from the available labor market. This subject is discussed in a separate section.

Pervading all personnel practices are the special considerations directed toward ensuring that equal opportunity for employment is afforded to all segments of society. This also is discussed in a separate section.

Lastly, personnel procedures involving company benefits, salary administration, and labor relations are not discussed, since they are functions of the industrial relations department, of which the personnel office forms only a part.

23-1. Employee Classification

Some of the technical terms used throughout the chapter to classify the various types of employees include:

Exempt—(meaning exempt from the Fair Labor Standards Act) Professional engineering and management personnel who are not required by law to be paid for their overtime services.

Nonexempt—clerical employees and technicians who are required by law to be paid for overtime services.

Manual Employee—skilled craftsman, generally a member of a local union.

Salaried Employee—Exempt and nonexempt employees, hired on a fixed salary and paid weekly, biweekly, or twice a month in most companies.

Hourly Employee—Manual employee, hired on an hourly basis of a fixed rate established by collective bargaining agreements.

23-2. Responsibility for Employment Functions

The personnel department of a construction company is typically a service activity to technical and administrative management providing the following:

- Manpower planning
- Recruitment
- Hiring and termination
- Orientation
- Upgrading and transfer
- Employee record keeping
- Personnel development

As shown in Figure 23-2-1, it forms a part of the industrial relations department, which also takes charge of labor relations, salary administration, employee benefits, employee safety, and project security.

The personnel department of a construction company, however, is also very different from that in any other business or industry. The difference lies in the fact that the construction company is the sum of its field projects. where most of the activities of a personnel office really take place. The field projects may be large or small, established locally or across the country, possibly even on the other side of the world. Because personnel activities exist at every project site and the size and location of the projects differ greatly, personnel functions may be executed by the home office, the field office, or, in many cases, other project departments.

On a very large project located remotely from the headquarters, it is common practice to set up a separate field industrial relations group to carry on the duties and responsibilities of the central offices, including personnel.

On smaller projects encountering no labor relations difficulties, there may not be a field industrial relations department. If the project is sufficiently large to support its own payroll office, the personnel duties are taken over by the accounting office manager. If not, the field project manager or his administrative designate normally assumes the responsibilities of personnel.

Personnel responsibilities in a construction company, therefore, are fragmented and widely distributed, depending on the size and location of the project. Consequently, it is important to distinguish between personnel practices at the central offices and those in the field, and the coordinated activities of both offices during the planning and construction phases of the project.

Manpower planning and personnel development are necessarily central office responsibilities since they involve the employees of the entire company, including all field projects. The other activities of recruitment, hiring, employee processing, and record keeping must be performed both at home and in the field.

The remaining sections of this chapter will consider manpower planning, personnel development, and Equal Employment Opportunity (EEO) as separate sections carried out by the central personnel department for the benefit of all company personnel. Other sections show the coordinated effort between the central and field personnel offices during the planning and construction phases of the project.

23-3. Resources for Manpower Planning

Finding the best qualified personnel for the start of a new project is the major task of a construction company's personnel department. Figure 23-3-1 shows the typical organization chart for a large project, indicating the various positions of key project personnel. Each position is filled by a technical specialist with years of experience in his field. A recruitment effort of this magnitude would be impossible, unless the personnel manager had earlier developed ready sources of manpower supply.

Outside of the company the manager's most important sources are personal contacts in the various public and private agencies, other companies, and the colleges and universities. Professional and trade associations offer another source of manpower and oftentimes provide a roster of available members whom they help to place.

In the category of the skilled craftsmen, the almost exclusive source of supply in the United States is the local union "hiring hall," whose business representative is the bargaining agent for the craft. This, however, is a special personnel function, handled by the Labor Relations Department of Industrial Relations.

Advertising is another major manpower source, especially for recruiting large numbers of personnel at the start of a project. A classified advertisement, placed in a local

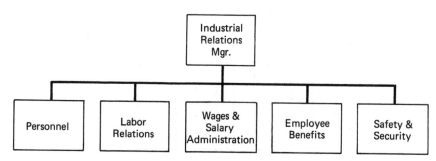

Figure 23–2–1 Personnel department within the organization structure of industrial relations

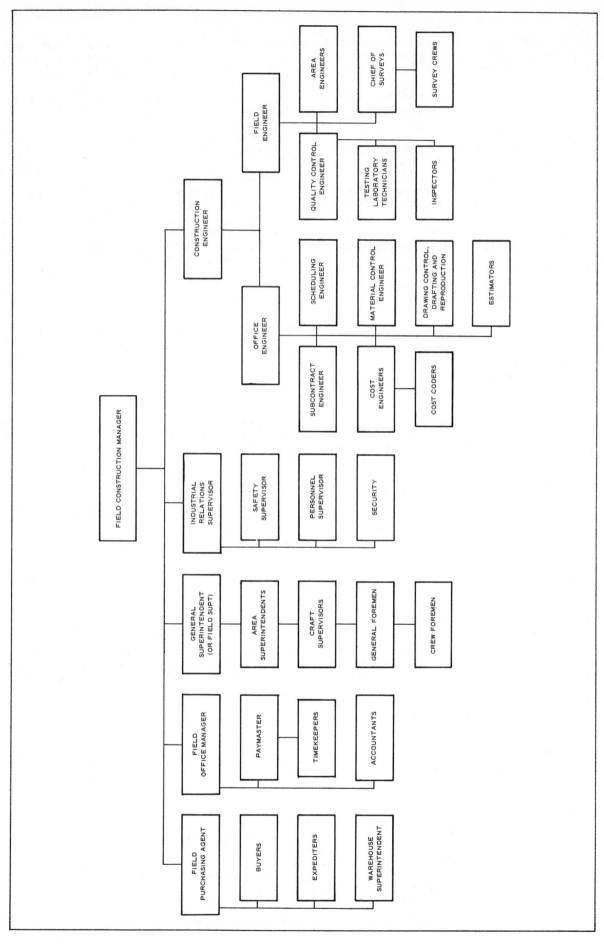

Figure 23-3-1 Typical personnel organization chart of a large construction project

newspaper, is the most practical means of recruiting the needed clerical and other nonexempt skills for the project. For professional management and engineering personnel, however, an area or nationwide advertising campaign is normally conducted, either through the various trade publications, local area daily newspapers, or nationally circulated newspapers, such as the *Wall Street Journal.*

One of the best sources for job-site recruiting of nonexempt personnel and a good source of exempt personnel in some areas is the local state employment office. These offices are usually very cooperative and in many cases will handle skills testing and preliminary screening for nonexempt applicants.

Although the above methods represent excellent sources for manpower recruitment, the most effective planning sources lie inside the company, within the personnel department's own records. The central office of personnel maintains records on all company employees, as well as on all individuals applying for employment. These records, when properly organized, can work for the company during early manpower planning of the project.

The remainder of this section will discuss three basic methods of manpower planning available within the company: the applicant file, skills inventory, and availability lists.

23-4. Applicant File

A medium-sized construction company may process up to 400 solicited and unsolicited applications every month. Over a three-year period, which is the legal requirement for maintaining an application, this means an accumulation of more than 10,000 applications. If these applications are filed in categories matching the general job positions of the projects, they can greatly facilitate the recruitment effort. One category, for example, may contain as few as ten or as many as 100 applicant folders, which are arranged alphabetically by the applicant's name and which contain his formal application, résumé, and any pertinent correspondence.

During the course of a new project or a project expansion, when a special job skill is called for, the personnel department can go immediately to the particular job category and retrieve all the applicant folders for the project

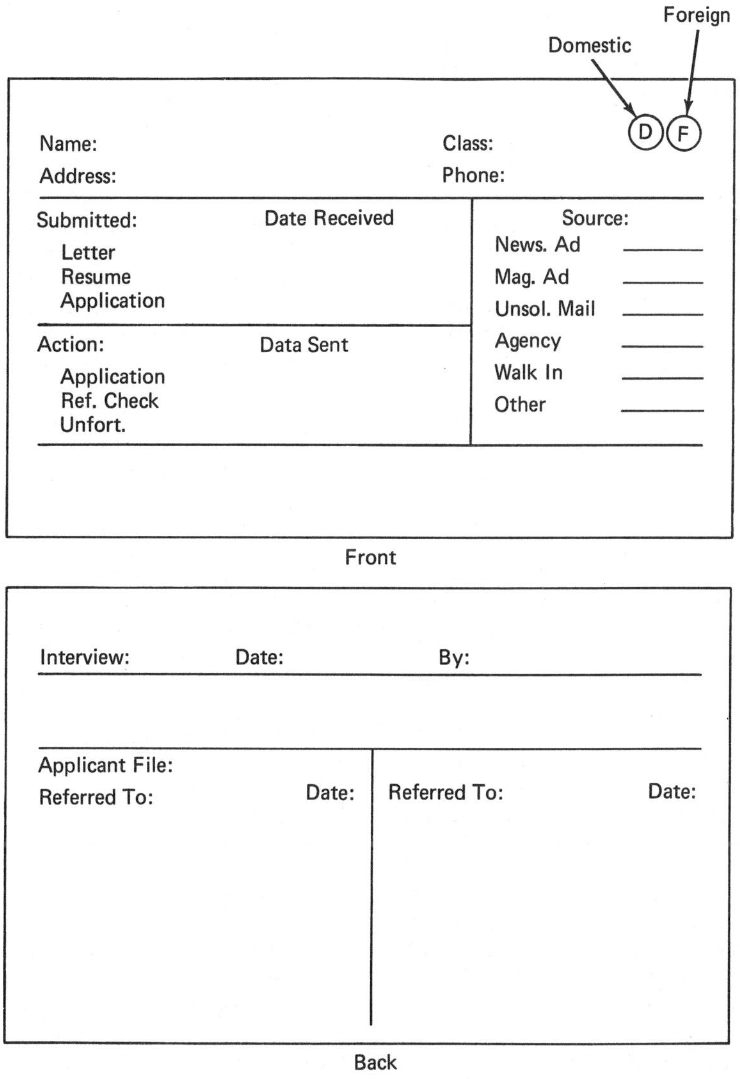

Figure 23–4–1 Locator card file alphabetically lists all job applicants and contains pertinent information on each individual for quick reference

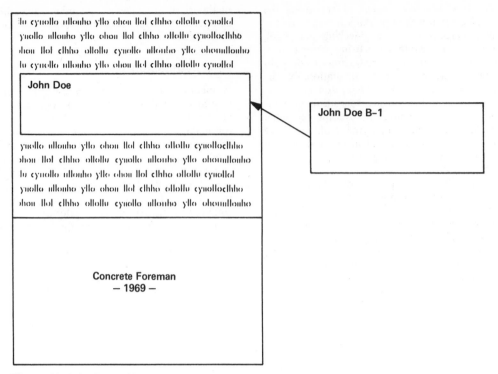

Figure 23–4–2 Desk-side locator card file is the key in retrieving applicant folders that are filed by job category

manager's examination and evaluation. Upon selecting the most suitable candidates the personnel department then contacts the individuals to determine their availability and interest in follow-up interviews.

In some cases the personnel director or project manager may have a particular individual in mind and may want to review his application. In this case, a locator card, prepared during the applicant's first interview, is used to determine in which category the applicant's folder is located.

The locator card, as shown in Figure 23-4-1, is a standard 3 by 5 inch reference card, containing pertinent information on the individual. This includes his name, address, and telephone number; the submission dates for his résumé and application, and the action taken. The designation D or F refers to whether the applicant is seeking domestic or foreign employment.

The locator cards are arranged alphabetically by name, and the complete card file index is kept by the personnel director or clerk on desk top for quick reference.

The class number shown on the card is the key in locating the applicant's file folder. In Figure 23-4-2, the locator card for the applicant shows that he is in category (class) B-1 for carpenter superintendent. Referring to the category B-1 in the applicant file, under the individual's name, the clerk finds the file folder.

To keep the system up to date, a new applicant file series and locator card index are started at the beginning of each year. Each applicant record is kept for a period of three years, whereupon it is declared inactive and sent to the records center for storage.

23-5. Skills Inventory for Present Staff

Perhaps the most practical system for planning the manpower organization and strength of a construction project

from within the company is the skills inventory. Under this system all pertinent information concerning the biography, education, work skills, and project experience of present employees is collected and stored in a computer for fast reference. This information is then used by the personnel department and project management in quickly assessing work skill capability for job assignment purposes or for entry into proposals—a field of constant activity in the construction business.

Another major use of the skills inventory, especially in a large international company, is to determine language capability either for a foreign project assignment or for translation of foreign documents and correspondence. Less often, the skills inventory is a valuable tool in compiling employee information for government and private surveys.

There are two basic forms of the skills inventory, each differing only in the method of access to the computer data:

Runoff Data System The runoff data sheet system, more conveniently used in smaller construction companies, involves a printout of all the data stored in the computer at periodic intervals. Typically, the input data, on standard punch cards, are processed every three months and information published in separate sections. Figure 23-5-1 shows example pages from this system. The master list, as shown in Figure 23-5-1a, is an alphabetical listing of all employees, containing complete background information and work experience on each individual. From left to right across the top, the headings include the employee number, department number, name (and in the case of an engineer, his registration), year of birth, degree (if any), job classification, present location, language capability and fluency grade, and the various skill combinations. The latter code numbers, in groups of two numbers, refer to

```
PERSONNEL SKILLS ZZZZ                        MASTER. LIST                         DATE 09-70   PAGE 152


EMP#   DEPT NAME AND REGSTRIN   YR   DEGREE      JOB CLASS LOCATION LANG ***   WORK SPEC/PROJ EXP COMBINATIONS ***


6340   2092 RA CASS             33               ASSOC ENG MIDDLETO         D042 240 D054 240 E069 750 E069 238 E069 240

                                                                            E123 238 E147 086 E147 240 E147 382 C010 370

                                                                            C010 240 C024 238 C024 240 C042 240 ZZZZ 500


5715   4090 CM CAVNESS          12               CRFT SUPV LOA              C040 006 C040 064 C040 094 C040 198 C048 006

                                                                            C048 006 C040 064 C040 094 C040 198 C058 066

                                                                            C058 198 C053 006 C052 064 C052 224 M006 248


5864   1300 C  CAYLOR           29               SR DESGR  OAKLAND          D002 252 E149 248 C016 248 0029 224 M006 500


5522   0250 CO CHANDLER         33               SR DES    PITTSBUR 26 C    D028 ]56 D028 178 D028 382 D028 412 D028 224

                                                                     70 C    D024 040 C010 150 C010 224 D030 156 D030 158

                                                                            D030 382 E130 412 E152 156 E152 534 :DN52 412

                                                                            ZZZZ 500


8347   0292 FR CHURCH           34               SR DRFTS  MIDDLETO         D052 220 ZZZZ 500


3153   0328 JT CHAN             25   ARCH-BA     ASSOC ARC OAKLAND 42 C     E053 004 E053 006 E053 016 E053 008 C010 174

                                                                            C012 286 C014 002 C014 414 C014 054 0963 002

                                                                            0001 096 0001 414 ZZZZ 500


5422   5390 BJ COLVIN           30   ARCH-BS     CH CON MR AUSTRALI 04 C    C034 285 C034 368 C010 154 D004 264 D004 248

                                                                     38 C    D003 306 C078 038 C078 034 C060 032 C060 034

                                                                     58 C    C060 036 C060 053 0007 268 ZZZZ 500
```

Figure 23–5–1 One variation of the Skills Inventory—All employee qualifications are periodically printed out on long runoff sheets: (a) Master runoff list contains complete background information and work experience; (b) project list contains the names and qualifications of only those with experience in a particular project (in this case nuclear), (c) foreign language list categorizes employees by a particular foreign language capability.

the combinations of work specialty and project experience obtained from a questionnaire given to each employee upon entry into the company. This questionnaire will be described in detail shortly. In the case of the circled example, CO10 079 indicates that he was a field engineer on a flood control project.

The separate list shown in Figure 23-5-1b groups all employees by project experience—in this case, under the general category of Nuclear.

Essentially the same information in the master list is shown here except that the skill combinations (specialty/ project experience) relate to the employee's work in the nuclear field only. Other project experience is indicated in the last column.

The final list shown in Figure 23-5-1c is the language skill list, which categorizes all personnel by their foreign language capability. The example sheet here shows those employees with a proficiency in Italian. Information in this list includes the department number, present location,

job classification, and fluency grade, which is weighted heavily on speaking ability.

Three separate lists are prepared because each is preferable in retrieving certain kinds of information. The master list, for example, since it is a total listing of employees and their complete backgrounds, is advantageous in looking up specific facts concerning an employee.

The computer listing by project category is extremely useful for project reassignment or for insertion of names in proposals. The project manager or vice-president in charge of the division can simply request a search by the personnel department, based on an accurate job description.

The language skill list, as the name implies, categorizes individuals by their language ability. This list is particularly useful for locating employees in other departments to translate letters and correspondence from a foreign country or to act as interpreters in foreign project assignments. Fluency, of course, is the major factor in the final selec-

EMP#	DEPT NAME AND REGSTRIN	YR	DEGREE	C	JOB CLASS	LOCATION	LANG	**** WORK SPEC/PROJ EXP COMB **	OTHER PROJECTS
1044	0101 MS LEWIS	24 ME	-BS		ASSOC ENG	OAKLAND		0007 114	AERO COMM HYDR MAR
									MIN MFG CHEM MISC
8616	0531 AL LIBBY	20 ECON	-BS		PRO CO SU	OAKLAND		0041 110	AERO MISC
1740	0723 HR LIGGINS	23 CE	-BS		PR ENG B	OAKLAND	26 C	0002 126 0033 114 E141 144	COMM PIPE HYDR AERO
			CHEM-MS					E127 120 D018 114 E149 126	MISC
4817	0130 EO LINNELL	33 ME	-BS		SR ENG	OAKLAND	26 B	D028 110 D044 110 D017 114	POW AERO TRAN MFG
							34 A		MISC
4486	0130 EO LOE	]9			SR DESGR	OAKLAND		D034 123 D010 123 D028 132	CHEM MIN MFG MISC
8525	0130 CE LOGAN	34			DES DRFTS	OAKLAND		D064 130	AERO MISC
4608	0130 JW LOHR	14 ME	-BS		SR ENGR	OAKLAND	18 C	C098 114 C098 132 C038 144	COMM HO M HYDR MFG
							22 C	0037 143 0039 137 0039 143	PIPE MISC
							40 C		
8339	0230 LB LORENZ	24			SR DESGR	CHICAGO	24 C	0042 130	MAR CHEM I ST MFG
							54 C		MISC
3785	0292 JP LUNDBERG	12			ASSOC PR	MIDDLETO	50 C	D040 120 D040 132	COMM MIN I ST AERO
							58 C		HYDR POW CHEM MISC
4138	0130 CH LUDMAW	24 CE	-BS	F	ASSOC ENG	OAKLAND	24 C	0024 112	TRAN CHEM MIN MISC
	BRAZIL SURV						26 A		
	BRAZIL CIV						40 B		

Figure 23-5-1   (Continued)

tion of an individual from the list. For translation purposes, however, a person from within the same department and with a lower salary classification would be preferred.

On-Demand System   In construction companies that have a large number of employees, the practice of manually searching the published runoff sheets can be time consuming and can lead to errors or oversights. The on-demand system, an alternate inventory system, retains all personnel data within the computer until needed and then prints only the names and qualifications of those individuals meeting the input requirements. This method therefore eliminates the manual search procedure but requires a special search request form to translate the project manager's personnel requirements into the language of the computer. The system also requires close cooperative work with the personnel department to detail the employee requirements sufficiently for an adequate com-

puter search. Although initially harder to work with and more expensive to maintain, the on-demand system, when properly applied in a large company setting, can sort through thousands of employee names and qualifications and obtain the appropriate list of names within minutes.

Skills Inventory Questionnaire   The source of all information stored in the skills inventory computer program is a questionnaire, first given to the employee during his orientation period, and updated by him at a minimum of 18-month intervals thereafter. Separate sections cover biography and education, foreign language capability, and work experience. The latter section is most important for job reassignment purposes and proposals, and a portion of this section is shown in Figure 23-5-2. This is the section that relates an employee's various work specialties to his project experience. For example, John Doe's work specialty is warehouse supervisor (Code C036) with seven

```
PERSONNEL SKILLS 40                        LANGUAGE SKILL ITALIAN
```

EMP#	DEPT	NAME	YR	LOCATION	JOB CLASS	FLUENCY
8136	0130	CH ROBERTSON	19	OAKLAND	ASST ENGR	C
0657	5592	JJ ROBINS	27	GURI	PWRHS SUP	C
5286	0322	EE ROQUE	15	OAKLAND	PR ENGR A	C
0843	5592	AW ROPER	23	GURI	COMC TECH	A
4101	0130	CO ROSSI	29	OAKLAND	ASSOC ENG	C
5984	0130	AC RYAN	36	OAKALND	SR ENG	A
5927	0292	HR SADLER	38	MIDDLETO	SR ENGR	C
8404	0250	BK SACYR	25	PITTSBUR	SR DESGR	B
1030	0516	M SANDERS	11	OAKLAND	MR TE REP	C
4113	0292	M ⌐SCARR	23	MIDDLETO	SR ENGR	A
4173	0326	BV SCHMIDLY	35	BALTIMOR	SR ENGR	B
0438	5501	H SCHRADER	13	OAKLAND	PR ENGR B	A
3924	0221	W SELLERS	23	CHICAGO	ASSOC ENG	C
4608	0130	PR SHAVER	14	OAKLAND	SR ENGR	C

Figure 23–5–1   (Continued)

years of project experience on diesel electric stations (Code 162) and five years of project experience on combined cycle station (Code 170). Mr. Doe would then graphically *relate* his work specialty to types of project experiences as shown in Figure 23-5-2. Examples of work specialties and project experience are given in Figure 23-5-3, together with their corresponding code numbers. Separate work specialty lists can also be prepared for other department functions, such as administration, business development, contract administration, and sales.

In filling out the form shown in Figure 23-5-2, the employee is given the complete list of work specialties and

	COLUMN 1		COLUMN 2	COLUMN 3
Code No.	Work Specialty (From List A)	Code No.	Types of Project Experience (From List B)	No. Yrs. Experience on Project

Figure 23–5–2   Positions of the Skills Inventory Questionnaire provide space for listing various work specialty/work experience combinations which are most important in determining an individual job capability

COLUMN 1—*Construction Work Specialties*

C002 RESIDENT (PROJECT) MANAGER
C004 ASST. RESIDENT (PROJECT) MANAGER
C006 ADMINISTRATIVE MANAGER
C008 CONST. ENGINEER (TOP ENGINEER ON JOB)
C010 FIELD ENGINEER
C012 OFFICE ENGINEER
C014 COST ENGINEER
C016 FIELD ENGINEER (General)
C018 Electrical
C020 Instrumentation
C022 Mechanical
C024 Piping
C026 Structural
C030 GENERAL SUPERINTENDENT
C032 ASST. SUPERINTENDENT
C034 AREA SUPERINTENDENT
C036 WAREHOUSE SUPERVISOR/SUPERINTENDENT
C038 EQUIPMENT REPAIR & MAINTENANCE SUPERVISOR
      (MASTER MECHANIC)
C040 CRAFT SUPERVISOR/SUPERINTENDENT (General)

COLUMN 2—*Project Experience*

AEROSPACE AND DEFENSE
008 Operations Support Facility and Space
     Ground Support Facilities
022 Communications Systems

HYDRO AND HYDROELECTRIC
075 Power Transmission Lines
079 Flood Control
085 Seawater Desalination
087 Water Quality Control (Water Pollution)

MARINE
097 Beach Stabilization
099 Breakwater/Jetty

NUCLEAR
126 Dual Purpose Power—Desalting

PIPELINE
157 Long Distance Slurry

CHEMICAL
251 Fertilizer Plants

253 MINERAL (General)
254 Beneficiation

Figure 23–5–3 Partial list of work specialties and project experience categories which are referred to when filling out the Skills Inventory Questionnaire

project experience titles prepared by the company (see Figure 23-5-3). After reviewing the list he fills in his work specialties and corresponding code numbers in column 1 of Figure 23-5-2. For each work specialty he would then review the list of project experience titles, filling in the appropriate names and code numbers in column 2 and the number of years experience in column 3.

Updating the Skills Inventory To keep the skills inventory current, the personnel department should update the computer data periodically, preferably every three months. Prior to revision, the department should review all employee personnel folders for changes in job classification, department number, project assignment, and location. Information on new hires and terminations, obtained from weekly lists published by the department, is also recorded. In addition to these changes, the work experience code combinations should be brought up to date at least every 18 months by sending the original questionnaire and the updating supplement to the employee who then adds any new work specialty or project experience acquired during the intervening period.

23-6. Availability Lists

The skills inventory is an excellent reference source for assessing the qualifications of present employees for new project assignments. However, it cannot predict an individual's availability for reassignment which can only be determined if the expected date of completing his present assignment is accurately known. There are many good methods of accurately predicting availability; the most practical one for a particular company depends basically on its size.

In a small company, the executive manager or vice-president in charge of all construction projects is the person best able to judge the availability of his men for new job positions. During the preparatory work for an upcoming project, he and his subordinate managers are normally involved in planning the general organization of the project, scheduling each task, and assigning the project management personnel. He, therefore, is in the best position to select personnel for a new project, using the skills inventory as a guide and basing each choice on his knowledge of the person's availability. When he comes across a position for which he knows there is no qualified person available at the time, he can reshuffle personnel from within existing projects in order to free an individual for the new assignment. He can also request a search of the applicant file and authorize outside recruitment when the needed personnel cannot be found from within the company.

This method of placing the responsibility for determining availability in the hands of those involved in the actual assignment of personnel has been found to be most efficient for the small company, since it minimizes paperwork and outside department interference. In the large company, however, there may be several construction divisions, each handling a particular type of construction project. It is impossible in this case for one person to keep track of all the various project personnel, and the task of determining availability is best handled by the personnel department in the form of periodic "personnel available" lists.

In the large company approach, the personnel department can make more effective use of the skills inventory by including in the data input the expected completion date of each employee's present project assignment. These dates can be revised and updated periodically, based on the latest progress reports and separate runoff lists of available personnel published at regular intervals (weekly, if necessary) and disseminated throughout the various project divisions. These availability lists would include each person's present location, department number, and coded qualifications. The availability list ensures maximum utilization of employee talent through an efficient system of collecting and distributing employee information.

23-7. A Typical Project Scene

The basic tools used by the Personnel Department for Manpower planning have been examined. Now it is important to see how the Personnel Department, in coordination with other departments, uses these tools and others during the life of a typical project: First, at its inception, then during the active construction period, and finally, during the closing stages when the project undergoes demobilization.

Early in the project during the planning phase, a conference is called to select and assign members of the new project's management and engineering staff. To this meeting the field manager and other leaders of the project bring a tentative organization chart containing the names of personnel already assigned. The immediate task of the personnel representative is to find the additional personnel through the three manpower techniques described earlier—namely, the applicant file, skills inventory, and availability lists.

Concurrently, a personnel representative conducts a survey of the area, collecting local area information on schools, housing, medical facilities, cost of living, and available transportation. This may be accomplished at home base by writing the local chamber of commerce and local civic groups or by visiting the site, in which case it is good practice to take photographs for pre-

paration into a slide presentation for proper orientation of employees to the project.

23-8. Schedule of Job Classifications and Salary Ranges

The preliminary organization chart brought to this first personnel meeting is the basis for preparation of a schedule of job classifications and salary grades for project personnel. Using this chart, the salary administration department of industrial relations establishes an appropriate position title and salary grade for each member. The organization chart in Figure 23-8-1 illustrates this procedure. The salary grade is the number indicated in each box and represents a U.S. base salary range established by the company for exempt employees only. In assigning a salary grade to a position title, the salary administrator considers the size of the project and the responsibilities attached to the individual position.

Besides the exempt positions on the organization chart, the salary administrator must establish the salary ranges for the nonexempt positions—the clerks, typists, secretaries, and stenographers of the project. These ranges are normally based on the established local rates available from the Department of Employment of the state and from surveys of local business firms.

After the complete schedule is prepared, it is sent for review and approval to the field project manager and the division vice-president or general manager. The schedule of job classifications and salary ranges becomes an im-

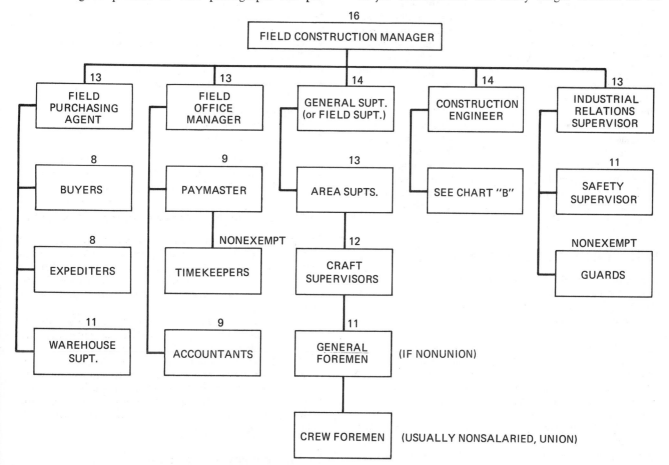

Figure 23–8–1 In establishing the salary for the project's exempt personnel, the salary administrator uses a number system that represents a country-wide base salary range

portant tool for management purposes in hiring, salary review, and dealings with the client. In some cases, a detailed schedule may have to be submitted prior to award of contract. Once approved by top management, the schedule of job classifications and salary ranges becomes the primary tool during recruitment.

23-9. Recruitment

Recruitment of nonexempt personnel is necessarily done at the project location from the residents of the local communities. It is also desirable, however, for the field office to recruit most exempt employees from areas as close to the project as possible to lessen the cost of travel and moving expenses provided to these people. Consequently, although the central personnel office provides assistance in recruiting critical hard-to-find skills, the primary responsibility for recruiting project personnel is in the hands of the field office.

With the schedule of job classifications and salary ranges already prepared, the only additional item necessary to begin recruitment is a list of job descriptions corresponding to the vacant positions. Job descriptions are normally prepared by the superintendent in charge of these personnel positions. However, they can also be taken from the records of similar prior projects. In many large companies job descriptions are written by a job analyst, a specialist in this activity working within the industrial relations department.

As soon as the salary schedule and the list of job descriptions become available, a local office is established at or near the project site. The recruitment effort begins with classified advertisements, and in some cases local radio and television commercials. For exempt personnel, classified ads are generally placed in the appropriate trade magazines, society publications, and major newspapers.

Personal contacts at employment agencies and other companies are also explored both at the central office and in the field. In addition, local college placement centers are also sought out for assistance in finding recent college graduates to fill some of the engineering positions. Upon successful interview and acceptance of a candidate, the field office is responsible for project orientation, scheduling a pre-employment physical examination, preparing all necessary employment records, and sending notice of the starting date.

23-10. Prejob Trades Conference

The bulk of the employees hired by a construction company for a field project comprise skilled craftsmen who are generally represented by the various craft unions. They are hired as needed on an hourly basis as set forth in collective bargaining agreements.

Prior to the beginning of construction it is desirable to hold a prejob conference with representatives of the various unions in order to review the general scope of work, to inform them of the manpower needed for the project and to determine their capabilities to supply the required skilled labor.

It is helpful to provide the union representatives with a booklet briefly describing the scope of work and listing approximate manpower requirements by craft. When there is a shortage of a particular craft in the area, a program for recruitment can be initiated with the local and international unions in order to obtain the needed craftsmen.

23-11. Employment Procedures Manual

On large construction projects it is frequently necessary for the field industrial relations department to prepare a separate procedures manual outlining the employment practices for the project, taking into consideration the applicable federal, state, and local laws, as well as provisions of the client as set forth in the contract. Detailed procedures are given on hiring, orientation, processing to job site, review, reclassification, transfer, and termination. Separate sections are devoted to the exempt, nonexempt, and the manual employee. Other sections of the manual include the safety and security programs and the employee benefits, including workmen's compensation.

The manual should also include copies of all the necessary forms, with explanations of their purpose and instructions on their use. Some of these forms may be the standard company forms used at the central office, those of the client in keeping with his employment records, or special forms newly created for the project.

Most important, the approval cycle for each personnel action must be explained, and the names or positions of the delegated approving authorities included.

Since the procedures manual is to be used as a guide by the local personnel office staff, it is expected to go into considerable detail on all personnel actions. For example, in the hiring of an exempt employee, the procedural information would cover:

- Formal application
- Interview technique and evaluation checklist
- Reference checking
- Preemployment physical examination
- Offer of employment (including form letters)
- Payroll notification
- Medical and insurance plans enrollment
- Skills inventory questionnaire
- Patent agreement
- Employee withholding
- Certification of citizenship
- Personnel record card

The employment procedures manual is intended both as a guide in the administration of a field personnel office and for the training of field personnel office staff.

23-12. Personnel Practices During Construction

The primary personnel responsibilities during the course of project are to set up and maintain employee records; supply labor; administer the employment function of hiring, transfer, upgrading, and termination; and provide for the demobilization of personnel at the completion of the project.

For purposes of discussion, this section will consider a large construction project in which the personnel functions are vested in an established personnel office which is part of the field industrial relations activity. As mentioned earlier, however, the personnel functions remain the same for any project and depending on the project's size, may be carried out by the accounting office, field office manager, or, in the case of an extremely small project, the field project manager himself.

23-13. Employment Records

The headquarters personnel records are models for records kept in the field. The basic employment record is the employee's personnel folder kept on all exempt and nonexempt employees of the company. The personnel folder contains the individual's complete employment history: formal application, résumé, medical examination certificate, patent agreement (if required), offer of employment and letter of confirmation, certificate of citizenship, and copies of all internal correspondence and personnel actions.

A register-like summary sheet is also attached to the inside front cover of the folder to record all changes in status by the effective date of change.

The personnel folder file is supplemented by a quick-reference card file. Pertinent information, such as present project assignment and location, present position, and address are transcribed from the folder to the card for quick checking during telephone requests for information.

At the inception of a project, copies of the records of all transferred personnel are sent to the field to establish the field office file. As new employees are hired in the field, the original records are sent to the headquarters files, and copies remain at the field office. The central personnel office therefore maintains a master file on all exempt and nonexempt employees for the entire company. The master file centralizes all personnel records for purposes of manpower planning, personnel development, reassignment, and benefits enrollment. It is also a safeguard against the loss or destruction of field files.

Because of the extremely transient nature of their employment and the difficulty of maintaining day-to-day records, it is general practice not to keep home-office records on manual employees. At the completion of a project, however, the field files on manual employees are returned to the headquarters office for permanent record.

23-14. Hiring Exempt and Nonexempt Employees

One of the most important functions of the field personnel office is to supply labor as required by the needs of the project. For exempt and nonexempt personnel, the process begins with a requisition form originated by the department supervisor. Figure 23-14-1 shows a typical personnel requisition form, which includes space for a description of the job position, experience needed, special qualifications, and the authorizing signature. Based on this information, the personnel office can assign a salary grade and begin recruitment.

Personnel Lists   In a large construction company it is sometimes necessary to prepare simplifying personnel lists to handle the large number of personnel requisitions passing through the personnel department daily. The two reports described here are used to facilitate hiring and reassignment of exempt employees when large numbers are involved:

1. *Personnel Available List*   This list, compiled by field personnel from project progress and activity reports, lists the names, position titles, and locations of individuals who will soon be completing their present jobs and ready for reassignment. The expected date of completion is recorded as far ahead as is possibly known, preferably as long as six months in advance. The list is sent to the home office monthly for their use in coordinating the hiring of exempt employees at all locations.

2. *Weekly Report of Open Positions*   This report compiled by the central personnel office from personnel requisitions sent from all field projects can be used as a quick reference in matching available employees and applicants to open positions at all locations. The report lists by project location the position title, number of open positions in this job category, and the name of the supervisor who has the hiring authority. By comparison of the Personnel Available List with the Weekly Report of Open Positions, it becomes a simple task to match qualified candidates to open positions.

Personnel Forms   Upon locating a qualified candidate and successful interview by the department supervisor, the personnel office completes the hiring procedure first by means of a Personnel Action Authorization form, and later by means of a Personnel Action form. These two forms which are also used to carry out other personnel actions, including hiring, transfer, upgrading, and termination, are described in detail as follows:

1. *Personnel Action Authorization Form*   Figure 23-14-2 shows a typical Personnel Action Authorization form with multiple copies. This form is used to record and gain approval of any action taken or change made regarding employment of each individual. Personnel actions involving the hiring, reclassification, salary adjustment, leave of absence, transfer, title change, promotion, standby leave, or termination of the employee are prepared on this form by the personnel department and forwarded to the appropriate department head and the project manager for approvals.

   Space is provided to indicate the nature of the proposed action and, if necessary, an explanation in the "Reason for Action" section. Space is also provided for the employee's name, present salary and effective date, proposed salary and effective date, amount of increase or decrease, and the percent of increase or decrease.

   "Present Status" indicates the position of the employee prior to the proposed action; "Proposed Status" indicates the employee's position after the proposed action.

   Actions involving headquarters approvals are sent to the appropriate offices for signature. Upon approval of the personnel action, both the home and field personnel offices retain copies, as well as the approving authorities.

2. *Personnel Action Form*   After the action has received the necessary authorization and has, in fact, been carried out, it is necessary for the office executing the action to send confirming notice to all parties concerned. This is done by means of a Personnel Action form, as shown in Figure 23-14-3. Copies of this form are sent to the headquarters personnel office for the master files and to all approving authorities for their records.

23-15. Hiring Union Craftsmen

Union craftsmen are hired daily, some for a four-hour concrete pour, others for month-long assignments. The hiring process must be swift and thorough, fulfilling union, legal, and safety requirements.

Since the approving authority for hiring skilled craftsmen rests in the hands of the field construction manager, there is no need to coordinate the hiring activity with the headquarters personnel office. In addition, no master file

PERSONNEL REQUISITION

REQ. NO._____   DATE_____

DEPARTMENT_____

CLASSIFICATION_____

RECOMMENDED SALARY RATE_____

NO. REQUIRED_____

NEW OR REPLACEMENT FOR_____

PERSON PREFERRED (if any)_____

DATE NEEDED_____

FULL TIME OR PART TIME_____

REGULAR OR TEMPORARY_____

EXPIRATION DATE, IF TEMPORARY_____

SEX_____ AGE PREFERRED_____

PREFERRED EDUCATIONAL ATTAINMENT_____

JOB SPECIFICATION_____

MINIMUM EXPERIENCE REQUIRED_____

SPECIAL QUALIFICATIONS_____

OTHER COMMENTS_____

APPLICANTS TO BE INTERVIEWED BY   1._____   2._____

ORIGINATOR_____   AUTHORIZING SIGNATURE_____

REFERRALS_____DATE_____

HIRED_____DATE_____

DISTRIBUTION — ORIGINAL - - - - - - - - - PERSONNEL OFFICE
FIRST COPY - - - - - - - - - PERSONNEL OFFICE
SECOND COPY - - - - - - - - ORIGINATOR

KEP-34
(Revised 3-63)

Figure 23–14–1 Typical personnel requisition form, initiated by the project manager or his authorized staff, starts the search for a needed project member

is kept on manual employees at the home office because of the extremely short-term nature of their work and the complexity of the task in transmitting records between the field and the home offices.

In contrast to exempt and nonexempt employees, the hiring of skilled craftsmen is greatly facilitated because of their availability through local unions. Similarly, however, the process begins with a written personnel requisi-

tion. The requisition, as illustrated in Figure 23-15-1, is prepared by the craft superintendent and signed by the field project manager. The simple form calls for the type and number of personnel required and the duration of employment.

The personnel office, upon receipt of the requisition, simply relays this information by phone to the local union. The requisition is then passed along to the badge register

PERSONNEL ACTION AUTHORIZATION

NAME

	AMOUNT	EFFECTIVE DATE
PRESENT SALARY	/	
PROPOSED SALARY	/	
INCREASE OR DECREASE		PERCENT

NATURE OF PERSONNEL ACTION:

☐ Hire ☐ Transfer

☐ Reclassification ☐ Title Change

☐ Salary Adjustment ☐ Promotion

☐ Leave of Absence ☐ Standby Leave ☐ Termination ☐ Other (Specify)

REASON FOR ACTION:

STATUS OF EMPLOYEE'S POSITION:

<u>PRESENT</u>   <u>PROPOSED</u>

LOCATION

OFFICE OR DEPARTMENT AND NUMBER

POSITION TITLE

EXEMPT OR NON-EXEMPT

DATE SUBMITTED   PROPOSED EFFECTIVE DATE OF ACTION

Originating Supervisor

Approving Supervisor   ☐ Field Industrial Relations Dept.
                                        ☐ Oakland Personnel Office

Position Number
Position Grade
Exempt or Non-Exempt

DATE PROCESSED_____EFFECTIVE DATE_____

Supervisor, Wage & Salary Administration

REVIEWED BY:

Name	Date		Name	Date

Name	Date		Name	Date

Approving Executive   Date

FORM KE3-25 Rev. 7-65

COPY 1 Oakland Personnel Office

Figure 23–14–2 Personnel Action Authorization form, initiated by the personnel department, is used to obtain management authorization for all personnel actions

PERSONNEL ACTION
(Non-union Employees)

EMPLOYEE ACTION NO.:_____ Field Office:_____ Date:_____

Employee's Name:_____ _____ _____ Social Security No.:_____|____|_____
LAST   FIRST   INITIAL

Marital Status:_____ No. of Exemptions Claimed:_____

EFFECTIVE DATE OF ACTION (TO BE USED FOR ALL ACTIONS):_____

Brief Explanation for Action:_____

TYPE and DETAILS of ACTION

☐ **HIRE**   Birth Date:_____

Classification:_____ Rate:_____ Previous Kaiser Employment: Yes ☐ No ☐

(FORWARD COMPLETE APPLICATION AND WORK HISTORY ON ALL EXEMPT PERSONNEL TO OAKLAND PERSONNEL OFFICE)

☐ **REHIRE**   Location Last Kaiser Job:_____ Approx. Date_____

Classification:_____ Rate:_____ Prev. Class:_____ Rate:_____

Date Previous Termination:_____ Reason:_____

☐ **PROMOTION**

Old Classification:_____ _____ New Classification:_____

Old Rate:_____ New Rate:_____ Date Last Increase:_____

☐ **MERIT INCREASE**

Original Date of Hire:_____ Rate:_____ Classification (Present):_____

Last Promotion or Increase: Date:_____ Rate:_____ New Rate:_____

☐ **TERMINATION**

Resignation ☐   Lay-off ☐   Discharge ☐   Explain:_____

Classification:_____   Eligible for Rehire   Yes ☐ No ☐

☐ **TRANSFER** Company_____ Company_____

From: Location _____ To: Location_____

Paid Thru:_____ Classification at time of Transfer:_____

Date Scheduled to Report to Point of Transfer:_____

Terms of Transfer:

Transportation Allowed   Self ☐ Family ☐   Leave of Absence without Pay of _____ days authorized

Travel Expense Allowed   Hotel ☐ Meals ☐ Mileage ☐   Vacation of _____ days granted en route

Moving Expense Allowed   Household Goods ☐ Personal ☐   Current year unused vacation at transfer _____ days

Relocation Expense Allowed_____weeks   Vacation credits accrued toward next May 1,_____ days

Advance Expense Allowance:_____ Total vacation credits transferred _____ days

Vacation funds to cover credits transferred to _____

_____ Days Travel Allowed Enroute

Group Life Insurance: Date Used to Establish Service:_____ Date Enrolled:_____

Class: Employee Premium:_____ Coverage Last Deduction:_____

Medical & Hospital Insurance: New York Life ☐ Kaiser Foundation ☐ Paid Thru:_____ Rate:_____

Major Medical ☐   Paid Thru:_____ Rate:_____

Disability Insurance: New York Life ☐ State ☐ Other ☐   (For transfers within California only)

KSRP ☐   Date used to establish service:_____ Date Enrolled:_____

Employee contribution:_____ Coverage Last Deduction:_____

SSRP ☐   Date used to establish service:_____ Date Enrolled:_____

Employee contribution:_____ Coverage Last Deduction:_____

Contributions each of last 3 months_____, _____, _____. Date of Last Contribution:_____

Gross Yearly Earnings thru:_____ Amount:_____ Total FICA Deduction:_____

☐ "Kaiser Companies Rate History" Attached.   Transfer Authority:_____

(See Reverse Side for Instructions)   _____
                                                                  Signature

FORM KEP-24 Rev. 7-65   _____
                                                                  Title

Original—To Oakland Personnel Office

Figure 23–14–3 Personnel Action form notifies all parties concerned, authorized by the form in Figure 23–14–2, that the action has officially taken place

```
┌─────────────────────────────────────────────┐
│  ᵇᵗˣ         LABOR REQUISITION               │
│                                              │
│                         NO. 450              │
│    Operators              1                  │
│  ──────────────────  ──────────────────      │
│      Craft           Number Required         │
│                                              │
│  Special Qualifications  Oiler Truck Crane (days) │
│                                              │
│  Temporary replacement 6-13-66               │
│  ──────────────────────────────────────      │
│  to 6-27-66                                  │
│  ──────────────────────────────────────      │
│                                              │
│  ──────────────────────────────────────      │
│  Called in 6-8-66  1:30PM                    │
│  ──────────────────────────────────────      │
│                                              │
│  ──────────────────────────────────────      │
│                                              │
│  ──────────────────────────────────────      │
│                                              │
│     6/13/66            Sudkins                │
│  ──────────────     ──────────────           │
│  Date Required       Craft Supt.             │
│                                              │
│  Approved  G. W. Zimmerman  J. K. Lockie      │
│          ──────────────   ──────────────      │
│            Gen'l Sup't.     Const. Mgr.      │
└─────────────────────────────────────────────┘
```

Figure 23-15-1 Labor Requisition form is conveniently used by a craft supervisor to authorize the field personnel office to send needed craftsmen. (Note: The duration of employment is included)

clerk who checks the requisition number against that on the referral slip carried by the entering craftsman.

Processing In   The processing-in procedure is swift and routine. Upon entry the craftsman registers for a badge number, completes the Employee Income Tax Withholding form and signs a Job Fitness Statement in which he indicates his physical capability to perform his job assignment. At the same time, the personnel clerk fills out a work record card, the original of which is kept on file, and a copy is sent to the field payroll department as notification of hire.

Following this procedure, the craft employee signs for company equipment, including a hard hat, and is issued a safety manual and his identification badge. Before starting work, he is also given a safety briefing by the safety supervisor or the craft superintendent.

If the employee has worked on the project previously, his record card is pulled from the Manual Terminations file. If eligible for rehire, he is issued the same badge number and must complete another Job Fitness Statement and an Employee Income Tax Withholding form.

Termination   The procedure for terminating craft employees is also designed for speed. The craft superintendent in charge of the man completes a termination slip and forwards it to the department head for signature. The employee then hand carries the slip to the warehouse to return equipment and gain clearance. Next, he proceeds to the payroll department, where he receives his paycheck, plus a copy of the termination slip for his records.

The paymaster, in turn, retains one copy and sends a duplicate to the personnel office for filing with the employee's work record card.

23-16. Demobilization of Project

As the project nears completion it is necessary to plan for the orderly withdrawal of all exempt and nonexempt employees of the project. This is accomplished through one of the following means:

- Transfer to client operations
- Transfer to another project
- Temporary assignment in home office
- Leave of absence
- Termination

Transfer to Client Operations   Depending on the type of facility being built, it is sometimes possible to transfer some of the locally hired clerical and technical personnel to operating positions within the facility after startup. Through prior negotiations with the client, it is determined what kinds of skills are needed and whether or not they may be recruited from present project personnel. In some cases, the client may specify such cross-training of personnel as part of the contract agreement.

As the project nears completion the field personnel office coordinates with the client's personnel office in effecting the transfer of records.

Transfer to Another Project   Reassignment of permanent project personnel is the responsibility of the headquarters construction division and personnel department. As early as possible in advance of actual project demobilization, the field personnel office assists the field project manager in developing a preliminary demobilization schedule showing the anticipated release date of all engineering and construction management personnel. The demobilization schedule is sent to the construction division and used for reassignment of employees to upcoming projects. When there is no open position for a particular job skill, the headquarters personnel office is notified and assists in finding reassignment in or through other divisions of the company.

During the preparation of the demobilization schedule, the field personnel office also coordinates the compilation of performance evaluations for all project employees. Figure 23-16-1 shows a sample job-site evaluation which provides the personnel office with the needed performance data to properly evaluate his qualifications for reassignment.

Temporary Assignment in Home Office   Certain key personnel, including project managers, key department heads, key superintendents, employees with special skill, employees with future potential for advancement to key positions, and other selected employees are often assigned to the home office on a temporary basis, pending reassignment to another field location or to an opening in the home-office staff. Both the permanent home-office staff and the field personnel can benefit from this association.

Leave of Absence   Individuals who wish to continue their employment even though there is no present job assignment available may request and be placed on temporary leave of absence. The method permits the individual to work elsewhere while waiting for a new job assignment and in some companies to continue coverage in the company's benefits plan. The company, if it anticipates using the man's services in the near future, benefits from this policy because it ensures the availability of the individual when needed, without having to carry him on the payroll.

JOBSITE EVALUATION   Date: _____

NAME _____ CLASSIFICATION _____ PROJECT _____
                          (at termination)

DATE _____ ARRIVED_____ DEPARTED _____ TERMINATION _____
(at termination)   JOBSITE   JOBSITE

RECOMMENDED FOR RE-EMPLOYMENT _____ FOREIGN _____ DOMESTIC _____

JOBSITE WORK RECORD

CLASSIFICATION AT ARRIVAL: _____

RECLASSIFICATIONS:

Title	Date	Rate

ATTENDANCE RECORD:

Good:_____

Fair:_____

Poor:_____

WORK ASSIGNMENTS: (List even though not covered by classification changes)

Type of Work	Appropriate Dates	Supervisor	Rating

ABILITY:

	Excellent	Good	Fair	Poor
As Craftsman				
Staff Supv's				
Non-Staff				

ADAPTABILITY:

	Excellent	Good	Fair	Poor
To Working Conditions				
To Social Conditions				
To Camp Life				

WILLINGNESS TO:

Work				
Accept Assgns				
Cooperate				
Accept Respblty				
Fulfill Assgns				

CHARACTER:

Personality				
Disposition				
Conduct				
Sobriety				

COMMENTS: (Special abilities or lack of ability affecting job performance, disciplinary, personality, or other problems and factors to be considered in regard to this employee, etc.)

_____
_____
_____

FINAL RATING: CONTRARY OPINION:
Excellent - would want on any job requiring this craft _____ _____
Good   - would be satisfied to have on future jobs   _____ _____
Fair   - would want only if better men unavailable   _____ _____
Poor   - would not want under any conditions   _____ _____

RATING BY: (If joint opinion, list all supervisors   By_____
            whose opinions are reflected.)   Title_____
NAME _____ TITLE _____
NAME _____ TITLE _____
VERIFIED & APPROVED BY (Name) _____ (Title)_____

Figure 23–16–1 Typical job-site evaluation form is used by personnel office to evaluate an individual's performance for upgrading, salary adjustment, and reassignment.

Termination Personnel not able to transfer to the client's operation, or to another project, and not wishing to be placed on leave of absence are terminated from company employment by written notice, as far in advance as possible. Those hired at a distant location from the project are normally provided travel and moving expenses to the point of original hire or an equal distance, as specified in the terms of the written offer of employment letters. The headquarters personnel office coordinates with the field office in making arrangements for moving household effects.

Like other personnel actions, all transfers, requests for leave of absence, and terminations must be properly recorded so that all parties concerned are aware of the change of status. The Personnel Action Authorization Form (Figure 23-14-2) can be used for such purposes.

23-17. Personnel Development

Construction companies, like other businesses, progress by keeping pace with technology. New materials and construction techniques are being tried and proven daily. As a result, engineers are continually having to update their knowledge and skills to maintain their proficiency.

The management profession is also advancing, and newer and better techniques are being devised to plan projects, direct large groups of people, and coordinate the diverse number of project activities.

Ensuring that present employees maintain excellence in technical skills and management techniques is a task and goal of the personnel department, or, in the case of a large company, a separate personnel development department attached to industrial relations. This group is responsible for establishing and administering policies and programs for the continued training of all employees.

One widely accepted policy provides tuition reimbursement to all employees who take formal college courses or recognized correspondence courses in subjects related to the construction business and the employees job position.

Another practice is to bring certain general-interest college courses in-house for wider participation by company employees. Courses in technical writing, computer programming, CPM scheduling, and estimating can be taught by visiting professors from nearby colleges.

Although both of the above practices are successful in promoting continued academic and vocational training of employees, the training imparted is of a general nature, often not directly applicable to the specific needs of the construction company. Accordingly, it is becoming essential that the construction company conduct in-house training programs for its employees. To this end, the personnel development department should be engaged in planning and implementing courses covering key aspects and activities of construction. Necessarily included in these programs should be the following:

- Specialist Training Program—Technical training for the technical specialist
- Professional Management Program—Management training for middle management
- Supervisor Training Program—Management training for supervisory personnel
- Management Trainee Program—On-the-job management training and orientation for the recent college graduate
- College Cooperative Program—On the job training for the college student.

Specialist Training Program The specialist training program strikes directly at the heart of employee stagnation and evolved as part of an effort toward improving the various specialty skills of the construction department. Under this program, technical courses in estimating,

critical path scheduling, computer programming, procurement, cost control, and forecasting are given by the heads of the respective departments to their staffs. The established theory is first covered and is followed by group discussions of the company's present practice on the various projects. Suggested improvements brought forth during these discussions are developed into detailed procedures for trial workout.

Classes are conducted at the department location during work hours depending on the work load. A floating two-hour session approximately every two weeks seems the optimum frequency interval, allowing both the department head and staff time to digest and mull over previous discussions.

A significant advantage of specialist training courses is that the top members of each specialty can pass along their years of practical experience to younger staff members, at the same time brushing up on the latest techniques and gleaning new ideas from the neophytes.

Professional Management Program The professional management program is intended for members of middle management to review the fundamental principles and techniques of management in order to apply these fundamentals in their thinking and planning on the job.

Conducted by a training expert of the personnel development department, the program reviews the basic management functions of planning, organizing, leading, and controlling. Separate sessions include discussions of organization structure, leadership skills, decision making, techniques and approaches to problem solving, personal development determinants, and effective communication.

The sessions are conducted in a discussion-type format in small groups of ten to fifteen people selected to represent a cross section of the company's various operating and staff divisions. The small groups provide ample opportunity for in-depth discussions, with emphasis directed toward applying the management fundamentals to actual job situations. The diversification of management people provides a useful stage for exchanging viewpoints and ideas. Introducing real project or administrative problems in an atmosphere of discussion allows the participants to cooperatively practice management.

For most effective results (i.e., application to actual job situations) management programs must be comprehensive, requiring 15 to 20 sessions and conducted on a weekly or biweekly schedule.

Supervisor Training Program Aligned with the professional management program for middle management is a condensed management training program for supervisory personnel both at the home offices and in the field. The supervisor, since he is the person in direct contact with project personnel, must be capable of motivating and guiding the behavior of subordinates in order that they may carry out the tasks and objectives of the organization.

The supervisor training program presents the principles and concepts of supervisory management. Segments of the program are devoted to planning fundamentals, organization relationships, effective communications, and promoting job productivity and employee satisfaction through motivation and selection techniques.

The program should be succinct and job-problem oriented. It is conducted in-house by a member of the personnel development department. However, on a small field project, correspondence courses may be used as an acceptable substitute.

Management Trainee Program   The continued existence and viability of a company depends on how well management recruits and prepares individual members to take over the reins of management as older members approach retirement.

The management trainee program is designed to introduce recent college graduates into the practice of construction management. Select candidates with the aptitude, imagination, and desire for advancement are offered the opportunity to work on-the-job in all aspects and activities of the construction business. Each individual is under the sponsorship of a divisional vice-president who is responsible for the overall guidance of the trainee during the program. The objective is to provide each candidate with practical experience in the integral workings of the construction business, with the ultimate aim of an assignment in the sponsoring division.

The program is divided into two groups—one technical and the other administrative. Technical trainees spend the first part of their time in the various project departments performing assignments in cost control, estimating, critical path scheduling, procurement, subcontract administration, and industrial relations, as well as others. Once exposed to the various home-office activities of the construction project, the trainee is given a field assignment directly related to his interests and potential. As he continues to gain experience and demonstrate ability, he moves into positions of greater responsibility, such as construction engineer, area superintendent, field construction manager, and resident manager. His rate of progress will then depend on his performance on the projects.

The administrative program is designed to acquaint the trainee with the people, operations, and policies of the various administrative departments of the company. He spends a total of 12 months in the accounting, procurement, legal industrial relations, technical reports, and business relations departments learning all facets of administrative management through on-the-job experience. In addition, he is versed in many of the latest technical skills, including mathematical systems, computer applications, cost analyses, and economic studies. Upon completion, the trainee receives an assignment from the vice-president of administration.

The objectives of the management trainee program are to stimulate interest and participation in company ways and practices. The long-range purposes is obvious: to provide a continuous supply of talented people with the initiative, imagination, and leadership ability to keep the company competitive in its field.

College Cooperative Program   Paralleling the management trainee program as a method of obtaining new management potential is the college cooperative program for recruiting the best technical talent. Under this program students attend formal college classes for the first six months of each year and then work for the company the remaining six months. The plan is of mutual benefit to both the company and the student. The student has the opportunity to practice what he has learned over an extended period, and the company has the opportunity to size up his technical capability for possible future permanent employment.

The personnel development department is involved in planning the work activities and job assignments with the project department heads and the college guidance counselor, and requesting periodic evaluations of his performance in school and on the job. This information is used for reference purposes after graduation when consideration is being made for permanent employment.

23-18.   Equal Employment Opportunity (EEO)

From an economic as well as a social standpoint, there is need for strong measures to eliminate the fact and the causes of minority discrimination in employment. "Every unemployed and underemployed American, whether or not he is from a minority group, is a liability to our society." That statement, made by Edgar F. Kaiser, President of Kaiser Industries Corporation, focuses on the vicious circle that affects the economic viability of urban areas. Poverty of ghetto residents causes increased welfare costs, which cause higher taxes, which cause industries to move out or expand in other less-taxed areas. This results in fewer jobs for the residents and hence greater poverty for the community, thus completing the circle. Although the short-term effects of discrimination on the nation's economy and social order may be disregarded, the effect over an extended period can only be disastrous as businesses find fewer and fewer unblighted areas for new sources of labor.

Although the trend to move industry from the urban centers to the suburbs has been accelerating in the past decade, there is a counterbalancing trend which will force a national re-evaluation of the adverse effects of discrimination in employment practices. With the continued growth of business and the resultant increasing shortage of skilled labor over the next decade, companies will come to recognize the minority groups as the last reservoir of manpower. Thus, economic considerations will combine with social pressures to accomplish assimilation of minorities into the work force.

Affirmative Action   Because of the accelerating economic and social changes relative to equal employment opportunity, it is impossible to prescribe any long-lasting formulas or rules. The reasons for the remedial action taken now may be nonexistent in a few years.

For the construction industry, however, the basic precept for affirmative action has been well established in Executive Order Number 11246 of the Code of Federal Regulations: "The Contractor will take affirmative action to ensure that applicants are employed and that employees are treated during employment without regard to their race, creed, color, or national origin." Affirmative action in the context of the order means a comprehensive program that is result-oriented and extends to all personnel functions, including recruitment, hiring, assignment, transfer, promotion, and termination. It necessarily involves a thorough appraisal of company policy, seeking out evidence of inadvertent or purposeful discrimination. A good starting point is a review of all job categories to determine actual numbers of minority employees versus white employees being utilized. Where deficiencies exist that are disproportionate to the minority representation with requisite skills in the community, there is cause for immediate corrective action. Detailed procedures involving all personnel functions should be prepared, in which targets and goals are set, individual responsibilities are assigned, and an audit system is set up to measure progress.

During recruitment, for example, it is important to check all labor sources to assure that minority candidates are not unknowingly being excluded. Frequent contact should be made with local minority agencies, civil rights organizations, training centers, and schools to provide maximum exposure of the company to minority applicants.

During the planning phase of a field project, a survey should be made to determine the ethnic makeup of the community and the percentage of skilled and unskilled minorities in the area from which the work force is drawn. In cooperation with the unions, the company can participate directly in recruiting the skilled minority forces and encourage unskilled members to join craft apprenticeship programs.

It is important to make the union hierarchy aware of the company's EEO policy at the very start of the project. During the prejob trades conference and other union-contractor meetings, the company's written policy statement should be read aloud to notify union representatives of the company's intent to ensure equal employment opportunity for minorities. Allegations made by an employee or applicant that hiring hall procedures and referral practices are being conducted in a discriminatory manner should be thoroughly investigated and, if justified, brought to the attention of the union representative through formal compliance audit or complaint review.

The recruitment and hiring of exempt and nonexempt personnel should be approached with equal vigor. Employment agencies should be made aware of the company's EEO policy.

A tally should be made of all walk-in applicants and referrals from agencies to compare numbers of minority applying for employment versus numbers actually hired. Personnel interviewers should be asked to document their reasons for rejecting an applicant, the aim being to explore ways to eliminate the causes for rejection. One reason may be unwarranted educational qualifications of unvalidated testing methods. For example, in the United States it is common practice to hire only high school graduates for nonexempt job positions, even though a particular position may not require such comparatively high educational qualifications. It should be the responsibility of the personnel office to measure accurately the qualifications needed for each job position and, when necessary, to promote EEO policy hire on the basis of only those qualifications necessary to perform the job.

Training Programs Although hiring based on bare minimum qualifications is a good starting point in implementing EEO policy, it can lead to serious problems when the time comes for promotion of these employees to positions of greater responsibility. Therefore, any effort made in the direction of improving opportunity for unskilled minorities must be complemented by a comprehensive plan of training, in cooperation with local schools, community training centers, and state employment agencies. Every effort should be made to encourage minority employees to continue their education, offering job advancement and salary increases as inducements.

Large companies can become more directly involved in training to meet company needs through formal in-house training programs. One such program successfully being employed by one company is the Draftsman Training Program. An initial seven weeks of full-time in-house classroom training in mechanical drawing techniques and basic mathematics is followed by another seven weeks of specialized training in one of the design disciplines.

During the second seven weeks, one-half day is spent in classroom instruction, and the remainder is spent in the design section performing small training assignments and observing other draftsmen. Upon completion of the classroom phase, the trainee is assigned full-time productive work as an apprentice draftsman until qualified to perform competitively as a junior draftsman. This final phase varies upon the individual performance of the trainee, but generally requires from six to nine months.

The salary structure for trainees is designed to provide incentive by giving quarterly salary increases based upon performance reviews. The quarterly review permits recognition of the trainees' rapid increase in productivity from initial classroom training phase to full production at the completion of the on-the-job training phase.

Another successful program is the Field Construction Aide Program. Similar to the college cooperative program described earlier, it consists of one-half-time on-the-job training in the estimating or cost engineering departments, and one-half-time classroom study at a local college. The college courses provide specialized training in construction engineering subjects so as to complement the on-the-job training. The program continues for two school years with full-time employment during the summer months. At the end of this period, the successful trainee will be qualified for assignment to a construction project at entry level.

Since trainees are paid only for time worked (approximatley 20 hours per week) and generally have a productivity level equivalent to their salary, the cost of this program to the company is insignificant.

The value of these and other similar on-the-job training programs is that they open the door for minority groups to meaningful and respected job positions, without undo burden on the company or the community.

Communicating EEO Policy Company policy, if it is to be effective, must have the participation and support of the excutive management. Either wholly originated in executive chambers of carefully planned through the guidance of an EEO expert, a written policy on EEO will carry the necessary authority for proper implementation at every level within the company.

Once EEO policy has been formulated, the task of implementation falls to the personnel department, or sometimes, in the case of a large organization, to a separate department of industrial relations. The EEO director must see to it that all company employees are made aware of the new policy and that its guidelines and procedures meet with intended results.

Besides the strictly legal requirement of posting government notifications, there are many opportunities for communicating EEO policy within the company. The company newspaper or magazine, for example, should include articles reporting on EEO developments. Orientation of new employees can include a review of the company's policies and procedures regarding this activity. The EEO director should also hold frequent meetings with supervisors and managers to explain the intent of policy and individual responsibilities.

To make communication two-way, an internal procedure should be set up so that employees may discuss any questions or problems either with their line supervisor or directly with the EEO director. In line with this, individual employee counseling should be an inherent part of job training programs.

Outside the company, publishers of periodicals and local newspapers should be given full assistance in the preparation of articles and stories. Management should also be encouraged to participate in community organizations.

The possibilities for promoting EEO within and outside a company are boundless once management and employees realize that EEO is not only law, not only good business, but also only fair.

23-19.   Foreign Project Aspects

The foreign project imposes some additional responsibilities on both the headquarters and field personnel offices, as determined by the manpower availability within the particular country. For example, there may or may not be a skilled labor force, comparable to the union craftsmen of the United States. If not, the project will entail considerable training of local-hire personnel. In some countries, the construction company may have to deal directly with the government in working out labor problems; in other countries, the local unions may be the authorized representative of labor. The living conditions of the country also pose a problem for the construction companies. In remotely located areas, the company may have to provide housing, schools, medical facilities, as well as a commissary and field laundry for the project employees and their families.

Hiring practices are particularly complicated on foreign projects, especially those in the developing nations. Normally, there is considerable pressure applied by the government to hire local personnel whenever possible. In some cases working permits for expatriates are strictly regulated to assure that local personnel have every advantage to work on the project. Although it is generally more economical to hire local personnel, many positions of the project require persons with specialized training and experience. Consequently, although it becomes a problem working out some of the labor-relations problems, it is also necessary to hire and transport to the project U.S. expatriates and citizens of a third country (called third-country nationals) to perform the specialized tasks required. Unless hired as permanent employees of the company, these individuals are hired on a contract basis for a stipulated period of time, with all terms and conditions of employment specified in a written employment agreement.

Headquarter's   Personnel   Responsibilities After   an organization chart has been approved the headquarter's personnel office is responsible for the recruitment and processing of all expatriates, including third-country nationals, until the time that the field office is established, whereupon the field personnel office may assume some or all of these responsibilities. During this time, the headquarters office processes the expatriate personnel to the job site, helps them to obtain the necessary passport and visa, and also, when required, the work permit. At the same time, the salary administration department prepares a complete schedule of job classification and salary ranges, based on the currency of the country.

Employment Contract All personnel not considered permanent employees of the company and who are hired outside of the country in which the project is located are employed on a contract basis. The employment agreement sets forth the terms and conditions applying to the employment of the individual for a particular project. These terms include position, working conditions, effective date of hire and duration of employment, salary, transportation, travel, moving expenses, housing, schooling, and taxes. Other information in the agreement includes the employee's permanent address, list of dependents, and person to be notified in case of an emergency.

The employee agreement delineates both the company's rsponsibilities for the employee and the employee's responsibilities to the company for the duration of the contract. The company normally pays the cost of transporting the employee to the work site and also for the return trip upon completion of the contract. In addition, the company may either provide housing or a monthly housing allowance. Family status employees are reimbursed for the cost of privately educating dependent children when adequate schooling is not available locally.

The employee, on the other hand, must abide by the position, working conditions, and salary terms for the duration of the agreement.

24 LABOR RELATIONS AND THEIR EFFECT ON EMPLOYMENT PROCEDURES

LEE E. KNACK

Director of Labor Relations
Morrison-Knudsen Company, Inc.
Boise, Idaho

THE FIELD of labor relations continues to grow in complexity and imposes an increasing demand for greater management expertise. One can be easily tempted into writing many broad statements and generalizations about labor relations; however, since we are dealing with construction, it is essential that we establish the fact that labor relations in the construction industry are not the same as labor relations in manufacturing, transportation, services, or other industries.

24-1. Uniqueness of Construction Labor Relations

The contractor faces labor relations from the moment he decides to enter into business. For many different reasons, he may personally favor to operate his business on an open shop or "nonunion" basis. For other reasons, he may favor operating his business "union" or under the terms of a collective bargaining agreement. At this point in his decision making, he is establishing his labor-relations policies. If it is his choice to operate open shop, his labor-relations policies and programs can and will be quite similar to employers in other industries who also operate under open-shop policy. This parallel condition will remain until the contractor finds himself at the bargaining table with one or more of the seventeen unions that comprise the Building and Construction Trade Crafts.

When the contractor begins to enter into one or more labor agreements, his labor-relations policies will begin to parallel those of the contractor who initially chose to commence his operations under union agreements.

Here the separation between construction labor relations and other labor relations begins to appear. In prac-tically all industry except construction, no employer may enter into a collective bargaining agreement with a union unless his employees have designated that union as their collective bargaining representative. An individual engaged in the construction industry can legally enter into collective bargaining agreements with construction craft unions even before he has any employees. Specifically, the employers and unions in the construction industry can legally enter into pre-hire contracts.

Once the contractor is at the bargaining table, either by his own choice or as a result of successful organizing tactics on the part of the Building Trades Crafts, he will find himself dealing with the specific craft unions on the basis of the type of contracting in which he engages. For example, if he is a general contractor, he undoubtedly will be dealing with what is known as the six basic crafts. These are laborers, teamsters, carpenters, cement finishers, operating engineers, and iron workers. If he engages in specialized contracting or is known as a specialty contractor, he will find himself dealing primarily with one union. For example, if he is in the plumbing and heating business, he will be dealing primarily with plumbers and fitters, possibly sheet-metal workers. A painting contractor will be dealing solely with the painters. An electrical contractor will be dealing with the electricians, and so forth.

24-2. Collective Bargaining Methods

The contractor, as he approaches the need to bargain collectively, may choose to do so in one of four ways: (1) he may decide to carry out his collective bargaining on his own and on an individual basis; (2) he may become affiliated with other contractors of like kind in a formal association and have his collective bargaining conducted on his behalf by the association; (3) he may decide to become signatory to one or more national agreements to suit his needs. (4) he may have a combination of any of the preceeding three methods.

Several factors will influence which of the four decisions he will make. If he is just commencing in the contracting business, is performing limited type of work, and does not engage in public works, he may seriously consider conducting his own labor relations directly with whatever unions he may deal. If he does a considerable volume of work in the area, if his work is commercial and industrial, and if he engages in public works, he will be more inclined to become associated in a formal organization with other like contractors. If he has grown to large size, if he engages in specialized industrial work in various locations throughout the country, if he has been established in business for some long period of time, and if his work is primarily with clients from the private sector, he is likely to handle his labor relations from a collective bargaining standpoint on the national agreement level. National agreements are not tendered to contractors on an indiscriminate basis, however, and it must be recognized that national agreements are restricted to a limited number of either general contractors or specialty contractors. If the contractor is long established, has a well coordinated organization including a labor-relations department, is diversified as to type and location of work, and works for a variety of clients, he may enter into a national agreement with one or two of the specialty crafts and he may then handle the remainder of his labor relations in the area where he is performing work by joining the local Contractor's Association, by individual agreement, or even by negotiating a special project agreement.

Each of the four labor-relations approaches that may be taken in operating a construction company under union agreement has its advantages and disadvantages. The following analysis of the four approaches will not necessarily be all inclusive, but the primary advantages and disadvantages will be highlighted.

24-3. Independent Action

Where the contractor elects to conduct his labor relations independently and on his own, he has the advantage of dealing directly with the union. Where decisions are involved, he can weigh involvement solely as it effects him. He is not confronted with the impact it may have on others. He has the advantage of negotiating only for those classifications of workmen he employs and he can insist that discussion of matters not pertaining to him are not proper subjects for negotiation. For example, if he works in a limited area around a metropolitan center, he does not have to concern himself with negotiating travel time or subsistence allowances for work performed outside of the metropolitan area where he does his contracting.

There are many other matters that relate only to him, and when he negotiates on his own he needs only to analyze and evaluate those specific instances that do relate to him. In preparing for his negotiations, therefore, he does not have to be concerned with extraneous matters, and his preparation, as well as the negotiations can, therefore, be much more simplified.

A further advantage to the contractor who conducts his own labor relations comes in the administration of the agreement after it has been negotiated. His administration can comply strictly to his own company practices and policies, and the absence of outside influence in administration can be as important to him as the absence of outside influence in negotiations. Self-determination of administration is of primary importance when jurisdictional disputes are concerned. In the case of conflicting claims which may be made for the same bit of work, the contractor who is independent does not have to observe the claims of any craft or union with whom he does not have an agreement. As an independent functionary, if and when a jurisdictional dispute arises, he can resort to the provisions of the National Labor Relations Act for the resolution of any work stoppage or threatened work stoppage. Because he is independent, he can more effectively plead economy and efficiency for the basis of making his particular work assignment. This is in contrast to when he is coupled with other contractors through voluntary association or by virtue of national agreements which bind him more to local area or industry practices in the making of work assignments.

When he retains the independent right to administer the labor agreement, he has the advantage of deciding the manner in which he will resolve grievances that might arise during the term of the agreement. He has no obligation to go through any other agency or association and, because he has no such obligation, he retains the determination of whether to resolve the grievance at any given stage or to force it through all steps, including arbitration.

The basic advantages of choosing to act independently are directly related to the management ability of the contractor's organization. When he has internal expertise to negotiate a successful labor agreement and where he has internal capability for the administration of such an agreement, the advantages of acting independently weigh heavily in his favor.

The disadvantages of acting independently are equally related to the individual's ability to handle his problems. He must be prepared to devote the necessary time to make his organization knowledgable and competent in the technical and legal aspects of labor relations. It will not take him long to find that ignorance of the law is no defense for committing an unfair labor practice, and his lack of knowledge can be very expensive. Because he cannot draw upon talent from among his associates when he is acting independently, he must match his organization's capabilities against those of the unions and their representatives, who are trained and who spend full time in the field of labor relations. As an independent operator, he does not have the advantage of benefiting from the experience of other contractors. He is more vulnerable to secondary boycotts for which (even though they are illegal) prompt and capable action is the only deterrent to a costly and irrecoverable loss.

As an independent operator, he must be prepared to face more attention from union business agents. They are fully aware of the kind of resistance they will meet when acting against an employer who is a member of a group. They know that their ability to accomplish "nuisance settlements" with the independent contractor are greater than it is with an association of contractors. This causes unions to single out the independent contractor as fertile territory for testing out contract provisions. It is also a fertile area for them to establish area practices or jurisdictional claims which they might not otherwise establish. Administratively, therefore, a disadvantage of acting independently is the excessive amount of time the contractor may have to devote to labor relations occurring solely because he is independent.

When weighing the advantages and disadvantages of being independent, obviously the contractor is going to examine the same circumstances should he become affiliated with other contractors of like kind in a formal association and grant that association the right to conduct collective bargaining in his behalf. The advantages of association are multiple. Through association, the contractor immediately can draw upon the vast storehouse of knowledge and experience that is compositely represented by his fellow contractors who also belong to the association. The long history of area collective bargaining is available to him. Area practices in the conducting and making of work assignments are readily available. Previous arbitration decisions can provide him with a shield against similar demands or in the handling of similar grievances which the union may make or bring up during the term of the agreement.

24-4. Affiliated Action

Through association, the contractors are often able to employ Association personnel who are specialized in the field of labor relations as are the union representatives. The preparation for negotiations which can involve the research of conditions, surrounding wage patterns, and other significant economic factors, can be more efficiently handled through an association than by the individual. In the association, a stronger and more unified balance of collective bargaining power on behalf of the contractor can be presented at the bargaining table. The union recognizes that action against one member of the association can readily bring into full force and play the economic power and force of the whole organization. The advantages of association with other contractors can best be reflected in the efficiency of negotiations, in the cost of negotiations, and in the time spent in negotiations. A contractor who belongs to an association will find himself having more time to devote to the other phases of his business.

The disadvantages of association with other contractors are those that are often identified with group action. It is almost impossible to obtain unanimity of thought and action; and when the moment of decision occurs and action must be taken, group relationship and the effort to obtain group concurrence can be a most frustrating experience. For example, if the moment of decision as to whether or not a strike should be taken occurs, those members of the group who would be least effected by the strike are, of course, most adamant in maintaining the contractor's position, which could result in the strike. Those who are affected the most and who would be hurt the most by a strike are the ones who are willing to "sweeten the pot" in order to avoid the strike. Nowhere is the phrase, "caught in the middle," more appropriate than it is for the contractor who, as a member of the association, is asked to cast his vote on a strike situation and he has just enough work to be affected but not so much that he couldn't tolerate a strike action.

There is also a disadvantage in association in the administration of agreements that have been negotiated. An individual who has a problem in the form of a grievance or a threatened work stoppage may want to take very strong action to resolve the problem area. However, because he must work through his association, he may find reluctance on the part of the association to pursue the action he desires. Even though his company may be of sufficient size to have a board of directors in the matter of decision making in negotiations and contract administration, he may find himself faced with the dilemma of a decision from the board of directors of the association which does not coincide with the decision of his own board of directors.

Indifferent and careless administrative practices on the part of fellow members of his association can and do often accrue to the disadvantage of the member of the Contractor's Association. For example, a member of an association may, under a given set of circumstances, agree to the featherbedding demand on the part of the union. By a process of insiduousness, such a practice might spread among the weaker members of the association until, ultimately, it has spread so far that the union takes it up as a matter of industry practice. Then all members of the association can find themselves saddled with a costly practice. Some of the most uneconomical and costly factors currently in the construction industry have found their inception in this manner.

24-5. National Agreements

To the contractor who grows sufficiently large and engages ultimately in diversified client business, there comes a time when he may weigh the advantages and disadvantages of becoming signatory to national agreements as a method of handling his labor relations. The advantages of national agreements can be extremely attractive because strong collective demands are not necessary to negotiate the agreement itself. National agreements in the construction industry contain boiler-plate language covering general conditions, but they are primarily designed as an agreement: the contractor agrees to be bound by all of the terms and conditions of the local agreements wherever he performs work, and the union agrees that there will be no strikes on that employer's work, including the period when the union might be striking the contractors at the local level. The national agreement might be described as one in which the employer agrees not to become involved and not to influence the local negotiations, and the union agrees not to involve the employer in any economic actions that might result from impasses that occur during the local negotiations. One can readily see that such an easy arrangement, providing what appears to be on the surface a kind of peace and harmony not contractually obtainable otherwise, makes the national agreement most advantageous.

Administratively, the national agreement offers further inducement in that grievances are usually handled at the local level initially, and if they are unresolved at that stage they are then referred to the general president of the international union involved and then to a designated high official of the contractor. The administrative capability of being able to resolve critical and serious grievances and disputes at the more mature level of the national union puts a large plus on the side of the national agreements.

In view of what appears to be strong advantages for the national agreement, one could contemplate as to why there are not more national agreements in effect. National agreements are much more one-sided than one may realize. They are strictly voluntary in nature, and international unions are careful in selecting those to whom they allow the privilege of signing a national agreement. One of the disadvantages to national agreements, therefore, is the fact that the contractor is chosen more than he chooses; and because he is chosen, he is expected to perform in the interests of the international union with whom he has signed a national agreement or suffer the consequences of having such a national agreement cancelled on as little as thirty days notice. This could mean that a contractor having invested a great deal of security in the peace and harmony of his organization could, if he fell into disfavor of a particular union, find his national agreement cancelled at a time when he had considerable work going on across the country. In one fell swoop, all of his work, regardless of location, could be put in jeopardy because of a dispute or grievance which occurred at a given single location.

Because contractors who have national agreements invest a great deal of security in their intent and purpose, they often will agree to costly conditions to preserve that security. For example, each union who grants a national agreement to a contractor expects that that contractor will make all of the work assignments favorable to that union. Thus, when a contractor has two national agreements with unions who are competing for the same work, he finds himself with an almost unsolvable dilemma. However, he finds the solution by putting an equal number of men from both unions on the same piece of work; thus, twice as many people are employed to do the job. The local representatives are satisfied and the international union or unions may or may not be aware of the problem and its solution. The constant threat of the loss of the national agreement weighs heavily over the heads of those who are signatory to it.

24-6. Combination of Methods

The fourth procedure which is available to the contractor in pursuing his labor relations is the combination of any of the preceding three methods. He may, for example, sign a local agreement with one or more crafts as an independent contractor; he may be bound to some local agreements with other crafts by virtue of being a member of the local Contractor's Association; and he may be signatory to one or more national agreements with still other crafts.

All of the advantages of each of the preceding methods of handling labor relations obviously do not accrue to the contractor who follows the combination method. He does enjoy some of the advantages of each of the other methods, but he must have much greater expertise to keep his organization in balance under such a conglomerate. It does provide him, however, with an opportunity to employ some unions and craftsmen directly without going through a subcontractor, thus avoiding subcontracting costs and avoiding the loss of direction over some of his work. Not all of the advantages of the preceding methods of handling labor relations are available to the contractor who follows the conglomerate action, but most certainly all of the disadvantages of each of the preceding methods exist not only in their original form, but in some circumstances they are multiplied.

Although a great deal of emphasis is on the expert capability of the contractor to follow an individual program of handling his labor relations, even more expertise and capability are required of the contractor who follows the conglomerate method.

His first responsibility is to be certain that his working organization is fully conversant with which of his agreements is in existence at a given location. One can easily imagine the confusion that can and does result when job supervision assumes that because a National Agreement existed on a job they recently did on one location, that same National Agreement is being applied to their current job. However, it is conceivable that the contractor under the conglomerate action has decided for his own benefit and at the particular job in question that he is bound by a local agreement negotiated by an association activity. The confusion that can exist, therefore, at the job level can cause labor-relation headaches. It is for this reason that the more complicated the labor-relations program of the company, the more need there is to see that the complexity is held to a minimum at the operating level. This requires that management establish clear and precise operating procedures and policies and then that such policies and procedures be carefully relayed and communicated to the operating segments of his company.

No matter what the contractual method of handling his labor relations, the contractor still feels the basic effect of increasing wages, fringe benefits, featherbedding practices, etc. As the decade of the seventies progresses, the impact of increasing costs in the construction industry is being felt as never before. Long-term agreements are being negotiated covering from one to five years. Wage settlements are running from as low as 50¢ per hour per year to as high as $1.50 per hour per year. The percentage increase is varying from 10 to 11 percent to as high as 35 percent, and in some isolated cases the percentage increase is even greater. The total cost packages are being represented in both wages and fringe benefits. The fringe benefit settlements are directed mostly toward improved health and welfare benefits, pension programs, training programs, and vacation plans. Although it would be illogical to charge the construction industry with being totally responsible for the escalating costs of medical care, it would be equally as illogical to believe that the ever-increasing demand for more and more hospital benefits and medical services to be provided through collective bargaining has not made a major contribution to the inflationary spiral that has hit the medical-care program.

Through the many years of past negotiations in the construction industry, no matter how bitter or lengthy the negotiations were, upon their completion one could always rationalize and understand the justification of increases that had been hard fought and hard won at the bargaining table. In the current settlements, however, there is apparently little rhyme or reason for the size and extent of the agreements being reached. Those who spend many hours at the bargaining table recognize that the demands being put forth by the unions in the construction industry are being dictated by the militant and vocal members of the local unions. Justification for demands are often predicated solely on the basis that somebody else got the same amount. As a matter of fact, the membership itself in sitting at the bargaining table, not being required to justify the demands, apparently picks numbers out of a hat or out of thin air sometimes to surpass another craft, sometimes to surpass a neighbor sister local, sometimes to "get even" with the local bargaining committee of contractors, and sometimes for a multitude of other unrealistic and undistinguishable characteristics that can often be categorized as being emotional rather than rational.

This same attitude is also resulting in less productivity at the same time that basic costs are increasing. Some studies show that, in the ten-year period from 1960 to 1970, wages in certain areas of the construction industry have increased by as much as 100 percent, while specific productivity has decreased as much as 35 to 40 percent. The increased costs coupled with the decreased productivity in such areas can result in a total increased cost for labor in the amount of 150 to 160 percent in the next ten-year span. As significant and as much of an impact that these figures might have when looked at on a ten-year basis, they have even a greater significance when one recognizes that the majority of the increase in wage costs and the majority of the productivity decreases have occurred in the last three-year period of the ten-year study.

The inflationary effect of increased wages coupled with lowered productivity has become a major concern of both government and the private purchasers of the services and products of the construction industry. Major private owners throughout the United States have organized and formed what is known as "Construction Users Anti-Inflation Roundtable." The list of participating companies and their chief executives who have concerned themselves over this problem would represent the major companies and the who's who of industry in the United States.

24-7. Influence of Inflation on Labor Negotiations

The effect of the inflationary cycle in the construction industry has been felt at the bargaining table by industrial leaders. Industrial unions are constantly pointing to the collective bargaining gains made by the building and construction tradesmen and are using these settlements as a measuring device against which to formulate their own demands with the industrial owners. Whenever the industrial owner points to the difference of employment in construction as contrasted to other industries, the industrial unions counter with the demand that the owner not subcontract or contract the construction work to construction contractors and construction unions but to do so with his own industrial forces. Heretofore, the industrial owner has purchased these services from the construction industry fully recognizing that it was to his advantage economically to do so. The skill capabilities of the construction worker under the direction of the construction contractor under the competitive bid or negotiated bid system provided him with a service and the purchasing of a facility at a cost cheaper than he himself could accomplish. It eliminated the necessity for him to enlarge his work force to develop the special skills or to recruit those that possess the special skills.

The margin of difference, however, between doing it himself and purchasing a contract is narrowing rapidly, due to increased costs in the construction industry coupled with lowered productivity. Both the contractor and construction laborer must face the reality that the owner is looking to other sources for the purchase of construction services and the physical facilities. He cannot afford to pay labor through a contractor at a much greater wage rate than he pays his own forces. He cannot tolerate excessive overtime conditions. He recognizes the insiduousness of tolerated featherbedding can spread throughout his own organization, and he has no comprehension or patience or understanding of the justification of the numerable work stoppages that occur among construction unions as a result of an ever-existing jurisdictional bickering. All of these costly conditions have brought him to the table for evaluation and discussion among his associates and contemporaries.

The house building industry has felt the impact of increasing costs and, as a result, the majority of the homes in the cities of the United States are being built "open shop" or by nonunion craftsmen and workers. In many instances in the home building industry, contractors pay union wages; but by operating through the open-shop method, they are not faced with jurisdictional disputes, restrictive work practices, featherbedding, and all the other conditions that contribute to lowered productivity.

The government has been studying the method of awarding contracts, which causes most of the construction activities to be concentrated in a seasonal period and often results in an erroneous and unnecessary shortage of craftsmen available to do the work. The government as an owner is seriously considering the disbursement of construction contracts over a wider calendar period of time, thus providing more employment for more workers and reducing costly overtime. The myth of seasonality in the construction industry is being carefully scrutinized.

With all of the problems relating to increasing costs and with the recognition of concern on the part of the purchasers of construction services and facilities, contractors and contractor groups are considering all avenues to improve the economic outlook for the industry. The first and foremost matter with which they must concern themselves is how does one slow down the rate of increase in wage costs in union negotiations, and it appears definitely to be a problem of slowing the rate down. There seems to be no single major answer or solution to slowing down the demands with which the industry is faced.

The enigma of the problem is highlighted when construction falls off in some areas for short periods of times and the union worker readily accepts employment from the nonunion contractor and performs for him better than he does when working where a union agreement exists. In the days of modern technology, considering the advances we have made one cannot accept the cynicism of Ambrose Bierce who, at the turn of the century, was quoted as saying that the answer to all labor problems was hunger. One can, however, accept the reality that the inflationary spiral in the construction industry will slow down and ultimately be halted when the construction worker moderates his wage demands and ultimately reaches the point where, out of necessity, he is willing to work for the same wage today that he received yesterday, and in the process he is willing to produce to the same degree of capability.

Contractor groups may contribute to a slowing-down process through better preparation and stronger positions at the bargaining table. A possible improvement in collective bargaining could occur if all construction labor agreements covering a given location were to have a common expiration date. More unification and association between the various segments of the contractors association might be able to accomplish more effective negotiations if they were to join in more concerted effort. At the present time, the fact is that the general contractors negotiate through their association and often independently and without consultation between themselves as employers. The plumbers and fitters, electricians, sheet-metal workers, painters and many other craft unions and contractor associations negotiate their own respective agreements. This separated negotiation practice has undoubtedly contributed to the inflationary spiral. Politically, the business agent of one craft must at least equal in accomplishment the negotiation attainments of other crafts with whom he works, and if possible he must surpass them.

Unification and possibly more moderate settlements would occur if all the contractor groups within a given area were to negotiate, if not jointly, at the same period of time and certainly with full communication between each other. Some experiments in this direction have been conducted in Pittsburgh and a few other areas: They have produced some effective results; yet it is realistic to state that unity alone has not provided the expected or hoped for answer. Contractor groups might be more effective in obtaining reasonable settlemens if they were to communicate the known and obvious losses that the industry has suffered as a result of inflationary conditions.

Conferences are taking place between large construction organizations and international unions, and there are some healthy attitudes developing out of these conferences; but it is a considered opinion here that more communications and more healthy attitudes must be developed at the local levels. Contractor groups can make invaluable contributions toward improving the productivity at their work sites. Some joint management-labor seminars have been held in various locations, and surprisingly effective discussions have been taking place on the need to improve skill productivity as well as skill development. These seminars appear to be more effective when conducted in association with contractor groups than when they are initiated by the individual contractor.

There is no single answer, but there is hope for improving the situation if contractor groups will consider the following courses of action:

1. Exchange information of common interest and have closer association.
2. Plan for unity and cohesiveness among all contractor groups and associations for collective bargaining purposes.
3. Negotiate their regional or area bargaining agreements during the same calendar period.
4. Prepare for negotiations with affirmative and definite information and assume more aggressive posture and less of the defensive posture at the bargaining table.
5. Prepare and develop carefully oriented seminars and work clinics between management and labor, preferably to be held at times other than under the heat of negotiating or renegotiating agreements.

24-8. Strike vs. Lockout

In the search for effective tools, the strike versus the lockout as a balance to economic power at the bargaining table has received considerable attention. The principle of a strike against one is a strike against all was widely embraced by employer's associations. In accepting this principle, associations eliminated the complaints that occurred during strikes from contractors who were shut down by a picket line as opposed to those who were working because there was no picket line. Unions in the construction industry have historically employed the tactic of selective striking against members of the associations. Naturally, they selected those they believed could least stand a strike and would capitulate the soonest to their demands. This type of hit-and-run tactic created dissension and a lack of unity among the contractors in the associations.

As was previously reported, the ability to obtain unanimity among contractors in an association is not an easy task, and it is compounded when some members of the association are working and some members are struck. Primarily because the lockout placed all of the members in the same category (i.e., of not progressing with any work during the economic crisis), contractor's associations studied the device as they saw it used in other industries and began to use it as their own.

There are many schools of thought concerning the effecitveness of a lockout.

First, when a lockout is contemplated, consideration must be given to its legality. This is particularly true when an association has existing and continuing agreements with some crafts and is contemplating a strike-lockout action in relation to one craft. It is possible that due to specific circumstances on a given job or project, a contractor locking out in joint effort with his associates might place himself in legal jeopardy in relation to some of his other existing contractual relationships. No firm rule of thumb can be drawn in this text, and only competent legal evaluation can determine the validity of any contractor engaging in a lockout.

Secondly, when a lockout is contemplated, its practical effectiveness must be carefully weighed. When unions picket, it is costly to them. They must pay the pickets a wage and they must physically supervise the picketing to be certain that it does not become involved in illegal activity. Under these conditions, a union can realize considerable economic benefits if, in striking one contractor of an association, the others all lock out, thus effectively eliminating the necessity of the unions to hire and supervise many pickets. It must be recognized that a strike-lockout situation is a matter of testing economic strength of the parties involved, and there is a school of thought that believes there is economic advantage to forcing the union to employ and supervise as many pickets and picket locations as possible. Indeed, there are instances in the past where treasuries of local unions have been effectively depleted in such a maneuver, thus generating considerable respect and reluctance toward any future economic actions.

Third, the attitude of workers should be considered, and proper understanding must be conveyed to them if a lockout is to be effective. If a strike action and a vote to take a strike are marginal within the local union, this condition should be given some evaluation in contemplatng a lockout on the part of the contractor. It is entirely possible that a hastily conceived and executed lockout could solidify the union members into a longer strike and holdout for a longer period of time toward a settlement than if a marginally voted strike action had been allowed to run its course. Simply put, it is rare that all workers are unanimously agreed and desirous of engaging in a strike.

Those who are forced to go on strike but who are not in agreement with it can have their dissatisfaction with their local union leadership transferred to dissatisfaction toward the local contractors if a strike is converted into a lockout without considering the attitude of the members of the local union. Each economic impasse must be approached on the reasonable assumption that there is an ultimate solution and resolution to the impasse sometime in the future. This reasonable assumption cannot be disregarded when a lockout is being weighed. It would be totally unreasonable and unrealistic for employers to engage in a lockout if the calculated aspects of the lockout were to indicate a prolonging of the impasse and the delaying of a settlement. It would be equally as unrealistic and impractical not to engage and instrument a lockout if, after all considerations of its legality have been resolved, it was calculated that the lockout would expedite a settlement and aid in resolving the impasse.

Lockouts should not be engaged in solely because they may have worked in some other area any more than they should be disregarded because they failed to work in some other area. As a tool or device toward reaching an ultimate resolution of a serious problem, all factors involved in the activity should be weighed and calculated. When the pluses of those calculations are favorable to the association, the effectiveness of a lockout is considerably enhanced. If the calculations are in the opposite direction, an indiscriminate lockout can complicate the situation rather than making an effective contribution to it.

In considering the question of economic strength and power, a strike is that device used by labor to withhold the services of the workmen from the contractor until an agreement has been reached. A lockout is that device of the employer whereby employment is withheld from the workers until a settlement is reached. The key to this important aspect of labor relations is the flow and supply of workmen to the place of employment.

In making a diagnosis of what appears to be a sick situation, the flow of workmen to the place of employment must be given serious consideration. Wherever and whenever a constriction of that flow exists, the end result will be an increased cost or reduced productivity, which in reality are one and the same thing.

24-9. Hiring Halls

In the construction industry, the permissiveness of negotiating exclusive hiring-hall provisions in labor agreements is one of the most significant and critical conditions in modern construction industry. It has always been the custom, where craft capability is required, for the employer to look to the union as his primary source for the skill capabilities of the craftsmen. Craftsmanship, however, as has been noted elsewhere herein, has been on a decline and the circumstances to improve it have not been meeting the need. To put it more simply, with craftsmen retiring at an earlier age, the various programs for training craftsmen, including apprenticeship, have been falling woefully behind the need. As a result, unions have not met, and are not able to meet, the demands for the numbers of skilled craftsmen. The seventeen unions that comprise the Building and Construction Trades crafts are registering workers as being available for work who do not possess the capability to perform the work involved. As contractors continue to call the union for their worker needs, depending upon the conditions which may exist in the location, more than 50 percent of the men referred to them can be lacking in the necessary skills. This is the price the construction industry is paying for the exclusive hiring-hall condition that exists in the construction industry.

Exclusive hiring became a part of the construction industry as a result of a worker filing a charge with the federal government against a contractor's association and a union claiming that he had been discriminated against because of his failure to obtain employment in the construction industry. In essence, his complaint was that the union and the contractor's association had by contract agreed to obligate him to obtain his employment in the construction industry through the union hall. In resolving that historical case, criteria were established by the National Labor Relations Board which, if the criteria were met by the bargaining parties, would make the exclusive hiring-hall practice a permissive and allowable condition in the construction industry.

After the rendering of this decision, negotiations for exclusive hiring halls became a pattern. They are commonplace and exist practically in all parts of the United States, including those states with Right-to-Work laws. Most unions now claim that under the registration requirements of maintaining an exclusive hiring hall, they must accept the registrants without discrimination and, therefore, they are obliged to "enroll" on the records many individuals who are incompetent.

Although the above indictment cannot be made of all unions, it is sufficiently broad and prevalent to cause a serious and continuing problem. Where and when construction activities are at a high point in a given area and when the labor supply is low, many contractors have individuals referred to them from the union hiring halls who are less capable than workers that might be available from other sources. The exclusive hiring-hall permissiveness in the industry has allowed the unions to maintain a tighter control over construction manpower; but in the process of maintaining that control, the unions have allowed their reputation of having craftsmanship available to become diluted.

In many instances, contractors would do better to train younger and totally inexperienced men than to be obligated to accept some of the workers being sent to them from the unions. For example in a study made in the mid-1960's, on one project in the middle-western United States, 17 out of 262 journeymen employed on that project were from the local union, 105 were permit holders (none fully qualified), and the rest were travelers. On another mid-western job, only one member of a particular craft was a local journeyman and the other 40 were permit men. In studies made in the Gulf Coast area, over a two-month period it was determined that the ratio of unskilled permit men to qualified journeymen ranged from 22 to 1 to 38 to 5. In still another job, 33 of 44 journeymen were, in fact, permit men.

At this point, a permit man should be defined. He is an individual who does not qualify as a journeyman because he has not had the proper training and experience. The local union does not choose to have him become a full-fledged voting member and so, for a fee, they issue him a permit to work in the trade or craft. Always he must be paid and is paid the full journeyman scale, together with all the fringes called for in the agreement. Many permit men referred by the union are less capable than young apprentices with less than one year's experience.

One union member in describing the difficulties stated:

"It has been found that Building Trades Unions unable to supply sufficient manpower have resorted to referring the moonlighter, the student, the chronically unemployed, and members of other trades. These bodies are unqualified job roamers, who contribute to the low productivity figures. These substitutes are not referred out as helpers, trainees, or probationary members at low wage rates but are sent to the job as mechanics at the negotiated wage. Only recently, to cite an example, three 'mechanics' had to leave a job in mid-day because the local metropolitan police department called in all off-duty patrolmen."

In another instance, a number of workmen on a project who had been referred out by a union as highly skilled craftsmen refused to work overtime since they had to go to their night-time work, thus revealing that the Building Trades Union had called up the bartenders to supply the construction job with "bodies."

A considerable amount of space and time could be devoted to this particular aspect of the problem, and certainly in its examination one comes to realize there is no single answer and solution to it. It is necessary, however, that the many facets of this problem be given attention without delay. Improved methods of training workmen must be adopted. Obsolete and unnecessarily long training programs must be revised. Younger workers must be sought and provided with adequate training, and certainly the disadvantaged and minority groups must be recognized as a source of trainable and usable workers for the construction industry. It is essential that management not look to government or unions or even to joint efforts as a sole solution to the problem.

24-10.   Turnover Policy

The administration responsibilities of management must be brought into play in approaching this condition. It is necessary to borrow administrative pages from other industry personnel approaches. The construction company that maintains adequate labor turnover and daily absentee reports can make a valuable contribution to the solving of their own problem in the manpower field.

In establishing a turnover policy, particular attention should be paid to establishing the accurate reasons for termination of employees. When job supervision terminates an individual as "laid off—reduction in force" as an easy way out when, in reality, the individual should have been discharged as incompetent or incapable, not only does that company suffer an unrealistic and unnecessary charge against its unemployment compensation rate, but it also suffers the consequence of deceiving itself as to what the real manpower problem is in a given area.

All job supervision, therefore, should be required to turn in written terminations and to be properly instructed in listing the termination in an honest and realistic manner. Once such accurate records are conveyed to management, it becomes a simple matter to translate them into labor turnover reports which, when examined, give a true reflection of the manpower problems that may exist in the locality where the work is being performed. Special emphasis should be given to "quits" and discharges, and sample exit interviews and reports thereof will help to evaluate a manpower problem in a given area.

24-11.   Absenteeism

Daily absenteeism is another management and administrative function in which there has been laxness in the construction industry. The high rate of absenteeism following payday and certain holidays as a normal condition on the construction job has for too long a period been passively accepted by the industry. Studies have shown that daily absenteeism as high as 25 to 30 percent of the work force does occur on construction projects and in many cases without concern or attention from management. Properly established corrective absentee programs have demonstrated that these rates can be reduced to as low as 1 and 2 percent. There is no reason why the construction absenteeism rate cannot compare favorably, and in some instances surpass, the absentee experiences of other industries.

In passing, it should be noted that because of the tendency for job roamers to be attracted to construction, many workers pull up stakes and just disappear without any notification to management. Companies that do not maintain absenteeism controls frequently are unaware that an individual or individuals have left their organization, and this lack of knowledge has been known to exist for as long

as two to three weeks at a stretch. The maintaining of daily absentee records and daily follow-up would eliminate this totally ridiculous situation.

The proper administration of labor turnover and attendance on the job is in reality taking steps to ensure the best of the labor productivity which is currently and daily available. But the manpower problem in the construction industry needs more attention than that. Management cannot wait for the manpower problem to be solved by someone else. They must begin to solve that problem through their own initiative. In looking at the problem, they must ask themselves: Where and how can adequate and qualified manpower be obtained? The solution to this problem will depend largely on the imagination and the initiative of each employer and how well he applies his option of selectivity.

24-12.   Selectivity in Hiring

Those engaged in the construction industry are often deprived of areas of selectivity that are open to employers in other industries. For example, the selection of the location of the site of a dam cannot be made by the employer. That location and its selection have already been determined by nature and there is no choice in the matter. Consequently, the scene of operations in the construction industry is frequently in some remote area and far from available labor markets. Conversely, employers engaged in manufacturing or processing have the flexibility and mobility to exercise selectivity in plant location and can thereby take advantage of available transportation systems, utility services, and most importantly, established labor markets where adequate and qualified manpower is available. Even so, the construction industry has the opportunity to exercise some degree in selection in solving its manpower problems. It has the selection of the people to be employed (in spite of a severe restriction of a union hiring hall); it has the selection of those to be retained in employment; it can and should have the selection of those who enter training programs; it can select the training programs that best fit the specific requirements; and it has and should exercise the selection as to the men who will serve as instructors in training.

Selectivity at the time of the original hiring is of utmost importance. Personnel departments should develop and follow criteria that will produce the best men from the available manpower pool. Because of the state we find ourselves in today, aside from the obvious considerations such as physical condition, experience, and qualifications, consideration must be given to the prospective employee's ability to accept training. Is he young enough to have a career potential in the industry? Is his interest and attitude compatible with the training environment? Has he shown or demonstrated an aptitude for the skills that must be learned? Personnel administrators will have to place more emphasis on the training potential of each individual employed if the manpower shortages faced by the industry are to be alleviated.

Management must regain control of its prerogative of selectivity in employment, and careful attention should be paid to establishing probationary clauses and wage rates for probationary employees in labor agreements. If an employee proves to be unsatisfactory or exhibits an unsatisfactory potential, he should be terminated as soon as this evaluation has been made and a decision reached.

The obvious answer to the manpower problem is training. The manpower market is such that it is unrealistic to expect a sufficient number of qualified men to make themselves available for the work that needs to be done. Consequently, a program to train the inexperienced and to upgrade the capabilities of the partially skilled must be adopted. If the selectivity of the employment applicant has been properly handled, the selection of men as trainees has been automatically achieved. In selecting men from the present work force whose skills are to be upgraded by training, consideration should be given to demonstrated performance, aptitude, and attitude. An evaluation by first-line supervisors and crew foremen is most valuable, and criteria for such evaluation should be developed and put into effect.

24-13. Training Personnel

Without doubt, the most important factor in a training program is the selection of the program itself. An evaluation must be made as to what specific and immediate need must be met, and a program should then be developed around that specific need. The most successful training program is one that teaches the individual how to perform the required job in the shortest possible time and in the most efficient manner. The breadth and scope of the program should be carefully calculated and caution should be exercised that extraneous social connotations do not become a major part of the program. Today it is popular to become involved in correcting the social ills of the deprived, the culturally disadvantaged, and the unemployable, but an educational or academic program is no substitute for a practical job-training program that efficiently prepares a man to perform a task in the shortest time.

Training programs must be looked at in the light of training people for jobs that need them rather than training people who need jobs. An example of a successful program was conducted on the west coast for work to be performed under compressed air. No compressed air work had been performed in this area for more than thirty years; therefore, experienced and qualified compressed air workers were not available. Some experienced workers drifted in from other sections of the country, but since this was a four-shift, around-the-clock operation it was impossible to rely on the influx of experienced workers. The nature of compressed air work being what it is, there was no way that a classroom could be set up which would realistically duplicate the conditions found at the heading of a compressed air tunnel. It became obvious that the only realistic training environment was actually in the tunnel itself.

A number of inexperienced young men were selected and were passed through a medical examination to qualify them for compressed air work. When a regular employee was absent, one of these preselected, inexperienced young men stepped in and filled out the crew. He was assigned cleanup work requiring the least experience, and the man that had been performing that cleanup work regularly moved on up to the job vacancy caused by the absentee. Thus a dual training function occurred. The preselected, inexperienced worker became exposed to working under compressed air in the crowded conditions that exist in a tunnel heading, and simultaneously the cleanup man in the regular crew was able to get experience in a more skilled classification, such as steel support ring erection. When a vacancy became permanent, the preselected trainee became a permanent member of the crew.

The selection of these young people was made on the basis of good health and an expressed desire to learn compressed air work. They included many applicants from minority groups, some of whom had extensive police

records and had been regarded as unemployable. The only attention given to them on the job, however, was to train them to become good compressed air workers and nothing more. The program proved to be so successful that when a new heading was started, a complete crew of inexperienced men was placed under an experienced foreman. Through practical exposure to work under a planned and selected training program, this crew soon learned the rudiments of tunnel driving and support erection and was performing to the same degree of efficiency as experienced compressed air workers.

The success of the example cited was attributed to the fact that it was designed to perform the function of training people for jobs that needed them. The training was programmed on a timed basis for skill acquisition as soon as possible. It was a program without extraneous frills. It was a simple, practical approach to the problem, the extent of which can be understood when it is realized that approximately 450 men were so trained.

Because of the importance of manpower and the training of manpower, training programs should be selected to fill a specific need and should then be streamlined and tailored to that need. The grab-bag approach to training, wherein an effort is made to take a universal program or one that has been developed to meet another need, should be avoided. The selection of instructors or training supervisors is also an important factor in developing specific programs. It is human nature that the instructor or supervisor will have the tendency to train people in the same way he was trained. This tendency can result not only in archaic methods, but it can also result in overtraining, a waste of training time, the failure to meet a production schedule, and the development of unneeded and irrelevant skills. Because of these hazards, it is imperative that the instructors be given a definite written program and schedule aimed directly at the desired training goal. This will prevent digression and overtraining in unnecessary areas.

Employers everywhere are facing changing concepts in manpower and manpower developments. The tendency in corporations is toward smaller, relatively independent, self-controlling, responsible, and flexible units. To obtain this flexibility, constant training and retraining will be necessary. This adaptability will depend upon management's ability to be selective in all areas relating to manpower, in hiring and placement, and above all in training.

24-14. Looking Ahead

In the field of labor relations with all of its attendant problems and complications, one cannot look backward only. There must be projections, and it is even necessary to indulge to some degree in prognostication. There is no question that this is more true in the construction industry than in many other industries. Long-term projects often extend beyond the time covered by labor agreements. Competitive bidding is an inherent part of the construction industry; and not only is it necessary to make accurate estimates as to material availability, costs, equipment usages, and so forth, but also fairly accurate estimates must be made as to the next round of wage increases, the availability of manpower, the productivity of manpower, and other related cost factors. This means that there must be effort to look into the future. There is no advantage in making the wrong guess and getting a job because one puts in a low bid which subsequently results in an award that produces only a loss. At the same time, it is of no value to the contractor to pad the bid with so many contingencies and so much protection as to price himself out of the award consideration.

Crystal-ball gazing is a constant requirement of the construction industry. One of its most nebulous areas is in the field of labor relations. In this presentation, no useful purpose would be served by attempting to forecast the actual terms or specific figures about wage increases or labor costs over any predictable period of the future. It seems obvious that wage increases will continue during the seventies and they will at least parallel increases in the cost of living and in some instances go beyond in the constant effort to provide improvements in the standard of living of American workers.

There will be continued effort toward establishing some degree of guaranteed employment, and coupled with that will be an effort to establish some form of minimum annual income. The pressures for these specific developments have already been indicated in some proposed government programs. For example, the Family Assistance Plan which has had attention in the halls of Congress would provide income maintenance payments to the underemployed.

Aside from wages, there will be increased effort in certain benefit areas. Recently, the National Labor Relations Board (NLRB) held that retired employees are considered to be employees under the National Labor Relations Act; therefore, changes in benefits for retirees is a proper subject for collective bargaining with the union. Although the Court of Appeals reversed the Board and although the final disposition of this case rests at this time with the U. S. Supreme Court, the original decision of the NLRB points to the fact that greater benefits will be demanded for employees upon retirement.

The first and most obvious form these benefits will take will be a dramatic increase in pension benefits and the attendant costs for such increased benefits. Pension benefits and Social Security payments during the last twenty years have risen from $190 a month to a current average of $400 a month. Indications are this trend will continue. With workers' increases being tied to the increased cost of living, it can be expected that efforts will be made to attach the pensioner's income to the reflected cost of living increases. Much study is being devoted to provisions for pension flexibility. With a multiplicity of local unions which exist in the construction industry, there are scores upon scores of pension programs, and the older a worker gets the less inclined he is to sacrifice pension credits that he may have acquired in one geographical area and which he would stand to lose if he moved to work in another geographical area. Unions will seek to negotiate transferability of pension credits and benefits. We can also expect there will be pressures for legislation to impose portability concepts on all types of pension plans. All of these factors in the areas of the fringe benefit of pension must be considered as increased labor costs.

Even in the face of rising medical costs and perhaps because of rising medical costs, it is reasonable to expect there will be continued efforts in the future to expand health and welfare benefits through increased hourly contributions. Special attention can be expected to be given to preventive health care procedures, such as annual physical checkups. There is considerable attention being given now to the inclusion of psychiatric care in medical programs. The emphasis currently being placed on mental health will contribute to this increased effort. We can also expect that the advent of medical programs in certain areas will result in the request or demand that such services be made available to the working man. This could include artificial hearts, transplants, aging controls, and some of the newer exotic medical procedures.

It is reasonable to expect that a demand will be made for an increase of women in the work force. Women flagmen, truck drivers, equipment operators, and other categories of classifications will no longer be viewed as novelties, and we can expect to see more women employed in the construction industry. This can result in the demand for child-care facilities and might even extend to absorbing some costs of preschool education. We should recognize that pressures for these particular increased cost factors may not only come from unions, but may well take the form of legislation.

There will be a demand not only for paid vacations within the construction industry but, where they already exist, for longer vacations. It is even contemplated in some circles now that the pay for vacation periods should be greater than the regular pay, since the regular pay is spent for the regular and continuing household expenses and thus there is no extra money to indulge in vacation activities.

The population increase is going to add to employment problems, and the push for a shorter workday and a shorter work week can be a real expectation. Much of the demand for the shorter work week and shorter workday will be advocated as a means of increasing employment and providing more jobs for more people. This philosophy should not be accepted as being accurate in the construction industry unless the unions are willing to relax exclusive hiring restrictions and become more realistic about skill training and development. Future employee benefits could also include the group purchasing in the fields of insurance, such as automobile liability and homeowners insurance.

As mentioned at the onset of the discussion on predicting the future, there is no intent here to attach specific figures to these cost conditions, but suffice it to say that they will indeed be significant. Not only will the significance of *how much* be prominent, but when will they become effective is most important for those who are preparing to bid jobs that may last for two or three years or may extend from one negotiating period into another. These benefits, wage increases, and the so-called improved standard of living of workers cannot be accomplished solely by passing the cost on to the consumer or the job as a whole. Labor must be prepared to improve productivity. The trend of todays wages running away from productivity must be halted.

As far back as 1958, a prominent union official stated:

"It seems to me that we have reached a point in our economy where it behooves both labor and management to do everything within their combined power to effect economies in our industry. We in the labor end of this partnership cannot continue to take from the industry without putting our fair share into it. In many instances, our members seem to think that any time is starting time after 8:00 a.m. and quitting time is any time before 4:30 p.m. Anything exceeding one-half hour is an acceptable lunch period, when our agreements call for one-half hour for lunch time. Coffee break, something which is not provided for in our agreements, is merely a matter of our "job condition" and, therefore, is of no concern to an employer."

That same individual carried forth an example whereby on a wage scale of $3.80 per hour, if each worker building a home took a fifteen minute coffee break each morning, the cost to the buyer would be $5.70 per day or $570 for the construction of the home. He went on to point out that some of the union members would no doubt take the position that anyone who can afford to buy a house can well afford to buy the luxuries of our member's coffee breaks. He then said,

"We have yet to bear in mind that most home builders need a mortgage to buy a home and somewhere along the line, at least a part of this $570 becomes a part of that mortgage. It means that he is paying interest on all or part of this luxury from fifteen to twenty-five years, depending on the length of time required to amortize the mortgage."

These observations were made in 1958. In today's wages the construction industry, the $570 would be closer to $1200, or more; and as one adds to the decline in productivity, this same figure could be close to $2000. This, more than any other example, clearly shows what effect increased costs can have in the future if there is no increase in productivity.

In looking at the future then, it would be less than realistic to believe that either wage costs or productivity will remain or attain a status quo level for any time in the foreseeable future. There is no single artery where one can feel the overall pulse beat and determine precisely what the facts and figures of the future will be. It is essential, however, that to be anywhere close in the ballpark, as the saying goes in the construction industry, the contractor will need to keep his finger on many pulses and, most of all, will have to have within the framework of his organization a sophisticated and calculated approach to labor relations.

24-15. Labor Laws

One pulse which the contractor must certainly know and continually monitor is that of the body of federal and state laws and regulations covering labor relations activities.

The basic federal labor law effecting the nation is the National Labor-Management Relations Act, commonly known as the Taft-Hartley Act. Although this law effects all industry involved in interstate commerce and covers the basic labor-management relationship, including representation, the rights of workers, collective bargaining, and unfair practices by employers and unions, there are important exclusions and provisions that relate directly to the construction industry.

As mentioned previously, construction is one industry where the pre-hire contract is legal under the Taft-Hartley Act, and the contractor may sign a labor agreement without going through the process of organization, employee representation election, and certification of the collective bargaining representative. This exception allows the contractor the flexibility of coming into a new area and becoming signatory to the existing local labor agreement without going through the time-consuming process required by union organization, certification, and the eventual negotiation of a labor agreement.

The Act also provides another exclusion which is directly applicable to the construction industry. In other industries in those states in which compulsory union membership is not prohibited by state law, employees may be required by the labor agreement to become union members after thirty days of employment or be subject to discharge upon the union's request. Because of the temporary and transitory nature of construction employment, the requirement for union membership, where legal, may

be imposed after the seventh day of employment. The liberalization of this provision in regard to the construction industry also contributes to the union's control of the labor force.

In the unfair labor practices section of the Taft-Hartley Act are two clauses which are of particular importance to the contractor. These are Section 8(b)(4)(B), which prohibits secondary boycotts, and Section (8)(b)(4)(D), which prohibits work stoppages or other forms of coersion to force an employer to assign work to employees represented by a particular labor organization rather than to employees represented by another labor organization.

Section 8(b)(4)(B) relating to secondary boycotts is paricularly important to contractors who may be utilizing subcontractors and to contractors who may be working on a large project where several different construction firms may be performing work. This section makes it an unfair labor practice for a union to strike or withhold labor from one contractor to exert pressure on another contractor with whom the union may have a primary dispute or whom the union may be trying to organize. For example, if the union were trying to organize the employees of a sheet-metal specialty subcontractor, it would be an unfair labor practice for the union to picket the general contractor and to stop his work in an effort to make the general contractor pressure the specialty subcontractor into signing a labor agreement on pain of losing the subcontract.

On a multi-employer construction site, it is also an unfair labor practice for a union to picket the work of neutral contractors who are not party to a primary labor dispute with the union. This is particularly true if the neutral contractors have clearly marked separate entrance and exit gates for the exclusive use of their own employees. Unions have attempted to legalize general or "situs" picketing of entire construction projects by having Section 8(b)-(4)(B) amended, but to date they have been unsuccessful in their efforts. Such an amendment would put a powerful weapon in the hands of organized labor, as it would allow them to embroil all contractors in labor disputes and would enable them to bring severe pressures on the contractor involved in the primary dispute.

Section 8(b)(4)(D) has a direct application to one of the industry's most persistent labor-relations problems—the jurisdictional dispute. On a typical construction job, there may be one or more craft unions employed ranging from laborers and carpenters on through equipment operators, teamsters, and ironworkers, to pipe fitters, electricians, and boilermakers. With this spectrum of skills and crafts, invariably a dispute will arise between two or more crafts as to which one should perform an item of work; and oftentimes a work stoppage results. Since the turn of the century, a great volume of material has been developed concerning the resolution of jurisdictional disputes, including labor conventions, agreements between crafts, decisions, and procedural rules by voluntary and impartial boards, and industry-wide and area-work practices. However, when all voluntary methods fail, an unfair labor practice complaint filed under Section 8(b)(4)(D) of the Taft-Hartley Act against the offending union will generally end the work stoppage. The NLRB will then make a determination on which craft should perform the disputed work after conducting a hearing into the matter under Section 10(k) of the Act.

There is no more important piece of labor legislation to the contractor than the Taft-Hartley Act, because it effects labor-relations activity on all of his work whether it be public or private, union, or open shop, and at all of his construction locations. It is the law of the land, and he cannot be too well versed in all of its ramifications.

Because of its universal application, the Fair Labor Standards Act, commonly called the Wage and Hour Act, is also of prime concern to the contractor. This law, which establishes minimum wage rates and overtime conditions, covers any employer whose operations affect interstate commerce. At this writing, the Fair Labor Standards Act requires the payment of a minimum wage not less than $1.60 per hour and overtime at the rate of one and one-half (1½) times the basic hourly wage rate for all work performed in excess of forty (40) hours in the established work week.

This law requires special attention by the contractor in what is construed to be the "basic hourly rate." For example, if a bonus is paid for superior production on a piecework basis above and beyond the regular hourly rate, this bonus must be taken into account in calculating the overtime rate; i.e., the proper overtime rate is not just one and one-half times the hourly wage rate, but is one and one-half times the hourly wage rate *plus* the performance bonus. The same factor must be applied to shift premiums or any other monetary factor which affects the weekly income of the employee.

The Fair Labor Standards Act also carries regulations concerning the employment of minors and hazardous occupations from which minors are excluded. Most construction activities involve the potential hazards of moving machinery, falls by materials or individuals, etc. Consequently, it is well if the contractor adopts a firm policy of employing no one under eighteen years of age at any of his work sites.

If the contractor is performing work for a federal agency or is building a project financed in part with federal funds, he is faced with additional regulations covering wages and hours. The Davis-Bacon Act requires the payment of prevailing wage rates and fringe benefits to laborers and mechanics employed on such a project. The prevailing rates are determined by the Department of Labor and are made a part of the contract specifications. Although the rates are intended to reflect what the majority of the construction laborers and craftsmen are being paid in a particular geographic area, they often are merely a restatement of the wage rates and fringe benefits of the current "prevailing" labor agreement in the area.

In the event the contractor is not signatory to the area labor agreements of that area and does not make fringe benefit contributions (health and welfare, pension, vacation, apprenticeship, etc.), he may comply with the law by paying his employees the value of the fringe benefits in cash as additional wages, or he may establish and contribute to equivalent benefit programs. The Act provides that the contracting agency may withhold funds from a contractor who fails to pay the proper rates and fringes and the agency may distribute monies directly to the underpaid laborers and/or mechanics. In case of willful and repeated violations, the contracting agency is empowered to cancel the contract of the offending contractor and the contractor may be banned from performing work on government projects for a three-year period.

Also, if the contractor is performing work on a federal project, he comes under the Work Hours Act of 1962. This law requires the payment of overtime at the rate of at least time and one-half (1½) for all hours of work in excess of eight hours in any one day. Whereas the Fair Labor Standards Act requires payment of overtime after forty hours per week, the contractor on the federal construction project must also pay overtime after eight hours per day.

The contractor also finds that he is confronted with a large body of laws and regulations relating to the civil rights of his employees and nondiscrimination in the employment relationship. The Civil Rights Act of 1964 applies to any employer employing twenty-five or more workers regardless of where he works or whether he performs public or private work. Of particular importance is Title VII of the Act, which prohibits discrimination by the employer in hiring, termination, upgrading, transfer, and promotion against any individual because of race, creed, color, sex, or national origin. Of special interest is the fact that this section also makes it an unfair employment practice for a union to cause, or attempt to cause, an employer to discriminate against any individual. Title II of the Act, known as the Public Accomodations Section, stipulates that there shall be no discrimination in the employee facilities which may be maintained by the contractor, including rest rooms, lunch rooms, and camp facilities.

Investigation of complaints and enforcement of Title VII is delegated to the Equal Employment Opportunity Commission, which was created by the Act. The Commission processes complaints and, in cases of alleged discrimination, may refer the case to Federal District Court if compliance is not secured by voluntary means. The Court then may issue injunctions or orders, and may require the employment or reemployment of the discriminee with or without back pay.

The Civil Rights Act is also augmented by a number of Presidential Executive Orders prohibiting discrimination, some of which predate the passage of the Act. Executive Order 11246 prohibits discrimination on federal or federally assisted contracts. This Order is enforced by the Office of Federal Contract Compliance in the Department of Labor. This office has the authority, in cases of willful or continuous violation, to cancel the contract and to bar the contractor from bidding further federal work.

In 1964, Executive Order 11141 was issued, which also prohibited discrimination because of age on federal work. However, the Age Discrimination in Employment Act of 1967 has the effect of extending the coverage of Order 11141 to all employers engaged in commerce who employ twenty-five or more people. The Act prohibits discrimination against people in the 40 to 65 year age bracket on matters relating to hire, discharge, rates of pay and terms, conditions, or privileges of employment. Exceptions to the law are permitted when age is a bona fide occupational qualification reasonably necessary to the normal operations of the business.

In addition to the foregoing federal laws and regulations which can be considered universal, the contractor must be aware of the applicable state laws and regulations which may effect his various operations at various locations. Virtually every state has a management labor-relations act, which covers employers who may not come under the purview of the Taft-Hartly Act. As mentioned previously, Section 14(b) of the Act makes it permissible to enact laws which prohibit compulsory union membership. In states where Right-to-Work laws have been enacted, the mandatory union membership or "union shop" clauses of the Taft-Hartley Act are invalidated. It is, therefore, highly important for the contractor to be aware of the state labor-relations law where he is performing work. As this text is written, nineteen states have Right-to-Work laws on the books.

The states also have their individual wage and hour

laws which may contain minimum wages and overtime conditions for specific occupations and often carry protective laws covering the employment of minors and women.

Almost every state has now enacted a Fair Employment Practice Law, Human Relations Law, or State Civil Rights Law, which provides for nondiscrimination in hiring and employment, and many states have also enacted age discrimination laws which are similar to the Federal Age Discrimination in Employment Act of 1967. Thus, almost any contractor, even those with fewer than twenty-five employees, are subject to laws providing for equal employment opportunity without regard to the applicants' or employees' race, creed, color, sex, national origin, or age.

In addition, most states have enacted laws making it mandatory for employers to allow qualified employees sufficient time to vote in either general, special, primary, or all elections. Some of these laws also provide for the payment of the time that the employee must be off work in order to have an opportunity to cast his ballot. In addition some state voting time laws have provisions requiring that notices be posted, that employees request time off in advance, and that there be penalties for violations of the law.

It can be seen from the foregoing that the contractor must continually keep abreast of a large volume of ever-changing federal and state laws and regulations. If the contractor does not maintain a knowledgeable and well-informed labor-relations department, he must be prepared to devote a considerable amount of his own time to maintaining accurate and current knowledge of the laws which effect the personnel and labor-relations activities of his business. Because of the complexity and changing nature of this vast body of laws and regulations, it is extremely easy for the contractor inadvertently to commit a violation because of his lack of adequate knowledge. In no area is the proverbial "an ounce of prevention is worth a pound of cure" more applicable than in the realm of law effecting construction labor relations.

25 SAFETY PROCEDURES AND PRACTICES

LEE E. KNACK

Director of Labor Relations
Morrison-Knudsen Company, Inc.
Boise, Idaho

SAFETY IS NO ACCIDENT!
SAFETY IS UP TO YOU!
DON'T BE HALF SAFE!
BE ALERT—STAY ALIVE!
TEAMWORK PREVENTS ACCIDENTS!
SAFETY PAYS!

You can see these slogans on posters, signs, or in magazines whenever you are around men working in the construction industry. All too frequently, many companies feel that by providing this visual lip service to accident prevention they have a viable safety program. Safety must be a basic component of the management philosophy, just as operating at a profit is, because the cost of accidents presents a serious drain of profit dollars. An aggressive company will examine each of its operations with a keen interest to see not only that work is done in the most efficient manner to ensure greateset potential profit, but that it is done as safely as possible for the very same reason.

In this chapter we will examine the factors that make safety pay for: (a) the employer; (b) the employee; and (c) the customer. To achieve this end, it will be necessary to discuss Workmen's Compensation provisions, costs of providing Workmen's Compensation protection, how safety records affect the cost of providing Workmen's Compensation protection, how safety affects employer-employee relations, and how safety affects production.

25-1. Safety Benefits the Employer

The significance of safety as it relates to the construction industry comes into full perspective when we realize that the construction industry is the largest industry in the United States. Because safety and profit have an integral relationship, the discussion of safety itself would become a moot question for discussion if construction companies were not making a profit. There is no lack of humanitarian concern if we view safety from a profit standpoint, providing we recognize that it is profitable not only to the employer but it is profitable to the worker as well. The construction worker who is injured suffers a financial loss as well as pain and discomfort. The construction employer who disregards safety suffers an indirect cost which could ultimately affect his survival also. The impact of this cost relationship is even more significant when we consider that the labor expense in construction today is equivalent to 45 to 50 percent of the total contract price. Thus, with at least half of the construction cost consisting of labor, any constructive type of safety program will result in economies, and be of great importance to the worker, to the contractor, and to the public as a whole.

It is not sufficient to merely say that an adequate safety or accident prevention program is an important way to reduce labor costs. It is not difficult to see that accidents and losses involving both people and equipment result in a waste of time and money. What is not so evident, however, is the extent of those losses. All too often an employer believes that as long as he has provided for insurance to protect himself against direct losses resulting from accidents, he has no longer any concern as to profit and loss once he has paid that insurance premium. This attitude can lead to disastrous results, and a simple evaluation of the relationship between insured costs and uninsured costs will reflect the importance of this relationship.

25-2. Insured vs. Uninsured Costs

As you can surmise from the foregoing, uninsured costs are approximately nine times greater than insured costs. Thus, for every dollar in direct loss or cost of an accident, nine additional dollars are spent indirectly. Those nine

INSURED COSTS

Injuries

1. Compensation for lost earnings
2. Medical and hospital cost
3. Awards for permanent disabilities
4. Rehabilitation costs
5. Funeral charges
6. Pensions for dependents

Property Damage

Insurance premiums or charges for:
   a. Fire
   b. Loss and damage
   c. Use and occupancy
   d. Public liability

Uninsured costs which result in indirect losses are as follows:

UNINSURED COSTS

Injuries

1. First-aid expenses
2. Transportation costs
3. Cost of investigations
4. Cost of processing reports

Associated Costs

1. Difference between actual losses and amount recovered
2. Rental of equipment to replace damaged equipment
3. Surplus workers for replacement of injured workmen
4. Wages or other benefits paid to disabled workers
5. Overhead costs while production is stopped
6. Loss of bonus or payment of forfeiture for delays

Wage Losses

1. Idle time of workers whose work is interrupted
2. Man-hours spent in cleaning up accident area
3. Time spent repairing damaged equipment
4. Time lost by workers receiving first aid

Off the Job Accidents

1. Cost of medical services
2. Time spent on injured workers' welfare
3. Loss of skill and experience
4. Training replacement worker
5. Decreased production of replacement
6. Benefits paid to injured worker or dependents

Production Losses

1. Product spoiled by accident
2. Loss of skill and experience
3. Lowered production of worker replacement
4. Idle machine time

Intangibles

1. Lowered employee morale
2. Increased labor conflict
3. Unfavorable public relations

dollars show up eventually on the profit and loss statements and are just as real a drain on company profits as any other whole dollar loss.

As we examine those items listed under insured costs, compensation for lost earnings stands out for item number one, and well it should, for no one suffers greater loss than the injured. Recognizing this, society rose to the occasion by enacting Workmen's Compensation Laws.

25-3. Workmen's Compensation Laws

These regulatory provisions to protect the worker were a direct outgrowth of the industrial revolution, which had its seat in England.

Just prior to the period of the so-called industrial revolution, a standard shop complement consisted of the craftsman and one or two apprentices, and the work was most often accomplished at the craftsman's home or an attachment thereto. Whenever one of the individuals so engaged in the home workshop was injured, neighborhood opinion and attitude would result in the employer providing the injured with assistance, and it was just a matter of good business on his part to prevent the injured employee from becoming a public charge.

As the advent of the factory system with mass production began to develop and displace the home-shop operator with his one or two apprentices, mass employment developed. The close personal relationship between the employer and one of his employees began to disintegrate; and because

the employer was not dependent upon the immediate locale for the sale of his products and his services, local attitudes and pressures were not as significant as they were with the smaller employer. As a result, the injured employee, with a break in the communication with his employer, was often replaced with a hale and hearty worker and then turned loose to overcome his difficulty as best he could, often becoming a public charge. In effect, therefore, the cost of compensating the worker for his injury was born by the public rather than by the employer. Those employers who might have been sympathetic to the employee's plight were often not aware of the hardships that existed. The foreman and supervisors who were expected to produce profits often deliberately withheld injury and accident information from their employers for selfish reasons.

Coupled with the communication gap that appeared directly between the employer and his employees was the change in the methods of performing work. Work that had been previously performed by hand with manual tools was rapidly being done by power-driven machinery. The power-driven machinery not only increased quantity of products, but it increased the speed with which the work was being accomplished and, in so doing, provided the factory environment with hazards and dangers that never were present in the home workshop. During this period, therefore, the worker's contact with his employer was becoming severed and his exposure to hazardous conditions was being increased.

When a worker was injured on the job at the factory, he

received little personal attention and was forced to seek restitution for his suffering and losses in common law courts based on the same common law conditions that the employer owed to any member of the general public.

Under the common law, an employer was required to meet these five responsibilities:

1. To provide a resonably safe work place.
2. To provide reasonably safe tools and equipment.
3. To use reasonable care in selecting employees.
4. To enforce reasonable safety rules.
5. To provide reasonable instruction regarding the dangers of employment.

It would seem then that the employee was "reasonably" protected in his employment. Unfortunately, enforcing those protections was almost an impossibility. If an injured employee felt the employer had failed to maintain his responsibilities, he could sue for damages.

Common law, while assigning the employer with the forementioned responsibilities, also provided the employer with three basic and crucially important defenses:

1. Fellow Servant Rule. An employer could not be held liable for injury caused by the negligence or carelessness of fellow employees.
2. Contributory Negligence. The employer was not responsible if negligence of the employee contributed to the cause of the injury.
3. Assumption of Risk. The employee, when he accepted the job, assumed all the obvious and customary associated risks.

It is estimated that in 40 percent of all industrial accidents that cause disability to workers, both the employer and employee have a degree or responsibility of contributory cause. The employee primarily, though not entirely, is responsible for an additional 30 percent of industrial accidents. Therefore, through the three defenses provided the employer, in approximately 70 percent of those industrial accidents the employee could not collect damages under the legal systems that prevailed.

It is readily evident that a successful suit required considerable legal preparation and the legal fees took a substantial portion of the monies awarded in damages, if any. The injured employee was faced with two choices— either a long drawn-out legal fight or a settlement out of court for the much needed cash, often in grossly inadequate amounts. Thus, injured employees suffered the balance of the losses themselves or became public charges. The ever-swelling list of disabled workers on public welfare became a major concern within society. To correct this situation, Workmen's Compensation regulations were enacted.

England started this evolution by limiting the use of the fellow servant rule in 1880. Germany created the first self-supporting and self-directing insurance plan with the enactment of their Workmen's Compensation Act in 1884. This Act provided for government supervision of the plan. This form of legislation spread throughout Europe, Austria passing the law in 1887 and England in 1897.

By now American labor was pressing for Workmen's Compensation, and in 1908 the federal government passed the first compensation law to protect its own employees. The first state laws were enacted by Washington and Kansas. These early compulsory laws conflicted with the court's interpretation of the 14th Amendment to the Constitution,

since compelling an employer to pay without fault was denying the employer due process of law. Subsequent laws were enacted that were either elective, where the employer and employees must select to be covered, or elective presumptive, where the employer and employees are presumed to be protected unless they stipulate otherwise. In 1917, the United States Supreme Court declared that states could enact and enforce compulsory Workmen's Compensation laws under its power to provide for the public health, safety, and welfare without violating the 14th Amendment.

As a result 29 states now have compulsory laws, two states have compulsory laws for hazardous occupations, and 22 states have elective laws. In those states with elective laws, if the employer refuses to participate he forgoes the right to claim the three basic defenses in common law heretofore mentioned. Those state laws that are compulsory usually contain a provision for a certain number (2 to 10) of employees before the employer is subject to the law. Additionally, certain occupations are exempt, such as domestic servants, farm laborers and casual employees. Approximately 20 percent or 18 million employed workers are outside the protection of Workmen's Compensation. Workmen's Compensation laws were enacted to assure the injured worker prompt compensation of wages lost; to limit the employer liability to the degree stipulated in the law without satisfying the question of fault or negligence; to reduce the cost of industrial injuries to society by minimizing the loss of the injured worker's productivity; and to reduce the burden on the community of supporting the injured worker and his family.

The economic justification of these laws was that liability for industrial accidents should be regarded as part of the cost of operating a business so that that cost would be transferred from the worker and the employer to the consumer.

State Workmen's Compensation acts generally cover accidental injuries arising out of and in the course of employment. Additionally, all states recognize occupational diseases, such as silicosis, asbestosis, radiation disability, and occupational loss of hearing due to prolonged exposure to noise; however, the amount of benefits vary between the states.

25-4. Amount of Benefits

The amount of benefits is based on the degree of disability. Occupational injuries or disease are divided into five classifications:

1. Death. A fatality resulting from an industrial injury.
2. Permanent Total Disability. An injury which permanently and totally incapacitates an employee from following any gainful occupation.
3. Permanent Partial Disability. The complete loss of a member or part of a member of the body, or permanent injury which partially incapacitates the employee and prevents him from following gainful employment.
4. Temporary Total Disability. An injury which does not result in permanent impairment but does prevent the injured person from performing any kind of work for a period of time, after which, having had proper medical care, he can return to work in as good health as he had before he was injured.
5. Temporary Partial Disability. An injury which does not result in permanent impairment or temporary total

disability but which temporarily impairs the injured employee's full earning power.

Additionally, the length and amount of payment for disabilities varies among the states:

1. Payment as long as disability lasts. This method is used primarily for permanent total disability or to extended payments to dependents in the case of the death of the worker.
2. Payment for a specific time limit. This method provides for a definite period of payment, usually stated in weeks of compensation benefits, such as 300 weeks of the weekly benefit amount.
3. Payment of a specified maximum amount. This method fixes a maximum sum of money, such as an award of $7500, with weekly payments terminating when that amount is reached (see Figure 25-4-1).

Injuries can be further defined as scheduled injuries. These scheduled injuries are losses of members or parts of members of the body where wage loss based on the impairment is presumed. The benefits for scheduled injuries are defined by law and may be awarded in addition to temporary disability benefits (see Figure 25-4-2).

This discussion of payment of benefits must include the statement that many states are without "Second Injury" funds. These funds have been established so that when an employee has already received an injury which created permanent partial disability and is again injured in such a manner as to cause permanent total injury, the employer pays for that portion of the disability related to an accident while in his employment. In a state without a "Second Injury" provision, when an employer hires an employee with only one eye he hires that employee as a whole man. Should that employee lose his remaining eye in an industrial accident, the employer is liable for permanent total disability. This obviously creates special problems in hiring handicapped persons. Efforts are being made to establish Second Injury funds in every state and, wherever possible, employers should assist those efforts.

25-5. Methods of Complying with the Law

In all but six states and a commonwealth, the employer has several choices in complying with the law. He may purchase insurance policies from private companies; if he meets the qualifications, he may elect to be self-insured; or he may participate in state insurance funds.

Nevada, North Dakota, Ohio, Washington, West Virginia, Wyoming, and Puerto Rico are monopolistic states; i.e., they have exclusive state insurance funds and the employer must provide Workmen's Compensation insurance by participating in those funds.

Penalties for failure to provide Workmen's Compensation vary among the states and may include fines as high as $5000, imprisonment, enjoinment from doing business, and/or loss of common law defenses.

Workmen's Compensation Premiums Workmen's Compensation insurance premiums are based on the injury experience that the state has had with a specific industry or operation classification. Each state maintains a Workmen's Compensation Board to recommend what risk is associated with each job classification. Their function would be similar to that of an insurance actuary. Although the states can establish varying rates, the National Council on Compensation Insurance published a manual listing between 600 and 700 classifications and their respective rates, which are used as a guide by the State Workmen's Compensation Boards. The risk is then translated into cost per $100 annual payroll per job classification. An employee's income in excess of $300 per week is not included in the annual payroll and hours worked as premium time can be recorded as straight time for this specific payroll determination.

Experience Rating Plan The annual premium is determined by the insured's annual payroll, and the rate for the classification to which the insured's work is assigned. Annual payrolls are estimated and a deposit premium is made at the beginning of the policy year. At the completion of that policy year, an audit is made to establish an experience rating. The experience ratings for the previous three years are averaged to establish the percent of the basic premium to be charged the employer for the following year. The experience rating for the year just completed will be used the following year to establish the three-year average. A yearly log is created by this method.

It is here that a good accident record makes its presence felt economically. Companies with a good average experience rating can receive a discount in the following policy year. Those companies with a poor accident record may be assessed a penalty rating. A company may be required to pay only 65 percent or less of the basic premium, whereas another company may be charged 125 percent or more of the basic premium for the same job classification. In some industries, the state's basic rate is in excess of $11 per $100 unit of payroll. The lumber industry in Oregon is one such example. In this case, if an employer had $150,000 annual payroll in that assigned job classification, the basic annual premium would be $16,500. A good experience rating such as described above could save that employer $5775. A bad experience rating could cost an employer $4125. With today's net profit after taxes percentage for most industries approximating 1½ to 2 percent, the employer with a poor experience rating would be required to increase his volume more than $495,000 to make up for the poor experience rate as compared with the other company. When the annual payrolls are expanded into the millions per job classification, it is readily apparent why a good experience rating is so vitally important.

Retrospective Rating Plan The retrospective rating plan is the most advantageous to the average employer. It overcomes two disadvantages of the experience rating plan: First, the experience rating plan does not count the current year's loss record; and, second, the improvements in loss records will not become felt to their fullest until the previous three years of poorer records have been removed from the average.

The retrospective rating plan begins with a standard premium determined from the estimated payroll multiplied by the current risk's experience rate. A basic premium is then figured, which is composed of a specific percentage of the standard premium necessary to cover the administrative expenses of servicing the risk and the insurance charge for excluding excess losses. The incurred losses are multiplied by a conversion factor which varies by state and which increases the losses to include the cost of handling claims. The Retrospective Premium is established by adding the converted losses to the basic premium, and this sum is multiplied by a tax factor to include the premium taxes charged by various states.

INCOME BENEFITS FOR PERMANENT
AND TEMPORARY TOTAL DISABILITIES

January 1, 1971

JURISDICTION	LIMITATIONS ON PERMANENT TOTAL					LIMITATIONS ON TEMPORARY TOTAL					NOTATIONS
	MAXIMUM PERCENT OF WAGES	MAXIMUM WEEKLY PAYMENT	MINIMUM WEEKLY PAYMENT	TIME LIMIT	AMOUNT LIMIT	MAXIMUM PERCENT OF WAGES*	MAXIMUM WEEKLY PAYMENT	MINIMUM WEEKLY PAYMENT	TIME LIMIT	AMOUNT LIMIT	
ALABAMA	65 [10]	$50.00	$15.00	550 weeks	$20,000	65	$50.00	$15.00[1]	300 weeks	$15,000	
ALASKA	65	82.55	25.00[1]	Life		65	127.00	25.00[1]	Disability	17,000	Disfigurement maximum, $3,500.00
ARIZONA	65	150.00[2]	32.50	Life		65	150.00[2]	32.50	433 weeks	65,000	
ARKANSAS	65 [26]	49.00	10.00			65 [26]	49.00	10.00	450 weeks	19,500	Disfigurement maximum, $3500
CALIFORNIA	65	52.50	20.00	Life*		61-3/4	87.50	25.00	240 weeks [20]		60% maximum after 400 weeks.
COLORADO [15]	66-2/3	59.50 [21]	13.00	Life*	18,623	66-2/3	59.50	13.00	Disability	18,623	50% increase in compensation where employer has failed to comply with insurance provisions. 50% decrease in compensation where injury results from failure to obey safety regulations or from intoxication.
CONNECTICUT	66-2/3	84.00[19]	20.00	Life	25	66-2/3	84.00[19]	20.00	Disability		
DELAWARE	66-2/3	75.00	25.00[1]	Life		66-2/3	75.00	25.00[1]	Disability		
DIST. OF COLUMBIA	66-2/3	70.00	18.00[1]	Life		66-2/3	70.00	18.00[1]	Disability	24,000[18]	
FLORIDA [33]	60	56.00	12.00[1]	Life		60	56.00	12.00[1]	350 weeks		
GEORGIA	60	50.00	15.00	400 weeks	18,000	60	50.00	15.00	400 weeks	18,000	
GUAM	66-2/3	56.00	28.00[1]	Life	20,000	66-2/3	56.00	28.00[1]	Disability	20,000	Disfigurement maximum, $3,500.
HAWAII	66-2/3	112.50*	18.00*	Life	35,100 [14]	66-2/3	112.50	18.00[1]	Disability	35,100 [13]	Director may order payment of $150 per month for attendant, paid from special fund.
IDAHO	60	43.00*	26.00*	Life[6]		60	43.00*	26.00*	Disability[6]		Maximum $51.00 with dependent spouse. Add $6.00 each child. Maximum $99.00.
ILLINOIS	(*)	71.00	31.50	Life	(*)	(*)	91.00	31.50	8 years		Limited to amount if death had resulted. Pension thereafter.
INDIANA	60	57.00	21.00[1]	500 weeks*	25,000	60	57.00	21.00[1]	500 weeks	25,000	Additional benefits from second injury fund.
IOWA	66-2/3	56.00	18.00[1]	500 weeks		66-2/3	61.00	18.00[1]	300 weeks		Weekly compensation based upon 50% of state average wage, annually.
KANSAS	60	56.00	7.00	415 weeks	23,240	60	56.00	7.00	415 weeks	23,240	
KENTUCKY [22]	66-2/3	56.00	22.31	425 weeks	23,400	66-2/3	56.00	22.31	425 weeks	23,400	Disfigurement benefits.
LOUISIANA [29]	65	49.00	10.00[1]	500 weeks	22,500	65	49.00	10.00	300 weeks	13,500	
MAINE [23]	66-2/3	73.00				66-2/3	73.00				Disfigurement benefits, $5,000 max.
MARYLAND	66-2/3	85.00	25.00[1]		45,000	66-2/3 [27]	81.50 after first 42 days @ $55.00	25.00[1]	208 weeks		If permanent disability exceeds 50% of the body as a whole, employee is entitled to additional compensation for the full disability from the "Subsequent Injury Fund" after completion of payments by the employer.
MASSACHUSETTS [16A]	66-2/3	70.00*	20.00[3]	Life		66-2/3	70.00*	20.00[3]	Disability	18,000	$6.00 additional each wholly dependent but not to exceed weekly wage. Combined total compensation for total and partial disability not to exceed $18,000.
MICHIGAN [24]	66-2/3	75.00*	27.00	Disability [16]	[16]	66-2/3	75.00*	27.00	Disability		$6 add. for ea. dependent up to 5, max. $104.
MINNESOTA	66-2/3	70.00	17.50	Life	[14]	66-2/3	70.00	17.50	350 weeks	21,000	Additional $5,000 allowable in certain cases. Disfigurement benefits.
MISSISSIPPI	66-2/3	40.00	10.00*	450 weeks [9]	15,000 [9]	66-2/3	40.00	10.00*	450 weeks [9]	15,000 [9]	Less in partially dependent cases. $2,000 disfigurement maximum.
MISSOURI	66-2/3	58.00	16.00	300 weeks [7]		66-2/3	63.50	16.00[1]	400 weeks	22,800	$2,000 disfigurement maximum.
MONTANA	66-2/3	60.00*	34.50	500 weeks [31]		66-2/3	65.00[32]	39.50	300 weeks		Reducing schedule if less than 5 children.
NEBRASKA	66-2/3*	55.00*	35.00[1]	Life [8]		66-2/3*	55.00	35.00[1]	300 weeks [8]	16,500	45% after 300 weeks, maximum $41.00 minimum $31.00 (or actual wages if less.)
NEVADA	90	66.46[12]		Life		90 [12]	79.96		100 months	29,250	Additional allowance for constant attendant if necessary. $50.00 a month.
NEW HAMPSHIRE	66-2/3	67.00	20.00[1]	(*)		66-2/3	67.00	20.00[1]	(*)		After six successive years of payment, additional payments may be made only on order of the commissioner upon application by the employee and to the employer. If employer objects, medical panel provided for.
NEW JERSEY	(17)	91.00*	15.00	450 weeks*		(17)	91.00	15.00	300 weeks		After 450 weeks at reduced rate, if employed; at full rate if not rehabilitable.
NEW MEXICO*	60	48.00	24.00[1]	500 weeks	24,000	60	48.00	24.00[1]	500 weeks	24,000	10% additional compensation payable by employer for failure to provide safety devices.
NEW YORK [28]	66-2/3	80.00	20.00[1]	Life		66-2/3	95.00	30.00[1]	Disability		Additional compensation for vocational rehabilitation.
NORTH CAROLINA	60	50.00	10.00	400 weeks*	18,000*	60	50.00	10.00	400 weeks	18,000	In cases of paralysis from a brain or spinal injury, payments may be extended for the life of the claimant and the total may exceed $18,000.

*See Notations column.
1 Actual wage if less.
2 No actual limit in computing average monthly wage. All wages in excess of $1,000 per month excluded.
3 Actual wage if less, but not under $10.00 for work week of 15 hours or over.
4 Disability extending beyond period compensated from second injury fund.
5 Actual wage if less, but in no case less than $20.00
6 400 weeks at maximum disability, reduced thereafter to $26 per week.
7 50% thereafter but not less than $20.00 or more than $40.00 for life.
8 Reduced amounts after 300 weeks.
9 Plus rehabilitation allowance, maximum $160 for 104 weeks.
10 Percentage increased 5% each, for dependent wife and children. Maximum 65%, wife and children.
11 May not exceed actual wage.
12 65 per cent of average monthly wage not in excess of $385 per month plus an additional 15% for each dependent not to exceed 90%.
13 Same rate of compensation thereafter from special fund. Disfigurement maximum $15,000.
14 Old age and survivors insurance benefits credited on compensation after $25,000 has been paid.
15 Disfigurement maximum $1,000.
16 Persons receiving less than benefits provided after 1955 receive difference in amounts from second injury fund.
16A If no benefits paid prior to final decision of claim, award shall be based upon benefits in effect at time of decision instead of date of injury.

17 Maximum not to exceed 66-2/3 per cent of average industrial wage determined annually (as of 1/1/70)
18 Does not include rehabilitation allowance.
19 60% of average production wage. To be determined annually by Labor Commissioner. Determined to be $84.00 as of Oct. 1, 1970.
20 Within period of 5 years from date of injury.
21 If employee is receiving social security benefits for disability, compensation may be reduced by 50% of such payments.
22 Maximum shall not exceed 55% of 85% of average weekly state wage; minimum shall be 25% of 85% of same, promulgated annually by Workmen's Compensation Board as of Jan. 1, 1971.
23 Maximum not to exceed 66-2.3% of state average weekly wage fixed by Maine Employment Security Commission, as of 6/1/70.
24 All benefits adjusted to average state wage.
25 Additional allowance of $5 per dependent child but not to exceed 50% of benefit or 75% of average weekly wage but may exceed 60% of annual average production wage. Retroactive benefit increases provided for cases prior to 1953 and 1969, and prospectively for cases after 1969. Benefits also adjusted annually based on cost of living.
26 Compensation increased 15% if disability due to employer's violation of safety regulations.
27 Based upon State's average weekly wage computed annually.
28 Supplemental retroactive benefits payable in permanent total disability cases before 1960, maximum weekly benefit $50. Payments made from Reopened Cases Fund.
29 After December 31, 1969 maximum weekly benefit is increased to $49.
31 Board may order further benefits in hardship cases where necessary.
32 Amount decreased by $5 after first 26 weeks.
33 During training and rehabilitation in use of artificial members, maximum of 40 weeks.

Figure 25-4-1

Income Benefits For Permanent and Temporary Total Disabilities—January 1, 1971 (Continued)

JURISDICTION	LIMITATIONS ON PERMANENT TOTAL					LIMITATIONS ON TEMPORARY TOTAL					NOTATIONS
	MAXIMUM PERCENT OF WAGES	MAXIMUM WEEKLY PAYMENT	MINIMUM WEEKLY PAYMENT	TIME LIMIT	AMOUNT LIMIT	MAXIMUM PERCENT OF WAGES*	MAXIMUM WEEKLY PAYMENT	MINIMUM WEEKLY PAYMENT	TIME LIMIT	AMOUNT LIMIT	
NORTH DAKOTA	55*	59.00*		Life		55	59.00*		Disability		55% of state's average weekly wage, determined annually. Plus $5.00 for each child under 18. Max. not to exceed weekly take-home pay after taxes.
OHIO	66-2/3	56.00	45.50 [1,9]	Life		66-2/3	56.00*	25.00 [1]	Disability	10,750	During first 12 weeks of temporary total disability, max. compensation is $63.00.
OKLAHOMA	66-2/3	43.00	15.00 [1]	500 weeks	20,000	66-2/3	49.00	15.00 [1]	300 weeks		Disfigurement $3,000 maximum.
OREGON	55	62.50	40.00	Life		90	80.00*	30.00	Disability		Reducing schedule if less than 6 children.
PENNSYLVANIA	66-2/3	60.00	35.00 [11]			66-2/3	60.00	35.00			
PUERTO RICO [19]	66-2/3	28.86	11.54	Life		66-2/3	45.00	10.00	312 weeks		Additional benefits in specific cases such as for vocational rehabilitation or constant companion at not more than $30 a month. Disfigurement $3,000 max.
RHODE ISLAND	66-2/3*	70.33	30.00	Life [12]	(12)	66-2/3	70.33	30.00	Duration [12]	(12)	Additional benefit of $6.00 per week each dependent but total shall not exceed his average weekly wage nor 60% of average state wage, computed annually (9/1/70)
SOUTH CAROLINA	60	50.00	5.00	500 weeks	12,500	60	50.00	5.00	500 weeks	12,500	
SOUTH DAKOTA	55	50.00	27.00 [1]	Life *	78,000	55	50.00	27.00 [1]	312 weeks	15,600	After 300 weeks, maximum $15.00 per week. Minimum $12.00
TENNESSEE	65	47.00	15.00 [4]	550 weeks	18,800	65	47.00	15.00 [4]			After 400 weeks $15.00 per week, or actual wage if less but not less than $12.00. Disfigurement benefits.
TEXAS	60	49.00	12.00	401 weeks	19,649	60	49.00	12.00	401 weeks	19,649	
UTAH	60*	47.00 [2]	27.00 [1]	Life *	20,280 [6]	60	47.00 [2]	27.00 [1]	312 weeks	20,280	After 260 weeks 45% plus $3.60 for a dependent wife and $3.60 for each dependent minor under 18 up to four such children. Disfigurement benefits.
VERMONT [20]	66-2/3	61.00 [5]	31.00 [1]	330 weeks	20,130	66-2/3	61.00 [5]	31.00 [1]	330 weeks	20,130	
VIRGINIA [26]	60	62.00	14.00	500 weeks	24,800	60	62.00	14.00	500 weeks	24,800	Disfigurement benefits.
WASHINGTON		81.23*	42.69	Life			81.23*	42.69	Disability		Additional allowance for constant attendant, if necessary $115.00 per month. Reducing schedule if less than 5 children.
WEST VIRGINIA [23]	66-2/3	65.50	26.00	Life		66-2/3	65.50	26.00	208 weeks	13,622	
WISCONSIN	70	79.00	14.00	Life		70	79.00	8.75	Disability		Additional compensation for vocational rehabilitation. [14]
WYOMING		34.61*	28.80	Life		66-2/3	63.46	33.46	Disability	12,000	Permanent—$34.61 plus $7.50 for each child (no limit)[3]. Aggregate sum for children $10,000.
FEDERAL EMPLOYEES' COMPENSATION ACT	75 [18]	455.66*	70.63 [1]	Life		75	454.66*	70.63 [1]	Disability		Additional allowance of $300.00 per month for constant attendant if necessary.
LONGSHOREMEN AND HARBOR WORKERS' ACT	66-2/3	70.00	18.00 [1]	Life		66-2/3	70.00	18.00 [1]	Disability	24,000 [7]	
ALBERTA	75	95.20	40.00 [1,25]	Life		75	95.20	40.00 [1]	Disability		75% of maximum earnings of $6,600 per year.
BRITISH COLUMBIA	75	109.62	38.22 [1]	Life		75	109.62	35.85 [1]	Disability		75% of maximum earnings of $7,600 per year. [15]
MANITOBA	75	95.20	35.00 [1,16]	Life		75	95.20	35.00 [1]	Disability		75% of maximum earnings of $6,500 per year.
NEW BRUNSWICK	75	86.54	25.00 [1]	Life		75	86.54	30.00 [1]	Disability		75% of maximum earnings of $6,000 per year.
NEWFOUNDLAND	75	86.54	28.84 [1]	Life		75	86.54	25.00 [1]	Disability		75% of maximum earnings of $6,000 per year.
NOVA SCOTIA	75	86.54 [21]	35.00 [1]	Life		75	86.54	35.00 [1]	Disability		75% of maximum earnings of $6,000 per year.
ONTARIO	75	101.00	40.00 [25]	Life		75	101.00	40.00	Disability		75% of maximum earnings of $7,000 per year. Disfigurement benefits.
PRINCE EDWARD ISLAND	75	86.54	25.00 [1]	Life		75	86.54	25.00 [1]	Disability		75% of maximum earnings of $6,000 per year.
QUEBEC [13]	75	86.54	35.00	Life		75	86.54	35.00	Disability		75% of maximum earnings of $6,000 per year. [24]
SASKATCHEWAN	75	95.20	36.00 [17]	Life		75	95.20	36.00 [1]	Disability		75% of maximum earnings of $6,600 per year. [8]
CANADIAN MERCHANT SEAMEN COMPENSATION ACT	75	64.90	12.50	Life		75	64.90	12.50	Disability		75% of maximum earnings of $4,500 per year.

* See notations column.

[1] Actual wage if less.

[2] $3.60 additional for dependent wife and $3.60 for each dependent child under 18, up to four such children.

[3] Court will supervise disbursement of fund for children.

[4] Actual wage if less but with a minimum of $12.00

[5] Maximum not to exceed 1/2 of average industrial wage determined annually. (as of 7/1/70) Additional amount of $3.50 per week for each dependent child under 21.

[6] Employees tentatively found permanently and totally disabled referred to rehabilitation program. If employee has cooperated, cannot be rehabilitated and has exhausted benefits, then maximum of $47.00 per week is paid by special fund upon termination of payments by employer and carrier until employee's death.

[7] Plus rehabilitation allowance.

[8] Board has discretion to choose the 12 months in the preceding 3 year period most advantageous to workmen for computation of his earnings.

[9] Actual wage if less, but not under $10.00 for work week of 15 hours or over.

[11] Actual wage if less but in no case less than $22.00.

[12] Disability extending beyond 500 weeks or $32,500 paid from second injury fund. Maximum weekly benefit not to exceed 60% of state average weekly wage computed annually on September 1. Added benefits to dependents excepted from 60% maximum. Employee paid compensation for 3 months or more shall be evaluated for rehabilitation services.

[13] Beginning September 30, 1965, benefit increases varying from 1.1-40% for awards made from September 1, 1931 and January 1, 1965 will be paid existing cases.

[14] Compensation reduced 15% for employee's failure to use safety devices.

[15] Applicable to all cases prior to January 1, 1965. Benefits to be increased annually by 2% increase in Consumer Price Index. Maximum wage rate to be adjusted according to annual gross earnings of workmen. Increased benefits payable prospectively.

[16] Minimum benefits of $150 per month increased retroactively to August 5, 1959.

[17] Minimum benefits increased retroactively as of July 1, 1965.

[18] Maximum is based upon grade 15 of Gen. Schedule Classification Act ($31,523) minimum upon grade 2 ($4,897) as of 1/10/71). Benefits to be increased annually by 3% increase in Consumer Price Index after 1967.

[19] Compensation doubled if disability due to employer's violation of safety or health law or regulation.

[20] Maximum benefit shall equal 50% of annual state average weekly wage. (7/1/70)

[21] Extra allowance of $60 monthly for attendant, if needed.

[23] Maximum not to exceed 50% of state's average weekly wage - 7/1/70.

[24] Beginning January 1, 1970, benefits shall be increased annually by 2% increase in Consumer Price Index. Increased benefits payable prospectively.

[25] Increase in minimum benefits made retroactive to existing cases. Alberta is paying same from Gen. Revenue Fund.

[26] Failure to pay compensation within 2 weeks after due, 20% penalty added.

Figure 25-4-1 (Continued)

MAXIMUM INCOME BENEFITS FOR SCHEDULED INJURIES

January 1, 1971

JURISDICTION	ARM AT SHOULDER	HAND	THUMB	FIRST FINGER	SECOND FINGER	THIRD FINGER	FOURTH FINGER	LEG AT HIP	FOOT	GREAT TOE	OTHER TOES	ONE EYE	HEARING ONE EAR	HEARING BOTH EARS
IN THIS GROUP OF STATES COMPENSATION FOR TEMPORARY DISABILITY IS ALLOWED IN ADDITION TO ALLOWANCE FOR SCHEDULED INJURY														
ALASKA f	$16,000	$12,000	$2,000	$1,200	$700	$700	$350	$14,200	$10,200	$1,000	$350	$8,000	$3,000	$10,000
ARIZONA 5,6	33,000	27,500	8,250	4,950	3,850	2,750	2,220	27,500	22,000	3,850	1,380	16,500	11,000	33,000
ARKANSAS11,30	9,800	7,350	2,940	1,715	1,470	980	735	8,575	6,125	1,470	490	4,900	1,960	7,350
COLORADO21	12,367	6,188	2,975	1,547	1,071	654	773	12,367	6,188	1,547	654	8,270	2,082	8,270
CONNECTICUT 20/d	26,208	21,168	7,980	4,536	3,696	2,604	2,184	19,992	15,792	3,528	1,092	19,740	4,368	13,104
DELAWARE a	18,750	16,500	5,625	3,750	3,000	2,250	1,500	18,750	12,000	3,000	1,125	17,625	5,625	13,125
DISTRICT OF COLUMBIA	21,840	17,080	5,250	3,220	2,100	1,750	1,050	20,160	14,350	2,660	1,120	11,200	3,640	14,000
FEDERAL EMPLOYEES' COMPENSATION ACT	141,960	111,020	34,125	20,930	13,650	11,375	6,825	131,040	93,275	17,290	7,280	72,800	23,660	91,000
GUAM	15,680	11,872	2,856	1,568	1,008	952	392	13,888	9,688	1,456	448	7,840	2,912	11,200
HAWAII22	35,100	27,450	8,437	5,175	3,375	2,812	1,687	32,400	23,062	4,275	1,800	18,000	5,850	22,500
IDAHO	10,320	8,600	3,010	1,720	1,720	1,290	860	7,740	5,375	1,290	516	6,020	1,505	6,450
LONGSHOREMEN AND HARBOR WORKERS' ACT	21,840	17,080	5,250	3,220	2,100	1,750	1,050	20,160	14,350	2,660	1,120	11,200	3,640	14,000
MAINE c	12,775	10,950	3,650	2,336	2,044	1,460	1,241	12,775	10,950	1,825	730	7,300	3,650	7,300
MARYLAND 28	10,500	8,750	3,500	1,400	1,225	1,050	875	10,500	8,750	1,400	350	8,750	4,375	8,750
MASSACHUSETTS 7	5,625	4,375	(7)	(7)	(7)	(7)	(7)	5,000	3,750	(7)	(7)	5,000	3,750	10,000
MICHIGAN e	24,479	19,565	5,915	3,458	3,003	2,002	1,456	19,565	14,742	3,003	1,001	14,742	–	–
MISSISSIPPI	8,000	6,000	2,400	1,400	1,200	800	600	7,000	5,000	1,200	400	4,000	1,600	6,000
NEVADA 5	11,999	9,968	3,000	1,800	1,385	1,015	785	9,968	7,984	1,385	508	9,968	4,015	11,999
NEW HAMPSHIRE	14,338	11,725	3,350	2,077	1,742	1,273	871	14,338	10,117	1,742	670	8,442	3,484	14,338
NEW JERSEY	12,000	9,200	3,000	1,600	1,600	1,200	800	11,000	8,000	1,600	600	9,000	2,400	8,000
NEW MEXICO 5,11	9,600	6,000	2,640	1,344	1,056	816	672	9,600	5,520	1,680	672	6,240	1,920	7,200
NORTH CAROLINA 15	11,000	8,500	3,250	2,000	1,750	1,100	800	10,000	7,200	1,750	500	6,000	3,500	7,500
NORTH DAKOTA 4	7,875	6,300	2,047	1,260	945	630	504	7,371	4,725	945	378	4,725	1,575	6,300
OHIO 24	12,600	9,800	3,360	1,960	1,680	1,120	840	11,200	8,400	1,680	560	7,000	1,400	7,000
OREGON 2	10,560	8,250	2,640	1,320	1,210	550	330	8,250	6,425	990	220	5,500	3,300	10,560
PUERTO RICO 29	10,000	9,000	3,375	1,800	1,350	1,125	675	10,000	7,875	1,350	675	–	2,250	9,000
RHODE ISLAND	14,040	10,980	3,375	2,070	1,350	1,125	900	14,040	9,225	1,710	450	7,200	2,700	9,000
SOUTH CAROLINA	10,000	7,500	3,000	1,750	1,500	1,000	750	8,750	6,250	1,500	500	5,000	3,500	7,500
SOUTH DAKOTA 31	10,000	7,500	2,500	1,750	1,500	1,000	750	10,000	6,250	1,500	500	5,000	–	7,500
TENNESSEE	9,400	7,050	2,820	1,645	1,410	940	705	9,400	5,875	1,410	470	4,700		7,050
UTAH 3,27	13,000	9,750	3,900	1,950	1,950	1,300	780	11,700	8,125	1,950	780	7,800	3,250	13,000
WASHINGTON 5	15,000	13,500	5,400	3,375	2,700	1,350	675	15,000	5,250	3,150	1,150	6,000	2,000	12,000
WISCONSIN 1,11	24,250	19,400	6,063	2,910	2,183	1,262	1,388	24,250	12,125	4,026	1,588	13,338	2,668	15,891
WYOMING	6,800	5,500	900	400	300	300	300	5,800	4,300	950	300	5,000		
IN THIS GROUP OF STATES COMPENSATION FOR TEMPORARY DISABILITY IS ALLOWED IN ADDITION TO SCHEDULED INJURY WITH CERTAIN LIMITATIONS AS TO PERIOD														
ALABAMA	11,100	8,500	3,100	2,150	1,550	1,100	800	10,000	6,950	1,600	550	6,200	2,650	8,150
CALIFORNIA 8	15,750	12,600	2,520	1,680	1,680	1,260	1,260	16,800	10,500	2,100	420	6,300	2,100	10,500
FLORIDA	11,200	9,800	3,360	1,960	1,680	1,120	840	11,200	9,800	1,680	560	9,800	2,240	8,400
GEORGIA 17	10,000	8,000	3,000	2,000	1,750	1,500	1,250	11,250	6,750	1,500	1,000	6,250	3,000	7,500
ILLINOIS 3	22,200	14,060	5,180	2,960	2,590	1,850	1,480	20,350	11,470	2,590	888	11,840	3,700	9,250
INDIANA 10	14,250	11,400	3,420	2,280	1,995	1,710	1,140	12,825	9,975	3,420		9,975	4,275	11,400
IOWA 11 /g	14,030	10,675	3,660	2,135	1,830	1,525	1,220	12,200	9,150	2,440	915	7,625	3,050	10,675
KANSAS 16	11,760	8,400	3,360	2,072	1,680	1,120	840	11,200	7,000	1,680	560	6,720	1,680	6,160
KENTUCKY b	10,200	7,650	3,570	2,805	2,040	1,530	1,275	10,200	6,375	1,530	510	6,120	3,825	7,650
MINNESOTA 9	17,010	13,860	4,095	2,520	2,205	1,575	1,260	13,860	10,395	2,205	945	10,080	3,465	10,710
MISSOURI 5	13,456	10,150	3,480	2,610	2,030	2,030	1,276	12,006	8,700	2,320	812	8,120	2,552	9,744
NEBRASKA	12,375	9,625	3,300	1,925	1,650	1,100	825	11,825	8,250	1,650	550	6,875	2,750	5,500
NEW YORK 11/32	24,960	19,520	6,000	3,680	2,400	2,000	1,200	23,040	16,400	3,040	1,280	12,800	4,800	12,000
OKLAHOMA	10,750	8,600	2,580	1,505	1,290	860	645	7,525	6,450	1,290	430	4,300	4,300	8,600
PENNSYLVANIA 11,19	12,900	10,500	3,600	2,100	1,800	1,200	900	12,900	9,000	2,400	960	9,000		10,800
VERMONT 11 /g	13,115	10,675	3,050	1,852	1,525	1,220	732	13,115	10,675	1,525	610	7,625	3,172	13,115
VIRGINIA	12,400	9,300	3,720	2,170	1,860	1,240	930	10,850	7,750	1,860	620	6,200	3,100	6,200
IN THIS GROUP OF STATES COMPENSATION FOR TEMPORARY DISABILITY IS DEDUCTED FROM THE ALLOWANCE FOR SCHEDULED INJURY														
LOUISIANA	9,800	7,350	2,450	1,470	980	980	980	8,575	6,125	980	490	4,900	(13)	(13)
MONTANA 3	15,000	12,000	4,500	4,500	2,220	1,500	900	18,000	10,800	2,220	960	9,900	2,400	12,000
TEXAS	9,800	7,350	3,430	2,695	1,960	1,519	1,225	9,400	6,125	1,470	490	4,900	3,675	7,350
WEST VIRGINIA	12,960	10,800	4,320	2,160	1,512	1,080	1,080	12,960	7,560	2,160	864	7,128	3,240	9,720

a Disfigurement awarded in addition to schedule but not to exceed 20% thereof, in certain cases.

b Additional $370 for the loss of a metacarpal bone for the corresponding thumb or finger. Maximum weekly benefit 50% of 85% of average state wage = $51.00 for 1971. Added benefits may be granted if there is wage loss.

c Maximum weekly benefit 2/3 of State's average weekly wage = $73.08 as of 6/1/70. Disfigurement benefits, $1500 maximum.

d Award for disfigurement not due to loss of a member may be up to 205 weeks benefits.

e Computed @ $91 weekly based upon 3 dependents. Range of benefits runs from $75.00 to $104.00 maximum for 5 or more.

f Aggregate maximum compensation for temporary disability and permanent partial disability not to exceed $20,000.

g Based upon 50% of state's average weekly wage, annually—computed as of 7/1/70 to be $61.

Figure 25-4-2

NOTE: The Canadian laws contain no provisions for scheduled injuries and the Board decides each case on its merits.

[1]Computation based on employees 50 years of age or less.

[2]Scheduled by degrees.

[3]Figures are maximum with children or dependents. Temporary total allowed to date of amputation if it occurs subsequent to accident.

[4]25% additional for loss of master hand or member thereof.

[5]Computations for major members and in Washington 25% less for disability not involving amputations. 10% more in Missouri if complete loss of use or loss by severance, may allow 30 weeks maximum healing period additional.

[6]Calculated on $1,000 a month maximum.

[7]Specified schedule in addition to all other compensation. Proportional benefits for functional loss of use of leg, foot, arm, or hand.

[8]Schedule variable. Figures given based on standard rating—a 39 year old laborer and figures based on major arm and parts thereof.

[9]350 weeks for back injury. Maximum healing period for any injury 104 weeks, maximum benefit during same $70.

[10]26 weeks total disability, without deduction.

[11]Additional weeks for healing periods. (In New York, the weeks for the temporary total disability period in excess of the statutory healing period.) (Vermont adds $3.50 per week for each dependent child.)

[13]Commission may provide benefits up to maximum of 100 weeks.

[15]$5,000 maximum for loss of or permanent injury to any important external or internal organ or part of the body for which no compensation is provided in any other section. Total loss of use of back $15,000. Disfigurement, maximum of $5,000.

[16]Additional healing period up to 15 weeks may be allowed.

[17]Not to exceed 10 weeks total disability may be paid in addition. Maximum of $18,000 for two permanent injuries.

[18]Special provision for occupational deafness.

[19]Special provision for concurrent injuries.

[20]60% of average production wage determined to be $84.00 as of Oct., 1970. Average production wage to be determined annually by Labor Commission. Commission may award additional compensation based upon loss of earnings.

[21]Compensation is based on 66-2/3% of the impairment of the employee's earning capacity rather than his average weekly wages.

[22]In cases in which the disability is determined as a percentage of total loss or impairment of physical or mental function of the whole man, the maximum compensation shall be computed on the basis of the corresponding percentage of $35,100.

[24]Compensation in cases of permanent partial disability may be paid for 200 weeks if the percentage of disability equals or exceeds 90%.

[26]Percent of total disability it represents.

[27]Maximum amount payable $13,000 or $20,280 for any combination of disabilities.

[28]Scheduled injuries and "other cases" of 250 weeks or more, except disfigurement, shall be considered "serious" cases and entitled to an extra award of 1/3 the number of weeks awarded, not to exceed $65 per week and not subject to $17,500 maximum total.

[29]Compensation doubled if disability due to employer's violation of safety or health law or regulation.

[30]Compensation increased 15% if disability due to employer's violation of safety regulations.

[31]Loss of use of back computed as percentage of 325 weeks benefit.

[32]Compensation for wage loss in addition to schedule if impairment due to loss of 50% or more of member.

Figure 25-4-2 (Continued)

RETROSPECTIVE RATING PLAN EXAMPLE

A	B	C	D	E
		Loss	Converted	Basic
Standard	Incurred	Conversion	Losses	Premium
Premium	Losses	Factor	$(B) \times (C)$	Percentage
$150,000	$25,000	1.29	$32,250	20.5%

F	G	H	I	J
	Minimum		Maximum	
Basic	Retrospective	Minimum	Retrospective	Maximum
Premium	Premium	Premium	Premium	Premium
$(A) \times (E)$	Percentage	$(A) \times (G)$	Percentage	$(A) \times (I)$
$30,750	24%	$36,000	132%	$198,000

K	L
State	Retrospective
Tax	Premium
Multiplier	$[(F) + (D)] \times (K)$
1.035	$63,000

Self-insured Employers may comply with the law if they meet the requirements within that state by becoming self-insured. Through this plan, the employer would not have to pay for the losses incurred by other employers which was used to establish the manual rate; but by the same token, the employer could not share the risk of his own losses. A serious accident taking the lives of several people could jeopardize the self-insured employer's chances of remaining solvent.

25-6. Labor Turnover

So far, we have discussed only the obvious costs of industrial accidents through the insurance premiums charged to provide Workmen's Compensation. Another expense not as easy to assess is labor turnover. A recent study has been made in which underground miners were subjected to a definite unavoidable safety hazard created by unstable underground geological conditions. The stresses in the rock in that area created the phenomenon known by the miners as "rock popping." Slabs of rock would explode from the sides of the tunnels doing serious damage to anything in their path (see Figure 25-6-1). Turnover for the miners averaged 23.27 percent, with some monthly rates reaching a maximum of 43.7 percent. Monthly miner turnover in heavy construction in general will average about 10 percent, plus or minus 2 percent. This project experienced a turnover rate as high as four times the normal.

The turnover rate for all other crafts ranged from 1.07

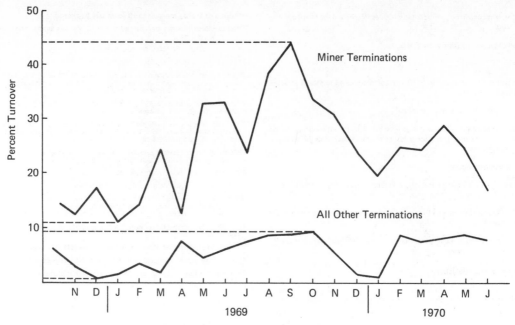

Figure 25–6–1 Percentage turnover

percent minimum to 9.27 percent maximum. The average turnover for these other crafts was 5.65 percent which fell very close to the accepted norm for similar projects.

The monthly miner turnover generally paralleled the number of monthly mining accidents which required a physician's attention. Thus, when the contractor experienced intensified adverse geological conditions, as in May 1969, the number of mining accidents increased, which in turn increased the rate of miner turnover (see Figure 25-6-2). The overall effect was lost efficiency in mining progress due to the continual influx of inexperienced miners. (Additional national surveys not associated with this project indicate that an overwhelming majority of accidents involve workers with less than three months of experience on that job. The work force was then composed of a majority of workers who fell within the category

of being most susceptible to industrial accidents.) The situation evolved to the point that the contractor's mining operations never escaped the learning curve for its miners.

As an example of the contractor's stalled position in the learning curve, consider Figure 25-6-3, for total round time vs. time period for a section of tunnel. Round time is the time required to complete a specific amount of excavation usually measured by the number of feet the tunnel is extended. Notice that after twelve weeks of excavation, total round time had still not stabilized.

In Figure 25-6-4, a direct correlation between miner turnover and total round time is established. It can be seen that as miner turnover decreases, so too does total round time for the excavation cycles.

Unable to control the geological conditions, the contractor in the spring of 1969 initiated a bonus plan in an

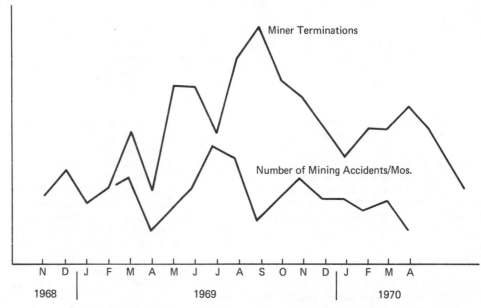

Figure 25–6–2 Terminations and accidents for miners

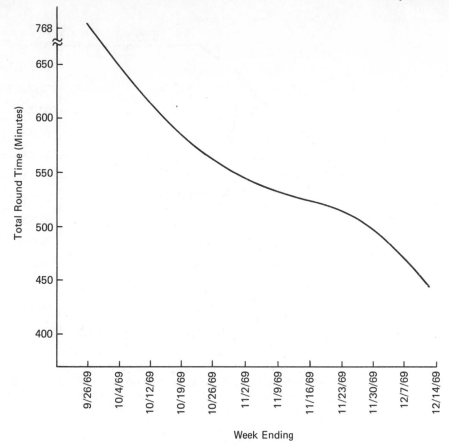

Figure 25–6–3 Typical total round time

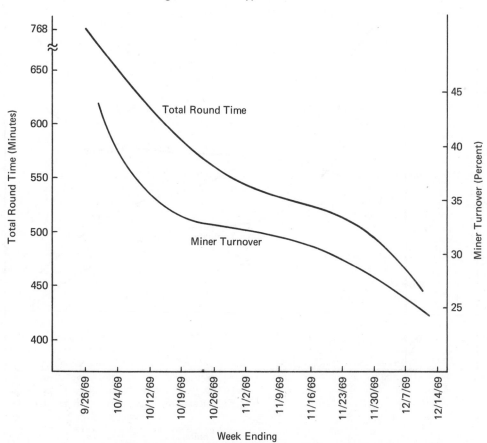

Figure 25–6–4 Typical total round time and miner turnover

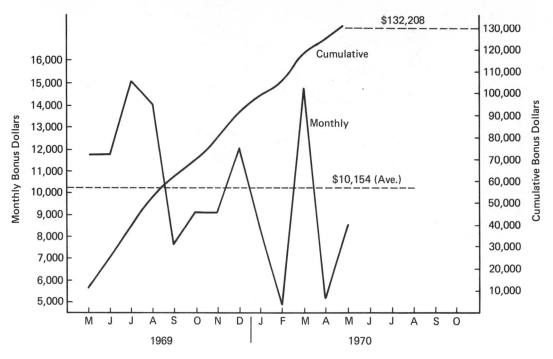

Figure 25-6-5   Typical monthly bonus for miners

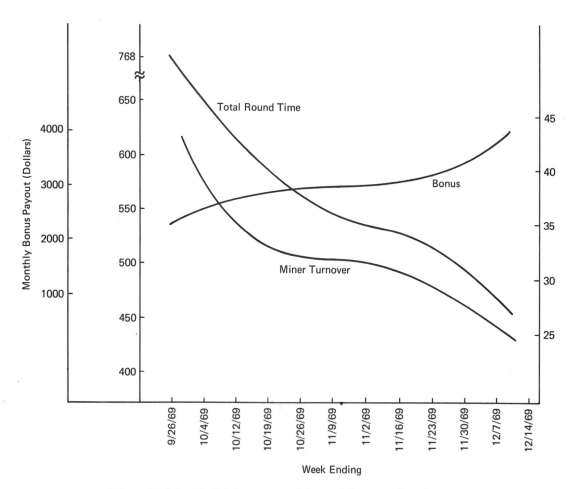

Figure 25-6-6   Monthly bonus vs. total round time on miner turnover

attempt to increase production rates in the area of underground excavation. Comparing the monthly bonus payments indicated in Figure 25-6-5 with miner terminations charted in Figure 25-6-1, you can see that bonus payments were only moderately successful in curtailing turnover.

Figure 25-6-6, indicating the twelve-week span for the end of 1969, does show that miner turnover and thus total round time did decrease, while for the same period bonus payout was generally increasing. For a contractor to be willing to pay out more than $130,000 in an attempt to curtail labor turnover and, therefore, increase productivity indicates what a serious expense such turnover must be. This example may be severe, but it does show a clear example of what unsafe, accident-ridden working conditions can do toward demoralizing a work force.

25-7. Safety Benefits the Employee

When a worker covered by Workmen's Compensation is injured on the job, he can expect compensation for wages lost and payment of medical expenses, unless in some jurisdictions he has been found guilty of intoxication, deliberate infliction of injury upon himself, violation of safety rules, violation of a law or an order, deviation from his work to attend to personal affairs, or other actions that can be described as intentional or willful misconduct. The amount of benefits vary by state, but rarely does the benefit exceed two thirds of his wages. The maximum weekly benefits, as illustrated in Figure 25-6-1, hardly compare with those wages normally earned by a construction worker. The loss in wages alone creates a heavy burden for the injured worker. If you couple that burden with the pain and anguish that the worker and his family experience, the loss is truly noncompensable.

When a dismemberment occurs, usually that employee can no longer apply his craftsmanship in the construction industry. The worker is normally forced to accept less skilled and lower paid labor than was his previous occupation. Even after the time has healed the physical wounds, mental discomfort may remain. A worker in such a physical industry as construction usually develops a pride in his physical capabilities. Any accident that results in reducing that physical power creates serious challenges to that worker's projected manhood. If he attempts to return to work in the construction industry with a permanent partial disability, he often resents the special concern other workers afford him. Whether or not it is true physically, often the injured worker considers himself a cripple.

Special care must be afforded the injured worker in both physical and mental rehabilitation. Due to necessity or desire, rehabilitation frequently requires retraining for employment in another industry. The worker's life style may be completely changed. He may be forced to relocate to find employment in his new vocation. In any event, no injury can be considered anything but an unfortunate occurrence. Every aspect is normally negative. If an accident prevention program prevented such an occurrence, that in itself would justify its existence; but safety pays in less obvious ways.

The worker who has learned to do his work efficiently and safely is a better employee. Employers continually seek out this type of employee; therefore, the employee increases his chances of steadier employment and higher lifetime earnings. His actions indicate that he can analyze the work to be performed and select the most efficient and safest methods. One could even generalize further to say that normally he could be expected to work well with other employees, for in the construction industry, teamwork prevents accidents. A safe, efficient employee is then an asset to himself and the industry.

25-8. Safety Benefits the Customer

As the expenses of industrial accidents are to be considered a cost of operation and as such are to be transferred from the worker and employer to the consumer, eventually the customer will bear the cost of poor accident prevention programs. It is very possible that the immediate cost will not affect the customer, particularly in the case of fixed bid contract construction, but the effect that losses due to injury creates within the cost of providing Workmen's Compensation Insurance will eventually be felt.

The cost of a project as viewed by the customer not only includes the price for construction but also the cost of nonproductive capital. High interest rates make it mandatory that money be used in the most productive manner possible. Accidents create time losses: the time a project is delayed while another piece of equipment is brought in to replace one damaged accidentally, the time spent replacing an injured worker, and time spent training him to bring that worker to full production. Serious accidents can create delays that may never be overcome. When that occurs, the on-line projection of the customer must be revised and all moving plans must be revamped. Often customers have new equipment in transit to be installed after construction is completed and delay in completion of construction can create a chain reaction affect, causing real hardship to the customer. If an employer can do his work efficiently and safely, everyone benefits, and particularly the customer.

25-9. Safety and Employer-Employee Relations

As indicated in the miner study, industrial injuries effect labor turnover. When accidents increase, workers quit; but very few workers will admit that they are scared or concerned for their safety. Most workers will find other reasons to look elsewhere for employment. In the field, whenever labor is upset about something they cannot resolve to their satisfaction, safety is used as a verbal football. Where a real safety problem exists, labor generally uses this vehicle to promote additional causes. When men begin to get agitated for whatever reasons, and unsafe conditions can be incorporated in that agitation, the resultant feelings can become extremely bitter. When a worker is concerned about his grievances, his production decreases, and concern about production can cause an employer to fail to perceive the real reasons for dissatisfaction.

Good accident records on a project usually indicate that management has carefully planned the work to be performed. While examining the operation plans, management has selected the safe and efficient way to get the job done. The careful approach to directing performance of work on the project is conducive to being aware of management's responsibility toward its workers. Such awareness can be the basis for cooperative employer-employee relations.

25-10. Safety Program Planning

Effective safety programs are the product of planning, coordination, and commitment by all employees of a company, from the worker lowest in the chain of command to the administrator in the highest office.

Safety programs have five basic units:

1. Management support and direction
2. Safety organization
3. Worker education
4. Hazard control
5. Medical and first-aid plans

25-11. Management Support and Directions

The company must issue a safety and accident prevention policy statement so comprehensive and decisive that no question remains that the program will become an on-going process. Management must delegate responsibilities and authority so as to place the enforcement of that policy in the main stream of operations. Management must plan personnel selection, placement, and indoctrination in such a fashion that safety concerns become a way of life. Management must establish safe, realistic work standards. It is management's responsibility to provide safe machinery and equipment for the workers to use. Management must also see that personal protective equipment is used where applicable. Management must assist in planning general pre-job safety. Above all, management must maintain an active and continuing participation in the safety program.

The modern trend of safety programing is toward an integration of safety into the productive system from the standpoint of materials, machines, and people, and getting away from being considered as an adjunct or afterthought if time and money permit. The forementioned responsibilities of management can bring such programing into reality.

25-12. Safety Organization

A safety organization's foundation is qualified safety personnel. Safety or accident prevention is rapidly approaching an applied science, and the role of safety officer for a company is no longer a position to fill with someone passing time until retirement. A dynamic, qualified person is the key to sound accident prevention programs. In addition to staff personnel in the safety organization, safety committees representing all levels of workers and administration should be formed. These people are more closely associated with the daily operation of the company and can provide valuable assistance in spotting potentially hazardous conditions or unsafe practices.

Accident-cause analysis records must be maintained and used to point out patterns of accident or injuries. If minor accidents continue to occur in a given area, eventually a major accident will occur.

Detailed remedial action procedures must be created and distributed to key personnel and the general work force. Prompt corrections of minor problems can prevent major damage.

A system must be designed to process safety suggestions, and merit awards should be issued for those suggestions that correct conditions which might result in injury to workers or damage to equipment.

Finally, the safety organization must periodically review the safety program, looking constantly for methods to improve the work environment and the processing of safety information.

25-13. Worker Education

As stated earlier, most injuries occur to workers who have had less than three months of experience on the job; therefore, the soundest safety program begins with individual job training. The new worker must be taught the correct way to do his job before it becomes necessary to correct bad work habits or procedures. Safety training courses should be required of all employees, and these courses should be reinforced with safety meetings of the tool box type for the construction industry. Every opportunity to utilize demonstrations, displays, posters, exhibits, and other visual aids must be taken to keep safety on the minds of all employees. Additional material, such as manuals, rule books, payroll inserts, periodicals, and handouts should be developed and disseminated where practicable to assure that everyone gets the word.

25-14. Hazard Control

Safety inspections should be the responsibility of not only the safety organization, but of every employee for his own work area. Formal safety inspections should be as frequent as possible, with follow-up inspections of potential hazards. Every production function should have a job hazard analysis. Production procedures would be analyzed, and improper or inefficient work methods would be highlighted.

The care of machinery and equipment must be included in hazard control. Defective equipment not only reduces efficiency but it presents a potential cause of accident or injury. In addition to proper maintenance of machinery, guards must be provided to prevent parts of the body or clothing from being caught in moving parts.

Sanitation and occupational disease control are another important part of hazard control. Care must be taken to provide the employee with a clean environment in which to work.

No amount of personal protective equipment will prevent damage or injury if that equipment has not been maintained and kept in proper working order. Instruction must be provided to each worker who is issued protective equipment on how to service and repair that equipment to provide the protection it was designed to afford.

Finally, hazard control is best realized by enforcing safe practices. The rules designed to protect life and property will never do the job unless they are enforced 100 percent of the time.

25-15. Medical and First-Aid Plans

The safety organization must make and inform the workers of arrangements for hospitals, doctors, nursing, and ambulance service. When an injury occurs, prompt treatment can help prevent aggravation of the possible damage caused by delay in receiving medical attention. First-aid facilities and competent attendants must be available to assist in the treatment of minor injuries and maintain records to supplement accident-cause analysis.

The foregoing description represents a model accident prevention program probably in an ideal state. Few, if any, companies could boast of such a program of this scope that reaches all of its operation-production forces. It should be remembered that no program can prevent all accidents, but a current definition of accident can shed light on management's need to press for comprehensive accident prevention. Webster's Seventh New Collegiate Dictionary defines "Accident" as: "An unfortunate event resulting from carelessness, unawareness, ignorance, or unavoidable causes." All but the last cause can be attributed to management's failure to educate and motivate the employee to the degree that safety is a way of life.

The construction company, in evaluating its approach to safety, must not overlook any available source of achieving a responsive and responsible safety program. All of the outside help that is available should be used; but more importantly, the organization must develop a constant and prevailing internal awareness of values of safety. Periodic reports of accident frequency and severity must be maintained and analyzed. Each and every accident must be thoroughly investigated to acquire information that will prevent a similar accident from occurring.

It is necessary that all standards, including prevailing laws and regulations, be known to the company and its operations. Much legislation is being enacted in the area of occupational safety and health. A prime example is the Occupational Safety and Health Act of 1970. Every employer engaged in any kind of enterprise should be certain that his organization is thoroughly familiar with this and other safety regulations.

26 PUBLIC RELATIONS FOR CONTRACTORS

PAUL W. CANE

Manager of Public Relations
Bechtel Corporation
San Francisco, California

"We don't need any public relations on this project," a midwestern construction executive commented recently. "The job is so big we'll get plenty of stories in the papers."

The gentleman, a seasoned and capable performer, will probably do an excellent technical job on the project. His work record establishes that. But his belief that public relations consists only of getting "plenty of stories in the paper"—implying that a number of randomly reported articles will contribute to public acceptance—labels him unschooled in the discipline of public relations.

The comment illustrates that the midwestern executive lacks understanding of the ingredients of a comprehensive public relations program, which would establish targets, encompass qualitative consideration, and seek an identity with the public interest.

Moreover, he exhibits a lack of cognizance of the need to develop employee, community, and political understanding of the job, its purpose, and the contribution it will make to the community. And he ignores the importance of two-way communication between his company and the several different publics who will have a legitimate interest in his work.

Because he is unaware, he may well find, as the job progresses, that he will be subject to annoyance, costly delays, harassment, embarrassment, possible ridicule, damage to his firm's reputation, and increased costs. That need not happen, of course, for with public support (or, at a minimum, the absence of opposition), he will be able to do his job and make his contribution to society.

To gain support, he must perform well—and then adequately communicate that performance to the various publics in which he has an interest and which have an interest in him. This means relating the job to the public interest. It means recognizing the impact of the job on people. And it means allowing people to communicate their views and impressions to him—as well as his to them.

Since public relations is inherent in every enterprise, the question facing a firm is not: "Shall we have public relations?" There really is no choice in the matter; public attitudes *exist* whether we want them or not. Instead, the only option is: "What *kind* of public relations will we have?" Effective, positive public relations can be achieved if the firm understands, accepts, and adheres to some basic principles.

26-1. Gaining a Reputation

Good performance, first, and then adequate communication of that performance are really what public relations is all about. I recall a conversation with an excutive of a large (and very good) eastern construction company whose opinions had been molded in a quieter decade. It was this executive's belief that communicating with many segments of the public—and the general public in particular—was not his concern. He reasoned that performance spoke for itself and that professionalism demanded that contractors, like doctors, avoid initiating publicity. It was also his belief that his firm had no real relationship with the general public.

These beliefs no longer have validity in today's world, because the world has changed and continues to change in a rapid, sometimes bewildering, manner. The rate of technological growth leaves us gasping. Each new development brings on, just as surely as summer follows spring, political, economic, social, and psychological shifts and stirrings, which in turn influence human needs and human desires.

Today, as never before, legislators, the press, and the general public are keeping a watchful eye on construction projects. They are deeply concerned about such problems as pollution, use or misuse of land and forest areas, the disposal of waste, the agonizing questions of how best to solve monumental transportation and urban problems. The

582

people expect to have—indeed, are demanding—a voice with respect to the siting of power plants, freeways, industrial plants, or mass transit systems.

When a project is built with public money, the people have an even greater right to be heard. This means that it is imperative for the contractor to earn a reputation of respect and acceptance for his organization *before* he moves into a public project. Such a reputation is earned by good performance and adequate communication of it. A climate of public acceptance is particularly valuable if the project in which he may become involved develops into a controversial one.

In such a case, it can be most helpful and effective if the contractor is attuned to public attitudes and moods. He then may counsel and confer with the client both on anticipated reaction to a proposed project and on a program to earn understanding and acceptance. An ability to think in public relations terms is required to assess these attitudes and moods. For that reason, a firm's public relations voice is part of management and should contribute to policy decisions on projects from their conception.

The public's good opinion of a particular construction company, or the industry as a whole, is important to its commercial success. President Abraham Lincoln recognized the importance of public opinion more than a century ago, when he noted: "Public sentiment is everything. Without it, nothing can succeed; with it, nothing can fail." This presidential appraisal of a century ago is magnified today in terms of fast-changing social attitudes and the communications explosion.

26-2. Public Relations Objectives

Simply stated, the objective of public relations is gaining and keeping the understanding, goodwill, and cooperation of the people who affect and are affected by a company.

That definition is a broad one. To be more specific about public relations objectives, one should first be more specific about what is meant by "people." There is no such thing as *a* public; there are many publics—and the interests of the various publics may not always be compatible.

Depending on the size or the organizational setup of a company, one may find himself dealing regularly with some or all of the following distinct publics:

1. Clients and potential clients
2. Employees and members of their families
3. The project and branch office community
4. The mass media
5. Government officials and political figures
6. Stockholders and the financial community
7. Competitors
8. Suppliers
9. Technical and professional societies
10. Faculty and students at engineering schools
11. The "general public"

26-3. Foundations of Good Public Relations

What one should hope to accomplish by way of public relations activities with each of his publics, and how to do it, is suggested below. But first it should be noted that there are four foundation blocks on which successful relationships with publics can be built. They are:

1. Sound public relations is based on performance, not words. The words—oral or written—are essential for

effective communication of achievements or policies, but they will deservedly fall on deaf ears if they are not supported by reality.
2. There is no substitute for face-to-face communication. Its importance cannot be overemphasized. Developing appropriate personal contacts—and then maintaining them on a friendly, professional, and cooperative basis —should be part of everyday responsibilities.
3. Public relations is a long-term, continuing job. Although occasionally it may be necessary to use public relations techniques to put out a fire, the public support that can be expected (there are times when it will be needed more than others) is based not only on what may be said or done at a moment of crisis, but on what may have been done and said over a long period of time.
4. Truth. To lie is to lose credibility and to undermine all that has gone before.

26-4. Relations with Clients and Potential Clients

Ralph Waldo Emerson said it well: "What you are thunders so that I cannot hear what you say to the contrary." It's a point the contractor should remember. For no matter how much expensive literature he publishes, no matter how many ads he sponsors, no matter how many press releases he issues, and how many interviews he grants the media, all the effort is in vain if his message is not backed up by solid performance. It is performance that earns and secures one's reputation and image, and they are developed only through integrity and by following through on promises.*

It could be a temptation, once a project has been completed, to steer a bulldozer off into that setting sun (which always signaled the end of the old Fitzpatrick travelogues) and forget about the client. There is a better way.

Once a project is finished, it is desirable to call on the client to make sure everything is working as it should. Chances are the only time the client will call the contractor is when something has gone wrong, and it's an unfortunate human trait to remember those things that went wrong more vividly and longer than those things which went right. Contact a past client occasionally even if another job from him is doubtful.

Offer to help the client gain recognition for his project. If the client has his own press-relations officer, offer to work with him to develop and coordinate information that can be mutually beneficial.

The contractor can provide a valuable service to a client by alerting him, before work begins and while the project is still in the planning stage, to anticipated public reaction to the job and by suggesting ways of avoiding negative attitudes. To provide this service, the contractor and his public relations officer must be sensitive to the public mood and aware of the vagaries of public attitudes.

Some companies publish magazines which are edited specifically for clients or potential clients. This is one way to identify past projects, to keep the contractor's name in

*Perhaps some comments on that popular but overworked word "image" are in order. The word came into the general public relations vocabulary about a dozen or so years ago. As originally used, it was a good word: one intended to describe valid impressions or reflections of a person's or a corporation's posture and reputation But through broad use, and misuse, it has come to connate only the surface impression—and not the reality—of a situation. In fact, to many in the general public, an "image" has little relationship to fact. Most annoyingly, it seems generally accepted to be a false reputation, rather than a reflection of what exists.

the mind of clients and potential clients, and to remind them of the scope of his capabilities. Copies of such a publication might also be sent to employees to satisfy their need to know what is going on in the company.

A client publication should be of high quality. It will be a reflection of the company's professionalism. It is not uncommon for companies to spend $90,000 or more a year on these publications. Smaller firms which do not have the budget to produce a regularly distributed publication might prepare a brochure which briefly and attractively sums up the company's capabilities. These brochures can be good door-openers when dealing with potential clients and can also be used to answer requests for general information on a firm.

Also helpful in communicating capabilities to potential clients are movies (or less-expensive slide films) which describe a particular project summarize overall operations. These films can also be shown to some service clubs (which frequently search for program material), schools, and other community groups to give additional exposure with other important community publics.

26-5. The "General Public"

With increasing public awareness of, interest in, and often concern about, many construction projects, the contractor should be sensitive to the attitudes of the general public It may be helpful to keep one public relations tenet in mind: Understanding begets acceptance.

To develop understanding, the contractor should begin his public relations activities long before work on the visible aspects of a project commences. During some projects (construction of an office building, for example) many persons will be inconvenienced. Sidewalks will disappear, streets may be closed off or parking on them banned, and the din of construction will annoy people working in neighboring buildings. On certain other projects (such as construction of a nuclear power plant) the contractor may run into opposition because of such factors as unfounded fear for public safety.

What can be done? How do you gain acceptance? Let us, for example, assume that you have won a contract to build a major office building. Once the contract has been signed, you may—working closely with the client, of course—arrange for a news conference to announce the project. The local political leaders and the Chamber of Commerce might be invited to participate.

An honest appraisal of anticipated disruptions should be made. Point out you will do everything possible (and then make sure you do) to keep distractions and disruptions to a practical minimum. Stress at this time the positive benefits which will result from the project (the employment it will generate, how the completed building will enhance the livability of the area, etc). As construction proceeds, invite the media to make periodical progress inspections.

At major urban project locations, bleachers can be provided for sidewalk superintendents and inexpensively printed progress reports can be made available at the site. On many projects, it will be necessary to erect barriers. They need not be an eyesore which offends the community. You might consider inviting local art students to create paintings or designs directly on them. Then go one step further and invite the public to vote on which painting or design they like best, award prizes, and publicize the contest.

Employees in nearby buildings will be distracted by construction, of course. Personal visits may be made to the new neighbors and meetings set up at which time you can outline your construction schedule and tell them what they can expect in the way of inconvenience. Again, stress the benefits which will result when the work is finished.

After the project is completed, offer to help the client arrange tours and open houses. One tour should be reserved for the media. The employees who will be working in the building deserve a special open house so they can show it off to their families. The general public can be invited to dedication ceremonies. And the people in the neighborhood who were inconvenienced by your work should have a special tour of the facilities at which light refreshments can be served as a gesture of thanks for their patience.

Successful public relations is always based on what has been done well and said effectively over a long-term period of time. A firm with a long-established reputation as a "good citizen" in a community finds it much easier to have its viewpoint heard when it becomes involved in a crisis (an accident or a strike, perhaps) or when there is opposition to one of its projects.

Active participation in civic affairs will enhance the contractor's reputation as a good citizen. He and his employees will undoubtedly be involved in working with the local chamber of commerce, school boards, the United Crusade and many other groups, and his firm and employees may also be making significant contributions to such organizations. He will gain recognition for these efforts and it is unnecessary to be blatant about his role.

26-6. Employee-Company Publications

Good public relations begins at home. What a company's employees—who make up one of its most important publics— think of their firm has a profound effect on morale and thus influences the quality and quantity of work performed, the turnover rate, what the general public and potential clients think of the company, and the ability of the company to attract new employees.

It has been said, with truth, that every individual on the payroll is to some extent a member of the company's public relations staff, because every time an employee talks to his wife or his neighbors about his company, he is dealing in public relations. Every time one of the employees accidentally bulldozes out somebody's prize begonias, a public attitude is being formed. Every time an employee serves on a local governmental body or takes on a job working with the local United Crusade, he is helping to form a public opinion of what his company is all about. Every time a secretary answers the telephone she is, by her language, manner, and tone of voice, creating in somebody's mind a favorable or unfavorable impression of the entire organization.

A vast potential for positive communication about a firm exists among its employees. This is illustrated by research conducted by Opinion Research Corporation of Princeton, New Jersey, which showed that the average adult in the United States is involved in sixteen different contacts each day. (Casual "good morning" encounters were not considered a "contact.") Each contact affords an opportunity for people to exchange ideas and to identify and discuss a firm. To give an example of the potential which exists: with 1000 employees, multiplied by sixteen daily contracts, you find there are *16,000 opportunities every day* to have the company identified and discussed— and hopefully reflected in a positive way.

Do employees talk about their jobs and employer? Other Opinion Research Corporation data show that, next to "family," the item of greatest interest among people in the United States is their jobs. Obviously, people talk about what interests them. And what they say gets around.

Of course, you can't expect an employee to speak of his firm in glowing terms if he regards the employer as a "slave driver" or if benefit plans or salary schedules are not competitive with other firms in the same industry and area. In relations with employees, as in relations with every other public, words alone will not do a job. Performance is the bedrock on which all successful public relations programs must be built.

Because employees are such an important public, concentrated attention should be given to establishing an effective employee communications program. A properly executed program will:

1. Provide a way of satisfying the individual's desire to have management recognize his role in the organization and the importance of his contributions to the over-all success of the firm.
2. Help employees identify themselves as part of the team (this is a deep psychological need).
3. Create an atmosphere in which, because morale is high, an employee can do a better job.
4. Assist in the development of an understanding of your firm. (Again, understanding begets acceptance.)
5. Encourage people, because they have the facts, to talk intelligently about the company, to speak of it positively and thus make personnel recruitment easier.
6. Provide employees with information which can be discussed with neighbors, friends, and business contacts. Through the dissemination of such information, an awareness of the organization is developed in the community and a certain amount of status emerges for the employee. Employees and members of their families want to be able to identify themselves—and identify themselves with *pride*—with an organization that is recognized publicly by their friends and neighbors.

Although certain information in a firm is confidential and should not be given general release, every attempt should be made to communicate as fully as possible with employees. Surveys have shown that employees especially want to know of their company's plans for the future, what problems the company is facing, and what the individual's opportunities are likely to be in the future and where he may fit into the whole picture. It would be tempting to give out only the good news, but credibility in communications will not be fully achieved unless management also discusses setbacks.

First-line supervisors play a key role in a successful employee communication program. Ideally, they make employees under their supervision aware that they can feel free to ask questions and that they can expect honest, authoritative answers. Since communication is a two-way street, employees must be assured that, in addition to getting their questions answered, they can make their views known to the supervisor as freely as the supervisor makes his views known to them. A supervisor should be highly visible on the job, and not hide out constantly in an office behind a closed door, to remind employees that he is available to help solve their problems and answer their questions.

As companies grow, it becomes increasingly difficult for any one supervisor to be aware of everything that is going on in the firm. It then becomes necessary to develop more formal—but only supplementary—channels of communication. These can include employee publications, bulletin boards, letters to the employees and their families, movies and slide films, staff meetings, award dinners, established telephone numbers which employees may call to hear recorded news summaries, and so on.

Because of the permanence and the authority of the written word, employee publications are an effective method of communicating with employees. Today, in fact, there are an estimated 1,000 company publications printed regularly in the United States. They run the gamut from newsletters to newspapers to magazines; are issued from daily to quarterly; are reproduced on the company mimeograph machine or are as lavish as any national magazine found on the corner newsstand. The kind of publication issued should be determined by the need and the budget.

No matter what format is selected, and no matter how much money you spent on the publication, the effort and expense will be wasted if the publication is not read. Management should not make the mistake of assuming that simply because an employee is vitally interested in his job (as noted earlier) he will read anything the company publishes. He won't. His reading time is as limited as management's, and when the company publication arrives it will compete for his attention with *Playboy, Life,* the TV set, and the coffee table he's building in the basement workshop.

To help ensure readership, two factors should be kept in mind. One is that the publication should be professionally edited to capture attention. The second is that the material should be relevant to the audience.

Small firms that feel they cannot afford a full-time editor on the staff might consider having their publication prepared by a public agency or (if the firm is located in a city large enough so these services are available) by a freelance editor who prepares publications for several companies. Whether the editor is on the staff or not, top management should temper the temptation to edit copy with too much vigor. Certainly all copy for a company publication should be cleared to assure accuracy, to ensure conformity with corporate policy, and to make sure confidential information is not released. But respect the editor's professionalism—or replace him. Employees, it must be remembered, are extremely adept at recognizing—and ignoring—the vague verbosities some managements have been known to try to inflict on corporate publications.

What is relevant to employees? As indicated earlier, employees have a deep interest in the issues that affect the progress of the company and, as a result, their own careers. The company publication should discuss these issues candidly and provide sufficient background so the issues are clearly understood.

There is a rule of writing which says: "Never overestimate your reader's information." There is another rule of public relations, also worth remembering: "Because you have told an important story once, don't assume it has been told."

In the years when company publications first began to appear on the scene, there was a mistaken assumption that employees were interested only in seeing their own names in print and wanted to read of nothing more significant than their coworkers' babies and fishing trips. This type of reporting soon became known as the "live babies and dead fish" school of journalism. Fortunately, many of these publications failed to survive. It's just as well. Some never earned their keep.

Let us assume you now have, or are about to begin, an employee publication. You may find it profitable to ask yourself this question: Now that we have made the investment in this publication, can I get more return or value from it?

The major expenses in publishing a newspaper or magazine are incurred up to the point where the publication actually comes off the press: for salaries, typesetting, page makeup, color, paper, press time, etc. Once the press run begins, each additional copy you print will cost very little.

Hence, at small additional cost, you can also send copies to clients and potential clients, stockholders (if you are a public owned company), government officials on various levels, editors of trade journals, local community leaders and opinion molders (newspaper editors, TV and radio stations, presidents of local service groups), libraries (high school, college, and public) in areas you serve or hope to serve, and to guidance or job placement counselors in schools in your area. The list can be as large as you want to make it.

It is a worthwhile practice to mail to those outsiders on your mailing list, about once a year, a self-addressed, prestamped post card on which the recipients can indicate whether they wish to continue receiving your publication.

Who on the outside will be interested in your employee publication will be determined by its content. If the publication heavily emphasizes personnel news or the social activities of employees, there will be limited outside interest in the publication. If its content is broader, and emphasizes company involvements, goals, problems, and accomplishments, outside interest will increase.

There will be times when, aware that your employee publication will be going outside the company, you might decide it is desirable to modify your information with either greater or lesser emphasis on certain points. There is one hazard here. You might end up succeeding in the communications process with a handful of outsiders, but lose the ear of many of your employees.

When publication of each issue is being planned, consider who—in addition to your employees—might be interested in a story in that issue. Increase your print order for that issue to cover your additional distribution needs and, when the publication appears, send a copy to the "outsider" accompanied by a brief note calling his attention to the story. If this note goes to an editor, offer reprint permission or invite the editor to call you if he wishes to explore the topic in greater depth. On certain issues you may end up mailing only a handful of copies; on others, you may mail several hundred.

This approach requires substantially more mental input than casually adding names to a mailing list. But it increases the possibilities of getting the additional exposure you want,—and you do get increased return on your investment in communication.

26-7. The Project or Branch Office Community

Let us suppose that, as contractor, your intent is to concentrate your public relations objectives on a local scale, whether you maintain a permanent branch office in a community or merely open one for the duration of a specific project.

Dealing with a smaller public, you will have a solid opportunity, as far as public relations activities are concerned.

But this self-same opportunity could create a problem for you.

The opportunity: in dealing with a relatively small public, you will find it easier to establish close personal relationships with leadership segments. You will, for example, find it easier to develop valuable political contacts. Because of your role in the community, you will find yourself invited to attend meetings of service organizations, to work with them on civic interests, and be asked to supply speakers to appear before community groups and schools. And you will be able to establish a close personal relationship with the local media. Through your activities, however, your responsibilities to the community will be greater—and so will be your vulnerability to criticism.

People, especially in small communities, can become resentful when somebody from the outside—especially if he represents a large or powerful organization which they know will be there for only a fairly brief time—moves in and appears to take over. Far too often, a representative of a large organization will move into a community, work hard on its behalf, then suddenly find himself the target of what he considers (often rightfully) unwarranted criticism and sniping.

The secret of successful public relations on the local level is to be active, but always gentle. Don't throw your weight around, and don't be afraid to let others take the spotlight. If you are sensitive to the feelings and pride of your fellow citizens in a branch or project office community, it can result in a sound platform upon which to build a good reputation for yourself and your firm.

26-8. The Mass Media

What the public thinks of a firm will be influenced by what it reads in newspapers and magazines or hears on radio and television. For that reason, it is essential to establish contacts and then a cooperative working relationship with representatives of the mass media.

Some people tend to view reporters and editors with a degree of distrust and suspicion. The direct, detailed, and sometimes pointed questions a reporter must ask to get his story may give people inexperienced in dealing with the media an uncomfortable impression that the reporter is out to "get" them. Generally, he isn't. He merely out to get his story.

What type of story will result will be determined not only by the news value of the event being covered, but also by responses to the reporter's questions. Suppose you are being interviewed in behalf of your firm. You need not fear a reporter. But unless you know him well, don't speak off the record. As a general practice, it is well to be responsive and answer inquiries fully. Do not, however, be gratuitous and voice any comment you would be uncomfortable seeing in print.

By being available and answering questions, your firm will develop a reputation as being honest, cooperative, and reputable. Once such a reputation has been established, you increase your chances—should news develop which might cast your firm in a bad light—of having reporters and editors want to get in touch with you to hear your side of the story.

The fact that they will want to call you does not mean you will be able to manipulate news or be in a position to ask reporters to slant a story in your favor. A responsible reporter has an obligation to his publication and his audience and cannot, if he is to maintain his own credibility, afford to distort a story.

There are times when, for competitive reasons, you may feel you cannot answer a reporter's question. Any good reporter understands this and indeed comes to expect a certain number of replies phrased along the line of "I'm sorry, but if I answered that question I would be giving out information which would be helpful to our competitors."

Unfortunately, some persons have shown a tendency to take what they erroneously consider the easy way out by answering all questions with "*no* comment" (a very bad response under *any* circumstance) whenever they have the slightest doubt about the advisability of revealing information. In the long run, such lack of cooperation builds up an aura of mystery and suspicion about your firm. And the response, "no comment," really is a comment, loaded with negative implications.

Be honest with a reporter. If you don't know the answer to a question, or if you think it would be advisable to check with somebody else in the firm before you release information, tell him. Promise him you'll check out the question and call him back. Then keep your promise, remembering always that the reporter is probably facing a deadline and that he needs a prompt response. Never try to cover anything up with a lie. You will be found out eventually. And the reporter will for a long time remember with bitterness any attempt to deceive him.

It would be delightful if the only time you got publicity was when it was favorable to your firm. But unfortunately there are times when situations arise which could reflect unfavorably on you: an accident or a disaster at your job site, for example. At such times you might be tempted to keep silent or under the pressure of questioning by reporters, make statements not based on fact. Either reaction can, in a few moments' time, destroy years of work in establishing good relationships with the media and building up the general public's confidence in your firm. If you keep silent, rumors will spread quickly. Rumors, by their very nature, generally tend to exaggerate what actually happened. If you make statements based on supposition, you can expect later to be accused of making false statements and attempting to mislead the public. The only thing you can do at the time of an accident or disaster is to get together what facts are immediately available, make them known to the media, and be prepared to say, if questioning continues, that you do not yet have all the facts but that you will let the media know as soon as the information is available.

There are times when, instead of waiting for a reporter to come to you, you will want to issue a press release on your own. Before you do, determine to your own satisfaction that the information you release is of genuine news value. Even fairly small publications receive floods of press releases each day and there is no point in wasting time and money by issuing barrages of news-less "news" releases. They find their way into the editor's wastebasket very rapidly. If you do have news to release, make sure your mailing list is complete. Don't play favorites by including only certain publications which might be interested in your release while purposely ignoring others. Such favoritism is resented by the publications which were not informed of your news.

things to say that the businessman may feel government is his "partner" (and not a silent one at that).

The contractor, like all businessmen, must abide by volumes of governmental regulations and laws, and be alert to the implications of proposed new legislation on the federal, state and local level which may affect his operations. (How much new legislation is considered each year? An idea can be gained from the fact that it is not uncommon for some 17,000 bills to be introduced during only one session of Congress alone.)

Today, also, the contractor is more than ever before involved in governmental relationships because government is often his client as well as his monitor. For these reasons, close contact with legislators, governmental officials, and political figures is increasingly important. To maintain such a contact, larger companies with branch offices, or with long-term projects underway or in the planning stage in several localities, find it helpful to designate one individual in each location to be responsible for governmental and legislative relations in each area.

A contractor, like anybody else, has every right to speak up and have his views aired on legislation which would affect his business. At first though, such action may sound merely self-serving. But the expression of the contractor's opinions can fulfill a public responsibility as well. His expertise in his field should be made available to legislators and the general public so that, especially in the case of legislation which deals with technically complicated matters or highly emotional issues, the public interest will truly be served in the enactment and administration of laws and regulations.

The public and lawmakers, however, pay little attention to the pleas of a company, or an entire industry for that matter, if over many years it has not shown a willingness to regulate itself so that the public and the industry are equally well served. The days of William H. Vanderbilt's "the public be damned" attitude are long past, and it is a fact of life that if one is not willing to keep his own house clean, somebody is going to be more than happy to come along and do it for him. And far too often, when that happens, emotion will rule the day and the broom will raise up more dust than is necessary or desirable. Several industries have learned this the painful way. Because so much of what the construction industry does relates to environment—a particularly emotional subject—the industry will do well to make sure it does nothing which could stimulate legislative overreaction which in the long run can do more harm than good.

Some companies have found it worthwhile to sponsor a nonpartisan training course in practical politics for their employees. An effective one has been prepared by the U.S. Chamber of Commerce and many local and state chambers have similar programs. These courses are devoted to the practical aspects of politics and explain how and why political campaigns are run as they are and show, through hypothetical situations, how a campaign (whether for an office or to assure passage or defeat of a bond issue) is conducted. The courses encourage employees to become more active in political and government affairs, and they have helped those whose positions require political contact to become more effective and knowledgeable spokesmen.

26-9. Government Officials and Political Figures

In any business venture undertaken today, government at one or more levels will have something to say about how the job can be done and sometimes will have so many

26-10. Stockholders and the Financial Community

A construction company which is publicly held will want to generate confidence by keeping its shareholders, as well as the financial community as a whole, fully informed on its financial position, progress, and future plans. This confidence, built on a continuing, long-term basis, is especially important when a company is negotiating acquisitions or seeking additional capital or supplementary financing.

Some shareholders are interested primarily in learning how much the firm netted as compared to the previous year, the earnings prospects for the next few years, and whether dividend rates will continue unchanged. Others are more interested in growth. No matter what the shareholder's interest, the information can be imparted through the annual report, quarterly reports distributed with dividend checks, letters to shareholders from the top executive, financial news stories, and published interviews with the firm's top executives. Information which goes to shareholders can also be effectively disseminated to employees.

One excellent forum for presenting a company's financial story is meetings of the local chapters of the Financial Analysts Federation, which can be found in almost every major American city. The job of the security analysts who make up the membership is to collect all possible information about a particular company, evaluate what they have learned, and advise investors about securities. The analysts will expect detailed information on the operations of the company and are interested in knowing not only what is happening in the company and industry, but *why* it is happening and what trends are foreseen. Meetings of security analysts generally are attended by reporters from financial, general circulation, and trade publications, which helps communicate the information to an even broader audience.

A word of warning on releasing financial information. Before doing so, check with experts to be certain such actions are consistent with Securities and Exchange Commission regulations.

26-11. Competitors

The magnitude of some construction projects—a major rapid transit system, for example—will sometimes require that the skills and talents of more than one company be combined in a joint venture to assure successful completion of the job or to minimize possible financial losses. The harmony necessary to make a joint venture work smoothly will be impossible to achieve if the participating companies have not over many years exhibited respect for each other's abilities and shown a willingness to work together on industry-wide problems.

It is often essential for individual contractors to enlist the aid of their competitors when a common industry front is necessary for a successful public information or government relations program. To provide coordination for such activities, trade associations have been formed on many geographical levels. Contractors, no matter how small their firms, will find it beneficial to support such associations to the greatest extent possible. In fact, smaller firms find that the benefits they derive—the opportunity to exchange ideas, and the ability to get their viewpoints known to legislators and the general public—often exceed their contributions in terms of money and time.

A handful of companies, because of their size, could dominate and control activities of a trade association. Representatives of these large companies usually are aware that lack of tact or consideration for the opinions and problems of smaller competitors can result in resentment which could frustrate positive cooperative efforts. No matter

what the size of a particular company, its representatives normally remember the need to maintain relations of mutual respect with competitors and to work with them in all suitable—and, of course, legally permissible—ways.

26-12. Suppliers

When dealing with suppliers, every firm wants to get the best possible quality, assurances that deliveries will be made on time, and courteous prompt treatment when adjustments are necessary—without, of course, paying a premium.

To get this kind of service requires that consideration and courtesy be part of all relationships with suppliers. Every employee who deals with vendors must be alert to the fact that maintaining good relations with suppliers will help get reasonable treatment when a particular material or service may be in short supply for some reason. It is a shortsighted contractor who refuses to give a few minutes to the sales representative of a potential supplier, or who makes salesmen cool their heels in the waiting room for an inordinate amount of time, because he feels "I don't need this guy." He may indeed not need him today. But he may tomorrow.

One effective way to help suppliers is to permit them to use the company name of the contractor in their advertising. This technique also offers additional exposure for that firm. Care should be taken, however, to avoid testimonials which claim—or even imply—that supplies of one vendor are used exclusively when that is not the case. Such a testimonial will merely harm relations with other suppliers. Endorsement of a product or a service which has been less than completely satisfactory should also be avoided. After all, some other contractor may well accept such an endorsement for the quality of the product or service (that's why testimonials are considered desirable).

26-13. Technical and Professional Societies

Involvement in the work of technical and professional societies can pay rich rewards. For management, membership provides the benefit of increasing knowledge of the industry and for broadening the company's reputation. For employees, participation—which should be strongly encouraged by top management—provides an opportunity to get together with peers and stimulates their own thinking.

Some companies have established formal technical information programs under which employees are encouraged to present papers at symposia and to write articles for trade publications. In those companies, there is often available a staff member whose job it is to help employees with editorial problems and to help find an audience for their papers and articles. A technical information program not only enhances the stature of the employees who participate, but can also result in increased business and a favorable employee climate. Exposure at symposia and in publications keeps the name of a company in the minds of potential clients and affords an opportunity to remind a large audience, especially when placing articles, of the varied capabilities of a firm.

To embark on such a program, one must be prepared to face the fact that he will be required to disclose some specific information about company activities. No editor will accept an article which is mere puffery. Nor will any professional society grant time at its meetings to speakers who have no information to share with your colleagues.

26-14. Faculty and Students at Engineering Schools

Many firms that hire new engineering school graduates hope to attract the pick of each year's crop. As a contractor, your chances of getting these students, who are your talent and management of tomorrow, can be enhanced if you have established, long-term relations at the schools where you recruit.

Because of the continual turnover in student population, "long-term" relations with individual students are difficult to establish. The sort of relationship can be achieved with the faculty, however—and indeed should be, since faculty members are in a position to transmit their good (or, we would hope not, bad) opinions of your firm to their students.

A number of techniques can be used by small as well as by large firms to increase a company's stature and reputation among both students and faculty. They include:

- Establishing scholarships for students and research grants for faculty members in your company name.
- Supporting engineering student publications (by advertisements, which offer an opportunity to stress career opportunities at your firm, or by offering to have members of your staff write articles for the publication).
- Setting up summer work programs (for faculty members as well as students so both can get to know more about your company).
- Offering to provide guest lecturers for classes. (If you can select somebody from your firm who is an alumnus of the college where he will be speaking, you increase the chance that the students will identify with the speaker and your firm.)
- Responding promptly and cooperatively (within reason) when faculty and students turn to you for help on research projects.

If you recruit regularly, you may want to prepare descriptive material about your firm to be available in college placement offices. It is advisable to have this material prepared professionally and by an individual or agency in tune with the aspirations of today's young people. Today's students, reared in an era of affluence, are looking for more than money on a job (though of course your opening salaries must be competitive). They are not, at least in their first few years of employment, overly interested in benefit plans and in fact tend to take them for granted. Most of today's young people are looking for challenge; they want to believe they are working toward some socially worthwhile goal. They are "itching" to get out of the classroom situation and they chafe when they are forced to endure inordinately long training or indoctrination programs when they begin work for you. Most of all, they want quickly to be given responsibility and an opportunity to prove what they can do.

26-15. Hazards of Increased Exposure

Increased involvement in public relations activities will bring positive benefits, but it must be pointed out that additional public exposure can have its irritations.

When dealing with the general media one accepts the hazards of news reporting. This means accepting someone else's judgment of what is important, someone else's skill at interpreting through writing, someone else's time pressures which can affect accuracy, and someone else's relative news judgment of whether the material should be published at all.

If you are a contractor, you have a personal subjective involvement in dealings with the media, and minor inaccuracies or an interpretation that does not jibe with your own may cause your blood pressure to go up a bit. But to individuals who are not so intimately involved with your firm, generally good exposure can be a plus despite a few relatively minor faults in accuracy or interpretation. Public attitudes are shaped by general impressions—not by detail. Then, too, you may profit from the knowledge you gain when you see how somebody from the outside evaluates your firm and your work.

You may wish that, in publications of general circulation, you had some control over precisely how a message is to be presented. The only way you can absolutely control your message in a newspaper is to buy advertising space. Or to purchase the publication.

There is another hazard of increased public exposure: as your firm becomes better known, you'll receive more and more letters and phone calls asking for donations, help, and information. Just because you are invited does not mean you have to say "yes" to every request. There is nothing wrong with a "no" response if you have been fair, consistent, and cordial in your consideration. But no matter what your reply to a request, it will take time, talent, and money to process it. That is a price you must be prepared to pay.

26-16. Staffing the Public Relations Function

Very small firms may find it impractical to have a full-time public relations officer on the staff. In such situations, it is desirable to have one high-level executive coordinate all public relations activities. Coordination under one individual tends to ensure that an integrated posture will be developed and maintained in all public relations activities. In addition, this arrangement makes it easier to develop personal relationships with those who may have—or should have—an interest in the firm's activities. The rank of the public relations officer allows for high-level consideration of public relations issues and permits others in management to gain accurate feedback directly from him—without multi-layered filtration.

A small company may consider hiring a public relations counseling firm on a retainer basis. The rates for counselors vary, depending on the types of service the agency is to offer and the amount of time to be devoted to the account. One should expect to pay a minimum of $15,000 to $20,000 a year for this service. Even large companies with their own public relations department frequently use outside counseling service, to extend their own effectiveness and gain the advantage of an impartial, outside viewpoint.

Most companies of size have found it essential to establish their own public relations department. Depending on the scope of the firm's activities, this department may consist of one high-level, full-time individual or—in the case of larger firms—several dozen individuals, under one head, whose duties are divided according to various specialties: media relations, employee communications, public affairs, community relations, audio-visual presentations, publications, advertising, urban affairs, and so on.

No matter how large or small the public relations department—and no matter whether public relations activities are handled on a part-time or a full-time basis—it is desirable that the public relations officer report directly to top management. The reason: the most important function a public relations officer will perform is to counsel top management on the public relations implications of every activity in which the company is involved. Without the status of high management rank, his voice will be muted and his contribution to policy-making decisions will be stunted. It is precisely at this time—when policy is first being formulated—that important public relations implications should be taken into account.

Let us assume that you have not had a full-time public relations officer on your staff in the past and are now considereing hiring one. What qualifications should you look for in this individual?

There is a misconception in some quarters that a public relations officer is a hail-fellow-well-met type who spends most of his time slapping people on the back and trying to influence their opinions by buying them four martinis at three-hour lunches. That might be a pleasant (albeit short) life, but a sound professional public relations officer knows his job is not entertainment. His job is creative but practical, imaginative but disciplined. If he is good, he is intelligent, socially aware, politically perceptive, economically knowledgeable, and has exceptional sensitivity to words, people, and trends.

Certainly, the professional you hire should display an ability to function effectively with people, although he does not necessarily—paradoxical though it may seem—have to "like" people. What is important is that he *understand* people and what motivates and influences them. Training in economics, the social sciences, business, politics, and psychology is beneficial.

Several colleges now offer majors in the field of public relations. In addition to academic training, it would be desirable if your potential public relations officer had some working experience in journalism, politics, and business.

Once you have hired your public relations officer, be prepared to consider and evaluate his advice. That is where he will earn his salary.

26-17. Evaluating Public Relations Activities

Back at the turn of the century, famed Philadelphia merchant John Wanamaker one day looked at his advertising budget and commented ruefully: "I know that half of the money I spend on advertising is wasted. The only problem is: I don't know which half."

In the same way, some people have puzzled over the money spent on some public relations activities. Some of the benefits can be difficult to measure in terms of dollars and cents. Also, because public relations is a long-term, continuing job, it is regarded difficult to measure progress specifically at any given moment. But there are techniques available to assess the value of the activity.

An ear sensitively attuned to feedback can help evaluate the effectiveness of public relations work. Comments and actions by political leaders, employees, and community leaders will give a clue as to what the various publics think of a firm. First-line supervisors are an especially valuable source of information as far as employee attitudes are concerned.

Public opinion surveys, if conducted properly, can tell at a given moment what a specific public thinks of several aspects of a firm, its activities, its management. When conducted on a regular basis, the trends these surveys reveal indicate whether public relations activities have been successful—and why or why not.

To measure effectiveness, of course, one must know precisely what he is trying to measure. And this means that these must be a goal to work toward, i.e., what is one trying to accomplish? Like all company goals, these must be established at the top management level.

Many firms find it helpful to subscribe to a press clipping service to give them an idea of what government leaders, the public, and the media think of their companies and their industry. There can be a tendency, however, when reviewing press clippings, to evaluate impact or warning signals solely by virtue of the *number* of stories. A public relations counsel should be able to evaluate more precisely than that—and then to counsel whatever action may be needed. Press releases from a company may result in dozens upon dozens or even occasionally hundreds of stories, but the sheer number of clippings will not necessarily tell anything too significant. One still won't know, for example, how many people actually saw or *read* the article. And the responses of those who did see it will not be known.

Because one may feel an article was favorable does not necessarily mean everyone else will react to it in the same way. A manager of a corporation producing consumer items once told me how delighted he was with a story about his firm which appeared in a major metropolitan newspaper. A few days after publication, he received in the same mail two letters commenting on the article. One writer said he was so impressed with what the company was doing that he was making it a point to buy its products exclusively thereafter. The second writer said she was so disgusted by the company's action she would never buy its products again.

The reason such divided opinion can result is that intellect and logic occasionally have little to do with public reaction. The latter can be largely a matter of emotion and unconscious impression.

As we have noted, a solid reputation is based on performance. Without it and adequate communication, you will not have the understanding and support of the publics with which you must deal. No firm can afford to forget—as public relations executive Richard W. Darrow has expressed it—"the public is the ultimate power."

As editor and essayist Charles Dudley Warner said: "Public opinion is stronger than the legislature, and nearly as strong as the Ten Commandments."

In the words of Woodrow Wilson, "Opinion ultimately governs the world."

As Mark Twain put it, "[Public opinion] is held in reverence. It settles everything. Some think it is the voice of God."

27 LEGAL AND CONTRACTUAL PROBLEMS

O. P. EASTERWOOD, JR.

Partner
McNutt, Dudley, Easterwood and Losch
Washington, D. C. 20006

To begin this chapter, let us take a hypothetical case. You have just received your Notice of Award of contract and you are busily engaged in furnishing your payment and performance bonds (if required); you are setting up your schedule of performance (an intelligent bar chart unless you are on CMP or PERT); you are making up your payment schedule; you are ordering your materials; you are contacting the labor unions for workmen; and you are negotiating with your prospective vendors and subcontractors.

Is it too early to anticipate possible claims on your part against the owner for changed or additional work; for delaying your work forces; for acceleration of your work, or for possible claims by your subcontractors and vendors against you? Certainly not! Now is the time to put little wedges in the dam of frenzied effort that you will be engaged in to mobilize the job and start the wheels turning that you hope will eventually grind out a profit for you. Don't ignore or postpone the vital and necessary safeguards at the critical, initial stages of the work, or you will be flooded with a tide of angry subcontractors and vendors and with the relentless drive of the owner to keep you on or ahead of your schedule, notwithstanding your problems. You will be left helpless unless you have made the proper records and have them readily available for your protection. But what are these safeguards and records, and why do you need them? Let us analyze your situation.

27-1. Subcontracts and Purchase Orders

Don't be a "slavish form follower" and pull out a dusty old document off the shelf, fill in a price, and send it to your subcontractor or vendor. Be meticulous and be prompt. Take the forms you have been using and match them up against your contract. With the present-day trend of owners—be it the federal government, a state-created corporation or agency, or a private corporation—all con-

tract forms have been modified and updated to include more controls over your operations, be it labor, your purchasing practices, or your quality control. Be sure that your proffered form to your subcontractor or vendor reflects completely the obligations you expect him to perform. If your company does not have an appropriate form to use, one can be obtained from the American Institute of Architects, the American Society of Civil Engineers, or the Associated General Contractors, and *modified to suit your needs.*

After your subcontract or purchase order form contains all the clauses to reflect properly your obligation under the prime contract, the complete agreement between you and your subcontractor or vendor must be clearly and unequivocally set forth. Always remember that if you are the prime contractor, the owner has a right to expect you to perform each and every obligation set forth in the contract you signed with him. Therefore, if you have inadvertently or negligently failed to incorporate some portion of the work that you expected your subcontractor or vendor to perform, you will be required to do the work yourself. Remember also, an owner has no privity of contract with a subcontractor or vendor and will not (or should not) deal with him directly. By the same token, your subcontractor or vendor may only deal with the owner *through* the prime contractor.

If you are a subcontractor or vendor and you receive a prime contractor's proposed subcontract form or purchase order, you should carefully analyze it to see if it describes exactly the scope of work you have agreed to perform. If you are given certain rights or placed under certain restrictions by an incorporation-by-reference of provisions of the prime contract, it is vitally important to you to know what such provisions are, and you should insist upon being furnished with a copy of the prime contract or pertinent extracts therefrom so that you will know your rights and obligations. It is common practice for the

591

subcontract to provide that the subcontractor receive no increases or decreases in price unless the owner grants such to the prime contractor. A provision such as this will restrict the right of a subcontractor to file suit against the prime until the prime contractor processes the subcontractor's or vendor's claim through such channels to the owner as the prime contract provides. In this connection see the case of *Humphreys & Harding, Inc.* v. *Pittsburgh-Des Moines Steel Co.,* 397 F 2d 227 (1968), where it was held that a subcontractor was bound by the decision under the general contract disputes clause and could not sue the general contractor until the entire appeal procedure was exhausted.

If the prime contractor does not have such a provision in the subcontract form, the parties might consider whether an arbitration clause referring all disputes to the procedures prescribed by the American Arbitration Association is desirable. Because of the general trend for arbitrators to seek a compromise settlement figure, it is the prevailing view among attorneys that an arbitration provision is more beneficial to a subcontractor or vendor than to a prime contractor. Many prime contractors, for this reason, may not include an arbitration provision in the subcontract form. When no provision such as "disputes clause procedure through the Owner" or a standard arbitration procedure is specified, a subcontractor is left to proceed before the courts to reach the prime contractor.

However, when federal government contracts are involved, a subcontractor or vendor has the benefit of the Miller Act under which to litigate. This Act, in effect, permits suit to be brought, after appropriate notice, against the prime contractor *and* his surety company. It does not permit protection beyond the third tier of subcontractor, and strict compliance with its notice provisions must be observed. Although this particular Act is applicable to federal government contracts only, most states have statutes permitting a subcontractor or vendor to proceed and attach property or have other remedies against a prime contractor where materials are furnished and labor is performed upon a project.

Care should be taken in the specific wording of the scope of work to make it clear which subcontractor has the obligation to "hook up" or "wire" or "test" certain items that they may install. When a prime contractor uses an incorporation-by-reference system of certain "sections" of an owner-prepared specification, he may have left the obligation of the subcontractor to perform these services in a vague or uncertain status. For example, who wires the starting motors to a large generator—is it under the electrical or mechanical subcontract? Much litigation has been caused by vague and indefinite references to sections of specifications, and in many instances a prime contractor will be punished for his carelessness by having to perform the work himself that he hoped he had subcontracted!

Further, the necessity for prompt submittal and execution of subcontracts and purchase orders should be obvious. During times of inflation a subcontractor or vendor may well have "second thoughts" about performing when he has quoted an extremely low competitive price and he has time to observe his materials or labor increasing beyond what he contemplated paying. If he has not furnished a bond, it is rather difficult to force him to sign a subcontract, and he can well claim that he made a mistake on his quotation. The prime contractor must then go on the market and obtain another subcontractor or vendor to perform the work at a higher price, or elect to perform the work himself. The longer the period between quotation

and executed subcontract and purchase order, the more the chances are that the agreement will not be consummated.

27-2. Bonding of Subcontractors

Should a prime contractor consider it desirable to insist upon a bid and performance bond from his prospective subcontractors? To do so, he must expect to pay the price of the bond in the subcontract price. He must also speculate as to whether the requirement of a bond will eliminate lower prices from other subcontractors, because many reliable and competent subcontractors are unable to furnish bond because of their size or because of their experience or credit rating. The obtaining of a bond does not protect against a possible default on the part of the subcontractor—but with protracted litigation, the damages might be recovered from the subcontractor's surety company.

It is believed that the contractor must carefully investigate the area by trades and capabilities before he makes a determination either to require bonds or to waive them. If there are capable and reliable electrical subcontractors, for example, it might be desirable to waive bonds. On the other hand, if the area is remote and the possibility of getting a replacement subcontractor (should such action be required) is not good, a bond may be desirable. This is an important executive decision that must be made—preferably during the prebidding period. In any event, whatever the decision as to requiring bond, appropriate subcontracts and purchase orders should be executed as soon as possible so that the possibility of delays in scheduling of work will be eliminated and reliance can be made on a firm agreement with a reliable subcontractor or vendor.

27-3. Documents and Records

Today is a day of mechanization of records. We have payrolls and other computations in computers; we have Xerox and other copying machines that turn out volumes of reproductions; and government, state, and large private corporations have multi-copy forms with instructions in fine print that challenge the most ingenious to fill them out properly. Further, the owner always has an abundance of inspection personnel to ensure compliance with every detail of the contract and specifications that is favorable to him. These inspectors generally have certain prescribed duties and daily inspection reports to fill out and turn over to the owner. These become the "official records" of an owner and will be accepted as such by any appeal board or court.

A. The Daily Diary   In addition, the inspectors are generally instructed to and do keep a "daily diary." In these neatly bound volumes are found the opinions and self-serving declarations of the individuals who daily observe the activities of a contractor. It must be remembered that the owner has only three basic obligations under a contract: (1) to inspect the work as it progresses to ensure that it meets all the terms and conditions of the contract; (2) to enforce the progress schedule to see that the work is performed within the time limits; and (3) to pay the contractor for the percentage of satisfactory work in place. These responsibilities leave ample time for meticulous record keeping on the part of the owner, and an ominous burden to meet on the part of a successful contractor. Therefore, like it or not, it is incumbent upon a contractor to keep sufficient records to equalize or offset the documents accumulated by the owner as the "official record."

B. Job Photographs It has almost become standard practice for all owners to take and retain "progressive photos" to record and document each stage of the work. This, in turn, requires a prudent contractor to have periodic job progress photograhs, *taken at the angle and from the location* the contractor selects. For this purpose, the camera should be as important a piece of equipment as the first typewriters and adding machines that are shipped to the job office!

C. Serial Letters But first things first. In starting a job, it is strongly recommended that a series of "serial numbered" letters be inaugurated so that all correspondence can be properly logged and located throughout the period of the work. It is also recommended that Serial Letter No. 1 set up the organization on the work site that you, as the contractor, propose to utilize and, most important, that it delineate the authority of your personnel. In other words, do you want to give your project engineer authority to sign change orders? Do you want your office manager committing the organization for all subcontracts and purchase orders? Under good management and organization contained in other chapters herein, a determination will be made as to organization.

The emphasis here is that the duties and responsibilities be definitely determined and delegated and properly described in this first letter. By doing this for your personnel, you have every right and you should *demand* that the owner furnish to you *in writing* the authority that has been delegated to his representatives on the job. This may save many arguments and misunderstandings as the job progresses. It may also save many hours in conferences with personnel who have no authority to act on matters where you must have a decision to proceed with your work.

D. Document Events in Writing Having a mutual understanding as to the duties and responsibilities of personnel on the job, it is next suggested that you, as contractor, relate to the owner's representatives on the job that it is customary on your work (and required by your board of directors, senior partners, or corporate officers) that you document various events that transpire on the job so that they will be in a better position to review all details of the job. In other words, there will be numerous occasions when "field changes" or "problem areas" are temporarily solved, but the owner's representative is reluctant to, or actually refuses, to reduce his verbal decision to writing. This is a situation where you must immediately write a letter confirming the oral decision—who made it and when—so that, *if necessary,* it can be used later to support your position.

Contentions based upon oral conversations frequently turn out to be just allegations, whereas a letter that should have and did document the happening on the occasion turns out to be the *statement of accepted fact.* To repeat, when any decision is made by an owner's representative that may cause a change in your work or in your schedule of work sequence, it should be documented immediately. It may never be used, but if it should become critical later in a matter of proof, it will be worth its weight in gold! So long as the owner's representatives are given to understand, *in advance,* that you intend to document in writing

the various events that occur, it will cause much less friction than if you wait until your collections on estimates vs. expenditures show a deficit and you suddenly pour in reams of correspondence attempting to establish claims!

E. Allocation of Work on Time Cards Once these preliminary matters are out of the way, what do you tell your office manager to set up in the way of files and records? Whether you follow the Dewey decimal system, an alphabetical system, or a numbered system, you must be able to allocate in some manner *categories of work to items of cost.* This is the basic key to costing out any contract recovery you may ultimately receive under the contract.

Perhaps the most reliable method is by the proper use of time cards for individual workmen engaged upon a specific item of work. For example, suppose you are building room X and when you are halfway through with your rough carpentry, the owner issues a change order to you to make a 45° angle and jut one wall out 3 feet from elevation 6 feet to the eave of the roof elevation 10 feet. Although this sounds like a simple change at this stage of the work, in order to estimate its cost initially and then to be able actually to price it out ultimately, it will necessitate the consideration of numerous trades and supplies.

Suppose that in attempting to negotiate out a contract change on an estimated basis the owner is either slow in accepting or does not like your estimating basis. You cannot stop the job while the paper work is being done, but you can start immediately to keep time cards on this particular segment of the work so that you will ultimately know what it has cost to perform.

You would start with your carpenters assigned to put in the changed arrangement of studs. They would probably have to put in additional bracing and blocking because of the change in weight-bearing timbers. If the owner does not furnish you a detailed drawing as to load-bearing timbers, you may have to allocate engineering time, or stop all work until an appropriate design is furnished.

Once the rough timbers are in, you may have a plastering subcontractor or a sheetrock installer from whom you will need a detailed quotation for the small added quantities of his work and an allocation of time of his workmen on the changed work vs. what the unchanged work was priced out on their estimate.

Further, this may have changed your electrical installation, your heating and cooling system and your painter's work. It may have changed the planned sequence of your operations. All of the time of any workman who worked upon this changed installation should be carefully allocated to it and all of these data placed in one file so that they can be quickly referred to. You may have to order more lumber, more plaster or sheetrock, and more cable or wiring. These added materials costs should also be included in this file.

If the alteration has changed the sequence of your operations (i.e., a delay in receipt of materials, or an argument over bearing timbers) and it occasions moving in and out of scaffolding, a crane, or other equipment, this should be included in the file. If you were delayed in this operation for any reason (other than your own fault or negligence), a notation to this effect should be contained in the file and a time extension sought in connection with your pricing of the item.

In other words, each and every item that is requested by the owner that varies in any respect from what is shown in your contract plans and specifications should be the subject of a separate file with time of labor and materials allocated to it. This same procedure should be followed by your subcontractors and vendors and their data promptly submitted to you.

Admittedly, this makes the cost and time records more voluminous and more difficult to keep, but it is absolutely essential if you expect to receive for yourself, your vendors, and subcontractors a complete recovery for additional work performed under the contract. You may say, "I have good cost records. I can account for every penny spent on the job." This may be quite true on a *total cost* basis, but can you state how much each change order cost you—how much each request for what you claim is additional work under your contract cost you? Probably not, and this is most important.

27-4.   Total Cost Less Bid Price

The contracting officers, the Courts, and the Contract Appeals Boards are most reluctant to accept as a pricing basis the difference between a contractor's estimate and his actual costs as the measure of payment for the work item. To quote from the headnote of a recent Court of Claims decision along this line: *(Turnbull, Inc.* v. *United States,* 180 Ct. Cl. 1010, 389 F. 2d 1007 (1967):

"The total cost theory of computing the amount due a contractor under a cause of action against the Government, i.e., the difference between the contractor's bid price and the actual cost of performing the entire contract as changed by order of the Government as distinguished from computing damages on the basis of the increased costs relative to specific or separate items, is not the most satisfactory method since it assumes that the contractor's costs were reasonable and that the contractor was not responsible for any of the increases in costs, and it also assumes that the contractor's bid was accurately computed which is not always the case. Accordingly, this means of computing damages is only used where other and better proof is lacking. *F. H. McGraw & Co.* v. *United States,* 131 Ct. Cl. 501, 511, 130 F. Supp. 394, 400 (1955)."

Thus, it is much better to have available a realistic estimate, and ultimately a file that contains completely detailed costs specifically allocated to this item. Estimates can generally be optimistic to forestall possible contingencies, but after-the-fact pricing must be supported by records; otherwise, an owner-audit might prove most embarrassing.

27-5.   Record of Negotiations and Schedules

The files must also contain all negotiations or "understandings" with vendors and subcontractors from the initial contact with them through all stages of the work and up to the time that all matters are settled and a final release is obtained. As protection for all concerned, all work schedules, all job changes, all owner caused delays that affect the vendor and subcontractor must be immediately communicated to him and such reference included in the file.

Much litigation is created by the failure of a prime contractor to currently advise a subcontractor, on an accurate progress schedule, just when his portion of the work is expected. If he is not on an accurate schedule, and he is forced to mobilize without proper notice, he may have a justifiable acceleration claim. By the same token, if the contractor is unduly delayed because of failure to furnish the subcontractor with proper scheduling, he may well claim a delay and rescheduling of his work, which in-

creases his costs of performance. This means that the "scheduling file" must be constantly revised to meet changing conditions, and notice of all changes promptly communicated to all interested parties with appropriate notation in the files that this has been done.

Further the simple transmission of a schedule change by the owner to the contractor and thence to the subcontractor and vendor is not sufficient. The contractor must ascertain whether the schedule change affects the work of his subcontractors or vendors and, if so, accumulate all of this material and appropriately advise the owner of the increase in costs and/or time extension, or both, if such is the case. To save some correspondence back and forth, when these matters are submitted to subcontractors or vendors, a phrase might well be included in the letter along these lines:

"Unless a written notice is received from you within _____ days it will be presumed that the attached _____ does not affect your price or time for performance and that you release _____ from any claim or demand for increases therein."

Although admittedly this is a self-serving type of communication, it nevertheless tends to crystallize problem areas before they become too serious and end up in litigation toward the end of the job. When a party decides to file a claim or a suit in court, he generally picks up every trifling matter he has, and this can prolong and increase the expenses of litigation.

27-6.   Daily Chronology of Progress

Last but not least, a contractor's project manager (or superintendent) should keep and maintain an accurate daily diary, setting forth in succinct terms the progress of work and the routine and unusual happenings of each day. All job conferences with owner's representatives should be covered, as well as any meeting with subcontractors and vendors. Problems of the job should be written up in sufficient detail to later form the basis of correspondence, if needed. Weather details and working conditions can be covered briefly, and equipment performance or problems covered in brief fashion.

Many times an owner will claim that equipment is not in good working condition and that excessive downtime is charged into a contractor's costs. If feasible, the diary should be reviewed by one other member of the contractor's personnel, such as the project engineer, the quality control representative, or the labor-relations officer to assure its completeness and to eliminate careless statements that might later reflect upon the contractor's good workmanship and ingenuity. With the current day discovery proceedings now in vogue with appeal boards and courts, you can expect your complete set of files and records to be examined and copied by the owner's personnel if you become engaged in an adversary proceeding of any nature. Therefore, it behooves you to keep them orderly and with content that reflects upon your good qualities and capabilities for performance.

Since many of a prime contractor's claims are in part based upon actions and activities of his subcontractors, they should likewise keep similar records. Further, it should be remembered that if the subcontract has a "disputes clause through the owner" the subcontractor's files should not reflect contentions against the prime—they should be primarily directed toward adverse actions of the owner affecting the subcontractor through the prime. If, after the disputes procedure is pursued against the owner, there remain issues between the prime and subcontractor, these can then be considered.

27-7. Trouble Flags

You ask, when does a contractor file a claim for himself or for his subcontractors and vendors against an owner? This is an easy question to ask but extremely difficult to answer. All home-office supervisors, all surety company executives, and all those individuals who have a financial interest in the ultimate profit or loss on a construction job have been attempting to solve this question, but variables are so numerous that the question defies a quick and easy solution.

For example, if the project manager wants to keep the supervisor and surety company "off his back" while he is running the job, he can very easily turn in optimistic weekly or monthly financial status reports. This can be done by underestimating his equipment hours, by allocating his overhead in a conservative manner, or by many other means best known to accountants. Further, he can overestimate his percentages of completion and, unless the owner is most conservative, he may not object because he knows that ultimately the pay quantities will balance out with the work in place—he can always hold back payments toward the end of a job to be sure he has not overpaid.

When a unit price contract is in effect, it is not unusual for a contractor to unbalance his bid so that the first pay items on the unit prices will bring in more money than their actual value. This helps in financing the job. Frequently, on a long-term contract, the owner will set forth a pay item for "mobilization." This helps materially in financing the work. Hence, between a project manager wanting his job to look good during the early stages to fend off criticism, and the owner being liberal by custom and practice on the early pay items, it is almost impossible to tell whether a job is making money or losing money in the early stages.

27-8. Personalities in Owner's Administration

A claim of any kind should ordinarily be filed just as quickly as a contractor can price it out with reasonable assurance that his estimates on the pricing are satisfactory. However, an exception to this may well be that to do so would cause the owner's inspection personnel to become obnoxious and overbearing and end up by causing the contractor more needless expense to get the work accepted than the value of a meritorious claim. Unfortunately, in the contracting business, we are dealing with individuals and, generally speaking, although they intend to be fair and impartial, some of them are inclined to bear grudges and it is not uncommon for them to "take it out on the contractor."

Of course, the proper thing to do in a situation of this kind is to go to the next higher echelon in the owner's organization and seek relief. Generally, if this matter is properly pursued, an echelon will be found that will be fair and equitable and will straighten out their inspection personnel. But if it is known in advance by reputation, or by past experience, that all echelons of an owner's organization will support their inspection personnel, no matter how unreasonable they may be, it behooves a contractor to think carefully whether he should currently

file his claims as they accrue, or whether he should wait until the end of the work and then file his justifiable claims.

In the event it should be determined to defer claims submission until toward the end of the work in order to satisfy the peculiarities of an owner, a thorough review of any of the contract articles requiring that "notice" be given prior to the filing of a claim should *be meticulously observed*. The giving of a proper notice within the time limits stated in a contract may, or may not, be waived. Particularly is this so in contract articles similar to the Changed Conditions Article (now termed the "Differing Site Conditions" article.) The reason for the strict construction is that, unless notice is given at the time the condition is being encountered, the owner would not be in a position to investigate it and determine its merits. When an owner is thus prejudiced, the boards and courts will not permit recovery under the contract, no matter how meritorious the claim may be. (See Appeal of *Carson Linebaugh, Inc.*, ASBCA No. 11384 (October 5, 1967) 67-2 BCA 6640, pages 30, 787.)

The written notice may be waived by the Owner *(General Casualty Co. of America* v. *U. S.*, 130 Ct. Cl. 520 (1955)), or may become supererogatory when the owner has in fact knowledge of the condition and of the difficulties encountered by the contractor. *(Peter Kiewit Sons' Co.*, ASBCA No. 5600, 60-1 BCA 2580.)

Many owners also place time limitations in their Changes Clauses and in their Suspension of Work clauses which may or may not be waived. The alleged purpose of these time limitations is so that the owner may be accurately apprised of the extent of funds that he may need to finance the work. In any event, a contractor should not expect an owner to waive the time limitations both parties have agreed to by entering into the contract. When a time limitation provision is placed in a contract for an appeal from a final decision, it becomes jurisdictional and cannot be waived. (See *Poloron Products, Inc.* v. *U.S.*, 126 Ct. Cl. 816 (1953.) Therefore, it is fatal not to give notice of appeal from a final, adverse decision, and many valid claims have been forfeited because of a contractor's failure to give a Notice of Appeal.

For whatever reason a contractor or subcontractor may have for not promptly filing and pursuing whatever valid claims he may have, it is absolutely essential that he give notices of claims to be filed, wherever the contract requires them. In fact, where an area of claim is even suspected, it might be well to file the notice for protection and then, if the claim does not materialize upon a further review, it can be abandoned. However, if no notice has been filed, a claim will either be seriously prejudiced, or completely lost, if jurisdictional.

27-9. Your Original Estimate and Schedule

Any contractor or subcontractor has carefully made an estimate and a schedule at the time he has bid upon work to be performed. These data should be carefully preserved as they may become his most important tool for locating his trouble areas. If his costs start to soar and his revenues become totally inadequate to balance his operation, either his problem is grossly underbidding his work, or he is performing work or suffering delays and rescheduling of his work forces not contemplated at the time of preparing his bid.

Examine the original estimate carefully and note the categories of work sequences where the costs are greatly exceeding the revenue. Try to pinpoint the reasons for the loss areas. If the reasons turn out to have been *caused by the owner,* then the contractor is in a position to match these factual causations against provisions within his contract which may afford him an opportunity to present his losses as a claim under the terms of the contract. By the terms of most contracts, the owner has obligated himself to the contractor to do certain things, and if he fails in these duties, there may be a provision for compensation for such failure.

27-10.   Breach of Contract vs. Administrative Remedies

Generally speaking, when the owner owes a duty to a contractor, and he fails or refuses to perform such duty, there is a "breach of contract" situation that could be actionable in any court of law. Although it is true that anyone with $10 can file a lawsuit against any party, a person would be foolish to do so without some probability of recovering damages. Therefore, the breach of contract by the owner must also be coupled with specific damage done to a contractor, because of such breach. One of the areas of breach of contract most frequently used and pleaded before the Court of Claims is:

"There is an implied obligation on each party to a contract to do nothing to hinder or prevent performance by the other party." (*WRB Corp.* v. *U.S.,* 183 Ct. Cl 409 (1968); *Clack* v. *U.S.,* 184 Ct. Cl. 40, 395 F 2d 773 (1968).)

This is a broad area and must be specifically proven to recover. *However,* before any consideration is made to "go into court," a contractor must carefully review the provisions of his contract under clauses generally entitled "disputes" or "arbitration." If the contract contains such a clause, it is tantamount to the parties agreeing that *before they go to court,* they will exhaust their "Administrative Remedies." These consist of following the article set forth in the contract which may either provide for an appeal from a contracting officer's adverse final decision, if it is a contract with a governmental agency; an arbitration procedure as specified in the contract article; or a decision of the owner's engineer (or hired architect-engineer). These clauses are *conditions precedent* to going into court. As the Court of Claims states (*Schlesinger* v. *U.S.,* 181 Ct. Cl. 21, 383 F. 2d 1044 (1967)):

"Where the contract offers an administrative remedy for the claim sued upon, it is a necessary prerequisite to suit that the contractor exhaust such contract remedy. Thus, where the contract provides for a decision on a dispute to be made by the contracting officer with the right of appeal by the contractor, if dissatisfied, within 30 days, failure to appeal or failure to appeal within the time specified in the contract is proper ground for the appellate board or officer to refuse to consider the claim and the claim is also not one on which the contractor may bring suit in the Court of Claims."

(See also the *Humphreys & Harding* case cited in Paragraph 1, supra.)

Therefore it is mandatory that the contractor exhaust his administrative remedies, if he has any, before he goes into a court. Also, it is generally desirable to do so in that the costs of processing a claim through the administrative procedure steps are ordinarily less and the time from filing the claim to the final decision is not as extended. Accordingly, it is strongly recommended that every attempt be made to bring the claim within one of the provisions of the contract permitting a settlement. Although this cannot always be done, the emphasis can be placed upon facts that fall under the contract adjustment provisions and facts that tend to support a breach of contract can be de-emphasized.

If a contractor presses for a breach of contract declaration from the owner, he is making it extremely easy for the owner to reject his claim in toto under the contract, because an owner cannot legally consider or pay for anything not covered by the contract and its adjustment provisions. Presentations to the owner should therefore be couched in terms of "requests for equitable adjustments" or "for consideration under Article _____ of the contract"; or, in case of doubt, present the claim in the alternative under two or three articles that may indicate that payment can be made thereunder. This type of presentation will at least require a decision from an owner under the provisions of the contract, whereas a protestation of "you've breached our contract" or "you've damaged me" will evoke a polite reply that breaches of contract cannot be settled under the terms of the contract.

27-11.   Various Contract Adjustment Clauses

If you are the contractor and you have determined that your reasonable costs are exceeding your revenue because of some action or inaction of the owner, or having had your "home-office supervisor" or your "surety company" upon examination of your profit-and-loss statement urge you, or having had the owner, either verbally or in writing direct you, to perform additional work, or work in a manner different than you scheduled it, you are now ready to consider filing a formal claim to attempt to recoup your losses.

Your first action should be to carefully review your entire contract and specifications to ascertain what provisions are contained therein that would permit you to increase the contract price. Most every contract, whether it be with a governmental agency, a private utility company, or a foreign government, will have articles such as: (1) a *Changes Article,* granting an increase or decrease in contract price where the owner wants a change in the work; (2) a *Changed Conditions* (or *Differing Site Conditions*) Article, granting an increase in the contract price where a condition is encountered that was not contemplated by the parties; (3) a *Suspension of Work Article,* whereby the owner reserves the right to suspend your work, or any part thereof, in consideration for paying you your standby costs; (4) a *Variation in Quantities Article,* where, in a unit-price contract, a contract adjustment may be made in which the actual quantities of work to be performed; (5) a *Termination for Default—Damages for Delay—Time Extensions—Force Majeure—Articles,* where generally all that may be obtained is a time extension for certain specified reasons; (6) an *Inspection and Acceptance article,* where, if work is alleged to be defective and it is found to meet the requirements of the contract after being "torn out" or "uncovered," an increase in contract price is in order; (7) a *Possession Prior to Completion* Article, where the owner reserves the right to occupy and use the premises prior to physical completion upon the payment of an equitable adjustment; (8) a *Termination for Convenience of the Government* Article, where a settlement on a cost basis may be made for all or any part of the work so terminated; (9) a *Government Furnished Property* Article, where certain provision is made for an equitable adjustment in the event such property is not furnished, or delivered in a timely manner; or where particular specification provisions provide for the cost of (10) *Escalation of Labor or Materials;* or, if it is a contract with a foreign government, a clause to protect against (11) *Currency Fluctuations,* or a guarantee to pay for loss of Equipment and Materials in the event of an uprising or hostile enemy action. In addition to specific contract articles and unique provisions in specifications, there is also for consideration beyond the four corners of the contract, (12) *"Constructive" Changes;* (13) *Acceleration of Work;* (14) *Ripple or Impact Effect,* and (15) *Cardinal Changes.* Each of these items will be separately reviewed so that the maximum usage may be made of each, depending upon your peculiar factual situation.

27-12. Changes Article

Most owners will put a changes clause of some type in a contract. Knowing that most larger construction projects will have changes desired by the owner during the course of construction running from 5 to 12 percent of the project, owners are anxious to reserve unto themselves the right to make such changes. In so doing, they generally write into this article time limitations and, in some cases, a formula on percentage of contract adjustment that may be realized. Many others simply provide for an equitable adjustment in contract price.

The Supreme Court of the United States put a limitation on the form of changes article the federal government used in the Standard Form 23-A, Construction Contract in the Rice and Foley cases *(U.S.* v. *Rice,* 317 U.S. 61 (1942) and *U.S.* v. *Foley,* 329 U.S. 64 (1946)) in that the contracting officer would not pay for any delay costs in connection with making a change or changed condition determination, nor would the increased costs in connection with the unchanged work be recognized. However, that has been changed by the more liberal Changes Article included in the October 1969 Revision of Form 23-A.

Inasmuch as many States and their Utility Commissions and many large corporations use rather restrictive language about what they will consider to be an equitable adjustment, the Changes Clause should be read carefully before a price is submitted for the changed work. Also, the time limitation on notifying the contracting officer as to the contractor's intent to file a claim for increased costs should be strictly observed. If the extent of the change cannot be estimated within the time limitation, a communication to the contracting officer to that effect should be sent in so that the record will indicate that you intend to request a contract modification as soon as the extent of the change and the reasonable cost thereof can be estimated. If the request is not received within the time limitation and the contracting officer still acts on the claim, he has waived the limitation.

As the contractor, it is also necessary that you familiarize yourself with the original work as specified under the contract to determine whether the change directed is within the general scope of the work under the contract. If it is not, the contractor has not agreed in advance to perform it, and a Supplemental Agreement may be insisted upon, or the contractor may refuse to perform the work. A supplemental agreement is a two-party agreement which requires the concurrence of the contractor, and it cannot be issued unilaterally by the contracting officer. Familiarity with the scope of work is also important because it will

assist in determining whether any unchanged work will be affected; whether rescheduling and/or remobilization will be required; whether winter weather conditions will be encountered; and what subcontractor or vendor problems may be created by the directed change. The estimate for performing the changed work should be carefully considered and all anticipated costs included therein, because once the estimate, including the request for whatever additional time will be needed to perform the work, as changed, is accepted and the change order accepted, the price and time granted are final and conclusive and may not be reopened.

Care should be exercised not to "volunteer" to perform extra or additional work suggested by an owner's representative who may have no authority to bind or commit the owner. If the owner is unwilling to have his representative request the contractor to perform the change *in writing,* the contractor should refuse to perform such work. In an emergency situation, an oral directive from the contracting officer would suffice, provided a confirmation in writing of the oral directive is expeditiously acknowledged.

Some owners may require a certificate of current cost or pricing data to accompany a change in excess of $100,-000, except where the price of the change is based upon adequate price competition, established catalog or market prices of commercial items sold in substantial quantities to the general public, or prices set by law or regulations.

Where a deductive change order substantially affects the scope of work in the original contract, it may be more desirable to treat it is a Partial Termination for Convenience and submit the pricing on an audited cost basis. If not, the contractor would be entitled to spread his original overhead proportionately among the units remaining in the contract as part of his estimated cost for performing the work.

27-13. Changed Conditions (Differing Site Conditions) Article

The Court of Claims has defined the purpose of the Changed Conditions Article as follows *(United Contractors* v. *U.S.,* 177 Ct. Cl. 151; 368 F. 2d 585 (1966)):

"The purpose of the Changed Conditions article is to prevent bidders from adding high contingency factors to protect themselves against unusual conditions while excavating, and is thus expressly designed to take some of the gamble out of subsurface operations. Its object is therefore to persuade contractors to calculate bids on the basis of the descriptions contained in the specifications, plans and drawings, including test or core borings and profiles."

If there is included a Changed Conditions clause in the contract, it will generally cover two types of situations. It will deal with the subsurface and latent physical conditions at the site of the work which differ materially from those stated in the contract. Essentially, what it means is that, if the data shown in the test borings or specifications are in error, and if the contractor has been misled by representations of the owner in the specifications, the contractor is entitled to an equitable adjustment of the contract price. The second part of the clause deals with a condition at the work site, of an unusual nature, differing materially from those which a contractor could ordinarily expect to meet. No misrepresentation on the part of the owner is required.

Should you, as contractor, believe you have encountered

either of these situations, you must give notice thereof so that the contracting officer may investigate the conditions. If you fail to give the notice, your claim can be easily jeopardized because the owner has a right to investigate the condition before it changes or is covered up; otherwise he will be prejudiced. There are exceptions to this, one of which is that, if the owner is already on notice of such condition, no further notice need be given. The other exception is that, if there is an emergency or urgent necessity, such as happened in the Bianchi case where steel ribs had to be installed to protect a tunnel from collapsing, no notice need be given. (*Bianchi* v. *U.S.,* 144 Ct. Cl. 500, 169 F. Supp 514 (1959).)

The conditions contemplated by the first part of the clause are physical in nature. Also, they are generally below the surface or latent, meaning hidden, dormant, or not manifest; they must be at the site and they must differ from what is described in the plans and specifications. The entitlement in the first part can result from errors in surveying or exploratory drilling of the site; from incomplete presentation of drilling logs, from omission of subsurface water information or omissions from the drawings. However, it is incumbent upon the contractor to examine the site of the work carefully, and if there are cores available for inspection, he should examine them. In fact, every representation of the owner must be carefully documented so that the actual conditions encountered may be shown to be materially different. Most of the problem areas turn around grossly inaccurate estimated quantities, inaccurate portrayal of subsurface water or rock conditions, misrepresentation of soil condition, borrow pits, contour lines, or failure to show accurately existing utility lines.

As for the second part of the clause, the condition must have been unknown when the parties entered into the contracts. Hence, if a reasonable investigation would have disclosed it, this could be fatal to the claim. For example, if a contractor is working near the rock mountains, his general experience should put him on knowledge that there might be subsurface rocks or boulders. Likewise, if he is working in some parts of Florida, he is certainly aware of the nearness of the water table to the surface. Thus, he cannot well claim that he has encountered an "unknown" condition.

Neither part of the clause covers situations such as labor problems, change in wage rates, inability to procure materials, or acts of God, such as weather, rainfall, flooding, etc. However, acts that are "man-made" or artificial have been considered to be Changed Conditions, such as when irrigation ditches are not closed contrary to representations, or scrap is dumped where excavation is to be performed.

Although numerous cases from appeal boards and courts could be cited as to what is and what is not considered to be a Changed Condition, it is believed that almost every set of facts regarding a changed condition claim is unique. When it is believed that the factual situation is covered by either the first or second portion of the article, reference should be made to any good law index of "changed condition" cases and comparison made with the situation prior to filing. This is probably one of the best indexed subjects in the contract administrative field.

27-14. Suspension of Work Article

This Article will probably be found only in federal government contracts because it was first put into construction specifications by the Corps of Engineers, U.S. Army. The development and purpose of the article is well set forth in a decision by Mr. Joel Shedd, Jr., of the Armed Services Board of Contract Appeals, as follows (*Appeal of Bateson Construction Co.,* 60-1 BCA 2552 (1960) at pages 12, 345-6):

"GC-10 and GC-11 (the Suspension of Work Article) were drafted during World War II as standard provisions to be added to Standard Form 23 for all Army Engineers fixed price construction contracts. Mr. Easterwood, one of appellant's attorneys, was a legal officer in the Office of the Chief of Engineers when GC-10 and GC-11 were drafted and took an active part in the drafting of these clauses. . . . Under the circumstances, we are able to take cognizance of the history and purpose of these two clauses. . . .

"The 'Suspension of Work' clause was in part an outgrowth of the agitation following the decision of the Supreme Court in *United States* v. *Rice,* 317 U.S. 61, decided 9 November 1942, in which the Supreme Court interpreted the time extention provisions of the standard construction contract 'Changes,' 'Changed Conditions' and 'Delays-Damages' clauses as constituting the contractor's exclusive remedy for delays in performance caused by change orders, changed conditions and delays due to 'acts of the Government.' The Court said in effect that by providing in the contract for a time extension, but nothing else, when the contractor's performance was delayed by an act of the Government, the parties had contracted that when the contractor was delayed by the Government the contractor would be entitled to a time extension and only a time extension. . . .

"GC-11 grants to the contractor the right to a price adjustment when the Government 'orders' a total or partial suspension of the work of unreasonable duration and such suspension causes additional expense or loss to the contractor. Although GC-11 refers to the suspension of work as being ordered by the contracting officer, in an early decision involving this clause, it was interpreted by the Army Board of Contract Appeals as applying not only to a suspension of work ordered in writing in advance by the contracting officer, but also to a *de facto* suspension of work by the Government where the contracting officer failed to issue the suspension order that should have been issued, and the Board said that it would treat the contractor's claim as if the proper suspension order has been issued. *Guerin Brothers,* BCA No. 1551 (1948).

". . . First, it repels any possible interpretation of the contract as making a time extension the contractor's exclusive remedy for delays caused by acts of the Government. Secondly, it provides an administrative remedy for the settlement and payment of claims for losses and increased costs incurred by the contractor as a result of suspensions of work, or parts thereof, caused by the Government under certain circumstances.

"In order for the Government to be liable in damages, it must be found that the Government breached the contract by either failing to do what it was contractually obligated to do or by doing what it had no right to do under the contract."

The Article has gone through many changes at the hands of government agencies and, as it now appears in Standard Form 23-A, which is mandatory for use by all governmental agencies performing construction services, it requires *notice* and it eliminates profit. It further limits relief by excluding the application of other articles, if it is utilized. The restrictive wording is as follows:

"23. Suspension of Work. . . . However, no adjustment shall be made under this clause for any suspension, delay, or interruption to the extent (1) that performance would have been so suspended, delayed, or interrupted by any other cause, including the fault or negligence of the Contractor or (2) for which an equitable adjustment is provided for or excluded under any other provision of this contract.

"(c) No claim under this clause shall be allowed (1) for any costs incurred more than 20 days before the Contractor shall have notified the Contracting Officer in writing of the act or failure to act involved (but this requirement shall not apply as to a claim resulting from a suspension order), and (2) unless the claim, in an amount stated, is asserted in 'writing as soon as practicable after the termination of such suspension, delay, or interruption, but not later than the date of final payment under the contract."

With its restrictive wording and elimination of profit, about all that can be successfully claimed under it would be costs in the nature of standby costs and extended overhead—perhaps some inefficiency percentage. Thus, where a choice exists as to whether to utilize this clause as against the Changes or Changed Conditions Article, the latter articles would be preferable. Further, since the contractor always has the burden of proof on a claim, it almost takes an audit approach to establish standby costs, and, when the government agency calls in the Department of Defense Audit Agency to verify costs, the time element in ultimate recovery can become quite lengthy.

The Court of Claims had occasion to comment recently on this article in connection with a case in which the delay was caused by an error in the specifications. The Court stated (*Chaney and James Construction Company, Inc.* v. *U.S.*, No. 150-67 (Decided February 20, 1970), page 6):

"If plaintiff had brought his present claim as a breach claim, under the rationale of *Laburnum* and *Luria,* it would not have had to prove the delay unreasonable, since all delays due to defective or erroneous Government specifications are *per se* unreasonable and hence compensable. The fact that the claim was brought under a contract clause does not affect the reasonableness of the delay. Here, since the delay was caused by erroneous or defective specifications, it too was unreasonable.

". . . It would thus follow that if plaintiff could recover for *all* of the delay caused by defective specifications in an action at law for breach, it should be able to obtain a similar recovery under the contract."

This opens up a fertile field for construction contractors because many problems are caused by errors in specifications. In fact, the *Engineering News-Record* of May 21, 1970, had part of their editorial page devoted to this subject. Quoting a paragraph from page 7:

"The GAO report, a follow-up to a 1967 effort, recommends that the Defense Department develop a uniform contract clause to define the A-E's responsibility for designing within construction cost estimates; and that Defense begin work with other federal agencies to make such a clause government-wide. GAO also suggests that Defense think about requiring documentation during processing of large construction change orders to show: whether the A-E is responsible for design errors, what the added cost to government is because of defective A-E work, and what actions the federal agency takes regarding the errors. The Pentagon says steps are already being taken to carry out the recommendations. . . ."

In examining the delay factors on the job, and in comparing the original estimate on items against those which seem to be costing an inordinate amount of money, the facts may well be covered by this article rather than any of the other articles. The contractor must be sure that the reasons for the delay cannot be attributed to his fault or negligence. They must be attributable to the owner (government).

27-15. Variation in Quantities Article

This article would be found only in unit-price-type contracts when the owner makes an estimate on the quantities for bidding purposes and the contractor bids a number of unit prices that collectively make up the total estimated contract price. It is really a series of small lump-sum contracts blended into one document. The contractor is then paid the unit price for the actual quantities produced. The government has also used this type of contract for the sale of surplus supplies and materials when it is not sure of the inventory count.

In order that bids may be evaluated, the owner places estimates on the quantities of each item. Although it is advisable that prospective bidders make a detailed site survey, many bidders will rely on the owner specified quantities. There may be situations where this will be considered a warranty as to correctness of quantities (such as dredging contracts where the bidder is not expected to take cores; or tree stumps where the ground is covered with snow); nevertheless, if a bidder can investigate the conditions for himself, he will be held to the unit price for the variation of an underrun or an overrun in quantities. For this reason, and to obtain more reasonable prices by eliminating contingencies for this type of situation, a Variations in Quantities Article will be placed in a contract. Generally the article will provide that there will be no contract adjustment in unit price unless the quantities are in excess of or less than 15 percent. Sometimes they will go as high as 25 percent of the estimated quantities. Some articles provide for a different percentage of overruns and underruns. Most articles limit the adjustment to the items exceeding the 15 or 25 percent figure necessary to bring the article into use.

When there has been a substantial underrun in quantities, consideration should be given to claim this as a possible Changed Condition. The leading cases on this matter are the *Chernus* and *Peter Kiewit* cases. (*Chernus* v. *U. S.*, 110 Ct. Cl. 264 (1948) and *Peter Kiewit Sons' Co.*, v. *U. S.* 109 Ct. Cl. 517 (1947).) Quoting from the *Chernus* case, page 266:

". . . The issue presented by the demurrer is similar to that of the case of *Peter Kiewit Sons' Co.* v. *U. S.*, 109 Ct. Cl. 517. In that case, as in this, there was a very large underrun of one kind of work below the amount estimated in the invitation for bids. The effect of the faulty estimate there was to cause the contractor to make a bid of a composite unit price which, because of the underrun of the one kind of work, turned out to be unreasonably low. The specifications in the *Kiewit* case included a cautionary paragraph to the same effect as paragraph 1-05 quoted above. We held that the petition stated a cause of action; that the parties in making their contract did not intend that the cautionary language of the specifications should turn the process of bidding on a Government contract into a pure speculation. We thought that such a literal interpretation of the contract, in a case where the parties were, obviously laboring under a mutual mistake as to vital facts when they made it, would, in the particular case, be unfair to the victim of the interpretation, and, in the long run, ruinously costly to the Government. We thought that such an interpretation would undo all the good which other provisions of the contract, particularly Article 4 were carefully worked out to accomplish; that is, to induce bidders not to increase their prices to cover possible misfortunes which might result from unforeseen developments."

When a changed condition is established, a contractor does not have to deduct the first 15 or 25 percent overrun.

When the owner, by a written change order, causes a substantial reduction in quantities, it may also be considered as a "Partial Termination." (See *Appeal of Nolan Bros., Inc.* ASBCA 4378, 58-2 BCA 1910 (1958).) In this event, the matter can be handled on a "cost basis" under the contract Termination Article. This may prove to be more beneficial to a contractor than proceeding under the Variations in Quantities Article.

It is comparatively easy for a contractor to recognize when an overrun or underrun in quantities occurs. A simple mathematical computation of the quantities shown on the bid estimate vs. the quantities shown on the pay voucher are the only supporting documents needed. Costing the matter out may be more difficult, however. Generally speaking, it is considered that an overrun may be profitable to a contractor; on an underrun, it is considered equitable to permit the contractor to spread his overhead over the remaining units, and to increase the unit prices by at least that amount. However, there may be extenuating circumstances that would justify an increase in unit prices. Suppose, for example, that the original estimated quantities exhausted the designated quarry, and that the overrun caused the contractor to have to open a more distant, less productive quarry. This would justify an increase in the unit prices.

As is customary under all contract Articles, the contractor must notify the owner in writing of the overrun or underrun and ask for an equitable adjustment and carry the burden of proof in showing that his costs were increased because of such overrun or underrun. The increase in costs must be directly related to the overrun or underrun.

27-16. Termination for Default, Damages for Delay, Time Extensions, and Force Majeure Articles

These articles are known as the Time Extension *Only* Articles. When a time extension is given under them, it is presumed to be complete consideration for any claims of money amounts. The articles are generally related to delays of a contractor, the delays are considered excusable, and the contract time will be extended by the amount of time the contractor is delayed because of:

"Unforseeable causes beyond the control and without the fault or negligence of the contractor,"

(and these words should always preface any request for time extension under articles similar to these)

"including but not limited to, acts of God, acts of the public enemy, acts of the Government in either its sovereign or contractual capacity, acts of another contractor in the performance of a contract with the Government, fires, floods, epidemics, quarantine restrictions, strikes, freight embargoes, unusually severe weather, or delays of subcontractors or suppliers arising from unforeseeable causes beyond the control and without the fault or negligence of both the Contractor and such subcontractors or suppliers."

This is the clause used in the Standard Form 23-A Government construction contract. This clause is more restrictive than a Force Majeure clause used by corporations in overseas work. This is probably so because of the ready insurability of the categories of situations encountered by our stateside bonding and surety companies on work in the United States. A typical Force Majeure clause used overseas would provide (see article, *"A Changing World Needs a Stronger Force Majeure,"* appearing in the November/December 1967 Worldwide *P & I Planning Journal,* page 25):

"Neither party shall be considered in default in performance of its obligations hereunder to the extent that performance of such obligations, or any of them, is delayed by Force Majeure.

"Force Majeure shall include, but not be limited to hostilities, restraint of rulers or peoples, revolution, civil commotion, strike, epidemic, accident, fire, flood, wind, earthquake, explosion, blockade or embargo, lack of or failure of transporation facilities, or any law, proclamation, regulation or ordinance, demand, or requirement of any Government or Governmental agency having or claiming to have jurisdiction over the work, or with respect to materials purchased for the work, or over the parties hereto, or any Act of God or other act of Government, or any cause, whether of the same or different nature, existing or future, which is beyond the control and without the fault or negligence of the parties hereto."

The article in the *Worldwide P & I Planning Journal* suggests that the Suspension of Work Article be amended to incorporate excusable causes in the Damages for Delay or Force Majeure article as suspensions of work for which an equitable adjustment would be granted. This would be a most desirable combination for any contractor and would eliminate a contractor's risks on Acts of God, the elements of nature, etc. In fact, this would eliminate practically all financial risks in the contract. However, it is most doubtful that owners will go this far to eliminate possible contingencies.

Generally, this is the article that affords relief against the assessment of liquidated damages if they are specified in a contract. It is also the article by which a contractor is defaulted for his failure to prosecute the work with such diligence as will ensure its completion within the time specified in the contract. It makes provision for the assessment of damages by the owner in case of default and should be carefully analyzed because it can be most helpful in the event a contractor falls behind schedule for reasons other than his fault or negligence. It is also important to the surety companies who furnish bonds, because they are given the first option in the event of default of taking over and completing the work, or to pay the difference between the contract price of a completion contractor and the amount paid to date of default of the defaulted contractor. Where no time extensions can be obtained from any of the other articles of the contract, it is essential that this article be carefully reviewed to see whether a bona fide time extension can be obtained under it. It should be remembered that no acceleration claim can be successfully prosecuted until the legal right to time extension is established, and then the owner's failure to grant it on the one hand, but insisting upon completion of the work under the originally established completion dates on the other hand.

This is a *must* for a successful contract administrator: never sign and accept a change order granting time under this article if you are hopeful of recovering money under any other article for the same general reasons. Time extensions accepted under this article and so recited in a change order will be construed as your having accepted the time in lieu of any other claim for the same or similar facts.

For example, suppose the owner delays you in the furnishing of some pieces of equipment that were agreed to be furnished, and issues a change order for the number of days delay that it changed your schedule of completion. If you accept this time with the recitation of its being granted under the Damages for Delay or Force Majeure article, you will have foreclosed any right to make a money claim for the delay under the Changes Article or under the Government Furnished Equipment Article, if there is one in your contract.

Watch carefully on any change order that the owner prepares to assure yourself that you are not "signing away your rights." You can always type on an "exception" or a "Provided that" above your signature and except any other claim that you might have for the same set of facts. Probably such a qualification will not be acceptable to the owner, but at least it will put him on notice of a possible claim, and if he accepts the qualification, your subsequent claim presentation will not be barred by your having executed a change order granted under a clause that gives time only for the facts recited as the justification for utilization of the article.

As in so many of the other articles, the owner may place a time limit upon which to claim a time extension for delay arising under this article, and, if so, your claim for time must be asserted within that time, or notice that such a claim will be made when all facts are established should be given within such time limit. Should a unilateral change order be issued under this Article and should the time granted not be sufficient, it must be appealed since it is considered to be in the same category of a "final decision" which must be appealed within the 30 day or other time limit set forth in the Disputes or Arbitration Articles.

The comparatively recent *Schweigert* case *(Schweigert, Inc., v. U. S., 181 Ct. Cl. 1184, 388 F 2d (1967))*, holds, under this Article, that:

"Where a contract provides that the contractor's right to proceed shall not be terminated nor shall the contractor be charged with damages arising because of delays caused by subcontractors but resulting from unforeseeable causes beyond the control and without the fault or negligence of both the contractor and the subcontractors, the contract is referring to subcontractors or suppliers who are in privity with the prime contractor, i.e., subcontractors whom the principal contractor can control or for whom it was contractually responsible. The term "subcontractor" in this context does not refer to second tier subcontractors, i.e., suppliers of the subcontractor. Delays by such second tier subcontractors, even where culpable should not occasion imposition of liquidated damages on the prime contractor unless the contract clearly so requires."

Because of this decision the government has amended the article to define "subcontractors and suppliers" to mean "at any tier" so that now a contractor must prove that any delay was beyond the fault or negligence of all tiers of subs and suppliers. However, if your article does not contain such a clause, a delay on the part of your second tier sub or supplier on down would be excusable if it actually delayed the prime contractor.

27-17. Inspection and Acceptance Article

There may arise an occasion where a difference of opinion exists between the contractor and the owner's inspection force as to whether some embedded item complies with the specifications. Standard Form 23-A has a clause in its Article 10, Inspection and Acceptance, that provides:

"(e) Should it be considered necessary or advisable by the Government at any time before acceptance of the entire work to make an examination of work already completed, by removing or tearing out same, the Contractor shall, on request, promptly furnish all necessary facilities, labor, and material. If such work is found to be defective or nonconforming in any material respect, due to the fault of the Contractor or his subcontractors, he shall defray all the expenses of such examination and of satisfactory reconstruction. If, however, such work is found to meet the requirements of the contract, an equitable adjustment shall be made in the contract price to compensate the Contractor for the additional services involved in such examination and reconstruction and, if completion of the work has been delayed thereby, he shall, in addition, be granted a suitable extension of time."

Thus, if it can be proven that the work did meet the terms and provisions of the contract, a contractor should immediately insist not only upon complete reimbursement for all direct and indirect expenses for tearing out the work, but for putting it back in place and for any delay in accommodating the inspectors for this added work and delay. Too often, a contractor will forget about this clause and generally will feel so proud of his work when it is determined that it does meet the specifications that he will overlook making a claim for the inconvenience, harrassment, and additional time it took him, as well as his costs of the examination. Before any work is torn out, it might be desirable to have the contracting officer make a determination that this action is "considered necessary or advisable" under this specific article so that a predicate can be laid for filing the claim under it afterward.

27-18. Possession Prior to Completion Article

Frequently, an owner will desire to occupy the premises prior to completion of all construction therein. In order that the rights and obligations of the parties may be spelled out in such an event, an article along the following lines is used:

"The Owner shall have the right to take possession of or use any completed or partially completed part of the work. Such possession or use shall not be deemed an acceptance of any work not completed in accordance with the contract. While the Owner is in such possession, the Contractor, notwithstanding the provisions of the clause of this contract entitled 'Permits and Responsibilities' shall be relieved of the responsibility for loss or damage to the work other than that resulting from the Contractor's fault or negligence. If such prior possession or use by the Owner delays the progress of the work or causes additional expense to the Contractor, an equitable adjustment in the contract price or the time of completion will be made and the contract shall be modified in writing accordingly."

The contractor should be extremely careful to document when and how the owner occupied each segment of the project. Particularly is this true when the contractor is completing a manufacturing or chemical plant of some type, and the owner's operating staff comes in and may abuse the equipment installed to such an extent that they will claim it was "faulty workmanship" or failure to meet specifications. Further, the presence of the owner's representatives on the site may make the work more difficult because of interference, or because of the necessity to reschedule certain portions of the work. If this becomes a necessity, the contractor should immediately put the owner on notice and make claim for the added costs and extended time under this article. Although the article may imply that possession does not mean "acceptance," it will be most difficult for an owner to claim deficiencies in the work after there has been an election to take over and occupy the premises, and the owner's representatives have, in fact, operated the equipment and utilities and taken over the buildings. Careful documentation will be extremely important in situations such as this to establish rights and obligations.

27-19.   Termination for Convenience of the Government Article

All government contracts contain a lengthy Termination for Convenience Article as a mandatory requirement. (See ASPR Sec. VII and FPR, Part 1-8.7.) Even when the article was inadvertently omitted, the Court of Claims in the celebrated Christian case (*Christian and Associates* v. *U. S.*, 160 Ct. Cl. 1, 320 F. 2d 345 (1963)) incorporated it by operation of law. Their wording was (312 F. 2d 418 (1963), page 427):

"For all of these reasons, we believe that it is both fitting and legally sound to read the termination article required by the Procurement Regulations as necessarily applicable to the present contract and therefore as incorporated into it by operation of law."

(See also *Chris Berg, Inc.* v. *U. S.*, decided 5/15/70, No. 235-68, page 6, by the Court of Claims.)

The primary purpose of the article, from the owner's standpoint, was to prevent a contractor from making a claim for anticipatory profits in the event of a termination initiated by the owner. Under the common law and the sanctity of contracts, a contractor would be entitled to recover his anticipated profits on the entire contract if it were terminated by the owner. The *Christian* case, supra, allowed profit on work actually done but disallowed anticipated but unearned profits.

Immediately after the end of World War II, the emphasis was upon the prompt settlement of defense contracts and plant clearance so that the industry could divert its production from munitions to peacetime products. Contract termination personnel were urged to be prompt and "courageous" in that, if the cost to the government was a bit more for a prompt settlement, it was money well spent to get a plant back into production of civilian products. Today, the regulations have not changed very much, but the settlements are much less accelerated and the negotiated price is scrutinized with great particularity. Most termination settlements are centralized to the extent that the Department of Defense Audit personnel handle the accounting and the Department of Defense utilizes a "Termination Contracting Officer" who assumes jurisdiction and proceeds with the termination proceedings. The plant clearance aspect (disposition of all specialized and other equipment) is extremely slow because the inventory must be circulated to all the government agencies to ascer-

tain whether any of them desire to take any of the items. If the first group circulated does not want the items, the inventory goes to a secondary list of agencies, and the time aspect is extensive.

However, attempts are made to achieve an equitable settlement, and contractors should have their books, records, and inventory in good shape for auditing. In all probability a fair settlement will be reached. Advantages under the article are that a negotiated settlement may be reached, and this figure does not have to be matched 100 percent by audited costs. Discretion is given the Termination Contracting Officer to consider many things in the settlement.

Although the lengthy termination article was basically designed for supply or manufacturing type contracts because they are the most numerous and have by far the largest dollar volume in the Department of Defense, their use has been made mandatory in the construction field. Consequently, when it is found that you are in a termination status, the language and practices of the manufacturing type of contract will have to be mastered to understand the numerous forms that will have to be prepared and the steps that will be required to get down to the settlement figures. It is well to remember that the regulations require the Termination Contracting Officer to assist contractors in every way to handle the forms, accounting, and other matters, whereas on any other claim under a contract, the contracting officer takes an adversary position ab initio.

Some points for you, as the contractor, to remember in the event of termination for convenience of the government are:

1. Settlement will be on a *cost* basis, although the last phase may be negotiated as a lump-sum settlement. To this end, your accountant should guide and assist the defense audit team and give them help and explanations that will be most favorable to your contentions. Remember, their report generally ends up in three columns: For Allowance, For Disallowance, and For Further Consideration. Probably you will be sitting across the table from the Termination Contracting Officer who is considering whether to accept some of the items considered not allowable by the auditors and certainly passing on the items for "further consideration." The more items you can persuade the auditors to place in the "allowance" column, the better off you will be. Section XV of ASPR is the Guide on all audits. Be familiar with this section of the regulations.

2. You will probably be asked to reach settlement with all of your subcontractors. To this end, your meetings and discussions with your subcontractors should be carefully documented so that you can justify a negotiated settlement that you achieve. These settlements will all be reviewed by a Termination Review Board and they are looking for documentation.

3. If you have obtained any pieces of unique equipment, or forms, or other items that are not readily salable or reusable by you, they should be charged off 100 percent to the work. If your books show a depreciated basis, you may have trouble getting the full value back from the investment.

4. If changes or changed conditions have been experienced up to the time of termination, be sure and press for their consideration because (a) if otherwise you are in a loss position on your contract work, no profit will be considered, and (b) no settlement will be made in excess of the contract price. You will be entitled to increase the contract price by the amount of your justified changes or changed conditions, but they must be stressed and justified.

5. Consult your attorney and accountant on any problem you have on your accounting system or in your dealings with subcontractors. A reasonable allowance will be made in the settlement agreement for these fees under your other expenses of conducting the termination procedures.

27-20. Government Furnished Property Article

An article of this type will ordinarily provide for time of delivery or performance dates, an equitable adjustment in contract price for delay in furnishing such property, and an extension of contract completion time. The article will also provide for the use, maintenance, and risk of loss of the property; for title provisions, and for disposition of the property upon completion.

In construction contracts it is customary for the owner to buy and furnish long-term procurement items, such as generators for hydroelectric power plants, or certain vessels and fittings for sophisticated complex chemical plants. In such cases, the construction schedule must be closely co-ordinated with and tied to the delivery dates specified.

When there is no article of this type and the owner fails to furnish the equipment as represented or warranted, or furnishes defective equipment, it will be considered to be a breach of contract, and the only remedy is in the Court of Claims if it is a federal government contract, or the state or United States District Court if the contract is with private parties. Court litigation is generally a longer and more costly procedure for ultimate relief than is a proceeding under the contract article.

If it is known in advance that the owner will furnish some of the critical equipment or supplies for the project, it is strongly recommended that a reasonable article covering owner-furnished items be included. This will allow you, as contractor, to proceed administratively under the contract in the event of failure to furnish such items, or in case the items are not furnished within the time stated and their absence from the work materially affects the completion schedule. If the contract does not contain such an article and the probable need for it is anticipated, the owner can be asked to put in such an article at the first time any other amendment or change is made to the contract. It can be included by a supplemental agreement and any other change in the contract can be considered as "consideration" for its inclusion in the contract.

27-21. Escalation of Labor or Materials Article

On rare occasions, an owner will recognize the shortage of certain trades in certain areas, or will recognize problems in an untested and unstable labor market in a specified area. In order to receive more competive bids and eliminate contingencies, he will place in the specifications an escalator clause relating to labor. Further, when certain material prices are drastically affected, such as a steel strike making re-bar in short supply, or if copper is critical, an owner may include a clause permitting an increase in price for materials. If an opportunity exists to insist on such articles, it makes the probability of contingencies less. When real concern is generated over shortages, or over rapidly increasing prices, either for labor or for materials, such a clause is not totally unusual, nor will an owner object where good cause for the utilization of such clauses is shown.

One of the main problems in connection with these escalation clauses is the definition of the base from which to escalate. It is easy enough to prove actual costs at the time they are incurred by payrolls or by paid vouchers, but establishing a base is more difficult. If the escalation clause is limited in scope, the current rate or value of labor and materials may be known as well as those upon which you based your bid price. However, if it is a clause of general, broad application, care should be taken to document the reasons why you were obliged to pay the wage rates or the material prices at the beginning of the contract so that actual costs may be subtracted therefrom as they are incurred to establish the extent of your equitable adjustment or time extension.

27-22. Currency Fluctuations; Loss of Equipment and Materials

If work is being performed beyond the continental limits of the United States, it is sometimes possible for a contractor to get protection against a radical fluctuation in the currency of the foreign government within whose territory he is working. He may also get insurance or an indemnity agreement against the loss, destruction or confiscation of his equipment or materials in the event of an insurrection or enemy action. When the Agency for International Development (AID) is the owner, or when this Agency places a loan or grant to a foreign government, it will enter into separate insurance or indemnity agreements directly with the U.S. contractor to protect these items, if requested. The reason for this is that its policy requires that the contract be written between the foreign government and the contractor, and AID does not become a contracting party to the arrangement. Its policy is to attempt to assure more 'independence" and American contracting experience for the foreign government. However, it does, in most cases, contract with the architect-engineer directly and this firm acts as the liaison with the foreign government and in assisting in the supervision of the work for both the governmental agencies.

If requested, many petroleum products corporations or companies, who intend later to operate large plants, will similarly protect the construction contractor. Because the workmen in the trades are mostly native labor, it is necessary that they be paid in local currency, and many items that are purchased for incorporation into the work and the operation of commissaries require local currency. Therefore, it is most helpful to have a currency-convertibility clause in the contract so that, upon completion of the work, local currency can be converted into dollars that may be taken out of the country.

Although these matters are major risks of a contractor performing work in countries beyond the continental limits of the United States, there are many other matters that should be carefully studied. such as normal hours of work and labor practices in the country; their tax structures (particularly income tax); duties and restrictions on importation of materials and labor; laws affecting injury to native personnel or property; and restrictions and controls over American personnel, such as visas, purchase of real estate, and other matters that affect day-to-day living overseas.

Many contractors find it advisable to take in a reputable and reliable local joint venturer to guide them in the customs and practices of the country, and to assist them in matters involving the local government and the native personnel with whom they must deal.

Because of the many unknown factors of performing construction work in overseas areas, it is suggested that this matter be given the greatest of study before entering into a firm commitment. In some countries, the economic or commercial attaché in their embassy can be most helpful in giving guidance and advice. Every source of information available in the country should be exhausted in the United States before proceeding overseas.

27-23. "Constructive" Changes

Webster's dictionary defines "constructive" as "Derived from, or depending on, construction or *interpretation;* as, a constructive fraud." Thus, by interpretation, the Courts have created an area of recovery for a contractor *under* the contract so that he does not have to claim the "breach-of-contract route" and go directly into a court. As the Court of Claims defined the term in a recent decision: (*Turnbull, Inc.* v. *U.S.*, 181 Ct. Cl. 29, 385 F. 2d 438 (1967)-Headnote:

"Where the contracting officer without issuing a change order requires the contractor to do work in a manner which is different than that required by the contract, the oral directive amounts to a constructive change order and the work so performed is a constructive change for which the contractor is entitled to demand compensation for the extra expense incurred. If the contractor and the contracting officer then fail to agree upon the amount of the equitable adjustment to which the contractor is entitled, the dispute concerning this amount is one which, under the language of the 'Changes' provision of the contract, is to be determined in accordance with the 'Disputes' provision."

As is stated in Commerce Clearing House, Government Contract Reports, Sec. 10,150.120, *Constructive Change Orders:*

"Access to borrow pit. A contractor was entitled to additional compensation for a constructive change order which resulted from the government's limiting the area of, and access to, a borrow pit which the contract expressly allowed the contractor to use. The Government's limitations on his use of the area forced him to use an uneconomic and unreasonable method of performance. . . ."

Thus it can be seen that even where a change order has not been issued, but should have been under all of the circumstances, the appeal boards or courts will take a *nunc pro tunc* approach and grant relief *under the contract.* It is therefore strongly recommended that, when a situation indicates that a change order should be issued and the contracting officer refuses to act, cases or board decisions be shown him to convince him that he should act under the contract in accordance with his duty as a contracting officer. If he is still adamant, it is suggested that his legal staff be consulted. In many instances advice can be given to the contracting officer by his own legal staff that will indicate to him that he has certain legal duties and he will not be criticized for exercising them.

27-24. Acceleration of Work

This is another area of recovery under the contract but there is no contract article covering the subject. It has grown up under board of appeals and courts' decisions to the general effect that an owner has no right to change or hasten a contractor's completion schedule without paying for it. The use of this device is different from what is known as the "Bonus-Penalty" provision. There, the project completion was geared to an estimated completion date, and the parties to the contract would agree that if the project was completed a certain number of days late, the owner would, in effect, suffer liquidated damages; by the same token, if the job was completed sooner, the owner would be benefited and should pay for the early completion by the specific sum per day agreed upon.

In the situation discussed here, most construction projects are scheduled with other phases of a project, and the owner is not particularly desirous of having them completed ahead of schedule—he simply wants them completed *on schedule.* Thus, if he demands that they be completed on the specified completion date and refuses to grant excusable contract completion time, he has in fact accelerated the work.

An excellent Briefing Paper was written on this subject by Sol Ribakoff, now Counsel for Peter Kiewit Sons', Inc., entitled *Acceleration.* It was published by the *Government Contractor,* No. 64-3, June 1964. He sets forth the Key Elements as follows:

"Based on the existing legal decisions, acceleration of performance will be held to have resulted, and you will have the right to a price adjustment therefor, if *all* of the following *five* elements are present:

(1) You have encountered *excusable delay* for which you are *entitled to a performance time extension;* and

(2) You have *specifically requested* such extension from the contracting officer and have timely presented that request; and

(3) The contracting officer *failed* or *refused* to *grant* the extension; and

(4) The contracting officer either (a) *expressly ordered* you to accelerate the work—i.e., to complete it within the original performance period, or (b) *acted* in such a way that it was clear he *required* you to complete within the original performance period (constructive acceleration); and

(5) You actually *did* accelerate performance, therein incurring excess costs."

Of course, in order to recover, it is up to the contractor to prove that there has been an acceleration and that his costs were increased thereby.

Care should be taken in presenting an acceleration claim that it is not confused with the provisions in most contracts requiring that a progress schedule be filed by the contractor and that, if the contractor falls behind such schedule, the owner has a right to demand overtime, and to require additional shifts of labor and equipment to keep up to schedule. Therefore, if the contractor has fallen behind schedule and has failed to submit a revised schedule and to notify the owner of the excusable causes that entitle him to a time extension, he may well fall into a situation where the owner has a right under the Progress Chart to require him to accelerate at his expense to keep up to schedule. In many instances, a job will wind up with two schedules—the one maintained by the owner showing no revisions therein, and the one maintained by the contractor showing revisions wherever excusable delay factors were encountered.

Examples of costs that may be proposed on an acceleration claim would be the premium time over straight time and taxes and insurance thereon; loss of efficiency where congestion, disruption of work, and fatigue are present; subsistence expenses of additional workmen, if any; extra shifts; extra equipment; premium costs for delivery of materials; employee training expenses; outside subcontracted work because of congestion; and any rescheduling costs that increased the total cost of the work. Should costs be saved in overhead, these should be credited to the owner on the theory that the supervisory personnel were not on the job as long.

Although the above costs are easy to relate, and they are all costs that are logical and recognizable, they require extensive bookkeeping entries and appropriate allocation. Should it be considered that the basic elements are present for the perfection of such a claim, time cards and allocations of costs should be initiated immediately so that it can be properly documented.

27-25. Ripple or Impact Effect

These have become "words of art" created by the appeal boards and courts to give a cost increase to contractors to work that has been "unchanged" by the operations of an acceleration claim, a change order, a changed conditions matter, or a rescheduling requirement. This was primarily because of the Supreme Court decision in the *Rice* and *Foley* cases as discussed under Paragraph 4A, supra.

In many instances an owner will give selective orders as to crafts to be accelerated when the owner agrees to pay for the overtime and extra shifts of those crafts. However, a contractor will find that by working one craft overtime, others will threaten to and may actually strike unless they also receive overtime payments. Therefore, the contractor has suffered a "ripple or impact" effect upon crafts and operations of his job that were unchanged. These costs are compensable today if it can be shown that they are directly attributable to the partial acceleration.

Further, where change orders have been issued so frequently making such drastic changes that it soon becomes a matter of "changes on changes," the Boards and Courts have considered that this has a "ripple or impact" effect upon the unchanged work and will allow an equitable adjustment both upon the changed and unchanged portions of the work.

Under Notes and Comment, McBride and Wachtel, Government Contracts, Sec. 37.90, have this to say regarding ripple effect (*Chalender* v. *United States,* 127 Ct. Cl. 557 (1954), page 562):

"This case exemplifies what might be called the *ripple effect* which follows an interruption of a carefully planned work sequence. When some event occurs which delays or brings to a halt a construction project or a production line, costs can be expected to accumulate rapidly, and all segments of the operation may feel its effects. Illustrating this point is a case in which a building alteration contract was divided into four operations or phases, with each phase of work to commence after the completion of the prior phase. A delay in any of the first three phases automatically delayed the start of the fourth, resulting in a change. *Appeal of Rosen & Son, Inc.,* BACAB 429, 65-2 BCA 4936 (1965)."

"The quotation above makes it clear that if the Government is responsible under circumstances amounting to a breach, it must respond in damages."

Thus it follows that if the contractor can demonstrate that his planned work sequence was disrupted by some extracontractual action of the owner, the delay caused as the result is excusable and the additional expense is compensable.

The problem of being compensated for Ripple or Impact effect has largely been eliminated in government contracts because of the use of the new Changes and Differing Site Conditions Articles now being utilized in Standard Form 23-A. Both articles now indicate that any change or changed condition affecting either the new or existing work may be compensable, if otherwise coming within the terms and conditions of the Article. However, so far, these revised articles have not yet been interpreted by the appeal boards or courts and time will tell whether their effect is as broad as the present language would seem to indicate.

27-26. Cardinal Change

The Court of Claims defines a "Cardinal Change" in the headnotes to the case of *Luria Bros. & Co.,* v. *U.S.,* 177 Ct. Cl. 676, 369 F. 2d 701 (1966):

"Where changes made by the Government as the result of the contractor having encountered unsatisfactory material at subgrade elevations were of such magnitude that they were not within the scope of the original contract, they amount to cardinal changes constituting a breach of the contract."

and in *Keco Industries, Inc.,* v. *U.S.,* 176 Ct. Cl. 983, 364F. 2d 838 (1966):

"The ordering of a change which is clearly outside the scope of the contract is a cardinal change and a breach of contract. Whether or not a change ordered is a cardinal change depends on the particular facts involved. Where a contract provided for the procurement by defendant from plaintiff of 100 electric and 100 gasoline driven refrigeration units differing only in the type of power unit used, the defendant's change order eliminating the gasoline driven and substituting therefor a like number of electric powered units was not a cardinal change or breach of contract."

A claim of this nature is most difficult to prove and is usually a sort of a "last resort" matter for consideration. To give some idea as to the thinking of a contracting officer in reviewing and evaluating the impact effect of changes and whether they might be extensive enough in cumulative scope to constitute a cardinal change, the following is quoted from the *Appeal of C. W. Regan, Inc., et al,* ASBCA No. 12064, 68-2 BCA 7313 (1968):

"After completion of the contract work, appellant presented to the contracting officer sixty-eight (68) claims for equitable adjustments in the contract price. The requested adjustments were based partly on the performance of extra work due to changes and the encountering of changed conditions, partly on disagreements as to the amount due appellant for the increased cost of admitted changes, and partly on the impact of all changes made and alleged changed conditions encountered during the accomplishment of the contract as a result of which the efficient performance of appellant's work force was reduced and its costs were increased. . . .

Assuming that in settling up 31 of appellant's claims the parties gave some positive weight to each of them, there was a total of 95 changes compensable under the changes as distinct from the changed conditions article. For these appellant was paid less than 5% of the contract and would receive altogether 8% of the contract price if its claims for additional monies under the contested contract modifications (Nos. 60 to 67) were allowed. Nearly all of the changes involved minor work only and there is no showing that they caused costs beyond those recognized by respondent or in dispute under claims Nos. 60 to 67 for the work actually performed. Nor is there any showing that they caused any unreasonable suspension of work under the suspension of work article. On the contrary, in contract Modifications Nos. 10, 11, 13, 14, 15 and 19, which allowed extensions of time for changed work, appellant agreed that no suspension of work had occurred."

It can readily be appreciated that the burden of proof is most difficult to carry and that your contracting officer will exert every effort to minimize every contention made. That is why it is so important that good records be kept, complete with time-card allocation of work to specific items so that every factor on the job can be shown in its true perspective.

27-27. Platitudes in the Claims Field

In the parlance of the owner's contract administrators many platitudes are bantered around, and a slight knowledge of the purport and scope of the phrases may prove to be most helpful. A few of the most common with a suggested meaning will be set forth for guidance purposes.

27-28. A "War Powers Case"

This has reference to the extraordinary relief available to contractors under Public Law 85-804, which succeeded Title 11 of the World War II First War Powers Act. (50 USC 1431-35 and 55 Stat. 838, respectively.) A contractor has no legal or equitable rights under the Act—it was enacted to "facilitate the national defense" and the Contract Adjustment Board (created for the purpose of hearing such matters) must so find. One of the principal uses of the Act has been to correct certain types of mistakes. It has been determined that relief in this area will expedite the procurement program by assuring contractors that mistakes will be corrected. Three types of mistakes are set forth in ASPR 17-204.3, the first being the failure of the contract to express the agreement, as both parties understood it; a mistake on the part of the contractor which should have been obvious to the contracting officer, and a mutual mistake as to a material fact.

A second area of relief under 85-804 is a contract amendment without consideration." This is available when the actual or threatened loss under a defense contract will impair the future productive ability of a contractor necessary and essential to the defense effort. The third area is when some government action causing a contractor a loss is directed specifically at the contractor, such as when the government delayed in furnishing tooling, or making a work site available, or in issuing necessary modifications, causing undue testing requirements. The relief is adjudged on a matter of "fairness." The Act may not be utilized if there is any other legal remedy under the contract. (Breach of contract is not considered an administrative remedy under the contract.) Further, the request must be made within six months after the contractor has furnished the property or services, and subsequent to final payment on the contract.

The limitations of being essential to the defense effort and being impaired financially from proceeding with any work eliminates most of the possible applicants for relief. However, the area of "mistakes" is frequently utilized and should be sought if the facts qualify for the relief desired. It is suggested that the regulations set out in the Armed Services Procurement Regulations be carefully studied before proceeding with a War Powers Request for relief.

27-29. Contract Interpreted against Drafter

If a contract specification is ambiguous and is susceptible to two or more interpretations, it will be interpreted against the owner who drafted it. (See *C. J. Montag & Sons, Inc.* v. *U. S.*, 348 F. 2d 954 at 960.) However, before it is considered ambiguous, all specifications and drawings which form a part of the contract must be read together, and no portion is to be rejected or treated as meaninglesss if any meaning which is reasonable and consistent with the rest of the contract can be given to it. (*Appeal of Lucarelli & Co., Inc.*, ASBCA 4768, 59-2 BCA 2343 (1959).) This leads into another rule of interpretation and that is that "parol" (or oral) evidence "will not be admitted to vary the terms of a written instrument" if the contract is clear and unambiguous on its face. In this connection, see *United States* v. *Bethlehem Steel Co.*, 215 F. Supp. 62, page 68 (1962). When it is not "clear on its face," you may use the rule of "course of conduct of the parties" up to the time of the dispute, as to how they interpreted the provision; or you may bring in the standard practice of the industry, or what may have been the interpretation of other parties to the identical contract provision, or the record of negotiations between the parties prior to execution of the contract.

27-30. Specifications vs. Drawings

Other areas of interpretation *where a conflict exists* in a contract are: the specific provisions control over general provisions; typed provisions control over printed provisions. When a general term is joined with a specific one, the general term will be deemed to include only those things which are similar to the specific term. Finally, specifications will supersede drawings.

27-31. Duty to Seek Clarification

Where a contractor knows that there is a major patent discrepancy, or obvious omission or a drastic conflict in provisions, he is under a duty to seek clarification prior to bidding. As stated by the Court of Claims in the case of *Blount Brothers Construction Co.*, v. *U. S.*, 171 Ct. Cl. 478, page 496:

"However, contractors are businessmen, and in the business of bidding on Government contracts they are usually pressed for time and are consciously seeking to underbid a number of competitors. Consequently, they estimate only on those costs which they feel the contract terms will permit the Government to insist upon in the way of performance. They are obligated to bring to the Government's attention major discrepancies or errors which they detect in the specifications or drawings, or else fail to do so at their peril. But they are not expected to exercise clairvoyance in spotting hidden ambiguities in the bid documents, and they are protected if they innocently construe in their own favor an ambiguity equally susceptible to another construction, for as in *Peter Kiewit Sons' Co.* v. *United States*, 109 Ct. Cl. 390, 418 (1947), the basic precept is that ambiguities in contracts drawn by the Government are construed against the drafter."

27-32. The Wunderlich Act

A brief piece of legislation enacted on May 11, 1954, has had a salutary effect on administrative proceedings and particularly upon procedures in the Court of Claims. It is commonly known as the Wunderlich Act. (Public Law 356—83rd Congress, 68 Stat. 81.) It is quoted as follows:

"That no provision of any contract entered into by the United States, relating to the finality or conclusiveness of any decision of the head of any department or agency or his duly authorized representative or board in a dispute involving a question arising under such contract, shall be pleaded in any suit now filed or to be filed as limiting judicial review of any such decision to cases where fraud by such official or his said representative or board is alleged: provided, however, that any such decision shall be final and conclusive unless the same is fraudulent or capricious or arbitrary or so grossly erroneous as necessarily to imply bad faith, or is not supported by substantial evidence.

"Sec. 2. No Government contract shall contain a provision making final on a question of law the decision of any administrative official, representative, or board."

The legislation was requested because of a harsh decision by the Supreme Court in the case of *United States v. Wunderlich*, 342 U. S. 99; 96 L ed 113 (1951). The Court of Claims had been holding hearings de novo (new trials) before their commissioners and, notwithstanding a decision by a head of a department (or an appeal board) on a question of fact, might determine that the decision should be set aside on the basis of a "gross mistake implying bad faith" which the Court equated to fraud. The Supreme Court held that there had to be *actual* fraud to set aside a factual determination. Therefore, the Congress was persuaded to pass the above legislation, which provides that all questions of law shall be determined by the Courts and sets up a standard *of review* of factual decisions for the Court of Claims.

This has been activated by Rule 168 of the Court's Rules and is tantamount to making the Court of Claims an appellate or review court to determine whether final decisions on questions of fact by heads of departments (appeal boards) are "fradulent or capricious or arbitrary or so grossly erroneous as necessarily to imply bad faith, or is not supported by substantial evidence." If it is so found, the decision may be reversed, and the matter remanded to the Board for determination of an equitable adjustment under the contract. The equitable adjustment then agreed upon will be entered into the Court's final judgment on the case.

Whenever an appeal is taken from an adverse Board of Appeals decision to the Court of Claims, and the decision is made on a question of fact, a contractor will not receive a trial before a Commissioner but, instead, will submit a review of the record before the Board, stating the reasons why the decision should not be final under the Wunderlich Act, and will ask the Court of Claims for a summary judgment. The Department of Justice will then review the record from the government's side and will contend for finality of the decision. The matter is then up to the Commissioner to review and make findings. Ultimately, a decision by the Court will be rendered. But gone are the days of a new trial in the Court of Claims on appeals from administrative agencies where their decisions are considered to be on questions of fact. As

stated in a recent Court of Claims case (J. A. Terteling & Sons, Inc. v. U. S., 390 F. 926 (1968), 931):

"The province of this court in considering the case in its present posture is, of course, to determine whether the decisions of the Board on questions of fact are final under the Wunderlich standards, and to resolve independently questions of law which are presented. *River Constr. Corp v. U.S.*, 159 Ct. Cl. 254 (1962)); *Morrison-Knudsen Co. v. U.S.*, 345 F. 2d 833, 170 Ct. Cl. 757 (1965). The standard to be used in determining whether an administrative decision is supported by substantial evidence goes to the reasonableness of what the agency did on the basis of the evidence before it. . . .' *U.S. v. Carlo Bianchi & Co.*, 373 U.S. 709, 83 S. Ct. 1408, 10 L. Ed. 2d 652 (1963)."

27-33. Equitable Adjustments

This is a subject in which all contractors are interested. Most articles in contracts providing for an increase in compensation to a contractor use the phrase "equitable adjustment" in contract price to be the ultimate solution to all of his problems. It reminds the author of a phrase used by a prominent Department of Navy negotiator that a "good settlement" is when both parties (the owner and the contractor) walk away from the table a little bit unhappy! The same must be true of an "equitable adjustment," because the final figure must be "equitable" or fair to both sides.

But how do the appeal boards and the courts look at the problem? It is interesting to note the reasoning of the Court of Claims on the subject in their opinion in *Nager Electric Company, Inc. et al. v. U. S.*, 194 Ct. Cl. 835; 442 F 2d 936, at page 851 et seq.:

"The formula by which an equitable adjustment is to be computed is a determination of law, one, therefore, decided independently by the court. *N. Fiorito Co. v. U.S.*, 189 Ct. Cl. 215, 416 F. 2d 1284 (1969); *Keco Industries v. U.S.*, 174 Ct. Cl. 983, 364 F. 2d 838 (1966), cert denied, 386 U.S. 958 (1967). The pertinent authorities compel rejection of the limited test applied in the administrative proceedings below. The measure for an equitable adjustment to the contract price resulting from a change order is, as was correctly indicated by the AEC hearing examiner, framed in terms of 'reasonable cost.' *Bruce Constr. Co. V. U.S.*, 163 Ct. Cl. 97, 324 F. 2d 516 (1963). The search for 'reasonability' however, is not limited to inquiry of such factors as 'fair market value' or 'historical cost.' In addition, the reasonable cost concept includes both 'objective' and 'subjective' elements.

"The objective focus is on the costs that would have been incurred by a prudent businessman placed in a similar overall competitive situation. 163 Ct. Cl. at 101, 324 F. 2d at 518-19. See Spector, Confusion in the Concept of the Equitable Adjustment in Government Contracts, 22 Fed. B. J. 5 (1962). . . .

"Further, the equitable adjustment should not increase the plaintiff's loss nor decrease it at the expense of the Government."

As can be determined from the above, there are many tests to be applied and it seems to end up pretty much on a matter of what is fair under all of the circumstances. The goal seems to be to leave a contractor in just about the same position he was in—that is to say, he should not profit by a windfall; but on the other hand, he should not suffer a loss because of some action on the part of the owner.

It would seem in the last analysis that there is no real substitute for carefully documented cost records, and the trend is to consider them as the best evidence of what is an equitable adjustment. In most cases the problem is separating the cost of the changed work from the reasonable value of the unchanged work so that the net can become the price for the equitable adjustment. Here again, the original estimate can be the best evidence of the anticipated cost of an item. It should be carefully preserved along with other contract records.

28 TAXES

JOSEPH J. HYDE

Partner
Lybrand, Ross Bros. and Montgomery
San Francisco, California

Construction contractors are like most other businesses —various tax collectors take a slightly larger piece of the profit pie than is left for the shareholders or proprietors. And like other businesses, they are subject to gross receipts taxes, sales and use taxes, payroll taxes, property taxes, and many others.

Taxation, in addition to being a substantial financial and administrative burden, is ever changing because of changes in laws, regulations, administrative policy, and court decisions. The reader is cautioned to check current status before relying on specifics of this chapter.

For example, in March of 1971 the Internal Revenue Service proposed new regulations dealing with the entire subject of taxation of long-term construction contracts. These proposed regulations were withdrawn because of criticism by the construction industry and its tax advisors, and revised proposed regulations were issued in December 1971. The changes proposed (which will be discussed in detail later) are substantive and pervasive. If adopted in the form proposed, they will be of great financial importance to many contractors.

28-1. Taxation Is Monumentally Complex

Taxation is also monumentally complex, and a satisfactory job can be done only by full-time specialists. Contractors who are not sufficiently large to afford their own staff should retain accountants or attorneys who are knowledgeable in taxation to advise them of the tax results of contemplated transactions and to assist them in structuring transactions in a manner which will minimize taxes. United States tax law and the tax laws of many of the states are highly codified and widely analyzed by tax reporting services and other commentaries. Consequently, reliable answers to most questions can be found by research of available literature. This is not the case in most foreign countries or in the smaller states. In those jurisdictions, it is wise to obtain advice from practitioners in the respective countries and states who know how the taxing systems actually work.

28-2. United States Income Tax

It probably is fair to assume that most contracts performed in the United States are performed by corporations or by joint ventures of two or more corporations. Minor contracts and subcontracts are often performed by sole proprietorships, or by partnerships, but since the gross volume involved is estimated to be a minor percentage of total volume, we will confine our discussion to corporations.

The current U.S. corporation income tax rate is in substance 48 percent of taxable ordinary income and 30 percent of net long-term capital gains. There presently are no substantial credits against U.S. tax other than the foreign tax credit, the investment tax credit, and the credit for nonhighway use of fuels and lubricants.

The United States levies taxes on the basis of place of incorporation—i.e., a corporation formed under the laws of one of the states or the District of Columbia is taxable on its worldwide income. A foreign corporation (one incorporated outside the United States) is taxable only on its income from sources in the United States.

28-3. Tax Accounting Methods to Determine Income from Long-Term Contracts

It is beyond the scope of this chapter to describe the manner in which ordinary taxable income and net long-term capital gains are determined. Volumes have been written on those subjects and are available in any business library. Rather, we will concentrate on tax accounting methods used to determine and report income derived from performance of long-term contracts. Except for this aspect, the taxable ordinary income and capital gains of a contractor are determined under rules applicable to taxpayers in general.

The first question that arises is: What is a long-term contract? Oddly, the Internal Revenue Code does not mention long-term contracts at all. The controlling rules are set forth in Regulations Section 1.451-3, which are regulations under the Code section dealing with the taxable year in which income is to be reported. The language of the present regulations was adopted originally in 1918 and has not been changed substantially since.

The regulations state that a long-term contract is a construction contract covering a period of more than one year from the date of execution to the date the contract is finally completed and accepted. Court decisions have modified this rule to the effect that long-term contracts are those which are started in one taxable year and completed in another taxable year. (*Daley* v. *U.S.*, 243 F-2d 466 (9th Circuit 1957).) The new regulations proposed in December of 1971 follow the view of the courts, and since this is a sensible rule it is relatively safe to assume that final regulations will include this proposed change.

Hereafter, reference to proposed regulations means Proposed Regulations 1.451-3 as published in the Federal Register December 15, 1971.

TAX ACCOUNTING METHODS

There are four methods which may be used to account for income from contracts. They are:

- Cash receipts and disbursements
- Accrual
- Percentage of completion
- Completed contract

Each of these methods will produce a different timing of the recognition of income, although under each method the amount of income finally reported for taxation will be the same. Under present rules, whichever method is used must be used for all long-term contracts, and there is no requirement that accounting records be kept on the basis used for tax reporting. The proposed regulations seek to modify both of these rules.

A taxpayer adopts an accounting method by using it in the first return in which a long-term contract is involved. Although no formal election or statement as to use of the accounting method is required, it is good practice to include a statement identifying the accounting method being used for the sake of clarity. The proposed regulations would require a statement to be included in each return where income from a contract or contracts is reported on the completed contract method setting forth facts which justify use of that method. Once a method is adopted, it may be changed only upon receiving consent of the Commissioner of Internal Revenue and upon agreeing to such terms for making adjustments required in the year of change and other years, as the Commissioner requires.

28-4. Cash Receipts and Disbursements Method

The *cash receipts and disbursements method* is the simplest. Income consists of amounts received as progress or final payments for work performed, and deductions are amounts paid for project costs, other than amounts paid for depreciable equipment. This method is often used by small contractors and subcontractors but, to the author's knowledge, by few major contractors. Besides simplicity, this method has the virtue of matching cash flow with taxation. The taxpayer doesn't owe any tax unless cash receipts have exceeded expenses. It also allows the taxpayer to "manage" his taxable income by accelerating or deferring receipts and payment of expenses.

28-5. The Accrual Method

Under the *accrual method*, a taxpayer takes into income all amounts which he is entitled to bill under the contract terms and deducts all costs incurred (whether or not paid for) except costs of materials and supplies that do not enter into computation of amounts which can be billed. The general effect of this method is to recognize income or loss over the life of the job more or less in relation to work performed. This result is somewhat the same as the percentage of completion method, but it can also be significantly different.

The accrual method permits current deduction for mobilization and setup costs even though the contract may not provide for any billing for these costs. On the other end of the job, dismantle and move-out costs cannot be deducted until incurred. They usually are incurred after all billings on the contract have been made. The accrual method accelerates recognition of income where the payment schedule is loaded on the front end of the job.

One interesting problem where the accrual method is used is whether amounts of billings retained by the customer (a common provision) are includable in income. It is the position of the Internal Revenue Service that retained amounts are not income. (Rev. Rul. 69-314, 1969-1 C.B. 139.) Consistent with this treatment, a contractor cannot deduct as a cost the amounts he has held back from payments to his subcontractors. Many contractors have included retained amounts in both income and costs, which under the rationale of Rev. Rul. 69-314 constitutes an erroneous accounting method. Since the Internal Revenue Service would ordinarily have nothing to gain by correcting this erroneous method, the taxpayer must take the initiative if he wishes to change by requesting a change in accounting method from the Commissioner of Internal Revenue.

28-6. The Percentage of Completion Method

The *percentage of completion method* is in concept fairly simple. Revenue is the contract price multiplied by percent complete. Deductions are all contract costs, excluding materials and supplies which have not entered into the determination of percent complete. Current regulations say that the percentage of completion will be determined on the basis of certificates of architects or engineers. A practice commonly used, and which is acceptable for tax purposes, is to compute the percentage of completion by dividing costs incurred by expected total costs. The proposed regulations permit either of these methods. The method used to determine percentage of completion must be used consistently.

All three of the methods discussed so far will result in recognition of income or loss as jobs progress. The completed contract method provides for recognition of the entire income or loss from a contract in the year in which it is completed. The attraction here is that payment of tax is posponed as compared with any of the other methods.

28-7. The Completed Contract Method

The *completed contract method* recognizes income or loss from a contract in the year in which the contract is finished. Although this simple rule would seem entirely clear, there have been several areas of dispute under present regulations. However, the courts have evolved rules which are fairly clear.

The first question is: When is a contract complete? The regulations say a contract is complete in the year in which it is finally completed and accepted. Despite this seemingly clear rule, the Internal Revenue Service developed the position that a contract is complete when it is substantially completed. This was thought necessary to prevent taxpayers from postponing payment of tax by deliberately failing to complete a minor portion of the work. The Tax Court accepted this rule, but the Courts of Appeals for the 6th and 9th Circuits have ruled that the regulations mean what they literally say. (Thompson-King-Tate, Inc., 296 F2d 290, E. E. Black, Ltd., 211 F2d 879).

The proposed regulations (Proposed regulations 1.451-3 (b) (2).) read as follows on this subject:

"The term 'completed' means finished at least to the point where—

(i) The remaining costs required to entirely finish the contract are insignificant in comparison with the amounts already expended with respect to such contract;

(ii) No substantial dispute exists as to the acceptability of the work performed on the portion finished; and

(iii) The contract has been completed in all respects which are essential for the basic utility of the subject matter of the contract."

28-8. Income from Cost-Plus Contracts

The Internal Revenue Service has contended that income from cost-plus contracts may not be reported on the completed contract method. The theory here is that there is not sufficient uncertainty as to the amount of profit to be earned to justify postponement of reporting. The Tax Court (Sam W. Emerson, 37 TC 1063) did not agree, although there had been earlier Tax Court decisions to the contrary.

In the author's opinion, the method should be allowed where there are substantial contingencies in forecasting net income, such as a guaranteed maximum cost, an increased fee if total costs are less than a target, or the amount allowable as overhead is subject to audit and adjustment. The proposed regulations adopt this approach.

28-9. What Constitutes a Single Contract?

Another problem that arises when the completed contract method is used is the question of what constitutes a contract. The general rule which has evolved is that a single project, even though split up into several contracts, is a single contract. (*Helvering* v. *National Contracting Co.* 69 F 2d 252.) There also have been cases where a single contract to deliver completed units at a stated price was divided into a contract for each unit. (*Grays Harbor Motorship* v. *U.S.*, 45 F 2d 259.) Here, income or loss was recognized as each unit was delivered. The proposed regulations follow these cases in substance.

As mentioned earlier in this chapter, under present rules a single tax accounting method must be used for all contracts of the taxpayer. Proposed regulations 1.451-3 (a) 2 modified this rule as follows:

(2) (i) Except as provided in subdivision (ii) of this subparagraph, if a taxpayer reports income from a long-term contract in accordance with one of the long-term contract methods, such taxpayer must use a long-term contract method (but, ex-cept as otherwise provided in this section, not necessarily the same long-term contract method) for each long-term contract within the same trade or business. Once a method of reporting income from a long-term contract has been used, such method must be used consistently with respect to such contract. Similarly, once a method of determining percentage of completion is used with respect to a long-term contract, such method must be used consistently with respect to such contract.

(ii) Notwithstanding the provisions of subdivision (i) of this subparagraph, a taxpayer who has long-term contracts of substantial duration and long-term contracts of less than substantial duration in the same trade or business may report the income from the contracts of substantial duration on the percentage of completion method or the completed contract method and report the income from the contracts of less than substantial duration pursuant to another proper method of accounting. For example, if a manufacturer of heavy machinery has contracts of a type that generally take 9 months to complete, and also has contracts of a type that generally take 3 months to complete, the manufacturer may use a long-term contract method for the 9-month contracts and use a proper inventory method pursuant to section 471 and the regulations thereunder for the 3-month contracts.

The proposed regulations permit use of the completed contract method only if estimates of progress toward completion are not reliable. They cite use of a different method for financial reporting (accrual or percentage of completion) as creating a presumption that reliable estimates are available. This presumption may be rebutted by reference to other facts and circumstances which indicate that reliable estimates of progress toward completion are not available.

28-10. Problem Common to Accrual, Percentage of Completion, and Completed Contract Methods

There is a problem common to accrual, percentage of completion, and completed contract methods and that is what overhead should be charged to job costs and what general expenses should be deducted currently. In general, most contractors include field office costs in contract costs, and currently deduct regional and home office costs.

Although not a construction contract case, there was a 1970 Tax Court Decision (All-Steel Equipment, Inc. 54 TC No. 176) regarding the kinds of expenses to be included in overhead for purposes of computing inventory costs. The problem there was essentially the same as that involved in allocating overhead to construction contracts in progress. The Court held that deductions for taxes, losses, and research are expressly authorized in the year incurred and consequently need not be considered in arriving at inventory costs. A similar determination, based on somewhat different grounds, was made for repairs. The Commissioner conceded that profit-sharing plan contributions were currently deductible. This rationale is primarily of interest to new taxpayers and to those who are already currently deducting such expenses, since it is the Commissioner's position that a change requires his permission. It is very doubtful if such permission would be granted.

Possibly with this case in mind, the proposed regulations provide, as a general rule, that all overhead type expenses must be allocated to jobs. This rule is modified in certain circumstances to permit current deduction for taxes, depreciation, employee benefit expenses, officers' salaries, and general overhead.

It is the author's view that if the regulations are adopted as proposed, and they would require allocation of more overhead items to contracts than has been the taxpayer's practice, past practices should be continued.

FOREIGN TAXES

28-11.   Taxation of Foreign Operations

The large U.S contractors do considerable work abroad and consequently are exposed to foreign income taxation. Foreign taxation generally has no impact on the total tax payments unless the foreign tax rate on foreign income taxable in the U.S. return is higher than the present U.S. rate of 48 percent. This is accomplished by the allowance of foreign income taxes paid as a reduction of U.S. income tax, subject to the limitation expressed as follows:

$$\frac{\text{Foreign source taxable income included in the U.S. return}}{\text{Total taxable income}} \times \frac{\text{U.S. tax}}{\text{before credit}} = \frac{\text{Maximum foreign}}{\text{tax credit}}$$

This limitation may be computed separately for each country or for all foreign countries combined. The latter is usually the better method since it permits offsetting low tax rate countries against high tax rate countries and tends to maximize the allowable credit.

28-12.   Foreign Tax Credits

A foreign tax credit is also allowed in connection with dividends received from a foreign corporation in which a U.S. corporation owns at least 10 percent of the voting stock, and it is allowed to a foreign corporation with respect to dividends it receives from another foreign corporation in which it owns at least 10 percent of the voting stock. This credit is usually referred to as a "deemed paid" credit and the general result is to place the U.S. corporation in the same position with respect to foreign tax credit as if it had earned the income and paid the foreign tax directly. There is a somewhat more advantageous treatment for dividends received from less developed country corporations and their less developed country subsidiaries.

28-13.   Less Developed Country Corporations— Tax Credits

A less developed country corporation is a foreign corporation 80 percent or more of the gross income of which for a taxable year is from sources in less developed countries, and 80 percent or more of whose assets on each day of the taxable year consists of property associated with activities and certain investments in less developed countries and money or deposits with banks. (Sec. 955 (c) (1) I.R.C.) Less developed countries are those other than:

Australia	France
Austria	Germany
Belgium	(Federal Republic)
Canada	Hong Kong
Denmark	Italy
Japan	Norway
Liechtenstein	Union of South Africa
Luxembourg	San Marino
Monaco	Sweden
Netherlands	Switzerland
New Zealand	United Kingdom

Conducting business in less developed countries by means of a less developed country corporation can potentially be advantageous at the time of liquidation. Gain realized on liquidation of such corporations which have been so for ten years or more is long-term capital gain taxable at 30 percent, but without reduction for foreign tax credit. Gain from liquidation of other foreign corporations is treated as a dividend as to earnings and profits accumulated after 1962, but the foreign tax credit is allowed. A brief experiment with certain assumed facts discloses that an advantage would result only where the rate of foreign tax is low, somewhat under 30 percent.

Taxes paid to possessions of the United States are treated as foreign taxes paid.

In lieu of claiming the foreign tax credit for taxes paid directly (but not for taxes deemed paid) the U.S. corporation may deduct such taxes in computing its taxable income. This generally is not as advantageous as claiming the credit, but occasionally there are circumstances when it is. Foreign taxes paid which exceed the allowable credit for a year may be carried back to the two preceding taxable years and forward to the five following taxable years. Amounts creditable in the carry-back or carryover years are determined under the limitation formula for the year to which carried.

28-14.   Foreign Jurisdictions That Do Not Allow Completed Contract Methods of Reporting

Some foreign jurisdictions do not allow the completed contract method of reporting income for taxation. When a U.S. contractor who uses the completed contract method here does a job in such a foreign country, the foreign tax is considered to be paid in the year in which income from the contract is reported in the U.S. return. (Rev. Rul. 288, 1953-2 CB 27.)

28-15.   Foreign Countries Levying Withholding on Dividends

Many foreign countries levy a withholding tax on dividends and interest paid by foreign corporations to U.S. persons. These are treated as foreign taxes paid by the U.S. recipient. In computing the foreign tax credit, there is special limitation on the credit allowable with respect to investment interest received from foreign sources. The effect here is to prevent a taxpayer from bringing down the average effective rate of foreign tax by earning foreign source investment interest.

As stated previously, foreign taxation is not as highly codified as U.S. taxation. There is typically greater latitude for the administrators to interpret the law than there is here. On large foreign projects, many of which are performed for the foreign government, it is at times possible to reach a favorable agreement regarding taxation when the contract is being negotiated. Knowledgeable foreign nationals are usually essential for this purpose and for the purpose of advising on tax matters.

28-16. Controlled Foreign Corporations

In 1962 Congress added a nightmare to the Internal Revenue Code called Subpart F dealing with controlled foreign corporations (foreign corporations where more than 50 percent of voting shares are owned or treated as owned by U.S. shareholders). The general thrust of these provisions was to tax currently certain types of undistributed income earned by controlled foreign corporations. Current taxation can be avoided by distribution of certain percentages of the earnings and profits of the controlled foreign corporation. The lower the effective rate of foreign tax, the higher the distribution required to avoid current taxation.

The kinds of foreign income to which these provisions apply are not typically encountered in construction activities. The rules apply to any U.S. corporation or individual who owns, or who under attribution rules is considered as owning, 10 percent or more of the voting power of a controlled foreign corporation.

Code Section 951, reads in part as follows:

"SEC. 951. AMOUNTS INCLUDED IN GROSS INCOME OF UNITED STATES SHAREHOLDERS
(Sec. 951(a))

"(a) *Amounts Included.*

"(1) *In General*—If a foreign corporation is a controlled foreign corporation for an uninterrupted period of 30 days or more during any taxable year beginning after December 31, 1962, every person who is a United States shareholder (as defined in subsection (b)) of such corporation and who owns (within the meaning of section 958(a)) stock in such corporation on the last day, in such year, on which such corporation is a controlled foreign corporation shall include in his gross income, for his taxable year in which or with which such taxable year of the corporation ends—

"(A) the sum of—

(i) except as provided in section 963, his pro rata share (determined under paragraph (2)) of the corporation's subpart F income for such year, and

(ii) his pro rata share (determined under section 955 (a) (3)) of the corporation's previously excluded subpart F income withdrawn from investment in less developed countries for such year; and

(B) his pro rata share (determined under section 956(a) (2)) of the corporation's increase in earnings invested in United States property for such year (but only to the extent not excluded from gross income under section 959(a) (2)).

"Subpart F income is further defined as:

"Income from insurance of United States risks
Foreign personal holding company income
Foreign base company sales income
Foreign base company service income."

Foreign personal holding company income can be loosely described as investment income not related to the active conduct of a business. Interest on foreign bank accounts would be a good example. Foreign base company sales income is highly complex. Construction activities are treated under the regulations as being foreign base company service income, consequently it is unlikely that the foreign base company sales income will be involved.

Foreign base company service income is income from services performed for a related person and performed outside the country in which the foreign corporation is organized. An example of Subpart F income arising from construction activities would be when a U.S. corporation enters into a contract to construct a dam in, say, Australia, and assigns the contract to a subsidiary organized under the laws of a foreign country other than Australia. This result could be avoided by assigning the contract to an Australian subsidiary for performance.

There is a saving provision—if subpart F income is less than 30 percent of the income of the foreign corporation, there is no subpart F income. The reader is cautioned that this is an extremely loose account of subpart F. It is not meant to be definitive or exhaustive but merely to point to a potentially troublesome area.

Not to be outdone by Congress, in 1968 the Commissioner of Internal Revenue produced his own nightmare in the guise of Regulation 1.482. Section 482 gives the Commissioner the power to allocate income, deductions, credits, and allowances among taxpayers under common control where necessary to prevent tax evasion or to clearly reflect taxable income of the several taxpayers. These regulations cover about 23 printed pages and supply just about everything except practical answers to practical questions.

It is particularly attractive for a Revenue agent to make adjustments which transfer deductions from the returns of the U.S. parent to a foreign subsidiary, because he can win two ways. First, he gets a deficiency as to the parent, and by decreasing the foreign source income may also decrease the allowable foreign tax credit. It can also happen that under foreign tax laws the amounts disallowed in the United States are not allowable on the return of the foreign corporation, making for the worst of all possible worlds.

Tax treaties which we have with most of the developed countries provide that where deductions are not allowed in either country or the same income is taxed by both countries the so-called "Competent Authorities" of each country will, upon request of the taxpayer, get together and resolve the problem of double taxation in some equitable manner. Where there are no tax treaties, this relief is not available.

28-17. Necessity of Making Fair Charges to Foreign Subsidiaries for Services Rendered

In the circumstances the prudent course is to make sure that foreign corporations receive a fair charge from domestic parents for services rendered, goods sold, and general overhead. The form which the charge takes is important also. Canada and Japan, for instance, do not take kindly to management fees. They will, however, accept a charge for home-office overhead allocated on a fair basis, such as sales volume. Since tax rates in many foreign countries are now as high or higher than in the United States, there may be a tax saving as the result of making a fair and careful allocation of costs and expenses to foreign corporations.

28-18. Western Hemisphere Trade Corporations

The income of a Western Hemisphere Trade Corporation is taxed at a preferential rate as compared with a regular domestic corporation. Taxable income is reduced by 14/48, or 29 1/6 percent. The effective tax rate is therefore 34 percent. A Western Hemisphere Trade Corporation is a domestic corporation all of whose business, except incidental purchases, is done in North, South, or Central America or in the West Indies, and which derived 95 percent or more of its gross income for the preceding three taxable years (or such shorter period as the corporation was in existence) from sources outside the United States. Further, 90 percent of its gross income for the three-year or less period must have been from the active conduct of business.

Thus, a contract to be performed in the Western Hemisphere, when no substantial purchases are made outside the Western Hemisphere, can be performed by a Western Hemisphere Trade Corporation, and a tax saving will result if the effective rate of foreign income is less than 48 percent.

The 1971 Revenue Act provided for a deferment of the taxation of income of a "Domestic International Sales Corporation," commonly called a DISC. (Secs. 991-994 I.R.C.) These provisions seemingly cannot be used with respect to performance of construction contracts, but may possibly be used for performance of allied activities, such as consulting engineering and supervision of construction.

28-19.   Dangers of Joint Ventures in Certain Countries

The form of doing business in foreign countries can be important. For example, one South American country would tax a joint venture as a corporation, and would again tax each venture partner on its share of income from the venture as a dividend. In such a country a joint operation should be carried out by means of a jointly owned corporation.

28-20.   Problems of Income. Taxation of American Employees in Foreign Countries

Income taxation of employees in foreign countries can also be important. The U.S. employer cannot expect employees, particularly managers and other highly compensated employees, to go to a foreign country and suffer a net loss of take-home pay because of high foreign income taxes. And as highly as we think we are taxed here, there are many foreign jurisdictions where personal income taxes, particularly at higher income levels, are materially higher than ours. If these taxes cannot be avoided, the additional pay required to compensate for the higher taxes must be considered in estimating job costs.

Sometimes they can be avoided. The United Kingdom, for example, taxes an individual on income earned and received there. A good strategy is to limit an employee's pay which he will actually receive in the U.K. to an amount which will permit the U.K. tax to be fully credited against the employee's U.S. tax as a foreign tax credit. The employer should lend the employee the additional amount which is expected to be required for living expenses in the U.K. and have him take that money with him to the U.K. The balance of his salary would be paid to the employee in the U.S. and could in part or in total be applied to reduce the loan. In this way the employee's income tax will be no more than if he had earned the money in the U.S. and the employer will not suffer the additional expense of defraying excessive foreign taxes. Similar arrangements may work in other countries.

Before bidding a foreign job in an unfamiliar country, the possibility of wage and salary increases to cover per-

sonal income taxes of employees should be thoroughly explored. Foreign payroll taxes and import duties can be significant in amounts and should be carefully checked out before bidding on a foreign project.

There are U.S. tax advantages for U.S. citizens who are physically present outside the United States for 510 days out of eighteen months, and for citizens who become resident abroad. These advantages should be exploited, where possible, as an aid to recruiting and retaining competent foreign service personnel, and to reduce the employer's costs where reimbursement of excess foreign income taxes are involved.

STATE AND INTERSTATE TAXES

28-21.   State Taxes

Forty-four of the fifty states and the District of Columbia levy taxes on or measured by income. It is fair to say that there are forty-four different systems for determining taxable income and, until recently, that many different systems for determining the amount of income of a multistate business taxable by each state.

The states which do not have taxes measured by income are Florida, Nevada, Ohio, South Dakota, Texas, Washington, and Wyoming. Proposals for enactment of an income tax have recently been made in Florida, Ohio, and Washington. All of the foregoing states except Nevada levy a capital stock tax or similar tax based on capital employed within the state.

Congress has been threatening to enact legislation which will prescrible and limit state taxation of interstate business. Several bills have passed the House, but they died for lack of action by the Senate.

28-22.   Interstate Tax Bills, Allocation of Income

In an effort to avoid federal legislation prescribing rules for state taxation of income of multi-state businesses, many of the states have recently adopted a "Uniform Division of Income for Tax Purposes Act," which is model legislation developed by tax administrators from several of the states. The purpose of the Act is to provide rules for apportioning income of a business operating in more than one state to the several states in which business is conducted in such a way that no more and no less than the total income of the business will be allocated to some state. Not all states adopted the Act verbatim; some adopted it in substance with minor modifications.

Many of the states have also adopted the "Multistate Tax Compact" which incorporates in substance the provisions of the Uniform Division of Income Act.

In drafting the material regarding multistate taxation, attention was centered on businesses which manufacture and sell, or buy and sell, property. Scant attention seems to have been devoted to service businesses or to the situation of construction contractors. In writing regulations construing legislation, state tax administrators may produce rules that do not fairly fit construction contractors. It may take some period of time, and perhaps litigation, to evolve firm and equitable rules for taxation of income of contractors doing business in several states.

Table 28-22-1 tabulates the states which had adopted the Uniform Act as of March 30, 1971.

The first problem which comes up in allocation of

TABLE 28-22-1. STATE ADOPTION OF THE UNIFORM DIVISION OF INCOME ACT PROVISIONS (as March 30, 1971)

State	Adopted the Provisions	Special Note
Alabama	Yes	2
Alaska	Yes	1
Arizona	No	2
Arkansas	Yes	1
California	Yes	2
Colorado	Yes	1
Connecticut	No	
Delaware	No	
District of Columbia	Yes	
Florida	Yes	1
Georgia	No	
Hawaii	Yes	1
Idaho	Yes	1
Illinois	Yes	1
Indiana	Yes	1
Iowa	No	
Kansas	Yes	1
Kentucky	Yes	
Louisiana	No	2
Maine	Yes	
Maryland	No	2
Massachusetts	No	2
Michigan	Yes	1, 3
Minnesota	No	2
Mississippi	No	
Missouri	Yes	1
Montana	Yes	1
Nebraska	Yes	1
Nevada	Yes	1
New Hampshire	No	
New Jersey	No	2
New Mexico	Yes	1
New York	No	
North Carolina	Yes	
North Dakota	Yes	1
Ohio	No	
Oklahoma	Yes	
Oregon	Yes	1
Pennsylvania	Yes	2
Rhode Island	No	
South Carolina	Yes	
South Dakota	No	2
Tennessee	No	2
Texas	Yes	1
Utah	Yes	1
Vermont	No	
Virginia	Yes	2
Washington	Yes	1
W. Virginia	No	2, 4
Wisconsin	No	
Wyoming	Yes	1

KEY:
1. State has adopted the more comprehensive *multistate compact* (which includes the *uniform division of income act* provisions).
2. State is only an "Associate" member of *multistate compact.*
3. Michigan also applies the provisions to franchise and intangibles tax.
4. Virginia only adopted the part of the provisions relating to nonbusiness income.

income of an interstate business among the various states is whether the income shall be apportioned directly and totally to the state in which the work was performed, or whether total income of the company shall be allocated among the various states by use of allocation formulas. The former is proper if the company is not considered to be engaged in a "Unitary Business," and the latter is proper if it is a unitary business.

A unitary business may be carried on by a single taxpayer or maybe conducted by a group of related companies. The test of unity is where there is a commonality of ownership and one part or parts contribute to or benefit from activities of other parts of the business. Such things as centralized management, purchasing, accounting, financing, insurance, selling and bidding, planning, common business systems, the flow of goods between units, etc., are indicia that the business is unitary. The State of California treated contractors as not unitary businesses before 1967, and as unitary thereafter. Kentucky, however, takes the opposite position. It is hoped that the states subscribing to the MultiState Compact will eventually develop a single rule.

In connection with preparation of this chapter, the author requested information regarding state taxation of contractors directly from the administrators for the states. A tabulation of responses is given in Table 28-22-2.

As these responses indicate, there is little uniformity among the states in the taxation of interstate construction activities. Even among the states which have adopted the Uniform Division of Income Act, there is disagreement as to the proper method of allocating income to the taxing state. The positions adopted run the spectrum of possible alternatives—Kentucky, even after adopting the Act, ordinarily requires separate accounting; Kansas, which, prior to the adoption of the allocation provisions of the Uniform Act, required separate accounting, now does not have a clear position; California, which, prior to the adoption of the Act, required separate accounting, now requires allocation under the Act's three factor method.

With numerous variations, the allocation and apportionment methods which predominate are "the three-factor formula" advanced by the Uniform Division of Income Act, and the "separate" or "direct accounting" method.

The approach of the Uniform Division of Income Act is to first require the taxpayer to segregate his "nonbusiness income" from his "business income" ("business income" being income arising from the regular course of the taxpayer's trade or business, and all other income being "nonbusiness income"). The significance of this characterization is that "nonbusiness" income is generally not subject to the apportionment rules, but is directly attributed to the state where generated. The taxpayer's "business income" is apportioned to the taxing state by multiplying his total income by a fraction, the numerator of which is the "property factor" plus the "payroll factor" plus the "sales factor," and the denominator of which is three. The individual "property," "payroll," and "sales" factors are the ratios of the taxpayer's in-state "property," "payroll," and "sales" as compared with total "property," "payroll," and "sales" everywhere.

The "separate accounting" method, on the other hand, would require the taxpayer to compute his taxable income by reducing his gross receipts from jobs in the taxing state by direct jobs costs attributable to jobs in taxing state, and then reducing this figure by the portion, of his indirect job costs everywhere, that his gross receipts in the taxing state bear to his gross receipts everywhere.

ANALYSIS OF STATE TREATMENT OF CONSTRUCTION CONTRACTORS
DOING BUSINESS INTERSTATE

STATE	Does the State consider construction contracting to be a "Unitary Business"? What method of apportionment is to be used?	Does the State allow construction contractors to report on the completed contract basis?	Does the State prescribe any special allocation method when the completed contract basis is used?	SPECIAL NOTES
ALABAMA	Depends on facts	YES	NO	
ALASKA	YES – 3 factor formula (property, payroll, sales)	YES	NO	Favor separate accounting – 3 factor formula should be used only as second resort.
ARIZONA	Depends on facts	YES – but favor percent of completion method as more clearly reflecting income	If completed contract method is allowed, separate accounting is ordinarily required for construction contractors not operating a Unitary business. If it is found to be Unitary, 3 factor formula is required.	Favor separate accounting for the construction industry.
ARKANSAS	YES	YES	NO – but see SPECIAL NOTES.	Historically, contractors could use the direct accounting method, but Arkansas adopted the Uniform Division of Income Act, and contractors are not specifically exempt -- State is trying to put all contractors on an apportionment basis. Waiting for Multi-State Tax Compact regulations.
CALIFORNIA	YES – 3 factor formula	YES	YES – see SPECIAL NOTES.	Used separate accounting until 1967.
COLORADO	YES – 3 factor formula	YES	YES – Director has power to prescribe special formulas wherever he thinks it is justified. See SPECIAL NOTES.	Generally, where completed contract is adopted, the contractor must recognize percent of completion payments, wages paid, property owned or leased (8 times annual rent).
CONNECTICUT	YES – 3 factor formula	YES	NO	
DELAWARE	YES – 3 factor formula	YES	NO	
DISTRICT OF COLUMBIA	YES – 3 factor formula	YES	NO – see SPECIAL NOTES.	In certain cases, where the contractor does business in the Washington, D.C. metropolitan area (i.e., District, Maryland, Virginia), a single apportionment factor (gross receipts) is allowed if it is required by Maryland and Virginia.

FLORIDA –
No corporate
income tax

TABLE 28-22-2

STATE	Does the State consider construction contracting to be a "Unitary Business"? What method of apportionment is to be used?	Does the State allow construction contractors to report on the completed contract basis?	Does the State prescribe any special allocation method when the completed contract basis is used?	SPECIAL NOTES
GEORGIA	YES - but prefer reporting by separate accounting and completed contract method	YES	YES - prefer separate accounting with general overhead and administrative expenses to be apportioned by ratio of gross receipts within Georgia to gross receipts everywhere.	
HAWAII	YES - 3 factor formula (income, property, payroll)	YES	NO (just allocate on basis of actual work completed)	
IDAHO - see SPECIAL NOTES				Does not say except that it will follow regulations of Multi-State Tax Compact, when issued.
ILLINOIS	YES - unless contractor can show that his contracts are separable into units each constituting a unitary business - 3 factor formula	YES	NO	
INDIANA	YES - 3 factor formula (see SPECIAL NOTES)	YES	YES - if completed contract method is used, the cash method must be used for gross income tax purposes.	Corporations doing business in Indiana are subject to the greater of gross income tax or adjusted gross income tax - allocation only applies to the adjusted gross income tax, not to the gross income tax.
IOWA	YES - apportion on basis of sales or business activity within Iowa	YES	NO	
KANSAS	Open - see SPECIAL NOTES	YES	See SPECIAL NOTES	Law provides for, and Director prefers, direct accounting. However, state has adopted Uniform Division of Income Act which allows apportionment on 3 factor formula... favor the use of direct accounting for construction contractors, i.e., gross Kansas income less Kansas expense, with overhead computed by using costs in Kansas as compared to total costs in all states.
KENTUCKY	Depends on facts - see SPECIAL NOTES	YES	YES - separate accounting is mandatory	Requires that separate accounting be used except where the Director determines that this is not practical. Note that even where separate accounting is used, the Director must first be notified.
LOUISIANA	NO	YES	NO	
MAINE	YES - 3 factor formula	YES	NO	

TABLE 28-22-2 (Continued)

STATE	Does the State consider construction contracting to be a "Unitary Business"? What method of apportionment is to be used?	Does the State allow construction contractors to report on the completed contract basis?	Does the State prescribe any special allocation method when the completed contract basis is used?	SPECIAL NOTES
MARYLAND	YES - single factor formula (see SPECIAL NOTES)	YES	NO	Business income allocable to Maryland is allocated by applying the factor of stipulated contract price for the work to be performed or gross receipts received by the taxpayer for work performed in Maryland, to total stipulated contract prices or total gross receipts, for all contracts, whichever result is less.
MASSACHUSETTS	Depends on facts - if found to be Unitary, corporation can use 3 factor formula	YES	YES - separate accounting required (see SPECIAL NOTES)	Reporting is on the basis of Massachusetts gross receipts less deductions directly attributable to the Massachusetts contract. Deductions for indirect overhead must be taken on the basis of a formula approved by the bureau.
MICHIGAN	YES - 3 factor formula	YES	NO	Although Director expects contractors to use the 3 factor formula, several contractors have been allowed to use separate accounting.
MINNESOTA	NO - use separate accounting	YES	NO	
MISSISSIPPI	NO - direct accounting (see SPECIAL NOTES)	YES	NO	Reporting is on basis of direct accounting with allocation of administrative and overhead expenses by the ratio that this state's direct job cost bears to total direct job costs.
MISSOURI	YES - apportion net income by receipts within Missouri to total receipts	YES	NO	
MONTANA	YES - 3 factor formula	YES	NO	
NEBRASKA	YES - 3 factor formula	YES	NO	
NEVADA - No corporate income tax				
NEW HAMPSHIRE	YES - 3 factor formula	YES	NO	
NEW JERSEY	YES - 3 factor formula	YES	NO	
NEW MEXICO	YES - 3 factor formula	YES	NO	
NEW YORK	YES - 3 factor formula	YES	NO	
NORTH CAROLINA	YES - 3 factor formula	YES	NO	

TABLE 28-22-2 (Continued)

STATE	Does the State consider construction contracting to be a "Unitary Business"? What method of apportionment is to be used?	Does the State allow construction contractors to report on the completed contract basis?	Does the State prescribe any special allocation method when the completed contract basis is used?	SPECIAL NOTES
NORTH DAKOTA.	Varies with situation, usually require separate accounting	YES	NO	
OHIO - No corporate income tax				
OKLAHOMA	YES - 3 factor formula	YES	NO	
OREGON	YES - 3 factor formula, but see SPECIAL NOTES	YES	NO	The Oregon Supreme Court, in Utah Construction and Mining Co. (2-28-70), 90 Adv. 467, allowed the construction contractor to use the segregated method of reporting to determine net income attributable to Oregon.
PENNSYLVANIA	YES - 3 factor formula (see SPECIAL NOTES)	NO	NOT APPLICABLE	Only that portion of gross receipts from Pennsylvania will be included in the sales factor as the direct and indirect costs incurred in the state under the contract bear to the total costs incurred under the contract.
RHODE ISLAND	YES - 3 factor formula	YES	NO	
SOUTH CAROLINA	YES - 3 factor formula	YES	YES - the Director requires that separate accounting be used. The computation would be to apportion indirect overhead to gross job profit realized in South Carolina by using the ratio of South Carolina gross receipts to gross receipts everywhere. The gross job profit less the apportioned overhead, results in the net taxable income to South Carolina.	
SOUTH DAKOTA - No corporate income tax				
TENNESSEE	NO - must use separate accounting	YES	NO	
TEXAS - No corporate income tax				

TABLE 28-22-2 (Continued)

STATE	Does the State consider construction contracting to be a "Unitary Business"? What method of apportionment is to be used?	Does the State allow construction contractors to report on the completed contract basis?	Does the State prescribe any special allocation method when the completed contract basis is used?	SPECIAL NOTES
UTAH	YES - 3 factor formula	YES - but see SPECIAL NOTES	NO	Since the Utah tax is a prepaid tax, corporations are permitted to use the completed contract method only if they first agree with the tax commission that they will report the income from all contracts in Utah which begin prior to their date of withdrawal from Utah.
VERMONT	YES - 3 factor formula	YES	NO	
VIRGINIA	YES - 3 factor formula	YES	NO	
WASHINGTON - No corporate income tax				
WEST VIRGINIA	YES - apportion using 2 factor (payroll and property) formula	YES	NO	
WISCONSIN	Varies - see SPECIAL NOTES	YES	NO	Will ordinarily allow either separate accounting or apportionment. If apportionment is chosen, the corporation can use either a two factor formula of revenue and job costs, or a three factor formula (including tangible property if it was a material income producing factor). If separate accounting is permitted, general overhead items are normally allocated to jobs within the state on the basis of direct job costs.
WYOMING - No corporate income tax				

TABLE 28-22-2 (Continued)

28-23. Other State and Local Taxes

In addition to income taxes there are numerous other state and local taxes about which a contractor must have knowledge in estimating costs to perform.

States which have some type of gross receipts tax which would apply to construction contracts are:

Alaska Business License Tax—based on the gross receipts of any person or corporation engaged in any business.

Arizona Occupational Gross Income Tax—every person engaging or continuing in business in Arizona pays a tax based on gross income.

Hawaii General Excise (Gross Income) Tax—a privilege tax is levied on the gross income on account of their business and other activities in Hawaii.

Indiana Gross Income Tax—gross income of all persons or corporations doing business in Indiana, derived from sources within Indiana, are subject to tax.

Mississippi Occupational Sales Tax—imposed on every person engaging in business.

New Mexico Gross Receipts Tax—on persons "engaging in business in New Mexico."

West Virginia Occupational Gross Income Tax—levied on all persons and corporations for the privilege of engaging in business; based on gross income.

Washington Business Occupation Tax—imposed for the act or privilege of engaging in business activities.

Some cities levy gross receipts taxes (San Francisco and Los Angeles, for example) and San Francisco recently enacted a payroll tax. It can be expected, in view of the dire financial plight of the major cities, that taxes of these sorts, as well as city income taxes, will become more common in the future.

Most states have sales or use taxes which will be levied on materials purchased, equipment purchased, and in some cases on equipment brought into the state for use on a contract. Property situated most anywhere, such as inventory and equipment, is subject to local property taxes. Care should be taken to make sure whether the contractor or the owner is taxable on the value of construction in progress.

Payroll taxes are levied by all states, but rates may vary as between states and within states depending on circumstances.

29 CONTRACTORS' INDUSTRIAL INSURANCE

JOHN C. MOORE

Executive Vice President
Marsh and McLennan
Chicago, Ill.

PROTECTION against loss which could impair the assets of a contractor is important. One of the ways is through the sound use of insurance. No prudent construction firm should attempt any construction activity without the advice and counsel of a knowledgeable, professional insurance advisor who is experienced in arranging the necessary insurance protection.

The main factors to be considered in arriving at a decision as to what insurance should be purchased, and for what amounts and limits, are as follows:

1. What are the obligations of the contractor under the contract for loss of or damage to the work and injury or damage to third parties?
2. What are the probabilities of loss inherent in the particular type of project? For example, the construction of a dam represents a far more serious exposure for loss or damage to the work than does the construction of an office building.
3. The geographical location of the project. Is it in the center of a highly populated or sparsely populated area?
4. What is the exposure to specific perils such as earthquake, flood, etc?
5. Will the major part of the work be done by subcontractors?
6. The comparative costs of a full insurance program vs. self-insurance, and/or deductible forms of insurance.

The principal forms of insurance protection fall within the following classes:

1. Workmen's Compensation and Employer's Liability—Injury or death to employees
2. Public Liability and Property Damage—Injury or death to members of the public and damage to their property
3. Automobile
4. Builders Risk—Loss or destruction to work in progress prior to owners acceptance
5. Equipment Floater—Loss or destruction to contractors equipment
6. Extra Expense and Business Interruption
7. Fidelity and Forgery

29-1. Workmen's Compensation and Employer's Liability

Each of the fifty states, the District of Columbia, and all the Canadian Provinces have Workmen's Compensation laws requiring payment for compensable death, lost time, and medical benefits sustained by employees injured on the job. With few exceptions, the most economical way to discharge these obligations is by transferring them to an insurance company, or to state or Provincial Fund where private insurance is not permitted. The purchase of insurance from one of the above sources discharges the obligations of the employer under the Workmen's Compensation Statutes.

Basic Coverge There are two basic coverages afforded by the insurance policy. These are: (1) Workmen's Compensation, and (2) Employer's Liability.

The Workmen's Compensation section agrees to assume the liability imposed upon the insured by the Workmen's Compensation Law. The insurance policy defines "Workmen's Compensation Law" as the law of any state designated in the policy declarations and includes any Occupational Disease law of that state. In most cases, filling in the name of the state or states in which coverage is to apply is all that is necessary.

The Employer's Liability section protects the insured against the liability imposed by law for injury to his employees which injury arises out of and in the course of their employment. The insuring clause of this section is similar to other liability insuring clauses. It states that the insurance company will pay damages which the employer is legally obligated to pay because of "bodily injury by accident or disease, including death at any time resulting therefrom . . . by any employee of the insured arising out of and in the course of his employment by the insured either in operations in . . . the declarations or in operations necessary or incidental thereto."

The Workmen's Compensation Section covers the statutory limits of the applicable states. The insurance company is simply required to take over the entire liability of the insured under the Workmen's Compensation Laws of State(s). . . .

The Employer's Liability section in most states is subject to a limit per accident. The basic limit is $100,000. This limit may be increased subject to an additional premium. Consideration should be given to a minimum limit of $500,000.

The insurance policy restricts protection to the state law or laws listed in the policy declarations. To some degree most state laws are extraterritorial, applying to injuries sustained outside the state, but having some connection with operations in that particular state.

Endorsements for Additional Coverage   Depending on the operations of the employer and the states or locations where such operations are conducted, there are a number of endorsements which may be issued to the insurace policy that extend additional protection:

All States   This is intended to provide automatic coverage against liability under the Workmen's Compensation laws of states not listed in the policy declaration, where such protection may be lawfully provided.

Voluntary Compensation   The primary purpose of this extension of coverage is to secure Workmen's Compensation benefits for injured employees who may be outside a particular law or whose status may be doubtful.

Extra-Legal Medical   The Workmen's Compensation laws of some states place limitations upon the amount of medical benefits which the employer must furnish to an injured or disabled worker. Others place a limit as to the length of time over which medical treatment may be extended. Some place limitations on both amount and time. Where limitations are imposed the policy covers only the statutory limits. These limited benefits will sometime fall short of providing an adequate amount to cover all costs. To provide for situations like this, it is possible to purchase additional medical benefits coverage called Extra-Legal Medical. This is intended to pay medical and allied expenses over and above those to which the workman is entitled under the laws.

Other Workmen's Compensation Acts   If the construction work involves maritime or special United States Government operations, certain endorsements are required to cover the employer's legal requirements. Subject to the exposure(s) involved, these would include protection for:

- Longshoremen and Harbor Workers Act
- Outer Continental Shelf Lands Act
- Defense Bases Act
- Jones Act and High Seas Death Act

Direct and Indirect Costs of Injury   The principal costs of an industrial injury include direct and indirect costs:

Direct costs include:
- Medical payments
- Indemnity payments

Indirect costs include:
- Time lost from work by the injured employee
- Lost time by fellow workmen
- Loss of efficiency resulting from breakup of work crew
- Loss in earning power
- Many others that may develop from an accident

Insurance companies, including the state or Provincial funds, have available certain premium rating plans from which a contractor can select, so he can purchase the protection on the plan best suited to his needs and that which will reflect the most favorable cost.

Premium Rate Modification   A contractor is entitled to a premium rate modification based on his own experience as distinguished from the experience of an entire class within his state. This may reflect a credit or debit from the manual or published rates.

There are four states that use intrastate (one state only) rating, but do not allow interstate (more than one) rating. The following minimum premium eligibility requirements are needed:

California	$3000 at manual rates
Delaware	$1500 at current manual rates
New Jersey	$1600 at current manual rates
Pennsylvania	$2250 at current manual rates

If eligible, the remaining states (excluding monopolistic fund states) permit interstate rating. To qualify, the risk must have a premium at manual rates in one state of at least $1500 in the last year, or an average of $750 for the last two or more years of the experience period. In Oregon and North Carolina the requirement is $1000 in the last year or an average of $500 for the last two or more years.

The experience period used to make the computation is not less than one year and not more than three years, commencing four years and ending one year before the experience rates are to be established.

The following is an example using the three-year period:

Policy period	1/1/70–1/1/71
Remove one year	1/1/69–1/1/70
These years are used to develop the	1/1/68–1/1/69
experience modification for the	1/1/67–1/1/68
policy year 1/1/70–1/1/71	1/1/66–1/1/67

The reason for removing the latest year is that, the reserves are not sufficiently seasoned to afford equitable treatment in the computation of the experience rate.

On-the-Job Injuries in Foreign Lands   Operations in foreign lands involve a whole new obligation as respects on-the-job injuries to employees, and these obligations vary by country. Questions of responsibility for injuries or death occurring during transportation to the job, extension of state-side coverage to American nationals hired for the job, compliance with all foreign government requirements for the nationals at the job site, including their transportation, cost of repatriation, and similar problems must be examined and a proper program of protection provided by a knowledgeable advisor in this field before work is commenced.

Safety   An effective way to reduce cost is through accident prevention. An efficient workable plan will materially assist in reducing loss-time accidents. It should be tailored to your operations. Your broker and insurance company should present the program to be used. Effective results from the program can only be achieved by the management of the contractor seeing that the program is installed with continuity of follow-up.

A number of insurance brokers and many insurance companies and state funds have trained safety engineers who are available to work with the contractor in establishing a practical accident prevention program.

One insurance company offers the following steps to an effective Accident Prevention Program:

1. Safety Committees—One of the most effective accident prevention measures is the organization of a safety committee. Safety committees normally
   - meet on a monthly basis
   - review and approve safety inspection reports
   - review and discuss all safety recommendations made to the commitee
   - study the cause of accidents
   - set up systems to educate employees in the hazards of their work and in safe practices
2. Machine Guarding—Over a period of time, safety standards have been developed for the most effective methods of guarding machines—thus preventing injuries to employees. Safety engineers will advise on the most effective way to guard machinery, while maintaining or improving the effectiveness of operations.
3. Preconstruction Planning—Sometimes safety can be built in. Preconstruction planning during the design and construction of a building can provide a program for
   - fire protection
   - workmen's safety
   - protection of the public
   Safety engineers can assist in the blueprint stage and follow each step of construction until the project is completed.
4. Material Handling and Work Processes—These are areas where numerous accidents occur, but through careful training of employees in the safe techniques of material handling, lifting, storage, fork-lift truck operation, work production, and other related fields, accidents can be avoided.
5. On-the-job Inspections—Safety engineers conduct inspections of plant facilities or construction sites to make suggestions to management on
   - housekeeping
   - machine guarding
   - use of personal protective devices
   - work hazards
6. Occupational Disease Hazards and Exposures—These special exposures and hazards are analyzed and recommendations are made for the prevention of
   - hearing losses
   - pneumoconiosis
   - radiation exposures
   - chemical exposures
   - toxic contacts

7. Safety Aids—Company accident prevention departments make available many safety materials at no cost to the policyholder. These safety aids include
   - supervisory safety material for those who supervise other workmen
   - safety instruction pamphlets for distribution to workmen in major industries
   - safety posters and warning signs for posting on bulletin boards and in work areas
   - safety films and slides
8. Personal Protective Devices—The use of personal protective devices is probably one of the simplest—and yet most effective—methods of preventing injuries. Safety engineers consult with supervisory personnel on the various items of protective equipment and the way to obtain maximum value from such items as gloves, hard-toed shoes, hard hats, ear protectors, masks, and safety glasses.
9. Accident/Loss Record—As a special aid in directing accident prevention programs, companies prepare an Accident/Loss Record which has been designed to help you determine the
   - frequency of accidents
   - severity of accidents
   - accident prone employees
   - recurrence of same type of accident
   - estimated loss experience
10. Supervisory Safety Training—Research has proven that 8 out of 10 job injuries could have been prevented through supervisory action. Available are special comprehensive instruction courses to your key men of the insured on the fundamentals of accident prevention.

29-2.   Public Liability—Injuries to Members of the Public or Damage to their Property

This type of insurance is most critical for anyone in the construction industry because the possible financial losses can be ruinous. Courts and juries are unpredictable and claimants are quick to sue.

It is essential that insurance contracts be properly written to cover the exposures involved with limits sufficiently substantial to protect adequately the assets of the contractor.

Liability   This form of coverage agrees to pay on behalf of the insured all sums which the insured becomes legally obligated to pay if the loss is because of bodily injury or property damage caused by an accident or occurrence. The policy also agrees to defend the insured in any suit even if groundless, false, or fraudulent. Basically, the policy applies to the following:

- Contracting operations
- The ownership, operation, or use of premises including elevators
- Liability of the insured arising out of operations conducted by subcontractors of the insured
- Liability assumed in contracts or agreements as defined in the policy
- Automobiles

Since no one can predict the source of an injury which can result in an award of substantial damages, and since awards of $500,000 or more for injuries or death to one person are no longer uncommon, the purchase of substantial limits of liability is recommended. The general practice is to have a policy arranged to cover the construction exposures, and the owned and non-owned automobile exposures, including all others involved in the project work. The premium cost is dependent upon the protection afforded and the limits of liability provided.

The selection of proper limits of liability is a matter of judgment. These may be arranged under one or two separate policies. If the two-policy arrangement is selected, the first policy is referred to as the primary; the second, an excess or excess umbrella contract.

Under the two-policy form, an adequate primary limit should be provided with the excess limit selected after a full review of the exposures and consultation with your insurance advisor.

Umbrella Form of Coverage A properly-written umbrella form not only provides excess limits over the primary (including replacing exhausted primary property damage limits because of an aggregate limitation) but also will step down to provide primary insurance with a reasonable deductible where coverage for some reason is not provided by the primary insurance. One of the main points to watch for is that the umbrella not only contains the obligation to pay for losses, but also provides for defense of such excess claims.

Bodily Injury The Bodily Injury Insuring Agreement of a liability policy should be amended by endorsement to cover claims arising from the following:

False Arrest Any unlawful physical restraint of a person's liberty, whether in prison or elsewhere.

Malicious Prosecution A judicial proceeding instituted against a person, out of the prosecutor's malice and ill will, with the intention of injuring him without probable cause, and which terminates in favor of the person prosecuted.

Willful Detention The act of keeping back or withholding unlawfully and intentionally a person or thing.

Willful Imprisonment Any forcible detention of a man's person or control over his movements, either unlawfully or intentionally.

Libel That which is written or printed and published, which is done for the purpose of injuring the character or the reputation of another by bringing him into ridicule, hatred, or contempt.

Slander The speaking of falsehoods concerning another which results in injury to his reputation.

Defamation of Character The taking of one's reputation or the offense of injuring a person's character, name, or reputation by false and malicious statements.

Privacy The right of an individual (or corporation) to keep himself and his property from public scrutiny if he chooses.

Wrongful Eviction The unlawful removal of a person from a premise for which he has lawful right to the use or occupancy thereof.

Wrongful Entry An unauthorized intrusion by a person to a premise which interferes with the rightful person's use thereof.

Special Considerations Certain sections of the liability policy merit special attention for the reason they are often misinterpreted. These are as follows:

Definition of Occurrence The insuring agreement states that the insured will pay on behalf of the insured all sums which the insured has become legally obligated to pay as damages because of bodily injury or property damage . . . caused by "occurrence." Occurrence is defined as an accident which results in bodily injury or property damage *neither expected nor intended from the standpoint of the insured*. Thus, if an accident occurs, failure to correct the work method or eliminate the condition which caused the accident might permit the insurer to deny coverage for a second accident on the grounds that bodily injury or property damage resulting from the second accident could logically have been expected.

In the Event of an Accident or Occurrence The insured is required, among other things, to promptly take at his own expense all reasonable steps to prevent other bodily damage arising out of the same or similar conditions.

Contractual Liability The standard comprehensive liability policy provides coverage with respect to liability for bodily injury and damage to property of others which liability is assumed in an incidental contract or agreement with another.

An "incidental" contract means any written:

1. Lease of premises
2. Easement agreement except in construction or demolition operations on or adjacent to a railroad
3. Indemnification of a municipality required by a municipal ordinance except in connection with work for the municipality
4. Side track agreement
5. Elevator maintenance agreement

It can clearly be seen that the definition of "incidental contract" does not embrace construction contracts, and for this reason it is necessary to broaden the definition to cover the liability assumed by the contractor in such contracts. Most insurance companies are agreeable to broadening the policy in this respect.

Indemnification or hold harmless agreements vary in scope depending upon a number of circumstances including the relative bargaining position of the parties to the contract. In general, these agreements fall into one of the following three categories:

1. Limited Form of Indemnification Under this form the contractor agrees to indemnify or hold harmless the other party to the contract with respect to claims or damages arising out of the negligence of the contractor or his subcontractors.

2. Intermediate Form of Indemnification This type requires the contractor to indemnify or hold the other party harmless with respect to claims or damages arising out of the operations of the contractor or subcontractor even though the other party may be jointly negligent.

Broad Form of Indemnification The "broad" form requires the contractor to indemnify or hold the other party harmless even though the injuries or damage was caused by the sole negligence of the other party.

It must be kept in mind that if the insurance company is agreeable to broadening the policy to cover with respect to liability assumed in all written agreements, this extension is subject to all other terms, conditions, and exclusions of the policy, and consequently it may not cover all of the liability assumed.

If the insurance company is not agreeable to a broad extension of the policy so that, for example, only the limited or intermediate forms are automatically covered, it would be advisable to submit *all* agreements to the insurance company. This procedure will eliminate a dispute as to whether or not an agreement falls into the category for which automatic coverage is provided.

XCU Hazards   The XCU hazards are exclusions in the general liability policy relating to excavating, blasting, underground work, etc. These are defined as follows:

X—The Property Damage Liability rate and minimum premium for the classification (work or job description) exclude coverage for injury to or destruction of any property arising out of blasting or explosion, other than the explosion of air or steam vessels, piping under pressure, prime movers, machinery, or power transmitting equipment.

C—The Property Damage Liability rate and minimum premium for the classification exclude coverage for injury to or desruction of any property arising out of the collapse of or structural injury to any building or structure due:

1. To grading of land, excavation, borrowing, filling, back filling, tunneling, pile driving, coffer dam work or caisson work.
2. To moving, shoring, underpinning, razing or demolition of any building or structure, or removal or rebuilding of any structural support thereof.

U—The Property Damage Liability rate and minimum premium for the classification exclude coverage for:

1. Injury to or destruction of wires, conduit, pipes, mains, sewers, or other similar property, or any apparatus in connection therewith, below the surface of the ground, if such injury or destruction is caused by and occurs during the use of mechanical equipment for the purpose of grading land, paving, excavating, or drilling.
2. Injury to or destruction of property at anytime resulting therefrom.

If your project involves or could involve the type of construction referred to, it is well to inquire about the deletion of these exclusions.

Territorial Definition   The insurance should cover losses during the policy period that occur within the United States of America, its territories or possessions, or Canada.

If foreign work is undertaken, care should be exercised to see that the territorial provision of the policy is sufficiently broad to cover all areas where the operations are being conducted.

29-3. Automobile

Liability   Obligates the insurance company to pay on behalf of the insured all sums that the insured shall become legally obligated to pay as damages because of bodily injury or property damage to which the insurance applies caused by an occurrence or accident.

There are minimum limits to comply with the financial responsibility laws of the various states. These limits are inadequate to meet the needs of a catastrophe settlement(s). More realistic minimum limits would be:

Bodily injury	$100,000 each person
	$300,000 each occurrence
Property damage	$100,000 each accident

Physical Damage—Fire, Theft, Comprehensive, and Collision   Collision is normally written with a deductible.

Medical Payments   Subject to the limit provided, covers all reasonable medical expense incurred within a given period of time from the date of accident to or for each person who sustains bodily injury, caused by accident, while occupying a designated automobile.

29-4. Completed Operations Coverage

Completed operations coverage protects the contractor with respect to claims for damages to bodily injury or property damage which allegedly arose from work completed by the contractor as opposed to those that might arise during the contract period. Operations are considered completed:

1. When all operations to be performed by or on behalf of the named insured have been completed.
2. When all operations to be performed by or on behalf of the named insured at the site of the operation have been completed.
3. When the portion of the work out of which the injury or damage arises has been put to its intended use by any person or organization other than another contractor or subcontractor engaged in performing for a principal as a part of the same project.

29-5. Railroad Protective Liability

Many times the contractor is required to provide a railroad with a special liability policy to pay for claims involving bodily injury or property damage arising out of, or resulting from, the contractor's obligations in and/or around the railroad tracks. To comply with the railroad requirements, this coverage is usually written on a special form. The cost of this insurance will vary, depending upon the exposure or hazards involved.

29-6. Builder's Risk

The contractor's exposure to loss during course of construction is determined by the terms and conditions of the owner's specifications. It is imperative that there be a complete understanding before the job is bid as to who assumes the risk of loss for materials, tools, equipment, and the work itself until final completion and acceptance by the owner.

There should be a complete understanding as to types of loss assumed, as well as what property is involved. Many times the owner will specify the insurance that is to be purchased by the contractor, or he will purchase the insurance for all interests. He may only specify certain types of risk, such as fire and other perils, and if the contract requires the bidder to complete the project and deliver the finished work, the contractor may have the risk of loss for his own account for such perils as collapse, earthquake, flood, and similar losses.

It is extremely important to have a clear agreement as to when the risk of loss passes from the contractor to the owner. Questions of partial payment, partial occupancy by the owner before the project is completed, and other such interim arrangements may affect the protection of the insurance and need to be watched closely.

This type of insurance can be arranged by the contractor on a blanket basis covering all his work at an agreed average premium rate for all jobs based on his volume, or it can be arranged on a specific job as the need arises.

Such insurance is available under two different arrangements. The first is a reporting form, wherein a monthly or progress premium rate is established and the contractor reports to the insurance company his progress and pays a premium on the then value times the rate, gradually building up the amount of insurance as the work progresses.

The other method is the so-called "completed value" form, wherein a policy is written at the start of the job in the amount of its completed value and a premium rate established based on such value for the term of the job. The rate is usually 50 percent of the full term rate on the theory that the loss exposure starts from zero and increases proportionately over the term to the completed value; therefore, the premium will measure the average exposure and no monthly reports of values are required. The limit of loss provided by the insurance does not necessarily have to be the full amount of the contract price unless the owner's specifications require it. The premium rate will usually reflect a lower limit of loss if this is used.

One point to insist on if you purchase the completed value form is that the policy contain a provision that if the insurance company cancels the coverage for any reason in midterm, the earned premium should be calculated pro rata. This would be pro rata on the rate times the value built into the job at the time of cancellation and not calculated on the completed value. The reason being it is not possible in midterm to buy a new policy for pro rata of the entire premium when the major value exists during the last half of the construction period.

This protection may be purchased subject to named perils or what is commonly referred to as all risk. The named peril policy recites specific perils only, such as fire, windstorm, and the like, which is limited protection depending on the type of construction. The all risk form includes loss or destruction from any physical cause but here also, depending on the premium rate, may include or exclude specific perils, such as earthquake and flood.

The wording of these policies differs by company. Care must be exercised to see that the wording of the selected policy form provides the desired protection.

These policies are usually written with a deductible clause, and the rate and premium reflects credits through the use thereof. The deductible amount(s) will vary, depending upon the insurance company and the wishes of the contractor.

A word of caution is suggested about the term of insurance you purchase covering a specific project. If you buy a policy to expire on the date you expect to complete and have the owner accept the project, it may be very difficult to get an extension of the policy for an additional term if the job is not accepted by the anticipated date. The reason is that, the full value of the work is at risk. A better plan is to have the insurance policy recite that it can be extended at the contractor's option on a pro rata basis if acceptance is not had by the normal completion date. Another suggested way would be to purchase the insurance with some leeway with respect to time for completion. This can usually be done at no or a nominal additional premium if arranged when the policy is first negotiated.

Be sure the insurance you purchase includes coverage for materials from the time they are purchased or first become at your risk, wherever that might be, including in transit by any means to the job site.

Even though you purchase the so-called All Risk insurance, there still are hazards which are excluded by the policy. These usually have to do with losses arising out of defective workmanship, faulty materials, and improper design.

During the construction of a project, there are other sources of loss against which insurance may be obtained. When a contract has a penalty for noncompletion on a certain date, this penalty may be insured if the delay is caused by physical damage. How much insurance to buy is a matter of judgment which only experience can provide.

In the case of joint ventures, separate insurance should be arranged. The policies may or may not be extended to include the interest of subcontractors. The circumstances controlling a decision on this are largely governed by the prime contractor and/or the work involved.

29-7. Contractor's Equipment

Construction involves a large investment in equipment and tools of all kinds that are subject to a variety of perils. It is usually mobile and values can fluctuate greatly. Leased equipment, financed equipment, or equipment turned over to a joint venture all have specific requirements with regard to insurance.

There are a variety of methods of insuring this kind of equipment. It can be insured on a blanket basis automatically covering all equipment, or on a specific basis listing the items of equipment and the values they are insured for. Both plans are acceptable, depending on the circumstances.

Equipment can be insured for its full replacement value at the time of loss without deduction for depreciation however caused. This is, in effect, new for old. It can also be insured and usually is on a depreciated value, i.e., the actual cash value at the time of loss after reflecting depreciation and obsolescence. Another method that may be used is the so-called "agreed value" basis. This is a specific amount agreed in advance to be paid for a total loss. It is most favorable for unique types of equipment when the market value or replacement value is not readily determinable.

It is important to understand that these types of insurable values have nothing to do, except incidentally, with cost or book values and are measured by rules of their own rather than determined by the usual accounting technique.

There is latitude in this kind of insurance to tailor the policy wording and economize on your premium through the use of various deductible plans. These can be negotiated subject to a deductible applicable to each and every loss or in the aggregate. The policy may be extended to cover certain consequential losses such as loss of use or extra expense to provide duplicate equipment. The ability to quickly secure alternate equipment or the availability of proper repair and parts facilities adjacent to the job will determine this exposure.

The protection afforded may be against named perils or all risks. Either policy form contains certain standard exclusions, which should be carefully reviewed.

It is important to arrange this coverage so that equipment is covered whether working or in dead storage. Usually the insurance company will grant a lower rate for the idle equipment. Depending on the exposure and values at risk, the policy may be broadened to include office contents at project sites and various miscellaneous materials.

Contractors who have marine equipment of any kind used in lakes, rivers, or offshore work have a different and more complex problem with regard to insuring such equipment.

29-8.   Transit Insurance

This form of insurance agrees to pay for direct loss or damage to property in transit. The coverage may be for named perils or all risks.

29-9.   Business Interruption and Extra Expense

Certain types of projects have serious risk exposures involving business interruption and/or extra expense, which could result in loss of profits, continuing fixed expenses, and/or extra expenses, depending on the severity of the loss.

Business Interruption Insurance   The purpose of this coverage is to do for the insured during a period of total or partial shutdown what the project would have done for him if the shutdown had not happened. The function of this protection is to replace the operating income of the project during the period when damage to the project or equipment prevents this from being earned.

It is from operating income that the contractor meets his expenses—light, heat, advertising, telephone, etc.—and from which is derived his profit. If the contractor has to shut down for a period of time or operate at a reduced scale because of fire, windstorm, or insured peril, that income will be cut off or down.

Extra Expense   When a loss occurs the practice is to try and continue operations no matter how serious the damage. In some cases there may not be a desire for Business Interruption insurance when the principal risk or potential loss involves only extra expense, i.e., the cost of renting equipment until the damaged equipment is back in service.

On certain types of jobs the contractor is subject to both exposures. A good rule to follow is: if the income of the job will be cut off by a fire or other disaster, then Business Interruption insurance is needed. If operations must continue regardless of cost and will maintain the same or nearly the same volume of work, then Extra Expense insurance should be considered.

A careful review of the exposures should be made in connection with each job to determine the needs.

29-10.   Fidelity and Forgery

Fidelity   Fidelity bonds or dishonesty policies, as they are commonly called, cover losses incurred by an insured due to fraudulent or dishonest acts of his employees. The loss may involve such property as money, securities, or other property, real or personal, due to such acts.

Depositor's Forgery   Forgery is the fraudulent making or altering of a writing to another's prejudice with intent to defraud. It provides indemnity to the contractor and to any bank in which he carries a checking or savings account against loss by reason of the forgery or alteration of any check, draft, promissory note, bill of exchange, or similar written promise, order, or direction to pay a sum certain in money made or drawn by or upon the insured, or made or drawn by one acting as agent of the insured.

The Fidelity and Forgery coverages may be written separately or included in one of several blanket forms. The more common blanket form is called the "Three D" policy or the comprehensive, dishonesty, destruction, and disappearance policy. The basic "Three D" form includes five insuring agreements:

- Fidelity
- Money and securities inside the premises
- Money and securities outside the premises
- Safe deposit box coverage
- Depositor's forgery (outgoing instruments)

Additional coverage may be added by endorsement:

- Income check forgery
- Open-stock burglary
- Open-stock theft
- Broad form payroll, inside and outside premises
- Broad form payroll, inside premises only
- Open-stock burglary and limited theft
- Forgery of warehouse receipts

Additional forms of protection which should be considered, depending on the risk exposures of the contractor. These are described in the following sections.

29-11.   Officers' and Directors' Liability Insurance

When an individual becomes a director or officer of a corporation, he exposes himself to substantial personal financial risk. He personally becomes accountable to stockholders, creditors, and government, and no one can predict today how the courts or government agencies will deal with allegations of misconduct or error on his part tomorrow. The personal liability to which he is exposed may be insured under Directors' and Officers' Liability insurance.

There are an increasing number of suits being brought against directors and officers by stockholders or others to recover damages for real or fancied wrongs. Although the circumstances vary by case, the claims made in suits for financial recovery are based largely upon negligent management, violation of statute or regulation, or breach of fiduciary duty. In some cases where judgments resulted, for the defendants, the courts recognized that, although the alleged cause of action existed, liability was not established by the evidence. In such cases, substantial costs were incurred in defending the action which, in most instances, would be covered by this insurance.

29-12.   Architects' or Engineers' Professional Liability

The purpose of this form of coverage is to protect against liability from professional services rendered.

29-13. Installation Floater

This form of protection is designed to cover the insured property against the hazards of transportation, including other specified hazards at the job premises of installation until the interest of the contractor in the property ceases.

29-14. Package Policies

Depending on the volume and type of construction work performed, various forms of package policies may be tailored for the contractor.

On large office buildings or heavy industrial construction, a wrap-up program may be arranged. Under this underwriting concept, the prime contractor and all subcontractors may be included for the coverages selected. The coverages normally included are:

- Workmen's Compensation and Employer's Liability
- Public Liability and Property Damage covering the construction operations, including the ownership and operation of owned and nonowned automobiles
- Builder's Risk
- Contractor's Equipment

In the case of large heavy construction and engineering contractors, package contracts may be arranged to cover practically any or all insurable exposures.

Some of the advantages of these plans are:

1. Lower net costs due to the larger concentration of premium dollars.
2. No disclosure of certain bid information prior to the project bid date
3. Known premium rates for your estimating engineers to work with

4. Broader protection than usually obtainable under separate policies for each contractor and/or project
5. Avoidance of gaps and overlaps
6. Considerably less time and expense required of office personnel who administor the detail for the insurance requirements

29-15. Important Matters to Watch for and the Action to Be Taken

1. Waiving rights of subrogation
2. Naming another party as an additional insured
3. Agreeing to save another party harmless or free of any liability arising out of your operations or those for which you are obligated under the construction contract

[*Note:* Where possible the above should always be avoided. If it is necessary, check with your agent or broker to be sure the applicable insurance policies cover such agreements or are acceptable to the insurance company(ies). If the insurance policies do not automatically cover the additional risk, secure the necessary endorsement.]

4. Where equipment is purchased for the project, be certain of any warranty provisions which may impose liability on you.
5. Tailored or manuscript policies vs. standard policies. The primary purpose of tailored or manuscript policies is to design the coverage to fully cover the exposures. Usually, they afford broader protection. Extreme care should be taken with respect to the wording of all tailored or manuscript policies.
6. Claim reporting. It is not only important but an obligation of the insurance policies that all losses be promptly reported to the insurance company(ies). Failure to do so could prejudice recovery or void the protection.

30 GROUP INSURANCE PLANS

JAMES A. ATTWOOD

Senior Vice President, Group Operations
Equitable Life Assurance Society of the U.S.
New York, New York

GROUP insurance is an important management tool. It is generally recognized that the company without a sound group insurance plan cannot compete successfully with others in its field in attracting and holding worthwhile personnel. Group insurance is basic to a positive and productive employer-employee relationship. It is no longer a fringe benefit but a vital part of employee compensation. Group insurance protection is highly valuable to employees. Regardless of the extent to which they may share the cost of the insurance with their employer, whatever they pay will normally be substantially less than the price to them as individuals outside of the employee group.

30-1. Designing a Group Insurance Program

The construction industry, unlike most, is characterized by the seasonal nature of its operations and the relatively temporary employment of its rank-and-file workers. The latter are frequently insured by group policies negotiated by their various unions, either under contracts written in the name of the union or issued to trustees representing both management and the union. Many firms, however, become involved in establishing or negotiating the terms of their own plans, and it is therefore important that management have a good basic understanding of the various benefits available. To get started in a study of employee benefits, management can:

a. Work with an insurance agent and his company's salaried specialists. Representatives of more than one company might be asked to submit bids.
b. Work with a broker or consultant who will represent the company. He will request bids from several insurance carriers, analyze them, and make his recommendations to management.

Because each employer's business and personnel situation has its own characteristics which make it at least a little different from others, a group insurance plan must be designed to fit it. This tailoring process is important because it decides who will be in the plan and how much the plan will pay in the event of death, disability, medical bills, and retirement. The succeeding pages describe the provisions of the various available benefit plans beginning with eligibility for membership.

30-2. Eligibility

Who is eligible for benefits? Eligibility must be defined in the master policy, and it must apply to all employees or all employees in a specified class or classes. Legally, eligibility must be based on some condition pertaining to employment, such as salary, hourly rate, occupation, location, or length of service, and it must preclude individual selection either by the employer or the employees. All the employees in one location could be made eligible, or eligibility could be limited to salaried or hourly paid workers.

Eligibility for plan membership should be set so as to eliminate most short-term or part-time workers because the administration involved makes it expensive to insure them. Therefore, it is advisable to include a probationary period in the definition of eligibility. This refers to the length of time an employee must be in service before he can become eligible for Group Insurance. A probationary period of three or six months might be sufficient to eliminate temporary workers, or perhaps as long as one year will be required. More than one year is not usually recommended. Further, to eliminate part-time workers, an additional requirement should be included limiting eligibility to those working at least a specified number of hours per week, such as 32 hours.

In an industry in which there is considerable labor mobility, some workers could miss being covered under single employer plans because they would not work long enough for any one company to stay beyond its probationary period. This problem was one of the factors that gave rise to the popularity of negotiated industry-wide or trade-wide plans which cover their members regardless of which member company they work for or for how long.

FORMS OF GROUP INSURANCE BENEFITS

30-3. Group Term Life Insurance for Employees

This was the first form of Group Insurance developed and offered to industry. Benefits are payable in the event of an employee's death from any cause, and at the option of the employee or beneficiary the insurance may be paid in a lump sum or in installments. The employee selects his own beneficiary, who may be a relative, friend, or institution (such as a college); the employer, however, may not be named as beneficiary.

The plan of benefits may provide one to two years' salary, or may be graded by earnings, occupation, or length of service, or it may offer the same amount to each employee.

Any plan providing benefits based on earnings levels, in effect, continues an employee's earnings for a given time. An occupational plan can achieve much the same result but is more cumbersome to administer because of the need to amend the policy each time there is change in job title or class. A service plan, under which benefits increase with each year of service completed, is not particularly popular today, because the benefit does not reflect the relative value of the employee. Also, the cost of such plans rises more quickly than that of salary or occupational class plans since the highest amounts of insurance are on the oldest lives. Flat benefit plans are adequate if only a nominal benefit is desired or if all employees have about the same income.

The available maximum benefit which may apply for a particular plan is determined by (a) the laws (if any) of the state where it is located, and (b) the internal underwriting rules of the insurance company.

Substantial amounts of insurance on older employees causes the cost of group life insurance to increase. Therefore, most employers reduce these benefits at retirement or a specified age, whichever is the earlier, to an amount equaling about 25 to 50 percent of their original coverage. The most usual method of doing this is to reduce benefits each year following retirement by a fixed percentage of the original amount (10 percent is common) to the final reduced amount which will then continue for his lifetime. Alternatively, the full reduction may be made in one step. The rationale for reduction (other than the cost reason) is that the employee has had full protection while his responsibilities were heaviest (during employment years) and receives a reduced amount only when those needs are less.

30-4. Group Term Life Insurance for Dependents

Dependent Group Life Insurance provides a limited benefit for the spouse and dependent children of an employee in the event of their death from any cause. Benefits usually range from $1000 to about $5000 on the life of a spouse, depending on state law, and from $100 to $1000 for children, depending on the child's age.

30-5. Survivor Income Benefits

A relatively new form of employee death benefit provides regular monthly income paid to an employee's surviving eligible spouse and/or eligible children. It is called a Survivor Income Benefit. It is never paid in a lump sum, but only in monthly installments and only to a spouse or children. The benefit terminates if a spouse remarries, dies, or attains a limiting age (such as 62) or if a child dies, marries or reaches the limiting age in the policy.

The Survivor Income Benefit is usually a percentage of the employee's monthly earnings, such as 15 or 25 percent. A monthly payment equal to a portion of the spouse's benefit is allowed for the eligible children of the employee if there is no eligible spouse when the employee dies or if the spouse dies while receiving Survivor Income Benefits. The period for which benefits are payable to the spouse may be a specific time (5 or 10 years), or until the spouse is age 62 to coincide with the beginning of Social Security payments, or it may be for the spouse's lifetime. There are other periods as well.

30-6. Group Accidental Death and Dismemberment Insurance (for Employees Only)

Group Accidental Death and Dismemberment Insurance is paid in the event of accidental death or dismemberment except in a few instances, such as death or disability resulting from an act of war, self-inflicted injuries, etc. Twenty-four hour coverage is usually provided; i.e., the death or disability may occur on or off the job, but, alternatively, coverage may be nonoccupational only.

Group Accidental Death and Dismemberment benefits usually are the same amounts as group life insurance. If the life insurance is reduced at a specified age, the Accidental Death and Dismemberment is also reduced, and it generally terminates when an employee retires.

30-7. Group Weekly Indemnity Insurance (for Employees Only)

Weekly Indemnity Insurance is written to provide partial reimbursement for an employee's earnings in case of short-term disability resulting from nonoccupational accident or sickness. Workmen's compensation pays benefits on disabilities resulting from on-the-job accident or sickness. In New York, New Jersey, Hawaii, Rhode Island, California, and Puerto Rico, disability plans must be provided by nearly all employers. Except in Rhode Island, these plans may be underwritten by private insurance companies.

Typically, the Weekly Indemnity plan pays from 50 to 75 percent of an employee's weekly income. A waiting period of three to seven days in case of sickness, and sometimes accident, is normally written into the plan so as to exclude minor claims. Benefits may continue for 13 weeks or some other period, but usually no longer than 52 weeks for any one disability. If plans allow payments for 26 weeks or more, benefits are usually integrated with Social Security disability benefits (which start after 26 weeks). This eliminates overpayment of benefits and maintains an incentive for the employee to return to work. Weekly Indemnity benefits are terminated when an employee retires.

30-8. Long Term Disability Income Insurance (for Employees Only)

Many employers have been supplementing their short-term disability plans with insurance that will protect employees in case of disabilities that might last for years. Long Term Disability Income Insurance benefits start after the short-term benefits are ended and are payable until the employee reaches age 65 (or for life, in some instances). This coverage usually pays 50 to 66⅔ percent of an employee's earnings. The standard plan is written on the twenty-four hour basis, i.e., benefits are payable regardless of whether the disability had an occupational or nonoccupational cause.

Long Term Disability benefits are generally integrated with, and reduced by, other disability benefits which an employee may receive, such as Social Security disability benefits, retirement plans, Workmen's Compensation, etc. Because of the liberality of Social Security benefits for employees earning less than the Social Security maximum tax base, it is inadvisable to offer Long Term Disability to these employees, since the integration with Social Security results in little or no payments under the plan. Long Term Disability Income Insurance terminates when an employee retires or reaches age 65, whichever is earlier.

30-9.   Hospital-Surgical-Medical Expense (for Employees and Dependents)

Dependents are considered to be an employee's spouse and children, ages between 14 days and 19 years. However, children may be covered from birth and up to age 21 or even 23, if desired, or only students may be covered beyond age 19. Children who are handicapped and dependent on the employee for full support may be insured for life.

Because of constantly increasing hospital and medical costs, most persons cannot begin to afford even limited health care unless they have some form of Group Insurance. A basic plan of health insurance consists of one or more of the following benefits which are payable for *nonoccupational* causes:

Hospital Expense

Hospital expense reimburses for room and board charges and for additional charges, such as those for the operating room, recovery room, X-rays, etc., which are incurred while an individual is an in-patient in a hospital. Certain types of out-patient charges are also covered in case of accidents.

One type of plan provides a fixed dollar amount per day for room and board benefits. The room and board benefit may be paid for any desired number of days up to 365 for any one confinement. Benefits for additional charges usually represent 10, 15, or 20 times (or some other multiple) of the daily room and board benefit for the charges incurred while confined in the hospital.

Another type of plan (sometimes known as a "service-type") offers benefits described as the average semiprivate charge in the hospital used. In these cases, the insurance company would pay the full semiprivate charge, or the full ward charge, if ward accommodations were used; private room accommodations would be reimbursed at the average semiprivate rate. This type of plan is of particular interest to companies with branches in several locations, when a flat dollar amount might mean inadequate insurance in one branch location and perhaps too much in another. Additional charges may be a multiple of the average semiprivate charge, or be set on some other basis.

There are innumerable variations of Hospital Expense plans since each plan is designed to meet the employer's needs. The "service-type" plan is generally the more expensive, since the benefits will automatically increase as hospital charges increase, whereas the fixed dollar plan does have built-in maximum benefits. As the benefits of the fixed dollar plan are increased to meet rising hospital costs, however, the costs must increase as well.

If desired, maternity benefits may be included under hospital expense. These benefits usually pay either a flat dollar amount or its equivalent, e.g., 10 times the daily room and board benefit, with the stipulation that this benefit is to include additional hospital charges as well.

Surgical Benefits

This coverage reimburses for doctors' fees for surgical procedures performed either in or outside of a hospital. There is usually a schedule of surgical procedures set up with a maximum amount of reimbursement for each operation. The schedule may allow a maximum of $300 or $400 or more, depending upon the employer's selection. Alternatively, plans providing for "reasonable and customary" benefits are also made available. On this basis, reimbursement for surgical expense is geared to the usual range of fees charged, in the locality where the expense is incurred, by most of the physicians who have had similar training and experience. Since many doctors have a sliding scale of fees, based on the income of the patient, this factor is considered in determining rates and payments. The "reasonable and customary" basis is more expensive than the scheduled plans, since it pays benefits which in many cases are higher than those which a scheduled plan would allow.

Medical Expense

Reimbursement for doctors' fees for treatment of nonsurgical and nonmaternity disabilities may be provided while the individual is hospitalized or may also cover treatment in home or doctor's office as well. (The exclusion of charges for surgery or in case of maternity disability is simply because the doctor normally combines his charges for patient visits with his fee for operating or delivery.)

The In-Hospital plan usually provides a maximum benefit for doctors' visits which is based upon a fixed dollar amount for each day of hospital confinement up to a designated number of days. The benefit may accrue, for example, at the rate of $5 per day for 50 days or a total benefit of $250. This maximum would permit payment of benefits for charges of $10 per visit for 25 visits or charges of $25 per visit for 10 visits, if the patient were confined for 50 days. It is the maximum accrued benefit which is significant and not the actual number of visits.

In a more comprehensive plan, doctor's home and office calls are also reimbursed. Generally, home visits are reimbursed at a higher level than hospital or office visits. In some plans, reimbursement is not made for the expense of the first two or three home and office calls in order to eliminate budgetable expenses and to reduce the cost of benefits. Unlike the in-hospital calls, those for home and office are limited to one call per day.

30-10.   Ancillary Health Care Benefit Plan

Although the Hospital, Surgical, and Medical Expense benefits are the keystones of any basic health care plan, any of several other forms of specialized protection could also be included. For example:

X-Ray and Laboratory Benefits These cover charges incurred as other than an in-patient of a hospital. Maximum payments may be listed in a schedule or the benefits may be paid without maximum limitations.

Dental Expense Benefits Covered are expenses for all forms of general dental care and orthodontia.

Radiation Therapy Benefits The cost of radiation treatment and drugs is covered on a scheduled basis. Charges may be incurred either in or outside of a hospital.

Prescription Drug Benefits Provides partial reimbursement for prescription drugs purchased when the patient is not hospitalized.

Nursing Home Care Benefits Ordinarily room and board charges are covered when a patient enters a nursing home within a required number of days (e.g., seven) after being released from a hospital where confinement lasted for a specified period (e.g., five days). This coverage does not provide custodial care for the aged.

30-11. Supplemental Major Medical Expense (for Employees and Dependents)

In cases of severe or prolonged disability, even a substantial basic plan of health care expense benefits is not adequate to meet an individual's needs. Such an illness may completely use up a family's savings and a heavy burden of debt may have to be assumed if extra benefits are not available. Major Medical Expense Benefits were developed to supplement those of a basic plan to help pay the costs of catastrophic illness. Major Medical's key elements are:

1. Major Medical is usually offered on a "per calendar year" or "per cause" basis. This description generally refers to the deductible which is the amount of expense the covered individual must pay before plan benefits are payable. As the name implies, it applies either once in a calendar year or for each cause. The calendar year basis is the more popular. The deductible may apply:
   a. after the base plan benefits have stopped, in which case it may be $100 or some other appropriate amount, or
   b. it may be integrated with the base plan benefits, in which case it is usually $300, $500 or more. In these situations, the Major Medical benefits commence after the basic benefits have been exhausted, assuming their value is at least equal to the stated deductible. If the value of the basic plan benefits paid is less than the stated deductible, an out-of-pocket expense is incurred until the value of the basic plan benefits plus out-of-pocket expenses equal the stated deductible.

   Although a deductible usually applies separately to each insured person in a family, it can be waived after the expenses of a certain number of family members (usually 3) have met it.

2. A coinsurance feature applies under which the plan pays 75 or 80 percent of the covered charges, and the individual pays the balance.
3. A high maximum of plan payments is available— $15,000, $25,000, $30,000, or even more, depending

on the number of covered employees, the other terms of the plan, and the underwriting rules of the insurance company.

Major Medical Expense plans normally provide payment for charges incurred while the insured is either in or outside of a hospital. Hospital, surgical, and medical expenses are included as eligible for reimbursement as well as those for out-of-hospital prescription drugs, private nursing care, and charges usually covered under the ancillary benefit plans. Maternity and obstetrical charges are not covered by Major Medical except in cases of complications of pregnancy.

30-12. Health Care Expense (for Employees and Dependents)

A modification of the base plan plus Major Medical concept is presented under Health Care (often called Comprehensive Major Medical Expense), which incorporates in one coverage all the features separately provided by a basic plan of hospital, surgical, medical expense and supplementary Major Medical.

A Health Care plan generally follows this pattern:

1. A small deductible, typically $50, is required generally in each calendar year. It is frequently waived for hospital charges. There may also be a "family" deductible as described above under the Major Medical coverage.
2. Full payment is made for hospital charges up to a specified limit, such as $1000.
3. For hospital expenses in excess of $1000, as in the example above, and for other charges, reimbursement is made on a coinsured basis with the plan paying 75 or 80 percent.
4. A maximum at the same level as described under a Major Medical plan may apply per lifetime or per cause.
5. Covered charges under Health Care and Major Medical are the same except that the former provides benefits for normal maternity-obstetrical procedures in addition to making payment for complications of pregnancy.

30-13. Medicare

When a policyholder has only a basic plan of hospital, surgical, and medical benefits in force, most insurance companies recommend that it be terminated as employees reach age 65, since Medicare is a satisfactory substitute.

If an employer has a basic plan plus Major Medical or Health Care on his active employees, the recommendation might be to terminate all coverage as employees reach age 65 but to offer a special plan for Medicare participants which would pay benefits in instances of expense which are not reimbursable under Medicare. In this way, comprehensive protection may be continued. If an employee retires before age 65, his insurance could be continued until he reaches that age.

* * *

The possible forms and combinations of health insurance plans are almost unlimited, and the actual plan adopted should depend on factors inherent in the employer's situation, i.e., location, whether unions are involved, preponderance of wage or salaried employees, labor turnover, available funds for planned costs, etc.

30-14. Costs of Group Insurance Benefits

Rates for determining the costs of particular plans of insurance are calculated for each employer group and are affected by such factors as employee age, sex, income, and location as well as plan structure, industry, etc. Since most forms of group insurance are on a yearly renewable term basis, rates and costs are usually redetermined annually.

The employer's share of Group Insurance premiums is a business expense for tax purposes and is not reportable as taxable income to covered employees except for higher amounts of group life insurance.

30-15. Foreign-Based Employees

Because of the many legal and administrative problems involving the inclusion of foreign nationals employed in overseas locations under a Group policy issued in the United States, many insurance companies prefer to arrange with foreign insurance companies to underwrite coverage for these employees.

Most insurance companies will include under the Group policy American nationals who are transferred overseas temporarily, or transferred foreign nationals living in the United States, provided that these employees represent only a small percentage of the entire group. Generally, premiums and claims are paid in United States currency.

*   *   *

This chapter has, of course, touched only upon the highlights of Group Insurance. An insurance advisor will be able to provide the necessary overall plan design and the intricacies of each form of employee benefit.

31 FUNDAMENTAL CONCEPTS UNDERLYING PENSION PLAN FINANCING AND COSTS

FREDERICK P. SLOAT

**Retired Principal and formerly
National Director of Actuarial and
Benefits Consulting Services
Lybrand, Ross Bros. and Montgomery**

DAVID V. BURGETT

**Partner
Lybrand, Ross Bros. and Montgomery
San Francisco, California**

EDITORS' NOTE:

Both the title and the text of this chapter have been adapted from a 1970 copyrighted publication by Lybrand, Ross Bros. and Montgomery, and it is included here in its presented form with permisison of the copyright owners. In addition to its outstanding value for accounting purposes, its basic information concerning pension plans will assist construction management in their consideration of such plans. Brief statements concerning the authors follow.

Frederick P. Sloat is a retired principal of Lybrand and formerly head of the Actuarial and Benefits Consulting Services. Mr. Sloat was actuarial adviser to the Accounting Principles Board during the development of Opinion 8 and worked with its subcommittee and drafting group. He is a Fellow of the Society of Actuaries, and a member of the Wharton School's Pension Research Council and of the Committee on Pension and Profit-Sharing Terminology of the American Risk and Insurance Association.

David V. Burgett is a partner in Lybrand's San Francisco office and a member of the Firm's SEC Review Board. Mr. Burgett started in the firm's Los Angeles office in 1958, and later transferred to the National Office in October 1966. Having gained auditing experience on two coasts, he returned to Los Angeles in 1969 and transferred to San Francisco in 1972.

This monograph discusses, for accountants and others, the considerations underlying the determination of funding contributions toward pension costs. It describes several commonly used types of pension plans and the related funding methods in a way designed to provide the internal accountant or the CPA with a working knowledge of the major cost elements of a pension program, and to help him better evaluate the relationships between a plan's basic costs, the related periodic funding contributions, and the appropriate periodic accounting accruals.

Opinion No. 8 of the Accounting Principles Board (APB), "Accounting for the Cost of Pension Plans," which was designed to stabilize the amounts of annual provision for pension costs, has had far-reaching effects on this entire area; it is analyzed in detail hereinafter. This monograph also discusses some tax aspects of pension plans. Its central purpose, however, is to give the accountant enough general actuarial knowledge to help him cooperate more fully with the actuary in solving client problems in pension areas where the skills of the two professions overlap.

Essentially, the accountant must start with an understanding of three basic areas: pension costs, funding contributions, and charges against operations.

Pension costs, because they encompass the entire work span of a group or groups of employees, are long-range by their very nature. Any recognition of the amount of costs accruing before actual disbursements are made during employees' retirement can only be an estimate. It is predicated on assumptions about future occurrences such as continuance in employment, mortality, and earnings on funds allocated for the future payment of pensions. Actuaries deal with these assumptions in order to evaluate the financial impact of pension proposals, and to suggest funding and/or accounting arrangements appropriate for providing for the ultimate expenditures required under the terms of a plan.

The actual *contributions made toward funding* have frequently formed the basis of accounting recognition of accruing pension costs. They may, in fact, provide an appropriate measurement of accruing costs, since the considerations underlying pension payments and accounting accruals are similar. Computations in both areas involve the same method, although a funding arrangement is primarily a financing device, and thus does not differ essentially from any arrangement to pay, over a period of years, for any goods or services a business may purchase. The people involved with the financial management of a company are responsible for decisions concerning the funding arrangements to be adopted, but the management, of course, utilizes both the technical skill and the advice of actuaries.

The accountant must decide which *charges to levy against operations,* in an accounting period, to represent accruing costs of pension obligations, chargeable against revenues. In carrying out this responsibility, he should not assume that currently accruing costs are necessarily the same as funding contributions. Rather, he must decide whether accruing costs may exceed contributions or whether a portion of contributions represents an amount which should be charged against future operations. Since pension costs are computed on the premise that they will be invested until needed to pay benefits, an adjustment must be made, whenever costs are funded either earlier or later than they are recognized, for the difference in the period of investment.

This monograph is written primarily from a funding standpoint. When costs are determined without regard to funding, as they are for APB 8, the same concepts are applicable, but adjustment must be made for the interest differentials between costs as recognized and costs as funded.

Familiarity with the terminology commonly used in discussing pensions, and especially pension costs, is obviously necessary to an understanding of the nature of such costs. Unfortunately, pension terminology has not always been uniform: The same term may be used for several different concepts, and a number of terms may be used synonymously to describe one concept.

A Committee on Pension and Profit Sharing Terminology established by the American Risk and Insurance Association has been working to establish uniform terms, and those that have already been decided on are used here. In addition, synonymous or previously used terms are included; where appropriate, other usages are given. When terms used by actuaries and other nonaccountants in discussing pensions may have different connotations for accountants, an explanation is given unless the context makes clear the sense in which that term is used.

TYPICAL PENSION PLANS AND THEIR PROVISIONS

31-1. Types of Pension Plans

A pension plan* is an arrangement whereby an employer can provide retirement benefits for employees in recognition of their service to the company before their retirement. The arrangement may be an *informal* pension plan calling for voluntary payments to the retired employee in amounts and under conditions more or less at the discretion of the employer, or it may be a *formal* plan with the benefit payments and other features explicitly stated

or readily computed or otherwise determined. A formal plan is usually set forth in a writen document, but an established policy with regard to pension payments may be sufficiently definitive to constitute a formal plan.

A pension plan may be unilaterally adopted by an employer. This type of plan is frequently referred to as a *conventional* pension plan or a *voluntary* pension plan. If, however, it results from collective bargaining between employer and employees, it is referred to as a *negotiated* pension plan. *Pattern* pension plan is a term sometimes used to describe a negotiated pension plan whose form has been established by bargaining between a labor union and one or more companies. It is usually adopted, perhaps with variations, by companies in the same or allied industries.

A qualified pension plan is a plan that conforms with Internal Revenue Code requirements. It provides certain tax advantages, though these are subject to specified restrictions, including the deductibility of employer contributions to the plan as a business expense, exemption of investment income from the pension fund, and exemption for the employee of those contributions and the income therefrom made before he actually receives any retirement payments.

A *contributory* pension plan is one to which both the employees and the employer make contributions. A *noncontributory* pension plan is one to which only the employer makes contributions.

31-2. Provisions

A pension plan's provisions specify the benefit payments* to be disbursed to retired employees and the conditions of entitlement and payment. The specifications form part of the data used to compute the employer's contributions. The following review of typical provisions of pension plans is designed to be background material for later discussions of these plans.

Eligibility Requirements   These refer to the earliest date or age when an employee can become a participant in a plan and begin to accumulate retirement benefits. In some plans, an employee may become eligible immediately upon being hired, regardless of age or period of service; however, most plans require a period of service or a combination of age and period of service (e.g., 30 years of age with at least two years of service). Frequently, only certain classes or categories of employees may be included in a particular plan.

Retirement Dates   A designated retirement age when an employee may receive the normal pension is termed the *normal retirement date* or *age* (for example, age 65). *Mandatory* retirement date also referred to as *compulsory* retirement date or *automatic* retirement date means that the employee must retire at a given age, which may be the same as, or later than, the normal retirement age. Most plans provide for an *early* retirement date prior to the normal retirement age; if disability is involved, this date may be referred to as *disability* retirement. Early retirement for other reasons may be at the election of the employee or the employer. Benefits paid upon early retirement are usually less than normal retirement benefits. A plan may

---

*The terms "pension plan" and "retirement plan" are now used interchangeably.

*The terms "benefit payments" and "benefits" are used synonymously, to describe payments to retired employees under a pension plan.

also permit an employee to postpone his retirement from the normal retirement date to a *deferred* retirement date, also referred to as a *delayed* retirement date or *late* retirement date. Such late retirement may be at the employer's or employee's election, and may not involve higher than normal benefits.

Vesting Employees usually have a vested right to receive benefits when they become eligible to retire. However, if an employee's entitlement to benefits arising from employer contributions is not contingent on his continuing in the service of the employer until he is eligible to retire, he is said to have a *vested interest*. The applicable provision in the plan is described as a *vesting provision*. The vested benefits to which the employee would be entitled upon termination of employment are usually in the form of a deferred annuity income, with payments commencing at the normal retirement date. If the vested benefits are in an amount equal to the accrued pension rights determined up to the date his employment ends, the plan is said to provide for *full vesting*. If a portion of the accrued benefits is forfeitable, the term *partial vesting* is used.

In addition to any vested interest in employer contributions, employees generally have the option, at termination of employment, to receive a refund of their own contributions under contributory plans, usually with accrued interest. If the employee chooses such a refund of his own contributions, however, he usually forfeits any vested interest that he would otherwise have. This kind of vesting is called *conditional vesting*.

Normal Retirement Benefits A variety of methods is used for determining the amount of retirement benefits. Most of these specify benefits in terms of percentages of each year's compensation, often with fixed minimum and maximum amounts. The total benefit so accumulated is said to be on a career average compensation basis. The compensation base may be determined, however, by average earnings during a certain specified period such as the final five or ten years of service (final average pay plan). Different percentages may be used for compensation before and after the adoption of a plan. Many plans correlate or integrate pension payments with federal Social Security retirement benefits. This is often provided for in the benefit formula by the application of a lower percentage to the part of earnings covered by Social Security.

In some cases, pension payments are determined as the amount of a retirement annuity purchasable by the contributions specified in the plan. This type of plan is referred to as a *money purchase plan*. All other things being equal, this method produces decreasing benefit units with advancing years of service because the period of investment of contributions diminishes and the probability of the employee's living to retirement age becomes greater. A *defined benefit* plan states the formula for determining the benefits. A *defined contribution* plan states the formula for determining the employer's contributions (e.g., money purchase plan).

Death Benefits Some retirement plans provide a *death benefit,* which is any benefit payable upon the death of an employee. In a contributory plan, the minimum death benefit would necessarily be the total contributions made by the employee, together with accumulated interest. In some plans, however, death benefits are related to the employee's anticipated pension benefits or the value of his accumulated pension credits, and hence may be substantial.

Duration of Retirement Benefits Most pension plans provide retirement benefits for the remaining lifetime of the annuitant (i.e., the retired employee). The annuity thus provided, which terminates upon death, is called a *life* annuity or a *straight life* annuity. A contributory plan will usually include a *cash refund* annuity provision, making the death benefit for a retired employee the amount that would have been payable at retirement date, less pension benefits paid between retirement date and death. Benefit payments in some plans are guaranteed for a minimum fixed period after retirement, with the annuitant's beneficiary or estate receiving the remaining payments should he die during the period. This type of annuity is called a *period certain life* annuity. If a plan calls for a five-year guaranteed period, for example, the annuity would be referred to as a five-year certain life annuity.

Some plans allow the annuitant an optional benefit to be elected prior to retirement. A *joint and survivorship* option is a common optional arrangement that permits the employee to elect to receive reduced retirement benefits with provision for a continuation of benefits, in full or in part, to a named beneficiary after his death. Such a beneficiary, usually a spouse, is referred to as the *joint annuitant* or the *contingent annuitant*. The benefit amount payable to the annuitant and to the joint annuitant depends on their ages, and the total is usually the actuarial equivalent of the normal benefit.

Federal Income Tax Considerations Most formal funded plans are designed to meet the Internal Revenue Code's qualification requirements, so that they may receive substantial tax advantages. These requirements are as follows:
1. A plan must not discriminate in favor of employees who are officers, shareholders or highly compensated.
2. Deductions under the plan must be reasonable in amount when considered with other forms of compensation.
3. The plan must be permanent.
4. Contributions to the fund must be exclusively for the benefit of employees.
5. Deductions cannot exceed actual contributions.
6. The amounts of deductions are subject to specified limitations.

If a plan qualifies, contributions by the employer are currently deductible, income on accumulated funds held for a qualified plan is not subject to tax, and the employees will be subject to income tax on the resulting benefits only as payments are received.

It is customary to obtain a determination from the Internal Revenue Service that a plan is qualified, to avoid later difficulties in interpreting the technical requirements.

PRELIMINARY ACTUARIAL CONSIDERATIONS

In most instances, a pension plan specifies the benefits to be paid to eligible employees (defined-benefit plan). In these cases, the amount of the employer's contributions must be ascertained. However, if a plan specifies the amount of the employer's contributions (defined-contribution plan), it is the type of benefit structure that can be established on the basis of the given contributions that must be ascertained.

Computations relating to pension plan costs, contributions, and benefits are made by an actuary. An actuary is an expert in pension, life insurance, and related matters that involve life contingencies. He employs mathematical, statistical, financial, and other techniques to compute costs or benefits, equate costs with benefits, and evaluate and project actuarial experience under a plan. Membership in the American Academy of Actuaries, or one of the other recognized actuarial organizations, identifies a person as a member of the actuarial profession.

31-3. The General Nature of Pension Costs

Some consideration should be given at this point to the nature of pension costs.* The ultimate cost of any pension plan is determined primarily by the total pension benefits that will actually be paid to retired employees over their lives. The choice of a funding agent affects total pension contributions only to the extent to which, under a particular type of plan, the portion of costs met from investment income increments varies, or the extent to which expense factors vary. However, the *actuarial cost method** adopted does influence the incidence of contributions or payments.

A general statement of a company's prospective pension obligation at any given time with respect to then covered employees, i.e., the *present value* (PV) of future contributions required to meet retirement payments, may be expressed by the following formula:

$$\frac{PV}{contributions} = \frac{PV}{benefits} + \frac{PV}{expenses} - \frac{Funds}{on\ hand}$$

This general statement provides a measurement of total cost to be met by contributions, but it does not furnish information for the important decision on allocating this amount to fiscal periods. The most rational and commonly used approach to reaching this decision recognizes that pension costs are a part of total employee compensation. Accordingly, they are usually regarded as a factor related to total current compensation, although past service costs involve considerations that frequently cause them to be dealt with separately.

Since direct compensation is readily allocable to fiscal periods, relating pension costs to it provides a basis for associating their incurrence with fiscal periods. Thus, contributions necessary to meet a plan's pension payment requirements may be assigned to (1) prior services, (2) current services, or (3) future services.† Pension costs assigned to prior services are said to have "accrued." Terms for the cost associated with prior, current, and future services necessarily vary because their treatment varies under the different actuarial cost methods.

31-4. Normal Cost

Basic to determining pension costs is normal cost or the level of periodic contributions appropriate under a particular actuarial cost method, whether or not there was a late start in recognizing the cost for benefits. Thus, *normal*

*The term "cost" is used here to refer to contributions for funding future benefit payments specified in a pension plan. Determinations of periodic costs for accounting purposes may involve the same or differing considerations.

*An actuarial cost method is a particular actuarial technique for establishing the amount and incidence of the accrued actuarial cost of pension plan benefits and the related actuarial liability.

†The terms "current services" and "future services" are often used interchangeably. A desirable distinction can be made, however, by using the former term to relate to a specific current period and the latter to relate to the entire period after a specified date.

cost is the cost assigned to a given year under an actuarial cost method, and is not dependent upon the cost of benefits relating to service before the inception of a funded plan. Such service is called past service. (Normal cost is also sometimes referred to as "current service cost" and "future service cost.")

The *prior service cost* (sometimes called "accrued liability"*) under a plan is the value assigned under an actuarial cost method to services up to any given date. It may or may not be equal to the present value of benefits for such services. (Prior service costs are sometimes referred to as "past service costs," but this term should be used to refer to *initial* past service cost.)

The term *supplemental cost* refers to the difference, arising with the establishment or amendment of a plan, between the actuarial present value of total projected benefits and the value of total expected normal costs. It may also be termed *initial past service cost*.

The term *current costs* usually refers to the total of normal costs and the portion of the supplemental cost provided for during a given period under the actuarial cost method in use. This term is also used at times as a synonym for normal costs.

Calculation of normal costs may be made on one of two bases: (a) the *accrued benefit cost method,* sometimes called "step-rate" or "single premium," or (b) the *projected benefit cost method,* sometimes called "level cost" or "level premium."

The accrued benefit cost method considers as cost for a period (which may be a definite time interval, such as a fiscal year) the value of an annuity for each employee that would provide for retirement payments applicable to pension credits related to that period. Other things being equal, costs so calculated will increase with the advancing age of employees because of the smaller time remaining for income accretions before retirement age and the greater likelihood of survival to retirement. It is because of this ascending characteristic, that this method is often referred to as *step-rate*.

Under the projected benefit cost method of calculating periodic pension costs, the effect of the ascending characteristics of the accrued benefit cost method is counteracted. Actuarial costs are based on providing for the present value of total projected benefits by level amounts or level percentages of compensation. Costs so calculated will, other things being equal, remain at a uniform rate and unchanged unless employee compensation changes occur or pension benefits are revised.

31-5. Actuarial Assumptions

Pension costs are predicated on assumptions concerning future occurrences, such as continuance of employment with the employer, earnings on funds set aside for the future payment of pensions, or mortality. Actuaries use these assumptions when evaluating the financial impact of pension plan benefits and suggesting appropriate funding (or costing) arrangements to provide for the expenditures a plan will require. Here are some significant factors for which actuarial assumptions must be made.

Mortality As pension benefits are not paid unless the employee lives to retirement, and usually cease with

*The term "liability" is used in this monograph to describe the actuarial cost of a category of benefits. This term has been adopted because of common usage, but such adoption is not intended to suggest that there is necessarily any legal liability.

the death of a retired employee, an actuarial assumption about mortality rates of covered employees is a major consideration in the actuarial determination of pension costs. Making allowances for future mortality is sometimes referred to as *discounting for mortality*. The value of any included death benefits also depends on the mortality assumptions.

Employee Turnover   An assumption may also be made about the rates of future employee turnover, since, except where there is vested interest, termination of employment before retirement age eliminates benefits that would otherwise accrue, thereby reducing pension costs. Studies made of turnover rates usually involve recognition of the effects of age, sex, length of employment, type of work, etc., on turnover. Making allowance for future employment severance is called *discounting for turnover*.

Retirement Age   When plans permit retirement at a date other than the normal retirement age, i.e., at either an early retirement date or a deferred retirement date, assumptions about the rates at which employees will retire at various ages may be needed. However, in many plans the benefits are adjusted for early retirement to amounts that are equivalent, actuarially, to those at normal age. It can then be assumed that all employees will retire at normal retirement age.

Salary Scales   When benefits are keyed to future salary rates, as in a percentage-of-compensation formula, assumptions may be made about future salary scales. This is essential in the case of a final-pay plan, where all benefits are related to earnings for a limited period of years immediately before retirement.

Interest Rate   Increases in a pension fund result not only from contributions but also from the income earned on investments in, and net realized gains or losses of, fund assets. The income may be in the form of interest, dividends, or gains or losses, and sometimes the term *income yield* is used to describe both types.

Valuation of Fund Assets   Although it has been general practice to value the investments in a pension fund at cost, some plans now use the market value of the fund; others use book values adjusted from cost values by some appropriate formula that gives gradual recognition to unrealized appreciation. In the latter case, any resulting unrealized gain or loss is an additional factor that may modify the basic income yield.

Other Assumptions   Actuarial assumptions in addition to those already discussed may be required, depending on the provisions of a specific plan. Thus where plans provide for disability retirement or death benefits, or contain features dependent on marital status, appropriate assumptions are needed as to future events or status changes relative to these conditions.

Obviously, all the assumptions used will be subject to occasional revision, as experience under a plan increases. Assumptions are tested by comparison with actual experience as part of the periodic *actuarial valuations*. Assumptions thus indicated to have been conservative give rise to "actuarial gains" and optimistic assumptions, conversely, accumulate actuarial losses. The general approach to actuarial valuation and the disposition of gains or losses will be analyzed in the section on Actuarial Gains and Losses.

31-6.   Present Value

Basic to all funding methods is the concept of *present value* (PV), sometimes referred to as capitalized value. The PV principle permits the value at any given point of time under one set of future conditions to be expressed as the equivalent value at the same point of time under any other set of future conditions. The principle is particularly useful in dealing with financial transactions involving a time series, such as periodic contributions, retirement annuities, etc. It permits the computation of an entire series of financial transactions over a period of time to be expressed as a single value at any point of time.

The term "present value" is frequently used in accounting literature. The present value of a debt of "Y" dollars payable "X" years hence is the amount that, with accumulated compound interest thereon for "X" years, would accrue to "Y" dollars. This illustration employs a single assumption of the interest increment because the continued existence of the debt is certain. Actuarial use of the concept of present value in pension plans, however, involves the probability of a payment's being made "X" years hence, based on a variety of assumptions, including mortality, severance of employment, etc. The use of previously considered factors, such as discounting for interest, mortality, and severance, is intrinsic to the actuary's concept of present value.

The principle of present value in a particular instance will reflect the relevant actuarial assumptions. For example, if the value of a retirement benefit commencing at a given date is equated to one commencing at a different date, only the assumptions concerning income yield and mortality usually need be considered. The new value will be referred to as the *actuarial equivalent*. In plans providing for early retirement, for instance, the value of reduced benefits resulting from such early retirement may be described as the actuarial equivalent of the value of normal retirement benefits. The actuarial equivalent concept is also used when the value of a retirement benefit under one set of conditions is equated to one under a different set of conditions, e.g., the conversion from a normal benefit to a joint and survivorship optional benefit.

FUNDING POLICIES AND METHODS

31-7.   Financing Policies

Financing policy may be defined as a company's general operational policy, which enables it to provide for pension payment obligations as they mature and become payable. Financing policies usually fall into one of three general categories:

- *Pay-as-you-go,* under which no funding is done for pensions except as payments are made to retired employees. (Reference to pay-as-you-go as a "funding" policy constitutes a misnomer.)
- *Terminal funding,* under which no funding is done for pensions until the date of an employee's retirement. At that time a fund is set aside that is actuarially equivalent to all payments to him expected to be called for under the pension plan.
- *Advance funding,* under which contributions are made to a pension fund during the active service lives of employees under one of several actuarial cost methods.

Practices under advance funding methods may involve either full or partial funding. *A fully funded* plan is one in which funding contributions are at least enough to cover pension credits accrued to the current applicable date for both credited past service and credited membership service. Since the amount applicable to credited past service at the inception of a plan may be substantial, a plan may be referred to as fully funded if it includes an orderly and systematic arrangement that contemplates future full funding of such past service costs, even though it may not be fully funded during its earlier stages. A plan that does not contemplate complete funding of past service credits would be a *partially funded* plan.

Reference to full funding versus partial funding suggests the concept of *actuarial soundness,* a term frequently used in connection with pension plans. This is a somewhat flexible concept. Although definitions vary, depending upon viewpoint and conservatism, in general a pension plan may be said to be actuarially sound if:

● A formal and substantial system of periodic contributions has been established and adhered to
● There is a reasonable certainty of the employer's intention and ability to continue such contributions
● Such contributions, after reflecting relevant actuarial factors, are fully expected to be enough to meet pension payment obligations under the plan as they arise.

There is some difference of opinion as to whether or not a partial funding policy is actuarially sound. From a strict viewpoint, a program contemplating only partial funding does not seem to fit the definition of actuarial soundness, since ability to meet obligations as they arise does not rest solely on the fund, but continues to depend to some extent on the employer. Some partially funded plans, nevertheless, may provide reasonable assurance that probable obligations will be met.

31-8.   Effect of Funding Policy on Periodic Pension Costs

Advance funding methods call for larger contributions during the earlier period of a plan than do pay-as-you-go and terminal funding methods, which defer any consideration of pension costs until retirement dates are reached. Similarly, those advance funding plans using the projected benefit cost (level-cost) method generally call for larger contributions initially than do those using the accrued benefit cost (step-rate) method.

In the determination of the impact upon contributions of a particular actuarial cost method in a specific case, consideration must be given to such factors as the average age or population maturity of the employee group and the period over which past service costs are expected to be funded. A young, growing company, for example, would have employees whose age distribution would lean heavily toward the younger side. An older company, particularly one whose size has stabilized, not only would have an older age distribution of employees but also would probably have a substantial number of employees already on a pension roll. Such a company could be described as one with a *mature population.* The funding requirements of the two groups would obviously be quite different, since much of the liability on the latter company would be for those beyond or close to retirement. Therefore, not as much of the cost of providing pensions could be obtained through investment income produced by funds under the plan.

One of the purposes of pre-funding a plan is to obtain tax-free investment income. The more rapidly the funding is done, accordingly, the smaller will be the total direct costs ultimately required, and the greater the portion of ultimate payments that comes from fund earnings.

31-9.   Past Service Costs (Supplemental Cost*)

In addition to choice of actuarial cost methods, the policy adopted for funding past service costs has an important bearing on the periodic incidence of pension contributions. The separate determination of initial past service cost and its differentiation from normal cost is significant because of the variety of methods available for funding such initial costs, and because of the income tax limitations on the amounts permitted to be deducted in any one year. Because the initial past service cost is usually substantial, and emerges suddenly upon the adoption of a pension plan, a greater degree of flexibility is permissible in its funding than would be deemed actuarially sound in funding normal cost.

When there is a separate determination of past service costs, several methods of funding or otherwise providing for these costs can be followed. Frequently, past service costs are funded in level amounts over a specified period of years—ranging from as low as 11 years to as high as 40. Under plans that do not differentiate between past service costs and normal cost, there is an implicit or built-in funding of past service credits over the remaining service lives of employees who were covered at the plan's inception.

In some cases, an employer may decide to delay the funding of initial past service costs indefinitely. In these circumstances it is usually necessary, for IRS qualified plans, to make minimum contributions equivalent to interest increments on the unfunded cost in order to prevent its increase. Some employes may elect to meet the initial past service cost by contributing a large sum to the pension fund at the inception of the plan, or in its early stages. However, this course may be discouraged by limitations on currently obtaining income tax deductions for amounts contributed in excess of those allowable under pertinent tax regulations.

In general, past service cost contributions to plans that qualify under Section 401 of the Internal Revenue Code are not deductible in any one year in an amount in excess of 10 percent of the initial past service cost. Contributions that exceed 10 percent in any year may be carried forward as deductions to succeeding years when contributions would be less than 10 percent. An exception to this 10 percent limitation applies to certain level-funding plans. The minimum funding requirements of the Internal Revenue Code are met if the total unfunded cost of the plan at any time does not exceed the initial past service cost. This is accomplished if the total of the payments to date, in addition to normal cost of the plan, is at least equal to interest (at the assumed rate) on the initial past service cost for each year of the plan.

31-10.   Pay-As-You-Go Method

Under a completely unfunded plan, the contributions are merely the pension payments as no advance funding

---

*Under the proposed standard terminology, supplemental cost is the term for any liability treated as an element of actuarial cost separate from normal cost. Past service cost would, therefore, be a supplemental cost.

has been undertaken. This is referred to as the pay-as-you-go method. The effect of pay-as-you-go financing will vary in proportion to the population maturity. The basic characteristic of the pay-as-you-go method is that payments initially may be low, but increase rapidly and unevenly until an employee group (including retired persons) achieves a mature composition, at which time the payments tend to become uniform.

31-11. Terminal Funding Method

Under terminal funding, contributions are made only upon the retirement of employees. At that time, the present value of future benefits is paid into the fund as a single contribution. Sometimes, however, this payment is made in installments, giving such terminal funding some aspects of a pay-as-you-go method.

Terminal funding resembles the pay-as-you-go method in that contributions initially may be low, but will increase unevenly until the employee group achieves a mature composition. Contributions then tend to become level. The ultimate normal costs of terminal funding will be somewhat less than those of pay-as-you-go funding, however, because terminal funding produces some tax-free income yield on contributions.

31-12. Advance Funding Methods

Accrued Benefit Cost Method This funding method employs the unit-cost (or step-rate) basis of measuring accruing pension costs. It is also referred to as the *single premium* or *unit purchase* method of funding. The accrued benefit cost method operates on the principle that the retirement pension is or may be divided into units of benefit, each unit being related to a year of employment service. The normal (i.e., the current service) cost for each year of service is the amount of the contribution needed to provide in full for the benefit unit or units applicable, or allocated, to that year.

The benefit units applicable or allocated to the various years of an employee's credited service are usually equal in amount or equal as a percentage of earnings. The normal cost applicable to an individual employee, however, will increase with each advancing year, since it is increasingly probable that he will live, and work for the employer, until retirement age, and there is less time to earn interest on the amount funded. The total annual contribution for the group as a whole usually does not reflect the same marked step-up effect in progressive years because of the effect of replacement, generally, of older employees upon their death or retirement by younger employees. For a mature population, therefore, the normal cost may be relatively uniform, while the normal cost for an initially immature group will rise before ultimately leveling off.

The supplemental cost (past service cost) is the single sum necessary at the inception of a plan to provide for retirement benefit units applicable or allocated to credited service for the years before inception. A supplemental cost can also arise when a plan is amended to provide increased benefits. The supplemental cost may, of course, be funded in a variety of ways. A popular method is to amortize (fund) it on the basis of a series of uniform payments, including an interest factor, over a specified term. Some insured plans using the unit credit funding method provide that the initial accrued liability for any individual must be fully funded by his normal retirement date.

The group annuity type of insured plan, which will be discussed at length in the section on Funding Agencies, is a classic example of the accrued benefit cost method of funding, but the method is also used in self-administered (trusteed or deposit administration) plans. Since current service costs are usually funded regularly, the unfunded accrued liability will probably be simply the unfunded supplemental cost.

Projected Benefit Cost (Level-Funding) Methods Some form of level funding can be used to eliminate the step-rate increases in normal costs as an employee gets older. This obviously requires higher initial contributions than does funding under the accrued benefit cost method, though contributions will be lower in later years. A level-cost method may use contribution rates that are level in dollar amounts or a level percentage of earnings. If level funding is established at an empoyee's current age to cover the benefits for all subsequent years, it is known as the *attained-age* normal cost method. If level funding is established as of the age the employee could have entered the plan had it been in existence, it is known as the *entry-age* normal cost method. These concepts will be discussed more fully in later sections of this chapter.

The usual level funding serves to smooth out only the increases resulting from advancing age. If the increases resulting from the higher benefits that occur when salary rates are raised are also to be leveled out, it is necessary to include assumed rates of expected future salary progression (salary scale) with the other actuarial assumptions used.

Entry-Age Normal Cost Method This cost method adopts the level-cost approach, wherein the uniform contribution rate is that which would have been needed to fund pension benefits over the entire service life or lives of an employee or group of employees had the current pension plan always been in effect.* Frequently, an average age for all employees is used. The supplemental cost is separately determined and funded.

This supplemental cost may be defined as the "fund" that would have been accumulated at the inception of a plan if annual contributions had, in fact, been made over the entire credited service lives of employees up to that time. This is a retrospective method.

The supplemental cost is usually determined, however, by a prospective method. The present value of the future normal contributions is subtracted from the present value of all expected benefit payments, and the difference is regarded as the initial supplemental cost. If all elements were in accordance with the actuarial assumptions, this would equal the amount that would have been on hand if the plan had always been in effect. For any subsequent valuation date, the value of all expected benefits would be determined by the same procedure. The amount so determined, less the value of future normal contributions and less the funds accumulated under the plan, would represent the then unfunded supplemental cost.

The initial supplemental cost, as has been described, may be funded in a variety of ways.

The entry-age normal cost method is used in self-administered trusteed plans and for insured plans such as the deposit administration group annuity. These types of plans will be discussed in the section on Funding Agencies.

*The term "entry-age" derives from the fact that the contribution rate is determined on the assumption that the plan had always been in effect, and contributions commenced when the employee entered the plan.

Individual Level-Cost Method This method uses the level funding approach, but fixes the level cost or level rate of contributions for each employee at an amount that would fund future pension payments over the remaining future service of each employee.

For those employed after the adoption of a plan, this method gives results similar to the entry-age normal cost method. For other employees, however, the individual level-cost method differs from the entry-age method, which relates the factor of normal cost to the entire period of service and makes separate provision for the initial unfunded cost of supplemental cost (treated as past service cost). The individual level-premium method computes the periodic cost on the basis of funding both past and future service costs, without distinction, from the date the employee becomes covered (which, for older employees, is the date the plan was established) to the normal retirement date.

This method is also termed the *individual funding to normal retirement age* cost method. The supplemental cost is funded (i.e., amortized) in this case over the remaining service lives of employees covered at the adoption of the plan. Accordingly, the first annual contribution will be especially high in order to fund fully the benefits for employees retiring in the first few years, unless they are excluded. The same pyramiding of costs in the early years would occur, but in declining amounts, for those retiring in those years.

The individual level-cost method is used in insured individual policy or group permanent plans and can be adopted for self-administered trusteed or insured plans.

Aggregate Method This method of funding is similar to the individual level-cost method, except that calculations are made on a collective basis. The present value of benefits expected to be paid after a given date, with respect to all employees or retired employees covered as of that date, is determined, to arrive at a rate of funding; this amount, less any accumulated funds, is divided by the present value of estimated future-compensation of all covered employees. No part of cost accruals is identified with the projected benefits of specific individuals.

The aggregate method is used in self-administered trusteed plans and may also be used in an insured deposit administration plan. As in the individual level-cost method of funding, the aggregate method does not provide for a separate method of funding the initial supplemental cost.

A combination method is often used, referred to as an "aggregate method with frozen initial liability." The first-year costs are determined as they are under the entry-age normal method. Initial supplemental cost is amortized in the same way past service cost is, but it is not adjusted at any time to recognize actual experience. It is thereby termed "frozen initial liability." In each year after the first, cost is computed as under the aggregate method, except that the unfunded part of the frozen initial liability is subtracted as well as the accumulated funds. Since the former is being amortized by separate payments, which can be flexible in amount, it is thus excluded from normal cost.

The aggregate method follows the level-cost approach in that an overall uniform contribution rate is established.

Attained-Age Normal Method This method of funding combines some features of the accrued benefit cost method and others of the aggregate method. As under the accrued benefit cost method, future benefits are divided into units applicable to past and future service. The initial supplemental cost, i.e., the single sum necessary to provide for retirement benefit units applicable to credited service for years prior to the inception of the plan, is computed as it is under the accrued benefit cost method. This initial supplemental cost may be funded separately from normal cost, and in a variety of ways.

The funding of future service benefits, however, follows the aggregate method. Whereas the aggregate method arrives at the over-all contribution rate by equating all future benefits to future compensation (on a present value basis), the attained-age normal method relates the present value of that portion of future benefits atributable to future service (i.e., service after the plan's effective date) to the present value of future compensation. The normal cost is the amount derived when this over-all rate is applied to total current annual compensation. The funding approach to normal costs therefore uses the level-cost approach. This method's name derives from the view of normal cost as the annual amount necessary to fund "future service" benefits over the period beginning with the age the employee has attained at he date of the plan's inception, or, if he was employed after its inception, the date of initial coverage under the plan.

FUNDING AGENCIES

31-13.   Trusteed Plans

A pension plan that uses a trust or pension trust as its funding agency is called a *trusteed* pension plan. Other terms used for such a plan are *self-administered* pension plan; *self-insured* pension plan; and *uninsured (or non-insured)* pension plan. Trusteed plans are generally set up under an agreement between an employer and a bank or trust company that is to act as trustee; the agreement prescribes the terms under which the trust is to be created and administered. Subject to the provisions of the trust agreement, investment decisions may be made by the bank or by trustees of the fund. Cash and securities and the collection of income and disbursements for trust purposes are usually handled by the bank.

Under a trusteed plan, contributions are made to the pension trust, and the fund so established is invested according to the terms of the trust agreement. Earnings, of course, serve to increase the fund, and the fund's assets are used to provide pensions or other benefits for employees. Retirement benefits may be paid out by the trust during the period of retirement, or an annuity may be purchased for the employee upon his retirement, thus transferring the obligation at that date to an insurance company (see "split funding" given subsequently).

Retirement benefits to be paid are based upon the stipulations in the plan. The trustees assume responsibility for the proper administration of the trust, but do not guarantee actuarial adequacy, rate of income, or availability of assets to pay benefits.

The majority of large- and medium-sized funded pension plans currently in effect are of the self-administered trusteed type.

31-14.   Insured Plans

An arrangement with an insurance company may be used as a medium for the receipt of pension contributions, administration of a pension fund, and payment of benefits. A plan that uses only an insurance company as a funding agency is called an *insured plan*. Under some insured plans, insurance companies contract to pay specific retirement benefits in return for certain set premium payments. Under others, the insurance companies accept and accumulate funds for use at subsequent dates, either to pay specified retirement benfits in return for certain set premium payments taken from the accumulated funds or to pay benefits directly from the accumulated funds. The retirement benefits and eligibility for such benefits are established, generally, on the basis of the provisions of the pension plan. Insured plans may be adopted that would provide either for individual policy contracts or for master group contracts. Under some insured plans, segregated accounts may be used which accumulate funds separately from the insurance company's general assets. This procedure differs little from funding through a trust fund. Some of the more common types of insured plans are described here.

Group Annuity Plan Under this type of insured plan, the employer enters with the insurance company into a master contract that covers the entire group of eligible employees. The contract calls for the purchase each year of a full paid-up deferred annuity in an amount equal to the benefit accrued in that year for each participating employee. This is the *accrued benefit cost* method for computing contributions to the plan. When an employee retires, the total amount of his benefits will be equal to the sum of the separately purchased units of deferred annuity. This plan is also referred to as a *deferred annuity group annuity plan.* Because there is a direct correspondence between the premium and the amount of deferred annuity that is purchased by a given single premium, the group annuity plan is also said to employ the *single premium* method of funding.

The premium rates applicable to the various ages under a group annuity plan are guaranteed by the insurance company for a fixed period, usually five years, after which a new premium structure may be established and guaranteed for a future fixed period.

Since group annuity contracts usually do not provide for death benefits to be paid only from employer contributions, premium rates related to these contributions reflect a discount for mortality. Premium rates for employee contributions are set at a higher level and because they do not involve a discount for mortality, the employee's contributions, usually with added interest, are paid as a death benefit if the employee dies before his retirement date. Group annuity plans do not discount for turnover; past employer's contributions which have been made for the account of those separated employees who do not receive vested benefits are credited against the employer's next premium payment, less a portion retained by the insurance company as a *surrender charge* to provide for applicable administrative expenses. Employees' contributions are returned, usually with interest.

The premium payment schedule for a given employee is usually of the *steprate type,* i.e., premiums increase each year with advancing age. However, the stability of the overall average cost for all employees depends upon the current and future age distribution of the group or its degree of maturity.

Although this is not common, some group annuity plans may use a money purchase basis for computing benefits (as described under Normal Retirement Benefits). In such instances, the annual contribution applicable to each employee will be determined by the plan. The amount of the deferred annuity that can be purchased for an individual each year with a constant amount of annual contribution will become progressively smaller as his age increases.

A group annuity plan ordinarily uses an explanatory booklet instead of a plan document. The provisions of the plan are incorporated in the master contract and individual employees receive *certificates* as evidence of participation.

At one time most insured plans were of the group annuity plan type. This type is now used primarily for small groups or special situations.

Individual Policy Plan As the name implies, this type of plan involves the purchase by the employer of a separate life insurance or annuity policy in the name of each covered employee. The premiums paid under such a policy are referred to as *level annual* premiums because they remain uniform for each employee through the period of coverage to retirement date. The premium is determined by reference to annuity benefits the employee will receive. When benefits are revised for any reason, such as changes in compensation, other policies will be issued and total premiums will change accordingly. An annuity policy is sometimes referred to as a *level premium deferred* annuity or an *annual premium retirement* annuity.

Individual policy plans frequently use insurance policies that provide, for an additional premium, life insurance protection in addition to the annuity benefits. A policy that combines annuity and life insurance features is called, variously, a *retirement income* policy, an *insurance income* policy, or an *income endowment* policy. Such an individual policy provides a death benefit before retirement equal to a specified sum related to monthly retirement income or to the cash value of the policy, whichever is greater. An employee who is uninsurable because of poor health may receive a policy with a death benefit limited to the cash value of the policy. Death benefits after retirement will vary with each plan, but a common practice is to provide a period certain life annuity, the period usually being 10 years.

Individual policy premiums do not involve discounting for either mortality or severance before retirement age. In the event of severance, the employer is entitled to a refund, but this is considerably less than the amount credited for severance under a group annuity plan.

Because of the possible hazards of direct ownership of the policies by either the employer or the employee, individual policy plans usually provide for a trust under which the trustee holds title to the individual policies and administers the plan in accordance with the formal plan document or a separate trust instrument.

Deposit Administration Group Annuity Plan Under this plan, as in the case of a group annuity plan, a master contract is drawn between the employer and the insurance company. This contract may be referred to as a *deposit administration* contract, a *deposit administration group annuity* contract, or, in abbreviated form, a *DA* contract.

No annuity units are purchased for the individual employee before the date he retires, as is done under a group annuity plan. Instead, the periodic contributions of the employer are deposited with the insurance company in an unallocated fund, to which the insurance company adds interest at a rate usually guaranteed for five years but subject to change thereafter. The rate, when changed, applies to funds deposited from the date of change, but the initial rate often continues to apply to previously deposited funds. When an employee retires, the amount required to buy an *immediate* annuity to provide for his pension is applied by the insurance company as a single premium from the unallocated fund. The premium rates for such annuities to be purchased in the future are also guaranteed by the insurance company on a five-year basis. The insurance company guarantees the pension payments on purchased annuities, but does not guarantee that the unallocated fund will necessarily be adequate to meet the cost of annuities to be purchased. The contract provides for periodic dividends (rate credits, if the company is a stock company) determined by the insurance company, at its discretion, on the basis of its experience under the contract. In effect, these dividends represent an adjustment of premiums. Contributions to the unallocated fund may be determined by the use of any one of several actuarial cost methods.

Immediate Participation Guarantee Contract   This contract is somewhat similar to the DA contract. The master contract, if it is with a mutual insurance company, is called an *immediate participation guarantee* contract or, in abbreviated form, an *IPG* contract. If it is with a stock insurance company, it is called a *direct rerating* contract.

As in the case of a DA contract, the employer makes periodic payments into a fund maintained by the insurance company, and the insurance company credits interest to the fund and pays annuity benefits on employees' retirement. However, unlike the DA contract, which guarantees interest, the IPG contract calls for interest credits to be based essentially on the rate of the insurance company's earnings for the year on its investment portfolio after investment profits and losses are reflected. The insurance company charges expenses directly to the fund, whereas under DA contracts expenses are charged only in the determination of dividends or rate credits. Essentially, the IPG contract is intended to give the employer the immediate effect of experience under the contract, including the insurance company's investment results; at the same time the insurance company makes fewer guarantees under such a contract.

The IPG contract is written in two forms with the same ultimate result. One form, like the DA plan, may provide for the actual purchase of single premium annuities as employees retire. There is an adjustment each year for actual experience under the annuities, based upon the insurance company's analysis of mortality, benefits paid, and earnings. The annual adjustment for favorable or unfavorable experience is reflected by an addition to, or a deduction from, the fund.

In the other form, the IPG contract may accomplish the same objective through a different technique. When an employee retires, the retirement income payments to him are made directly out of the fund without the purchase of an annuity. However, the fund on deposit with the insurance company must be maintained at the amount required, according to a premium schedule in the contract, to provide for the remaining retirement benefits for all those on retirement at any time. Thus, if necessary, the fund could always be applied to buy all annuities in force. Under either form of IPG contract, the insurance company guarantees lifetime benefit payments to retired employees.

Separate Accounts   Legislation permitting an insurance company to offer separate accounts as an adjunct to one of the other insured plan contracts is now in effect in most states. The investments in these accounts are not restricted by insurance laws; they are used primarily for investing pension plan funds in equities. Such an account may be established solely for use in conjunction with the contract for one policyholder. Usually, however, such an account is pooled with funds of various policyholders. The availability of separate accounts with DA or IPG contracts

has resulted in a considerable increase in the number of plans using insurance company funding facilities.

The separate account facility is now sometimes used as a basis for the insurance company to offer investment services only without necessarily being involved in the payment of plan benefits. This is the same as the investment service of a bank under a trusteed plan.

Split Funding   Many employers use both a trust fund and an insurance contract in an effort to get some of the expected advantages of each method. If this is done, the annuities to retired employees would probably be paid under the insurance contract, and all or part of the funds for active employees would be accumulated in the trust fund. The DA or IPG insurance contract is usually used for this purpose. If the contract calls only for the purchase of immediate annuities as employees retire, with all funds for active employees being accumulated in a trust fund, it is similar to the portion of the DA contract used to purchase annuities as employees retire.

ACTUARIAL GAINS AND LOSSES

To the extent that actual experience after actuarial valuation differs from the actuarial assumptions used in the valuation, actuarial gains or losses will arise. For example, if the actual experience is such that the contributions to the plan, which were previously determined on the basis of such assumptions at a particular valuation date turn out to be larger than necessary, an actuarial gain would result. If the contributions turn out to be inadequate, the result would be an actuarial loss. In other words, differences in the actual rates of mortality, turnover, and income yield from those originally assumed would result in actuarial gains or losses. If, for example, more employees die prior to retirement, or soon after retiring, than had been anticipated, less retirement benefits will have to be paid, since the contributions will have been too great, and an actuarial gain results. The opposite situation would involve an actuarial loss.

When the actuarial costs of a plan are being computed, which is normally done annually, the effect of actual experience to date as it differs from the actuarial assumptions will automatically be reflected in the new computation of costs. However, the present value of projected benefits and other elements in the valuation, from the new date on, will again be based on the actuarial assumptions. It is necessary, therefore, to determine when the experience indicates that a revision should be made in the actuarial assumptions used for projecting into the future.

Actual experience may differ from the actuarial assumptions merely because of fluctuations occurring when the group is not extensive enough for averages to work out. On the other hand, the differences may arise because the assumptions used are not applicable to the group, or have now ceased to be applicable. For example, many plans that were started in the 1950's used an interest assumption of 2½ percent, which at that time seemed as high as one could reasonably expect on a conservative basis. The investment returns have been so well above this for many years that this rate is no longer applicable into the foreseeable future.

It was pointed out previously (see Preliminary Actuarial Considerations) that the following formula must be used for pension plan funding:

$$\begin{matrix} PV \\ contributions \end{matrix} = \begin{matrix} PV \\ benefits \end{matrix} + \begin{matrix} PV \\ expenses \end{matrix} - \begin{matrix} Funds \\ on\ hand \end{matrix}$$

Anything that reduces the present value of contributions can be considered a gain. The formula shows that such a gain would result if there were a reduction in the present value of benefits, a reduction in the present value of expenses, or an increase in the funds on hand. It could also result if the actuarial factors for determining the present value of contributions should cause a reduction. For example, if the investment income to date has been greater than the rate assumed, the funds on hand will have been augmented and the present value (PV) of contributions will be reduced. If mortality to date has been greater than expected, fewer employees will survive to retirement and the present value of benefits will be reduced. To the extent that employee turnover is different from that expected, there will be a different number of employees included in the present value of benefits. Conversely, anything that would increase the present value of contributions could be looked upon as a loss. Changes that occur from causes outside the experience under the plan, however, would not be considered a gain or loss. An example would be a change in the present value of benefits arising from a change in the benefit formula.

When a change is made in the actuarial assumptions, it will affect both the present value of units of contributions in the future and the present value of benefits. The net gain or loss is therefore determined by this interaction.

31-15.   Computation of Actuarial Gain or Loss

It is harder to analyze and compute the amount of gain or loss attributable to each actuarial assumption than it is to calculate aggregate net gain or loss on an overall basis. As part of his annual actuarial valuation of a plan, however, the actuary will frequently make the technical computations necessary to analyze the gains or losses by source to determine the effect with respect to each of the actuarial assumptions. It is necessary for him to make these analyses in order to decide when adjustment should be made in the actuarial assumptions for use in the future.

The aggregate gain or loss can be generally determined by a comparison of the actual fund on hand at the valuation date with the fund that would have been on hand if experience had followed the actuarial assumptions used, taking into account the actual contributions made to the fund. If the actual fund exceeds the "expected" fund, the difference is the actuarial gain; a deficiency means an actuarial loss. If there has not been full funding of the supplemental cost (initial past service cost) by the valuation date as of which the actuarial gain or loss is calculated, the expected fund would consist of the accrued liability at the end of the period less the expected unamortized supplemental cost as of that date.

The following schedule shows a computational technique which corresponds to the description just given but reflects the inclusion of additional details:

Present value of future benefits payments to employees for credited service to end of period	$300,000
Less expected unfunded supplemental cost at end of period	95,250
Expected fund	204,750
Less actual fund at end of period	206,000
Actual loss (gain)	($ 1,250)

For the above computation, the expected unfunded supplemental cost at the end of the period, e.g., a year, may be computed as follows:

Unfunded supplemental cost at beginning of period	$100,000
Plus interest on above to end of period (at the assumed interest rate)	5,000
Total	105,000
Less contributions during period toward supplemental cost	10,000
Plus interest on above from contribution dates to end of period at assumed rate	250
Expected unfunded supplemental cost at end of period	$ 95,250

This general procedure for calculating actuarial gains or losses is easily adaptable for some funding methods, but requires modification for others. In terminal funding, which fully funds the accrued liability only at retirement date, the gain or loss calculation involves no more than a comparison of the actual fund with the present value of the future benefits for employees already retired.

Because the individual level cost and aggregate cost methods do not differentiate between normal and past service costs, the calculation of actuarial gain or loss for a given period is performed on an overall basis. The actual unfunded cost at the end of the period (the actual accrued liability less the actual fund on hand) is compared with the expected unfunded cost (the expected accrued liability less the expected fund on hand). Any excess of the actual unfunded accrued liability over the expected amount constitutes an actuarial loss, while the reverse would constitute a gain. The difference between this procedure and the general method previously described should be noted. The general method compares actual and expected funds, while this procedure compares the actual with the expected unfunded accrued liabilities.

31-16.   Actuarial Gain or Loss Adjustment

There are two ways of adjusting for actuarial gains or losses. The *immediate* method entails the immediate addition of an amount equal to the loss, or subtraction of an amount equal to a gain, to the next normal cost contribution. Where there is a loss, however, it can be added to the unfunded supplemental cost; this is almost always done with the accrued benefit (step-rate) cost method.

The second adjustment method, called the *spread* method, spreads the adjustment over future periods and is almost always used with the projected benefit cost method (level funding). The gain or loss is spread over the same future period used by the basic funding method. If the actuarial assumptions are being modified, the adjustment of prior actuarial gains and losses may be added to the effects of the new actuarial assumptions in order to arrive at a modified contribution rate for future periods. Although the adjustment usually affects only future normal costs, in some situations it may be apportioned to the unfunded supplemental cost.

In the *aggregate cost* method, the spread technique is automatically used because the funding percentage is based on the aggregate unfunded present value of all future benefits. Thus, a revised funding percentage at any valuation date automatically spreads the adjustment over future periods. Since the initial supplemental cost is not separately determined in the computation of the funding percentage, the aggregate method, in effect, applies the actuarial gain or loss to both normal and past service contributions.

31-17. Valuation of Fund Assets

One of the factors that enters into the calculation of actuarial gains and losses is the amount of the pension fund. It is also a factor in the calculation of any underfunding or overfunding of the pension plan. Thus it is important to use an acceptable method of valuing the assets of the fund.

In arriving at actuarial valuations, investment portfolios of pension funds traditionally have been valued at cost; asset gains or losses, therefore, have not entered into the calculation until they have been realized through sales. Use of cost rather than current market value stems from the consideration that securities are held for long periods; pension fund valuations should not reflect cyclical market conditions. The use of a cost basis can provide unrealized gains that have not been taken into account to act as a buffer against future investment losses. Alternating such gains may be considered desirable to help adjust benefits in future years that become inadequate through loss of purchasing power.

APB Opinion No. 8 specifies that unrealized appreciation and depreciation should be recognized in the determination of pension cost for accounting purposes. This has caused employers to depart from the use of the cost basis where the effect would be material. As a result, several techniques are in use for the gradual recognition of unrealized appreciation, since many funds have developed substantial gains of this type. The additional amounts so recognized become actuarial gains. The most direct method would be to use market value in valuing fund assets, but the effects of short-term fluctuations often make this undesirable. The various techniques now used seek to smooth out these short-term effects.

THE DEVELOPMENT OF ACCOUNTING FOR PENSION COSTS

31-18. Historical Background

Payments to retired employees have probably been made ever since commerce and industry began. Early payments were granted on a purely informal, almost casual, basis; their amounts and the persons to whom they were given depended almost entirely upon the will of the employer, as did continuance of payment after retirement. Early pensions were probably largely gratuitous, granted in recognition of past employee loyalty. Undoubtedly, though, an equally important consideration was the removal of superannuated employees from the payroll.

This informal basis of dealing with pensions carried into the present century, in the early part of which formal pension plans were largely confined to government employees and employees of railroads and utilities. During this time, pensions invariably were accounted for on a pay-as-you-go basis; even formalized plans did not generally provide for advance funding. The prevailing opinion was that the granting of a pension to a particular employee was a specific decision related to the circumstances of that employee at the time of the grant, rather than an established general policy.

As time went on, however, the granting of pensions gradually became a permanent part of employment policy. A pension grant came to imply that payments would continue to the employee throughout his retirement, and that similar consideration would be given to other employees. At this point, managements began to appreciate the desirability of making some advance provisions, during the active service lives of employees, for pensions they expected to pay those employees upon retirement. Such provisions often took the form of contractual arrangements with insurance companies. Self-administered funding arrangements were also instituted—a form of self-insurance. Enactment of Internal Revenue Code provisions under which contributions to a qualified pension plan fund were made immediately deductible by an employer (but with the resultant benefits not becoming taxable income to the employees until actual pension payments were made) gave considerable impetus to the formation of self-administered funding plans.

The increasing adoption of formalized funded pension plans and the consequent need for determining the amounts of contributions to the funds appropriate under the plans led to recognition of the concept of past service costs as an element in determining the amount of funding contributions, particularly after the tax deductibility of contribution amounts applicable to past service became subject to separate treatment. This gave rise to considerable uncertainty, and some difference of opinion, among accountants about the nature of contributions made for past service credits. Since these contributions were for past services, the question arose whether they should be charged to surplus at the time they were made, or to current expenses, which is the generally accepted procedure for contributions for current services.

31-19. Statements of the American Institute of Certified Public Accountants (AICPA)

The AICPA's first pronouncement on accounting for pension plan costs was set forth by the Committee on Accounting Procedure in Accounting Research Bulletin (ARB) No. 36, published in 1948. In this bulletin, entitled "Accounting for Annuity Costs Based on Past Service," the Committee expressed the belief that costs of annuities based upon past services were generally incurred in contemplation of present and future services, not necessarily of the individual affected, but of the organization as a whole, and that such costs should be allocated to current and future services.

As expressed by the Committee, this belief was based on the assumption that:

". . . although the benefits flowing from pension plans are intangible they are nevertheless real. The element of past services is one of the important considerations of most pension plans and costs incurred on account of such services contribute to the benefits gained by the adoption of a plan. It is usually expected that such benefits will include better employee morale, the removal of superannuated employees from the payroll, and the attraction and retention of more desirable personnel, all of which should result in improved operations."

The clear-cut opinion, therefore, was that so-called past service costs should be charged to current and future periods, not to surplus.

The Committee did not offer an opinion, however, on the question of how accruing pension costs should be recognized in the accounts. It may be inferred that, aside from clarifying the accounting position with regard to contributions for past services, the Committee was probably satisfied with the existing practice of according accounting recognition to accruing pension costs in amounts equivalent to pension fund contributions. This was also generally in line with Internal Revenue Code requirements governing the deductibility for tax purposes of contributions to qualified pension plans.

The Internal Revenue Code limited deductions for pension costs to actual payments or contributions to qualified pension funds. As was noted in the section on Funding Policies and Methods, deductible contributions in any one year were required as a minimum, to be in such an amount as to prevent the total unfunded cost of the plan from becoming greater than initial past service costs at any time. As a maximum, they could not, except in the case of certain level funding plans, exceed current costs plus 10 percent of the initial past service cost.

Although these requirements undoubtedly were entirely appropriate from the financial and tax administrative viewpoint, they did not necessarily form a rational basis for the recognition of accruing pension costs for accounting purposes. ARB No. 36 was, in effect, reaffirmed without substantial change in 1953 by its inclusion as Chapter 13(a) in ARB No. 43—Restatement and Revision of Accounting Research Bulletins.

In September 1956, ARB No. 47 was issued. In this bulletin the Committee was more specific about how past service cost should be accounted for, and also introduced the concept of vested benefits. The Committee, in ARB No. 47, expressed a preference for full accrual of pension costs over the remaining service lives of employees covered by a plan. However, it regarded as being acceptable 'for the present," minimum accruals to the extent that:

". . . the accounts and financial statements should reflect the accruals which equal the present worth, actuarially calculated, of pension commitments to employees to the extent that pension rights have vested in the employees. . . ."

The Committee expressly recognized the desirability of systematically accruing pension costs over the expected period of active service of covered employees, generally on the basis of actuarial calculations. It specifically stated that these accruals should not necessarily depend on funding arrangements, or on strict legal interpretations of a plan. It indicated a view that past service costs should be charged off over a reasonable period on a systematic and rational basis that would not distort the operating results of any one year.

Despite the Committee's expressed preference for this full accrual basis as a desirable procedure in accounting for pension costs, it recognized that opinion had not yet sufficiently crystallized to assure agreement on any one method of dealing with the subject. Also, it observed that a substantial body of opinion considered full accrual of pension costs to be unnecessary because:

- Amounts accrued might differ materially from payments to retired employees.
- Such accrual might result in recording pension costs in excess of legal liability.
- The company might never use the actuarially calculated full accrual.
- In the event of liquidation, amounts accrued in excess of vested rights would revert to surplus.

Although it recognized minimum accruals, the Committee did not define the term "vested," or make any recommendation concerning appropriate actuarial cost methods or recognition of actuarial gains or losses.

Following the issuance of ARB No. 47, then, widely divergent accounting practices continued. Furthermore, certain companies seized upon the Committee's recognition of minimum accrual to determine the amount of pension costs charged to income. During 1958, a few well-known companies that had previously followed the practice of accruing the full amount of current service costs (which coincided with contributions to the funds) either eliminated or drastically reduced pension costs charged to income. This generally resulted in a material increase in net income. The supporters of these actions justified them on the grounds that funds provided in the past were enough to afford reasonable assurance of the continuance of pension payments, and more than enough to meet the company's liabilities for the then vested rights of employees; the minimum requirements of ARB 47 were satisfied.

There was considerable question over whether these reductions conformed with the intent and spirit of ARB 47. Carman Blough, then Director of Research of the AICPA, held that the Committee on Accounting Procedure intended the minimum accrual requirements in ARB 47 to be only a temporary provision to avoid drastic changes in existing arrangements. He did not consider the "minimum accrual method" an alternate and equally acceptable basis of accounting to which a company could shift at any time.

It was against this background that the Accounting Principles Board (APB), which succeeded the Committee on Accounting Procedure, decided that the subject needed further study, and authorized an accounting research study. This study, published in 1965, detailed the accounting complexities of pension plans. In November 1966, after lengthy considerations, the APB promulgated its Opinion No. 8, "Accounting for the Cost of Pension Plans," whose major objective was to eliminate inappropriate fluctuations in the amount of annual provision for pension costs.

31-20. Present Practice

Effective with fiscal periods beginning after December 31, 1966, costs charged to income are to be determined in accordance with the guidelines expressed in APB Opinion No. 8. The conclusions of the Board can be broadly stated:

"The provision for pension cost should be based on an actuarial cost method that gives effect, in a consistent manner, to employee group data, pension benefits, pension-fund earnings, investment gains or losses, and other assumptions regarding future events. The method selected should result in a systematic and rational allocation of the total cost of pensions among the employees' years of active service. If the method selected includes past service cost as an integral part of normal cost, the provision for pension cost should be normal cost adjusted for the effect on pension fund earnings of differences between amounts accrued and amounts funded. If the actuarial cost method deals with past service cost separately from normal cost, the provision for pension costs should include normal cost, an amount for past service cost, and an adjustment for the effect on pension fund earnings of differences between amounts accrued and amounts funded.

"Provisions for pension cost should not necessarily be based on contributions to the pension fund, be limited to the amounts for which the company has a legal liability, or fluctuate widely as a result of pension fund investment gains and losses or other causes unrelated to the employee group."

The Opinion narrowed the limits within which the annual provision for pension costs must fall by:

- Eliminating pay-as-you-go and terminal funding as acceptable methods of computing the annual provision for pension costs, except in the rare instances where their application would not result in amounts materially different from amounts obtained by the application of acceptable actuarial cost methods.
- Increasing the minimum required annual provision for pension costs to include a supplementary provision for vested benefits, if applicable.
- Requiring that actuarial gains and losses and unrealized appreciation and depreciation be recognized in the computation of the annual provision for pension costs in a consistent manner that reflects the long-range nature of pension cost and avoids large fluctuations in the year-to-year charges to the income statement.

The Board concluded that all employees who can reasonably be expected to receive benefits under a pension plan should be included in pension cost determination. It also concluded that any change made in the method of accounting for pension costs should not be applied retroactively. Disclosure recommendations, for accounting method changes as well as for other pertinent pension cost data, are set forth in the Opinion.

31-21. Lybrand's Comment on APB Opinion No. 8

The APB Opinion No. 8 applies "both to written plans and to plans whose existence may be implied from a well-defined, although perhaps unwritten, company policy." Where a company has been making regular payments to retired employees, there is a presumption that a pension plan exists within the meaning contemplated by the Opinion.

Defined-Contribution Plan The distinguishing characteristic of the defined-contribution plan is the requirement that the employer contribute to the plan pursuant to a stated formula (for example, a fixed amount for each ton produced, so many cents per hour worked, or a fixed percentage of compensation). Under a plan where the benefits for each employee are the amount that can be provided by the sums contributed for him (as in the case of a money purchase plan), it is clear that the periodic cost of the plan is the amount contributed by the employer. However, some plans that call for a contribution by formula also provide for a contemplated scale of benefits. This is true of some union-negotiated agreements. In these cases, careful analysis is required to ascertain whether such a plan should be considered a defined-benefit plan rather than a defined-contribution plan.

The Opinion states that the periodic cost of a defined-contribution plan is usually appropriately measured by the amount of contribution specified in the plan. In addition, we believe that where the employer's liability is limited to the amount of the pension fund, the amount of the defined contribution is presumed to be the proper amount of the current charge to expense. This would be the case with bilateral plans negotiated by either an employer or a group of employers (such as an industry group) with a union or

other employee representatives. If a plan provides both a formula for plan contributions and a scale for plan benefits, and the contributions are found to be inadequate or excessive for the purpose of funding the scale of benefits, either the contributions or the benefits, or both, may be subsequently adjusted. If the plan history indicates that only the scale of benefits is adjusted, the plan presumably should be treated as a defined-contribution plan. However, if the plan history indicates and/or the current employer policy contemplates the maintenance of benefit levels regardless of the amount of defined contribution or legal limitation of the employer's liability for such benefits, we believe that the plan should be treated as a defined-benefit plan, with the current charge to expense computed actuarially. If a company's liability for benefit levels is not limited by the amount of the pension fund, minimum current charges to expense should generally not be less than an amount actuarially computed on the basis of specified benefit levels.

Deferred Compensation Contracts Accounting for deferred compensation contracts is generally governed by the provisions of paragraphs 6, 7, and 8 of APB Opinion No. 12, which specifically relate to these contracts. Opinion No. 8, however, also refers to these contracts, indicating that its provisions are applicable to them if the contracts taken together are equivalent to a pension plan; however, the circumstances indicative of such equivalence are not specified. We believe that this equivalence would be most likely when a company's policy is to enter into deferred compensation contracts with an entire group or class of employees (such as officers of a certain level), instead of with only certain individuals in that group or class. The principal difference in the accounting method called for under the two opinions is that Opinion No. 12 requires accrual of compensation over the remaining service lives of individual employees (or, from its effective date, up to 10 years if remaining service lives are less), whereas Opinion No. 8 is more flexible in this respect.

Minimum Provision Opinion No. 8 provides for a minimum annual provision for pension costs equal to the total of (1) normal cost, (2) an amount equavalent to interest on unfunded prior service cost, and (3) a supplementary provision for vested benefits if required. If a supplementary provision for vested benefits is required, it may be the lesser of:

- An amount, if any, by which 5 percent of the unfunded or unprovided-for value of vested benefits at the beginning of the year is more than the reduction in the comparable unprovided-for value of vested benefits at the end of the year.
- An amount sufficient to make the aggregate annual provision for pension cost equal to the normal cost, amortization of prior service cost on a 40-year basis (including interest), and interest equivalents on differences between provisions and amounts funded.

Within the general framework of these "minimum" provisions, a company may adopt, as a single accounting policy, one of three procedures regarding unfunded or otherwise unprovided-for prior service cost. It may provide for such cost on the basis of (1) the interest only plus vesting method, (2) the 40-year amortization method, or (3) the annual lower of the two, even though either method may call for a larger amount than the other in a particular year. The 40-year amortization procedure has the practical advantage of avoiding the need of separately determining the annual change in unprovided-for vested benefits.

The current amount of such unprovided-for vested benefits, however, will have to be determined for the disclosure requirements of the Opinion, unless it can be actuarially determined that there are none. The 40-year basis of amortization is computed as a level annual amount, including the equivalent of interest.

Maximum Provision   The maximum annual provision for pension cost is stated as being the total of normal cost, 10 percent of past service cost at inception of the plan and of increases or decreases in prior service cost arising from plan amendments (in each case until fully amortized), and interest equivalents on differences between provisions and amounts funded. The 10 percent includes an interest factor and the maximum provision therefore requires (assuming interest rates of between 3 and 4 percent) from about 12 to slightly over 13 years in order to fully amortize past and prior service costs.

Supplementary Vesting Provision   This provision is required if the unfunded or otherwise unprovided-for value of vested benefits at the end of the year is not at least 5 percent less than the comparable amount at the beginning of the year. The provision, if required, is equal to 5 percent of the unfunded or otherwise unprovided-for value of vested benefits at the beginning of the year, to the extent that such 5 percent is not covered by the net reduction, if any, in such unfunded or unprovided-for amounts resulting from plan experience during the year. Increases will arise from additional vested benefits, reductions from contributions to the fund, fund earnings, and actuarial gains. The 5 percent is a declining balance type of provision; assuming no change in unprovided-for vested benefits other than through the provision, it would take about 45 years to reduce the unprovided-for vested benefits to 10 percent of the original amount.

Vested Benefits   Vested benefits are defined as benefits "that are not contingent on the employee's continuing in the service of the employer." The value* of vested benefits at a particular date includes not only the value of the amounts of pensions payable or to be payable to retired employees and to ex-employees who retain vested benefits, but also the value of benefits that would remain to the credit of those employees in active service who would become entitled to vesting if they were to terminate service as of the valuation date. Vested benefits do not accrue ratably between valuation dates, but reflect the vesting percentage in effect for the individual employee on the valuation date. The accrued benefit cost (unit credit) method of calculation should be used to determine the value of vested benefits, even though a different method may be used for other purposes.

In the determination of unfunded or unprovided-for value of vested benefits, we believe that the method of valuing the pension fund assets should preferably be consistent with that employed in periodic actuarial valuations under the actuarial cost method in use for accounting purposes. The use of full market value is undesirable, however, for the same reason that immediate recognition of full unrealized appreciation is considered undesirable.

*That is, the present value, at a particular actuarial valuation date, of future expected payments as determined actuarially.

Actuarial Cost Method   Opinion 8 defines five actuarial cost methods whose application would result in the appropriate annual provision for pension costs. In addition, variations of those methods may be used (e.g., the aggregate method with frozen, separately amortized, past service cost). For accounting purposes we prefer the entry-age normal cost (or entry-age level cost) method, the aggregate method, or the variation referred to in the example above. For unfunded plans, we consider level-cost methods especially preferable to the accrued benefit cost method. Generally, the accrued benefit cost method has cost ascendancy tendencies that are accentuated when applied to unfunded plans, because of accretions from interest equivalents inherent in the application of conventional actuarial methods to unfunded plans.

Actuarial Gains and Losses   Actuarial gains and losses arise from the need to use assumptions concerning future events. Under Opinion No. 8, adjustments required from time to time to reflect actual experience must be recognized in a consistent manner that reflects the long-range nature of pension cost. Except in specified circumstances relative to plant closings, acquisitions, etc., actuarial gains and losses are to be spread or averaged rather than accorded immediate recognition; the latter course is considered undesirable because of the possibility of fluctuations in annual pension expense. From 10 to 20 years is considered a reasonable period over which to spread actuarial gains and losses when spreading is accomplished by separate adjustment rather than by the routine application of the actuarial cost method used.

The average method of dealing with actuarial gains and losses entails arriving at an estimated average annual amount of such gains or losses that occurred in the past, taking into consideration those expected to occur in the future, e.g., a five-year moving average could be used. We consider averaging adaptable to situations where gains or losses tend to be repetitive, but it would probably not be feasible in the case of an occasional (as to nature or size) gain or loss. In some cases it may be necessary to use estimates of future actuarial gains and losses that may cause the computation of the average to be disadvantageous or impractical. In certain instances, a combination of methods may be desirable; for example, spreading might be applied to gains or losses that are not expected to recur frequently, while averaging may be applied to recurring items. When actuarial assumptions have been changed, past experience regarding actuarial gains and losses may not be a reliable basis for averaging. The averaging method, when suitable, has the advantage of avoiding a possible cumulative effect that could result from use of the spreading method.

Opinion No. 8 recognized that "an effect similar to spreading or averaging may be obtained by applying net actuarial gains as a reduction of prior service cost in a manner that reduces the annual amount equivalent to interest on, or the annual amount of amortization of, such prior service cost and does not reduce the period of amortization." For example, the application of a $100,000 gain to prior service cost would, assuming a 4 percent rate, reduce the "interest" charge by $4,000 annually. As a period of 10 to 20 years is considered reasonable for spreading, we believe that net actuarial gains should not be applied to prior service cost if the remaining amortization period of such prior service cost is less than 10 years. The Opinion does not discuss applying a net actuarial loss to prior service; however, we believe that this is acceptable, provided that prior service costs are being amortized over a remaining period of between 10 and 20 years, the period over which actuarial losses could otherwise be separately spread.

Actuarial Valuation Date   The date as of which the actuarial valuation is made will not generally coincide with the end of a company's fiscal year. In the absence of any substantially changed conditions (e.g., acquisition or disposal of a significant division, product line or segment of the business) that would render obsolete the data on which the actuarial valuation was made, valuations made as of a date within 12 months from the end of the fiscal year should be suitable for accounting determinations. A valuation made as of an earlier date may also be suitable, but its use could present problems in some areas—value of vested rights, for example. Technical problems concerning the use of the valuation date should be discussed with actuaries.

Unrealized Investment Appreciation and Depreciation Opinion 8 requires recognition of unrealized appreciation or depreciation in the determination of pension cost on a rational and systematic basis that avoids giving undue weight to short-term market fluctuations. Implicit in the Opinion's recommendations, we believe, is that the method of recognition used should be conservative enough to minimize the possibility that a reduction of pension expense in one year, in order to recognize appreciation, will result in the need to increase pension expense in the following year in order to recognize depreciation. Also, we believe

that it is particularly important that a consistent method be adopted for according recognition to unrealized appreciation and depreciation.

Interest on Unfunded Plans   Costs under unfunded pension plans should be determined under an accepted actuarial method including the use of interest equivalents. Possible bases for determination of an appropriate interest rate include rates used by funded pension plans of the company or generally used by other companies, or the borrowing rate paid by the company. We do not believe that use of an assumed rate of return on the employment of additional capital would generally be appropriate.

31-22.   Future Changes

Accounting for the cost of pension plans took a significant step forward with the promulgation of APB Opinion No. 8. We believe that further improvement can be attained through continuing efforts toward the further reduction of certain alternatives permitted under the Opinion. Under the Opinion, for example, accounting for past service cost may range from a minimum recognition consisting of interest on past service cost (plus a supplemental provision for vested benefits, if required) to a maximum recognition consisting of a 10 percent annual amortization. The very broad range of available alternatives the Opinion provides for accounting for pension costs can result in substantially differing amounts of annual provision in otherwise similar situations. Thus, the next logical step may well be the narrowing of the presently available alternatives.

INDEX